Principles
of Cell Biology

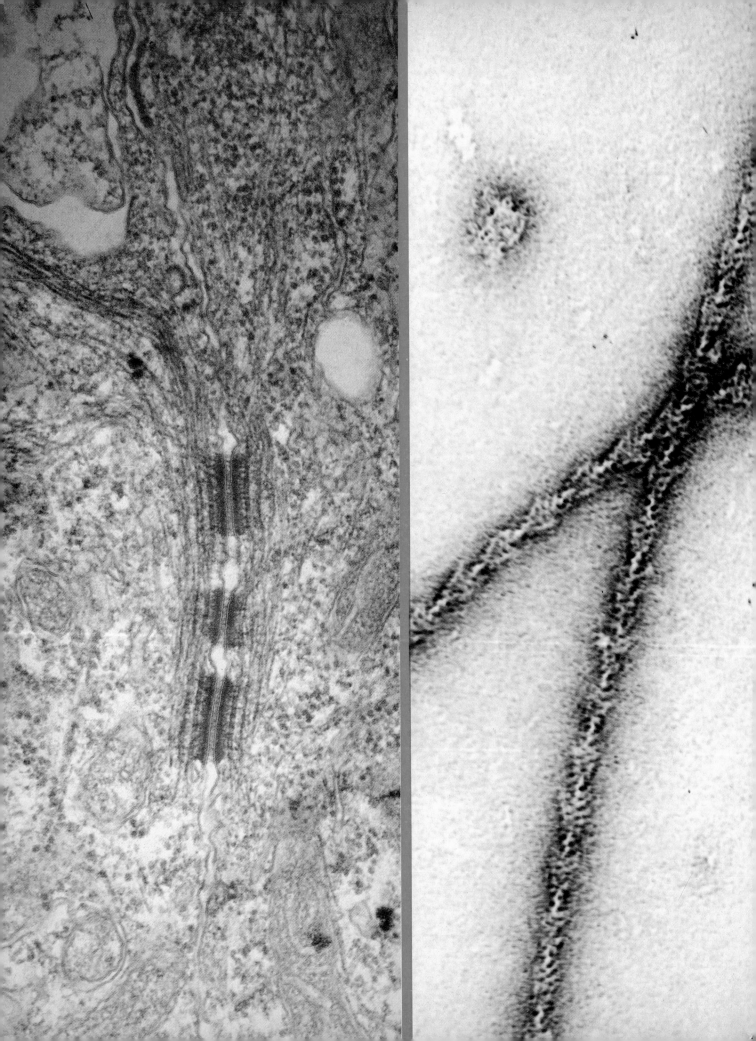

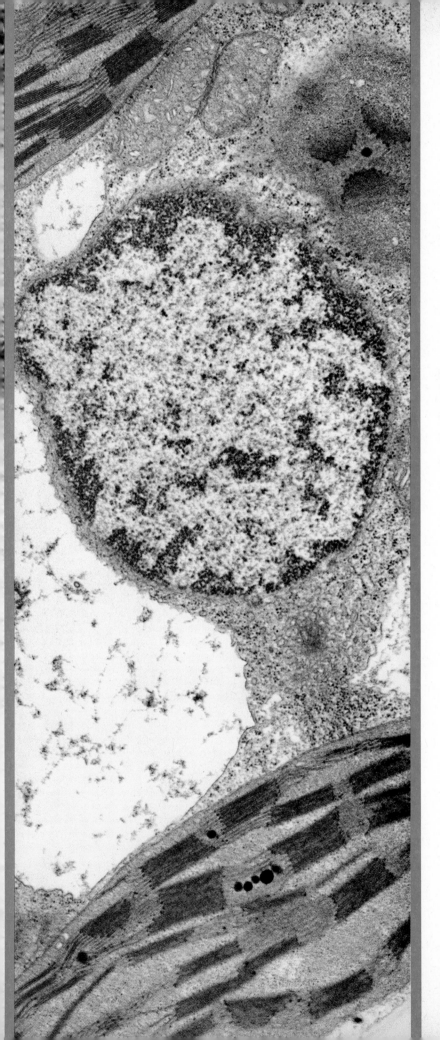

PRINCIPLES OF
CELL
BIOLOGY

Lewis J. Kleinsmith
The University of Michigan

Valerie M. Kish
Hobart and William Smith Colleges

 HarperCollins*Publishers*

For permission to use copyrighted materials,
grateful acknowledgment is made to copyright holders
listed on pages 761–772.

Sponsoring Editor: Claudia M. Wilson
Project Editor: Joan Gregory
Text and Cover Design: Maria Carella
Text Art: Vantage Art, Inc.
Production Manager: Kewal Sharma
Compositor: Progressive Typographers
Printer and Binder: Murray Printing
Cover Photographs: (Top): Mitosis; Reschke, © Peter Arnold.
(Far left): Fertilized sea urchin eggs; © Edwards/Sea Studios,
Peter Arnold. (Large photo): Living human endothelial cells; ©
1986, Siu, Peter Arnold. (Bottom): Spruce bud, showing needles
in cross-section; © Kage, Peter Arnold.
Part Opening Photographs: (Left): Courtesy of R. S. Decker.
(Middle): Reproduced from J. A. Spudich, H. E. Huxley, and J. T.
Finch (1972). J. Mol. Biol. 72:619–632. © Academic Press, Ltd.,
London. (Right): Courtesy of E. L. Vigil.

Principles of Cell Biology

Library of Congress Cataloging in Publication Data

Kleinsmith, Lewis J.
    Principles of cell biology.
    Includes bibliographies and index.
    1. Cytology.   I. Title.   II. Kish, Valerie M.
[DNLM: 1. Cells—physiology.   2. Molecular Biology.
QH 631 K64c]
QH581.2.K54   1988   574.87'6   87-8356

ISBN 0-06-043712-X

93 94 95 MPC 10 9 8 7

Brief Contents

PREFACE   XV

PART 1
Introduction
1

1
Cells, Viruses, and the
Origins of Life   3

2
Enzymes and Catalysis   35

3
Experimental Approaches
for Studying Cells   57

PART 2
Cell Organization
91

4
Molecular Organization
of Membranes and Cell Walls   93

5
Functions of the Cell
Surface   129

6
Functions of Cytoplasmic
Membranes   160

7
Mitochondria and the
Conservation of Energy
Derived from Foodstuffs   208

8

Chloroplasts and the
Conservation of Energy
Derived from Sunlight   248

9

The Cytoskeleton: Its Role
in Cell Organization
and Motility   293

10

Information Flow in the
Cell: General Principles   338

11

The Nucleus and Transcription
of Genetic Information   385

12

The Ribosome: Translation
of Genetic Information   447

13

The Cell Division Cycle   490

14

Biogenesis of Mitochondria
and Chloroplasts   536

**PART 3
Formation of
Specialized Cells**
579

15

Germ Cells, Early Development,
and Cell Differentiation   581

16

Hormones and Their Role in
Intercellular Communication   619

17

The Cellular Basis
of the Immune Response   643

18

The Neuron: Functional Unit
of the Nervous System   667

19

The Organization and Function
of Vertebrate Muscle Cells   701

20

The Cancer Cell   727

APPENDIX: PREFIXES, SYMBOLS,
AND ABBREVIATIONS   757

PHOTOGRAPH AND ILLUSTRATION CREDITS 761

INDEX   773

Detailed Contents

PREFACE   XV

PART 1
Introduction
1

1

Cells, Viruses, and the Origins of Life   3

WHAT ARE CELLS?   3
    Cell Functions, 4
    Cell Architecture, 4
    Comparison of Prokaryotes and Eukaryotes, 10

MOLECULAR COMPOSITION OF CELLS   11
    Water, 11
    Organic Building Blocks, 12
    Polysaccharides, 12
    Lipids, 13
    Proteins, 15
    Nucleic Acids, 22

INVESTIGATING THE ORIGIN OF LIFE   24
    Spontaneous Generation, 24
    The Primitive Earth, 24
    Formation of Organic Molecules, 25
    Intrinsic Molecular Functions, 26
    Emergence of the First Cells, 27

WHAT ARE VIRUSES?   30
    Viral Reproduction and Life Cycles, 30
    Evolutionary Origin of Viruses, 33

SUMMARY   33

SUGGESTED READINGS   34

2

Enzymes and Catalysis   35

DISCOVERY OF BIOLOGICAL CATALYSTS   35

THERMODYNAMICS OF CHEMICAL REACTIONS   37
    Determination of Free-Energy Changes, 37

KINETICS OF CHEMICAL REACTIONS   39

PROPERTIES OF ENZYME-CATALYZED
REACTIONS 40
The Michaelis–Menten Equation, 40
Effects of pH and Temperature on Enzyme
Activity, 42
Enzyme Cofactors, 43
Enzyme Inhibitors, 45
Regulation of Enzyme Activity, 47
Mechanism of Enzyme Action, 51

COUPLED REACTIONS AND THE ROLE
OF ATP 53

SUMMARY 55

SUGGESTED READINGS 55

3
Experimental Approaches
for Studying Cells 57

GROWING CELLS IN THE
LABORATORY 58

MICROSCOPY 60
The Light Microscope, 60
The Electron Microscope, 63
X-ray Diffraction, 66
Cytochemical Techniques, 69

ISOLATION OF CELL ORGANELLES 73
Principles of Centrifugation, 73
Subcellular Fractionation, 77

ANALYSIS OF ISOLATED
MACROMOLECULES 80
Spectrophotometric Analysis, 80
Isotopic Tracers, 81
Dialysis, 82
Precipitation, 82
Electrophoresis, 83
Chromatography, 85
Nucleic Acid Hybridization, 88

SUMMARY 88

SUGGESTED READINGS 89

PART 2
Cell Organization
91

4
Molecular Organization of
Membranes and Cell Walls 93

MODELS OF MEMBRANE STRUCTURE 93

Early Observations, 94
The Danielli–Davson Model, 95
The Unit Membrane Hypothesis, 95
The Fluid Mosaic Model, 98

MOLECULAR ORGANIZATION OF THE
PLASMA MEMBRANE 105
The Red Cell Membrane, 106
The Purple Membrane, 112
Relationship of the Plasma Membrane to
Other Membranes, 113

THE CELL WALL AND RELATED
STRUCTURES IN PROKARYOTES 113
The Capsule Layer, 113
The Bacterial Cell Wall, 114
The Periplasmic Space, 119

THE CELL WALL AND RELATED
STRUCTURES IN EUKARYOTES 120
The Plant Cell Wall, 120
The Animal Cell Glycocalyx, 123

THE EXTRACELLULAR MATRIX 123
Collagens, 124
Proteoglycan Networks, 125
Noncollagenous Matrix Glycoproteins:
Fibronectin and Laminin, 126

SUMMARY 127

SUGGESTED READINGS 128

5
Functions of the
Cell Surface 129

MOVEMENT OF MATERIALS ACROSS
THE PLASMA MEMBRANE 129
Simple Diffusion, 130
Facilitated Diffusion, 133
Active Transport, 139
Endocytosis/Exocytosis, 146
The Membrane Potential, 146

INTERCELLULAR COMMUNICATION
AND THE CELL SURFACE 147
Membrane Receptors, 147
Cell-Cell Recognition and Adhesion, 149

METABOLIC FUNCTIONS OF THE
CELL SURFACE 157

SUMMARY 158

SUGGESTED READINGS 159

6
Functions of
Cytoplasmic Membranes 160

THE ENDOPLASMIC RETICULUM 160

Metabolism of Carbohydrates and Lipids, *163*
Detoxification of Drugs, *165*
Transport and Processing of Proteins, *167*

THE GOLGI COMPLEX AND CELL
SECRETION   *171*
Processing of Proteins for Secretion, *175*
Processing of Proteins for Intracellular Use, *180*
Polysaccharide Synthesis and Cell Wall
Formation, *182*

LYSOSOMES   *182*
Formation of Lysosomes, *186*
Endocytosis, *187*
Autophagy and Autolysis, *192*
Extracellular Digestion by Hydrolytic
Enzymes, *193*
Lysosomal Storage Diseases, *193*
Spherosomes and Related Vacuoles in Plants,
195

PEROXISOMES, GLYOXYSOMES, AND
OTHER MICROBODIES   *195*
Metabolic Functions of Peroxisomes, *197*
Glyoxysomes and the Glyoxylate Cycle, *199*
Biogenesis of Peroxisomes, *200*
Evolutionary Origin of Peroxisomes, *202*
Other Microbodies, *202*

MEMBRANE BIOGENESIS   *203*

SUMMARY   *205*

SUGGESTED READINGS   *206*

7

Mitochondria and the Conservation of Energy Derived from Foodstuffs   208

OVERVIEW OF GLUCOSE
METABOLISM   *208*
Oxidation-Reduction and Redox Potentials, *209*
Role of Coenzymes in Oxidation-
Reduction, *210*
Glycolysis, *211*
The Krebs Citric Acid Cycle, *213*
Electron Transfer and Oxidative
Phosphorylation, *215*
Energy Yield of Glucose Oxidation, *215*

ANATOMY OF THE MITOCHONDRION   *218*
Arrangement of Mitochondria Within the
Cell, *223*
Isolation of Submitochondrial Components, *225*
The Outer Mitochondrial Membrane, *226*
The Intermembrane Space, *227*
The Mitochondrial Matrix, *228*
The Inner Mitochondrial Membrane, *228*

HOW MITOCHONDRIA MAKE ATP   *229*
Components of the Respiratory Chain, *229*
Sequential Arrangement of Carriers, *231*
Coupling Sites for Oxidative Phosphorylation,
235
Reconstitution of the Inner Membrane, *235*
The Chemiosmotic Theory, *237*
Inner Membrane Transport Reactions, *241*
Regulation of ATP Formation, *243*

HOW BACTERIA MAKE ATP   *243*

SUMMARY   *245*

SUGGESTED READINGS   *247*

8

Chloroplasts and the Conservation of Energy Derived from Sunlight   248

OVERVIEW OF PHOTOSYNTHESIS   *249*
Elucidation of the Basic Equation of
Photosynthesis, *249*
Discovery of the Light and Dark Reactions, *249*

ANATOMY OF THE CHLOROPLAST   *251*
Arrangement of Chloroplasts Within the
Cell, *255*
Isolation and Subfractionation of
Chloroplasts, *257*
The Chloroplast Envelope, *258*
The Chloroplast Stroma, *260*
Thylakoid Membranes, *262*

THE LIGHT REACTIONS OF THE
THYLAKOID MEMBRANES   *262*
Light-absorbing Pigments and the
Photosynthetic Unit, *262*
The Photosynthetic Electron Transfer
Chain, *266*
Photophosphorylation and the Chemiosmotic
Theory, *269*
Substructure of Thylakoid Membranes, *276*
Origins and Significance of Membrane
Stacking, *278*

THE DARK REACTIONS OF THE
STROMA   *279*
The Calvin Cycle, *279*
Overall Efficiency of Photosynthesis, *283*
C_4 Pathways, *283*
Photorespiration, *286*

PHOTOSYNTHESIS IN PROKARYOTES   *287*

ARTIFICIAL LIGHT-TRAPPING
SYSTEMS   *289*

SUMMARY   *290*

SUGGESTED READINGS   *292*

9

The Cytoskeleton: Its Role In Cell Organization and Motility   293

MICROTUBULES   *293*

Assembly and Disassembly of Microtubules, *295*

Eukaryotic Cilia and Flagella, *301*

Other Axonemal Structures, *311*

Cytoplasmic Microtubules and the Movement of Subcellular Components, *312*

Cytoplasmic Microtubules and the Cell Surface, *314*

Cytoplasmic Microtubules and Cell Shape, *315*

MICROFILAMENTS   *316*

Microfilaments and Cytoplasmic Streaming, *319*

Microfilaments and Amoeboid Movement, *320*

Microfilaments and Cell Division, *324*

Microfilaments, Cell Shape, and the Cell Surface, *324*

Microfilaments and Endocytosis/Exocytosis, *327*

Microfilament/Microtubule Interaction, *328*

INTERMEDIATE FILAMENTS   *328*

CYTOSKELETAL INTERCONNECTIONS   *329*

BACTERIAL FLAGELLA   *330*

SUMMARY   *335*

SUGGESTED READINGS   *337*

10

Information Flow in the Cell: General Principles   338

IDENTIFICATION OF DNA AS THE GENETIC MATERIAL   *338*

Discovery of DNA and Early Views Concerning Its Significance, *338*

Genetic Transformation and the Role of DNA, *340*

Verification of the Genetic Role of DNA, *341*

HOW IS DNA REPLICATED?   *343*

The DNA Polymerase Reaction, *345*

Reverse Transcriptase, *350*

HOW IS DNA UTILIZED?   *350*

The One Gene–One Polypeptide Chain Theory, *350*

The Role of Messenger RNA, *354*

The Genetic Code, *359*

The Role of Transfer RNA, *362*

HOW IS GENETIC INFORMATION ALTERED?   *365*

TECHNICAL ADVANCES IN THE STUDY OF INFORMATION FLOW   *370*

Nucleic Acid Hybridization, *371*

Recombinant DNA Technology and Gene Cloning, *375*

DNA Sequencing, *380*

Chemical Synthesis of DNA Sequences, *381*

Recombinant RNA, *381*

Implications of Recent Advances in Nucleic Acid Technology, *381*

SUMMARY   *383*

SUGGESTED READINGS   *384*

11

The Nucleus and Transcription of Genetic Information   385

ULTRASTRUCTURAL ORGANIZATION OF THE NUCLEUS   *385*

The Nuclear Envelope, *386*

The Nucleoplasm, *391*

The Nucleolus, *392*

Chromatin Fibers, *393*

The Nuclear Matrix, *402*

GENE TRANSCRIPTION   *404*

Microscopic Visualization of Gene Activity, *404*

The RNA Polymerase Reaction, *409*

Processing of the Primary Transcript, *414*

REGULATION OF GENE EXPRESSION   *424*

Cytoplasmic Influence over Nuclear Activity, *425*

DNA Organization in Prokaryotic and Eukaryotic Cells, *426*

Gene Regulation: The Role of Chromatin Structure, *429*

Gene Regulation: The Role of Changes in DNA Structure, *431*

Gene Regulation: The Role of DNA-associated Proteins, *436*

Gene Regulation: The Role of RNA Processing and Export, *442*

SUMMARY   *443*

SUGGESTED READINGS   *445*

12

The Ribosome: Translation of Genetic Information 447

INTRODUCTION TO THE RIBOSOME 447

RIBOSOME BIOGENESIS 450
Role of the Nucleolus, 450
Amplification of Ribosomal Genes, 452
Synthesis and Processing of Ribosomal RNA, 453
Joining of RNA with Ribosomal Proteins, 455
Visualization of Ribosome Formation, 456
Ribosome Formation in Prokaryotic Cells, 457
Regulation of Ribosome Formation, 460

THREE-DIMENSIONAL MODELS OF RIBOSOMAL SUBSTRUCTURE 460
Ribosome Reconstitution, 461
Biochemical and Microscopic Probes of Ribosome Substructure, 461
General Conclusions Concerning Ribosome Substructure, 463

THE RIBOSOME AND PROTEIN SYNTHESIS 464
Ingredients Required for Protein Synthesis, 464
The Direction of Messenger RNA Translation, 465
The Role of Ribosome Dissociation, 465
Initiation of Protein Synthesis, 466
Elongation Phase of Protein Synthesis, 470
Termination of Protein Synthesis, 474
Polysomes, 476
Inhibitors of Protein Synthesis, 479

TRANSLATIONAL AND POSTTRANSLATIONAL CONTROL MECHANISMS 481
Messenger RNA-Binding Proteins, 481
Alterations in Protein Synthesis Factors, 482
Messenger RNA Stability, 482
Regulatory Role of Transfer RNA, 484
Posttranslational Modifications, 485

SUMMARY 486

SUGGESTED READINGS 488

13

The Cell Division Cycle 490

INTRODUCTION TO THE CELL CYCLE 490

DNA REPLICATION AND THE CELL CYCLE 493
Prokaryotic Chromosome Replication, 493
Eukaryotic Chromosome Replication, 495

PROKARYOTIC CELL DIVISION 499

EUKARYOTIC CELL DIVISION: MITOSIS 501
Summary of Mitotic Stages, 501
Structure of Mitotic Chromosomes, 504
Structure and Assembly of the Mitotic Spindle, 507
The Mechanism of Chromosome Movement, 509
The Mechanism of Cytokinesis, 510

EUKARYOTIC CELL DIVISION: MEIOSIS 514
Summary of Meiotic Stages, 514
The Molecular Basis of Genetic Recombination, 518
Role of the Synaptonemal Complex, 523
Lampbrush Chromosomes, 526

REGULATION OF CELL DIVISION 528
Control of the Cell Cycle in Dividing Cells, 528
External Factors That Stimulate Cell Division, 529

SUMMARY 533

SUGGESTED READINGS 535

14

Biogenesis of Mitochondria and Chloroplasts 536

GENETIC EVIDENCE FOR CYTOPLASMIC INHERITANCE 536
Chloroplast Genetics, 536
Mitochondrial Genetics, 538

MOLECULAR BASIS OF CYTOPLASMIC INHERITANCE 540
The Discovery of DNA in Mitochondria and Chloroplasts, 540
Properties of DNA from Mitochondria and Chloroplasts, 543
Mechanism of Organellar DNA Replication, 548
Mechanism of Maternal Inheritance, 551
Transcription in Mitochondria and Chloroplasts, 552
Protein Synthesis Within Mitochondria and Chloroplasts, 554
Translocation of Polypeptides into Mitochondria and Chloroplasts, 560

DEVELOPMENTAL ORIGINS OF MITOCHONDRIA AND CHLOROPLASTS 561
Mitochondrial Development, 561
Chloroplast Development, 564

EVOLUTIONARY HISTORY OF
MITOCHONDRIA AND CHLOROPLASTS 568
The Endosymbiont Theory, *569*
Direct Filiation Theories, *572*
Cellular Evolution and Organelle
Biogenesis, *575*

SUMMARY   575

SUGGESTED READINGS   577

PART 3
Formation
of Specialized Cells
579

15

Germ Cells, Early Development,
and Cell Differentiation   581

OVERVIEW OF DEVELOPMENT   581

GAMETOGENESIS   584
Oogenesis, *584*
Spermatogenesis, *589*

FERTILIZATION AND EARLY
DEVELOPMENT   593
Early Responses of Fertilization, *593*
Late Responses of Fertilization, *597*
Cleavage, *600*
Unique Features of Plant Development, *600*

CELL DIFFERENTIATION   602
Cytoplasmic Influence on Cell
Differentiation, *602*
The Concept of Nuclear Totipotency, *603*
The Role of Gene Expression in
Development, *605*
Gene Injection Techniques, *606*
Protein Turnover and Cell Differentiation, *607*
Differentiation and Cell Division, *608*

CELL-CELL INTERACTIONS AND
MORPHOGENESIS   608
Cell Movements and Cell Shape Changes, *608*
Cell Adhesion, *609*
Embryonic Induction, *610*
Hormone-mediated Interactions, *613*

AGING   613

SUMMARY   615

SUGGESTED READINGS   617

16

Hormones and Their Role in
Intercellular Communication 619

INTRODUCTION TO THE VERTEBRATE
ENDOCRINE SYSTEM   619
The Discovery of Hormones, *619*
Endocrine Glands and Their Hormone
Products, *620*
The Neuroendocrine System and Control of
Hormone Release, *622*

HORMONES ACTING ON THE PLASMA
MEMBRANE   623
Plasma Membrane Hormone Receptors, *624*
Cyclic Nucleotides and the Second Messenger
Concept, *625*
Calcium Ion as an Intracellular Messenger, *631*
Termination of Hormone Action, *632*

HORMONES ACTING DIRECTLY ON
THE NUCLEUS   632
Cytosol and Nuclear Receptors, *633*
Activation of Transcription of Specific
Genes, *636*

THE EVOLUTION OF INTERCELLULAR
COMMUNICATION   637
Plant Growth Substances, *638*
Hormones Produced by Nonendocrine
Cells, *638*
Other Mediators of Intercellular
Communication, *639*
Cell Specialization and Hormone
Evolution, *640*

SUMMARY   640

SUGGESTED READINGS   642

17

The Cellular Basis of the
Immune Response   643

GENERAL PROPERTIES OF THE
IMMUNE SYSTEM   643

THE B-LYMPHOCYTE AND ANTIBODY
SYNTHESIS   645
Evidence That B-Cells Synthesize
Antibodies, *645*
Cellular Cooperation in Antibody
Synthesis, *647*
Antibody Structure, *648*
The Origin of Antibody Diversity, *650*

Genetic Control of Antibody Organization, *652*
Activation of Antibody Synthesis in B-Cells, *655*
Monoclonal Antibodies, *658*
How Do Antibodies Recognize and Inactivate
Foreign Antigens? *659*

THE CELL-MEDIATED IMMUNE
RESPONSE   *660*
The Discovery of T-Cell Involvement in
Cell-mediated Immunity, *661*
How Do T-Cells Recognize Antigen and
Confer Immunity? *661*

IMMUNOLOGICAL TOLERANCE   *662*

INNATE VERSUS ACQUIRED
IMMUNITY   *663*

SUMMARY   *664*

SUGGESTED READINGS   *665*

18

The Neuron: Functional Unit of the Nervous System   667

STRUCTURE AND DEVELOPMENT OF
THE NEURON   *667*
The Cell Theory and Its Relationship to
Nervous Tissue, *667*
Structure of Neurons, *668*
Myelin and Glial Cells, *671*
Axonal Transport, *672*
The Synapse, *675*
Growth and Development of Neurons, *678*

ELECTRICAL PROPERTIES OF THE
NEURON   *681*
The Action Potential, *681*
Propagation of the Nerve Impulse, *684*
Initiation of Electrical Signals by Sensory
Stimuli, *685*

SYNAPTIC TRANSMISSION OF NERVE
IMPULSES   *689*
Neurotransmitter Release, *689*
Neurotransmitter-Receptor Interaction, *692*
Neurotransmitter Inactivation, *697*

TRIGGERING A PHYSIOLOGICAL
RESPONSE   *697*

SUMMARY   *698*

SUGGESTED READINGS   *699*

19

The Organization and Function of Vertebrate Muscle Cells   701

STRUCTURE AND DEVELOPMENT OF
SKELETAL MUSCLE CELLS   *701*
Structure of Skeletal Muscle Cells, *701*
Developmental Origin of Skeletal Muscle, *706*

BIOCHEMISTRY OF CONTRACTILE
PROTEINS   *708*
Myosin: Major Constituent of the Thick
Filament, *709*
Other Thick Filament Proteins, *710*
Actin: Major Constituent of the Thin
Filament, *711*
Other Thin Filament Proteins, *712*

MECHANISM OF CONTRACTION IN
SKELETAL MUSCLE   *713*
The Sliding Filament Hypothesis, *713*
Excitation-Contraction Coupling, *715*
Energy Sources for Contraction, *719*
Abnormalities in Skeletal Muscle Function, *720*

SMOOTH MUSCLE   *721*
Structure of Smooth Muscle Cells, *721*
The Contractile Machinery, *721*

CARDIAC MUSCLE   *723*

SUMMARY   *724*

SUGGESTED READINGS   *725*

20

The Cancer Cell   727

WHAT IS CANCER?   *727*
Characteristics of Malignant Tumors, *728*
Profile of the Cancer Cell, *729*

WHAT CAUSES CANCER?   *736*
Chemical Carcinogens, *736*
Radiation-induced Cancer, *739*
Oncogenic Viruses, *739*
Heredity and Cancer, *746*
Genetic Versus Epigenetic Views of Cancer, *747*

CAN CANCER BE PREVENTED OR
CURED?   *748*
Prevention of Cancer, *748*
Surgery, Chemotherapy, and Radiation, *752*
Future Prospects, *754*

SUMMARY   *754*

SUGGESTED READINGS   *756*

APPENDIX: PREFIXES, SYMBOLS,
AND ABBREVIATIONS 757

PHOTOGRAPH AND ILLUSTRATION CREDITS 761

INDEX 773

Preface

In the last two decades the field of cell biology has undergone a veritable revolution, leading to major advances in our understanding of cell structure and function. The convergence of cytological, genetic, and biochemical approaches has generated a rich panorama of detail, the significance of which we are still attempting to unravel. *Principles of Cell Biology* is written as an introduction to this rapidly growing field. Our goal is to acquaint the undergraduate student who is encountering the subject for the first time with the fundamental principles that characterize cellular organization and function. This textbook has been drawn from our collective teaching experiences, both at a large university and at a small liberal-arts college. Our primary concern has been to write a book that is readable.

PLAN OF THE BOOK

Principles of Cell Biology is organized into three sections, with the first two sections making up the core, involving organelle structure and function. The remaining section allows the student to see how these general principles of cellular organization apply to various differentiated cell types. Because of this arrangement the book can easily serve a full year course, one that is a semester in length, or courses of shorter duration.

In Part 1, we present background information essential to the understanding of concepts discussed later in the book. Although some students may be able to use this section as a review of material already learned, we have assumed only a basic understanding of biological and chemical principles and have included these first three chapters as a means of introducing certain elements fundamental to the subject matter. Chapter 1 presents an overview of the various organelles found in eukaryotic and prokaryotic cells and discusses in detail the structure of important cellular molecules. The chapter concludes with a short discussion of the origin of cells and a description of how cells differ from viruses. This chapter sets the stage for the rest of the book and should be read as a prelude to the remaining material. Because enzymes are critically involved in many, if not most, cellular events, we have devoted an entire chapter to a discussion of their properties. Chapter 2 focuses on enzymes as biological catalysts, discussing thermodynamic and kinetic features of these important cellular proteins. The last chapter in Part 1 serves as a repository for the discussion of experimental techniques commonly used in the study of cells. Chapter 3 should be used as a reference that the student can consult when necessary. The single exception to this approach involves the discussion of techniques used to dissect and study genetic systems. Because of the difficult vocabulary and necessity for having a well-developed understanding of genetics, we have placed the discussion of these techniques within the chapter to which they apply (Chapter 10). After reading Chapters 1 and 2, and becoming familiar with Chapter 3 as a reference, students should be prepared to begin a study of the core material presented in Part 2.

There are many ways to organize a discussion of cellular organelles. We have chosen to begin by concentrating on the architecture and function of membranes, since this organelle plays such an important role in a diverse spectrum of cellular processes. Chapters 4 and 5 deal with the plasma membrane, cell walls, and the extracellular material covering the cell surface, with Chapter 4 devoted to structural considerations and Chapter 5 concerned primarily with function. The cytoplasmic membrane system is the subject of Chapter 6. Following a discussion of the membranes of the endoplasmic reticulum and Golgi complex, coverage is expanded to include the biology of lysosomes, since they are involved in membrane recycling events. Theories of how membranes are formed and cycled throughout the cell are also a part of this material. Finally, we discuss the biogenesis and function of microbodies such as peroxisomes and glyoxysomes.

The next major subject to be introduced is cellular energetics. The mitochondrion has historically served as a model system illustrating the essential ingredients of structure-function relationships, revealing to cell biologists some fundamental principles of subcellular organization. Following a description of how the cell is able to transform energy in preformed organic molecules into a usable form (involving a study of glycolysis and mitochondrial respiration pathways), we turn to the chloroplast in Chapter 8. Using Chapter 7 as a framework, we build on that knowledge, showing why the evolution of cells that could make their own organic molecules from inorganic precursors was such an important event in the history of living things. The study of photosynthesis illustrates the complexities involved in transducing radiant energy into energy of the chemical bond, and provides the student with a solid basis on which to evaluate the relationship between organelle architecture and function.

The study of the cytoskeleton is a rapidly expanding area in cell biology. We have included an overview of the principles of cytoskeletal organization in Chapter 9, focusing on the structural and functional differences encountered in comparing microtubules, microfilaments, and intermediate filaments. The rate at which experimental data are accumulating in this area has often exceeded our abilities to formulate conceptual models about the array of functions the cytoskeleton provides the cell. Nevertheless we have tried to present the basic principles that ultimately may lead to a fuller understanding of the diverse roles of the cytoskeletal framework.

Chapters 10 through 13 are devoted to an analysis of the flow of information from nucleus to cytoplasm. We begin this section with Chapter 10, which provides the conceptual groundwork concerning genetic principles of information transfer. Chapter 11 describes how RNA synthesis is achieved and regulated, and relates this information to the morphology of the nucleus. Chapter 12 is devoted to a discussion of the biogenesis and structure of the ribosome, as well as to its role in protein synthesis. Chapter 13 concludes this sequence with an analysis of cell division. The final chapter in Part 2, Chapter 14, focuses on the biogenesis of mitochondria and chloroplasts; in it topics such as cytoplasmic inheritance, the genetics of these organelles, and their evolutionary origins are considered. These topics are more typically discussed in units on mitochondria and chloroplasts, but because they require an understanding of the principles of genetic information flow, we chose to offset the material from its usual location. This eliminates any interruption in the flow of dialog between Chapters 7 and 8, and allows a direct comparison of the biogenesis of mitochondria and chloroplasts within the same chapter.

Whereas Chapters 1 through 14 present principles of organelle structure and function in the "typical" cell, very often it is easy to lose sight of the fact that many cells of multicellular organisms are programmed to perform a limited array of highly specialized functions. This differentiation is revealed by unique shapes, sizes, and molecular functions of the cells involved. In Part 3 we show how specific cells have modified their basic architectural plans in order to achieve these specialized functions. Chapter 15 introduces this section with a brief overview of developmental processes. In Chapter 16 the molecular events set in motion by hormones acting at the levels of the plasma membrane and nucleus are brought into focus. Chapter 17 deals with the cells of the immune system, some of which are responsible for antibody synthesis, while others are involved in cell-mediated immune reactions. The role of the neuron as the functional unit of the nervous system is the topic of Chapter 18. In this chapter the unique electrical properties of this cell type are highlighted. The unusual molecular composition of muscle cells is the subject of Chapter 19. Here we discuss the biochemistry of contractile proteins and the way in which these proteins must work together to achieve contraction. In the final chapter of the book we introduce the cancer cell, which can be considered a specialized cell in the sense that it represents a unique, though aberrant, phenotype. Chapter 20 examines the causes of cancer, discusses some properties of cancer cells, and, finally, offers prospects for the control of this disease.

AIDS TO THE STUDENT

One of the major concerns of this book is to provide students with a basic understanding of what a cell is and why the cell is the fundamental unit of life. Our approach throughout has been to focus on the major questions involved and the experimental approaches utilized in addressing these questions. This emphasis on experimental design should allow students to understand not only current theories and models regarding cell biology, but also the complex thought processes required to arrive at these ideas. To aid students in identifying key terms, we have highlighted those that we feel most relevant to the ideas presented by placing them in boldface type. Other, somewhat less important words appear in italics. The illustrations in this text are of a variety of types. Some are composites of data drawn from the literature, while others are original illustrations that interpret new findings that have appeared. Others are taken directly from published sources. We have also included a large number of micrographs that help to round out discussions of morphology and also aid in tying together biochemical data with structural observations. As an added feature, the magnification of each micrograph has been transposed to a bar with the length noted above it, thereby allowing size comparisons to be made more easily. Each illustration or micrograph is cited in a special credits section at the end of the book. Finally, each chapter is concluded by a substantial summary, which can be used as a general introduction to the chapter as well as a review when study of the material is completed.

Because the principles of cell biology are constantly being reevaluated in the light of current scientific information, the scientific literature is the place to search for the most up-to-date information on a given subject. We have included as suggested readings a representative listing of recent books and articles that can be used by students as a bridge to the literature. Because of space constraints, we have decided not to include all of the classical papers and other references used in preparing the text. Some information along these lines is available in the credits section at the end of the book. Finally, a word about the index. A properly designed index is absolutely essential for the efficient use of a text. We have compiled the index with the following features in mind: It should be pertinent, detailed without being cumbersome, and easily read. As an added feature, pages on which key terms are defined appear in boldface type. We believe that such an approach is more useful than a simple glossary, since it allows the meanings of terms to be looked up within their appropriate context. We have also included an appendix that serves as an easy guide to symbols, prefixes, and abbreviations in common usage.

ACKNOWLEDGMENTS

We are also pleased to thank those colleagues who graciously consented to contribute micrographs to this endeavor, as well as the authors and publishers who have kindly granted permission to reproduce copyrighted material. Additionally, we are indebted to those students and reviewers whose efforts to improve our manuscript have resulted in a much clearer and more concisely written text. In particular we acknowledge the following: Max Alfert, University of California, Berkeley; Ross Johnson, University of Minnesota; Arthur H. Whiteley, University of Washington; Robert E. Lee Black, College of William and Mary; Milton Saier, University of California, San Diego; Edwin V. Gaffney, The Pennsylvania State University; Carol Jeanne Muster, University of Illinois; William B. Busa, University of California, Davis; Peter J. Rizzo, Texas A & M University; Richard Nuccitelli, University of California, Davis; Hal Krider, University of Connecticut; Kenneth R. Miller, Brown University; Fred Wilt, University of California, Berkeley; David R. McClay, Duke University; William R. Jeffery, University of Texas, Austin; and Sidney Fox, University of Miami. We are especially obliged to all those at Harper & Row whose unflagging concern and attentiveness have made the book possible. We wish to thank Maria Carella, who was responsible for the design and the cover of the book; Joan Gregory, whose demanding job of checking and rechecking the manuscript and art program was carried out with consideration, good humor, and a remarkable attention to detail; and in particular Claudia M. Wilson, Senior Editor, who with unstinting support cheerfully nurtured the manuscript through its various stages to completion. Finally, we would like to acknowledge our families, whose patience, support, and encouragement during these long days and nights have been paramount in making this book a reality.

Lewis J. Kleinsmith
Valerie M. Kish

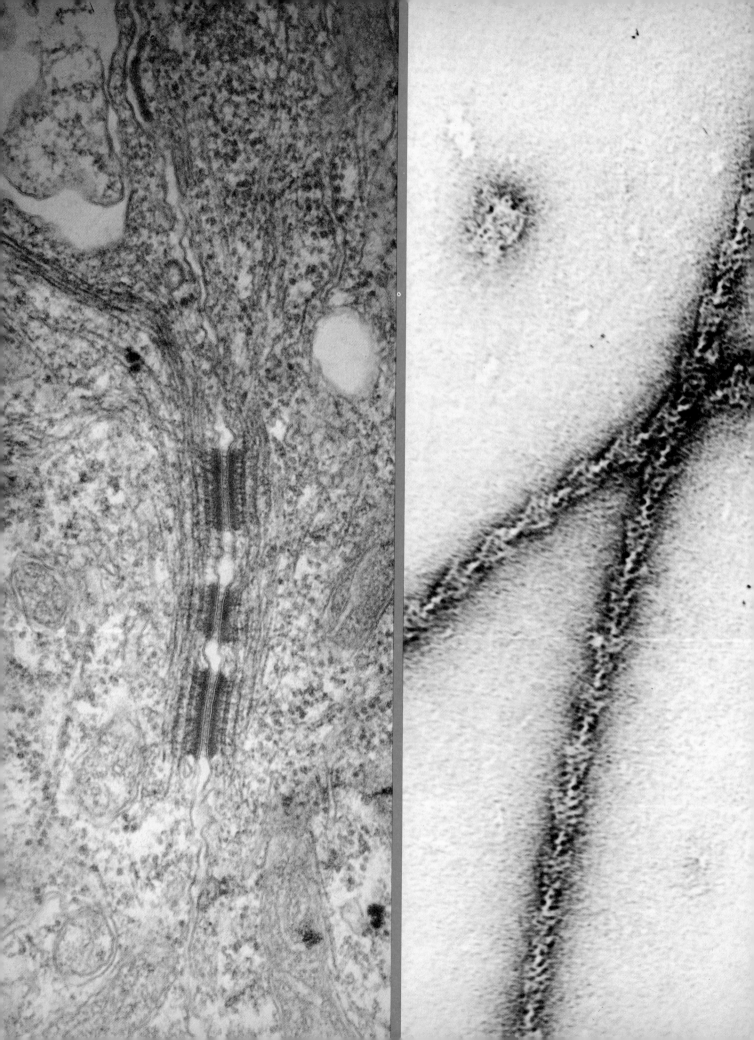

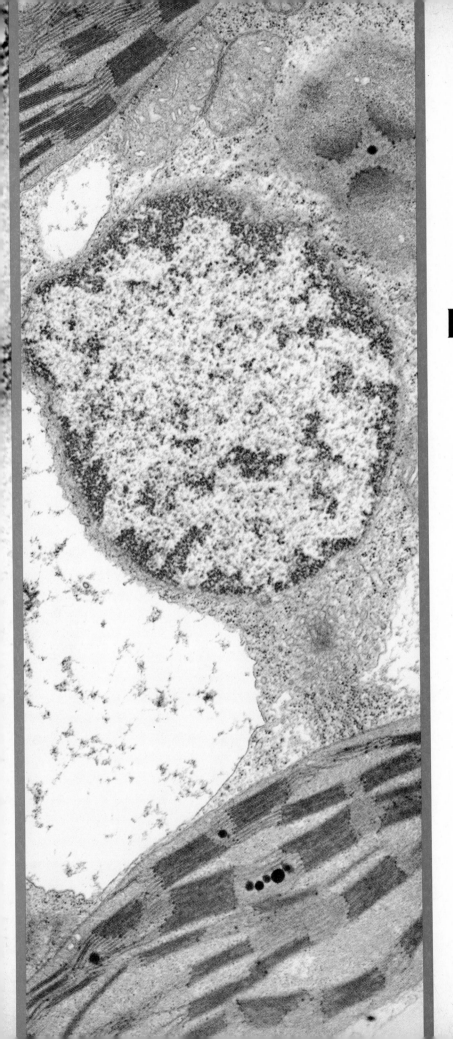

1
Introduction

1

Cells, Viruses, and the Origins of Life

The cell is the fundamental unit of life, the basic building block from which all organisms are constructed. The properties of the cell, this smallest unit manifesting the characteristics of living matter, define both the potential capabilities and the inherent limitations of all forms of life. In the past few decades extensive laboratory investigation has led to tremendous advances in our understanding of how cells are constructed and how they carry out the activities required for the maintenance of the living state. These experimental approaches and the information obtained from them in essence comprise the subject matter of this book.

As a general introduction to this undertaking, the present chapter will consider the following basic questions: What are cells? How are they organized? Of what are they composed? How did they originate? and How do they differ from viruses? Though a detailed discussion of such questions will occupy us throughout the book, a general consideration of these fundamental issues at the outset will help to set the stage for what is to follow.

WHAT ARE CELLS?

Because cells are generally invisible to the naked eye, their existence was unknown to scientists prior to the invention of the microscope. When the ability to cast and grind magnifying lenses was first perfected in the early seventeenth century, it triggered a scientific and intellectual revolution. Pointed at the sky, such lenses opened the universe to our vision; pointed at small objects, these lenses opened the microscopic world to our sight. Robert Hooke, one of the first scientists to build his own microscope, used it to examine thin slices of cork cut with a razor blade. Under the microscope, these slices of cork appeared to be made up of "little boxes," which Hooke termed **cells.**

What Hooke had in fact seen were no more than dead cell walls, with little evidence of internal cell structure (Figure 1-1). But these observations stimulated other scientists to examine biological materials under the microscope, leading to the discovery of an incredible variety of cell types. In 1839 the German biologists Mathias Schleiden and Theodor Schwann integrated the growing body of information on the universal occurrence of cells into one of the first great unifying theories of biology, the so-called **cell theory.** This theory had two important facets. First, it stated that in spite of the great diversity of living organisms on earth, all

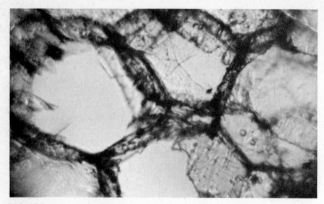

FIGURE 1-1 Original section of elder pith cut by van Leeuwenhoek in 1674, viewed through a surviving Leeuwenhoek microscope. The cell walls are clearly visible but no internal cell structure can be seen. Courtesy of B. J. Ford.

organisms are composed of cells. And second, it went on to propose that all living cells are similar to one another in structure and function.

Considering that these generalizations were based on relatively crude microscopic observations and virtually no biochemical data, it is remarkable how accurate they have turned out to be. In the century since the cell theory was first proposed, our tools for examining both cell architecture and the chemical activities occurring within cells have been revolutionized, leading to an understanding of the cell whose extraordinary detail would have astounded Schleiden and Schwann. Within this detail we have certainly discovered differences among cells, but in their basic functions and in the kinds of intracellular structures employed in carrying out these functions, cells are remarkably similar to each other.

Cell Functions

Although the cell theory was originally proposed on the basis of microscopic observations, the fundamental similarities among cells are even more pronounced at the functional level than at the structural level. If one examines the properties of a diverse spectrum of cells, it soon becomes apparent that the following four functions are virtually universal:

1. *Cells maintain a selective barrier* that separates the cell's internal contents from the extracellular environment. This barrier maintains an optimum intracellular environment by rigorously regulating the passage of materials into and out of the cell. Similar barriers are used to subdivide the cell interior into separate compartments specialized for particular activities.
2. *Cells inherit and transmit genetic material* containing encoded instructions used for directing the synthesis of most of the cell's components. The genetic material is duplicated prior to cell division so that each newly formed cell inherits a complete set of instructions.
3. *Cells carry out catalyzed chemical reactions* utilized for the synthesis and breakdown of organic molecules. Grouped together under the term **metabolism,** these reactions convert foodstuffs into molecules needed by the cell, trap energy in chemical forms that can be used to drive energy-requiring activities, and degrade molecules that are no longer needed.
4. *Cells manifest several types of motility* that result in locomotion of the cell as a whole as well as movement of individual components within the cell. Such movements are an essential ingredient of many cellular activities.

Each of the four functions listed above is associated with a particular set of morphologically specialized structures, termed **organelles,** which exist inside cells. In the following section, devoted to an introduction to cell architecture, the major organelles will be briefly summarized. Each organelle will be described in depth in an appropriate chapter later in the book.

Cell Architecture

One of the most striking features of a cell is its ability to create an extraordinarily complex substructure in the confines of a very tiny space. Figure 1-2 summarizes the dimensions of typical cells, organelles, and molecules. The major constraint on maximum cell size appears to be the surface area available for exchange of nutrients and wastes. As the size of a sphere increases, its internal volume increases proportionally to the cube of its radius. Hence smaller cells have a higher ratio of surface area to volume than do larger cells, so a point is eventually reached in larger cells where the surface area available is no longer sufficient to accommodate the exchange of materials required by the cell's internal machinery. For this reason most cells fall in the range of $1-30 \mu m$ in diameter.

Some large cells have extended this limit by modifying the outer membrane with multiple invaginations or extrusions, thereby increasing the available surface area. Another problem faced by large cells is the need for enough genetic material to guide all the events occurring in the large cell volume. An increase in the number or size of the chromosomes may therefore be required. Such modifications allow certain unicellular organisms, such as the *Amoeba*, to attain diameters of up to several millimeters, ten times greater than that of more typical cells. Even larger sizes are reached in extreme cases, such as the egg of a hen or ostrich, but these are atypical situations because such eggs consist largely of stored food. The metabolically active portion of the egg, destined to become the future embryo, is no more than a tiny speck on the surface of the yolk.

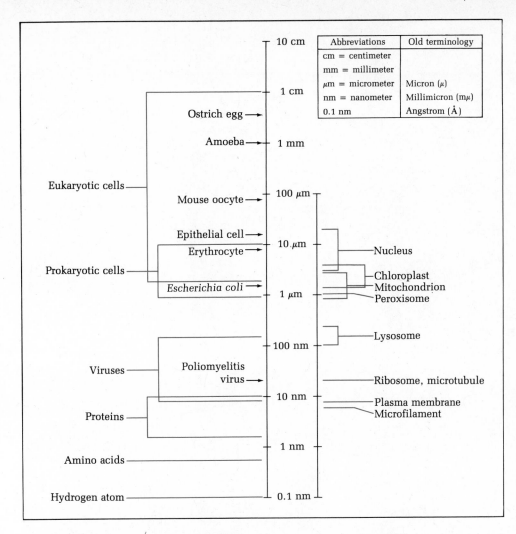

Abbreviations	Old terminology
cm = centimeter	
mm = millimeter	
μm = micrometer	Micron (μ)
nm = nanometer	Millimicron (mμ)
0.1 nm	Angstrom (Å)

FIGURE 1-2 Dimensions of typical cells, subcellular organelles, viruses, and molecules. The units of measurement in cell biology are in a state of transition, with proper metric units replacing some of the older terms such as the micron and the Ångstrom unit.

Restrictions are also placed on the minimum size that cells can attain. On the basis of our current understanding of the biochemical events occurring within cells, it can be calculated that a minimum of about 10^{-15} g of protein are needed to carry out essential metabolic functions. This is about the amount of protein found in *Mycoplasma* bacteria, the smallest organisms known. These cells are only about 0.1 μm in diameter, yet are capable of a completely independent existence. Although it is conceivable that smaller cells will yet be discovered, it is clear that the lower limits of cell size are being approached in *Mycoplasma*.

In spite of basic similarities in the organization of all cells, detailed structural comparisons have led to subdivision of the cellular world into two major groups. **Prokaryotes,** which include bacteria and blue-green algae, are single-celled organisms of relatively small size and simple structure. **Eukaryotes,** which include all other unicellular organisms (e.g., algae, fungi, and protozoa) and the cells of multicellular plants and animals, are generally larger, more complex cells. The most obvious feature distinguishing these two groups of cells

from each other is that the genetic material is surrounded by a membrane envelope in eukaryotes, but not in prokaryotes. There are also other, more subtle differences that will become apparent as we move on to describe the basic organelles involved in carrying out the four universal cell functions.

Organelles Functioning as Barriers: Membranes and Cell Walls All living cells are enclosed by a **plasma membrane** 8–10 nm thick (Figures 1-3, 1-4, and 1-5). Though minor chemical and structural differences distinguish the plasma membranes of prokaryotic and eukaryotic cells, the major functions performed in both cases are the same. The most important of these involves regulating the exchange of materials between the cell and its external environment, resulting in an intracellular milieu that differs substantially from that outside the cell. In order to maximize the surface area available for exchange of materials, the cell membrane is sometimes thrown into a series of extensive folds or invaginations.

The plasma membrane uses at least three different

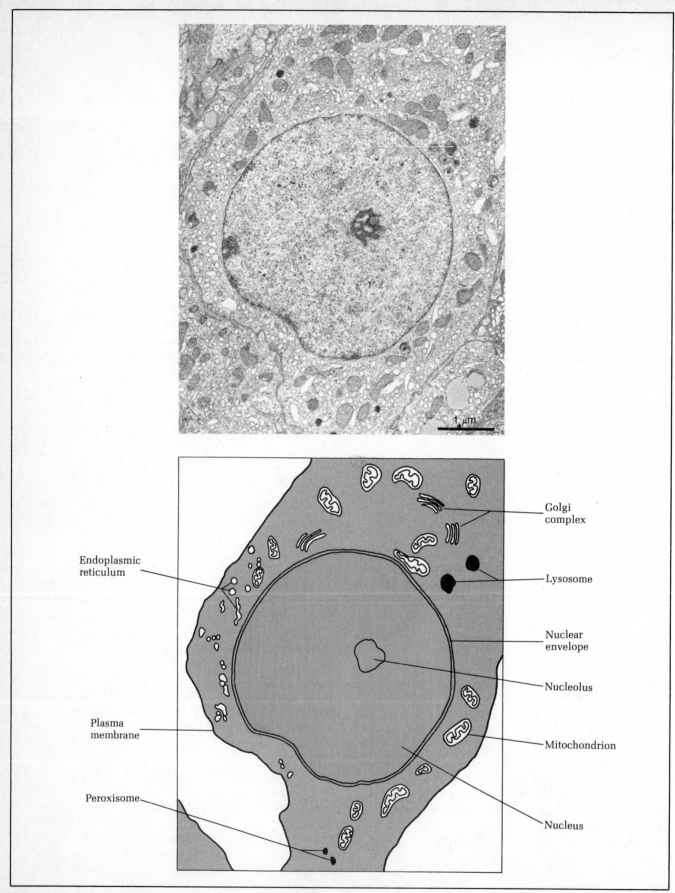

FIGURE 1-3 The general organization of an animal cell. *(Top)* Electron micrograph of a granulosa cell in the corpus luteum. Courtesy of E. L. Vigil. *(Bottom)* Line drawing of the cell shown above indicating some important features of cellular organization in animal cells.

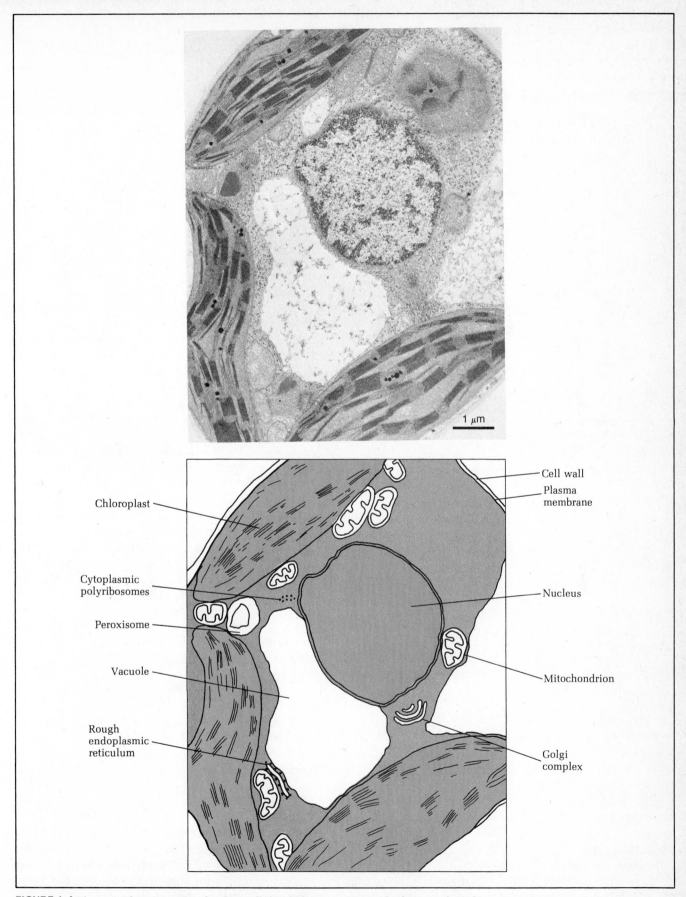

FIGURE 1-4 The general organization of a plant cell. *(Top)* Electron micrograph of a parenchymal cell of a growing leaf. Courtesy of E. L. Vigil. *(Bottom)* Line drawing of the cell shown above illustrating some important features of cellular organization in plant cells.

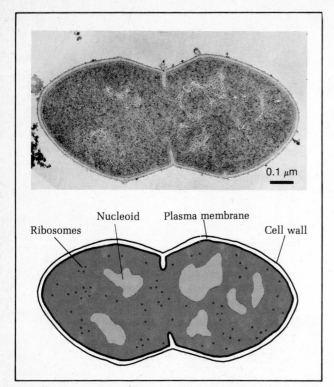

FIGURE 1-5 The general organization of a bacterial cell. *(Top)* Electron micrograph of a bacterial cell *(Streptococcus)* in the act of dividing. Courtesy of M. L. Higgins. *(Bottom)* Line drawing of the cell shown above with important features labeled.

Ribosomes Nucleoid Plasma membrane Cell wall

mechanisms for regulating passage of materials into and out of the cell. First, all plasma membranes are *selectively permeable,* meaning that some molecules move through the membrane more readily than others. This movement may involve either an energy-independent or an energy-facilitated diffusion of molecules. Second, membranes can selectively move specific molecules against a concentration gradient, maintaining either a lower or higher concentration inside the cell than out. This process requires an expenditure of energy and is therefore called *active transport.* Finally, large amounts of material are moved across the plasma membrane by a process of vesiculation. In this case pieces of membrane are "pinched off," forming spherical vesicles with various substances trapped within them. Movement of materials into the cell in this fashion is referred to as *endocytosis,* while movement out of the cell is termed *exocytosis* (see Figure 6-33).

In addition to its main function in regulating the flow of materials into and out of the cell, the plasma membrane contains specialized enzymes and proteins responsible for particular metabolic functions. The plasma membrane also serves a structural role in maintaining the physical integrity of the cell, although this property is less important in plant and prokaryotic cells, which are supported by rigid **cell walls** (see Figures 1-4 and 1-5). The cell wall is located immediately outside the plasma membrane and, by virtue of its rigid struc-

ture, molds the shape of the cell it encloses. Because of its great mechanical strength, the presence of a cell wall permits the cell to withstand extreme environmental conditions. Cell walls are not essential features of cell structure, however, for they can be removed artificially without destroying the enclosed cells. Such wall-less cells, termed **protoplasts,** are more fragile and susceptible to breakage than are walled cells.

Membranous structures are not restricted to the outer surface of the cell. The interior of most cells is partitioned into compartments by a series of complex infoldings or derivatives of the plasma membrane. This compartmentalization is most extensive in eukaryotic cells, though some internal membrane organization also occurs in prokaryotes. For example, **chromatophores** or **photosynthetic lamellae** in prokaryotes are single or parallel sheets of membranes deemed to be involved in photosynthesis, while **chondroids** are similar configurations believed to play a role in respiration. Some bacterial cells also contain an easily recognizable cluster of closely aligned concentric membranes referred to as a **mesosome.** Although the specific function of this unique organelle is unknown, it has been implicated in replication of the genetic material, as well as in energy metabolism.

The interior of eukaryotic cells is compartmentalized at a level of complexity far exceeding that of prokaryotes. The principal internal membrane system consists of an interconnected series of tubules, flattened sacs, and spherical vesicles. This network, termed the **endoplasmic reticulum (ER),** is a series of membrane-enclosed spaces whose interrelationships are in a state of constant flux. Points of connection between the endoplasmic reticulum and the outer nuclear membrane are commonly observed, and the internal space enclosed by the membranes of the endoplasmic reticulum, termed the **cisterna** (cavity), is also continuous with a stack of flattened membranous sacs termed the **Golgi complex** (see Figures 1-3 and 1-4). This organelle is involved in processing newly synthesized macromolecules and in packaging them into granules for subsequent storage or secretion.

A variety of other membrane-enclosed vesicles exhibit dynamic interrelationships with the endoplasmic reticulum and Golgi complex. **Lysosomes** act as an intracellular digestive system, breaking down foreign materials brought into the cell as well as aiding in the recycling of intracellular organelles. **Microbodies** are a heterogenous group of small vesicles involved in complex oxidative pathways. Membrane **vacuoles,** which are usually larger than lysosomes or microbodies, are especially prominent in plant cells. Such vacuoles usually function in the storage of synthetic products or in the excretion of waste materials.

In the following sections we shall see that, in addition to the membrane-bound structures described above, eukaryotic cells have membrane envelopes surrounding their genetic material and their energy-transforming organelles.

Organelles Associated with the Genetic Material: Nucleus and Ribosomes Most of the genetic material present in eukaryotic cells is contained within a large membrane-enclosed **nucleus** (see Figures 1-3 and 1-4). The membrane system surrounding the nucleus consists of two concentric membranes, together termed the **nuclear envelope.** Occasional connections between the nuclear envelope and the endoplasmic reticulum have been observed, but these two membrane systems are distinctly different from each other. The most striking difference is the presence in the nuclear envelope of numerous specialized structures, termed **nuclear pores,** that measure 50–70 nm in diameter and that interrupt the continuity of the envelope at various points. The nuclear envelope regulates the flow of materials between the nucleus and the **cytoplasm** (the region outside the nuclear envelope), serves as an anchoring point for fibers of genetic material, and carries out certain metabolic reactions, especially those involving oxidation-reduction.

The bulk of the nucleus consists of a mass of intertwined **chromatin** fibers containing the molecules of **DNA** in which the cell's genetic information is stored. There are numerous points of attachment between chromatin fibers and the nuclear envelope, most often in the regions of the nuclear pores. During the process of cell division the chromatin fibers condense into compact structures known as **chromosomes.** Also present in the eukaryotic nucleus is a small spherical structure referred to as the **nucleolus** (see Figure 1-3). There are usually several nucleoli per cell, functioning primarily in the manufacture and assembly of cytoplasmic ribosomes (see below). An amorphous fluidlike material termed the **nuclear sap** fills the spaces between the other nuclear structures. The nuclear sap contains the soluble molecules of the nucleus along with a variety of granular particles, and contains within it a fibrous network called the **nuclear matrix.**

In prokaryotic cells the genetic material is packaged into one or more fibers that are folded into a compact structure referred to as the **nucleoid** (see Figure 1-5). A fairly sharp boundary exists between the nucleoid and the rest of the cell, but no membrane encloses the nucleoid, nor are any nucleoli present. Occasional connections between the nucleoid and the plasma membrane are observed, often mediated by a mesosome.

In both prokaryotes and eukaryotes the genetic information encoded in the cell's DNA is ultimately utilized to guide the synthesis of specific protein molecules. This process of protein synthesis occurs on small cytoplasmic granules referred to as **ribosomes** (see Figures 1-4 and 1-5). Ribosomes measure approximately 25 nm in diameter and consist of roughly equal amounts of *protein* and *RNA*. Prokaryotic ribosomes are slightly smaller than eukaryotic ribosomes, but the two are basically similar to each other in structure and function. Eukaryotic ribosomes occur free in the cytoplasm as well as attached to membranes of the endoplasmic reticulum. Small numbers of ribosomes are also present in the mitochondria and chloroplasts (see below).

Organelles Involved in Metabolism: Mitochondria and Chloroplasts Chemical reactions occur virtually everywhere in the cell, but those most central to the metabolism of foodstuffs are localized predominantly in the cytoplasm. The biochemical reactions involved in the initial stages of the degradation of foodstuffs occur in the soluble phase of the cytoplasm, or **cytosol.** The latter stages, where most of the energy is released and used to drive the formation of the energy-rich molecule *ATP*, occur in specialized cytoplasmic organelles termed **mitochondria** (see Figures 1-3 and 1-4). Present only in eukaryotic cells, mitochondria are double-membrane enclosed structures as large as typical bacteria. The inner of the two mitochondrial membranes is folded into a series of **cristae** that project into the internal cavity, or **matrix,** of the mitochondrion. In addition to being the site of many chemical reactions, the matrix space also contains some DNA and ribosomes, suggesting that mitochondria exhibit a certain degree of genetic autonomy. Projecting from the inner surface of the cristae are tiny particles that are the main sites of mitochondrial ATP formation.

In prokaryotic cells, where mitochondria are not present, the analogous metabolic reactions occur in the cytosol and plasma membrane. Particles resembling those on the inner membrane of mitochondria project from the inner surface of the plasma membrane, suggesting that the plasma membrane plays a role in ATP synthesis analogous to that of the mitochondrial inner membrane. It is also possible that infoldings of the plasma membrane, such as the mesosomes or chondroids mentioned earlier, perform roles in ATP formation.

In addition to obtaining energy from the degradation of foodstuffs, some organisms are capable of trapping energy from sunlight and converting it to a useful chemical form. These **photosynthetic** organisms include both unicellular and multicellular plants (eukaryotes), as well as photosynthetic bacteria and blue-green algae (prokaryotes). In photosynthetic eukaryotes, the reactions of photosynthesis take place in specialized organelles termed **chloroplasts** (see Figure 1-4). Although chloroplasts are usually a bit larger than mitochondria, the structure of the two organelles is quite similar. Like mitochondria, chloroplasts are enclosed by two membranes surrounding an internal compartment, in this case designated the **stroma.** Within the stroma are the **thylakoid membranes,** which contain the light-absorbing pigments of photosynthesis, such as chlorophyll. The thylakoid membranes are thought to be derived from invaginations of the chloroplast inner membrane, though connections between the two are not seen as often as one sees connections between the inner mitochondrial membrane and cristae. As in the case of the mitochondrial inner membrane, thylakoid membranes exhibit protruding particles where ATP synthesis occurs.

The stroma is the site of many metabolic reactions associated with photosynthesis. As in the analogous matrix space of mitochondria, the stroma of chloroplasts contains both DNA and ribosomes, indicating that chloroplasts too exhibit a certain degree of genetic autonomy.

In photosynthetic prokaryotes, where chloroplasts are not present, the reactions involved in capturing energy from sunlight and converting it to a useful chemical form are also membrane associated. In this case the special infoldings of the plasma membrane mentioned earlier, namely chromatophores and photosynthetic lamellae, appear to be the major sites involved.

Organelles Involved in Motility: Microtubules, Microfilaments, and Related Structures Motility of some type or other is characteristic of virtually all cells. The movement of a cell from one place to another or the movement of fluids across the surface of a stationary cell are both accomplished by cellular appendages known as **cilia** and **flagella.** The major feature distinguishing the two is that cilia are shorter and present in larger numbers. Although such motile appendages occur in both prokaryotes and eukaryotes, their structures are quite different in the two cell types. Prokaryotic flagella are only 10–20 nm in diameter and exhibit a very simple internal structure. Eukaryotic flagella, on the other hand, are 25–50 times thicker and exhibit a complex inner structure consisting of nine pairs of hollow tubules surrounding two inner tubules (the "9 + 2" arrangement). In spite of the structural differences between the motile appendages in prokaryotic and eukaryotic cells, both generate movement by mechanisms utilizing energy derived from ATP.

In addition to motile external appendages, cells also contain internal mechanisms for producing movement. This is especially true for eukaryotic cells, whose larger size often poses serious problems of intracellular transport. **Microtubules** are relatively rigid tubules 25 nm in diameter; they contain the protein *tubulin* and occur widely in eukaryotic cells. They are involved in the construction of the *spindle apparatus* that moves the chromosomes during cell division and also form the backbone responsible for the motility of both cilia and flagella. In addition, microtubules contribute to the **cytoskeleton** that provides structural support for the cell as well as a framework upon which other components can be transported throughout the cell.

Smaller fibers called **microfilaments** also play a role in generating movement within the cytoplasm of eukaryotic cells. These fibers measure about 6 nm in cross-sectional diameter and are constructed from *actin,* a protein that also occurs in the contractile fibers of muscle cells. Microfilaments are involved in producing intracellular movements of cytoplasm known as *cytoplasmic streaming,* as well as amoeboid-type locomotion of the cell as a whole. Microfilaments have also been implicated in movements of the plasma membrane, especially those occurring when the cytoplasm is divided during cell division. The presence of the protein actin in microfilaments suggests that the contractile proteins of muscle cells may simply represent an elaborate specialization of a more primitive and widespread contractile system based on microfilaments.

The third type of cytoskeletal element is the **intermediate filament,** so named because its diameter of 7–11 nm is in between that of microfilaments and microtubules. These filaments have different protein compositions depending on the cell type in which each is found, and they play an important role in maintaining the cytoskeletal framework of the cell.

Comparison of Prokaryotes and Eukaryotes

Although the basic functions carried out by prokaryotic and eukaryotic cells are remarkably similar, we have now noted a number of structural differences between these two types of cells. In essence, prokaryotes are smaller and simpler than eukaryotes. The major difference is the great elaboration of internal membranes that characterizes the typical eukaryotic cell, leading to the production of multiple intracellular compartments. Membranes involved in this process include the nuclear envelope, endoplasmic reticulum, Golgi complex, lysosomes, and microbodies. Other features that distinguish eukaryotes are the presence of nucleoli, microtubules, microfilaments, intermediate filaments, and the size

TABLE 1-1
Structural features of prokaryotic and eukaryotic cells compared

	Prokaryotes	Eukaryotes	
		Animal	Plant
Plasma membrane	+	+	+
Cell wall	+	0	+
Ribosomes	+	+	+
Endoplasmic reticulum	0	+	+
Golgi complex	0	+	+
Lysosomes	0	+	+
Microbodies	0	+	+
Nuclear envelope	0	+	+
Nucleolus	0	+	+
Mitochondria	0	+	+
Chloroplasts	0	0	+
"9 + 2" cilia/flagella	0	+	+
Microtubules	0	+	+
Microfilaments	0	+	+
Intermediate filaments	0	+	0

Note: + = structure present; 0 = structure absent.

TABLE 1-2
Elemental composition of living matter

CONSTITUENTS OF MACROMOLECULES

Hydrogen	Constituent of water and organic compounds
Carbon	Forms backbone of all organic compounds
Nitrogen	Required for many organic compounds, including all proteins and nucleic acids
Oxygen	Constituent of water, organic compounds, and all macromolecules
Phosphorus	Constituent of nucleic acids, phospholipids; key role in energy transfer
Sulfur	Constituent of almost all proteins; used in some other organic compounds

OTHER MAJOR ELEMENTS

Sodium	Principal extracellular cation
Potassium	Principal cellular cation
Chlorine	Principal cellular and extracellular anion
Calcium	Major component of bone; required for activity of some enzymes
Magnesium	Required for chlorophyll and many enzymes
Manganese	Required for activity of several enzymes
Zinc	Required for activity of many enzymes
Copper	Essential in some oxidative and other enzymes
Iron	Essential for activity of many enzymes, hemoglobin, respiratory proteins

TRACE ELEMENTS

Boron	Essential in some plants; function unknown
Iodine	Essential for thyroid hormones
Fluorine	Growth factor in rats; possible constituent of teeth and bone
Silicon	Possible structural unit of diatoms; essential in chicks; function unknown
Cobalt	Required for activity of several enzymes
Molybdenum	Required for activity of several enzymes
Chromium	Essential in higher animals; required for insulin action
Vanadium	Essential in lower plants, certain marine animals and rats
Selenium	Essential for liver function
Tin	Essential in rats; function unknown
Nickel	Requirement under study

and 9 + 2 arrangement of their cilia and flagella. An overall summary of the major structural differences between prokaryotic and eukaryotic cells is provided in Table 1-1.

A detailed description of the structure and function of the organelles introduced in this section will be presented later in the text. At that time, side-by-side comparisons will be given to show how prokaryotic and eukaryotic cells accomplish the same basic functions through a somewhat different architecture.

MOLECULAR COMPOSITION OF CELLS

All the properties manifested by living cells ultimately reflect the characteristics of the organic molecules from which cells are constructed. In order to understand how cells carry out their essential functions, it is therefore necessary to briefly describe the types of molecules inhabiting the cell. Of the 92 natural elements potentially available for use in constructing these biological molecules, a relatively small number are of major importance. **Carbon, hydrogen, oxygen, nitrogen, phosphorus,** and **sulfur** predominate in the large **macromolecules** that make up most of the cell's dry mass. A variety of other elements are required in

smaller quantities for the performance of particular specialized functions (Table 1-2).

Water

The most abundant compound in the cell is water, accounting for 75–90 percent of the total mass in most cells (Figure 1-6). Hence life occurs in an aqueous envi-

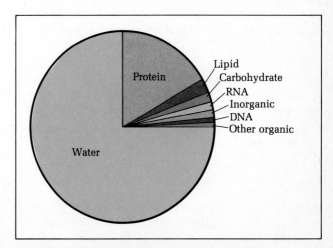

FIGURE 1-6 Molecular composition of a typical cell.

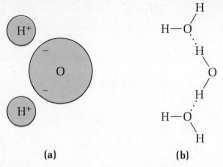

FIGURE 1-7 Polarity of the water molecule. (a) The electrons of the two covalent bonds in water are closer to the oxygen atom than to the hydrogens, imparting a partial negative charge to the oxygen and a partial positive charge to the hydrogens. (b) These partial charges result in weak electrostatic interactions between adjacent water molecules.

ronment, with the properties of water dictating to a large extent the activities that can take place. Water plays at least two critical roles. It is the *medium* in which all cellular reactions occur, as well as a direct *reactant* in the case of hydrolysis and dehydration reactions.

One of the key features determining the properties of the water molecule is the fact that its electrons are not equally shared between the oxygen and the hydrogens, but are instead more closely associated with the oxygen (Figure 1-7). One end of the molecule therefore exhibits a partial negative charge and the other a partial positive charge. This inherent **polarity** permits weak binding of water molecules to each other, as well as to any charged substance.

This attraction between water molecules leads to several properties important to living systems. For example, the intermolecular attraction between water molecules causes water to exhibit a high heat capacity (i.e., it takes a relatively large amount of heat to change the temperature of water). Cells are therefore protected from rapid temperature fluctuations in the external environment. The polarity of water also contributes to its great solvent power and its ability to dissociate other molecules into ions.

It has long been assumed that intracellular water behaves like a typical liquid. However, certain kinds of physical measurements suggest that at least some intracellular water molecules are induced to form a partially ordered arrangement by interaction with the cell's macromolecular structures. If this is true, the existence of such partially organized regions of water may exert a significant influence on reactions taking place within these areas of the cell.

Organic Building Blocks

All cellular molecules other than water exhibit one basic feature in common, the presence of a **carbon skeleton.** About 150 years ago it was thought that car-

bon-containing or **organic substances** are too complex to be understood, and that living systems perform some type of transcendental chemistry beyond the reach of laboratory experimentation. This belief was finally destroyed in 1828, when the German chemist Friedrich Wöhler achieved the first synthesis of an organic compound (urea) outside of a living organism. After this pioneering discovery thousands of organic compounds were synthesized in the laboratory, and it has gradually become apparent that the chemical viewpoint can greatly enhance our understanding of living cells. Contrary to the expectations of the early nineteenth-century scientists, the chemistry of living cells turned out not to be staggeringly complex, at least from the point of view of the organic molecules from which cells are assembled. In fact, in spite of the tremendous diversity and complexity of living organisms, most cellular constituents are constructed from only four simple kinds of chemical building blocks: **sugars, fatty acids, amino acids,** and **nitrogenous bases.**

Although these same four building blocks are utilized by all cells, prokaryotic as well as eukaryotic, their major use is not in the form of free sugars, fatty acids, amino acids, and nitrogenous bases. Rather they are assembled into four types of larger macromolecules called **polysaccharides, lipids, proteins,** and **nucleic acids.** These polymers of the basic building blocks form the major structural and functional components of living cells.

Polysaccharides

Polysaccharides belong to a class of organic molecules, referred to as **carbohydrates,** whose members are characterized by a $1:2:1$ ratio of carbon, hydrogen, and oxygen. Simple sugars (or **monosaccharides**) are the smallest carbohydrates. They may contain as few as three carbons (*trioses*), but five- and six-carbon sugars (*pentoses* and *hexoses*) are the most important for biological systems.

The most widely occurring six-carbon sugar in nature is **glucose,** a key source of energy and a building block for larger carbohydrates. Glucose occurs as both a ring and an open chain, a pattern common for most sugars containing five or more carbons. As shown in Figure 1-8, the ring form of glucose occurs in two configurations, α and β, which differ in the orientation of the hydroxyl group attached to the first carbon. The α and β forms of glucose are in equilibrium with each other through the open-chain intermediate, with the ring forms greatly predominating.

Other hexoses, such as galactose and fructose, have the same number of carbon atoms as glucose, but the positions of the hydroxyl (—OH) or aldehyde (C=O) groups are different. Although hexoses play a key role in metabolism, they are not the only biologically impor-

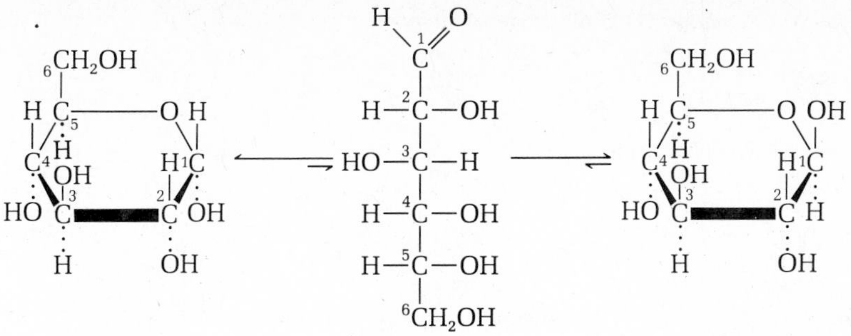

α-Glucose
(ring form)

Glucose
(chain form)

β-Glucose
(ring form)

FIGURE 1-8 The structure of glucose. The open-chain and ring forms are in equilibrium with each other. The numbers used in referring to the six carbon atoms are shown.

tant monosaccharides. As will be seen later, the pentoses *ribose* and *deoxyribose* are especially important because they are utilized in the construction of nucleic acids, the polymers in which genetic information is stored.

Monosaccharides serve as basic building blocks for the construction of a variety of larger molecules. Two monosaccharides can be joined together to produce a **disaccharide,** splitting out a molecule of water in the process. Subsequent addition of more monosaccharide units produces **oligosaccharides** (oligo = few), and eventually very long chains referred to as **polysaccharides** (poly = many). The most commonly encountered polysaccharides in living organisms are cellulose, starch, and glycogen (Figure 1-9). **Starch** and **glycogen,** long polymers constructed from α-glucose subunits, are used for energy storage in plant and animal cells, respectively. **Cellulose,** on the other hand, is a polymer of β-glucose and the major constituent of plant cell walls. The great mechanical strength of cellulose is responsible for many of the physical properties of wood. The difference between the orientation of a single hydroxyl group in α- and β-glucose is thus seen to generate polysaccharides with profoundly different chemical and physical properties.

Some polysaccharides are composed of mixtures of several different kinds of sugar derivatives. Amino- and acetyl-substituted sugars are extremely important in this regard, as are sugars containing sulfate or phosphate. Such **complex polysaccharides** are usually formed by alternating two different sugar derivatives. Such "mixed" polysaccharides serve a variety of important <u>functions</u>, especially as constituents of extracellular fluids and materials.

Lipids

The lipids comprise a chemically heterogenous group of molecules that share the common property of being insoluble in water. They are therefore readily extracted from cells by organic solvents such as acetone, chloroform, and ether. The lipids most commonly encountered in biological systems can be subdivided into five

principal categories: *neutral fats, phospholipids, glycolipids, steroids,* and *terpenes.*

Neutral fats are constructed from two types of building blocks, glycerol and fatty acids (Figure 1-10). **Glycerol** is a reduced alternative of a three-carbon sugar, while **fatty acids** are long carbon chains with an acidic (carboxyl) group at one end. Most biologically important fatty acids have an even number of carbon atoms, ranging up to a maximum of around 24. The formation of a neutral fat involves the linkage of the carboxyl group of a fatty acid to one of the hydroxyl groups of glycerol, splitting out a molecule of water in the process. This reaction removes the charge originally present on the fatty acid carboxyl group, and so yields a nonpolar neutral fat exhibiting very low water solubility.

Neutral fats may be constructed by joining one, two, or three fatty acids to a molecule of glycerol, producing **monoglycerides, diglycerides,** and **triglycerides,** respectively. The fatty acids attached to the three carbons of glycerol may be of the same or different lengths, and may be *saturated* (all single bonds) or *unsaturated* (containing one or more double bonds). Because many different kinds of fatty acids can be joined to glycerol, a large variety of neutral fats exist. Neutral fats function predominantly in storing energy because the fatty acids released during the breakdown of neutral fats are an extremely concentrated energy source.

In contrast to the extremely low water solubility of neutral fats, **phospholipids** contain ionic phosphate groups incorporated into the lipid backbone. This produces an *amphipathic* molecule, that is, a molecule with both *hydrophilic* (water-loving) and *hydrophobic* (water-hating) properties. **Phosphoglycerides** are constructed by linking a phosphate group to the first carbon of glycerol. Fatty acids are then joined to the two remaining carbons, while various alcohols can be joined to the phosphate (Figure 1-11). In the other major type of phospholipid, called **sphingophospholipids,** the phosphate group is esterified not to glycerol but to a complex amino alcohol called *sphingosine* (see Figure 1-11). Phospholipids of both types play important roles in the structure and function of cellular membranes.

Glycolipids are derivatives of sphingosine charac-

FIGURE 1-9 Diagrammatic representation of polysaccharide formation. (a) Joining together molecules of α-glucose into long chains creates starch and glycogen. (b) Joining together of β-glucose molecules creates cellulose. The actual metabolic pathways used by cells to synthesize such polysaccharides are more complex than illustrated because of the need to provide an energy input to drive the reactions.

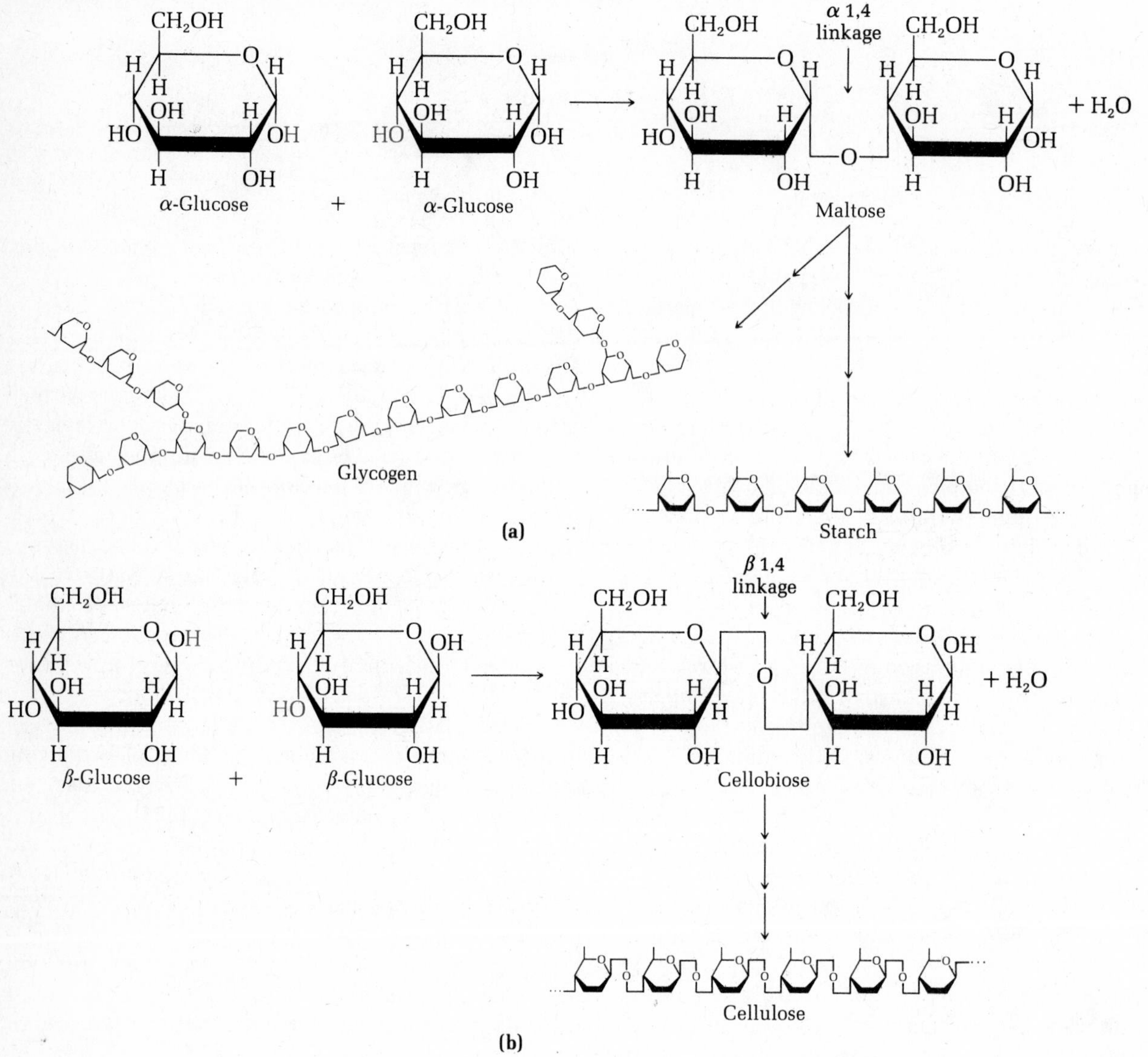

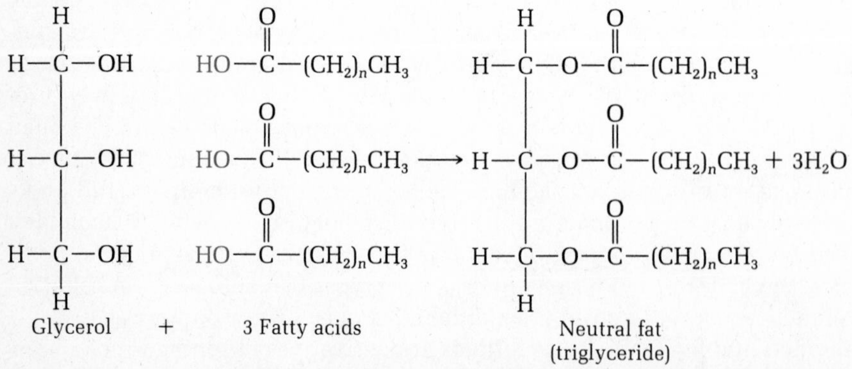

α-Glucose + α-Glucose → Maltose + H_2O (α 1,4 linkage)

Glycogen

Starch

(a)

β-Glucose + β-Glucose → Cellobiose + H_2O (β 1,4 linkage)

Cellulose

(b)

Glycerol + 3 Fatty acids → Neutral fat (triglyceride) + $3H_2O$

FIGURE 1-10 Diagrammatic representation of the formation of a typical neutral fat (triglyceride). The actual pathway used by cells for synthesizing neutral fats is more complex than illustrated because of the need to provide an energy input to drive the reaction.

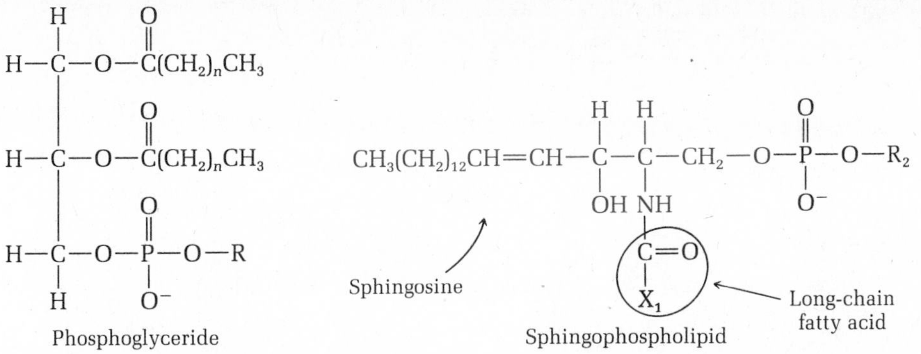

H—C—O—C(CH₂)ₙCH₃

Phosphoglyceride   Sphingosine   Sphingophospholipid   Long-chain fatty acid

FIGURE 1-11 Structure of the two major types of phospholipids. Abbreviations: R = alcohol group, X = fatty acid chain.

terized by the presence of a carbohydrate group. They also contain fatty acid chains but have no phosphate groups and so are less water soluble than the sphingophospholipids. However, one class, the *sulfatides*, contain an ionic sulfate group and therefore exhibit a hydrophobic-hydrophilic character similar to that of phospholipids. Glycolipids are important constituents of membranes, and are found in especially high concentration in the central nervous system.

Steroids are derivatives of the four-membered phenanthrene ring system and so are structurally distinct from the other lipids discussed so far (Figure 1-12a). Although some steroids may be linked to fatty acids, this is not generally the case. Most steroids are regulatory molecules, for example, the hormones *cortisone*, *testosterone*, *estrogen*, and *progesterone*. *Vitamin D*, another important steroid, regulates the absorption of calcium from the intestine. *Cholesterol*, a steroid serving a nonregulatory function, is a structural constituent

of cell membranes and an important metabolic precursor for the synthesis of other steroids.

Terpenes are lipids constructed from the five-carbon compound **isoprene** (see Figure 1-12b). This building block and its derivatives are joined together in various combinations to produce substances such as *vitamin A*, *coenzyme Q*, and *carotenoids*. Despite similarities in their construction, such molecules differ significantly in the functions they carry out.

As is evident from the foregoing discussion, lipids exhibit a variety of chemical structures, each suited for a particular purpose. The principal feature unifying this diverse group of compounds is their hydrophobic nature, a property that does impose a certain set of attributes. Probably the most important in terms of the origin and evolution of cells is that hydrophobic molecules are ideally suited for creating barriers between aqueous environments. Hence lipids have been intimately involved in the creation of the membrane systems required for the evolution of cells as we know them today.

FIGURE 1-12 (a) Structure of the steroid ring. (b) Structure of the isoprene unit from which terpenes such as vitamin A are constructed.

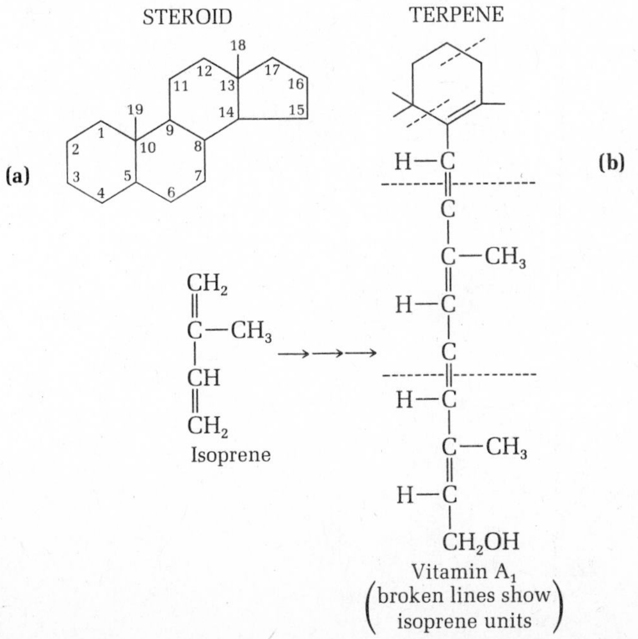

STEROID   TERPENE

Isoprene

Vitamin A₁
(broken lines show isoprene units)

Proteins

Proteins are the most structurally and functionally diverse macromolecules in the cell, accounting for two-thirds or more of the cell's total dry weight. Each type of protein is assembled by the joining together of **amino acids** in a specific linear sequence. Twenty different amino acids, each characterized by a unique type of side chain, are employed in the process (Figure 1-13). The sequential joining of amino acids involves a reaction in which the carboxyl group of one amino acid is linked to the amino group of another amino acid, splitting out a molecule of water and forming a covalent **peptide bond** (Figure 1-14). Two amino acids joined in this way form a *dipeptide*, three a *tripeptide*, and many amino acids a **polypeptide**. A typical **protein** consists of one or more polypeptide chains, each containing from a few dozen to several hundred amino acids.

Since the 20 different amino acids can be incorporated in any combination and sequence, a vast number of different kinds of proteins can be formed. A polypeptide chain 200 amino acids long, for example, could

FIGURE 1-13 Structures of the 20 amino acids commonly occurring in natural proteins. The amino acids are divided into four groups whose members have similar side chains. For each amino acid the side chain (R) is shown in color.

NONPOLAR (HYDROPHOBIC) AMINO ACIDS

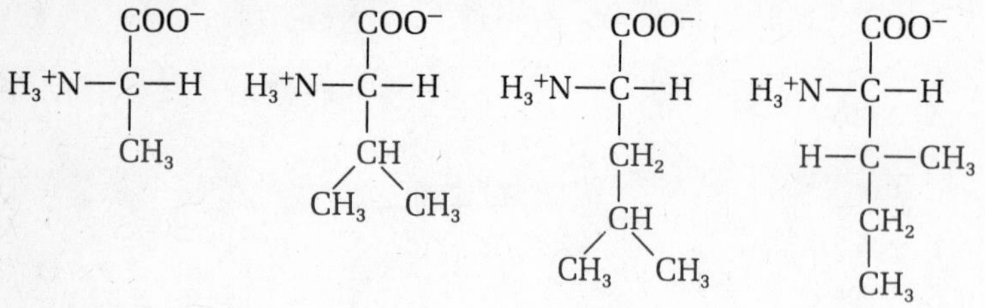

Alanine (Ala) Valine (Val) Leucine (Leu) Isoleucine (Ile)

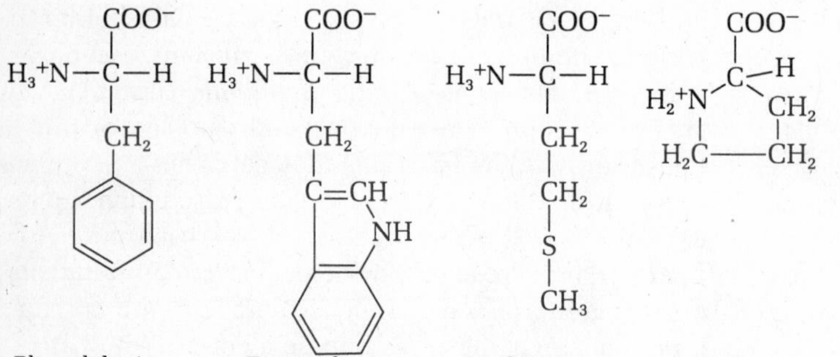

Phenylalanine (Phe) Tryptophan (Trp) Methionine (Met) Proline (Pro)

POLAR UNCHARGED AMINO ACIDS

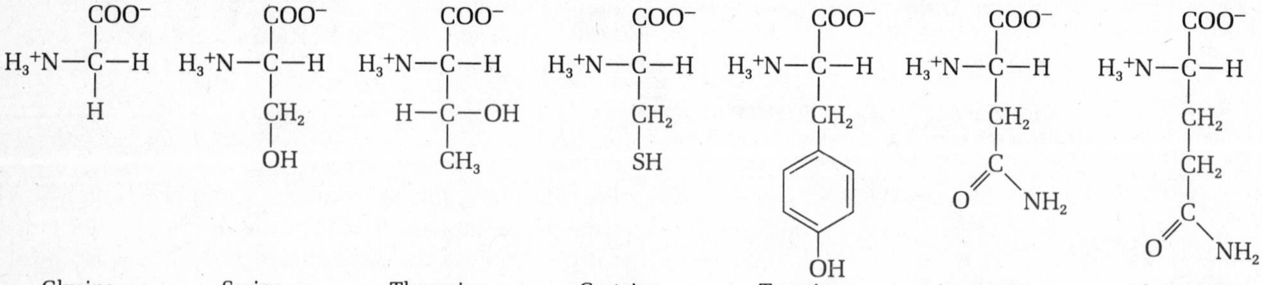

Glycine (Gly) Serine (Ser) Threonine (Thr) Cysteine (Cys) Tyrosine (Tyr) Asparagine (Asn) Glutamine (Gln)

NEGATIVELY CHARGED AMINO ACIDS   POSITIVELY CHARGED AMINO ACIDS

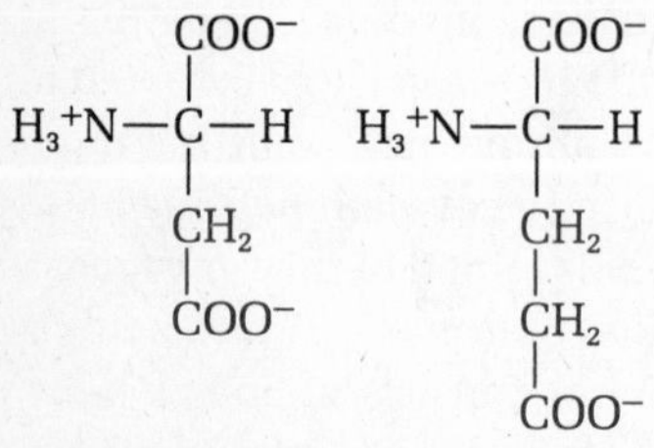

Aspartic acid (Asp) Glutamic acid (Glu) Lysine (Lys) Arginine (Arg) Histidine (His)

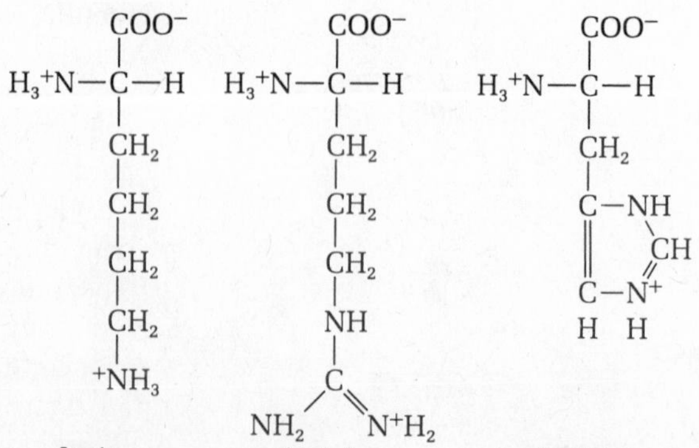

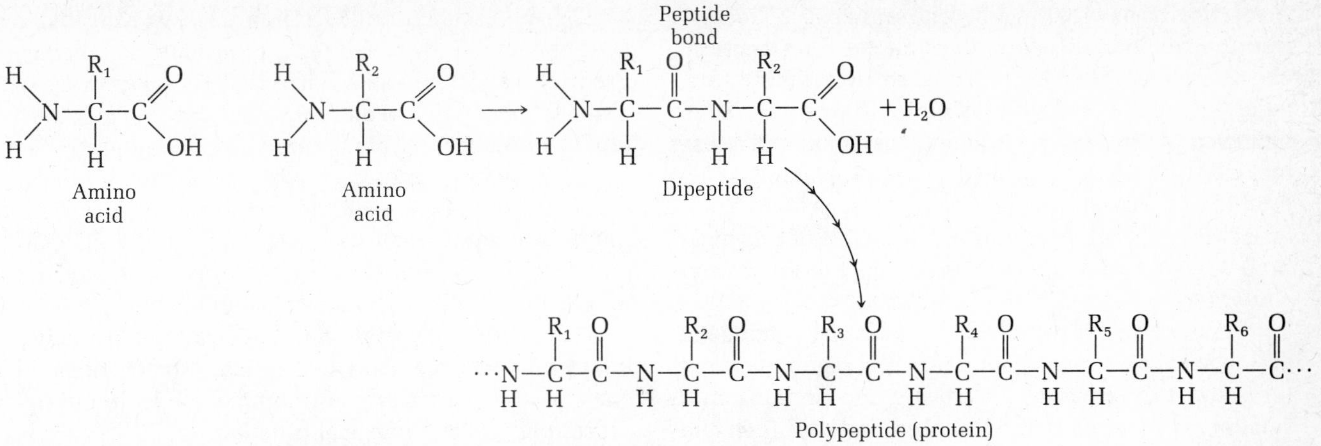

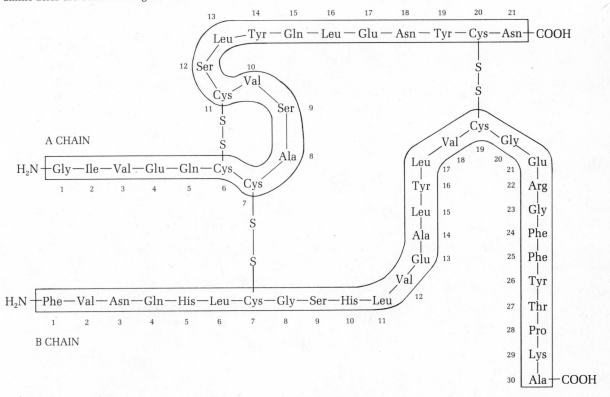

FIGURE 1-14 Diagrammatic representation of peptide bond formation between amino acids. The actual process used by cells to form peptide bonds is much more complex, owing to the need both for an input of energy to drive the reaction, and an input of information to specify the linear order of the amino acids being joined.

FIGURE 1-15 Complete primary structure of bovine insulin. Differences in the primary structure of insulin from several organisms are summarized in the lower box. Abbreviations used to refer to amino acids are defined in Figure 1-13.

	A Chain				B Chain		
Position:	4	8	9	10	3	29	30
Bovine (shown above)	Glu	Ala	Ser	Val	Asn	Lys	Ala
Human	—	Thr	—	Ile	—	—	Thr
Sheep	—	—	Gly	—	—	—	—
Dog	—	Thr	—	Ile	—	—	—
Rat₁	Asp	Thr	—	Ile	Lys	—	Ser
Rat₂	Asp	Thr	—	Ile	Lys	Met	Ser

theoretically have 20^{200} different sequences. This extraordinary flexibility is responsible for the structural and functional diversity manifested by protein molecules, and is the principal basis for their central role in biological systems. This tremendous complexity also makes it difficult to describe the structure of individual proteins. In order to simplify the situation protein structure is usually subdivided into four levels of organization, designated *primary*, *secondary*, *tertiary*, and *quaternary*.

Primary structure refers to the linear sequence of amino acids comprising a protein. This sequence is determined experimentally by degrading proteins into smaller fragments (peptides) and determining the amino acid sequence of each fragment. Insulin, a pro-

tein hormone containing 51 amino acids, was the first to be sequenced in this manner, a milestone achieved by Fredrick Sanger in 1956. Since that time hundreds of other proteins have been sequenced, leading to important generalizations concerning the relationship between protein structure and biological function. For example, it is now known that each kind of protein exhibits its own characteristic sequence of amino acids, but the same protein taken from different organisms may exhibit a slightly different sequence (Figure 1-15, p. 17). This finding reveals that though a protein's activities ultimately depend upon its primary structure, small changes in sequence can be accommodated without significant effects on biological function.

The fact that the primary structure of a protein mol-

FIGURE 1-16 The pleated-sheet or β-configuration type of protein secondary structure. Polypeptide chains may run either in the same (parallel) or opposite (antiparallel) direction. A three-dimensional representation is given for the parallel configuration. Some of the atoms below the plane have been omitted for clarity.

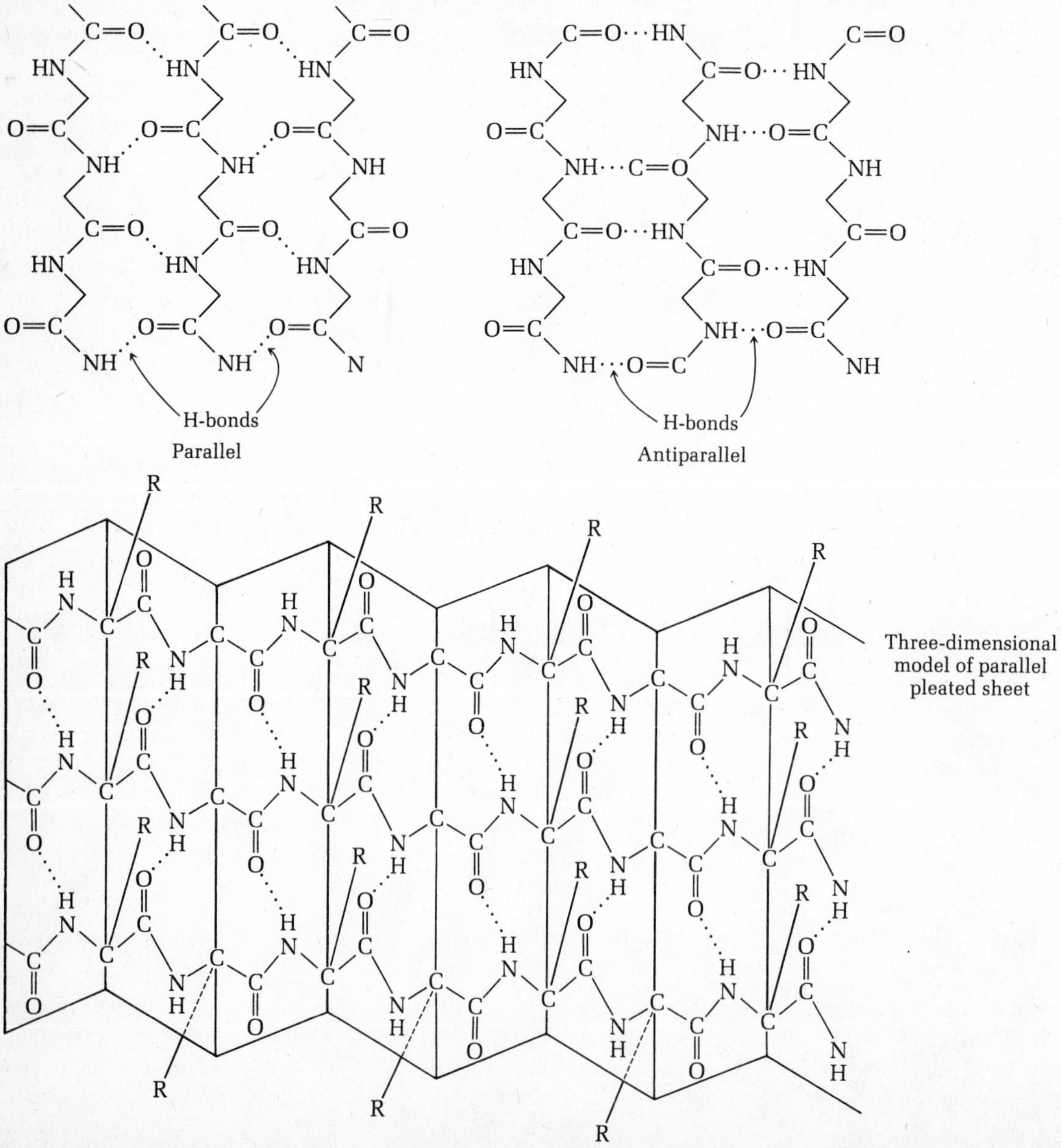

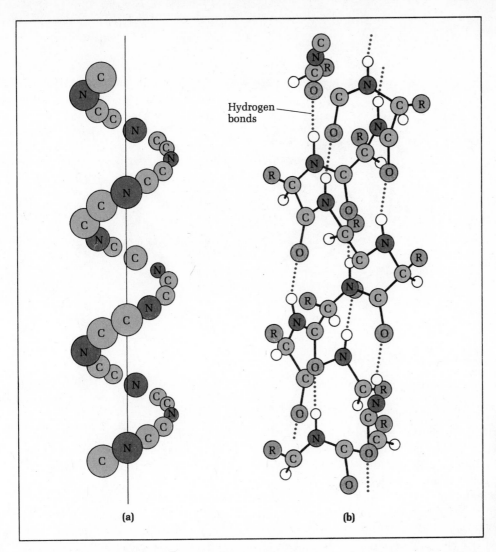

Hydrogen
bonds

(a)   (b)

FIGURE 1-17 The α-helical form of protein secondary structure. (a) Diagrammatic representation of the position of the backbone atoms. (b) Ball-and-stick model of the α-helix. Note that the C=O and N—H groups form the hydrogen bonds that hold the structure together and that the amino acid side chains (R) project outward from the helix.

ecule dictates its functional properties does not necessarily mean that proteins exist as simple extended chains of amino acids. Interactions among amino acid groups result in the folding of protein molecules into more complex structures. This folding, which is the net result of secondary, tertiary, and quaternary structure, is required in order for proteins to exhibit their normal biological activities.

Secondary structure is the result of hydrogen bonding between the —C=O group of one peptide bond and the —NH group of another. On the basis of X-ray diffraction studies, the chemist Linus Pauling has determined that proteins exhibit two principal types of secondary structure. The first is the **pleated sheet** or **β-configuration,** exhibited by insoluble "fibrous" proteins such as silk and keratin. In this structure two separate polypeptide chains (or two different regions of the same polypeptide chain) are held together by hydrogen bonds to form an extended, sheetlike structure (Figure 1-16). In the other type of secondary structure, which occurs in the water-soluble "globular" proteins that predominate in most cells, the hydrogen bonds cause

the polypeptide chain to coil into a spiral structure known as an **alpha (α)-helix** (Figure 1-17). Globular proteins usually contain several regions of α-helix intermixed with nonhelical regions. In some proteins over 90 percent of the chain is helical, while in others little if any helix occurs. Nonhelical regions are often imposed by the presence of certain amino acids (especially proline) that prevent the helix from forming. Another reason for the presence of nonhelical areas is the existence of forces involved in generating tertiary and quaternary structure, which may produce configurations more stable than the α-helix.

Tertiary structure refers to the folding of polypeptide chains caused by the formation of bonds or weak associations between amino acid side chains located within a single polypeptide chain. At least four different types of side-chain interactions are involved: **covalent** bonds, **ionic** interactions, **hydrogen** bonds, and **hydrophobic** interactions. Examples are the formation of covalent disulfide bonds between the sulfhydryl groups of cysteine residues; ionic attractions between positively charged side chains (present in lysine, arginine, and

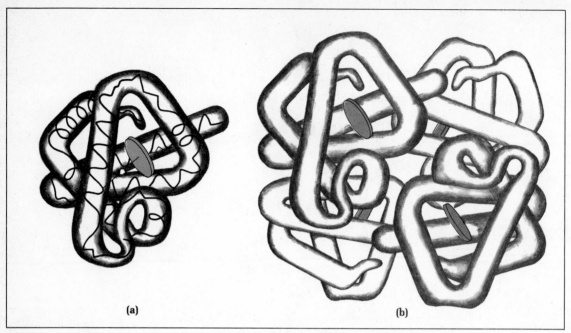

FIGURE 1-18 The three-dimensional structure of the protein hemoglobin as deduced from X-ray crystallography. (a) Structure of a single polypeptide chain. The disk-shaped structure represents the heme prosthetic group. (b) The quaternary structure of hemoglobin, showing how the protein is made of four polypeptide chain subunits. Differences in shading are used to indicate that two slightly different types of subunits are involved.

histidine) and negatively charged side chains (present in glutamic and aspartic acids); hydrogen bonding between various side chains, including tyrosine-histidine and serine-aspartic acid combinations; and hydrophobic interactions between nonpolar amino acids, such as valine, leucine, isoleucine, tryptophan, and phenylalanine. In combination, these various kinds of bonds cause a twisting and folding of the polypeptide chain into a specific three-dimensional shape (Figure 1-18).

Quaternary structure represents the most complex level of protein organization. The bonds are of the same type as those occurring at the tertiary level of organization, except that they occur between amino acids located in different polypeptide chains and therefore function to hold multiple chains or **subunits** together. Although not all proteins are composed of multiple subunits, the majority with molecular weights over 50,000 exhibit such organization. The individual subunits may be identical to one another or of different types. Typical examples of proteins exhibiting quaternary structure are hemoglobin, containing four subunits of two different types (see Figure 1-18), and bacterial RNA polymerase, an enzyme comprised of five polypeptide chains.

The overall shape of protein molecules thus involves interactions at several levels of organization (Figure 1-19). The final three-dimensional shape generated by these forces is referred to as protein **conformation.** With highly purified proteins, X-ray crystallography can be used to obtain rather detailed pictures of the native conformation (see Figure 1-18). Maintenance of

this proper conformation is essential for the biological activity of most proteins. Agents that disrupt any of the bonds involved in secondary, tertiary, or quaternary structure may alter the proper conformation and thereby cause a loss of functional activity. Such disruption of a protein's normal conformation is referred to as **denaturation.** Conditions or substances that interfere with hydrogen bonds (e.g., urea), ionic bonds (e.g., salt), or hydrophobic bonds (e.g., organic solvents) can all denature proteins. Likewise moderate increases in temperature, which disrupt all these weak kinds of interactions, also cause protein denaturation.

The question of how proteins normally acquire their proper conformation has been extensively investigated by the biochemist Christian Anfinsen. Ribonuclease, a protein that catalyzes the degradation of RNA, was utilized as a model in these studies because its conformation is maintained by four disulfide bridges that can be experimentally altered, and it exhibits a biological activity that can be readily measured. Anfinsen first demonstrated that treating ribonuclease with urea (which breaks hydrogen bonds) and mercaptoethanol (which disrupts disulfide bonds) causes a loss of normal activity. Upon removal of these denaturing agents, however, the ribonuclease spontaneously regains both its proper conformation and normal biological activity (Figure 1-20). Such observations suggest that under appropriate conditions, proper protein conformation is achieved spontaneously because it is the most stable configuration.

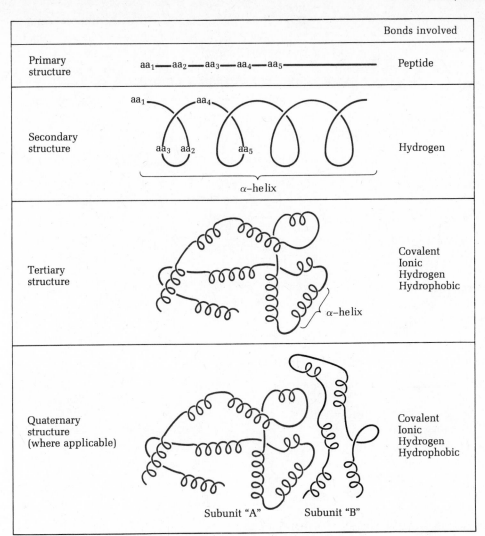

		Bonds involved
Primary structure	aa₁—aa₂—aa₃—aa₄—aa₅	Peptide
Secondary structure	α-helix	Hydrogen
Tertiary structure	α-helix	Covalent Ionic Hydrogen Hydrophobic
Quaternary structure (where applicable)	Subunit "A" Subunit "B"	Covalent Ionic Hydrogen Hydrophobic

FIGURE 1-19 Diagrammatic representation of the four levels of forces involved in generating protein conformation.

FIGURE 1-20 Summary of the Anfinsen experiment establishing that denatured proteins may refold spontaneously back into their active conformation.

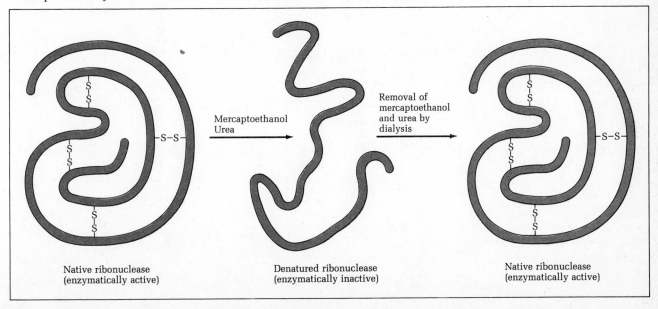

Native ribonuclease (enzymatically active) → Mercaptoethanol Urea → Denatured ribonuclease (enzymatically inactive) → Removal of mercaptoethanol and urea by dialysis → Native ribonuclease (enzymatically active)

TABLE 1-3
Functions of protein molecules

Function	Examples
1. Structural materials	Collagen (tendons, connective tissue)
	Keratin (skin, hair, nails)
2. Hormones	Insulin
	Gonadotropins
3. Contractile proteins	Actin
	Myosin
4. Transport proteins	Hemoglobin
	Transferrin
5. Storage proteins	Casein
	Ferritin
6. Osmotic proteins	Serum albumin
7. Respiratory proteins	Cytochromes
8. Antibodies	Immunoglobulins
9. Toxins	Diphtheria toxin
	Botulinum toxin
10. Enzymes	Oxidoreductases
	Transferases
	Hydrolases ⎫ To be defined
	Lyases ⎬ in Chapter 2
	Isomerases
	Ligases ⎭

The vast potential for diversity in protein structure and conformation allows proteins to carry out a wide spectrum of cellular functions (Table 1-3). Among these functions enzymatic activity, which refers to the ability to catalyze chemical reactions, is the most widespread and crucial. Without enzymes, life as we know it would be impossible. The properties of enzymes are so basic to our understanding of how cells operate that Chapter 2 will be devoted entirely to their study.

Nucleic Acids

Nucleic acids serve as vehicles for the storage and transmission of genetic information. This critically impor-

FIGURE 1-21 Structure of a typical nucleotide, adenosine monophosphate (AMP).

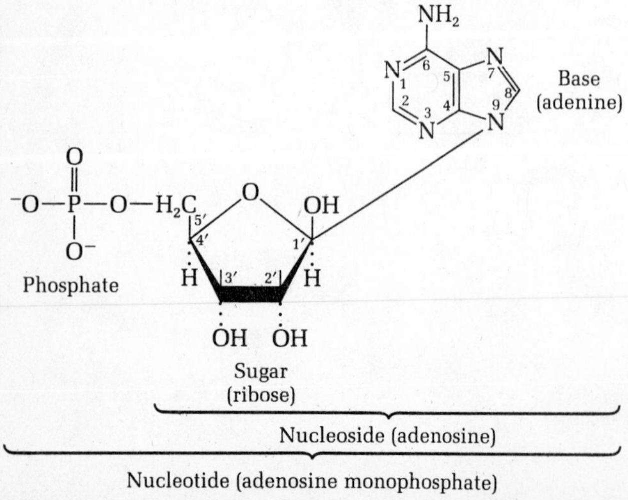

tant information is encoded within the linear sequence of monomer building blocks, or **nucleotides,** strung together to form a nucleic acid molecule. A nucleotide is composed of three components: a **nitrogenous base,** a **pentose sugar,** and a **phosphate group** (Figure 1-21). Nucleotides contain from one to three phosphate

FIGURE 1-22 Structures of the common bases and sugars found in nucleic acids. Note that the numbering systems for the purine and pyrimidine rings run in opposite directions.

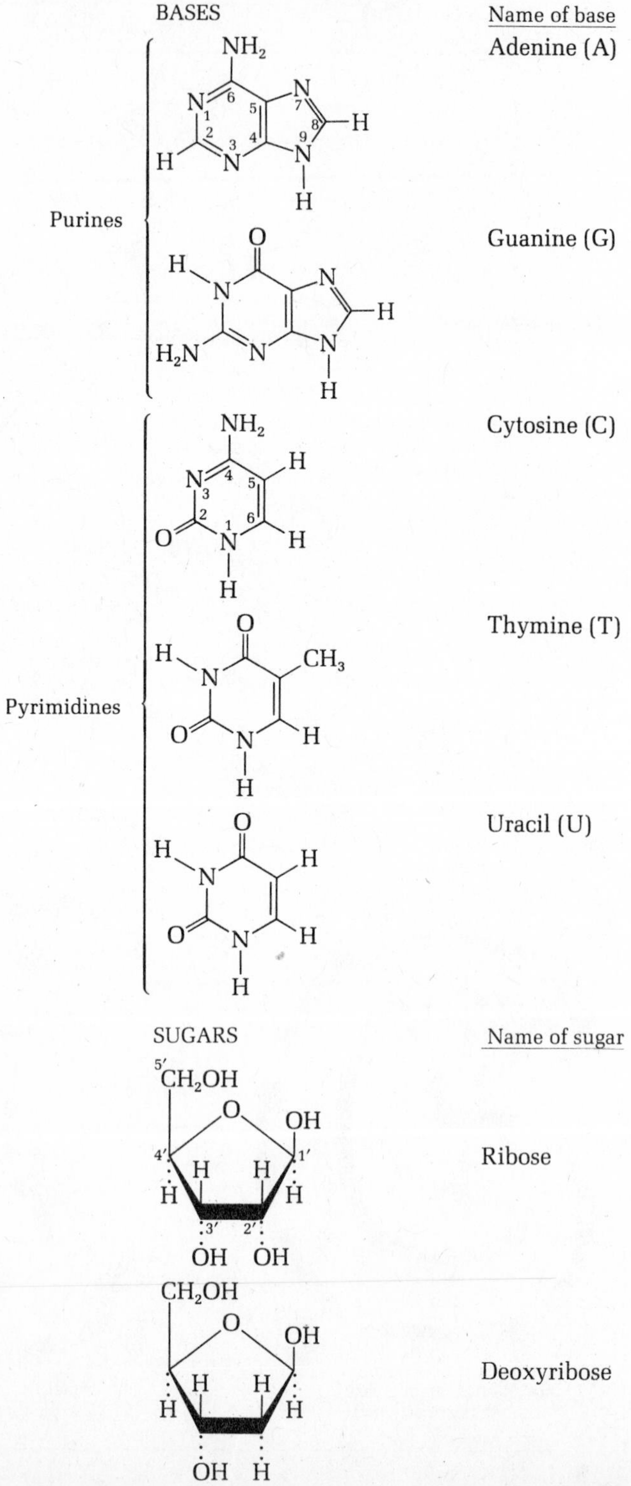

groups, and are thus designated as mono-, di-, or tri-phosphates. When the phosphate group is absent, the term **nucleoside** is employed.

Cells contain two major classes of nucleic acids, **deoxyribonucleic acid (DNA)** and **ribonucleic acid (RNA),** which differ in both the sugar and nitrogenous bases employed in their construction. DNA contains the five-carbon sugar **deoxyribose,** while RNA utilizes **ribose** instead. The only difference between these two sugars is the absence of the oxygen on the 2'-carbon atom of the deoxyribose molecule (Figure 1-22).

The nitrogenous bases present in DNA and RNA are also illustrated in Figure 1-22. **Adenine** and **guanine** are derivatives of the two-membered **purine** ring, while **cytosine, thymine,** and **uracil** are derived from the single-ringed **pyrimidine.** The capital letters A, G, C, T, and U are commonly used to refer to adenine, guanine, cyto-sine, thymine, and uracil, respectively. DNA and RNA each contain four of the five bases. A, G, and C are present in both types of nucleic acid, but the fourth base is T in DNA and U in RNA. Chemically modified forms of these bases, as well as other more unusual bases, also occur occasionally in DNA and RNA.

The formation of a nucleic acid from nucleotide building blocks involves the joining of the phosphate group on the 5'-carbon of one sugar to the hydroxyl group on the 3'-carbon of another sugar, forming a **3',5'-phosphodiester linkage** (Figure 1-23). Successive addition of nucleotides yields an extended polynucleotide chain containing a repeated sugar-phosphate backbone with the bases projecting outward. The chains produced in this way may be thousands of bases long for RNA, and up to hundreds of thousands or even millions of bases long for DNA. The final three-dimensional

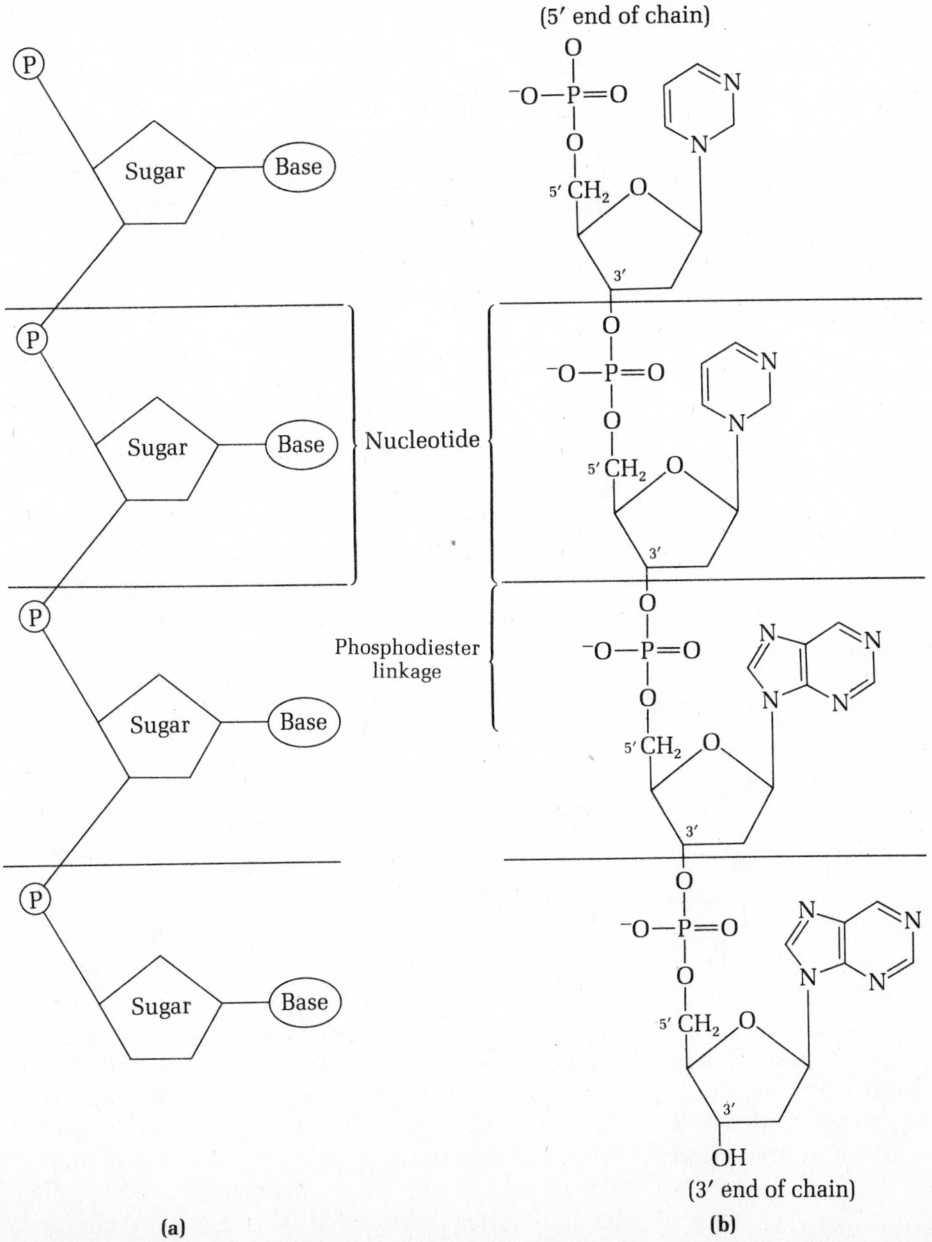

FIGURE 1-23 General structure of nucleic acids. (a) Skeleton diagram showing the sugar-phosphate backbone with projecting bases. (b) A more detailed chemical formula, showing how adjacent nucleotides are joined by a phosphate group linking the 5'-carbon of one sugar with the 3'-carbon of the next sugar.

structure of DNA and RNA molecules involves interactions between nitrogenous bases both within and between polynucleotide chains. However, such interactions are so intimately associated with the process of information transfer that we will delay further consideration of this topic until Chapters 10 through 12, when cellular information flow will be discussed in detail.

INVESTIGATING THE ORIGIN OF LIFE

We have now seen that cells are composed of several different kinds of organic molecules that interact with one another to generate both the architectural and functional features of this basic unit of life. One of the most profound questions one can pose about this arrangement is how did it originate? The question of the ultimate origin of living cells is not an easy one to deal with experimentally. Much of what we know about present-day cells is derived from the investigative approach: Cells are broken open, specific organelles are isolated for study, and conclusions are drawn concerning the functioning of these organelles in the intact cell. But we do not have examples of primitive cells to work with, so this type of approach cannot be applied to studies on the origin of life. However, one may use information derived from contemporary experiments to generate models for what could theoretically have occurred under primordial conditions. This approach cannot tell us with certainty what did take place, but it does provide us with significant insights into the underlying organization and possible evolution of biological systems.

Spontaneous Generation

Up until the mid-nineteenth century, it was widely believed that cells routinely arise spontaneously from the self-organization of inanimate matter. The main "experimental" basis for this concept was the observation that nutrient solutions left standing around are soon found to be teeming with unicellular organisms. However, in 1861 the French microbiologist Louis Pasteur dealt this theory a mortal blow by showing that if the air and nutrient broth in a culture flask are carefully sterilized to destroy all microorganisms, no living cells appear in the broth as long as the neck of the flask is constricted. As soon as the neck of the flask is opened to permit entry of unsterile air, growth of microorganisms ensues. This clearly suggested to Pasteur that living cells arise only from previously existing cells. A similar conclusion was reached by cell biologists who examined developing organisms under the microscope and concluded that new cells arise from the division of pre-existing cells. In 1855 the German pathologist Rudolph Virchow summarized these observations in his famous statement *omnis cellula e cellula* (all cells come from cells).

The basic validity of the above conclusion is now unquestioned, but it is ironic that the development of these ideas may have hindered our understanding of the evolutionary origins of life. For Pasteur's observation that spontaneous self-organization of inanimate matter into living cells fails to occur under defined laboratory conditions does not necessarily prove that it could not have occurred under primitive earth conditions. The biggest obstacle to creating a scientific explanation for how living cells might have arisen on earth is the need to invoke the occurrence of such a self-organization step. Pasteur's experiments made this concept fairly disreputable, and in the absence of an experimental demonstration that spontaneous self-organization occurs, models on the origin of life based on such events could be little more than metaphysical. In the 1960s, however, an era of experimentation evolved that demonstrated that **self-assembly** of macromolecules into supramolecular functional entities can be observed in the laboratory. Self-organization is now appreciated to be a biologically important phenomenon occurring routinely in nature. Among the many examples of self-organization known to take place in nature are the spontaneous assembly of protein subunits into microtubules and flagella, RNA and protein subunits into ribosomes, and nucleic acids and proteins into viruses.

Now that the concept of spontaneous self-assembly has a firm experimental basis, it is possible to construct plausible models for the evolutionary origin of living cells. Hence we have come almost full circle, from the pre-nineteenth-century view that cells arise spontaneously through self-organization of inanimate matter, through Pasteur's refutation of this concept, and back to spontaneous self-organization as an experimentally viable notion. If we are to apply this concept to the question of how cells originated on earth, however, some information regarding conditions on the primitive earth is obviously essential.

The Primitive Earth

The disciplines of astronomy, astrophysics, nuclear physics, geology, geochemistry, and geophysics have together accumulated an enormous mass of data bearing on the age of the earth and the conditions prevailing at the time of its origin. These independent approaches, whose results buttress one another, are in general agreement that the earth is about 4.5 billion years old. During its first few billion years, the earth is thought to have been largely submerged under a cover of water that boiled and steamed beneath an atmosphere of toxic gases. Fossil imprints of early unicellular organisms have been discovered in sedimentary rocks estimated to be 3.5 billion years old (Figure 1-24). Hence formation of the first cells must have occurred sometime during this first billion years under conditions that at first glance do

10 μm

FIGURE 1-24 Photomicrograph showing the fossilized remains of a cluster of unicellular plankton estimated to be 1.4 billion years old. Courtesy of G. Vidal.

not appear to be particularly hospitable to the formation of living cells.

As it turns out, it is easier to envision the origin of cells in such a context than it would be under current earth conditions. The principal approach used to study this problem has been to analyze the behavior of various molecules under laboratory conditions designed to simulate the primitive earth. In the process of conceiving such experiments it has become apparent that three phenomena must be accounted for in any model purporting to explain the origin of the first living cells: first, the formation of organic molecules; second, the performance of appropriate biological functions by these molecules; and third, the organization of these functioning molecules into the first cells. This last step—the transition from the inanimate to the animate, from nonliving to living—is difficult to investigate experimentally. In the following sections we will see what insights the experimental approach has provided concerning each of these events.

Formation of Organic Molecules

The first insight into the question of how organic molecules could have formed on the prebiotic earth was pro-

vided in the late 1920s by Aleksandr Oparin, a Russian biochemist to whom we owe many of our current ideas on the evolutionary origin of cells. Oparin suggested on purely theoretical grounds that the chemical and physical conditions of the primitive atmosphere were favorable for the spontaneous formation of simple organic building blocks. According to his theory, gases in the primitive atmosphere reacted to form small organic molecules in the presence of an energy source such as lightning, heat, or radiant energy from the sun.

Evidence in support of this theory was not available until 1953, when Stanley Miller performed a classic experiment on the formation of organic molecules under simulated primitive earth conditions. Because organic compounds are unstable in the presence of molecular oxygen, it has long been thought that the primitive earth contained a reducing (oxygen-free) atmosphere. Miller attempted to recreate this atmosphere by mixing together methane (CH_4), ammonia (NH_3), hydrogen, and water vapor. When he subjected this mixture to electric sparking (comparable to the lightning that occurred in the primitive atmosphere), a variety of organic acids were formed. Among these were significant quantities of physiologically important compounds, including amino acids and fatty acids.

These experiments seemed to support Oparin's theory that organic molecules arose spontaneously on the primitive earth. However, the results remained somewhat controversial because the exact composition of the primitive earth atmosphere is unknown. Many scientists do not believe that reduced gases such as methane and ammonia were present in high concentrations. In fact it has been suggested that a more oxidizing atmosphere prevailed, such as a mixture of nitrogen, hydrogen, carbon monoxide (CO), carbon dioxide (CO_2), and water vapor. Subsequent experiments of the Miller type have therefore been carried out on this, as well as many other, gas mixtures, using a variety of energy sources; in all cases biologically important, simple organic molecules have been formed, including virtually all the small molecules utilized by living cells (Table 1-4).

All the chemical building blocks needed for the evolution of living systems could thus have formed spontaneously on the primitive earth. But simple building blocks are not enough. Cells are constructed predominantly from polymeric macromolecules, which

TABLE 1-4
Results of some typical primitive earth simulation experiments

Gas Mixture	Energy Source	Products
CH_4, NH_3, H_2, N_2O	Electric discharge	Amino acids, urea, organic acids, fatty acids
CO_2, CO, N_2, NH_3, H_2, H_2O	Electric discharge	Amino acids
CH_4, CO_2, CO, NH_3, N_2, H_2, H_2O	X-rays	Amino acids
HCHO, CH_3CHO, CH_2OH—CHOH—CHO, $Ca(OH)_2$	Heat	Sugars
CH_4, NH_3, H_2O	β-rays	Adenine
CH_4, NH_3, H_2O	UV-radiation	Amino acids, fatty acids

are synthesized by the joining together of these monomeric building blocks. The question therefore arises as to how large macromolecules formed on the primitive earth. **Condensation reactions,** in which two smaller molecules are joined together with the splitting out of water, are not energetically favorable. Consider, for example, the formation of protein molecules, which involves successive condensation reactions between individual amino acids:

Amino acid$_1$ + amino acid$_2$ $\rightleftharpoons$
$$\text{amino acid}_1 - \text{amino acid}_2 + H_2O$$

The equilibrium for this reaction lies far to the left, indicating that the joining together of amino acids will not yield much product. This situation is typical for condensation reactions. In essence, the reverse reaction involving the hydrolytic breakdown of macromolecules is energetically favored over the forward reaction by which the larger molecules are formed.

There are two potential ways of overcoming this difficulty. One is to carry out the reaction under anhydrous conditions, thus removing the water required for the reverse hydrolysis reaction. Sidney Fox has shown that under such conditions, amino acids are readily polymerized into proteinlike molecules by moderate heating. The second alternative is to use a **condensing agent,** a compound whose free energy can be used to shift the equilibrium in the direction of macromolecule formation. *Polyphosphate*, a linear chain of phosphate groups covalently joined together, is an example of a condensing agent formed under simulated primitive earth conditions. The energy released during the breakdown of polyphosphate has been used to facilitate the formation of nucleic acid- and proteinlike polymers. It is interesting to note that present-day cells utilize nucleoside triphosphates, such as ATP (see Figure 2-22), for this purpose. Such triphosphates contain within their structures a "polyphosphate" sequence that is three phosphate groups long.

Intrinsic Molecular Functions

We have now seen how simple organic molecules, as well as more complex macromolecules, could have formed spontaneously under primitive earth conditions. Our next concern is the question of how these molecules acquired the functions required for the development of living cells. Two functions are particularly important in this regard. One is the ability to enhance the rates of chemical reactions, a property referred to as **catalysis.** The other is the ability to *store, reproduce, and transmit* **information.** Recent experiments suggest that both these functions are intrinsic properties of macromolecules synthesized under simulated primitive earth conditions, and that no special process needs to be invoked to explain how they arose.

Catalysis It is known from organic chemistry that simple organic acids and bases catalyze reactions involving the uptake or loss of protons. It follows that simple amino acid polymers containing acidic or basic side chains could have readily served as simple proton catalysts on the primitive earth. As evolution progressed, selection for amino acid polymers with enhanced catalytic activity and reaction specificity could have gradually led to the formation of enzymes as we know them today.

The likelihood of this progression of events has been experimentally investigated by Sidney Fox and his associates. As a model system they have studied the properties of proteinlike polymers, or **proteinoids,** synthesized in the laboratory by heating mixtures of amino acids in the absence of water. Proteinoids have molecular weights of many thousands and contain all the commonly occurring amino acids. Instead of a random mixture of amino acid polymers, a relatively small number of proteinoids predominate. Their amino acid compositions resemble those of natural proteins and the distribution of amino acid residues follows specific patterns, suggesting that the polymerization reaction spontaneously generates a certain degree of order.

The most interesting property of proteinoids, however, is their ability to enhance the rate of specific chemical reactions. Among the various kinds of chemical reactions found to be specifically catalyzed by proteinoids are hydrolysis, decarboxylation, oxidation-reduction, and amination. These catalytic effects exhibit many of the properties typical of present-day enzymatic catalysis. Furthermore, the catalytic activity of proteinoids increases in rough proportion to their size, perhaps explaining why enzymes have evolved into such large macromolecules.

Information Transfer Nucleic acids are the carriers of hereditary information in present-day cells. The instructions for guiding most cellular activities are stored in the base sequence of DNA molecules, which is reproduced every time a cell divides. But the reproduction of DNA base sequence information, as well as its transfer to RNA, requires the participation of highly specific protein catalysts. The question therefore arises as to how the ability to transfer base sequence information between nucleic acids arose on the primitive earth prior to the advent of such specific catalysts.

Leslie Orgel and his colleagues have carried out experiments suggesting that enzymatic catalysts are not an absolute requirement for the transfer of base sequence information. If one mixes together the building blocks from which RNA is constructed and adds a condensing agent to drive their polymerization, RNA chains are synthesized containing a random mixture of the four bases (A, G, C, and U). But if poly-A (an RNA molecule containing only the base A) is first added to the system, RNA chains containing the base U will be

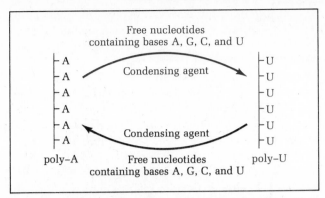

Free nucleotides
containing bases A, G, C, and U

poly-A

Condensing agent

Condensing agent

poly-U

Free nucleotides
containing bases A, G, C, and U

FIGURE 1-25 Diagram showing how the ability of nucleic acids to transfer information in the absence of enzymatic catalysts could be exploited to allow nucleic acid self-replication under primitive earth conditions. See text for experimental details.

selectively synthesized. The resulting poly-U, in turn, promotes the preferential synthesis of poly-A. In this way a crude replicating system can be established, where the initial presence of poly-A ultimately promotes the synthesis of more poly-A (Figure 1-25). In a similar fashion, the presence of poly-C causes the preferential formation of RNA chains containing the base G, and vice versa. Because the above phenomenon occurs under simulated primitive earth conditions with no enzymatic catalyst present, Orgel's experiments lead to the conclusion that it is possible for nucleic acids to transmit and duplicate base sequence information in the absence of enzymatic catalysts.

Emergence of the First Cells

Having established plausible mechanisms for the synthesis of functioning macromolecules on the primitive earth, we now arrive at the crucial question of how such molecules were assembled to form the first living cells. In order for a group of molecules to work together as a functional unit, some type of boundary is needed to separate these molecules from their surroundings. At least two experimental models have been developed to show how such boundaries could have formed spontaneously on the primitive earth.

The first model system is the formation of coacervate droplets. **Coacervation** is a well-known phenomenon that occurs in concentrated solutions of large polymer molecules. In such solutions, spontaneous separation into two phases often occurs. The phase separating out consists of small droplets ranging from 1 to 100 μm in diameter. The concentration of polymer in these coacervate droplets is several orders of magnitude higher than the concentration of polymer in the surrounding solution.

Oparin and his colleagues have studied coacervates extensively as models of primitive cells. Although coacervation requires a relatively high concentration of molecules, it is possible that organic molecules on the

primitive earth became concentrated by evaporation of water in the small lagoons that formed at the peripheries of the primitive seas. Oparin has shown that coacervates are able to trap and metabolize other molecules within them. For example, coacervates made from concentrated solutions of phosphorylase, an enzyme involved in polysaccharide metabolism, will take up glucose-phosphate and convert it to starch. Accumulation of the newly synthesized starch causes the coacervate droplets to grow and divide, thus providing a useful model for how a primitive metabolizing cell might have arisen.

The coacervate model of primitive cells is not without its limitations, however. One problem is that coacervates are generally unstable, tending to break up upon standing. Furthermore the boundary of coacervate droplets is not a selective barrier, so substances concentrated within the droplet eventually leak back into the external environment. This behavior contrasts sharply with that of the plasma membrane, which selectively regulates the passage of material into and out of the cell. Yet another limitation of the coacervate experiments is that the use of biological molecules such as phosphorylase to make coacervate droplets may not necessarily be applicable to conditions on the primitive earth, when such complex molecules did not yet exist. If one wants to create a laboratory model for the formation of primitive cells, it would be more realistic to utilize molecules formed under primitive earth conditions.

Such a model system, using proteinoids synthesized by heating amino acids in the absence of water, has been worked out in Sidney Fox's laboratory. When proteinoids are dissolved in water and allowed to cool slowly without disturbance, small vesicles a few micrometers in diameter appear (Figure 1-26). In contrast to coacervates, these **microspheres** are of uniform size and shape, and are stable for long periods of time. Microspheres are bounded by a semipermeable barrier that selectively influences the passage of materials into and out of the vesicle. Microscopically this boundary bears a superficial resemblance to the plasma membrane of contemporary cells, although it lacks the lipid component present in current biological membranes.

Microspheres exhibit a number of functional attributes superficially resembling those of living cells. For example, microspheres exhibit nonrandom movements that are enhanced by the addition of high-energy compounds such as ATP. Microspheres formed from solutions of proteinoids exhibiting catalytic activity are able to carry out metabolic reactions, including the polymerization of amino acids and nucleotides. When exposed to certain ions or changes in pH, microspheres divide to form more microspheres [Figure 1-27 (*top*)]. If a solution of microspheres is allowed to sit for several weeks, new microspheres form by a budding process resembling that which occurs in yeast [see Figure 1-27 (*bottom*)]. Hence many properties we currently associate with liv-

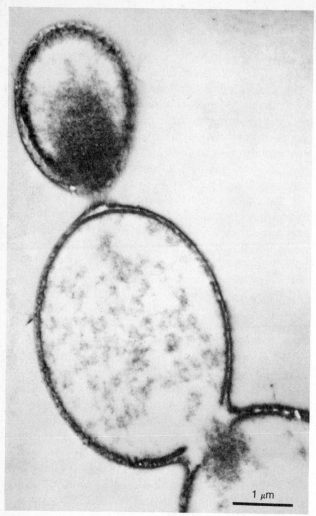

FIGURE 1-26 Electron micrograph of a proteinoid microsphere showing the double-layered outer boundary. Courtesy of S. Fox.

ing cells are in fact exhibited by these unique vesicles formed in the complete absence of biological material. It is also intriguing to note that ancient fossils resembling microspheres in structure have been detected in geological formations estimated to be billions of years old.

Although the study of microspheres has shed light on the question of how certain biological properties might have first arisen, it is a long way from a microsphere to a contemporary cell. Microspheres are no more than small vesicles containing mediocre catalysts and therefore capable of a primitive type of metabolism, growth, and division. Microspheres lack the structural and metabolic complexity of present-day cells, and have no genetic system for encoding and accurately transmitting hereditary information. Of course there is no reason for believing that the first cells arising on earth had to resemble those occurring today, so the term **protocell** has been introduced to refer to a primitive cell-like structure that might have been the evolutionary precursor of contemporary cells. Microspheres are certainly a plausible model for such protocells.

In order for protocells to evolve into cells as we know them today, a genetic system capable of accurately reproducing information is needed. Hence if a particular protocell happened, by chance, to make a new protein whose properties offered a significant advance over previously existing proteins, such a genetic system would allow the information for making this new protein to be replicated and passed on to future cell generations. In this way, gradual evolutionary improvement would be possible. In contemporary cells this genetic function is carried out by nucleic acid molecules, whose base sequences store information coding for the amino acid sequences of protein molecules. This base sequence information can be readily reproduced and transmitted from nucleic acid to nucleic acid, allowing it to be passed on from cell to cell. But how did the ability of nucleic acids to encode information for amino acid sequences arise in the first place? Since contemporary cells utilize nucleic acids to guide the synthesis of proteins, but require proteins (enzymes) for the synthesis of nucleic acids, this question is like asking which came first, the chicken or the egg? As one might imagine, there are two schools of thought on the matter.

Nucleic Acids Before Proteins The viewpoint espoused by many biologists, most notably Leslie Orgel and Francis Crick, is that the nucleic acid system arose

FIGURE 1-27 Proteinoid microspheres that exhibit some features of living cells. *(Top)* Microspheres appear to be dividing. *(Bottom)* Budding microspheres. Courtesy of S. Fox.

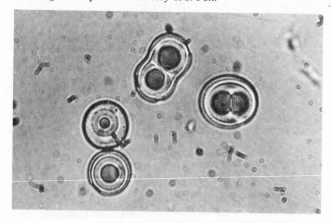

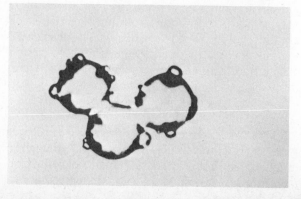

before proteins. We have already seen that macromolecules resembling nucleic acids could have formed spontaneously on the primitive earth, and that nucleic acids are capable of transmitting base sequence information and therefore replicating themselves in the absence of enzymatic proteins (see Figure 1-25). Let us now suppose that one such spontaneously arising, self-replicating nucleic acid folded up into a particle that in turn catalyzed the condensation of amino acids into proteins. Since the particles involved in protein synthesis in contemporary cells are called ribosomes, let us call this particle a "protoribosome." Although protoribosomes, unlike present-day ribosomes, would contain no protein, the notion that they might have some catalytic properties is not inconceivable. As we shall see later in the text, recent studies have revealed that RNA molecules do carry out certain catalytic functions in the absence of proteins (page 424).

We can also postulate the spontaneous formation of a second class of self-replicating nucleic acids, in this case consisting of molecules capable of specifically binding to amino acids. Let us call these molecules "transfer nucleic acids," and assume that different transfer nucleic acids arose for each of the commonly occurring amino acids. Finally, all we need for the creation of a primitive genetic system for guiding protein synthesis is to utilize other spontaneously arising nucleic acids as crude "messages." The operation of such a primitive system is illustrated in Figure 1-28, where it is seen that the base sequence in the "message" nucleic acids is recognized by base sequences in the "transfer nucleic acids," causing their attached amino acids to be lined up in a specific order prior to being joined together by a condensing agent. The net result is that the sequence of amino acid residues in the protein being synthesized is determined by the base sequence of the nucleic acid "message," which in turn is capable of self-replication. The information coding for any particular protein could therefore be passed on from generation to generation. If a randomly arising nucleic acid happened by chance to code for a protein that facilitated the operation of such a system, it would enhance the system's chances for survival and would therefore be selected for as evolution proceeded. In this way the various proteins and enzymes utilized by contemporary cells could have gradually evolved, converting the protocell with its primitive genetic system into a cell as we know it today.

Although the above scenario seems to be at least theoretically feasible, it must be emphasized that there is little experimental support for such speculations. Nucleic acid systems capable of guiding protein synthesis have not been created in the laboratory under simulated primitive earth conditions, nor have model protocells capable of metabolic growth and division been created in the laboratory starting solely from nucleic acidlike polymers. Although it has been suggested that a primitive metabolism might have been based on the use

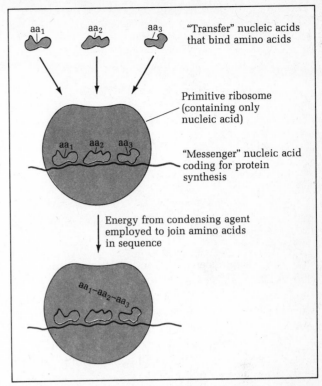

FIGURE 1-28 Hypothetical model of a primitive protein-synthesizing system in which the information contained in nucleic acids is employed to specify the linear sequence of amino acid incorporation, without the participation of specific enzymatic catalysts.

of nucleic acids as catalysts, this has not been demonstrated and it is hard to imagine a primitive protocell without significant metabolic activity. For these reasons there are drawbacks to the idea that cells arose using nucleic acids prior to proteins.

Proteins Before Nucleic Acids The behavior of laboratory-produced proteinoid microspheres has led Sidney Fox to advocate the idea that proteins evolved before nucleic acids. According to Fox, protocells contained proteinoids with catalytic activity capable of joining amino acids into proteinlike polymers and nucleotides into nucleic acidlike polymers. Like microspheres created in the laboratory, these protocells would also be capable of growth and division. Because proteinoids with certain amino acid sequences tend to form preferentially under experimental conditions, one would predict that proteinoids that were synthesized as protocells grew and divided would tend to resemble the previously existing proteinoids. Once they arose, proteinoid-based protocells would dominate evolution because of their metabolic and physical advantages over noncompartmentalized organic molecules. According to this model, the original unit of biological replication was thus the intact protocell rather than individual nucleic acids. Only later did such protocells evolve a genetic system in which the information coding for the

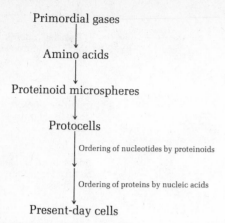

Primordial gases

↓

Amino acids

↓

Proteinoid microspheres

↓

Protocells

↓ *Ordering of nucleotides by proteinoids*

↓ *Ordering of proteins by nucleic acids*

Present-day cells

FIGURE 1-29 A flow diagram for the evolution of cells based on the idea that proteins came first.

amino acid sequences of protein molecules came to be encoded in the base sequence of nucleic acids.

A hint as to how this step might have occurred is provided by Fox's discovery that certain types of proteinoids tend to selectively bind to specific nucleic acids. For example, proteinoids enriched in the amino acid lysine bind to poly-C much better than to poly-G, while the opposite holds for proteinoids enriched in the amino acid arginine. This ability of proteinoids to bind specific nucleotide base sequences supports the theory that primitive proteins might have specifically facilitated the synthesis of nucleic acids with particular base sequences. Later in evolution the process could have been reversed, with the nucleic acids generated in this way being used to direct the synthesis of the proteins that had originally coded for their particular base sequences. This scenario fits nicely into a flowsheet with no awkward gaps (Figure 1-29) and is consistent with laboratory experimentation in many of its steps.

In spite of these advantages, the protein-directed synthesis of specific nucleic acids postulated above is yet to be convincingly demonstrated in the laboratory. Even if it were, one could not state with certainty that a similar event occurred on the primordial earth. Although such experimentation has deepened our understanding of the inherent properties of biological molecules, no one has yet to cook up a "primitive soup" and let it simmer away until living cells appear. In fact, when one considers the myriad of conditions that would have to be met in order to be able to generate life in this fashion, the origin of life appears to be quite miraculous.

WHAT ARE VIRUSES?

Closely related to any discussion of the evolutionary origin of cells is a consideration of the nature and origins of viruses. The discovery of viruses in the late nineteenth century was initially based on their ability to cause infectious diseases. It was already known that bacteria are responsible for many illnesses in animals and plants, but the infectious agents for certain diseases (e.g., smallpox in humans and mosaic disease of tobacco) would not grow or reproduce in the laboratory outside of living cells, nor could they be seen with the light microscope as could bacteria. In 1892 the Russian biologist Dmitri Iwanowsky showed that the infectious agent responsible for tobacco mosaic disease will pass through a porcelain filter with pores too small to permit the passage of bacterial cells. The term **virus** was introduced to refer to such infectious agents, with the term **bacteriophage** being introduced later to refer to viruses that selectively attack bacterial cells.

As experimentation progressed in the early twentieth century, it gradually became apparent that viruses resemble complex chemicals more than they do living cells. In 1935 Wendell Stanley showed that tobacco mosaic virus can be crystallized, and that such crystals retain the ability to cause disease and reproduce when injected into tobacco plants. The development of electron microscopy in the early 1950s finally confirmed that viruses are quite different in structure from living cells.

In general, virus particles consist of a core of DNA or RNA surrounded by a protein coat. Most virus particles are either polyhedral (solids with multiple plane faces) or helical, or some combination of the two (Figure 1-30). One of the most commonly encountered shapes is a 20-sided polyhedron called an *icosahedron*, in which each of the faces is an equilateral triangle. The inherent stability and efficiency of such a structure has been dramatically demonstrated by the American architect and philosopher Buckminster Fuller, who applied this principle to the construction of his famous geodesic domes. Helical viruses also exhibit a regular symmetry, although in this case the protein subunits surrounding the nucleic acid core usually form a cylinder composed of protein subunits arranged in a spiral configuration.

Some viruses have a more elaborate structure than these simple polyhedral or helical configurations. In the *enveloped* viruses, a polyhedral or helical protein coat is enclosed by a membranous envelope consisting of a mixture of protein, lipid, and/or carbohydrate. Unlike the membranes of true cells, however, this structure is not capable of regulating the flow of materials into and out of the virus. Other viruses may exhibit a complex combination of polyhedral and helical symmetry. The **T-even bacteriophages,** for example, consist of a polyhedral head attached to a helical tail. The remarkable efficiency with which DNA is packed into the small head of such bacteriophages is dramatically illustrated by electron micrographs of ruptured virus particles (Figure 1-31).

Viral Reproduction and Life Cycles

Viruses appear to be adapted for growth in specific cell types. In some cases the interaction is so specific that

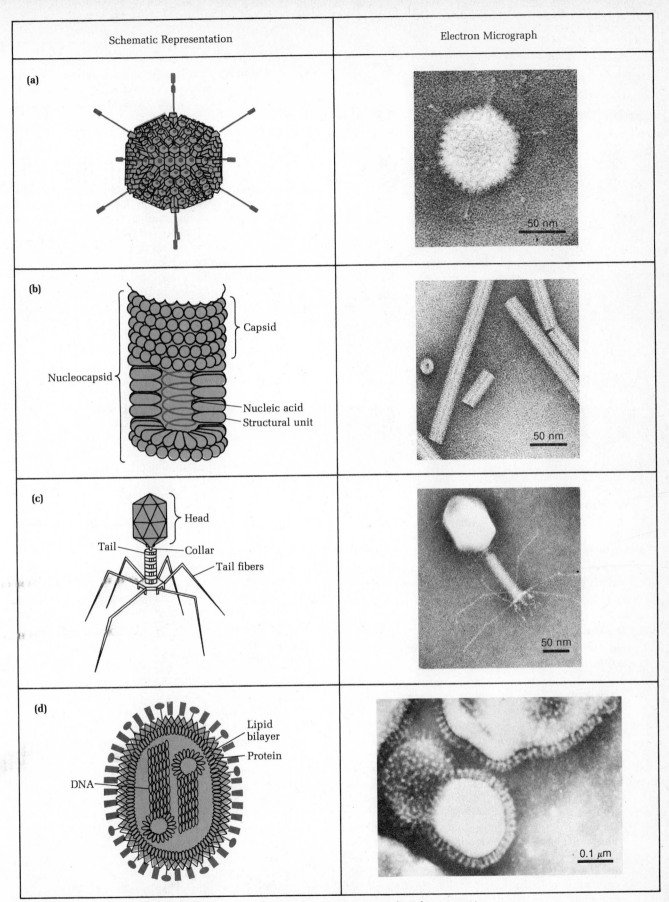

Schematic Representation	Electron Micrograph
(a)	50 nm
(b) Nucleocapsid — Capsid, Nucleic acid, Structural unit	50 nm
(c) Head, Tail, Collar, Tail fibers	50 nm
(d) DNA, Lipid bilayer, Protein	0.1 μm

FIGURE 1-30 Examples of DNA- and RNA-containing viruses. (a) Adenovirus. (b) Tobacco mosaic virus. (c) Bacteriophage T4. (a–c) Courtesy of R. C. Williams. (d) Influenza virus. Courtesy of W. G. Laver.

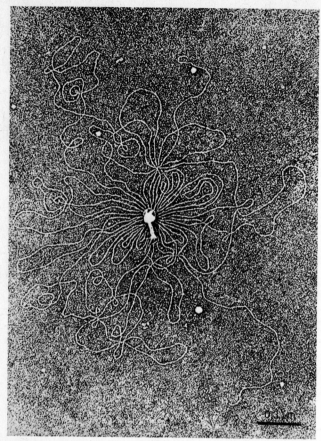

FIGURE 1-31 Release of DNA from a T2 bacteriophage ruptured by osmotic shock. Courtesy of A. K. Kleinschmidt.

only certain genetic strains of an organism are susceptible to infection by a particular virus. This specificity is governed by a number of factors, including viral recognition of the host cell surface and the ability of a virus to utilize intracellular components to aid in its own replication. Although the details of the infection process vary among viruses, five basic steps can be distinguished: *adsorption, penetration, replication, maturation,* and *release.*

1. **Adsorption.** Viral infections are generally initiated by the binding of a viral coat protein to a specific chemical receptor site on the surface of the host cell. Such receptor sites generally consist of macromolecules that are normal components of the plasma membrane or cell wall. Although the binding of a virus to the cell surface is important in aiding the efficiency and determining the host cell specificity of a viral infection, this step is not an absolute requirement. It has been possible in the laboratory to infect cells with viral nucleic acid with its protein coat removed and to show that this naked nucleic acid can enter cells and lead to the production of complete new virus particles.
2. **Penetration.** After attachment of a virus particle

to the host cell surface, the next step is the movement of its nucleic acid into the cell. In the case of the well-studied T-even bacteriophages, an enzyme located in the virus tail digests a hole in the cell wall. The tail then contracts and injects the viral DNA through the hole into the cell. Animal viruses, on the other hand, are usually engulfed intact by an infolding of the cell membrane, a process called *endocytosis*. In this case both the viral nucleic acid and protein are brought into the host cell. Once inside the host cell, however, the viral protein is usually degraded, so in both cases the final result is the introduction of a naked molecule of viral nucleic acid into the cell.

3. **Replication.** There are two possible fates for the viral nucleic acid once it has entered the cell. In the case of a **virulent virus,** the nucleic acid directs the host cell to make viral nucleic acid and protein molecules. From these are assembled new virus particles, which are released as the cell is destroyed. In the case of **temperate** or **lysogenic viruses,** on the other hand, the information from the viral nucleic acid becomes incorporated into the DNA of the host cell, where it is called a **provirus** or **prophage.** This arrangement can be stable for indefinite periods of time, although certain stimuli are known to trigger release of viral nucleic acid and an ensuing virulent infection.

   When the provirus DNA is released under such conditions, it often takes a bit of the host's DNA as a replacement for some of its own. If such a virus then infects another cell, the newly infected cell receives some of the genetic information obtained from the first host cell. Such a mechanism for the transfer of genetic information between cells, referred to as **transduction,** is an important mechanism for the exchange of genetic information in bacterial cells. Since transducing viruses have lost some of their own genes, they are often incapable of infecting other cells by themselves and are therefore referred to as being *defective*. In order for such defective viruses to function, the host cell must be simultaneously infected by a related virus that can provide the missing functions.
4. **Maturation.** The process by which newly synthesized viral nucleic acid and protein are assembled into intact virus particles is referred to as maturation. For many viruses this event appears to proceed by a simple self-assembly process. It was shown in the early 1950s by Heinz Fraenkel-Conrat that a mixture of RNA and protein purified from tobacco mosaic virus will spontaneously assemble into reconstituted virus particles that are infective. Exceptionally interesting is the finding that the RNA from one strain of virus can be mixed together with the protein component from another to form a hybrid virus

that is also infective, though as expected, it is the source of the RNA and not the protein that determines the type of virus that will be made by the infected cells.

Such experiments suggest that all the information required for the correct assembly of tobacco mosaic virus particles is inherent in the RNA and protein components themselves. It is not clear how far one can generalize this conclusion, especially in the case of more complex viruses. The maturation of T-even bacteriophages, for example, is a complex multistep process involving a sequential series of events. Thus formation of the head, insertion of DNA, assembly of the tail, addition of the tail fibers, and attachment of the head and tail all occur in a highly integrated sequence. Genetic mutants with this process blocked at various points have been identified, pointing to the existence of genetically determined molecules that guide the process of assembly. It is not known whether all these molecules are actually part of the virus, or whether some guide assembly without themselves becoming part of the developing particle.

5. **Release.** After new virus particles have been assembled, they are released from the infected cell and move on to infect other cells. This release is usually accomplished by **lysis,** in which the host cell membrane is broken open and the cell destroyed. As an example we may again use the T-even bacteriophages, which code for the production of an enzyme called *lysozyme*. This enzyme digests the bacterial cell wall, breaking open the cell and permitting the release of newly formed virus particles. It is interesting to note that the bacteriophage genes do not all function simultaneously, but that some are employed early in the infection process while others are used later. Since it is obvious that early production of lysozyme would destroy the host cell before the virus particles had sufficient time to be replicated, it is not surprising that the gene coding for this enzyme turns out to be one of the "late" genes.

Although enzymatically induced cell lysis is a common mode of virus release, such destruction of the host cell does not always occur. In certain types of bacteriophages, and in many animal viruses, the newly formed virus particles are continuously released from the infected cell by budding off from the cell surface. Part of the cell membrane may be taken along with the virus in the process, contributing to the viral envelope. Viruses released in this fashion do not dominate cellular metabolism the way those causing lysis do, and the normal functioning of the host cell may continue during the period of infection. Because such cells are not destroyed during virus release, they act as perpetual carriers of the infecting virus.

Evolutionary Origin of Viruses

It should be clear from the above description that viruses are quite different from living cells. The question therefore arises as to whether or not viruses are "alive." This question has no absolute answer since it depends to a great extent on how one defines life. But some insight into the relationship between viruses and living organisms might be gained if we had a better understanding of the evolutionary origin of viruses.

Two main hypotheses have been proposed in this regard. One states that viruses are a precellular form of life that first occurred on the primordial earth prior to the evolution of cells. This hypothesis appears somewhat unlikely, for it is hard to picture how viruses could have evolved before living cells when they require the existence of cells for their replication. The other viewpoint considers viruses to be degenerate cells or cell fragments. The most extreme form of this hypothesis is the view that viruses are simply pieces of DNA or RNA that have escaped from cells and have sufficient genetic information to be able to replicate in other cells. This theory could explain why viruses tend to preferentially infect certain cell types, since such nucleic acid fragments might be expected to function only in cells that are similar to the one from which they were originally derived.

If viruses are in fact cellular derivatives, it is not surprising to find that their functioning involves the same types of biological principles as apply to cells. But virus particles have lost most of the basic features of cells. Of the four defining properties of living cells mentioned earlier, viruses exhibit but one, the storage and replication of genetic information. Even this property depends upon living cells, in that viral replication occurs within a host cell using the host's synthetic machinery. The complete and integrated performance of all four functions essential to the living state clearly involves a great deal more structural complexity and compartmentalization than occurs in virus particles.

SUMMARY

All organisms are composed of cells that perform the same basic functions: maintenance of a selective barrier for regulating the flow of materials into and out of the cell, reproduction and utilization of genetic material for guiding cellular activities, carrying out of metabolic pathways for converting matter and energy, and movement of components within the cell as well as locomotion of the cell as a whole.

Most cells also resemble one another in structure, although eukaryotic cells are larger and have a more extensive internal membrane system than prokaryotic cells. All cells are enclosed by a plasma membrane that regulates the passage of materials by selective permeability, active transport, and endocytosis/exocytosis. In eukaryotic cells similar membranes form a series of cytoplasmic compartments, including the endoplasmic reticulum, Golgi complex, lysosomes, microbodies, and vacuoles. The genetic material consists of DNA-containing fibers that in eukaryotes are also enclosed by a double-membrane, the nuclear envelope. The information contained in this DNA is ultimately used to guide the synthesis of specific proteins on small cytoplasmic granules called ribosomes. Metabolism occurs predominantly in the cytosol, although in eukaryotic cells many of the important metabolic reactions involved in energy conversions occur in double-membrane enclosed mitochondria. In plant cells another double-membrane enclosed structure, the chloroplast, carries out photosynthesis. Motility is made possible by external appendages called cilia and flagella, and in eukaryotic cells by an internal cytoskeleton containing microtubules, microfilaments, and intermediate filaments.

Water, which accounts for up to 90 percent of the cell mass, is the medium in which all reactions take place, and is a direct reactant in some cases. The major building blocks of the cell are sugars, fatty acids, amino acids, and nitrogenous bases. These small molecules are assembled into four types of macromolecules: polysaccharides, lipids, proteins, and nucleic acids.

Polysaccharides are polymers of sugar molecules. Major examples are glycogen and starch, used as energy storage molecules, and cellulose, whose mechanical strength contributes to the rigidity of the cell wall. Lipids are a heterogeneous group of molecules that are hydrophobic in nature. Among them are the neutral fats used in energy storage, the phospholipids and glycolipids employed in membrane construction, and the terpenes found in certain vitamins, coenzymes, and carotenoids. Proteins are extremely complex molecules made by joining together amino acids in various sequences. In addition to the vast number of different sequences possible, protein structure is further complicated by the existence of secondary, tertiary, and quaternary structure. Together these elements create a specific three-dimensional conformation required for a protein's biological activity. Among the many functions carried out by proteins, enzymatic catalysis is the most widespread and crucial. Nucleic acids are polymers of nucleotides (nitrogenous-base–pentose-sugar–phosphate) that function in the storage, reproduction, and transmission of genetic information. The two major types are DNA (pentose = deoxyribose, bases = G, C, A, T) and RNA (pentose = ribose, bases = G, C, A, U).

Although spontaneous generation of cells from inanimate matter does not occur under defined laboratory conditions, recent experimentation suggests the possibility that cells might have arisen spontaneously on the primitive earth. The organic building blocks required by living cells, as well as macromolecules with catalytic and information-transferring properties, have been shown to form in the laboratory under simulated primitive earth conditions. Proteinoids made in this way are capable of spontaneously assembling into small vesicles, called microspheres, which carry out primitive cell functions. The origin of the genetic system in which nucleic acids code for specific protein molecules is an enigma that has led to two schools of thought, one emphasizing the primacy of nucleic acids and the other emphasizing the primacy of proteins.

Viruses are macromolecular complexes consisting of a nucleic acid core surrounded by a coat of protein (and occasionally carbohydrate and/or lipid). Viruses can only reproduce when they infect a living cell. The infection process involves five stages: adsorption, penetration, replication, maturation, and release. In the case of lysogenic viruses, viral DNA may be stably incorporated into the host cell for an indefinite period of time. Viruses have been proposed to represent either a precellular form of life or degenerate cells or cell fragments.

SUGGESTED READINGS

Books

Crick, F. (1982). *Life Itself: Its Origin and Nature*, Simon and Schuster, New York.

Day, W. (1984). *Genesis on Planet Earth*, 2nd Ed., Yale University Press, New Haven, CT.

Dillon, L. S. (1978). *The Genetic Mechanism and the Origin of Life*, Plenum Press, New York.

Fox, S. W., and K. Dose (1977). *Molecular Evolution and the Origin of Life*, Marcel Dekker, New York.

Kuchera, L. S. (1985). *Fundamentals of Medical Virology*, 2nd Ed., Lea and Febiger, Philadelphia.

Lehninger, A. L. (1982). *Biochemistry: The Molecular Basis of Cell Structure and Function*, Worth, New York, Chs. 1–7, 10–12, 37.

Miller, S. L., and L. E. Orgel (1974). *The Origins of Life on the Earth*, Prentice-Hall, Englewood Cliffs, NJ.

Norton, C. F. (1986). *Microbiology*, 2nd Ed., Addison-Wesley, Reading, MA.

Reid, R. A., and R. M. Leech (1980). *Biochemistry and Structure of Cell Organelles*, Blackie, Glasgow.

Tolbert, N. E., ed. (1980). *The Biochemistry of Plants*, Vol. 1, Academic Press, New York.

Wolfe, S. L. (1985). *Cell Ultrastructure*, Wadsworth, Belmont, CA.

Articles

Chothia, C. (1984). Principles that determine the structure of proteins, *Ann. Rev. Biochem.* 53:537–572.

Dickerson, R. E. (1978). Chemical evolution and the origin of life, *Sci. Amer.* 239 (Sept.):70–86.

Eigen, M., W. Gardiner, P. Schuster, and R. Winkler-Oswatitsch (1981). The origin of genetic information, *Sci. Amer.* 244 (Apr.):88–118.

Karplus, M., and J. A. McCammon (1986). The dynamics of proteins, *Sci. Amer.* 254 (Apr.):42–51.

Vidal, G. (1984). The oldest eukaryotic cells, *Sci. Amer.* 244 (Feb.):48–57.

CHAPTER
2
Enzymes and Catalysis

Thousands of chemical reactions occur in the typical cell. These reactions underlie such diverse activities as synthesis and degradation of chemical building blocks, trapping and utilization of chemical energy, synthesis of macromolecules, transmission of genetic information, transport of materials across membranes, and motility. Thus every important cell function introduced in Chapter 1 is ultimately dependent upon chemical reactions.

Most reactions upon which living organisms depend do not, by themselves, occur fast enough at moderate temperatures to sustain life without catalysts to speed these reactions. Enzymes are specialized proteins that have evolved to carry out this catalytic function. There are thousands and thousands of different kinds of enzymes, each designed to enhance the rate of a specific reaction. As a result of a century of intensive laboratory investigation, much is now known about what enzymes are, how they work, and how they are regulated. Because of the ubiquitous role of enzymes in the cell, it is important for us to consider these general issues at this early stage in the book.

DISCOVERY OF BIOLOGICAL CATALYSTS

In the early nineteenth century the Swedish chemist Jon Berzelius discovered that an extract of potatoes is more effective than concentrated sulfuric acid in promoting the breakdown of starch. Berzelius concluded that this effect is due to the presence of biological catalysts and with remarkable insight went on to predict that all materials found in living organisms are synthesized under the influence of such catalysts. This observation stimulated interest in the biochemical identity of these catalysts, and Louis Pasteur soon postulated that the catalytic effect is intimately associated with cell structure and so cannot be separated from living cells. If true, this meant that the catalysts could not be isolated and studied in purified form. It was therefore a great milestone when, in 1897, Edward Büchner demonstrated that an extract of yeast, from which all intact cells had been removed, is capable of catalyzing the breakdown of glucose. This was an especially significant observation because glucose degradation is a major metabolic pathway in cells. This finding firmly established that biological catalysts, in the absence of any cell organization, are capable of enhancing the rates of metabolic reactions.

The increasing importance of this class of molecules soon led to suggestions for a uniform nomenclature system to facilitate communication among biochemists. The term **enzyme** was adopted as a general designation for biological catalysts, and it was decided that individual enzymes would be named by adding the suffix "ase" to the name of the **substrate** (the substance upon which the enzyme acts). For example, the enzyme catalyzing the degradation of urea was termed *urease,* enzymes catalyzing hydrolysis of nucleic acids were designated *nucleases,* the enzyme catalyzing glucose oxidation was called *glucose oxidase,* and so forth. This basic approach to enzyme nomenclature is still in use today, although some older names introduced prior to the adoption of this system (e.g., trypsin, pepsin) have been retained.

In the years immediately following the initial demonstration of cell-free enzyme activity by Büchner, a large number of other enzymes were identified by biochemists. But little progress was made in understanding the structure or mechanism of these cellular catalysts. An example of the difficulties experienced during this period is the classic story of Richard Willstätter's encounters with the enzyme *peroxidase,* one of the first enzymes to be obtained in a highly purified form. When Willstätter chemically analyzed his purified enzyme preparation, he found no lipid, no sugar (and therefore no nucleic acid), and no protein. These results led to the prevailing opinion in the early twentieth century that enzymes belong to none of the commonly recognized classes of biological molecules.

In 1926, however, James Sumner reported the first crystallization of an enzyme, in this case urease. He found that the crystals consisted of pure protein, even after repeated cycles of recrystallization. Since crystallization is generally recognized as the ultimate criterion of purity, these results suggested to Sumner that the enzyme urease is a protein. Although this conclusion initially met with skepticism, similar experiments by John Northrop and his colleagues on other enzymes soon led to general acceptance of the idea that enzymes are made of protein. Willstätter, it turns out, had been misled by the fact that enzymes are active in extremely small quantities. Thus his peroxidase preparation exhibited measurable enzyme activity even though the protein content of the sample was too low to be detected by the analytical techniques then available.

In the years since the pioneering work of Sumner and Northrop, over a thousand enzymes have been purified and identified as proteins. Though the number of different reactions catalyzed by these enzymes is very large, one can subdivide this group into a relatively small number of categories (Table 2-1). All cellular reactions are catalyzed by enzymes that fit into one of these categories.

As Berzelius predicted many years ago, enzymes act as extraordinarily efficient chemical catalysts. Enzymes therefore exhibit the four defining characteristics shared by all catalysts: (1) They modify the rates of chemical reactions. (2) They are not consumed in the process. (3) They are effective in extremely small quantities. (4) They do not alter the equilibrium of the reaction being catalyzed.

In addition to these four attributes common to all catalysts, enzymes exhibit several unique features that

TABLE 2-1

Current nomenclature of six types of biochemical reactions catalyzed by enzymes

Enzyme Class	Type of Reaction Catalyzed	Example
Oxidoreductases	Oxidation-reduction	Dehydrogenases, oxidases, peroxidases, hydroxylases, reductases, oxygenases
Transferases	Transfer of a functional group from one molecule to another (AX + B → A + BX)	Transaminases, transmethylases
Hydrolases	Hydrolytic cleavage	Esterases, amidases, glycosidases, peptidases, phosphatases
Lyases	Cleavage of C—C, C—O, C—N, etc. by elimination, leaving a double bond; or addition to double bonds	Aldolases, synthases, deaminases, hydrases, decarboxylases, nucleotide cyclases
Isomerases	Intramolecular rearrangements	Isomerases, racemases, epimerases, mutases
Ligases	Joining together of two molecules coupled to the cleavage of a high-energy bond	Synthetases

distinguish them from the catalysts routinely encountered in organic or inorganic chemistry. First, enzymes are exceedingly efficient compared to other catalysts. A typical enzymatic reaction may proceed as much as 10^8–10^{11} times faster than the same reaction catalyzed by a nonenzymatic catalyst. Also, unlike the nonbiological catalysts routinely encountered in chemistry, the catalytic activity of enzymes can be regulated to suit specific cellular needs. Finally, one of the most striking properties of enzymes is their selectivity. Unlike other catalysts enzymes are highly specific, both in the types of reactions they catalyze and in the structures of the substances upon which they act. For example, the enzyme glucose oxidase is specific not just for the oxidation of glucose versus other six-carbon sugars but also for a particular three-dimensional form of the sugar, namely, α-glucose.

The unique properties of enzymes (efficiency, regulation, and specificity) all stem from the fact that enzymes are protein molecules and therefore exhibit all the remarkable characteristics of proteins. The study of how enzymes work is therefore intimately tied to the study of protein chemistry.

THERMODYNAMICS OF CHEMICAL REACTIONS

In order to discuss the question of how enzymes work, one must first have a basic familiarity with the concepts of thermodynamics and kinetics. **Thermodynamics** refers to the study of the changes in energy that occur during chemical reactions. Such information determines the extent to which equilibrium conditions favor the formation of product, but tells us nothing about reaction rates. (Kinetics, which deals exclusively with this latter issue of rates, will be discussed in the next section.)

Energy can be transferred into and out of systems in various forms, such as heat, light, electricity, mechanical energy, and chemical energy. However, the total energy content of any system plus its surroundings must remain constant, a principle known as the **first law of thermodynamics.** In addition to this overall conservation of energy, there are limitations on the types of energy transformations that can occur. The **second law of thermodynamics** states that every event proceeds in such a way that a system plus its surroundings exhibit a net increase in disorder, or **entropy.** This propensity to maximize entropy in the universe is the ultimate driving force for all chemical reactions. The portion of the total energy used to increase the entropy is not available to do work. This amount of energy can be calculated by multiplying the change in entropy (ΔS) times the absolute temperature (T), yielding $T\Delta S$. The energy available to do work, on the other hand, is known as the **free energy** (G). The changes in free energy and entropy

occurring in a system undergoing transformation are related by the equation

$$\Delta G + T\Delta S = \Delta H$$

where H equals the **enthalpy** or heat content. In biological systems, where pressure and volume are constant, enthalpy is equivalent to the total energy. Hence the above equation, which combines both the first and second laws of thermodynamics, tells us that in any system undergoing transformation, the total change in energy is equal to the sum of the changes in (1) free energy and (2) the energy involved in altering the entropy.

Although the second law of thermodynamics tells us that the entropy of a system plus its surroundings always proceeds toward a maximum, this does not necessarily mean that the entropy of an individual system always increases. It may either increase, decrease, or stay the same. If the system decreases in entropy, however, the entropy of the surroundings must increase by a sufficient amount so that the total entropy of system *plus* surroundings increases. This is exactly what happens during the development of living organisms. As highly organized living cells are produced, there is a decrease in entropy of the system, but only at the expense of an even greater increase in entropy of the surroundings.

The fact that a net increase in entropy accompanies all events proceeding toward equilibrium means that there must be a comparable decrease in free energy (due in essence to the loss of the free energy used to increase the entropy). Because of this tendency of free energy to decrease to a minimum, free-energy calculations are useful indicators of the direction in which reactions tend to proceed. They are especially useful in this regard because changes in free energy are much easier to determine than changes in entropy.

Determination of Free-Energy Changes

For the generalized chemical reaction

$$\text{Reactants} \longrightarrow \text{products}$$

the change in free energy (ΔG) can be calculated from the equation

$$\Delta G = \Delta G^\circ + 2.303RT \log_{10} \frac{[\text{products}]}{[\text{reactants}]} \qquad (2\text{-}1)$$

where R is the gas constant, T is the absolute temperature, [reactants] is the mathematical product obtained by multiplying together the initial molar concentrations of each of the reactants, [products] is the mathematical product obtained by multiplying together the initial molar concentrations of each of the products of the reaction, and ΔG° is the standard free-energy change. This **standard free-energy change** (ΔG°) is a measure of the amount of free energy released during conversion of reactants to products under "standard conditions" (defined as all reactants and products present at an initial concentration of 1.0M). It is critical that the distinction between ΔG and $\Delta G°$ be clearly understood. On the one hand ΔG is a measure of the *actual* free-energy change that occurs given a mixture of reactants and products at any particular concentrations; the value of ΔG thus varies, depending on the concentrations involved. On the other hand $\Delta G°$ is a constant for any given reaction at a given temperature. It is the free-energy change that occurs under standard conditions of reactant and product concentration.

When equilibrium is achieved, it means by definition that no further net change in free energy can occur, that is, $\Delta G = 0$. Substituting $\Delta G = 0$ into Equation 2-1, we obtain the equilibrium expression

$$0 = \Delta G° + 2.303RT \log_{10} \frac{[\text{products}_{eq}]}{[\text{reactants}_{eq}]}$$

where $[\text{reactants}_{eq}]$ and $[\text{products}_{eq}]$ are the molar concentrations of reactants and products at equilibrium, respectively. Rearranging terms we obtain

$$\Delta G° = -2.303RT \log_{10} \frac{[\text{products}]}{[\text{reactants}]} \quad (2\text{-}2)$$

Because the **equilibrium constant (K'_{eq})** for any chemical reaction is defined as

$$K'_{eq} = \frac{[\text{products}]}{[\text{reactants}]} \quad (2\text{-}3)$$

K'_{eq} can be substituted into Equation 2-2 to obtain the general relationship

$$\Delta G° = -2.303RT \log_{10} K'_{eq} \quad (2\text{-}4)$$

This final equation provides us with a relatively easy way of determining the standard free-energy change for any particular chemical reaction. One simply measures the concentrations of reactants and products after equilibrium has been achieved, uses this information to calculate K'_{eq} (Equation 2-3), and then substitutes this value of K'_{eq} into Equation 2-4 to obtain $\Delta G°$.

It can be readily calculated from Equation 2-4 that reactions with an equilibrium constant of 1.0 (equal concentrations of reactants and products at equilibrium) exhibit a $\Delta G°$ of 0. This means that no free-energy change occurs during conversion of reactants to products under standard conditions. For reactions with an equilibrium constant greater than 1.0 (equilibrium favoring products), $\Delta G°$ will be negative and free energy will therefore be released during conversion of reactants to products. When the equilibrium constant is less than 1.0 (equilibrium favoring reactants), $\Delta G°$ will be positive and free energy will therefore be consumed during conversion of reactants to products.

Reactions with a negative $\Delta G°$, termed **exergonic** reactions, can proceed spontaneously in the direction written. However, it is important to emphasize that in this particular context, the term *spontaneous* does not signify anything about the speed at which the reaction will occur. A spontaneous reaction may take seconds or years to achieve equilibrium; its spontaneity refers only to the fact that no net input of energy is required in order to make the reaction proceed. Reactions in which $\Delta G°$ is positive are termed **endergonic,** and cannot proceed spontaneously in the direction written. Instead, they will proceed in the reverse direction.

In a living cell, of course, one does not generally start with 1.0M concentrations of all ingredients, the standard conditions that apply to the above discussion of $\Delta G°$. Under other conditions it is ΔG, the actual free-energy change, that determines the direction in which a reaction proceeds. Depending on the particular conditions involved, ΔG may be smaller, larger, or the same as $\Delta G°$.

As an example to help summarize the above points, let us consider the following reaction that occurs as part of the pathway for metabolizing glucose:

Dihydroxyacetone phosphate   Glyceraldehyde 3-phosphate

No matter what the starting conditions, the ratio of glyceraldehyde 3-phosphate to dihydroxyacetone phosphate is always found to be 0.0475 once equilibrium has been attained. By substituting this value of K'_{eq} into Equation 2-4, we obtain

$$\Delta G° = -2.303RT \log_{10}(0.0475)$$

Substituting appropriate values for the gas constant R (0.00198 kcal/mol/K) and absolute temperature T (25°C = 298 K), we can calculate

$$\Delta G° = -2.303 \times 0.00198 \times 298 \times \log_{10}(0.0475)$$
$$= +1.8 \text{ kcal/mol}$$

Hence for every mole of dihydroxyacetone phosphate converted to glyceraldehyde 3-phosphate *under standard conditions*, 1.8 kcal of energy is consumed.

Now let us observe what would happen to this reaction under an arbitrary set of initial conditions such as might exist in the cell, for example, a dihydroxyacetone phosphate concentration of 10^{-4}M and a glyceraldehyde 3-phosphate concentration of 10^{-6}M. The actual free-energy change ΔG can be calculated for these conditions by substituting into Equation 2-1 as follows:

$$\Delta G = \Delta G° + 2.303RT \log_{10} \frac{[\text{products}]}{[\text{reactants}]}$$

$$= 1.8 \text{ kcal/mol} + 2.303RT \log_{10} \frac{10^{-6}}{10^{-4}}$$

$$= 1.8 \text{ kcal/mol} + (2.303 \times 0.00198 \times 298 \times -2)$$
$$= 1.8 \text{ kcal/mol} - 2.7 \text{ kcal/mol}$$
$$= -0.9 \text{ kcal/mol}$$

Thus in spite of the positive value of $\Delta G°$ previously calculated for the reaction, ΔG turns out to be negative under these particular conditions. This negative value of ΔG means that under the above specified conditions, the reaction will proceed spontaneously with a release of free energy, even though such would not be the case under standard conditions. Since ΔG is calculated on the basis of the prevailing initial concentrations of reactants and products rather than arbitrary standard conditions, ΔG rather than $\Delta G°$ determines the direction in which reactions proceed in the cell.

KINETICS OF CHEMICAL REACTIONS

The thermodynamic approach is useful in defining the directions in which reactions tend to proceed, but it tells us nothing about the rates at which they occur. Analysis of the factors involved in determining how fast a reaction will occur is referred to as the study of **kinetics.** Reaction kinetics are ultimately based on the fact that the conversion of reactants to products always involves an intermediate stage, called the **transition state,** whose energy level is higher than that of either the reactants or the products (Figure 2-1, black curve). The term **free energy of activation,** or **$\Delta G\ddagger$,** is used to refer to the difference between the free-energy level of the reactants and the free-energy level of the transition state. The rate at which a chemical reaction proceeds is directly proportional to the number of molecules attaining the high-energy transition state. As a result, any factor that increases the number of molecules having sufficient energy to attain the transition state will increase the rate of the reaction.

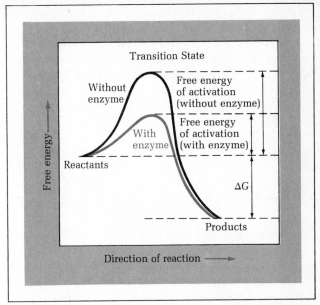

FIGURE 2-1 Energy diagram for a general chemical reaction in the presence or absence of an enzymatic catalyst.

There are two general ways in which this can be accomplished. The first involves temperature. At moderate temperatures a typical reaction mixture consists of a population of molecules with varying energy levels. Although the *average* free-energy level of the mixture is lower than that of the transition state, a small fraction of the molecules has sufficient energy to achieve this activated state. The relative number of molecules in this latter category determines how fast the reaction will proceed. Raising the temperature increases the overall thermal energy of the population as a whole, thereby increasing the number of molecules with free-energy levels high enough to enter the transition state (Figure 2-2a). This higher proportion of molecules attaining the

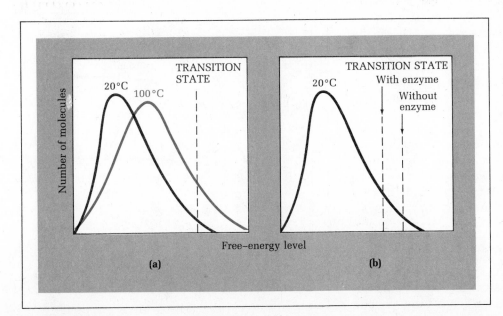

FIGURE 2-2 Energy distribution diagrams illustrating two ways in which reaction rates can be increased. (a) Raising the temperature increases the proportion of the molecules in a population having sufficient free energy to achieve the transition state. (b) The presence of a catalyst creates a transition state with a lower free energy, so at any given temperature more molecules will possess sufficient free energy to attain this state.

transition state in turn increases the rate of product formation.

The second general method for increasing reaction rates involves the use of catalysts. A catalyst combines with one or more reactants to generate a new transition state whose $\Delta G\ddagger$ is lower than the $\Delta G\ddagger$ of the transition state for the uncatalyzed reaction (see Figure 2-1, color curve). In other words, the catalyst provides an alternative reaction pathway with a transition state of lower free energy. Hence at any given temperature, the number of molecules possessing sufficient free energy to attain the transition state will be higher for the catalyzed reaction than for the uncatalyzed reaction (see Figure 2-2b).

Because the new transition state created by a catalyst is part of both the forward and reverse reactions, catalysts increase reaction rates in both directions rather than selectively enhancing the rates of either the forward or backward reaction. A catalyst therefore can only speed up the rate at which equilibrium is achieved. The equilibrium concentrations of reactants and products are solely determined by the thermodynamic calculation of ΔG, which is not influenced by the presence or absence of a catalyst.

PROPERTIES OF ENZYME-CATALYZED REACTIONS

Enzyme-catalyzed reactions obey all the basic principles of thermodynamics and kinetics described above. Like other catalysts, enzymes provide reactions with a new transition state of lower free energy, thereby speeding up the rate at which equilibrium is attained. And like other catalysts, enzymes do not affect the relative concentrations of reactants and products that prevail once equilibrium is attained. These equilibrium conditions are determined solely by the difference in free-energy levels (ΔG) between reactants and products, a factor upon which enzymes have no effect. Thus in the presence of enzymes, reactants need less of an energy input to be able to react to form products. This is of paramount biological significance for it permits metabolic reactions to occur at the relatively low temperatures compatible with living cells.

In addition to the above properties shared by all catalyzed reactions, enzymatic reactions exhibit several unusual characteristics not commonly encountered elsewhere. In the following sections we shall elaborate upon some of these characteristics and discuss their implications for the mechanism of enzyme action.

The Michaelis–Menten Equation

One of the first distinctive properties of enzyme-catalyzed reactions to be observed historically is an unusual effect of substrate (reactant) concentration upon reaction rates. In general, the rate of product formation in chemical reactions is directly proportional to the concentration of reactants because an increase in concentration increases the probability of collisions occurring between the reacting molecules. For enzymatic reactions this general principle holds true at low substrate concentrations, where the rate of product formation is almost directly proportional to substrate concentration. As the substrate concentration is raised, however, the reaction velocity does not continue to increase proportionally, and a point is eventually reached where further increases in substrate concentration cause no further change in rate. At this point the reaction has reached a **maximal velocity** (V_{max}), and the enzyme is said to be *saturated* with substrate. The actual maximal velocity attained varies with the enzyme concentration being employed. The more enzyme present, the faster the reaction will go before the enzyme becomes saturated (Figure 2-3).

In 1913 Leonor Michaelis and Maud Menten formulated a theory of enzyme action to explain this unique saturation effect. This theory is easiest to describe for the simple case of a reaction involving only one substrate. Under such conditions, Michaelis and Menten proposed that the enzyme (E) binds directly to its substrate (S), forming a transient **enzyme-substrate complex** (ES). This enzyme-substrate complex in turn breaks down to generate free enzyme plus product (P):

$$E + S \rightleftharpoons ES \rightleftharpoons E + P \qquad (2-5)$$

Because this equation indicates that the rate of product

FIGURE 2-3 Effect of varying substrate concentration on the rate of an enzyme-catalyzed reaction, illustrated for two different concentrations of enzyme. If one doubles the enzyme concentration, the maximal velocity (V_{max}) is also doubled. However, the concentration of substrate at which the reaction proceeds at one-half the maximal rate does not vary with enzyme concentration.

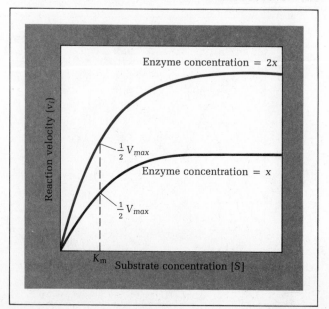

formation is directly related to the concentration of the enzyme-substrate complex, one can envision a simple explanation for the saturation effect described above. Enzymes are present in small quantities compared to substrate, so at high concentrations of substrate virtually all the enzyme molecules will be converted to the form of the enzyme-substrate complex. At this point raising the substrate concentration even higher cannot increase the concentration of the enzyme-substrate complex any further, so the reaction velocity has reached its maximum value.

Another attractive feature of the Michaelis–Menten model is that it allows one to visualize more directly how the presence of an enzyme can lower the free energy of activation. It can simply be postulated that the transition state of the enzyme-substrate complex has a lower free energy than the transition state of the substrate alone. In essence, the presence of the enzyme creates a new reaction pathway whose transition state can be produced at a faster rate because of its lower energy requirements.

In spite of the attractive features of the Michaelis–Menten theory, it was not until 25 years later that the existence of the postulated enzyme-substrate complex was verified experimentally. In the late 1930s David Keilin carried out an elegent set of studies on the enzyme *peroxidase*. This enzyme, which catalyzes the degradation of hydrogen peroxide, is normally a reddish-brown color in solution. Keilin discovered that when peroxidase is mixed with hydrogen peroxide in the absence of the second substrate needed to complete the reaction, the color of the enzyme solution changes dramatically. Such a change in color indicates an alteration in the chemical species present. Since no product formation was possible under these conditions, Keilin concluded that the color change must be due to the formation of a complex between the enzyme peroxidase and its substrate hydrogen peroxide, the only two ingredients present.

In the years since Keilin's pioneering work, the existence of enzyme-substrate complexes has been confirmed by a variety of other techniques. In some cases it has even been possible to directly isolate such complexes, permitting analysis of their three-dimensional structure. This information has provided solid support for the Michaelis–Menten notion that enzymes act by combining, however transiently, with their specific substrates.

In addition to predicting the existence of the enzyme-substrate complex, the Michaelis–Menten equation can be used to define two basic properties of enzymatic reactions. The first of these properties is the maximal velocity or V_{max}, a reflection of how quickly individual enzyme molecules can catalyze the conversion of substrate to product. Some enzymes are exceedingly efficient, with each enzyme molecule catalyzing

the conversion of several hundred thousand molecules of substrate to product per second; other enzymes may convert only a few molecules per second. The maximal velocity a reaction can achieve is therefore dependent on this basic efficiency of the enzyme involved. For any given enzyme the value of V_{max} will vary with enzyme concentration, since the more enzyme molecules present, the faster the reaction will proceed.

The other basic property of enzymatic reactions determined by the Michaelis–Menten equation is the **Michaelis constant** or K_m, which is defined as the substrate concentration at which the reaction velocity is equal to one-half the maximal velocity V_{max}. Unlike V_{max}, the value of K_m is independent of enzyme concentration. In other words the substrate concentration at which a reaction reaches half its maximal velocity is always the same, regardless of the value of the maximal velocity itself (see Figure 2-3). The Michaelis constant is a useful indicator of the relative affinity of an enzyme for its substrate. Thus, if the K_m for a given reaction is very high, a high substrate concentration is required for the reaction to proceed, and the enzyme must therefore have a relatively low binding affinity for its substrate. A low K_m, on the other hand, indicates that the reaction will proceed readily at low substrate concentrations, so the enzyme must have a high binding affinity for its substrate.

Therefore, V_{max} and K_m are useful parameters because they provide information about an enzyme's catalytic efficiency and affinity for substrate, respectively. Although approximate values for V_{max} and K_m can be obtained by examining a curve of reaction velocity versus substrate concentration (see Figure 2-3), simpler and more accurate approaches are based on mathematical manipulations of the Michaelis–Menten equation. The derivations of these equations are the province of biochemistry textbooks and will not be described in detail here. For our purposes it is adequate to begin with the fact that starting from the original formulation of the enzyme-substrate complex (Equation 2-5), one can mathematically derive the following relationship between V_{max} and K_m:

$$v_i = \frac{V_{max}\,[S]}{K_m + [S]} \qquad (2\text{-}6)$$

In this equation v_i stands for the **initial velocity,** which is the rate of product formation during the early stages of an enzyme reaction before product accumulation or other secondary events have caused the reaction to slow down.

There are several points worth emphasizing about Equation 2-6, which is commonly referred to as the **Michaelis–Menten equation.** The first is that it accurately predicts the shape of the curves obtained when one plots experimental data for reaction velocity versus substrate concentration (see Figure 2-3). This empirical verification of the Michaelis–Menten equation provides further support for the validity of the concept of the enzyme-substrate complex.

A second point worth noting about the Michaelis–Menten equation (2-6) is what happens when the substrate concentration is set equal to K_m:

$$v_i = \frac{V_{max}K_m}{K_m + K_m}$$

$$= \frac{V_{max}K_m}{2K_m}$$

$$= \frac{V_{max}}{2}$$

This is simply the mathematical derivation of our original definition of the Michaelis constant, that is, the substrate concentration at which the reaction proceeds at one-half maximal velocity.

The final point to emphasize about the Michaelis–Menten equation is that it can be easily used to determine V_{max} and K_m. By inverting Equation 2-6, we obtain the following relationship:

$$\frac{1}{v_i} = \frac{K_m + [S]}{V_{max}\,[S]}$$

$$= \frac{K_m}{V_{max}} \cdot \frac{1}{[S]} + \frac{1}{V_{max}} \tag{2-7}$$

This equation tells us that by plotting the reciprocal of reaction velocity ($1/v_i$) against the reciprocal of the substrate concentration ($1/S$), one can graphically determine V_{max} and K_m; the intercept with the y axis is the reciprocal of V_{max}, while the intercept with the x axis is the *negative* reciprocal of K_m. Such a graphic representation, called a **Lineweaver–Burk plot,** has the advantage of making the data fall on a straight line (Figure 2-4).

It is important to note that in the range of substrate concentrations near the K_m value, enzymes are serving most efficiently as biological catalysts. In this range an enzyme is sensitive to changes in substrate concentration, and yet is utilizing a substantial fraction of its catalytic power. The evolutionary development of enzyme molecules with such attributes has been an important step in permitting precise control of cellular reaction rates. The intracellular concentrations of most substrates appear to be within an order of magnitude of the K_m values for their respective enzymes, indicating that enzymes have evolved with substrate binding affinities in the optimal range for biological efficiency and regulation.

The determination of V_{max} and K_m is one of the first steps in examining the kinetic properties of an enzymatic reaction. As we shall see in the following sections, the ultimate rate of an enzyme-catalyzed reaction is also influenced by other factors such as environmental conditions and the presence or absence of various molecules that influence enzymatic activity.

Effects of pH and Temperature on Enzyme Activity

Although changes in pH or temperature do not alter the covalent structure of enzyme molecules, they may exert a dramatic effect on enzymatic activity. It is often stated that a given enzyme has an "optimum" pH and temperature range. The pH effect is due to changes in the ionization of the substrate and/or enzyme molecule caused by altering the pH. Changing the ionization of amino acid side chains may disrupt ionic bonds involved in tertiary and quaternary structure, thereby altering an enzyme's conformation or even denaturing it completely. This is why extreme changes in pH often inactivate enzymes. Most enzymes exhibit a sharp optimum of activity at pH values close to those occurring inside cells, that is, a pH near 7. Enzymes whose normal environments are more acidic or basic, on the other hand, usually exhibit optimal conformation at some other pH. For example, pepsin, the protein-digesting enzyme found in the highly acidic gastric juices of the stomach, functions optimally at a pH of 2 (Figure 2-5).

As in the case of pH, changes in temperature affect the rate of enzyme-catalyzed reactions in a complex manner. We have already seen that increasing the temperature enhances the rate of chemical reactions by increasing the frequency of collisions between reacting molecules. At some point, however, a temperature is reached that begins to disrupt the weak bonds involved in maintaining the enzyme's secondary, tertiary, and quaternary structure. At this stage, denaturation and

FIGURE 2-4 A Lineweaver–Burk plot of the data in Figure 2-3. When one graphs the reciprocal of reaction velocity against the reciprocal of substrate concentration, the data fall on a straight line. The *y* intercept corresponds to the reciprocal of V_{max}, while the x intercept equals the negative reciprocal of K_m.

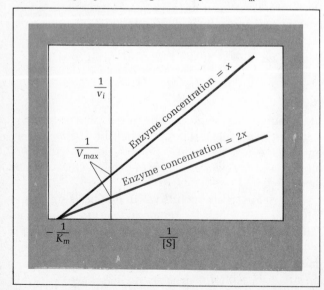

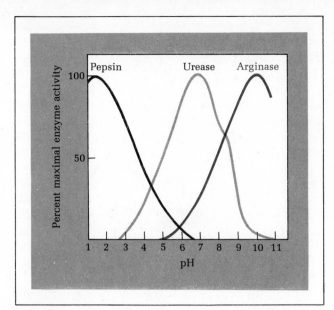

FIGURE 2-5 Effect of varying pH on the catalytic activity of three different enzymes.

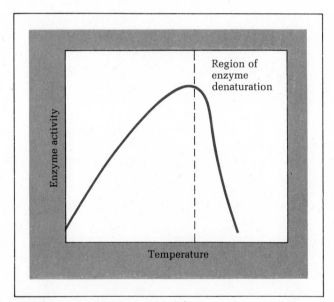

FIGURE 2-6 Effect of increasing temperature on the rate of an enzyme-catalyzed reaction.

loss of enzyme activity ensue. The effect of temperature on enzyme-catalyzed reactions is therefore biphasic, exhibiting an increase in reaction rate with moderate increases in temperature and a subsequent decline in reaction rate as further increases lead to enzyme denaturation (Figure 2-6).

The optimum temperature for mammalian enzymes is usually near the normal body temperature of 37°C. As with pH, however, considerable variation exists among enzymes. Enzymes occurring in bacteria that live in hot springs function well at temperatures over 90°C, for example, while enzymes from arctic bacteria work efficiently near freezing temperatures.

The optimum conditions of pH and temperature

determined for enzymatic reactions under artificial test-tube conditions may bear little relationship to optimal conditions inside a living cell. First of all, enzymes often exist in the cell as part of complex structural frameworks. The resulting intermolecular relationships no doubt affect the optimum activity an enzyme will display under given conditions of pH and temperature. Furthermore, the conditions for obtaining maximal catalytic activity in the test tube may in some instances begin to denature enzyme molecules after the reaction is in progress. In thinking about optimal conditions for enzyme activity inside the cell, one must balance the need for conditions that foster fast catalytic rates with the need for conditions that cause minimal enzyme denaturation.

Enzyme Cofactors

Although some enzymes are simple protein molecules, most require the presence of additional **cofactors** in order to manifest catalytic activity. Three principal classes of cofactors can be distinguished: *coenzymes*, *prosthetic groups*, and *metal ions*.

Coenzymes are low-molecular-weight organic molecules required in small amounts for the activity of certain enzymes. Such compounds are usually derived from vitamins rather than amino acids (Table 2-2). Examples are *nicotinamide adenine dinucleotide* (NAD) and *flavin adenine dinucleotide* (FAD), derivatives of niacin and vitamin B_2, respectively. NAD and FAD function as electron carriers for many enzymatically catalyzed oxidation-reduction reactions. Coenzymes bind reversibly to the enzymes with which they interact, participating directly in the reaction being catalyzed.

Prosthetic groups are similar to coenzymes but are attached to enzyme molecules by covalent instead of noncovalent bonds. An example of such an interaction occurs in the enzyme catalase. This reddish-brown molecule can be dissociated in acid to yield a colorless protein molecule and a red ferriprotoporphyrin prosthetic group. Such a protein missing its normal prosthetic group is referred to as an **apoprotein.** Like coenzymes, prosthetic groups are generally an essential component of the catalytic mechanism in enzymes where they occur.

The final class of cofactors consists of **metal ions** such as iron, copper, manganese, magnesium, calcium, and zinc. These ions can function in one of two basic ways. They may be necessary to stabilize the proper conformation of the enzyme and/or substrate, or they may participate directly in the chemical reaction being catalyzed. The iron-containing enzymes involved in mitochondrial electron transfer are classic illustrations of the direct participation of metal ions in enzymatic reactions.

TABLE 2-2
Some vitamins that function as cofactors for enzymatic reactions

Vitamin	Coenzyme or Prosthetic Group Derived from Vitamin	Function
Nicotinic acid (niacin)	Nicotinamide adenine dinucleotide (NAD)	Oxidation-reduction reactions
	Nicotinamide adenine dinucleotide phosphate (NADP)	
Riboflavin (vitamin B_2)	Flavin mononucleotide (FMN)	Oxidation-reduction reactions
	Flavin adenine dinucleotide (FAD)	
Pantothenic acid	Coenzyme A (CoA)	Acyl group transfer reactions
Thiamine (vitamin B_1)	Thiamine pyrophosphate	Oxidative decarboxylations
Pyridoxine (vitamin B_6)	Pyridoxal phosphate	Amino group transfer reactions
Biotin	Biocytin	CO_2 transfer reactions
Folic acid	Tetrahydrofolate coenzymes	One-carbon transfer reactions
Cobalamine (vitamin B_{12})	Cobamide coenzymes	Alkyl group transfer reactions

FIGURE 2-7 Two irreversible inhibitors of enzyme activity. (a) The reaction of diisopropylphosphofluoridate (DIPF) with serine hydroxyl groups. (b) The reaction of the alkylating agent iodoacetate with cysteine sulfhydryl groups.

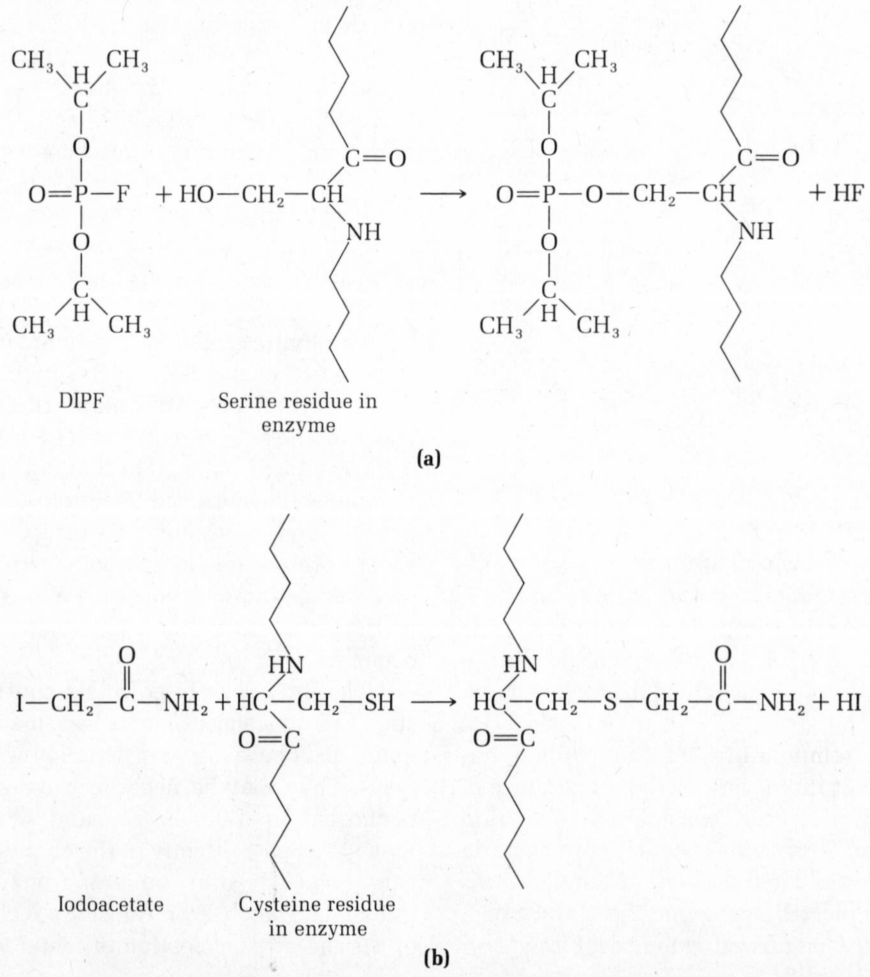

(a)

(b)

Enzyme Inhibitors

Many naturally occurring and synthetic chemicals are known to interfere with enzyme activity. Laboratory studies on enzyme inhibition have provided us with insights into basic questions such as the mechanism of enzyme action, the nature of catalytic sites, and the role of enzyme conformation in enzyme activity. The study of enzyme inhibitors has also had its practical benefits, helping us to deal more effectively with the effects of toxins and poisons that act on enzymes, as well as aiding in the design of specific enzyme inhibitors to be used as therapeutic drugs.

Irreversible Inhibitors Inhibitors of enzymatic reactions are divided into two general categories, irreversible and reversible, depending on whether or not the inhibition is permanent. *Irreversible inhibitors* bind to enzymes very tightly, usually by covalent linkage. The result is a relatively permanent disruption of enzyme structure and an associated inhibition of catalytic activity. Agents in this category are therefore highly toxic and are cause for extreme concern when encountered in the environment.

A classic example of irreversible inhibition occurs with *diisopropylfluorophosphate* (DIPF), a prominent component of insecticides and nerve gas. DIPF combines covalently with the hydroxyl group of the amino acid serine (Figure 2-7a). Since serine hydroxyl groups play a critical role in the catalytic sites of many enzymes, this reaction often results in enzyme inactivation. Enzymes involved in the transmission of nerve impulses are quickly inactivated in this way, explaining the highly toxic effects of DIPF and its analogs on the nervous system.

Alkylating agents, such as *iodoacetate* and *iodoacetamide*, are another class of irreversible enzyme inhibitors. In this case the sulfhydryl (—SH) group of the amino acid cysteine is the target (see Figure 2-7b). Modification of sulfhydryl groups by alkylating agents causes a virtually permanent alteration in enzyme structure. The degree to which enzymatic activity is inhibited depends upon the location of the cysteine group and the extent to which it is critical for proper protein conformation and enzyme function.

Reversible Inhibitors The *reversible inhibitors* of enzyme activity are a diverse group of substances with varying mechanisms of action. For convenience they are subdivided into two classes, competitive and noncompetitive, based on whether or not the inhibition can be reversed by increasing the concentration of substrate. **Competitive inhibitors** generally resemble one of the substrates of the reaction being inhibited, and are therefore capable of competing with the substrate for binding to the catalytic site of the enzyme. A classic

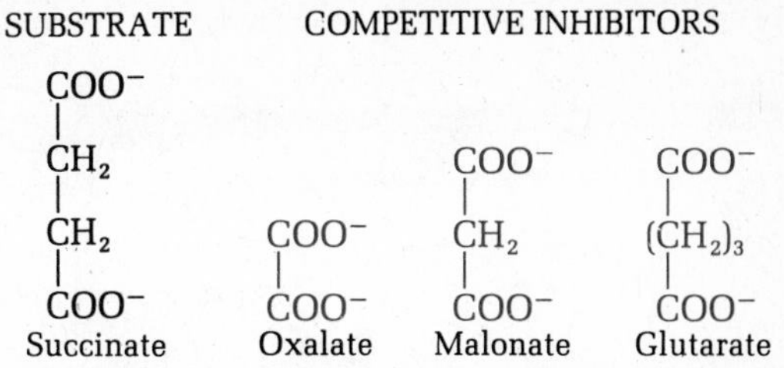

FIGURE 2-8 Competitive inhibitors of the enzyme succinate dehydrogenase. Malonate, oxalate, and glutarate resemble the normal substrate succinate in structure, and therefore compete for binding to the enzyme's catalytic site.

example involves *succinate dehydrogenase*, a mitochondrial enzyme that catalyzes the oxidation of succinate. Compounds resembling succinate in structure bind to the enzyme in its place, but do not undergo reaction (Figure 2-8). Enzyme molecules to which such compounds have bound are of course unavailable for binding the normal substrate, causing a net decrease in reaction rate.

The binding of competitive inhibitors to their respective enzymes is noncovalent and reversible, leading to an equilibrium between bound and unbound inhibitor molecules. During the period when inhibitor is not bound to a given enzyme molecule, the free enzyme is potentially available to bind to either a molecule of substrate or to another molecule of inhibitor. Hence the rate of the enzymatic reaction can be increased by raising the concentration of substrate relative to inhibitor, since a high concentration of substrate will effectively compete with the inhibitor for binding to the enzyme. At high concentrations of substrate the inhibition may be overcome entirely, since the probability becomes very great that any free enzyme molecule will encounter a molecule of substrate rather than inhibitor.

It is therefore characteristic of competitive inhibition that the effect of the inhibitor can be overcome by adding excess substrate; in other words, the maximal velocity (V_{max}) of a reaction is unaffected by competitive inhibitors. However, because of the binding of inhibitor molecules to the enzyme, it takes a higher substrate concentration to achieve V_{max} or one-half V_{max}; hence the K_m will be higher. An unaltered V_{max} and an increased K_m are therefore the hallmarks of competitive inhibition [Figure 2-9 *(top)*].

The sulfa drugs represent a striking example of how the above principles of competitive inhibition can be put to beneficial medical use. The structure of *sulfanilamide* closely resembles that of p-aminobenzoic acid, a substance from which bacteria manufacture the essential cofactor, folic acid (Figure 2-10). By competitively inhibiting the bacterial enzyme responsible for metabolizing p-aminobenzoic acid, sulfanilamide eventually leads to bacterial death. Human cells, on the other hand,

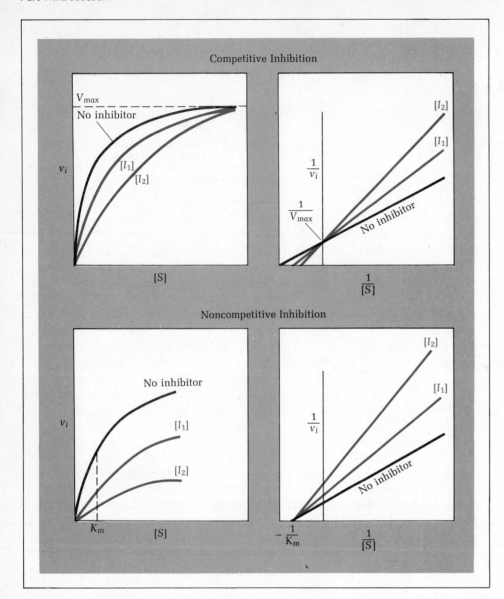

FIGURE 2-9 Graphic comparison of the effects of increasing concentrations of competitive and noncompetitive inhibitors on enzyme reaction rates. The two graphs on the right are plotted by the Lineweaver–Burk method. $[I_1]$ and $[I_2]$ represent data obtained in the presence of lower and higher concentrations of inhibitor, respectively.

do not use the metabolic pathway for converting p-aminobenzoic acid to folic acid, and so are not adversely affected.

In contrast to competitive inhibitors, the binding of **noncompetitive inhibitors** to enzymes cannot be reversed by adding excess substrate. The reason for this difference is that noncompetitive inhibitors bind to regions of the enzyme molecule other than the catalytic site where the substrate binds. Hence there is no direct competition between substrate and inhibitor for binding to the enzyme. Noncompetitive inhibitors instead cause alterations in enzyme conformation that lead to a decreased catalytic efficiency of the enzyme rather than to a decreased proportion of enzyme molecules having bound substrate. The value of V_{max} is therefore reduced in the presence of a noncompetitive inhibitor, and cannot be restored regardless of substrate concentration. A reduced V_{max} and an unaltered K_m are there-

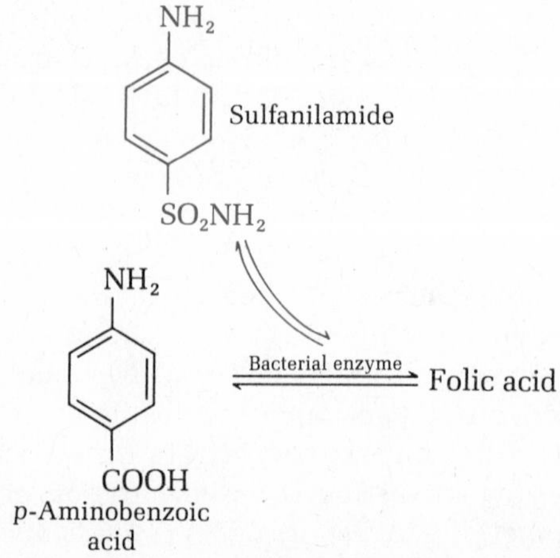

FIGURE 2-10 Competitive inhibition by sulfanilamide. Human cells are not adversely affected by this drug because they do not contain the enzyme that converts p-aminobenzoic acid to folic acid.

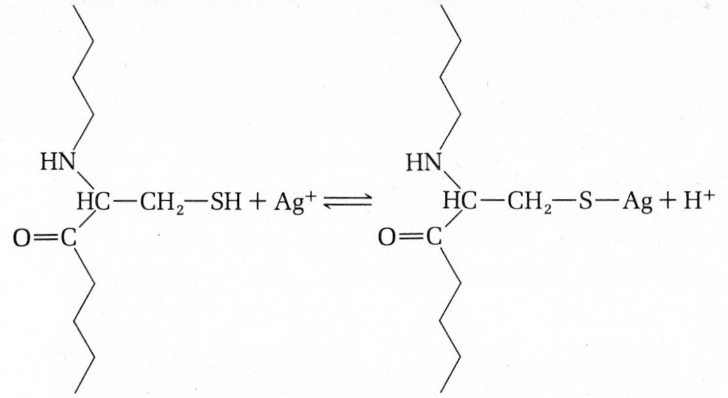

$$HN \diagdown \atop O=C \diagup HC-CH_2-SH + Ag^+ \rightleftharpoons \quad HN \diagdown \atop O=C \diagup HC-CH_2-S-Ag + H^+$$

Cysteine residue in enzyme

FIGURE 2-11 Noncompetitive inhibition of enzymes by heavy metals. A reversible reaction with the sulfhydryl group of the amino acid cysteine leads to an altered enzyme conformation and a resulting loss of catalytic activity.

fore the hallmarks of noncompetitive inhibition [see Figure 2-9 *(bottom)*].

Heavy-metal ions are an important class of substances known to act as noncompetitive inhibitors. Compounds in this group, such as *mercury, lead,* and *silver,* form reversible covalent bonds with sulfhydryl groups of the amino acid cysteine (Figure 2-11). The net result is a change in protein conformation and an associated decrease in enzymatic activity. The accumulation of heavy-metal pollutants in our environment is therefore a cause for growing concern.

Another type of noncompetitive inhibition occurs in the case of enzymes requiring metal-ion cofactors. Enzymes in this category are noncompetitively inhibited by agents capable of reversibly binding to the required ion. Examples are the binding of *cyanide* to iron, and the binding of *ethylenediaminetetraacetate* (EDTA) to magnesium. Because iron-containing proteins play essential roles in mitochondrial oxidation-reduction reactions, and magnesium ions are required for all reactions involving ATP, the effects of such inhibitors on the cell can be extremely widespread.

Analysis of enzyme kinetics in the presence of unknown inhibitors occasionally yields results that do not fit the typical pattern of either competitive (K_m altered) or noncompetitive (V_{max} altered) inhibition. Such inhibition kinetics involve an alteration in both K_m and V_{max}, indicating that the binding of the substrate to its enzyme as well as the enzyme's catalytic efficiency have been altered. In the cell, such mixed effects may be the rule rather than the exception.

Yet another type of kinetics is observed when enzyme inhibition is mediated by interactions between the various subunits of an enzyme. Such inhibition differs fundamentally from the examples described above in that it is part of a mechanism used by cells in the normal regulation of enzyme activity. It will therefore be discussed in the following section on enzyme regulation.

Regulation of Enzyme Activity

One of the key features that distinguishes enzymes from nonbiological catalysts is the fact that enzyme activity can be controlled. Regulation of enzyme activity is clearly required if cells are to maintain their metabolic pathways in proper balance, with each pathway generating neither too much nor too little of its particular product. Although long-term regulation can be achieved by altering the rates at which specific enzymes are synthesized or degraded, other mechanisms are available for inducing short-term increases or decreases in enzyme activity. In general, these short-term mechanisms involve altering the structure or conformation of preexisting enzyme molecules in such a way as to increase or decrease catalytic efficiency.

Allosteric Regulation of Enzyme Activity Short-term modulations of enzyme activity are often mediated by molecules that are intermediates or products of the metabolic pathways involved. When a metabolic product functions to inhibit one or more of the enzymes involved in the pathway by which it is synthesized, the phenomenon is termed **feedback inhibition.** This mechanism ensures that an oversupply of any particular product is not produced, for as more product is formed, the pathway producing that product becomes progressively more inhibited (Figure 2-12).

The molecular mechanism underlying feedback inhibition was not comprehended until 1963, when Jacques Monod, Jean-Pierre Changeux, and Francois Jacob first proposed their general theory of enzyme regulation. It was known at the time that feedback inhibitors do not structurally resemble the substrates of the enzymes they inhibit. These investigators therefore proposed that feedback inhibitors bind to a region of an enzyme distinct from the catalytic site. This region, which they termed the **allosteric site** (another site), is generally located on a different enzyme subunit than the catalytic site. According to this model, the reversible noncovalent binding of an inhibitor to its allosteric site causes an overall change in enzyme conformation and a resulting decrease in catalytic activity. It was also proposed that some enzymes contain allosteric sites where substances that increase enzyme activity may bind. Binding of such allosteric activators would lead to a change in enzyme conformation that

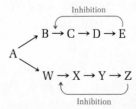

FIGURE 2-12 The principle of feedback inhibition. Each end product regulates the rate of its own formation by selectively inhibiting the activity of the appropriate enzyme(s).

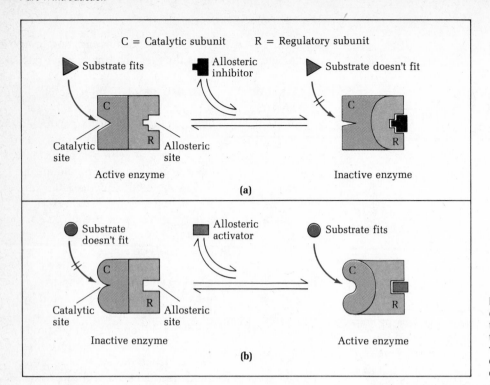

C = Catalytic subunit   R = Regulatory subunit

Substrate fits

Catalytic
site

Allosteric
inhibitor

Allosteric
site

Active enzyme

Substrate doesn't fit

Inactive enzyme

(a)

Substrate
doesn't fit

Catalytic
site

Allosteric
activator

Allosteric
site

Inactive enzyme

Substrate fits

Active enzyme

(b)

FIGURE 2-13 Allosteric control of enzyme activity. Allosteric inhibitors induce conformational changes that decrease enzyme activity (a), while allosteric activators induce conformational changes that increase enzyme activity (b).

increases rather than decreases catalytic activity (Figure 2-13). Each allosteric site, like a catalytic site, was hypothesized to be very specific for binding a particular substance.

There was little direct evidence to support the above model of allosteric control when it was first formulated. A few years later, however, John Gerhart and Howard Schachman found the behavior of the enzyme *aspartate transcarbamoylase* to closely conform to the allosteric model. Aspartate transcarbamoylase catalyzes the first step in the biosynthesis of the pyrimidine nucleotides cytidine triphosphate (CTP) and uridine triphosphate (UTP), and is subject to feedback inhibition by one of the end products of this pathway, CTP. When Gerhart and Schachman treated aspartate transcarbamoylase with the denaturing agent p-chloromercuribenzoate (PCMB), the enzyme was found to maintain its catalytic activity but lose its susceptibility to inhibition by CTP. Analysis of the denatured enzyme by moving zone centrifugation revealed the presence of two types of polypeptide subunits instead of the intact enzyme (Figure 2-14). One type of subunit possessed enzymatic activity but could not be inhibited by CTP. The other had no enzymatic activity but was capable of binding CTP. When the two types of subunits were recombined, normal enzyme activity susceptible to inhibition by CTP was regenerated. These results led to the conclusion that aspartate transcarbamoylase consists of at least two different types of subunits, **catalytic subunits** responsible for carrying out the reaction and **regulatory subunits** that bind the allosteric inhibitor, CTP. The multi-subunit makeup of this enzyme has since been confirmed by electron microscopy (Figure 2-15). It is

FIGURE 2-14 Experimental demonstration of the existence of separate catalytic and regulatory subunits in the allosteric enzyme aspartate transcarbamoylase. *(Top)* Analysis of control enzyme by moving-zone centrifugation reveals the presence of one component, susceptible to inhibition by CTP. *(Bottom)* Analysis of enzyme denatured by PCMB treatment reveals the presence of two components, a larger subunit with enzymatic activity but not inhibitable by CTP, and a smaller subunit with no enzyme activity but capable of binding CTP.

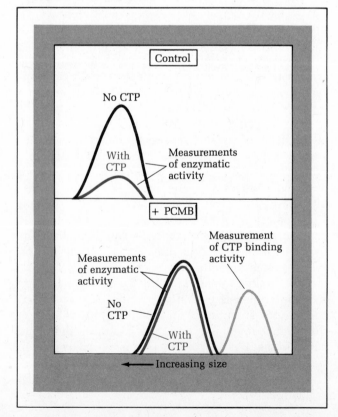

Control

No CTP

With
CTP

Measurements
of enzymatic
activity

+ PCMB

Measurements
of enzymatic
activity

Measurement
of CTP binding
activity

No
CTP

With
CTP

Increasing size

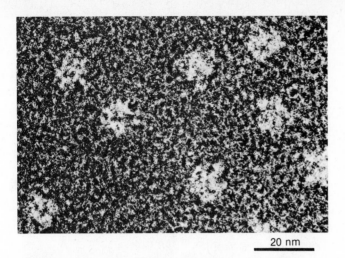

FIGURE 2-15 Electron micrograph of aspartate transcarbamoylase molecule, showing the existence of multiple subunits. Courtesy of R. C. Williams.

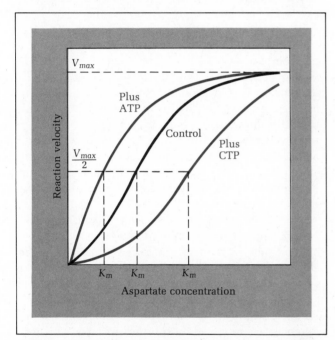

FIGURE 2-16 Allosteric regulation of aspartate transcarbamoylase. The allosteric inhibitor CTP increases the K_m of the reaction, while the allosteric activator decreases the K_m. The sigmoidal shape of the curve obtained for the control enzyme indicates the existence of cooperativity.

for several different types of allosteric inhibitors and activators, making allosteric control an extremely powerful mechanism for the control of enzyme activity.

In the case of both allosteric inhibitors and activators, an equilibrium exists between molecules bound to the enzyme and those free in solution. Hence a decrease in the intracellular concentration of an allosteric regulator will cause this molecule to dissociate from the allosteric site, while an increase in intracellular concentration will have the opposite effect. In this way, enzymatic activity can be quickly increased or decreased in response to changes in intracellular conditions.

Allosteric enzymes generally respond to allosteric regulators with an alteration in K_m, although sometimes the predominant effect is on V_{max}. The observed inhibition kinetics do not usually conform to those observed with typical competitive or noncompetitive inhibitors, however. This divergent behavior is usually caused by the presence of more than one catalytic or allosteric site, leading to cooperative effects between sites of the type described below.

Cooperativity Enzymes composed of multiple subunits often do not exhibit typical Michaelis–Menten kinetics. When one plots reaction velocity against substrate concentration for the allosteric enzyme aspartate transcarbamoylase, for example, an S-shaped or sigmoidal curve is obtained instead of the normally expected rectangular hyperbola (compare Figures 2-3 and 2-16). A sigmoidal shape means that the enzyme is becoming catalytically more efficient as the substrate concentration is increased. In this phenomenon, termed **positive cooperativity,** the binding of the first molecule of substrate is believed to induce a conformational change in the enzyme that causes its other catalytic sites to exhibit an increased affinity for binding substrate. Some enzymes, on the other hand, exhibit **negative cooperativity,** in which the binding of the first substrate molecule changes the protein's conformation so that the other catalytic sites bind substrate less readily. In this case the curve of reaction velocity against substrate concentration looks superficially like a hyperbola, but in fact it is significantly distorted (Figure 2-17).

Cooperativity provides a mechanism for producing enzymes that are more or less sensitive to changes in substrate concentration. In the case of positive cooperativity, an enzyme's catalytic activity will increase faster than normal as the substrate concentration is increased, while with negative cooperativity the opposite effect will occur. Under physiological conditions it is appropriate for some enzymes to be extremely sensitive to changes in substrate concentration, while for others a decreased sensitivity is more appropriate. The existence of positive and negative cooperativity gives cells a way of making both types of enzymes. Cooperativity can

now known that aspartate transcarbamoylase consists of two catalytic and three regulatory subunits, and that in addition to allosteric inhibition by CTP, it is susceptible to allosteric activation by ATP (Figure 2-16).

Many other examples of allosteric control have been discovered since this pioneering work with aspartate transcarbamoylase. Like aspartate transcarbamoylase, most allosteric enzymes consist of multiple polypeptide chains with catalytic and allosteric sites present on different subunits. Both types of sites can be on the same subunit, however, and in a few cases allosteric enzymes appear to consist of only a single polypeptide chain. A given enzyme may have specific sites

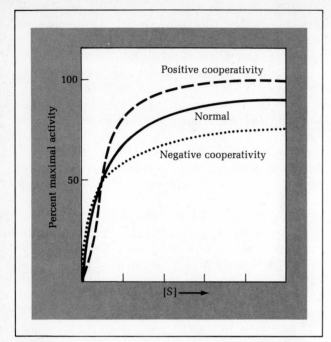

FIGURE 2-17 Effects of negative and positive cooperativity on enzyme kinetics. With positive cooperativity, enzyme activity is more sensitive to changes in substrate concentration and achieves maximal velocity at lower substrate concentrations than in the absence of cooperativity. Negative cooperativity achieves the opposite effect.

also affect the binding of allosteric regulators to allosteric sites. In this way enzymes with greater or lesser sensitivity to fluctuations in allosteric regulator concentrations can be produced.

Covalent Modification of Enzyme Structure Conformational changes leading to alterations in enzyme activity can also be induced by direct covalent modification of enzyme structure. One common mechanism for modifying the structure of preexisting enzyme molecules is by the addition or removal of chemical groups such as methyl, acetyl, adenyl, and especially phosphate. Among the numerous enzymes known to be subject to modification by phosphate groups is *phosphorylase,* an enzyme catalyzing the breakdown of glycogen to glucose-1-phosphate. Phosphorylase exists in two forms that differ from each other in structure and activity. The more active form, designated phosphorylase *a,* is a tetramer consisting of four identical polypeptide chains, while the less active phosphorylase *b* is a dimer containing only two chains. Each of the four subunits of phosphorylase *a* contains a serine residue to which a phosphate group is covalently attached. When these four phosphate groups are removed (by another enzyme designated phosphorylase phosphatase), the tetramer phosphorylase *a* dissociates into two molecules of the dimer, phosphorylase *b.* The relatively inactive phosphorylase *b,* in turn, can have these same serine residues rephosphorylated (by an enzyme

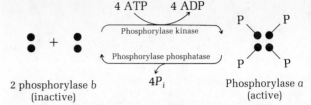

2 phosphorylase *b* $4P_i$ Phosphorylase *a*
(inactive)   (active)

FIGURE 2-18 Regulation of glycogen phosphorylase activity by covalent modification.

designated phosphorylase kinase), reconverting it to the more active phosphorylase *a* (Figure 2-18). The regulation of this enzyme is therefore seen to involve both the covalent modification of amino acid side chains and the association-dissociation of subunits.

Phosphorylation-dephosphorylation is also involved in the regulation of *pyruvate dehydrogenase,* an enzyme that catalyzes the conversion of pyruvate to acetyl-CoA. Pyruvate dehydrogenase is part of a larger **multienzyme complex,** a term referring to the physical clustering of several enzymes that function in sequence. Such complexes are fairly common in intermediary metabolism because they limit the distance each substrate molecule must diffuse as it is acted upon by subsequent enzymes in a reaction pathway. Among the enzymes present in the pyruvate dehydrogenase complex are a protein kinase and a protein phosphatase, which add and remove phosphate groups from pyruvate dehydrogenase respectively. Addition of phosphate groups inhibits pyruvate dehydrogenase activity, while their removal activates the enzyme.

Another way in which the covalent structure of enzyme molecules can be altered is by the cleavage of peptide bonds. Certain enzymes are synthesized by the cell as inactive precursors known as **proenzymes** or **zymogens.** Zymogens are catalytically inactive because of the presence of extra amino acids in their primary sequence. At the appropriate time and place, the extra sequences are removed by specific enzymes. This pattern of regulation is exemplified by the digestive enzymes *pepsin, trypsin,* and *chymotrypsin,* which are synthesized as the inactive zymogens *pepsinogen, trypsinogen,* and *chymotrypsinogen,* respectively. Only after these zymogens have been secreted by the cell into the digestive tract are they converted into their active forms. In the case of trypsinogen this activation involves removal of a hexapeptide from one end of the molecule, a reaction catalyzed by an enzyme called *enterokinase.* Similar reactions involving the removal of small peptides are involved in the activation of the other zymogens. Enzymes manufactured in the form of inactive zymogens are predominantly those whose catalytic activities would have deleterious effects if exerted within the cell. For example, the proteases pepsin, trypsin, and chymotrypsin would obviously destroy many essential proteins if they were allowed to act inside the cell.

Isozymes Regulation of enzymatic pathways can also be influenced by using different forms of the same enzyme, or **isozymes.** Isozymes catalyze the same reaction but are coded for by separate genes and manifest subtle differences in structural, physical, and catalytic properties. Although the presence of more than one enzyme catalyzing the same reaction may at first glance appear inefficient, the evolution of isozymes apparently reflects an adaptation to the needs of different cell types. A classic example is *lactate dehydrogenase* (LDH), the enzyme catalyzing conversion of pyruvate to lactate. Lactate dehydrogenase contains four subunits, each of which occurs in two types (designated H and M). A total of five different forms of lactate dehydrogenase can therefore be created. Two consist of four identical subunits and are designated H_4 and M_4, respectively. The other three are composed of varying mixtures of the two types of subunits, and are designated H_3M, H_2M_2, and HM_3, respectively.

Although these five isozymes all catalyze the same reaction, kinetic analyses have revealed that they differ in both K_m and V_{max}. The isozyme M_4 has a relatively low K_m and a high V_{max}, while the opposite is true for the isozyme H_4. This difference means that M_4 is more efficient at converting pyruvate to lactate than is H_4. It is interesting to note that the more efficient enzyme M_4 is enriched in skeletal muscle and embryonic tissue, where conditions often require the formation of lactate. The less efficient isozyme H_4, on the other hand, is enriched in heart muscle, where lactate production is appropriate only under certain emergency conditions.

The existence of isozymes thus allows the same enzymatic activity to be specifically adapted to the needs of different cell types. Isozymes can also be adapted for functioning in different subcellular locations, and may exhibit varying sensitivities to allosteric regulation. In addition to isozymes, proteins other than enzymes are also known to exist in multiple forms.

Mechanism of Enzyme Action

The mechanisms responsible for the extraordinary specificity and efficiency of enzymatic catalysts have long interested biologists and chemists. In the late nineteenth century Emil Fischer first noted that enzymes can distinguish between closely related three-dimensional configurations of the same substrate, leading him to propose that each enzyme has an **active site** whose three-dimensional configuration is exactly complementary to the shape of the substrate molecule upon which it acts (Figure 2-19). If the shape of the substrate isn't exactly complementary to that of the active site, no binding can occur and hence the reaction will not proceed.

In spite of its usefulness in explaining enzyme specificity, this **lock-and-key theory** is somewhat of an oversimplification. Kinetic data and physical measurements suggest that enzymes are dynamic structures that undergo changes in conformation upon binding of substrate. According to the **induced-fit model** proposed by Daniel Koshland, the proper three-dimensional fit between enzyme and substrate is triggered by a change in enzyme conformation that occurs after the substrate binds to the active site (Figure 2-20). This emphasis on the dynamic nature of enzyme-substrate interaction has implications for the mechanism of enzyme action to be discussed shortly.

In most enzymatic reactions the substrate is considerably smaller than the enzyme. For example the enzyme catalase, with a molecular weight of 250,000 daltons, catalyzes the breakdown of hydrogen peroxide, having a molecular weight of 34. It therefore follows that only a small portion of an enzyme's amino acid backbone can be involved in the active site that physically associates with the substrate. A variety of techniques have been devised for mapping the arrangement of the involved amino acids. One such approach is to

FIGURE 2-19 The lock-and-key model of enzyme-substrate interaction.

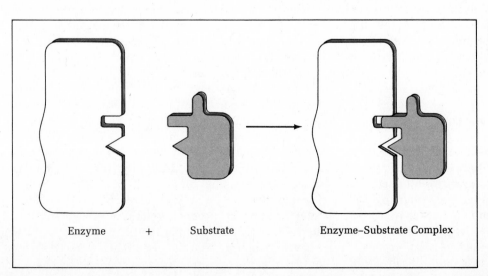

Enzyme + Substrate   Enzyme–Substrate Complex

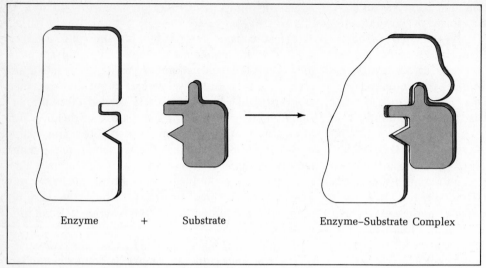

Enzyme   +   Substrate   Enzyme–Substrate Complex

FIGURE 2-20 The induced-fit model of enzyme-substrate interaction, showing how the interaction of substrate with the active site causes the enzyme to undergo a conformation change that enhances the complementary fit between enzyme and substrate.

treat enzymes with molecules that form covalent bonds with certain amino acid side chains. When the enzyme ribonuclease is reacted with the alkylating agent *iodoacetate*, for example, two histidines become selectively alkylated (amino acids 12 and 119). Alkylation of either of these histidines destroys the catalytic activity of the enzyme, suggesting that they are part of the active site. Such observations have made it clear that amino acids that are far removed from one another in terms of linear amino acid sequence come together to form the active site. Hence disruptions in protein conformation that alter the folding of the polypeptide chain can change the spatial relationship among amino acids forming the active site, leading to a loss in enzymatic activity.

The phosphorylating agent *DIPF*, which inactivates certain enzymes by attaching to the hydroxyl group of serine residues, has also been used to map active sites. This compound selectively interacts with a class of enzymes that share the presence of an essential serine residue at the active site. By analyzing the amino acid sequences adjacent to the serine with which the DIPF has reacted, considerable information about the arrangement of amino acids at the active site can be obtained. In this way it has been ascertained that certain regular patterns tend to recur in the active sites of different enzymes (Table 2-3).

The large acceleration in reaction rate occurring in enzymatically catalyzed reactions can be explained by at least four distinct phenomena involving the active site. First, the active site brings reacting molecules in close proximity to one another, precisely oriented so as to optimize the chances of the reaction occurring. Second, some enzymes act by forming an unstable covalent intermediate between enzyme and substrate. Such covalent intermediates generate the final reaction product more readily than would the substrate by itself. Third,

TABLE 2-3

Amino acid sequence of active sites of various enzymes

Enzyme	Sequence of Active Site[a]
Trypsin	Gly — Asp — Ser — Gly — Gly — Pro
Chymotrypsin	Gly — Asp — Ser — Gly — Gly — Pro
Elastase	Gly — Asp — Ser — Gly — Gly — Pro
Thrombin	Asp — Ser — Gly
Alkaline phosphatase	Asp — Ser — Ala
Butrylcholine esterase	Gly — Glu — Ser — Ala — Gly
Acetylcholine esterase	Glu — Ser — Ala
Aliesterase	Glu — Ser — Ala

[a] Note that Asp and Glu are similar amino acids (both acidic), as are Gly and Ala (both aliphatic).

some active sites contain functional groups capable of acting as acids or bases, permitting them to facilitate reactions by acting as general proton donors or acceptors. Finally, once an enzyme has bound to its substrate, it may induce a strain or distortion in the substrate molecule that enhances its reactivity. The fact that enzymes undergo an induced change in conformation upon binding substrate is at least consistent with this notion, since the resulting conformational change may distort the enzyme and/or substrate in such a way as to make them more reactive.

For any given enzyme, catalytic activity is most likely due to a combination of the above factors. In the case of the extensively studied enzyme chymotrypsin, for example, serine 195 is known to form a covalent intermediate with the acyl group of the peptide bond being hydrolyzed (Figure 2-21). Formation of this intermediate is facilitated by histidine 57, which acts as a proton acceptor. This property of histidine 57 also facilitates the release of product from the enzyme in the final step of the reaction. By themselves, the participation of a covalent intermediate and proton catalysis account for only part of the rate enhancement achieved by chymo-

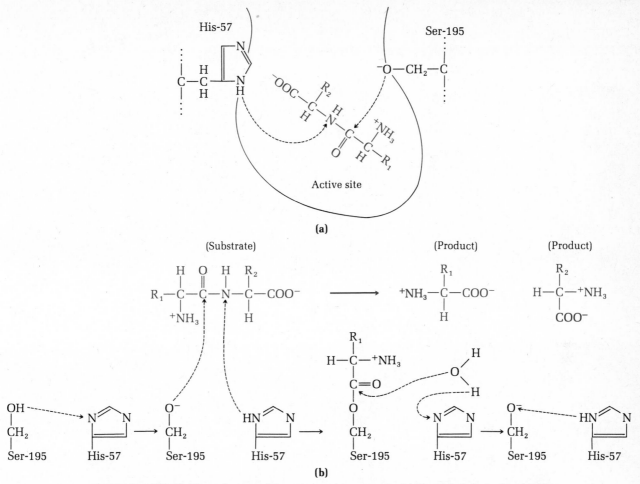

FIGURE 2-21 Mechanism of action of the enzyme chymotrypsin. (a) The active site is shown with substrate in place. (b) Steps involved in the reaction mechanism. The covalent bond formed with the substrate involves the hydroxyl group of serine 195 pictured in the drawing of the active site. In this step histidine 57 acts as a proton acceptor. Release of the covalent intermediate also involves the proton-accepting ability of histidine 57.

trypsin. Proximity, orientation, and strain effects must also be postulated in order to account for the extraordinary efficiency of this enzyme.

COUPLED REACTIONS AND THE ROLE OF ATP

Many important cellular events involve endergonic reactions that require an input of energy. The required energy is obtained from the oxidation of foodstuffs (or directly from the sun in the case of photosynthetic organisms). The transfer of energy from energy-requiring (endergonic) to energy-yielding (exergonic) processes involves the utilization of common chemical intermediates for coupling reactions to one another. The most prominent example of such an intermediate is **ATP** (adenosine triphosphate), a nucleoside triphosphate whose pivotal role in the cell was first perceived by Fritz Lipmann in 1940. As is summarized in Figure 2-22, the terminal phosphate group of ATP can be removed in an exergonic reaction, yielding **ADP** (adenosine diphos-

phate) and inorganic phosphate. The terminal phosphate of ADP can in turn be removed by another exergonic reaction, yielding **AMP** (adenosine monophosphate) and inorganic phosphate.

On the basis of the above free-energy changes, Lipmann postulated the existence of an "ATP cycle" in which the energy released from the oxidation of nutrients is used to drive the formation of ATP. When the energy stored in this ATP is needed, it is broken down to ADP and the free energy released. In this way ATP serves to couple energy-yielding reactions with energy-requiring ones (Figure 2-23).

ATP is not the only compound that functions to couple enzymatic reactions with one another. In oxidation-reduction reactions, cells employ a series of coenzymes to carry electrons from one reaction to another. Hence during the oxidation reaction A → B, electrons removed from substance A might be utilized to reduce a coenzyme. The electrons present in the reduced coenzyme might later be utilized in the reduction reaction X → Y, in which electrons are added to substance X to

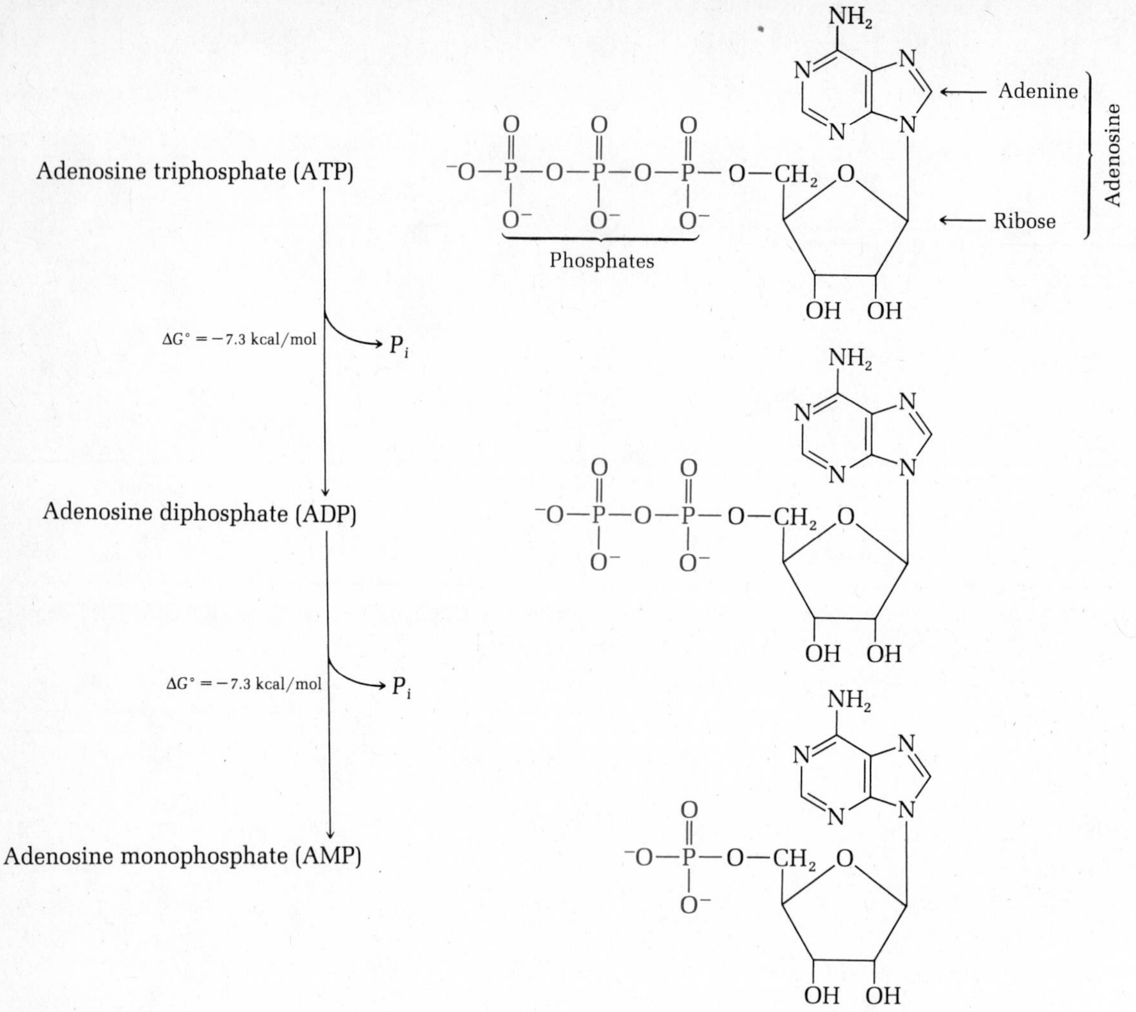

Adenosine triphosphate (ATP)
Adenine
Adenosine
Ribose
Phosphates
ΔG° = −7.3 kcal/mol
P_i
Adenosine diphosphate (ADP)
ΔG° = −7.3 kcal/mol
P_i
Adenosine monophosphate (AMP)

FIGURE 2-22 Summary of the free-energy changes accompanying the interconversion of ATP, ADP, and AMP.

CO2
+
H2O
Solar energy
(photosynthesis)
Energy
ATP
Energy for:
Biosynthesis
Active transport
Mechanical work
(movement)
Organic fuels
+
O2
ADP + P_i

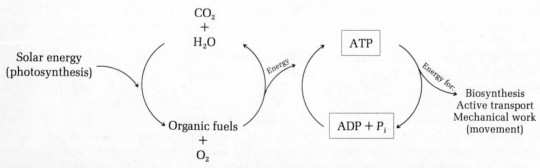

FIGURE 2-23 The ATP cycle. Energy released during the oxidation of foodstuffs is conserved by utilizing it to drive the formation of ATP. This energy stored in ATP can be later used by the cell to drive energy-requiring processes such as biosynthesis, active transport, and mechanical work.

convert it to Y. In this way, the coenzyme serves to couple the oxidation reaction A → B to the reduction reaction X → Y. Like ATP, such coenzymes couple many reactions involved in the mainstream of cellular metabolism.

SUMMARY

Enzymes are specialized proteins designed to speed up the rate of chemical reactions required by living organisms. Like all catalysts, enzymes modify reaction rates without affecting the final equilibrium conditions, and act in small quantities without being consumed. Enzymes can be distinguished from nonbiological catalysts, however, by their extraordinary efficiency and specificity, and by their ability to be regulated.

All catalyzed as well as noncatalyzed reactions are governed by a basic set of thermodynamic and kinetic principles. Thermodynamics refers to the study of energy changes that accompany chemical reactions. Reactions that give off free energy favor the formation of products at equilibrium, while reactions that consume free energy will have reactants predominate at equilibrium. Kinetics refers to the study of how fast reactions proceed toward equilibrium, rather than the equilibrium conditions themselves. Reaction kinetics are based on the amount of free energy required to achieve the transition state. Catalysts increase reaction rates by creating a new transition state of lower free energy, thereby increasing the rate at which equilibrium is achieved without affecting the equilibrium conditions themselves.

Kinetic analysis of enzyme-catalyzed reactions led Michaelis and Menten to propose that enzyme and substrate interact to form a transient enzyme-substrate complex. From this model one can mathematically derive two constants, V_{max} and K_m, which are useful in characterizing enzymes. V_{max} is the maximal reaction velocity obtainable with a given amount of enzyme, and hence is a measure of an enzyme's inherent catalytic efficiency. The Michaelis constant K_m, on the other hand, is the substrate concentration at which half maximal velocity is achieved. K_m, which is independent of enzyme concentration, is a measure of the binding affinity of an enzyme for its substrate.

In addition to V_{max} and K_m, other factors influencing reaction kinetics include pH, temperature, cofactors, inhibitors, and regulators. The effects of pH are due to alterations in ionization of the substrate and/or enzyme, which in turn influence enzyme conformation and substrate recognition. Increases in temperature cause an increase in reaction rate until the point of enzyme denaturation is reached. Cofactors are small nonprotein molecules, such as coenzymes, prosthetic groups, and metal ions, required for the activity of certain enzymes.

Enzyme inhibitors are divided into two categories, irreversible and reversible. Irreversible inhibitors usually bind to enzymes by covalent linkage, permanently disrupting proper conformation. Reversible inhibitors can be of either the competitive or noncompetitive type. Competitive inhibitors resemble an enzyme's normal substrate and therefore compete with the substrate for binding to the active site. Noncompetitive inhibitors bind to regions of an enzyme other than the active site, causing an alteration in conformation that decreases enzymatic activity. Competitive, but not noncompetitive, inhibitors can be overcome by adding excess substrate.

Enzyme activity can be regulated by a variety of physiological mechanisms. Allosteric inhibitors and activators bind to special regulatory sites, altering conformation in such a way as to inhibit or enhance catalytic activity. Enzymes composed of multiple subunits may exhibit cooperativity, in which the binding of one molecule of substrate influences the binding of subsequent substrate molecules. Covalent alterations of amino acid side chains, association-dissociation of polypeptide subunits, and cleavage of peptide bonds may be employed to alter an enzyme's structure and hence activity. Finally, enzymes can be manufactured in multiple isozymic forms characterized by slightly different properties.

The specificity of enzyme action is based on the active site, whose three-dimensional shape is complementary to that of its substrate. Amino acids far removed from each other in terms of primary sequence are brought together to form the active site. The large acceleration in reaction rates observed during enzymatic catalysis is due to a variety of factors, including proximity and orientation effects, formation of covalent intermediates, general proton catalysis, and strain.

Transfer of energy between enzymatic reactions involves common chemical intermediates that couple reactions to one another. The most widespread example of this phenomenon involves ATP, which stores energy released from exergonic reactions and uses it to drive endergonic reactions. Likewise, coenzymes capable of undergoing oxidation-reduction are used to carry electrons from one reaction to another.

SUGGESTED READINGS

Books

Lehninger, A. L. (1982). *Biochemistry: The Molecular Basis of Cell Structure and Function*, Worth, New York, Chs. 8 and 9.

Stryer, L. (1981). *Biochemistry*, 2nd Ed., Freeman, San Francisco, Ch. 6.

Articles

Gerhart, J. C., and H. K. Schachman (1965). Distinct subunits for the regulation and catalytic activity of aspartate transcarbamoylase, *Biochemistry* 4:1054–1062.

Lipmann, F. (1941). Metabolic regulation and utilization of phosphate bond energy, *Adv. Enzymol.* 1:99–162.

Monod, J., J. P. Changeux, and F. Jacob (1963). Allosteric proteins and cellular control systems, *J. Mol. Biol.* 6:306–309.

Sumner, J. B. (1926). The isolation and crystallization of the enzyme urease, *J. Biol. Chem.* 69:435–441.

CHAPTER
3

Experimental Approaches for Studying Cells

The purpose of the introductory section of this book has been to set the stage for a detailed analysis of the contributions each organelle makes to the life of the cell. We have therefore sought to provide basic information necessary for understanding the coming discussions on cell organelles, without getting too enmeshed in biochemical or structural details for the moment. One more background topic needs to be covered in the same general manner, and that concerns the experimental techniques used by cell biologists in investigating cell structure and function.

An early discussion of methodology is especially important because the remainder of the book places a heavy emphasis on the analysis of experimental data. This experimental focus has been chosen because our concepts about the cell are in a state of continual flux, changing in either subtle or significant ways as each major new set of experiments is carried out. To try to teach nothing but a set of "facts" about the cell would therefore be a disservice, for most "facts" will sooner or later be superseded by newer information. What is more significant for the student to understand is the way in which our ideas about the cell have developed. In practice, the science of cell biology is the process of asking questions, carrying out experiments designed to investigate these questions, drawing conclusions and formulating new questions based on the results, and repeating this cycle again and again. Our conclusions are always tentative, subject to further investigation and refinement. Disconcertingly few nonscientists understand what the scientist means by the word "truth." As Lynn White, Jr., author of *Machina ex Deo*, so aptly put it, to the scholar truth "is not a citadel of certainty to be defended against error: it is a shady spot where one eats lunch before tramping on" (1968:3).

The experimental techniques used by cell biologists encompass a wide spectrum of approaches, from the microscopic examination of cells and tissues to the isolation of organelles and macromolecules for biochemical analysis. The purpose of this chapter is to provide an overview of these approaches as general background material for later discussions of specific experiments. Instead of reading this chapter in intimate detail at this point, some students may prefer a more cursory examination designed to provide a general acquaintance with the kinds of techniques available. Later in the book, as specific procedures are mentioned in the context of individual experiments, one can refer back to this chapter for more detail. The advantage of putting these techniques together in one chapter near the beginning of the text, rather than spreading them throughout the book, is that it provides a more readily accessible reference source and overview of the procedures used by cell biologists.

GROWING CELLS IN THE LABORATORY

Ideally cells should be studied under conditions that are as close to natural as possible. Experiments carried out while cells are still in their native environment within intact animals or plants are said to be carried out *in vivo* (in life). Unfortunately, the types of experiments that can be carried out *in vivo*, especially in the case of complex multicellular organisms, are quite limited. It is possible to take cells from living organisms for microscopic analysis or to administer radioisotopes to intact organisms for monitoring metabolic activities occurring *in vivo*. But many variables cannot be controlled *in vivo* and the kinds of experimental manipulation possible are severely restricted. For this reason cell biologists often study the properties of cells or organelles that have been removed from their normal environment. Such experiments are said to be carried out *in vitro* (in glass). Methods for growing and maintaining cells *in vitro* will be discussed below, while techniques for isolating organelles will be described later in the chapter.

The science of cell culture dates back to the turn of the century, when pioneers such as Ross Harrison and Alexis Carrel first showed that tissues and cells can survive outside the body in an artificial culture medium containing blood serum or plasma. Carrel's discovery that extracts of chicken embryos cause cultured cells to grow and divide led to the establishment of permanent cell lines, and focused attention on tissue culture as an important tool for studying the regulation of cell growth, division, and differentiation. Many culture techniques have subsequently been developed, and hundreds of different cell lines are now routinely maintained in the laboratory.

Cells of animal or plant origin are usually grown either in **suspension culture** or as a **monolayer.** In suspension culture cells are dispersed throughout a liquid medium and can therefore be grown in relatively large quantities. Cells that do not grow well under these conditions can be cultivated as monolayers on glass or plastic surfaces, to which they tightly adhere. In this technique cells are placed in a tube or flask and covered with a thin layer of nutrient medium. The cells become attached to the glass or plastic surfaces and grow until they form a confluent layer of cells covering the entire surface of the container. This "monolayer" of cells can be detached and a portion transferred to another container to keep the culture constantly growing and dividing (Figure 3-1).

Bacteria can also be maintained in culture, either in suspension or on agar plates. *Agar* is a carbohydrate material derived from seaweed that can be made into semisolid gels containing various nutrients. Bacteria are grown on agar by streaking a sample of cells across the gel surface. From this initial inoculation individual colonies, or *clones*, that have grown from single cells can be selected for further analysis.

The composition of the nutrient medium has a critical influence on the properties of cells grown in culture. Prokaryotic cells require only a source of carbon, usually provided by a carbohydrate like glucose, and inorganic salts, which provide essential elements such as N, P, S, Mg, Na, and K. Trace elements required for cell growth are generally present in sufficient quantities as impurities in the other ingredients.

The nutritional requirements of eukaryotic cells are generally more complex than those of bacteria. For this reason early tissue culture work utilized natural biological substances, such as serum or plasma, as nutrient sources. Harry Eagle spent many years in defining the nutritional needs of mammalian cells and formulating media to optimize their growth. As more details about nutrient requirements gradually became available, attempts were made to design synthetic media composed of a completely defined mixture of substances such as salts, amino acids, vitamins, and carbohydrates. Although it is generally a requirement to supplement such concoctions with serum in order to obtain optimal cell growth, a few cell types do grow well in completely synthetic media.

FIGURE 3-1 Micrograph of fibroblasts grown as a monolayer in tissue culture. Courtesy of K. K. Sanford.

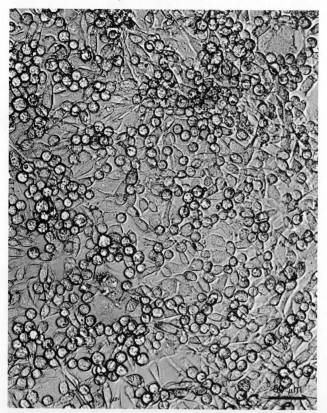

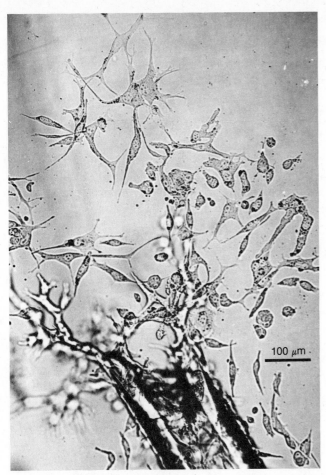

100 μm

FIGURE 3-2 A clone of fibroblasts that originated from a single cell in a capillary tube. The cells have migrated out through one end of the tube. Courtesy of K. K. Sanford.

Cells that normally exist in nature as independent entities adapt to growth in tissue culture with relative ease. Cells from multicellular organisms, on the other hand, are not accustomed to growing as individual cells and often do not adjust well to culture. And those cells that do adjust to growing in culture often lose their capacity to divide after a certain period of time. This period varies among cell types, averaging about 50 rounds of division for human cells. Such cell populations capable of only limited multiplication in culture are referred to as **primary cultures.**

In contrast, some cells acquire the ability to undergo unlimited division in culture, and are therefore said to be **transformed.** Transformation may occur spontaneously or may be induced by certain chemicals or viruses. Because of their unlimited capacity to divide, transformed cells can be used to set up permanently dividing **cell lines.** It is usually desirable to work with pure cell lines consisting of one cell type. Since the original tissues from which cultures are established generally consist of heterogeneous populations of cells, one must be able to isolate a single cell and allow it to grow by successive divisions into a homogeneous cell population called a **clone** (Figure 3-2). Although the rapid growth and unlimited lifespans of permanent cell lines make them an ideal source of material for many experiments, it should be emphasized that such transformed cells exhibit many aberrant properties. They usually possess an abnormal number of chromosomes, and produce tumors when injected back into intact organisms.

In many instances cell lines have been established from tumors rather than from normal tissue. These tumor cell lines behave just like cells transformed in culture. The classic example of a permanent cell line established in this way is the *HeLa* line. The original source of HeLa cells was a malignant uterine tumor surgically removed from a woman named Henrietta Lacks in 1951. Although she died a year later, the cells from her tumor have been maintained in culture to this day. HeLa cells are now grown in hundreds of laboratories around the world, and their use has contributed to the solution of a large number of important biological and medical problems.

Although the vigorous growth of HeLa cells in culture has contributed to their usefulness in scientific research, it has recently become evident that it may also pose some serious problems. If another tissue culture cell line is accidentally contaminated by even a single HeLa cell, that culture may eventually become overgrown with HeLa cells. Because most cell lines look alike, this process may occur without any accompanying visual changes, so it is important to make certain that the actual identity of cells maintained in tissue culture is known. One way to identify cells is to check the number and appearance of their chromosomes, for at this level differences between cells types can be readily detected.

The major advantage of using tissue culture is that it allows cells to be studied under carefully controlled conditions. Cellular responses to various external agents, such as hormones or metabolic inhibitors, can be gauged without the complicating effects of secondary factors present *in vivo.* Because they have been selected for rapid growth characteristics, tissue culture cells are also excellent models for studying the regulation of cell growth and division. Finally, certain cell lines retain differentiated properties characteristic of their native state, permitting detailed analysis of specialized cell functions.

The development of a technique called **cell fusion** has significantly enhanced our ability to probe the functional activities of cultured cells. In 1960 a research team in Paris headed by Georges Barski discovered that two cells of different types can fuse together to form a single hybrid cell. Barski had been studying two permanent mouse cell lines that differed in both morphology and chromosome content. When the two cell lines were grown together in the same culture, a third cell type eventually appeared. This third cell type contained genetic material from both original cell lines combined together in a single nucleus, and was therefore termed a **synkaryon** (*syn* = together; *karyon* = nucleus).

Shortly thereafter Henry Harris discovered that the rate of cell fusion can be dramatically increased by treating cells with Sendai virus that have been previously inactivated by exposure to ultraviolet light. Sendai virus apparently facilitates fusion between the plasma membranes of adjacent cells, producing synkaryons as well as **heterokaryons** (hybrid cells in which the nuclei of the fused cells retain their separate identities). The remarkable thing about cell fusion is that in addition to generating hybrids between cells of different types and from different species, the hybrids produced are capable of functioning and, in the case of synkaryons, reproducing as permanent cell lines.

The cell fusion technique has played an important role in many areas of cell biology. Mechanisms underlying specialized cell functions have been investigated by fusing together cells with and without the function in question. The nature of the defect present in cancer cells has been studied in hybrids of normal and malignant (cancerous) cells. Cell fusion has also been used to assign specific genes to individual human chromosomes. In this chromosome-mapping process, advantage is taken of the fact that hybrids formed between human cells and certain rodent cells selectively lose human chromosomes. By correlating the presence or absence of various genetic traits in the hybrid cells with the presence or absence of particular chromosomes, human genes can be assigned to individual chromosomes. The significance of these and other applications of the cell fusion technique will become more apparent in the appropriate contexts later in the text.

MICROSCOPY

Our current appreciation of the complexity of cell ultrastructure and its relationship to cell function has been made possible by a combination of powerful microscopic and biochemical techniques. It was the microscope, of course, that first opened our eyes to the existence of cells. Historically the use of curved surfaces for magnifying objects was reported as early as A.D. 127, although it was not until 1235 that this principle led the Englishman Roger Bacon to the invention of eyeglasses. By 1590 Jans and Zacharias Janssen had combined two convex lenses within a tube to construct the forerunner of the light microscope. As we saw in Chapter 1, the invention of the microscope in turn led to the discovery of cells and eventually to the formulation of the cell theory.

The Light Microscope

Typical prokaryotic and eukaryotic cells fall within the size range of 1–100 μm (see Figure 1-2). Because the unaided human eye cannot resolve objects smaller than 100 μm in size, microscopes are needed for the visualization of subcellular architecture. Although microscopes create a magnified image of the object being observed, magnification per se is not the issue. One can magnify a photograph, for example, to any size one wants, but eventually nothing more is gained because the image simply becomes blurred. The important issue is not magnification but **resolution,** which refers to the ability to distinguish closely adjacent objects as separate entities. The greater the ability to distinguish two objects from one another, the greater the clarity of the image produced.

The lower limit of resolution for any optical system can be calculated from the following relationship:

$$r = \frac{0.61\lambda}{n \sin \alpha} \tag{3-1}$$

where r, or **resolving power,** is the minimum distance between two points that can be recognized as separate, λ is the wavelength of light (or other radiation) used to illuminate the object, n is the refractive index of the medium in which the object is placed, and $\sin \alpha$ is the sine of half the angle between the specimen and the objective lens (Figure 3-3). The entire term $n \sin \alpha$ is often referred to as the **numerical aperture.**

It is apparent from this equation that only a small number of variables affect the resolving power of a microscope. The refractive index can be increased by immersing the sample in oil ($n = 1.5$) rather than air ($n = 1.0$), and moving the lens closer to the specimen to increase α. The upper theoretical limit of α is 90°, meaning that the value of $\sin \alpha$ cannot exceed 1. Hence the maximum numerical aperture of an optical system employing an oil immersion lens will be $1.5 \times 1 = 1.5$. A microscope using white light, which has an average wavelength of about 550 nm, will therefore have a resolving power of 550/1.5, or about 220 nm. This means that objects closer to one another or smaller than 220 nm cannot be distinguished. A resolving power of 220 nm is adequate to see some details of subcellular structure, but many organelles, such as ribosomes, cellular membranes, microtubules, microfilaments, intermediate filaments, and chromatin fibers, cannot be resolved at this level.

In order to take maximum advantage of the inherent resolving power of the light microscope, samples must be prepared in a way that produces **contrast,** that is, a difference in darkness or color of the various structures being examined. Without special treatment to increase contrast, little discernible structure will be visible because most cell structures will appear either opaque or transparent. Contrast is usually enhanced by the application of specific *stains* that color or otherwise alter the light-transmitting properties of cell components. As will be seen in a moment, certain modifica-

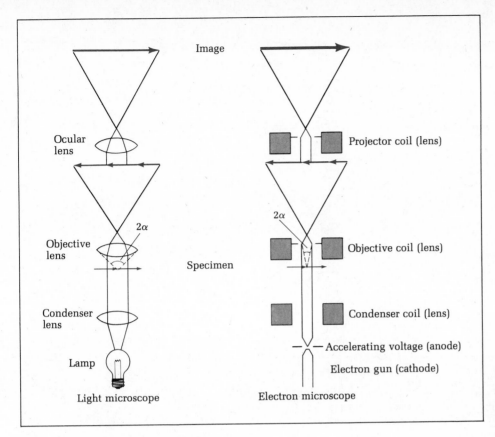

Image

Ocular lens

Objective lens

2α

Specimen

Condenser lens

Lamp

Light microscope

Projector coil (lens)

2α

Objective coil (lens)

Condenser coil (lens)

Accelerating voltage (anode)

Electron gun (cathode)

Electron microscope

FIGURE 3-3 Image formation in the light and electron microscopes. The aperture angle (2α) in the electron microscope is usually no more than a few tenths of a degree, but is drawn larger to make it visible. To facilitate the side-by-side comparison, the electron microscope is drawn upside down.

tions of the optical system can also enhance contrast.

The most common approach taken for the preparation of tissues for light microscopy involves slicing the material into sections thin enough to be transparent under the microscope. The first step in this process is treatment of the tissue with a fixative. Fixation kills cells while largely preserving their native structural appearance. The most widely employed fixatives are acids and aldehydes such as acetic acid, picric acid, formaldehyde, and glutaraldehyde. To provide the physical support needed for cutting thin sections, tissues are either embedded in paraffin wax or quick-frozen. Sections are cut on a special instrument called a *microtome*, after which they are spread out on glass microscope slides and stained. Although a fairly reliable picture of cell structure can be obtained from the examination of stained thin sections, artifacts due to the fixation, sectioning, and staining procedures can introduce distortions.

Because fixation destroys all behavioral and metabolic activity, examination of fixed tissue sections provides only a static view of the cell. As an alternative it is possible to visualize living cells directly under the light microscope, but the thickness and opacity of intact cells create problems of resolution and contrast. In response to these problems, several modifications of the light microscope have been introduced with the purpose of enhancing contrast by exploiting differences in thickness and refractive index. The **refractive index** of a sub-

stance is a measure of the velocity with which light waves pass through it. A beam of light passing through an object of high refractive index will be slowed down during its passage, resulting in a change in phase relative to light waves that have not passed through the object. Many cellular structures have high refractive indices and thus alter the phase of light waves, but unfortunately the human eye cannot detect such phase changes.

The **phase contrast microscope** overcomes this problem by converting phase differences into alterations in brightness. This conversion is accomplished by taking waves that have been refracted and using them to interfere with the unrefracted waves. **Interference** refers to the process by which two or more light waves combine to reinforce or cancel one another, producing a wave equal to the sum of the two combining waves. The refracted and unrefracted waves are easily separated from each other because waves passing obliquely into an area of changing refractive index not only have their phase altered, but also have the angle of their paths deflected. Thus the illumination source in a phase microscope is a cone of light whose rays pass obliquely through the specimen, causing refracted waves to deviate away from the unrefracted rays. When the unrefracted and refracted rays are focused back into the image plane, they undergo interference. For most biological materials, however, the phase difference between the unrefracted and refracted waves is only

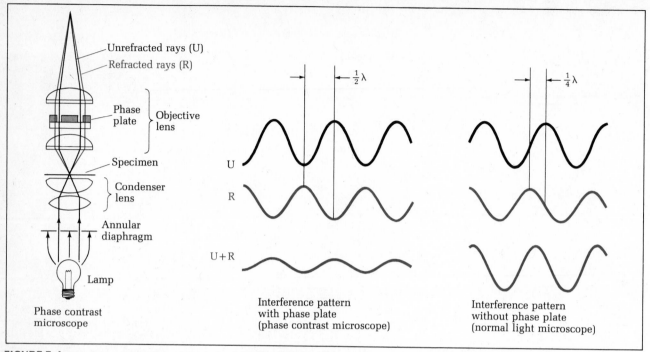

FIGURE 3-4 Operation of the phase contrast microscope. Refracted rays, which are delayed about one-fourth wavelength in the normal light microscope, are delayed an extra one-fourth wavelength by the phase plate. A comparison of the wave diagrams on the right illustrates how this extra delay enhances the interference obtained when the unrefracted and refracted rays are focused back together (U + R).

about one-fourth wavelength, producing only minimal interference. In an ordinary light microscope, this effect would be undetectable to the eye.

In a phase contrast microscope, however, a *phase plate* of refracting material designed to retard the refracted waves by an additional one-fourth wavelength is placed in the path of the refracted waves. When the refracted and unrefracted waves are now refocused in the image plane, they again undergo interference. But the additional one-fourth wavelength delay in the refracted waves increases the interference between the two to such a degree that a significant decrease in the amplitude of the resulting wave occurs (Figure 3-4). Since changes in amplitude are detected by the human eye as changes in brightness, the reduced amplitude appears as a darkened area in the specimen. In this way the phase microscope converts differences in refractive index and thickness to contrasting degrees of brightness (Figure 3-5b).

The **interference microscope** is similar to the phase contrast microscope in principle, but is more sophisticated in operation because it employs special mirrors to split the light beam into two separate rays. One beam passes through the specimen, while the other is routed through a control slide or medium. When the two beams are recombined, any changes that have occurred in the phase of the specimen beam cause it to interfere with the control beam. Complete separation of the two beams enhances sensitivity to small differences in refractive

index. Interference microscopy can be used to convert phase changes into a vivid spectrum of colors somewhat reminiscent of stained preparations. The magnitude of the phase changes can even be quantitated and related to the dry weight of cellular components. A special type of interference microscope, known as the **differential interference microscope** with Nomarski optics, enhances contrast by recombining the split beam in polarized light. The remarkable enhancement in resolution obtained in this way produces a striking three-dimensional image of the specimen being examined (see Figure 3-5c).

Yet another way of increasing contrast with living material is the use of **darkfield microscopy,** which employs a special condenser to illuminate the specimen at an angle such that no direct light enters the lens. Under these conditions the only light reaching the observer is that which has been scattered by interacting with the specimen. Hence in darkfield microscopy the background appears dark, while cell structures are seen as areas whose brightness is directly proportional to the amount of light scattered (see Figure 3-5d).

One of the oldest methods for improving contrast with the light microscope involves the use of *polarized light,* whose waves travel in a single plane. **Polarization microscopy** is most useful for visualizing *birefringent* structures, which differ in refractive index in different planes. In polarization microscopy a special *polarizer* is used to transmit polarized light through the specimen,

FIGURE 3-5 Examples of various types of microscopy. (*Top to bottom*) (a) Living fibroblast viewed by light microscopy. (b) The same cell viewed by phase contrast microscopy. (c) The same cell viewed by differential interference microscopy. (d) The same cell viewed by dark-field microscopy. (e) Fixed and stained fibroblasts grown in tissue culture and viewed with the light microscope. (f) Thin-section electron micrograph of fibroblast in connective tissue (arrow). Top four photographs courtesy of K. Roberts. Bottom two photographs courtesy of D. W. Fawcett.

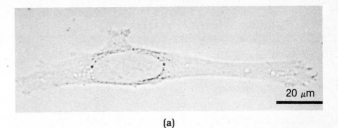

(a)

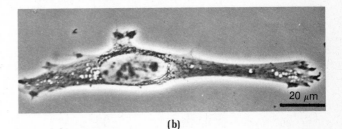

(b)

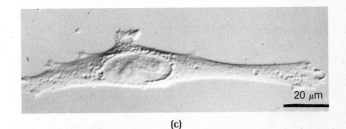

(c)

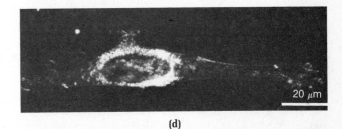

(d)

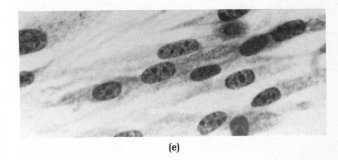

(e)

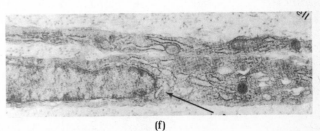

(f)

while a similar device called an *analyzer* is placed above the objective. By measuring how much one must rotate the analyzer to achieve maximum and minimum brightness of the specimen, differences in refractive index between planes can be determined. Since these differences in refractive index are determined by an object's molecular organization, polarization microscopy provides useful information about the molecular substructure of cell organelles. This approach has been useful in the study of crystalline and fibrous materials, such as the mitotic spindle and muscle fibers, but its general utility as an analytical tool for investigating cell ultrastructure has been superseded by advances in electron microscopy.

The Electron Microscope

In spite of the numerous modifications introduced to increase the usefulness of the light microscope, one cannot overcome its inherently limited resolving power. The basic factor underlying this limitation is the wavelength of the light involved (Equation 3-1). By switching the source of illumination from light to electrons, however, an enormous improvement can be obtained. The development of the electron microscope and its first application to biological materials in the late 1940s therefore ushered in a new era, opening our eyes to an exquisite subcellular architecture never before seen.

The discovery that electrons have wavelike properties and can be focused on objects using an electromagnetic field was the major impetus to the construction of such a microscope. By the late 1930s a system of magnetic coils capable of focusing an electron beam had been designed, and it was shown that the effective magnification of images created by illuminating objects with such a beam can be regulated by adjusting the current flowing through the coils. In this way the first electron microscope was created, forming an image with a focused electron beam in much the same way as a light microscope uses a focused light beam (see Figure 3-3).

Because the effective wavelength of a typical electron beam is 100,000 times shorter than the wavelength of light, an enormous enhancement in resolving power is theoretically possible. But practical limitations in the design of the magnetic coils or "lenses" used to focus the electron beam have prevented the electron microscope from achieving its full theoretical potential. The underlying problem is that magnetic lenses produce considerable distortion when the angle of illumination (2α) is more than a few tenths of a degree. Such a small angle is orders of magnitude less than that of a good glass lens, giving the electron microscope a numerical aperture that is considerably smaller than that of the light microscope. The practical resolving power for the typical electron microscope is therefore only about 0.5 nm, far from the theoretical limit of 0.002 nm. Of course 0.5 nm is still nearly a thousand times better than the capabilities of a typical light microscope.

The most commonly employed type of electron microscope is termed the **transmission** electron microscope because it forms an image from electrons that have passed through the specimen being examined. These electrons are not visible to the human eye, however, so they must be detected by interaction with a fluorescent screen or photographic plate. Components in the specimen that scatter electrons, preventing them from passing straight through, appear dark in the photographic image and are referred to as being *electron dense*. Regions with less electron-scattering ability, on the other hand, appear lighter. The final image obtained thus depends on differences in the electron-scattering abilities of the various structures that make up the cell.

The electron-scattering ability of these cell structures in turn depends on the atomic number of the elements from which they are constructed. The higher their atomic number, the greater the electron-scattering ability. The most common elements occurring in biological molecules (C, H, O, N, P, S) have relatively low atomic numbers and so scatter electrons poorly, resulting in an image with little contrast. To counteract this problem *electron stains* have been introduced to enhance contrast. Such stains include metals of high atomic number such as osmium, uranium, and lead, which form complexes with biological molecules and thereby enhance their electron-scattering properties.

In spite of the enhanced resolution made possible by use of the electron microscope, this approach is not without its inherent limitations. An electron beam is too weak to pass an appreciable distance through air, so the internal chamber of an electron microscope must be kept under vacuum. This lack of penetrating power also limits specimen thickness to no more than a few hundred nanometers. Such restrictions create many technical problems in preparing biological material for observation, and make examination of living cells virtually impossible.

Pioneering studies by Keith Porter, George Palade, Fritiof Sjøstrand, and Humberto Fernández-Morán carried out in the early 1950s led to the first suitable techniques for fixing and sectioning tissues for electron microscopy. The fixatives chosen were mainly oxidized metallic compounds such as osmium tetroxide, potassium permanganate, and phosphotungstic acid. Such

agents cause minimal distortion of cell structure and simultaneously function as electron stains. For gentler fixation, formaldehyde and glutaraldehyde have subsequently become popular. After fixation, tissues are embedded in a hard plastic resin and sectioned with a special instrument called an *ultramicrotome* to a thickness of no more than 50–100 nm. Examination of the resulting *thin sections* with the electron microscope provided our first views of the complexity of subcellular organization, and this approach remains the most widely used electron microscopic technique to this day.

In addition to thin sectioning, several other methods for preparing tissues for electron microscopy have been devised. For example, the shape and surface appearance of small particles, such as intact organelles or viruses, can be examined without slicing these specimens into thin sections. In the **negative staining** technique such particles are suspended in a heavy-metal solution that does not stain the specimen and are then visualized in relief against the stained background (Figure 3-6). In the closely related **positive staining**

FIGURE 3-6 *(Top)* Diagram of the negative-staining technique in which the sample is treated with a heavy metal that stains only the background. *(Bottom)* Electron micrograph of negatively stained wound-tumor virus. Courtesy of R. C. Williams.

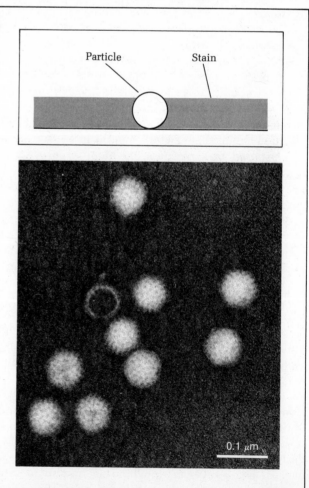

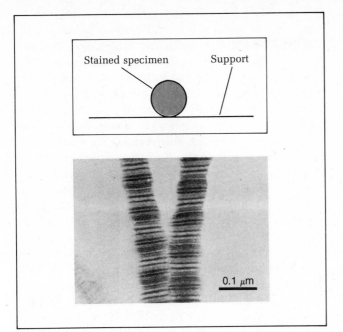

FIGURE 3-7 *(Top)* Diagram illustrating the positive-staining method in which the specimen is stained and the background stain is then removed. *(Bottom)* Electron micrograph of positively stained collagen fibers. Courtesy of J. Woodhead-Galloway.

technique, a specimen is first reacted with the stain and the stain then is removed, producing a stained sample visible against an unstained background (Figure 3-7).

Isolated particles can also be visualized by placing them in an evacuated chamber and spraying heavy metal across their surfaces. This **shadow-casting** process causes metal to be deposited on one side of the specimen, creating a "shadow" and a resulting three-dimensional appearance (Figure 3-8). A related procedure has been developed by Albrecht Kleinschmidt for visualizing purified macromolecules such as DNA and RNA. In this technique a solution of DNA and/or RNA is spread on an air-water interface, creating a molecular monolayer that is collected on a thin film and visualized by uniformly depositing heavy metal on all sides.

Yet another way of preparing unsectioned material for the electron microscope involves direct viewing of specimens without any staining or heavy-metal deposition. The inherent thickness of such **whole-mount** preparations provides sufficient electron scattering to produce an image. The relative darkness of various regions of the image can even be measured and used to calculate the overall mass of the specimen. This technique of whole-mount microscopy was especially helpful in early studies of chromatin and chromosome structure.

The **freeze-etch** technique of sample preparation is fundamentally different from the approaches described thus far. Instead of employing fixatives to preserve cell structure, specimens are rapidly frozen in liquid Freon, placed in a vacuum, and struck with a sharp knife edge.

At this temperature biological samples are too hard to be cut, and instead fracture along lines of natural weakness. These weak areas are generally associated with biological membranes. Brief exposure of the broken tissue to vacuum results in sublimation of water from the fractured surfaces. This removal of water produces an "etching" effect, that is, an accentuation of surface detail. A replica of the freeze-etched specimen is made by shadowing the fractured surface with a heavy metal, such as platinum, and then backing it with a carbon film. After dissolving the tissue in strong acid, the remaining metal replica can be viewed with the electron microscope.

Such preparations provide a unique picture of cells, particularly where membranes are involved (Figures 3-9 and 4-7). At least four different views of a membrane can be generated by fracture planes: two surface views and two images of the fracture plane. Figure 3-10 illustrates the origins of these four views and summarizes a

FIGURE 3-8 *(Top)* Diagram illustrating how the spraying of heavy metal at an angle causes an accumulation of metal on one side of a specimen and a "shadow" area lacking metal on the other side. *(Bottom)* Electron micrograph of tobacco mosaic virus shadowed with uranium. Courtesy of R. C. Williams.

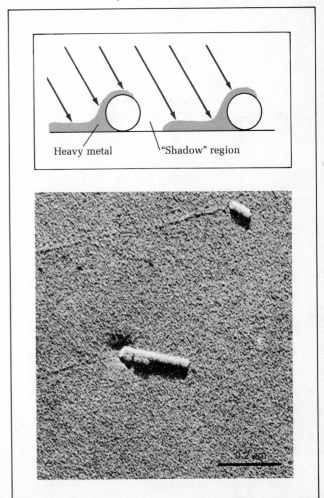

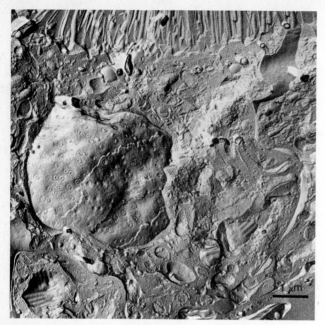

FIGURE 3-9 Freeze-etch micrograph of rat kidney cell showing folded plasma membrane at the top and the large nucleus in the center. Courtesy of H. D. Fahimi.

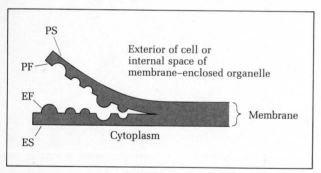

FIGURE 3-10 Four views of biological membranes can be generated by freeze-fracturing. The internal faces revealed by splitting the membrane down the middle are referred to as protoplasmic and exoplasmic fracture faces (PF and EF), while the membrane surfaces are referred to as the protoplasmic and exoplasmic surfaces (PS and ES).

system of nomenclature employed to facilitate reference to the various surfaces.

Freeze-etching is particularly useful because it avoids exposure to fixatives, embedding agents, and stains, all of which may deform cell ultrastructure. Unlike such treatments, rapid freezing causes minimal tissue distortion and permits instantaneous arrest of cell function. In general, the picture of cell architecture obtained by the freeze-etch technique supports that derived in other ways. The fact that fracture planes either follow membrane surfaces or bisect them internally has also made this procedure useful for the analysis of membrane structure.

The electron microscopic techniques described thus far all involve formation of an image by electrons transmitted through a specimen. Another type of electron microscope has been developed, known as the **scanning electron microscope,** which forms images from electrons deflected off a specimen's surface. The scanning electron microscope uses a fine beam of electrons that is moved back and forth across the specimen in much the same way as a moving electron beam is used to form a television image. As the beam traverses the surface of the object, electrons are deflected to varying degrees. In some cases, secondary electrons are also induced and emitted. These deflected and emitted electrons are detected by a photomultiplier tube and used to form a three-dimensional image of the object's surface features (Figures 3-11 and 3-12). The 10-nm resolving power of present-day scanning electron microscopes is considerably less than that of the transmission electron microscope, although projected technological advances should ultimately permit a resolution of 0.5 nm. Transmission and scanning electron microscopes could have similar resolving power, permitting two- and three-dimensional analysis of cellular ultrastructure.

The preparation of samples for scanning electron microscopy usually involves fixation or freeze-drying to preserve the specimen's shape, though simple air drying is suitable if shape changes can be tolerated. The sample is then shadowed with a heavy metal and viewed directly with the microscope. Since the scanning microscope utilizes deflected electrons rather than electrons transmitted through the sample, the image depicts only surface features. Neither light microscopy nor transmission electron microscopy are particularly well suited for this purpose, so the development of the scanning microscope has been of major importance in unraveling the three-dimensional surface architecture of cells and organelles.

X-Ray Diffraction

Although the resolving power of the best transmission electron microscopes now approaches a few tenths of a nanometer (about the diameter of an amino acid), there are several reasons why this instrument cannot be used to discern the shape and internal structure of individual macromolecules. First of all, the low electron-scattering ability of biological molecules requires the use of electron stains that reduce resolvable detail by coating the molecules with which they react. Furthermore, most electrons are scattered at or near the surface of the object being examined, thus precluding any view of a molecule's internal structure. Finally electron beams are destructive, altering the structure of the object being examined.

Due to these inherent limitations, biologists have turned to X-rays as an alternative source of illuminating radiation. X-rays have greater penetrating power and

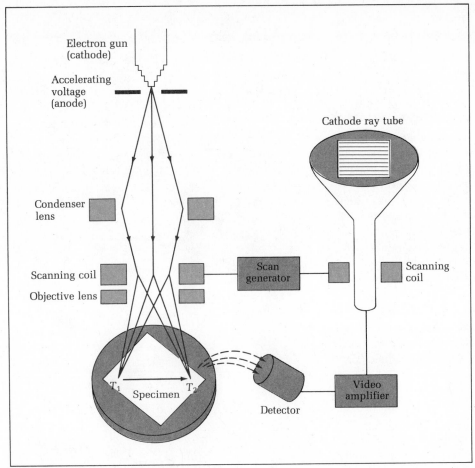

FIGURE 3-11 Image formation in the scanning electron microscope. The electron beam is moved back and forth across the specimen by a scanning coil, illuminating different points in the sample at different times (T_1, T_2). The scan generator synchronizes the movement of this beam with the beam in a cathode ray tube (television tube). Electrons scattered or emitted from the specimen are picked up by a detector that modulates the beam in the cathode ray tube, thereby forming an image of the specimen.

FIGURE 3-12 Scanning electron micrograph of a rounded HeLa cell. Courtesy of R. G. Kessel.

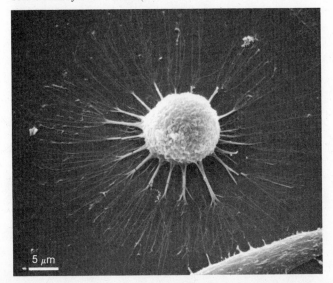

are less destructive than electrons, but unfortunately no suitable lenses are available for focusing X-ray beams. Neither conventional glass lenses nor magnetic lenses will work, so until an appropriate lens has been devised, the potential resolving power of X-rays cannot be readily applied to an examination of cell architecture.

Nonetheless, important information about molecular structure can be obtained with X-rays without the need for a finely focused beam. A narrow unfocused beam of X-rays passing through a specimen encounters atoms that are opaque to X-rays. When a wave passes by an opaque edge its path is altered, a phenomenon known as **diffraction.** Since the wavelength of X-rays is very short (~ 0.1 nm), all atoms larger than hydrogen act as separate diffraction edges. By placing a photographic plate on the opposite side of the specimen, the pattern of X-ray diffraction can be recorded and used to determine the position of individual atoms within a molecule.

Interpretation of such diffraction patterns requires

FIGURE 3-13 A comparison of the X-ray diffraction patterns of various molecules. *(Top)* Myoglobin. Courtesy of J. C. Kendrew. *(Middle)* DNA. Courtesy of M. H. F. Wilkins. *(Bottom)* Random crystal. Courtesy of P. Luger.

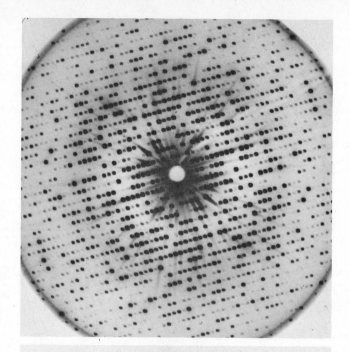

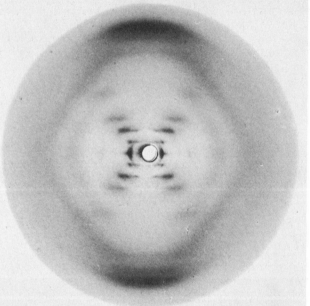

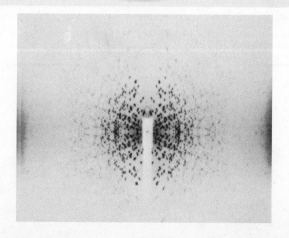

specimens with considerable molecular order. Diffraction patterns produced by randomly distributed molecules in dilute solution are unintelligible because of interference between individual molecules. Crystals or fibers with repeating structures, on the other hand, give good diffraction patterns because the patterns generated by individual atoms reinforce one another. Since every atom other than hydrogen contributes to the diffraction pattern, X-ray analysis of macromolecules containing thousands of atoms is a singularly complex task. The situation can be simplified somewhat by analyzing diffraction produced by arrays of atoms rather than by each atom itself. This type of analysis is carried out by rotating the specimen in the X-ray beam. If the sample is highly ordered, a distinct pattern of spots is obtained. Spreading of the spots occurs as disorder increases, until eventually a concentric pattern of spots is obtained (Figure 3-13). This latter pattern represents an analysis of randomly oriented crystals.

Determining the three-dimensional structure of macromolecules from X-ray diffraction patterns requires complicated mathematical analysis. One of the key developments in the application of this type of analysis to protein structure has been the introduction of heavy-metal atoms into specific regions of the molecule being examined. The resulting alteration in the diffraction pattern permits one to identify the particular region of the molecule to which the metal is bound. This information in turn provides a point of departure from which the rest of the structure can be deduced. Pioneers such as Linus Pauling, John Kendrew, Max Perutz, and Maurice Wilkins have played crucial roles in the application of X-ray analysis to the structure of biological macromolecules, and have provided information that has been crucial to recent developments in cell and molecular biology.

X-ray diffraction has been especially important in determining the three-dimensional structure of DNA, which has in turn had a dramatic impact on our understanding of how DNA stores and transmits genetic information. X-ray diffraction has also been used to unravel the three-dimensional structure of many important proteins, yielding crucial information about protein specificity and the mechanism of enzyme action. Finally, this approach has been useful in analyzing the structure of membranes and muscle fibers, and promises to continue being an indispensable tool for studying the structure and interactions of biological macromolecules.

Cytochemical Techniques

Our discussion of microscopy and X-ray diffraction has thus far focused mainly on the analysis of cell structure, with little consideration given to functional activities. Although biochemical and cell fractionation techniques have been primarily responsible for unraveling the details of cell function, microscopy has also provided crucial information. The field of **cytochemistry,** which encompasses all the microscopic techniques designed to provide information about the localization of chemical substances and biochemical activities, has been most important in this regard. Some of the more commonly utilized cytochemical procedures will be briefly described in the following sections.

Staining Procedures Selective staining reactions have been used to provide important information about the subcellular localization of particular kinds of molecules. Some staining reactions, for example, are selective for protein molecules: the *Millon reaction,* in which a mercury salt reacts with tyrosine and tryptophan residues in protein molecules to form a red precipitate; the *Sakaguchi reaction,* which employs α-naphthol and hypochlorite to produce a red color in the presence of arginine residues; and lastly, a large group of dyes that react with the sulfhydryl group of cysteine residues.

The high concentration of acidic phosphate groups in DNA and RNA permits them to be readily stained with basic dyes, some of which can distinguish between DNA and RNA. *Methyl green,* for example, selectively stains DNA a green color, while *pyronin G* stains both DNA and RNA red. These two stains together color chromatin blue, and nucleoli and cytoplasm red, thereby differentiating these components in the same specimen. Nucleic acids can also be stained by reagents that bind to nitrogenous bases. A classic example is the dye *acridine orange,* which imparts a yellow-green fluorescence to DNA and a reddish fluorescence to RNA. This particular example illustrates how the same dye can produce different colors, depending on the particular molecule with which it interacts. This property of certain dyes makes it possible to determine the cellular location of different kinds of molecules using the same stain.

One of the most widely used cytochemical tests for nucleic acids is the specific reaction for DNA known as the **Feulgen reaction.** In this technique tissue sections are first treated with dilute hydrochloric acid to remove purine residues from DNA, liberate free aldehyde groups on the adjacent deoxyribose moieties, and degrade any RNA present. The free aldehyde groups on the depurinated DNA are then reacted with the colorless *Schiff reagent* (leucofuchsin), producing a vivid red color. Cells treated in this way exhibit intense nuclear

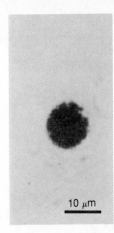

FIGURE 3-14 Photomicrograph of a Feulgen-stained pea-root cell. The cytoplasm surrounding the heavily stained nucleus is only faintly visible.

staining because the bulk of the cell's DNA is located there (Figure 3-14). The Feulgen reaction is so selective that the staining intensity can be quantitated with a microspectrophotometer attached to the microscope and the information used to determine the amount of DNA present.

The cytochemical localization of polysaccharides involves a related procedure, called the **periodic acid– Schiff** technique, in which the adjacent hydroxyl groups present in sugar molecules are selectively oxidized to aldehydes by treatment with periodic acid. The resulting aldehyde groups are then treated with the Schiff reagent to produce a red color. Any molecule containing sugars with adjacent hydroxyl groups can be identified in this manner.

Cytochemical identification of lipids is based on the use of hydrophobic dyes whose preferential solubility in lipid causes them to accumulate within intracellular fat droplets. *Sudan red* and *Sudan black* are commonly employed stains in this category. Unsaturated lipids may also be stained by reagents that react selectively with double bonds.

The specificity of cytochemical staining reactions can often be enhanced by coordinated use of appropriate enzymes. One might observe, for example, that the staining of a particular region of a cell can be prevented by first treating the specimen with the enzyme ribonuclease. Such an observation would suggest that the stained region contains RNA. Other hydrolytic enzymes, such as deoxyribonucleases or proteases, can also be used to help clarify the nature of stained components. Enzymes may also be used to detect molecules that are not stained. In cells viewed with ultraviolet light, for example, nucleic acids absorb the most light and will therefore appear darkest. Precise identification of the UV-absorbing material can be accomplished by treating the specimen with the enzymes ribonuclease or deoxyribonuclease. By determining where in the cell the UV absorption has been altered, one can ascertain where DNA and RNA are normally present.

Unfortunately the dyes used for light microscopy are not generally visible in the electron microscope because they do not scatter electrons very well. Their color per se cannot be detected in this instrument because of the use of electrons rather than light for illumination. The heavy-metal stains normally employed in electron microscopy are not very specific, although salts of uranium, iridium, and bismuth are somewhat selective for nucleic acids while phosphotungstic acid and osmium are more effective for staining proteins. Potassium permanganate, which acts as a fixative as well as a general stain, provides special contrast for cellular membranes, while osmium reacts well with lipids. Despite these observations, electron stains by themselves do not generally exhibit the specificity characteristic of the staining reactions used in light microscopy.

Immunocytochemical Techniques The inherent specificity of cytochemical reactions can be greatly enhanced by utilizing the discriminatory properties of **antibody** molecules, which are naturally occurring proteins whose differing three-dimensional shapes allow each antibody to specifically recognize and bind to a particular type of molecule recognized as being "foreign." The term **antigen** is employed to refer to the specific molecule to which a given antibody will bind. Specific antibodies are obtained in the laboratory by injecting an appropriate antigen into an experimental animal (usually a rabbit or mouse), and then isolating the newly formed antibodies from the animal's bloodstream after a suitable waiting period. In recent years a highly selective approach, called the *monoclonal antibody technique*, has been devised for obtaining large amounts of purified antibody molecules exhibiting identical specificity. This procedure, which has greatly facilitated the utilization of antibodies in biological and medical research, is described in detail in Chapter 17.

Because antibodies are not readily seen under the microscope, cytochemical techniques require that an antibody first be linked to a visible substance. For light microscopy the most commonly employed substance is *fluorescein*, a dye that fluoresces yellow-green when exposed to ultraviolet light. Fluorescein-labeled antibody can therefore be applied to a cell or tissue section, and its localization then determined by illumination with ultraviolet light. For example, if an antibody specific for a plasma membrane protein is used in such an experiment, only the plasma membrane will be fluorescent. Similarly when a fluorescein-labeled antibody specific for the protein tubulin is employed, only the microtubules containing tubulin become fluorescent (Figure 3-15).

In addition to the preceding direct approach for using fluorescent antibodies, an *indirect* fluorescent-

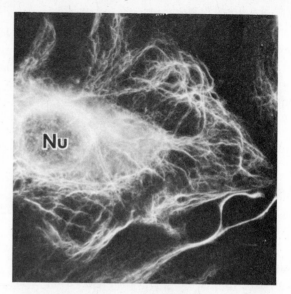

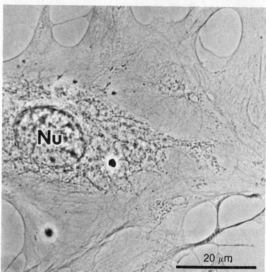

FIGURE 3-15 *(Top)* Fluorescence micrograph of a cell stained with fluorescent-labeled antibody against tubulin, a microtubule protein. *(Bottom)* Phase contrast micrograph of the same cell. The nucleus is labeled in each case. Courtesy of S. H. Blose.

labeled antibody technique is even more widely used. In this indirect approach, a cell or tissue section is first reacted with a nonfluorescent antibody directed against a particular cellular component. After removing the excess unbound antibodies, the tissue section is then reacted with a fluorescent antibody directed against the first antibody. In other words, if the initial antibody had been made in rabbits, the second (fluorescent) antibody might be a goat antibody directed against rabbit antibodies. The major advantage of this approach is that a single type of fluorescent antibody (e.g., goat antirabbit antibody) can be used to localize any cellular component that has first been reacted with a specific rabbit antibody.

Antibodies can also be employed to localize cellular components at the electron microscopic level. In this case specific enzymes rather than fluorescent dyes are employed to visualize the antibodies. For example, in one common procedure the enzyme *horseradish peroxidase* is linked to a specific antibody prior to incubation with tissue sections. After the antibody-peroxidase complex has bound to the cell, diaminobenzidine is added. This substance is converted to a brown insoluble product at the site of peroxidase activity, thereby localizing the antigen to which the antibody in question is directed. Because this reaction product is electron dense, it can be readily seen with both the electron and light microscopes (Figure 3-16). Coupling antibodies to *ferritin*, a protein whose high iron content makes it electron dense, is an alternative way of visualizing antibodies for use in electron microscopy.

Enzyme Cytochemistry Using a slightly different approach, cytochemical techniques can be employed to localize specific enzymes in the cell. The basic principle involved is quite simple. A tissue section is first incubated with a substrate of the enzyme one wishes to localize. The reaction product is then detected by converting it to a colored or electron-dense substance. Because harsh fixatives inactivate enzymes, sections must be cut either from unfixed frozen tissue or from tissue treated with mild fixatives, such as cold formaldehyde or glutaraldehyde.

This general approach is well illustrated by the enzyme acid phosphatase, which hydrolyzes phosphate ester bonds in a wide variety of substrates. In the *Gomori technique* for localizing acid phosphatase, tissue sections are incubated at low pH in a medium containing a phosphate ester, such as β-glycerophosphate, and lead nitrate. Acid phosphatase catalyzes hydrolysis of the β-glycerophosphate, releasing inorganic phosphate. The released inorganic phosphate in turn reacts with the lead nitrate to form an insoluble precipitate of lead

phosphate. This precipitate, which marks the site of acid phosphatase localization in the cell, is electron dense and therefore visible in the electron microscope (Figure 3-17). In adapting this technique for light microscopy, the lead phosphate is treated with ammonium sulfide to produce lead sulfide, a readily visible brown pigment.

Similar principles have been used in the microscopic localization of many other enzymes, including esterases, oxidoreductases, glycosidases, and proteases. The information obtained has added a great deal to our understanding of where biochemical activities occur in the cell.

Microscopic Autoradiography A related approach for microscopic localization of specific molecules and biochemical activities employs radioactive isotopes as tracers and is termed **autoradiography.** In this case radioactive precursors are incubated with tissue sections or administered to intact cells or organisms. After sufficient time has elapsed for the radioactive precursors to become incorporated into newly synthesized molecules, the remaining unincorporated radioactivity is removed, and the radioactive tissues or cells are sectioned and coated with a layer of photographic emulsion. The radioactive particles emitted from the tissue interact with silver bromide crystals in the emulsion, creating silver grains when the slide is photographically developed. The location of these silver grains, which are readily visible with both the light and electron microscope, can be used to pinpoint the regions of the cell containing the radioisotope (Figure 3-18). Counting the number of grains even provides a quantitative measure of the amount of isotope incorporated.

In practice, only isotopes emitting relatively weak forms of radiation are useful for microscopic autoradi-

FIGURE 3-16 Light micrograph of pancreatic cells in the islets of Langerhans showing the localization of specific hormones using peroxidase-labeled antibodies. *(Left)* Peroxidase-labeled antibody specific for the hormone insulin. *(Middle)* Peroxidase-labeled antibody specific for the hormone glucagon. *(Right)* Control. Courtesy of S. L. Erlandsen.

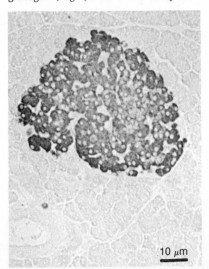
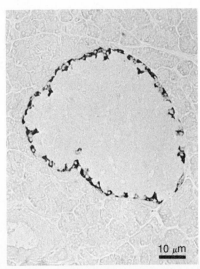
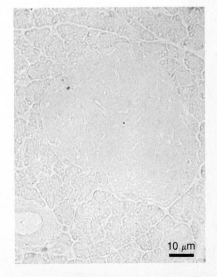

10 μm   10 μm   10 μm

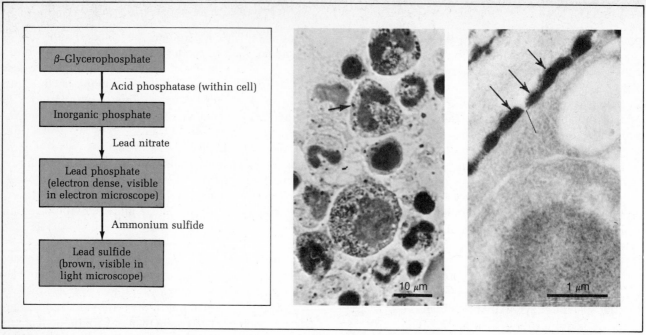

FIGURE 3-17 *(Left)* Outline of the Gomori procedure for the cytochemical localization of acid phosphatase. *(Middle)* Light micrograph of white blood cells stained to reveal Gomori-positive granules in the cytoplasm (arrow). Courtesy of D. F. Bainton. *(Right)* Cytochemical localization of acid phosphatase at the electron microscopic level. Courtesy of J.-C. Roland.

FIGURE 3-18 *(Left)* Outline of the principle underlying microscopic autoradiography. The silver grains formed when the photographic emulsion is developed are readily visible in light and electron micrographs. *(Top right)* Light micrograph of *Allium* nucleus showing incorporated radioactive DNA precursor. *(Bottom right)* Electron micrograph of *Allium* nucleus under the same conditions. Courtesy of W. Nagl.

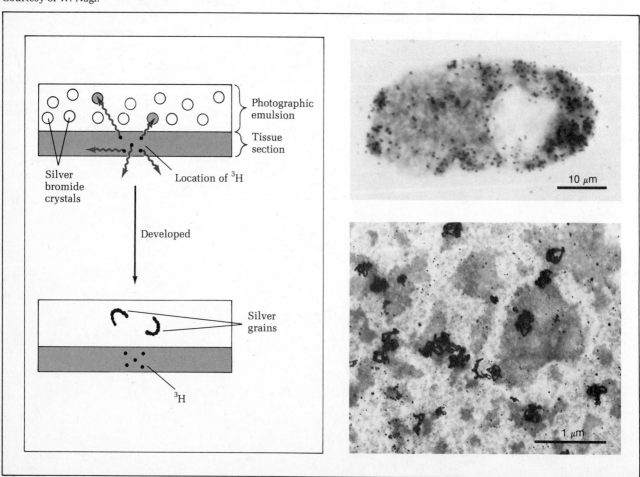

ography because stronger radiation penetrates the emulsion too far to permit accurate localization. For this reason the most widely used isotope in autoradiography is tritium (^{3}H), whose low-energy radiation permits a resolution of about 1 μm with the light microscope and close to 0.1 μm with the electron microscope. Since hydrogen is ubiquitous in biological molecules, a wide range of ^{3}H-labeled compounds are potentially available for use in autoradiography.

ISOLATION OF CELL ORGANELLES

Although microscopic analysis has led to an appreciation of the complexity of cell ultrastructure, the amount of functional information that can be obtained in this way is quite limited. Without question the most powerful tool for investigating the functional organization of cells is the isolation of individual organelles by subcellular fractionation. Once they have been isolated, organelles can be subjected to experimental manipulations and biochemical analyses that would be inconceivable with the microscope. Subcellular fractionation is probably the most widely used technique in cell biology laboratories today, and is responsible for much of our current understanding of how cells function. Because almost all subcellular fractionation procedures are based on centrifugation, we will begin this section with a discussion of the principles governing the behavior of particles in a centrifugal field.

Principles of Centrifugation

The centrifuge is probably the most important tool available to cell biologists for the separation and analysis of subcellular organelles and their constituent macromolecules. Centrifugal force is generated by placing tubes containing samples to be fractionated in a holder, or *rotor*, which can be rotated at high speeds in a machine called a *centrifuge*. The *ultracentrifuge*, a special high-speed instrument developed by the Swedish physical chemist Theodor Svedberg during the period 1920–1940, is designed to generate centrifugal forces up to several hundred thousand times gravity. Such high forces have made it possible not only to isolate and purify subcellular organelles but also to define the physical characteristics of proteins, nucleic acids, and other macromolecules. The ultracentrifuge is the key instrument in every cell biology laboratory, without which we would lack much of our current appreciation of the functions carried out by subcellular organelles.

Several types of centrifugation are commonly employed for subcellular fractionation. In order to understand their underlying rationales, it is necessary to know something about the basic principles governing the movement of a particle or molecule in a centrifugal field. The rate of movement of a spherical particle in a centrifugal field is given by the Stokes formula:

$$\frac{dx}{dt} = \frac{2r^2(\rho_P - \rho_M)}{9\eta} \cdot g \qquad (3\text{-}2)$$

where dx/dt is the velocity at which the particle moves toward the bottom of the tube, r is the particle radius, ρ_P and ρ_M are the densities of the particle and the suspending medium, respectively, η is the viscosity of the suspending medium, and g is the centrifugal force exerted on the particle. This formula tells us that when the density of a particle exceeds that of the medium, its velocity will be positive and the particle will migrate in the direction of the force field (toward the bottom of the tube). If the particle is less dense than the medium, its velocity will be negative, meaning that it will move toward the top of the tube. It is also apparent that particles whose densities are equal to that of the suspending medium will not move, because the term $\rho_P - \rho_M$ will equal zero.

The centrifugal force created by a centrifuge is directed outward from the axis of rotation, and is measured in gravity units. Its value can be calculated from the formula

$$g = \frac{4\pi^2 \, (\text{rpm})^2}{3600} \cdot x \qquad (3\text{-}3)$$

where rpm is the number of revolutions per minute at which the rotor is spinning and x is the distance between the central axis of the centrifuge and the point at which the centrifugal force is being measured in the rotor. Thus the strength of the centrifugal field is determined by the speed of the centrifuge and the size of the rotor, both of which can be easily determined in the laboratory.

The rate of movement (dx/dt) of a particle in a centrifugal field can also be determined in the laboratory by measuring the distance a particle has migrated in a centrifuge tube as a function of time. Special analytical ultracentrifuges have even been designed to make such measurements automatically. Thus we see that for Equation 3-2 above, values for both g and dx/dt can be readily measured experimentally.

If two particles are centrifuged under conditions of equal centrifugal force, Equation 3-2 tells us that their rates of migration (dx/dt) will depend on four factors: r, ρ_P, ρ_M, and η. Rather than attempting to measure each of these variables independently, they are grouped together to yield a **sedimentation coefficient (s):**

$$s = \frac{2r^2 \, (\rho_P - \rho_M)}{9\eta} \qquad (3\text{-}4)$$

Hence s can be substituted in Equation 3-2 to give

$$\frac{dx}{dt} = sg$$

which can be rearranged to

$$s = \frac{dx/dt}{g} \tag{3-5}$$

Because values of dx/dt and g are easy to measure experimentally, one can calculate values of s for various particles and molecules. The importance of knowing the sedimentation coefficient s is that it is a crude estimate of the size and shape of a molecule or organelle. Although s was seen in Equation 3-4 to depend on r, ρ_P, ρ_M, and η, in most cases the latter two factors are held constant. In other words, sedimentation of different particles is compared in a medium of the same density (ρ_M) and viscosity (η). With these two factors held constant, any differences observed in the s values for different particles must be due to differences in their size (r) and density (ρ_P). Differences in density are usually less important because density does not differ much between biological particles and molecules. Their respective sizes and hence r, on the other hand, vary over many orders of magnitude. Since the value of s is proportional to the square of r, size differences have the greatest impact on sedimentation rate. It should be noted that the shape of a particle also affects its sedimentation rate (s). Although the equations above apply only to the idealized spherical particle with radius r, it is possible to introduce corrections that account for nonspherical shape. Only in the case of a pronounced asymmetry will the sedimentation rate of a molecule be modified appreciably.

The value of s for most biological particles lies in the range of $1 - 200 \times 10^{-13}$ sec. In order to avoid repetition of the term "10^{-13} sec," a sedimentation coefficient of 1×10^{-13} sec has been defined as a **Svedberg unit,** abbreviated **S.** Thus a particle with a sedimentation coefficient of 18×10^{-13} sec would have a sedimentation coefficient of 18S. The convenience of this nomenclature is that the sedimentation coefficients of biological molecules can be referred to as simple integers.

Because the sedimentation coefficient is an approximate measure of size, it is not surprising to find that the values obtained for various particles and molecules correlate roughly with their molecular weights (Table 3-1). The relationship between the two is not directly proportional, however, because of the influence of shape and density. In order to determine molecular weight from such data, one needs to measure the diffusion constant (D) as well as the sedimentation coefficient. Molecular weight can then be calculated using the *Svedberg equation:*

$$M = \frac{RTs}{D(1 - \bar{v}\rho_M)} \tag{3-6}$$

where M is the molecular weight, R the gas constant, T the absolute temperature, s the sedimentation coefficient, D the diffusion coefficient, $\bar{v}$ the partial specific volume of the particle (the volume of fluid displaced per gram of particle), and ρ_M the density of the medium.

TABLE 3-1
Molecular weights and sedimentation coefficients of selected macromolecules

	Molecular Weight	Sedimentation Coefficient (S)
Myoglobin	16,900	2.0
Lysozyme	17,200	2.2
Transfer RNA	25,000	4.0
Hemoglobin	64,500	4.5
Catalase	248,000	11.3
Urease	483,000	19.0
Small ribosomal RNA (bacterial)	550,000	16.0
Ribosome (mammalian)	4,500,000	80.0
Tobacco mosaic virus	40,600,000	198.0

The major application of centrifugation to cell biology is not in the calculation of molecular weights, however, but in the fractionation of organelles and macromolecules. Two types of centrifugation are commonly employed for these purposes, *velocity centrifugation* and *isodensity centrifugation* (also called isopycnic or equilibrium centrifugation). These two techniques exploit different molecular properties to achieve fractionation, and so both have useful biological applications.

Velocity Centrifugation When the density of a particle is greater than the density of the medium, the particle will move toward the bottom of the tube when centrifuged. The rate of migration of different particles under these conditions will vary according to size, with larger particles sedimenting more rapidly than smaller particles (Equation 3-2). This difference in velocity can be exploited in one of two ways to separate particles from one another.

The first way, known as **differential** centrifugation, involves a successive series of centrifugations at increasing speeds. The initial centrifugation step is designed to be fast enough to sediment the largest particles into a pellet at the bottom of the tube, leaving the smaller particles suspended in the remaining solution, or *supernatant.* This supernatant is removed and centrifuged again at a higher force to sediment the smaller particles (Figure 3-19). As will be seen in a moment, this approach is widely employed in subcellular fractionation experiments, but its usefulness is restricted to the separation of particles exhibiting relatively large differences in size.

When separating particles with small differences in size, a second type of velocity centrifugation can be employed. This technique, termed **moving-zone** centrifugation, involves applying the sample as a narrow, discrete zone on the surface of the solution in a centrifuge tube. Centrifugation is carried out long enough for the particles to move part way down the tube. Because the migration velocity of different particles will vary, depending on their respective sizes, the particles will

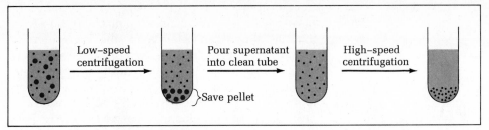

FIGURE 3-19 The principle of differential centrifugation. The tube initially contains a mixture of small and large particles. After an initial low-speed centrifugation to pellet the larger particles, the supernatant is removed and centrifuged at higher speed to sediment the smaller particles.

separate from one another as they move down the tube. Centrifugation is therefore stopped during the middle of the run, and the various components are recovered by puncturing the tube and collecting sequential fractions (Figure 3-20). Care must be taken not to centrifuge too long, or all the material will end up at the bottom of the tube. In order to minimize diffusion of the particle zones as they migrate down the tube, the solution through which the particles are moving is modified by including a shallow gradient of sucrose or glycerol (denser at the bottom of the tube and gradually less dense toward the top). Such gradients stabilize the fluid environment in the centrifuge tube and sharpen the resolution of the separating bands of particles. Gradients of this type are made with a mechanical device like the one illustrated in Figure 3-21. It is important to emphasize that even

when such gradients are used, the density of the particles being separated is greater than the density of the separating medium everywhere in the tube.

Isodensity Centrifugation In contrast to velocity centrifugation, which separates particles on the basis of differences in size, isodensity centrifugation fractionates on the basis of differences in density. We have already noted from our examination of Equation 3-2 that no movement occurs when the density of a particle (ρ_P) equals the density of the medium (ρ_M). Isodensity centrifugation exploits this fact by utilizing a gradient whose density overlaps that of the components being separated. In such a gradient, particles will migrate until they reach the region in which their density is equal to that of the medium. At this point they reach

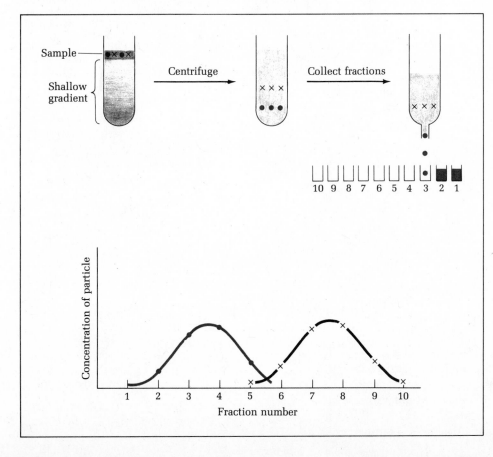

FIGURE 3-20 The principle of moving-zone centrifugation. A sample containing two types of particles of slightly different size (● and X) is layered on the surface of a shallow density gradient of sucrose or glycerol. After centrifuging for an appropriate amount of time, centrifugation is stopped and the bottom of tube punctured to permit collection of fractions. Each fraction is assayed for the presence of the two particles, yielding a graph like the one illustrated.

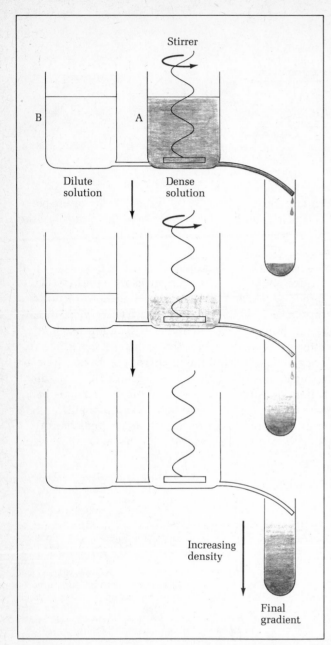

FIGURE 3-21 Mechanical apparatus for making density gradients. As the dense solution leaves chamber A, dilute solution from chamber B enters chamber A to equalize the fluid levels. In this way the density of the solution in chamber A gradually falls, generating a continuous gradient of decreasing density.

from a variety of materials. Sucrose and glycerol solutions have a maximum density of about 1.3 g/cm³, limiting their applicability to the separation of membranous organelles of relatively low density (Golgi complex, endoplasmic reticulum, lysosomes, mitochondria, etc.). It should be emphasized that the function of such sucrose or glycerol gradients differs in principle from the moving-zone gradients discussed above. In moving-zone centrifugation the only purpose of the gradient is to minimize diffusion; even at the bottom of the gradient, the density of the particles is greater than that of the medium and they continue to move. In isodensity centrifugation, on the other hand, the gradient is of sufficient density to cause the particles to stop migrating when they reach the region in which the density of the medium equals their own.

Fractionation of molecules with densities greater than 1.3 g/cm³, such as DNA, RNA, and nucleoproteins, requires gradients formed from substances denser than sucrose or glycerol. Heavy-metal salts, such as cesium chloride and cesium sulfate, are especially useful for this purpose. The density of such salts is so great that gradients need not be premade with a mechanical device. If a uniform solution of cesium chloride is simply centrifuged at high gravitational forces, the centrifugal field itself will cause the cesium salt to become more

FIGURE 3-22 The principle of isodensity centrifugation. (a) A density gradient is preformed in the centrifuge tube and the sample is then layered on top. During centrifugation particles migrate downward until they reach the region of the gradient corresponding to their own density, at which point they stop. (b) Density gradient is generated by centrifugal force rather than being preformed. The suspended particles move either upward or downward until each reaches the region corresponding to its own density.

equilibrium and cannot move in either direction, no matter how long centrifugation is continued. After equilibrium is achieved, centrifugation is stopped and fractions are collected (Figure 3-22a). Because a particle's density is the only factor determining its final equilibrium position in such a gradient, this method is useful for separating components that differ in density. In practice, with appropriate density gradients it is possible to separate components whose densities differ from each other by less than one percent.

Gradients for isodensity centrifugation are formed

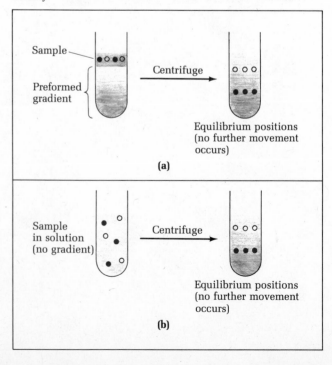

concentrated at the bottom of the tube and less concentrated toward the top. Because such gradients form spontaneously during centrifugation, the material to be fractionated can be mixed directly with the cesium chloride (or other heavy-metal salt) solution, and the entire mixture centrifuged. As the density gradient forms during centrifugation, particles or molecules dispersed throughout the tube will either sediment toward the bottom or move upward until they reach their isodensity points (see Figure 3-22b). The high density that can be obtained with cesium salts (up to 1.9 g/cm³) makes them useful in the fractionation of very dense molecules. In a later chapter we shall describe landmark experiments on DNA structure and replication that were made possible by this property of cesium gradients.

Two major approaches are thus available for the fractionation of subcellular components by centrifugation. Particles or molecules differing in size can be separated from each other by velocity centrifugation (either differential or moving zone), regardless of whether their densities are the same or different. Particles or molecules differing in density, on the other hand, can be separated by isodensity centrifugation, regardless of whether they are of the same or different sizes. When used in appropriate combination, these two approaches make centrifugation an extremely powerful tool for separating, isolating, and purifying cellular constituents (Figure 3-23).

Subcellular Fractionation

Although the first attempts at isolating organelles were reported over a century ago, only recently have schemes been devised for the comprehensive fractionation of cells into their constituent organelles. The invention of the ultracentrifuge has played a key role in these developments, but it is not the only factor involved. Cells must first be broken open to release their organelles into solution, and the type of medium used for this purpose is of critical importance. Pioneering studies carried out by Albert Claude in the early 1940s revealed that disrupting cells in buffered salt solutions of physiological ionic strength permits the subsequent isolation of several discrete organelle fractions by centrifugation. But it was also noted that this medium causes considerable aggregation of cell components, with an accompanying loss in their morphological integrity. In 1948 George Hogeboom, Walter Schneider, and George Palade reported that sucrose is a better medium for cell breakage and fractionation because it causes minimal aggregation and structural alterations. This technically simple change has revolutionized cell biology. Virtually every important subcellular fractionation experiment performed since that time has employed a sucrose-containing medium for the isolation of subcellular organelles.

The use of sucrose is not entirely without its drawbacks, however. Many biological substances are water soluble and therefore easily extracted from organelles during isolation. To counteract this problem nonpolar solvents, like carbon tetrachloride and benzene, have also been used in cell fractionation studies. Balancing the advantage of this approach in minimizing the loss of water-soluble molecules is the disadvantage that nonpolar solvents extract lipids and denature many proteins. Although nonpolar solvents have found certain applications, such as in isolating nuclei that retain

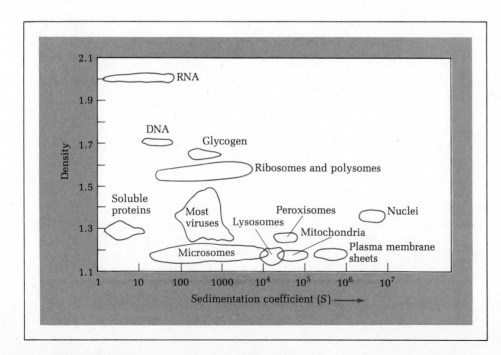

FIGURE 3-23 Sedimentation coefficients and densities of various organelles, macromolecules, and viruses. Because each component exhibits a relatively unique combination of size and density, velocity and isodensity centrifugation can be used in combination to separate most of these materials from one another.

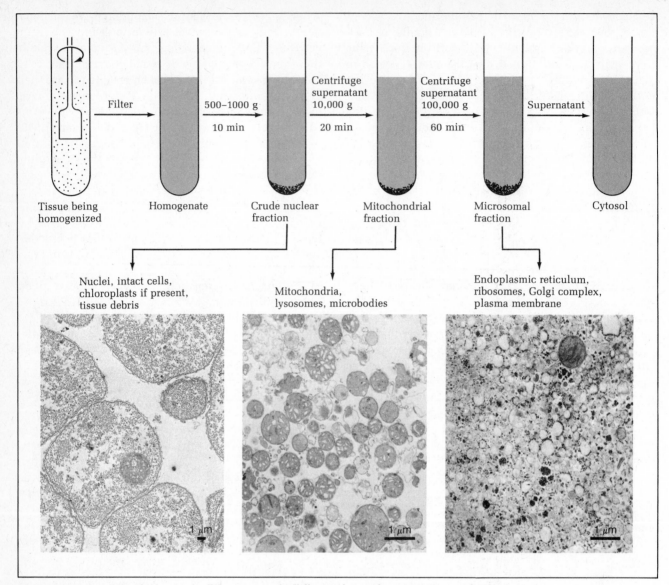

FIGURE 3-24 Generalized scheme of cell fractionation by differential centrifugation. Micrographs show the appearance of each fraction. *(Left)* Courtesy of D. J. Morré. *(Middle, Right)* Courtesy of P. Baudhuin.

all their water-soluble constituents, the inherent disadvantages of this approach have severely limited its general usefulness.

Another factor critical to the success of subcellular fractionation studies is choosing a reliable means for breaking cells open. Ideally the plasma membrane should be disrupted under conditions that cause minimal damage, releasing organelles into solution with their morphological features and functional integrity intact. This goal is difficult to achieve in practice, however, and the method chosen for breaking cells usually involves some compromises based on optimizing conditions for the organelle in which one is most interested. Organelles such as mitochondria, chloroplasts, and nuclei are readily released intact, while membranes such as the endoplasmic reticulum, the Golgi complex, and the plasma membrane are usually broken into smaller fragments.

Probably the most common procedure for disrupting cells involves *homogenization* of a cell or tissue suspension in a tube fitted with a glass or Teflon pestle. By rotating the tightly fitting pestle, cells are disrupted by shearing forces. Other methods achieving a similar end include blending in a high-speed mixer, freezing and thawing, grinding with a mortar and pestle, forcing material through a narrow orifice at high velocity, osmotic lysis, and subjecting cells to sonic or ultrasonic vibrations (sonication). The decision concerning the most appropriate technique to use is generally based on the specific objectives of the experiment being performed.

The suspension of cell organelles and tissue debris obtained from such procedures is referred to as an **homogenate.** Differential centrifugation is usually used to then fractionate the homogenate into its various constituent organelles (Figure 3-24). Following removal of large pieces of tissue by filtration, the homogenate is centrifuged at 500–1000g to sediment the largest cell components. This fraction, referred to as the **crude nuclear fraction,** is enriched in nuclei but also contains unbroken cells, tissue debris, and some chloroplasts if present. The supernatant from this first centrifugation step is recentrifuged at 10,000–20,000g to sediment intermediate-sized organelles such as mitochondria, lysosomes, and microbodies. The pellet obtained is usually designated the **mitochondrial fraction,** in spite of the other organelles present. The remaining supernatant is centrifuged at high speeds (~100,000g) to sediment the smallest cell components. This **microsomal fraction** contains fragments of smooth and rough endoplasmic reticulum, as well as free ribosomes. The final supernatant contains the soluble molecules of the cytoplasm, and is therefore designated the **cytosol fraction.**

The distribution of organelles among the various fractions can be altered to a certain extent by modifying the speed and time of each centrifugation step. However, none of the fractions obtained from such a differential centrifugation scheme are ever completely pure. To obtain a homogeneous preparation of a single organelle, subsequent purification of these crude fractions by other types of centrifugation is necessary. For example, isodensity gradients are useful for purifying mitochondria, lysosomes, and microbodies because these organelles differ from one another in density. Likewise,

TABLE 3-2

Markers commonly employed for identifying organelle fractions during subcellular fractionation

Organelle	Marker
Nuclei	NAD synthetase, DNA
Mitochondria	Cytochrome oxidase, succinate dehydrogenase, monoamine oxidase
Chloroplasts	Ribulose-bisphosphate carboxylase, photosystems I and II
Lysosomes	Acid phosphatase, acid deoxyribonuclease
Microbodies	Catalase, urate oxidase, D-amino acid oxidase
Microsomes: (endoplasmic reticulum)	Glucose-6-phosphatase (liver), esterase, cytochrome b_5, cytochrome P-450
(ribosomes)	RNA
Golgi complex	Nucleoside diphosphatase, N-acetyl-glucosamine: β-galactosyltransferase
Plasma membrane	5'-nucleotidase, alkaline phosphodiesterase I, nucleotide pyrophosphatase
Cytosol	Enzymes of glycolysis (glyceraldehyde-3-phosphate dehydrogenase, phosphoglycerate kinase, etc.), phosphoglucomutase

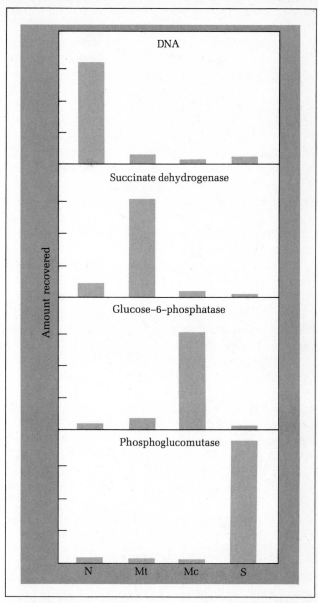

FIGURE 3-25 Distribution of marker substances for nuclei (DNA), mitochondria (succinate dehydrogenase), endoplasmic reticulum (glucose-6-phosphatase), and cytosol (phosphoglucomutase) after fractionation of a rat liver homogenate according to the general scheme outlined in Figure 3-24. N = nuclear fraction, Mt = mitochondrial fraction, Mc = microsomal fraction, and S = the final supernatant (cytosol fraction).

velocity centrifugation of the crude nuclear fraction through a solution of dense sucrose (~2.0M) results in a relatively pure nuclear pellet.

It is essential to independently verify the purity of cell fractions obtained by centrifugation. Electron microscopic examination of each fraction is clearly helpful, but biochemical analyses can be equally important because certain molecules are preferentially localized in particular organelles. Table 3-2 summarizes some of the more important molecular markers used to identify the major organelles. The data summarized in Figure 3-25 illustrate how such markers can be employed to confirm the identity of the fractions obtained in a classic differential centrifugation experiment.

In spite of the great power of the subcellular fractionation technique, the possibility of artifacts should always be kept in mind. First of all, solubilization of water-soluble molecules in aqueous media can result in a significant loss of molecules from their native organelles and their concomitant appearance in the cytosol fraction. Substances normally dissolved in the cytoplasm, on the other hand, may stick to organelles under the nonphysiological conditions of cell fractionation. Aggregation of organelles may also affect their sedimentation properties. Membranes of the endoplasmic reticulum, for example, normally sediment at high speed in the microsomal fraction. But, if the pH of the medium is too low, the membranes aggregate together, forming large masses that sediment at low speed in the crude nuclear fraction. Regardless of the precautions taken during subcellular fractionation experiments, no fraction is ever completely pure. Although each one may be highly enriched for a particular organelle, there is always some overlap between fractions.

In spite of these problems, subcellular fractionation is an extraordinarily powerful tool for the dissection of cell function. Studies on the chemical composition, enzymatic activities, and metabolic properties of isolated organelles have provided us with a wealth of information about how the cell is organized and how it operates. As each of the cell's organelles is discussed in detail later in the book, it will become increasingly obvious that much of our current understanding of the cell is intimately tied in with subcellular fractionation experiments of one type or another.

ANALYSIS OF ISOLATED MACROMOLECULES

In the course of studying the properties of individual organelles, it is often necessary to isolate and characterize the molecular components from which they are constructed. Much of our understanding of how nucleic acids and proteins function in the cell, for example, has depended on the development of appropriate biochemical procedures for the purification and analysis of these macromolecules. In this section we shall briefly review the biochemical methods that have had the most important applications in cell biology.

Spectrophotometric Analysis

Most biological compounds either absorb light in the ultraviolet region of the electromagnetic spectrum or react with certain chemical reagents to form colored compounds that absorb visible light. The exact amount of light absorbed by a solution can be quantitated in an instrument called a *spectrophotometer*, and the result-

ing value used to calculate the concentration of the absorbing material. By altering the wavelength of the illuminating light beam, the presence of compounds with different light-absorbing properties can be detected. One can also measure absorption at various wavelengths, generating an **absorption spectrum** that can be used to identify the type of material present.

Both proteins and nucleic acids absorb ultraviolet light. Though the absorption spectra for these two substances overlap, proteins exhibit a characteristic peak at 280 nm while nucleic acids peak at 260 nm (Figure 3-26). It is therefore common practice to measure absorption at either 260 or 280 nm when analyzing for nucleic acids or proteins, respectively. Caution is needed when using this approach, however, because nucleic acids absorb more strongly than proteins on a per weight basis. This means that small amounts of protein can be easily overshadowed by the presence of contaminating nucleic acids. The difference in absorption intensity between proteins and nucleic acids is caused by the fact that unsaturated rings are responsible for the

FIGURE 3-26 Ultraviolet absorption spectra of proteins and nucleic acids. Note that the protein concentration must be 40-fold higher than that of the nucleic acid to achieve absorption of comparable magnitude.

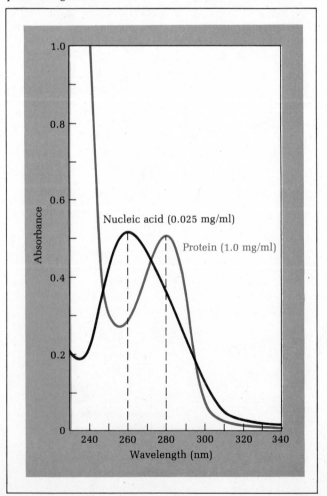

TABLE 3-3
Colorimetric reactions used to identify biological molecules

Molecule Detected	Name of Reaction	Reagents	Color Produced
Protein (peptide bond)	Biuret	Copper sulfate (alkaline)	Purple
Protein (tyrosine; tryptophan)	Lowry	Copper sulfate (alkaline); phosphomolybdate/ phosphotungstate	Blue
Amino acid, amines	Ninhydrin	Triketohydrindene hydrate	Purple
DNA (deoxyribose)	Dische	Diphenylamine	Blue
DNA (deoxyribose)	Indole	Indole, cupric sulfate	Yellow
RNA (ribose)	Cysteine	Cysteine-HCl	Yellow
Polysaccharide	Iodine	Iodine, iodide	Blue-black
Sugars (and polysaccharides)		Concentrated acid, various phenols	Different colors
Sugars (and polysaccharides)	Molisch	α-Naphthol	Red-violet
Sugars (pentoses)	Phloroglucinol	Phloroglucinol-HCl	Red

TABLE 3-4
Properties of isotopes commonly employed in cell biology

Element	Predominant Natural Isotope	Useful Isotopic Tracers	Radioactive	Half-life	Energy of Radiation (meV)[a]
Hydrogen	^{1}H	^{2}H (deuterium)	No	—	—
		^{3}H (tritium)	Yes	12.3 years	0.018 (β^-)
Carbon	^{12}C	^{13}C	No	—	—
		^{14}C	Yes	5,570 years	0.16 (β^-)
Nitrogen	^{14}N	^{15}N	No	—	—
Oxygen	^{16}O	^{18}O	No	—	—
Sodium	^{23}Na	^{24}Na	Yes	15 hours	1.39 (β^-)
					2.75 (γ)
Phosphorus	^{31}P	^{32}P	Yes	14.3 days	1.71 (β^-)
		^{33}P	Yes	25.2 days	0.248 (β^-)
Sulfur	^{32}S	^{35}S	Yes	87.1 days	0.17 (β^-)
Chlorine	^{35}Cl	^{36}Cl	Yes	310,000 years	0.71 (β^-)
Potassium	^{39}K	^{42}K	Yes	12.5 hours	3.55 (β^-)
Iodine	^{127}I	^{125}I	Yes	59.7 days	0.035 (γ)
		^{131}I	Yes	8 days	0.61 (β^-)
					0.36 (γ)

[a] Million electron volts.

absorption peaks at 260 and 280 nm. In nucleic acids every nucleotide contains a nitrogenous base with one or two such rings, while only a few of the amino acids occurring in proteins contain unsaturated rings.

In addition to the absorption peaks at 260 and 280 nm generated by the presence of unsaturated rings, peptide bonds absorb ultraviolet light at 220–230 nm. Because nucleic acid absorption is minimal at these lower wavelengths, measurements at 220–230 nm are a more sensitive way of quantitating protein.

Measurement of ultraviolet absorption is a quick and simple way of detecting proteins or nucleic acids in solution, particularly if the sample is relatively pure. For increased specificity this method can be supplemented by treating with chemical reagents that yield characteristic colored compounds in the presence of certain types of macromolecules. Table 3-3 summarizes the more commonly used color reactions of this type and the macromolecules identified by each.

Isotopic Tracers

The use of chemical isotopes as tracers for detecting the synthesis of biological molecules has had a tremendous impact on cell biology in recent years. Isotopic forms of chemical elements can be readily monitored because they are either radioactive or differ in density from the naturally predominant form of the element in question. Both types of isotopes have had important roles in the history of cell biology, although radioactive isotopes are the more commonly employed. Table 3-4 summarizes the properties of the isotopic tracers most widely used in cell biology today.

One of the major advantages of using radioactive

isotopes is that they permit the detection of very small amounts of material. Radioactivity can be detected either by autoradiography, which utilizes a photographic film to locate the isotope, or with specialized radiation detectors. With such detectors one can measure radioactivity levels as low as 5–10 counts per minute above background, which may correspond to as little as 10^{-13} or 10^{-14} g of the labeled compound being studied. Thus one can monitor the presence of substances in amounts too small to be detected by ordinary chemical means. Most radioisotopes employed as biological tracers emit β-particles, a relatively low-energy form of radiation. The most efficient method for detecting β-emitting radioisotopes involves placing the radioactive material into vials containing a special solution that gives off light (fluorescence) each time it is excited by an emitted β-particle. The flashes of light (scintillations) are detected in an instrument called a *liquid scintillation counter*, and expressed as counts per minute (cpm).

In liquid scintillation counting, the intensity of the light flashes varies with the energy of the radioactive emission. Thus ^{32}P, which emits higher intensity β-particles than ^{14}C or ^3H, causes brighter flashes of light. The light detection device in a scintillation counter is able to detect these differences and can therefore be used to distinguish between ^{32}P, ^{14}C, and ^3H. In this way two or more radioisotopes can be counted simultaneously. "Double-label" experiments, in which the fates of two different isotopes are followed simultaneously, are very useful in monitoring the behavior of biological molecules.

By choosing the appropriate radioisotopes, one can obtain information about specific metabolic processes. If cells are labeled with ^{32}P-phosphate, for example, all molecules containing phosphorus (i.e., DNA, RNA, phospholipids, nucleotides, etc.) will incorporate radioactivity during their synthesis. If one wishes to specifically follow the synthesis of DNA, on the other hand, labeling can be carried out with ^3H- or ^{14}C-thymidine because thymidine does not occur in RNA or any other macromolecule. In a similar way radioactive uridine can be used to monitor the synthesis of RNA; radioactive amino acids can be used to follow protein formation; and radioactive monosaccharides can be employed for studies on polysaccharide biosynthesis.

In addition to their usefulness in studying the synthesis of specific molecules, radioisotopes can be employed in **pulse-chase experiments** to determine what happens to molecules after they have been synthesized. In this type of experiment, cells are incubated with a particular radioactive compound for a brief period (the pulse), followed by removal of the radioisotope and further incubation (the chase). During the chase period no further radioactivity can be incorporated by the cell, so any changes in radioactivity that do occur must reflect changes in previously synthesized material. This approach has helped elucidate the existence of precursor-product relationships in various metabolic pathways, and has led to the important discovery that most cellular macromolecules are subject to continual *turnover* (degradation and resynthesis).

In contrast to radioactive isotopes, nonradioactive isotopes such as ^{15}N, ^{18}O, and ^2H do not emit radiation and so can be detected only by their characteristic densities, which differ from the normally predominant forms of these elements. The use of such isotopes is practical only when the element involved accounts for a substantial portion of a particular biological molecule. For example, nucleic acids, which contain about 15 percent nitrogen by weight, have their densities appreciably altered when ^{15}N is substituted for the common isotope ^{14}N. DNA synthesized in the presence of ^{15}N can be separated from normal DNA by isodensity centrifugation, an approach that has been extremely useful in the study of DNA replication.

Dialysis

It was pointed out in Chapter 1 that biological molecules can be divided into two general categories: large macromolecules and small building blocks. Molecules in these two categories can be readily separated from one another by using artificial membranes with small pores that permit passage of water and small molecules, but not large molecules. This technique, called **dialysis,** is carried out by placing a mixture of small and large molecules in a membrane bag and then placing the bag in a large volume of solution. Small molecules pass through the pores in the membrane until their concentrations outside and inside the bag become equal, at which point equilibrium is attained. Large molecules, on the other hand, are selectively retained inside the bag (Figure 3-27). Dialysis is generally used to remove and change salts, buffers, and other small molecules occurring in macromolecule preparations. Membranes with pores that permit the passage of substances with molecular weights less than 10,000–15,000 are most often employed, but membranes with cutoffs ranging from as small as 3,500 to greater than 50,000 are available for appropriate applications.

Precipitation

Small and large molecules can also be separated from one another by precipitating with *trichloroacetic acid* (TCA) or *perchloric acid* (PCA). The anions of TCA and PCA form insoluble salts with proteins and nucleic acids, causing them to aggregate and precipitate out of solution. Any lipid present in the precipitate can be extracted with a nonpolar solvent, and the precipitated nucleic acids can be resolubilized by mild heating.

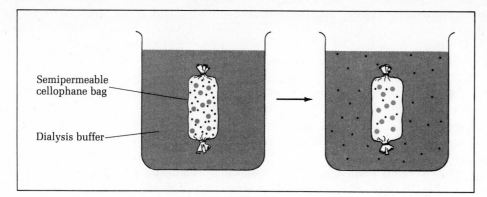

FIGURE 3-27 The use of dialysis to remove small molecules (black dots) from a solution of macromolecules (color dots). By replacing the dialysis buffer several times in succession, the concentration of the small molecules within the dialysis bag can be lowered to trace amounts.

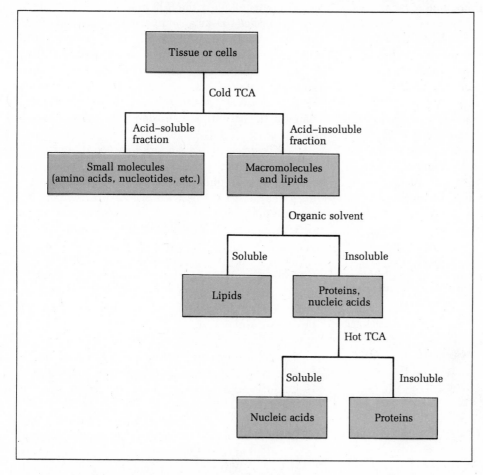

FIGURE 3-28 Fractionation of the major classes of macromolecules based on differential solubility in trichloroacetic acid (TCA) and organic solvents.

Combining these procedures in sequence permits a crude fractionation of cellular macromolecules (Figure 3-28).

Acid precipitation is often utilized in experiments where the synthetic rates of macromolecules are being studied. In cells incubated with radioactive amino acids, for example, the incorporation of radioactivity into protein can be determined by precipitating with TCA or PCA. The labeled amino acids employed as precursors are soluble in acid, so any radioactivity present in the precipitate represents newly synthesized protein. A similar approach employing precursors of nucleic acid biosynthesis can be utilized for studying nucleic acid synthesis.

Electrophoresis

Most biological molecules contain ionizable groups that exhibit a net negative or positive charge at appropriate pH. The technique of **electrophoresis** utilizes an electric field to separate such charged molecules from each other. Positively charged molecules move toward the negative pole or cathode, and are hence called *cations*, while negatively charged molecules migrate toward the positive pole or anode, and are designated *anions*. The rate at which a given molecule moves during electrophoresis depends on the ratio of its charge to its mass. The higher this ratio, the more rapid the electrophoretic mobility.

Samples to be fractionated by electrophoresis are usually applied as a small spot or zone on a supporting medium such as paper, cellulose acetate, starch, or polyacrylamide gel. Of these media, polyacrylamide gives the best resolution and is most commonly employed for the separation of proteins and nucleic acids. The material being analyzed is layered on top of the polyacrylamide gel and a voltage applied to induce migration of the charged molecules into the gel. Depending on their relative charges and sizes, molecules move through the polyacrylamide matrix at differing rates. The resulting bands of material can be visualized by appropriate staining or autoradiographic techniques (Figure 3-29).

In polyacrylamide gel electrophoresis, the charge-to-mass ratio is not the only factor affecting separation. Polyacrylamide gels are composed of cross-linked polymers that create a series of "pores" through which the electrophoresing molecules must pass. Because the diameter of many macromolecules approaches that of the pores themselves, the larger the molecule the more hindered its passage through the pores, and hence the slower its progress. This principle is particularly important in electrophoretic fractionation of nucleic acids, which are generally similar to one another in terms of charge-to-mass ratio, but differ in size. Polyacrylamide

gel electrophoresis separates nucleic acids principally on the basis of size because smaller nucleic acids migrate through the pores of the gel more rapidly than do larger ones.

By employing the detergent *sodium dodecyl sulfate* (SDS), polyacrylamide gel electrophoresis can also be used to fractionate proteins on the basis of size. In aqueous solution SDS forms an anion that coats the surfaces of protein molecules. This coating disrupts the weak bonds involved in maintaining protein conformation and dissociates proteins into polypeptide subunits if quaternary structure is present. The large negative charge contributed by the SDS molecules bound to the protein surface masks any inherent charge present in the protein itself. Since all protein molecules are coated with such a negative charge in the presence of SDS, protein size is the only variable affecting electrophoretic mobility. For this reason SDS-polyacrylamide gel electrophoresis is useful for separating polypeptide chains on the basis of size, and determining their molecular weights.

The basic principle underlying electrophoresis is applied in a somewhat different way in **isoelectric focusing,** a technique that exploits the fact that a molecule's net charge varies with pH. As pH is increased, molecules tend to lose positive charge and/or increase negative charge, while a decrease in pH results in the opposite effect. By varying the pH, a point can be reached where the negative and positive charges in a molecule exactly balance one another. At this pH, called the *isoelectric point,* the molecule's net charge is zero. Proteins differ from one another in isoelectric point because of inherent differences in the number and type of ionizable side chains present.

Isoelectric focusing exploits these inherent differences in isoelectric point by establishing a gradient of pH, either in solution or in a support medium such as polyacrylamide. When a protein sample is introduced and an electric potential applied, the protein molecules will begin to migrate (Figure 3-30). Those protein molecules present at the cathode end of the gradient where the pH is high will have a net negative charge, and will migrate toward the anode end of the gradient where the pH is lower. But as the migrating proteins move through regions of progressively lower pH, their negative charge will gradually decrease. When they finally reach the region of the gradient corresponding to their isoelectric pH, net charge will become zero and migration will stop. Proteins that are initially positively charged at the anode end of the gradient where the pH is low will move in the opposite direction toward the cathode until they reach their particular isoelectric pH, at which point they will also stop migrating. Eventually each protein molecule becomes localized in a sharp band corresponding to the position in the pH gradient of its isoelectric point. This technique is an extremely sensitive way

FIGURE 3-29 Separation of proteins by polyacrylamide gel electrophoresis. *(Left)* Marker proteins with molecular weights indicated in kilodaltons. *(Right)* Cell extract separated by this technique showing the diversity of polypeptides present. Courtesy of S. H. Blose.

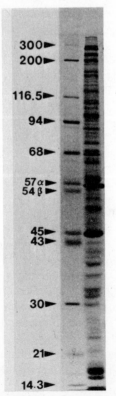

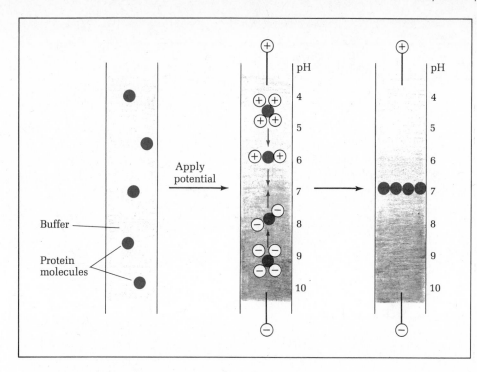

FIGURE 3-30 The principle of isoelectric focusing. The buffer contains special molecules that form a gradient of pH when an electric potential is applied. Protein molecules migrate until they reach the areas where the pH corresponds to their respective isoelectric points. Each protein therefore becomes sharply focused into a band in the region of the gradient where the pH equals that of its own isoelectric point.

of fractionating protein molecules because it is so sensitive to small differences in isoelectric point.

In 1975 Patrick O'Farrell designed a procedure for carrying out polyacrylamide gel electrophoresis and isoelectric focusing on the same sample, generating a two-dimensional separation that greatly increases the number of individual components that can be resolved from one another (Figure 3-31). Such two-dimensional gels, which are capable of resolving a thousand or more proteins in the same sample, are currently being used to map the entire protein makeup of bacterial as well as human cells.

FIGURE 3-31 Separation of bacterial cell proteins by a two-dimensional separation technique involving polyacrylamide gel electrophoresis in one direction, and isoelectric focusing in the other. Each spot represents a different polypeptide. Courtesy of P. O'Farrell.

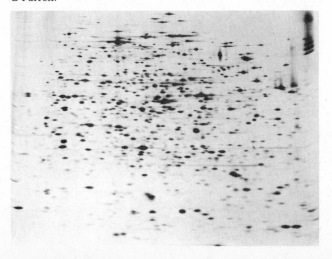

Chromatography

The term **chromatography** encompasses a broad spectrum of techniques that separate molecules from each other on the basis of differences in partitioning between two phases. In essence a sample containing the molecules to be fractionated, in either a liquid or gas phase, is passed through or across a bed of material comprising the second phase. Depending on their relative affinities for the two phases, molecules move at different rates and so can be physically separated from one another.

Liquid chromatography involves the passage of a liquid solution of molecules through a column filled with an appropriate packing material. If the substances to be fractionated are charged, one can employ packing materials with either positively or negatively charged side chains (Table 3-5). Molecules with positive charge will bind to negatively charged column materials, and vice versa. By changing the pH and ionic strength, the relative binding of molecules to the column material can be altered. In a typical experiment the sample to be fractionated is applied at the top of the column and washed in with an appropriate buffer. Molecules retained by the column are then eluted by changing the pH and/or ionic strength of the washing buffer, either in discrete steps or in a continuous gradient. Molecules with different charge will emerge from the column at different rates, and can be collected as they come off the bottom (Figure 3-32). Because this technique involves the reversible binding of charged ions to the column matrix, it is referred to as **ion-exchange** chromatography.

Gel filtration chromatography is an alternate type

TABLE 3-5
Selected column materials used in liquid chromatography

Ion-Exchange Chromatography

Material	Reactive Group	Substances Fractionated
Bio-Rex 70	$-COO^-Na^+$	Positively charged molecules
Carboxymethylcellulose	$-O-CH_2-COO^-Na^+$	Positively charged molecules
Phosphocellulose	$-O-P(O)^-(O^-Na^+)_2$	Positively charged molecules
Dowex-50 or Bio-Rex AG-50	$-O-SO_3^-H^+$	Positively charged molecules
DEAE-cellulose or DEAE-Sephadex	$-O-C_2H_4-N^+H-(C_2H_5)_2Cl^-$	Negatively charged molecules
Dowex-1 or Bio-Rex AG-1	$-\varnothing-CH_2N^+(CH_3)_3Cl^-$	Negatively charged molecules
QAE-Sephadex	$-O-C_2H_4-N^+(C_2H_5)_2-CH_2-CHOH-CH_3Br^-$	Negatively charged molecules

Gel Filtration Chromatography

Material	Chemical Makeup	Substances Fractionated
Bio-Gel P	Polyacrylamide beads	Any molecules by size
Bio-Gel A	Agarose beads	Any molecules by size
Sephadex G	Dextran beads	Any molecules by size

Adsorption Chromatography

Material	Substances Fractionated
Hydroxylapatite (hydrated calcium phosphate)	Double- and single-stranded DNA; proteins
Calcium phosphate gel	Proteins
Alumina	Steroids

of column chromatography designed to fractionate molecules on the basis of size rather than charge. In this case the column matrix consists of small beads of polymer or glass containing numerous microscopic pores. Molecules too large to enter the pores flow around the beads and therefore pass through the column more quickly than do small molecules, which pass into the pores of the beads and are temporarily retarded (Figure 3-33). A molecule's rate of passage through such a column is therefore a direct function of its molecular size. Beads in a variety of pore sizes are available for the separation of molecules with widely differing molecular weights (see Table 3-5).

Adsorption chromatography is a variation of liquid chromatography that separates molecules utilizing binding forces other than charge and size alone. A com-

FIGURE 3-32 Fractionation by ion exchange chromatography. Separation is dependent on differences in electrostatic interaction between the charged column matrix and the molecules being fractionated. When the sample enters the column, molecules with charge opposite to that of the column matrix stick to the column, while other molecules pass through. A change in the ionic strength or pH of the buffer allows the bound molecules to be subsequently eluted.

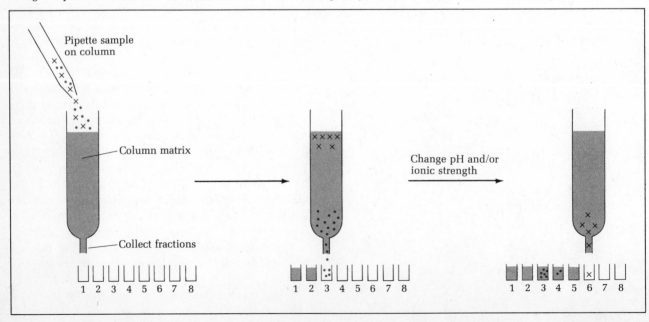

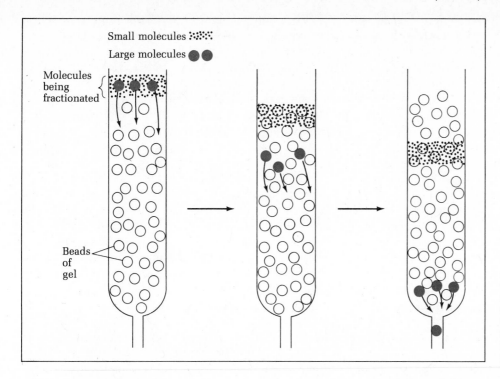

Small molecules ⠒⠶⠒

Large molecules ●●

Molecules being fractionated

Beads of gel

FIGURE 3-33 Gel filtration chromatography. Larger molecules, which cannot pass into the beads of the column matrix, move around them and through the column more quickly than smaller molecules, which are momentarily retained in the beads during passage.

FIGURE 3-33 Gel filtration chromatography. Larger molecules, which cannot pass into the beads of the column matrix, move around them and through the column more quickly than smaller molecules, which are momentarily retained in the beads during passage.

mon example involves the use of hydroxylapatite, a hydrated form of calcium phosphate, for fractionating proteins and nucleic acids. Hydroxylapatite chromatography is especially useful in fractionating DNA because double-stranded DNA binds to such columns more tightly than does single-stranded DNA.

Yet another type of column chromatography involves the prior attachment of specific molecules to an inert column matrix. In this technique, termed **affinity chromatography,** the attached substance is generally a compound expected to interact selectively with particular types of molecules. As an example, antibodies can be covalently linked to a packing material to form columns to be used for the isolation of specific antigens. Likewise enzyme substrates can be attached to the matrix to form columns useful for the isolation of specific enzymes. The inherent selectivity of this approach has made it a powerful tool with a wide variety of applications.

Although the above examples of liquid chromatography all involve a solid phase packed into the form of a column, the solid phase can also be spread out as a sheet. In **paper chromatography,** for example, the sample is applied as a small spot to a piece of filter paper and the solvent is then allowed to flow across the paper by capillary action. Molecules in the sample will migrate at different rates, depending on their relative affinities for the solvent and the paper. The separated compounds are then visualized by spraying the sheet with an appropriate staining reagent or by autoradiography. In the closely related technique of **thin-layer chromatography (TLC),** the solid phase is spread out as a thin layer on a rigid support of glass or plastic (Figure 3-34). The

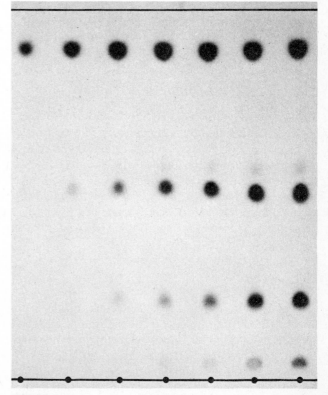

FIGURE 3-34 Separation of lipids by thin-layer chromatography. Samples are applied as separate spots at the bottom and the lipids are separated from one another as the solvent moves upward. The leading edge of the solvent migration is indicated by the line at the top. Courtesy of R. Douce.

thinner the layer, the better the resolution of separated components. By varying the chemical nature of the solid phase, it is possible to separate molecules by the principles of ion exchange, gel filtration, or absorption chromatography. Paper and thin-layer chromatography are most commonly employed in the fractionation of relatively small molecules because macromolecules do not migrate sufficiently to produce good separation.

In addition to chromatographic techniques designed for the separation of molecules in liquid solution, chromatography has also been adapted for the separation of molecules in a gaseous phase. This type of chromatography is limited to molecules that are readily volatilized, such as fatty acids, steroids, and other lipids. The basic principle is similar to liquid chromatography, with the molecules to be fractionated flowing across a bed of material for which they have differing affinities. In **gas chromatography,** however, the column bed is contained in a long coil through which the gas phase containing the sample is passed. The solid material of the bed is inert, but is coated with a heavy liquid with which the sample molecules can interact. The molecules to be separated therefore pass through the coil at different rates, depending on their relative affinities for the heavy liquid phase. Because the molecules undergoing fractionation are partitioned between the gas phase and the heavy liquid, this technique is most properly termed **gas-liquid chromatography (GLC).**

Nucleic Acid Hybridization

The technique of nucleic acid hybridization, which exploits specific interactions between nucleic acid molecules of similar base sequence, is one of the most powerful tools available for the isolation and characterization of nucleic acids. Because an understanding of this technique requires an intimate acquaintance with the structure and function of nucleic acids, we shall delay its discussion to the appropriate place later in the text (page 371).

SUMMARY

Cell biology is an experimental science based on the execution and interpretation of experiments designed to provide information about cell structure and function. Cells to be utilized for morphological or biochemical analysis can be obtained either directly from intact organisms or from cell cultures maintained in the laboratory. The major advantage of cell culture is that in spite of the artificial environment involved, conditions can be carefully controlled and a wide variety of experimental manipulations are possible.

Our current appreciation of the relationship between cell structure and function has been made possible by a combination of microscopic and biochemical techniques. The light microscope was historically responsible for the discovery of cells and the formulation of the cell theory, but its limited resolving power of about 200 nm severely restricts its usefulness for studying the details of cell architecture. Even with modifications such as phase contrast, interference, and polarization microscopy, organelles in the size range of ribosomes, membranes, microtubules, microfilaments, and chromatin fibers cannot be resolved.

By switching the source of illumination from light to electrons, resolving power can be enhanced several orders of magnitude from 200 nm to about 0.5 nm. The invention of the transmission electron microscope therefore revolutionized our view of cell architecture. The development of a diverse set of procedures for specimen preparation, such as thin sectioning, negative staining, positive staining, shadow casting, whole mounting, and freeze-etching, has opened our eyes to the existence of an exquisite subcellular architecture. And the more recent development of the scanning electron microscope has provided us with a direct three-dimensional view of the cell surface. Cytochemical techniques employing specific stains, antibodies, enzyme substrates, and radioisotopes have permitted both the light and electron microscopes to be used for studying the subcellular localization of chemical substances and biochemical activities.

By switching the illumination source to X-rays, yet another order of magnitude improvement in resolution is possible. Though no lenses are currently available for focusing X-ray beams, the diffraction pattern obtained by passing narrow beams of unfocused X-rays through purified protein or nucleic acid preparations has provided considerable information about macromolecular structure. This approach has played a crucial role in unraveling the structure of DNA and its role in the storage and transmission of genetic information.

Although the structural approach has provided us with a great deal of information about how cells are organized, the amount of functional information that can be obtained in this way is relatively limited. The most powerful tool for exploring cell function has been the isolation of individual organelles from cell homogenates for subsequent biochemical and morphological analysis. The historical development of techniques for subcellular fractionation was intimately associated with the invention of the ultracentrifuge, which generates forces large enough to sediment the small organelles and molecules present in cells.

Two major types of centrifugation are commonly employed in cell biology, velocity centrifugation and isodensity centrifugation. In velocity centrifugation the density of the particles or molecules being fractionated is greater than the density of the medium, so they migrate toward the bottom of the tube at a rate directly related to their size. Velocity centrifugation can be carried out on a sample randomly suspended in solution (differential centrifugation), or on a sample that has been applied to the top of a tube as a discrete zone (moving-zone centrifugation). In isodensity centrifugation the density of the particles or molecules being fractionated is within the same range as the density of the medium. In such a gradient substances stop migrating when they reach the region in which their density is equal to that of the medium. Because isodensity centrifugation fractionates on the basis of differences in density and velocity centrifugation fractionates on the basis of differences in size, a combination of these two approaches makes centrifugation an extremely powerful tool for separating, isolating, and purifying cell components.

In the course of studying the properties of individual organelles, it is often necessary to isolate and characterize the macromolecules from which they are constructed. Spectrophotometric analysis, isotopic tracers, dialysis, precipitation, electrophoresis, chromatography, and nucleic acid hybridization are all important experimental tools in the cell biologist's arsenal for characterizing macromolecules. Using information obtained in this way, great strides have been made in relating the properties of macromolecules to the structure and function of the organelles in which they occur.

SUGGESTED READINGS

Books

Chase, G. D., and J. L. Rabinowitz (1967). *Principles of Radioisotope Methodology*, Burgess, Minneapolis.

Davidson, R. L., and F. de la Cruz (1974). *Somatic Cell Hybridization*, Raven Press, New York.

Freifelder, D. (1982). *Physical Biochemistry. Application to Biochemistry and Molecular Biology*, Freeman, San Francisco.

Morris, C. J. O. R., and P. Morris (1973). *Separation Methods in Biochemistry*, 2nd Ed., Halsted Press, Wiley, New York.

Paul, J. (1970). *Cell and Tissue Culture*, 4th Ed., Williams and Wilkins, Baltimore.

Shaw, D. J. (1969). *Electrophoresis*, Academic Press, New York.

Sheeler, P. (1981). *Centrifugation in Biology and Medicine*, Wiley, New York.

White, L., Jr. (1968). *Machina ex Deo: Essays in the Dynamism of Western Culture*, MIT Press, Cambridge, MA.

Articles

de Duve, C., and H. Beaufay (1981). A short history of tissue fractionation, *J. Cell Biol.* 91:293s–299s.

Pease, D. C., and K. R. Porter (1981). Electron microscopy and ultramicrotomy, *J. Cell Biol.* 91:287s–292s.

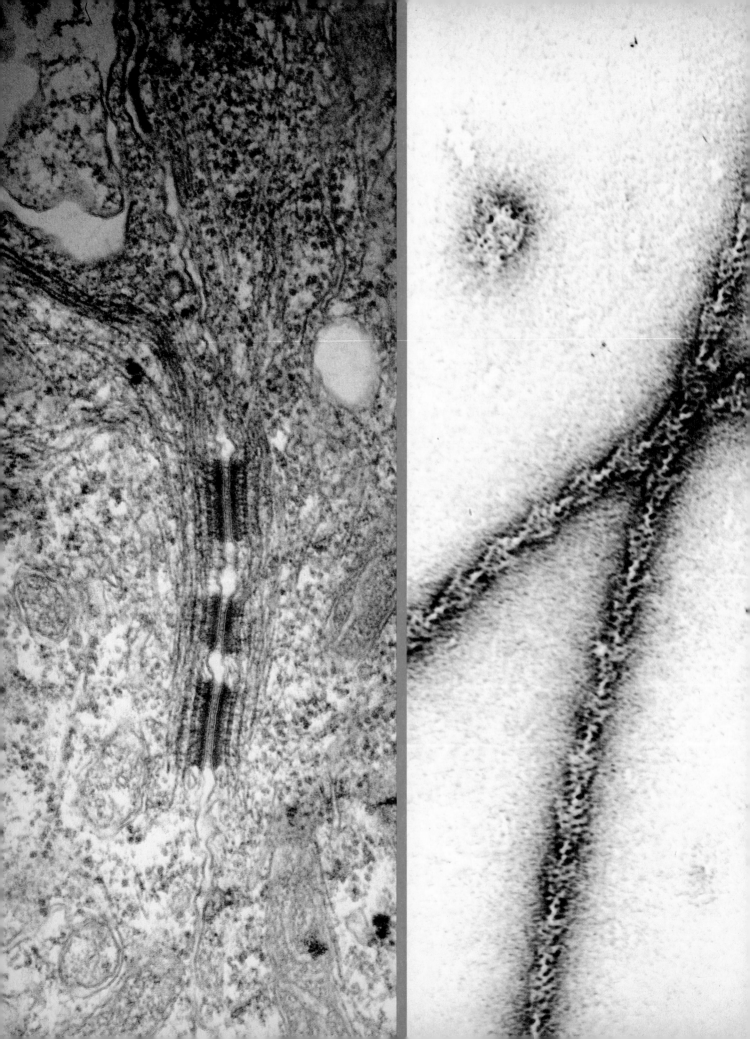

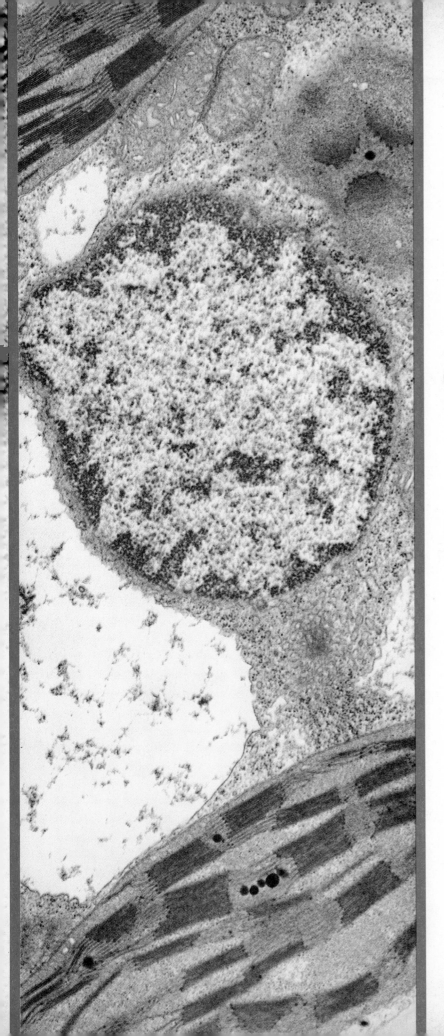

2

Cell
Organization

CHAPTER
4

Molecular Organization of Membranes and Cell Walls

At the surface of every cell is a thin and fragile barrier, the plasma membrane, that separates the cell from its external environment. The creation of a membrane boundary designed to protect and regulate the intracellular environment from the potentially disruptive conditions of the outer world must have been one of the first steps in the evolutionary origin of cells. Later in evolution this external membrane apparently proliferated, giving rise to a series of intracellular membranes designed to subdivide the cell into multiple compartments. Although such membranes were once viewed as static structures, functioning solely to separate the various regions of the cell from one another, it has gradually become apparent that membranes exhibit a multitude of functional capabilities. Virtually every important activity of the cell involves events that occur within, on, attached to, or through membranes. The question of how membranes are constructed is therefore crucial to our overall understanding of cell function.

In this chapter we shall explore the principles underlying the organization of the plasma membrane and its associated structures. Data obtained from many decades of experimentation on membrane behavior have finally led to a widely accepted molecular model of membrane organization. As will become apparent in later chapters, this model has important implications for our understanding of the functions carried out by all biological membranes.

MODELS OF MEMBRANE STRUCTURE

Biological membranes are constructed from lipids, proteins, and small amounts of covalently attached carbohydrate. Although membranes derived from different organelles and cell types resemble one another in basic morphology, there is considerable variability in the amounts and kinds of proteins and lipids employed in their construction. The protein-to-lipid ratio of most membranes is about 50:50, although it ranges from as high as 80:20 in some bacterial plasma membranes to as low as 20:80 in certain nerve cell membranes. The lipid component usually consists of a mixture of various phospholipids, but in some membranes glycolipids and/or steroids may be major components of the lipid fraction. The protein constituents of biological membranes may range anywhere from one or a few simple polypeptides to heterogeneous mixtures of dozens or even hundreds of different proteins.

In view of these marked differences in membrane

composition, the question arises as to whether there are any common principles upon which the construction of all membranes is based. This problem is analogous to trying to understand the basic principles of automobile construction by examining the materials from which cars are made. One finds varying proportions of different ingredients (steel, aluminum, plastic, cloth, rubber, glass, chrome) assembled in diverse ways to produce cars with superficially different appearances. But in spite of this diversity, the underlying principles of organization and operation are the same for all automobiles. The problem is to distinguish between the basic invariant features of design and those features that provide for superficial differences in appearance. This is the problem membrane biologists have faced over the years in trying to ascertain which, if any, features are fundamental to membrane organization and which are responsible for membrane diversity and specialization.

Early Observations

Because they are too thin to be resolved with the light microscope, no one had seen a biological membrane prior to the advent of electron microscopy. Yet indirect evidence led biologists to postulate the existence of such membranes long before they could be directly visualized. In the late nineteenth century C. Nageli,

W. Pfeffer, and C. E. Overton studied rates at which various molecules pass into and out of cells. Their discovery that different compounds enter and leave cells at significantly different rates suggested the existence of a surface membrane that regulates the passage of materials into and out of the cell. Overton made the especially important discovery that the rate at which a given substance passes into cells is directly related to its solubility in lipid; the more soluble it is in lipid, the more readily a substance passes into cells. This correlation led Overton to propose that cells are covered by a membrane containing a thin film of lipid.

The next important contribution to our understanding of membrane structure was made by Irving Langmuir, who discovered that phospholipids spread on a water surface spontaneously form a film whose dimensions suggest it to be one molecule thick. Because phospholipids are amphipathic molecules containing both hydrophilic and hydrophobic regions, Langmuir theorized that such monomolecular films involve an organized arrangement in which the hydrophilic or "head" groups of the lipid molecules are aligned next to the water surface while their hydrophobic "tails" extend out toward the air (Figure 4-1).

Shortly thereafter E. Gorter and F. Grendel exploited this phenomenon in their studies on the behavior of lipids extracted from red blood cells. When these

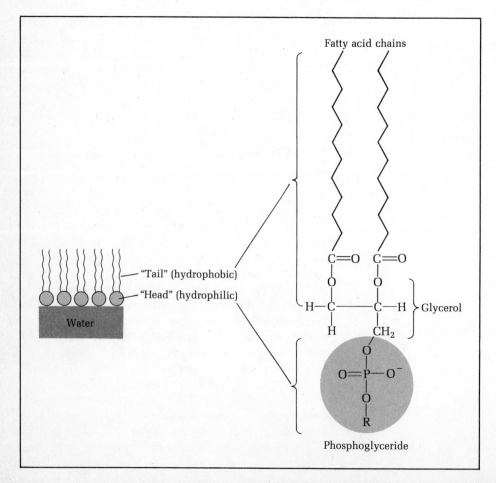

FIGURE 4-1 Langmuir's interpretation of the experiment in which phospholipids were found to form a layer one molecule thick when spread on a water surface. The chemical formula of a typical phospholipid is drawn to illustrate the typical "head-tail" arrangement characteristic of amphipathic lipids.

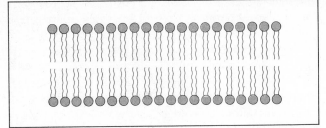

FIGURE 4-2 The lipid bilayer model of membrane organization proposed by Gorter and Grendel.

extracted lipids were spread as a monolayer on water, the film was found to cover an area twice that of the calculated surface area of the red cells from which the lipids had been extracted. Gorter and Grendel therefore concluded that the red cell is surrounded by a lipid membrane two molecules thick; such a **lipid bilayer** would be most stable if the hydrophilic head groups were exposed to the aqueous environments at the two membrane surfaces, and the hydrophobic tails were sequestered away from water in the membrane interior (Figure 4-2). In retrospect, it is clear that these experiments contained two major sources of error. The lipid extraction procedures were incomplete, leading to an underestimate of the total lipid present; and the erythrocyte was assumed to be spherical rather than biconcave in shape, causing an underestimate in the cell's surface area. But these two factors canceled one another, resulting in a conclusion about the existence of a lipid bilayer that has turned out to be basically sound. The lipid bilayer model of Gorter and Grendel is an important milestone in membrane research because it was the first attempt to describe the organization of a biological membrane in molecular terms, and because it has had a profound influence on subsequent models of membrane organization.

The Danielli–Davson Model

Shortly after Gorter and Grendel proposed the lipid bilayer model of the plasma membrane in 1925, its general validity was cast in doubt by surface tension experiments carried out by E. Harvey and J. Danielli. The data obtained by these investigators revealed that the surface tension of biological membranes is considerably lower than that of pure lipid droplets, implying that natural membranes cannot be made solely of lipid. It was subsequently discovered that the addition of protein to pure lipid droplets causes the surface tension to fall to levels comparable to those observed with natural membranes, implicating protein as a component of membrane structure.

Such observations, in combination with theoretical considerations on the permeability characteristics of protein-lipid films, led Danielli and Hugh Davson to propose a detailed molecular model of the plasma membrane in which an inner bilayer of lipid molecules, oriented with their hydrophilic head groups toward the membrane surfaces, is covered on both sides by layers of protein (Figure 4-3). The protein was presumed to be bound to the hydrophilic head groups of the lipid bilayer through ionic bonds. In order to account for the observed ability of some water-soluble molecules to diffuse through membranes, intermittent protein-lined pores were believed to interrupt the lipid bilayer. The Danielli–Davson model, first elaborated in detail in 1934, was to dominate the thinking of cell biologists for many decades to come.

The Unit Membrane Hypothesis

One of the more remarkable aspects of the Danielli–Davson model was that it was proposed before anyone had ever "seen" a membrane. During the early 1950s the application of electron microscopy to cell ultrastructure permitted the first direct visualization of biological membranes. In addition to confirming the expected existence of a plasma membrane around all cells, electron microscopy revealed for the first time the occurrence of elaborate intracellular membrane systems in eukaryotes. Upon close examination, all membranes were found to measure 7–8 nm in thickness and to exhibit the same staining appearance: two electron-dense lines separated by a more lightly stained central zone (Figure 4-4). Because of the presence of these three layers,

FIGURE 4-3 The Danielli–Davson model of membrane organization.

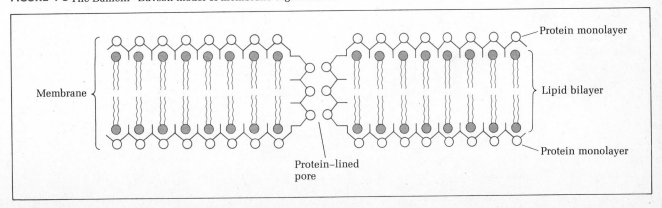

FIGURE 4-4 Electron micrograph showing the trilaminar appearance of the red blood cell plasma membrane. Courtesy of J. D. Robertson.

ance led J. D. Robertson to postulate in his **unit membrane hypothesis** that all biological membranes share a common underlying structure.

To investigate the relationship between the trilaminar appearance observed in electron micrographs and the underlying molecular organization of membranes, Robertson chose myelin as a model system for study. *Myelin* surrounds the axons of certain nerve cells and is composed of multiple layers of plasma membrane successively wound around one another. This stack of membranes is derived from specialized *Schwann cells* that are intimately associated with the axons (Figure 4-5). One may wonder why it was necessary to use myelin, rather than a more typical membrane, as a model for study. The answer is that the repeating layers of membrane present in myelin create a quasi-crystalline structure amenable to analysis by X-ray diffraction. Electron microscopic observations could therefore be correlated with X-ray diffraction data to determine the arrangement of the lipid and protein molecules within the myelin membranes. The X-ray diffraction analyses, carried out by F. O. Schmitt and later by J. B. Finean, indicated that myelin consists of lipid bilayers alternating with layers of protein. The data were not sophisticated enough, however, to distinguish between the three different arrangements of protein and lipid that would fit such a general pattern (Figure 4-6).

In order to distinguish among these three alternatives, Robertson carried out an extensive series of studies on the staining properties of lipid and protein molecules. These investigations led him to conclude that in electron micrographs of biological membranes, protein molecules as well as the hydrophilic head groups of lipid molecules are stained. Given this information, only one of the three possible arrangements of lipid and protein mentioned above would impart to

membranes were said to have a "trilaminar" appearance. This trilaminar arrangement, most readily observed in tissues fixed in potassium permanganate, occurs in a broad spectrum of membranes including eukaryotic and prokaryotic plasma membranes, the endoplasmic reticulum and its derivatives, and mitochondrial, chloroplast, and nuclear membranes. The widespread occurrence of this morphological appearance

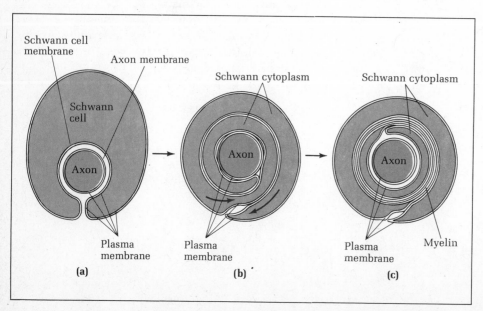

FIGURE 4-5 Developmental origin of myelin membranes. (a) In the early stages of myelin formation, nonmyelinated axons become enveloped by Schwann cells. (b) The Schwann cell plasma membrane begins to wrap around the axon in a spiral sheath. (c) Eventually this sheath becomes quite extensive and the layers of Schwann cell membrane become closely stacked upon each other, eliminating all cytoplasmic material from the spaces between the membranes. This final stack of closely packed membranes is termed myelin.

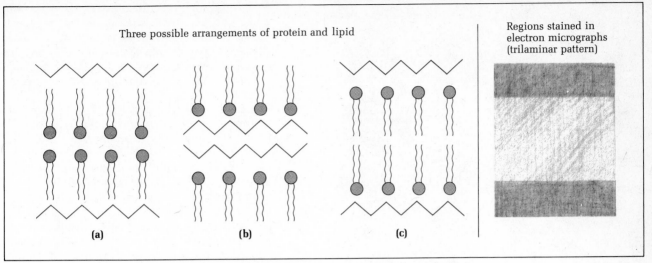

Three possible arrangements of protein and lipid

(a)   (b)   (c)

Regions stained in
electron micrographs
(trilaminar pattern)

FIGURE 4-6 X-ray diffraction analysis of myelin membranes yielded data compatible with all of the above possible arrangements of membrane proteins (zigzag line) and lipids. When Robertson subsequently showed that electron stains interact with proteins and the hydrophilic head groups of lipid molecules, it was concluded that only pattern (c) is compatible with the trilaminar appearance of biological membranes observed in electron micrographs.

membranes the typical trilaminar appearance observed in electron micrographs. This arrangement consists of an internal bilayer of lipid molecules covered by protein, with the lipid head groups projecting outward toward both membrane surfaces (see Figure 4-6c). In essence, this is the type of membrane organization originally proposed by Danielli and Davson.

The combination of electron microscopic and X-ray diffraction analyses therefore provided strong experimental support for the Danielli–Davson model of plasma membrane organization, and extended it to other membranes as well. The total body of evidence seemed so convincing that it led Danielli to state in 1963 that the basic structure of the plasma membrane "is that which I suggested in 1934, and it is also highly probable that the same structure is present in many other intracellular membranes. So far as it is possible to predict at the present time, it is unlikely that this general picture will be substantially disturbed, and the focus of attention is likely to shift to other fields" (1963:172).

Within a few years of this statement, however, isolated reports began to appear in the literature that were hard to reconcile with the Danielli–Davson–Robertson view of membrane organization. Among the more significant issues raised were the following:

1. The use of myelin as a general model for other biological membranes was questioned because myelin is atypical in its lack of enzymatic and metabolic activities, low protein content, and abundance of unusual glycolipids.
2. The dimensions of the Danielli–Davson–Robertson model of membrane organization place severe constraints on the amount and type of protein that can be accommodated in mem-

branes. Of the total membrane thickness of 7–8 nm typically observed in electron micrographs, about 4–5 nm is required for the lipid bilayer. The 3 nm remaining for the protein components must be divided between the two membrane surfaces, giving only about 1.5 nm per membrane surface. Only a thin monolayer of protein in the pleated-sheet or β-configuration would fit in such a small space. Yet physical measurements using circular dichroism (absorption of circularly polarized light) and infrared spectroscopy have revealed that most membrane proteins are not in the β-configuration, but are globular proteins with extensive regions of α-helix. Such globular proteins would be too large to form a monolayer 1.5 nm thick.

3. According to the Danielli–Davson model, membranes are held together by electrostatic attraction between the hydrophilic head groups of the lipid molecules and ionic side chains in the protein molecules. It therefore follows that increasing the ionic strength should disrupt membrane structure, releasing into solution protein molecules whose ionic side chains make them readily soluble in water. These premises are not supported by the available experimental data. Most membrane proteins are not removed by increases in ionic strength and, once isolated by more drastic extraction techniques (organic solvents or detergents), they are found to be relatively insoluble in water. Such observations suggest that membrane proteins have extensive hydrophobic regions that may be involved in binding them to membrane lipids.
4. Freeze-etch micrographs reveal the presence of protein particles in the interior of many membranes (Figure 4-7), indicating that the lipid bi-

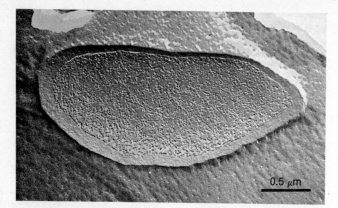

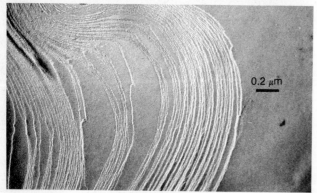

FIGURE 4-7 Freeze-etch micrographs showing the presence of distinct protein particles embedded within the inner regions of some, but not all, biological membranes. *(Top)* Red blood cell plasma membrane. *(Middle)* Chloroplast membranes. *(Bottom)* Myelin sheath. *(Top* and *Bottom)* Courtesy of D. Branton. *(Middle)* Courtesy of R. B. Park.

layer is interspersed with proteins rather than existing as a continuous sheet.

5. The accuracy of calculations indicating that membranes contain exactly enough lipid for a complete layer two molecules thick has been questioned. The data are not precise enough to rule out the possibility that the amount of lipid present is less than that required for a complete bilayer.

6. According to the Danielli–Davson model, the hydrophilic head groups of the lipid bilayer are covered over by a protein monolayer. Yet

treatment of intact membranes with the enzyme phospholipase, which selectively cleaves phospholipid head groups, leads to degradation of up to 75 percent of the membrane lipids. This susceptibility to enzymatic digestion suggests that many of the phospholipid head groups are exposed at the membrane surface rather than being covered by a protective layer of protein.

7. Optimum thermodynamic stability requires that hydrophobic regions of membrane lipids and proteins be sequestered from contact with water; hydrophilic regions of membrane lipids and proteins, on the other hand, should be exposed to the aqueous environment. The Danielli–Davson model does not meet these criteria for thermodynamic stability because the hydrophilic lipid head groups are sequestered from aqueous contact while the membrane proteins, which contain extensive hydrophobic regions (see issue 3 above), are exposed to the aqueous environment at the membrane surface.

In all fairness to the Danielli–Davson model, it should be pointed out that many of the above objections can be overcome by relatively minor modifications to the model. For example the presence of proteins interrupting the lipid bilayer, which is implied by several of the criticisms listed above, was already implicit in the protein-lined "pores" included in one of the early versions of the Danielli–Davson model (see Figure 4-3). Nonetheless the sheer number and variety of criticisms directed at the Danielli–Davson model was enough to stimulate considerable interest in the design of new models of membrane organization. This surge in model building culminated with the emergence of the fluid mosaic model in 1972.

The Fluid Mosaic Model

Of the various models of membrane organization proposed in recent decades, the **fluid mosaic model** of S. J. Singer and G. Nicholson has had the greatest impact on our thinking. According to this more recent model, illustrated in Figure 4-8, three principles guide the organization of all biological membranes. (1) Membrane lipids are arranged predominantly in the form of a bilayer, but this bilayer may be frequently interrupted by the presence of embedded proteins. (2) Membrane proteins exist in two classes, integral proteins embedded in the lipid bilayer and peripheral proteins bound to the bilayer surface. (3) The lipid bilayer is fluid, thereby permitting lateral movement of both membrane proteins and lipids. In the following sections, each of these points will be elaborated.

The Lipid Bilayer As mentioned earlier, the calculations that originally led Gorter and Grendel to propose

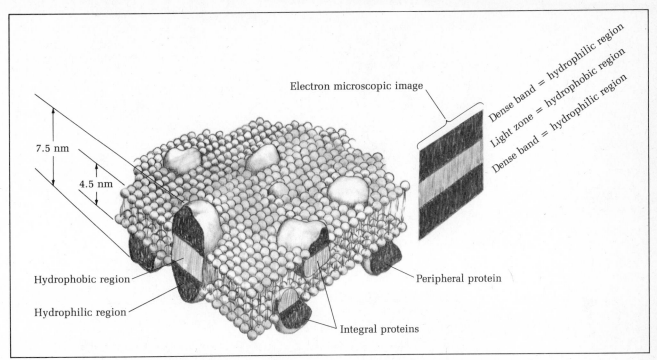

FIGURE 4-8 The fluid mosaic model of membrane organization. Such a model is compatible with the trilaminar appearance of membranes in electron micrographs if one postulates that the hydrophilic head groups of the lipid bilayer and the hydrophilic regions of the membrane proteins have a higher affinity for electron stains than the hydrophobic regions of these lipids and proteins.

the existence of a lipid bilayer were not precise enough to be completely convincing. But as more sophisticated techniques have gradually been developed, the data have continued to support the theory that membrane lipids are organized predominantly in the form of a bilayer. The most widespread approach to this problem has been to compare the properties of natural membranes to those of artificially created phospholipid bilayers. Such artificial bilayers can be generated in the laboratory by either placing a drop containing amphipathic lipids in a small hole separating two aqueous compartments or by exposing an amphipathic lipid-water suspension to ultrasonic vibrations. The first procedure generates planar bilayer membranes, while the second produces enclosed bilayer vesicles, or **liposomes** (Figure 4-9). Among the physical and chemical properties studied in such artificial lipid bilayers are thickness, electrical properties, permeability to water and solutes, temperature-dependent changes in state, electron-spin-resonance spectra, and X-ray diffraction patterns. In all cases the data obtained are similar to those derived from studies of natural membranes, suggesting that the majority of the lipid in natural membranes is arranged in the form of a bilayer. This does not rule out the possibility, however, that the continuity of the bilayer is occasionally interrupted by the presence of embedded proteins.

The universal presence of lipid bilayers in biological membranes occurs in spite of widespread differences in the particular kinds of lipids involved (Table 4-1 and Figure 4-10). In most membranes the majority of the lipid is phospholipid, although glycolipids predominate in myelin and chloroplast membranes. Even among membranes containing mostly phospholipid, the kinds of phospholipid vary significantly. Phosphoglycerides are generally the most abundant, but the various types of phosphoglyceride (phosphatidylcholine, phosphatidylethanolamine, etc.) occur in differing proportions. *Cardiolipin*, which contains two phosphoglyceride molecules linked together by a glycerol bridge, occurs in mitochondrial inner membranes, lysosomal membranes, and bacterial plasma membranes. Sphingomyelin is an important phospholipid of animal plasma membranes, but is scarce or absent in mitochondrial, chloroplast, and bacterial plasma membranes. In addition to phospholipids, some membranes contain glycolipids and steroids. Glycolipids are abundant in myelin and chloroplast membranes, while the steroid cholesterol is a significant component of animal plasma membranes and myelin. The steroids sitosterol and stigmasterol are restricted to the plasma membranes of plant cells.

The fatty acid chains employed in the construction of membrane lipids also vary among membranes. Less than 10 percent of the fatty acid chains in myelin membrane lipids are unsaturated, while in mitochondrial

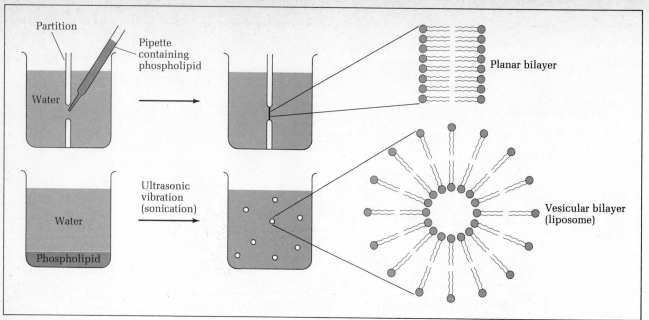

FIGURE 4-9 Procedures for creating artificial lipid bilayers in the laboratory.

TABLE 4-1

Principal lipids employed in the construction of biological membranes

Category and Name	Structure
I. Phospholipids	
A. Phosphoglycerides	Derivatives of phosphatidic acid (see Figure 1-11) in which $X =$
Phosphatidylcholine (lecithin)	$-(CH_2)_2-\overset{+}{N}-(CH_3)_3$
Phosphatidylethanolamine (cephalin)	$-(CH_2)_2-NH_3^+$
Phosphatidylserine	$-CH_2-CH(COO^-)-NH_3^+$
Phosphatidylthreonine	$-CH(CH_3)-CH(COO^-)-NH_3^+$
Phosphatidylglycerol	$-CH_2-CHOH-CH_2OH$
Phosphatidylinositol	
Cardiolipin (diphosphatidylglycerol)	
B. Sphingophospholipids	Derivatives of sphingosine (see Figure 1-11) in which $X =$
Sphingomyelin	$-(PO_3^-)-(CH_2)_2-NH_3^+$ or $-(PO_3^-)-(CH_2)_2-\overset{+}{N}-(CH_3)_3$
II. Glycolipids	Derivatives of sphingosine (Figure 1-11) in which $X =$
Cerebrosides	Galactose or glucose
Sulfatides	Galactose-sulfate
Gangliosides	Oligosaccharide chain plus a sialic acid
III. Steroids	
Cholesterol	
Stigmasterol	

FIGURE 4-10 Comparison of the protein and lipid contents of some typical membranes. Abbreviations: P-lipid = phospholipid, Chol = cholesterol, Glyco = glycolipid, PE = phosphatidylethanolamine, PS = phosphatidylserine, PC = phosphatidylcholine, PI = phosphatidylinositol, PG = phosphatidylglycerol, CL = cardiolipin, SM = sphingomyelin.

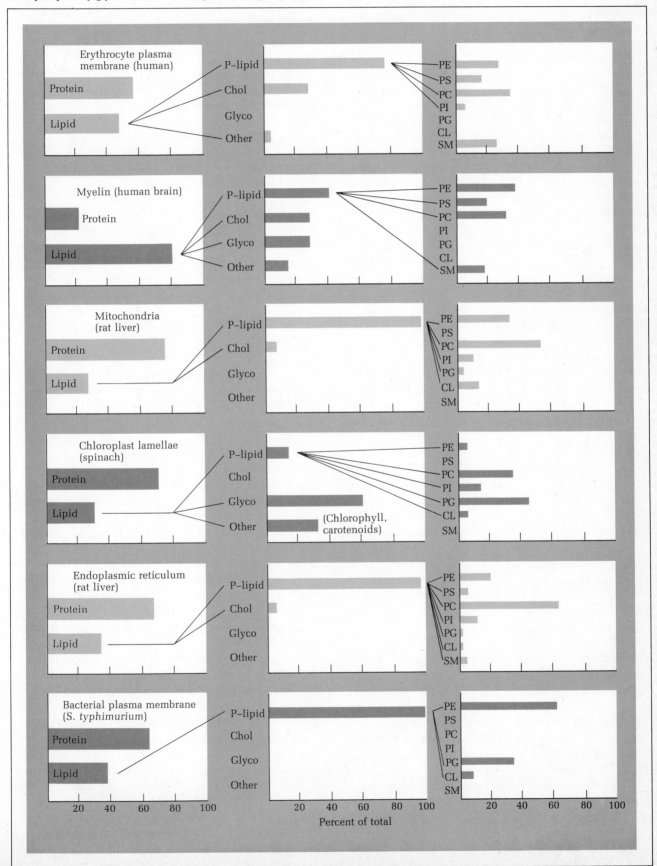

and chloroplast membranes the value is in excess of 50 percent. Chloroplast membranes are rich in branched-chain fatty acids, while bacterial membranes often contain unusual fatty acids with branched chains or cyclopropane rings. Changes in dietary intake or physiological conditions can induce significant alterations in the spectrum of fatty acid chains occurring in membrane lipids, indicating that their fatty acid makeup is not rigidly specified.

In view of the above differences in lipid composition, it may seem surprising that all membranes exhibit a similar bilayer organization. This common bilayer structure is possible, even in the face of such diversity among the lipids involved, because virtually all membrane lipids are amphipathic molecules exhibiting a hydrophilic "head" and a hydrophobic "tail." Hence in spite of their chemical differences, membrane lipids can all be arranged in the form of a bilayer. This does not mean, of course, that the differences in lipid composition that occur among membranes are of no significance. Many of the physical and biological properties of membranes, such as permeability, fluidity, and enzymatic activities, may be significantly influenced by the nature of the lipids present.

Arrangement of Membrane Proteins The fluid mosaic model divides membrane proteins into two general categories, integral and peripheral. **Integral proteins** are embedded in the lipid bilayer and can only be removed by relatively harsh treatments, such as exposure to detergents or organic solvents. Integral proteins are bound to the membrane primarily by hydrophobic interactions with the tails of the lipid bilayer, and may either span the bilayer completely (transmembrane proteins) or be embedded in one side of the bilayer or the other. In order for such an arrangement to be thermodynamically stable, integral proteins must be amphipathic molecules whose hydrophobic regions are buried in the membrane interior and hydrophilic regions are exposed at the membrane surfaces. Such molecules account for the bulk of the membrane proteins, including most membrane-associated enzymes, receptors, and antigens.

In contrast to integral proteins, **peripheral proteins** are bound to membranes by relatively weak ionic interactions with the hydrophilic head groups of the lipid bilayer. Such proteins are therefore easily removed from membrane surfaces by raising the ionic strength. Peripheral proteins do not cover the entire surface of the bilayer, leaving many vacant areas where the lipid head groups are exposed at the membrane surface (see Figure 4-8).

The arrangement of integral and peripheral proteins in the fluid mosaic model overcomes many of the criticisms of the Danielli–Davson model mentioned earlier. For example, the presence of integral proteins embedded in the lipid bilayer is consistent with the protein particles observed within the membrane interior in freeze-fracture micrographs; it is also compatible with the observation that most membrane proteins are not readily extracted by increasing the ionic strength. The presence of proteins embedded in the bilayer is also consistent with the known existence of globular membrane proteins containing α-helical structure, while the exposure of lipid head groups at the membrane surface is compatible with their observed susceptibility to phospholipase digestion. Finally, the fluid mosaic model is thermodynamically reasonable because the hydrophilic regions of the membrane lipids and proteins are exposed to the aqueous environment, while their hydrophobic regions are buried in the membrane interior away from contact with water.

Membrane Fluidity The first indication that membranes may be fluid rather than rigid structures came from studies employing the technique of **electron-spin-resonance (ESR)** spectroscopy to monitor the freedom of movement of membrane lipids. In this technique a chemical group containing an unpaired electron, usually a nitroxide group, is attached to the fatty acid tail of a phospholipid molecule. The term *spin-label* is often used to refer to such substances. The presence of the unpaired electron in the spin-label causes it to absorb and emit energy when exposed to an external magnetic field of appropriate intensity. If one plots the amount of energy absorbed as a function of the intensity of the magnetic field, an ESR spectrum is obtained. The shape of such a spectrum is influenced by the mobility of the molecule containing the unpaired electron. Thus one can introduce phospholipid containing a nitroxide spin-label into biological membranes and use ESR spectroscopy to monitor the mobility of the phospholipid molecules within the bilayer.

Such experiments, carried out in the laboratories of Harden McConnell and O. Hayes Griffith in the late 1960s, provided the first direct evidence for the fluidity of the lipid bilayer in biological membranes. In these studies spin-labels inserted into biological membranes were found to exhibit an ESR spectrum whose shape is intermediate between that of a spectrum produced by a rigidly fixed molecule and that produced by a completely mobile molecule (Figure 4-11). Such an intermediate pattern indicates that the lipid molecules in the bilayer are neither fixed in position, as in a crystal, nor do they have complete freedom of movement, as in a liquid. This state is referred to as a **liquid crystal,** a condition in which individual lipid molecules are capable of lateral diffusion within the bilayer but always retain the same orientation, with their hydrophilic head groups pointed toward the membrane surface and their hydrophobic tails projecting toward the membrane interior.

The fluidity of the lipid bilayer has also been studied by **differential scanning calorimetry,** a technique

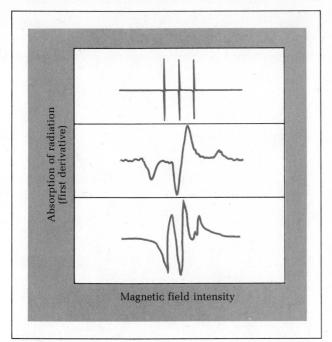

FIGURE 4-11 Electron spin resonance spectra of a spin-labeled phospholipid under various conditions. *(Top)* Spin-label in solution at room temperature yields a spectrum typical of a completely mobile molecule. *(Middle)* The same phospholipid spin-labeled at −65°C yields a spectrum typical of an immobilized molecule. *(Bottom)* When inserted into a biological membrane, the spectrum of the spin-label is intermediate between the above two extremes.

FIGURE 4-12 Differential scanning calorimetry of membrane preparations obtained from *Mycoplasma* cells grown under different conditions. The transition temperatures, which correspond to the point at which each membrane melts from a gel into a liquid crystal, are designated by arrows. Membranes obtained from cells grown in media enriched in oleate, an unsaturated fatty acid, have a much lower transition temperature than normal; membranes from cells grown in media enriched in stearate, a saturated fatty acid, have a higher transition temperature than normal.

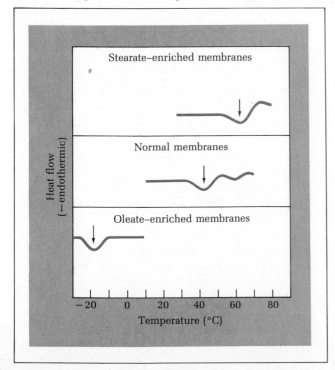

that exploits the fact that the transition between different physical states (e.g., from solid to liquid or liquid to gas) is accompanied by the uptake of heat. At relatively low temperatures lipid bilayers are frozen into a solid or gel state, while at higher temperatures they "melt" into the liquid-crystalline state. The *transition temperature* at which this melting occurs can be determined by placing a membrane or artificial bilayer into a sealed chamber (calorimeter) in which the uptake of heat can be measured as the temperature is raised. The point of maximum heat absorption corresponds to this transition temperature (Figure 4-12). Such experiments have revealed that the transition temperature for most biological membranes is below the physiological temperatures at which they normally function, indicating that under physiological conditions membranes are in a liquid-crystalline state. By quantitating the amount of heat uptake that occurs at the transition temperature, one can even calculate the percentage of the total lipid that has assumed the fluid state. For most membranes, this value is between 70 and 90 percent. The remaining lipid is thought to have its mobility restricted because it is tightly associated with membrane proteins. The term *boundary lipid* has been given to this relatively immobile lipid fraction.

Differential scanning colorimetry has also shown that the transition temperature varies with the type of lipid present in the bilayer. An increase in unsaturated fatty acids, a decrease in fatty acid chain length, or a decrease in cholesterol content all cause a lowering of the transition temperature and hence an increase in membrane fluidity. This effect is not a trivial one; membranes enriched in lipids containing the unsaturated fatty acid oleate, for example, have transition temperatures that are as much as 80°C lower than membranes enriched in lipids containing the saturated fatty acid stearate (see Figure 4-12).

Such differences in fluidity may have profound effects on membrane function. It has been shown in *Mycoplasma* cells, for example, that growth in a medium enriched in saturated fatty acids results in the formation of membranes with an abnormally high ratio of saturated to unsaturated fatty acids. As a result, the transition temperature of such membranes is raised to almost 60°C. Because this temperature is considerably higher than that at which these cells normally grow, the lipid bilayer in their membranes will be maintained in the gel state rather than in its usual liquid-crystalline state. *Mycoplasma* containing such membranes tend to swell and rupture, indicating that the normal permeability and transport properties of the plasma membrane depend on a liquid-crystalline state of the lipid bilayer. Such observations are especially intriguing because in higher organisms the fatty acid composition of cellular membranes, and hence their relative fluidity, is known to be influenced by diet. The nutritional state of an organism can thus have significant effects on the structural and functional properties of its cellular membranes.

The conclusion that biological membranes contain a lipid bilayer that is relatively fluid raises the question of whether membrane proteins are fixed in location or whether they too are free to diffuse laterally within the plane of the membrane. In a classic experiment reported in 1970, David Frye and Michael Edidin employed the technique of fluorescence microscopy to follow the fate of membrane proteins in human and mouse cells fused together by treatment with Sendai virus. The location of mouse membrane proteins was detected with mouse-specific antibodies linked to the dye fluorescein, which fluoresces green, while human membrane proteins were located with human-specific antibodies linked to the dye rhodamine, which fluoresces red. Although it was found upon microscopic examination that the plasma membrane of newly fused cells contained distinct regions fluorescing either green or red, within an hour these separate regions had completely intermixed. This intermixing could not be prevented by metabolic inhibitors, suggesting that it is the result of passive diffusion of membrane proteins through a fluid lipid matrix. This interpretation was further supported by the discovery that lowering the temperature of the culture below the transition temperature of the lipid bilayer prevented mixing of the two types of fluorescence.

Similar conclusions about the lateral mobility of membrane proteins have been reached from experiments involving treatment of white blood cells with antibodies directed against specific membrane antigens. When visualized by fluorescence microscopy these protein antigens appear to be diffusely spread over the entire cell surface in normal white cells; within an hour of antibody treatment, however, the protein antigens become localized at one end of the cell, forming a structure known as a "cap" (Figure 4-13). This sequence of events again suggests that membrane proteins are capable of relatively rapid migration through a fluid lipid matrix.

If membrane proteins are free to diffuse laterally through the lipid bilayer, then their distribution pattern should be random. To test this prediction Singer and his collaborators employed ferritin-labeled antibodies to monitor the distribution of several different proteins within the plasma membrane. The results of these studies confirmed the expected random distribution of at least some membrane proteins. Intramembrane protein particles observed in freeze-fracture micrographs are also randomly arranged in most membranes. The distribution pattern of such particles can be altered, however, by changing the fluidity of the lipid bilayer (Figure 4-14), confirming the fact that membrane fluidity affects the behavior of membrane proteins.

Although many membrane proteins are free to diffuse laterally within the lipid bilayer, some proteins appear to have their movement restricted. One widely used technique for quantitating the rates at which

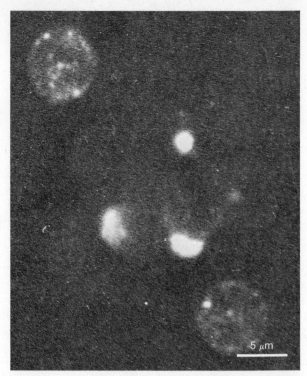

FIGURE 4-13 Fluorescence photomicrograph of lymphocytes treated with fluorescein-labeled antibodies that specifically bind to cell surface antigens. Cells in the upper left and lower right exhibit a patched distribution of fluorescein-labeled antibody, while the three cells in the center illustrate aggregation of the labeled molecule at a single point on the cell surface referred to as a "cap." Courtesy of P. Sheterline.

membrane proteins (and lipids) diffuse is **fluorescence photobleaching recovery.** In this procedure fluorescent chemical groups are attached to membrane proteins or lipids, and a narrow beam of laser light is then focused on the membrane to bleach the fluorescent label in a small spot on the membrane. The diffusion of unbleached molecules from adjacent parts of the membrane into the bleached region will cause the fluorescence of the bleached area to gradually recover. The rate at which this recovery occurs can be used to calculate the diffusion rate of the fluorescent protein or lipid.

Diffusion measurements employing fluorescence photobleaching recovery have led to the conclusion that typical membrane lipids diffuse at rates approaching several micrometers per second. This means that an individual lipid molecule can diffuse from one end of a cell to the other in a matter of seconds. Membrane proteins, on the other hand, exhibit considerable variability in diffusion rates. Some diffuse almost as rapidly as membrane lipids, but most appear to move more slowly than would be expected if they were completely free to diffuse at random within the lipid bilayer. In some cases the reason for restricting the mobility of a particular membrane protein is clear. The enzyme acetylcholinesterase, for example, is preferentially localized at the tip of the axon in nerve cells because it is required there for

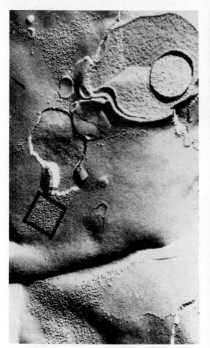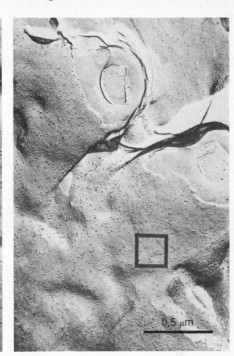

0.5 μm

FIGURE 4-14 The fluidity of the plasma membrane is illustrated in this sequence of freeze-fracture micrographs of membranes incubated at several different temperatures. As the temperature is increased (left to right), the membrane particles change from a densely packed configuration to a more random array (see areas enclosed in boxes). Courtesy of G. Thompson, Jr.

the breakdown of the neurotransmitter acetylcholine. In cells lining the intestinal tract certain membrane proteins involved in solute transport are located only on the side of the cell where such transport occurs. Such observations make it clear that mechanisms for controlling the mobility and distribution of specific membrane proteins exist. Such mechanisms, which include cytoskeletal attachments and intermolecular associations between membrane proteins, will be discussed shortly when we examine the structure of the red cell and purple membranes in detail.

The fluid mosaic model of membrane organization has several advantages over earlier models of membrane organization. It provides answers for criticisms raised against the Danielli–Davson model, it incorporates recent information on lipid and protein mobility, it offers a dynamic rather than a static picture of the membrane, and it is broad enough to accommodate membranes with differing lipid and protein compositions. The ultimate test of this model, of course, is its applicability to specific membranes where detailed information about molecular organization is available. Two such examples will be described in the following section.

MOLECULAR ORGANIZATION OF THE PLASMA MEMBRANE

A detailed understanding of the molecular organization of the plasma membrane ultimately depends on the availability of techniques for isolating the plasma membrane from other cell components. When cells are homogenized in a sucrose-containing medium, the plasma membrane is broken into fragments that spontaneously reseal to form tiny vesicles. Depending on the conditions these vesicles may either be right-side-out, reflecting the normal *in vivo* situation, or they may be inside-out, exposing their cytoplasmic surfaces to the exterior. The plasma membrane vesicles generated during homogenization can be separated from other cellular organelles by a combination of differential and isodensity centrifugation, with the exact conditions of centrifugation depending on the cell type involved. For example with mammalian red blood cells, which have no internal membranes or organelles, plasma membrane vesicles can be isolated in a single centrifugation step. More complex tissues may require an elaborate sequence of centrifugation steps before a "pure" plasma membrane fraction is obtained. Even then, such fractions are almost always contaminated to some extent with membranes derived from other organelles.

A combination of microscopic and biochemical approaches is required for the identification of subcellular fractions containing plasma membrane. With the electron microscope one would obviously expect to see membrane vesicles and fragments. The presence of specialized structures characteristic of the plasma membrane, like the cell junctions to be discussed in the next chapter, can help to confirm that isolated membranes were derived from the cell surface. Biochemical

TABLE 4-2
Molecular markers employed for the identification of isolated plasma membrane fractions

ANIMAL CELLS	PLANT CELLS
5'-Nucleotidase	K$^+$-ATPase
Na$^+$,K$^+$-ATPase	Glucan synthetase II
Adenylate cyclase	
Alkaline phosphodiesterase I	**BACTERIAL CELLS**
Aminopeptidase	NADH dehydrogenase
Hormone receptors	ATPase
Lectin receptors	Malate dehydrogenase
Neurotransmitter receptors	Succinate dehydrogenase
Toxin receptors	
Membrane glycoproteins	

analyses for markers such as plasma membrane enzymes, glycoproteins, and drug or hormone receptors are also useful in identifying vesicles derived from the plasma membrane (Table 4-2). Once such approaches have been used to identify plasma membrane-containing fractions, these fractions can be subjected to various extraction and labeling techniques designed to ascertain the identity and orientation of the molecules from which these membranes are constructed.

The Red Cell Membrane

The above approach has been most successfully applied to the mammalian red blood cell (erythrocyte), where the absence of intracellular membranes and organelles has facilitated purification of large quantities of plasma membrane. The red cell membrane is simple enough to permit isolation and analysis of its major molecular components, yet it exhibits functional activities that are potentially representative of other plasma membranes. The biochemistry of the red cell membrane has accordingly received a great deal of attention in recent years, and promises to be one of the first plasma membranes whose molecular structure is understood in complete detail.

FIGURE 4-15 Scanning electron micrograph of a red cell ghost showing the interior of the empty cell. Courtesy of T. L. Steck.

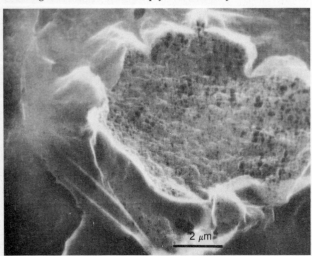

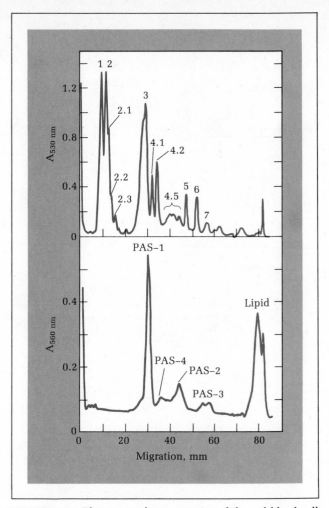

FIGURE 4-16 Plasma membrane proteins of the red blood cell fractionated by SDS-polyacrylamide gel electrophoresis. The gels have been stained with Coomassie-blue dye to detect protein *(top)* and with the PAS reagent to stain carbohydrate *(bottom)*. The intensity of each stained polypeptide band has been quantitated by measuring the amount of light it absorbs. Courtesy of T. L. Steck.

The plasma membrane can be isolated intact from red blood cells by simply using hypotonic conditions to make the cells leaky. Under these conditions the intracellular contents leak out, leaving behind an empty membrane sac or **ghost** (Figure 4-15). Such membrane ghosts consist of protein, lipid, and carbohydrate in a ratio of about 50:40:10 by weight. The lipid can be removed by extracting isolated membranes with organic solvents, leaving the protein and most of the carbohydrate behind. Membrane proteins, on the other hand, can be extracted with the detergent sodium dodecyl sulfate (SDS), which simultaneously dissociates protein molecules into their constituent polypeptide chains.

When membrane polypeptides isolated in this manner are fractionated by SDS-polyacrylamide gel electrophoresis and stained with a protein-specific dye such as Coomassie blue, a relatively small number of different components are seen. Prominent among these are **spectrin, ankyrin, band 3, bands 4.1 and 4.2, actin,** and **glyceraldehyde-3-phosphate dehydrogenase**

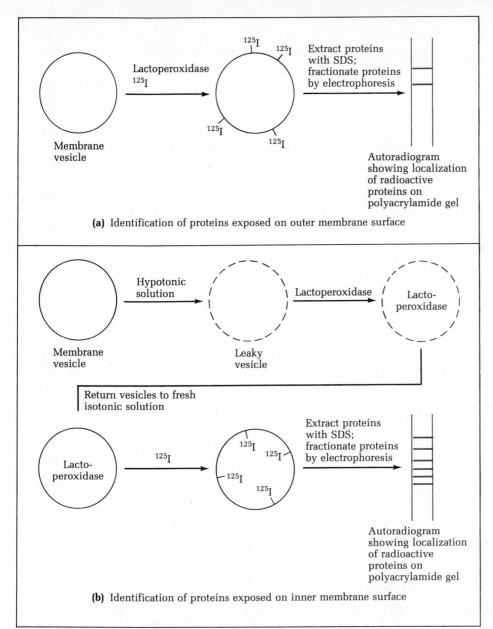

(a) Identification of proteins exposed on outer membrane surface

(b) Identification of proteins exposed on inner membrane surface

FIGURE 4-17 Outline of one common approach for analyzing the distribution of proteins on the inner and outer surfaces of membrane vesicles.

(Figure 4-16). Staining such electrophoretic gels for the presence of carbohydrate with the periodic acid–Schiff (PAS) reagent reveals that some of the red cell membrane proteins contain carbohydrate, and are hence glycoproteins. The most prominent glycoprotein, termed **glycophorin,** is two-thirds carbohydrate by weight. This high carbohydrate content hinders the staining of glycophorin by protein-specific dyes such as Coomassie blue. Among the other glycoproteins detected by PAS staining is the band-3 protein.

Identification of its major protein components is only the first step in unraveling the molecular oganization of the red cell membrane. Next one must ascertain where these proteins are located within the membrane. Spectrin, ankyrin, bands 4.1 and 4.2, actin, and glyceraldehyde-3-phosphate dehydrogenase are readily extracted by increasing the ionic strength, indicating that they are peripheral rather than integral proteins. Radioactive-labeling procedures have been useful in elucidating which membrane proteins are located on the exterior membrane surface and which are located on the inner (cytoplasmic) membrane surface. The most common labeling procedures employ either the enzyme *lactoperoxidase* to catalyze labeling of membrane proteins with radioactive iodine (^{131}I or ^{125}I) or the enzyme *galactose oxidase* to catalyze labeling of membrane-associated carbohydrate groups with ^{3}H-borohydride. Because the enzymes used in these labeling techniques are too large to pass through the plasma membrane, only proteins exposed on the outer membrane surface will be labeled. Fractionation of the labeled proteins on SDS-polyacrylamide gels will therefore indicate which erythrocyte membrane proteins are exposed on the exterior membrane surface (Figure 4-17a).

To ascertain which proteins are exposed on the inner (cytoplasmic) membrane surface, vesicles can be made "leaky" by exposing them to a hypotonic medium. Lactoperoxidase will now freely enter such ghosts, which are then resealed by transferring them back to an isotonic medium. After washing away any remaining external enzyme, the resealed ghosts are labeled by incubation with the appropriate radioactive substrate. Only membrane proteins originally exposed on the inner membrane surface become radioactive under such conditions (see Figure 4-17b).

An alternative approach for localizing membrane proteins, pioneered by Theodore Steck and his colleagues, involves exposing red cell ghosts to conditions designed to produce a mixture of right-side-out and inside-out vesicles. These two types of vesicles are then separated from each other by centrifugation. Comparison of the proteins labeled in right-side-out versus inside-out vesicles again provides information concerning the identity of the proteins exposed on the exterior and cytoplasmic surfaces of the red cell membrane.

Data obtained from the above approaches have revealed considerable asymmetry in the localization of red cell membrane proteins. Glycophorin and the band-3 protein are the only major proteins labeled from the outer membrane surface. Both are glycoproteins whose attached carbohydrate groups are released as glycopeptides when right-side-out ghosts are treated with proteolytic enzymes. Since such enzymatic treatment affects only molecules exposed at the outer membrane surface, it can be concluded that the carbohydrate portions of these membrane glycoproteins are exposed at the cell surface.

In contrast to the outer membrane surface, where glycophorin and the band-3 protein predominate, all the major proteins of the red cell membrane are exposed at the inner membrane surface, including glycophorin and band 3. The fact that every membrane protein can be labeled from one membrane surface and/or the other means that none are completely buried within the lipid bilayer. The exposure of glycophorin and band 3 on both the inner and outer membrane surfaces raises two possibilities. Either separate molecules of glycophorin and band 3 are present on both the external and cytoplasmic surfaces of the membrane, or else each glycophorin and band-3 molecule is a transmembrane protein spanning from one side of the membrane to the other.

To distinguish between these two possibilities, experiments have been carried out in which radioactive protein fragments derived from membranes labeled from either the inner or outer surface were compared. When the red cell is labeled from the outside, one set of fragments from glycophorin and band-3 protein are found to be radioactive, while labeling from the inside produces a different set of radioactive fragments. These

results indicate that each molecule of glycophorin and band 3 is a transmembrane protein spanning the membrane in an asymmetric fashion and hence exposing a unique portion at each membrane surface. The fact that radioactive peptides produced by labeling membrane vesicles from the inside are not observed when vesicles have been labeled from the outside (and vice versa) indicates that glycophorin and the band-3 protein cannot rotate their ends back and forth between the cytoplasmic and external membrane surfaces. Because they span the membrane, such transmembrane proteins are in a position to transfer materials or information across the plasma membrane. The possible involvement of transmembrane proteins in events of this type will be discussed in the following chapter.

A great deal is now known about the structures of glycophorin and the band-3 protein. Glycophorin consists of a polypeptide chain 131 amino acids long to which 16 oligosaccharide chains are covalently attached. The oligosaccharide chains are joined to the amino acids serine, threonine, or asparagine and are exposed on the exterior surface of the membrane. As will be discussed in the following chapter, such cell surface oligosaccharide chains play important functional roles as recognition and attachment sites for other cells and molecules. The amino acid sequence of glycophorin can be divided into three regions. By convention, the beginning of a protein chain is referred to as its *amino-terminal* end because the first amino acid of the sequence retains a free amino group; the other end of a protein chain is referred to as the *carboxy-terminal* end because the last amino acid in the chain retains a free carboxyl group. In the case of glycophorin, the amino-terminal region of the polypeptide chain contains the attached oligosaccharide chains and is located at the outer membrane surface. The middle of the molecule is enriched in hydrophobic amino acids and is buried in the membrane interior. Finally, the carboxy-terminal end is enriched in hydrophilic amino acids and is localized on the cytoplasmic surface of the membrane. Glycophorin is thus seen to be an amphipathic protein with its hydrophobic region buried in the membrane interior and its hydrophilic ends associated with the membrane surfaces, an arrangement consistent with the basic principles of the fluid mosaic model.

In contrast to glycophorin, the carboxy-terminal rather than the amino-terminal end of the band-3 protein contains the covalently attached carbohydrate chains and is exposed at the exterior surface of the plasma membrane. The middle segment of the band-3 protein is buried in the membrane interior, while the amino-terminal end is exposed on the cytoplasmic surface of the membrane. The band-3 protein is therefore amphipathic like glycophorin but is oriented across the membrane in the opposite direction.

Information concerning the localization of red cell

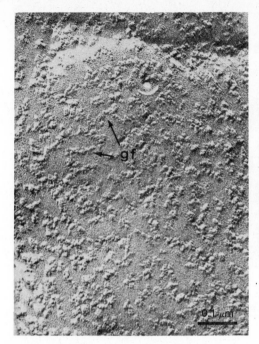

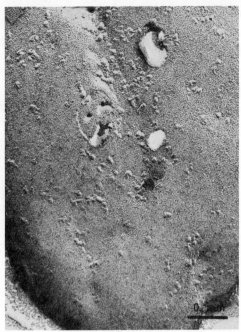

FIGURE 4-18 Localization of band-3 protein by freeze-fracture analysis of inside-out membrane vesicles derived from red blood cells. *(Left)* In untreated vesicles the granulofibrillar component (gf) can be seen. *(Right)* Incubation of the vesicles with trypsin, which cleaves off the amino-terminal end of the band-3 polypeptide, causes the granulofibrillar component to disappear. Courtesy of T. L. Steck.

membrane proteins has also been obtained from electron microscopy. In freeze-fracture micrographs, numerous particles can be seen randomly distributed throughout the membrane interior (see Figure 4-7). The size and frequency of these particles suggest that they are predominantly composed of band-3 protein. On the cytoplasmic surface of the membrane one can observe a granulofibrillar component believed to represent the amino-terminal fragment of band 3 (Figure 4-18). This conclusion is supported by the finding that proteolytic digestion of red cell membranes under conditions known to remove the amino-terminal fragment of the band-3 protein causes the granulofibrillar component to disappear.

In addition to the granulofibrillar component, a weblike or fuzzy meshwork has been observed on the cytoplasmic surface of the red cell membrane (Figure 4-19). The only protein present in sufficient quantity to account for this picture is spectrin, a molecule whose easy removal from the membrane defines it as a peripheral rather than an integral membrane protein. Isolated spectrin has been found to bind *in vitro* to two other membrane proteins, actin and band 4.1, forming a meshwork whose morphological appearance closely resembles that of the meshwork present on the cytoplasmic surface of the red cell membrane (Figure 4-20). Binding of isolated spectrin back to the red cell membrane can be prevented by prior removal of the membrane protein ankyrin, suggesting that spectrin binds to ankyrin. Ankyrin in turn has been shown to bind to the

FIGURE 4-19 The internal surface of the red cell, showing the meshwork present on the cytoplasmic surface of the plasma membrane. Courtesy of T. L. Steck.

FIGURE 4-20 Electron micrograph of two actin filaments cross-linked by the actin-binding protein spectrin in the presence of band-4.1 protein. Courtesy of C. M. Cohen.

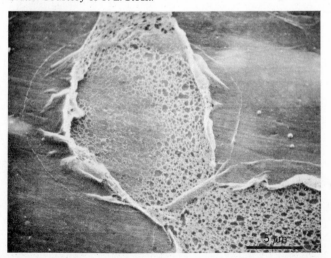

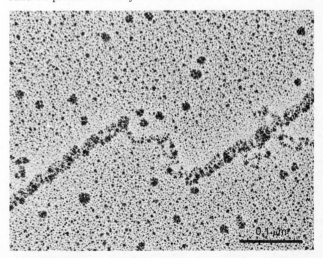

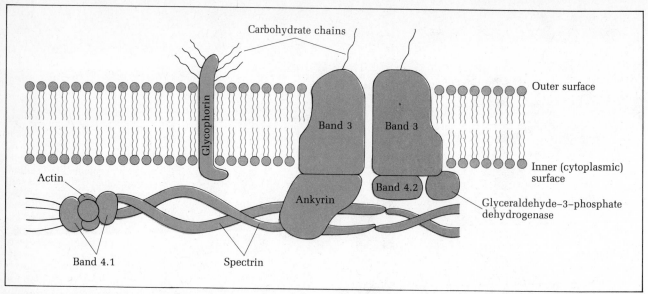

FIGURE 4-21 A schematic representation of the red cell membrane illustrating one possible arrangement of the major membrane proteins.

band-3 protein. These complex interrelationships, summarized in Figure 4-21, provide evidence for a considerable degree of protein-protein interaction in the red cell membrane.

A variety of observations have led to the conclusion that the spectrin–actin–band-4.1 meshwork influences the mobility of integral membrane proteins such as band 3 and glycophorin. It has been shown, for example, that removal of spectrin from red cell membranes increases the lateral mobility of band 3. Likewise, addition of spectrin to artificial lipid vesicles containing band 3 causes the band-3 particles observed in freeze-fracture micrographs to disperse or aggregate, depending on the particular conditions employed. Finally, the integral proteins have been shown by fluorescence photobleaching recovery to diffuse 50 times more rapidly in spectrin-deficient mutant red cells than in control red cells containing a normal spectrin network. It is therefore clear that the spectrin-containing meshwork present on the inner surface of the red cell membrane influences the distribution of integral proteins embedded within the membrane. This phenomenon has been proposed as a model for the way in which cytoskeletal elements influence the distribution of membrane proteins in other cell types.

Although this discussion of the red cell membrane has focused on the description of proteins present in sufficient quantity to be readily detectable by polyacrylamide gel electrophoresis, it should be emphasized that these are not the only proteins present in this plasma membrane. The distribution of proteins present in smaller quantities can be investigated if one has a way of detecting their functional activities. In essence, one simply assays for the activity in question in right-side-out and in inside-out vesicles and compares the results obtained. If, for example, both kinds of vesicles are incubated with ATP to test for ATPase activity, the values obtained are much higher for the inside-out vesicles. Hence the enzyme ATPase must be preferentially located on the inner membrane surface. This approach only works, of course, if the vesicles are relatively impermeable to the testing substrate. Such experiments have permitted the localization of numerous proteins or activities that cannot be directly detected with polyacrylamide gel electrophoresis (Table 4-3).

The data presented thus far indicate that proteins are asymmetrically distributed within the red cell membrane. Similar techniques have been used to study the distribution of phospholipids within the lipid bi-

TABLE 4-3
Localization of proteins, enzymatic activities, and binding sites on the two surfaces of the red cell membrane

OUTER SURFACE

Glycophorin (including carbohydrate chains)
Band 3 (including carbohydrate chains)
Acetylcholinesterase
Nicotinamide adenine dinucleotidase
Oubain binding sites
Lectin binding sites

INNER (CYTOPLASMIC) SURFACE

Glycophorin
Bands 3, 4.1, and 4.2
Spectrin
Ankyrin
Actin
Glyceraldehyde-3-phosphate dehydrogenase
ATPase
Cyclic-AMP binding sites
Protein kinase

layer. Phosphatidylcholine and sphingomyelin have been found to be more accessible to enzymatic digestion or radioactive labeling in right-side-out ghosts, while phosphatidylethanolamine and phosphatidylserine are more accessible in broken or inside-out ghosts. This asymmetric distribution of lipids within the bilayer indicates that the red cell membrane has an inherent asymmetry beyond that imparted to it by the distribution of membrane proteins.

The physiological significance of this asymmetry is not yet clear. It is known, however, that such asymmetry is also exhibited by artificial bilayer vesicles that form spontaneously in the laboratory when purified phospholipid-water suspensions are exposed to ultrasonic vibrations. In these artificial bilayers phosphoglycerides predominate in the inner half of the bilayer while sphingomyelin predominates in the outer half.

Such observations suggest that the lipid asymmetry observed in biological membranes may be due at least in part to inherent differences in the properties of the lipid molecules themselves. This conclusion is supported by the discovery that lipid asymmetry patterns tend to be similar in the various kinds of intracellular membranes (Figure 4-22).

The overall picture of the red cell membrane generated to date has provided strong support for the fluid mosaic model of membrane organization. As predicted by this model, the lipid bilayer is interrupted by embedded proteins whose amphipathic structures are consistent with the principles of the fluid mosaic model. The distinction between integral proteins, such as glycophorin and band 3, and peripheral proteins, such as spectrin and actin, is also evident. The random distribution of the intramembranous protein particles and the

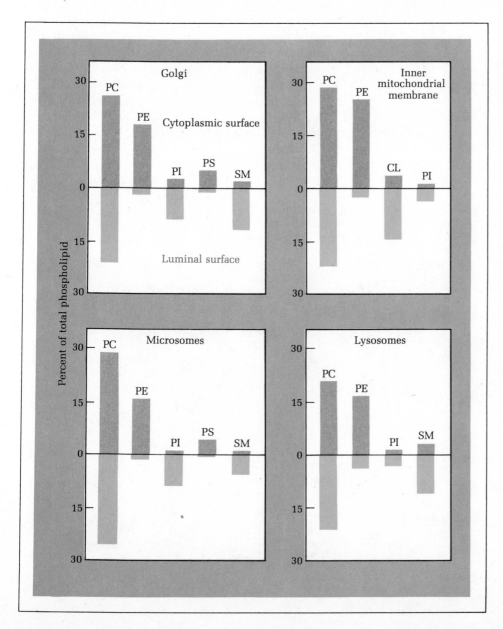

FIGURE 4-22 Patterns of phospholipid asymmetry in several types of biological membranes. Abbreviations: PC = phosphatidylcholine, PE = phosphatidylethanolamine, PI = phosphatidylinositol, PS = phosphatidylserine, SM = sphingomyelin, CL = cardiolipin.

ability of spectrin to influence their distribution are both consistent with the notion of a fluid membrane in which protein components can diffuse laterally. In terms of molecular structure and organization, this membrane therefore conforms to the basic principles of the fluid mosaic model. The functional ramifications of some of these features of the red cell membrane will be discussed in the following chapter.

The Purple Membrane

Because they do not possess the extensive repertoire of intracellular membranes present in eukaryotic cells, prokaryotes use their plasma membranes to carry out specialized functions relegated to the cytoplasmic membrane systems of eukaryotes. In this section we will spotlight a particularly striking case where a bacterial plasma membrane has assumed an unusual molecular organization in order to carry out a specialized function. This membrane therefore illustrates the point that structure and function are inextricably intertwined in biological systems, and that structural appearance is simply a visual reflection of the functional activities being carried out.

Halobacterium halobium is an oxygen-requiring bacterium specifically adapted to environments in which the salt concentration is high. When these cells are grown under oxygen-deficient conditions, patches containing a purple pigment appear within the plasma membrane. In freeze-fracture micrographs these patches of **purple membrane** exhibit a distinctive pattern of hexagonally arranged granules, quite unlike anything seen in the rest of the plasma membrane (Figure 4-23). The fact that these purple membrane patches retain separate identities as distinct regions of the plasma membrane implies that membrane fluidity must be restricted in these regions; otherwise the purple color would diffuse throughout the entire plasma membrane. Protein-protein interactions between the membrane proteins comprising the purple patches, rather than attachment to a cytoskeletal network, appear to be responsible for this restriction of mobility. Such protein-protein interactions produce the hexagonal lattice of granules observed in electron micrographs.

Purple patches can be isolated for biochemical analysis by first lowering the salt concentration of the medium to disrupt the plasma membrane. Under these conditions the purple patches retain their structural integrity and can be isolated free of other membranous material by centrifugation. Over two-thirds of the mass of isolated purple membranes is accounted for by a single protein, **bacteriorhodopsin.** Associated with this protein is the prosthetic group **retinal,** which imparts to it a purple color. Bacteriorhodopsin contains seven regions of α-helix connected to one another by short nonhelical regions. Each helical segment measures about 4 nm in length and fits precisely across the width of the lipid bilayer (Figure 4-24). The retinal groups face the

FIGURE 4-23 Freeze-fracture electron micrograph of the plasma membrane of the bacterium *H. halobium.* Two distinct areas can be seen: The rough-textured regions comprise the major part of the membrane, while the fine-grained regions with the hexagonal array of particles represent patches of purple membrane. Courtesy of W. Stoeckenius.

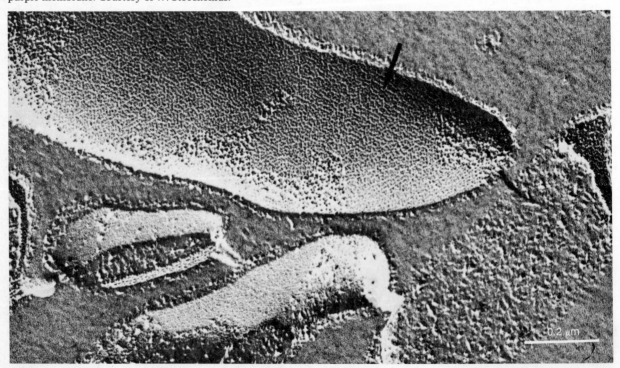

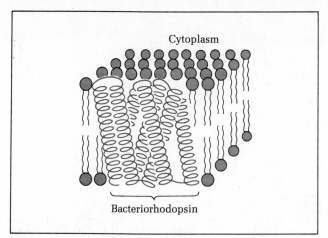

FIGURE 4-24 Arrangement of the bacteriorhodopsin molecule within the purple membrane. Seven regions of α-helix are connected to each other by nonhelical bends in the amino acid chain. The retinal group, located on the exterior surface of the membrane, is not shown.

extracellular surface of the membrane, where they function to absorb light. The energy trapped in this process is utilized by bacteriorhodopsin to generate a gradient of protons across the plasma membrane. As will be discussed in Chapters 7 and 8, such proton gradients can be used by cells as a source of energy. The formation of purple membrane patches under oxygen-poor conditions therefore serves to provide *H. halobium* with an alternative source of energy.

The example of the purple membrane indicates that a single type of protein embedded in a lipid bilayer is sufficient to form a functionally active membrane. The simplicity of this membrane system has made it a popular candidate for study, and we shall return to a more detailed discussion of its function in the appropriate context later.

Relationship of the Plasma Membrane to Other Membranes

The red cell membrane and the purple membrane have been chosen by cell biologists for extensive study because of their relative simplicity. Such studies have revealed that in spite of the marked differences between these two types of membranes, both are organized according to the basic principles of the fluid mosaic model. The plasma membranes of other cells contain a more heterogeneous collection of proteins that have been more difficult to study. Nevertheless, each membrane appears to be organized according to the same underlying principles.

The above generalization holds for other cellular membranes as well. Membranes of the nuclear envelope, endoplasmic reticulum, Golgi complex, lysosomes, microbodies, mitochondria, and chloroplasts certainly differ from one another in structure and function; these differences, however, are due to the particular set of proteins and lipids employed in the construction of each type of membrane. In all cases the proteins and lipids are assembled according to the principles of the fluid mosaic model. In coming chapters we shall see how the different kinds of membranes generated in this way have become specialized for particular functions. Although we shall tend to emphasize the specialized attributes of these different membranes, the common elements of design described in this chapter should always be kept in mind.

Another basic issue arising in the present context is the question of how these various kinds of membranes are manufactured by cells. Because the process of membrane formation is intimately tied in with events occurring in the endoplasmic reticulum and associated cytoplasmic membranes, this issue will be addressed when these intracellular membranes are described in detail in Chapter 6.

THE CELL WALL AND RELATED STRUCTURES IN PROKARYOTES

The plasma membrane is not the only structure located at the cell surface. Most cells contain one or more layers of material external to the plasma membrane, such layers being quite substantial in some cell types. The organization of this region is especially complex in prokaryotic cells, where it involves structures and molecules present nowhere else in nature. In bacteria, where these cell surface layers have been most extensively studied, three distinct regions can be distinguished: an outermost capsule, a cell wall, and a periplasmic space separating the cell wall from the plasma membrane.

The Capsule Layer

The **capsule** is a transparent zone of gelatinous material of varying thickness surrounding the cell wall in some, but not all, strains of bacteria. The capsule layer is usually well organized and tightly associated with the cell wall; when it is less organized and more loosely associated with the cell surface, the term **slime layer** is employed. Because of the viscous nature of the capsule layer, bacteria with capsules tend to form moist, glistening colonies on agar, while those lacking a capsule form colonies with a dry appearance (Figure 4-25). The presence of a capsule can also be detected by suspending bacteria in India ink. Under these conditions, the capsule appears as a clear zone surrounding the cell (Figure 4-26).

Capsules are generally made of polysaccharides containing the sugars glucose, galactose, mannose, ribose, fucose, and their derivatives. In some bacteria the capsule polysaccharide is constructed from a single type of sugar, while in others more than one type of sugar is employed. Capsule layers may also contain unusual

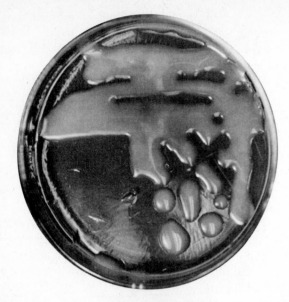

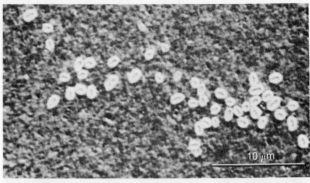

FIGURE 4-26 Bacterial cells stained with India ink and photographed by oil immersion light microscopy. *(Top)* Encapsulated bacteria. *(Bottom)* Nonencapsulated bacteria. Courtesy of R. N. Goodman.

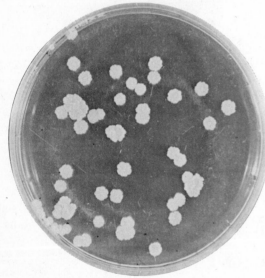

FIGURE 4-25 Illustration of the difference in colony morphology of encapsulated and nonencapsulated bacteria. *(Top)* Sticky mucoid colonies of capsule-forming bacteria. *(Bottom)* Rough coarse colonies of non-capsule-forming bacteria. Courtesy of R. E. Levin.

polypeptides constructed from one or two kinds of amino acids repeated many times in succession. In these polypeptides the amino acids are in the unusual D-configuration rather than the L-configuration characteristic of the amino acids present in virtually all proteins (Figure 4-27).

Cells that have had their capsules removed experimentally suffer no loss in viability, indicating that the capsule layer is not essential for survival. The presence of a capsule does, however, influence the ability of bacteria to cause disease. For example *Pneumococcus*, the organism responsible for bacterial pneumonia, only causes illness when a capsule layer is present. Strains of *Pneumococcus* that have lost their capsules lose their pathogenic status as well. The apparent explanation for this phenomenon is that host cells responsible for en-

gulfing and destroying invading bacteria are ineffective against the encapsulated forms of *Pneumococcus*. Examples of other encapsulated bacteria responsible for serious disease are *Streptococcus pneumoniae* (lobar pneumonia), *Bacillus anthracis* (anthrax), and *Clostridium perfringens* (gas gangrene).

The Bacterial Cell Wall

Bacterial cell walls are divided into two general categories based on their reaction to a staining procedure developed in the late nineteenth century by the Danish physician, Hans Christian Gram. In this technique bacteria are stained with a basic dye such as crystal violet, followed by a brief extraction with alcohol. Cells that still retain the purple dye after the alcohol extraction step are referred to as *gram-positive*, while those from which the dye is extracted are referred to as *gram-*

$$\text{H} \!-\! \overset{\overset{\displaystyle \text{COO}^-}{\vdots}}{\underset{\underset{\displaystyle \text{R}}{\vdots}}{\text{C}}} \!-\! \text{NH}_3^+ \qquad\qquad {}^+\text{H}_3\text{N} \!-\! \overset{\overset{\displaystyle \text{COO}^-}{\vdots}}{\underset{\underset{\displaystyle \text{R}}{\vdots}}{\text{C}}} \!-\! \text{H}$$

D-Amino acid   L-Amino acid

FIGURE 4-27 Structures of the D- and L- forms of a generalized amino acid. The wedge-shaped bonds project above the plane of the page, while the dotted bonds project behind the page. The L-form is used for the construction of virtually all proteins, but certain D-amino acids are occasionally employed in the construction of smaller peptides.

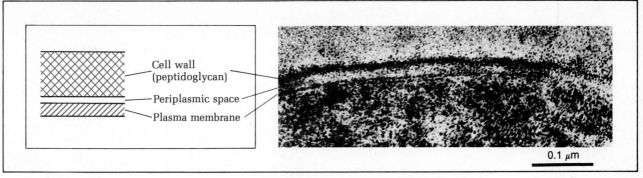

FIGURE 4-28 Electron micrograph of the cell wall of a gram-positive bacterium *(Clostridium).* Courtesy of J. W. Costerton.

negative. This difference in staining behavior reflects an underlying difference in the construction of the cell wall in gram-negative and gram-positive organisms.

The cell wall of gram-positive bacteria is a 30- to 100-nm-thick structure located directly outside the plasma membrane (Figure 4-28). Chemical analyses carried out on isolated cell walls have revealed that the predominant molecular component is a polysaccharide-peptide complex termed **peptidoglycan** or **murein.** The polysaccharide component of this complex contains alternating residues of N-acetylglucosamine and N-acetylmuramic acid (a lactic acid derivative of glucosamine), with short peptide bridges connecting the N-acetylmuramic acid groups of adjacent chains to each other (Figure 4-29). These peptide bridges contain only a few different kinds of amino acids, most commonly D- and L-alanine, D-glutamic acid, L-lysine, L-serine, glycine, and the unusual amino acid diaminopimelic acid (a close relative of lysine). The extensive cross-linking of polysaccharide chains by the peptide cross-bridges produces an intertwined three-dimensional network that imparts great strength to the cell wall.

Although peptidoglycan accounts for the vast bulk of the cell wall in gram-positive bacteria, other molecules are occasionally woven into its framework. Among these are polysaccharides, polypeptides, and teichoic acids. **Teichoic acids** are anionic polymers consisting of glycerol or ribitol derivatives joined into long chains by phosphodiester bridges (Figure 4-30). Such molecules, which are often covalently linked to the peptidoglycan skeleton, have been implicated in the regulation of certain enzymes and in the binding of specific cations required for plasma membrane stability and function.

The cell wall of gram-negative bacteria (Figure 4-31) differs in several ways from the wall of gram-positive cells. First, the peptidoglycan layer of the gram-negative wall is thinner ($\sim 3-8$ nm) and less extensively cross-linked, making it significantly weaker than the gram-positive peptidoglycan layer. This difference is structure is responsible for the differing behavior of gram-positive and gram-negative bacteria in the Gram-

staining procedure. The thinner wall of gram-negative bacteria readily permits removal of the Gram stain during the alcohol extraction step, while the thicker and stronger wall of gram-positive bacteria prevents the dye from being extracted.

The other major difference is the presence in gram-negative cell walls of an extra structure outside the peptidoglycan layer. Based on its electron microscopic appearance and physical properties, this extra layer appears to be a membrane composed of a lipid bilayer and associated proteins. It is termed the **outer membrane** to distinguish it from the plasma membrane lying beneath the cell wall. The protein and lipid composition of the outer membrane is significantly different from that of the plasma membrane. Especially unusual is the presence of **lipopolysaccharides,** which may account for up to half the mass of this membrane. These lipopolysaccharides consist of complex sugar chains to which fatty acid residues are covalently attached at one end (Figure 4-32, p. 118). Because the polysaccharide chains are hydrophilic and the fatty acid residues are hydrophobic, lipopolysaccharides are amphipathic molecules that can be readily inserted into the lipid bilayer with their hydrophobic regions buried in the membrane interior and their polysaccharide chains exposed at the membrane surfaces.

Lipopolysaccharides are of special medical interest because many disease symptoms associated with infection by gram-negative bacteria are no more than host inflammatory responses to the presence of lipopolysaccharide. Cell wall lipopolysaccharides that trigger such responses are termed *endotoxins* to distinguish them from toxins actively secreted by bacteria, which are termed *exotoxins.* The ironic aspect of endotoxin behavior is that these substances are not inherently toxic, that is, they do not directly interfere with any normal activities occurring in the infected host. Yet when cells of the infected organism detect the presence of lipopolysaccharide endotoxins they undergo what might aptly be termed a panic reaction, triggering a set of inflammatory responses that produce more damage than that directly caused by the bacteria.

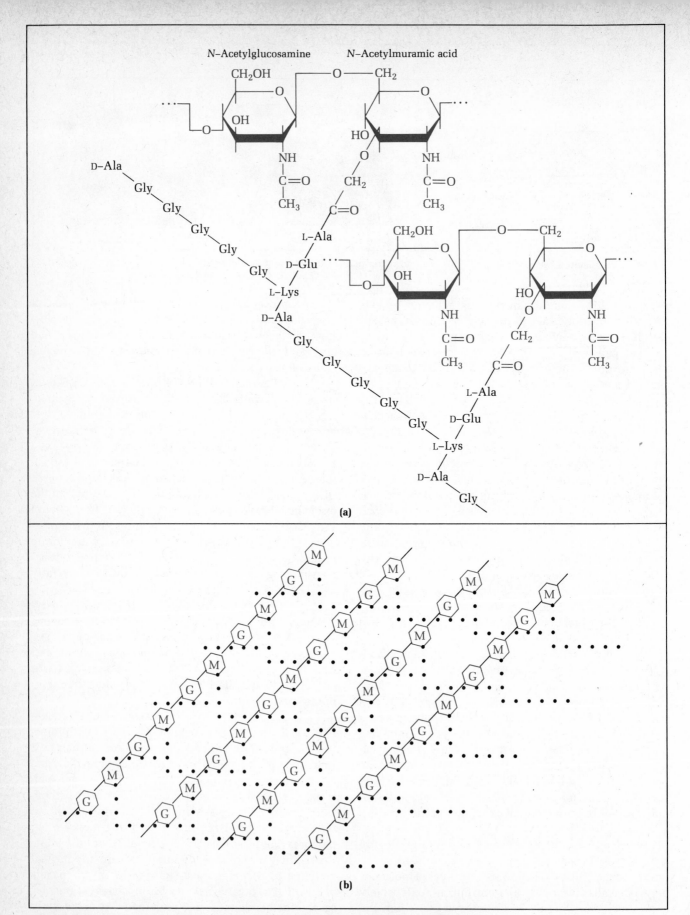

FIGURE 4-29 Structure of peptidoglycans. (a) Diagram of the two alternating sugars that make up the polysaccharide chains of the peptidoglycan, showing how adjacent chains are joined together by small peptide sequences. (b) Overview of the way in which such linkages form an intertwined peptidoglycan network in a gram-positive bacterium. The cross-linking of this network is less extensive in gram-negative bacteria. Abbreviations: G = N-acetylglucosamine, M = N-acetylmuramic acid, ● = an amino acid.

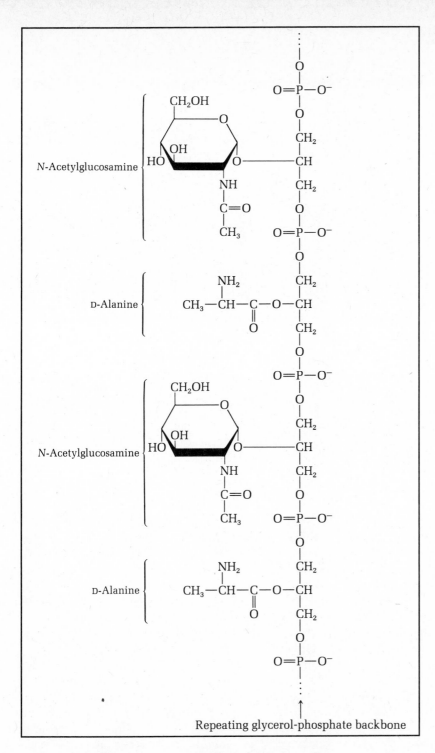

FIGURE 4-30 Structure of a typical teichoic acid.

Repeating glycerol-phosphate backbone

FIGURE 4-31 Electron micrograph of the cell wall of a gram-negative bacterium (*Bacteriodes*). Courtesy of J. W. Costerton.

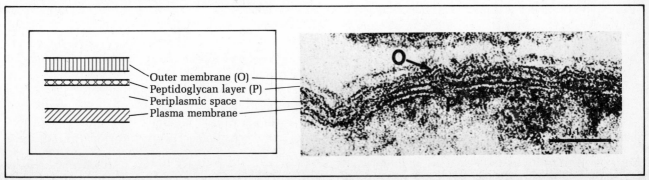

Outer membrane (O)
Peptidoglycan layer (P)
Periplasmic space
Plasma membrane

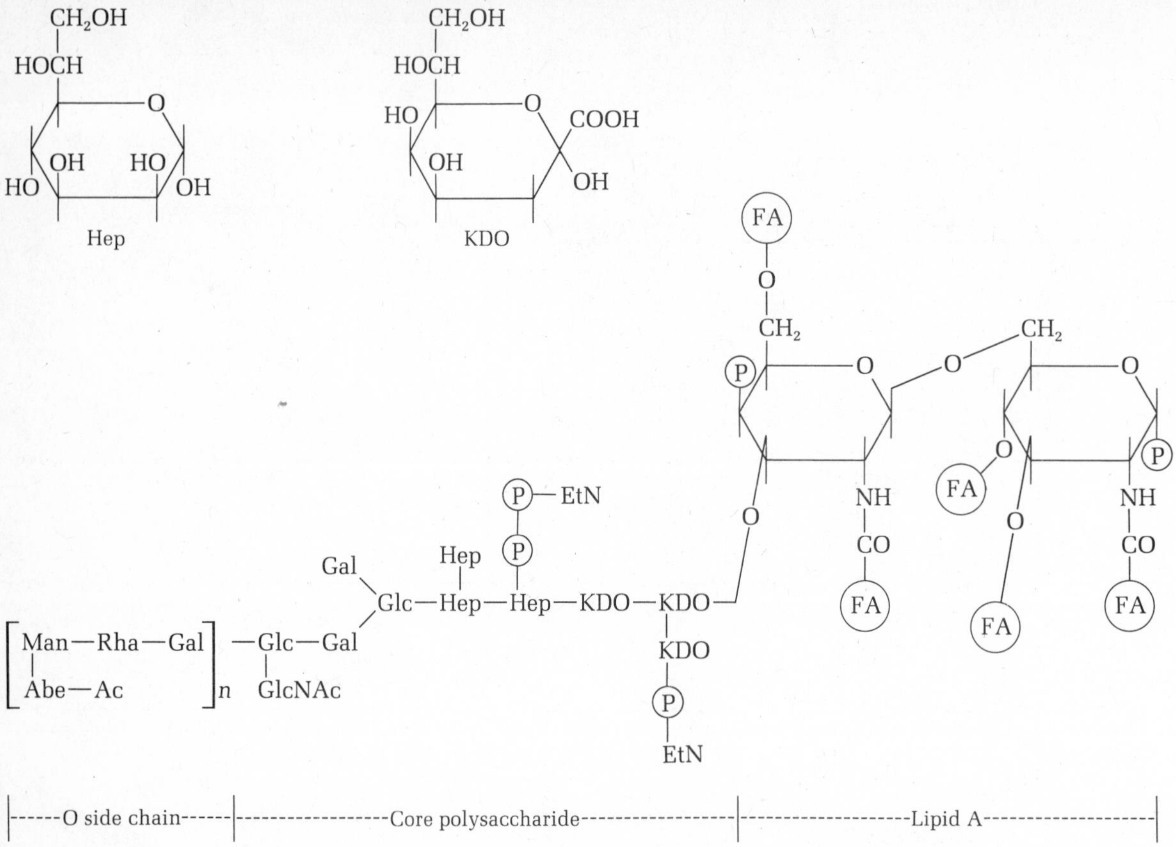

FIGURE 4-32 Structure of a typical lipopolysaccharide. Abbreviations: KDO = 2-keto-3-deoxyoctonic acid, Hep = L-glycero-D-mannoheptose, EtN = ethanolamine, Gal = galactose, Rha = rhamnose, Man = mannose, Abe = abequose, Ac = acetate, FA = fatty acid, P = phosphate.

In addition to lipopolysaccharide, several protein components are present in the outer membrane of the gram-negative cell wall. Among these are **matrix porins,** which function as general pores for facilitating the movement of hydrophilic solutes across the outer membrane, and the **Omp A protein,** which is important in maintaining cell morphology and outer membrane integrity. The major approach for elucidating the functions of such proteins has been to study the properties of mutant strains of bacteria in which the protein of interest is defective.

The cell wall in both gram-negative and gram-positive bacteria functions to protect the cell from injury. Because most bacteria live under conditions where the concentration of dissolved solutes in the external environment is low compared to that within the cell cytoplasm, water tends to flow into the cell. Such a flow of water through a semipermeable membrane from an area of lower solute concentration to an area of higher solute concentration is referred to as **osmosis.** The osmotic flow of water into bacteria generates a massive intracellular pressure that would cause cells to burst if a rigid cell wall were not present. The presence of the cell wall permits pressures as high as 5 atm in gram-negative cells and 25 atm in gram-positive bacteria to be tolerated without cell rupture.

The rigidity of the cell wall also gives this structure a role in determining cell shape. Although bacteria occur in a wide variety of different shapes, most of them can be subdivided into three general categories: spherical cells called **cocci,** rod-shaped cells termed **bacilli,** and spiral-shaped cells known as **spirilla** or **spirochetes** (Figure 4-33). Two lines of evidence suggest that these distinctive shapes are determined by the cell wall. First, isolated cell walls tend to retain the shape of the organism from which they were obtained (Figure 4-34). Second, cells that have had their cell walls removed all assume the same spherical shape, regardless of what their original shape had been. Such wall-less cells, or **protoplasts,** can be generated by treating bacteria with **lysozyme,** an enzyme that hydrolyzes the polysaccharide chains of peptidoglycan molecules. Lysozyme is a normal constituent of tears, saliva, mucous, and egg whites, where it serves a protective function by destroying the cell walls of invading bacteria.

The mechanism by which the bacterial cell wall is synthesized presents an interesting spatial problem because the wall is located outside the plasma membrane and is therefore separated from the cell interior by a permeability barrier. Current evidence suggests that the synthesis of the sugars and peptides involved in peptidoglycan formation occurs in the cytoplasm, but these

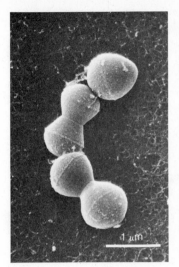
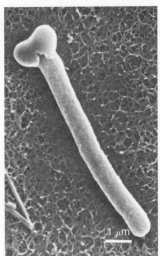
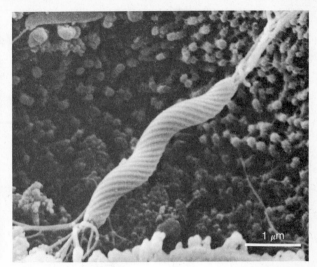

FIGURE 4-33 Scanning electron micrographs revealing the three principal bacterial shapes. *(Left)* Cocci. *(Middle)* Bacilli (emerging from the spore coat). Courtesy of K. Amako. *(Right)* Spirilla. Courtesy of S. E. Erlandsen.

building blocks are joined together into long chains by enzymes present in the plasma membrane. The resulting chains are then cross-linked in a *transpeptidation* reaction catalyzed by an enzyme localized on the outer surface of the plasma membrane. The plasma membrane thus plays a key role in the formation of the cell wall.

The synthesis of the cell wall is of special interest because the transpeptidation reaction is susceptible to inhibition by the antibacterial drug *penicillin* (Figure

FIGURE 4-34 Electron micrographs showing that the bacterial cell wall retains its shape upon isolation. *(Left)* The bacterial protoplast (P) is partially released from the constraining cell wall (W), which retains its elongated shape. *(Right)* A portion of an isolated bacterial cell wall. Courtesy of D. Abram.

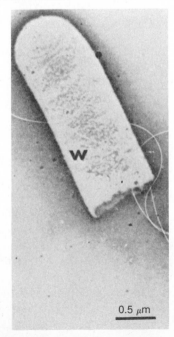

4-35). Since no other cells in nature have a peptidoglycan wall, the effects of penicillin are directed solely against bacteria. Inhibition of the transpeptidation reaction by penicillin hinders cell wall formation and results in the production of fragile cells that tend to burst and die. Gram-negative bacteria are less susceptible to the effects of penicillin than gram-positive cells because the outer membrane, which is not present in gram-positive cells, hinders access of the penicillin to its site of action.

Although most bacteria are extremely vulnerable in the absence of a cell wall, some wall-less forms do exist in nature. The most prominent are the *Mycoplasmas*, whose unusually strong and flexible plasma membranes permit them to swell to enormous sizes without bursting.

The Periplasmic Space

In electron micrographs a space is often seen separating the cell wall from the underlying plasma membrane, especially in gram-negative organisms (see Figure 4-31). Although it is difficult to rule out the possibility that this space is a shrinkage artifact, independent evidence for its existence has come from the finding that disruption of the cell wall leads to the release of a specific group of proteins into the external medium. This observation suggests that bacteria contain a discrete **periplasmic space**, located between the cell wall and the plasma membrane, where these molecules are normally trapped. The periplasmic space is most prominent in gram-negative organisms because the outer membrane, which is not present in gram-positive organisms, serves as a permeability barrier preventing the loss of molecules from this zone to the external medium.

The enzymes located in the periplasmic space are

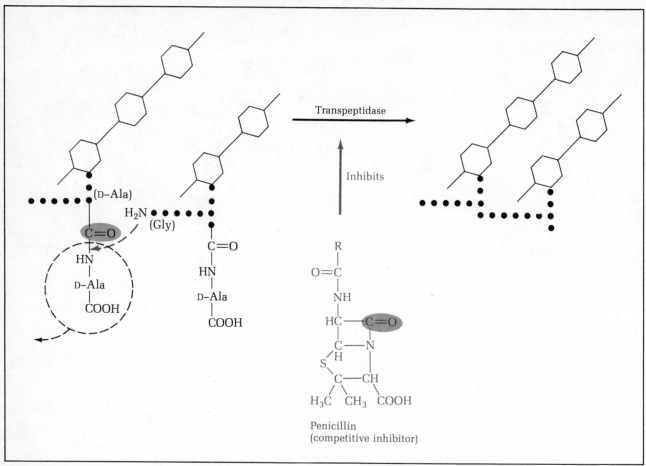

FIGURE 4-35 Mechanism of action of penicillin. The transpeptidase reaction of peptidoglycan formation is competitively inhibited by penicillin, which contains a C=O group that binds to the enzyme in place of the C=O group of the peptidoglycan chain.

mainly hydrolytic in nature; included in this category are alkaline phosphatase, 5′-nucleotidase, phosphodiesterase, ribonuclease, and deoxyribonuclease. In gram-positive organisms, where the periplasmic space is not as well defined, these same enzymes are either bound to the cell surface or are secreted into the medium. Hence in both gram-negative and gram-positive organisms these enzymes are associated with the cell surface, presumably because they function there to degrade macromolecules that are too large to pass through the plasma membrane and into the cell. This degradative function is similar to that performed by lysosomal enzymes in eukaryotes, suggesting that lysosomes may owe their evolutionary origins to infoldings of the plasma membrane containing trapped periplasmic enzymes.

THE CELL WALL AND RELATED STRUCTURES IN EUKARYOTES

Two major patterns of organization occur in the outer surface layers of eukaryotic cells. Eukaryotic plant cells, like bacteria, contain a rigid cell wall external to the plasma membrane, although this wall is quite different in composition from those of bacteria. Eukaryotic animal cells, on the other hand, are not surrounded by a rigid wall but do contain an external coat that is in some ways analogous to a cell wall.

The Plant Cell Wall

The cell wall of most plants is a rigid multilayered structure consisting of a cellulose framework upon which other molecules are superimposed. Depending upon the cell type and its stage of development, the cell wall may vary from ten to several hundred nanometers in thickness. In multicellular plants the outermost layer, termed the **middle lamella,** is enriched in polygalacturonic acid derivatives known as **pectins.** The middle lamella is shared between neighboring cell walls and functions to hold adjacent cells together. The remaining layers of the wall are designated primary, secondary, and tertiary walls, named in the order in which they are formed (Figure 4-36).

The **primary wall** is the first zone to appear, and in some plants represents the complete extent of wall development. Its framework consists of a loosely organized network of cellulose, the polysaccharide formed

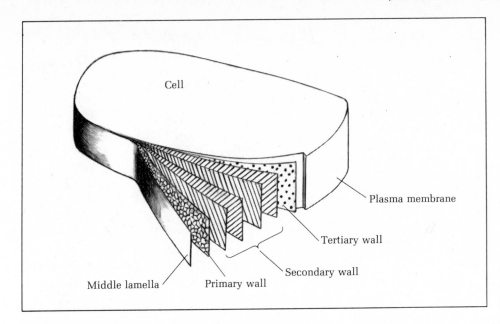

FIGURE 4-36 Organization of a generalized plant cell wall. Not every plant cell contains all the layers shown.

by linking β-glucose units into long chains (see Figure 1-9). Such chains have molecular weights of 500,000 or more and contain several thousand glucose residues. These individual cellulose chains are organized into **microfibrils** that measure 10–25 nm in diameter and consist of several thousand cellulose molecules ar-

ranged in parallel. In the primary wall the cellulose microfibrils are randomly distributed in an amorphous matrix (Figure 4-37 *left*). This loosely textured organization creates a relatively flexible structure capable of expansion during cell growth. It has recently been shown that primary walls contain glycoprotein mole-

FIGURE 4-37 Electron micrograph showing the arrangement of cellulose fibrils in the cell wall of a higher plant. *(Left)* Primary wall of loosely organized cellulose. *(Right)* Secondary wall of densely packed cellulose bundles oriented in parallel. Courtesy of K. Mühlenthaler.

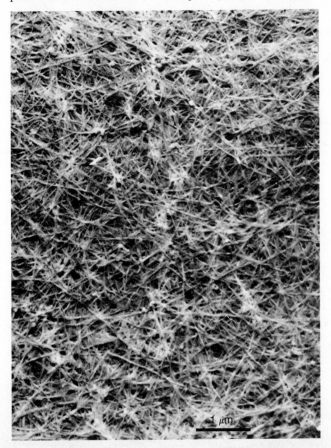

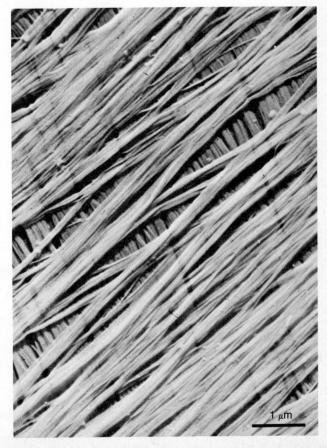

cules closely associated with the cellulose framework, but the functional significance of such molecules is yet to be ascertained.

The **secondary wall** is a multilayered structure superimposed on the inner surface of the primary wall after cell growth has ceased. Each layer of the secondary wall consists of densely packed cellulose microfibrils arranged in parallel, with each layer oriented so that its microfibrils lie at an angle to those of the adjacent layers (see Figure 4-37 *right*). This organization imparts great mechanical strength and rigidity to the secondary cell wall and is ultimately responsible for the characteristic strength of woody tissues. In some plants a final innermost layer is added to the cell wall. This **tertiary wall** (or lamella) is relatively thin and is lacking in cellulose microfibrils.

Although the cell wall provides a thick encasement for plant cells, this barrier does not necessarily seal the cell completely from its surroundings. Plant cell walls usually contain small openings or **plasmodesmata** through which cytoplasmic extensions of adjacent cells may come in direct contact (Figure 4-38). In some cases the individual plasma membranes of the two adjacent cells are even interrupted in these regions, resulting in direct cytoplasmic continuity between neighboring cells. Although such cytoplasmic continuity does not always occur in plasmodesmata, these openings still permit close physical contact between adjacent plant cells that would otherwise be separated by thick cell walls.

Despite the fundamental importance of cellulose in providing the basic framework of the plant cell wall, the relative content of cellulose varies among cell types. In some plants, such as cotton, the cell wall consists almost entirely of cellulose. In woody plants, on the other hand, the cell wall is about 50 percent cellulose, while in some seedlings the wall is less than 1 percent cellulose. Among the other macromolecules contributing to the cell wall are hemicelluloses, pectins, and lignins. The **hemicelluloses** are a complex group of carbohydrate polymers constructed from a variety of five- and six-carbon sugars, including xylose, arabinose, mannose, and galactose. The hemicellulose xylan, which utilizes the pentose xylose as its main building block, accounts for as much as 50 percent of the cell wall in woody tissues. **Pectins** are polymers of the carbohydrate galacturonic acid and its derivatives. They occur predominantly in the middle lamella and adjacent primary wall, where they aid in binding adjacent cell walls together. These same pectins are primarily responsible for the gelation that occurs during the process of making fruit jams and jellies. **Lignins** are a group of polymerized aromatic alcohols that occur mainly in the secondary cell walls of woody tissues, where they contribute to the hardening of the matrix.

In addition to the above macromolecules, cell walls

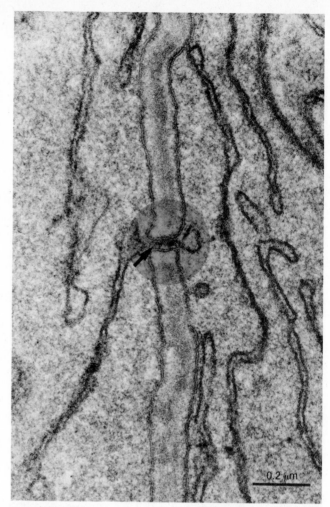

FIGURE 4-38 Electron micrograph of plasmodesmata connecting two corn plant cells. In this region (arrow) the plasma membranes of the cells are continuous and form a channel connecting the cytoplasmic compartments of each cell. Courtesy of H. H. Mollenhauer.

contain small amounts of protein, lipid, and minerals. The predominant protein, termed **extensin,** may account for as much as 10 percent of the wall in higher plants. This protein resembles collagen, the extracellular protein of animal connective tissues, in its high content of the unusual amino acid hydroxyproline. Extensin is thought to be covalently linked to cellulose microfibrils, creating a reinforced protein-polysaccharide complex analogous to the peptidoglycan of bacterial cell walls. Lipids, including waxes and other complex polymers, typically account for only a few percent of the mass of the cell wall. They occur mainly on the external surface of the wall, where they form a **cuticle** layer that protects the cell against injury and dessication. Finally certain minerals, especially calcium and potassium, are present in plant cell walls in the form of carbonate or silicate salts.

As in bacteria, the major function of the cell wall in plants is to provide a semirigid enclosure that protects cells from osmotic and mechanical injury. In higher plants the strength of the cell wall is also responsible for physically supporting the entire plant, a task that can reach mammoth proportions in the case, for example, of a large redwood tree. Because of its rigidity, the wall is a prime factor in determining the characteristic shapes of plant cells, which all become spherical when their walls are removed. The plant cell wall also functions as a permeability barrier, especially when a waxy cuticle is present, and as the site of certain enzymatic reactions. These reactions are mainly hydrolytic in nature, suggesting that as in bacteria, enzymes located at the plant cell surface function to degrade extracellular nutrients into smaller compounds capable of passing through the plasma membrane and into the cell.

The synthesis and assembly of the plant cell wall is a complex event in which the Golgi complex and related cytoplasmic membrane systems play a prominent role. In essence, cellulose and other cell wall polysaccharides are synthesized in the Golgi complex and transported via membrane vesicles to the cell surface, where they are secreted. The subsequent organization of these newly synthesized polysaccharide chains into microfibrils then takes place outside the plasma membrane. A discussion of the role played by the Golgi complex and related cytoplasmic membranes in such events will be elaborated upon in Chapter 6.

The Animal Cell Glycocalyx

Although animal cells do not contain a rigid wall, many are covered by polysaccharide coats that are in some ways analogous to a cell wall. The existence of such carbohydrate-rich layers exterior to the plasma membrane has been demonstrated mainly by cytochemical techniques. In spite of variations in the chemistry and morphology of these cell surface polysaccharide layers, the term **glycocalyx** (sweet husk) has been proposed to encompass all such coatings.

Glycocalyces can be subdivided into two general categories, depending on the degree of connection to the cell surface. An **attached glycocalyx** (or surface coat) is an inherent part of the cell surface that cannot be removed by mechanical means without simultaneously removing a portion of the plasma membrane itself. These coatings often appear in electron micrographs as fuzzy layers of filamentous material covering the cell surface (Figure 4-39). Since the carbohydrate chains of plasma membrane glycoproteins and glycolipids are major constituents of this layer, it may be more proper to consider it as part of the plasma membrane.

In contrast, an **unattached glycocalyx** (or extraneous coat) consists of material located external to the plasma membrane that can be readily removed without affecting the viability of the cell or disrupting the plasma membrane. Included in this category are the membranes surrounding most animal eggs, the outer

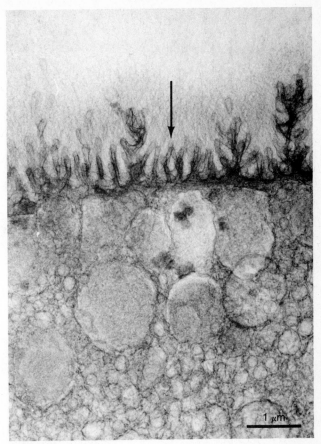

FIGURE 4-39 Electron micrograph of an egg cell showing branched microvilli and the rich feltwork of polysaccharides called the glycocalyx (arrow). Courtesy of J.-C. Roland.

coat of amoebas, and the sarcolemma of muscle fibers. Although the various types of glycocalyx do not provide a rigid enclosure like plant and prokaryotic cell walls, they have been implicated in functions such as cell recognition and adhesion, protection of the cell surface, and the creation of permeability barriers.

THE EXTRACELLULAR MATRIX

Some materials secreted from the cell surface accumulate in the spaces between cells, forming an **extracellular matrix.** Massive aggregations of extracellular matrix occur in connective tissues such as bone, tendon, cartilage, and dermis (the supporting layer of the skin). Beneath some cell layers the extracellular matrix is organized into a thin sheet of material, termed the **basal lamina** (Figure 4-40), which supports the overlying cells and separates them from the underlying connective tissue. In spite of differences in the physical and mechanical properties of the extracellular matrix in different tissues, the same kinds of molecules are often involved: *collagens, proteoglycans,* and *noncollagenous glycoproteins.* Depending on the relative contents and arrangements of these ingredients, the extracellular matrix may be watery, gelatinous, elastic, or rigid.

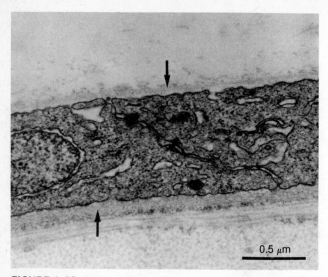

FIGURE 4-40 Electron micrograph of the basal laminae (arrows) present adjacent to both surfaces of this single layer of cells in the cornea. Courtesy of D. H. Dickson.

Collagens

The collagens are the most abundant type of protein present in higher vertebrates, accounting for a quarter or more of the total body protein. The amino acid composition of the collagen proteins is unusual in that as many as 25 percent of the amino acid residues are *glycine* and another 25 percent are the modified amino acids *hydroxyproline* and *hydroxylysine*, which rarely occur in other proteins. Some of the hydroxylysines have glucose and galactose residues attached to them, although the total amount of carbohydrate involved varies considerably among the different types of collagen. The major structural feature shared by the various kinds of collagen molecules is that they all consist of three intertwined polypeptide chains, termed **α-chains,** wound together to generate a relatively rigid, triple-helical structure. At least seven genetically different kinds of α-chains occur; these are in turn assembled in various combinations to form at least ten distinct types of collagen (designated types I through X).

Of the various collagens, types I, II, and III are by far the most abundant in connective tissue. These three forms are referred to as "fiber-forming" collagens because the individual collagen molecules line up adjacent to one another to generate organized structures termed **collagen fibrils** (Figure 4-41). These fibrils, containing thousands of individual collagen molecules and measuring up to several hundred nanometers in diameter, in turn aggregate into even larger structures, known as **collagen fibers,** which measure several micrometers in diameter and hence are visible with the light microscope. One of the most striking properties of collagen fibers is their enormous physical strength. It has been reported, for example, that it takes a load of more than 20 lb to break a collagen fiber measuring only a milli-

meter in diameter. Thus these fiber-forming collagens are the major contributors to the mechanical strength of protective and supporting tissues such as skin, bone, tendon, and cartilage.

Of the remaining types of collagen (types IV through X), a few are capable of forming fibrils and others are not. The best characterized nonfibrous form of collagen is type IV, the collagen component of the basal lamina. Instead of forming a rigid fibrous structure, the molecules of this type of collagen assemble into an open, nonfibrillar network created through end-to-end association of adjacent molecules.

In all types of collagen, the three α-chains of the triple helix are held together by hydrogen bonds involving both glycine and hydroxyproline residues. The importance of the hydroxyproline residues in maintaining the stability of the triple helix is dramatically illustrated by the disease *scurvy*, which is caused by an inadequate dietary intake of ascorbic acid (vitamin C). Ascorbic acid is a reducing agent responsible for maintaining the activity of prolyl hydroxylase, the enzyme that catalyzes proline hydroxylation. In the absence of ascorbic acid, collagen is inadequately hydroxylated and therefore cannot form a stable triple helix. This defective collagen causes individuals with scurvy to suffer from a variety of connective tissue disorders, including extensive bruising, hemorrhages, and breakdown of supporting tissues.

The collagen triple helix does not readily form

FIGURE 4-41 Electron micrograph illustrating the cross-striations of the collagen fibrils. Courtesy of N. Simionescu.

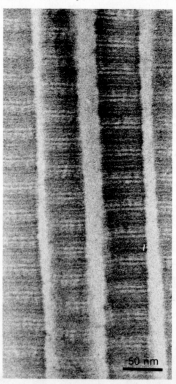

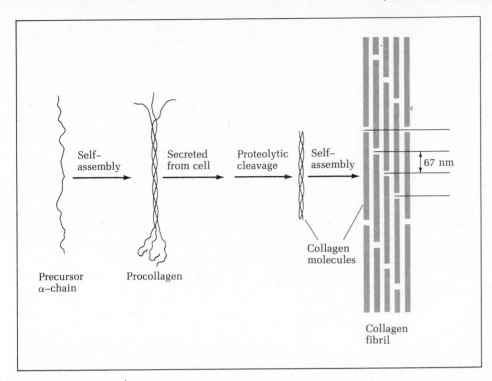

Self–
assembly

Secreted
from cell

Proteolytic
cleavage

Self–
assembly

67 nm

Precursor
α–chain

Procollagen

Collagen
molecules

Collagen
fibril

FIGURE 4-42 Schematic diagram summarizing the steps in the formation of collagen fibrils.

when isolated α-chains are simply mixed together *in vitro,* suggesting that the helical structure is generated by some mechanism other than self-assembly of individual α-chains. Insight into the mechanism involved has been obtained from experiments in which the steps in collagen formation were monitored by labeling fibroblast cells with radioactive amino acids. Such studies have revealed that the α-chains are initially synthesized as larger precursor molecules containing extra amino acids at both their amino- and carboxyl-terminal ends. Interactions between these extra segments are thought to play an important role in the initial formation of the triple-helical structure.

The triple helix of **procollagen** resulting from the assembly of these precursor chains is formed in the lumen of the endoplasmic reticulum. Procollagen molecules are subsequently transported out of the cell by a general pathway common to the handling of all secretory proteins (see Chapter 6). After its secretion into the extracellular space, procollagen is converted to collagen by a specific proteolytic enzyme that removes the appropriate terminal peptides. The resulting free collagen molecules can then aggregate with one another to form mature collagen fibrils. Collagen fibrils exhibit a characteristic pattern of cross-striations every 67 nm (see Figure 4-41), a distance corresponding to about one-quarter of the length of an individual collagen molecule. It has therefore been proposed that collagen fibers are constructed from overlapping rows of collagen molecules in which each row is displaced by one-fourth the length of an individual molecule (Figure 4-42). The stability of the final collagen fiber is reinforced by the for-

mation of chemical cross-links both within and between individual collagen molecules.

It should be clear from the preceding discussion that the existence of multiple forms of collagen differing in both chemical composition and structural organization permits the properties of this molecule to be adapted to the needs of the various types of connective tissues in which it occurs. In addition to this diversity inherent to collagen itself, we shall see in the following section that the properties of this molecule are influenced by the nature of the proteoglycan network in which the collagen is embedded.

Proteoglycan Networks

Proteoglycans are specialized glycoproteins characterized by two unique features. One is their exceptionally high content of carbohydrate, which may account for as much as 95 percent of the total mass. Second is the fact that the carbohydrate is composed of repeating disaccharides known as **glycosaminoglycans** or **mucopolysaccharides.** Each disaccharide repeating unit contains one amino sugar, and at least one acid group (sulfate or carboxyl). The chemical structures of the repeating disaccharides present in some of the more common glycosaminoglycans, such as hyaluronic acid, chondroitin sulfate, keratin sulfate, and heparin, are illustrated in Figure 4-43.

The joining together of glycosaminoglycans and proteins results in some rather elaborate and complex structures. In cartilage, for example, the proteoglycan component of the extracellular matrix contains multi-

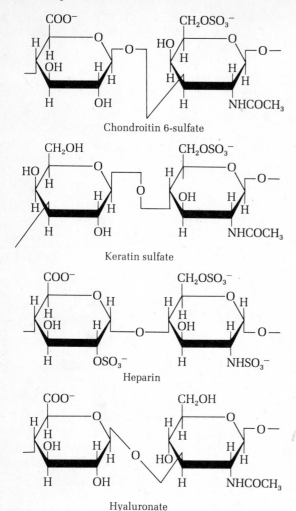

Chondroitin 6-sulfate

Keratin sulfate

Heparin

Hyaluronate

FIGURE 4-43 Chemical structures of the repeating disaccharide units of some common glycosaminoglycans.

ple "core proteins" attached to a long backbone of hyaluronic acid. In addition, each core protein has multiple chains of keratin sulfate and chondroitin sulfate protruding from it (Figure 4-44). The function of such massive proteoglycans appears to be to form a ground substance that traps water and small solutes, thereby creating a gelatinous matrix in which the collagen fibers can be embedded. The overall physical properties of the matrix material are therefore determined to a large extent by the relative proportions of collagen and proteoglycan present. Cartilage, for example, is relatively soft and pliable because it has a high content of proteoglycans. Tendons, on the other hand, are tough and strong because they contain a large number of collagen fibers.

Noncollagenous Matrix Glycoproteins: Fibronectin and Laminin

In addition to collagen, several other kinds of glycoproteins are present within the extracellular matrix. Among these, two are especially abundant and have been the focus of considerable attention in recent years: fibronectin and laminin. **Fibronectin,** an extracellular protein present in connective tissues, blood and other body fluids, and loosely associated with cell surfaces, is a large glycoprotein containing about 5 percent carbohydrate by weight and composed of two subunits with a combined molecular weight approaching 450,000. The fibronectin molecule contains separate sites for binding to collagen and to cell surfaces, suggesting that it functions to anchor cells to the extracellular matrix. If cultured cells are introduced into a culture flask whose surface has been coated with fibronectin, the cells readily attach to the surface and flatten out. Moreover the

FIGURE 4-44 Schematic model of a proteoglycan found in cartilage.

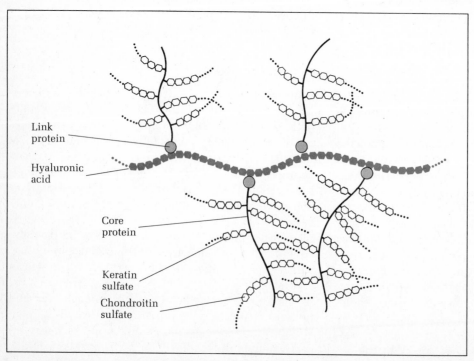

Link protein

Hyaluronic acid

Core protein

Keratin sulfate

Chondroitin sulfate

microfilaments within the cell become aligned with the fibronectin molecules present on the outside, indicating that the organization of the extracellular matrix can influence the organization of components within the cell. The discovery that the fibronectin content of tumor cells is often reduced compared to that of normal cells has led the speculation that the loss of fibronectin is responsible for some of the abnormal properties of cancer cells (see Chapter 20).

In contrast to fibronectin, which is widely dispersed throughout connective tissues, bodily fluids, and on cell surfaces, the glycoprotein **laminin** is localized predominantly within the basal lamina. Basal laminae are specialized sheets of matrix material that separate epithelial cell layers (cell layers covering external and internal body surfaces) from underlying connective tissue; in addition, basal laminae often surround nerve, muscle, and fat cells. In addition to laminin, the other major components of the basal lamina are type IV collagen and the proteoglycan, heparin sulfate. The basal lamina functions as a structural support for maintaining tissue organization as well as a selective barrier regulating the movement of both molecules and cells. In the kidney, for example, the basal lamina serves as a filter that allows small molecules but not blood proteins to be excreted into the urine. The basal lamina beneath epithelial cell layers prevents the passage of connective tissue cells into the epithelium, but not the migration of white blood cells needed to fight infection. The influence of the basal lamina on cell migration is of special interest because it has recently been discovered that the cell surface of some cancer cells is enriched in binding sites for laminin, suggesting that the resulting increase in the ability of cancer cells to bind to the basal lamina may facilitate their movement through this structure and hence allow them to migrate from one region of the body to another.

SUMMARY

All cells are separated from their external environment by a plasma membrane constructed from protein, lipid, and small amounts of covalently attached carbohydrate. Although the lipid and protein composition varies among membranes, the underlying principles of organization are thought to be similar. According to the fluid mosaic model, these principles can be divided into three basic categories. (1) Membrane lipids are arranged predominantly as a bilayer in which the hydrophobic tails of the lipid molecules are buried in the membrane interior and the hydrophilic head groups face the membrane surfaces. Such a bilayer can be generated by phospholipids, which account for the bulk of the lipid in most membranes, as well as by other amphipathic lipids occurring in membranes. (2) Membrane proteins exist in two classes, integral proteins embedded in the lipid bilayer and peripheral proteins bound to the bilayer surface. Integral proteins, which are difficult to extract from membranes, have a globular shape and are often visible as particles in freeze-fracture micrographs. Such proteins are amphipathic, with their hydrophobic regions imbedded in the lipid bilayer and their hydrophilic regions exposed at the membrane surface. Peripheral proteins are bound to the lipid head groups by ionic bonds and are readily extracted from membranes by increasing the ionic strength of the medium. (3) The lipid bilayer is fluid, permitting lateral mobility of both protein and lipid molecules. The extent of lipid and protein mobility is influenced by the type of lipids present in the bilayer, interactions between membrane proteins and cytoskeletal elements, and protein-protein interactions among membrane proteins themselves.

Biochemical investigations on the red cell membrane have provided important support for the fluid mosaic model. From labeling studies on isolated membrane vesicles it has become clear that the red cell membrane has two major transmembrane proteins, glycophorin and the band-3 protein, that are capable of diffusing laterally within the plane of the lipid bilayer. The carbohydrate portions of these glycoproteins are exposed on the exterior cell surface. Several peripheral proteins, including spectrin, actin, ankyrin, and the band-4.1 protein, form an interconnected meshwork on the cytoplasmic surface of the red cell membrane. The two surfaces of the erythrocyte membrane are thus quite different in terms of the proteins present; this asymmetry also extends to the lipid bilayer, with different types of phospholipid preferentially located on the two sides of the bilayer.

In contrast to the relative complexity of the red cell membrane, the purple membrane of *H. halobium* is remarkably simple. A single transmembrane protein, bacteriorhodopsin, spans the lipid bilayer. Attached to its exterior surface is the prosthetic group retinal, which gives the membrane its purple color and is intimately involved in the membrane's function in absorbing and transforming light energy.

Most cells contain one or more layers of material external to the membrane. Prokaryotic cells exhibit three distinct layers: an outer capsule, a cell wall, and a periplasmic space. The capsule, which consists predominantly of polysaccharides, is not essential for cell survival but may influence bacterial pathogenicity. Cell wall structure differs in gram-positive and gram-negative bacteria. In gram-positive bacteria the wall is a relatively thick structure containing a rigid peptidoglycan framework. In gram-negative bacteria the cell wall is thinner, weaker, and contains an outer membrane not present in gram-positive walls. The lipopolysaccharide component of the outer membrane can act as an endotoxin, triggering acute inflammatory reactions during bacterial infections. The periplasmic space, located between the cell wall and the plasma membrane, is more prominent in gram-negative bacteria and contains hydrolytic enzymes involved in the degradation of nutrient macromolecules.

Two major patterns of organization occur in the external layers of eukaryotic cells. Plant cells are surrounded by a rigid wall containing cellulose and other complex macromolecules. As in bacteria, the plant cell wall is a semirigid enclosure that protects the cell from osmotic and mechanical injury. Animal cells do not contain a rigid cell wall but are often covered by a polysaccharide coat known as a glycocalyx. Such coatings, which may be either tightly or loosely associated with the cell surface, have been implicated in cell recognition, adhesion, permeability, and protection.

The extracellular matrix is composed of a mixture of collagens, proteoglycans, and noncollagenous glycoproteins. Collagens exist in a variety of types, some of which are capable of forming fibers characterized by great mechanical strength. Individual collagen molecules are initially synthesized as larger precursor polypeptides that spontaneously assemble into a triple-helical structure called procollagen. After secretion into the extracellular space, procollagen is converted by proteolytic cleavage into individual collagen molecules that in turn may be assembled into collagen fibrils and fibers.

Proteoglycans are glycoproteins consisting of a relatively small amount of protein combined with large quantities of repeating disaccharides known as glycosaminoglycans or mucopolysaccharides. Proteoglycans readily trap water and small solutes, creating a gelatinous matrix in which the collagen fibers are embedded. The physical properties of the extracellular matrix are largely determined by the relative proportions of collagen and proteoglycan present.

The main noncollagenous proteins of the extracellular matrix are fibronectin and laminin. Fibronectin, which has been implicated in anchoring cells to the extracellular matrix, is a large glycoprotein widely distributed in connective tissues, blood and other bodily fluids, and on cell surfaces. Laminin, on the other hand, is concentrated predominantly in the basal lamina, a thin sheet of matrix material found beneath epithelial cell layers and surrounding nerve, muscle, and fat cells. In addition to laminin, the main constituents of the basal lamina are type IV collagen and the proteoglycan, heparin sulfate. The basal lamina functions as a structural support for maintaining tissue organization as well as a selective barrier regulating the movement of both molecules and cells.

SUGGESTED READINGS

Books

Branton, D., and R. B. Park, eds. (1968). *Papers on Biological Membrane Structure*, Little, Brown, Boston.

Finean, J. B., R. Coleman, and R. H. Mitchell (1984). *Membranes and Their Cellular Functions*, Blackwell Scientific, Oxford.

Hay, E. D. (1982). *Cell Biology of Extracellular Matrix*, Plenum, New York.

Piez, K. A., and A. H. Reddi, eds. (1984). *Extracellular Matrix Biochemistry*, Elsevier, New York.

Preston, R. D. (1974). *The Physical Biology of Plant Cell Walls*, Chapman and Hall, Methuen, NY.

Robertson, R. N. (1983). *The Lively Membranes*, Cambridge University Press, New York.

Articles

Bretscher, M. S. (1985). The molecules of the cell membrane, *Sci. Amer.*, 253 (Oct.):100–108.

Danielli, J. F. (1963). In: *The Structure and Function of the Membranes and Surfaces of Cells*, Biochem. Soc. Symp., No. 22, Cambridge University Press, London.

Fowler, V. M. (1986). New views of the red cell network, *Nature*, 322:777–778.

Hakomori, S. (1986). Glycosphingolipids, *Sci. Amer.*, 254 (May):44–53.

Lazarides, E., and R. T. Moon (1984). Assembly and topogenesis of the spectrin-based membrane skeleton in erythroid development, *Cell* 37:354–356.

Robertson, J. D. (1981). Membrane structure, *J. Cell Biol.* 91:189s–204s.

Shockman, G. D., and J. F. Barrett (1983). Structure, assembly, and function of cell walls of gram-positive bacteria, *Ann. Rev. Microbiol.* 37:501–527.

Singer, S. J., and G. L. Nicolson (1972). The fluid mosaic model of the structure of cell membranes, *Science* 175:720–731.

Stoeckenius, W. (1976). The purple membrane of salt-loving bacteria, *Sci. Amer.* 234 (June):38–46.

Storch, J., and A. M. Kleinfeld (1985). The lipid structure of biological membranes, *Trends Biochem. Sci.* 10:418–420.

Sutherland, I. W. (1985). Biosynthesis and composition of gram-negative bacterial extracellular and wall polysaccharides, *Ann. Rev. Microbiol.* 39:243–270.

Unwin, N., and R. Henderson (1984). The structure of proteins in biological membranes, *Sci. Amer.* 250 (Feb.):78–94.

Viitala, J., and J. Järnefelt (1985). The red cell surface revisited, *Trends Biochem. Sci.* 10:392–395.

Yamada, K. M. (1983). Cell surface interactions with extracellular materials, *Ann. Rev. Biochem.* 52:761–799.

5

Functions of the Cell Surface

In Chapter 4 the general principles underlying the organization of the plasma membrane and its associated layers were examined. With that background in hand, we can now relate this information to the functional activities carried out by these cell surface components. Perhaps the most obvious contribution of the cell surface is that it creates a physical barrier that surrounds, protects, and holds the cell together. We have already discussed the role played by the rigid cell wall in this mechanical function of maintaining cell integrity. Cells lacking a wall must rely on the less rigid plasma membrane for this purpose, although we shall see later in the text that the cytoskeleton is also important in defining cell shape and maintaining cell integrity.

In addition to its relatively static role as a boundary that surrounds and protects the cell, the cell surface is involved in at least three other kinds of activities: It regulates the flow of materials into and out of the cell, is involved in cell-cell communication and adhesion, and is the site of specific enzymatic reactions and metabolic pathways. The plasma membrane is the only component of the cell surface present in all cell types, so it must be centrally involved in each of these three events. This chapter will therefore be devoted to a discussion of these three activities, focusing especially on the role played by specific components of the plasma membrane.

MOVEMENT OF MATERIALS ACROSS THE PLASMA MEMBRANE

The interior of a typical cell contains an intricately organized set of components whose environmental conditions must be carefully regulated, even in the face of widespread fluctuations in the external environment. By controlling the passage of materials into and out of the cell, the cell surface plays a crucial role in maintaining proper intracellular conditions. Although cell surface layers such as the cell wall or glycocalyx play some role in this endeavor, it is clearly the plasma membrane that is most intimately involved.

In cells where passage of materials across the cell surface is a major function, such as in cells lining the intestinal tract or the absorptive tubules of the kidney, the plasma membrane is thrown into thousands of microscopic fingerlike projections called *microvilli* (Figure 5-1). The presence of microvilli, which typically measure about 100 nm in diameter and up to a micrometer in length, greatly increases the total surface area

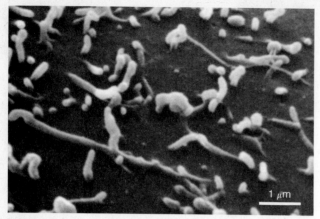

FIGURE 5-1 Scanning electron micrograph of microvilli on the surface of a HeLa cell. Courtesy of R. G. Kessel.

of membrane available for transporting molecules into and out of the cell. In cells where transport is not as extensive, microvilli are usually fewer and smaller, and may even be absent entirely.

Several mechanisms exist for controlling the passage of materials across the plasma membrane. For convenience in discussing the underlying principles involved, these mechanisms will be subdivided into four general categories: simple diffusion, facilitated diffusion, active transport, and endocytosis/exocytosis. As we discuss these four processes in the following sections, it will gradually become apparent that the plasma membrane has a powerful arsenal of tools for regulating the conditions prevailing within the cell interior.

Simple Diffusion

The most straightforward mechanism for moving materials from one side of a membrane to the other is *simple diffusion*, a process involving the unaided net movement of a substance from a region of higher to a region of lower concentration. This net movement is ultimately based on the random motion of individual molecules driven by their inherent thermal energies. No matter how a population of molecules is initially distributed, the eventual result of the random movements of individual molecules must be a random solution, that is, a solution in which the concentration everywhere is the same. To illustrate this point, let us consider an apparatus consisting of two chambers containing different concentrations of the same solute separated by a membrane through which the solute can freely pass. Random movements of individual solute molecules back and forth through the membrane will eventually cause the concentrations in the two chambers to become equal. Hence the random movements of individual molecules leads to a *net* movement of solute from the chamber where its initial concentration was higher to the chamber where its initial concentration was lower. After the concentration has been equalized on both

sides of the membrane, random movements of individual molecules back and forth continues, but no further *net* movement of diffusing molecules occurs. At this point equilibrium is said to be achieved.

Although diffusion generally occurs from regions of higher to regions of lower solute concentration, in strict thermodynamic terms it is differences in free energy rather than differences in concentration that determine the net direction of diffusion. As we learned in Chapter 2, the laws of thermodynamics dictate that all events proceed in the direction of decreasing free energy. Although the free energy of any solution is directly related to the concentration of its constituent molecules, other factors such as heat, pressure, and entropy also contribute to the total free energy. To illustrate this point let us consider an example in which two compartments, A and B, are separated by a membrane permeable to substance X. If the concentration of substance X is slightly lower in compartment A than in compartment B, but the heat or pressure is higher in compartment A, the free energy added by the heat or pressure may cause diffusion to occur in the direction A → B rather than B → A. It is therefore more proper to state that diffusion proceeds from areas of higher to lower free energy rather than from areas of higher to lower concentration. At thermodynamic equilibrium no further *net* movement occurs because the free energy of the system is at a minimum.

The above principle applies to the behavior of water as well as to dissolved solutes. The free energy of water molecules decreases as increasing concentrations of other substances are dissolved in the water. This decrease in free energy results from the fact that the dissolved solute molecules interrupt the ordered three-dimensional interactions that normally occur between individual water molecules. Hence individual water molecules will tend to diffuse from areas where the solute concentration is lower (and hence the free energy of the water molecules is higher) to areas where the solute concentration is higher (and hence the free energy of the water molecules is lower). This prediction can be tested by placing solutions of different solute concentration in two compartments separated by a *semipermeable membrane*, that is, a membrane permeable to water but not to the dissolved solute. As predicted, a net movement of water molecules from the chamber of lower solute concentration to the chamber of higher solute concentration will occur (Figure 5-2). Such movement of water molecules through a semipermeable membrane is referred to as **osmosis.**

The above thermodynamic principle explains why cells shrink or swell when the solute concentration of the extracellular medium is changed. Mammalian cells placed in sucrose solutions exceeding $0.25M$ in concentration, for example, lose water and shrink, a process known as **plasmolysis.** Sucrose solutions less concen-

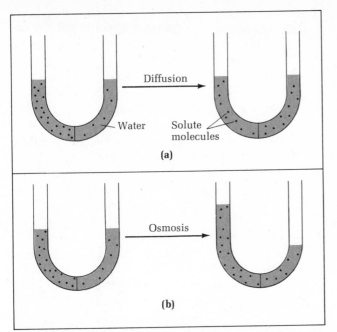

FIGURE 5-2 Comparison of simple diffusion and osmosis. (a) In simple diffusion the membrane is permeable to the dissolved solute, which diffuses from the chamber where its concentration is higher to the chamber where its concentration is lower. (b) In osmosis the membrane is not permeable to the dissolved solute, so water diffuses from the chamber where the solute concentration is lower to the chamber where the solute concentration is higher. See text for the thermodynamic explanation of such movements.

trated than 0.25M, on the other hand, cause cells to take up water and swell (Figure 5-3). The simplest explanation for this phenomenon is that cells are impermeable to sucrose, and that the observed changes in cell volume result from movement of water molecules into or out of cells driven by the difference in solute concentration on the two sides of the plasma membrane. According to this explanation, the sucrose solution, which causes no change in cell volume, must have the same solute concentration as the cell interior; such a solution is said to be **isotonic.** Solutions whose solute concentration is higher than that of the cell and that therefore cause water to diffuse out are termed **hypertonic;** those whose solute concentration is lower than that of the cell and thus cause water to diffuse inward are termed **hypotonic.**

The isotonic concentration is not 0.25M for all solutes. In the case of NaCl solutions, for example, the isotonic concentration is approximately 0.14M. The reason this value is so much lower than that for sucrose derives

FIGURE 5-3 *(Below)* Behavior of animal and plant cells placed in sucrose solutions of differing concentrations. In plant cells the rigid cell wall prevents much swelling from taking place in hypotonic solutions, although the internal pressure is increased by the water that does enter and the cell therefore becomes turgid. With hypertonic solutions the plant cell cytoplasm shrinks away from the cell wall (plasmolysis).

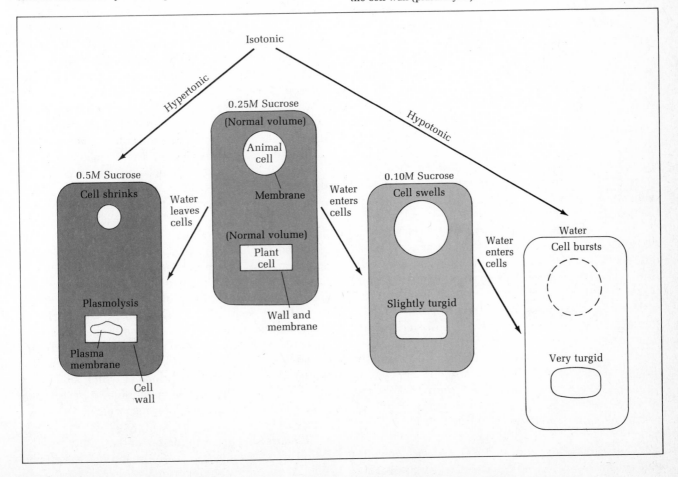

from the fact that NaCl dissociates into two separate ions, Na^+ and Cl^-, when dissolved in water. Sucrose, on the other hand, remains as a single intact molecule in solution. Thus at the same molar concentration, NaCl yields twice as many dissolved particles as sucrose, and so only about half as much NaCl is needed to achieve the same effective concentration of dissolved particles. The difference between 0.14M and 0.25M is not exactly twofold because the osmotic effect of ions such as Na^+ and Cl^- differs slightly from that of uncharged molecules such as sucrose.

The discovery that cells shrink or swell when placed in solutions containing molecules to which the plasma membrane is impermeable formed the basis for one of the first experimental assays for measuring the permeability of the plasma membrane to different substances. This approach, pioneered by C. E. Overton in the late nineteenth century, involves placing cells in concentrated solutions of various substances and then examining them under the microscope to determine whether any change in cell volume has occurred. Plant cells are especially well suited for this purpose because shrinkage of the cytoplasm causes it to pull away from the cell wall, an event that can be readily visualized. Overton noted that when plant cells are exposed to a medium containing concentrated sucrose, rapid plasmolysis (shrinkage) occurs. Plasmolysis does not occur, however, when plant cells are exposed to alcohols such as ethanol. From this observation Overton concluded that ethanol, unlike sucrose, must be capable of quickly diffusing into the cell. Such diffusion causes the intracellular and extracellular concentrations of ethanol to quickly equalize, leaving no difference in solute concentration to drive the flow of water out of the cell.

When experiments of this type were repeated with other substances, it became apparent that some solutes resemble ethanol in being able to readily diffuse through the plasma membrane, others resemble sucrose in not being able to penetrate the membrane at all, and many fall somewhere in between. In order to compare the penetration rates of these intermediary substances, Overton introduced a refinement to the plasmolysis assay based on the following rationale. If a molecule diffuses through the plasma membrane at a relatively slow rate, cells exposed to a concentrated solution of this material will initially undergo plasmolysis because of the high extracellular concentration of solute. As the solute gradually diffuses into the cell, however, the difference in solute concentration between the extracellular and intracellular compartments will gradually disappear and plasmolysis will therefore be reversed. By measuring the amount of time required for the reversal of plasmolysis, the relative permeability of the plasma membrane to different substances can be estimated.

The results of such studies led Overton to conclude that cells are surrounded by a **selectively permeable** membrane, that is, a membrane through which some solutes diffuse more readily than others. While investigating the factors that might be responsible for this differential permeability, Overton came upon the discovery that lipid solubility is a major factor determining a solute's ability to diffuse through membranes. Lipid solubility can be readily measured experimentally by shaking the solute under study in an oil/water mixture, allowing the two phases to separate, and then measuring the concentration of solute in the oil and water phases, respectively. The ratio of these two values, termed the oil/water **partition coefficient,** is a measure of the lipid solubility of the solute in question. Overton discovered that in general, the higher the oil/water partition coefficient, the more readily a substance will diffuse through the plasma membrane. As mentioned in the previous chapter, the resulting conclusion that lipid solubility determines a molecule's ability to diffuse through the plasma membrane first led to the idea that the plasma membrane contains a thin film of lipid.

Although subsequent studies have tended to support Overton's conclusion that a solute's ability to diffuse through biological membranes is directly related to its lipid solubility, other factors have also been implicated. Investigations on cell permeability initiated in the laboratory of R. Collander in the early 1930s led to the realization that size influences a solute's diffusion rate as well. Most notably, small hydrophilic molecules such as water and methanol appear to enter into cells more readily than would be expected on the basis of their low lipid solubilities alone (Figure 5-4). Some large molecules, on the other hand, diffuse into cells more slowly than would be predicted on the basis of their lipid solubility. Such observations suggest that the plasma membrane acts as a molecular sieve, differentially permitting the passage of smaller molecules.

The discovery that some small molecules pass through membranes readily in spite of being insoluble in lipid has led to the suggestion that membranes contain tiny channels or "pores" through which small hydrophilic substances can pass. Calculations based on the observed permeability of cells to small hydrophilic molecules suggest that such pores, if they exist, must be no more than 0.5–1.0 nm in diameter. The major substance of physiological importance small enough to traverse these postulated pores is water, which clearly passes in and out of cells more readily than would be expected on the basis of its negligible lipid solubility. Certain ions might also pass through such pores, although the widely differing rates at which small ions migrate through the plasma membrane suggests that a more selective mechanism of transport is generally involved.

Electric charge is another factor that influences the ability of solute molecules to diffuse through the plasma membrane. Highly charged molecules or ions generally

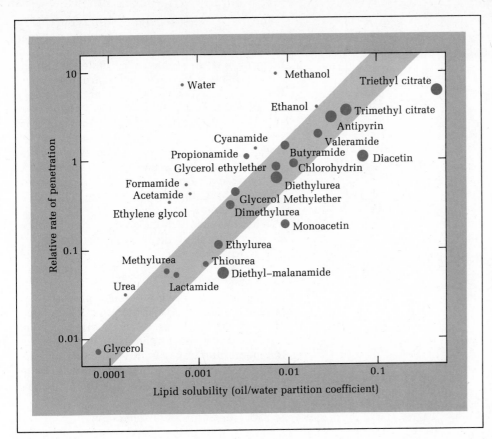

FIGURE 5-4 Penetration of various solutes into cells as a function of their lipid solubility. The diameter of each point is roughly proportional to the size of the molecule it represents. Note that in spite of the general correlation between penetration rate and lipid solubility, significant deviations occur. Those molecules penetrating faster than expected (lying above the shaded area) are relatively small, while those penetrating more slowly than expected (lying below the shaded area) are relatively large.

pass through membranes less readily than do weakly charged or uncharged substances. This phenomenon most likely results from a combination of the lipid solubility and size effects discussed above: Charged molecules tend to be insoluble in lipid and to attract a shell of water molecules around themselves, thereby increasing their effective size. Both factors would therefore tend to decrease their rate of diffusion through the plasma membrane.

It is clear from the above considerations that solutes differ widely in their ability to diffuse into and out of cells. The plasma membrane is most permeable to small molecules, such as water and dissolved gases, and to lipid-soluble substances. At the other extreme are large and/or electrically charged substances that penetrate slowly or not at all. Within this general framework, however, considerable variation is possible. Different cell types possess membranes that differ significantly in permeability, and the membrane of a given cell may undergo changes in permeability as part of its normal behavior. A dramatic example of this latter phenomenon will be discussed later in the text, when the change in sodium ion permeability that accompanies the transmission of nerve impulses is discussed (see Chapter 18).

Facilitated Diffusion

Simple diffusion of molecules through the plasma membrane often does not proceed fast enough to accommo-

date the needs of the cell. This is especially true for sugars and amino acids, which are relatively insoluble in lipid and so diffuse poorly through membranes. Because such hydrophilic substances are essential for many cellular activities, several mechanisms have evolved for facilitating their entry into cells.

The first of these mechanisms, termed **facilitated diffusion,** resembles simple diffusion in that net movement of solute occurs from regions of higher to lower concentration until equilibrium is achieved. However, four features distinguish facilitated diffusion from simple diffusion:

1. The rate of solute movement in facilitated diffusion is usually several orders of magnitude greater than in simple diffusion. If allowed to reach final equilibrium, however, the net result would be the same in both cases; the solute concentration on both sides of the membrane would become equal. Facilitated diffusion is thus analogous to enzymatic catalysis in that the time required for equilibrium to be attained is reduced, but the final equilibrium conditions remain unchanged.

2. Facilitated diffusion also resembles enzymatic catalysis in that it exhibits saturation kinetics. This means that in contrast to simple diffusion, whose rate is directly proportional to solute concentration at all concentration values, the rate of solute movement in facilitated diffusion approaches a maximum value as the solute con-

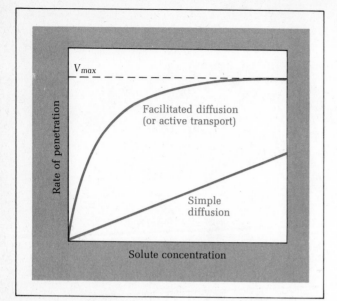

FIGURE 5-5 Comparison of the kinetics of simple diffusion with those exhibited by transporter-mediated processes (facilitated diffusion and active transport). Only transporter-mediated events exhibit saturation kinetics.

centration is increased (Figure 5-5). The saturation curve obtained when one plots the rate of facilitated diffusion as a function of solute concentration has the same hyperbolic shape as the curve obtained when one plots the rate of an enzyme-catalyzed reaction as a function of substrate concentration (compare Figures 5-5 and 2-3). This similarity indicates that facilitated diffusion can be described mathematically using the Michaelis–Menten equation for enzyme kinetics (Equation 2-6):

$$v_i = \frac{V_{max}[S]}{K_m + [S]}$$

In the case of facilitated diffusion, however, v_i stands for the rate of solute diffusion, [S] the solute concentration, V_{max} the maximal rate of solute diffusion observed at saturation, and K_m the solute concentration at which the rate of diffusion is half-maximal. The fact that facilitated diffusion exhibits saturation kinetics suggests that this type of transport process involves interaction of solute molecules with specific membrane components present in limited numbers (just as enzymatic catalysis involves interaction of substrate with a limited number of enzyme molecules). At saturation all such plasma membrane components would be occupied with solute, so further increases in solute concentration have no effect on transport rate. We will generally refer to such plasma membrane components as solute **transporters,** though the terms *pumps, porters, translocases, permeases,* and *carriers* are also employed for this purpose.

3. Unlike simple diffusion, in which structurally similar molecules diffuse through membranes at comparable rates, facilitated diffusion can be highly selective in distinguishing between closely related molecules. The diffusion of glucose, for example, can be facilitated while the passage of other hexoses, such as fructose, is not. Five-carbon sugars, though smaller than glucose, do not readily enter cells in which the diffusion of glucose is facilitated, indicating that glucose is not simply passing through small pores. The solute specificity of facilitated diffusion is comparable to the substrate specificity of enzymatic catalysis, suggesting that membrane transporters possess solute binding sites analogous to the active sites of enzyme molecules.

4. The final feature distinguishing facilitated diffusion from simple diffusion is its susceptibility to competitive and noncompetitive inhibitors. Molecules closely resembling the solute in structure inhibit facilitated, but not simple, diffusion. This type of inhibition, whose kinetic properties resemble competitive inhibition of enzyme activity, is presumably due to competition between inhibitor and solute for the solute binding site on the membrane transporter. Facilitated diffusion can also be inhibited by agents known to denature enzymes, again reinforcing the analogy to enzymatic catalysis.

Taken together, these four characteristics suggest that facilitated diffusion involves interaction of solute molecules with specific membrane transport systems. The resemblance of such an interaction to the binding of an enzyme to its substrate implies that membrane transporters may be protein molecules. In several cases this conclusion has been verified by the isolation of membrane proteins exhibiting the properties of putative transporters.

Two of the most thoroughly studied examples in this category occur in the human red blood cell, where the facilitated diffusion of both sugars and anions has been extensively studied. Glucose diffuses into human red cells at least a hundred times faster than it does into cells in which only simple diffusion is involved. This uptake exhibits the classic features of a transporter-mediated process: saturation kinetics, susceptibility to competitive and noncompetitive inhibition, and solute specificity. The specificity of the glucose transport system is broad enough to permit transport of a few structurally related sugars, such as galactose and mannose, but for most other sugars it is relatively ineffective (Figure 5-6).

Two basic approaches have found widespread application in attempts to isolate membrane transporters such as the one involved in glucose uptake (Figure 5-7). One approach, termed **reconstitution,** entails isolating membrane proteins and assaying individual components for their ability to confer a specific transport activity on artificial bilayer vesicles (liposomes). The alternative approach, **affinity labeling,** involves label-

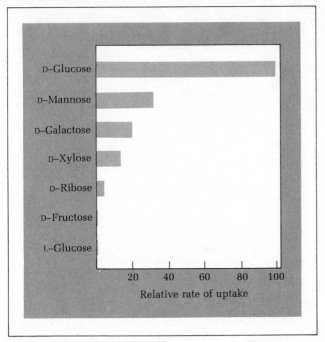

FIGURE 5-6 Solute specificity of the facilitated diffusion system for glucose uptake in human red blood cells.

ing the membrane with radioactive substances expected to selectively interact with the particular transport system under study. In the case of the glucose transporter, a potent inhibitor of glucose uptake called *cytochalasin B* has been used for this purpose.

Both experimental approaches have led to the identification of a 55,000 molecular weight glycoprotein as the glucose transporter. The carbohydrate moiety of

this glycoprotein can be labeled by incubating intact red cells or ghosts with ^{3}H-borohydride and galactose oxidase, indicating that the glucose transporter is exposed at the outer membrane surface. Treatment of inside-out vesicles with the protease trypsin, on the other hand, abolishes the facilitated diffusion of glucose; part of the glucose transport protein must therefore be exposed on the inner (cytoplasmic) membrane surface. Taken together, these observations indicate that the glucose transporter is a transmembrane protein. Kinetic analyses of the effects of various inhibitors on glucose transport have led to the conclusion that the glucose transporter contains one glucose binding site, and that facilitation of glucose diffusion involves conformational changes in which this site alternately faces the interior and exterior of the cell (Figure 5-8).

Another system for facilitated diffusion that has been extensively studied in red cells promotes the movement of anions. The major physiological role for this anion transport system is to facilitate removal of CO_2 from tissues and its delivery to the lung. The CO_2 produced in metabolically active tissues diffuses into red blood cells, where it is converted to HCO_3^- (bicarbonate) and H^+. As intracellular levels of bicarbonate rise, this ion diffuses out of the cell. To prevent an electric charge imbalance from occurring, the expulsion of each bicarbonate ion is accompanied by the uptake of one chloride ion from the extracellular fluid. At the lungs, this entire process is reversed. The net result of this overall pathway is to greatly increase the CO_2-carrying capacity of the blood.

Several lines of evidence have implicated the

FIGURE 5-7 Two general approaches for identifying membrane transporters. In the above example protein B is implicated in the transport of substance X by both the reconstitution and affinity-labeling approaches.

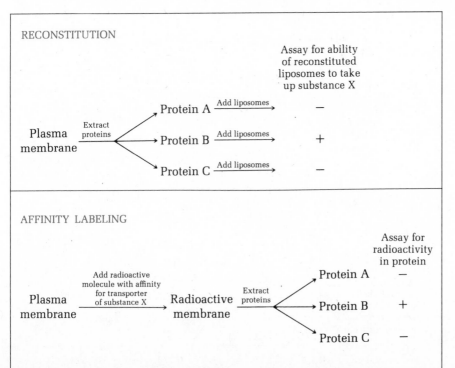

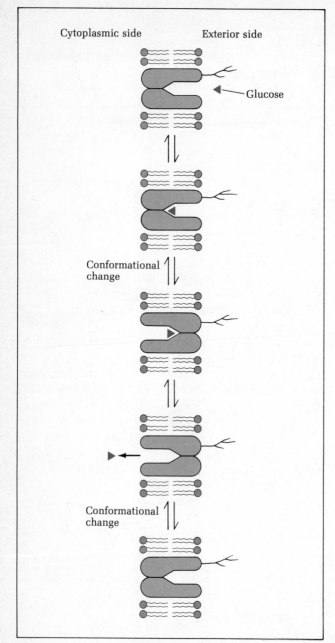

FIGURE 5-8 The alternating conformation model for facilitated diffusion by the red cell glucose transporter.

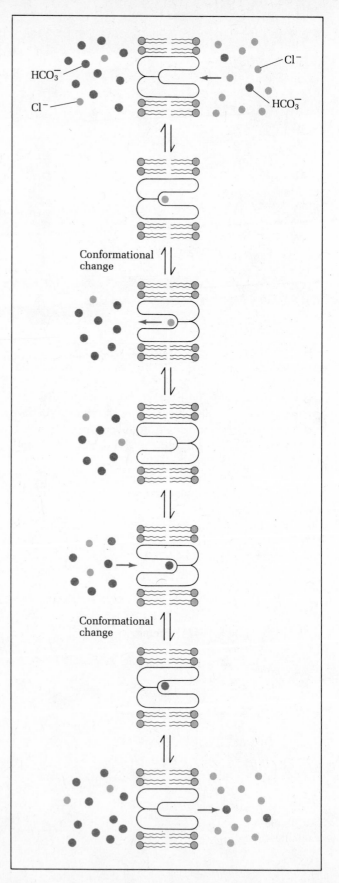

FIGURE 5-9 The alternating conformation model for exchange transport by the red cell anion transporter. According to this model, the anion binding site will accept either bicarbonate or chloride ions, depending on which is present in higher concentration. The diagram illustrates the predominant direction of transport in tissues where CO_2 levels are high. Bicarbonate and chloride would move in opposite directions in the lung.

band-3 protein of the red cell membrane in these movements of bicarbonate and chloride. First, isolated band-3 protein preparations are capable of conferring anion transport capacity on artificial bilayer vesicles. Second, incubation of red cells with radioactive inhibitors of anion transport results in selective labeling of the band-3 protein. Studies on isolated band-3 preparations indicate that this polypeptide readily forms dimers in solution, and electron microscopic analyses of intact membranes suggest that the band-3 protein may even be a tetramer. Although the role played by such quaternary structures in the transport process is not clear, kinetic evidence suggests that the band-3 protein alternately transports individual anions from one side of the membrane to the other. As is illustrated in Figure 5-9, a conformational model similar to the one described for glucose transport could generate such movements. The major difference is that in the case of anion transport, the binding site interacts with different substances on opposite sides of the membrane. In tissues where CO_2 levels are high, the binding site picks up bicarbonate ions from the cell interior and releases them to the exterior; chloride ions are then taken up from the exterior and released to the interior. The reciprocal process would occur in the lung, where CO_2 levels are low. This type of process, in which the uptake of one substance is coupled to the expulsion of another, is termed **exchange diffusion.**

An alternative approach for obtaining insight into the mechanism of facilitated diffusion has evolved from the discovery that certain microorganisms manufacture substances, termed **ionophores,** that increase the diffusion rate of specific ions across membranes. One of the more thoroughly investigated ionophores is *valinomycin,* a small cyclic molecule consisting of 12 alternating hydroxy and amino acids (Figure 5-10). When valinomycin is added to artificial phospholipid bilayers, which are normally quite impermeable to ions, the rate of potassium ion diffusion through the bilayer increases almost 100,000-fold. The change in permeability to sodium ions, on the other hand, is only slight, indicating that valinomycin meets at least two criteria of a membrane transporter: It accelerates the rate of solute penetration and it is solute specific. The solute specificity of valinomycin derives from the fact that it is a donut-shaped molecule containing a central cavity in which an ion such as K^+ can fit snugly, coordinately bonded to six oxygen atoms (see Figure 5-10b). Smaller ions, such as sodium and lithium, do not fit as snugly and so are bound less efficiently.

One can envision at least two mechanisms by which ionophores might increase the rate of ion movement across membranes. First, ionophores might act as reversible **carriers** that bind ions at one membrane surface, diffuse across the lipid bilayer, and then release the bound ions on the other side (Figure 5-11a). Such

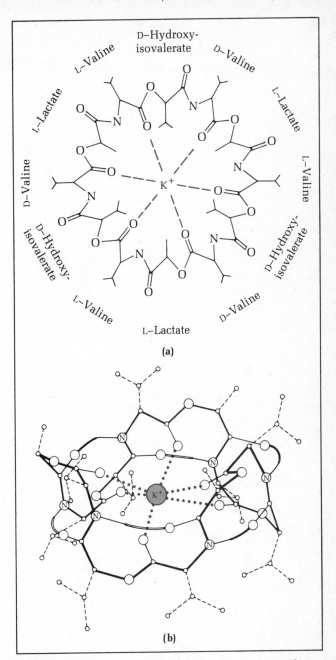

FIGURE 5-10 Structure of the ionophore valinomycin. (a) Chemical formula for valinomycin-K^+ complex. (b) Molecular model of a valinomycin-potassium complex, illustrating how the potassium ion is bonded in the central cavity by ion-dipole interactions with internally oriented oxygen atoms of the valinomycin backbone.

movements back and forth across the lipid bilayer are referred to as *transverse diffusion,* or "flip-flop," to distinguish them from lateral diffusion occurring in the plane of the membrane. The alternative possibility is that ionophores form **channels** that span from one side of the membrane to the other. Ions could enter either end of such a channel, diffuse through it, and exit at the other end (see Figure 5-11b).

A simple experiment has been devised to distinguish between these two possibilities. This test is based

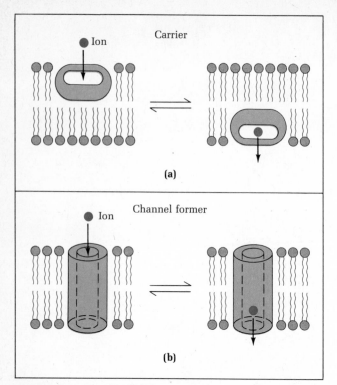

(a)

(b)

FIGURE 5-11 The two alternative mechanisms of ion transport by ionophores. (a) Carriers, such as valinomycin, transport ions by diffusing back and forth through the lipid bilayer. (b) Channel formers, such as gramicidin, create relatively fixed transmembrane channels through which ions can diffuse.

FIGURE 5-12 Influence of temperature on the rate of ion transport by carrier and channel-forming ionophores. The sharp transition in the curve for carrier ionophores is caused by melting of the lipid bilayer, which converts it from a gel to a liquid crystal.

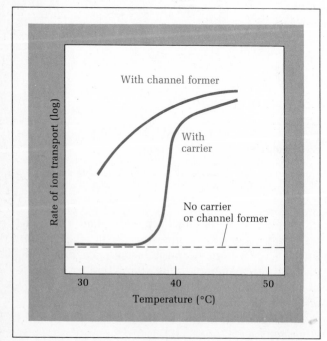

on the simple rationale that the diffusion rate of an ionophore across a lipid bilayer should be influenced by the fluidity of the bilayer. Transport by a channel-forming ionophore, on the other hand, does not require the ionophore to move, and so should be relatively unaffected by the fluidity of the bilayer. When experiments were carried out in which the fluidity of an artificial lipid bilayer was altered by raising the temperature, the rate of ion transport by valinomycin was found to be markedly increased at the gel to liquid-crystal transition (Figure 5-12). This observation suggests that valinomycin acts as a carrier that diffuses back and forth across the lipid bilayer. The conformation of the valinomycin-K^+ complex is ideally suited for this purpose because the charge of the bound potassium ion is masked in the molecule's central cavity while the hydrophobic groups of the valinomycin molecule face the exterior, enhancing the lipid solubility of the complex.

Not all ionophores act in this same way, however. *Gramicidin*, an ionophore consisting of a linear chain of 15 amino acids, transports cations across membranes in a manner that is virtually unaffected by the fluidity of the lipid bilayer (see Figure 5-12). This finding implies that gramicidin does not physically diffuse across the lipid bilayer but rather forms a relatively fixed channel through which ions pass. Structural studies suggest that the channel is created by two gramicidin molecules hydrogen bonded end to end. This dimer is in turn twisted into a helix to form a passageway through which the ions pass (Figure 5-13). The ion selectivity of the gramicidin channel follows the order: $H^+ > NH_4 > K^+ > Na^+ > Li^+$.

The realization that ionophores can function as either diffusible carriers or channel formers raises the question of whether these same two mechanisms are employed by normal membrane transport systems as well. The glucose and anion transporters discussed earlier clearly fit into the channel-forming category, as do all other naturally occurring membrane transporters identified to date. Although the possible existence of diffusible carriers in normal membranes cannot be ruled out entirely, thermodynamic considerations make such a mechanism appear unlikely. The rationale underlying this statement is that membrane proteins contain hydrophilic regions that would have to pass through the hydrophobic lipid bilayer in order to move from one side of the membrane to the other. A large free-energy barrier therefore restricts such movements.

Even for membrane lipids, whose hydrophilic regions are less extensive than those of membrane proteins, transverse diffusion across the membrane appears to be a rare event. In an elegant experiment designed to investigate this point, Roger Kornberg and Harden McConnell generated artificial phospholipid vesicles containing a nitroxide spin-label, and then incubated these vesicles with the reducing agent ascorbate, to which the vesicles are impermeable. Ascorbate con-

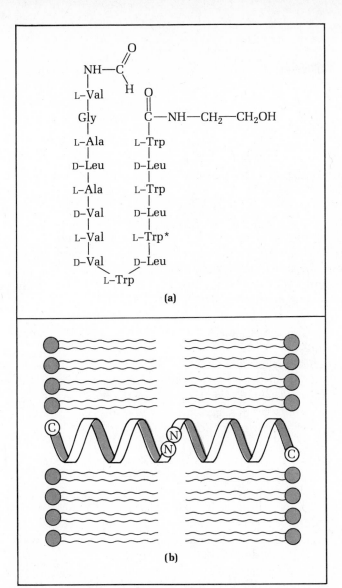

FIGURE 5-13 Structure of the ionophore gramicidin. (a) Amino acid sequence of gramicidin A. The asterisk indicates that other amino acids can replace the L-Trp at position 11, creating a family of related gramicidin molecules. (b) A current model of the gramicidin transmembrane channel, involving two gramicidin molecules in helical conformation joined end to end.

verts the nitroxide group to an amine, thereby abolishing the characteristic ESR spectrum of the spin-label. Within a few minutes of ascorbate treatment the magnitude of the ESR signal was found to decrease by half, indicating that all the spin-label groups exposed on the exterior of the vesicle had been degraded. The remaining signal, however, took many hours to decay in intensity. The gradual loss of the remaining spin-label apparently results from the occasional flip-flop of molecules from the inner half of the bilayer to the outer half, where they are reduced by the ascorbate present in the external medium. From the rate of decay of the ESR signal, it was calculated that flip-flop occurs once every several hours for a typical membrane lipid. This rate of transverse diffusion is orders of magnitude too low to be

involved in any type of transport reaction. It is therefore highly unlikely that diffusion of membrane lipids across the lipid bilayer plays any role in solute transport.

Facilitated diffusion is a powerful mechanism for increasing the rates at which specific substances move across the plasma membrane. Not only are the individual transporters selective in the substances they transport, but the numbers and kinds of transporters present in the plasma membrane of any particular cell type are subject to dynamic regulation in response to changing conditions. Exposure of fat cells to the hormone insulin, for example, causes the rate of glucose uptake by these cells to increase by an order of magnitude. This increased diffusion rate is produced by increasing the effective number of glucose transporters present in the plasma membrane.

The major limitation to the general usefulness of facilitated diffusion is that it can only enhance net movement of solute molecules from areas of higher to areas of lower concentration. Thus, at best, facilitated diffusion can only produce a condition in which the solute concentration is roughly equivalent inside and outside the cell. In a significant number of situations this final result is inadequate because substances need to be moved into or out of cells against significant concentration gradients. Under such conditions a mechanism termed **active transport** is required.

Active Transport

The main feature distinguishing active transport from facilitated diffusion is that active transport is unidirectional, selectively moving substances from regions of lower to higher concentration. Because movement of a substance against its concentration gradient is an energy-requiring process, active transport must be linked to a source of free energy. In its remaining characteristics, active transport closely resembles facilitated diffusion. Both processes accelerate the rate of solute movement, exhibit Michaelis–Menten saturation kinetics, are solute specific and susceptible to competitive and noncompetitive inhibition, and involve the interaction of solutes with membrane transporters (Table 5-1). Hence the only new issue to be considered in discussing active transport is the question of how a free-energy source is utilized to selectively transport solutes against their concentration gradients.

The exact amount of energy required to transport an uncharged substance, under conditions of constant temperature and pressure, from compartment A to compartment B can be calculated from the equation

$$\Delta G = 2.303 RT \log_{10} \frac{[C_B]}{[C_A]} \tag{5-1}$$

where R is the gas constant, T is the absolute temperature, $[C_A]$ is the initial concentration of solute in compartment A, and $[C_B]$ is the initial concentration of

TABLE 5-1
Comparison of simple diffusion, facilitated diffusion, and active transport

Property	Simple Diffusion	Facilitated Diffusion	Active Transport
Accelerated rate of solute movement	No	Yes	Yes
Saturation kinetics	No	Yes	Yes
Solute specificity	Low	High	High
Subject to competitive and noncompetitive inhibition	No	Yes	Yes
Direction	Bidirectional	Bidirectional	Against concentration gradient
Energy requirement	No	No	Yes

solute in compartment B. This equation is a simple variation of the basic expression describing the change in free energy for any chemical reaction (compare Equation 2-4).

A brief examination of the above equation reveals that ΔG will be negative when the concentration of solute is higher in compartment A than in compartment B. In other words movement of solute down a concentration gradient releases free energy, and can therefore proceed spontaneously with no energy input. But if the initial concentration of solute is lower in compartment A than in compartment B, the calculated value of ΔG will be positive. Hence transport of a solute from a region where its concentration is low to a region where its concentration is high will not occur without an input of free energy, that is, "active" transport is required.

If the solute undergoing transport is electrically charged, the charge gradient as well as the concentration gradient must be taken into account. The sum of the concentration and charge gradients is referred to as the **electrochemical gradient.** The free-energy content of an electrochemical gradient can be determined by modifying the above equation as follows:

$$\Delta G = 2.303RT \log_{10} \frac{[C_B]}{[C_A]} + ZF\,\Delta V \qquad (5\text{-}2)$$

where Z is the number of charges per molecule, F is the faraday (23.062 kcal/V/mol), and ΔV is the difference in electrical potential between the two compartments in volts. This equation tells us that during transport of an electrically charged molecule, it is the sum of the free-energy gradients contributed by the concentration gradient ($[C_B]/[C_A]$) and the charge gradient ($ZF\,\Delta V$) that determines whether or not active transport is required.

The energy required to drive active transport can be provided in two basic ways. One is to couple active transport to an energy-yielding chemical reaction, such as the hydrolysis of ATP. The alternative is to couple the active transport of one solute to the movement of another substance down its concentration gradient. The way in which such mechanisms operate in practice will become apparent in the following sections when we discuss the properties of several active transport systems in detail.

Active Transport of Sodium and Potassium Ions The active transport of sodium and potassium ions is one of the most thoroughly studied examples of a transport process driven by the hydrolysis of ATP. In most cells the concentration of K^+ is higher and the concentration of Na^+ is lower than the concentration of these same ions in the extracellular fluid (Figure 5-14). Potassium ions are required by cells for protein synthesis and the activity of certain enzymes, while the sodium ion gradient is critical to the transmission of nerve impulses, the contraction of muscle, and, as we shall see shortly, the transport of sugars and amino acids.

The first indication that transport of sodium and potassium ions might be related to each other was provided in 1955 by Alan Hodgkin and Richard Keynes, who reported that efficient transport of Na^+ out of squid nerve cells requires the presence of K^+ in the external medium. Subsequent studies carried out in other laboratories revealed a similar phenomenon in red blood cells. In these latter experiments the temperature was lowered to slow the transport of Na^+ out of red cells, thereby causing Na^+ to accumulate intracellularly. When the temperature was subsequently returned to normal, the accumulated Na^+ was found to be pumped out of the cell only when K^+ was present in the extracellular medium. The transport of K^+ from the external medium into the cells was likewise found to be dependent on the presence of intracellular Na^+. Quantitative measurements later established that two potassium ions are taken up by the red cell for every three sodium ions extruded. Transport systems of this type, which carry different substances in opposite directions, are referred to as **antiports.** Because three positively charged ions are expelled for every two positive ions taken up, an electric potential is created across the plasma membrane by this transport system. Pumps that create such a charge imbalance are said to be **electrogenic.**

A major breakthrough in our understanding of the molecular mechanism underlying Na^+ and K^+ transport occurred in 1957, when Jens Skou discovered that isolated nerve cell membranes contain an enzyme that breaks down ATP and whose activity requires the presence of both Na^+ and K^+ (Figure 5-15). This enzyme, termed **Na^+,K^+-ATPase,** was found to be inhibited by

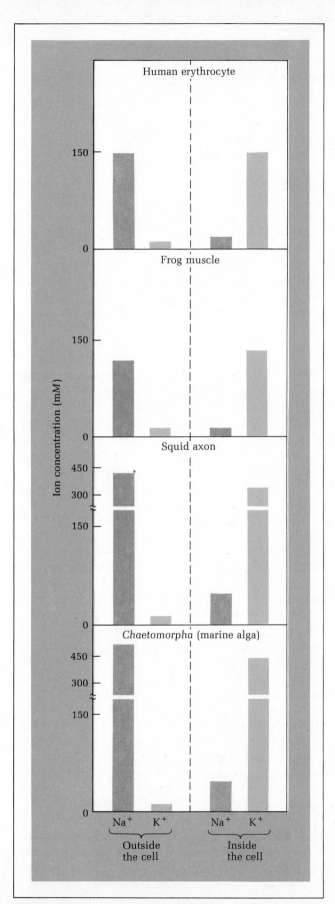

FIGURE 5-14 Internal and external sodium and potassium ion concentrations in several different cell types.

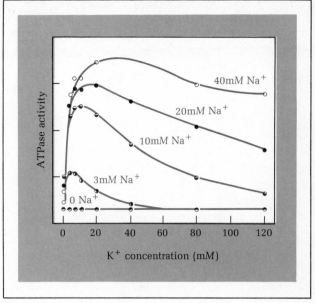

FIGURE 5-15 Membrane ATPase activity as a function of varying sodium and potassium ion concentrations. Note that both Na$^+$ and K$^+$ are required for optimal activity.

ouabain, a drug that inhibits transport of sodium and potassium ions. Similar Na$^+$,K$^+$-ATPases were soon discovered in other cells, and the amount of this particular enzyme present in any given cell type appeared to correlate with the cell's Na$^+$-K$^+$ pumping capabilities. Taken together, the above information strongly suggested that the Na$^+$,K$^+$-ATPase is involved in the transport of sodium and potassium ions.

Experiments designed to investigate the orientation of the Na$^+$,K$^+$-ATPase within the plasma membrane have been carried out in red cell ghosts. These studies revealed that sodium ions must be inside the ghost to activate membrane ATPase activity, while potassium ions only stimulate the ATPase from the outside. It was also discovered that ATP must be inside the ghosts for transport of Na$^+$ and K$^+$ to occur. Taken together, this information has led to the conclusion that the Na$^+$ and ATP binding sites of the membrane ATPase are located on the inner membrane surface, while the K$^+$ binding site faces the outer surface.

But how does interaction of ATP with its binding site drive transport of Na$^+$ and K$^+$? A significant insight into this question has emerged from the discovery that incubation of membranes with ^{32}P-labeled ATP leads to the incorporation of radioactive phosphate into the Na$^+$,K$^+$-ATPase. This incorporation of phosphate groups (phosphorylation) is enhanced in the presence of sodium ions, while the subsequent removal of the incorporated phosphate groups (dephosphorylation) is stimulated by the presence of potassium ions. This information has provided the basis for a relatively straightforward model of Na$^+$-K$^+$ transport based on the existence of two alternate conformations of the Na$^+$,K$^+$-ATPase. This general model, illustrated in Figure 5-16, postulates the following three stages: (1) In the initial

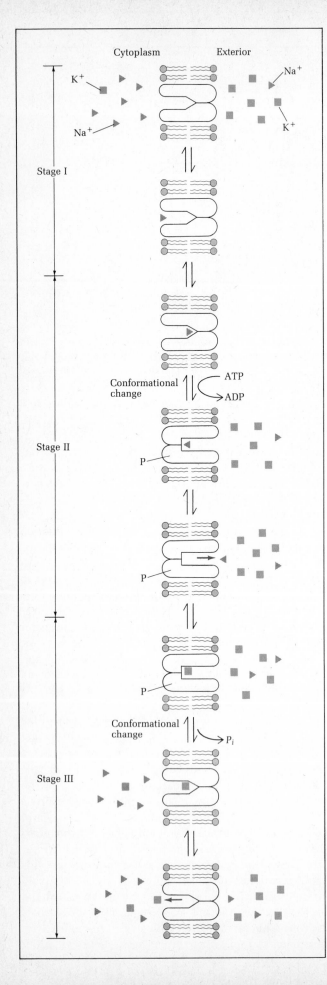

FIGURE 5-16 The three-stage model of Na$^+$-K$^+$ transport.

conformation of the Na$^+$,K$^+$-ATPase molecule, its Na$^+$ binding sites are exposed on the inner membrane surface where they efficiently bind Na$^+$. In this conformation the K$^+$ binding sites are inactive. (2) Binding of sodium ions triggers phosphorylation of the Na$^+$,K$^+$-ATPase molecule by ATP, resulting in a change in conformation. In this altered conformation the Na$^+$ binding sites of the Na$^+$,K$^+$-ATPase undergo two changes: They become exposed to the exterior rather than the interior of the cell, and they lose their binding affinity for sodium ions. Thus Na$^+$ is released to the external environment. In this new conformation the K$^+$ binding sites of the Na$^+$,K$^+$-ATPase become capable of binding potassium ions present in the external medium. (3) Binding of potassium ions stimulates dephosphorylation of the Na$^+$,K$^+$-ATPase and thus reestablishes the initial conformation. In this conformation the K$^+$ binding sites of the Na$^+$,K$^+$-ATPase are exposed to the interior rather than the exterior of the cell, and lose their binding affinity for potassium ions. Hence K$^+$ is released within the cell. The Na$^+$ binding sites, on the other hand, are again exposed to the cell interior and capable of efficiently binding sodium ions. The net effect of repeating the above three steps in sequence is that free energy stored in ATP is used to produce a cycle of conformational changes in the membrane-bound Na$^+$,K$^+$-ATPase, which in turn mediate the transport of sodium and potassium ions against their respective concentration gradients.

Rigorous testing of the above model will obviously require detailed information concerning the structure and conformation of the Na$^+$,K$^+$-ATPase. Purified Na$^+$,K$^+$-ATPase preparations consist of a glycoprotein containing two large (α) and two small (β) subunits. The α subunits, which span the plasma membrane, contain ATP-binding sites facing the cell interior and ouabain binding sites facing the cell exterior (Figure 5-17). Also present on the α subunits are the binding sites for sodium and potassium ions. The smaller β subunits, which contain covalently attached carbohydrate groups, are exposed only at the cell exterior. The function of the carbohydrate chains is unclear because they can be removed with no obvious effects on ion transport. As additional information concerning the detailed structure and conformation of this protein becomes available, it should eventually be possible to rigorously test and refine the three-step model of Na$^+$ and K$^+$ transport described above.

The Phosphotransferase System for Sugar Transport
ATP is not the only compound whose free energy can be employed to drive the process of active transport. Some bacteria take up sugars by an active transport mechanism whose energy is derived instead from the hydrol-

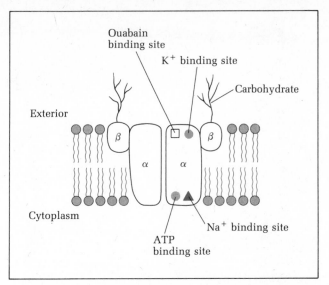

FIGURE 5-17 Schematic model showing arrangement of the large and small subunits of the Na$^+$, K$^+$-ATPase.

ysis of phosphoenolpyruvate, an intermediate in carbohydrate metabolism. The first insights into this type of transport were provided in 1964 by Saul Roseman, who reported that bacteria contain several enzymes and a small protein, designated **HPr,** which together catalyze the transfer of a phosphate group from phosphoenolpyruvate to various sugar molecules. A role for this **phosphotransferase** pathway in sugar

transport was suggested by several subsequent observations. First, it was discovered that bacterial cells defective in sugar transport contain a reduced amount of HPr, and that the addition of purified HPr back to such cells is capable of restoring normal transport. Other bacteria defective in sugar transport were found to lack one of the enzymatic components of the phosphotransferase pathway. It is now clear that there are at least three such enzymes involved, designated enzymes I, II, and III. Cells deficient in enzyme I and/or HPr lose the ability to transport a wide variety of sugars, indicating that these components are not sugar specific. Defects in enzymes II and III, on the other hand, selectively inhibit the transport of certain sugars; this specificity occurs because multiple forms of enzymes II and III exist, each specific for a particular type of sugar.

Biochemical studies on the individual reactions catalyzed by the three enzymes of the phosphotransferase pathway have led to the formulation of the sugar transport model depicted in Figure 5-18. According to this scheme, a high-energy phosphate group is first transferred from the initial phosphate donor, phosphoenolpyruvate, to enzyme I molecules situated in the cytoplasm. Phosphorylated enzyme I molecules in turn transfer their phosphate groups to the protein, HPr. These initial steps of the phosphotransferase pathway are common to the transport of a wide variety of sugars.

The next steps in the pathway depend on the partic-

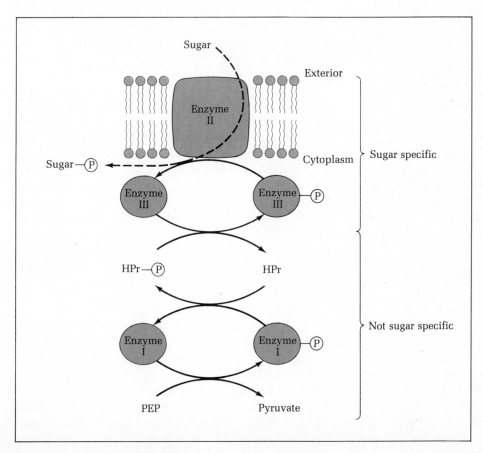

FIGURE 5-18 The phosphotransferase system for sugar transport in bacteria.

ular sugar being transported. In the case of glucose transport, the high-energy phosphate group is transferred from HPr to a glucose-specific form of enzyme III, which in turn transfers this phosphate group to a glucose-specific form of enzyme II, which is an integral plasma membrane protein. Finally, the phosphorylated enzyme II molecule donates its phosphate group to a glucose molecule being translocated through the membrane, leading to the release of glucose-6-phosphate within the cytoplasm. Although differing forms of enzymes II and III catalyze a similar process for other sugars, enzyme III is not always required. During transport of the sugar mannitol, for example, HPr transfers its phosphate group directly to a mannitol-specific form of enzyme II localized within the plasma membrane.

One of the unique features of the phosphotransferase transport system is that the sugars being transported are chemically altered as they pass through the membrane. At least one reason for this additional step appears to be that the newly acquired phosphate group, with its negative charge, traps the transported sugar molecule within the cell because the plasma membrane is relatively impermeable to charged substances. In this way high concentrations of sugar-phosphate can be accumulated within the cell under conditions where the external concentration of free sugar is relatively low.

Na$^+$-linked Transport of Sugars and Amino Acids As an alternative to using energy-rich compounds such as ATP or phosphoenolpyruvate to drive active transport, the immediate driving force can be provided by the movement of another substance down its concentration gradient. Many of the transport pathways specific for sugars and amino acids fall into this latter category. Studies carried out in the laboratory of Halvor Christensen during the early 1950s were the first to show that active uptake of amino acids by the cells of higher organisms is directly related to the sodium ion concentration of the external medium, an observation soon extended by others to sugar molecules as well (Figure 5-19). It was subsequently discovered that in such cases the uptake of amino acids or sugars is accompanied by the simultaneous uptake of sodium ions.

The use of competitive inhibitors to investigate the specificity of Na$^+$-dependent transport has led to the conclusion that there are several different transporters in this category, each specific for a related group of amino acids or sugars. Na$^+$-dependent transporters presumably contain two binding sites, one specific for sodium ions and the other specific for the particular sugars or amino acids being transported. Only when both sites are occupied can transfer of solute across the plasma membrane occur. Transporters of this type, which simultaneously carry two different substances in the same direction across the plasma membrane, are called **symports.** Although Na$^+$-dependent symports

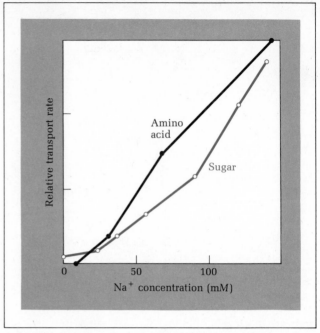

FIGURE 5-19 Data illustrating the Na$^+$ dependence of the uptake of the amino acid glycine into red blood cells, and the sugar 7-deoxy-D-glucoheptose into intestinal lining cells.

are theoretically capable of moving solutes in both directions across the plasma membrane, the high concentration of Na$^+$ outside the cell relative to that inside the cell preferentially drives the process inward. Hence the sugar or amino acid being transported by Na$^+$-dependent symports is selectively moved into the cell, even though it is usually moving against its own concentration gradient.

Although the Na$^+$ gradient provides the immediate driving force for Na$^+$-dependent transport, the existence of this gradient in turn depends on active expulsion of Na$^+$ from cells by the Na$^+$,K$^+$-ATPase. The cooperation that exists between these two types of transport systems is especially well illustrated in the cells that line the intestinal tract of higher organisms. These cells, which are specialized for absorbing nutrients from the intestinal lumen and transferring them to the bloodstream, exhibit a marked polarity (Figure 5-20). The plasma membrane on the side of the cell facing the intestinal lumen contains Na$^+$-dependent transporters specific for the active transport of glucose; because the Na$^+$ concentration is higher outside the cell than inside, this transport system selectively moves glucose into the cell against its own concentration gradient. The plasma membrane on the opposite side of the cell contains a facilitated diffusion system that speeds the passage of glucose out of the cell and into the bloodstream. This opposite side of the cell also contains the Na$^+$,K$^+$-ATPase system, which actively expels Na$^+$ from the cell. Without this system, the sodium ion concentration within the cell would gradually rise until the Na$^+$

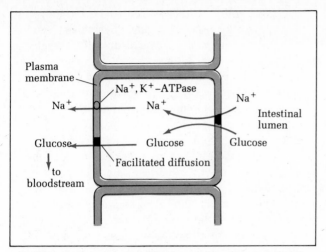

FIGURE 5-20 Arrangement of the three transport systems in intestinal lining cells that permit glucose to be transported from the intestinal lumen to the bloodstream.

gradient from outside to inside was abolished. At this point the Na^+-dependent glucose carrier would no longer be able to selectively transport glucose from the intestinal lumen to the cell interior. Hence the movement of glucose from the intestinal tract to the bloodstream ultimately depends on interactions among several different transport systems localized in different regions of the cell.

Proton-linked Active Transport: Lactose Permease In the early 1950s Georges Cohen, Howard Rickenberg, and Jacques Monod discovered that the bacterium *Escherichia coli* contains a dynamically regulated system for the active uptake of lactose and closely related sugars. Their studies revealed that bacteria grown in the absence of lactose are not capable of actively transporting this sugar, but they quickly develop the ability to do so when lactose is added to the growth medium. If inhibitors of protein synthesis are present, however, formation of the lactose transport system does not occur upon addition of lactose to the medium, indicating that synthesis of a new membrane protein must be involved.

The first major breakthrough in our understanding of how this protein, designated **lactose permease,** functions occurred in the early 1970s when Ian West and Peter Mitchell discovered that transport of lactose into bacterial cells is accompanied by an increase in the pH (i.e., a decrease in proton concentration) of the external medium. This observation suggested that the lactose permease might simultaneously transport protons and lactose molecules into the cell. Such a notion fits nicely with Mitchell's *chemiosmotic theory* of energy transformation, whose features will be spelled out in detail in Chapters 7 and 8. In brief, this theory states that energy released during the oxidation of nutrients is employed to drive the transport of protons across biological membranes. The free energy stored in the resulting proton

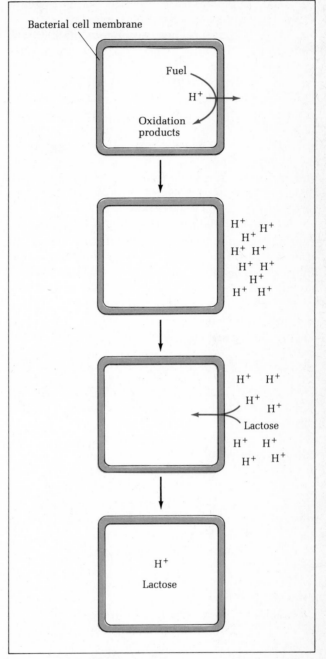

FIGURE 5-21 Model illustrating how a proton gradient established during the oxidation of nutrients can subsequently drive the active uptake of lactose by bacterial cells.

gradient is then employed to drive other energy-requiring processes, such as active transport or the synthesis of ATP. In the case of the lactose transport system described above, diffusion of protons down this free-energy gradient would drive the uptake of lactose (Figure 5-21). This model has been tested by Howard Kaback and his associates by exposing isolated bacterial membrane vesicles to artificially created proton gradients. Under such conditions the transport of lactose is significantly enhanced, supporting the conclusion that movement of protons down their free-energy gradient drives the accompanying transport of lactose.

Recent biochemical investigations on the structure and function of lactose permease have begun to shed some light on how the preceding type of transport is carried out at the molecular level. Lactose permease is a highly hydrophobic, integral membrane protein that contains a single lactose binding site and a single proton binding site. This protein spans the plasma membrane and behaves as if it exists in two conformational states, one in which these two binding sites face the cell exterior and one in which they face the inside of the cell. Transport is thought to be initiated by the binding of lactose and a proton to their respective sites on lactose permease when these sites are facing the cell exterior. This binding then induces a conformational change in lactose permease that reorients the binding sites so that they face the cell interior, allowing the lactose and proton to be released within the cell. This dissociation triggers a conformational change that reorients the binding sites so that they again face the outside of the cell, allowing the cycle to be repeated.

Endocytosis/Exocytosis

One basic feature shared by simple diffusion, facilitated diffusion, and active transport is that they all involve direct passage of materials through the lipid bilayer of the plasma membrane. In some instances, however, substances enter and leave cells enclosed within membrane vesicles. Uptake of material within membrane vesicles formed by invagination and budding off from the plasma membrane is termed **endocytosis;** the reverse process, involving the fusion of cytoplasmic vesicles with the plasma membrane and the discharge of their contents into the extracellular space, is called **exocytosis** (see Figure 6-33).

The functional significance of endocytosis and exocytosis varies among different cell types. In *Amoebae*, for example, endocytosis is a major route for the uptake of nutrients, while white blood cells employ this same process to capture and destroy invading bacteria. Exocytosis, on the other hand, mediates the discharge of secretory proteins such as hormones and digestive enzymes, as well as smaller substances such as neurotransmitters. Exocytosis can also be used to rid the cell of undigestible waste materials. Because endocytosis and exocytosis involve interaction of the plasma membrane with internal cytoplasmic membranes, we shall discuss these processes when the relevant cytoplasmic membrane systems are described in Chapter 6.

The Membrane Potential

Several mechanisms for regulating the passage of materials through the plasma membrane have now been described. Taken together these various processes, many of which are solute specific and subject to dynamic regulation in response to changing physiological conditions, provide cells with a powerful arsenal of means for regulating their internal environments. In some cases interactions between two or more of these processes are required. A case in point is the creation of the electric charge gradient, or **membrane potential,** that occurs across the plasma membranes of most cells.

Membrane potentials are created by diffusion and transport events in which the uptake and expulsion of charged solutes are not precisely balanced. In practice, two such processes are primarily responsible for the membrane potential. The first is active transport by the Na^+,K^+-ATPase system, which creates a high intracellular concentration of potassium ions and a low intracellular concentration of sodium ions. The second is the subsequent diffusion of these same two ions down their respective concentration gradients. If the plasma membrane were equally permeable to Na^+ and K^+, the tendency of positively charged potassium ions to leak out of the cell by simple diffusion would be balanced by the tendency of positively charged sodium ions to enter the cell by simple diffusion. Under such conditions the overall balance of charges inside and outside the cell would not be affected.

In reality, however, the plasma membrane is generally more permeable to K^+ than to Na^+. The rate of loss of potassium ions from the cell therefore exceeds the rate of entry of sodium ions, and an electric potential is gradually established in which the cell exterior is more positively charged than the cell interior. As the magnitude of this electric potential increases, the rate of K^+ diffusion out of the cell is gradually decreased by the electrostatic repulsion generated when the positively charged potassium ions move out of the cell into an area that is becoming positively charged. Eventually the exterior of the cell becomes so positive with respect to the cell interior that the rate of K^+ leakage decreases to the point where it equals the rate of Na^+ entry. At this stage, which occurs when the membrane potential reaches a value of about -50 to -100 mV, **electrochemical equilibrium** is said to be achieved and no further change in potential occurs.

Because the rate at which K^+ diffuses out of the cell is directly related to the steepness of its concentration gradient, the final value of the membrane potential is a function of the initial concentration of potassium ions inside and outside the cell (Figure 5-22). The magnitude of the expected membrane potential can in fact be calculated from the intracellular and extracellular K^+ concentrations by using the Nernst equation for electrochemical equilibrium:

$$E = 58 \log_{10} \frac{[K_{out}^+]}{[K_{in}^+]} \qquad (5\text{-}3)$$

where $[K_{out}^+]$ and $[K_{in}^+]$ are the extracellular and intracellular concentrations of potassium ions, and E is the

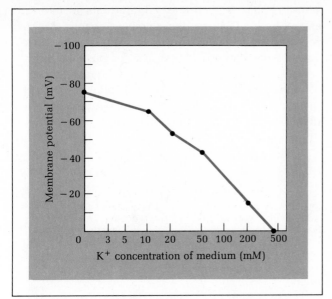

FIGURE 5-22 Dependence of the membrane potential on external potassium ion concentration. As the K^+ concentration of the medium is increased, the membrane potential in squid nerve cells is seen to decrease.

membrane potential in millivolts. One can calculate from this equation that under conditions where the potassium ion concentration is 100mM inside the cell and 10mM outside the cell, the membrane potential will equal -58 mV (by convention the minus sign indicates that the inside of the cell is negative with respect to the outside).

The major physiological importance of the membrane potential in higher organisms is that it underlies the mechanism by which nerve and muscle cells communicate, a topic to be discussed in detail in Chapters 18 and 19. Membrane potentials also provide part of the driving force for certain transport reactions, such as the proton-linked transport of lactose discussed earlier.

INTERCELLULAR COMMUNICATION AND THE CELL SURFACE

Because the plasma membrane is located at the cell exterior, it is natural for this structure to play a role in transmitting signals from the external environment to the cell interior. Such communication generally involves either the interaction of soluble molecules with the cell surface or direct physical contact between adjacent cells. In both cases special molecular components of the cell surface are required. In the following sections the general properties of these two types of communication will be described.

Membrane Receptors

One of the most important functions of the cell surface is its capacity to receive molecular signals from the extra-

cellular environment and to pass the information on to the rest of the cell. This function of the cell surface is made possible by the presence of special plasma membrane components, termed **receptors,** that specifically bind to other substances, or **ligands,** in much the same way that enzymes specifically bind to their respective substrates. Binding of an external ligand to its appropriate membrane receptor triggers changes in membrane-localized functions, such as enzymatic activity, permeability, or transport, that may in turn lead to major alterations in cell behavior. Individual cell types differ from each other in the kinds of receptors they contain, and therefore in the external signals to which they can respond. This arrangement permits highly specific networks of cell–cell communication to be established.

In recent years a great deal of progress has been made in identifying and isolating membrane receptors. The most widely used approach for receptor identification involves incubation of cells or isolated membrane preparations with a radioactive form of the substance with which the receptor normally reacts. For example, one would use radioactive insulin to identify the receptor to which the hormone insulin binds, or radioactive acetylcholine to identify the membrane receptor to which the neurotransmitter acetylcholine binds. After incubating cells or membrane fragments with such radioactive ligands, one can fractionate membrane components and monitor individual fractions for the presence of radioactivity. The demonstration that a radioactive ligand binds to a particular membrane constituent does not, however, constitute proof that a true receptor has been identified. At least three additional criteria must be met by the binding reaction. These criteria concern the *specificity, affinity,* and *number* of the binding sites involved.

Binding specificity is assessed by measuring the ability of nonradioactive substances to compete with the radioactive ligand for binding to the putative receptor. If binding to a true receptor is being observed, competition by substances unrelated to the ligand should not occur. The data from such an analysis of insulin receptors, for example, reveals that binding of radioactive insulin to its receptor is not inhibited by unrelated hormones such as glucagon, but is inhibited by nonradioactive insulin or insulin derivatives (Figure 5-23). Hence radioactive insulin must be binding to a receptor site specific for insulin.

Information concerning binding affinity and number of binding sites present can be obtained by analyzing the kinetics of receptor-ligand interaction. If the concentration of ligand bound to receptor is plotted as a function of total ligand concentration, a hyperbolic curve is obtained that is similar in shape to the curve generated when enzyme activity is plotted against substrate concentration (Figure 5-24a). The ligand concen-

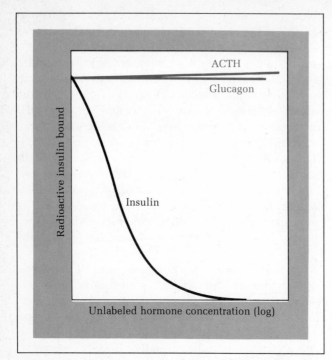

FIGURE 5-23 Experimental data illustrating the specificity of the insulin receptor. Radioactive insulin was incubated with isolated plasma membranes in the presence of varying concentrations of unlabeled insulin, glucagon, or ACTH. Note that only unlabeled insulin competes with radioactive insulin for binding to the plasma membrane receptor.

tration at which binding to receptor is half-maximal is referred to as the **dissociation constant** (K_d). This dissociation constant (analogous to the Michaelis constant of enzymatic reactions) is inversely related to the affinity of receptor for ligand. A high value of K_d indicates that a high concentration of ligand is required for binding to occur, and the receptor must therefore have a relatively low affinity for this ligand. Conversely, a low value of K_d indicates that binding to receptor occurs at low ligand concentration, so the affinity of receptor for ligand must be high.

Measurements of K_d are important in characterizing putative receptor preparations. Suppose, for example, one wished to identify the membrane receptor for insulin, a hormone whose normal concentration in the bloodstream is about $10^{-9}M$. If an isolated membrane component were found to bind insulin with a K_d of $10^{-6}M$, this could not be the true receptor for insulin because it does not have sufficient affinity to bind insulin at its normally prevailing concentration of $10^{-9}M$. Most isolated hormone receptors exhibit K_d values in the range of 10^{-8} to $10^{-11}M$, corresponding well with the low concentrations at which most hormones act.

The total number of binding sites present in an isolated receptor preparation can be determined by measuring the total number of ligand molecules bound to the receptor after saturation has been achieved. Because it is generally difficult to know when complete

saturation has been attained, binding data are often analyzed by a transformation of the Michaelis–Menten equation termed a **Scatchard plot.** This type of analysis converts the data to a linear form by plotting the concentration of ligand bound to receptor against the ratio of bound to free ligand (see Figure 5-24b). In such a plot the x intercept corresponds to the total number of binding sites, or B_{max}, while the slope of the line corresponds to the negative reciprocal of K_d.

FIGURE 5-24 Graphic representation of the kinetics of receptor-ligand interaction. (a) A standard Michaelis–Menten plot in which the concentration of bound ligand (B) is plotted as a function of free ligand concentration (F). Note that the receptor becomes saturated with bound ligand at high concentrations of free ligand. (b) A Scatchard plot of the same data in which the ratio of bound to free ligand (B/F) is plotted as a function of the free ligand concentration (F).

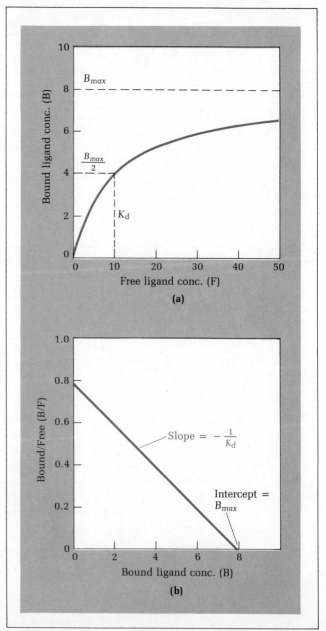

TABLE 5-2
Examples of polypeptide hormones, neurotransmitters, and other ligands for which specific plasma membrane receptors have been identified

POLYPEPTIDE HORMONES	NEUROTRANSMITTERS AND RELATED LIGANDS
Insulin	Acetylcholine
Glucagon	Serotonin
Thyrotropin (TSH)	Norepinephrine
Angiotensin	Epinephrine
Adrenocorticotropin (ACTH)	Opiates
Calcitonin	**OTHER LIGANDS**
Follicle stimulating hormone (FSH)	
Leuteinizing hormone releasing factor	Auxins
Prolactin	Prostaglandins
Lactogenic hormone	Sugars
Antidiuretic hormone	Lectins
Chorionic gonadotropin	Asialoglycoproteins
Thryotropin releasing hormone	Cholera toxin
Vasoactive intestinal polypeptide	Cyclic AMP
Growth hormone	Immunoglobulins
Parathyroid hormone	Complement
Vasopressin	Viruses
Epidermal growth factor	
Nerve growth factor	
Leuteinizing hormone (LH)	

Utilization of the above kinetic methods has led to the identification and characterization of a multitude of membrane receptors (Table 5-2). Many of these receptors are specific for protein hormones or growth factors, and are involved in mediating changes in cell metabolism, growth, and differentiation induced by these agents. The mechanisms by which the binding of such agents to their respective membrane receptors induces such profound changes in cell behavior will be the subject of a later chapter (Chapter 16). Another important category of membrane receptors are those specific for neurotransmitters such as acetylcholine and norepinephrine. In this case binding of ligand to receptor induces alterations in membrane permeability that are involved in transmitting nerve impulses from one cell to another. The behavior of such receptors will be discussed in detail in Chapters 18 and 19, devoted to nerve and muscle cells, respectively. Yet another important class of receptors are the antigen receptors of lymphocytes. Their role in triggering the formation of specific antibodies will be described in detail in Chapter 17.

It should be apparent from this brief introduction that membrane receptors are capable of triggering a variety of changes in cell behavior in response to differing external signals. The mechanisms by which such changes are induced will become clearer when we discuss specific receptors in the appropriate contexts elsewhere in the book. One generalization concerning receptor structure worth mentioning at this point, however, relates to the role of carbohydrate groups in determining receptor specificity. As it turns out, most receptors that have been isolated and characterized contain a carbohydrate component, usually attached to a backbone of protein (or occasionally lipid). In several instances exposing membranes or isolated receptor preparations to enzymes that degrade carbohydrate groups has been shown to abolish specific binding of ligand to receptor, suggesting that membrane carbohydrate groups may be involved in determining receptor specificity.

Cell–Cell Recognition and Adhesion

The successful functioning of multicellular organisms requires that individual cells become associated with each other in precise patterns to form tissues, organs, and organ systems. Such ordered interactions are ultimately based on the ability of individual cells to recognize, adhere, and directly communicate with one another. The existence of specific recognition between cells was first demonstrated experimentally in 1907 by H. V. Wilson, who investigated the behavior of cells obtained from two differently colored species of marine sponge. Wilson isolated individual cells from a yellow-pigmented sponge and from a red-pigmented sponge, and then mixed the two cell populations together. Under these conditions the cells selectively bound together into clumps composed exclusively of either red or yellow cells, but not a mixture of the two. These results were one of the first indications that cells are capable of recognizing and selectively binding to other cells of the same type.

In subsequent years many other examples of selective cell–cell recognition and adhesion have been uncovered. The importance of this phenomenon for the formation of tissues and organs during embryonic development will be discussed in detail in Chapter 15, which deals with the topics of early development and differentiation. In the present context we shall focus our attention upon the general mechanisms underlying cell recognition and adhesion.

A variety of experimental approaches have implicated cell surface carbohydrate groups in cell–cell adhesion. Removal of galactose residues from cell surface macromolecules, for example, renders cells less capable of adhering to one another. The role played by carbohydrate groups in promoting adhesion appears to be quite selective, for cells have been shown to bind to synthetic beads coated with the sugar galactose, but not to beads coated with glucose. In several instances glycoproteins isolated from the cell surface have been found to selectively enhance the binding together of the cells from which they were isolated. Such proteins, which promote cell–cell adhesion by binding to specific cell-surface carbohydrate residues, are referred to as **lectins.**

Some glycoproteins appear to play a less selective role in promoting cell adhesion. One such agent is **fibronectin,** a glycoprotein in plasma, extracellular fluids, and on the surface of a variety of cell types. This protein, which appears to act by a mechanism that is less selective than recognition of specific carbohydrate groups, promotes adhesion of cells to each other, to collagen, and to artificial surfaces such as the walls of tissue culture vessels. As we shall discuss in Chapter 20, changes in fibronectin appear to underlie alterations in cell morphology, mobility, and cell–cell interaction that occur in cancer cells.

One plasma membrane glycoprotein whose role in cell–cell recognition has been extensively investigated is the red cell protein glycophorin. One of the carbohydrate chains present in glycophorin is responsible for determining the standard blood types (A, B, AB, and O). Individuals with blood type A have N-acetylgalactosamine residues at the terminal ends of this branched carbohydrate chain, while individuals with blood type B have galactose instead. In blood type AB both configurations are present, while in type O these terminal sugars are missing entirely. Transfusions between individuals with differing blood types are hindered by the fact that people with blood type B have antibodies that recognize glycophorin chains containing terminal N-acetylgalactosamine residues, individuals with type A have antibodies that recognize glycophorin chains con-

taining terminal galactose residues, and individuals with type O have both types of antibodies (Table 5-3).

Blood group compatibility is not the only phenomenon in which the carbohydrate chains of glycophorin play an important role. Glycophorin is an unusual glycoprotein in that it contains large amounts of the carbohydrate sialic acid. These sialic acid residues, which are selectively localized at the terminal ends of carbohydrate chains, have been implicated in the mechanism by which aging red cells are recognized and targeted for destruction. Red blood cells generally have a lifespan of 3–4 months, after which they are destroyed by cells residing in the spleen. The role of sialic acid in the process by which aging cells are marked for destruction has been studied by treating isolated red blood cells with the enzyme *sialidase* (neuraminidase), which catalyzes the removal of sialic acid groups. When cells treated in this manner are injected back into the animals from which they were originally obtained, they are rapidly destroyed by the host's spleen cells. Normal red cells that have not had their sialic acid groups removed, on the other hand, survive for many weeks after being injected. Because removal of sialic acid exposes underlying galactose residues in the carbohydrate chains of glycophorin, it has been postulated that it is recognition of these newly exposed galactose residues by spleen cells that targets red cells for destruction. Since the sialic acid content of the red cell membrane decreases as the cell becomes older, a model can be envisioned in which the loss of sialic acid groups in aging red cells eventually exposes the galactose residues that permit the red cell to be recognized by the spleen cells responsible for its destruction.

The molecular mechanism by which cell-surface carbohydrate groups permit recognition and adhesion between cells is yet to be completely resolved. In the most straightforward model of cell–cell adhesion, cells with similar carbohydrate groups on their surfaces are bound together by lectins that recognize particular carbohydrate configurations. This binding of cells together by lectins is possible because each lectin molecule has more than one carbohydrate binding site and can therefore join adjacent cells to one another (Figure 5-25a). A more elaborate model, suggested by Saul Roseman, postulates that cell–cell adhesion is mediated by enzymes that catalyze the addition of carbohydrate groups to oligosaccharide chains. Although such enzymes, termed

TABLE 5-3
Molecular basis of the ABO blood group compatibilities

Blood type of individual	Sugar Present at Terminal End of Glycophorin Carbohydrate Chains		Serum Antibodies Directed Against Glycophorin Chains Terminating In	
	N-Acetylgalactosamine	Galactose	N-Acetylgalactosamine	Galactose
A	+	−	−	+
B	−	+	+	−
AB	+	+	−	−
O	−	−	+	+

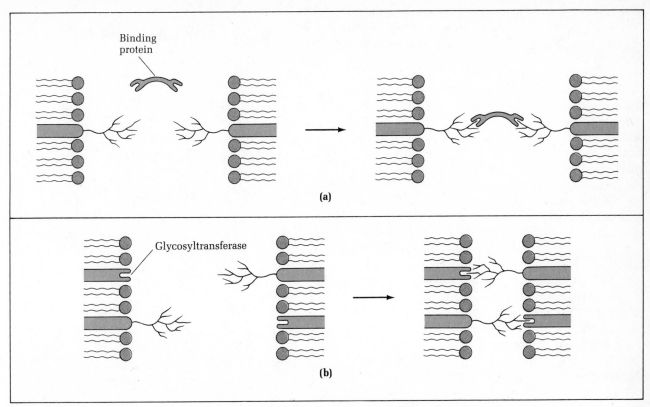

FIGURE 5-25 Two molecular models of cell-cell adhesion, both involving the carbohydrate residues of membrane glycoproteins (or glycolipids). (a) Model in which adjacent cells are bound together by an extracellular carbohydrate-binding protein, or lectin, having multiple carbohydrate-binding sites. (b) Model in which a glycosyltransferase on one cell recognizes its substrate (a carbohydrate chain) on the opposing cell.

glycosyltransferases, are predominantly localized in internal cytoplasmic membrane systems such as the Golgi complex and endoplasmic reticulum, they have also been identified in the plasma membrane. Because glycosyltransferases and the oligosaccharide chains upon which they act are both present on the outer surface of plasma membranes, a specific interaction between a glycosyltransferase on one cell and its specific substrate on another cell could cause the two cells to bind together (see Figure 5-25b). In such cases the simple binding of enzyme to oligosaccharide chain would be sufficient to promote cell–cell adhesion, and the glycosyltransferase need not necessarily catalyze the transfer of any new carbohydrate group to the substrate to which it is bound. If such a catalytic transfer of carbohydrate to the substrate were to occur, however, the glycosyltransferase would then be released. Hence an added feature of this model is that it provides a dynamic view of cell adhesion in which cell–cell contacts can be actively formed and broken.

Although the role of cell surface carbohydrate groups in cell–cell recognition and adhesion is well established, other molecules may also be involved. The divalent cation Ca^{2+} appears to be important, for example, because removing it from the external medium causes adhering cells to dissociate from one another. As

we shall see in a moment, Ca^{2+} has also been implicated in controlling cell–cell communication mediated by certain types of intercellular junctions.

Molecular interactions involving cell surface carbohydrates are clearly important in the initial events by which cells recognize and adhere to one another, but the creation of long-term stable connections between cells requires the formation of more complex structures known as **intercellular junctions.** These structures, which are unique to multicellular animals, perform three basic functions: *sealing, adhesion,* and *communication.* In the following sections we shall see that these functions are carried out by three types of junctions designated tight junctions, desmosomes (adhering junctions), and gap junctions, respectively. Because such junctions involve an intimate association between the plasma membranes of adjacent cells, the presence of a cell wall prevents the formation of these junctions in plants. However, the plant cell wall and its associated plasmodesmata (page 122) appear to carry out comparable functions for plant tissues.

Tight Junctions Cells that line the inner or outer surfaces of organs or body cavities are often linked together by intimate physical connections referred to as **tight junctions.** By injecting electron-dense tracer molecules into animals and then monitoring the distribution of such substances in electron micrographs, it has been shown that tight junctions establish leakproof barriers that prevent the movement of molecules through the spaces located between adjacent cells (Figure 5-26). Thus in cell layers sealed together by tight junctions, molecules cannot pass from one side of the layer to the other by diffusing through the spaces between cells. Movement of molecules across such a cell layer must instead occur through the plasma membranes of the constituent cells themselves, and is therefore subject to the precise control mechanisms inherent to transport through the plasma membrane. This sealing together of cells to form a leakproof barrier is especially important in organs such as the bladder, where seepage of urine back into the body tissues must be prevented, and the intestinal tract, where the passage of ingested materials into the body fluids must be carefully regulated.

In thin-section electron micrographs, tight junctions appear as regions in which the space between adjacent cells has been completely obliterated (Figure 5-27). The overall thickness of a typical tight junction is slightly less than the 15 nm that would be expected from two 7.5-nm membranes tightly joined together. Although the molecular basis for this reduced thickness is not completely understood, freeze-fracture micrographs reveal that the membrane surfaces involved in

FIGURE 5-26 Electron micrograph of a tight junction between two cells of the pancreas. The lanthanum tracer (black) can penetrate the intercellular space only as far as the tight junction, at which point its passage is blocked. Courtesy of D. S. Friend.

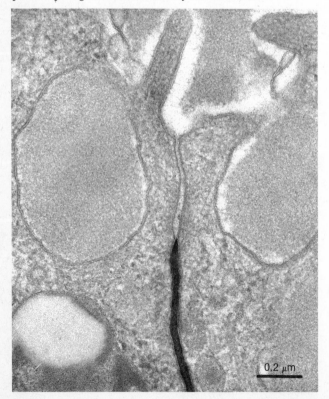

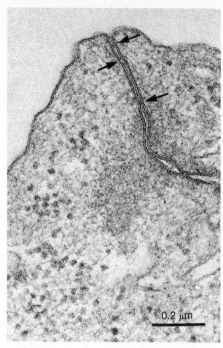

FIGURE 5-27 Thin-section electron micrograph showing the structure of the tight junction. Areas where membrane fusion occur are indicated by arrows. Courtesy of R. S. Decker.

the tight junction contain a meshwork of raised ridges and depressed furrows (Figure 5-28). Based on this picture, it has been postulated that the two adjacent membranes are fused together in the region of the tight junction by insertion of the ridges (sealing strands or junctional elements) from one membrane surface into the furrows on the opposing membrane. The ability of tight junctions to prevent diffusion of small molecules through the intercellular space is directly related to the number of ridges and furrows present. In cell layers where the tight junctions are composed of a relatively small number of ridges and furrows, the seal may be relatively leaky to small molecules. When larger numbers of ridges and furrows exist, on the other hand, the seal may be effective enough to prevent the passage of substances as small as ions.

Desmosomes and Other Adhering Junctions Many tissues and organs are exposed to mechanical forces that subject the constituent cells to considerable stretching and distortion. In order to maintain tissue integrity under such conditions, cells must be held together by junctions exhibiting significant mechanical strength. Of the several kinds of "adhering" junctions that function in this way, the most common is the **desmosome.** Desmosomes are characterized by the presence of dense plaques of fibrous material just beneath the plasma membranes of the two cells joined by the desmosome (Figure 5-29). From these plaques small fibers called **tonofilaments** radiate into the underlying cytoplasm.

Tonofilaments measure about 10 nm in thickness and function to anchor the desmosome to the cell interior. They belong to a general class of cytoplasmic fibers, known as *intermediate filaments*, whose properties will be described in detail in Chapter 9. The plasma membranes of the adjoining cells lie parallel to each other in the region of the desmosome, separated by an ~30-nm space in which thin filaments can be seen running perpendicular to the two membranes. These filaments react with cytochemical stains for carbohydrates and can be digested by brief treatment with the protease trypsin, which causes cells held together by desmosomes to separate from each other. Taken together, such observations suggest that the intercellular filaments consist of glycoprotein molecules responsible for holding the desmosome together. Several other components of the desmosome have been studied in isolated plasma membrane fractions enriched in desmosomes. In this way a 200,000-dalton polypeptide has been tentatively identified as the major constituent of the **plaque,** a re-

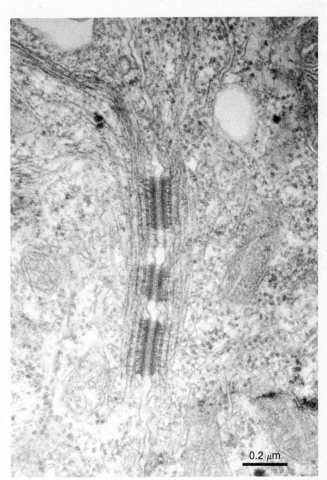

FIGURE 5-29 Electron micrograph of three desmosomes between two cells. The dense plaques directly beneath the plasma membrane are connected to tonofilaments that extend into the surrounding cytoplasm. Courtesy of R. S. Decker.

FIGURE 5-28 Freeze-fracture electron micrograph of a tight junction. Microvilli appear as bullet-shaped depressions and raised areas. Tight junction ridges and the complementary grooves are indicated by arrows. Courtesy of L. A. Staehelin.

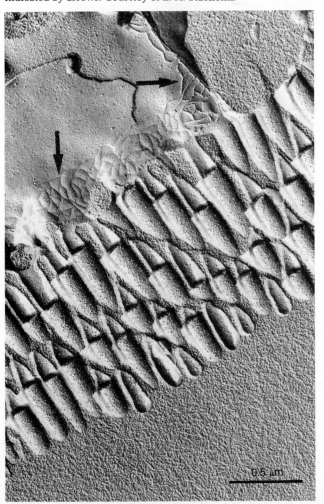

gion just under the plasma membrane, and a group of smaller polypeptides have been identified as components of the tonofilaments.

Desmosomes occur in most cells covering or lining the organs of multicellular animals, and are especially frequent in tissues where considerable mechanical stress is encountered, such as the skin and intestines. Desmosomelike structures also function in anchoring cells to extracellular structures, as occurs in the attachment of epithelial cells to the basal lamina, and in attaching cultured cells to the glass or plastic walls of the container in which they are growing. In these latter instances the plaque and tonofilaments occur in a single cell (Figure 5-30) rather than in two adjoining cells, so the structure is referred to as a **hemi-desmosome** rather than a desmosome.

Typical desmosomes involve small circular patches of plasma membrane measuring a few hundred nanometers in diameter. A closely related structure, termed the **fascia adherens** (belt desmosome), forms zones or belts around cells that are more extensive. The ultrastructure of the fascia adherens (Figure 5-31) differs

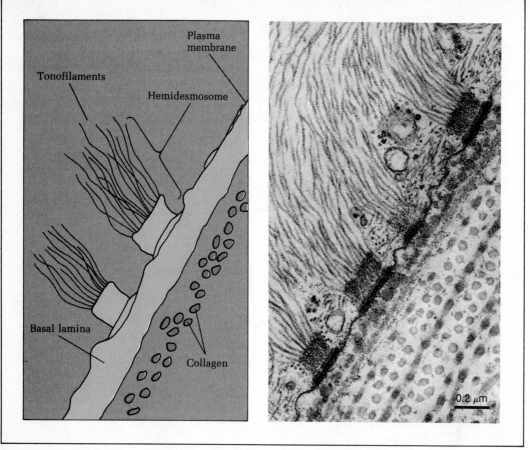

FIGURE 5-30 Electron micrograph and schematic diagram of hemidesmosomes showing tonofilaments extending into the cytoplasm on only one side, while the other side abuts directly on the basal lamina. Cross sections of collagen fibers also appear on this side of the hemidesmosome. Courtesy of D. E. Kelly.

FIGURE 5-31 Thin-section electron micrograph showing the fascia adherens (arrow) between two cells. Courtesy of S. Eichenberger-Glinz.

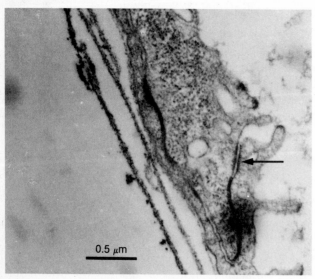

somewhat from that of classic desmosomes. The space between the adjacent membranes is reduced to 15–20 nm, the dense plaques are not present, and the fibers anchoring the junction to the underlying cytoplasm are thinner (7 versus 10 nm) than the tonofilaments underlying desmosomes. In morphology and biochemical properties, the 7-nm fibers associated with the fascia adherens resemble the actin microfilaments whose role in cell motility will be discussed later (Chapters 9 and 19). This presence of actinlike filaments raises the possibility that in addition to its mechanical role in holding cells together, the fascia adherens may also mediate the coordinated movement of adjacent cells. In the case of heart muscle, where adjoining cells are linked together by a special type of fascia adherens referred to as an **end plate,** such a role has been clearly established.

A third type of junction in the desmosome category is the **septate junction** (or septate desmosome). The characteristic feature of septate junctions is the presence of regularly spaced cross-bands or *septa* running across the 15- to 20-nm space separating the two opposed membranes (Figure 5-32). Although septate junctions occur most often in invertebrates, recent reports

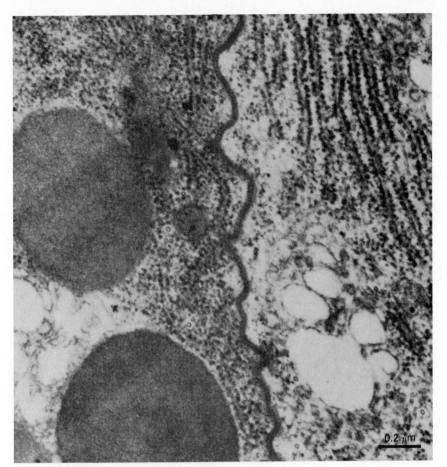

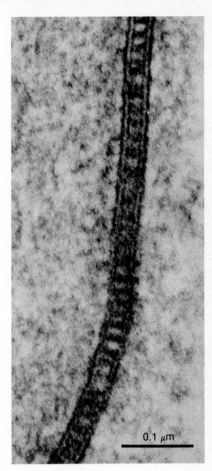

FIGURE 5-32 *(Left)* Electron micrograph of a septate junction. Courtesy of W. R. Loewenstein. *(Right)* Higher magnification of a septate junction showing the cross-bands, or septa. Courtesy of B. K. Filshie.

suggest that similar structures are present in the cells of higher organisms as well. Like desmosomes and the fascia adherens, septate junctions are most prominent in tissues subject to mechanical stress, where they function predominantly as an adhesive mechanism. It has also been claimed, however, that septate junctions allow direct molecular communication between cells in a fashion similar to the gap junction to be discussed below. But the cells upon which such claims have been made contain gap junctions as well as septate junctions, so the evidence implicating septate junctions in cell-cell communication is not conclusive.

Gap Junctions Communication between individual cells is essential for the coordinated functioning of the tissues and organs of multicellular organisms. One means for establishing direct communication between adjacent cells is the **gap junction,** which permits small molecules to move from one cell to another without passing through the plasma membrane itself. The idea that such a special type of channel facilitates the movement of molecules between adjoining cells owes much to the pioneering work of Werner Loewenstein and his

colleagues. By employing microelectrodes to measure the electrical resistance between adjacent cells, these investigators were the first to demonstrate that in some tissues, such as the insect salivary gland, the resistance to the flow of current between adjacent cells is several orders of magnitude lower than that normally measured across an intact plasma membrane. It was also noted that current does not readily pass from an electrode placed in one cell to another electrode positioned in the external medium. Based on this information, Loewenstein proposed that channels exist that permit conducting ions to pass directly from cell to cell without first appearing in the extracellular space.

In order to determine the effective size of these channels, studies were later carried out in which cells of various tissues were injected with fluorescent molecules of differing molecular weights. By examining the injected tissues with an ultraviolet microscope, movement of these fluorescent molecules into adjacent cells could be monitored (Figure 5-33). Such experiments revealed that molecules up to about 1000 daltons in molecular weight readily pass between cells, and in some tissues even somewhat larger molecules can do so. Al-

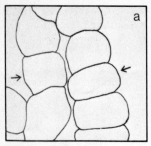
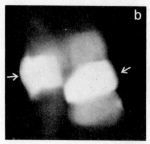
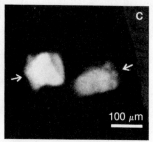

100 μm

FIGURE 5-33 Probing the permeability of junctional membrane channels with fluorescent molecules of differing sizes. (a) Arrows point to the two cells into which the fluorescent molecules were injected. (b) Fluorescence light micrograph showing distribution of smaller fluorescent molecule (molecular weight 1158) after 40 min. Note the passage of these fluorescent molecules into adjacent cells. (c) Distribution of larger fluorescent molecule (molecular weight 1926) 45 min after injection. Note that this larger molecule has been retained within the two injected cells. Courtesy of W. R. Loewenstein.

though some early confusion existed as to which type of intercellular junction is responsible for this phenomenon, it eventually became clear that the gap junction is the only junction whose presence always correlates with the existence of direct molecular coupling between cells.

Gap junctions occur in vertebrates as well as invertebrates and are especially abundant in tissues where extremely rapid communication between cells is required, such as nerve and muscle. In thin-section electron micrographs, gap junctions appear as regions in which the plasma membranes of the two adjacent cells are aligned in parallel and separated by a small gap of 2–3 nm (Figure 5-34). When examined in surface view by negative staining, the "gap" region is seen to be occu-

FIGURE 5-34 Electron micrograph of a gap junction. Courtesy of R. S. Decker.

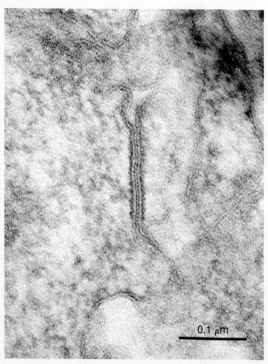

0.1 μm

0.1 μm

FIGURE 5-35 Electron micrograph showing the surface of a negatively stained gap junction. Note the hexagonal arrangement of the connexons. Courtesy of E. L. Benedetti.

pied by cylindrical structures, termed **connexons,** packed in a hexagonal array (Figure 5-35). Data obtained from a combination of X-ray diffraction and electron microscopy have led to the conclusion that connexons consist of six dumbbell-shaped protein subunits surrounding a central channel through which molecules can pass from cell to cell (Figure 5-36). The protein subunits that comprise the individual connexons are thought to be arranged with their hydrophilic residues facing the central channel and their hydrophobic residues on the outer surface exposed to the lipid bilayer. With such an arrangement, the central channel presumably provides a hydrophilic environment for the passage of water-soluble molecules. It is

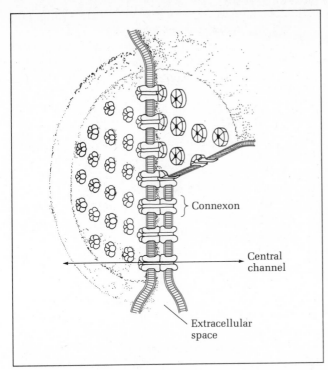

FIGURE 5-36 Model of the gap junction.

important to note that even though individual connexons provide direct passageways from one cell to the next, the spaces between connexons are in continuity with the extracellular medium. Thus gap junctions, unlike tight junctions, do not provide an impenetrable seal between cells, but rather consist of small cell-to-cell channels surrounded by open extracellular spaces.

An elegant series of experiments carried out by Werner Loewenstein and his colleagues have lead to the proposal that the permeability of gap junctions may be regulated under some conditions by calcium ions. In these studies the effect of Ca^{2+} on cell–cell communication was investigated by monitoring the migration of fluorescent molecules between cells microinjected with Ca^{2+}-containing solutions. The results revealed

that when the intracellular Ca^{2+} concentration is significantly increased above its normal levels of $< 10^{-7}M$, the movement of fluorescent molecules from cell to cell is inhibited (Figure 5-37). This closure of gap junctions is not an all-or-none event, for varying degrees of permeability were observed as the Ca^{2+} concentration was gradually increased from $10^{-7}M$ (maximum permeability) to $10^{-5}M$ (minimum permeability). Although the extent to which calcium ions exert a similar regulatory effect on gap junction permeability under normal physiological conditions is not clear, the preceding experimental observations suggest that gap junctions provide a means for cell–cell communication that is potentially subject to dynamic regulation.

METABOLIC FUNCTIONS OF THE CELL SURFACE

The final function of the cell surface to be mentioned in this chapter involves its role in metabolism. Several examples of metabolic enzymes located at the cell surface are discussed elsewhere in the text, including hydrolytic enzymes involved in nutrient breakdown (page 120), enzymes associated with cell wall biosynthesis (page 182), the enzyme responsible for phosphorylating sugars in the phosphotransferase system for sugar transport (page 143), and the glycosyltransferases postulated to play a role in cell-cell recognition and adhesion (page 151). In addition, the plasma membranes of higher organisms are often enriched in the enzymes alkaline phosphodiesterase, 5'-nucleotidase, and adenylate cyclase. Alkaline phosphodiesterase and 5'-nucleotidase catalyze removal of charged phosphate groups from nutrients such as nucleotides and sugar-phosphates, reducing their ionic character and hence facilitating their uptake through the plasma membrane. Adenylate cyclase, on the other hand, is intimately associated with the process by which hormones acting at the cell surface ultimately influence events occurring in the rest of the

FIGURE 5-37 Experiments demonstrating the effect of calcium ions on the movement of molecules through gap junctions. *(Left)* In calcium-containing medium, a fluorescent dye injected into one cell (3) is seen to diffuse into adjacent cells. *(Right)* In the same calcium-containing medium, holes are punched in cells 2 and 4 to allow the calcium to enter, thereby raising the intracellular calcium ion concentration. Under these conditions fluorescent dye injected into cell 3 does not diffuse into the adjacent cells. Courtesy of W. R. Loewenstein.

cell, a topic to be discussed in considerable detail in Chapter 16.

Due to the paucity of internal membranes in prokaryotic cells, some metabolic events that occur elsewhere in eukaryotes are associated with the plasma membrane in prokaryotes. One such event is the replication of chromosomal DNA, a process linked to the plasma membrane in bacteria but not in higher organisms. A more general category of reactions occurring in the prokaryotic plasma membrane are those involved in the oxidative degradation of nutrients and the associated formation of ATP. Because these reactions closely resemble events occurring in the mitochondria of eukaryotes, these processes will be discussed later in the chapter on mitochondria.

SUMMARY

In addition to providing a physical barrier that surrounds and protects the cell, the plasma membrane and its associated layers are involved in three major activities: regulating the flow of materials into and out of the cell, mediating cell-cell communication and adhesion, and serving as the site of specific enzymatic reactions and metabolic pathways.

Materials pass through the plasma membrane in at least four different ways: simple diffusion, facilitated diffusion, active transport, and endocytosis/exocytosis. Only the first three are discussed in the present chapter because the fourth (endocytosis/exocytosis) involves internal cytoplasmic membranes and so cannot be adequately discussed until such membranes are described in Chapter 6. Simple diffusion refers to the unaided net movement of a substance from regions of higher to lower concentration. In strict thermodynamic terms, differences in free energy rather than concentration are responsible for diffusion. Diffusion drives the net movement of dissolved solutes as well as water molecules; in the latter case the process is termed osmosis. The plasma membrane is said to be selectively permeable because some molecules, most notably those that are lipid soluble or small, tend to diffuse through more readily than others.

Like simple diffusion, facilitated diffusion involves the net movement of solute from regions of higher to lower concentration. The rate of solute movement is enhanced, however, by interaction with specific membrane transporters that are solute specific, exhibit Michaelis–Menten saturation kinetics, and are susceptible to competitive and noncompetitive inhibitors. The membrane transporters that mediate facilitated diffusion of glucose and anions in red blood cells have been tentatively identified. The glucose transporter is a transmembrane protein postulated to undergo conformational changes in which its glucose binding site alternatively faces inward and outward. The anion transporter corresponds to the band-3 protein and mediates exchange diffusion of bicarbonate and chloride ions by a similar conformational mechanism.

Ionophores have been employed as models for studying the mechanism by which transporters enhance the diffusion of ions across biological membranes. Some ionophores, such as valinomycin, act as carriers that diffuse back and forth across the lipid bilayer. Others, like gramicidin, form a relatively fixed channel across the membrane through which ions can diffuse. All naturally occurring membrane transporters identified to date appear to act as channel formers.

Active transport is the mechanism by which substances move into and out of cells against significant concentration gradients. The required energy is provided either by hydrolysis of an energy-rich compound such as ATP, or by coupling active transport of one solute to the movement of another substance down its concentration gradient. Active uptake of K^+ and expulsion of Na^+ are mediated by Na^+,K^+-ATPase, a transmembrane protein that contains binding sites for Na^+, K^+, and ATP. Transport of Na^+ and K^+ in opposite directions is thought to be mediated by conformational changes in the Na^+,K^+-ATPase induced by successive cycles of phosphorylation and dephosphorylation.

Some bacteria take up sugars by an active transport mechanism whose energy is derived from the hydrolysis of phosphoenolpyruvate. The sugar being transported becomes phosphorylated as it passes through the membrane, adding a negative charge that helps trap the molecule inside the cell. Among the active transport systems that use gradients as energy sources are the Na^+-dependent transporters for sugars and amino acids found in higher organisms, and the proton-linked transporter for lactose in bacteria.

A gradient of electric charge, or membrane potential, occurs across the plasma membrane of most cells. This electric potential is caused by leakage of K^+ from the cell driven by simple diffusion down its concentration gradient. Electrochemical equilibrium is usually achieved when the membrane potential reaches a value of -50 to -100 mV. The important role of the membrane potential in nerve and muscle cell communication will be described later in the text.

The plasma membrane contains receptors designed to receive signals from external agents such as hormones, neurotransmitters, and antigens. Receptors can be isolated and identified by studying the binding of radioactive ligands to isolated membrane fractions. Such binding must exhibit the appropriate specificity, affinity, and number of binding sites before one can be confident that a genuine receptor has been identified. Binding of ligands to their appropriate plasma membrane receptors triggers changes in membrane-localized functions (e.g., enzymatic activity, permeability, or transport) that lead to further alterations in cell behavior. The mechanisms by which specific receptors induce such changes will be discussed in appropriate chapters elsewhere in the text.

The successful functioning of multicellular organisms requires that individual cells adhere to each other in precise patterns to form tissues, organs, and organ systems. Cell surface glycoproteins play an important role in mediating cell–cell recognition and adhesion. In multicellular animals long-term stable connections between cells require the formation of membrane junctions. Tight junctions function to seal adjacent cells together in such a way that the diffusion of molecules through the spaces between adjacent cells is prevented. Desmosomes, fascia adherens, and septate junctions exhibit great mechanical strength and function to hold cells together in tissues subject to mechanical forces that tend to tear cell layers apart. Gap junctions permit small molecules to pass directly between adjacent cells, thereby establishing a virtually instantaneous mode of cell–cell communication.

The final function of the cell surface is its role in metabolism. Cell surface enzymes are involved in processes as diverse as the breakdown of potential nutrients, cell wall biosynthesis, enzymatic modification of solutes during transport, hormone action, and perhaps cell–cell recognition and adhesion. Furthermore, the plasma membrane of prokaryotic cells is the site of DNA replication, as well as the oxidative pathways for ATP formation that occur in the mitochondria of eukaryotes.

SUGGESTED READINGS

Books

Finean, J. B., R. Coleman, and R. H. Mitchell (1984). *Membranes and Their Cellular Functions*, Blackwell Scientific, Oxford.

Saier, M. H., Jr. (1985). *Mechanisms and Regulation of Carbohydrate Transport in Bacteria*, Academic Press, Orlando.

Stein, W. D. (1986). *Transport and Diffusion Across Cell Membranes*, Academic Press, Orlando.

Tonomura, Y. (1986). *Energy-Transducing ATPases—Structure and Function*, Cambridge University Press, New York.

Articles

Kaback, H. R. (1982). Membrane vesicles, electrochemical ion gradients, and active transport, *Curr. Topics Membrane Transport* 16:393–404.

Loewenstein, W. R. (1980). Cell-to-cell communication: permeability, formation, genetics, and functions of the cell–cell membrane channel. In: *Membrane Physiology* (Andreoli, T. E., J. F. Hoffman, and D. D. Fanestil, eds.), Plenum, New York, pp. 335–356.

———. (1987). The cell-to-cell channel of gap junctions, *Cell* 48:725–726.

Macdonald, C. (1985). Gap junctions and cell–cell communication, *Essays Biochem.* 21:86–118.

Overath, P., and J. K. Wright (1983). Lactose permease: a carrier on the move, *Trends Biochem. Sci.* 8:404–408.

Staehelin, L. A., and B. E. Hull (1978). Junctions between living cells, *Sci. Amer.* 238 (May):141–152.

Unwin, N. (1986). Is there a common design for cell membrane channels, *Nature*, 323:12–13.

CHAPTER
6
Functions of Cytoplasmic Membranes

One of the most striking morphological differences between prokaryotic and eukaryotic cells is the presence, in eukaryotes, of an elaborate series of membranes that divide the cytoplasm into multiple compartments. These membranes, which are thought to have arisen long ago in evolution by proliferation and invagination of the plasma membrane, are organized into several interconnected systems. In this chapter we shall focus on one of these systems, generally referred to as the *cytoplasmic membrane system*, whose major constituents are the endoplasmic reticulum, Golgi complex, lysosomes, and peroxisomes. The other internal membranes of eukaryotes, namely those enveloping mitochondria, chloroplasts, and the nucleus, are sufficiently independent to warrant their discussion in separate chapters later in the text (Chapters 7, 8, and 10).

Our current appreciation of the organization of the cytoplasmic membrane system of eukaryotes is a tribute to the power of the side-by-side use of subcellular fractionation and electron microscopy. The awarding of the Nobel Prize in 1974 to Albert Claude, George Palade, and Christian de Duve recognized the contributions of three pioneers who were especially instrumental in fusing the study of subcellular morphology and subcellular biochemistry. Claude devised the first systematic procedures for isolating organelles by subcellular fractionation. Palade applied subcellular fractionation and electron microscopic analysis to the study of the endoplasmic reticulum and Golgi complex, allowing him to uncover the significance of these membrane systems for the synthesis, transport, storage, and secretion of proteins. De Duve's studies on subcellular fractions, on the other hand, led him to predict the existence of lysosomes and peroxisomes before these structures had been recognized electron microscopically, a remarkable testimonial to the power of the subcellular fractionation approach. In this chapter we shall examine the experiments of these and other investigators who have investigated the roles carried out by the elaborate cytoplasmic membrane system of eukaryotes.

THE ENDOPLASMIC RETICULUM

In the late nineteenth century light microscopists first noted that certain areas of eukaryotic cytoplasm stain intensely with basic dyes. These regions, termed *ergastoplasm*, were found to be especially prominent in cells involved in secretion. Due to the limited resolving power of the light microscope, however, the reality of the ergastoplasm as a legitimate subcellular component remained in doubt until the advent of the electron microscope in the early 1940s.

In 1945 Keith Porter, Albert Claude, and Ernest Fullam discovered that cells grown in tissue culture become spread so thin that they can be examined by electron microscopy without the need for sectioning. This realization permitted the first electron micrographs to be taken of eukaryotic cells. These micrographs revealed that such cells contain a lacelike network of strands extending throughout their cytoplasm (Figure 6-1). When thin-sectioning techniques were developed shortly thereafter, it became clear that

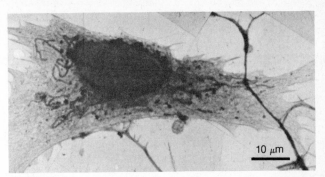

FIGURE 6-1 A series of electron micrographs spliced together to show an intact fibroblast along with three nerve fibers. Note the delicate lacework present throughout the cytoplasm. Courtesy of K. R. Porter.

FIGURE 6-2 Three-dimensional representation of the endoplasmic reticulum and related membrane systems, illustrating the profiles observed when cells are thin sectioned for electron microscopy.

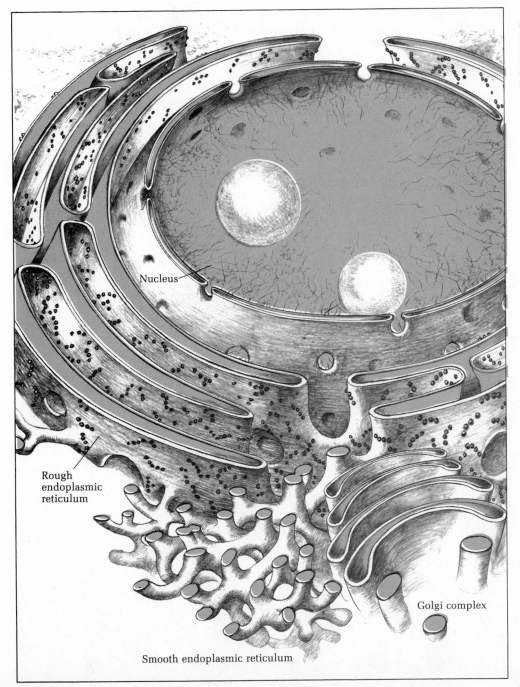

Nucleus

Rough endoplasmic reticulum

Smooth endoplasmic reticulum

Golgi complex

this network, referred to as the **endoplasmic reticulum (ER),** is composed of a heterogeneous collection of membranous tubules, vesicles, and sacs. Because tissue sections generally employed for electron microscopy are extremely thin, only slices through these structures are visible in routine electron micrographs. The observed profiles therefore look like separate membrane-enclosed spaces rather than interconnected tubules and vesicles (compare Figures 6-2, p. 161, and 6-3, below). It is fortunate that the endoplasmic reticulum was first noted in nonsectioned cells, for it would have been difficult to comprehend from thin-sectioned material alone that this system is composed of a single, interconnected network of membrane-enclosed channels.

The relationship between the endoplasmic reticulum and the ergastoplasm of classical light microscopy first became apparent when George Palade discovered that the endoplasmic reticulum exists in two forms: a "granular" or **rough endoplasmic reticulum (RER)** containing attached ribosomes, and an "agranular" or **smooth endoplasmic reticulum (SER)** lacking attached ribosomes (see Figure 6-3). Ribosomes contain large amounts of RNA, an acidic molecule exhibiting a strong affinity for basic dyes. It is therefore the presence of attached ribosomes in the rough endoplasmic reticulum that causes the cytoplasm to stain with basic dyes.

The rough and smooth endoplasmic reticulum differ in ways other than the presence or absence of attached ribosomes. Rough ER is generally arranged as large flattened sheets of membrane, while smooth ER more typically consists of an interconnected series of convoluted tubules (see Figures 6-2 and 6-3). The relative abundance of smooth and rough ER varies among cell types, with rough ER predominating in cells actively synthesizing protein for export, and an extensive smooth ER associated with cells involved in the metabolism of steroid hormones, drugs, and toxic substances. It has been estimated that in rat liver cells, the endoplas-

FIGURE 6-3 Comparison of the structure of the smooth and rough endoplasmic reticulum viewed by transmission and scanning electron microscopy. *(Top left)* Scanning electron micrograph of the rough endoplasmic reticulum of a pancreatic cell showing the membrane-enclosed sacs. Courtesy of K. Tanaka. *(Top right)* Thin-section electron micrograph of the rough endoplasmic reticulum. Courtesy of D. W. Fawcett. *(Bottom left)* Scanning electron micrograph of the smooth endoplasmic reticulum showing the branched and tubular membrane system. Courtesy of K. Tanaka. *(Bottom right)* Thin-section electron micrograph of the smooth endoplasmic reticulum. Courtesy of M. Bielinska.

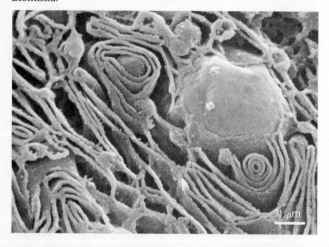

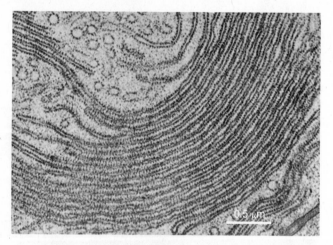

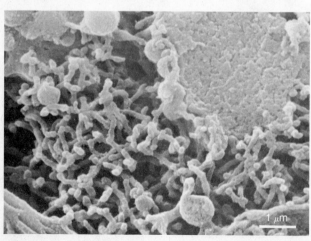

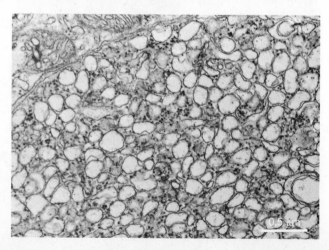

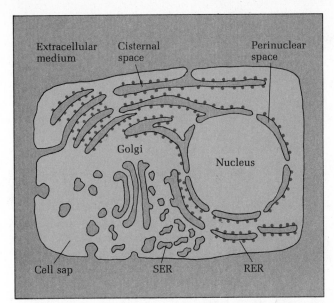

FIGURE 6-4 Relationship between the cisternal space (gray) and the cell sap (color). Note the close relationship between the cisternal space, the perinuclear space, and the external environment. Abbreviations: RER = rough endoplasmic reticulum, SER = smooth endoplasmic reticulum.

mic reticulum accounts for 19 percent of the total protein, 48 percent of the total lipid, and 58 percent of the total RNA of the cell. This organelle can therefore be a substantial contributor to the overall mass of the cell.

The endoplasmic reticulum divides the cytoplasm into two compartments, termed the **cell sap** and the **cisternal space** (Figure 6-4). The cell sap contains soluble enzymes involved in intermediary metabolism, transfer RNAs and other factors required for protein synthesis, and free ribosomes. Ribosomes attached to the endoplasmic reticulum are located on the side of the membranes facing the cell sap. The cisternal space of the endoplasmic reticulum, on the other hand, is functionally continuous with the internal cavities of the Golgi complex, lysosomes, and peroxisomes, the perinuclear space between the two membranes of the nuclear envelope, and the outside of the cell. Instead of being maintained by permanent membrane connections, this continuity is established by shuttling membrane vesicles that bud off from one membrane system, travel some distance, and then fuse with another membrane system.

The advantage of dividing the cytoplasm into multiple membrane-bound compartments is that separate environments, individually adapted to different activities, can be maintained. Regulation of the environmental conditions prevailing on opposite sides of the endoplasmic reticulum is facilitated by the fact that the ER membranes are selectively permeable and capable of carrying out active transport. The ability to transport and maintain gradients of Ca^{2+}, for example, appears to be especially well developed in the smooth ER, playing important roles in cell secretion, nerve excitation, and

muscle contraction. Another advantage to dividing the cytoplasm with membrane barriers is that a series of channels is created that provide a route for movement of various substances. As will be discussed later, this function has been most clearly established for newly made protein molecules. There is reason to believe, however, that other materials are also routed to various parts of the cell via the endoplasmic reticulum. Triglycerides present in the extracellular milieu, for example, are hydrolyzed to fatty acids in the cytoplasm, taken up into the cisternae of the ER, and quickly distributed throughout the cell through the channels of both the smooth and rough ER.

Our current understanding of the biochemical functions performed by the endoplasmic reticulum is based largely on studies carried out on isolated subcellular fractions. During homogenization the continuity of the endoplasmic reticulum is disrupted, creating membrane fragments that spontaneously reseal into vesicles called **microsomes.** Microsomes can be collected by high-speed centrifugation and further separated into smooth and rough ER fractions (Figure 6-5). Isolated microsomal vesicles, in spite of being artificially generated derivatives of a larger interconnected system of membranes, are useful model systems for studying the functional properties of the two types of endoplasmic reticulum. Biochemical studies carried out on such vesicles have revealed the presence of a wide variety of enzymatic activities (Table 6-1). Included in this group are enzymes involved in (1) the metabolism of carbohydrates and lipids, (2) the detoxification of drugs and other toxic chemicals, and (3) the processing and transport of protein molecules. These three important activities of the endoplasmic reticulum are discussed below.

Metabolism of Carbohydrates and Lipids

The enzyme **glucose-6-phosphatase,** which catalyzes the hydrolysis of glucose-6-phosphate to glucose and inorganic phosphate, is often employed as a *marker* for identifying microsome fractions because it is localized almost exclusively in ER membranes. This enzyme functions in the metabolism of glucose-6-phosphate released from glycogen granules located adjacent to the smooth ER in intact cells. The overall pathway involved in this process is summarized in Figure 6-6. Glycogen is first hydrolyzed to glucose-1-phosphate and this product converted to glucose-6-phosphate by enzymes present in the cell sap. The glucose-6-phosphate generated in this way is then transported across the smooth ER membrane, where it is broken down to glucose and inorganic phosphate by the glucose-6-phosphatase present on the luminal surface of the ER membrane. The free glucose released into the cisternal space can then diffuse through the channels of the endoplasmic

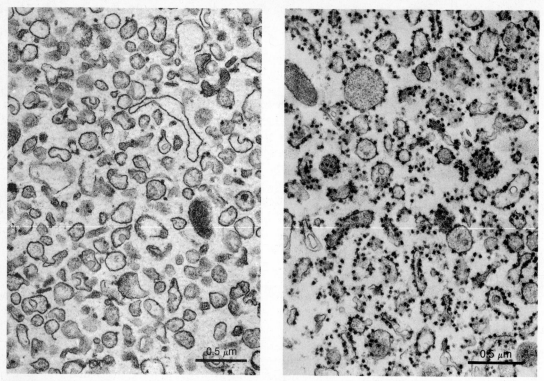

FIGURE 6-5 The appearance of microsomal subfractions separated according to density by centrifugation in a sucrose gradient. *(Left)* Smooth endoplasmic reticulum. *(Right)* Rough endoplasmic reticulum. Note the absence of ribosomes on the membranes of the smooth endoplasmic reticulum. Courtesy of C. de Duve.

TABLE 6-1

Major microsomal enzymes and their topological orientations

Metabolic Class	Enzyme Systems	Surface Localization
Carbohydrate metabolism	Glucose-6-phosphatase	Cisternal
	β-Glucuronidase	Cisternal
	Glucuronyl transferase	Cisternal
	GDP-mannosyl transferase	Cell sap
	Amylase	—
	L-Gulonolactone-ascorbate enzyme	—
Lipid metabolism	Fatty acid CoA ligase	Cell sap
	Phosphoglyceride formation	—
	Phosphatidic acid phosphatase	Cell sap
	Acyl-CoA synthetase	—
	Cholesterol hydroxylase	Cell sap
	Fatty acid reduction	—
	β-Hydroxysteroid dehydrogenase	—
	Steroid aromatization and hydroxylation	—
Detoxification of drugs and related oxidases	Cytochrome P-450	Both
	Cytochrome P-448	—
	NADPH-cytochrome P-450 reductase	Cell sap
	Cytochrome b_5	Cell sap
	NADH-cytochrome b_5 reductase	Cell sap
	Acetanilide-hydrolyzing esterase	Cisternal
	Monoamine oxidase	—
Other enzymes	ATPase	Cell sap
	5'-Nucleotidase	Cell sap
	Nucleoside pyrophosphatase	Cell sap
	Nucleoside diphosphatase	Cisternal
	Signal peptidase	Cisternal

Note: Dash indicates that surface localization is unknown.

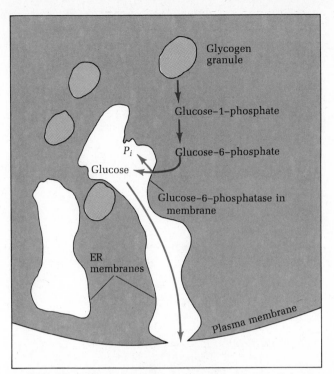

FIGURE 6-6 Diagram illustrating how the localization of the enzyme glucose-6-phosphatase in ER membranes facilitates the export of glucose derived from the breakdown of glycogen.

reticulum and eventually be released into the extracellular space. This overall pathway is particularly important in enabling liver cells to make energy stored in glycogen available to other cells of the body by releasing glucose into the bloodstream.

In addition to glucose-6-phosphatase, several other enzymes of carbohydrate metabolism are present in microsomal membranes. Among these are *amylase* and *β-glucuronidase*, involved in the hydrolysis of starch and mucopolysaccharides, respectively, the L-*gulono-lactone-ascorbate* enzyme, which catalyzes the formation of vitamin C (ascorbic acid), and *glucuronyl transferase*, which conjugates glucuronic acid to lipids and other hydrophobic compounds in order to render them water soluble and hence more readily excretable.

In addition to its role in carbohydrate metabolism, the endoplasmic reticulum is the major site of lipid biosynthesis. Neutral fats (triglycerides) synthesized in the endoplasmic reticulum are often stored in the cisternal space, forming fat droplets that serve as reservoirs of energy. As we shall see later in the chapter, phospholipids synthesized by microsomal enzymes are employed in the construction of cellular membranes. And finally, synthesis of cholesterol and subsequent oxidations, reductions, and hydroxylations involved in the formation of steroid hormones also occur in the endoplasmic reticulum. For this reason the smooth ER is especially well developed in tissues that produce steroid hormones, such as the adrenal cortex or the interstitial region of the testis (see Figure 6-3).

Detoxification of Drugs

One of the more important functions carried out by the endoplasmic reticulum of liver cells is the detoxification of foreign substances like drugs, pesticides, toxins, and other pollutants. Although most of these detoxification reactions involve oxidation, some entail reduction, hydrolysis, and conjugation as well. In all cases the net result is the conversion of a hydrophobic substance into a water-soluble one. This type of alteration is important because hydrophobic compounds tend to accumulate in body fats, while water-soluble materials are more easily excreted.

The first clear indication that cells contain a special enzyme system for oxidizing drugs was provided in 1953 by Gerald Mueller and James Miller, who reported that cancer-causing aminoazo dyes are oxidized by liver homogenates in a reaction requiring molecular oxygen and NADPH. Soon thereafter Bernard Brodie and his associates found that a vast number of other drugs and chemicals can be oxidized by isolated liver microsomes under similar conditions. The oxidation of these compounds involves the same common mechanism in which one molecule of oxygen is consumed per molecule of substrate oxidized:

$$RH + O_2 \xrightarrow{\quad NADPH + H^+ \quad NADP^+ \quad} ROH + H_2O$$

Because one atom of oxygen appears in the product while the other appears in water, enzymes catalyzing this type of reaction are referred to as **mixed-function oxidases.**

The mixed-function oxidase system of microsomes consists of several elements. The central component is **cytochrome P-450,** an iron-containing protein whose name is derived from the ability of its reduced form to absorb light at 450 nm. Cytochrome P-450 is a major constituent of microsomal membranes, accounting for up to 20 percent of the microsomal protein and 2–3 percent of the total protein of liver cells.

The role played by cytochrome P-450 in the mixed-function oxidase reaction is summarized in Figure 6-7. This reaction sequence involves four basic steps: (1) binding of an oxidizable substrate to cytochrome P-450, (2) reduction of the iron atom by NADPH, (3) binding of oxygen to cytochrome P-450, and (4) utilization of one atom of the bound oxygen to oxidize the substrate and the other to form water, with the iron atom becoming reoxidized in the process. In order for this reaction to occur with purified cytochrome P-450, two additional components must be present. One is **NADPH-dependent cytochrome P-450 reductase,** an enzyme catalyzing step 2 above. The other required molecule is a phospholipid such as phosphatidylcholine (lecithin). This phospholipid requirement may reflect the fact that in intact microsomes, the mixed-function oxidase sys-

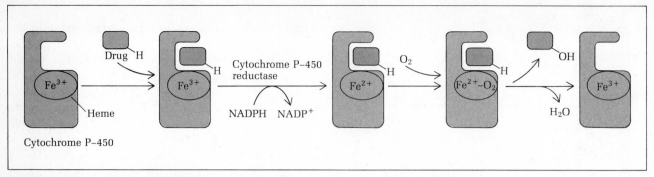

FIGURE 6-7 Mechanism of drug oxidation by a mixed-function oxidase system. After binding to the substrate, the Fe^{3+} present in the cytochrome P-450 is reduced to Fe^{2+} by cytochrome P-450 reductase. In this reduced form cytochrome P-450 can bind oxygen. One atom of the bound oxygen is employed to oxidize the substrate and the other is employed to form a molecule of water. During this process, the Fe^{2+} is oxidized back to Fe^{3+}.

tem functions within the lipid bilayer of the membrane. In fact, it has been suggested on the basis of transition temperature studies that cytochrome P-450 is enclosed in a phospholipid halo that is more rigid than the bulk of the membrane lipids. This halo may influence the behavior of lipid-soluble substrates and products in the vicinity of the cytochrome P-450 molecule.

In addition to its role in drug detoxification, the mixed-function oxidase reaction is employed in normal metabolic pathways for the oxidation of steroids and fatty acids. This broad utilization of a single enzyme system for catalyzing the oxidation of many diverse substrates means that alterations in this system may have widespread effects on cellular metabolism. To take a case with important practical implications, ingestion of the sedative phenobarbital causes a pronounced increase in the amount of mixed-function oxidase present in liver as well as a proliferation of the smooth ER membranes housing this enzyme system (Figure 6-8). The resulting increase in the ability of liver cells to degrade phenobarbital explains why habitual users of this drug require higher and higher doses to achieve the same effect. In addition, the broad substrate specificity of the induced mixed-function oxidase means that the liver acquires not only an enhanced ability to metabolize phenobarbital but an enhanced capacity to degrade other substances as well. Hence chronic use of barbituates increases the destruction rate and therefore decreases the effectiveness of other drugs, including therapeutically useful agents such as antibiotics, steroids, anticoagulants, and narcotics. The ability to induce drug-metabolizing enzymes is not restricted to phenobarbital but is shared by hundreds of other chemical substances.

The proliferation of endoplasmic reticulum that occurs in response to drug ingestion does not involve an equivalent increase in all microsomal enzymes. Phenobarbital, for example, induces cytochrome P-450 and NADPH-dependent cytochrome P-450 reductase, but not glucose-6-phosphatase (Figure 6-9a). The selectivity of microsomal enzyme induction is further demon-

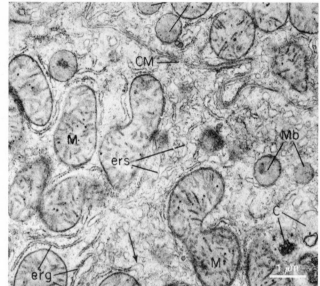

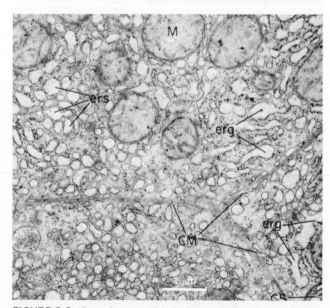

FIGURE 6-8 The induction of endoplasmic reticulum membrane proliferation by phenobarbital in rats. *(Top)* Electron micrograph of liver cell of control rat. *(Bottom)* Liver cell of phenobarbital-treated rat, showing the increase in smooth endoplasmic reticulum. Abbreviations: CM = cell membrane; erg = rough endoplasmic reticulum; ers = smooth endoplasmic reticulum; M = mitochondria; Mb = microbody. Courtesy of S. Orrenius.

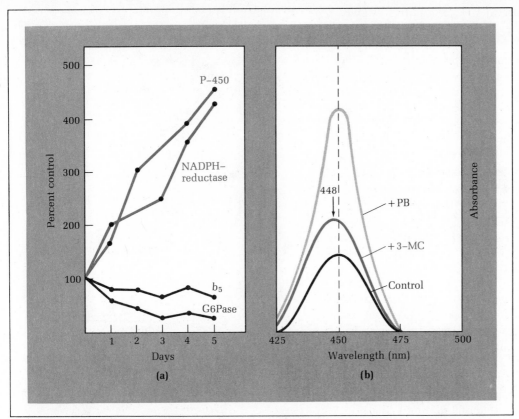

FIGURE 6-9 Experimental data illustrating selectivity of microsomal enzyme induction in rat liver. (a) Daily injections of phenobarbital cause an increase in cytochrome P-450 and NADPH-dependent cytochrome P-450 reductase, but not glucose-6-phosphatase or cytochrome b_5. (b) Comparison of the absorption spectrum of microsomal cytochromes after injections of phenobarbital or 3-methylcholanthrene. After 3-methylcholanthrene treatment the absorption maximum shifts from 450 to 448 nm, indicating the formation of a new cytochrome. Abbreviations: G6Pase = glucose-6-phosphatase; b_5 = cytochrome b_5; NADPH-reductase = NADPH-dependent cytochrome P-450 reductase; P-450 = cytochrome P-450; PB = phenobarbital; and 3-MC = 3-methylcholanthrene.

strated by comparing the properties of the mixed-function oxidase systems induced by different drugs. The cancer-causing agent 3-methylcholanthrene, for example, stimulates formation of a mixed-function oxidase system whose cytochrome has an absorption spectrum peaking at 448 rather than 450 nm (see Figure 6-9b). This **cytochrome P-448** metabolizes different substrates from those handled by cytochrome P-450. Because the mixed-function oxidase system employing cytochrome P-448 is most efficient in metabolizing polycyclic hydrocarbons, it is often referred to as **aryl hydrocarbon hydroxylase.**

Although the oxidation of polycyclic hydrocarbons by aryl hydrocarbon hydroxylase might be assumed to be a detoxification process, Harry Gelboin and his collaborators have suggested that in some cases the products generated may be more harmful than the original chemicals. Such a view is supported by the finding that mice with high levels of aryl hydrocarbon hydroxylase exhibit a greater incidence of spontaneous tumors than normal, and that animals given inhibitors of this enzyme acquire fewer tumors than normal when exposed to cancer-causing chemicals. Taken together, such ob-

servations clearly implicate aryl hydrocarbon hydroxylase in the causation of cancer. It is worth noting in this context that cigarette smoke is a potent inducer of aryl hydrocarbon hydroxylase.

The realization that cytochromes P-450 and P-448 differ in substrate specificity has spurred the search for other microsomal cytochromes. This search has led to the realization that cytochrome P-450 is not a single protein but a family of related molecules, each induced by and active upon a different class of substrates. In addition to cytochromes P-450 and P-448, another iron-containing protein selectively localized in the endoplasmic reticulum is **cytochrome b_5.** In contrast to the other microsomal cytochromes, cytochrome b_5 employs NADH rather than NADPH as a coenzyme. Cytochrome b_5 has been implicated in the desaturation of fatty acids and in NADH-dependent microsomal hydroxylation reactions.

Transport and Processing of Proteins

Unlike the smooth ER, whose main functions center around the metabolic events described above, the rough

ER is involved in the transport and processing of newly formed protein molecules synthesized on ribosomes attached to its cytoplasmic surfaces. The relative proportion of a cell's ribosomes found to be attached to the ER or free in the cytoplasm varies with both cell type and physiological state. In general, cells secreting large amounts of protein have an abundant supply of rough ER, while rapidly growing cells tend to possess large numbers of free ribosomes.

The mechanisms involved in the synthesis of protein molecules by ribosomes will be described in considerable detail in Chapter 12, and therefore need not be elaborated upon in the present context. For the moment it can simply be stated that the information specifying the amino acid sequence of newly forming proteins is encoded in molecules of messenger RNA, which bind to ribosomes and direct the specific order in which amino acids are added during the synthesis of polypeptide chains. Shortly after the discovery of membrane-bound ribosomes, Palade postulated that polypeptides synthesized on these ribosomes selectively pass into the cisternal space of the endoplasmic reticulum and are exported, while proteins synthesized on free cytoplasmic ribosomes are retained for use within the cell. The hypothesis that different classes of proteins are synthesized on free and membrane-bound ribosomes has now been amply verified. In liver cells, for example, proteins to be retained by the cells, such as ferritin, have been shown to be preferentially synthesized on free ribosomes. In contrast those destined for secretion into the bloodstream, such as albumin and gamma globulin, are predominantly synthesized on membrane-bound ribosomes. A similar phenomenon occurs in bacteria, even though an endoplasmic reticulum is not present. In this case proteins destined for secretion from the bacterial cell are synthesized on ribosomes attached to the plasma membrane, while proteins to be retained by the cell are made on ribosomes free in the cytoplasm. It can therefore be concluded that some type of signal must cause certain messenger RNAs to function on membrane-associated ribosomes, and others to function on free cytoplasmic ribosomes.

The first direct proof that proteins synthesized on membrane-attached ribosomes can pass directly into the cisternal space of the endoplasmic reticulum was provided by Colvin Redman and David Sabatini, who studied protein synthesis in isolated microsomes incubated with radioactive amino acids and other necessary factors. After a few minutes of incubation, the antibiotic *puromycin*, which stops protein synthesis and induces premature release of partially completed polypeptide chains, was added to the microsomal system. In order to determine whether radioactive polypeptides had been released into the lumen of the membrane vesicles, the detergent deoxycholate was added to disrupt the membranes and thereby release any radioactive material

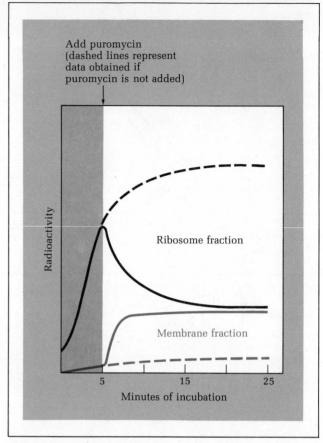

FIGURE 6-10 Experimental data demonstrating that proteins synthesized on membrane-attached ribosomes pass through the membrane as they are being synthesized. An *in vitro* protein-synthesizing system containing membrane-attached ribosomes was first incubated for 5 min with radioactive amino acids to label newly made polypeptide chains. Puromycin was then added, which stops protein synthesis and releases polypeptide chains that were in the process of being synthesized. The ribosomes were then separated from the membranes, and each fraction analyzed for radioactivity. The data show that the polypeptide chains released by puromycin treatment are trapped in the membrane fraction, indicating that these chains are being inserted through the ER membrane as they are synthesized.

trapped inside the vesicles. In this way, it was discovered that a large amount of radioactivity is released inside the microsomal vesicle after puromycin treatment (Figure 6-10). Since this radioactivity is present in polypeptides whose synthesis was interrupted by puromycin, the results indicate that polypeptides synthesized on membrane-attached ribosomes begin to pass through the membrane and into the lumen of the endoplasmic reticulum while they are still being synthesized.

A possible explanation for the mechanism by which certain messenger RNAs are selectively chosen to guide protein synthesis on membrane-bound ribosomes, while others function on free ribosomes, was proposed by Günter Blobel and David Sabatini in 1971. In essence, they postulated that attachment of particular messenger RNA–ribosome complexes to the endoplasmic reticulum is mediated by special "signal" sequences located near the beginning (amino-terminal end) of the proteins whose synthesis is being directed. Direct evidence for the existence of such signal sequences was obtained shortly thereafter by Cesar Milstein and his associates, who were studying the synthesis of the small polypeptide subunit, or "light chain," of immunoglobulin G. In a cell-free protein-synthesizing system containing free ribosomes but no microsomal membranes, the messenger RNA coding for this light chain was found to direct the synthesis of a polypeptide product that is 20 amino acids longer at its amino-terminal end than the authentic immunoglobulin light chain itself. When microsomal membranes were added to the protein-synthesizing system, however, the newly synthesized polypeptide was the same size as the normal light chain, suggesting that the microsomal membranes contain an enzyme capable of cleaving the extra 20 amino acids from the amino terminus. Subsequent studies have revealed that other proteins destined for secretion from the cell, as well as integral membrane proteins, are also synthesized as precursors containing extra amino acids at their amino-terminal ends. Proteins containing such leader sequences are referred to as "pre-" proteins (e.g., prelysozyme, preproinsulin, pretrypsinogen). Leader sequences tend to be enriched in hydrophobic amino acids (Table 6-2), suggesting that they facilitate passage of the polypeptide chain through the lipid bilayer.

On the basis of the preceding information, the **signal hypothesis** summarized in Figure 6-11 has been proposed. According to this model, protein synthesis is initiated by attachment of messenger RNA to free ribosomes. If a particular messenger RNA happens to code for a protein containing a signal sequence at its amino-terminal end, polypeptide synthesis ceases shortly after this signal sequence has been formed. This temporary interruption in protein synthesis is triggered by the binding to the ribosome of a specialized protein com-

TABLE 6-2
Hydrophobic nature of the signal sequence of several secreted polypeptides

Polypeptide	Sequence of Hydrophobic (x) and Nonhydrophobic (o) Amino Acids
Pretrypsinogen II (dog)	x o x x x x x x x x x x x x^{a}x
Prelysozyme (chicken)	x o o x x x x x x x x x x x x x x^{a}o
Preproinsulin (rat)	? x o o ? x x x x x o x x x x x x x o o x o x^{a}x
Preproparathyroid hormone (ox)	x x o o o o x x o x x x x x x x x x x x o o o o^{a}o

^{a} Represents site where signal peptide joins protein.

plex, termed the **signal recognition particle (SRP),** that specifically recognizes ribosomes synthesizing polypeptide chains containing a signal sequence. SRP, composed of a mixture of six polypeptides and a molecule of 7S RNA, is in turn responsible for the initial binding of ribosomes to membranes of the endoplasmic reticulum. Such membranes contain a special integral membrane protein, termed the **docking protein,** to which the SRP-ribosome complex specifically binds. Once this binding has occurred, the SRP is released and polypeptide synthesis resumes. During this stage, binding of the ribosome to the endoplasmic reticulum is thought to be stabilized by an interaction between ribosomes and other integral membrane proteins referred to as **ribophorins.**

Once polypeptide synthesis has been resumed by the membrane-bound ribosome, the signal sequence guides the insertion of the newly forming polypeptide chain through the ER membrane. The hydrophobic nature of the amino acid residues in the signal sequence presumably expedites this process. After the polypeptide chain passes through the membrane and begins to emerge into the cisternal space, the signal sequence has served its function and can be excised. This excision is

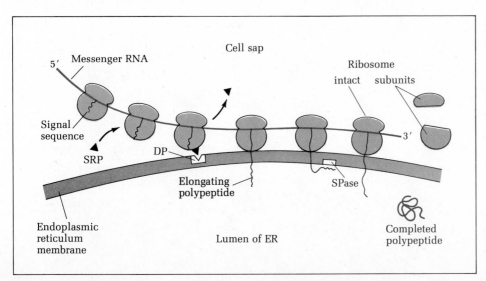

FIGURE 6-11 Schematic representation of the mechanism by which signal sequences function to guide the association of ribosomes with the endoplasmic reticulum. Abbreviations: SRP = signal recognition particle; DP = docking protein; SPase = signal peptidase.

catalyzed by a proteolytic enzyme, termed **signal peptidase,** which is located on the cisternal surface of the endoplasmic reticulum.

The major elements of the above model have been verified by experiments in which various types of messenger RNAs, ribosomes, and microsomal membranes have been mixed together in different combinations. If, for example, one takes messenger RNAs coding for proteins destined for secretion and mixes them with free ribosomes and microsomal membranes stripped of their ribosomes, the newly synthesized proteins become associated with the microsomal vesicles. If messenger RNAs coding for nonsecretory proteins are employed in the same system, however, the newly synthesized proteins do not become associated with the microsomal vesicles. Such findings clearly reveal that the information specifying that proteins are to be synthesized in association with the endoplasmic reticulum is ultimately contained in the messenger RNAs coding for those proteins. Several categories of protein molecules are now known to contain signal sequences that cause them to become associated with the endoplasmic reticulum while they are being synthesized. These include proteins destined for secretion from the cell, proteins destined to become localized within cytoplasmic vesicles such as lysosomes, and proteins destined to become integral membrane proteins. Of these three groups, the integral membrane proteins are somewhat unique in that they are not released into the lumen of the endoplasmic reticulum; the synthesis and processing of this special group of proteins will therefore be discussed separately in the section on membrane biogenesis later in the chapter.

The preceding scenario clearly illustrates how the endoplasmic reticulum functions to separate proteins destined for export, intracellular vesicles, or insertion into membranes from proteins destined to remain in the cell sap. Once newly synthesized proteins have been segregated within the endoplasmic reticulum, they can be transported through this system of membrane channels to other regions of the cell. During this transport process, the protein chains are often modified by enzymes present in the ER membranes. One such modification is the hydroxylation of proline and lysine residues, which takes place in fibroblasts making collagen (Chapter 4). Partial proteolysis of proteins, which converts single polypeptide chains into multiple subunits, may also take place in the ER, although this type of reaction appears to occur more typically in the Golgi complex.

One feature shared by many proteins destined for secretion is that they are glycoproteins. Included in this category are digestive enzymes such as pepsin and ribonuclease, plasma proteins like the immunoglobulins, hormones such as the gonadotropins, extracellular matrix materials such as collagen, and mucous secretions such as the salivary glycoproteins. The widespread occurrence of carbohydrate in secreted proteins has spawned the speculation that saccharide chains may help to identify these polypeptides for export from the cell, though it must be noted that some secretory proteins lack carbohydrate.

The covalent attachment of carbohydrate side chains to polypeptide chains occurs to a large extent in the lumen of the endoplasmic reticulum in reactions catalyzed by membrane-associated enzymes known as **glycosyltransferases.** The glycosylation reactions catalyzed by this family of enzymes involve the transfer of

FIGURE 6-12 Diagram summarizing the mechanism by which core oligosaccharide chains are transferred to polypeptide chains synthesized on ribosomes bound to the endoplasmic reticulum.

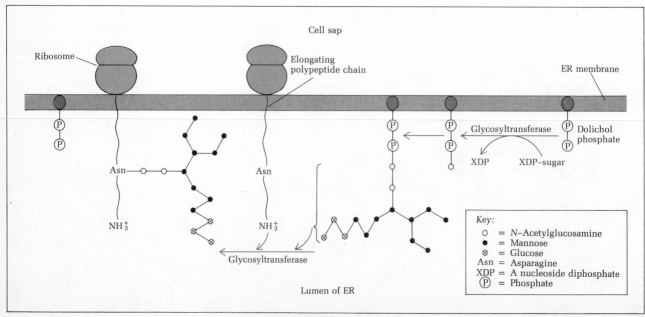

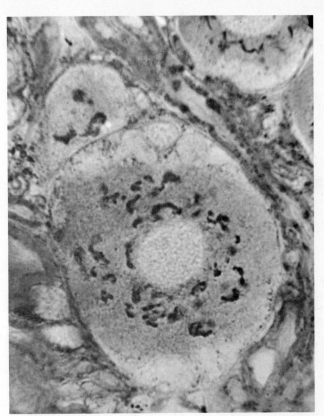

sugar residues to growing oligosaccharide chains, using energy-rich nucleoside-diphosphate sugar derivatives such as UDP-glucose as sugar donors. The oligosaccharide chains formed by these glycosyltransferases have precisely defined sequences generated by the substrate specificities of the enzymes involved. The main type of carbohydrate added to proteins in the endoplasmic reticulum is a **core oligosaccharide chain** consisting of a mixture of N-acetylglucosamine, mannose, and glucose. This core is initially assembled by sequentially adding sugars to **dolichol phosphate,** a long-chain alcohol whose massive hydrophobic tail, 80–100 carbon atoms in length, anchors it firmly within the ER membrane. After this oligosaccharide has been assembled on the dolichol carrier, it is transferred intact to an asparagine residue of a newly forming polypeptide as the growing chain emerges on the luminal side of the endoplasmic reticulum (Figure 6-12). After protein synthesis has been completed, the finished polypeptide chain is released into the lumen and travels to the Golgi complex, where the core oligosaccharide chains are modified by the removal and addition of specific carbohydrate residues. Although the bulk of the carbohydrate component is generally added while proteins still reside within the endoplasmic reticulum, some carbohydrate additions do occur in the Golgi complex, especially those involving the attachment of carbohydrates to the hydroxyl groups of the amino acids serine, threonine, and tyrosine.

THE GOLGI COMPLEX AND CELL SECRETION

In 1898 the Italian cytologist Camillo Golgi discovered that when cells of the nervous system are exposed to osmium tetroxide, this metal stain is deposited in the cytoplasm as a threadlike network surrounding the nucleus. Similar results were soon obtained with silver nitrate and other heavy-metal salts, and the stained region soon came to be known as the *Golgi body* (Figure 6-13). Although this staining reaction was observed in a variety of cell types, the method often failed to give consistent results and marked variations in staining pattern were evident. Since living cells viewed by phase microscopy exhibit no structure resembling the material stained by heavy metals, many cytologists viewed the region as an artifact of the staining process. In fact, the well-known Spanish microscopist Santiago Ramón y Cajal had seen the silver-stained material several years prior to Golgi, but had not reported the finding because, as he was to write later, "the confounded reaction never appeared again!"

The controversy surrounding the Golgi body was not settled until the advent of the electron microscope. The enhanced resolution afforded by this instrument revealed that wherever light microscopists had seen heavy-metal staining, a characteristic set of flattened

FIGURE 6-13 Light micrograph of a silver-stained Golgi complex in nerve cells of the rat. Courtesy of R. G. Kessel.

membranous vesicles is present (Figure 6-14). The authenticity of these membranes, now referred to as the **Golgi complex** (**dictyosome** in plant cells), was therefore verified. The variation in heavy-metal staining that had confused light microscopists can be explained by the fact that the number, shape, and location of the Golgi membranes vary considerably among cell types. The Golgi complex can take the form of multiple independent units, a single mass localized near the nucleus, or a widely dispersed, interconnected network.

In all cases, however, the Golgi complex can be recognized by one constant morphological feature: the presence of a stack of flattened, smooth-membrane-bounded sacs (see Figure 6-2). These tiny "saccules" are separated from each other by spaces of a few dozen nanometers and vary in number from a few to 20 or more. The membranes of the saccules are often curved, giving an overall cuplike shape to the organelle. In addition to the central stack of flattened saccules, Golgi complexes have a variable number of vesicles and membranous channels associated with them. Small vesicles are often clustered around the saccules, appearing as though they were being pinched off from the Golgi membranes. These vesicles usually contain a faintly granular material that becomes denser as the vesicles enlarge and become farther removed from the central stack. Occasionally, the saccules are dilated with a material of similar granular composition, reinforcing the

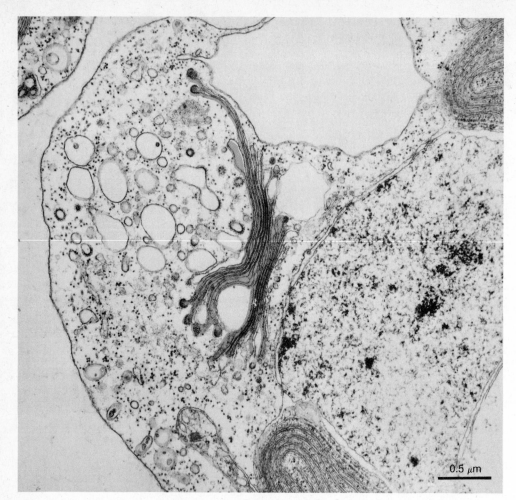

FIGURE 6-14 Thin-section electron micrograph of a golden-brown alga showing the prominent Golgi complex (color) with flattened membranous vesicles lying close to the nuclear envelope. Courtesy of M. J. Wynne.

0.5 μm

notion that the vesicles arise by budding from the Golgi region. The dilation of the saccules and proliferation of vesicles may become so extensive under some physiological conditions that the regular stacked appearance of the Golgi complex is largely obliterated.

The Golgi complex exhibits an inherent polarity in that saccules located at opposite ends of the stack differ from each other in size, shape, content, number of associated vesicles, and enzymatic activity. For convenience, the end of the stack receiving newly synthesized proteins from the ER is termed the **cis** region, while the opposite face is termed the **trans** region. In cells where the Golgi membranes are curved the *cis* face is easily identified because of its location on the convex, outer surface of the complex. For this reason the *cis* face is often referred to as the convex or outer surface, while the *trans* side is called the inner or concave surface (Figure 6-15). When the shape of the complex is not so distinctive, other criteria may be used. The saccules located at the inner surface, for example, generally contain a greater quantity of granular material and are associated with a larger number of vesicles than the saccules of the outer face.

The enzymatic composition of Golgi membranes has been studied by both cytochemistry and subcellular fractionation. The most reliable cytochemical marker for the Golgi complex is the enzyme *thiamine pyrophosphatase* (see Figure 6-15). Despite the fact that the physiological function of this enzyme is unknown, it is an especially useful marker for Golgi membranes because it is rarely found in other organelles. *Nucleoside diphosphatase*, another fairly consistent component of Golgi membranes, has also been employed as a marker, though it may be present in the endoplasmic reticulum as well.

Golgi membranes can be isolated and purified by a combination of moving-zone and isodensity centrifugation, using enzymatic assays as well as direct electron microscope observation to confirm the identity of Golgi-containing fractions (Figure 6-16). Biochemical analyses of isolated Golgi fractions have revealed the presence of several enzymes whose presence had not been suspected from cytochemical tests. Most prominent of these are *glycosyltransferases*, which catalyze the attachment of sugar residues to protein chains, and *glucan synthetases*, which catalyze polysaccharide biosynthesis. As we shall see shortly, the presence of these enzymes in the Golgi complex is related to an important role played by this organelle in carbohydrate metabolism.

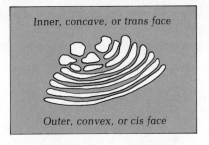

Inner, concave, or trans face

Outer, convex, or cis face

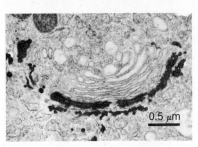

0.5 μm

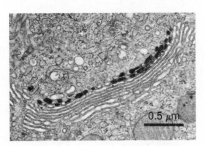

0.5 μm

FIGURE 6-15 The polarity of the membranes of the Golgi complex is shown in these electron micrographs of tissue taken from the mouse epididymis. *(Middle)* Tissue stained with osmium, which is preferentially deposited in the *cis* saccules. *(Bottom)* Tissue stained cytochemically for the Golgi enzyme thiamine pyrophosphatase, which is preferentially localized in the *trans* saccules. Courtesy of D. S. Friend.

FIGURE 6-16 Electron micrographs of Golgi complexes isolated from plant and animal cells. *(Left)* Thin-section electron micrograph of Golgi complexes isolated from onion tissue. Courtesy of H. H. Mollenhauer. *(Right)* Golgi membranes isolated from rat liver cells. These complexes are composed of a series of platelike structures each containing a central discoid cisterna (C) surrounded by a network of branching and anastomosing tubules (T). Depressions at the periphery of the cisternae (arrows) represent the site of fusion of tubules with the cisternal membranes. Courtesy of J. M. Sturgess.

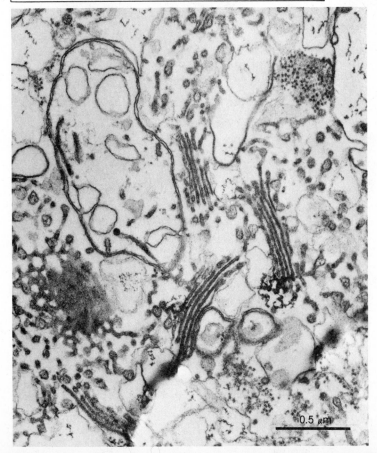

0.5 μm

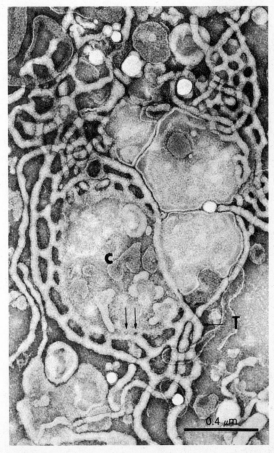

C

T

0.4 μm

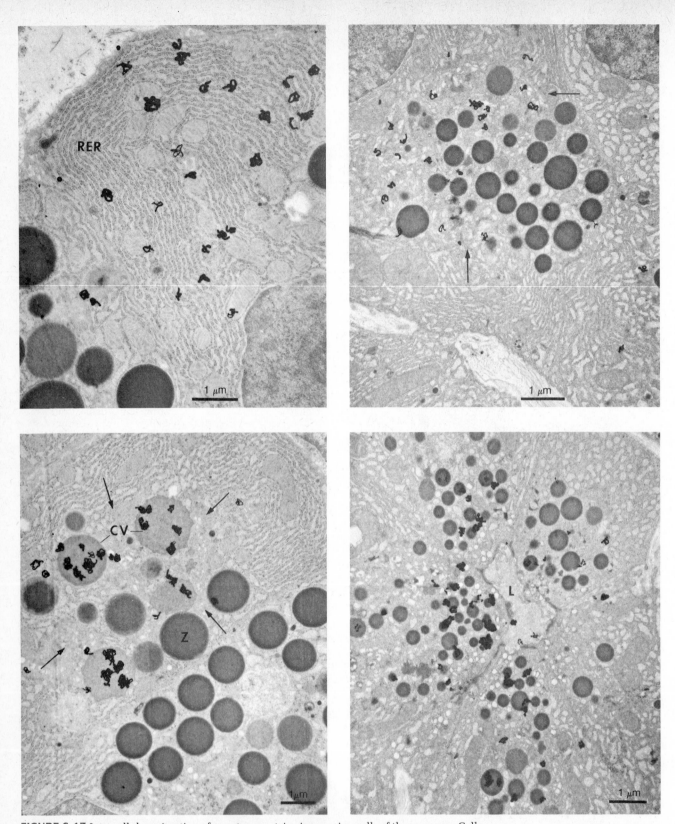

FIGURE 6-17 Intracellular migration of secretory proteins in exocrine cells of the pancreas. Cells were incubated for varying periods with the radioactive amino acid, leucine, and were then prepared for electron microscopic autoradiography. *(Top left)* After 3 min of incubation the silver grains, representing newly incorporated radioactivity, are localized almost exclusively over the rough endoplasmic reticulum. *(Top right)* After 7 min the majority of the newly synthesized radioactive protein has moved to the periphery of the Golgi complex. *(Bottom left)* After 37 min the silver grains are concentrated over the condensing vacuoles (CV). Arrows indicate the periphery of the Golgi complex. Zymogen granules (Z) are still unlabeled at this point. *(Bottom right)* After 117 min the radioactivity is localized primarily over zymogen granules situated near the apex of the cell, and some radioactive protein has even been secreted out of the cell into the lumen (L) of the gland. Courtesy of J. D. Jamieson.

Processing of Proteins for Secretion

The idea that the Golgi complex is intimately involved in the process of cell secretion originated with the early light microscopists, who observed both that vacuoles destined for expulsion from the cell originate in the Golgi region, and that cells with intense secretory activity stain strongly with the heavy-metal impregnation procedure first used to identify the Golgi complex. The increased resolution afforded by the electron microscope provided further support for this view by revealing that vesicles associated with the *trans* face of the Golgi complex contain a granular material similar to that seen in vesicles destined for secretion. Electron microscopy alone, however, gives a static picture of the cell at any given instant, and so cannot be used to follow the dynamic flow of materials. Direct proof that substances destined for secretion from the cell pass through the Golgi complex has therefore required the use of radioactive substrates to trace the movement of macromolecules through the various cell compartments.

In a classic series of investigations of this type, George Palade and his collaborators Lucien Caro and James Jamieson investigated the secretion of digestive enzymes by exocrine cells of the pancreas. Intact guinea pigs or isolated slices of pancreas from these animals were briefly exposed to the radioactive amino acid ^{3}H-leucine to label newly synthesized proteins. Electron microscopic autoradiographs revealed that in the first few minutes after exposure to ^{3}H-leucine, radioactive protein is associated with the rough endoplasmic reticulum. Within 10 min, however, it appears in the peripheral regions of the Golgi complex, at 30 min it is present in large Golgi vesicles, and after an hour, most of the radioactivity is localized in the large zymogen granules that contain protein destined for secretion (Figure 6-17). Shortly thereafter, radioactive protein is released from the cell. By counting the number of silver grains localized over various regions of the cell at successive time intervals after labeling, the passage of newly synthesized protein through the various cellular compartments can be viewed graphically. As illustrated in Figure 6-18, such data support the conclusion that secretory proteins are synthesized on the rough endoplasmic reticulum and then pass through the Golgi complex, into mature secretory granules, and finally out of the cell.

This conclusion, based on the use of autoradiography, has been reinforced by data obtained from subcellular fractionation experiments. In such studies cells are briefly exposed to radioactive amino acids, and are then homogenized and fractionated rather than examined microscopically. Results from experiments of this type have revealed that radioactive proteins appear first in the rough ER fraction, then in the smooth membrane fraction (which contains the Golgi membranes), and finally in large secretory granules (Figure 6-19).

FIGURE 6-18 Experimental data revealing movement of newly synthesized protein molecules through intracellular compartments in (a) guinea pig pancreas and (b) rabbit salivary gland. The number of autoradiographic silver grains found in various cell regions was counted at selected times after administration of radioactive amino acids to label newly synthesized proteins. The data reveal a general pattern in which newly made proteins pass from the rough endoplasmic reticulum to the Golgi complex, and finally to secretion granules.

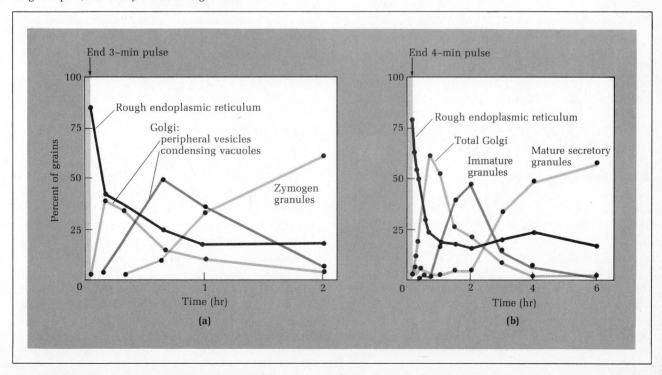

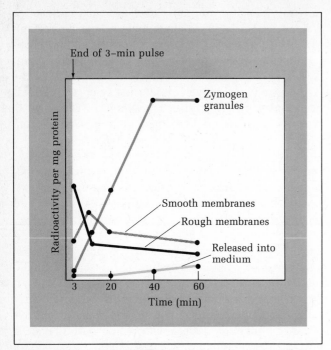

FIGURE 6-19 Experimental data from a subcellular fractionation experiment showing movement of newly made proteins through various cellular compartments. Isolated guinea pig pancreas was incubated with radioactive amino acids for 3 min to label newly made protein molecules. The radioactive isotope was then removed and incubation continued for various time periods. Analysis of the radioactivity present in isolated subcellular fractions reveals that radioactivity first appears in the rough ER membranes, followed by the smooth membrane fraction (which contains the Golgi complex), and finally zymogen granules. Release of radioactive protein from the cells occurs only after the zymogen granules have become radioactive.

On the basis of experiments such as those described above, it has been proposed that six basic steps are involved in cell secretion: (1) synthesis of secretory polypeptides on ribosomes attached to the rough endoplasmic reticulum, (2) segregation and modification of secretory product inside the cisternal space, (3) transport of product to the Golgi complex, (4) sorting, concentrating, and modifying the secretory products in the Golgi complex, (5) storage of the final product in secretion granules, and (6) discharge of the secretory product from the cell.

The first two steps in this scheme have already been discussed in the previous section devoted to the endoplasmic reticulum, although the molecular mechanisms involved in the process of protein synthesis (step 1) will not be described in detail until Chapter 12. The third step in the secretory pathway, namely, transport of product to the Golgi complex, appears to be mediated by **transitional vesicles** that bud off from the ER, move through the cytoplasm, and fuse with the *cis* face of the Golgi complex (Figure 6-20). The net result is transfer of the secretory protein from the cisternal space of the ER to the cisternal space of the Golgi complex. This transfer step is not blocked by inhibitors of protein synthesis and

so does not depend on protein accumulating in the cisternal space and generating a concentration gradient that drives the passage of material out of the endoplasmic reticulum. If ATP synthesis is inhibited, however, movement of proteins out of the endoplasmic reticulum stops, indicating the existence of an energy-requiring step.

The fourth step in the secretory pathway involves sorting, concentrating, and modifying the protein products present within the Golgi complex. A variety of different proteins with differing destinations are transported from the ER to the *cis* face of the Golgi complex, including many ER membrane proteins that are not destined for secretion at all. The first task of the Golgi is therefore to return these latter proteins to the ER. The recent finding that such ER proteins are present in large concentration in the *cis,* but not in the *trans,* Golgi regions suggests that one reason for the stacked arrangement of membranes within the Golgi complex may be to facilitate the gradual removal of ER proteins. A stacked arrangement of membranes would be expected to improve the overall efficiency of this process by making it a multistage event, successively repeated on purer and purer protein fractions. The material reaching the *trans* face would therefore be enriched in protein destined for secretion, though we shall see later that proteins destined for membranous organelles like lysosomes and the plasma membrane also follow this same route.

As proteins destined for secretion pass through the Golgi complex, they are concentrated and condensed into a highly compact form suitable for secretion. The particular region of the Golgi complex involved in concentrating secretions varies among cell types and organisms. In guinea pig pancreas, for example, large vesicles at the Golgi periphery, called **condensing vacuoles,** are the usual site for concentrating protein into granules. Other species employ the distal flattened saccules and/or the dilated margins of the Golgi stack for the same purpose (Figure 6-21). In either case, the net result is a highly concentrated secretory product contained within membrane-enclosed vesicles emerging on the *trans* side of the Golgi complex. These small secretory granules in turn fuse with one another to form larger particles referred to as **zymogen granules.**

Conditions that alter active transport by ion pumps have not been found to alter the concentration process, suggesting that a mechanism other than active transport is responsible for the gradual dehydration that occurs in condensing vacuoles. An alternative possibility is that large sulfated peptidoglycans, known to be formed in the Golgi complex, bind to secretory proteins and trigger the formation of insoluble aggregates. The net effect of forming such aggregates would be to lower the effective concentration of osmotically active particles (i.e., individual protein molecules) within the condensing vacuoles. This lowered osmotic concentration would be

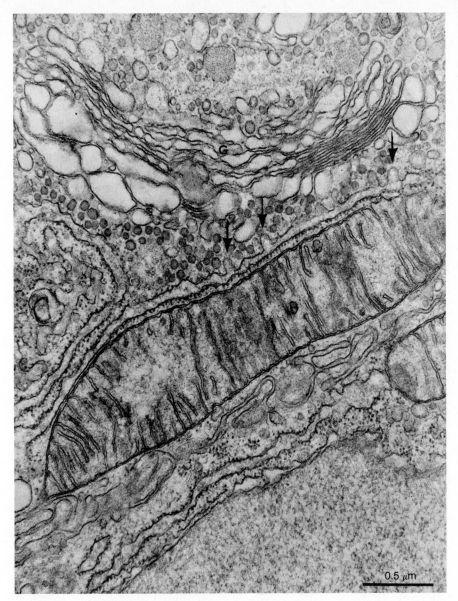

FIGURE 6-20 Electron micrograph showing transitional vesicles (color) budding from the endoplasmic reticulum (arrows) and moving to the *cis* face of the Golgi complex (G). Note that the surface of the endoplasmic reticulum is studded with ribosomes on the side next to the mitochondrion but lacks ribosomes on the side facing the Golgi membranes. Courtesy of D. S. Friend.

0.5 μm

expected to cause water molecules to flow from the vacuoles into the surrounding cell sap, effectively concentrating the secretory proteins present within the vacuole.

Besides its role in sorting and concentrating protein products, the Golgi complex carries out modification reactions on the molecules passing through it. For example, fatty acid residues are added to certain protein molecules to convert them to lipoproteins, and specific carbohydrate groups are added to and removed from other proteins. That the Golgi complex plays an important role in the covalent attachment of carbohydrate side chains to protein chains was first clearly established by studies carried out by Marian Neutra and C. P. Leblond on mucous-secreting cells of the intestine. The predominant constituent of intestinal mucous is a glycoprotein whose extensive carbohydrate component accounts for the viscosity of this secretion. To determine the subcellular site where attachment of the carbohydrate occurs, rats were injected with ³H-glucose to label carbohydrate chains, and the fate of the radioactive chains was then traced autoradiographically. In the first few minutes most of the radioactive label was found in the Golgi region. By 40 min it had become predominantly localized in mucous granules adjacent to the Golgi complex, and by 4 hr the labeled granules had migrated to the cell apex for secretion (Figure 6-22).

These results provided the first indication that the Golgi complex is involved in adding carbohydrate residues to proteins synthesized in the rough endoplasmic reticulum. Further support for this conclusion was provided later by the discovery of glycosyltransferases in isolated Golgi membranes. The relationship between the protein glycosylation reactions occurring in the Golgi complex and those taking place in the ER has been clarified by studies carried out on the synthesis of glyco-

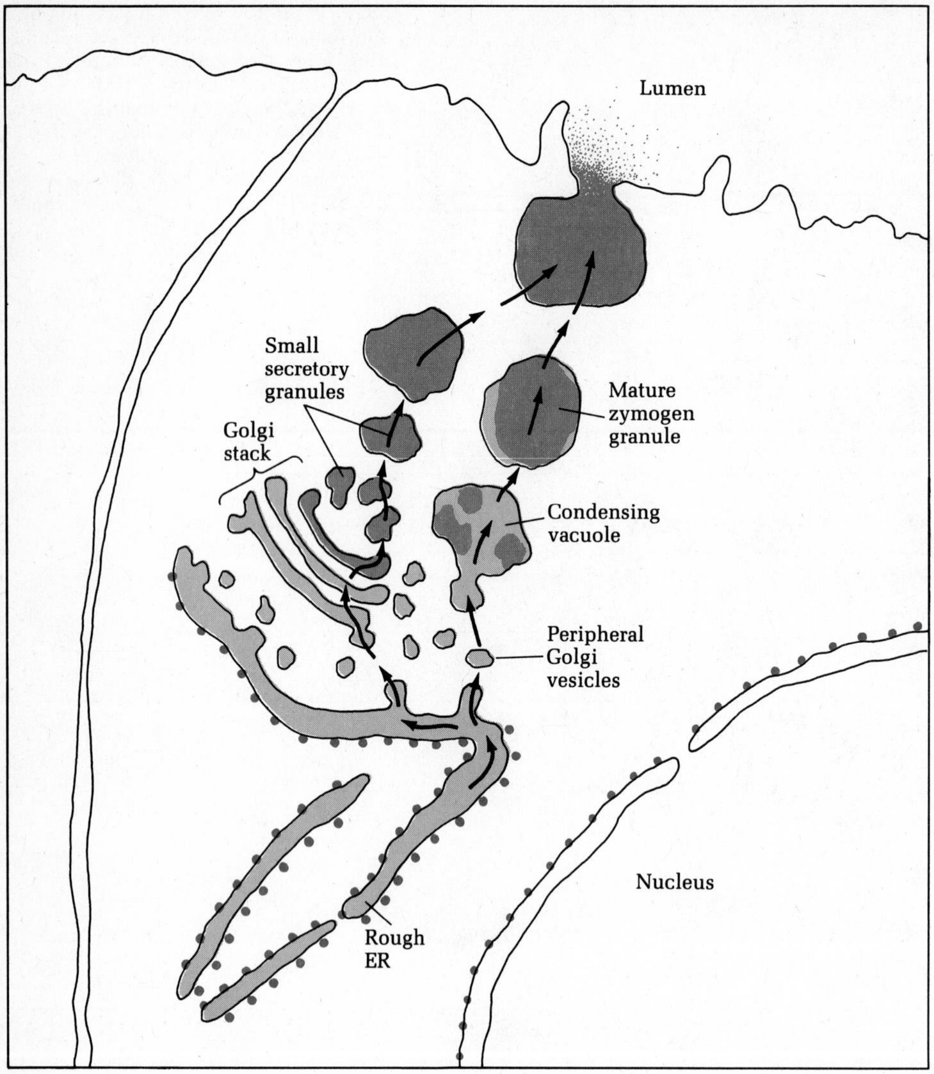

FIGURE 6-21 Major routes of intracellular transport of proteins destined for secretion.

proteins whose carbohydrate sequences are known. In the glycoprotein secreted by thyroid cells, for example, radioactive mannose has been found to be added in the ER while radioactive galactose is incorporated in the Golgi complex. It is known from structural studies that the mannose residues are located close to the polypeptide backbone, while the galactose residues are situated at the terminal ends of the oligosaccharide chains. Hence glycosylation appears to occur by a stepwise pathway in which sugar residues initiating the carbohydrate chains are attached to the polypeptide backbone in the ER, while the more terminal sugars are added after transport of the protein to the Golgi complex.

After emerging from the Golgi complex, secretory products enter the fifth stage in the process of cell secretion: storage in secretion (zymogen) granules. Secretion granules appear in electron micrographs as membrane-enclosed vesicles of varying sizes with granular contents of differing densities. This variability suggests that fusion of vesicles and concentration of protein products continues after emergence from the Golgi complex, finally giving rise to mature zymogen granules. These mature granules are either stored in the cytoplasm until an appropriate signal triggers their release from the cell, or they are discharged immediately.

The sixth and final stage in the secretion process involves discharge of the contents of secretion granules from the cell. Although a variety of different neurotransmitters and hormones can stimulate granule discharge, they all share one feature: the ability to increase the intracellular concentration of calcium ions. This increase in intracellular Ca^{2+} may result either from an increased permeability of the plasma membrane to Ca^{2+} or the release of calcium ions from a sequestered form within the cell. In either case the mechanism by which increased levels of free intracellular Ca^{2+} trigger the subsequent discharge of secretion granules is uncertain.

Granule discharge usually occurs at a particular location on the cell surface. In glandular cells, for example, discharge occurs only at the region of the cell facing the lumen. The way in which granules are guided to the appropriate region of the cell surface is not clear, but a clue comes from the observation that agents that disrupt microtubules can also inhibit secretion. Whether

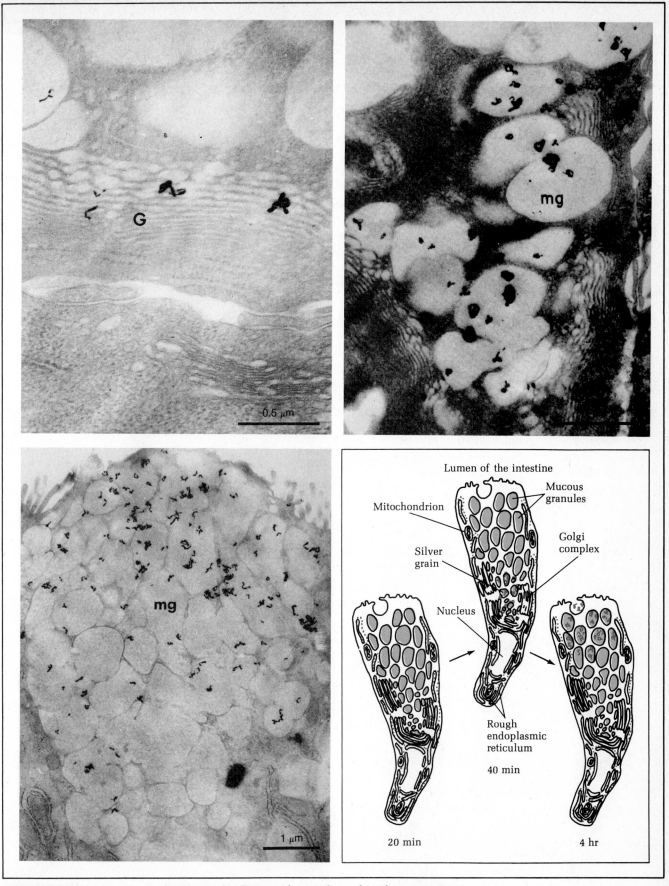

FIGURE 6-22 Electron micrographs showing localization of autoradiographic silver grains in rat intestinal cells at varying times after injection of ³H-glucose. *(Top left)* Twenty minutes after injection. *(Top right)* Forty minutes after injection. *(Bottom left)* Four hours after injection. These pictures reveal that glucose is initially added to secretory proteins in the Golgi complex (G), and that these proteins then appear in the mucous granules (mg) prior to secretion from the cell. Courtesy of C. P. LeBlond.

microtubules passively guide secretion granules along a specific pathway or actively propel the vesicles to the cell surface is yet to be determined.

When the secretion granule reaches the cell surface, its internal contents are expelled into the extracellular space by **exocytosis** (see Figure 6-33), a process involving the fusion of the granule membrane with the plasma membrane. The two conjoined membranes break open at the point of fusion, releasing the contents of the granule into the external medium without disrupting the continuity of the membrane barrier surrounding the cell (Figure 6-23). Because the membrane of the original secretion granule becomes incorporated into the plasma membrane during this process, the question arises as to what prevents uncontrolled expansion of the plasma membrane in secretory cells. At least part of the answer is to be found in the existence of a recycling process in which membrane vesicles invaginate and bud off from the cell surface. These internalized membrane vesicles often migrate back to the region of the Golgi complex, where they fuse with the Golgi saccules or their associated membranes. Cell secretion can therefore be viewed as a continuous cycle in which membranous elements are translocated back and forth between the Golgi complex and the cell surface, all the while maintaining the individuality of the various membranes involved (Figure 6-24).

The six steps that comprise the process of cell secretion (synthesis, segregation, transport, sorting/concentration/modification, storage, and discharge) provide a basic framework for understanding secretion in a broad spectrum of cell types. This does not mean, however, that variations in certain steps do not occur. In fibro-

FIGURE 6-23 Electron micrograph of rat pancreas cell showing secretory granules at different stages extruding their contents into the extracellular space (arrows). Courtesy of L. Orci.

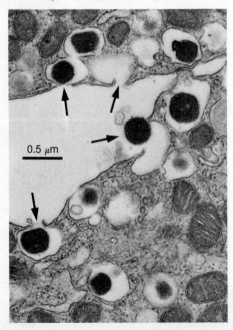

0.5 µm

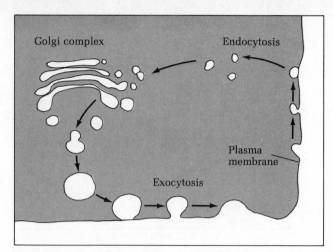

FIGURE 6-24 An overview of the role of endocytosis and exocytosis in cellular membrane flow.

blasts and plasma cells, for example, the secretion products are not concentrated as extensively as in other cell types. Phagocytic cells, on the other hand, modify the discharge step, emptying the contents of cytoplasmic granules into phagocytic vacuoles rather than into the extracellular environment. In spite of such differences, however, the underlying sequence of events is remarkably similar among different cell types.

Because most eukaryotic cells contain at least one Golgi complex, the question arises as to whether all cells are secretory. It is gradually becoming apparent that secretion of macromolecules is a more widespread activity of cells than was once thought. All plant cells are secretory in the sense that they discharge polysaccharides and proteins outside the plasma membrane to form the cell wall. Multicellular animals contain cells that release digestive enzymes, hormones, neurotransmitters, plasma proteins, antibodies, and extracellular matrix materials such as collagen and elastin. In those cells where obvious secretion does not occur, we shall see below that the Golgi complex may play a role in the production of lysosomes or other intracellular vesicles and granules. Such organelles are formed by the same basic pathway as are secretory granules, but their contents are utilized internally rather than externally. When secretion is viewed in this general way, the various activities of the Golgi complex can be seen to have a common unifying basis.

Processing of Proteins for Intracellular Use

In addition to processing and packaging of protein products destined for secretion, the Golgi complex handles two kinds of proteins retained by the cell. One category encompasses proteins that ultimately become incorporated into cellular membranes, most notably the plasma membrane. Because the process of membrane biogenesis involves interactions between several cytoplasmic

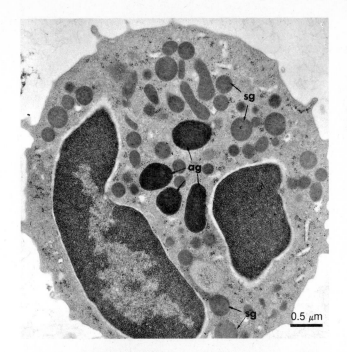

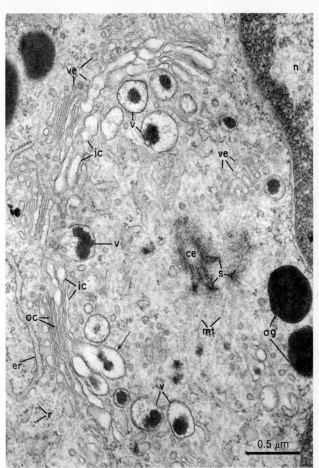

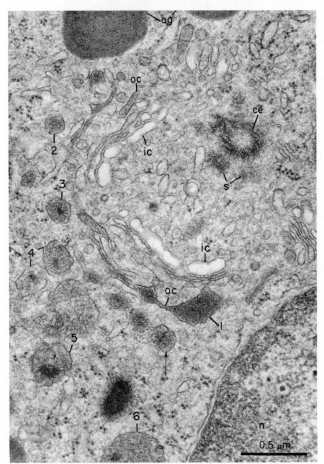

FIGURE 6-25 The production of azurophil and specific granules in polymorphonuclear leukocytes. *(Top left)* Mature polymorphonuclear leukocyte showing four azurophil granules (ag) and many smaller specific granules (sg) in the cytoplasm. *(Top right)* Immature polymorphonuclear leukocyte showing the outer cisternae (oc) and inner cisternae (ic) of the Golgi complex. Vacuoles (v) destined to form azurophil granules are seen budding from the *cis* face of the complex. Several mature azurophil granules are also present outside the Golgi zone. *(Bottom left)* Formation of specific granules from the *trans* face of the Golgi complex. Small granules first accumulate along the *trans* membranes (labeled 1, 2, and 3), and then merge to form larger granules (designated 4, 5, and 6). Courtesy of D. F. Bainton.

membrane systems, it will be discussed as a separate topic later in the chapter. The other category involves proteins destined for incorporation into granules that function intracellularly. The lysosome is the major organelle of this type, but also included in this group are several other kinds of enzyme-, pigment-, and yolk-containing granules.

One of the more striking examples of intracellular granule formation occurs in *polymorphonuclear leukocytes* (*polymorphs* or *neutrophils*), the white blood cells that ingest and destroy invading bacteria. From the extensive studies of Dorothy Bainton and Marilyn Farquhar, we know that mature polymorphs contain two types of cytoplasmic granules that can be distinguished from each other on the basis of differences in size, density, staining characteristics, and mode of origin (Figure 6-25, *top left*). The **azurophil granules,** so named because of their affinity for the basic dye methylene azure, are large dense granules, roughly 800 nm in diameter, formed relatively early in polymorph development. In contrast **specific granules** are smaller (~500 nm), less dense, and form late in development.

Cytochemical tests for individual enzymes have revealed that azurophil granules are filled with hydrolytic enzymes characteristic of lysosomes. Specific granules, on the other hand, contain alkaline phosphatase and lysozyme. During phagocytosis both kinds of granules contribute to the destruction of ingested bacteria by discharging their enzymes into phagocytic vacuoles containing ingested organisms.

Electron micrographs of developing polymorphs reveal an elegant picture of the specificity of the Golgi complex in producing these two types of granules. The first granules to appear in immature cells are the azurophils, which form from large vacuoles budding off the *cis* face of the Golgi saccules (see Figure 6-25 *top right*). These vacuoles characteristically have a dense central core of material destined to become the contents of the azurophil granules. After the newly developing azurophil granules have matured, the granular condensation disappears from the *cis* face of the Golgi complex. Smaller vesicles, containing a homogeneous granular matrix, then begin to bud off the *trans* Golgi saccules and evolve into specific granules (see Figure 6-25 *bottom left*). The fact that the azurophil granules develop from the *cis* saccules while the specific granules are formed in the *trans* saccules shows the Golgi complex to be a highly specific organelle, discriminating between the various macromolecules it processes, and sorting and packaging them in distinctly different ways.

Polysaccharide Synthesis and Cell Wall Formation

In addition to its role in catalyzing the addition of carbohydrate residues to glycoprotein molecules, the Golgi complex is also involved in the formation of certain polysaccharides. In animal cells hyaluronic acid, chondroitin sulfate, and other sulfated glycosaminoglycans destined for secretion into the extracellular matrix are the major components in this category. In plants, on the other hand, cell wall polysaccharides of the hemicellulosic and pectic types appear to be principally involved. The site of synthesis of cellulose itself has been the source of some controversy. Autoradiographic studies suggest that cellulose synthesis occurs at the plasma membrane – cell wall interface, but the enzyme responsible for cellulose synthesis, β-1,4-*glucan synthetase*, is located principally in the Golgi fraction of cell homogenates. These two apparently conflicting observations can be reconciled by the theory that the Golgi complex concentrates and packages this enzyme into vesicles, which then migrate to the cell surface and fuse with the plasma membrane. In intact cells, this enzyme is apparently active in cellulose synthesis only after it has been incorporated into the plasma membrane.

The relatively high concentration of β-1,4-glucan synthetase in isolated Golgi membranes may provide an explanation for the nucleoside diphosphatase activity routinely detected in the Golgi complex. If isolated Golgi preparations are permitted to age for several days or are treated with detergents, their glucan-synthetase activity decreases while nucleoside diphosphatase activity simultaneously appears. It has therefore been proposed that the nucleoside diphosphatase activity of Golgi membrane preparations is a by-product of the inactivation of glucan synthetase, and has no functional significance of its own.

LYSOSOMES

The identification of the lysosome as a distinct class of membrane-enclosed cytoplasmic organelle was originally derived from an extensive series of subcellular fractionation experiments carried out by Christian de Duve and his associates in the early 1950s. This accomplishment represents a dramatic example of the exquisite power of cell fractionation techniques, for this approach permitted lysosomes to be identified on the basis of their biochemical properties before they had been recognized electron microscopically. It was only after subcellular fractionation experiments provided electron microscopists with information as to what to look for, and where to look for it, that lysosomes came to be identified in tissue sections.

Serendipity, often responsible for important new advances in science, played a significant role in the discovery of lysosomes. De Duve had begun his work with an interest in the effect of insulin on carbohydrate metabolism. He therefore chose to investigate the subcellular localization of glucose-6-phosphatase, the enzyme responsible for the release of glucose into the bloodstream. As a control, he decided to assay for an enzyme not directly involved in carbohydrate metabolism. For this purpose he chose the enzyme acid phosphatase. After homogenization of liver tissue in 0.25M sucrose and differential centrifugation into four fractions enriched in nuclei, mitochondria, microsomes, and cytosol, respectively, de Duve found glucose-6-phosphatase activity to be associated predominantly with the microsomal fraction. This observation was of fundamental importance, for at the time it was widely believed that microsomes were simply fragments of disrupted mitochondria. The fact that glucose-6-phosphatase occurred in microsomes, but not mitochondria, established the unique biochemical identity of the microsomal fraction.

The results of the distribution of acid phosphatase in the subcellular fractions were initially less impressive. The total recovery of acid phosphatase activity in the homogenate was only about 10 percent of that previously known to be present in liver from experiments involving harsher extraction conditions. De Duve's initial reaction was that his assay was not working properly. Since it was too late in the day to repeat the measurements, the homogenate was put in the refrigerator for storage. When the acid phosphatase activity of this stored homogenate was again measured a few days later, the results were tenfold higher than when originally assayed.

Rather than dismissing the initial data as a technical error, as some might have done, de Duve considered the possibility that the acid phosphatase had been somehow "masked" in the original homogenate, and that the enzyme was subsequently activated during storage. A clue as to how this might have occurred emerged from the discovery that mitochondrial fractions stored for several days and then assayed for acid phosphatase exhibited a much greater increase in enzymatic activity than did supernatant fractions assayed under the same conditions. De Duve reasoned that the "masked" form of the enzyme might be physically bound to mitochondria in such a way that its activity could not be detected, but that upon storage the mitochondria became disrupted and released the acid phosphatase into the supernatant in an active form.

This hypothesis was easily tested by recentrifuging the stored mitochondrial preparations that had exhibited the increased enzyme activity. Instead of the acid phosphatase activity sedimenting with the mitochondria, as had occurred when the mitochondria were first prepared, the enzyme activity now appeared in the supernatant. Hence the tenfold increase in enzyme activity observed upon storage was accompanied by conversion of the acid phosphatase from a form that sedimented with mitochondria to a freely soluble form appearing in the supernatant.

Though this discovery was unrelated to de Duve's original interest in insulin (recall that the acid phosphatase assay had only been included as a control), he decided to deviate from his original research because of a feeling that he had stumbled upon something of general importance. This decision proved to be a fruitful one, for he soon discovered that a variety of different conditions cause acid phosphatase activation comparable to that observed with homogenate storage. For example, homogenization in a Waring blender, freezing and thawing, warming to 37°C, and exposure to detergents or hypotonic solutions (Figure 6-26) all appeared to stimulate the acid phosphatase activity present in liver homogenates. Since each of these treatments can result in membrane disruption, de Duve postulated that acid phosphatase is localized within membrane-bound vesicles, and that disruption of vesicle integrity leads to acid phosphatase release. According to this view, the lack of enzyme activity in gently prepared fractions can be explained by the impermeability of the vesicle membrane to substrates used in the assay (Figure 6-27).

De Duve at first assumed that the membrane-enclosed vesicle containing acid phosphatase was the

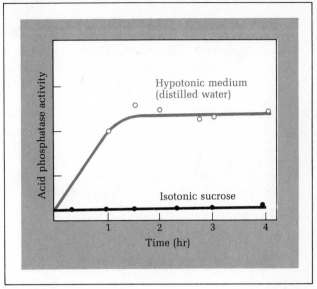

FIGURE 6-26 Experimental data illustrating the ability of hypotonic conditions to promote activation of acid phosphatase in a crude mitochondrial fraction prepared from rat liver.

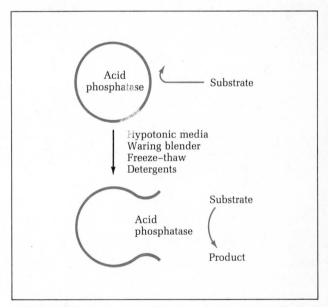

FIGURE 6-27 Model proposed by de Duve to explain acid phosphatase activation by treatments that disrupt membrane vesicles.

mitochondrion, although he was disturbed by the observation that a sizable portion of the acid phosphatase appeared in the microsomal fraction. This problem was soon resolved in an unexpected way when the breakdown of the ultracentrifuge routinely used to prepare the mitochondrial fraction forced one of de Duve's students to use a slower centrifuge. Under these slower centrifugation conditions large quantities of mitochondria could still be obtained, but they were almost totally devoid of acid phosphatase activity. This chance observation suggested to de Duve that the "mitochondrial" fraction as normally prepared includes more than one

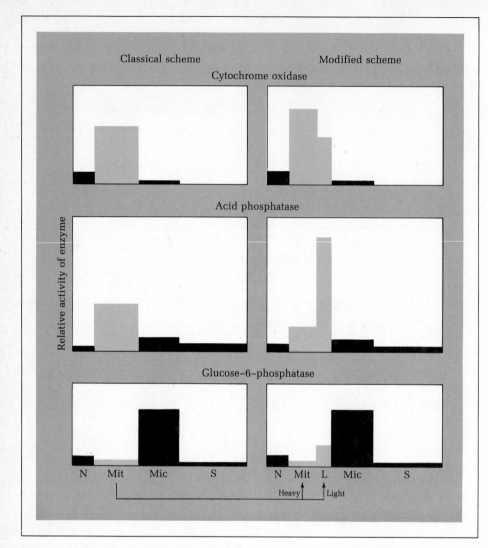

Classical scheme | Modified scheme

Cytochrome oxidase

Acid phosphatase

Glucose–6–phosphatase

Relative activity of enzyme

N   Mit   Mic   S   N   Mit   L   Mic   S

Heavy   Light

FIGURE 6-28 Comparison of the classical four-fraction scheme of subcellular fractionation with the modified five-fraction scheme introduced by de Duve. The major difference is the subdivision of the crude mitochondrial fraction into "heavy" and "light" components. Note that the modified scheme enhances the separation of particles containing cytochrome oxidase (a mitochondrial enzyme) from those containing acid phosphatase (a lysosomal enzyme). Abbreviations: N = nuclear fraction; Mit = mitochondrial fraction; L = lysosomal fraction; Mic = microsomal fraction, and S = supernatant (cytosol or cell sap fraction).

type of organelle that can be separated from one another by altering the centrifugation conditions.

To test this idea, de Duve revised the standard subcellular fractionation scheme to subdivide the mitochondrial fraction into "heavy" and "light" components based on sedimentation at lower and higher centrifugal forces, respectively (Figure 6-28). Under these conditions the "heavy" fraction was found to be enriched in typical mitochondrial enzymes (e.g., cytochrome oxidase), so it was concluded that this fraction consists of purified mitochondria. The light fraction, on the other hand, was found to be enriched in acid phosphatase and four other enzymes (ribonuclease, deoxyribonuclease, β-glucuronidase, and cathepsin) that can be activated by treatments similar to those that activate acid phosphatase. Because these five enzymes are all hydrolytic, de Duve concluded that they are enclosed within a special type of vesicle designed to perform some type of intracellular digestive function. In 1955 he introduced the term **lysosome** (hydrolytic particle) to describe this new cytoplasmic organelle.

Although the biochemical evidence for the existence of lysosomes was quite impressive, the nagging problem remained that no one had ever identified such an organelle in electron micrographs. These remaining qualms were soon alleviated when Alex Novikoff examined acid-phosphatase-containing subcellular fractions with the electron microscope. Besides a few mitochondria, he saw large numbers of smaller vesicles, several hundred nanometers in diameter, that were absent from mitochondrial preparations devoid of acid phosphatase activity (Figure 6-29). Membrane-enclosed particles with similar morphological features were subsequently identified in thin sections of liver cells. Although unequivocal identification of these particles was hindered by the fact that lysosomes lack a distinctive morphology, this obstacle was soon overcome with the development of cytochemical stains specifically directed toward the localization of acid phosphatase and other lysosomal enzymes (Figure 6-30).

Although de Duve's original list of five lysosomal enzymes has been substantially increased by subse-

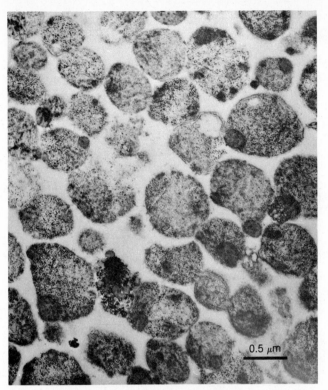

FIGURE 6-29 Electron micrograph of an isolated subcellular fraction enriched in lysosomes. Courtesy of P. Baudhuin.

quent investigations, all lysosomal enzymes still appear to be hydrolases exhibiting optimal activity at acid pH (Table 6-3). Taken together, these enzymes are capable of degrading all major categories of biological macromolecules (proteins, nucleic acids, polysaccharides, and lipids). In order to avoid self-hydrolysis, the enzymes possess a specific structure that renders them relatively resistant to the proteases present in the lysosome.

All lysosomes do not contain the same spectrum of enzymes. Enzymatic analyses carried out on lysosomal subfractions separated by isodensity centrifugation has revealed the presence of a heterogeneous array of particles differing from one another in enzyme content (Figure 6-31). That all lysosomes are not the same is reinforced by the observation that lysosomal enzymes occur in vesicles exhibiting a wide variety of sizes, shapes, and staining characteristics. We will see shortly that at least some of this diversity reflects differences in functional state and stage of development.

De Duve's discovery that lysosomes are a unique class of cytoplasmic vesicles containing a specialized set of hydrolytic enzymes raised two fundamental questions: How are these specialized vesicles formed, and what are their biological functions? These two issues will be addressed in the following sections.

FIGURE 6-30 Localization of lysosomes by cytochemical staining for acid phosphatase activity. *(Left)* Light micrograph showing localization of acid phosphatase-containing lysosomes in mature polymorphonuclear leukocytes. Courtesy of D. F. Bainton. *(Right)* Electron micrograph showing how the acid phosphatase reaction can be used to distinguish lysosomes (arrows) from surrounding mitochondria. Courtesy of P. Baudhuin.

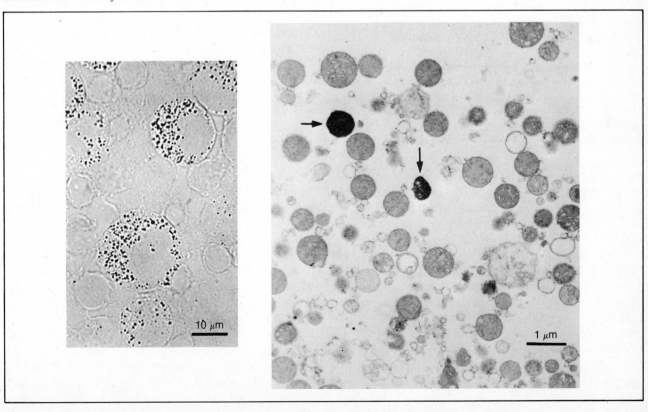

TABLE 6-3

Selected lysosomal enzymes

Enzyme	Natural Substrates
PHOSPHATASES	
Acid phosphatase	Phosphomonoesters
Acid pyrophosphatase	ATP, FAD
Phosphodiesterase	Phosphodiesters
Phosphoprotein phosphatase	Phosphoproteins
Phosphatidic acid phosphatase	Phosphatidic acid
SULFATASES	
Arylsulfatase	Aryl sulfates
PROTEASES AND PEPTIDASES	
Cathepsins (several)	Proteins
Collagenase	Collagen
Arylamidase	Amino acid arylamides
Peptidase	Dipeptides
NUCLEASES	
Acid ribonuclease	RNA
Acid deoxyribonuclease	DNA
LIPASES	
Triglyceride lipase	Triglycerides
Phospholipase	Phospholipids
Esterase	Fatty acid esters
Glucocerebrosidase	Glucocerebrosides
Galactocerebrosidase	Galactocerebrosides
Sphingomyelinase	Sphingomyelin
GLYCOSIDASES	
α-Glucosidase	Glycogen
β-Glucosidase	Glycoproteins
β-Galactosidase	Glycolipids, glycoproteins
α-Mannosidase	Glycoproteins
α-Fucosidase	Glycoproteins
β-Xylosidase	Glycoproteins
α-N-Acetylhexosaminidase	Heparin
β-N-Acetylhexosaminidase	Glycoproteins, glycolipids
Sialidase	Glycoproteins, glycolipids
Lysozyme	Bacterial cell walls
Hyaluronidase	Hyaluronic acid, chondroitin sulfate
β-Glucuronidase	Polysaccharides, mucopoly-saccharides, steroid glucuronides

Formation of Lysosomes

Most of the hydrolytic enzymes present in lysosomes are glycoproteins synthesized on ribosomes bound to the endoplasmic reticulum. Like other proteins synthesized by this mechanism, insertion of the newly forming polypeptide chain through the endoplasmic reticulum is guided by a signal sequence that is subsequently removed, and oligosaccharide chains are transferred from the dolichol-phosphate carrier system to the polypeptide chain as it emerges into the lumen of the endoplasmic reticulum. The major distinction between lysosomal enzymes and other glycoproteins manufac-

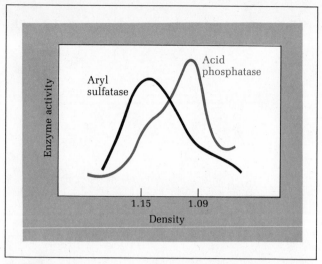

FIGURE 6-31 Isodensity centrifugation of homogenate prepared from fibroblast cells provides evidence for lysosomal heterogeneity. Peak activity of the enzymes acid phosphatase and aryl sulfatase is obtained in particles that differ slightly in density, indicating that these two enzymes are contained in different populations of lysosomes.

tured in this way is that mannose residues situated within the core oligosaccharide chains of lysosomal enzymes become phosphorylated shortly after the oligosaccharide chain has been introduced. The resulting **mannose-6-phosphate** groups serve to distinguish presumptive lysosomal enzymes from other glycoproteins as the various proteins synthesized within the rough endoplasmic reticulum migrate to the Golgi complex.

This recognition of lysosomal enzymes by virtue of their phosphorylated mannose groups is made possible by the fact that some Golgi membranes contain specific receptors that recognize phosphorylated mannose, causing the lysosomal enzymes to become selectively bound to those regions of the Golgi complex. The unique presence of mannose-6-phosphate in lysosomal enzymes thus allows the Golgi complex to distinguish these molecules from other glycoproteins and package them into specialized vesicles destined to become lysosomes. After these vesicles have budded off from the Golgi complex, the internal pH of the vesicles is lowered by the action of an active transport system that pumps protons into the interior of the vesicle. As the pH within the lysosome is lowered, the hydrolytic enzymes are released from their binding sites on the membrane because the interaction between the phosphorylated mannose groups and their membrane receptors is disrupted at low pH. After release into the lysosome interior, the free hydrolytic enzymes tend to lose their phosphorylated mannose groups.

At this point the stage is set for the normal functioning of the lysosome. The realization that all lysosomal enzymes catalyze hydrolysis reactions has naturally fostered the theory that lysosomes serve a degradative or digestive role within the cell. As we shall now see,

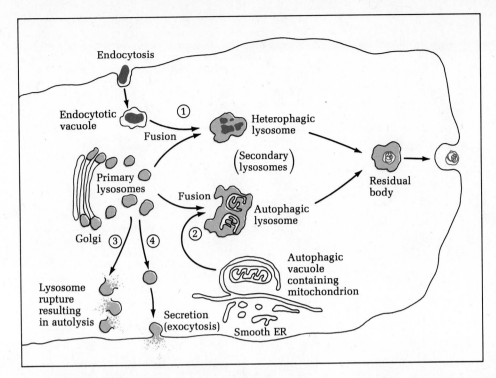

FIGURE 6-32 Schematic representation of the four basic digestive functions of lysosomes. 1 = degradation of material taken up by endocytosis, 2 = autophagy, 3 = autolysis, 4 = digestion of extracellular materials.

there are at least four distinct ways in which this digestive function is utilized by cells (Figure 6-32). These include (1) degradation of foreign matter taken up by endocytosis, (2) destruction of worn-out organelles (*autophagy*), (3) breakdown of cellular structures associated with cell death (*autolysis*), and (4) digestion of extracellular materials.

Endocytosis

One of the most important functions of the lysosome is its ability to catalyze the breakdown of complex macromolecules brought into the cell by the process of endocytosis. Endocytosis refers to the uptake of extracellular materials trapped in membrane vesicles that pinch off from the plasma membrane (Figure 6-33). When the

materials taken up by such vesicles consist of large particles (i.e., those visible by light microscopy), the term **phagocytosis** is applied. **Pinocytosis** refers to the uptake of all other matter, including small particles and water-soluble macromolecules such as antibodies, enzymes, hormones, and toxins.

Endocytosis usually involves an interaction between the substance being ingested and plasma membrane binding sites. Some of the earliest evidence for the importance of such binding was provided by Ralph Steinman and Zanvil Cohn, who investigated the uptake of soluble and insoluble forms of the protein horseradish peroxidase. These studies were carried out using specialized scavenger cells, termed **macrophages,** which are highly active in carrying out phagocytosis. The advantage of using horseradish peroxidase is that

FIGURE 6-33 Diagrammatic representation of endocytosis and exocytosis.

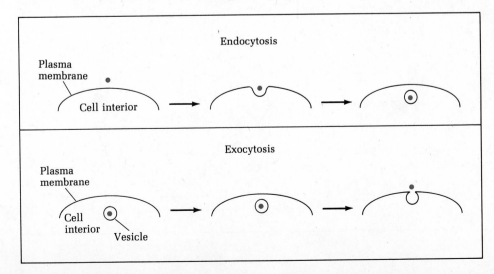

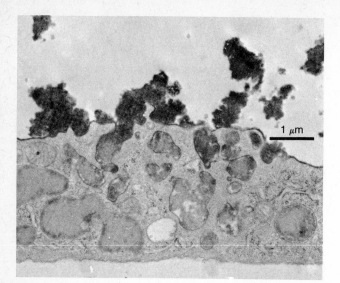

FIGURE 6-34 Electron micrograph showing phagocytosis of horse-radish peroxidase by a mouse macrophage grown in tissue culture. The enzyme produces an electron-dense reaction product that is visible in the electron microscope. Uptake of the electron-dense material into invaginations of the plasma membrane, which then form intracellular membrane-enclosed vesicles, can be clearly seen. Courtesy of R. M. Steinman.

FIGURE 6-35 The formation of coated vesicles. *(Top left)* Electron micrograph of a coated pit in the process of invaginating. *(Top right)* Electron micrograph of an internalized coated vesicle containing gold particles taken up from the extracellular space. The fuzzy edge of the coated vesicle is indicated by an arrow. Courtesy of M. C. Willingham. *(Bottom left)* Isolated cages from calf brain. Courtesy of S. C. Harrison. *(Bottom right)* Diagram of clathrin molecules (color) making up a cage.

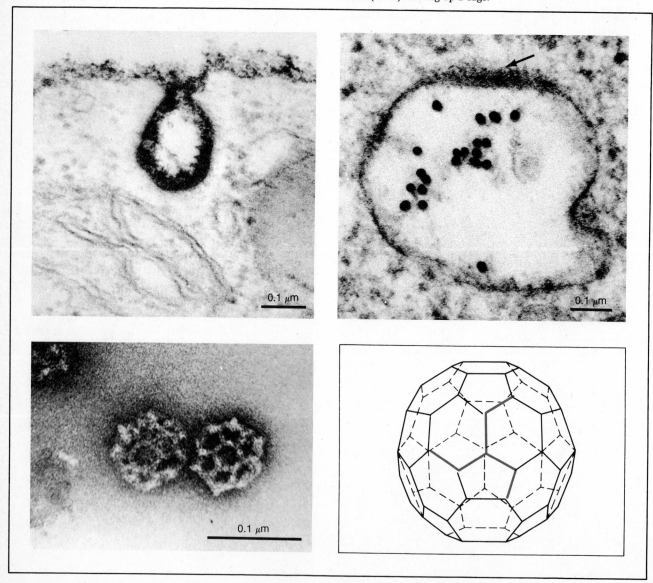

this protein can be readily visualized electron microscopically. When macrophages were incubated with insoluble peroxidase complexes, which bind to the plasma membrane surface, the peroxidase was found to be taken up thousands of times faster than when cells are exposed to soluble peroxidase, which does not bind to the macrophage surface (Figure 6-34). Thus, binding of substances to the exterior surface of the plasma membrane appears to stimulate their uptake by endocytosis.

The conclusion that membrane binding sites play an important role in endocytosis has found further support in the discovery that the binding of various protein hormones, toxins, and other ligands to specific membrane receptors can trigger their subsequent internalization into the cell by endocytosis. In this process of **receptor-mediated endocytosis,** the interaction of ligand with its appropriate plasma membrane receptor is followed by a lateral movement of the receptor-ligand complexes toward specific regions of the plasma membrane known as **coated pits.** The cytoplasmic surface of these pits is coated with a network of molecules consisting predominantly of the protein **clathrin** (Figure 6-35). Clathrin interacts with several other proteins to form a polyhedral cage surrounding the coated pit. After the ligand-receptor complexes have become clustered within the coated pit, it invaginates and pinches off from the plasma membrane, releasing a **coated vesicle** into the interior of the cell. This internalized vesicle then sheds its clathrin coat, forming an uncoated vesicle known as an **endosome.**

The endosome is subsequently processed in one of several different ways, depending on the particular ligand that has been taken up. As illustrated in Figure 6-36, the ligand and receptor may both be degraded, the ligand may be degraded while the receptor is recycled back into the plasma membrane, the ligand and receptor may be transported across the cell and the ligand released on the opposite side, or the ligand may be released on the side of the cell where uptake initially occurred. In the first two pathways, which involve degradation of the internalized ligand, lysosomes play a central role in the degradation process. The role of lysosomes in mediating the breakdown of materials taken up by endocytosis was first shown by Werner Strauss in 1964 in experiments utilizing the enzyme horseradish peroxidase. The location of this enzyme in tissue sections was determined by using a cytochemical procedure that yields a blue reaction product when it reacts with peroxidase. The same tissue sections were also stained by a procedure that gives a red color in the presence of acid phosphatase. When cells were examined by these staining procedures half an hour after exposure to horseradish peroxidase, separate red and blue granules could be visualized within the cell. This indicates that the peroxidase had been taken up by the cell and was present in vesicles (blue) that are distinct from lysosomes. When cells were examined at later times, however, purple granules were seen instead of separate red and blue granules, suggesting that the lysosomes and peroxidase-containing vesicles (endosomes) had fused

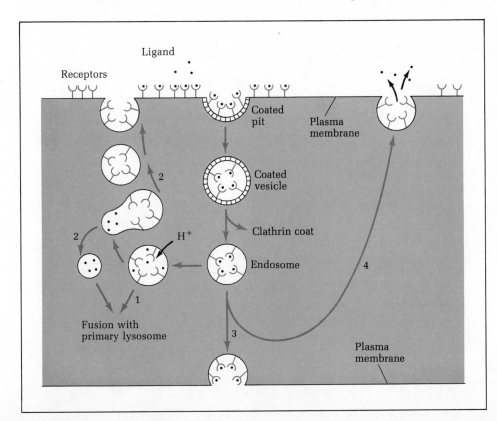

FIGURE 6-36 Alternative pathways of receptor-mediated endocytosis: (1) The incoming ligand and its membrane receptor are both degraded; (2) the incoming ligand is degraded while its membrane receptor is recycled back to the plasma membrane; (3) the incoming ligand and its membrane receptor are both transported across the cell, where the ligand is released from the cell and its membrane receptor is incorporated back into the plasma membrane; and (4) the incoming ligand is released from the cell and its membrane receptor is incorporated back into the plasma membrane on the same side of the cell where uptake originally occurred.

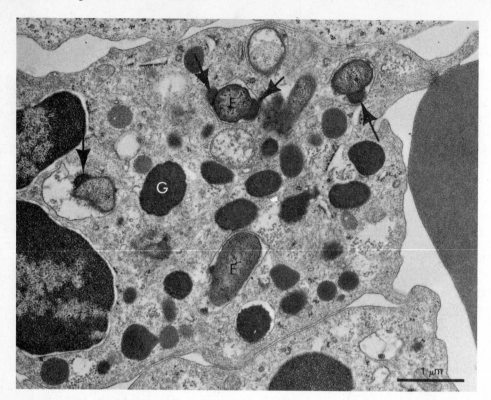

FIGURE 6-37 Formation of a secondary lysosome. A bacterium (E) is enveloped by the plasma membrane of a white blood cell to produce a phagocytic vacuole. The surrounding primary lysosomes (G) then fuse with this phagocytic vacuole to generate a secondary lysosome (arrows). Courtesy of D. Zucker-Franklin.

together. A day later the color of the granules reverted back to red, indicating that the lysosomal enzymes had broken down the peroxidase responsible for the blue color.

The mechanism guiding the fusion of endosomes and lysosomes is not completely clear. Movement of endosomes toward the Golgi region of the cell, where lysosomes are formed and localized, is believed to be guided by microtubules because drugs such as *colchicine*, which disrupt microtubules, cause the movement of endosomes to become disorganized and chaotic rather than selectively directed toward the Golgi complex. Endosomes nearing the Golgi region first encounter **primary lysosomes,** defined as those lysosomes not yet involved in digestive activity. In order for digestion of the contents of the endosome to proceed, the hydrolytic enzymes of the lysosome must be introduced. Since any uncontrolled release of these enzymes into the cytoplasm would result in unwanted digestion of the cell components, some mechanism must ensure the selective fusion of lysosomes with the appropriate vesicles. This interaction is thought to be guided by specific recognition between lysosomal and endosome membranes, leading to membrane fusion and a mixing of the contents of the two vesicles. The new, larger vesicle formed by this fusion is referred to as a **secondary lysosome** (Figure 6-37). The lysosomal enzymes can now hydrolyze the foreign matter originally present in the endosome, releasing small molecules that diffuse through the vesicle membrane and into the cell sap. Undigested substances are retained and accumulate in the secondary lysosome, eventually converting it to a structure

known as a **residual body** (Figure 6-38). Residual bodies ultimately fuse with the plasma membrane and expel their contents from the cell by exocytosis. When this discharge is slow or absent, the resulting accumulation of residual bodies is believed to contribute to cellular aging.

FIGURE 6-38 Electron micrograph showing the formation of residual bodies in a lung cell. Undigestible deposits of colloidal silver initially taken up by pinocytic vesicles (arrows) are seen to accumulate within residual bodies (RB). Courtesy of E. Essner.

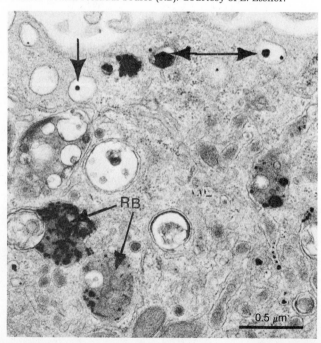

FIGURE 6-39 Scanning electron micrograph of phagocytic macrophages adhering to flattened cells. Courtesy of G. Maul.

Endocytosis can be an exceptionally impressive event, permitting the uptake and degradation of particles whose sizes approach that of the cell itself (Figures 6-39 and 6-40). A dramatic example is the ingestion and destruction of invading bacteria by polymorphonuclear leukocytes. As was discussed in the section on the Golgi complex, the cytoplasm of this cell type contains two kinds of granules, specific granules and azurophil granules (lysosomes). Although "specific granules" are not lysosomes, they contain hydrolytic enzymes active at neutral or alkaline pH that are also useful in degrading foreign matter. When bacteria are taken up by phagocytosis, the specific granules discharge their contents into the vacuoles containing the ingested bacteria to initiate the destruction process. A few minutes later the azurophil granules (lysosomes) discharge their contents into the same vacuoles, releasing massive amounts of hydrolytic enzymes that quickly complete degradation of the invading bacteria. The destruction that occurs is often so massive that the leukocytes themselves die in the

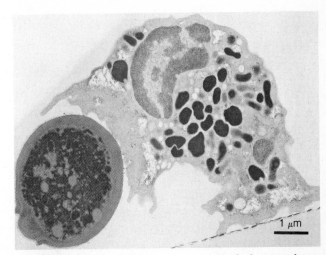

FIGURE 6-40 Thin-section electron micrograph showing phagocytosis of a yeast cell (color) by a polymorphonuclear leukocyte. Courtesy of D. F. Bainton.

process. Although the preceding mechanism is generally quite effective, some bacteria are able to survive and even multiply inside phagocytic cells. In such cases a factor produced by the bacterial cells appears to inhibit fusion of primary lysosomes with phagocytic vacuoles. One suspected candidate for this agent is 3′,5′-cyclic AMP, a compound known to inhibit lysosomal enzyme discharge and to be produced by bacteria capable of growing unharmed in phagocytic cells.

In addition to mediating the degradation of foreign organisms and molecules, endocytosis is also utilized in the uptake of normal extracellular substances. Developing egg cells, for example, employ receptor-mediated endocytosis to accumulate yolk proteins from the bloodstream, storing these proteins for use as nutrients later in embryonic development (see Chapter 15). Another example involves the **low-density lipoproteins (LDL)**, which are cholesterol-containing complexes utilized to transport cholesterol through the bloodstream. The LDL complex binds to specific plasma membrane receptors and is taken up in a process mediated by the binding of LDL to specific LDL receptors present on the cell surface. After the membrane-bound LDL has been taken up into coated vesicles, the clathrin coat is shed and endosomes are formed. The LDL receptors are then recycled to the plasma membrane for reutilization (see Figure 6-36, pathway 2), while the cholesterol component of LDL is released within the cell and utilized for appropriate metabolic purposes, including the feedback inhibition of excessive cholesterol synthesis by the cell itself. An unusual genetic disease, termed *familial hypercholesterolemia*, is caused by an inherited defect in cell surface LDL receptors. Individuals with such a deficiency cannot take up the cholesterol needed to regulate their own synthesis of this substance. As a result excessive amounts of cholesterol are formed and secreted into the bloodstream, leading to severe circulatory disease.

Autophagy and Autolysis

One obvious reason for packaging hydrolytic enzymes within lysosomes is to protect the cell from digestion by a group of enzymes that in combination could degrade every macromolecule and cell structure present. Situations do occur, however, that call for the destruction of cellular constituents. If the need for destruction is selective, such as occurs with aging or damaged organelles, lysosomes can be called upon to selectively degrade these structures. Such selective destruction, or **autophagy,** is useful to the cell because it removes unwanted organelles and because it permits the chemical building blocks generated during the organelle's destruction to be recycled for other purposes. Although the molecular signaling mechanism by which organ-

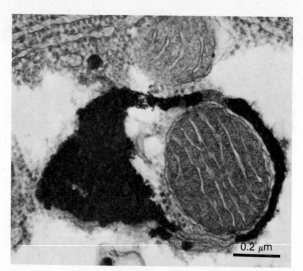

FIGURE 6-41 Electron micrograph showing a mitochondrion encapsulated within an autophagic lysosome that has been identified by cytochemical staining for acid phosphatase (dark material). Courtesy of M. Locke.

elles are marked for destruction is not well understood, the process is known to be initiated by wrapping the organelle targeted for destruction in one or more layers of smooth ER membrane. The resulting vesicle, termed an **autophagic vacuole,** then fuses with primary lysosomes to form a secondary lysosome in which digestion of the organelle can take place. In order to distinguish the resulting secondary lysosomes from those formed during endocytosis, the terms *autophagic* and *heterophagic* lysosome have been introduced. The term *heterophagic* refers to secondary lysosomes containing materials taken up by endocytosis, while the term *autophagic* refers to secondary lysosomes containing materials of intracellular origin. Autophagic lysosomes can often be recognized by the identifiable remnants of cellular organelles contained within them (Figure 6-41).

In some situations complete dissolution of cell structure, rather than the selective elimination of a few organelles, is required. Such self-inflicted cell destruction, or **autolysis,** plays an important role in the development of certain organ systems and body structures. For example, embryonic formation of fingers or toes from undifferentiated blocks of tissue requires selective destruction of the cells located in between the newly forming digits. Likewise, metamorphosis of a tadpole into a frog requires resorption of the tadpole's tail. In both cases tissue destruction is mediated by lysosomal digestion of the constituent cells. Although the mechanism controlling lysosomal enzyme release under such conditions is yet to be unraveled, lysosomes appear to rupture and release their hydrolytic enzymes into the cell sap, precipitating a massive breakdown of cell constituents.

Extracellular Digestion by Hydrolytic Enzymes

In addition to digesting materials contained within the cell, hydrolytic enzymes may in some instances be secreted into the extracellular space. During fertilization, for example, penetration of the sperm head through the outer layers surrounding the egg surface is aided by the release of hydrolytic enzymes derived from the Golgi complex of the sperm cell. Situations also occur, however, where release of hydrolytic enzymes from the cell has detrimental effects. A case in point is the lysosomal enzyme release that often follows cell damage caused by physical trauma or microbial infection. In such instances the released enzymes may trigger additional tissue damage and inflammation. The idea that release of lysosomal enzymes can contribute to tissue inflammation is supported by the discovery that an arthritislike disease can be produced in rabbits by injecting drugs known to disrupt lysosomes. This finding explains why drugs that inhibit lysosomal enzyme release, such as the steroid hormones cortisone and hydrocortisone, are effective antiinflammatory agents. Vitamin A, on the other hand, increases the lability of the lysosomal membrane and promotes discharge of hydrolases from the cell. This explains why connective tissue damage and spontaneous bone fractures often occur in individuals consuming excess quantities of vitamin A.

Lysosomal Storage Diseases

Several dozen genetic diseases afflicting young children are now known to be caused by an excessive intracellular accumulation of polysaccharides or lipids. The quantity of polysaccharide or lipid stored is often massive enough to interfere with and even destroy the cells involved. Depending on the particular cell types affected, symptoms such as muscle weakness, skeletal de-

TABLE 6-4
Selected lysosomal storage diseases

Disease	Symptoms	Substance Accumulated	Enzyme Defect
Type II glycogenosis	Muscle weakness	Glycogen Gluc–α–Gluc	α-Glucosidase
Niemann–Pick disease	Liver-spleen enlargement Mental retardation	Sphingomyelin Cer—P-choline	Sphingomyelinase
Gaucher's disease	Liver-spleen enlargement Bone erosion	Glucocerebroside Cer—β—Gluc	β-Glucosidase
Metachromatic leukodystrophy	Mental retardation	Sulfatide Cer—β—OSO$_3^-$	Sulfatidase
Fabry's disease	Skin rash Kidney failure	Ceramide trihexoside Cer—β—Gluc—β—Gal—α—Gal	α-Galactosidase
Pseudo-Hurler disease	Liver enlargement Skeletal deformities	Ganglioside (GM$_1$) Cer—β—Gluc—β—Gal—β—Gal—Gal NeuAcN NAc β	β-Galactosidase
Tay-Sachs disease (infantile amaurotic idiocy)	Mental retardation Blindness Muscular weakness	Ganglioside (GM$_2$) Cer—β—Gluc—β—Gal—Gal NeuAcN NAc	β-N-Acetylhexosaminidase

Note: Abbreviations: Gluc = glucose; Cer = ceramide (N-acylsphingosine); Gal = galactose; NeuAcN = N-acetylneuraminic acid; NAc = N-acetyl; P = choline-phosphatidylcholine. Dotted lines represent sites of enzymatic cleavage.

formities, and mental retardation may result. The first of these so-called *lysosomal storage diseases* to have its underlying mechanism unraveled was *type II glycogenosis*, an illness whose victims die at an early age with abnormally large amounts of glycogen in the liver, heart, and muscles. In 1963 H. Hers discovered that type II glycogenosis is caused by a severe deficiency of the lysosomal enzyme β-glucosidase, which catalyzes hydrolysis of glycogen to oligosaccharides and glucose. In the absence of this enzyme, undigested glycogen accumulates within lysosomes.

Not only did this discovery provide an explanation for glycogen storage disease, but it also led Hers to postulate the existence of a wide spectrum of other diseases corresponding to genetic defects in particular lysosomal enzymes. He predicted that in each case, the disease symptoms are caused by an abnormal accumulation of the undigested substrates of the defective enzyme. This unifying theory has led to an understanding of the causes of what were once a bewildering array of mysterious diseases (Table 6-4). Although in some instances excessive storage of simple polysaccharides or mucopolysaccharides is involved, most of these storage diseases result from an abnormal accumulation of glycolipids. Because glycolipids are highly concentrated in brain tissue and are important constituents of the myelin sheath surrounding the axon, the symptoms of

these storage diseases usually include severe mental retardation. The glycolipid accumulated in brain cell lysosomes can often be recognized morphologically by its tendency to form unusual layered structures known as *zebra bodies* (Figure 6-42).

The discovery of the molecular basis of the lysosomal storage diseases has led to the suggestion that these illnesses might be alleviated by replacing the defective enzymes. Direct administration of missing lysosomal enzymes, however, is impractical for several reasons. To begin with, destruction of the administered enzymes by serum proteases or by the individual's immune system would minimize the effectiveness of treatment. Furthermore, many cells do not take up foreign molecules efficiently. An alternative approach is to encapsulate the required enzyme in artificial lipid vesicles (liposomes). This encapsulation process protects the enzymes from destruction and has the added advantage that liposomes are known to be actively taken up by cells and fused with lysosomes. In this way a missing lysosomal enzyme might be delivered directly to its appropriate site of action. Encouraging results from animal studies suggest that this approach may ultimately be useful for the treatment of inherited enzyme deficiencies in humans. An alternative therapeutic approach is to attempt to correct or replace the gene coding for the defective lysosomal enzyme. Although such

FIGURE 6-42 Electron micrograph of zebra bodies in a nerve cell showing the distinct layered arrangement of the glycolipid that accumulates in these defective lysosomes. Courtesy of H. Loeb.

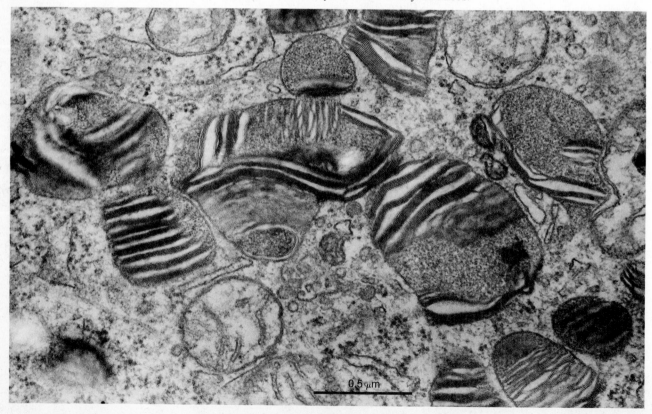

0.5 μm

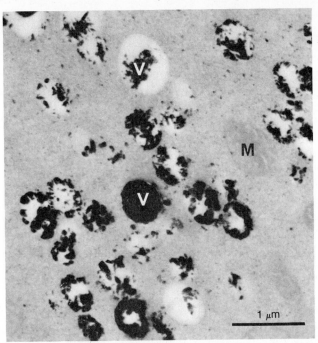

FIGURE 6-43 Demonstration of acid phosphatase activity in spherosomes and larger vacuoles (V) of a plant cell. Note that lead deposits representing the localization of acid phosphatase are absent from mitochondria (M). Courtesy of N. Poux.

ideas are quite speculative at present, recent advances in recombinant DNA technology may ultimately make such treatment feasible.

Spherosomes and Related Vacuoles in Plants

The isolation of lysosomes from plant tissues has been a difficult undertaking because methods developed for animal tissues cannot be directly applied to plant lysosome isolation. Using modified procedures, however, it is now possible to recover acid hydrolase activity in particulate fractions of plant cells. The purified fractions containing these hydrolases are enriched in small, membrane-enclosed vesicles. These structures resemble **spherosomes,** which are highly refractive spherical particles, approximately 0.5–1.0 μm in diameter, originally identified by light microscopists in the plant cell cytoplasm. Cytochemical staining for acid phosphatase in tissue sections has confirmed that spherosomes, as well as larger cytoplasmic vacuoles, contain lysosomal enzymes (Figure 6-43).

A principal feature distinguishing animal lysosomes from plant spherosomes is that the latter stain intensely with fat-soluble dyes, indicating a high lipid content. The accumulation of lipid in vesicles containing hydrolytic enzymes suggests that the spherosome functions in storing and mobilizing reserve lipid. In addition to its role in lipid metabolism, the spherosome and other hydrolase-containing plant vacuoles are thought to be involved in digesting and recycling intracellular constituents in a manner analogous to the animal cell lysosome. Digestion of foreign particulate

matter is probably not a major function of the spherosome, however, because the presence of the plant cell wall generally prevents phagocytosis from bringing in foreign particulate matter.

PEROXISOMES, GLYOXYSOMES, AND OTHER MICROBODIES

Among the enzymes monitored by de Duve in his early studies on lysosomes was the purine-catabolizing enzyme, **urate oxidase.** In spite of its high concentration in lysosomal fractions isolated by differential centrifugation, some minor differences in its distribution pattern relative to that of other lysosomal enzymes suggested that it might be localized in some other organelle. When isodensity gradient centrifugation was employed to enhance the resolution obtained, it became evident that urate oxidase is contained in a particle whose density differs from that of lysosomes (Figure 6-44 *top*). Shortly

FIGURE 6-44 Data illustrating separation of peroxisomes, lysosomes, and mitochondria by isodensity centrifugation. These three organelles are identified by the presence of urate oxidase, acid phosphatase, and cytochrome oxidase, respectively. *(Top)* Small inherent differences between the densities of these three organelles permit their partial separation by isodensity centrifugation of partially purified homogenates. *(Bottom)* Injection of animals with the detergent Triton WR-1339 enhances the separation that can be obtained because this substance is taken up by lysosomes, thereby lowering their density.

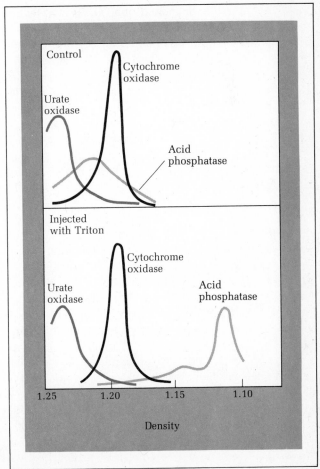

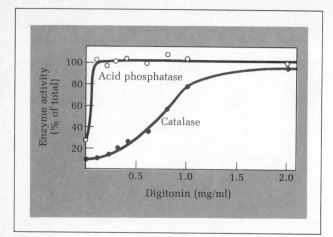

FIGURE 6-45 Differential effects of the detergent digitonin on the release of acid phosphatase from lysosomes, and catalase from peroxisomes. The fact that these two enzymes are not released simultaneously verifies that they are contained within different organelles.

thereafter the enzymes **D-amino acid oxidase** and **catalase,** involved in the formation and breakdown of hydrogen peroxide, respectively, were found to behave the same way as urate oxidase. Because of the involvement of these latter enzymes with hydrogen peroxide metabolism, the organelle containing them came to be called the **peroxisome.**

The conclusion that peroxisomes and lysosomes are distinct organelles received subsequent support from experiments in which the rate of release of acid phosphatase and catalase from detergent-treated subcellular fractions was compared. If these two enzymes were located in the same particle, one would expect them to be released simultaneously as one adds increasing concentrations of detergent to rupture the vesicles in which they are contained. In fact ten times as much detergent is required to liberate catalase as compared to acid phosphatase, indicating that these two enzymes are localized in particles with differing properties (Figure 6-45).

Because of the relatively small difference in density between peroxisomes and lysosomes, isodensity centrifugation does not normally achieve a complete separation of these particles from each other. Better separation can be accomplished, however, by isolating subcellular fractions from animals that have been first injected with the detergent Triton WR-1339. This detergent is selectively accumulated within lysosomes and causes their density to decrease dramatically (see Figure 6-44 *bottom*), making possible the large-scale isolation of peroxisomes for structural and functional studies.

Electron microscopic examination of isolated subcellular fractions enriched in peroxisomes has confirmed the unique identity of these organelles. The small, roughly spherical vesicles present in such fractions resemble structures seen in tissue sections by

early electron microscopists and given the general name **microbodies.** Microbodies vary from 0.2 to 2 μm in diameter and consist of a finely granular matrix surrounded by a single limiting membrane. These structures often contain elaborate crystalloid *cores* (Figure 6-46) containing the enzyme urate oxidase. Microbodies in which such cores are present can clearly be identified as peroxisomes, but in cases where distinctive cores are not present, peroxisomes may be difficult to distinguish from lysosomes and other cytoplasmic vesicles. Unequivocal identification can be made, however, by using the cytochemical staining reaction for catalase. This reaction, based on the ability of catalase to oxidize *diaminobenzidine (DAB)* to an electron-dense reaction product (Figure 6-47), permits the unequivocal identification of peroxisomes. It should be emphasized that the term *microbody* is only a morphological label employed by electron microscopists when they see a "small body." Unless these structures have a urate oxidase core or are stained for catalase, they cannot be clearly stated to be peroxisomes. As shall be seen later, other types of microbodies exist in addition to peroxisomes.

Peroxisomes occur primarily in the liver and kidney cells of vertebrates, in the leaves and seeds of plants, and in eukaryotic microorganisms such as yeast, protozoa, and fungi. The spectrum of enzymes present in peroxisomes varies considerably among tissue sources (Table 6-5). The only enzyme common to all peroxi-

FIGURE 6-46 Electron micrograph of a microbody with crystalloid inclusions in a rat kidney cell. Both rough endoplasmic reticulum and mitochondria are closely associated with the microbody membrane. Courtesy of J. M. Barrett.

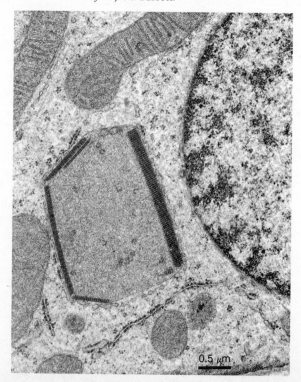

0.5 μm

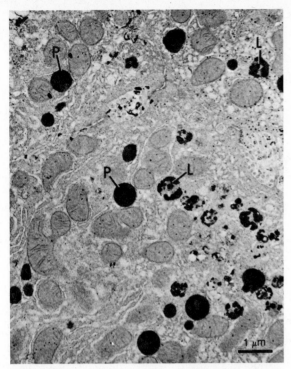

FIGURE 6-47 Thin-section electron micrograph of rat liver stained for both acid phosphatase to identify lysosomes (L), and for catalase to identify peroxisomes (P). Because the cytochemical reaction for acid phosphatase generates a more granular deposit than that produced by the reaction for catalase, lysosomes and peroxisomes are readily distinguished from one another. Courtesy of S. Goldfischer.

somes is catalase, which accounts for up to 15 percent of its protein. In addition to catalase, hydrogen-peroxide-producing oxidases such as urate oxidase, amino acid oxidase, and glycolate oxidase are usually present. These enzymes employ molecular oxygen as oxidizing agent and generate hydrogen peroxide (H_2O_2) according to the following general reaction:

$$RH_2 + O_2 \longrightarrow R + H_2O_2 \qquad (6\text{-}1)$$

TABLE 6-5
Major enzyme activities of peroxisomes

| | Source of Peroxisomes | | |
Enzymatic Activity	Rat Liver	Plant Seedling	Plant Leaf
1. Catalase	+	+	+
2. Urate oxidase	+	+	+
3. D-Amino acid oxidase	+	+	−
4. L-Amino acid oxidase	±	+	−
5. L-α-Hydroxy acid oxidase	+	+	+
6. Glycolate oxidase	+	+	+
7. Fatty acid β-oxidation	+	+	−
8. Citrate synthase	−	+	−
9. Aconitase		+	−
10. Isocitrate lyase	−	+	−
11. Malate synthase	−	+	−
12. Malate dehydrogenase	−	+	+
13. Glyoxylate reductase	+	−	+
14. Glyoxylate transaminase	−	−	+

Note: (+) = present, (−) = absent, (blank) = not looked for.

The H_2O_2 formed by reactions of this type can subsequently be hydrolyzed by catalase acting in one of two alternative ways. In the so-called *catalatic mode*, one molecule of H_2O_2 donates its electrons to another molecule of H_2O_2, forming a molecule of water and a molecule of oxygen:

$$H_2O_2 + H_2O_2 \xrightarrow{\text{Catalase}} O_2 + 2H_2O \qquad (6\text{-}2)$$

In the *peroxidatic mode* the electron donor is a substrate other than H_2O_2:

$$RH_2 + H_2O_2 \xrightarrow{\text{Catalase}} R + 2H_2O \qquad (6\text{-}3)$$

In either case, the net result is the breakdown of hydrogen peroxide.

The hallmark of peroxisomes is therefore the association of H_2O_2-producing oxidases with the H_2O_2-degrading enzyme, catalase. Although the exact functional significance of this association is not certain, some of the more likely possibilities will be discussed below.

Metabolic Functions of Peroxisomes

Because hydrogen peroxide is toxic, it might seem logical to conclude that the peroxisome functions to protect cells from exposure to this substance by compartmentalizing the enzymes that produce and destroy it. The flaw in this argument, however, is that many H_2O_2-producing oxidases occur in the cell sap and in mitochondria, suggesting that the peroxisome contributes little to the protection of the cell from the effects of H_2O_2. This conclusion is further reinforced by the fact that many cells producing H_2O_2 do not even contain peroxisomes. It therefore appears that the need for peroxisomes must be related to the metabolic pathways these organelles carry out rather than with the simple need to protect cells from H_2O_2. Examination of the various enzymes present in peroxisomes has led to consideration of the following possibilities.

1. *Inactivation of toxic substances.* One possible role of peroxisomes may involve the coupling of H_2O_2 degradation to the inactivation of toxic compounds. Catalase is the most active enzyme present in the peroxisome, and therefore degrades hydrogen peroxide at a much faster rate than it is formed by other peroxisomal reactions. The resulting scarcity of hydrogen peroxide favors the peroxidatic mode of catalase action, in which an alternative electron donor is substituted for the second molecule of H_2O_2 (Equation 6-3). The electron donors employed for this purpose include methanol, ethanol, phenols, nitrites, formaldehyde, and formate. Oxidation of these compounds driven by the breakdown of H_2O_2 results in the detoxification of what are otherwise noxious substances.

2. *Regulation of oxygen tension.* Because peroxi-

somal oxidases employ molecular oxygen as an oxidizing agent (Equation 6-1), the reactions catalyzed by these enzymes may have a significant effect on oxygen levels within the cell. In liver cells, for example, as much as 20 percent of the total oxygen consumption is accounted for by peroxisomes. Most of the remaining oxygen consumption occurs in mitochondria, where energy released during the oxidation of organic fuels by molecular oxygen is used to drive formation of the energy-conserving molecule, ATP. Hence **respiration,** a term referring to the oxidation of organic fuels by molecular oxygen, occurs in both mitochondria and peroxisomes. The major difference is that a portion of the energy released during respiration in mitochondria is trapped as ATP, while in peroxisomes the energy is all lost as heat.

Peroxisomal respiration also differs from mitochondrial respiration in its sensitivity to oxygen concentration. Mitochondrial respiration occurs at a maximal rate when the oxygen concentration is around 2 percent, with further increases in oxygen producing no further increase in respiratory rate. The rate of peroxisomal respiration, on the other hand, increases in roughly direct proportion to oxygen tension (Figure 6-48). This disparity gives mitochondria an advantage over peroxisomes in utilizing small amounts of oxygen, but also allows peroxisomes to exceed the respiratory ability of mitochondria at high oxygen concentrations. This property of peroxisomes may help to protect cells from the toxic effects of high concentrations of oxygen. Support for this idea has been obtained from studies carried out on plant cells exposed to bright light to enhance photosynthetic activity. The resulting increase in oxygen production stimulates peroxisomal respiration, thereby helping to consume the excess oxygen. This phenomenon, known as **photorespiration,** will be discussed in more detail in Chapter 8 in the context of our discussion of photosynthesis.

FIGURE 6-48 Graph showing the relationship between oxygen concentration and respiration rate in peroxisomes and mitochondria.

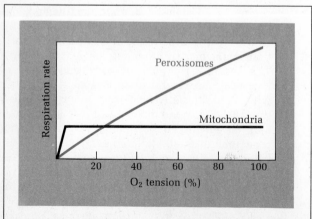

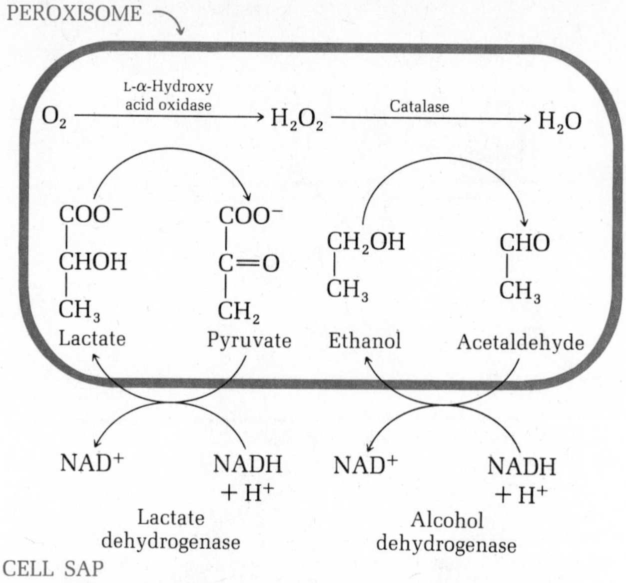

FIGURE 6-49 Mechanism by which peroxisomal oxidations might facilitate the regeneration of cytoplasmic NAD^+ from NADH.

Although photorespiration per se does not occur in animal tissues, an analogous enhancement of peroxisomal respiration takes place when oxygen tension rises in the cell because of a reduction in the mitochondrial respiration rate. If the depression in mitochondrial respiration rate has been caused by a lack of oxidizable substrates, some of these substrates may be regenerated by the peroxisomal oxidation reactions stimulated by the rise in oxygen tension. Such a feedback system permits regulation of mitochondrial respiration by peroxisomal respiration, and vice versa.

3. *Regeneration of cytoplasmic NAD^+.* Many of the oxidized compounds generated by peroxisomal respiration can be converted back to their reduced forms by NADH-dependent enzymes present in the cell sap. For example, pyruvate produced in peroxisomes by the oxidation of lactate can be reduced back to lactate in the cell sap, accompanied by the conversion of NADH to NAD^+ (Figure 6-49). As will be described in Chapter 7, this conversion of NADH to NAD^+ is useful in cases where NADH is being produced in the cell sap by glycolysis faster than it can be regenerated back to NAD^+.

4. *Metabolism of nitrogenous bases, lipids, and carbohydrates.* Because uric acid is a degradation product of the purine bases present in DNA and RNA, the localization of urate oxidase and other enzymes of purine metabolism in peroxisomes suggests that this organelle plays a role in the breakdown of nitrogenous bases derived from nucleic acids. Enzymes involved in the breakdown of fatty acids, and in the metabolism of various carbohydrates and organic acids, have also been identified in peroxisomes. The particular combination of enzymes present suggests the possibility that peroxisomes are involved in

gluconeogenesis, the synthesis of carbohydrate from fats and other noncarbohydrate materials. This speculation is reinforced by the fact that the only vertebrate cell types with significant numbers of peroxisomes are liver and kidney, tissues known to be major sites of gluconeogenesis. However, the role of peroxisomes in gluconeogenesis has been most firmly established in plant seedlings, where a special type of peroxisome known as a *glyoxysome* is involved.

Glyoxysomes and the Glyoxylate Cycle

Many plants store large amounts of lipid in their seeds for subsequent use as an energy source during germination. At the appropriate time, this lipid is converted to carbohydrate by a pathway that can be divided into two major parts. The first part involves degradation of fatty acids by *β-oxidation,* a process in which two-carbon fragments derived from the fatty acid chain are successively released in the form of a molecule called acetyl-CoA. These two-carbon units are then converted to carbohydrate by the **glyoxylate cycle,** a modified version of the Krebs tricarboxylic acid cycle (Chapter 7). In essence, the glyoxylate cycle by-passes the CO_2-evolving steps of the Krebs cycle with the aid of two enzymes not present in animal cells, isocitrate lyase and malate synthase. As shown in Figure 6-50, the net result of the glyoxylate cycle is the conversion of two molecules of

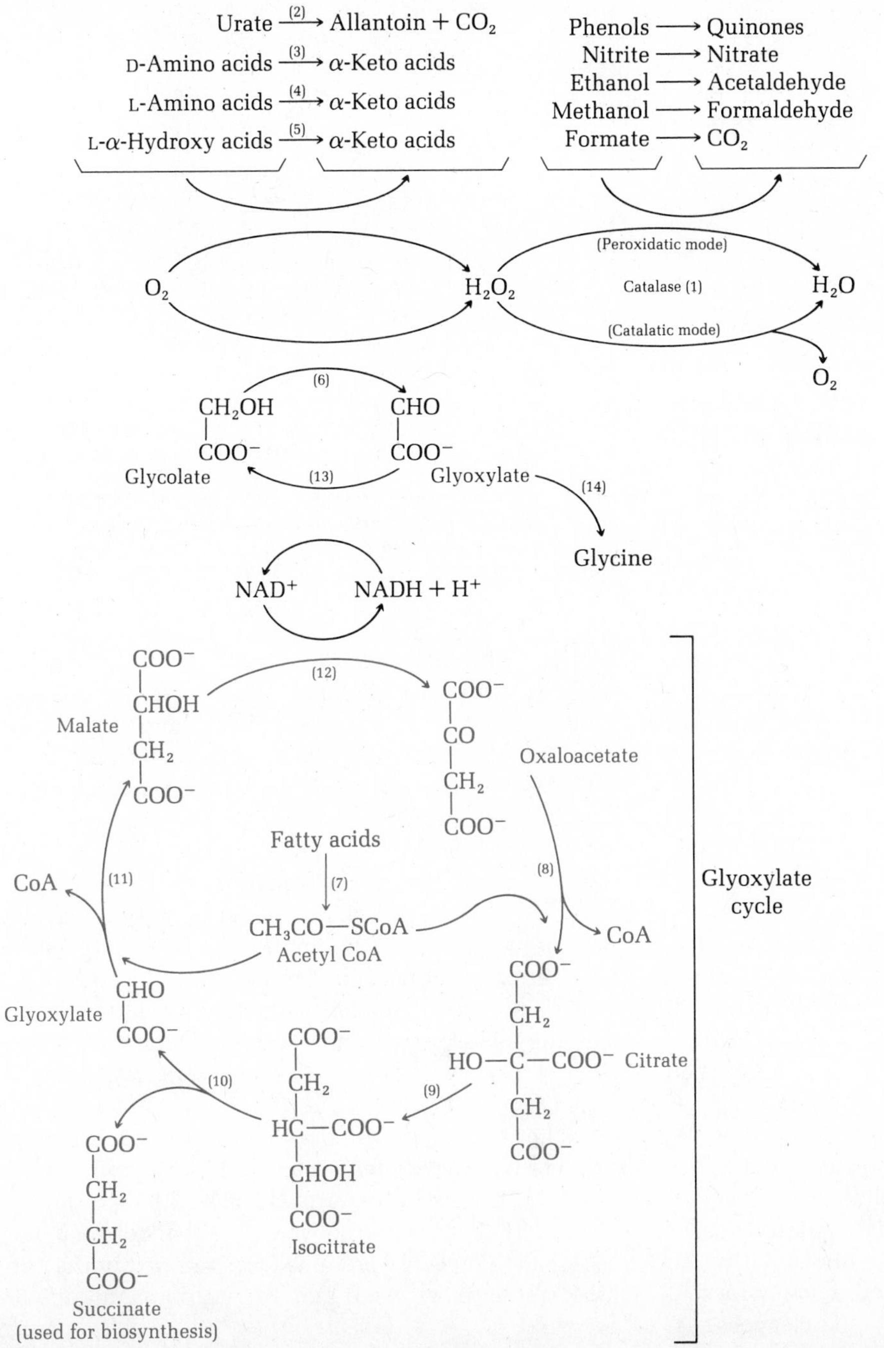

FIGURE 6-50 Composite view of the major metabolic reactions occurring in peroxisomes. The numbers refer to enzymes listed in Table 6-5. The complete glyoxylate cycle is restricted to the peroxisomes (glyoxysomes) of plant seedlings.

acetyl-CoA to one molecule of succinate, a precursor for the synthesis of various carbohydrates.

In the late 1960s Harry Beevers and his associates discovered that in plant seedlings, the enzymes of the glyoxylate cycle and fatty acid β-oxidation are localized in particles that can be separated by isodensity centrifugation from both mitochondria and chloroplasts. These particles, logically named **glyoxysomes,** also contain catalase and several H_2O_2-producing oxidases. Hence they represent a special type of peroxisome in which the complete set of glyoxylate cycle enzymes is present along with the normal H_2O_2-associated enzymes.

True glyoxysomes bearing the complete set of enzymes required for gluconeogenesis via the glyoxylate cycle have thus far been identified only in plant seedlings. Protozoa and yeast have peroxisomes containing some, but not all, of the glyoxylate cycle enzymes. In such cases the glyoxylate cycle may be carried out by cooperation between mitochondria and peroxisomes, each executing a portion of the cycle. In higher animals fatty acid β-oxidation has been observed in peroxisomes, but the products of this pathway cannot be used for gluconeogenesis by the glyoxylate cycle because this cycle does not occur in higher animals.

Biogenesis of Peroxisomes

Two theories have been proposed to explain how peroxisomes are manufactured by cells (Figure 6-51). The first model suggested that peroxisomal proteins, like secretory proteins, are synthesized on ribosomes bound to the endoplasmic reticulum; these newly formed peroxisomal proteins were thought to pass into the cisternae of the ER and into outpocketings that pinch off to form peroxisomes. A more recent model proposes that peroxisomal proteins are synthesized on free ribosomes, released into the cell sap, and then taken up by peroxisomes. According to this latter view, new peroxisomes are formed by expansion and budding of existing peroxisomes.

Early support for the first model was obtained from electron micrographs in which peroxisomes were seen to exhibit "tails" believed to represent connections to the endoplasmic reticulum (Figure 6-52). However, peroxisomal and ER membranes are difficult to distinguish from one another in electron micrographs, and it can be just as easily argued that the "tails" represent connections between an interconnected network of peroxisomes. Hence in order to clearly distinguish between the alternative models of peroxisome biogenesis, it has been necessary to carry out biochemical studies designed to determine where peroxisomal enzymes are synthesized within the cell. Subcellular fractionation studies have therefore been carried out on tissues briefly incubated in the presence of radioactive amino acids. According to the first model, the newly formed radioactive peroxisomal proteins should first appear as-

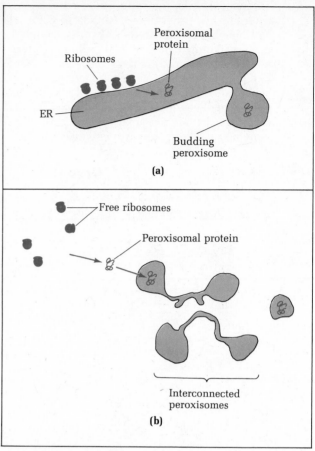

FIGURE 6-51 Two alternative models of peroxisome biogenesis. (a) Classical model in which peroxisomal proteins are synthesized on membrane-attached ribosomes and sequestered in the cisternae of the ER for transport into newly forming peroxisomes. (b) Model in which peroxisomal proteins are synthesized on free ribosomes and released into the cell sap prior to uptake by peroxisomes.

sociated with the rough ER of the microsomal fraction. The experimental data have failed to support this prediction, however. Instead, peroxisomal proteins such as catalase first appear in the cytosol fraction and only later become concentrated in peroxisomes, suggesting that they are being synthesized on free ribosomes. Additional support for this conclusion has come from the demonstration that peroxisomal enzymes are synthesized by messenger RNAs isolated from free, but not membrane-bound, ribosomes (Figure 6-53). The notion that peroxisomal proteins are not synthesized like secretory proteins is further supported by the discovery that the major peroxisomal proteins are neither glycosylated nor synthesized as precursors containing a signal sequence.

The above findings raise the significant question of how peroxisomal proteins made in the cell sap are selectively transported into peroxisomes. Although the answer to this question is not completely understood, it is thought that specific proteins localized within the peroxisomal membrane aid in the recognition and uptake process.

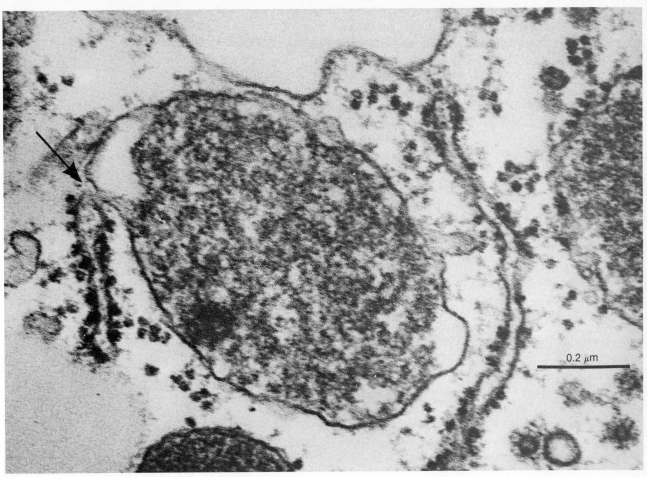

FIGURE 6-52 Electron micrograph showing a connection between a microbody and membranes of the rough endoplasmic reticulum (arrow). Courtesy of E. L. Vigil.

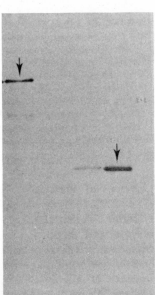

FIGURE 6-53 Determination of the sites of synthesis of the peroxisomal enzymes catalase and uricase. The proteins synthesized by messenger RNA extracted from free and membrane-bound ribosomes were purified by precipitation with antibodies specific for catalase and uricase, and the purified products were then analyzed by polyacrylamide gel electrophoresis. As can be seen, these two peroxisomal enzymes are synthesized by the messenger RNAs present on free, but not membrane-bound, ribosomes. Courtesy of G. Blobel.

Evolutionary Origin of Peroxisomes

Although several possible functions of the peroxisome have been mentioned, there are reasons for questioning whether any of these are important enough to justify the existence of such an organelle, especially in the cells of higher animals. To begin with, most vertebrate cells get by perfectly well without peroxisomes, and even when they are present their necessity is questionable. Humans inheriting genetic deficiencies of the enzyme catalase, for example, exhibit no obvious disease symptoms. The peroxisomal enzyme urate oxidase is absent in many organisms, including humans, and the function of the peroxisomal enzyme that oxidizes D-amino acids is somewhat puzzling because it is the L-amino acids that are normally found in protein molecules. Finally, most of the metabolic events supposedly occurring in peroxisomes can take place elsewhere in the cell.

The above considerations have spawned the speculation that the present-day peroxisome is a fossil organelle derived from an ancestral particle that performed critical functions hundreds of millions of years ago, but is no longer needed and is in the process of dying out. According to this theory, the original peroxisome formed when oxygen first appeared on earth. Oxygen can be toxic because it tends to react with biological molecules to form hydrogen peroxide. Adaptation to the appearance of oxygen in the atmosphere therefore required the evolution of a system for disposing of hydrogen peroxide. The peroxisome was especially well suited to this task because it uses the oxygen-induced production of hydrogen peroxide to facilitate the oxidation of intermediates involved in cellular carbohydrate metabolism. The major disadvantage of peroxisomal respiration, however, is that the energy released during this carbohydrate oxidation is not conserved in a useful chemical form, but is dissipated as heat. When mitochondria appeared later in evolution, their ability to link respiration to the formation of ATP permitted them to prevail over peroxisomes. Hence the peroxisome gradually lost many of its enzymes and became less and less important.

As we shall see in Chapter 14, some biologists believe that mitochondria evolved from ancient bacteria that were engulfed by primitive nonbacterial cells hundreds of millions of years ago. Such a notion can be combined with our picture of the ancestral peroxisome to form an integrated theory of the evolution of respiration (Figure 6-54). This theory asserts that when oxygen first appeared on the earth, cellular life diverged in two directions. One resulted in relatively small cells that linked respiration to ATP formation, while the other produced larger cells that depended on peroxisomes for respiration. At a later time some of the smaller, bacteria-like cells were engulfed by the larger, peroxisome-containing cells. The engulfed bacteria ultimately evolved into mitochondria, displacing the peroxisomes in importance.

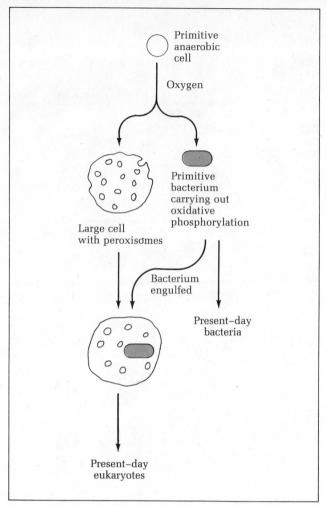

FIGURE 6-54 A hypothetical model proposed to explain the evolutionary origins of peroxisomes and mitochondria.

In spite of the apparent superiority of mitochondria in respiration, one should not overlook the fact that peroxisomes have persisted in the presence of mitochondria for hundreds of millions of years, albeit only in certain cell types. This persistence must reflect some useful function, though we are not certain what it may be. In plant cells the glyoxylate cycle certainly falls in this category, but the enzymes of this cycle were lost during the course of animal evolution, and are even lacking in the leaves of green plants. Thus the reason for the evolutionary persistence of peroxisomes in most cell types remains somewhat of a mystery.

Other Microbodies

The riddle of peroxisome function has recently become even more complicated with the discovery of two new types of microbodies. One, termed the **hydrogenosome,** is involved in the oxidation of pyruvate; the other, called a **glycosome,** appears to catalyze the anaerobic breakdown of glucose. Each of these organelles is restricted to a single group of protozoa. Hydrogenosomes and glycosomes appear to have little overlap with each other, or with peroxisomes or glyoxysomes, but their very existence raises the question of whether there are other undiscovered types of microbodies whose properties will help us to better understand the interrelationships and functional properties of this general family of organelles.

MEMBRANE BIOGENESIS

Although we have seen how certain proteins come to be compartmentalized within the luminal spaces of the ER, Golgi complex, lysosomes, and peroxisomes, there is still a question about how the membranes of these organelles are formed. This is a problem of considerable magnitude because each type of cellular membrane has its own characteristic spectrum of proteins and lipids. Two alternative models of membrane assembly can be envisioned. One is that membranes are formed *de novo* by the spontaneous self-assembly of lipids and proteins. The experimental basis for this idea is that isolated phospholipids can spontaneously form lipid bilayer vesicles that incorporate membrane proteins within them. The only problem with such self-assembled membranes as a model for membrane assembly *in vivo* is that the protein arrangement in self-assembled membranes almost always lacks the asymmetry inherent in natural biological membranes. Presumably the reason for this lack of asymmetry in artificial membranes is that proteins can be randomly inserted on either side of the lipid bilayer as it forms.

The alternative to self-assembly is that membranes grow by insertion of lipids and proteins into existing membranes. According to this view, selective insertion of molecules from one side of the membrane or the other is possible, allowing an asymmetrical arrangement of membrane components to be established and maintained. If the membrane were broken open, however, then proteins or lipids could be inserted from either side and the asymmetry would be lost. The existence of membrane asymmetry therefore argues strongly for the conclusion that membranes grow by expansion of intact membranes rather than by self-assembly.

Biochemical studies on membrane synthesis have provided strong support for this conclusion. In eukaryotic cells, for example, the synthesis of membrane phospholipids has been shown to be catalyzed by enzymes located within the membranes of the endoplasmic reticulum. The phospholipids formed by these enzymes are immediately incorporated into the lipid bilayer, causing it to expand. This new membrane material is ultimately passed on to other organelles, such as the Golgi complex, lysosomes, peroxisomes, and the plasma membrane, by the process of membrane budding and fusion. Thus the endoplasmic reticulum is the ultimate source of most new membrane material for the eukaryotic cell.

In prokaryotic cells the comparable enzymes of phospholipid synthesis are localized within the plasma membrane, again indicating that lipids are synthesized within membranes and immediately added to the lipid bilayer. Important insight into the question of how lipid asymmetry is generated has been obtained from studies on phospholipid synthesis by prokaryotic plasma membranes. In these experiments, carried out on the bacterium *Bacillus megaterium* by James Rothman and Eugene Kennedy, two types of labels were used to monitor synthesis of the membrane phospholipid, phosphatidylethanolamine (PE). The first, ^{32}P-phosphate, was used to label newly synthesized molecules of PE. Following a 1-min incubation of cells with ^{32}P-phosphate to label newly made PE, the reagent trinitrobenzenesulfonic acid (TNBS) was added to the cell suspension. TNBS reacts chemically with PE to form a stable, covalent complex. Because TNBS cannot cross the plasma membrane, it will only combine with PE molecules exposed on the exterior surface of the plasma membrane. After labeling cells with ^{32}P-phosphate and TNBS, the plasma membrane fraction was isolated and dissolved, and the membrane lipids separated from each other by thin-layer chromatography. Because the PE-TNBS complex migrates faster than unreacted PE, the PE present on the outer surface of the plasma membrane can be separated from the PE present on the inner side (Figure 6-55). When the separated PE fractions were subsequently assayed for radioactivity, only the PE from the cytoplasmic surface (not combined with TNBS) was found to contain ^{32}P. Hence it could be concluded that

FIGURE 6-55 Data obtained from an experiment in which phospholipid synthesis was investigated in bacterial cells incubated with ^{32}P-phosphate for 1 or 30 min, and then reacted with TNBS. Lipids were extracted from the membrane and fractionated by thin-layer chromatography. Autoradiography of the thin-layer chromatogram was carried out to detect the presence of radioactive phospholipid. Refer to the text for details concerning the interpretation of this experiment.

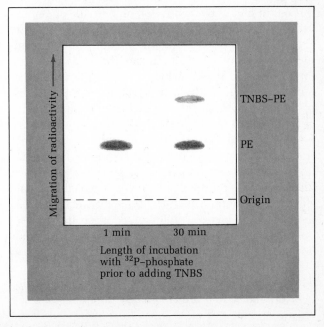

the synthesis of PE is confined to the cytoplasmic side of the plasma membrane.

The only problem with this conclusion is that it fails to explain how PE ever gets to the outer surface of the plasma membrane bilayer. We noted earlier that transverse diffusion, or flip-flop, of membrane lipids across the bilayer is an extremely rare event. Yet PE clearly must reach the outer half of the bilayer because 30 percent of the total PE of the bacterial plasma membrane is known to be located there. Although the mechanism by which the PE molecules synthesized on the cytoplasmic side of the plasma membrane migrate to the exterior half of the lipid bilayer is not known with certainty, further studies employing ^{32}P-labeling have clearly shown that this must occur. If, for example, one incubates bacterial cells with ^{32}P-phosphate for half an hour instead of only a minute and then adds TNBS, 30 percent of the total radioactivity is now found in the PE-TNBS complex and must therefore be associated with PE molecules present on the exterior surface of the plasma membrane. It has been calculated from these data that this rapid movement of PE to the exterior surface is 100,000 faster than expected from flip-flop. A similar rapid movement of lipids across the bilayer has recently been discovered in the endoplasmic reticulum of eukaryotic cells. It has therefore been argued that areas of new membrane formation contain special protein channels that facilitate the passage of phospholipid from one side of the lipid bilayer to the other. The general absence of such channels in other membrane areas would explain the low rate of lipid flip-flop observed elsewhere.

The above model raises a new question, however, and that concerns the mechanism by which lipid asymmetry is established. We noted in an earlier chapter that phospholipids tend to be asymmetrically distributed in biological membranes, with certain phospholipids present in higher concentration on one side of a membrane and other phospholipids enriched on the opposite side. If regions of membrane formation contain channels that permit the free diffusion of lipids back and forth across the bilayer, why isn't an equilibrium distribution achieved with each phospholipid equally distributed on both sides? Although the answer to this question is not known with certainty, many scientists now believe that the observed phospholipid asymmetries are not incompatible with a thermodynamic equilibrium. This view simply postulates that each phospholipid is thermodynamically more stable on one side of a membrane than the other because environmental conditions are different on opposite sides of membranes. For this reason the thermodynamic equilibrium for each particular phospholipid may cause it to favor one side of the bilayer over the other.

The asymmetric arrangement of membrane proteins originates in a fundamentally different way. The mechanism by which membrane proteins are synthesized and assembled into membranes has been extensively investigated by James Rothman and Harvey Lodish using vesicular stomatis virus (VSV) as a model system. VSV, which causes an influenzalike disease in farm animals, directs the cells it infects to synthesize several virus-specific proteins. Among them are two proteins that become incorporated into the host cell plasma membrane. One, designated the *M protein*, is a peripheral membrane protein associated with the cytoplasmic surface of the plasma membrane. The other, designated the *G protein*, is an integral plasma membrane glycoprotein. Studies in which radioactive amino acids were used to label and follow the fate of newly synthesized viral proteins revealed that the M protein is synthesized on free cytoplasmic ribosomes and released in a soluble form into the cell sap. Only later does it become bound to the cytoplasmic surface of the plasma membrane. Although the mechanism that guides this protein to the appropriate cellular membrane is unclear, this appears to be a general phenomenon because other peripheral membrane proteins have also been shown to be synthesized as soluble cell sap proteins.

In contrast to the M protein, the G protein of VSV is synthesized on membrane-attached ribosomes. Like secretory proteins, it contains a signal sequence that guides its binding to the ER membrane. As the G protein is being synthesized, it is gradually inserted through the ER membrane. As it begins to emerge on the luminal side, its signal sequence is removed and carbohydrate chains are added. Although these events are all similar to the behavior of secretory proteins, one major difference occurs. In this case the G protein remains anchored in the ER membrane, rather than passing completely through it and being released into the lumen.

The G protein, now embedded within the ER membrane, is directed to the plasma membrane by a series of steps similar to those occurring in cell secretion. ER membrane containing the G protein forms small transitional vesicles that bud off and fuse with the Golgi complex, where the carbohydrate chains of the G protein are further modified. Membrane vesicles containing the G protein bud off from the Golgi complex and migrate to the cell surface. Here they fuse with the plasma membrane, bringing the G protein to the cell surface. The important point to note about this process is that the portion of the G protein containing the carbohydrate chains always faces the lumen of the membrane system in which it is present. When a membrane vesicle containing the G protein fuses with the plasma membrane, this orientation causes the carbohydrate chains to face the cell exterior (Figure 6-56). From the results of these, as well as related experiments involving other membrane proteins, it has been concluded that integral membrane proteins are synthesized on membrane-attached ribosomes, and that their final asymmetric

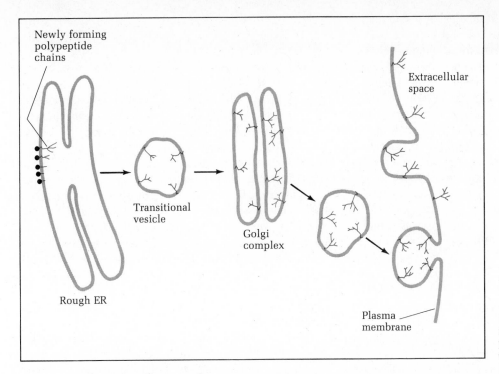

Newly forming
polypeptide
chains

Transitional
vesicle

Golgi
complex

Rough ER

Extracellular
space

Plasma
membrane

FIGURE 6-56 Summary diagram illustrating how the asymmetry of plasma membrane glycoproteins is established.

orientation is established at the time of synthesis by the direction in which the signal sequence inserts them into the ER membrane. In prokaryotic cells, where no endoplasmic reticulum is present, a comparable mechanism exists involving ribosomes bound to the plasma membrane rather than to the ER.

Although experiments carried out on simple systems such as those described above have led to several general conclusions about membrane synthesis and assembly, important problems still remain to be solved. One of the most crucial relates to the question of how each type of membrane acquires and retains its own characteristic set of proteins and lipids. If all integral membrane proteins are synthesized in the rough ER, some obviously remain as ER proteins while others are distributed to the Golgi, lysosomal, peroxisomal, or plasma membranes. The molecular mechanism responsible for this sorting and distribution of membrane components remains to be unraveled by future investigations.

SUMMARY

Eukaryotic cells have evolved an elaborate system of membranes for dividing the cytoplasm into multiple compartments. The endoplasmic reticulum (ER) exists in two forms, a "rough" ER containing attached ribosomes and a "smooth" ER lacking ribosomes. Rough ER is usually arranged as flattened membranous sheets, while smooth ER consists of an interconnected complex of membranous tubules. The endoplasmic reticulum divides the cytoplasm into two compartments, the cell sap

and the cisternal space. The cell sap contains enzymes involved in many metabolic pathways, while the cisternal space provides a route for the intracellular movement of materials, especially proteins destined for export.

During subcellular fractionation, the endoplasmic reticulum breaks into small fragments that reseal to form vesicles called microsomes. Isolated microsomal membranes contain enzymes involved in the metabolism of carbohydrates, lipids, and drugs. Among these are glucose-6-phosphatase, involved in releasing glucose from glycogen stores, and enzymes for the synthesis of neutral fats, steroids, and phospholipids. Also present are mixed-function oxidases containing cytochrome P-450 and related hemoproteins; these enzyme systems are involved in the detoxification of foreign substances and in normal metabolic pathways for the oxidation of steroids and fatty acids.

The rough ER plays a crucial role in the processing of secretory proteins, proteins destined for incorporation into cytoplasmic granules such as lysosomes, and integral membrane proteins. Proteins of this type contain a special signal sequence at their amino-terminal end that guides their attachment and insertion through the ER membrane while they are still in the process of being synthesized. The signal sequence is removed on the luminal side of the ER, after which protein synthesis is completed and the protein is released into the lumen for transport to the Golgi complex. Many secretory proteins have carbohydrate chains added to them while they are in the process of being synthesized in the ER.

The Golgi complex consists of a stack of flattened,

smooth-membrane-bounded saccules and associated vesicles. The membrane stack has an inherent polarity, with saccules at opposite ends exhibiting differences in structure, composition, and function. One end is referred to as the *cis* face, while the other is called the *trans* face. The most reliable enzyme marker for Golgi membranes is thiamine pyrophosphatase, although nucleoside diphosphatase is also used. The Golgi complex is involved in concentrating, sorting, chemically modifying, and packaging proteins destined for secretion or for incorporation into intracellular vesicles such as lysosomes. Secretory proteins leaving the Golgi complex are stored in secretion granules prior to their eventual expulsion from the cell by exocytosis, a process involving fusion of the secretory granule membrane with the plasma membrane.

Lysosomes are small, membrane-enclosed vesicles containing hydrolytic enzymes active at acid pH. Together this group of enzymes is capable of degrading virtually every type of macromolecule cells contain. The existence of lysosomes was originally predicted on the basis of enzymatic analyses of subcellular fractions. Only later, with the aid of the cytochemical staining reaction for acid phosphatase, were these organelles identified electron microscopically. During lysosome formation, hydrolytic enzymes destined for incorporation into the lysosome are identified by the presence of phosphorylated mannose groups that are recognized by specific membrane receptors present in the Golgi complex. Lysosomes serve four digestive functions for the cell: (1) degradation of foreign matter taken up by endocytosis, (2) destruction of worn-out organelles (autophagy), (3) breakdown of cell structure associated with cell death (autolysis), and (4) digestion of extracellular materials. Genetic deficiencies in individual lysosomal enzymes can cause "storage diseases" in which the undigested substrates of the missing enzymes accumulate and interfere with cellular activities. Experiments aimed at replacing defective enzymes provide hope that these debilitating and often fatal diseases may one day be conquered.

Peroxisomes are a family of cytoplasmic particles characterized by the presence of catalase, the enzyme that destroys hydrogen peroxide, and various oxidases that generate hydrogen peroxide. Peroxisomes occur in vertebrate liver and kidney, in eukaryotic microorganisms, and in fatty seedlings and green leaves of higher plants. Peroxisomes are often characterized by the presence of a crystalloid core and can be unequivocally identified with the DAB-staining reaction for catalase. Peroxisomal enzymes appear to be made on free ribosomes and released free in the cell sap, after which they are taken up by peroxisomes.

Several possible roles have been considered for peroxisomal metabolism, including inactivation of toxic substances, regulation of oxygen tension, NAD^+ regeneration, metabolism of purines, fatty acid oxidation, and gluconeogenesis. The glyoxylate cycle, which permits synthesis of carbohydrate from the products of fatty acid oxidation, is present in plant seedling peroxisomes, leading to their designation as glyoxysomes. Whether any of the above functions is adequate to explain the evolutionary persistence of peroxisomes, especially in animal cells where the glyoxylate cycle does not occur, is yet to be ascertained.

Membrane biogenesis occurs by expansion of existing membranes. Phospholipids are synthesized within the ER membrane (plasma membrane in prokaryotic cells) and incorporated directly into the lipid bilayer. Areas of new membrane formation have been postulated to contain special protein channels that facilitate the equilibration of newly formed lipid across both sides of the bilayer. The thermodynamic equilibrium for each particular phospholipid may cause it to favor one side of the bilayer over the other, explaining the observed phospholipid asymmetries.

Peripheral membrane proteins are synthesized on free cytoplasmic ribosomes and released in a soluble form into the cell sap. Only later do they become associated with membrane surfaces. Integral proteins, in contrast, are formed on membrane-attached ribosomes and inserted directly into the membrane while they are being synthesized. The question of how the appropriate proteins and lipids are directed to each type of subcellular membrane remains to be determined.

SUGGESTED READINGS

Books

Dingle, J. T., and R. T. Dean, eds. (1976). *Lysosomes in Biology and Pathology*, Vol. 5, North-Holland, Amsterdam.

Gething, M.-J., ed. (1985). *Protein Transport and Secretion*, Cold Spring Harbor Laboratory, Cold Spring Harbor, N.Y.

Pastan, I., and M. C. Willingham, eds. (1985). *Endocytosis*, Plenum, New York.

Watts, R. W. E., and D. A. Gibbs (1986). *Lysosomal Storage Diseases*, Taylor and Francis, Philadelphia.

Articles

Farquhar, M. G. (1985). Progress in unraveling pathways of Golgi traffic, *Ann. Rev. Cell Biol.* 1:447–488.

Goldstein, J. L., M. S. Brown, R. G. W. Anderson, D. W. Russell, and W. J. Schneider (1985). Receptor-mediated endocytosis: concepts emerging from the LDL receptor system, *Ann. Rev. Cell Biol.* 1:1–39.

Jessup, W., P. Leoni, and R. T. Dean (1985). Constitutive and triggered lysosomal enzyme secretion. *In: Developments in Cell Biology*, Vol. 1, Secretory Processes, R. T. Dean and P. Stahl, eds., Butterworth, New York.

Lazarow, P. B., and Y. Fujiki (1985). Biogenesis of peroxisomes, *Ann. Rev. Cell Biol.* 1:489–530.

Rothman, J. E. (1985). The compartmental organization of the Golgi apparatus, *Sci. Amer.* 253 (Sept.):74–89.

Walter, P., and V. R. Lingappa (1986). Mechanism of protein translocation across the endoplasmic reticulum, *Ann. Rev. Cell Biol.* 2:499–516.

Wickner, W. T., and H. F. Lodish (1985). Multiple mechanisms of protein insertion into and across membranes, *Science* 230:400–407.

7

Mitochondria and the Conservation of Energy Derived from Foodstuffs

Organisms cannot survive without an input of both chemical building blocks and energy from the external environment. Chemical building blocks are required for the synthesis of the small and large molecules from which all cellular constituents are made, while an input of energy is necessary for active transport, motility, and all the energy-requiring chemical reactions of the cell (most of which are involved in the biosynthesis of macromolecules). Chemical building blocks and energy are interrelated topics in that both are present in the nutrients that organisms and cells ingest, and both are simultaneously transformed every time a chemical reaction occurs. The overall set of chemical reactions and pathways used by cells in the transformation of chemical building blocks and energy are collectively referred to as **metabolism.**

Although metabolic reactions occur in various cellular locations, reactions central to the process of transforming energy and conserving it in a useful chemical form are localized predominantly within membranes. In eukaryotic cells mitochondrial and chloroplast membranes function in conserving energy derived from foodstuffs and sunlight, respectively, while in prokaryotic cells the plasma membrane and its derivatives perform comparable functions. In each of these cases the energy released is utilized to drive the synthesis of ATP molecules, which in turn transfer the energy to a variety of energy-requiring processes. Much is now known concerning both the metabolic pathways that release the energy stored in foodstuffs and the mechanism by which this released energy is utilized to drive ATP formation. One of the more pleasant surprises to emerge in recent years is the extent to which mitochondrial, chloroplast, and prokaryotic membranes employ similar mechanisms for carrying out this synthesis of ATP. In this chapter we shall focus on the role of membranes in the transformation of energy derived from the oxidation of foodstuffs, while Chapter 8 considers the related role of membranes in the transformation of energy derived from light.

OVERVIEW OF GLUCOSE METABOLISM

Although cells obtain chemical building blocks and energy from the metabolism of many different kinds of organic nutrients, most of these foodstuffs are broken down into molecules that in one way or another enter the pathway for glucose degradation. This pathway for glucose metabolism performs two basic functions for

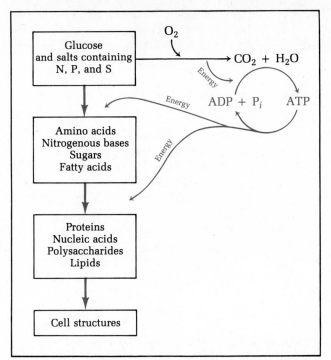

FIGURE 7-1 Diagram summarizing the central role played by glucose in the flow of chemical building blocks and the flow of energy.

the cell. First, energy released during the oxidation of glucose is utilized to drive the formation of ATP. Second, chemical intermediates produced during the process of glucose breakdown can be utilized as carbon skeletons for the synthesis of other molecules needed by the cell, such as amino acids, nitrogenous bases, fatty acids, and various sugars. These compounds are in turn assembled into macromolecules by biosynthetic pathways that utilize energy stored in ATP. Thus glucose plays a central role in the metabolism of both energy and chemical building blocks in living cells (Figure 7-1). Because the key energy-transforming steps in the metabolism of glucose involve oxidation of its carbon skeleton, we shall begin our overview of this pathway by reviewing the fundamental principles of oxidation-reduction reactions.

Oxidation-Reduction and Redox Potentials

By definition, an oxidation-reduction reaction is any chemical reaction involving the transfer of electrons. The substance losing electrons is said to become **oxidized,** while the substance receiving electrons becomes **reduced.** Several basic kinds of oxidation-reduction reactions are summarized in Figure 7-2. Consideration of these examples reveals that electrons can be transferred either alone or in association with hydrogen atoms. The term *oxidation* was originally coined because oxygen commonly acts as an electron acceptor in such reactions, but it is now clear that many substances other than oxygen also function as electron acceptors.

Oxidation-reduction reactions involving transfer of an electron pair:

1. $\quad X\!:\!H_2 \;+\; Y \;\rightarrow\; X \;+\; Y\!:\!H_2$

2. $\quad X\!:\!H_2 \;+\; Y \;\rightarrow\; X \;+\; Y\!:\!H \;+\; H^+$

3. $\quad X\!:\!H_2 \;+\; \tfrac{1}{2}O_2 \;\rightarrow\; X \;+\; H_2\!:\!O$

Oxidation-reduction reaction involving transfer of a single electron:

4. $\quad X^{2+} \;+\; Y^{3+} \;\rightarrow\; X^{3+} \;+\; Y^{2+}$

FIGURE 7-2 Examples of several different kinds of oxidation-reduction reactions. In examples 1 and 2 substance X is oxidized by donating a pair of electrons, while substance Y is reduced by gaining a pair of electrons. In example 3 molecular oxygen is the substance that undergoes reduction by accepting electrons. In example 4 substance X undergoes oxidation by transferring a single electron to substance Y, which becomes reduced.

The relative ability of a given substance to act as an electron donor or acceptor can be determined experimentally using a simple electrical cell like the one pictured in Figure 7-3. To illustrate how such a cell operates, let us consider a substance that exists in both a reduced state A and an oxidized state A⁺. Such a pair, consisting of the oxidized and reduced forms of the same substance, is referred to as a **redox couple.** In the apparatus illustrated in Figure 7-3, a mixture of A and A⁺ is placed in the *sample half-cell,* and a mixture of H⁺ and H₂ is placed in the *reference half-cell.* Electrons are then permitted to flow from one half-cell to the other through an agar bridge. If substance A has a lower affinity for electrons than does H₂, electrons will flow from the sample half-cell to the reference half-cell. If substance A has a higher affinity for electrons than does H₂, current will flow in the opposite direction. In either case the magnitude of the electron flow can be measured using a voltmeter connected to electrodes immersed in the two half-cells. When the above experiment is performed under standard conditions (1M concentration of A, A⁺, and H⁺, and 1 atm H₂), the resulting voltage is referred to as the standard oxidation-reduction poten-

FIGURE 7-3 Diagram of an electrical cell designed to measure standard redox potentials.

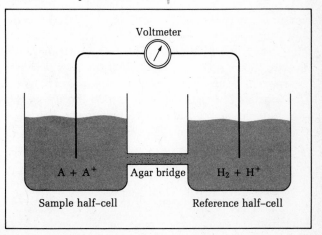

TABLE 7-1
Standard oxidation-reduction (redox) potentials of selected electron carriers

Oxidant	Reductant	n	E_0' (Volts)
NAD^+	$NADH + H^+$	2	-0.32
FAD	$FADH_2$	2	-0.22
FMN	$FMNH_2$	2	-0.22
Pyruvate	Lactate	2	-0.19
Cyt b^{3+}	Cyt b^{2+}	1	0.00
Ubiquinone	Ubiquinone-H_2	2	$+0.10$
Cyt c_1^{3+}	Cyt c_1^{2+}	1	$+0.22$
Cyt c^{3+}	Cyt c^{2+}	1	$+0.25$
Cyt a^{3+}	Cyt a^{2+}	1	$+0.25$
Cyt a_3^{3+}	Cyt a_3^{2+}	1	$+0.38$
$\frac{1}{2}O_2 + 2H^+$	H_2O	2	$+0.82$

Note: E_0' is the standard redox potential (pH 7, 25°C) for the partial reaction: oxidant + $ne^- \rightarrow$ reductant, where n = the number of electrons (e^-).

tial, or **redox potential (E_0').** A negative value of E_0' means that electrons flow from the sample half-cell to the reference half-cell, and the sample therefore has a lower affinity for electrons than does H_2. A positive value of E_0', on the other hand, occurs when electrons flow from the reference half-cell to the sample half-cell, meaning that the sample has a higher affinity for electrons than does H_2.

The redox potentials for many biologically important redox couples have been measured in this way (Table 7-1). Such information permits the relative strengths of various oxidizing and reducing agents to be easily compared. Substances whose redox potentials are very negative will be strong reducing agents, and will therefore tend to donate electrons to substances with more positive redox potentials. Conversely, substances whose redox potentials are very positive will be potent oxidizing agents, and will tend to accept electrons from substances with more negative redox potentials.

Information concerning redox potentials is also valuable because it can be used to calculate free-energy differences. The following equation relates these two forms of energy:

$$\Delta G° = -nF \, \Delta E_0'$$

where n is the number of electrons transferred, F is the equivalent of the Faraday in calories (23.062 kcal/V/mol), $\Delta E_0'$ is the difference in redox potential between the two sets of reactants, and $\Delta G°$ is the change in free energy that occurs when electrons are passed between the two sets of reactants. To illustrate the use of this equation, let us consider the oxidation of pyruvate ($E_0' = -0.19$) by oxygen ($E_0' = +0.82$). Substituting in the above equation

$$\Delta G° = -2 \times 23.062 \times [0.82 - (-0.19)]$$
$$\Delta G° = -2 \times 23.062 \times 1.01$$
$$\Delta G° = -46.6 \text{ kcal/mol}$$

Hence for any two redox couples, one can use the above equation to calculate the change in free energy that occurs when electrons are passed from one couple to the other.

Role of Coenzymes in Oxidation-Reduction

Enzymes that catalyze biological oxidation-reduction reactions are almost always associated with coenzymes that function as the immediate electron donors or acceptors for the reactions being catalyzed. The most prominent coenzyme of this type is **nicotinamide adenine dinucleotide** or **NAD,** a derivative of the vitamin niacin. In its oxidized state this coenzyme is generally written NAD^+ because the nicotinamide ring carries a positive charge. Upon accepting two electrons and one hydrogen atom, NAD^+ is converted to its reduced form, NADH. Because they are only loosely bound to the enzymes with which they function, NAD^+ and NADH can be employed as mobile carriers to transfer electrons from one reaction to another.

To illustrate how this occurs, let us consider a hypothetical reaction in which an organic molecule (XH_2) is oxidized by an enzyme (A) that utilizes NAD^+ as a coenzyme:

$$XH_2 \xrightarrow[\text{Enzyme A}]{NAD^+ \quad NADH} X + H^+$$

After becoming reduced, the NADH dissociates from enzyme A and diffuses to another enzyme B, where its electrons reduce a second substance (Y):

$$Y + H^+ \xrightarrow[\text{Enzyme B}]{NADH \quad NAD^+} YH_2$$

Summing the above two reactions yields

$$XH_2 + Y \longrightarrow X + YH_2$$

This summary equation reveals that the NAD^+/NADH redox couple has simply acted as a carrier system for transferring electrons from substance X to substance Y, and that the NAD^+/NADH couple itself undergoes no net change in the process.

Several other coenzymes play roles similar to that of NAD^+/NADH. A phosphorylated derivative of NAD^+, termed **nicotinamide adenine dinucleotide phosphate ($NADP^+$),** transfers electrons in much the same way as NAD^+. **Flavin adenine dinucleotide (FAD)** and **flavin mononucleotide (FMN),** which are coenzyme derivatives of the vitamin riboflavin, behave somewhat differently from NAD^+ and $NADP^+$. The first difference is that FAD and FMN are more tightly bound to the enzymes with which they interact, and are therefore classified as prosthetic groups rather than coenzymes. Proteins containing FAD or FMN as prosthetic groups are referred to as **flavoproteins.** The other important difference is that the reduction of FAD or FMN is asso-

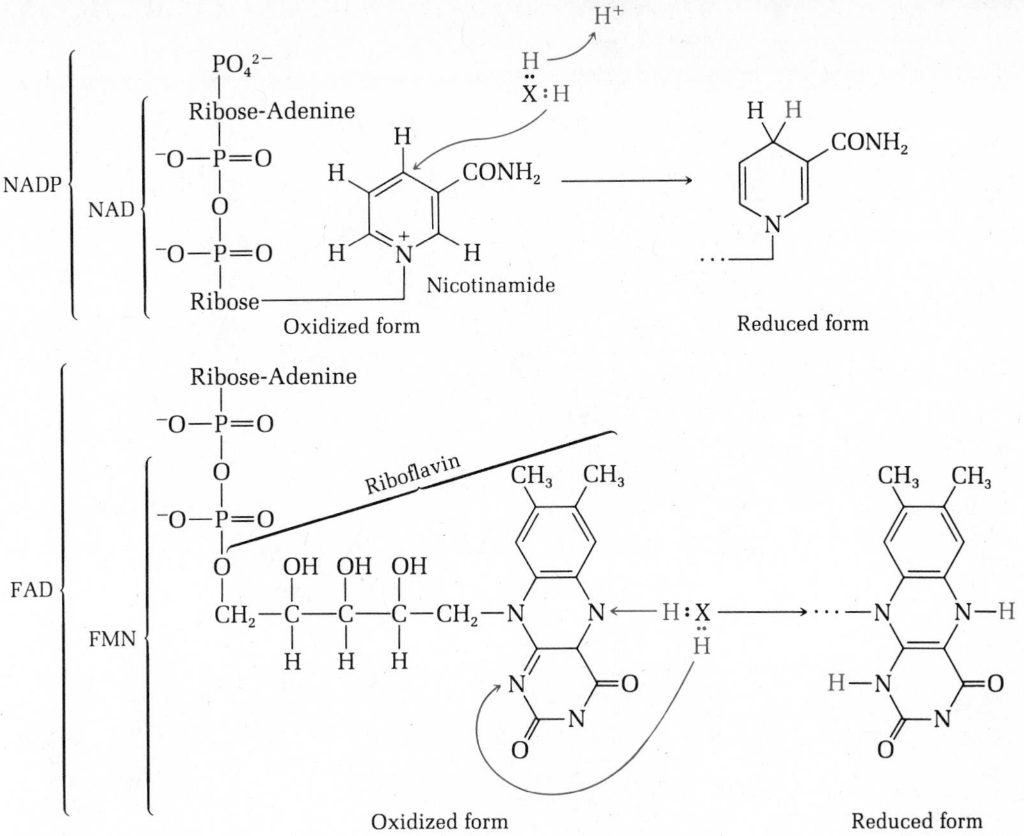

FIGURE 7-4 Structures of the coenzymes NAD, NADP, FMN, and FAD in the oxidized and reduced states.

ciated with the addition of two hydrogen atoms instead of one:

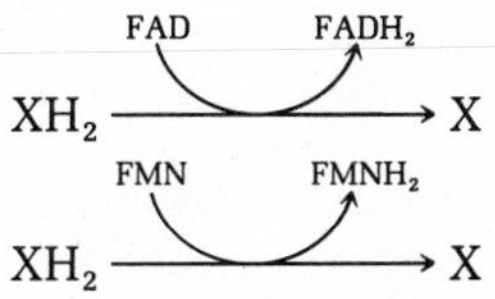

However, in each of the above reactions two electrons are accepted by the coenzyme, which is the same as the number of electrons involved in the reduction of NAD^+ or $NADP^+$ (Figure 7-4).

Glycolysis

With the above background information concerning oxidation-reduction reactions and coenzymes in hand, we can now summarize the major steps that occur during the metabolic breakdown of glucose. In the first portion of this pathway, which is referred to as **glycolysis,** glucose is broken down to two molecules of pyruvate. Several features of this glycolytic pathway, summarized in Figure 7-5, deserve special emphasis:

1. The six-carbon skeleton of glucose is cleaved to generate two molecules of the three-carbon compound, pyruvate.
2. Two of the early reactions in glycolysis are accompanied by the degradation of ATP to ADP; a

total of two ATPs is therefore broken down per molecule of glucose metabolized. Two reactions occurring later in glycolysis are accompanied by the synthesis of ATP; each of these steps produces two molecules of ATP per initial molecule of glucose metabolized, for a total of four. Since two molecules of ATP are broken down and four are synthesized, the *net* result is the formation of two new molecules of ATP.

3. One oxidation-reduction step also occurs in glycolysis. This reaction involves the oxidation of glyceraldehyde-3-phosphate, which is coupled to the reduction of NAD^+ to NADH. Because two molecules of glyceraldehyde-3-phosphate are produced per molecule of glucose, the total yield is two molecules of NADH.

The net result of glycolysis is therefore the conversion of one molecule of glucose to two molecules of pyruvate, accompanied by the net formation of two molecules of ATP and the conversion of two molecules of NAD^+ to NADH. Under normal physiological conditions glycolysis does not stop at this point, however, because the conversion of NAD^+ to NADH would quickly cause the cell to become depleted of NAD^+. Because it is the electron acceptor in the reaction in which glyceraldehyde-3-phosphate is oxidized, NAD^+ is an essential ingredient for glycolysis. A lack of NAD^+ would cause this reaction to stop, in turn bringing glycolysis to

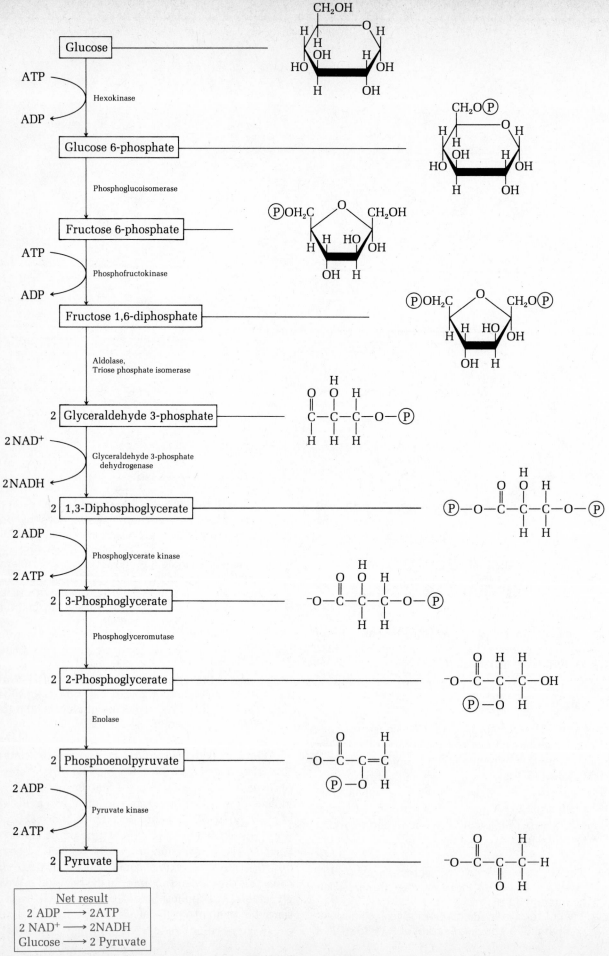

FIGURE 7-5 Summary of glycolysis. Abbreviations: Ⓟ = phosphate group.

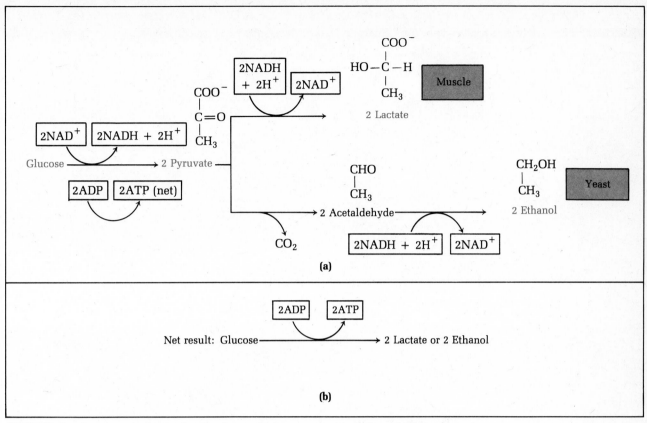

FIGURE 7-6 (a) Comparison of anaerobic glycolysis in higher organisms (muscle) and in yeast. (b) In both cases glucose breakdown is accompanied by the net production of two molecules of ATP.

a halt. The NADH generated during glycolysis must therefore be oxidized back to NAD⁺.

Two alternative mechanisms are employed for this purpose, depending on whether or not oxygen is present. Under *anaerobic* conditions (lack of oxygen), the pyruvate generated during glycolysis is reduced in a reaction that is coupled to the oxidation of NADH to NAD⁺. This reduction of pyruvate yields lactate in animal cells, but in microorganisms other substances may be produced (Figure 7-6). Yeast, for example, convert pyruvate to ethanol, a reaction that has been extensively exploited by the beer- and wine-making industries. No matter which reaction product happens to be formed, however, the important result is that the associated reoxidation of NADH to NAD⁺ regenerates the NAD⁺ needed for glycolysis.

This overall pathway for the anaerobic conversion of glucose to a reduced derivative of pyruvate is referred to as **anaerobic glycolysis** or **fermentation.** Anaerobic glycolysis has no net effect on NAD⁺/NADH levels because the conversion of NAD⁺ to NADH that accompanies the breakdown of glucose to pyruvate is reversed when pyruvate is reduced to lactate (or some other reduced compound). Hence the only net change accompanying the conversion of glucose to lactate is the formation of two molecules of ATP. The energy conserved in this ATP represents only about 2 percent of the total free energy potentially available in a molecule

of glucose. In spite of this low energy yield, however, the two molecules of ATP produced by anaerobic glycolysis can be the sole source of energy for microorganisms that live in an oxygen-free environment.

Under *aerobic* conditions (oxygen present), additional energy is liberated by the combustion of pyruvate to carbon dioxide and water. Not only does this pathway provide an alternative way of replenishing the NAD⁺ required for glycolysis, but it also traps a much larger fraction of the total energy available in the glucose molecule. This aerobic pathway, referred to as **respiration,** consists of several separate components known as the Krebs citric acid cycle, electron transfer, and oxidative phosphorylation.

The Krebs Citric Acid Cycle

The steps involved in the aerobic breakdown of pyruvate are summarized in Figure 7-7. In the first step of this sequence, catalyzed by the enzyme *pyruvate dehydrogenase*, pyruvate is oxidized in a reaction coupled to the reduction of NAD⁺ to NADH. In the process one of the three carbons present in pyruvate is released as CO_2. The remaining two carbons are joined to coenzyme A (CoA), a derivative of the vitamin pantothenic acid, to form a compound called **acetyl-CoA.** Acetyl-CoA, which is a central intermediate in the oxidative metabolism carried out by mitochondria, is produced by the breakdown of fatty acids as well as carbohy-

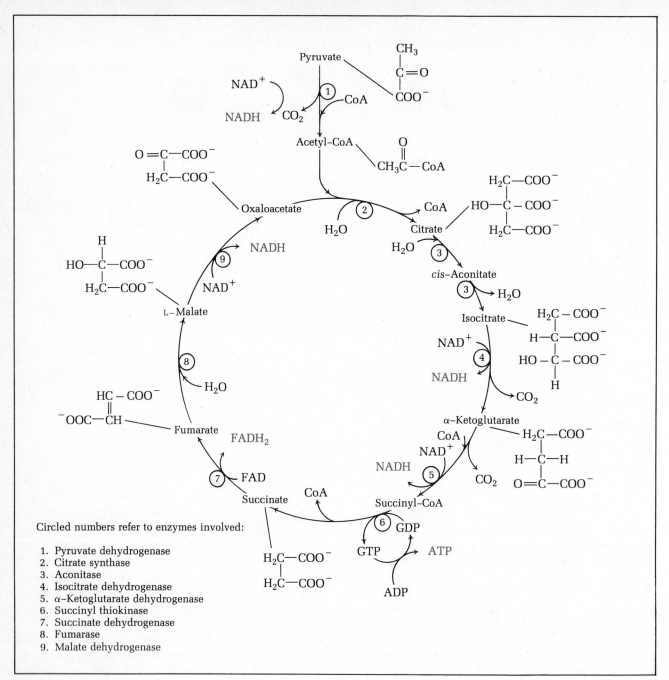

FIGURE 7-7 The Krebs citric acid cycle.

Circled numbers refer to enzymes involved:

1. Pyruvate dehydrogenase
2. Citrate synthase
3. Aconitase
4. Isocitrate dehydrogenase
5. α–Ketoglutarate dehydrogenase
6. Succinyl thiokinase
7. Succinate dehydrogenase
8. Fumarase
9. Malate dehydrogenase

drates. The degradation of fatty acids to acetyl-CoA is catalyzed by the **β-oxidation pathway**, which also occurs within mitochondria. Fatty acids liberated from stored fats are a particularly important source of acetyl-CoA during starvation, when a general deficiency of glucose and other carbohydrates exists.

The main oxidative pathway for the breakdown of acetyl-CoA begins with a reaction in which acetyl-CoA combines with the four-carbon compound **oxaloacetate** to generate the six-carbon compound **citrate** (see Figure 7-7). Citrate in turn undergoes a series of reactions that

include oxidations (accompanied by the reduction of NAD^+ to NADH or FAD to $FADH_2$), cleavages (which release CO_2), and the synthesis of ATP. This complete reaction sequence generates a total of three molecules of NADH, one molecule of $FADH_2$, and one molecule of ATP; two carbons are also released as CO_2, resulting in the conversion of citrate (six carbons) to oxaloacetate (four carbons). The oxaloacetate produced can then react with another molecule of acetyl-CoA to form a new molecule of citrate, creating a cycle that is repeated again and again.

The discovery of this cyclic set of reactions by Sir Hans Krebs in the 1930s is a classic example of elegant biochemical reasoning. By adding various four-, five-, and six-carbon compounds to homogenates of muscle tissue and analyzing the resulting products, Krebs was able to detect each of the enzymatic reactions involved in the conversion of citrate to oxaloacetate. The crucial discovery that led him to suspect that these reactions are tied together in a cycle involved use of the enzyme inhibitor *malonate*. Malonate is a competitive inhibitor of the enzyme succinate dehydrogenase, which catalyzes the reversible reaction by which succinate is converted to fumarate (see Figure 7-7). When malonate is added to muscle tissue, however, an unexpected result is obtained. The conversion of succinate to fumarate is found to be depressed, but the conversion of fumarate to succinate is not. Since the enzyme succinate dehydrogenase is inhibited when malonate is added, the conversion of fumarate to succinate must occur by some mechanism other than the simple reversal of the succinate dehydrogenase reaction. Krebs suggested that this mechanism involves a circular set of reactions by which fumarate is first converted to oxaloacetate, which is in turn converted to citrate and then back to succinate. The cyclic nature of this pathway, first predicted on the basis of these relatively simple observations, has since been confirmed by a variety of experimental approaches. In honor of Krebs the pathway is often referred to as the *Krebs cycle*, although the terms *citric acid cycle* and *tricarboxylic acid cycle* are also in common usage.

The glycolytic pathway degrades glucose to two molecules of pyruvate, which are in turn converted to two molecules of acetyl-CoA; hence two turns of the Krebs cycle are required for the metabolism of the acetyl-CoA produced by the breakdown of glucose. After two turns of the Krebs cycle are completed, it can be calculated that a total of six molecules of CO_2 have been produced (four from the two turns of the Krebs cycle and two from the reaction in which two molecules of pyruvate are converted to two molecules of acetyl-CoA). Since glucose has only six carbons, the net effect is equivalent to the complete combustion of glucose to CO_2.

If glucose has been completely oxidized to CO_2, what has become of the free energy originally present in this molecule? Some of the released energy is trapped in the two molecules of ATP produced by glycolysis and the two additional molecules of ATP synthesized during two turns of the Krebs cycle, but these four ATP molecules contain only a small fraction of the energy liberated by the oxidation of glucose. The answer is that most of the free energy becomes trapped in the form of the reduced coenzymes NADH and $FADH_2$. This energy is subsequently transformed into the chemical energy of ATP by a set of reactions collectively referred to as electron transfer and oxidative phosphorylation.

Electron Transfer and Oxidative Phosphorylation

The energy trapped in the reduced coenzymes NADH and $FADH_2$ is released by reoxidizing these substances back to NAD^+ and FAD, respectively; the energy liberated during this oxidation step is in turn utilized to drive the formation of ATP from ADP and inorganic phosphate. In an overall sense this oxidation process involves the transfer of electrons from NADH or $FADH_2$ to oxygen, thereby forming water:

$$NADH + \tfrac{1}{2}O_2 + H^+ \longrightarrow NAD^+ + H_2O$$
$$FADH_2 + \tfrac{1}{2}O_2 \longrightarrow FAD + H_2O$$

If carried out as written, however, these reactions would release an unmanageably large amount of energy. In order to capture the released energy in a more controlled fashion, cells utilize a sequential series of carriers designed to transfer electrons in a stepwise sequence from NADH or $FADH_2$ to oxygen. In this way free energy is released in a more gradual fashion and can be more readily utilized to drive the synthesis of ATP.

The sequence of carrier compounds employed for this purpose, known as the **electron transfer chain** or **respiratory chain,** is summarized in Figure 7-8. As electrons donated by NADH and $FADH_2$ are transferred through this series of carriers, the released energy is utilized to drive the synthesis of ATP. This type of ATP formation, known as **oxidative phosphorylation,** generates approximately three molecules of ATP for every NADH that donates its pair of electrons through the respiratory chain to oxygen, and approximately two molecules of ATP for every $FADH_2$ that passes its pair of electrons through the chain to oxygen. In addition to releasing energy that can be used to drive the formation of ATP, transfer of electrons from NADH and $FADH_2$ to the respiratory chain regenerates the oxidized coenzymes NAD^+ and FAD. These oxidized coenzymes can be utilized again in the Krebs cycle or in other reactions that require them.

The components of the respiratory chain and the mechanism by which they function in oxidative phosphorylation will be described in considerable detail later in the chapter. For the moment we shall conclude our general overview of glucose metabolism by summarizing the net changes in energy that occur.

Energy Yield of Glucose Oxidation

The overall result of glycolysis, the Krebs cycle, electron transfer, and oxidative phosphorylation is the oxidation of the energy-rich glucose molecule to six molecules of CO_2; six molecules of oxygen are consumed in the process, six molecules of water are produced, and energy is released. The net equation for glucose combustion can therefore be written:

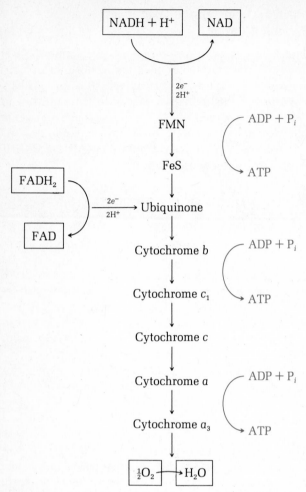

FIGURE 7-8 An overview of the mitochondrial electron transfer chain *(black)* and the associated reactions of oxidative phosphorylation *(color)*, showing the pathway followed by a pair of electrons donated to the chain by NADH or FADH$_2$. A more detailed description of the organization and operation of this chain will be presented later in the chapter (e.g., Figure 7-36).

$$C_6H_{12}O_6 + 6O_2 \longrightarrow 6CO_2 + 6H_2O + \text{Energy}$$

The $\Delta G°$ of this reaction has been experimentally determined to be -686 kcal/mol, that is, 686 kcal of energy is released per mole of glucose oxidized. Although some of this energy is simply lost as heat, a significant portion is utilized to drive the formation of ATP and the energy-rich reduced coenzymes, NADH and FADH$_2$ (Table 7-2). The energy conserved in these reduced coenzymes is subsequently used to drive the synthesis of additional ATP, though we shall see later in the chapter that some uncertainty exists as to the exact amount of ATP generated during the oxidation of these reduced coenzymes.

Determining the exact amount of ATP synthesized during glucose breakdown is further complicated by the existence of several mechanisms for oxidizing the NADH molecules generated by glycolysis. In eukaryotic cells the oxidation of this NADH is complicated by the fact that the electron transfer pathway for oxidizing NADH does not occur in the same cellular compartment as does glycolysis. Glycolysis takes place in the cell sap, while the ensuing conversion of pyruvate to acetyl-CoA, the Krebs cycle, electron transfer, and oxidative phosphorylation all occur inside mitochondria. The problem created by this arrangement is that the cytoplasmic NADH molecules generated during glycolysis cannot pass their electrons directly to the electron transfer chain because mitochondria are impermeable to NADH. As a substitute, cytoplasmic NADH passes its electrons to carrier compounds that, unlike NADH, are capable of passing through mitochondrial membranes; these carrier compounds in turn pass the electrons to the electron transfer chain. Figure 7-9 summarizes the operation of two such carrier systems, the **malate-aspartate shuttle** and the **α-glycerophosphate shuttle**. It should be noted that the malate-aspartate shuttle de-

TABLE 7-2
Summary of the production of CO$_2$ and energy-rich molecules during the oxidation of glucose in eukaryotes

Source	Yield of CO$_2$	Net ATP Yield From Substrate-Level Phosphorylation	Yield of Reduced Coenzyme	Number of ATP Molecules Formed Per Reduced Coenzyme	Total ATP Molecules Produced
Glycolysis (glucose → 2 pyruvate)			2NADH ×	1.5–3[a]	= 3–6
		2ATP			2
2 Pyruvate → 2 acetyl-CoA	2CO$_2$				
			2NADH ×	2.5–3[b]	= 5–6
Krebs cycle (2 turns)	4CO$_2$				
			6NADH ×	2.5–3[b]	= 15–18
			2FADH$_2$ ×	1.5–2[b]	= 3–4
		2ATP			2
				Total yield =	30–38

[a] The number of ATP molecules produced from the oxidation of cytoplasmic NADH depends on the particular shuttle employed for transferring electrons from NADH to the mitochondrial respiratory chain.

[b] Laboratory measurements on the number of ATP molecules produced during the oxidation of NADH and FADH$_2$ by the respiratory chain have yielded somewhat variable results; hence a range of values is provided.

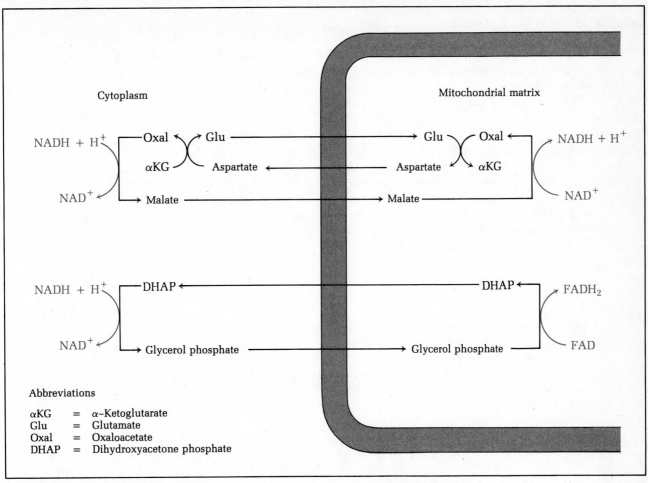

FIGURE 7-9 Diagram illustrating how the malate-asparate *(top)* and α-glycerophosphate *(bottom)* shuttles transport electrons from cytoplasmic NADH to the mitochondrial respiratory chain.

livers electrons to intramitochondrial NAD^+, while the α-glycerophosphate shuttle delivers electrons to intramitochondrial FAD. Since the oxidation of NADH by the respiratory chain yields approximately three molecules of ATP while the oxidation of $FADH_2$ generates closer to two ATPs, the malate-aspartate shuttle is inherently more efficient than the α-glycerophosphate shuttle. In addition to having their electrons transported into mitochondria by shuttle mechanisms, NADH molecules generated by glycolysis may also be oxidized to NAD^+ by peroxisomal pathways that yield no ATP at all (page 198).

Because of the preceding variations in the amount of ATP produced by the oxidation of cytoplasmic NADH, as well as current uncertainties concerning the exact yield of ATP when electrons from NADH and $FADH_2$ are passed through the electron transfer chain, the overall efficiency of energy conservation during glucose oxidation cannot be precisely determined. However, a rough approximation can be made based upon a few reasonable assumptions. First, the total amount of ATP formed per molecule of glucose oxidized

is presently thought to lie in the range of 30–38 molecules of ATP (see Table 7-2). Second, the ΔG for the reaction by which ATP is synthesized from ADP and inorganic phosphate under conditions prevailing within the cell appears to be about 11 kcal/mol, that is, the formation of each mole of ATP utilizes roughly 11 kcal. Given these two assumptions, the total energy conserved in the form of ATP can be calculated by multiplying the 30–38 mol of ATP made per mole of glucose times the 11 kcal/mol of energy utilized in making ATP, yielding a total of 330–418 kcal of energy conserved in the form of ATP. It has already been noted that the complete oxidation of one mole of glucose liberates 686 kcal; hence the overall efficiency of energy conservation can be calculated by dividing the energy utilized in making ATP (330–418 kcal) by the total energy released by glucose degradation (686 kcal), yielding an overall efficiency of energy conservation in the range of 50–60 percent. Though the remaining 40–50 percent of the energy released during the oxidation of glucose is lost as heat rather than being conserved in a useful chemical form such as ATP, the overall efficiency of this process is still quite high, especially when compared to the efficiency of most engines and other energy-transducing devices.

Because oxidative phosphorylation is responsible for generating the vast majority of the ATP molecules synthesized by eukaryotic cells under aerobic conditions, this pathway is the crucial component of the energy-conserving process. In eukaryotic cells the reactions involved in oxidative phosphorylation occur in mitochondria, while in prokaryotic cells they are associated with the plasma membrane. In spite of this difference in location, however, the underlying mechanisms appear to be quite similar. In the remainder of the chapter we shall focus primarily on the events occurring in mitochondria, where electron transfer and oxidative phosphorylation have been more extensively investigated than in bacteria. At the end of the chapter, however, we shall briefly describe the comparable phenomena that take place in bacteria.

ANATOMY OF THE MITOCHONDRION

It is difficult to say who first discovered the mitochondrion. Many nineteenth-century light microscopists noted the existence of small particles in the cytoplasm, but it is not clear how many of these were actually mitochondria. One of the first systematic studies of this organelle was initiated in 1850 by A. Kölliker, who described the existence of regularly spaced particles in the cytoplasm of muscle cells. He subsequently isolated these particles by mechanical teasing and discovered that they swell when placed in water, suggesting the existence of a surrounding semipermeable membrane. Further studies by others revealed that such particles are not restricted to muscle tissue. Although a variety of names were given to these particles as they came to be identified in various tissues, the term **mitochondrion** (threadlike granule) has come to be universally accepted.

In 1900 Leonore Michaelis made a remarkable discovery while in the process of staining live cells with the dye Janus green B. Initially, this dye stains mitochondria green, but as cells consume oxygen the color gradually disappears. Such alterations in color are known to indicate a change in the oxidation-reduction state of the dye, suggesting that mitochondria function in oxidation-reduction reactions. Unfortunately, this clue to the physiological role of mitochondria was not generally appreciated because most biologists of that era believed mitochondria to be involved in transmitting hereditary characteristics to differentiating cells. Even Otto Warburg's discovery in 1913 that granules isolated from cell homogenates are able to consume oxygen did not trigger the realization that mitochondria were responsible.

The uncertainty concerning the functional role of mitochondria was not resolved until methods were developed for isolating and purifying this organelle in a

functionally competent state. In the early 1940s Albert Claude pioneered the development of the first subcellular fractionation techniques, which involved the use of salt solutions as an homogenization medium. In spite of the damage caused by the salt solution, biochemical analysis of mitochondrial fractions prepared in this way revealed the presence of cytochrome oxidase and succinate dehydrogenase, which are components of the electron transfer chain and the Krebs cycle, respectively. In 1948 the goal of isolating metabolically active mitochondria was finally reached by George Hogeboom, Walter Schneider, and George Palade. The critical contribution of these investigators was the use of a fractionation medium containing sucrose rather than salt. This breakthrough led to the discovery by Eugene Kennedy and Albert Lehninger that the Krebs cycle, electron transfer, and oxidative phosphorylation all occur in isolated mitochondria. Because of this association of the mitochondrion with the main energy-conserving pathways, it has come to be known as the "power plant" of the cell.

A few years later George Palade and Fritiof Sjöstrand independently published the first high-resolution electron micrographs of mitochondria. In place of the relatively amorphous granule familiar to light microscopists, the mitochondrion suddenly appeared as an organelle exhibiting a complex membranous organization (Figure 7-10). Its architectural plan was seen to involve two membranes, an **outer membrane** that defines the external boundary of the organelle and an **inner membrane** exhibiting a series of infoldings or invaginations, known as **cristae,** that project into the interior cavity of the mitochondrion. The cristae most commonly assume the form of flat sheets projecting into the mitochondrial interior, but in certain organisms they may also take the form of tubules, vesicles, prisms, or concentric whorls (Figure 7-11).

The outer and inner mitochondrial membranes divide the mitochondrion into two distinct compartments (Figure 7-12). The outer compartment includes the outer mitochondrial membrane and the **intermembrane space** that separates the outer from the inner membrane. The inner compartment consists of the inner mitochondrial membrane, with its invaginating cristae, and the interior of the mitochondrion, or **matrix space.** In negatively stained preparations the inner mitochondrial membrane can be seen to contain small spherical particles protruding into the matrix space (Figure 7-13). These inner membrane **spheres** or **knobs,** which measure 8–9 nm in diameter, are more difficult to detect in thin-section electron micrographs, though proper fixation does render them visible (Figure 7-14). We shall see later that the spheres contain a protein complex, called F_1, which plays a key role in mitochondrial ATP formation. Membrane spheres comparable to those protruding from the surface of the inner

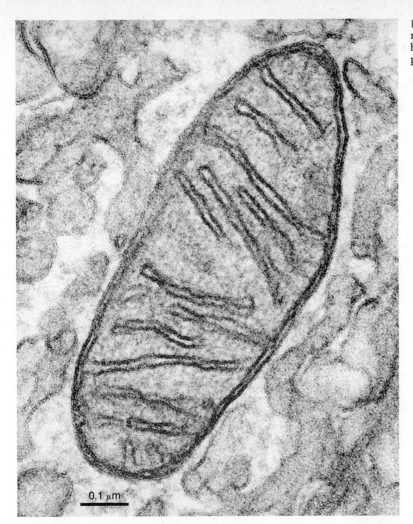

0.1 μm

FIGURE 7-10 Thin-section electron micrograph of a mitochondrion showing the outer and inner membranes and the cristae extending into the inner compartment. Courtesy of B. Tandler.

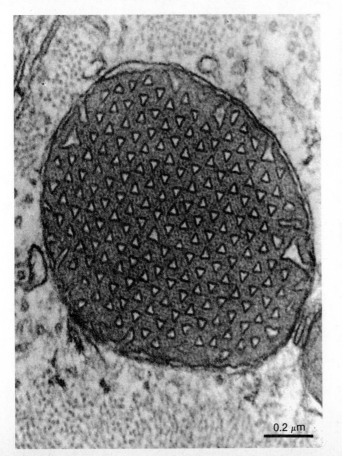

0.2 μm

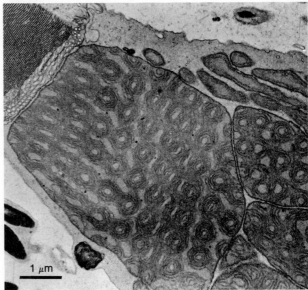

1 μm

FIGURE 7-11 Two arrangements of mitochondrial cristae that differ from the common tubular pattern. (Left) Prismatic cristae present in mitochondria of nerve cells in the hamster brain. Courtesy of N. B. Rewcastle and A. P. Anzil. (Right) Concentric cristae in giant mitochondria of photoreceptor cone cells in the eye of the tree shrew. Courtesy of T. Samorajski, J. M. Ordy, and J. R. Keefe.

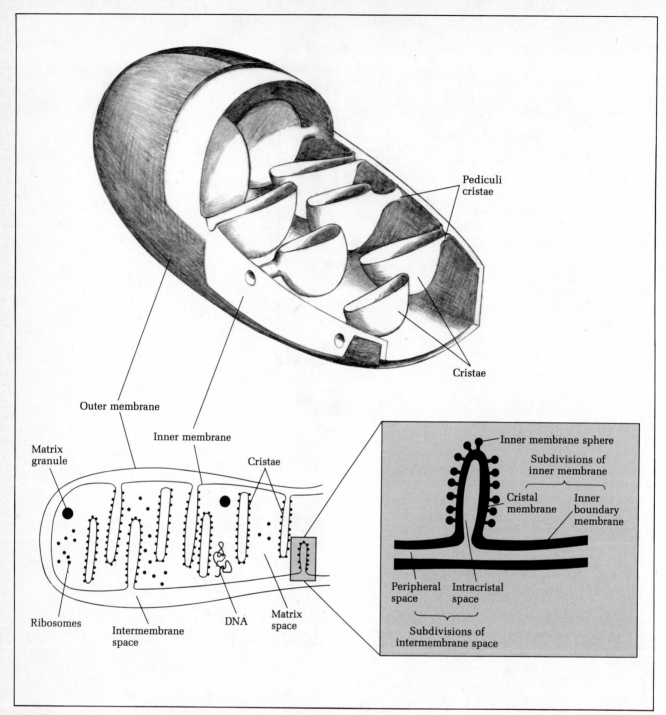

FIGURE 7-12 Diagrammatic representation of mitochondrial membranes and compartments.

mitochondrial membrane are not seen on the outer mitochondrial membrane.

Mitochondria undergo dynamic changes in morphology in response to alterations in their metabolic state. In cells where there is a large continued demand for ATP, for example, the cristae may increase in number. Morphological changes have also been observed in isolated mitochondria subjected to changing experimental conditions. Mitochondria incubated in the presence of the substrates required for electron transfer and

oxidative phosphorylation normally exhibit a *condensed* configuration in which the matrix space is reduced and the intermembrane space is expanded. If oxidative phosphorylation is prevented by omitting ADP from the medium, however, mitochondria assume the *orthodox* configuration in which the intermembrane space is small and the matrix space is distended (Figure 7-15). Such observations suggest that certain aspects of mitochondrial morphology reflect the functional state of this organelle.

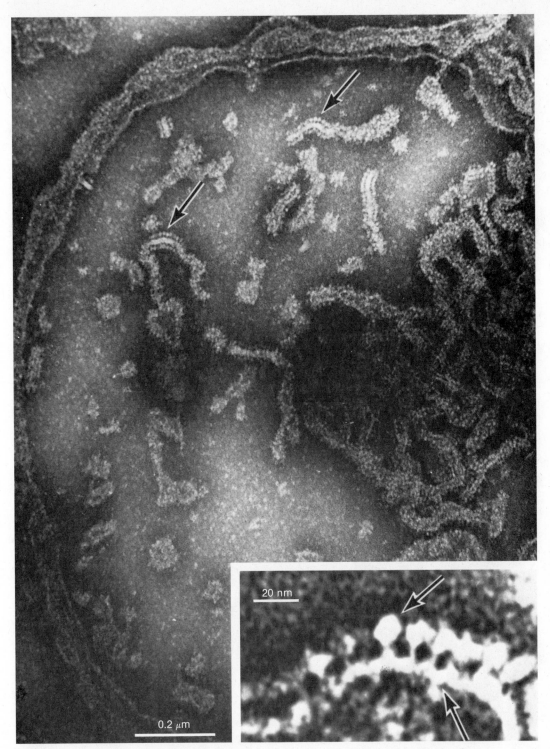

FIGURE 7-13 Electron micrograph of negatively stained fragments of beef heart mitochondria showing the spherical particles (arrows) protruding from the surfaces of the cristae. The inset shows these inner membrane spheres at higher magnification. Courtesy of H. Fernández-Morán.

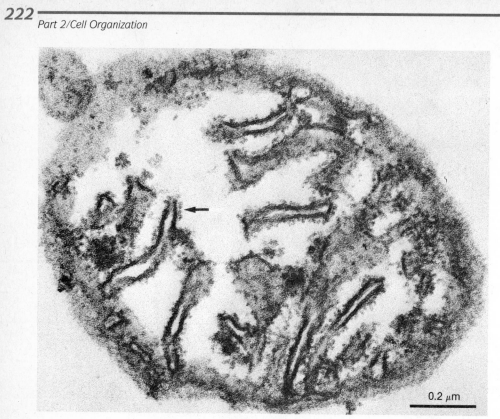

FIGURE 7-14 Thin-section electron
micrograph of brain mitochondrion
showing inner membrane spheres
projecting from the cristae (arrow).
Courtesy of E. Racker.

0.2 μm

FIGURE 7-15 *(below)* Electron micrographs illustrating the orthodox *(left)* and condensed *(right)* configurations of mitochondria. Courtesy of C. R. Hackenbrock.

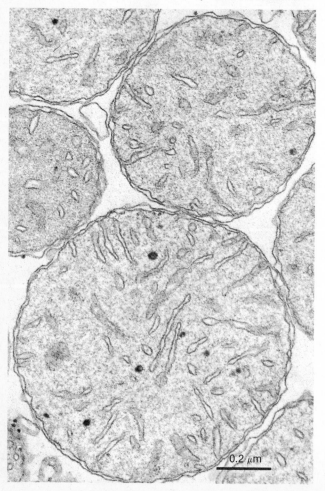

0.2 μm

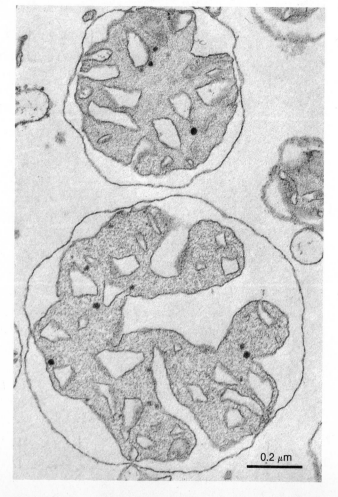

0.2 μm

Arrangement of Mitochondria Within the Cell

Mitochondria often occur in close association with structures that either utilize the ATP they produce or provide the mitochondria with oxidizable substrates. Two striking examples of such arrangements are presented in Figure 7-16. In one, muscle cell mitochondria are seen lined up adjacent to the fibrils that utilize ATP during muscle contraction. In the other, a mitochondrion is pictured surrounding a lipid droplet that contains fatty acids destined for mitochondrial oxidation.

In thin-section electron micrographs mitochondria typically appear as ellipsoid or oval profiles measuring several micrometers in length and 0.5–1.0 µm across. The widespread occurrence of the ellipsoid profile has led to the belief that most mitochondria are sausage shaped, and that typical cells contain from several hundred to a few thousand such structures. However, this commonly held view has been challenged by Hans-Peter Hoffman and Charlotte Avers, whose work on yeast cells has revealed that the three-dimensional shape of mitochondria within intact cells cannot be inferred from the examination of a few individual thin-section micrographs. After examining a consecutive series of micrographs obtained by thin-sectioning an entire yeast cell from one end to the other, these investigators were able to construct a three-dimensional model of yeast mitochondria by assembling all the mitochondrial profiles observed in the individual sections. The rather surprising result was that the ellipsoid profiles observed in individual thin sections were all found to be derived from cross sections through a single large, extensively branched mitochondrion.

When this experimental approach was applied to other eukaryotic cells, it was again found that many mitochondrial profiles seen in thin-section micrographs represent portions of larger, interconnected mitochondrial systems (Figure 7-17). Such results suggest that the number of mitochondria per cell is considerably smaller than once believed. The conclusion that mitochondria form large interconnected systems is further supported by phase contrast microscopy of living cells, where the problem of sectioning is eliminated (Figure 7-18). Because the interconnected segments of these large, branched mitochondria appear to be in a state of flux, continually pinching off and re-fusing with one another, the concept of the "number" of mitochondria per cell may in fact be meaningless.

In isolated subcellular fractions, mitochondria appear as small separate structures of relatively uniform size (Figure 7-19). It seems likely, however, that such individual "mitochondria" are artifacts caused by disruption of the branched, interconnected mitochondrial network during homogenization. This type of membrane rearrangement is similar to the fragmentation process that causes endoplasmic reticulum membranes to be broken up into microsomal vesicles during subcellular fractionation.

FIGURE 7-16 Electron micrographs showing the proximity of mitochondria to structures requiring ATP or providing substrates for ATP synthesis. *(Left)* Cross section through insect flight muscle showing myofibrils (M_f) surrounded by mitochondria (M_{it}). Courtesy of B. Sacktor. *(Right)* Section of a pancreatic exocrine cell showing a lipid droplet surrounded by a mitochondrion (arrows). Courtesy of J. D. Jamieson.

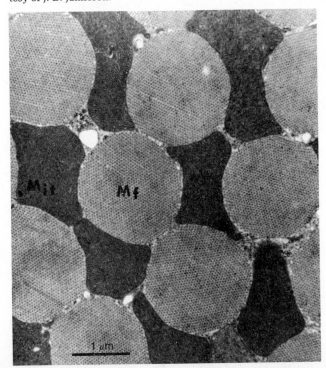

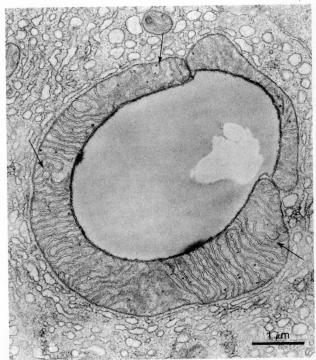

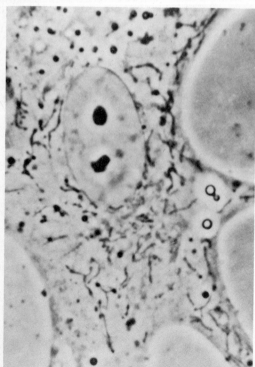

FIGURE 7-17 *(above)* Photograph of a three-dimensional model of interconnected mitochondria reconstructed from serial sections of a pig melanocyte (pigmented skin cell). The reconstruction is overlaid on a two-dimensional representation of the cell. Courtesy of M. L. Vorbeck.

FIGURE 7-18 *(left)* Phase contrast micrograph of a living fibroblast showing interconnected threadlike mitochondria in the cytoplasm and a nucleus containing two prominent nucleoli. Courtesy of D. W. Fawcett.

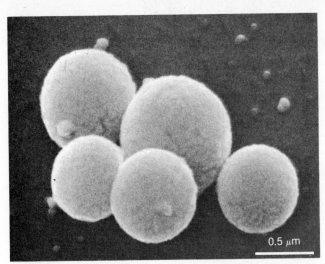

FIGURE 7-19 Scanning electron micrograph of isolated mitochondria. Although they appear as separate spherical structures, such individual mitochondria may be produced by disruption of the branched, interconnected mitochondrial network during homogenization and subcellular fractionation. Courtesy of C. R. Hackenbrock.

Isolation of Submitochondrial Components

The complex structural organization of the mitochondrion raises many questions concerning the functional significance of the various components that make up this organelle. The development of methods for separating and isolating these mitochondrial components has played an important role in advancing our knowledge in this area. The first successful technique for separating inner and outer mitochondrial membranes was developed by Donald Parsons and his colleagues in the 1960s. In this procedure mitochondria are placed in a hypotonic solution until the outer membrane ruptures, and the inner and outer membranes are then separated from each other by isodensity centrifugation. These two fractions can be readily distinguished from one another by electron microscopy; the isolated outer membranes look like empty sacs, while the isolated inner membranes form vesicles, called **mitoplasts,** which contain trapped matrix material within them (Figure 7-20). The detergent digitonin has also been useful in isolating submitochondrial fractions because it selectively disrupts the outer mitochondrial membrane and thus allows the outer and inner membranes to be separated from each other by centrifugation.

In both of the above procedures the contents of the intermembrane space are released into solution when the outer membrane is disrupted. Hence any material appearing in the supernatant fraction after the initial centrifugation can be ascribed to the intermembrane space. Once the mitoplast fraction has been isolated, it can be further separated into its membrane and matrix components by treating it with the detergent Lubrol, which disrupts the inner membrane, and recentrifuging the resulting mixture. Using this combination of techniques, the four major components of the mitochondrion can be separated from each other for biochemical analysis (Figure 7-21). It should be pointed out, however, that purified inner membrane preparations freed from matrix material generally reseal to form small ves-

FIGURE 7-20 Electron micrographs of isolated inner and outer mitochondrial membrane fractions. *(Left)* Outer membranes showing characteristic "folded bag" appearance. *(Right)* Negatively stained preparation of inner membranes (IM) with some outer membrane material (OM) still attached. Courtesy of B. Chance.

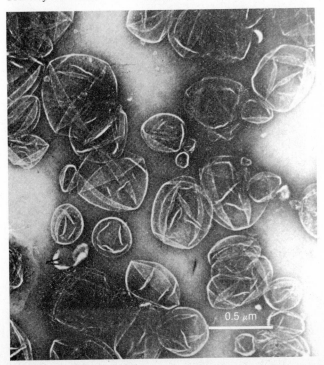

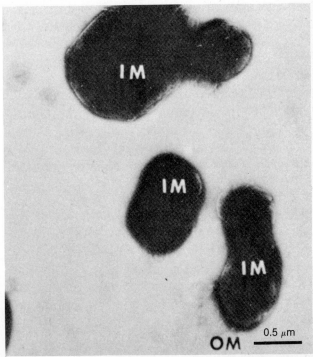

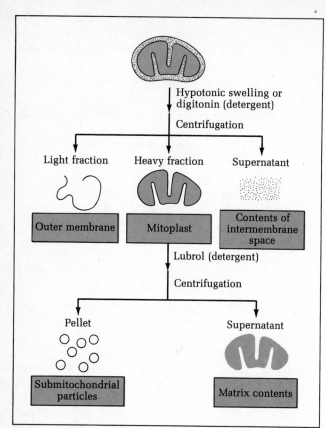

FIGURE 7-21 Flow diagram for mitochondrial subfractionation.

icles, termed **submitochondrial particles,** whose orientation is opposite to that of mitoplasts and intact mitochondria. In submitochondrial particles the inner membrane spheres protrude from the exterior surface of the vesicle instead of facing the interior cavity.

Analysis of the composition and properties of submitochondrial fractions has provided considerable insight into the location of particular metabolic activities

within this organelle (Table 7-3). The enzymes of the Krebs cycle and fatty acid β-oxidation are found mainly in the matrix fraction, while components involved in electron transfer and oxidative phosphorylation are localized in the inner membrane fraction. Although this accounts for the major mitochondrial activities discussed in the present chapter, the presence in submitochondrial fractions of enzymes involved in lipid breakdown, the urea cycle, the synthesis of fatty acids and certain amino acids, and the oxidation of substrates not directly related to oxidative phosphorylation points to the conclusion that the mitochondrion is a metabolically multifunctional organelle.

Because each submitochondrial component exhibits its own unique composition and function, we shall now examine each of their properties in more detail.

The Outer Mitochondrial Membrane

Striking similarities exist between the outer mitochondrial membrane and the membranes of the endoplasmic reticulum. A constant exchange of lipids between the two membrane systems causes them to closely resemble each other in lipid composition. Direct physical connections between these two types of membranes have occasionally been observed, and cytochrome b_5, an unusual protein considered to be a characteristic marker for endoplasmic reticulum membranes, has been identified in the outer mitochondrial membrane as well. In spite of these basic similarities, however, differences are also apparent. The enzyme monoamine oxidase, for example, is present in the outer mitochondrial membrane but not in the endoplasmic reticulum, while glucose-6-phosphatase occurs in the endoplasmic reticulum but not the outer mitochondrial membrane.

TABLE 7-3
Submitochondrial distribution of selected components

Outer Membrane	Intermembrane Space	Inner Membrane	Matrix Space
Cytochrome b_5	Adenylate kinase	NADH dehydrogenase	Fatty acid β-oxidation enzymes
NADH-cytochrome b_5 reductase	Nucleoside	Succinate dehydrogenase	Pyruvate dehydrogenase
Monoamine oxidase	diphosphokinase	Cytochrome b-c_1 complex	Krebs cycle enzymes
Fatty acyl-CoA synthetase	Nucleoside	Cytochrome oxidase	Citrate synthase
Glycerolphosphate-acyl	monophosphokinase	Cytochrome c	Aconitase
transferase		F_0-F_1 complex	Isocitrate dehydrogenase
Nucleoside diphosphokinase		Transport carriers (e.g., for	α-Ketoglutarate dehydrogenase
		adenine nucleotides, phosphate,	Fumarase
		Ca^{2+}, Na^+, etc.)	Malate dehydrogenase
			DNA polymerase
			RNA polymerase
			Ribosomes, protein synthesis
			factors, and transfer RNAs
Lipid content		Lipid content	
Phospholipid/protein = 0.9		Phospholipid/protein = 0.3	
Cardiolipin/phospholipid = 0.03		Cardiolipin/phospholipid = 0.22	
		Ubiquinone	

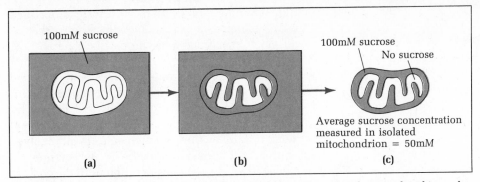

FIGURE 7-22 Measurement of mitochondrial permeability. (a) Mitochondria are placed in a solution containing 100mM sucrose. (b) The sucrose equilibrates across the outer mitochondrial membrane. (c) Mitochondria are removed from the sucrose solution and their overall sucrose content is measured. If the average sucrose concentration of the isolated mitochondria is 50mM instead of 100mM, it can be concluded that the sucrose attains access to only half the total mitochondrial volume.

The enzymes detected in outer mitochondrial membrane fractions represent a curious collection (see Table 7-3). It is not clear whether those that are similar to enzymes present in the endoplasmic reticulum play a legitimate function in the mitochondrion, or whether they are simply contaminants carried along during isolation. The presence of enzymes involved in lipid synthesis, such as the glycerolphosphate-acyl transferases, suggests a possible involvement in the biosynthesis of membrane phospholipids. Monoamine oxidase is involved in the metabolism of the neurotransmitter epinephrine and its derivatives (Chapter 18), but the physiological significance of its localization in the outer mitochondrial membrane is obscure.

The permeability of the outer mitochondrial membrane has been studied by incubating isolated mitochondria in the presence of various solutes and then measuring the amount of solute taken up. Most small molecules, such as salts, sugars, nucleotides, and coenzymes, penetrate rapidly. Their final concentration within the mitochondrion, however, is usually less than their concentration in the incubation medium, suggesting that only a portion of the mitochondrial fluid volume can equilibrate its solutes with the external environment (Figure 7-22). The volume of this readily permeable compartment, which varies from 20 to 80 percent of the total mitochondrial volume, correlates well with the size of the intermembrane space as visualized by electron microscopy. It has therefore been concluded that the outer membrane is readily permeable to small molecules, while the inner membrane is not.

When one considers the fact that the inner mitochondrial compartment is the site of the major metabolic reactions occurring in mitochondria, the highly permeable nature of the outer membrane seems appropriate because it allows the inner compartment more direct access to metabolites present in the cell sap. This picture is also compatible with theories that proclaim that mitochondria evolved hundreds of millions of years ago from parasitic bacteria (Chapter 14). According to this viewpoint, the inner membrane-matrix complex has descended from the bacterial plasma membrane and its enclosed cytoplasm, while the outer mitochondrial membrane is either derived from the bacterial outer membrane or represents an addition contributed by the membranes of the host cell. The latter possibility might explain the close resemblance between the outer mitochondrial membrane and membranes of the endoplasmic reticulum.

The Intermembrane Space

Although it is commonly assumed that the intermembrane space is a single continuous chamber bounded by the inner and outer membranes, some reservations have been expressed as to whether this is always the case. In thin-section electron micrographs the cristae often do not appear to be attached to the rest of the inner membrane, suggesting that the "intracristal space" situated between the folds of the cristal membranes may not be in complete continuity with the "peripheral space" located in the regions where the outer and inner membranes are closely opposed. (For clarification concerning the exact regions of the intermembrane space designated by these terms, refer to Figure 7-12.) Three-dimensional reconstructions generated from examination of consecutive thin sections through intact mitochondria have provided a more precise picture of how these membranes and spaces are arranged. Such analyses reveal that the large flattened cristae are connected to the rest of the inner membrane by relatively small tubular channels known as **pediculi cristae.** These channels measure as small as a few dozen nanometers in diameter, suggesting that they may be small enough to at least partially restrict continuity between the intracristal and peripheral regions of the intermembrane space. It is nonetheless clear that some continuity exists between these two areas because small solutes penetrate into both regions when added to suspensions of isolated mitochondria.

The highly permeable nature of the outer mitochondrial membrane causes the solute composition of the intermembrane space to closely mirror that of the cell sap. The number of enzymes present in the intermembrane space appears to be relatively small. Prominent among these is adenylate kinase, an enzyme that catalyzes transfer of the terminal phosphate group of ATP to AMP, forming two molecules of ADP.

The Mitochondrial Matrix

The matrix space contains all the enzymes and cofactors involved in the Krebs cycle with the single exception of succinate dehydrogenase, which is located in the inner membrane because it catalyzes the direct transfer of electrons from the Krebs cycle intermediate, succinate, to the electron transfer chain. Also present in the matrix are pyruvate dehydrogenase (which catalyzes the conversion of pyruvate to acetyl-CoA), and the enzymes involved in fatty acid β-oxidation, which degrade fatty acids into acetyl-CoA units that enter the Krebs cycle. Analyses carried out on isolated matrix preparations have revealed that the protein concentration is too high for all the matrix proteins to be in true solution. It has therefore been proposed that the enzymes of the Krebs cycle and fatty acid β-oxidation are anchored within a structural framework. This idea has received support from experiments carried out on mitochondria that have been artificially swollen by suspending them in a hypotonic medium. Under such conditions the matrix does not randomly disperse, but instead openings form in what appears to be an organized network.

In electron micrographs the matrix generally exhibits a finely granular appearance, although large matrix granules ranging from 30 to several hundred nanometers in diameter also occur. In addition DNA, RNA, ribosomes, and other enzymes and factors involved in nucleic acid and protein synthesis are present in the matrix. Since these components are all involved in the process of mitochondrial growth and division, we shall delay their further consideration until the chapter on mitochondrial biogenesis (Chapter 14).

The Inner Mitochondrial Membrane

The inner mitochondrial membrane is comprised of two distinct regions, the cristae and the inner boundary membrane (see Figure 7-12). Although these two regions of the inner membrane exhibit points of continuity with each other, the existence of at least some functional differences is suggested by the fact that the inner membrane spheres or knobs mentioned earlier appear to reside predominantly on the cristae. Because the cristae account for the bulk of the inner membrane fraction, most discussions concerning the functional properties of the inner membrane are in fact references to the cristae.

The inner mitochondrial membrane is unique in both chemical composition and morphological appearance. Its morphological uniqueness stems primarily

FIGURE 7-23 Electron micrograph showing mitochondria that have been stained with the cytochemical reaction for cytochrome oxidase. Note that the black reaction product, which indicates the presence of cytochrome oxidase, is localized on the cristae but not in the matrix space or on the adjacent muscle fibers. Courtesy of W. A. Anderson.

from the spherical particles that protrude from the matrix side of the cristae and thereby impart an asymmetry to the membrane. The inner membrane is also unusual in its high protein-to-lipid ratio, its high ratio of unsaturated to saturated phospholipids, the virtual absence of cholesterol, and the presence of large amounts of the unusual lipid, cardiolipin. The large protein content of the inner membrane results from the presence of an extensive number of proteins (50 or more) involved in electron transfer and oxidative phosphorylation. The conclusion that the inner membrane is the major site of electron transfer and oxidative phosphorylation rests not just on biochemical analyses of isolated membrane preparations but on cytochemical staining reactions as well (Figure 7-23).

Although the role of the inner membrane in electron transfer and oxidative phosphorylation will be our primary focus for the rest of the chapter, it is important to note that the inner membrane also performs other functions. In contrast to the outer membrane, which is freely permeable to small molecules, the inner membrane is an impermeable barrier to most nucleotides, sugars, and small ions. Small molecules in this category that need to be taken up or expelled by mitochondria are transported by specific carrier proteins situated within the inner membrane. The operation of these carriers will be described in more detail after we have discussed the mechanism by which the inner membrane conserves the energy released during electron transfer.

The inner mitochondrial membrane is also the site of certain enzymatic pathways not directly related to energy metabolism. Among the more important are the pathways of steroid metabolism, which involve cooperation between mitochondria and the endoplasmic reticulum. The first step in the synthesis of steroid hormones, for example, involves cleavage of the side chain of cholesterol by an enzyme present in the inner mitochondrial membrane. The product of this reaction, pregnenolone, is then released to the endoplasmic reticulum for subsequent processing. The resulting products are again taken up by mitochondria, where they are hydroxylated by NADPH-dependent enzymes. Both mitochondrial reactions involve a membrane-bound cytochrome P-450.

HOW MITOCHONDRIA MAKE ATP

Cells synthesize ATP in two fundamentally different ways: by substrate-level phosphorylation and by oxidative phosphorylation. **Substrate-level phosphorylation** refers to any chemical reaction in which the synthesis of ATP is an integral part of a reaction by which a substrate is converted to product. Such reactions commonly take the following form:

$$X\text{-}P + ADP \longrightarrow X + ATP$$

Reactions of this type include two steps in glycolysis

(see Figure 7-5), as well as the succinyl thiokinase reaction of the Krebs cycle, which involves the conversion of succinyl-CoA to free succinate and coenzyme A (see Figure 7-7).

In contrast **oxidative phosphorylation,** a membrane-bound reaction responsible for generating most of the ATP formed under aerobic conditions, involves the coupling of ATP synthesis to oxidation reactions:

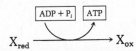

where X_{ox} and X_{red} are the oxidized and reduced forms of X, and P_i is inorganic phosphate. In reactions of this type, formation of a proton gradient across the inner mitochondrial membrane underlies the mechanism by which ATP synthesis is coupled to the oxidation of substrate. In the following sections we shall discuss the intricate series of steps by which this overall process is accomplished.

Components of the Respiratory Chain

Oxidative phosphorylation is driven by the energy released when electrons, initially derived from the oxidation of foodstuffs, flow through the respiratory chain to oxygen. To understand the mechanism underlying oxidative phosphorylation, we must therefore begin by examining the organization of the respiratory chain. At least five different kinds of molecules are constituents of the chain: pyridine nucleotides, flavoproteins, cytochromes, iron-sulfur proteins, and quinones.

1. The **pyridine nucleotide** NAD^+ serves as a mobile carrier that accepts electrons from three Krebs cycle reactions and the pyruvate dehydrogenase reaction, and transports them to the respiratory chain. The study of this coenzyme has been greatly facilitated by the discovery that its reduced form, NADH, exhibits a markedly enhanced absorbance at 340 nm (Figure 7-24). The appearance and disappearance of this absorption band can be used to monitor the oxidation state of the coenzyme and hence the progress of the reaction in which it is involved.

2. Flavoproteins containing either FMN or FAD as their prosthetic group function in the respiratory chain as well as in the Krebs cycle. In contrast to pyridine nucleotides, flavins absorb light in the oxidized rather than in the reduced state. Oxidized flavoproteins typically exhibit several absorption peaks in the region of the spectrum between 360 and 450 nm (see Figure 7-24).

3. Electron transfer proteins containing a heme prosthetic group (Figure 7-25) are called **cytochromes.** The heme group contains a central iron atom capable of accepting and donating electrons. In contrast to pyridine and flavin nucleotides, oxidation and reduction of the iron atom of the heme group is accomplished by the

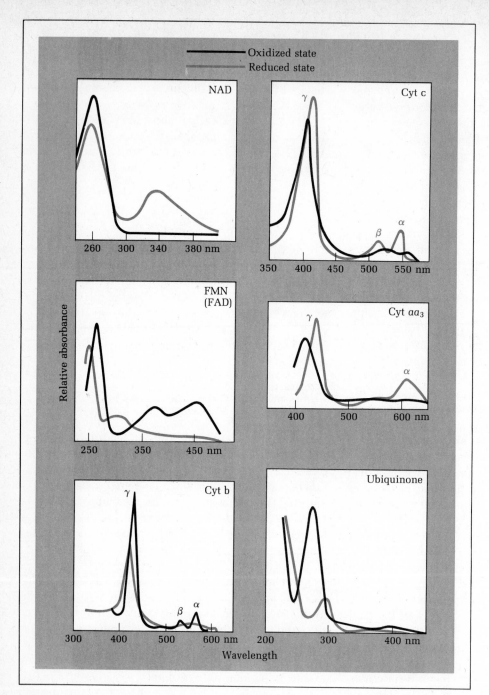

FIGURE 7-24 Absorption spectra of some respiratory chain components in the oxidized (black) and reduced (color) states. Abbreviation: Cyt = cytochrome.

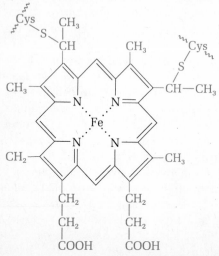

FIGURE 7-25 Structure of the heme group. The side chains (color) attached to the porphyrin ring (black) vary among different cytochromes. The particular heme illustrated is present in cytochrome c. Most other cytochromes do not have their heme groups covalently attached to the polypeptide chain. Abbreviation: Cys = cysteine residue of protein.

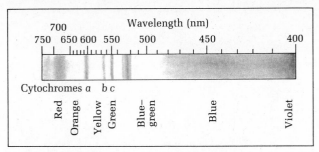

FIGURE 7-26 Absorption spectrum of intact muscle as seen by Keilin with a spectroscope. The three major bands are the α-bands of cytochromes *a*, *b*, and *c*.

transfer of a single electron rather than an electron pair. Reduction involves conversion of Fe^{3+} to Fe^{2+} by addition of a single electron, while oxidation involves the reverse reaction.

Cytochromes were first discovered in a series of elegant experiments carried out in the 1920s by David Keilin. In these classic studies light was first passed through a thin piece of insect muscle or a culture of yeast cells and was then put through a prism to divide the light into the spectrum of colors of which it is composed. Keilin discovered that the muscle or yeast cells caused three dark bands to appear in the spectrum (Figure 7-26), indicating the presence of molecules that absorb light at these particular wavelengths. The three major bands, occurring at approximately 605, 566, and 550 nm, were named cytochromes *a*, *b*, and *c*, respectively. The most significant discovery made by Keilin was that bubbling oxygen through a suspension of yeast causes the absorption bands to disappear. With remarkable insight, he concluded that cytochromes function in cellular oxidative pathways in which electrons are transferred from foodstuffs to oxygen.

Subsequent work by Keilin and others led to the isolation and characterization of the proteins responsible for each of the absorption bands. The individual cytochromes were found to differ from each other in size, absorption spectra, and the type of heme ring they contain. In the intact mitochondrion most of the cytochromes, with the exception of cytochrome *c*, were found to be so firmly attached to the inner membrane that they are difficult to extract and purify. Refined spectroscopic analysis revealed that each cytochrome band is a mixture of at least two components: cytochrome *b* = a mixture of cytochromes b_T and b_K, cytochrome *c* = a mixture of cytochromes *c* and c_1, and cytochrome *a* = a mixture of cytochromes *a* and a_3 (the cytochrome a-a_3 complex is alternatively referred to as **cytochrome oxidase**). In the reduced state most of the cytochromes exhibit three absorption bands, designated α, β, and γ. The γ-bands are the most intense; but it is the α-bands that Keilin originally discovered and that are most useful because they are more clearly separated from each other, and they undergo more dramatic changes upon oxidation-reduction (see Figure 7-24).

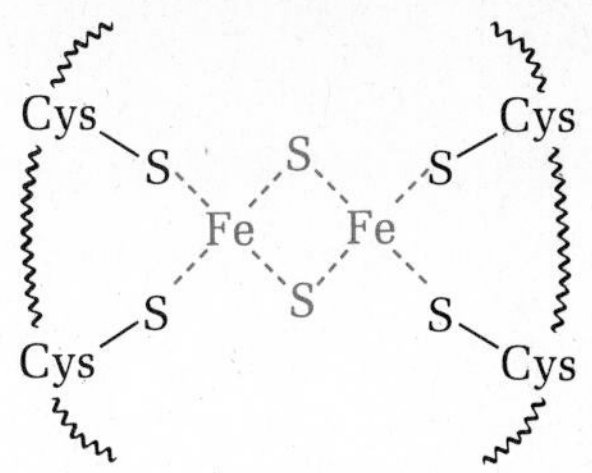

FIGURE 7-27 Configuration of the active site of iron-sulfur proteins. Abbreviation: Cys = cysteine residue of protein.

In addition to the iron atoms contained within its heme groups, the cytochrome a-a_3 complex contains two atoms of bound copper. These copper atoms are thought to transfer single electrons by $Cu^{2+} \rightleftharpoons Cu^+$ transitions in the same way that iron atoms transfer electrons by $Fe^{3+} \rightleftharpoons Fe^{2+}$ transitions.

4. Not all the iron present in the respiratory chain occurs in the heme groups of cytochromes. The more recently discovered **iron-sulfur proteins** lack the heme group, but contain two, four, or eight iron atoms per molecule and exhibit the rather unusual property of evolving hydrogen sulfide when exposed to acid. The amount of hydrogen sulfide liberated is equimolar to the amount of iron present, suggesting that the iron and sulfur atoms are part of the same active site. The organization of these "iron-sulfur centers" has been found to involve two iron atoms, each one attached to the protein chain by bonds to a pair of cysteine residues and linked to the opposing iron atom by two sulfur cross-bridges (Figure 7-27). It is the sulfur cross-bridges that are released as sulfide by acid treatment. Iron-sulfur centers function in oxidation-reduction reactions by $Fe^{3+} \rightleftharpoons Fe^{2+}$ transitions of its iron atoms. For each pair of iron atoms present in an iron-sulfur center only one can transfer electrons, so the total number of electrons carried per active site is one, just as in cytochromes.

5. Ubiquinone (also called **coenzyme Q** or simply "Q") is an unusual member of the respiratory chain in that it is a lipid rather than a protein. Several forms of ubiquinone exist that differ in the lengths of their isoprenoid side chains (Figure 7-28). Reduction of ubiquinone involves the addition of two hydrogens and their associated pair of electrons; oxidation occurs by the reverse process. The oxidation-reduction state of ubiquinone can be monitored by measuring its absorbance at 275 nm, which is greatly enhanced in the oxidized state (see Figure 7-24). In certain microorganisms, quinones other than ubiquinone may function in the respiratory chain.

Sequential Arrangement of Carriers

The sequence of the respiratory chain, which was illustrated earlier in the chapter (see Figure 7-8), has been unraveled through a combination of experimental approaches. First, examination of the standard redox po-

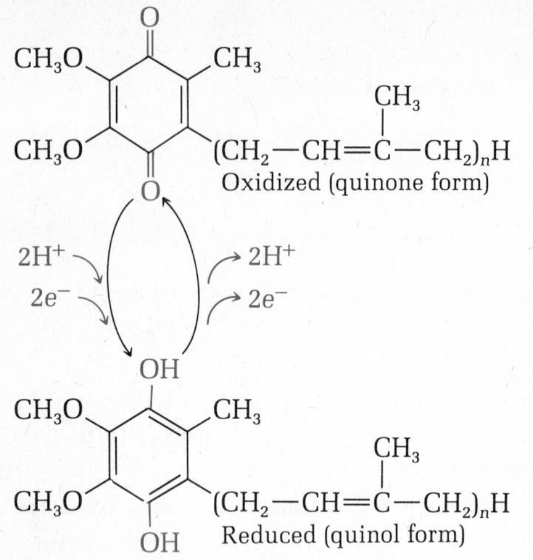

O

CH₃O ⎯⎯⎯⎯ CH₃

CH₃O ⎯⎯⎯⎯ (CH₂—CH=C—CH₂)ₙH
                          |
                          CH₃
O
Oxidized (quinone form)

$2H^+$ ⇄ $2H^+$
$2e^-$ ⇄ $2e^-$

OH

CH₃O ⎯⎯⎯⎯ CH₃

CH₃O ⎯⎯⎯⎯ (CH₂—CH=C—CH₂)ₙH
                          |
                          CH₃
OH   Reduced (quinol form)

FIGURE 7-28 Structure of the oxidized and reduced forms of ubiquinone. Although the length of the side chain is variable, in most mammalian mitochondria n = 10.

tentials of the carriers allows one to list them in order beginning with those that have the greatest tendency to give up electrons and ending with those that have the greatest tendency to accept electrons (see Table 7-1). Such information yields a sequence of carriers that is for the most part consistent with that deduced from other approaches, though it should be pointed out that a consideration of redox potentials only provides information as to the energetically most favorable sequence of reactions. It does not necessarily indicate that this identical sequence is utilized by the cell.

An alternative approach for determining the sequence of the respiratory chain is to examine the ability of isolated carriers to oxidize and reduce each other. It has been shown in this way, for example, that NADH will pass its electrons to the FMN group of NADH dehydrogenase but not to any of the cytochromes. Under similar conditions the transfer of electrons from NADH to cytochromes has been found to require the presence of both NADH dehydrogenase and ubiquinone. These data give rise to the overall sequence: NAD → FMN group of NADH dehydrogenase → ubiquinone → cytochromes.

In spite of the usefulness of the above approaches, they do not provide indisputable evidence concerning the order of the carriers within intact mitochondria. The most direct way for studying the behavior of individual carriers while they still reside in their normal environment is to monitor their light-absorbing properties, which vary with oxidation and reduction. Such measurements are usually carried out by taking separate samples of mitochondria in the reduced and oxidized states and measuring the difference in the light-absorbing properties of one versus the other. The result, called a **difference spectrum,** exhibits a series of peaks and valleys that correspond to the wavelengths at which the absorbance of the respiratory carriers differs between the oxidized and reduced state (Figure 7-29).

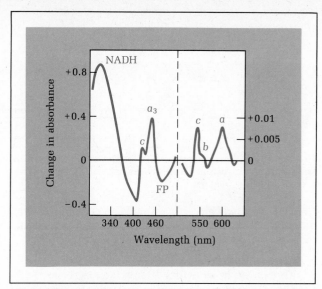

FIGURE 7-29 Difference spectrum of isolated rat liver mitochondria. The colored line represents the difference in absorbance between mitochondria in a reduced state (excess substrate, no oxygen) and those in an oxidized state (excess oxygen). Most of the respiratory carriers appear as peaks because they absorb more strongly when reduced. One exception is the FMN group of NADH dehydrogenase, which absorbs less in its reduced state and therefore appears as a valley. The prominence of each peak or valley can be used to determine the oxidation-reduction state of the individual carriers. The small letters (c, a₃, c, b, and a) refer to the various cytochromes.

The amplitude of these peaks and valleys can therefore be used to calculate the relative degree of oxidation of each of the individual carriers. In order to increase the sensitivity and reliability of this method, Britton Chance has measured the absorption of each carrier at two wavelengths, a wavelength at which its absorption change is maximal, and a nearby reference wavelength where its absorption does not change upon oxidation-reduction. This *dual-wavelength* approach permits nonspecific changes in the light-absorbing properties of mitochondria to be canceled out by subtracting the absorbance reading obtained at the reference wavelength from the reading obtained at the point of maximum absorption.

Difference spectra have been used in several ways to provide information concerning the sequence of the respiratory chain. In one approach, oxygen is added to oxygen-depleted mitochondria and the resulting changes in the difference spectra are recorded with time. Under such conditions the first component to become oxidized is cytochrome a₃, followed sequentially by cytochromes a, c, and b, the FMN group of NADH dehydrogenase, and NADH. These results indicate that cytochrome a₃ is located closest to oxygen in the respiratory chain, and that each of the remaining components is located progressively further from oxygen in the sequence.

In an alternative approach difference spectra have been employed in conjunction with the use of inhibitors that block the flow of electrons at specific points in the respiratory chain. All carriers located prior to the site where electron flow is blocked will remain in the reduced state because their electrons cannot be passed on to oxygen; in contrast, carriers located beyond the block will remain oxidized after they have passed on their electrons because the blockage earlier in the chain prevents them from receiving any more electrons. Using difference spectra to measure which carriers remain reduced and which are oxidized in the presence of various inhibitors therefore allows one to deduce the order in which these carriers are arranged. For example, examination of difference spectra obtained from mitochondria exposed to the inhibitor *antimycin* reveals that NADH, the FMN group of NADH dehydrogenase, ubiquinone, and cytochrome *b* are all fully reduced, while cytochromes *c*, *a*, and a_3 become fully oxidized. This pattern is compatible with the interpretation that antimycin blocks the flow of electrons between cytochromes *b* and *c*. In this experiment the region between cytochromes *b* and *c* is referred to as a crossover point because all the carriers on one side are reduced and all the carriers on the other side are oxidized. Analysis of the crossover points generated in the presence of other inhibitors of electron transfer (Table 7-4) has helped to elucidate the overall sequence of the respiratory chain.

The sequence of the respiratory chain presented in Figure 7-8 has been derived from data generated by a combination of the above approaches. The arrangement of certain components, especially in the ubiquinone–cytochrome *b* region, is still somewhat tentative and may be more complicated than indicated. One feature

of this particular sequence that deserves special mention is that several steps involve either the release or uptake of protons. For example, the transfer of a pair of electrons from the FMN group of NADH dehydrogenase to its iron-sulfur centers is accompanied by the release of protons, while the subsequent transfer of electrons to ubiquinone involves proton uptake. The oxidation of ubiquinone again involves the loss of protons, while the transfer of electrons to oxygen to form water is associated with the uptake of protons. As we shall discuss later, these proton fluxes are believed to play an important role in the process by which the energy released during electron transfer is utilized to drive the formation of ATP.

Another aspect of the respiratory chain that requires comment concerns the number of electrons transferred in each step. The first two carriers, NADH and the FMN group of NADH dehydrogenase, transfer electron pairs. The iron-sulfur centers following FMN, however, each carry a single electron. Ubiquinone again transfers electron pairs, while cytochromes carry single electrons. Finally, the reduction of oxygen to water, which is the last step in the sequence, requires four electrons: $O_2 + 4e^- + 4H^+ \rightarrow 2H_2O$. This terminal step is especially crucial because the reduction of oxygen by less than four electrons would generate toxic substances such as hydrogen peroxide and the superoxide anion (O_2^-). Some mechanism must therefore ensure that electron flow is organized in such a way that the terminal oxygen receives four electrons.

One factor that may assist in this process is the grouping together of the individual components of the respiratory chain into organized protein-lipid com-

TABLE 7-4
Selected inhibitors of mitochondrial ATP synthesis

Agent	Mechanism of Action
Inhibitors of electron transfer	Inhibit electron transfer between
Rotenone	FMN → Q
Amytal	FMN → Q
Piericidin	FMN → Q
Antimycin A	$b \rightarrow c$
Carbon monoxide	$aa_3 \rightarrow O_2$
Cyanide	$aa_3 \rightarrow O_2$
Uncoupling agents	Inhibit ATP synthesis, but not electron transfer, by acting as
2,4-Dinitrophenol	Proton ionophore that abolishes $\Delta pH + \Delta \psi$
Gramicidin	Proton ionophore that abolishes $\Delta pH + \Delta \psi$
Carbonyl cyanide *m*-chlorophenyl hydrazone (CCCP)	Proton ionophore that abolishes $\Delta pH + \Delta \psi$
Valinomycin[a]	K^+ ionophore that abolishes $\Delta \psi$
+	
Nigericin[a]	H^+/K^+ exchange ionophore that abolishes ΔpH
Inhibitors of oxidative phosphorylation	Inhibit ATP synthesis, and secondarily inhibit electron transfer, by binding to
Oligomycin	F_0
Aurovertin	F_1

[a] Valinomycin and nigericin uncouple oxidative phosphorylation only when added in combination.

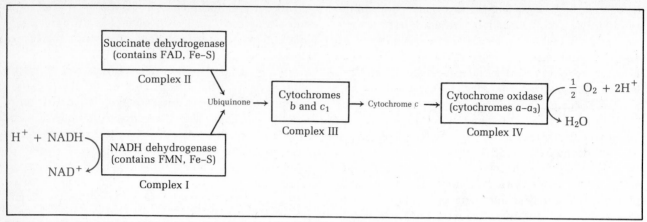

FIGURE 7-30 Organization of the respiratory chain into four complexes linked by soluble carriers.

plexes. If one extracts mitochondrial membranes under gentle conditions, only ubiquinone and cytochrome *c* are readily released. The remaining electron carriers stay tightly bound to the membrane, and are not released until the membrane is disrupted by harsh detergent or salt extractions. When such harsh procedures are used, the carriers are released into solution not as individual molecules but as four distinct complexes. Each complex contains carriers responsible for electron flow through a particular region of the respiratory chain: **Complex I** catalyzes the transfer of electrons from NADH to ubiquinone, **complex II** from succinate to ubiquinone, **complex III** from ubiquinone to cytochrome *c*, and **complex IV** from cytochrome *c* to oxygen. The respiratory chain can thus be viewed as a series of protein-lipid complexes that pass electrons from one end of the chain to the other (Figure 7-30).

Freeze-fracture electron microscopy has provided some important insights into the topological relationships that exist between these complexes in intact mitochondrial membranes. Because the inner mitochondrial membrane is virtually free of cholesterol and

has an unusually high ratio of unsaturated to saturated phospholipids, its fluidity is exceptionally high. One can therefore ask the question: Are the four respiratory complexes free to diffuse randomly within the plane of the membrane, or are they linked together to form a true "chain"? This question has been probed in the laboratory of Charles Hackenbrock, where artificial phospholipid vesicles were fused with the inner mitochondrial membrane to increase its phospholipid content. Freeze-fracture micrographs of the resulting membranes revealed that the average distance between intramembrane particles increases as membrane phospholipid is added (Figure 7-31), indicating that these particles are free to diffuse within the membrane rather than being joined together as a latticed network. Biochemical analyses of such phospholipid-enriched membranes further revealed that the rate of electron transfer between complexes is depressed, presumably because the average distance separating the complexes has been increased. These experiments indicate that the respiratory chain is not an ordered structural chain within the membrane, but consists of several independent complexes that are

FIGURE 7-31 Freeze-fracture electron micrographs of mitochondrial inner membrane showing the arrangement of intramembrane protein particles in membranes with differing phospholipid content. *(Left)* Control membranes with no added phospholipid. *(Middle* and *Right)* As increasing amounts of phospholipid are added to the membrane, note the increasing distance between the intramembrane particles. Such observations indicate that the membrane particles are free to diffuse within the membrane. Courtesy of C. R. Hackenbrock.

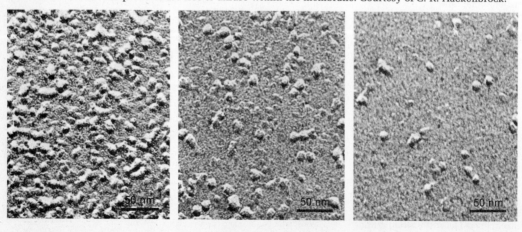

free to diffuse randomly within the lateral plane of the membrane. Such an arrangement suggests that factors that influence the diffusional movement of the respiratory chain complexes may play a role in regulating the rate of electron transfer through the chain.

Coupling Sites for Oxidative Phosphorylation

The conclusion that ATP synthesis accompanies the flow of electrons through the respiratory chain was first proposed in the late 1930s by Vladimir Belitzer, whose measurements of the rate of ATP synthesis and oxygen consumption in minced muscle preparations led him to conclude that at least two ATPs are formed per pair of electrons passed to oxygen. Subsequent measurements by other investigators have led to the conclusion that this **P/O ratio** (number of ATPs formed per oxygen atom reduced) is actually closer to three, though disagreement still exists about its exact value. Calculations on the amount of free energy released during the various stages of electron transfer have led to the conclusion that the reactions catalyzed by complexes I, III, and IV each yield sufficient energy to drive ATP formation. Based on this and other experimental observations to be described later, it has been proposed that each of these three complexes contains a **coupling site** where the energy released during electron transfer is utilized to drive ATP formation. Coupling site 1 is associated with complex I (NADH → ubiquinone), coupling site 2 is associated with complex III (ubiquinone → cytochrome *c*), and coupling site 3 is associated with complex IV (cytochrome *c* → oxygen). In contrast the reaction catalyzed by complex II, which involves the transfer of electrons from the Krebs cycle intermediate succinate to ubiquinone, does not yield sufficient energy to drive ATP formation.

This tentative assignment of ATP coupling sites has been experimentally confirmed using "site-specific" assays that utilize electron transfer inhibitors in combination with added electron donors and/or acceptors. For example, in mitochondrial vesicles incubated in the presence of antimycin to prevent the flow of electrons past cytochrome *b*, NADH can be added to act as electron donor and ubiquinone can act as electron acceptor. Under such conditions roughly one molecule of ATP is synthesized per pair of electrons transferred from NADH to ubiquinone. It can therefore be concluded that a coupling site for ATP synthesis occurs between NADH and ubiquinone. If an artificial electron donor known to reduce cytochrome *c* is employed in place of NADH, one ATP molecule is produced per pair of electrons transferred from cytochrome *c* to oxygen. Hence a coupling site for ATP synthesis must occur between cytochrome *c* and oxygen. Using this general approach, the presence of coupling sites associated with complexes I, III, and IV has been experimentally verified.

The above arrangement of coupling sites means that a pair of electrons donated to the respiratory chain from NADH will encounter three coupling sites for ATP synthesis because the electrons pass through complexes I, III, and IV, all of which have coupling sites for ATP formation. In contrast electron pairs donated by succinate (from the Krebs cycle) encounter only two coupling sites for ATP synthesis because the electrons pass through complexes II, III, and IV, only two of which have coupling sites for ATP formation (complexes III and IV).

Reconstitution of the Inner Membrane

Another powerful approach for dissecting the role of individual mitochondrial constituents in electron transfer and oxidative phosphorylation involves taking the inner membrane apart and reassembling it from selected components. Such experiments usually begin with submitochondrial particles, which consist of inner membrane vesicles freed from associated matrix materials (see Figure 7-21). In the presence of NADH, ADP, and inorganic phosphate, submitochondrial particles are capable of carrying out both electron transfer and oxidative phosphorylation. The first successful disassembly and reconstitution of this system was accomplished in the laboratory of Efraim Racker in the 1960s. In these experiments submitochondrial particles were disrupted by treatment with urea or shaking with glass beads, and were then separated into two fractions by centrifugation. The pellet produced in this way was found to carry out electron transfer but not oxidative phosphorylation, while the supernatant was capable of neither electron transfer nor oxidative phosphorylation. Adding the supernatant back to the pellet was found to restore the ability of the pellet to carry out oxidative phosphorylation (Figure 7-32).

Such observations suggested that the supernatant contains a factor responsible for coupling ATP synthesis to the flow of electrons through the respiratory chain. Although examination of the supernatant did not reveal the presence of any molecule capable of directly catalyzing the synthesis of ATP, an enzyme catalyzing ATP hydrolysis was identified. It was proposed that this enzyme, termed **F_1-ATPase** or **F_1**, is responsible for catalyzing the synthesis of ATP by intact mitochondrial membranes. F_1 alone does not catalyze ATP synthesis, however, because the respiratory chain must be present to provide the energy that drives the ATP-forming reaction.

Electron microscopic examination of the various fractions generated in the above experiments provided significant insights into the architectural arrangement of the mitochondrial inner membrane (see Figure 7-32). The original fraction, consisting of intact submitochondrial particles, was found to contain membranous vesi-

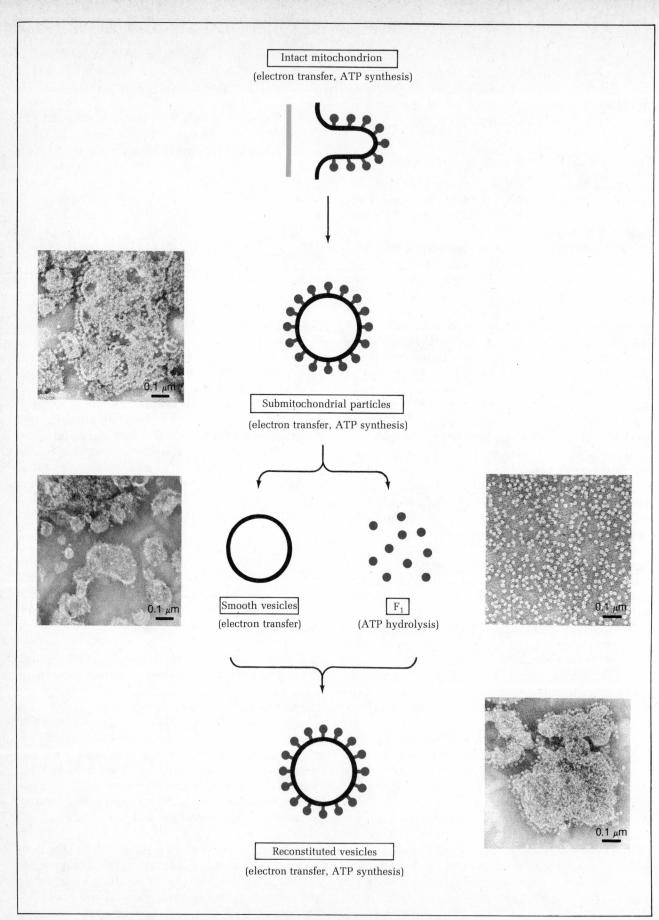

FIGURE 7-32 A diagrammatic summary of the steps involved in the disassembly and reconstitution of the mitochondrial inner membrane, indicating the metabolic capabilities and morphological appearance of the various fractions. Micrographs courtesy of E. Racker.

cles with inner membrane spheres protruding from their external surfaces. After treatment under conditions that remove the F_1-ATPase, the membrane vesicles no longer exhibited these attached spheres; the spheres were instead found floating free in the F_1-containing supernatant fraction. Addition of the F_1-containing supernatant back to the F_1-depleted membranes yielded vesicles in which the spheres were again seen protruding from the external surface. Since F_1-depleted vesicles lose their ability to perform oxidative phosphorylation but retain the capacity to carry out electron transfer via the respiratory chain, these results suggest that the respiratory chain is localized within the smooth portion of the membrane and that ATP synthesis occurs in the spherical particles protruding from the membrane surface.

The reconstitution approach has also been extended to the analysis of individual coupling sites. In one such study Racker's group reconstituted membrane vesicles from a mixture of isolated phospholipids, complex I, F_1, and a group of hydrophobic membrane proteins, termed **F_0**, that normally function to anchor F_1 to the mitochondrial membrane. Membrane vesicles generated in this way are capable of coupling the synthesis of ATP to the flow of electrons from NADH to ubiquinone; coupling site I has thus been reconstituted in an artificially constructed membrane. Comparable results have since been obtained with the other coupling sites. Because ATP synthesis in such reconstituted systems requires the presence of both F_1 and F_0, the F_1/F_0 complex is sometimes referred to as **ATP synthase**.

The Chemiosmotic Theory

The major reason for dissecting apart the components of the inner mitochondrial membrane is to try to unravel the mechanism by which the energy released during electron transfer is employed to drive the synthesis of ATP. Early work on this problem produced a vast torrent of complex literature and ideas, prompting an expert in the field to comment that anyone who is not confused about oxidative phosphorylation just does not understand the situation. In recent years, however, an innovative theory proposed in the early 1960s by Peter Mitchell has begun to dominate this field of research, providing us with a unifying framework for understanding the basic mechanisms underlying energy conservation not just in mitochondrial membranes, but in chloroplast and bacterial membranes as well.

Prior to the development of Mitchell's ideas, oxidative phosphorylation was widely believed to proceed by the formation of high-energy chemical intermediates. According to this viewpoint, the free energy released during an oxidation-reduction reaction can be conserved by linking one of the substrates of the reaction to some unidentified substance to form a high-energy intermediate:

$$A_{red} + B_{ox} + X \longrightarrow X \sim A_{ox} + B_{red}$$

where A and B are components of the respiratory chain, and X is the unidentified substance employed to activate the chemical intermediate. In a subsequent reaction the high-energy intermediate ($X \sim A$) would be broken down in a reaction that makes ATP:

$$X \sim A_{ox} + ADP + P_i \longrightarrow X + A_{ox} + ATP$$

The above theory, which was first clearly formulated in 1953 by E. C. Slater, led to a vigorous search for the hypothetical high-energy intermediates. Many claims were made to the effect that the elusive intermediate had been discovered, but all proved to be unfounded. While the biochemists of the time were feverishly occupied in the hunt for such intermediates, the alarming suggestion was made by Peter Mitchell that they, like the Emperor's clothes, might not exist at all. In 1961 Mitchell proposed that the flow of electrons through the respiratory chain drives the transport of protons from one side of the mitochondrial inner membrane to the other. The result would be an electrochemical gradient consisting of two components: a difference in electric charge or membrane potential ($\Delta\psi$), and a difference in concentration of protons (ΔpH). Mitchell further postulated that the synthesis of ATP is driven by the reverse flow of protons down this electrochemical gradient.

This **chemiosmotic theory** originally met with considerable skepticism, especially since it was proposed in the virtual absence of experimental data. As years have passed, however, a considerable body of supporting data has been amassed, and it now appears almost certain that the basic tenets of this theory are correct, though its details are still to be refined. Some of the more crucial data, and their implications for the chemiosmotic model, are summarized below.

1. Shortly after the chemiosmotic model was first proposed, Mitchell and his colleague Jennifer Moyle experimentally verified that proton transport across the mitochondrial membrane occurs in conjunction with the flow of electrons through the respiratory chain. In these experiments mitochondria were first suspended in a medium in which electron transfer could not occur because oxygen was lacking. The proton concentration (pH) of the medium was then monitored as electron transfer was stimulated by the addition of known quantities of oxygen. Under such conditions the pH of the medium declined rapidly. Because a decline in pH reflects an increase in proton concentration, it can be concluded that electron transfer causes the mitochondrial inner membrane to pump protons from the mitochondrial matrix into the external medium (the presence of the outer mitochondrial membrane can be ignored because it is freely permeable to small molecules and ions). The magnitude of the observed pH change is

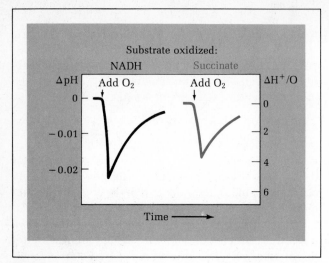

FIGURE 7-33 Data obtained from an experiment in which the pH of the external medium was monitored after exposure of anaerobic mitochondria to a brief pulse of oxygen. Under such conditions oxidation of NADH by the electron transfer chain is accompanied by the expulsion of roughly six protons from the mitochondrial matrix per pair of electrons passing from NADH to oxygen; during the oxidation of succinate, roughly four protons are expelled per pair of electrons transferred from succinate to oxygen.

greater when NADH is oxidized by the respiratory chain than when succinate is oxidized (Figure 7-33), as would be expected on the basis of the fact that the oxidation of NADH by the respiratory chain yields more ATP than the oxidation of succinate. Quantitation of this difference in pH revealed that roughly six protons are pumped out of the mitochondrion per NADH oxidized, and four protons per succinate oxidized. Taken together with the fact that NADH oxidation involves three coupling sites for ATP synthesis and succinate involves two sites, it was postulated that each coupling site pumps a pair of protons from the mitochondrial matrix to the intermembrane space, and that the electrochemical proton gradient established in this way is eventually utilized to drive ATP synthesis (Fig-

ure 7-34). Although subsequent studies have led to some differences in opinion concerning the exact number of protons transported at each site, such investigations have generally supported the conclusion that each coupling site establishes a proton gradient that in turn drives ATP formation.

2. Thermodynamic calculations indicate that the magnitude of the proton gradient is sufficient to account for the amount of ATP synthesized. In metabolically active mitochondria, the gradient consists of a pH difference of about 1.4 pH units (lower pH outside) and an electrical potential or $\Delta\psi$ of approximately 140 mV (more positive outside). Using the previously derived equation for determining the free-energy content of an electrochemical gradient (page 140), one can calculate that the $\Delta G°$ for such a gradient is approximately 5.3 kcal/mol of protons. If two protons are employed per molecule of ATP synthesized, then $2 \times 5.3 = 10.6$ kcal of energy are available. Current estimates suggest that the ΔG for ATP synthesis under conditions prevailing within the mitochondrion is around 11 kcal/mol, which is quite close to the 10.6 kcal provided by the proton gradient.

Although the preceding calculations indicate that the free-energy content of the proton gradient is within the general range required for driving ATP formation, the absolute values of the numbers involved should not be taken too literally. First of all, it was mentioned in the previous section that there is a difference of opinion as to how many protons are actually transported per coupling site. Mitchell's proposal of two protons per site appears to be a minimum value, and if the real value is higher, the free-energy content of the proton gradient would be correspondingly higher. Differences of opinion also exist as to exactly how many molecules of ATP are synthesized per coupling site. Before the advent of the chemiosmotic model, it was believed that each coupling site synthesizes one molecule of ATP per pair of electrons transferred. Hence the oxidation of NADH,

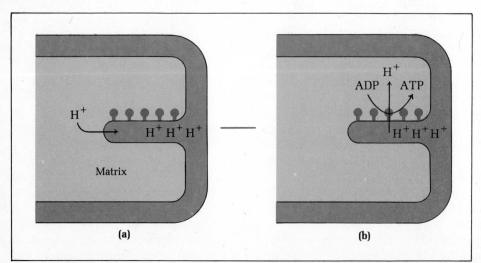

FIGURE 7-34 Orientation of proton flow across the inner mitochondrial membrane. (a) During passage of electrons through the respiratory chain, protons are pumped from the mitochondrial matrix to the intermembrane space (because of the permeability of the outer mitochondrial membrane, these protons can readily pass from the intermembrane space to the external environment). (b) The subsequent diffusion of protons down their electrochemical gradient drives ATP synthesis.

whose electrons pass through three coupling sites, was thought to be accompanied by the formation of exactly three molecules of ATP. With the chemiosmotic model, however, there is no need to be constrained to integer numbers such as 3.0, and experimental measurements have suggested that the number of ATP molecules formed per molecule of NADH oxidized may actually lie somewhere between 2 and 3 (e.g., 2.5).

3. The fact that a proton gradient is established during electron transfer and that it contains sufficient free energy to drive ATP formation do not prove that it actually does so. More direct evidence has been obtained by the use of agents that abolish the electrochemical gradient by making the inner membrane freely permeable to protons. The first such substance to be studied in mitochondrial systems was *dinitrophenol*, a compound known for many years to inhibit oxidative phosphorylation but not electron transfer. Agents of this type, which allow electrons to flow through the respiratory chain but prevent the released energy from being used to drive ATP formation, are known as **uncoupling agents.** In 1963 Mitchell demonstrated that dinitrophenol is a proton ionophore that makes biological membranes freely permeable to protons. Hence in the presence of dinitrophenol, membranes are incapable of maintaining an electrochemical proton gradient. Because dinitrophenol abolishes the ability to make ATP, this observation suggests that the existence of an electrochemical gradient is a necessity for ATP formation.

Other agents that enhance membrane permeability to protons have also been found to uncouple oxidative phosphorylation. Two of the more commonly employed agents of this type are carbonyl cyanide *m*-chlorophenylhydrazone (CCCP) and the peptide antibiotic gramicidin. Such substances completely uncouple oxidative phosphorylation from electron transfer because they abolish both the pH and membrane potential components of the electrochemical gradient. Ionophores that promote the diffusion of ions other than protons often influence only one of these two constituents of the electrochemical gradient. Valinomycin, for example, dissipates the membrane potential by making membranes freely permeable to potassium ions. Hence in the presence of valinomycin, the membrane potential created by the extra protons outside the mitochondrion is abolished by the valinomycin-mediated flow of K^+ into the mitochondrion. In contrast, the ionophore nigericin facilitates a neutral exchange of protons for potassium ions. Because this exchange is electrically neutral, the membrane potential is unaffected. But by allowing protons to equilibrate across the mitochondrial membrane, this ionophore abolishes the pH gradient. In the presence of either valinomycin or nigericin alone, oxidative phosphorylation can continue because either a pH gradient or a membrane potential is still present. In combination, however, both components of the electro-

chemical gradient are dissipated and ATP synthesis is uncoupled from electron transfer.

When mitochondria are uncoupled, the energy released during electron transfer is lost as heat instead of being employed to drive ATP formation. In hibernating animals, the mitochondria present in brown fat are deliberately uncoupled so that this released heat can help to maintain body temperature. In such cases special proteins or fatty acids inserted into the mitochondrial inner membrane may serve as proton ionophores.

4. The question of whether proton gradients are responsible for driving ATP formation has also been addressed by exposing mitochondria or submitochondrial particles to artificial pH gradients. If mitochondria are suspended in a solution in which the external pH is suddenly lowered by the addition of acid, ATP will be formed in response to the artificially created proton gradient. Such artificial pH gradients induce ATP formation even in the absence of oxidizable substrates capable of passing electrons to the respiratory chain, indicating that ATP synthesis can be induced by proton gradients even in the absence of electron transfer.

5. An obvious prediction of the chemiosmotic model is that oxidative phosphorylation requires an intact mitochondrial membrane enclosing a defined compartment. Otherwise, the proton gradient that drives oxidative phosphorylation could not be maintained. This prediction has been verified by the discovery that electron transfer carried out by isolated respiratory carriers cannot be coupled to ATP synthesis unless the carriers are incorporated into membranes that form intact vesicles.

6. Because it requires that protons be selectively pumped in one direction (from inside the mitochondrion to outside), the chemiosmotic model predicts that the electron carriers involved in the three coupling sites are asymmetrically oriented within the mitochondrial inner membrane. The topographical arrangement of carriers within the mitochondrial inner membrane has been studied using specific antibodies, proteolytic enzymes, and nonpenetrating labeling reagents as probes. Such experiments have revealed that the components of the respiratory chain are asymmetrically organized across the mitochondrial inner membrane.

7. Experiments involving the reconstitution of membrane vesicles from mixtures of isolated components have provided considerable support for the chemiosmotic theory. The predicted ability of each of the three coupling sites to transport protons, for example, has been directly verified by reconstituting artificial phospholipid vesicles containing either complex I, III, or IV. When provided with appropriate oxidizable substrates, such vesicles establish proton gradients corresponding to approximately two protons pumped per pair of electrons transferred (Figure 7-35).

The source of protons pumped by the three cou-

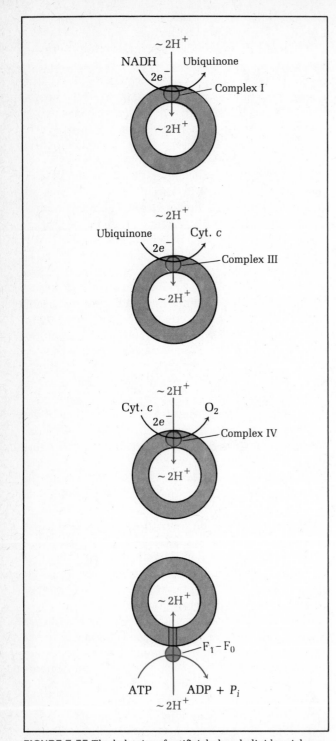

FIGURE 7-35 The behavior of artificial phospholipid vesicles reconstituted in the presence of various mitochondrial protein complexes. Vesicles reconstituted with complex I, III, or IV transport two protons per electron pair transferred from an appropriate substrate to an appropriate electron donor. Vesicles reconstituted in the presence of F_1-F_0 take up two protons per molecule of ATP hydrolyzed. This latter observation suggests that the reverse reaction (ATP synthesis) is driven by the flow of two protons in the opposite direction.

pling sites is easiest to envision for site 1. Within intact mitochondria, NADH present in the mitochondrial matrix donates a pair of electrons and a proton to the FMN group of NADH dehydrogenase. After taking up another proton from the matrix space, this FMN group transfers the two protons to the opposite side of the membrane; when the FMN group subsequently passes its electrons to the iron-sulfur centers, the two protons are released into the intermembrane space (Figure 7-36). Proton transport by sites 2 and 3 involves a more complicated series of steps whose details are yet to be fully elucidated.

8. The reconstitution approach has also been employed to demonstrate that two protons are sufficient to drive the synthesis of a molecule of ATP. In these experiments, artificial phospholipid vesicles were reconstituted with F_1-ATPase and a group of inner membrane proteins, designated F_0, which normally anchor the F_1-ATPase to the mitochondrial membrane. The resulting vesicles contained the F_1 spheres on their outer surfaces. Since the respiratory chain complexes are not present, such vesicles cannot carry out ATP synthesis. However, they will hydrolyze ATP to ADP and inorganic phosphate; at the same time, approximately two protons are pumped into the vesicle (see Figure 7-35). It can therefore be concluded that in the reverse reaction, which occurs during mitochondrial oxidative phosphorylation, synthesis of ATP by the membrane spheres is accompanied by the movement of protons in the opposite direction.

9. The mechanism by which the proton gradient drives ATP formation is yet to be clearly elucidated, but several experimental observations have begun to shed light on this issue. We have already seen that the F_1-containing spherical particles attached to the membrane surface are thought to be the normal site of ATP synthesis. F_1 is anchored to the inner membrane by attachment to a group of integral membrane proteins, collectively referred to as F_0. Certain inhibitors of oxidative phosphorylation, such as **oligomycin,** act by binding to F_0.

A clue to the role of F_0 has emerged from the discovery that removal of F_1 causes the mitochondrial membrane to become leaky to protons. These leaks can be sealed by treating the membranes with oligomycin or by rebinding of F_1, suggesting that F_0 serves as a transmembrane channel through which protons flow down their electrochemical gradient on the way to F_1. According to this view, the ability of oligomycin to inhibit oxidative phosphorylation is based on its capacity to prevent the flow of protons through F_0; this inhibitory mechanism is quite different from that of uncoupling agents, which inhibit oxidative phosphorylation by making the membrane more permeable to protons.

Both direct and indirect mechanisms have been proposed to explain how protons interacting with F_1

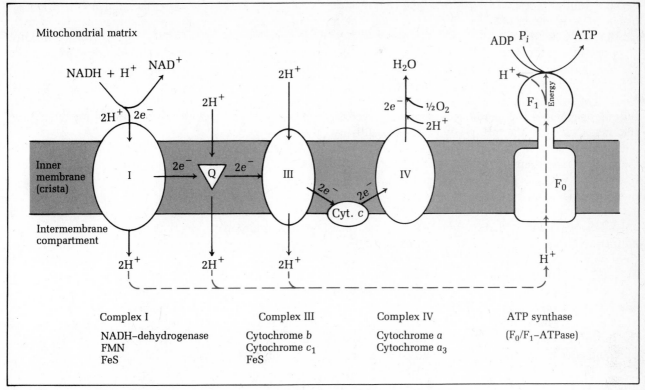

Complex I — NADH–dehydrogenase, FMN, FeS
Complex III — Cytochrome b, Cytochrome c_1, FeS
Complex IV — Cytochrome a, Cytochrome a_3
ATP synthase — (F_0/F_1–ATPase)

FIGURE 7-36 A highly stylized model of the inner mitochondrial membrane indicating the asymmetric arrangement of carriers and their possible roles in proton transport. The organization of the carriers in certain regions of the sequence, especially in the ubiquinone–cytochrome b region, is somewhat tentative and may be more complicated than pictured. Abbreviations: FeS = iron-sulfur center, Q = ubiquinone.

might drive the synthesis of ATP. Mechanisms of the first type involve the direct participation of protons in the reaction by which ATP is made, while mechanisms of the second type are based on the idea that the passage of protons through F_1 induces conformational changes that promote ATP synthesis. It has been experimentally verified that F_1 undergoes changes in conformation when mitochondrial membranes are exposed to a pH gradient, but it is difficult to ascertain whether such changes are responsible for triggering ATP synthesis or whether they are simply an incidental by-product of the catalytic mechanism.

Inner Membrane Transport Reactions

Although the main function of the proton gradient generated by the respiratory chain is to drive ATP synthesis, several inner membrane transport systems for moving ions and small molecules into and out of the mitochondrion also appear to be driven by the proton gradient (Figure 7-37). One such system involves a membrane carrier that expels sodium ions from the mitochondrion in exchange for protons. The high external proton concentration generated by activity of the respiratory chain drives this exchange process. The mem-

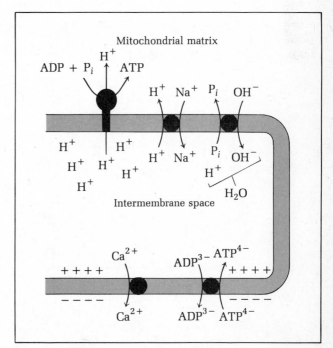

FIGURE 7-37 Diagrammatic representation of some of the events driven by the electrochemical proton gradient in mitochondria.

brane potential plays no role in this type of transport because the exchange of one sodium ion for one proton is electrically neutral.

For some mitochondrial transport systems, the membrane potential rather than the difference in proton concentration is the driving force. The uptake of calcium ions, for example, is dependent on the magnitude of the membrane potential, but is not influenced by the magnitude of the pH gradient. The positively charged calcium ions appear to be pulled across the mitochondrial membrane by electrostatic attraction to the matrix side, which is negatively charged relative to the mitochondrial exterior.

This same membrane potential appears to be involved in the transport of ADP and ATP. In the presence of uncoupling agents, which prevent a proton gradient from being established, ADP and ATP are transported into and out of mitochondria at equal rates. Under normal conditions, however, ADP is selectively taken up and ATP is selectively expelled, an appropriate arrangement for an organelle designed to manufacture ATP for the rest of the cell. This selectivity is ultimately based upon the effect of the membrane potential on the nega-

tively charged molecules, ADP and ATP. ADP has a charge of -3, while ATP has a charge of -4. Hence the uptake of a molecule of ADP coupled to the expulsion of a molecule of ATP is equivalent to the net expulsion of one negative charge. Because the membrane potential causes the exterior of the mitochondrion to be positively charged relative to the interior, the membrane potential can serve as a driving force for this type of exchange.

The above considerations suggest that uptake of ADP and export of ATP by mitochondria utilize some of the free energy stored in the electrochemical proton gradient. The uptake of inorganic phosphate, which is the other substrate required for mitochondrial ATP synthesis, is also driven by the proton gradient. In this case phosphate is taken up in exchange for hydroxyl ions; an expelled hydroxyl ion will combine with a proton to form water, thus decreasing the external proton concentration. The combined effect of these transport events is equivalent to the uptake of one proton per ATP molecule exported from the mitochondrion. Any protons employed in this way will not, of course, be available for driving ATP synthesis.

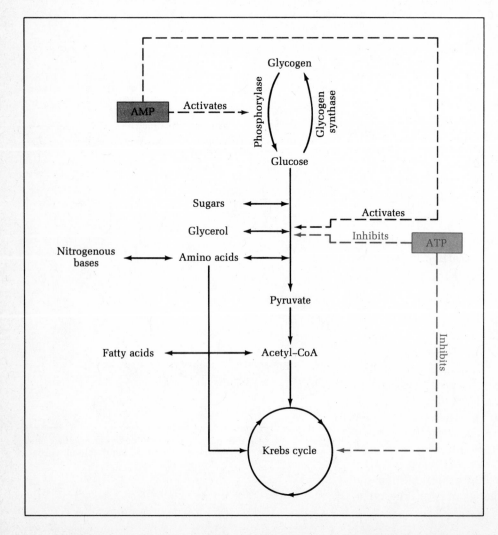

FIGURE 7-38 Summary of the pathways involved in glucose metabolism, emphasizing the points at which ATP and AMP regulate key allosteric enzymes.

Regulation of ATP Formation

The oxidation of glucose and the utilization of the released energy to drive ATP formation involves a complex set of interrelated cytoplasmic and mitochondrial pathways whose activities must be precisely coordinated and regulated. One obvious situation that calls for such regulation involves alterations in oxygen availability. When oxygen is absent, only two molecules of ATP can be formed per molecule of glucose metabolized. Under aerobic conditions, on the other hand, mitochondrial oxidative phosphorylation permits a maximum of 30–38 ATPs to be synthesized per molecule of glucose oxidized. It therefore follows that glycolysis must proceed more rapidly in anaerobic cells than in aerobic cells to maintain the same level of ATP production. This type of regulation was first observed over a century ago by Louis Pasteur, who discovered that exposure of anaerobic cells to oxygen causes the rate of glucose consumption to decrease dramatically.

The mechanism of this so-called **Pasteur effect,** however, was not elucidated until more recently. The enzyme *phosphofructokinase*, which catalyzes a key rate-limiting step in glycolysis, has been found to be allosterically controlled by substances such as ATP and AMP. Under aerobic conditions, the large amount of ATP produced by mitochondrial oxidative phosphorylation acts to inhibit phosphofructokinase, thus slowing down glycolysis. Under anaerobic conditions, on the other hand, the intracellular concentration of ATP falls because of the lack of oxidative phosphorylation. The utilization of ATP by energy-requiring activities causes it to be broken down to ADP, which is subsequently degraded to AMP. This AMP then acts as an allosteric activator of phosphofructokinase, stimulating glycolysis and causing more ATP to be made. By exerting similar allosteric effects on several other key enzymes, ATP and AMP help to coordinate and regulate the activities of the various metabolic pathways involved in glucose degradation and ATP formation (Figure 7-38).

In addition to ATP and AMP, ADP also plays a role in the regulation of energy-transforming pathways. When most of the cell's ADP has been converted to ATP by oxidative phosphorylation, the resulting fall in ADP concentration somehow causes the rate of electron transfer through the respiratory chain to decrease. This phenomenon, termed **respiratory control,** prevents the continued oxidation of NADH and FADH$_2$ under conditions where mitochondria are unable to synthesize ATP because of a lack of the required substrate, ADP.

HOW BACTERIA MAKE ATP

Although bacteria lack mitochondria, the metabolic pathways by which they synthesize ATP using energy released during the oxidation of foodstuffs are remarkably similar to those occurring in eukaryotes. Because

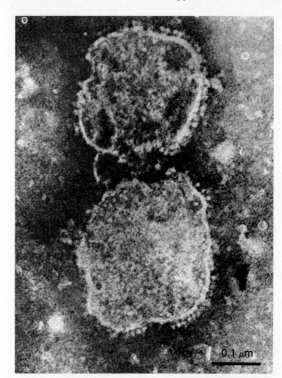

FIGURE 7-39 Electron micrograph of a negatively stained plasma membrane fragment from the bacterium *Mycobacterium* showing spherical particles protruding from the membrane surface. Courtesy of A. F. Brodie.

electron transfer and oxidative phosphorylation are membrane-bound reactions in eukaryotes, it might be expected that membranes are involved in prokaryotic cells as well. This idea has been verified by isolating plasma membrane vesicles from bacteria and demonstrating that they are capable of oxidizing NADH, using the energy released to drive the formation of ATP. In addition, electron micrographs of negatively stained bacterial membranes reveal the presence of spherical particles protruding from the inner surface that resemble the spheres present on the mitochondrial inner membrane (Figure 7-39). Isolation of these spherical particles has led to the discovery that they contain an ATPase activity comparable to that of mitochondrial F$_1$-ATPase.

The detailed organization of bacterial respiratory chains differs somewhat from that of mitochondrial chains. In the bacterium *E. coli* the chain begins with the transfer of electrons from NADH to the FAD (rather than FMN) group of a membrane-bound flavoprotein. The electrons next pass through a series of iron-sulfur proteins to ubiquinone, from which they are transferred through cytochromes *b* and *o* to oxygen (Figure 7-40). For each pair of electrons passing through this sequence of carriers, roughly four protons are exported from the bacterial cell.

As in mitochondria, the synthesis of each molecule of ATP appears to be associated with approximately two protons flowing down their electrochemical gradient

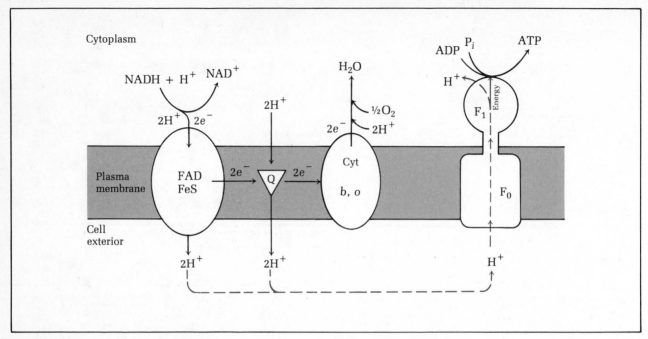

FIGURE 7-40 A highly stylized model of the arrangement of the electron transfer chain in the plasma membrane of the bacterium, *E. coli.* Though this chain contains one less coupling site for ATP synthesis than does the mitochondrial electron transfer chain, the basic similarity between the two respiratory systems is apparent. Abbreviations: FeS = iron-sulfur center, Q = ubiquinone.

through F_1. Thus the four protons exported per pair of electrons passing through the respiratory chain are sufficient to drive the synthesis of a maximum of two molecules of ATP. The first coupling site for ATP synthesis is associated with the transfer of electrons from NADH to ubiquinone, while the second site is associated with the transfer of electrons from ubiquinone to oxygen. The respiratory chains of other bacteria can exhibit significant variations from the sequence present in *E. coli.* As a consequence bacterial respiration is more flexible than mitochondrial respiration, oxidizing a wider range of substrates and in some cases employing other electron acceptors, such as sulfate or nitrate, in place of oxygen.

Studies on the mechanism by which ATP synthesis is coupled to electron transfer in bacterial cells have provided additional support for the chemiosmotic theory. Not only are a pH gradient and membrane potential established during electron transfer, but it has been demonstrated that creation of an artificial membrane potential, in the absence of electron transport, triggers the formation of ATP. The conclusion that a proton gradient drives ATP formation is also supported by the discovery that the hydrolysis of ATP by bacterial membranes causes protons to move in a direction opposite to the one in which they move during ATP synthesis.

A relatively new approach that is beginning to provide detailed information about the molecules involved in oxidative phosphorylation is based upon the study of bacterial mutants defective in ATP synthesis. In the early 1970s the first *E. coli* mutants were described in which oxidative phosphorylation is uncoupled from electron transfer. Characterization of such *unc* mutants has led to the identification of several genes and gene products involved in oxidative phosphorylation. For example, the *unc A* gene has been found to code for a polypeptide subunit of F_1, while the *unc B* gene codes for a subunit of F_0. Future studies on such genes and the proteins they code for should provide new insights into the components involved in the coupling of ATP synthesis to electron transfer.

As in mitochondria, the proton gradient established during bacterial electron transfer is capable of driving active transport as well as ATP synthesis. The proton-driven carrier that mediates the uptake of lactose across the bacterial plasma membrane has already been described (Chapter 5). Similar systems mediate the uptake of some amino acids. The involvement of the electrochemical proton gradient in solute transport may explain the action of certain toxic proteins secreted by *E. coli.* These toxins, termed **colicins,** kill susceptible bacteria by several different mechanisms. Colicins of one particular class, which includes colicins E1 and K, cause a decrease in the intracellular concentration of ATP and inhibit active transport (including active transport by carriers that do not utilize ATP as an energy source). These particular colicins have been shown to disrupt the membrane potential, probably by virtue of their ability to promote the diffusion of potassium ions. A decrease in the membrane potential component of the electrochemical gradient would of course be expected to inhibit ATP formation as well as active transport systems driven by the gradient.

In addition to its role in active transport, we shall see in Chapter 9 that the electrochemical proton gradient provides the energy for driving the rotation of bacterial flagella. When free energy stored in the electrochemical gradient is employed to promote flagellar rotation and/or active transport, the amount of ATP synthesis that can be driven by the gradient of course undergoes a comparable reduction.

SUMMARY

Although cells obtain chemical building blocks and energy from the metabolism of various kinds of organic foodstuffs, most of these are broken down into molecules that enter at one point or another into the pathway for glucose degradation. The key energy-releasing steps in this pathway involve oxidation-reduction. The relative tendencies of different substances to become oxidized (donate electrons) or reduced (accept electrons) is determined experimentally by measuring their redox potentials, which can in turn be employed to calculate the change in free energy that occurs when electrons are transferred from one substance to another.

Under anaerobic conditions, glucose is degraded by the glycolytic pathway to a reduced derivative of pyruvate (e.g., lactate, ethanol, etc.) accompanied by the net production of two molecules of ATP. Under aerobic conditions the pyruvate produced during glycolysis is subsequently degraded to acetyl-CoA. Acetyl-CoA, which is also generated by the breakdown of fatty acids, is then oxidized by a cyclic series of reactions known as the Krebs citric acid cycle. After two turns of the Krebs cycle, the equivalent of the entire carbon skeleton of glucose has been oxidized to carbon dioxide. The energy released during this process is trapped mainly in the form of reduced coenzymes (NADH and $FADH_2$), which are subsequently reoxidized by passing their electrons to oxygen through a sequential chain of carriers known as the electron transfer or respiratory chain. The energy released during electron transfer is utilized to drive ATP formation by a process termed oxidative phosphorylation.

In eukaryotic cells the conversion of pyruvate (and fatty acids) to acetyl-CoA, as well as the Krebs cycle, electron transfer, and oxidative phosphorylation all occur within the mitochondrion, earning this organelle the designation "power plant" of the cell. Although mitochondria are commonly pictured as independent sausage-shaped structures, three-dimensional reconstructions suggest that they may in some cases form interconnected branched systems within the cell. Mitochondria are enveloped by two membranes, a smooth outer membrane and an inner membrane thrown into a series of invaginations known as cristae. The area between the outer and inner membranes is termed the intermembrane space, while the region enclosed by the inner membrane is termed the matrix. Spherical particles containing the F_1 protein protrude from the inner surface of the cristae. Much of our understanding concerning the functional roles of the various mitochondrial membranes and compartments has been made possible by the development of methods for separating and isolating these components from each other.

The outer mitochondrial membrane, whose chemical composition bears a striking resemblance to that of the endoplasmic reticulum, is highly permeable to small molecules. As a result, the solute composition of the intermembrane space closely resembles that of the cell sap. The intermembrane space is composed of two regions, the intracristal space located between the folds of the cristae and the peripheral space located in the regions where the outer and inner membranes are closely opposed. Although the two areas are connected to each other by small channels (pediculi cristae), these channels may be narrow enough to hinder continuity between the two regions. Among the small number of enzymes present in the intermembrane space is adenylate kinase, which synthesizes ADP by transferring a phosphate group from ATP to AMP.

The matrix space contains most of the enzymes of the Krebs cycle and fatty acid β-oxidation. Because the protein content of the matrix is too high for all the matrix proteins to be in true solution, it is believed that the enzymes catalyzing the Krebs cycle and fatty acid β-oxidation are anchored within a structural network. In addition to its role in oxidative metabolism, the matrix contains DNA, ribosomes, and related components whose role in mitochondrial growth and division will be discussed in Chapter 14.

The inner mitochondrial membrane is the site of the respiratory chain and oxidative phosphorylation. The morphological uniqueness of the inner membrane stems from the presence of the small spherical particles that protrude from its matrix side, while its biochemical uniqueness stems from its high protein-to-lipid ratio, its high ratio of unsaturated to saturated phospholipids, the virtual absence of cholesterol, and the presence of large amounts of cardiolipin. The respiratory chain, which is also localized within the inner mitochondrial membrane, consists of pyridine nucleotides, flavoproteins, cytochromes, iron-sulfur proteins, and quinones. The sequential organization of these carriers has been elucidated by a combination of experimental approaches, including measuring their standard redox potentials, assaying their abilities to oxidize and reduce each other in test-tube reactions, and using difference spectra to examine the sequence in which they are oxidized in the presence and absence of various inhibitors. Such data have led to the conclusion that the respiratory chain is arranged in the following order: NADH → FMN group of NADH dehydrogenase → iron-sulfur centers → ubiquinone → cytochromes $b \rightarrow c_1 \rightarrow c \rightarrow a \rightarrow a_3 \rightarrow$ oxygen.

Within the membrane these carriers are organized into three complexes: complex I (NADH → ubiquinone), complex III (ubiquinone → cytochrome c), and complex IV (cytochrome c → oxygen). A fourth complex, designated complex II, transfers electrons from succinate (produced by the Krebs cycle) to ubiquinone. Thermodynamic considerations and site-specific assays have led to the conclusion that complexes I, III, and IV have coupling sites for ATP synthesis, while complex II does not. A pair of electrons donated to the respiratory chain from NADH will therefore generate a maximum of three molecules of ATP because all three coupling sites are encountered, while electron pairs donated by succinate yield a maximum of two molecules of ATP because only two coupling sites are encountered.

Membrane reconstitution experiments have provided important insights into the functional organization of the inner mitochondrial membrane. For example, removing the spherical particles from isolated inner membrane fractions (submitochondrial particles) causes the membrane to lose its ability to synthesize ATP, although electron transfer still occurs. The isolated spheres contain a protein complex, designated F_1, which by itself catalyzes ATP hydrolysis. A membrane capable of carrying out both electron transfer and ATP synthesis can be reconstituted by adding the spherical particles back to F_1-depleted membranes. Such observations indicate that the respiratory chain is localized within the smooth, particle-free membrane and that ATP synthesis occurs in the spherical particles protruding from the membrane surface.

Many years of experimental work on the mechanism that couples ATP synthesis to electron transfer has gradually led to the conclusion that the chemiosmotic theory is most compatible with the available data. According to this theory, the flow of electrons through the respiratory chain drives the expulsion of protons from the mitochondrial matrix. The result is an electrochemical gradient consisting of a membrane potential ($\Delta\psi$) and a difference in proton concentration (ΔpH). The synthesis of ATP is subsequently driven by the reverse flow of protons down this electrochemical gradient. Among the experimental observations that support this theory are the following. (1) Measuring the pH of the external medium has demonstrated that protons are expelled from mitochondria during flow of electrons through the respiratory chain. (2) The free energy of this proton gradient is sufficient to account for the amount of ATP synthesized. (3) Ionophores that abolish the electrochemical proton gradient also uncouple ATP synthesis from electron transfer. (4) Artificially created pH gradients will drive ATP synthesis in the absence of electron transfer. (5) Coupling of ATP synthesis to electron transfer requires an intact mitochondrial membrane enclosing a defined compartment. (6) The components of the respiratory chain are asymmetrically oriented within the mitochondrial inner membrane. (7) Reconstituted membrane vesicles containing either complex I, III, or IV establish proton gradients corresponding to approximately two protons pumped per pair of electrons transferred. (8) Reconstituted membrane vesicles containing F_1 but no respiratory carriers break down ATP, using the released energy to pump two protons into the vesicle. It follows that synthesis of an ATP molecule by F_1 is driven by the movement of two protons in the opposite direction. (9) Removal of F_1 causes the inner membrane to become leaky to protons. It therefore appears that F_1 is anchored to a membrane attachment site (designated F_0) that serves as a transmembrane channel through which protons flow down their electrochemical gradient to F_1.

In addition to driving ATP synthesis, the electrochemical proton gradient powers several mitochondrial transport systems. The pH component of the gradient drives the uptake of inorganic phosphate in exchange for hydroxyl ions, as well as the expulsion of sodium ions in exchange for protons. The membrane potential ($\Delta\psi$) component of the gradient is the driving force for the uptake of calcium ions, as well as for the uptake of ADP in exchange for ATP. Because the uptake of ADP and inorganic phosphate, as well as the export of ATP, utilize some of the free energy stored in the electrochemical proton gradient, such transport processes lower the net yield of ATP produced by mitochondria.

Coordination and regulation of the pathways involved in glucose metabolism is facilitated by allosteric control mechanisms. One example of this type of regulation involves the effects of ATP and AMP on the enzyme phosphofructokinase, which catalyzes the rate-limiting step of glycolysis. When cellular levels of ATP are high, glycolysis slows down because ATP acts as an allosteric inhibitor of phosphofructokinase. When ATP levels fall, glycolysis is stimulated because AMP, which is generated as the ultimate breakdown product of ATP, serves as an allosteric activator of phosphofructokinase.

Although bacteria lack mitochondria, the metabolic pathways by which they synthesize ATP using energy released during the oxidation of foodstuffs are remarkably similar to those occurring in eukaryotes. The bacterial plasma membrane, which exhibits F_1-containing spherical particles protruding from its inner surface, serves in lieu of the mitochondrial membrane as the site of electron transfer and oxidative phosphorylation. The respiratory chain in *E. coli* is somewhat simpler than in higher organisms, containing only two coupling sites instead of three. During electron transfer protons are pumped out of the cell, creating an electrochemical proton gradient that drives ATP synthesis as well as certain active transport reactions and flagellar rotation.

SUGGESTED READINGS

Books

Fiskum, G., ed. (1986). *Mitochondrial Physiology and Pathology*, Van Nostrand Reinhold, New York.

Stein, W. D. (1986). *Transport and Diffusion Across Cell Membranes*, Academic Press, Orlando.

Tzagoloff, A. (1982). *Mitochondria*, Plenum Press, New York.

Articles

Al-Awqati, Q. (1986). Proton-translocating ATPases, *Ann. Rev. Cell Biol.* 2:179–199.

Chernyak, B. V., and I. A. Kozlov (1986). Regulation of H^+-ATPases in oxidative- and photophosphorylation, *Trends Biochem. Sci.* 11:32–35.

Ernster, L., and G. Schatz (1981). Mitochondria: a historical review, *J. Cell Biol.* 91:227s–255s.

Ferguson, S. J. (1986). The ups and downs of P/O ratios, *Trends Biochem. Sci.* 11:351–353.

Hatefi, Y. (1985). The mitochondrial electron transport and oxidative phosphorylation system, *Ann. Rev. Biochem.* 54:1015–1069.

Hinkle, P. C., and R. E. McCarty (1978). How cells make ATP, *Sci. Amer.* 238 (March):104–123.

Mitchell, P. (1981). Biochemical mechanism of proton-motivated phosphorylation in F_0F_1 adenosine triphosphatase molecules. In: *Mitochondria and Microsomes* (Lee, C. P., G. Schatz, and G. Dallner, eds.), Addison-Wesley, Reading, MA, pp. 427–457.

Palade, G. E. (1953). An electron microscope study of the mitochondrial structure, *J. Histochem. Cytochem.* 1:188–211.

Racker, E. (1980). From Pasteur to Mitchell: a hundred years of bioenergetics, *Fed. Proc.* 39:210–215.

Saraste, M. (1983). How complex is a respiratory complex?, *Trends Biochem. Sci.* 8:139–142.

Schneider, E., and K. Altendorf (1984). The proton-translocating portion (F_0) of the *E. coli* ATP synthase, *Trends Biochem. Sci.* 9:51–53.

Srere, P. A. (1980). The infrastructure of the mitochondrial matrix, *Trends Biochem. Sci.* 4:120–121.

Wainio, W. W. (1985). An assessment of the chemiosmotic hypothesis of mitochondrial energy transduction, *Int. Rev. Cytol.* 96:29–51.

CHAPTER
8

Chloroplasts and the Conservation of Energy Derived from Sunlight

In Chapter 7 we examined the process by which cells obtain useful chemical energy from the degradation of foodstuffs such as carbohydrates and fats. Molecules of this type are excellent sources of nutrition because they contain reduced carbon atoms whose oxidation to carbon dioxide releases free energy. The continued renewal of the earth's food supply therefore requires a mechanism for reducing the carbon atoms that are continually being oxidized by cellular respiration. This reduction is ultimately accomplished by photosynthesis, a metabolic process that allows certain organisms to trap energy from the sun and use it to drive the energy-requiring reduction of carbon. As a by-product the photosynthetic pathway generates oxygen molecules whose release is largely responsible for the oxygen-rich atmosphere characteristic of our planet today.

Cells that generate their own reduced carbon by photosynthesis are referred to as *autotrophs* (self-feeding), while cells that cannot photosynthesize and must therefore consume foodstuffs containing reduced carbon are termed *heterotrophs* (feeding on others). Although multicellular plants are the most obvious photosynthetic organisms, more than half of the roughly 40 billion tons of carbon reduced annually by photosynthesis is generated by microorganisms. Among the microorganisms carrying out photosynthesis are both eukaryotes (algae, diatoms, and dinoflagellates) and prokaryotes (blue-green algae and photosynthetic bacteria). A great deal of effort has been expended in recent years in trying to unravel the mechanism by which photosynthetic organisms trap solar energy and utilize it to drive the synthesis of organic molecules containing reduced carbon. In eukaryotic autotrophs the light-trapping reactions occur within specialized cytoplasmic organelles known as chloroplasts, while in prokaryotic autotrophs the comparable reactions take place within membranes dispersed throughout the cytoplasm. But in spite of this superficial difference, the mechanisms underlying the photosynthetic pathway are basically similar in the two classes of organisms. In addition, these underlying mechanisms are remarkably similar to the principles governing mitochondrial respiration. As we shall see in the present chapter, in both cases the crucial energy-transforming steps take place within membranes, involve electron transfer reactions, and synthesize ATP by a mechanism that involves the formation of an electrochemical gradient of protons.

OVERVIEW OF PHOTOSYNTHESIS

Before embarking upon a detailed description of photosynthetic membranes and the mechanisms underlying their ability to trap and conserve solar energy, an initial overview of photosynthesis will be helpful. In this introductory section we shall briefly review the experimental evidence that first led to the formulation of the basic equation of photosynthesis, and to the idea that photosynthesis consists of separate "light" and "dark" reactions.

Elucidation of the Basic Equation of Photosynthesis

The original formulation of the chemical equation summarizing photosynthesis required the work of several scientists spanning a period of nearly 200 years. The first major contributor was the Flemish physician Jan Baptista van Helmont, who carried out a simple experiment in which he measured the increase in weight of a willow tree grown in a bucket of soil to which only water was added. After five years the tree was found to have gained 160 pounds, while the soil in which it was planted weighed only 2 ounces less than it had at the beginning of the experiment. Since water was the only ingredient added, van Helmont concluded that the organic matter synthesized by plants is derived from water molecules.

It was not until 100 years later that the English minister Joseph Priestly implicated the gaseous components of air in this process. In 1771 Priestly reported an experiment in which a candle was burned in an enclosed chamber until it spontaneously extinguished. Mice subsequently placed in such a chamber were found to suffocate, leading to the conclusion that the burning candle had "injured" the air. When Priestly placed a small green plant in the chamber, however, he observed that the mice could breathe and survive normally. His conclusion that plants can "restore" air to a normal state was a monumental realization, but its significance was not fully appreciated at the time because it preceded the discovery of the gases oxygen and carbon dioxide.

News of Priestly's observations stimulated Jan Ingenhousz, a Dutchman serving as court physician to the Austrian empress, to investigate this phenomenon in more detail. A few years later Ingenhousz made two crucial discoveries: Only the green portion of a plant can "restore" air, and light is required. In 1782 a Swiss minister, Jean Senebrier, further observed that the "restoration" of air is accompanied by the uptake of carbon dioxide. Around this same time oxygen was also discovered, leading to the realization that the restoration of air by green plants is based upon their ability to produce oxygen. Taken together the above observations suggested that photosynthesis involves the uptake of carbon dioxide and the production of organic matter and oxygen. But careful quantitation of the uptake and production of these substances during photosynthesis led the Swiss chemist Theodore de Saussure to conclude that the amount of organic matter and oxygen a plant produces is greater than the weight of the carbon dioxide it consumes. This discrepancy led de Saussure to rediscover van Helmont's old idea that the synthesis of organic matter by plants also requires the uptake of water.

Thus by the end of the eighteenth century it was clear that photosynthesis converts water and carbon dioxide to organic matter and oxygen. In 1845 the German surgeon Julius Robert Mayer, who formulated the law of conservation of energy, suggested that light provides the energy needed to drive this energy-requiring process. With this final insight, the overall equation describing photosynthesis could be written as follows:

$$CO_2 + H_2O \xrightarrow[\text{Green plant}]{\text{Light}} (CH_2O)_n + O_2$$

The principal type of organic matter produced by this pathway is glucose, in which case n equals 6. The equation can therefore be rewritten in its more common form:

$$6CO_2 + 6H_2O \xrightarrow[\text{Green plant}]{\text{Light}} C_6H_{12}O_6 + 6O_2$$

The elucidation of this basic equation of photosynthesis illustrates how the independent contributions of many individuals may be required for an important scientific breakthrough. In this particular case the observations of individuals of differing eras, nationalities (Dutch, English, Swiss, German), and occupations (physicians, ministers, chemists) were involved. But in spite of the magnitude of this accomplishment, it was only the first of many steps in unraveling the basic mechanism of photosynthesis. As biochemical techniques gradually improved, it became apparent that the reaction summarized by the above equation does not take place as a single step, but is rather the net result of the interactions of dozens of metabolic steps organized into two discrete pathways known as the "light" and "dark" reactions, respectively.

Discovery of the Light and Dark Reactions

The idea that photosynthesis occurs in two discrete stages first emerged from studies reported in 1905 by F. Blackman. His work revealed that the rate of oxygen production by photosynthetic cells exposed to intense illumination and low CO_2 concentration increases with increasing temperature. The rate of oxygen production by cells maintained under conditions of low light intensity and high CO_2 concentration, on the other hand, does not vary with changing temperature. On the basis of these observations Blackman proposed that photosynthesis involves two separate pathways: a temperature-insensitive pathway that requires light and a temperature-sensitive process requiring CO_2. These two pathways were designated the **light** and **dark** reactions of photosynthesis, respectively. The light reactions are responsible for absorbing light and producing oxygen, while the dark reactions are involved in CO_2 "fixation" (taking up carbon dioxide from the atmosphere and converting it to organic matter).

Additional support for the existence of separate light and dark reactions was provided by the discovery that photosynthesis is stimulated to a greater extent by flashing light than by continuous light. As the length of the dark period between light flashes is increased, the quantity of CO_2 incorporated per light flash rises (Figure 8-1). The most straightforward interpretation of this phenomenon is that the reactions involved in CO_2 fixation continue during the dark periods, driven by energy absorbed by the light reactions during the preceding light period.

A series of classic experiments carried out by Robert Hill in the 1930s provided yet more support for the concept of separate light and dark reactions. Hill was the first to show that isolated chloroplasts evolve oxygen when illuminated. This reaction only occurs, however, when an artificial electron acceptor is added in place of the normal electron acceptor, CO_2. For instance, addition of the electron acceptor Fe^{3+} permits the following reaction to take place:

$$4Fe^{3+} + 2H_2O \longrightarrow 4Fe^{2+} + 4H^+ + O_2$$

This production of oxygen accompanied by the reduction of an artificial electron acceptor such as Fe^{3+} has come to be known as the **Hill reaction.** The existence of

the Hill reaction indicates that the light-requiring steps of photosynthesis can occur in the absence of the dark reactions of CO_2 fixation. It also reveals that of the two inputs to the photosynthetic pathway (water and CO_2), water must be the source of the oxygen atoms liberated during photosynthesis because oxygen can be produced in the absence of CO_2 fixation.

In the early 1950s Daniel Arnon and his colleagues showed that chloroplasts isolated under gentler conditions than those employed by Hill are capable of executing the dark reactions as well as the light reactions of photosynthesis. With such chloroplast preparations an artificial electron acceptor need not be added because carbon dioxide acts as the natural electron acceptor. In order to investigate the localization of the light and dark reactions within the chloroplast, these workers removed the soluble components of the chloroplast interior, or **stroma,** by extracting chloroplasts with water. The remaining chloroplast membranes were found to carry out the light, but not the dark, reactions of photosynthesis, while the enzymes required for the dark reactions were found to be present in the water-soluble extract. It was therefore concluded that the light reactions are catalyzed by chloroplast membranes while the dark reactions occur within the stroma.

The question of how such physically separated light and dark reactions are linked to each other was at first a puzzle. The first clue to the nature of the intermediates that join these reactions was obtained in 1951, when Wolf Vishniac and Severo Ochoa discovered that illuminated chloroplasts reduce the coenzyme $NADP^+$ to NADPH. A few years later Arnon reported that illuminated chloroplasts also synthesize ATP. Since ATP and NADPH are both high-energy compounds, it was proposed that they function as intermediates that transfer energy from the light reactions to the energy-requiring steps of the dark reactions.

To test this idea, Arnon added ATP and NADPH to soluble chloroplast extracts known to contain the enzymes that carry out the dark reactions. The result was a marked stimulation of CO_2 fixation. Because maximal stimulation of CO_2 fixation was found to require the presence of NADPH as well as ATP, Arnon concluded that both these substances are involved in transferring energy from the light reactions to the dark reactions. At about the same time that Arnon was carrying out these experiments, the enzymatic steps involved in the dark reactions were being investigated by Melvin Calvin and his associates. This work led to the discovery that the carbon-fixing pathway involves a cyclic set of reactions that has come to be known as the **Calvin cycle.** Most interesting was the discovery that the Calvin cycle requires an input of both NADPH and ATP. Taken together the above observations suggest that solar energy trapped by the light reactions of photosynthesis is utilized to drive the formation of NADPH and ATP, and that these high-energy compounds are subsequently

FIGURE 8-1 Relationship between the length of the dark interval between light flashes and the amount of CO_2 fixed by the green alga *Chlorella*.

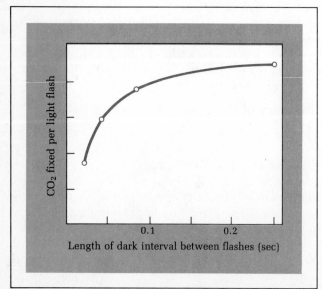

CO$_2$ fixed per light flash

0.1   0.2

Length of dark interval between flashes (sec)

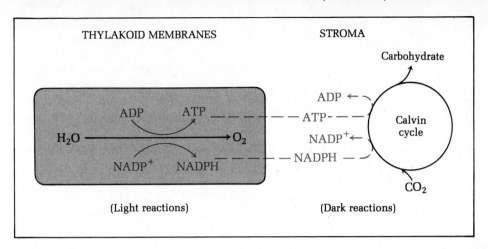

FIGURE 8-2 Relationship between the light and dark reactions of photosynthesis. The ATP and NADPH generated by the light reactions is utilized to drive the energy-requiring dark reactions.

employed to drive the energy-requiring steps of the Calvin cycle (Figure 8-2).

The main question left unanswered is how does the energy derived from sunlight drive the formation of NADPH and ATP? By analogy to mitochondrial respiration, where ATP synthesis is driven by energy released as electrons flow from reduced coenzymes to oxygen, the idea arose in the late 1950s that ATP synthesis in chloroplasts is mediated by the light-induced flow of electrons from water to the coenzyme, NADP⁺. According to this viewpoint the removal of electrons from water generates oxygen, while transfer of these electrons to NADP⁺ produces the NADPH required by the Calvin cycle. Although this model has turned out to be basically correct, many questions arise as soon as one begins to probe its details. Among the more crucial questions are: How is solar energy absorbed? What is the nature of the carriers that transfer electrons from water to NADP⁺? How is the released energy trapped to drive ATP formation? How are the relevant pathways organized within the cell? Answering such questions requires us to first examine the structure of the major organelle in which photosynthesis occurs, the chloroplast. Later in the chapter we shall see that in prokaryotic cells, where chloroplasts are not present, photosynthesis nonetheless involves similar kinds of membrane-localized phenomena.

ANATOMY OF THE CHLOROPLAST

Because of their relatively large size and conspicuous green color, chloroplasts were one of the first intracellular structures to be discovered by light microscopists. Shortly after the invention of the compound microscope in the seventeenth century the presence of chloroplasts in algal cells was noted by van Leeuwenhoek, and by 1800 chloroplasts had been identified in a wide variety of different plant cells. In the mid-nineteenth century it was discovered that chloroplasts rupture when placed in a hypotonic medium, suggesting that they are bounded by a semipermeable membrane.

The first significant insight into the function of this membrane-enclosed organelle was provided in 1894 by T. Engelmann. By that time it was already known that photosynthesis generates oxygen. To investigate the question of where photosynthesis occurs within plant cells, Engelmann utilized a special strain of bacteria known to be attracted to oxygen. Such bacteria were placed in a medium containing *Spirogyra*, a large unicellular alga that contains a single large chloroplast. When these algal cells were illuminated with a finely focused beam of light, the oxygen-seeking bacteria were found to congregate near those regions of the algal cells where the chloroplast had been illuminated (Figure 8-3). It was therefore concluded that the light-absorbing

FIGURE 8-3 Diagram of Engelmann's experiment with *Spirogyra*, a large algal cell containing a single spiral-shaped chloroplast. Oxygen-seeking bacteria are attracted to the region of the cell where the chloroplast has been illuminated, indicating that light-induced oxygen formation occurs within the chloroplast.

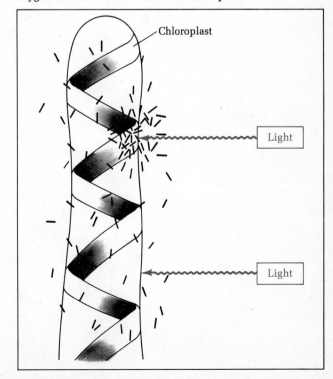

and oxygen-producing steps of photosynthesis are housed within the chloroplast. This conclusion was an important milestone in cell biology because it was the first time a specific metabolic pathway had been assigned to a particular subcellular organelle.

The conclusion that photosynthesis occurs within chloroplasts raises the question of what role the architectural organization of this organelle plays in the photosynthetic functions it carries out. Under the light microscope chloroplasts are seen to contain small green granules, or **grana,** suspended within an amorphous **stroma** (Figure 8-4). Fifty or more grana, each measuring 0.5–2.0 μm in diameter, may be present within a single chloroplast. The true architectural complexity of the chloroplast, however, is only evident in electron micrographs. When thin-sectioning techniques were first developed in the early 1950s it became apparent that the chloroplast has an extremely complex membranous organization. As expected from the ability of hypotonic solutions to rupture chloroplasts, electron

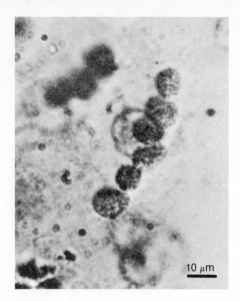

10 μm

FIGURE 8-4 Light micrograph of chloroplasts in which grana can be seen as dark granular areas. Courtesy of Y. Ben-Shaul.

FIGURE 8-5 Electron micrograph of a spinach leaf chloroplast lying near the periphery of the cell, directly adjacent to the cell wall. Within the chloroplast both stacked and unstacked thylakoids are visible. Courtesy of J. V. Possingham.

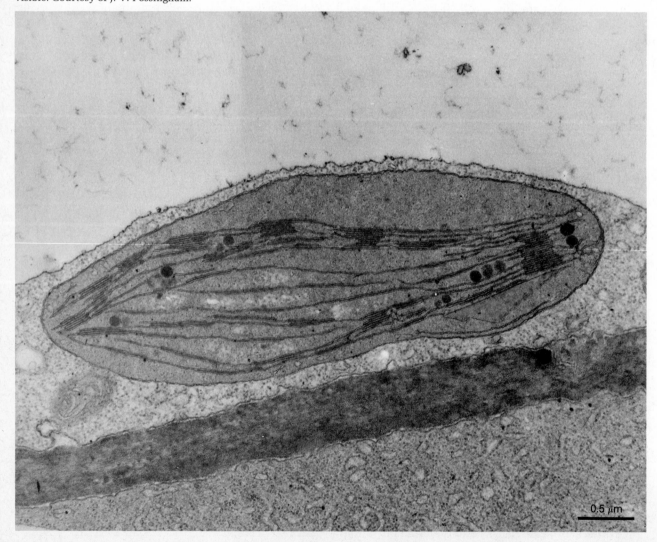

0.5 μm

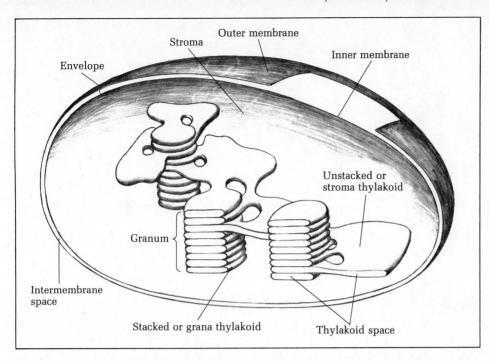

FIGURE 8-6 Diagrammatic representation of the arrangement of chloroplast membranes and compartments.

FIGURE 8-7 Electron micrograph of negatively stained thylakoid membrane showing CF₁-containing particles protruding from the surface. Courtesy of M. P. Garber.

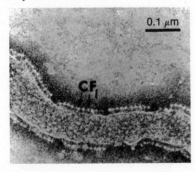

micrographs revealed that this organelle is surrounded by a membranous envelope. The chloroplast envelope consists of two closely opposed membranes, termed the **outer** and **inner membranes,** separated by an **intermembrane space.**

Electron microscopy has also revealed that the grana observed with the light microscope correspond to clusters of flattened membranous sacs, known as **thylakoids,** which are stacked upon each other like coins (Figure 8-5). The membranes that make up these

internal stacks are referred to as **stacked** thylakoids, while the cavities within the sacs are termed **thylakoid spaces.** A typical granum consists of several dozen thylakoids tightly stacked upon each other (Figure 8-6). Grana stacks are often connected to one another by membranous channels referred to as **unstacked** thylakoids. Negative staining has revealed that spherical particles resembling those that occur on mitochondrial cristae protrude from the surfaces of the thylakoid membranes into the stromal space (Figure 8-7). We shall see later that these spheres contain a protein complex, designated CF_1, which plays a central role in photosynthetic ATP formation.

The exact three-dimensional relationship of the unstacked thylakoids to the stacked thylakoids has been difficult to ascertain because thin-section electron micrographs only permit the visualization of thin slices through this interconnected membrane system. At one time it was thought that unstacked thylakoids were simple tubules connecting to single thylakoid membranes within a granum stack, but more detailed examinations suggest that each unstacked thylakoid is a large flattened sheet that makes connections to many or all of the individual thylakoids of a given granum (see Figure 8-6).

FIGURE 8-8 Electron micrograph of a corn cell chloroplast showing points of continuity (arrows) between the inner membrane of the chloroplast envelope and the thylakoid membranes. Courtesy of J. Rosado-Alberio.

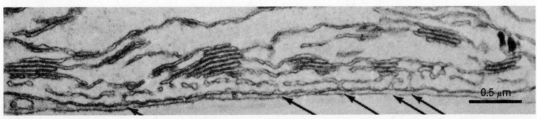

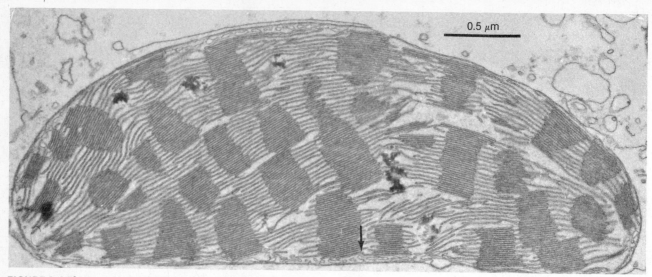

0.5 μm

FIGURE 8-9 Electron micrograph of a chloroplast showing the peripheral reticulum (arrow) located between the thylakoid membranes and the chloroplast envelope. Courtesy of J. Rosado-Alberio.

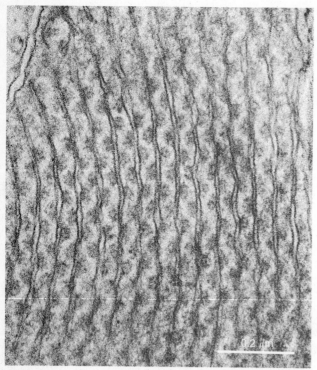

0.2 μm

FIGURE 8-10 High magnification electron micrograph of the thylakoid membranes of a chloroplast from red alga. Note the parallel arrangement of the individual thylakoid membranes. The granules attached to the membranes are phycobilisomes. Courtesy of E. Gantt.

As a consequence of this arrangement, the individual thylakoid spaces of each granum stack become interconnected both with each other and with the thylakoid spaces of other granum stacks. Hence the entire internal membrane system of the chloroplast may constitute a single, enormously complex, membrane-enclosed compartment.

During chloroplast development points of continu-

ity are occasionally observed between thylakoid membranes and the inner membrane of the chloroplast envelope (Figure 8-8, p. 253). Connections between thylakoids and the chloroplast envelope are rarely seen in mature chloroplasts, though a series of interconnected vesicles and tubules termed the **peripheral reticulum** may form bridges between the thylakoids and the inner membrane in certain kinds of chloroplasts (Figure 8-9). The organization of the thylakoids differs appreciably in different kinds of plant cells. While the stacking of thylakoid membranes into grana is typical of chloroplasts of almost all higher plants and green algae, it does not occur in lower algae. The chloroplasts of such organisms usually exhibit a series of separated thylakoids running parallel to one another. In red algae these parallel thylakoids are completely separate from each other (Figure 8-10), but in other algae they may be grouped into sets of two or three. Even when grouped in this way, however, the adjacent thylakoids do not appear to be fused to each other as they are in the grana stacks of higher plants. In red and blue-green algae, pigment-containing granules known as **phycobilisomes** are attached to the surfaces of the thylakoid membranes.

The overall organization of the chloroplast resembles that of the mitochondrion in several ways. Both organelles are surrounded by double-membrane envelopes and both contain a complex internal system of membranes from which protrude spherical particles involved in ATP synthesis. The major difference is that the thylakoid membranes are not generally continuous with the inner membrane of the envelope, as is the case for the mitochondrial cristae. Hence the thylakoid spaces represent a distinct compartment separate from the intermembrane space of the envelope, while the mitochondrial cristae do not enclose such a separate compartment (Figure 8-11).

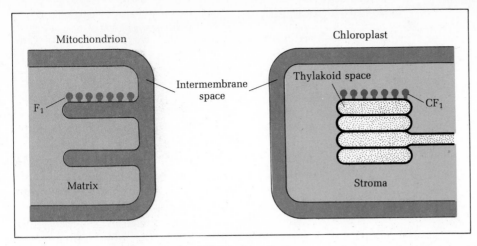

FIGURE 8-11 Diagram comparing the relationships between the compartments of mitochondria and chloroplasts.

Arrangement of Chloroplasts Within the Cell

Because the metabolic activities of the chloroplast are interrelated with events occurring in mitochondria and peroxisomes, these three organelles are often closely associated with one another in plant cells (Figure 8-12).

In some instances, mitochondria are even located within invaginations of the chloroplast surface (Figure 8-13). Relationships of this type are especially prominent when plants are grown in dim light, suggesting that mitochondria may provide ATP to the chloroplast under such conditions.

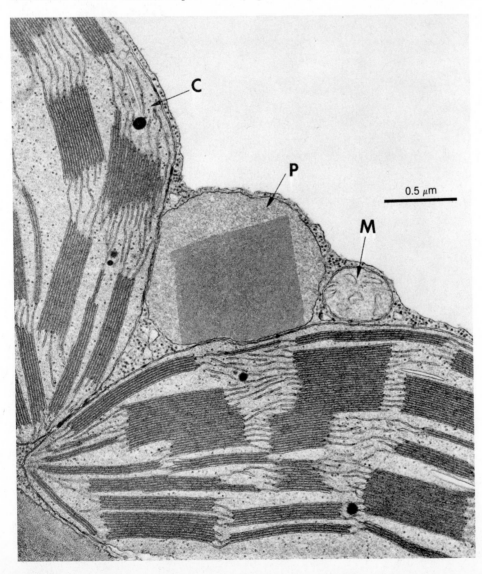

FIGURE 8-12 Electron micrograph of a tobacco leaf cell showing two chloroplasts (C), a peroxisome (P), and a mitochondrion (M) in close proximity to one another. The peroxisome contains a large crystalline inclusion. Courtesy of E. H. Newcomb.

0.5 μm

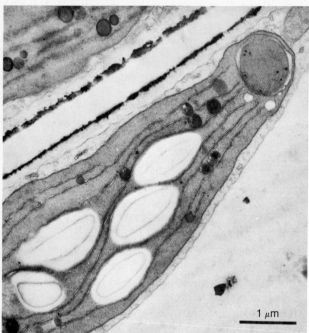

FIGURE 8-13 Electron micrograph of a leaf cell showing a mitochondrion (color) lying within an invagination of the chloroplast. The large white granules are starch grains. Courtesy of B. J. Ford.

In thin-section electron micrographs the chloroplasts of higher plants appear as disk-shaped or ellipsoid structures 2–4 μm in diameter and up to 10 μm in length (roughly twice the size of typical mitochondrial profiles). The chloroplasts of algae may even be larger and often assume unusual shapes such as spirals, cups, and circular bands around the cell. In algae as few as one or two large chloroplasts may be present, but higher plant cells typically contain several dozen or more. These multiple chloroplasts appear to be largely independent of each other, rather than forming an interconnected network as is often the case for mitochondria.

Chloroplasts undergo dynamic morphological changes in response to alterations in their metabolic state. If cells that had been kept in the dark are suddenly exposed to light, their chloroplasts shrink in volume by as much as 50 percent. At the same time a measurable

FIGURE 8-14 Electron micrographs of chloroplasts in spinach leaves incubated in the dark (*left*) and in the light (*right*). Note the difference in chloroplast shape. Courtesy of K. Shibata.

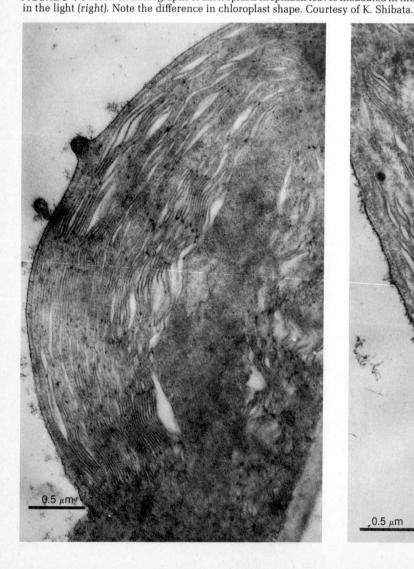

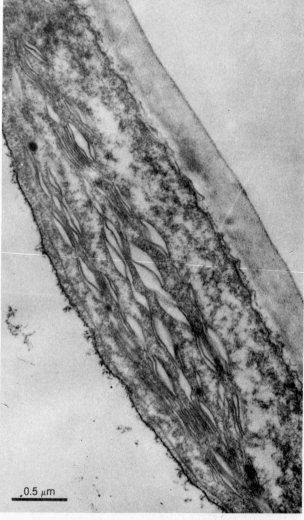

increase in the length to width ratio occurs (Figure 8-14). Upon removal of light, the volume and shape of the chloroplast gradually return to normal. The significance of these light-induced structural alterations is not well understood, but the magnitude of the volume change indicates that movement of water, perhaps secondary to ion movements, is involved.

Isolation and Subfractionation of Chloroplasts

The development of techniques for isolating and subfractionating chloroplasts has played an important part in advancing our understanding of the roles played by their various membranes and compartments. Chloroplasts can be isolated from plant cells in several different ways. The older procedures are based on the use of harsh homogenization techniques, such as grinding plant cells with sand in a mortar and pestle to break the cell wall and release the chloroplasts into solution. Following removal of nuclei by centrifugation at low speed, chloroplasts are collected by centrifuging at higher speeds. A more recent approach to chloroplast isolation employs the enzymes cellulase and pectinase to remove the cell wall. The advantage of this approach is that the resulting wall-less cells, or **protoplasts,** can be disrupted under milder conditions prior to isolation of the chloroplasts by centrifugation.

Chloroplasts isolated under harsh conditions are usually capable of carrying out the light reactions of photosynthesis (i.e., the light-induced formation of oxygen, ATP, and NADPH), but not the CO_2-fixing dark reactions. Electron microscopic observation of such defective *class II* chloroplasts reveals little or no stroma and broken or missing outer envelopes. In contrast, gentler disruption techniques yield completely intact *class I* chloroplasts capable of carrying out the light as well as the dark reactions of photosynthesis. Class I and II chloroplast preparations have both been useful as starting material for subfractionation experiments (Figures 8-15 and 8-16). In one commonly used procedure, class I chloroplasts are exposed to a hypotonic medium to rupture the envelope, followed by isodensity centrifugation to separate the stromal material, outer enve-

FIGURE 8-15 Outline of two general procedures for isolation and subfractionation of chloroplasts.

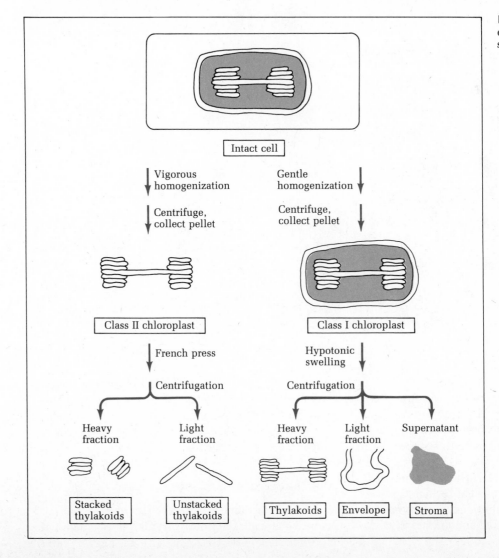

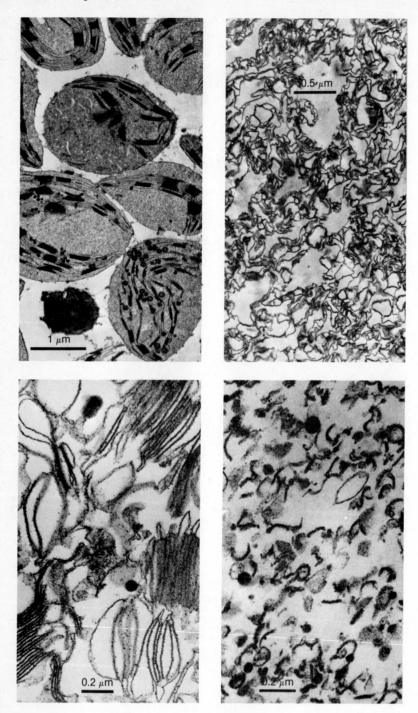

FIGURE 8-16 Electron micrographs of fractions obtained during chloroplast subfractionation. *(Top left)* Intact isolated chloroplasts. *(Top right)* Isolated chloroplast envelopes. Courtesy of A. A. Benson. *(Bottom left)* Isolated stacked thylakoids (grana). *(Bottom right)* Isolated unstacked thylakoids and end membranes of stacked thylakoids. Courtesy of R. B. Park.

lope, and thylakoids from one another. As an alternative, class II chloroplasts lacking the outer envelope and stroma can be used as starting material for subfractionation of the thylakoid membranes into stacked and unstacked thylakoids. In this approach the chloroplasts are disrupted in a special apparatus, called a French pressure cell, that shears the thylakoid membranes by forcing them through a small orifice following rapid decompression. These shearing forces break the connections between the stacked and unstacked thylakoids, thereby allowing the two membrane fractions to

be separated from one another by differential centrifugation. Biochemical and metabolic studies on chloroplast subfractions isolated in the above ways have provided a great deal of information concerning the functional organization of this organelle.

The Chloroplast Envelope

The chloroplast envelope serves as a barrier that separates the cytoplasm from the interior regions of the chloroplast where the reactions of photosynthesis

TABLE 8-1
Comparison of lipid content of chloroplast envelope, thylakoid, and microsomal membranes of spinach leaves

Lipid	Percent of Total Lipid		
	Envelope	Thylakoids	Microsomes
Glycolipids			
Monogalactosyldiglyceride	22	38	12
Digalactosyldiglyceride	32	19	8
Sulfoquinovosyldiglyceride	5	5	2
Phospholipids			
Phosphatidylglycerol	8	7	3
Phosphatidylcholine	27	2	35
Phosphatidylinositol	1	1	7
Phosphatidylethanolamine	0	0	30
Phosphatidylserine	0	0	3
Light-absorbing pigments			
Chlorophyll		20	
Carotenoids	0.4	6	
Quinones		3	
Protein/lipid ratio	0.5	1.5	

occur. Isolated envelope membranes are characterized by a lower protein-to-lipid ratio and a lower glycolipid content than thylakoid membranes (Table 8-1). Although electrophoretic analysis of chloroplast envelope extracts has revealed the presence of dozens of different proteins, the functional significance of most of these molecules is yet to be determined. It has been ascertained, however, that isolated envelopes contain enzymes that catalyze the synthesis of many of the lipids present within the chloroplast, such as galactolipids, carotenoids, and phenylquinones. Freeze-fracture micrographs of the inner envelope membrane reveal the presence of 7-nm particles on its stromal surface and 9-nm particles on the side of the membrane facing the intermembrane space. The outer membrane of the envelope is characterized by a much lower particle density, with 9-nm particles predominating.

The permeability of the chloroplast envelope has been studied by incubating intact chloroplasts in the presence of various solutes and then measuring the amount of solute taken up. Although most small molecules penetrate into the chloroplast readily, their final concentration within the chloroplast is generally less than in the surrounding medium. As was the case for comparable experiments carried out on mitochondria, results of this type indicate that the chloroplast consists of two compartments, one that readily equilibrates its solutes with the external environment and one that does not. Since the volume of the readily permeable compartment calculated from such experiments has been found to correlate with the relative size of the intermembrane space as visualized by electron microscopy, it can be concluded that chloroplasts resemble mitochondria in having an outer membrane that is readily permeable to small molecules and an inner membrane that is not.

The inner membrane of the envelope is relatively

impermeable to many substances that play important roles in photosynthesis, such as ATP, ADP, inorganic phosphate, and $NADP^+$/NADPH. The movement of these substances into and out of the chloroplast is accomplished by transport and shuttle mechanisms analogous to those occurring in mitochondria. Among the transport systems localized in the inner membrane are the ATP/ADP exchange carrier, whose primary function is to bring ATP into the chloroplast in the absence of light, and the phosphate-exchange carrier, which promotes the uptake of inorganic phosphate in exchange for products of photosynthesis such as 3-phosphoglycerate.

The impermeability of the chloroplast inner membrane to coenzymes has led to the development of shuttle systems that transport electrons into and out of the organelle just as analogous shuttle mechanisms transport electrons into and out of mitochondria. For example, reduced NADPH generated within the chloroplast can pass its electrons to 3-phosphoglycerate in an ATP-requiring reaction sequence that generates dihydroxyacetone phosphate. The dihydroxyacetone phosphate is then transported by the phosphate exchange carrier to the cytoplasm, where it is converted back to 3-phosphoglycerate in a reaction sequence involving the reduction of NAD^+ to NADH and the regeneration of ATP. The net result is the transfer of both reducing power and ATP from the chloroplast to the cytoplasm (Figure 8-17).

FIGURE 8-17 Two shuttle systems that mediate the transfer of reducing power from the chloroplast to the cytoplasm.

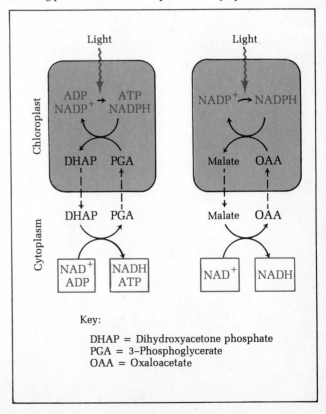

Key:

DHAP = Dihydroxyacetone phosphate
PGA = 3-Phosphoglycerate
OAA = Oxaloacetate

An alternative shuttle system utilizing oxaloacetate and malate operates in a similar fashion, although ATP is not involved.

The Chloroplast Stroma

The chloroplast stroma, which is a compartment distinct from both the internal thylakoid spaces and the intermembrane space of the envelope, exhibits a finely granular appearance in electron micrographs. Within this generally amorphous, viscous region of the chloroplast, which is analogous to the mitochondrial matrix, several kinds of structures may occur. Prominent among these are **starch grains,** which store some of the carbohydrate products produced by the dark reactions of photosynthesis (Figure 8-18). Also present are lipid-containing deposits, known as **plastoglobuli,** which accumulate in association with the breakdown of thylakoid membranes (e.g., in aging leaves) and decrease in size and number under conditions where new thylakoid membranes are being synthesized (e.g., in dark-adapted cells newly exposed to light). Such correlations suggest that plastoglobuli serve as reservoirs of lipid for thylakoid membrane formation. The stroma of algal chloroplasts often contains large granular deposits referred to as **pyrenoids** (Figure 8-19). The major constituent of these structures is crystallized ribulose-bisphosphate carboxylase, an enzyme whose role in CO_2 fixation will be discussed later in the chapter. The chloroplasts of higher plants contain related structures, termed **stroma centers,** which are comprised of tightly packed fibrils 8–9 nm in diameter also thought to contain ribulose-bisphosphate carboxylase (Figure 8-20).

In addition to the above components, the stroma contains DNA, RNA, ribosomes, and other enzymes and factors involved in nucleic acid and protein synthesis. Since these components are all involved in chloroplast growth and division, we shall delay their further consideration until the chapter on chloroplast biogenesis (Chapter 14). Biochemical analyses have revealed that the stroma also contains all the enzymes required for carrying out the dark reactions of photosynthesis. A detailed description of this pathway will be provided later in the chapter.

FIGURE 8-18 Electron micrograph of a chloroplast showing two elongated starch grains (S) and many spherical plastoglobuli (P). Courtesy of J. V. Possingham.

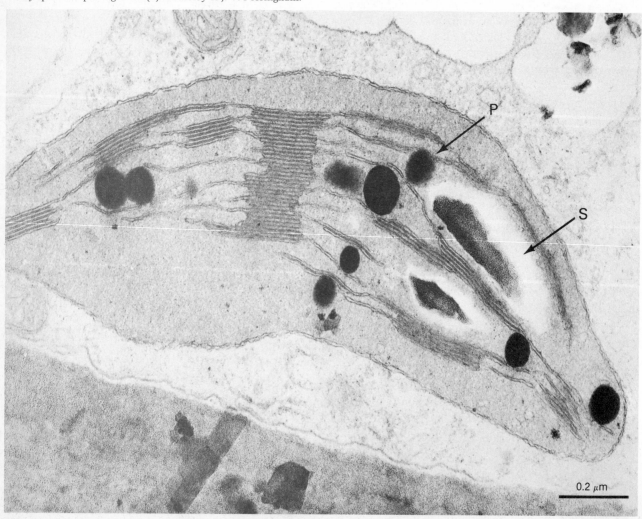

0.2 μm

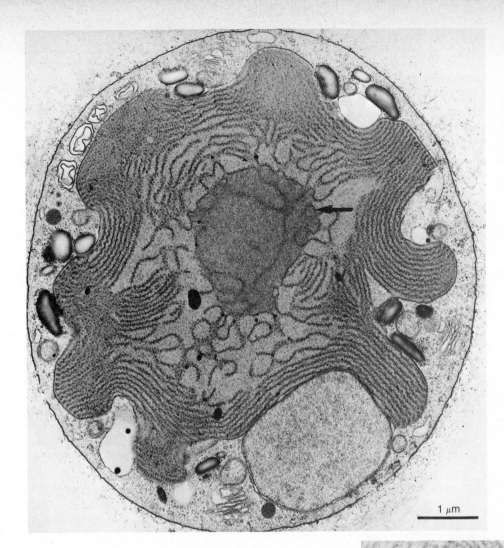

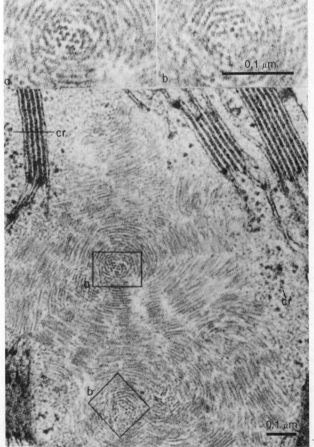

FIGURE 8-19 Thin-section electron micrograph of the red alga *Porphyridium*. The majority of the cell is occupied by a large chloroplast (color) containing parallel thylakoids and a central granular deposit called the pyrenoid (arrow). Courtesy of E. Gantt.

FIGURE 8-20 Electron microscopic appearance of the stroma center present in the chloroplasts of higher plants. *(Below)* A chloroplast exhibiting a fibrillar stroma center (S) surrounded by dark-staining granules (og). The chloroplast envelope (ce) is also evident. *(Right)* A higher magnification view of a stroma center. The boxed region contains a circular arrangement of fibers that at even higher magnification (inset) are seen to be composed of granular subunits. Chloroplast ribosomes (cr) are also evident. Courtesy of B. E. S. Gunning.

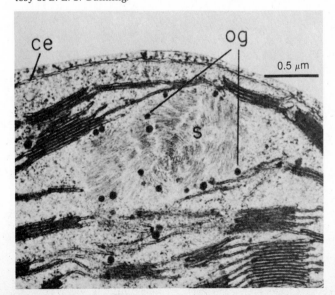

Thylakoid Membranes

Thylakoid membranes are unique in both morphological appearance and chemical composition. Their morphological uniqueness stems from the unusual stacked arrangement exhibited by the grana thylakoids and from the presence of the CF_1-containing spherical particles that protrude from the stromal surfaces of the unstacked thylakoids. In terms of chemical composition thylakoids are unusual in that phospholipid accounts for only about 10–20 percent of the total lipid. This contrasts markedly with mitochondrial membranes, where more than 90 percent of the membrane lipid is phospholipid, or endoplasmic reticulum membranes, where about 80 percent of the lipid is phospholipid. The bulk of the lipid in thylakoid membranes is accounted for by glycolipids, mainly in the form of galactosylglycerides and sulfate-containing glycerides (Figure 8-21). A significant portion of the lipid is also accounted for by light-absorbing pigments such as chlorophyll and carotenoids (see Table 8-1). In addition to their unusual lipid composition, thylakoid membranes are characterized by an exceptionally high protein content. About 60 percent of the membrane mass is generally accounted for by protein molecules whose roles in the light reactions of photosynthesis will be discussed in the following sections.

THE LIGHT REACTIONS
OF THE THYLAKOID MEMBRANES

When isolated thylakoid membranes are illuminated, they produce oxygen, reduce NADP+ to NADPH, and synthesize ATP from ADP and inorganic phosphate. The ability of thylakoid membranes to carry out the complete set of light reactions in the absence of other chloroplast components means that the thylakoid membranes hold the key to the mechanism by which solar energy is captured and converted to useful chemical forms. The light reactions depend on the existence of three classes of molecules present within the thylakoid membrane: First, light-absorbing pigments that contain electrons and that become activated upon absorption of light; second, electron carriers that transfer these activated electrons to NADP+; and third, proteins involved in coupling ATP synthesis to this electron transfer pathway. Below we shall describe each of these three classes of molecules and the events in which they are involved.

Light-absorbing Pigments
and the Photosynthetic Unit

The initial event in photosynthesis is the absorption of solar energy by pigments localized within the thylakoid membranes. For this purpose all photosynthetic cells contain some type of **chlorophyll,** a term referring to a family of lipid pigments responsible for the green color of plants. Evidence that chlorophyll plays a crucial role

in the light-absorbing step of photosynthesis has been provided by quantitating the relative abilities of light of varying wavelengths to stimulate photosynthesis. Such studies show that blue or red light is much more effective than green light in promoting photosynthesis in higher plants (Figure 8-22 *top*). This pattern matches the absorption spectrum of isolated chlorophyll molecules, which absorb red and blue, but not green, light. The realization that the wavelengths of light most strongly absorbed by chlorophyll are also the most effective in stimulating photosynthesis suggests that chlorophyll is primarily responsible for absorbing the light energy that drives photosynthesis.

Chlorophyll and other lipid pigments can be readily isolated from plant tissue by extraction with organic solvents. In the early 1900s the Russian botanist Mikhail Tswett discovered that the colored pigments present in such an extract can be separated from each other by passing the mixture through columns containing inert adsorbents. Tswett coined the term **chromatography** (color-writing) for this general type of fractionation procedure, which has subsequently been adapted to the purification of many other types of molecules.

The chemical and physical properties of chloro-

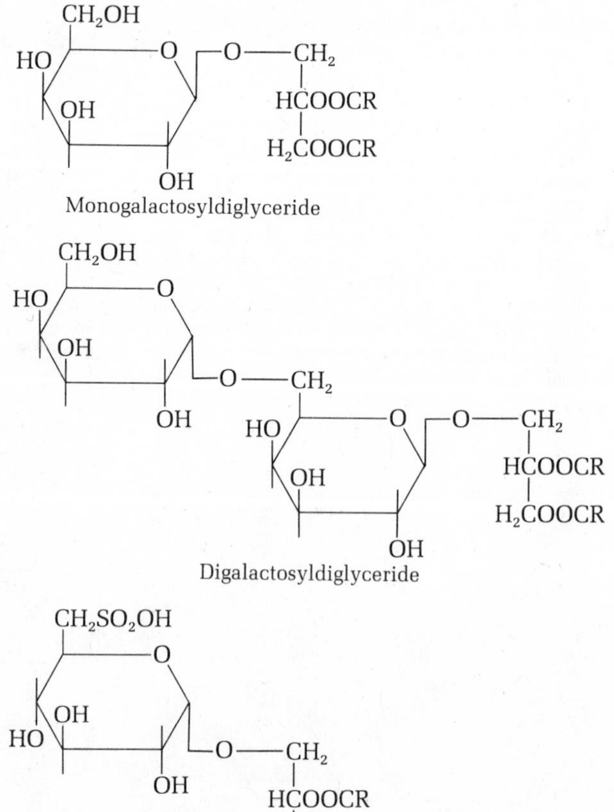

FIGURE 8-21 Structures of the major glycolipids occurring in chloroplast membranes. R = fatty acid side chains (hydrophobic), usually 16 or 18 carbon atoms in length.

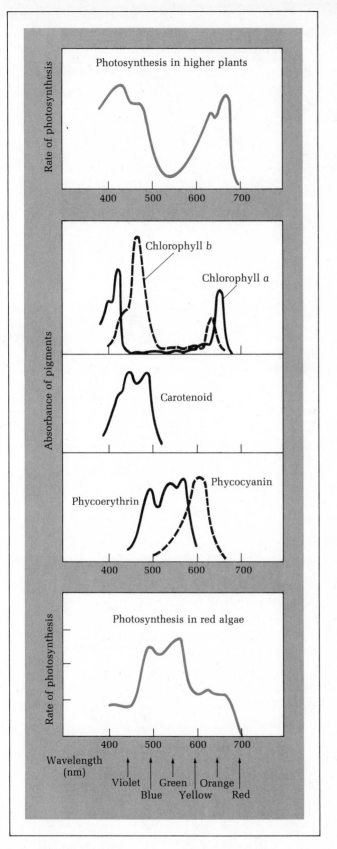

FIGURE 8-22 Comparison between the light-absorbing properties of various photosynthetic pigments (middle three graphs) and the ability of light of varying wavelengths to stimulate photosynthesis in higher plants (top graph) and red algae (bottom graph). The data suggest that chlorophyll is the major light-absorbing pigment for photosynthesis in higher plants, and that the accessory pigments play a more prominent role in red algae.

phyll molecules are ideally suited for a membrane-localized, light-absorbing substance. The basic skeleton of the chlorophyll molecule consists of a central *porphyrin ring* to which a long hydrocarbon side chain, called *phytol*, is attached (Figure 8-23). The hydrophobic phytol chain permits the chlorophyll molecule to be inserted within the lipid bilayer of the thylakoid membranes, while the alternating pattern of double bonds in the porphyrin ring facilitates the absorption of light. Although porphyrin rings also occur in cytochromes, hemoglobin, and myoglobin, the porphyrin ring of

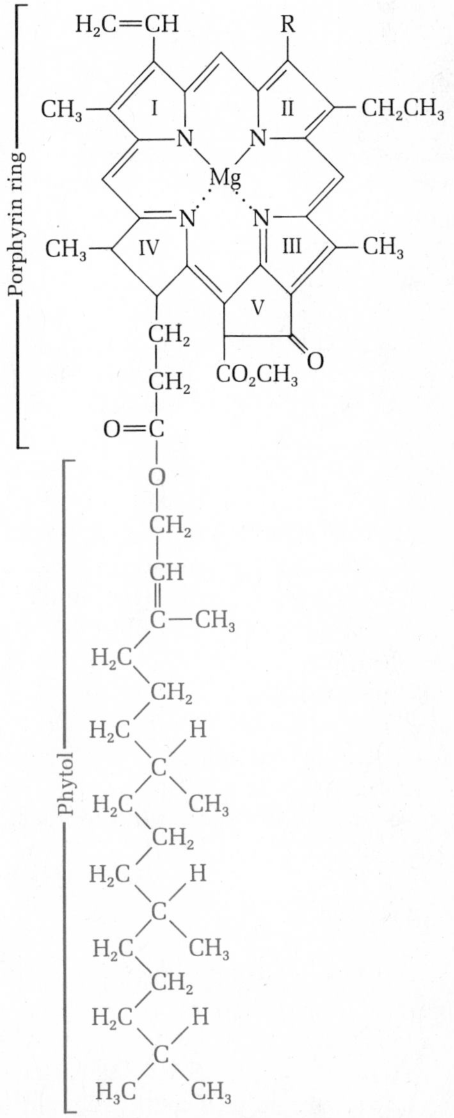

FIGURE 8-23 Structure of chlorophyll. In chlorophyll *a*, R = —CH$_3$, while in chlorophyll *b*, R = —CHO. Ring V is the fused cyclopentanone ring. In bacteriochlorophylls, the double bonds in ring II are not present.

TABLE 8-2
Species distributions of photosynthetic pigments

	Eukaryotes				Prokaryotes		
	Higher plants	Green algae	Diatoms, brown algae	Red algae	Blue-green algae	Purple bacteria	Green bacteria
Chlorophylls							
Chlorophyll *a*	+	+	+	+	+		
Chlorophyll *b*	+	+					
Chlorophyll *c*			+				
Chlorophyll *d*				+			
Bacteriochlorophyll *a*						+	+
Bacteriochlorophyll *b*						+	
Chlorobium chlorophyll							+
Carotenoids	+	+	+	+	+	+	+
Phycobilins				+	+		

chlorophyll is distinguished by several unique features: A *magnesium* atom, rather than an iron atom, is bound to the center of the ring; a cyclopentanone ring is fused to the basic porphyrin structure; and some of the side chains attached to the porphyrin ring differ from those occurring in other porphyrins.

Several types of chlorophyll, differing in their porphyrin side chains, occur in plants. The predominant form, designated **chlorophyll *a*,** is present in all photosynthetic eukaryotes and in blue-green algae. In addition to chlorophyll *a*, most photosynthetic cells contain a second type of chlorophyll (Table 8-2). In higher plants and in green algae this additional molecule is **chlorophyll *b*;** in brown algae, diatoms, and dinoflagellates it is **chlorophyll *c*;** and in red algae it has tentatively been identified as **chlorophyll *d*.** Slightly different types of chlorophyll, designated **bacteriochlorophylls** and ***Chlorobium* chlorophyll,** are utilized by photosynthetic bacteria. The light-absorbing properties vary somewhat among the different types of chlorophyll. Chlorophyll *a*, for example, absorbs light maximally at 420 and 663 nm, while chlorophyll *b* absorbs most strongly at 460 and 645 nm (see Figure 8-22). When chlorophyll molecules are situated within chloroplast membranes, their close association with specific membrane proteins causes small shifts in the position of these absorption peaks and the formation of several additional peaks.

In addition to chlorophyll, thylakoid membranes contain several **accessory pigments** that aid in the light-absorbing process. Such pigments absorb light maximally in those regions of the spectrum where chlorophyll absorbs light inefficiently, thereby increasing the overall efficiency of light absorption by the thylakoid membrane. The most widely occurring accessory pigments are the **carotenoids,** a family of lipid molecules constructed from long hydrocarbon chains containing alternating double bonds and ending in substituted cyclohexene rings (Figure 8-24). Carotenoid

FIGURE 8-24 Structure of the accessory pigments β-carotene and phycoerythrobilin. In phycocyanobilin, the $CH_2{=}CH-$ group shown in color is replaced by CH_3CH_2-. Note the resemblance of the four pyrrole rings in phycobilins to the four pyrrole rings that are joined together to form the porphyrin ring of chlorophylls (Figure 8-23).

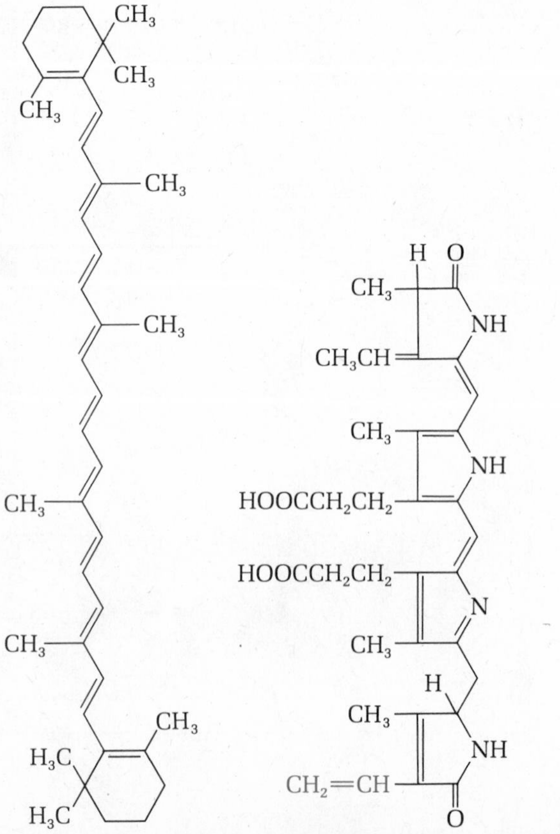

β-Carotene   Phycoerythrobilin

molecules absorb light in the violet/blue-green region of the spectrum (400–500 nm), giving them brilliant yellow, orange, and red colors. Carotenoids are responsible for the colors of many vegetables, such as carrots and tomatoes, but in actively photosynthesizing tissues their presence is generally masked by the green color of the more abundant chlorophyll molecules. At the end of the growing season, however, chlorophyll molecules often break down before carotenoids, giving many plants their brilliant fall colors.

A special class of accessory pigments, known as **phycobilins,** occurs in red and blue-green algae. The typical color of red algae is caused by the presence of a phycobilin pigment called phycoerythrobilin, while the color of a blue-green algae is imparted by a related substance, phycocyanobilin. Phycobilins are constructed from the same four rings present in porphyrins, but in phycobilins these rings are assembled into a linear chain rather than a circular structure (see Figure 8-24). Because of this linear arrangment, the centrally bound magnesium atom characteristic of the chlorophylls does not occur. Within intact cells, phycobilin pigments are complexed with specific proteins to form particles called **phycobilisome granules** (see Figure 8-10).

Although accessory pigments broaden the range of visible light that can be absorbed by photosynthetic cells, they cannot substitute for chlorophyll because the light energy absorbed by these pigments must first be transferred to chlorophyll a before it can enter the photosynthetic pathway. This qualification in no way minimizes the importance of the accessory pigments for photosynthesis. In red and blue-green algae, for example, maximum rates of photosynthesis occur when cells are illuminated with light of wavelengths maximally absorbed by the accessory pigments rather than by chlorophyll (see Figure 8-22 *bottom*). In such cases the accessory pigments are more important than chlorophyll in the *initial* light-absorbing step, even though the absorbed energy is subsequently channeled to chlorophyll.

In addition to the interaction between accessory pigments and chlorophyll, individual chlorophyll molecules also interact with each other in the light-absorbing step of photosynthesis. This conclusion first emerged in the early 1930s from experiments in which Robert Emerson and William Arnold exposed algae to brief flashes of light. It was expected that light of sufficient intensity to drive photosynthesis at its maximum rate would cause one molecule of carbon dioxide to be fixed per molecule of chlorophyll per light flash. The data revealed, however, that a maximum of 0.0004 mol of CO_2 is fixed per mole of chlorophyll present (Figure 8-25). This number indicates that the fixation of one molecule of CO_2 requires roughly $1 \div 0.0004 = 2500$ molecules of chlorophyll.

The above observations led Emerson and Arnold to

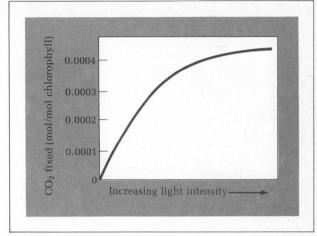

FIGURE 8-25 The effect of increasing light intensity on the rate of CO_2 fixation in algae. These data indicate that the fixation of one molecule of CO_2 requires several thousand chlorophyll molecules.

conclude that chlorophyll molecules are organized into functional groups of several thousand chlorophyll molecules. When light is absorbed by a chlorophyll molecule contained within such a **photosynthetic unit,** one of its electrons is activated to an excited state. This excited state is then transferred from chlorophyll to chlorophyll until it reaches a special chlorophyll a–protein complex known as the **reaction center.** The bulk of the chlorophyll molecules within a given photosynthetic unit thus function as "antennae," funneling their activation energy to the reaction-center chlorophyll. Excitation energy is transferred among the various antennae pigments in a defined sequence that serves to direct the pattern of energy migration toward the reaction center (Figure 8-26). As we shall see shortly, the reaction-center chlorophyll in turn passes its excited electron to the photosynthetic electron transfer chain.

Although the work of Emerson and Arnold permitted an estimate of the number of chlorophyll molecules required for the fixation of a single molecule of CO_2, it did not reveal how much solar energy is utilized in the process. Light is composed of discrete units of energy, termed **quanta** or **photons,** whose free-energy content is about 40 kcal/mol of quanta (for red light). The conversion of CO_2 to reduced organic matter, on the other hand, has a ΔG of roughly 115 kcal/mol of CO_2 reduced. Hence the increase in free energy that accompanies the fixation and reduction of a single molecule of CO_2 is roughly three times greater than the energy content of a quantum of light. A minimum of three quanta of light is therefore required per molecule of CO_2 fixed. In practice, the amount of light absorbed by actively photosynthesizing cells is closer to eight quanta per molecule of carbon dioxide converted to organic matter. Since the data of Emerson and Arnold indicated that roughly 2500 chlorophyll molecules are utilized per molecule of CO_2 fixed, each of the eight quanta required for the fixation

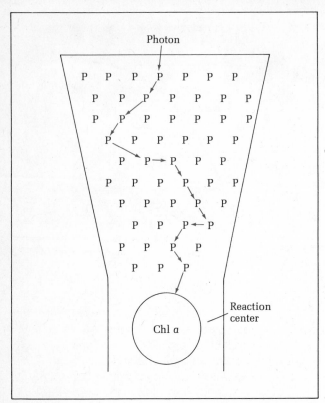

Photon

Chl *a*

Reaction
center

FIGURE 8-26 Schematic representation of energy transfer from antennae pigments (P) to the reaction-center chlorophyll *a* molecule (Chl *a*).

of a single molecule of CO_2 must have a target of $2500 \div 8 = \sim 300$ chlorophyll molecules. The term *photosynthetic unit,* originally defined as the number of chlorophylls per CO_2 fixed, is now commonly used to refer to this number of chlorophyll molecules per quantum of light absorbed.

The Photosynthetic Electron Transfer Chain

We have now seen how the absorption of light by chlorophyll and its accessory pigments creates an excited state that is ultimately funneled to the reaction-center chlorophyll, causing an electron in the reaction-center chlorophyll to be elevated to a state of high free energy. This activated electron is next transferred through a chain of carriers whose function is analogous to that of the respiratory chain of mitochondria. The molecules that comprise this photosynthetic electron transfer chain fall into categories that closely resemble those of the mitochondrial electron transfer chain: pyridine nucleotides, flavoproteins, cytochromes, iron-sulfur proteins, quinones, and copper-containing proteins. For a review of the general properties of such molecules, the reader is referred to Chapter 7. Below we shall briefly list the specific carriers of these types that have been identified in chloroplasts.

1. The major *pyridine nucleotide* employed by the photosynthetic electron transfer chain of eukaryotes is **nicotinamide adenine dinucleotide phosphate (NADP⁺),** a molecule differing from NAD^+ by the presence of an additional phosphate group (see Figure 7-4). In mitochondrial respiration electrons are transferred to the respiratory chain from the reduced coenzyme NADH (thereby oxidizing the NADH to NAD^+), while the light reactions of photosynthesis involve transfer of electrons from the photosynthetic electron transfer chain to $NADP^+$ (thereby reducing the $NADP^+$ to NADPH).

2. The major *flavoprotein* taking part in photosynthetic electron transfer is the enzyme **NADP reductase.** This FAD-containing protein catalyzes the transfer of electrons from ferredoxin (see number 4 below) to $NADP^+$.

3. Several *cytochromes* unique to plant cells are involved in photosynthetic electron transfer. Those whose roles have been most clearly defined are cytochromes *f* and *b*, which exhibit α-absorption bands at 553 and 563 nm, respectively.

4. In the late 1950s an *iron-sulfur* protein was isolated from plant leaves that enhances the rate of $NADP^+$ reduction when added to isolated chloroplasts. This protein, termed **ferredoxin,** contains a single active site composed of two iron atoms bound to two sulfur atoms (see Figure 7-27). Like other iron-sulfur proteins, ferredoxin transfers electrons by $Fe^{3+} \rightleftharpoons Fe^{2+}$ transitions. In its oxidized state ferredoxin exhibits strong absorption bands at 420 and 463 nm.

5. The participation of lipid molecules in the photosynthetic electron transfer chain was first suggested by the observation that extraction of chloroplasts with organic solvents inhibits the Hill reaction. The essential lipid was later identified as **plastoquinone** (Figure 8-27), a small molecule whose structure and absorption spectrum closely resembles that of ubiquinone. Like ubiquinone, plastoquinone transfers electrons by cycling between its quinone and quinol forms.

6. A copper-containing protein called **plastocyanin** has also been implicated in the photosynthetic electron transfer chain. In its oxidized state this protein absorbs light in the 600-nm region of the spectrum, giving the molecule a blue color. Upon reduction the absorption band and blue color disappear.

These six types of electron carriers are arranged into a chain designed to transfer activated electrons from reaction-center chlorophyll molecules to $NADP^+$. Observations made by Emerson in the early 1940s first suggested that light energy is absorbed at more than one point in this photosynthetic electron transfer chain. These studies were triggered by the observation that oxygen production declines dramatically when chloro-

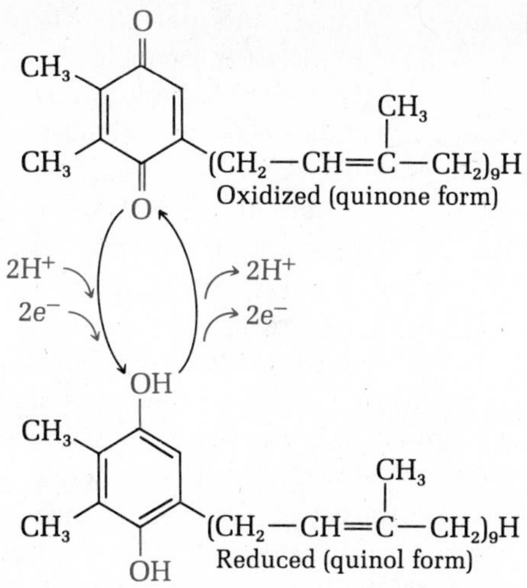

FIGURE 8-27 The structure of plastoquinone A. Note the close structural resemblance to mitochondrial ubiquinone (see Figure 7-28).

chlorophyll associated with a small number of additional chlorophyll molecules, pigment-binding proteins, and electron donors and acceptors, and (2) a **light-harvesting complex** containing chlorophyll molecules, accessory pigments, and their associated proteins. Because isolated photosystem II preparations tend to be characterized by a higher ratio of chlorophyll b to a than photosystem I preparations, it was once thought that chlorophyll b is the essential pigment of photosystem II. However, it was subsequently discovered that mutant strains of barley lacking chlorophyll b are able to photosynthesize at normal rates and efficiencies. On this basis

plasts are illuminated with light in the far red region of the spectrum, even though chloroplast pigments still absorb light of these wavelengths (Figure 8-28). Emerson discovered that this decrease in photosynthetic activity at long wavelengths, termed the *red drop*, can be reversed by simultaneous illumination with a second beam of light of a shorter wavelength. For example, simultaneous illumination with beams of 650 and 700 nm causes oxygen to be produced at a rate that is considerably faster than that obtained when each beam is used separately and the two rates of oxygen production are added together. This enhancement effect suggests that two independent light-absorbing events, with differing absorption optima, are involved in the light reactions of photosynthesis. The light-absorbing system that is more efficient at longer wavelengths is called **photosystem I,** while the system active at shorter wavelengths is referred to as **photosystem II.** The chlorophyll-protein complex that comprises the reaction center of photosystem I is referred to as **P700** because it absorbs light maximally at 700 nm, while the chlorophyll-protein complex that comprises the reaction center of photosystem II is designated **P680** because it absorbs maximally at 680 nm. The difference in the absorption maxima of the reaction-center chlorophylls is determined by the microenvironments created by the proteins with which they are associated rather than by any inherent differences in the molecules of chlorophyll a themselves.

Subsequent investigations have revealed that the two photosystems can be physically separated from each other by disrupting chloroplasts in the presence of detergents and fractionating the resulting suspension by differential centrifugation. Each isolated photosystem can be further subfractionated into two components: (1) a **core complex** containing the reaction center

FIGURE 8-28 Experimental data revealing the existence of the "red drop." The top curve is the absorption spectrum of the green alga *Chlorella*. The bottom graph shows the photosynthetic activity of this organism when illuminated with light of varying wavelengths. Under control conditions the rate of oxygen production declines dramatically with wavelengths of light near 700 nm, even though chlorophyll still absorbs appreciably at this wavelength. This so-called red drop is partially reversed in the presence of a supplemental light beam of 650 nm (colored line). See text for a more detailed interpretation of this phenomenon.

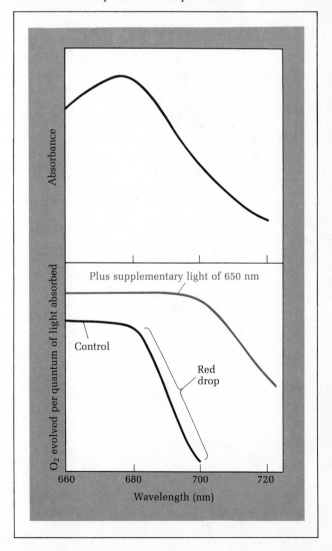

it has been concluded that chlorophyll *b* is an antennae pigment that simply happens to be present in higher concentration in photosystem II than in photosystem I, while chlorophyll *a* is essential for the reaction centers of both photosystems.

Shortly after the existence of two separate photosystems was first suggested by Emerson's experiments, Robert Hill and Fay Bendall proposed that the two photosystems are arranged in series rather than parallel. In an arrangement of this type one light-absorbing event triggers an initial activation of a chlorophyll electron, followed by a further activation of the electron by a second light-absorbing event. This proposal was supported by simple calculations that indicated that a single electron requires an energy input equivalent to at least two quanta of light in order to achieve an activated state whose free energy is sufficient to reduce $NADP^+$ to NADPH.

If the two photosystems are arranged in series, the question arises as to the location of these two steps within the overall pathway of photosynthetic electron transfer. Many of the experimental approaches employed to investigate the sequential organization of the mitochondrial respiratory chain have also been applied to the photosynthetic electron transfer chain. These approaches, which were described in detail in Chapter 7, include (1) determining the standard redox potentials of isolated carriers, (2) testing the ability of isolated carriers to oxidize and reduce each other, (3) examining light-induced changes in difference spectra, (4) using specific inhibitors to block electron flow at particular points, (5) utilizing artificial electron donors and acceptors to study individual portions of the chain, and

(6) analyzing genetic mutants defective in particular regions of the photosynthetic chain.

Data obtained from all the above approaches have been utilized in formulating the model of the photosynthetic electron transfer chain summarized in Figure 8-29. Because the diagram of this pathway resembles a sideways "Z," it is commonly referred to as the *Z-scheme*. The arrangement of carriers in the Z-scheme is still tentative in several respects and certain intermediates remain to be clearly identified, but the general framework of the pathway is fairly well established. As can be seen from examining Figure 8-29, the first light-absorbing step is mediated by photosystem II. Light energy absorbed by the accessory and antenna pigments of photosystem II is funneled into the reaction center (P680), where an electron associated with the reaction center chlorophyll is boosted to an excited state. Upon activation this electron is transferred to a protein-bound plastoquinone molecule designated "Q." The loss of this electron leaves the reaction center with a net positive charge, which serves as a driving force for attracting electrons from water. As electrons are removed from water, the water molecules are split into oxygen and free protons. The molecular mechanism underlying this water-splitting reaction is not well understood, but it is thought to involve a manganese-containing protein whose manganese atom serves as an intermediate carrier of the electrons between water and the P680 reaction center.

After transfer of an activated electron from P680 to the primary electron acceptor Q, the electron passes through another acceptor, designated B, to a pool of mobile plastoquinone molecules. The exact sequence of

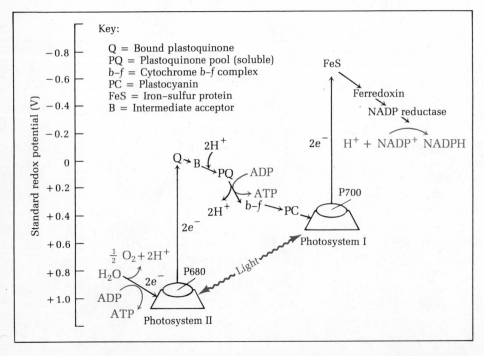

FIGURE 8-29 Proposed sequence of carriers in the photosynthetic electron transfer chain, plotted according to their standard redox potentials (E_0'). Absorption of light by the two photosystems elevates the oxidation-reduction potential, while the remainder of the reactions flow spontaneously down the potential gradient. Because of the resemblance of this diagram to the letter "Z," it is sometimes referred to as the "Z-scheme."

carriers beyond the plastoquinone pool has not been unequivocally established, but it is currently thought to involve the general arrangement: cytochrome $b \rightarrow$ cytochrome $f \rightarrow$ plastocyanin. Beyond plastocyanin, the electron transfer sequence encounters photosystem I. Absorption of light by photosystem I causes electrons associated with its reaction center (P700) to become activated. These activated electrons are passed to a tightly bound complex of at least three iron-sulfur proteins. From here the electrons are transferred to ferredoxin, which in turn passes them to $NADP^+$ in a reaction catalyzed by the enzyme NADP-reductase. Because the reduction of $NADP^+$ to NADPH involves two electrons, two molecules of reduced ferredoxin are required to effect the reduction of $NADP^+$. The "electron hole" left in photosystem I each time an electron is passed from P700 to $NADP^+$ is filled by electrons obtained from plastocyanin.

The net effect of the Z-scheme pathway is the light-induced flow of electrons from water to $NADP^+$. Although the entire electron transfer pathway from water to $NADP^+$ is collectively referred to as the "light reactions" of photosynthesis, it is only in the light-absorbing steps mediated by photosystems II and I that light is directly involved. The absorption of light energy by the two photosystems is sufficient to increase the redox potential from $+0.82$ V in water to -0.32 V in NADPH (see Figure 8-29), which corresponds to an increase in free energy of approximately 50 kcal. In addition, some of the energy absorbed by the photosystems is used to drive the formation of ATP as described below.

Photophosphorylation and the Chemiosmotic Theory

It was mentioned earlier in the chapter that the light-induced flow of electrons from water to $NADP^+$ is accompanied by the formation of ATP. This type of ATP synthesis, known as **photophosphorylation,** is analogous to mitochondrial oxidative phosphorylation in that both are driven by energy released when electrons flow through a chain of carriers. In mitochondrial respiration, however, the energy driving this electron flow is derived from the oxidation of foodstuffs, while in photosynthesis the energy is obtained from sunlight.

The main pathway of photophosphorylation involves the coupling of ATP synthesis to the light-induced flow of electrons from water to $NADP^+$. Because this process links ATP synthesis to the one-way passage of electrons through the Z-scheme carriers, it is referred to as **noncyclic photophosphorylation.** By definition, noncyclic photophosphorylation is accompanied by the production of oxygen and the reduction of $NADP^+$ to NADPH.

The alternative pathway for light-induced ATP formation, termed **cyclic photophosphorylation,** is based

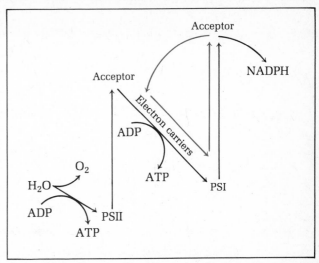

FIGURE 8-30 A diagrammatic representation of the distinction between cyclic photophosphorylation (color) and noncyclic photophosphorylation (black). Note that cyclic photophosphorylation involves only one of the two coupling sites for ATP synthesis.

upon a cyclic flow of electrons centered around photosystem I. In cyclic photophosphorylation light induces the activation of electrons from photosystem I as usual, but instead of being transferred to $NADP^+$, these activated electrons are returned to photosystem I (Figure 8-30). The energy released as the activated electrons return to the lower free-energy level of photosystem I is used to drive the synthesis of ATP. In such a pathway there is obviously no reduction of $NADP^+$ and no production of oxygen. It has been estimated that cyclic photophosphorylation normally accounts for 10–20 percent of the total ATP yield of photophosphorylation, although its contribution can be larger when there is insufficient $NADP^+$ available to function as the terminal electron acceptor of the Z-scheme pathway.

Coupling Sites for Photophosphorylation The conclusion that ATP formation accompanies both the linear and cyclic flow of electrons through the photosynthetic electron transfer chain raises the question of the number and location of the coupling sites for ATP synthesis. Definitive evidence for the existence of two separate coupling sites was first obtained in the laboratories of Achim Trebst and Norman Good, who employed artificial electron donors and acceptors to probe the ATP-synthesizing capabilities of individual regions of the photosynthetic electron transfer chain. Such experiments revealed that chloroplasts incubated in the presence of artificial electron acceptors such as ferricyanide or methylviologen, which take up electrons after photosystem I, exhibit a P/O ratio of about 1.2; that is, about 1.2 molecules of ATP are synthesized per atom of oxygen produced. In contrast electron acceptors such as phenylenediamine, which pick up electrons prior to photosystem I, support a P/O ratio closer to 0.6. This

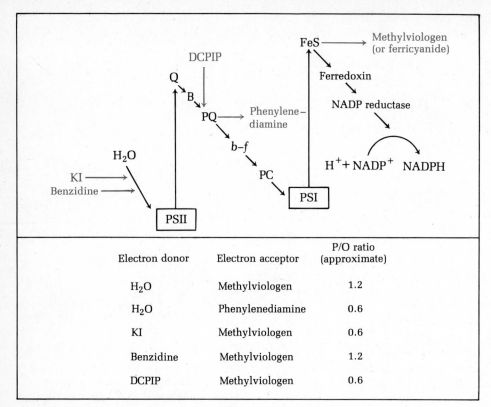

Electron donor	Electron acceptor	P/O ratio (approximate)
H_2O	Methylviologen	1.2
H_2O	Phenylenediamine	0.6
KI	Methylviologen	0.6
Benzidine	Methylviologen	1.2
DCPIP	Methylviologen	0.6

FIGURE 8-31 Summary of experimental data that led to the conclusion that photophosphorylation involves two coupling sites. See text for interpretation. The abbreviations for the carriers are defined in Figure 8-29.

halving of the ATP yield suggests that there are two coupling sites for ATP synthesis in the photosynthetic electron transfer chain, one located before and one located after the site of phenylenediamine action (Figure 8-31). The above measurements suggest that each of these coupling sites synthesizes roughly 0.6 ATP molecules per pair of electrons passing through the site.

To localize these two sites more specifically, ATP yields in the presence of other electron donors have been quantitated. For example reduced dichlorophenolindophenol (DCPIP), which bypasses photosystem II by donating electrons directly to plastoquinone, has been found to exhibit a P/O ratio of only 0.6, indicating that one coupling site must be situated *prior* to plastoquinone. Localization of this site has been facilitated by the use of artificial electron donors such as potassium iodide and benzidine, which can substitute for water as a source of electrons feeding the electron transfer chain. As is the case with water, electrons derived from such artificial electron donors flow through the entire sequence of carriers and are eventually employed to reduce $NADP^+$ to NADPH. In terms of the amount of ATP synthesis induced, however, artificial electron donors do not all behave the same as water. When potassium iodide is substituted for water, the P/O ratio declines to roughly half its normal value, indicating that water itself must participate in one of the coupling sites for ATP synthesis. An insight into the role played by water has been provided by the observation that unlike potassium iodide, benzidine produces the same amount of ATP as when water is the electron donor. Benzidine resembles water in that it releases protons upon transferring its electrons to the photosynthetic electron transfer chain, while potassium iodide does not. Taken together, the above observations suggest that the protons released during the splitting of water play an important role in the first coupling site for photophosphorylation.

If one examines the photosynthetic electron transfer chain in the region implicated by the phenylenediamine experiments to contain the second coupling site for ATP synthesis, another reaction involving proton fluxes is discovered. This reaction is the reduction and oxidation of plastoquinone, which is accompanied by both the uptake and release of protons (see Figure 8-27). Given the well-established role of proton gradients in mitochondrial oxidative phosphorylation (Chapter 7), it is perhaps not unreasonable to expect proton gradients to play a comparable role in photophosphorylation. In the following section, we shall review the major evidence supporting this idea.

Evidence that Proton Gradients Mediate Photophosphorylation

1. Shortly after Peter Mitchell proposed in his chemiosmotic theory that electrochemical proton gradients drive ATP synthesis coupled to electron transfer reactions, experiments carried out in the laboratory of André Jagendorf supported the proposed role of proton

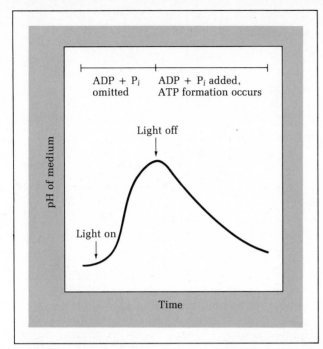

FIGURE 8-32 Data summarizing the changes that occur in the pH of the external medium when chloroplasts are exposed to light and darkness. The discovery that ATP formation can occur in the dark after a pH gradient has been established in the light is compatible with the theory that electrochemical proton gradients drive ATP synthesis.

matically. Since an increase in pH reflects a decline in proton concentration, it was concluded that protons are pumped from the medium into the chloroplast during the light reactions of photosynthesis. An involvement of this proton gradient in driving ATP formation was suggested by a subsequent experiment in which ADP and P_i were added back to the chloroplast suspension after the proton gradient had been established and the light had been turned off. Under these conditions ATP synthesis was observed to occur as the pH of the medium decreased (Figure 8-32). It was therefore concluded that ATP synthesis is associated with a flow of protons back out the chloroplast (down their concentration gradient).

2. Although the above observations demonstrate that a proton gradient is established in illuminated chloroplasts and dissipated in chloroplasts synthesizing ATP in the dark, such data do not prove that the proton gradient actually drives ATP formation. To directly test the idea that the flow of protons down an electrochemical gradient can induce ATP formation, Jagendorf created an artificial pH gradient by soaking chloroplasts for several hours in the dark in a pH 4 buffer, and then placing them in a medium of pH 8. In response the pH of the stroma quickly rose to 8, while the pH within the thylakoid sacs remained at 4. This artificially created pH gradient triggered a rapid burst of ATP synthesis that continued until the pH gradient was dissipated (Figure 8-33). Since the entire experiment was carried out in the dark, it was concluded that an electrochemical proton gradient can drive ATP synthesis even in the absence of light-induced electron flow. This experiment, reported in 1966, was the first direct demonstration in any experimental system that a pH gradient can drive ATP formation.

gradients in photophosphorylation. In these experiments chloroplasts were first placed in a medium in which ATP synthesis could not occur because the necessary substrates, ADP and P_i, were lacking. When such chloroplasts were subsequently illuminated with bright light, the pH of the medium was found to increase dramatically. Since an increase in pH reflects a decline in

FIGURE 8-33 Summary of experiment revealing that an artificially formed pH gradient can drive ATP synthesis in chloroplasts in the absence of light-induced electron transfer.

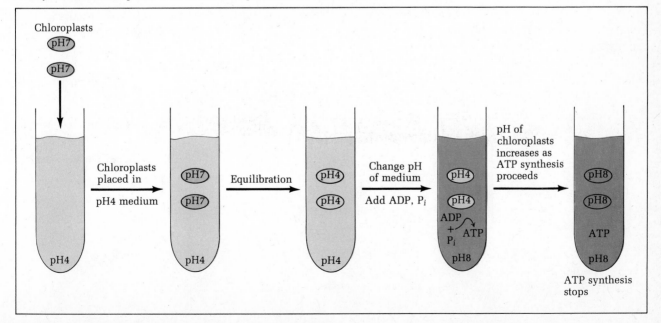

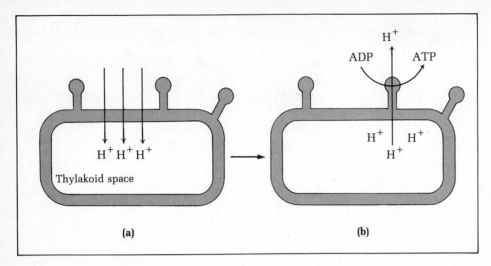

FIGURE 8-34 Orientation of proton flow in thylakoid sacs. (a) During light-induced electron transfer, protons are pumped into the thylakoid space. (b) The subsequent outward flow of protons down their electrochemical gradient drives ATP formation.

3. Taken together the above two sets of experiments suggest that light-induced photosynthetic electron transfer induces the transport of protons into the thylakoid sacs, generating a gradient of pH (interior acid) as well as a membrane potential (interior positive). The diffusion of protons back down this electrochemical gradient then drives ATP formation (Figure 8-34). Although the artificial gradient of 4 pH units utilized in Jagendorf's experiments might seem drastic when compared to physiological conditions, subsequent measurements have revealed that normally the internal pH within the thylakoid sacs declines to about 4.5 and the pH of the stroma rises to 8 during photosynthesis in chloroplasts stimulated with saturating light intensities. Compartmentalization of the strongly acidic side of the gradient within the thylakoid sacs protects the remaining components of the chloroplast, including the enzymes of CO_2 fixation present in the stroma, from this harsh environment. Measurements of the membrane potential ($\Delta\psi$) of illuminated thylakoids has generally yielded values less than 10 mV, an order of magnitude smaller than the membrane potential in respiring mitochondria. It has therefore been concluded that the driving force for ATP synthesis in chloroplasts derives almost entirely from ΔpH, whereas in mitochondria both ΔpH and $\Delta\psi$ make significant contributions. The

reason for the low value of $\Delta\psi$ in chloroplasts is that thylakoid membranes are more permeable than mitochondrial cristae to ions such as Mg^{2+} and Cl^-. Hence passive diffusion of ions through the thylakoid membrane tends to dissipate the membrane potential component of the electrochemical proton gradient (Figure 8-35).

4. Comparison of the magnitude of the pH change with the amount of oxygen produced when chloroplasts are exposed to brief flashes of light has led to the conclusion that roughly four protons are transported into the thylakoid sac for every molecule of oxygen produced (the production of one atom of oxygen corresponds to the removal of one pair of electrons from water). Since there are two coupling sites for ATP synthesis, it can therefore be surmised that two protons are pumped across the thylakoid membrane per coupling site per pair of electrons passing through the chain. In this regard chloroplasts resemble mitochondria, where each coupling site also pumps two protons per pair of electrons passing through the chain. As discussed previously, one of these proton pairs is derived from water and the other is transported across the membrane by the successive reduction and oxidation of plastoquinone.

5. Thermodynamic calculations based on the measured values of ΔpH and $\Delta\psi$ have been carried out to

FIGURE 8-35 Diagram illustrating how the permeability of the thylakoid membrane to Mg^{2+} and Cl^- allows the membrane potential component of the electrochemical gradient to dissipate.

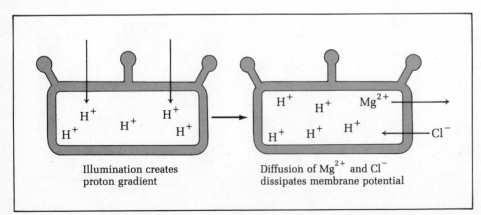

Illumination creates proton gradient

Diffusion of Mg^{2+} and Cl^- dissipates membrane potential

determine whether the free-energy content of the electrochemical proton gradient is sufficient to account for the amount of ATP synthesized. Using the previously defined equation for determining the ΔG of an electrochemical gradient (page 140), it can be calculated that the free-energy content of the proton gradient in illuminated chloroplasts is approximately 4.8 kcal/mol of protons. The energy required to synthesize ATP under conditions prevailing within the chloroplast has been estimated to be as high as 14.5 kcal/mol, which is significantly higher than the requirement of 11 kcal/mol for mitochondrial ATP formation (Chapter 7). The reason for this difference is that the concentration of ADP and P_i is lower in chloroplasts than in mitochondria, shifting the equilibrium point of the reaction and effectively increasing the free energy required. Since the free energy of the protons crossing the chloroplast membrane is only 4.8 kcal/mol, thermodynamic considerations suggest that at least 3 protons must flow back through the thylakoid membrane per molecule of ATP synthesized ($3 \times 4.8 = 14.4$ kcal, which is very close to the 14.5 kcal required). Although experimental determinations of the number of protons flowing through the thylakoid membrane per molecule of ATP synthesized have yielded values ranging between 2 and 3, the above thermodynamic considerations suggest that the correct value is probably 3.

We saw earlier that a total of four protons are transported across the thylakoid membrane during the light-induced flow of one electron pair from water to $NADP^+$ (two protons per coupling site). If three protons are needed to drive the synthesis of one molecule of ATP, then these four protons are thermodynamically sufficient for the synthesis of $4 \div 3 = 1.3$ molecules of ATP. According to this rationale, the P/O ratio of intact chloroplasts should be no higher than 1.3. Although experimental determinations of the actual P/O ratio under normal conditions yield variable results, values as high as 1.5 are commonly reported; that is, a total of 1.5 ATPs is formed per pair of electrons passing from water to $NADP^+$. If this discrepancy between the predicted and experimentally determined values is not due to experimental error, it suggests that the additional ATP (~ 0.2 molecules) must be formed by cyclic photophosphorylation, a pathway that allows additional protons to be contributed to the electrochemical gradient without the transfer of electrons from water to $NADP^+$ or the associated production of oxygen.

6. Additional support for the role of proton gradients in photophosphorylation has been provided by the use of ionophores that disrupt the electrochemical gradient. Agents such as gramicidin and CCCP, which make membranes freely permeable to protons and thereby collapse both ΔpH and $\Delta\psi$, uncouple ATP synthesis from electron transfer in chloroplasts as they do in mitochondria. Ionophores that eradicate only ΔpH

(e.g., nigericin) or $\Delta\psi$ (e.g., valinomycin) can also be used in combination to uncouple photophosphorylation. When ionophores are employed to collapse the electrochemical gradient, the light-induced flow of electrons through the photosynthetic chain continues, accompanied by the production of oxygen and the reduction of $NADP^+$ to NADPH. Photophosphorylation, however, is abolished. Such observations indicate that photophosphorylation, like oxidative phosphorylation, is driven by an electrochemical proton gradient.

7. Further evidence for the involvement of proton gradients in photophosphorylation has been provided by kinetic experiments in which the relationship between the magnitude of the proton gradient and the capacity to synthesize ATP has been measured as a function of time after illumination. This approach has revealed that the two processes have identical kinetics, that is, the rate at which the proton gradient forms is closely correlated with the rate at which ATP synthesis increases (Figure 8-36).

8. The chemiosmotic theory predicts that the coupling of ATP synthesis to light-induced electron transfer requires an intact thylakoid membrane enclosing a defined compartment, for a proton gradient could not otherwise be maintained. This prediction is consistent with the observation that chloroplast subfractions capable of carrying out photophosphorylation always contain intact membrane vesicles.

9. Reconstitution of membrane vesicles from isolated thylakoid components has yielded results that in many ways parallel those obtained with mitochondrial systems. Racker's laboratory has isolated a protein complex consisting of two components, designated CF_1 and CF_0, that closely resemble the F_1 and F_0 constituents of the mitochondrial ATP-synthesizing complex. CF_1 is a hydrophilic protein that is relatively easy to remove

FIGURE 8-36 Comparison of the rate at which the pH gradient is established (color) with the rate of ATP synthesis (black) in chloroplasts exposed to light. Note the close correlation in the timing of these two events.

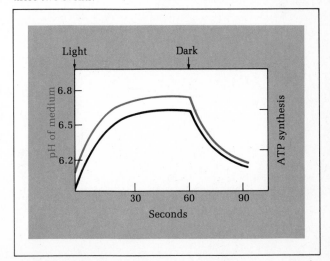

from thylakoid membranes. Electron microscopy of such isolated CF$_1$ preparations reveal the presence of spherical particles 9–11 nm in diameter that resemble the spheres that normally protrude from the outer surface of the thylakoid membrane. Like F$_1$, these isolated CF$_1$ preparations exhibit ATPase activity but cannot synthesize ATP by themselves because the photosynthetic electron transfer chain must be present to provide the energy that drives the ATP-forming reaction. Thylakoid membranes that have had CF$_1$ removed cannot carry out ATP synthesis either, although the light-induced transfer of electrons from water to NADP$^+$ proceeds as usual. Addition of purified CF$_1$ back to CF$_1$-depleted membranes is associated with the reappearance of the ability to synthesize ATP. Such observations suggest that the photosynthetic electron transfer chain is localized within the smooth portion of the thylakoid membrane, and that ATP formation occurs within the spherical particles that protrude from the membrane surface (Figure 8-37). The resemblance of this organization to that of the mitochondrial inner membrane is striking.

CF$_1$ is anchored to the thylakoid membrane by CF$_0$, a set of hydrophobic proteins spanning the membrane. Removal of CF$_1$ causes the thylakoid membrane to become leaky to protons, suggesting that CF$_0$ serves as a

FIGURE 8-37 A diagrammatic summary of the steps involved in the disassembly and reconstitution of the thylakoid membrane, indicating the metabolic capabilities and morphological appearance of the various fractions produced. Note the general similarity of these reconstitution experiments to those carried out on mitochondrial membranes (Figure 7-32). Micrographs courtesy of M. P. Garber.

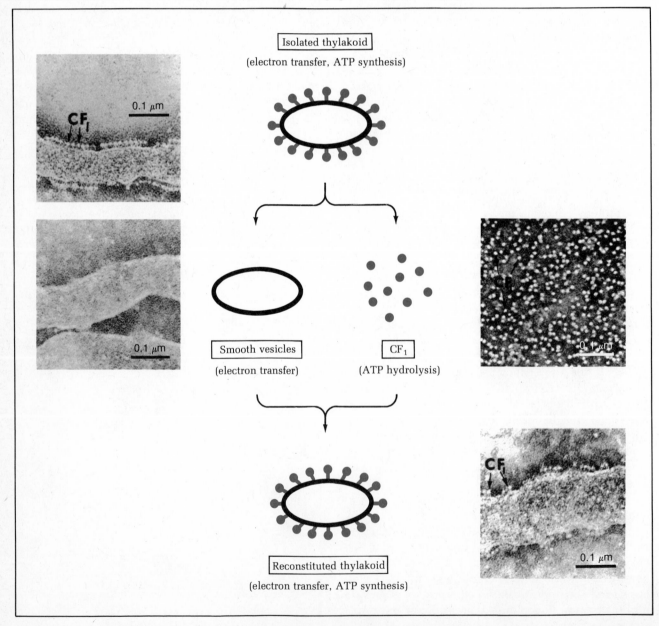

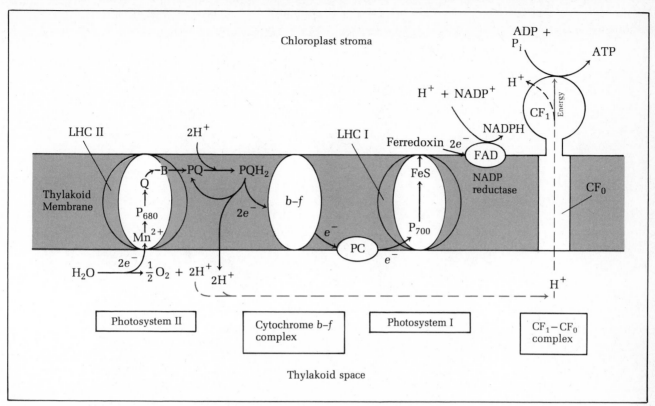

FIGURE 8-38 A highly stylized model of the thylakoid membrane, emphasizing the existence of four major structural units: photosystems I and II, the cytochrome *b-f* complex, and the CF_0-CF_1 complex. Abbreviations: LHC = light-harvesting complex.

transmembrane channel through which protons flow on their way to CF_1 (Figure 8-38). This overall arrangement, in which protons flow through CF_0 and drive ATP synthesis by interacting with CF_1, is basically the same as occurs in mitochondria with the analogous proteins, F_1 and F_0.

10. In addition to the information it has provided concerning the role of the CF_1-CF_0 complex, the reconstitution approach has also been used to study the behavior of isolated photosystems. By mixing purified photosystem preparations with isolated phospholipids and sonicating the mixture, it is possible to construct artificial membrane vesicles capable of carrying out the light-induced transport of protons. If the CF_1-CF_0 complex is included in the reconstitution mixture, the proton gradient can be utilized to drive ATP formation, providing yet additional support for the idea that photophosphorylation is mediated by formation of an electrochemical proton gradient (Figure 8-39).

11. Because it requires that protons be selectively pumped in one direction (from the stroma into the thylakoid spaces), the chemiosmotic model predicts that the components of the photosynthetic electron transfer chain must be asymmetrically oriented within the thylakoid membrane. This prediction has been tested by determining the accessibility of these membrane components to various probes. Two classes of probes have

been especially useful in this regard: antibodies and artificial electron donors/acceptors. Incubation of thylakoid vesicles with antibodies directed against various components of the electron transfer chain has been employed to determine whether a given constituent is present on the outer or inner thylakoid surface. In this way it has been discovered, for example, that antibodies against CF_1, ferredoxin, and NADP-reductase readily bind to intact thylakoid vesicles, implying that these components are exposed on the outer membrane surface. The ability of artificial electron donors and acceptors to interact with intact thylakoid vesicles has also been used to assess the orientation of membrane constituents. In general, lipid-soluble electron donors have been found to be capable of reducing photosystems I and II when added to intact membrane vesicles, while lipid-insoluble electron donors have not. Since only lipid-soluble electron donors can pass through the membrane to react with components exposed on the inner membrane surface, such observations suggest that the electron-accepting sites of photosystems I and II are located on the inner surface of the thylakoid membrane.

On the basis of these and related approaches, the highly schematic model of the thylakoid membrane presented in Figure 8-38 has been developed. Although the details of this model will undoubtedly be subject to

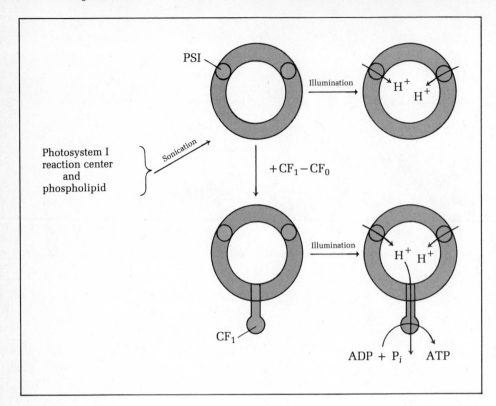

FIGURE 8-39 A diagrammatic summary of the experimental approach utilized for reconstituting artificial membrane vesicles capable of carrying out light-induced electron transfer and ATP synthesis. Experiments of this type provide strong evidence that light-induced electron transfer is associated with the formation of an electrochemical proton gradient, and that the CF_0-CF_1 complex utilizes this proton gradient to drive ATP formation.

future revision, the overall picture is clearly consistent with the chemiosmotic model in that the electron carriers are asymmetrically distributed within the membrane in such a way that two pairs of protons are released into the thylakoid space each time a pair of electrons flows through the chain.

Substructure of Thylakoid Membranes

In recent years electron microscopists have begun to probe the ultrastructural organization of the thylakoid membrane in the hopes of further advancing our understanding of the light reactions of photosynthesis. The most useful approach for analyzing the internal substructure of the thylakoid membrane is freeze-fracture microscopy. This technique has revealed the existence

FIGURE 8-40 Schematic representation of the two surfaces and two fracture faces observed in freeze-fracture micrographs of thylakoid membranes.

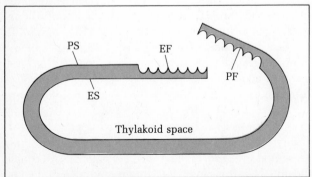

of such a bewildering array of particles within the thylakoid membrane, however, that it has taken a number of years for biologists to agree upon what they are seeing. Although agreement in certain cases is still incomplete, a general picture is gradually beginning to emerge that correlates well with the biochemical features of the photosynthetic pathway.

Freeze-fracturing tends to split membranes along the plane of the lipid bilayer, creating two membrane halves. In fractured thylakoid membranes the surface view and fracture face of the membrane-half adjacent to the stroma are referred to as PS and PF, while the surface view and fracture face of the membrane-half adjacent to the thylakoid space are referred to as ES and EF (Figure 8-40; see also Figure 3-10). Particles protruding from the membrane exterior are restricted primarily to the outer or PS surface of the unstacked thylakoids (Figure 8-41). These particles, which measure roughly 11 nm in diameter, disappear under conditions known to extract CF_1 activity from the membrane and are therefore thought to correspond to CF_1. Although CF_1 particles are usually absent from stacked thylakoids, exposing a mixture of stacked and unstacked thylakoids to hypotonic solutions causes the membranes to become unstacked and the CF_1 particles to assume a random distribution over the surface of the newly unstacked membranes. This observation indicates that CF_1 is free to diffuse within the lateral plane of the thylakoid membranes. The absence of CF_1 from the stacked thylakoids, where most of the light reaction activity occurs, is further support for the chemiosmotic theory because it

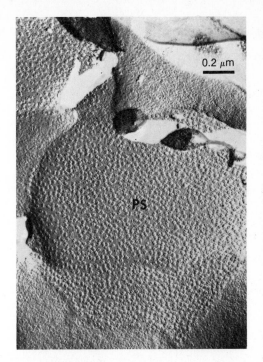

FIGURE 8-41 Freeze-fracture micrographs showing the various surfaces and fracture faces of stacked and unstacked thylakoids. *(Left)* Outer surface (PS) of unstacked thylakoids revealing particles thought to be CF_1. *(Right)* Fracture face of the membrane half exposed to the thylakoid interior in stacked (EF_s) and unstacked (EF_u) thylakoids. The fracture faces of the opposite membrane half are also shown for stacked (PF_s) and unstacked (PF_u) thylakoid regions. Courtesy of K. R. Miller.

shows that ATP synthesis need not be physically tied to electron transfer. Instead, ATP synthesis occurs primarily in the unstacked stroma thylakoids utilizing the proton gradient initially established within the stacked thylakoids (Figure 8-42).

The fracture faces of the thylakoid membranes exhibit several kinds of particles. The predominant component of the PF face is a family of small particles 8–11 nm in diameter, while the EF face contains a group of relatively large particles 10–18 nm in diameter (see Figure 8-41). The small PF particles occur in roughly equal numbers in stacked and unstacked thylakoids, while the larger EF particles are present in greater numbers in stacked than in unstacked membranes. An important clue to the identity of these protein particles has been provided by studies involving

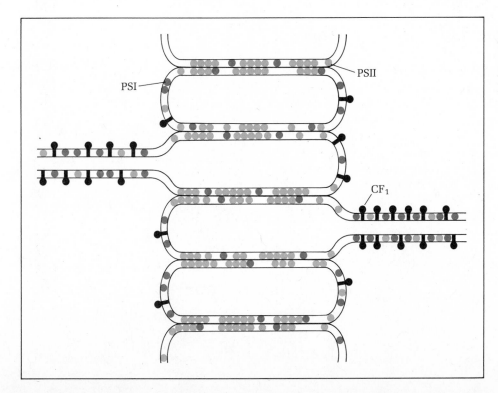

FIGURE 8-42 Schematic representation of the distribution of photosystem I, photosystem II, and CF_1 in stacked and unstacked thylakoids. Interactions between the light-harvesting complexes of photosystem II are generally believed to induce membrane stacking.

disruption of thylakoid membranes with detergents or concentrated salt solutions. Such treatments release the components of the photosynthetic electron transfer chain into solution not as individual molecules but as four distinct protein complexes: photosystem I, photosystem II, a cytochrome *b-f* complex, and CF₁-CF₀. The two photosystem complexes are involved in light harvesting, the cytochrome *b-f* complex functions in electron transfer between the two photosystems, and the CF₁-CF₀ complex mediates ATP synthesis (see Figure 8-38).

Several experimental approaches have been utilized for investigating the relationship between these protein complexes and the particles seen in freeze-fracture micrographs. Observations such as the following have led to the conclusion that the large and small particles correspond to photosystem II and photosystem I, respectively. (1) Stacked thylakoid fractions isolated by chloroplast subfractionation techniques are enriched in photosystem II activity and the large EF particles. Unstacked thylakoid fractions, on the other hand, are more concentrated in photosystem I activity and small PF particles. (2) Plants grown in the dark for extended periods of time contain modified chloroplasts, called **etioplasts,** which lack chlorophyll, thylakoids, and photosynthetic activity. Upon exposure to light, etioplasts evolve photosynthetic activity by a stepwise series of events. One of the first steps to occur in this sequence is the development of photosystem I activity, which appears at a time when only unstacked thylakoids and small PF particles are present. Later photosystem II activity appears, accompanied by the formation of grana stacking and the appearance of large freeze-fracture particles. (3) Mutant strains of tobacco exist whose leaves contain large yellow patches deficient in chlorophyll. The chloroplasts present in these yellow areas have only unstacked thylakoids, small freeze-fracture particles, and photosystem I activity. The lack of photosystem II activity thus correlates with the absence of the large freeze-fracture particles. (4) Barley mutants deficient in photosystem II activity have been found to contain reduced numbers of large EF particles, although the smaller PF particles are present in normal amounts.

Although the above lines of evidence strongly support the conclusion that photosystems I and II correspond to small and large freeze-fracture particles, respectively, examination of the number of particles of each type present has uncovered a discrepancy: There is a threefold excess of small versus large particles, even though photosystem I and II complexes are present in approximately equal numbers. Some of the small PF particles must therefore represent protein complexes other than photosystem I. Both the cytochrome *b-f* complex and the CF₀ portion of the CF₀-CF₁ complex are of the appropriate dimensions, and are believed to contribute to the PF particle population.

Origins and Significance of Membrane Stacking

The preferential association of photosystem II with stacked thylakoids raises questions concerning both the origins and functional significance of such an arrangement. Several observations suggest that this correlation occurs because the mechanism underlying membrane stacking involves the light-harvesting complex (page 267) associated with photosystem II. This complex, called **LHCII,** has been found to be present in reduced amounts in mutant organisms deficient in stacked thylakoids. Moreover, treatment of isolated thylakoid membranes with proteolytic enzymes known to degrade the protein constituents of LHCII abolishes the

FIGURE 8-43 Data from two experiments suggesting the existence of a relationship between grana stacking and photosynthetic efficiency. (a) Effect of varying light intensity on photosynthesis in a mutant strain of *Chlamydomonas* lacking stacked thylakoids. In comparison to normal organisms containing stacked thylakoids, a higher light intensity is required for maximal photosynthetic activity. (b) A summary of the effects of illumination on pea seedlings that had been germinated and kept in the dark for one week prior to exposure to light. Note that light induces the synthesis of LHC and the formation of grana stacks. At the same time the number of light quanta required per pair of electrons transferred through the photosynthetic electron transfer chain decreases, indicating an increased efficiency of the light-trapping process.

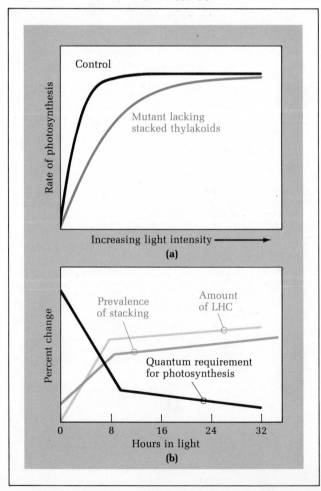

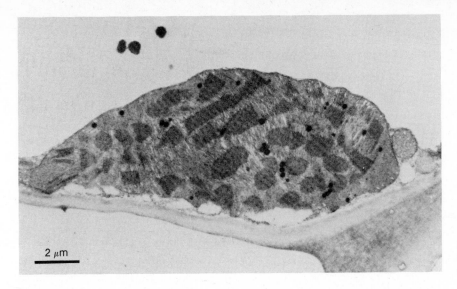

FIGURE 8-44 Electron micrograph of a chloroplast from a shade-adapted plant showing enlarged grana stacks. Courtesy of A. Melis.

2 μm

ability of the thylakoids to form stacks. Finally, artificial phospholipid vesicles reconstituted in the presence of LHCII form membrane stacks, while reconstituted vesicles lacking LHCII do not.

The conclusion that thylakoid stacking is induced by the presence of LHCII does not in itself provide much insight into the functional significance of grana stacking. An important clue has emerged, however, from the discovery that a mutant strain of *Chlamydomonas* lacking stacked thylakoids requires a higher than normal light intensity to achieve maximal rates of photosynthesis (Figure 8-43a). This difference cannot be accounted for by differences in chlorophyll content since the mutant cells actually contain more chlorophyll than normal cells. Comparable results have been obtained in studies in which dark-adapted pea seedlings were exposed to light to trigger chloroplast development. Under such conditions the formation of grana stacks occurs as LHCII is formed, and both events correlate with an increased photosynthetic efficiency as measured by the number of light quanta required per pair of electrons transferred (see Figure 8-43b). Taken together the preceding observations suggest that grana stacking is mediated by interaction between LHCII complexes across adjacent membranes, creating a topographical organization of light-absorbing pigments that increases photosynthetic efficiency. The conclusion that grana stacking is involved in regulating light-harvesting efficiency is compatible with the finding that plants grown at low light intensities contain enlarged grana stacks (Figure 8-44).

THE DARK REACTIONS OF THE STROMA

The Calvin Cycle

The discovery that the light reactions of photosynthesis are accompanied by the synthesis of ATP and the

reduction of $NADP^+$ to NADPH raises the question of how energy trapped in ATP and NADPH is subsequently utilized to drive the dark reactions of CO_2 fixation. When the radioactive isotope of carbon, ^{14}C, first became available to the scientific community after the end of World War II, it provided biologists with a powerful new tool for analyzing the pathway by which CO_2 is fixed during photosynthesis. This opportunity was quickly seized by Melvin Calvin and his collaborators, who exposed photosynthesizing cells to an atmosphere containing $^{14}CO_2$ and then analyzed the radioactive products by the technique of paper chromatography. Such experiments revealed the presence of a large number of different radioactive products in plant cells that had been exposed to radioactive CO_2 for several minutes or more. When the length of exposure to $^{14}CO_2$ was reduced to a few seconds, however, the vast majority of the radioactivity was found in a single compound, **3-phosphoglycerate** (Figure 8-45).

Because chemical analyses revealed that only one of the three carbon atoms in 3-phosphoglycerate is radioactive, Calvin first guessed that the dark reactions of photosynthesis involve linkage of CO_2 to a two-carbon acceptor molecule to form 3-phosphoglycerate. After several years of searching in vain for such a two-carbon precursor, a new observation revealed the error in this approach; it was discovered that longer exposures to $^{14}CO_2$ cause all the carbon atoms in 3-phosphoglycerate to become radioactive instead of just one. The most straightforward explanation of this unexpected finding was that the acceptor molecule to which $^{14}CO_2$ is joined is itself gradually becoming radioactive. Since all radioactivity initially appears in 3-phosphoglycerate, it was concluded that the acceptor with which 3-phosphoglycerate combines is itself a product of 3-phosphoglycerate metabolism. In other words the CO_2-fixing pathway must be circular. This realization eventually allowed Calvin to work out the details of the circular reaction

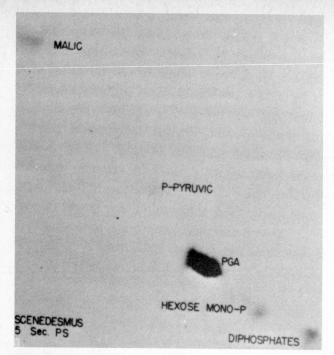
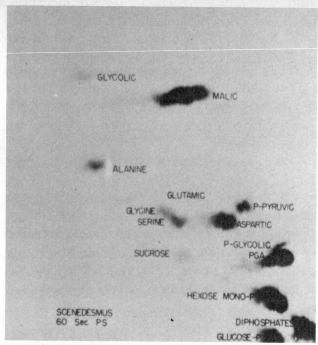

FIGURE 8-45 Autoradiogram of photosynthetic products extracted from algal cells and separated by paper chromatography. *(Left)* After 5 sec of incubation in the presence of $^{14}CO_2$, most of the radioactivity is in 3-phosphoglycerate (PGA). *(Right)* After 60 sec of incubation, radioactivity is detected within a large number of products. Taken together, the two experiments suggested that the initial product of CO_2 fixation during photosynthesis is 3-phosphoglycerate. Photographs courtesy of M. Calvin.

FIGURE 8-46 The Calvin cycle. The numbers in parentheses indicate how many molecules of each type are involved in the synthesis of one molecule of glucose. The substances within the boxes are the inputs and outputs of the cycle. Color is used to indicate the energy inputs derived from the light reactions of photosynthesis.

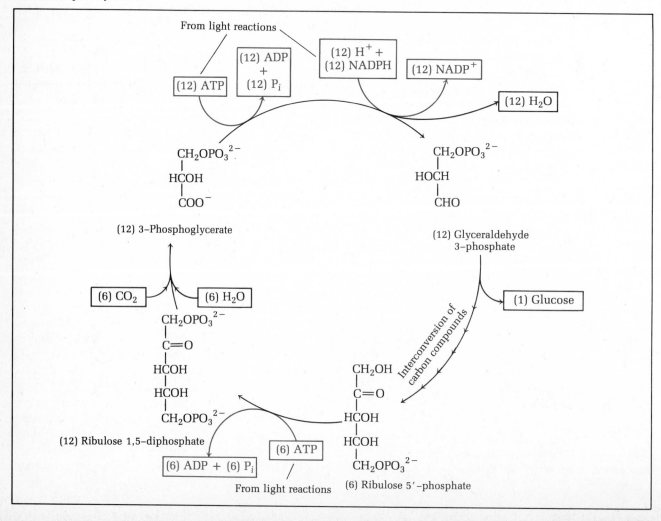

pathway, or **Calvin cycle,** by which photosynthetic cells fix and reduce CO_2 (Figure 8-46).

One of the crucial insights that allowed Calvin to unravel the organization of this pathway was the realization that the acceptor molecule to which CO_2 is initially joined might contain five carbon atoms rather than two. According to this hypothesis, the reaction of CO_2 with a five-carbon acceptor molecule produces a transient six-carbon compound that rapidly breaks down to two molecules of 3-phosphoglycerate. One five-carbon molecule capable of generating two molecules of 3-phosphoglycerate upon addition of CO_2 and subsequent cleavage is **ribulose 1,5-bisphosphate (RuBP).** If RuBP is the true acceptor molecule, one would expect its concentration to increase if CO_2 is suddenly removed from the atmosphere since there would no longer be any CO_2 available with which the RuBP could react. At the same time the concentration of 3-phosphoglycerate would be expected to fall, for this substance is the predicted product of the reaction between CO_2 and RuBP. Not only did Calvin confirm these predictions experimentally, but he went on to show that the opposite occurs in plants switched from light to darkness, namely RuBP decreases and 3-phosphoglycerate increases in concentration (Figure 8-47).

In addition to providing evidence for the role of RuBP as initial CO_2 acceptor, the discovery that 3-phosphoglycerate accumulates in the dark implies that energy derived from the light reactions of photosynthesis is not required for the initial CO_2-fixing reaction. The major energy requirement of the Calvin cycle occurs in the next step, where the NADPH and ATP generated by the light reactions are utilized to drive the reduction of 3-phosphoglycerate to 3-phosphoglyceraldehyde. It is in this reaction that the carbon atom incorporated during CO_2 fixation is actually reduced. As discussed earlier in the chapter, the presence of reduced carbon atoms gives molecules a high free-energy content, and it is therefore this reduction step that holds the key to the photosynthetic formation of foodstuffs.

Once the newly fixed carbon atom has been reduced, the remaining steps of the Calvin cycle are designed to accomplish two major objectives: To convert the newly fixed carbon into more complex carbohydrates and other energy-rich molecules, and to regenerate the acceptor molecule RuBP, which is needed for subsequent turns of the cycle. Both objectives are achieved by a complex set of pathways catalyzed by two enzymes, *transketolase* and *aldolase*, which transfer two- and three-carbon fragments back and forth between carbohydrates. The net result of this pathway is that some molecules of glyceraldehyde-3-phosphate are transformed into carbohydrates such as glucose, while others are converted back to RuBP. In order to synthesize one molecule of glucose, which contains six carbon atoms, one must start with six molecules of CO_2. The

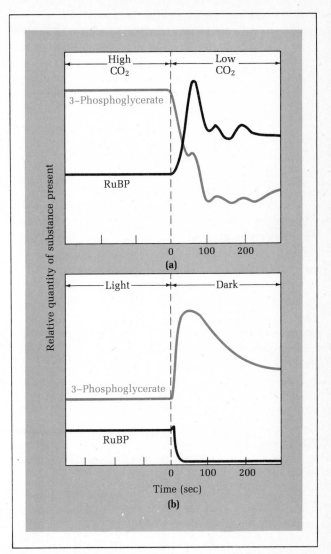

FIGURE 8-47 Effects of CO_2 concentration (a) and illumination (b) on the concentration of 3-phosphoglycerate (color) and RuBP within intact chloroplasts. See text for interpretation.

reaction of 6 molecules of CO_2 with 6 molecules of RuBP generates 12 molecules of 3-phosphoglycerate, which are in turn converted to 12 molecules of glyceraldehyde-3-phosphate. The carbon atoms from two of these glyceraldehyde-3-phosphate molecules are subsequently joined together to form one molecule of glucose, while the ten remaining molecules of 3-phosphoglycerate are used to regenerate the six molecules of RuBP that started the cycle. These six RuBP molecules can again react with six molecules of CO_2, and the entire cycle repeated. When all the inputs and products associated with the fixation of six molecules of CO_2 are added together, the following equation is obtained:

$$6RuBP + 6CO_2 + 6H_2O \xrightarrow[\substack{12NADPH + 12H^+ \quad 12NADP^+}]{\substack{18ATP \quad 18ADP + 18P_i}} 6RuBP$$
$$+ \; C_6H_{12}O_6 + 12H_2O$$

Water and RuBP are present on both sides of the equation, indicating that they are inputs as well as outputs of the cycle. Subtracting these components from each side of the equation yields the following *net* reaction of the Calvin cycle:

$$6CO_2 \xrightarrow[\substack{\text{12NADPH + 12H}^+ \quad \text{12NADP}^+}]{\substack{\text{18ATP} \quad \text{18ADP + 18P}_i}} C_6H_{12}O_6 + 6H_2O$$

Of the 18 molecules of ATP included in this summary equation, 12 are utilized during the conversion of 12 molecules of 3-phosphoglycerate to glyceraldehyde-3-phosphate, and the remaining 6 are used in the final step of the cycle to generate 6 molecules of RuBP. The 12 NADPH molecules are all employed in the conversion of 3-phosphoglycerate to glyceraldehyde-3-phosphate.

The ultimate source of the 12 NADPH and 18 ATP molecules used by the Calvin cycle is of course the light reactions occurring in the thylakoid membranes. In order to generate 12 molecules of NADPH, the light reactions must transfer 12 pairs of electrons from water to NADP$^+$, yielding 12 atoms of oxygen. The precise number of ATP molecules synthesized by photophosphorylation as 12 electron pairs pass through the photosynthetic electron transfer chain has been difficult to determine experimentally, but earlier in the chapter we cited experimentally determined values of 1.5 for the P/O ratio (page 273), meaning that roughly 1.5 molecules of ATP are synthesized per atom of oxygen released. If this value is correct, then it can be calculated that the release of 12 atoms of oxygen would be accompanied by the production of $12 \times 1.5 = 18$ ATP molecules, the exact number of ATPs required by the Calvin cycle. Hence current experimental estimates of the ATP yield of the light reactions match reasonably well with the ATP requirements of the dark reactions.

The step that appears to be most important in regulating the rate at which the Calvin cycle operates is the initial reaction in which CO_2 and RuBP are joined together to form two molecules of 3-phosphoglycerate. The enzyme that catalyzes this reaction, **RuBP carbox-ylase,** accounts for roughly half the total protein of the chloroplast, and as much as 20 percent of the entire protein content of typical plant cells. It is present in all photosynthetic organisms, including photosynthetic bacteria and blue-green algae, and is easily the single most abundant protein in nature. The RuBP carboxylase molecule of higher plants is composed of eight catalytic subunits of 56,000 molecular weight, and eight regulatory subunits of 12,000 molecular weight. The large size of this molecule renders it readily visible under the electron microscope, where it appears as a regular cube 10–15 nm across. Although normally found in the chloroplast stroma, RuBP carboxylase tends to stick to thylakoid membranes and can occasionally be seen as large particles located on the outer thylakoid surface. Unlike CF$_1$, which is also located on the thylakoid surface, RuBP carboxylase is readily removed from the membrane surface by extraction with water and is therefore not considered to be a genuine constituent of the thylakoid membrane.

Purified preparations of RuBP carboxylase require much higher CO_2 concentrations for maximal activity than is required for the operation of the Calvin cycle within intact chloroplasts. This discrepancy initially led to skepticism concerning the physiological significance of this enzyme, but doubts were subsequently dispelled when it was demonstrated that mutant organisms lacking RuBP carboxylase are unable to fix CO_2. It was later shown that purified RuBP carboxylase does not function as efficiently as expected because it is an allosteric enzyme whose activity is stimulated by factors normally present in the chloroplast. Two factors especially important in this regard are magnesium ions and pH. We have already seen that the pH of the stroma increases as a result of the light-induced transport of protons into the thylakoid space. Illumination of chloroplasts also stimulates transport of magnesium ions from the thylakoid spaces into the stroma. RuBP carboxylase is activated by both the increase in pH and the increase in Mg^{2+}. In addition to explaining why RuBP carboxylase is less active when removed from the chloroplast, the sensitivity of this enzyme to pH and magnesium ions

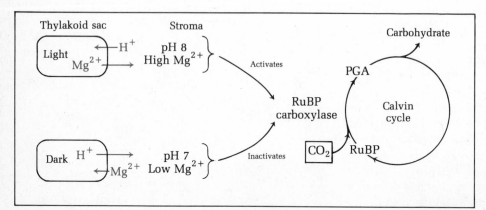

FIGURE 8-48 Regulation of RuBP carboxylase activity by light-induced transport of protons and magnesium ions. Abbreviations: PGA = 3-phosphoglycerate, RuBP = ribulose 1,5-bisphosphate.

provides a mechanism by which the activity of the dark reactions can be adjusted within the chloroplast in response to changes in the light reactions (Figure 8-48).

Although carbohydrate is the major synthetic product of the Calvin cycle, the reduced carbon atoms generated by this pathway can also be incorporated into amino acids and/or fats. The formation of these alternative products appears to be controlled by environmental conditions. For example, bright illumination stimulates the formation of carbohydrates, while dim illumination enhances the production of amino acids. It is worth noting in this general context that the reducing power generated by the light reactions is not used solely for carbon reduction. Most higher plants have alternate metabolic pathways that use this reducing power for other purposes, such as the reduction of nitrate to ammonia.

Overall Efficiency of Photosynthesis

Now that the light and dark reactions have been described, we shall see how the equations summarizing these two pathways can be combined to generate the overall equation of photosynthesis. The light reactions, which involve the transfer of electrons from water to $NADP^+$ accompanied by the production of ATP and oxygen, can be summarized as follows:

$$12H_2O + 18ADP + 18P_i + 12NADP^+ \longrightarrow$$
$$6O_2 + 18ATP + 12NADPH + 12H^+$$

In the above equation the yield of 18ATPs is based on the assumption that 1.5 ATPs are formed per pair of electrons passing from water to $NADP^+$ (page 273).

The dark reactions, which utilize the ATP and NADPH generated in the light reactions to drive the fixation and reduction of CO_2, can be written

$$6CO_2 + 18ATP + 12NADPH + 12H^+ \longrightarrow$$
$$C_6H_{12}O_6 + 18ADP + 18P_i + 12NADP^+ + 6H_2O$$

Adding the above two equations together yields

$$6CO_2 + 6H_2O \longrightarrow 6O_2 + C_6H_{12}O_6$$

which is of course the net reaction of photosynthesis described earlier in the chapter.

The $\Delta G°$ for this overall equation is 686 kcal/mol, meaning that the light reactions must provide at least 686 kcal of free energy (in the form of ATP and NADPH) to the dark reactions for every mole of glucose to be synthesized. As discussed earlier, the light reactions absorb 8 quanta of light per molecule of CO_2 converted to organic matter. Because the formation of one molecule of glucose requires the fixation of six molecules of CO_2, a total of $8 \times 6 = 48$ quanta of light are needed. Although the free energy of light quanta varies with wavelength, we can use 41 kcal/mol of quanta, which corresponds to light of 700 nm, as an approximation. The total energy input per mole of glucose formed is

therefore

$$48 \text{ mol of quanta} \times 41 \text{ kcal/mol} = 1968 \text{ kcal}$$

Since the synthesis of one mole of glucose traps 686 of these kilocalories, the overall efficiency of photosynthesis is $686 \div 1968 = \sim 35$ percent. At lower wavelengths the energy content of light quanta is higher, resulting in a corresponding reduction in efficiency. For light of 400 nm, for example, the calculated efficiency drops to about 20 percent.

C₄ Pathways

Although the Calvin cycle is the principal pathway for the photosynthetic fixation and reduction of CO_2, some plants have an additional pathway that also fixes CO_2. The existence of such alternative pathways was first suggested by experiments in which plants such as sugarcane, maize, and sorghum were exposed to an atmosphere containing $^{14}CO_2$. In contrast to the results originally obtained by Calvin, where the three-carbon compound 3-phosphoglycerate was the first radioactive compound to be observed, radioactivity was detected in the four-carbon acids oxaloacetate and malate before it appeared in 3-phosphoglycerate. To ascertain whether these four-carbon, or C_4, acids are intermediates in a unique carbon-fixing pathway, M. Hatch and C. Slack carried out experiments in which sorghum leaves were exposed to a $^{14}CO_2$-containing atmosphere for 15 sec, followed by transfer to a CO_2-free atmosphere. Under these conditions radioactivity was first detected in C_4 acids, but at later times it was found to sequentially appear in 3-phosphoglycerate, hexose-phosphates, and finally sucrose (Figure 8-49). These results suggested the existence of a pathway in which CO_2 is incorporated into C_4 acids prior to its passage into the Calvin cycle.

The first such C_4 pathway to be unraveled employs the enzyme **phosphoenolpyruvate carboxylase** to catalyze the initial CO_2-fixing step. This enzyme joins CO_2 to phosphoenolpyruvate to form the four-carbon acid, oxaloacetate. Oxaloacetate is then converted to malate, which is in turn degraded to CO_2 and pyruvate. The net result is therefore a cycle in which CO_2 is first joined to a small organic molecule and then released again as CO_2 (Figure 8-50). Although at first glance the value of fixing CO_2 in an organic form only to release it again as free CO_2 appears questionable, closer examination reveals that this process functions as a CO_2-concentrating mechanism because of the high efficiency of the enzyme phosphoenolpyruvate carboxylase. Unlike the Calvin cycle enzyme RuBP carboxylase, which requires relatively high concentrations of CO_2 for optimal activity, phosphoenolpyruvate carboxylase is active with low CO_2 concentrations. When CO_2 concentrations are too low to drive the Calvin cycle, the C_4 pathway can still convert CO_2 to C_4 acids. These C_4 acids are in turn concentrated within plant cells and later broken down

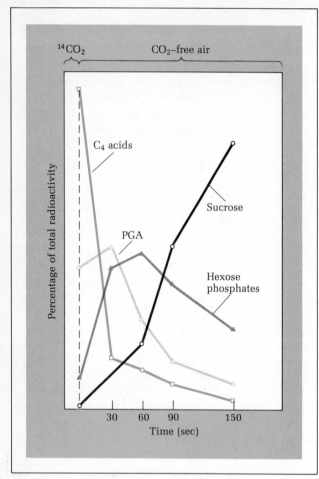

FIGURE 8-49 Data obtained from experiment in which sorghum leaves were briefly exposed to $^{14}CO_2$ and then transferred to a CO_2-free atmosphere. The results indicate that newly fixed carbon appears first in C_4 acids, and then sequentially in 3-phosphoglycerate, hexose-phosphate, and sucrose.

to liberate CO_2 in high enough concentrations to drive the RuBP carboxylase reaction of the Calvin cycle. The C_4 pathway thus functions as a CO_2-concentrating mechanism that feeds the Calvin cycle rather than as an alternative pathway. The ability of this pathway to concentrate CO_2 is not without its costs, however; as is shown in Figure 8-50, the C_4 pathway is accompanied by the breakdown of ATP.

Plants that incorporate CO_2 into four-carbon acids prior to its entrance into the Calvin cycle are commonly referred to as *C_4 plants*, while those that depend solely on the Calvin cycle are referred to as *C_3 plants* because the initial product of CO_2 fixation is always the three-carbon molecule, 3-phosphoglycerate. Several versions of the C_4 pathway have evolved in different plants. In some cases, for example, aspartate rather than malate is used to transport newly fixed carbon atoms to the Calvin cycle (see Figure 8-50). But in spite of such superficial differences, C_4 pathways all serve the same function: to fix CO_2 more efficiently than the Calvin cycle can do alone.

Leaves of C_4 plants are characterized by the presence of a special layer of large, thick-walled cells around their vascular bundles (Figure 8-51). These **bundle sheath cells,** which are the major site of photosynthesis, contain chloroplasts enriched in Calvin cycle enzymes. The exterior of the leaf, on the other hand, is covered by a layer of **mesophyll cells** containing the enzymes of the C_4 pathway. Hence the mesophyll cells function as the initial sites of CO_2 fixation, producing C_4 acids destined for transport to the bundle sheath cells. Upon arrival in the bundle sheath cells the C_4 acids are degraded, releasing the CO_2 which drives the Calvin cycle.

This arrangement for trapping and delivering CO_2

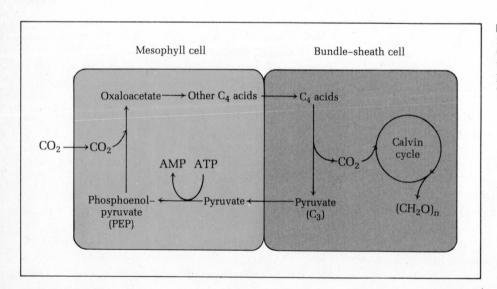

FIGURE 8-50 The C_4 pathway for fixing CO_2 and delivering it to the Calvin cycle. Several variations of this pathway exist, differing in the particular sequence of C_4 acids employed.

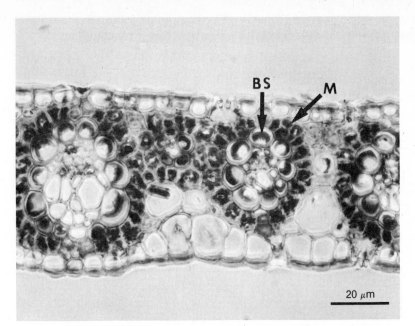

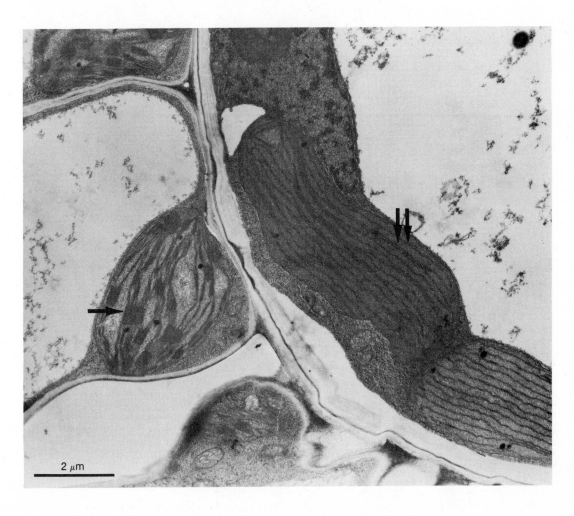

FIGURE 8-51 Leaf anatomy in C₄ plants. *(Top)* Low-magnification cross section of a sugarcane leaf. The bundle sheath cells (BS) lie adjacent to the vascular bundle, while the mesophyll cells (M) lie on the exterior. *(Bottom)* Electron micrograph showing the junction between mesophyll cells on the left and bundle sheath cells on the right. Note the stacked thylakoids (single arrow) in the chloroplast of the mesophyll cell and the unstacked thylakoids (double arrow) in the chloroplast of the bundle sheath cell. Courtesy of W. M. Laetsch.

has clear adaptive advantage in hot arid environments, where most C_4 plants flourish. In hot dry weather the leaves of such plants are protected from dessication by the closing of their surface pores, or **stomata,** which normally function to permit gas exchange with the atmosphere. Although closing these pores reduces the loss of water by evaporation, it also limits access of atmospheric CO_2 to the leaf cells. As a result, the concentration of CO_2 within the leaf falls to levels that are too low to effectively drive the Calvin cycle. However, the more efficient C_4 pathway can still fix CO_2 under such conditions and deliver it in concentrated form to the Calvin cycle, thereby allowing photosynthesis to proceed under conditions where it otherwise could not.

Photorespiration

In addition to triggering photosynthesis, light stimulates a pathway that consumes oxygen and evolves CO_2 (the opposite of what occurs in photosynthesis). This process, termed **photorespiration,** reduces the overall efficiency of photosynthesis because it oxidizes the energy-rich reduced carbon atoms produced by the photosynthetic pathway without capturing the released

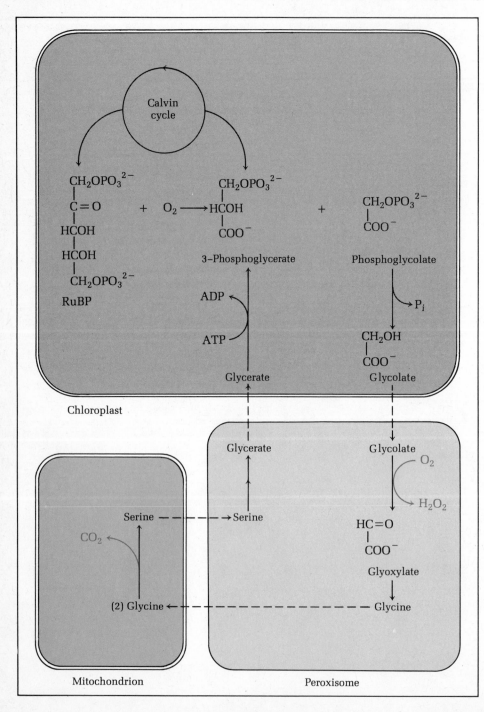

FIGURE 8-52 A summary of photorespiration. Oxygen-mediated oxidation reactions occur in chloroplasts and peroxisomes, while CO_2 production takes place in mitochondria. Organelles are not drawn to scale.

energy in a useful chemical form such as ATP or reduced coenzymes. Photorespiration is mediated by the oxidation of at least two different substrates by molecular oxygen (Figure 8-52). The first of these is RuBP, the normal CO_2 acceptor of the Calvin cycle. When oxygen is present in high concentration it competes with CO_2 in the RuBP carboxylase reaction, causing RuBP to react with oxygen rather than carbon dioxide. As a result, RuBP is oxidized to phosphoglycolate and 3-phosphoglycerate. Phosphoglycolate is then converted to glycolate, which is transferred to peroxisomes (glyoxysomes) for further oxidation by molecular oxygen. The final product of peroxisomal metabolism is the amino acid glycine, which is transported to mitochondria for degradation to CO_2 and the amino acid serine. Serine is returned to the peroxisome for conversion to glycerate, which then returns to the chloroplast to enter the Calvin cycle. Photorespiration is thus a cyclic pathway involving the cooperation of three organelles: chloroplasts, mitochondria, and peroxisomes. This metabolic interaction is one of the reasons these three organelles often occur in close association with one another in plant cells (see Figure 8-12).

The rate of photorespiration in C_3 plants can be exceptionally high, wasting as much as 50 percent of the reduced carbon generated by photosynthesis. The reason for the existence of photorespiration is unclear, but the discovery that this pathway is stimulated by high oxygen tensions raises the possibility that it may function to protect plants from the toxic effects of high concentrations of oxygen. In contrast to C_3 plants, photorespiration is absent or barely detectable in C_4 plants, presumably because the high concentration of CO_2 delivered to the Calvin cycle via the C_4 pathway prevents oxygen from effectively competing with CO_2 for the enzyme RuBP carboxylase. The absence of photorespiration in C_4 plants makes their overall efficiency in harvesting solar energy considerably greater than that of C_3 plants. In fact, it is likely that the C_4 pathway evolved at least in part to overcome the inefficiency of photorespiration.

PHOTOSYNTHESIS IN PROKARYOTES

Although photosynthetic prokaryotes do not contain chloroplasts, the light reactions are still membrane localized. In blue-green algae these reactions take place within flattened membranous sacs resembling the thylakoid membranes of higher plants (Figure 8-53). The shape and arrangement of these thylakoid sacs, which

FIGURE 8-53 Electron micrograph of blue-green algae showing photosynthetic membranes. *(Left)* A cell exhibiting thylakoid membranes distributed through the cell interior. *(Right)* In this algal cell several thylakoids are lined up directly beneath the plasma membrane. Courtesy of N. J. Lang and B. A. Whitton.

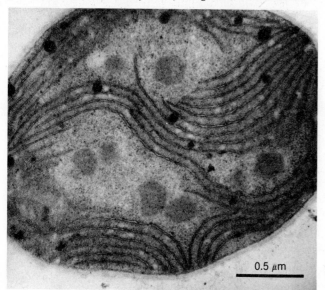

0.5 μm

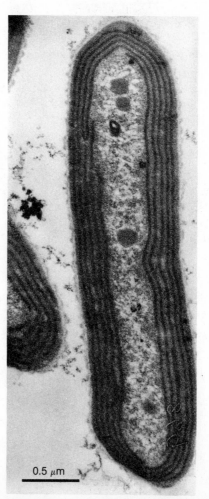

0.5 μm

are distributed throughout the cytoplasm, vary significantly among the various kinds of blue-green algae. Most of the light energy that drives photosynthesis in such cells is absorbed by phycobilin pigments contained within the phycobilisome granules attached to the thylakoid membranes (see Figure 8-10). The absorbed energy is then passed on to chlorophyll molecules situated within the thylakoid membranes to which the phycobilisomes are attached. The subsequent light and dark reactions are fundamentally similar to the analogous pathways of eukaryotes.

In addition to blue-green algae, two other kinds of prokaryotes are photosynthetic: purple bacteria and green bacteria. These organisms carry out the light reactions in specialized intracellular membranes termed **chromatophores** or **photosynthetic lamellae** (Figure 8-54). In purple bacteria the photosynthetic lamellae are created by invagination of the plasma membrane into vesicular or tubular channels that occasionally form stacks resembling grana. In green bacteria the photosynthetic membranes form spherical cytoplasmic vesicles that appear to be independent of the plasma membrane.

The basic mechanisms underlying bacterial photosynthesis are similar in principle to those occurring in higher plants, although there are many differences in detail. In terms of the light reactions, bacteria utilize light of longer wavelength (840–1000 nm), a single photosystem instead of two, a somewhat different sequence of carriers, and NAD^+ instead of $NADP^+$ as the terminal electron acceptor (Figure 8-55). Another unique feature of the bacterial light reactions is the use of substances other than water as electron donors. The most common electron donor is hydrogen sulfide (H_2S), a gas that generates elemental sulfur upon donating electrons to the photosynthetic electron transfer chain:

$$12H_2S + 6CO_2 \longrightarrow C_6H_{12}O_6 + 6H_2O + 12S$$

The sulfur produced in this way accumulates in large cytoplasmic globules that are eventually excreted from the cell. When substances such as H_2S are substituted for water as electron donors, oxygen is not produced by the light reactions because water molecules are not being broken down.

Bacterial photosynthesis also differs from the analogous process in higher plants in the electron acceptors it employs for the dark reactions. Besides reducing the classic electron acceptor, CO_2, bacteria employ electrons derived from the light reactions of photosynthesis to reduce hydrogen ions to H_2, and atmospheric nitrogen to ammonia. This latter reaction is especially important because it provides a mechanism for fixing atmospheric nitrogen in a form that is useful to living organisms.

In addition to the preceding cell types, another bacterium capable of capturing solar energy is *Halobacterium halobium*. This bacterial species is not photosynthetic in the classic sense because it does not utilize light energy to drive the reduction of carbon or other electron acceptors. *Halobacterium* does, however, trap solar energy and utilize it to power ATP formation. The crucial component in this process is a light-sensitive protein, termed **bacteriorhodopsin,** which resembles the light-absorbing pigment rhodopsin used by higher animals to make vision possible (Chapter 18). Bacteriorhodopsin absorbs light maximally at 560 nm, giving it a vivid purple color. Within the plasma mem-

FIGURE 8-54 Electron micrograph of a photosynthetic bacterium showing the presence of photosynthetic membranes within the cytoplasm. Courtesy of A. R. Varga.

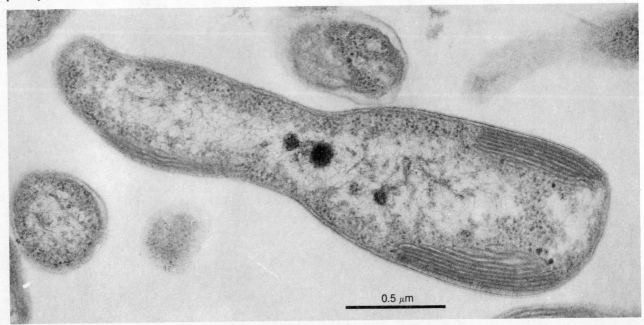

0.5 μm

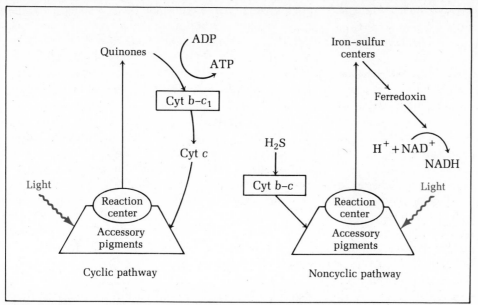

FIGURE 8-55 General organization of cyclic and noncyclic pathways for light-induced electron transfer in photosynthetic bacteria. Although the detailed arrangement of these pathways is yet to be elucidated, the underlying resemblance to the comparable pathways of higher plants is apparent.

brane of *Halobacterium* these bacteriorhodopsin molecules congregate into large masses, forming "purple patches" that occupy up to 50 percent of the membrane surface area.

In order to investigate how this molecule functions, Walther Stoeckenius and Efraim Racker incorporated bacteriorhodopsin and the ATP-synthesizing protein, F_1, into the membranes of artificial phospholipid vesicles. Subsequent illumination of these vesicles was found to trigger formation of a proton gradient and the ensuing synthesis of ATP. Since no electron transfer carriers are present in this simple system, it was concluded that light causes bacteriorhodopsin to act as a proton pump, generating a proton gradient that can be used to drive the synthesis of ATP by F_1. These results provide yet another example of the widespread applicability of the chemiosmotic theory, and reveal that electrochemical proton gradients can be established and utilized as energy sources even in the absence of an electron transfer chain.

ARTIFICIAL LIGHT-TRAPPING SYSTEMS

We have seen in this chapter that photosynthetic organisms are extremely efficient at trapping and utilizing solar energy. As society has become aware in recent years that the earth's supply of oil, gas, and coal is rapidly diminishing, the question has arisen as to whether solar energy can be directly utilized for human needs as it is by plants. The most attractive features of sunlight as an energy source are that it is in no danger of being used up and it is virtually pollution free. Although these con-

siderations have led to the construction of various types of solar cells, the products developed to date are both expensive to manufacture and relatively inefficient in terms of energy production. The question therefore

FIGURE 8-56 Model of an artificial solar cell based upon the principles of the photosynthetic membrane.

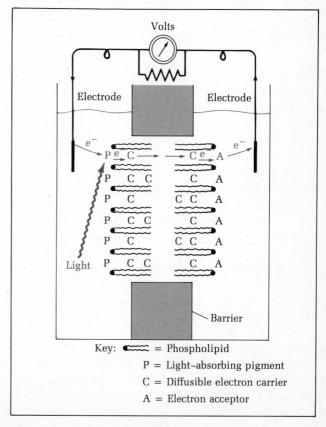

Key: 〰〰 = Phospholipid
P = Light–absorbing pigment
C = Diffusible electron carrier
A = Electron acceptor

arises as to whether our knowledge of the mechanisms underlying photosynthesis might provide any clues as to how to devise more efficient and compact systems for trapping solar energy.

Although such technology is still in its early stages, our current understanding of photosynthesis suggests that artificial membranes containing light-absorbing molecules may one day be developed to serve as a practical source of electricity. One way in which this might be accomplished is to place a light-absorbing pigment on one side of an artificial membrane, an electron acceptor on the other, and a diffusible electron carrier within the hydrophobic membrane interior (Figure 8-56, p. 289). Illumination of such a membrane should cause electrons in the light-absorbing pigment to become activated. The activated electrons will tend to pass to the electron carrier, which can diffuse to the other side of the membrane and transfer the electrons to the electron acceptor. In this way an electric potential would be established that could subsequently be used to generate an electric current. Although the properties of model membranes of this type are just beginning to be explored for energy-generating purposes, it is conceivable that in the not too distant future our knowledge of photosynthesis will help us to create such artificial systems capable of capturing the energy of the sun.

SUMMARY

All free energy consumed by living organisms is ultimately derived from solar energy captured by the process of photosynthesis. The overall reaction of photosynthesis is summarized by the equation: $6CO_2 + 6H_2O \rightarrow C_6H_{12}O_6 + 6O_2$. This reaction is the net result of dozens of metabolic steps organized into two discrete pathways known as the light and dark reactions. The light reactions involve the light-induced transfer of electrons from water to $NADP^+$, accompanied by the synthesis of ATP and the production of oxygen. In the dark reactions energy stored in the NADPH and ATP molecules generated by the light reactions is used to drive the fixation and reduction of CO_2.

In eukaryotic cells photosynthesis occurs within the chloroplast, a large cytoplasmic organelle bounded by an outer double-membrane envelope. The outer membrane of the envelope is relatively permeable to small molecules, while the inner membrane is not. The movement of materials into and out of the chloroplast is controlled by transport systems localized within this inner membrane. The inside of the chloroplast contains a mixture of stacked and unstacked thylakoid membranes that together enclose a complex compartment known as the thylakoid space. Protruding from the outer surfaces of the unstacked thylakoids are small spherical particles containing CF_1, the protein complex responsible for ATP formation. Thylakoid membranes

are unusual in that most of their lipid is glycolipid rather than phospholipid. The compartment between the thylakoids and the chloroplast envelope is referred to as the stroma. Studies on isolated chloroplasts and chloroplast subfractions have led to the conclusion that the light reactions of photosynthesis occur within the thylakoid membranes, while the dark reactions take place in the stroma.

The light reactions depend on the presence of three classes of molecules within the thylakoid membranes: light-absorbing pigments, electron carriers, and proteins involved in ATP formation. Together these components form a pathway that transfers electrons from water to $NADP^+$, coupled to the synthesis of ATP. The order of the steps comprising this pathway has been ascertained from a combination of data, including the oxidation-reduction potentials of the isolated electron transfer carriers, the ability of isolated carriers to oxidize and reduce each other, measurements of difference spectra in intact chloroplasts, the behavior of plants containing mutations in the electron transfer chain, and the effects of artificial electron donors, acceptors, and electron transfer inhibitors on photosynthetic electron transfer.

The energy driving the flow of electrons through the electron transfer chain is provided by two light-absorbing photosystems. The bulk of the chlorophyll and accessory pigment molecules associated with each photosystem serve as "antennae," absorbing solar energy and funneling it to a special chlorophyll-protein complex termed the reaction center. The first light-absorbing event in the chain is associated with photosystem II, which passes its activated electrons to a protein-bound plastoquinone designated Q. This loss of electrons from photosystem II leaves it with a net positive charge, which serves as a driving force for pulling electrons from water. This removal of electrons converts water molecules to oxygen and free protons. The electrons transferred from photosystem II to Q next flow through acceptor B to a pool of freely mobile plastoquinone molecules, and then through a series of carriers arranged in the following sequence: cytochrome $b \rightarrow$ cytochrome $f \rightarrow$ plastocyanin. Beyond plastocyanin lies photosystem I, where the second light-absorbing event occurs. Absorption of light by photosystem I causes an electron associated with its reaction center to be passed to a tightly bound complex of iron-sulfur proteins. From this, complex electrons are transferred to the final electron acceptor via the following sequence: ferredoxin $\rightarrow$ NADP-reductase $\rightarrow NADP^+$.

Some of the energy released as electrons flow through the above series of carriers is used to drive ATP formation, a process termed photophosphorylation. Noncyclic photophosphorylation, which accompanies the net flow of electrons from water to $NADP^+$, involves two coupling sites for ATP formation: one associated with the oxidation of water, and the other associated with the oxidation of plastoquinone. Cyclic photophosphorylation, which utilizes only the latter of these two coupling sites, is based on a circular flow of electrons around photosystem I, and therefore involves neither the production of oxygen nor the reduction of NADP$^+$. Current experimental measurements suggest that each coupling site has a P/O ratio of roughly 0.6, allowing a total of 1.2 molecules of ATP to be synthesized by noncyclic photophosphorylation for each pair of electrons passing through both sites on their way from water to NADP$^+$. In addition, cyclic photophosphorylation contributes some additional ATP synthesis, bringing the net yield of ATP closer to 1.5 molecules per atom of oxygen produced.

The two coupling sites for photophosphorylation are associated with the splitting of water and the oxidation-reduction of plastoquinone, respectively. Since both these steps involve the uptake and/or release of protons, it has been suggested that the formation of proton gradients in these two regions of the photosynthetic electron transfer chain drives ATP synthesis. Among the various lines of evidence that support this chemiosmotic view of photophosphorylation are the following: (1) Illumination of isolated chloroplasts induces the uptake of protons from the external medium. ATP formation is accompanied by the flow of protons in the opposite direction. (2) Artificially created pH gradients can drive ATP synthesis by chloroplasts in the absence of light. (3) The pH of the thylakoid space falls to about 4.5 and the pH of the stroma rises to 8 in illuminated chloroplasts. In addition to this gradient of 3.5 pH units, a small membrane potential of <10 mV is established. Hence in chloroplasts the driving force for ATP synthesis comes almost entirely from ΔpH, while in mitochondria both ΔpH and $\Delta\psi$ make significant contributions. (4) Quantitation of proton fluxes has revealed that two protons are transported into the thylakoid space per coupling site per pair of electrons transferred, just as occurs with mitochondrial coupling sites. (5) The free energy content of the electrochemical proton gradient is sufficient to account for the amount of ATP synthesized. (6) Ionophores that abolish the electrochemical proton gradient also uncouple photophosphorylation. (7) Kinetic analyses reveal that the rate of ATP formation correlates closely with the rate of formation of the proton gradient. (8) Coupling of ATP formation to light-induced electron transfer requires intact membrane vesicles. (9) Removal of CF$_1$ causes thylakoid membranes to become leaky to protons and to lose the ability to make ATP. (10) Artificial membranes reconstituted in the presence of isolated photosystems carry out light-induced proton transport. Inclusion of the CF$_1$-CF$_0$ complex permits these reconstituted vesicles to use the electrochemical proton gradient to drive ATP formation. (11) The components of the photosynthetic elec-

tron transfer chain are asymmetrically oriented within the thylakoid membrane. Taken together, the above 11 observations suggest that light-induced electron transfer is associated with the uptake of protons into the thylakoid space and that the subsequent diffusion of protons back down this electrochemical gradient drives ATP formation.

Freeze-fracture electron microscopy has revealed that several types of particles are associated with the thylakoid membrane. Particles protruding from the membrane exterior are restricted primarily to the outer surface of the unstacked thylakoids and are thought to correspond to CF$_1$. Small 8- to 11-nm particles are present within the PF face of unstacked and stacked thylakoids, while larger 10- to 18-nm particles are restricted primarily to the stacked membranes. The large freeze-fracture particles correspond to the photosystem II complex, while the smaller particles are thought to represent several different protein complexes: photosystem I, the cytochrome b-f complex, and the CF$_0$ portion of the CF$_0$-CF$_1$ complex. Grana stacking, which is mediated by membrane-membrane interactions involving the light-harvesting complex LHCII, creates a topographical arrangement of pigments that is more efficient in trapping light.

The dark reactions of photosynthesis occur within the chloroplast stroma and utilize the NADPH and ATP generated by the light reactions to drive the fixation and reduction of CO$_2$. In the Calvin cycle the initial product of CO$_2$ fixation is the three-carbon compound 3-phosphoglycerate. The energy-requiring steps of the Calvin cycle, which make possible the conversion of 3-phosphoglycerate to glucose and other foodstuffs, are driven by the ATP and NADPH provided by the dark reactions. In addition to the Calvin cycle, some plants contain a C$_4$ pathway for fixing CO$_2$ in mesophyll cells and delivering it in high concentration to the Calvin cycle in bundle sheath cells. This arrangement permits plants to adapt to hot arid environments, where the closing of leaf pores to minimize water loss hinders gas exchange with the external environment.

The overall efficiency of photosynthesis can be severely limited by photorespiration, a light-induced pathway that uses molecular oxygen to oxidize the energy-rich reduced carbon atoms generated by photosynthesis. Photorespiration involves the oxidation of RuBP by chloroplasts, a subsequent oxidation occurring in peroxisomes, and a CO$_2$-producing step in mitochondria. Because this process is not coupled to the synthesis of ATP or any other high-energy compound, it wastes energy that has been trapped by photosynthesis in the form of reduced carbon.

In photosynthetic prokaryotes the light reactions are associated with specialized internal membranes analogous to thylakoids. The photosynthetic pathway of blue-green algae closely resembles that of higher plants, but in purple and green bacteria the pathway exhibits several significant differences. The light reactions in these bacteria utilize a different sequence of electron carriers and light-absorbing pigments, a single photosystem instead of two, and NAD^+ rather than $NADP^+$ as terminal electron acceptor. In addition, oxygen is not produced because electron donors such as hydrogen sulfide are employed in place of water. The dark reactions of bacterial photosynthesis are also unusual in that electrons derived from the light reactions are used to reduce hydrogen or nitrogen in addition to the typical electron acceptor, CO_2. In *H. halobium* solar energy is trapped by the membrane pigment bacteriorhodopsin, generating a proton gradient that can be used to drive ATP formation in the absence of electron transfer or photosynthesis.

Because of the remarkable efficiency of photosynthetic organisms in capturing solar energy and converting it to a useful form, it is hoped that our deepening understanding of the photosynthetic process will one day permit us to design better artificial systems for trapping the energy of the sun.

SUGGESTED READINGS

Books

Danks, S. M., E. H. Evans, and P. A. Whittaker (1983). *Photosynthetic Systems. Structure, Function and Assembly*, Wiley, New York.

Foyer, C. H. (1984). *Photosynthesis*, Wiley, New York.

Govindjee, ed. (1982). *Photosynthesis*, Vols. 1 and 2, Academic Press, New York.

Hatch, M. D., and N. K. Broadman, eds. (1981). *The Biochemistry of Plants, Photosynthesis*, Vol. 8, Academic Press, New York.

Hoober, J. K. (1984). *Chloroplasts*, Plenum Press, New York.

Steinback, K. E., S. Bonitz, C. J. Arntzen, and L. Bogorad, eds. (1985). *Molecular Biology of the Photosynthetic Apparatus*, Cold Spring Harbor Laboratory, Cold Spring Harbor, NY.

Articles

Anderson, J. M. (1986). Photoregulation of the composition, function, and structure of thylakoid membranes, *Ann. Rev. Plant Physiol.* 37:93–186.

Bogorad, L. (1981). Chloroplasts, *J. Cell Biol.* 91:256s–270s.

Gounaris, K., J. Barber, and J. L. Harwood (1986). The thylakoid membranes of higher plant chloroplasts, *Biochem. J.* 237:313–326.

Miller, K. R. (1979). The photosynthetic membrane, *Sci. Amer.* 241 (Oct.):102–113.

Miller, K. R., and M. K. Lyon (1985). Do we really know why chloroplast membranes stack? *Trends Biochem. Sci.* 10:219–222.

Nugent, J. A. (1984). Photosynthetic electron transport in plants and bacteria, *Trends Biochem. Sci.* 9:354–357.

Zuber, H. (1986). Structure of light-harvesting antenna complexes of photosynthetic bacteria, cyanobacteria, and red alga, *Trends Biochem. Sci.* 11:414–419.

9

The Cytoskeleton: Its Role in Cell Organization and Motility

During the early part of this century the cell was regarded as a collection of physically independent organelles suspended in an amorphous fluid or cell sap. However, the advent of electron microscopy and the development of sophisticated cytochemical techniques has led to the realization that the eukaryotic cell sap contains a network of interconnected fibers and filaments that together impart an architectural framework to the cell. This elaborate and dynamically changing framework, or *cytoskeleton*, is constructed from an interconnected network of microtubules, microfilaments, and intermediate filaments.

The cytoskeleton performs at least two basic functions for the cell. The first, implied by the term "skeleton," is the creation of an internal scaffolding that supports and organizes the cell interior. This framework permits cells to assume elaborate shapes that would be unstable in the absence of a supporting lattice and also organizes and facilitates communication among intracellular organelles. The other general function of the cytoskeleton is its ability to generate movement. Just as the skeleton of the body is associated with muscles that implement movement of the body, the cytoskeleton contains motile elements that permit movement of the cell as a whole, as well as motion of intracellular components such as chromosomes, membranes, and granules.

In this chapter we shall investigate the structural organization of the cytoskeleton and the molecular basis for the functions it performs. In the process it will become apparent that the activities of this skeletal network are related to a broad spectrum of cellular events.

MICROTUBULES

The first component of the cytoskeleton to be clearly identified by electron microscopists was the **microtubule**, a long rodlike structure that measures 25 nm in diameter and up to several millimeters in length. Microtubules are involved in a diverse array of cellular activities, including ciliary and flagellar motion, movement of chromosomes and other subcellular components, modulation of plasma membrane topography, and the determination and maintenance of cell shape. This diversity of functions is made possible by two underlying properties of the microtubule: (1) its long rigid shape, which facilitates structural roles such as anchoring, guiding, orienting, and supporting other cellular constituents, and (2) its ability to generate movement, which is utilized in moving both subcellular components and the cell as a whole.

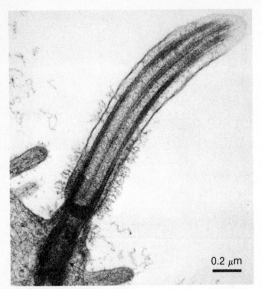
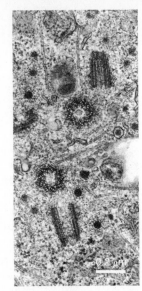
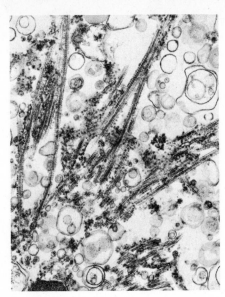

FIGURE 9-1 Electron micrographs showing the presence of microtubules in various parts of the cell. *(Left)* Longitudinal section showing microtubules running the length of a cilium. The basal body from which the cilium is derived is located directly beneath the plasma membrane (arrow). Courtesy of D. Sandoz. *(Middle)* Microtubules present within centrioles are shown in longitudinal and cross section. Courtesy of M. McGill. *(Right)* High-power view of microtubules present within the mitotic spindle of a HeLa cell. Courtesy of B. R. Telzer.

Because individual microtubules are too thin to be observed with the light microscope, their widespread occurrence in eukaryotic cells was unknown prior to the development of electron microscopy. Early electron microscopic studies, which employed osmium fixation, revealed the presence of microtubules within several subcellular structures, including cilia, flagella, centrioles, basal bodies, and the mitotic spindle (Figure 9-1). However, the extensive network of microtubules that pervades the cytoplasm of most eukaryotic cells was not detected until the early 1960s, when the introduction of gentler fixation techniques using glutaraldehyde permitted visualization of cytoplasmic microtubules that had been disrupted by the harsher osmium fixation procedures employed earlier (Figure 9-2).

FIGURE 9-2 Electron micrograph showing a network of free cytoplasmic microtubules. Courtesy of S. H. Blose.

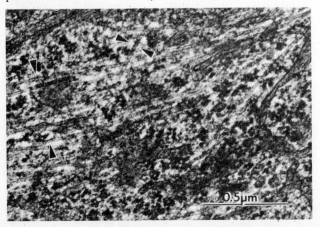

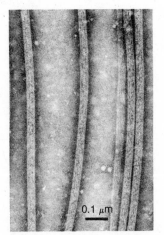
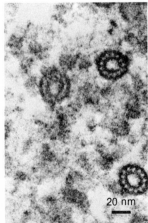

FIGURE 9-3 Microtubule fine structure as seen in high-power electron micrographs. *(Left)* Longitudinal section of a microtubule showing the protofilaments present within the microtubule wall. Courtesy of A. Klug. *(Right)* Cross section of a microtubule showing the 13 individual protofilaments that comprise the wall. Courtesy of P. R. Burton.

Cytoplasmic microtubules closely resemble the microtubules that form the backbone of larger structures such as cilia, flagella, basal bodies, centrioles, and the mitotic spindle. Both types of microtubules contain a hollow core 15 nm in diameter surrounded by a wall constructed of longitudinally arranged **protofilaments** (Figure 9-3). The microtubule wall generally contains 13 protofilaments, though in some cases this number may be slightly larger or smaller. The main building block of the protofilament is the protein **tubulin,** a dimer of approximately 110,000 molecular weight composed of nonidentical polypeptide chains desig-

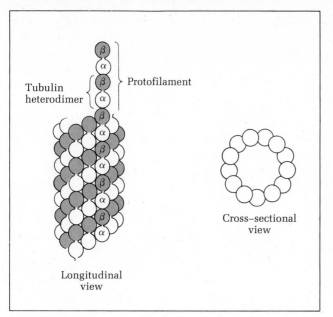

FIGURE 9-4 Schematic model of microtubule substructure.

nated α and β tubulin. Since microtubules can be easily dispersed into individual protofilaments, the forces that bind together the tubulin molecules of an individual protofilament must be stronger than the bonds between adjacent protofilaments. It follows that the α-β bond, which is only disrupted by harsh treatments, is located within individual protofilaments rather than between adjacent protofilaments. Additional information concerning the substructural organization of the microtubule has been obtained from X-ray and optical diffraction analyses, which indicate that the subunits of the microtubule wall are packed together with helical geometry. A model of the microtubule wall satisfying the requirements for both helical arrangement of tubulin subunits and the presence of the α-β bond within, rather than between, protofilaments is presented in Figure 9-4.

Assembly and Disassembly of Microtubules

One of the earliest insights into the question of how cells make microtubules emerged in the early 1960s from the laboratory of Shinya Inoué, where polarization light microscopy was employed to monitor the behavior of the microtubules present within the mitotic spindle. Inoué found that exposing cells to low temperature, high pressure, or the drug **colchicine** causes the birefringence of the spindle to decrease, indicating the breakdown of microtubular structure. As soon as the temperature is raised, the pressure lowered, or the colchicine removed, however, the spindle spontaneously reforms. This rapid recovery occurs even in the presence of inhibitors of protein synthesis, indicating that microtubule assembly does not require the synthesis of new molecules of tubulin. On the basis of these findings it was proposed that

microtubules are polymers of tubulin that exist in equilibrium with nonpolymerized tubulin, and that microtubule assembly-disassembly is regulated by changes in this equilibrium rather than by synthesis and degradation of tubulin subunits (Figure 9-5).

Subsequent insights into the process of microtubule assembly have emerged in large part from studies utilizing cell-free conditions. The first successful experiments of this type were carried out in the early 1970s by Richard Weisenberg, who discovered that microtubules spontaneously form in cell extracts that have been warmed to 37°C in the presence of GTP and the absence of calcium ions. Shortly after these experiments were reported, it was discovered that microtubules assembled under such conditions contain several proteins in addition to tubulin. These nontubulin proteins, which account for 10–15 percent of the microtubule mass, fall into two major classes: **tau proteins** of about 60,000 molecular weight, and **microtubule-associated proteins** (e.g., MAP-1 and MAP-2) of roughly 300,000 molecular weight. An early clue to the significance of these accessory proteins was provided by the observation that microtubules induced to undergo disassembly by lowering the temperature release two forms of tubulin into solution: The α-β tubulin dimer and larger tubulin aggregates that look like rings and spirals when viewed with the electron microscope (Figure 9-6). Although these isolated "ring forms" readily reassemble into microtubules upon warming, the dimers cannot reassemble unless the ring forms or microtubule fragments are also present to act as nucleation sites for tubulin polymerization. Subsequent analyses revealed that the isolated ring forms, but not the dimers, have accessory proteins associated with them. Taken together the above observations suggest that accessory proteins are required for

FIGURE 9-5 Microtubule assembly is governed by an equilibrium between polymerized microtubules and nonpolymerized tubulin molecules. Low temperature, high pressure, and the drug colchicine all promote microtubule breakdown by shifting the equilibrium in the direction of free tubulin.

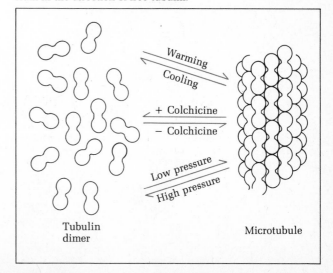

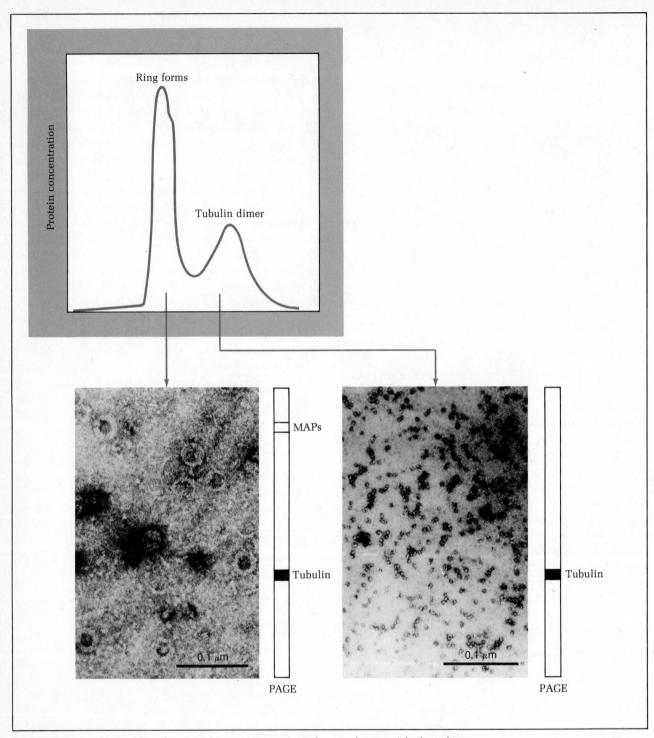

FIGURE 9-6 Data obtained from fractionation experiments carried out on the material released into solution when microtubules are disassembled by exposure to low temperatures. Gel filtration chromatography (top graph) size-fractionates the solubilized material into two fractions, one containing tubulin ring forms and the other containing tubulin dimers. Polyacrylamide gel electrophoresis (PAGE) of each fraction reveals that microtubule-associated proteins (MAPs) are restricted to the fraction containing the ring forms. Courtesy of J. L. Rosenbaum.

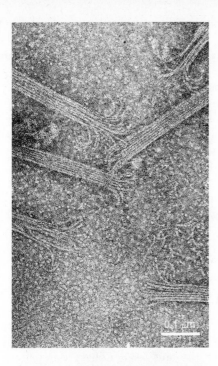

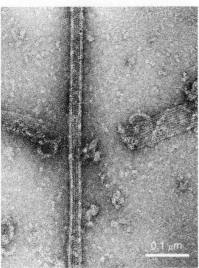

FIGURE 9-7 Electron micrographs of structures seen during microtubule disassembly and assembly. *(Top)* In a depolymerizing microtubule, the protofilaments at one end of the microtubule are seen to separate from one another and coil up. Courtesy of R. C. Williams. *(Bottom)* Micrograph of assembling microtubules, showing a long intact microtubule and several examples of incomplete walls with attached rings that appear to be continuous with the protofilaments of the wall. Courtesy of M. W. Kirschner.

the conversion of tubulin dimers to ring forms, and that the ring forms serve as nucleation intermediates in microtubule assembly. Direct support for this conclusion has been obtained by showing that the addition of microtubule-associated proteins to isolated dimers promotes their conversion to ring forms, and that this conversion in turn enhances the rate of tubulin polymerization into microtubules.

Electron microscopy has provided significant insights into the question of how the ring forms facilitate the polymerization of tubulin into microtubules. Micrographs of microtubules undergoing disassembly, for example, show individual protofilaments separating from each other at the end of the microtubule and curling into coiled structures that resemble partially formed rings (Figure 9-7 *top*). During microtubule assembly, on the other hand, the protofilaments look like they are being formed by uncoiling of the ring forms (see Figure 9-7 *bottom*). Such observations suggest that the ring forms are curved protofilaments and that microtubule assembly involves the uncoiling of these rings into protofilaments, the side-by-side joining of protofilaments into sheets, and the closing of these sheets into tubes when the number of protofilaments reaches 13 (Figure 9-8). After this nucleation stage is complete, elongation of the microtubule proceeds by addition of tubulin dimers to the microtubule tip.

One important question raised by the above model is whether tubulin dimers are added to one or both ends of the growing microtubule. To investigate this issue Joel Rosenbaum and his associates incubated fragments

FIGURE 9-8 Schematic representation of the steps involved in microtubule assembly under cell-free conditions.

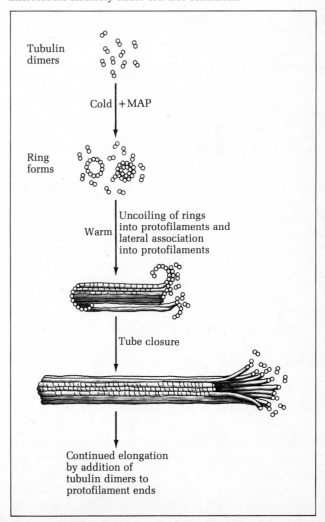

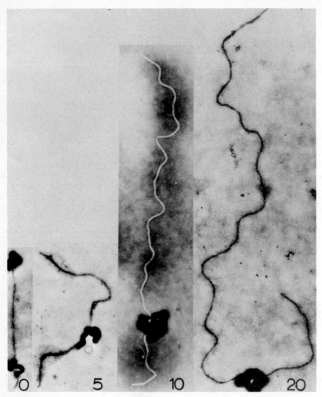

FIGURE 9-9 Electron microscopic autoradiographs showing unidirectional growth of microtubules. Elongating microtubules were mixed with radioactive microtubule fragments and unlabeled tubulin subunits for varying periods of time (as indicated by the number of minutes shown in the lower right of each micrograph). The results indicate that as the microtubules increase in length, the silver grains (dark spots) consistently appear near one end of the microtubule, in the same relative position as in the original preparation of radioactively labeled microtubule pieces. This pattern suggests that the unlabeled tubulin subunits are being added to only one end of the growing microtubule. The wavy appearance of the microtubules is caused by the fixation procedure employed. Courtesy of J. L. Rosenbaum.

of radioactively labeled microtubules with unlabeled tubulin under conditions designed to promote tubulin polymerization, and then examined the mixture by electron microscopic autoradiography. The resulting pictures revealed that most of the newly added, nonradioactive tubulin had been selectively added to one end of the growing microtubule (Figure 9-9). It was therefore concluded that microtubule assembly is a unidirectional process.

The conclusion that microtubule assembly generally proceeds in one direction is also supported by the behavior of microtubules within intact cells. Since microtubules tend to form within cells at specific locations and at specific times, it has been postulated that localized nucleation sites control the assembly of microtubules inside cells as they do under cell-free conditions. Such intracellular nucleation sites, or **microtubule-organizing centers,** occur in a variety of forms, including basal bodies, chromosomal kinetochores, and the granular material termed the **centrosome** that surrounds centrioles. Even microtubules lying "free" in

the cytoplasm generally have one end embedded in a diffuse granular matrix that might represent a microtubule-organizing center (Figure 9-10). Evidence that these various structures function as organizing centers for microtubule assembly has been obtained from several different approaches. First, isolated centrosomes, basal bodies, and chromosomal kinetochores have been shown to nucleate the polymerization of tubulin into microtubules under cell-free conditions (Figure 9-11). Second, the use of fluorescent tubulin-specific antibodies to visualize the microtubule networks of intact cells has yielded pictures of cytoplasmic microtubules emerging from organizing centers (Figure 9-12). Taken together, the preceding observations suggest that microtubule formation within intact cells occurs by the unidirectional addition of tubulin subunits to the ends of microtubules that are growing outward from microtubule-organizing centers.

Although considerable evidence suggests that microtubule assembly is a unidirectional process, there is less certainty about the nature of microtubule disassembly. In the late 1970s Robert Margolis and Leslie Wilson proposed that microtubules are disassembled at the end opposite from the one at which assembly occurs. This model was based on data obtained from a cell-free microtubule polymerizing system that had been briefly exposed to radioactive GTP. During microtubule assembly, **GTP** is known to bind to the growing end of the microtubule in association with the newly

FIGURE 9-10 Electron micrograph showing a single microtubule adjacent to the nuclear envelope with its end embedded in a diffuse, granular matrix (arrow). Courtesy of J. D. Pickett-Heaps.

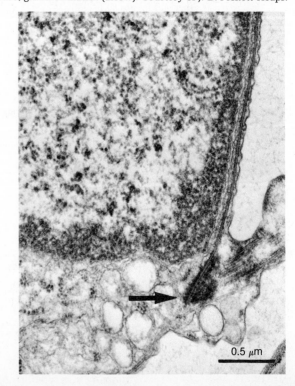

0.5 μm

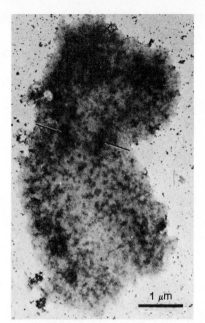
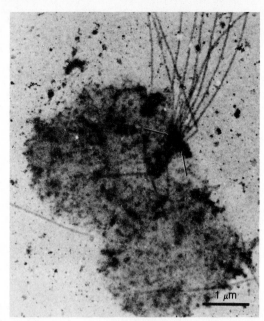

FIGURE 9-11 Electron micrographs showing the growth of microtubules from kinetochores in a cell-free system incubated in the presence of tubulin subunits. *(Left)* An isolated HeLa cell mitotic chromosome with the kinetochore appearing as two dark structures (arrows). *(Right)* Micrograph showing the elongating microtubules originating from the kinetochore region (arrows). Courtesy of J. L. Rosenbaum.

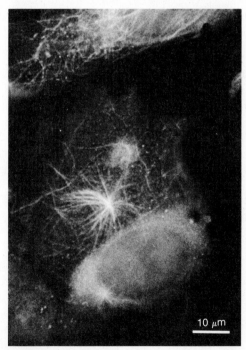

FIGURE 9-12 Light micrograph of a mouse fibroblast stained with fluorescent antitubulin showing microtubules emerging from an organizing center. Microtubule growth was stimulated by treating the cells with a drug that breaks down microtubules, and then removing the drug to permit the microtubules to re-form. Courtesy of B. R. Brinkley.

incorporated tubulin subunits. This GTP is subsequently hydrolyzed to GDP molecules that are later released from the microtubule. Because a delay of up to 90 min was found to occur between the uptake of radio-

active GTP and its release from the microtubule as radioactive GDP, Margolis and Wilson proposed that tubulin polymerization occurs at one end of the microtubule and depolymerization occurs at the opposite end, creating a unidirectional flux of tubulin molecules from one end of the microtubule to the other. According to this "treadmilling" model, the delay in the release of radioactive GDP is caused by the time it takes for the GDP to gradually move from one end of the microtubule to the opposite end (Figure 9-13a).

Subsequent studies on the behavior of microtubules assembled under cell-free conditions using isolated centrosomes as nucleation sites have led Tim Mitchison and Marc Kirschner to propose an alternative model of microtubule disassembly. These investigators discovered that the number of microtubules assembled per centrosome increases with an increasing concentration of free tubulin; as might be expected, lowering the tubulin concentration reverses the process and causes the number of microtubules per centrosome to decrease. More surprising, however, was the discovery that while some microtubules disappear when the tubulin concentration is lowered, the remaining ones get longer. To explain these observations Mitchison and Kirschner proposed that microtubules exist in two distinct populations, one growing and the other shrinking. According to this model (see Figure 9-13b), the presence of GTP bound to the tubulin at the end of a growing microtubule creates a stable microtubule tip to which further GTP-tubulin subunits can be added. Once all the GTP at the end of a microtubule has been hydrolyzed to GDP, however, an unstable tip is created and

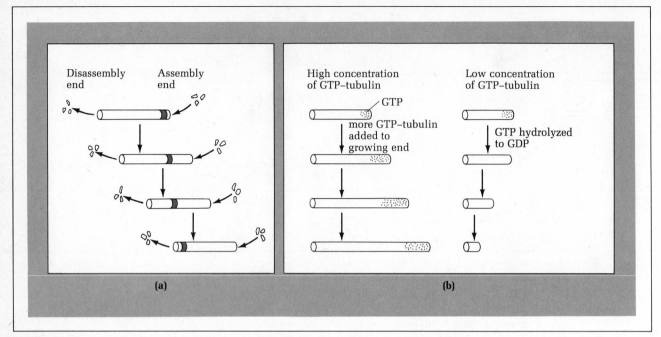

FIGURE 9-13 Two hypothetical models for microtubule disassembly. (a) According to the "tread-milling" concept, one end of the microtubule is specialized for assembly and the other for disassembly. The shaded region is employed to illustrate how individual tubulin or GTP molecules incorporated at one end would gradually move to the other end. (b) An alternative model of microtubule disassembly based on the existence of two populations of microtubules, a growing population containing GTP bound to the tip and a shrinking population in which the bound GTP has been hydrolyzed to GDP. The relative amounts of the two types of microtubules present will be determined by the concentration of free GTP-tubulin present.

rapid depolymerization occurs. When the concentrations of free tubulin and GTP are high, the microtubule tip will always have a sufficient amount of newly incorporated GTP-tubulin to maintain the growing state. If the concentration of free tubulin and/or GTP in the medium falls, however, the rate at which new GTP-tubulin is added to the microtubule tip will decrease. Under such conditions hydrolysis of the GTP already present at the microtubule tip will tend to predominate, creating an unstable microtubule that rapidly depolymerizes.

The preceding discussion suggests that several factors are potentially capable of regulating microtubule assembly/disassembly. First, we have seen how the availability of free tubulin and GTP influences the relative balance between microtubule assembly and disassembly. Another potential regulator of microtubule assembly is the availability of microtubule-organizing centers, a concept supported by experiments carried out on clam eggs by Richard Weisenberg. In these studies Weisenberg compared the ability of components isolated from dividing and nondividing clam eggs to induce the polymerization of tubulin into microtubules. Homogenates from dividing eggs were found to contain a sedimentable structure that, when incubated with non-polymerized tubulin, can function as an organizing structure for microtubule formation. Electron micrographs reveal that this microtubule-organizing center

consists of the amorphous granular material, or centrosome, that surrounds centrioles (Figure 9-14). Homogenates obtained from nondividing eggs, in contrast, lack such material and are unable to trigger the polymerization of tubulin into microtubules. However, tubulin isolated from nondividing eggs polymerizes into microtubules just as well as tubulin from dividing eggs when exposed to organizing centers isolated from dividing cells. It can therefore be concluded that the presence or absence of the organizing center regulates polymerization of microtubules into the mitotic spindle.

In addition to tubulin, GTP, and microtubule-organizing centers, other factors that are potentially capable of influencing microtubule formation within cells include the availability of microtubule accessory proteins (which promote tubulin polymerization) and the presence of calcium ions (which inhibit tubulin polymerization). The influence of Ca^{2+} on microtubule assembly requires the presence of *calmodulin*, a calcium-binding protein that has been implicated in the control of a wide variety of cellular activities (Chapter 16). Yet another possible mechanism for regulating microtubule assembly involves covalent modification of microtubule proteins. There is evidence, for example, that phosphorylation of two of the microtubule-associated proteins (especially MAP-2) by the enzyme cyclic AMP-dependent protein kinase causes fewer microtubules to be formed by shifting the balance between microtubule

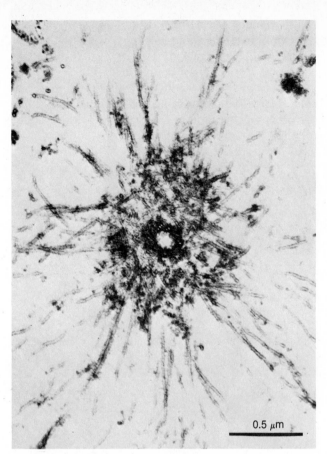

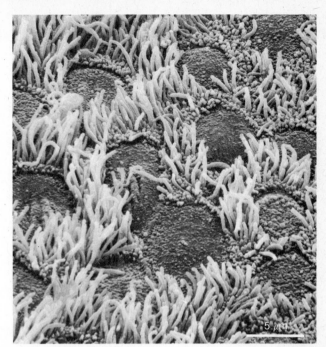

FIGURE 9-16 Scanning electron micrograph of the surface of epithelial cells lining the bronchioles of the respiratory system. Numerous cilia can be seen extending from the cells. Courtesy of A. T. Mariassy.

FIGURE 9-14 Electron micrograph showing microtubules growing from the granular material (centrosome) surrounding the centriole. Courtesy of R. W. Weisenberg.

FIGURE 9-15 Experiment showing that microtubule-associated proteins (MAPs) stimulate the polymerization of tubulin into microtubules under cell-free conditions. The data also reveal that phosphorylation of microtubule-associated proteins renders them less effective in stimulating microtubule formation.

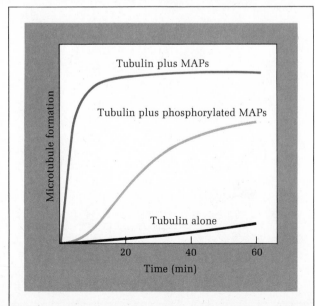

assembly/disassembly (Figure 9-15). Since cyclic AMP levels and the activity of this protein kinase are subject to control by a variety of hormones (Chapter 16), phosphorylation of microtubule proteins provides a potential mechanism for rendering microtubule assembly amenable to hormonal regulation.

Eukaryotic Cilia and Flagella

Among the various structures of the cell containing microtubules, cilia and flagella are probably the best understood in terms of microtubule organization and function. Cilia and flagella are large motile organelles that project from the surface of eukaryotic cells (Figure 9-16). Depending on the cell type, these organelles perform one of two alternative functions. Cells that are firmly anchored in place employ ciliary motion to move fluids across their surfaces; cells that are not firmly anchored, such as sperm cells or unicellular organisms, employ ciliary or flagellar movement to propel themselves through the liquid medium in which they are suspended.

Although cilia and flagella are closely related structures, the two can be distinguished from each other on the basis of differences in size, number, and pattern of movement. Flagella are relatively large organelles, measuring 150 μm or more in overall length; they are present in small numbers per cell and beat in a regular planar wave that propels liquid parallel to the flagellar axis. Cilia are shorter (5–10 μm average length), more numerous, and beat in a more complex pattern whose

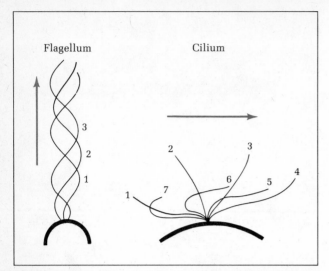

FIGURE 9-17 Comparison of the movement patterns of cilia and flagella. Numbers indicate successive steps in the movement of a single flagellum or cilium. The direction of liquid flow induced by these two types of movement is indicated by the arrows.

net result is to move fluid across the cell surface (Figure 9-17). In spite of these differences, however, the underlying substructure and mechanism of movement of these two organelles is quite similar, and most generalizations concerning the behavior of one apply to the other. This similarity does not extend, however, to bacterial flagella, whose unique properties will be described later in the chapter.

Architectural Organization of Cilia and Flagella Cilia and flagella are both constructed from a regularly arranged backbone of microtubules, termed an **axoneme,** surrounded by an extension of the plasma membrane. Axonemes typically consist of nine **outer doublet tubules** surrounding a pair of **central tubules,** an arrangement referred to as the *9 + 2 pattern* (Figure 9-18). The walls of the two central microtubules contain the typical arrangement of 13 protofilaments, but the outer doublet tubules are more complicated because each doublet consists of 2 microtubules physically joined to each other. One of these, termed the **A-tubule,** has a complete wall of 13 protofilaments; the adjoining **B-tubule,** in contrast, contains only 10 protofilaments because it shares a portion of the A-tubule wall (Figure 9-19). Projecting from each A-tubule are sets of **arms** that recur every 16–22 nm along the axoneme's length.

The orderly 9 + 2 arrangement of the axoneme is maintained by an intricate system of connections among the various tubules (Figure 9-20). The central microtubules are linked together by a **central microtubular bridge** and surrounded by a **central sheath.** The A-tubules of the outer doublets are connected to the central region by prominent **radial spokes** that terminate in thickened knobs lying adjacent to the central sheath. Adjacent doublets are connected to each other

by **interdoublet nexin links.** In addition, some cilia contain a prominent bridge connecting the arms of one particular doublet (designated "doublet 5") to the B-tubule of the adjacent doublet ("doublet 6").

Freeze-etch electron micrographs reveal the presence of longitudinal rows of particles embedded within the plasma membrane in the region over the doublet tubules of the axoneme and horizontal rows of particles, termed the **ciliary necklace,** located within the membrane surrounding the base of the cilium (Figure 9-21).

FIGURE 9-18 Electron microscopic appearance of cilia in cross and longitudinal section. (*Left*) Cross sections at three different levels. The 9 + 2 arrangement of microtubules is clearly visible in the bottom cross section. As the tip of the cilium is approached, the 9 + 2 arrangement of the microtubules disappears. (*Right*) Longitudinal section of the cilium. Courtesy of W. L. Dentler.

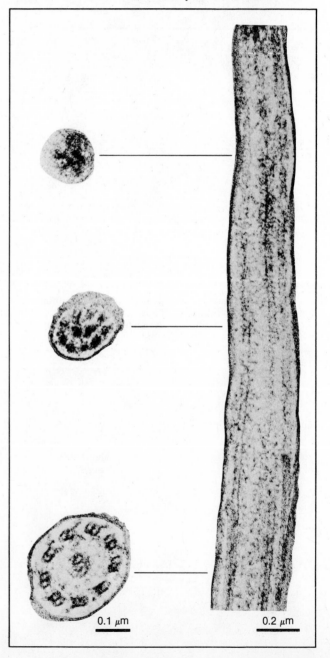

0.1 μm

0.2 μm

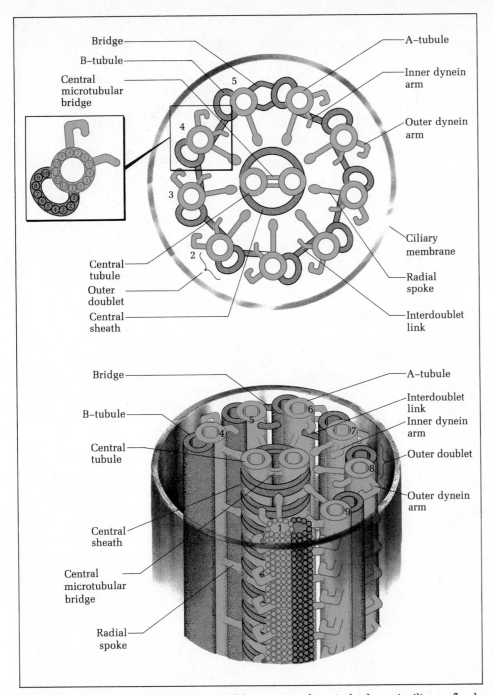

FIGURE 9-19 Diagrammatic representation of the structure of a typical eukaryotic cilium or flagellum. To enhance the visibility of the internal regions, several tubules, arms, and bridges have been omitted from the lower diagram. Only one doublet (1) has its tubulin molecules and protofilaments illustrated in detail.

Thin connections exist between the doublet tubules and some of these particles, raising the possibility that the axoneme may be attached to the overlying plasma membrane. Connections between the axoneme and the plasma membrane also occur at the tip of cilia and flagella. Although the organization of the ciliary tip varies somewhat among organisms and cell types, the end of the central pair of microtubules is generally covered with a *cap* structure that attaches to the plasma membrane, while the outer doublet tubules are often connected by small filaments to the plasma membrane (Figure 9-22).

Identification of the molecules that comprise the various substructures of the axoneme has been facilitated by the development of techniques for the stepwise degradation of cilia. This general approach, pioneered by Ian Gibbons, begins with the removal of cilia or flagella from the cell surface by gentle shearing or

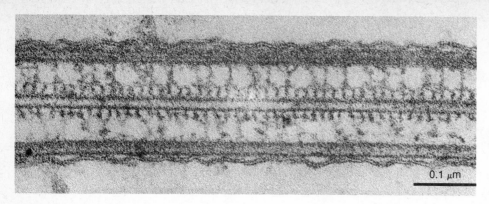

FIGURE 9-20 Fine structure of the cilium. *(Top)* Longitudinal section of a cilium showing the radial spokes that join the A-tubule of the doublet with the projections of the central sheath. *(Bottom left)* Cross section of a cilium with its membrane removed, showing a bridge between the two central microtubules (arrow). *(Bottom right)* Cross section of an intact cilium showing interdoublet links (arrow) connecting adjacent A- and B-tubules in the region of the dynein arms. Spokes connecting to the central sheath are also visible. Courtesy of F. D. Warner.

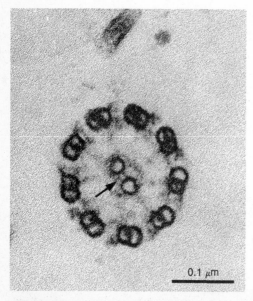

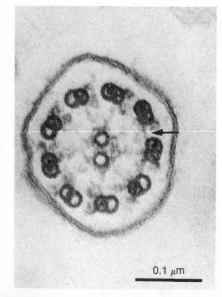

FIGURE 9-21 *(below, left)* Freeze-fracture electron micrograph showing the ciliary necklace (arrow) located in the transition region beneath the basal plate. Courtesy of P. Satir.

FIGURE 9-22 *(below, right)* Thin-section electron micrograph of the ciliary tip in *Tetrahymena.* The central microtubule cap is linked to the plasma membrane by small bridges (large arrowhead), while filaments link the A microtubules to the membrane (small arrowhead). Courtesy of W. L. Dentler.

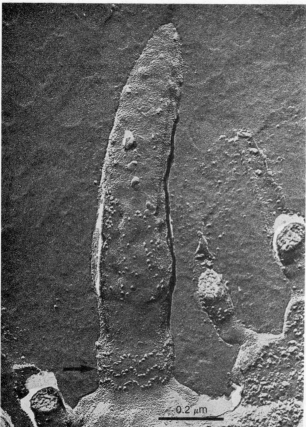

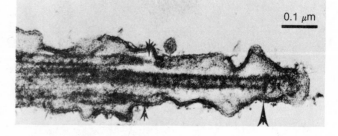

an appropriate chemical treatment. After low-speed centrifugation to remove intact cells, the isolated cilia are treated with detergent to remove their enclosing plasma membranes. Gibbons discovered that subsequent treatment of the exposed axonemes with the chelating agent ethylenediamine tetraacetic acid (EDTA) results in the loss of a protein, called **dynein,** that catalyzes the breakdown of ATP. Axonemes freed of dynein by EDTA treatment lack both the arms of the doublets and the pair of central tubules, suggesting that one or both of these structures corresponds to the protein dynein. To investigate this issue, Gibbons added dynein back to EDTA-extracted axonemes and then examined the resulting material with the electron microscope. Under these conditions the arms, but not the central microtubules, were found to reappear, indicating that the arms of the outer doublets correspond to the protein dynein (Figure 9-23).

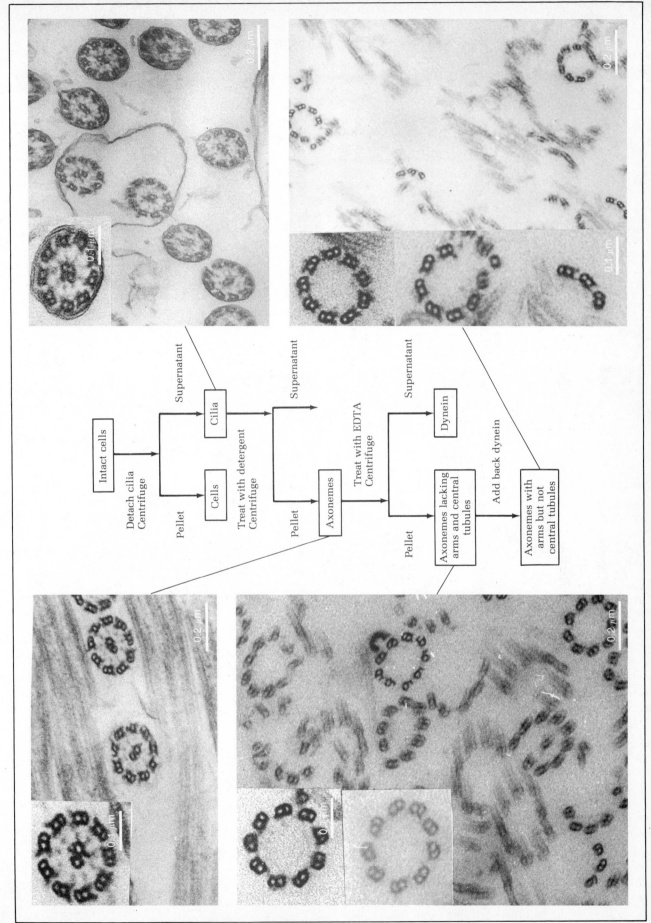

FIGURE 9–23 Schematic outline of ciliary fractionation experiments demonstrating that the arms on the outer doublets are composed of the protein dynein. Micrographs courtesy of I. R. Gibbons.

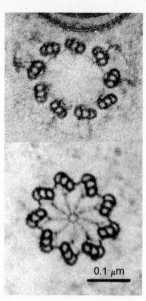

0.2 μm   0.1 μm

FIGURE 9-24 Appearance of the basal body in electron micrographs. *(Left)* Longitudinal section of a row of cilia showing the basal bodies (arrow). *(Right)* Cross section through a basal body at two levels, the top one closer to the cilium and the lower one closer to the point where the basal body is attached to the cell interior. In addition to the characteristic triplet arrangement of the microtubules, note the cartwheel spokes that occur in the lower section. Courtesy of F. D. Warner.

At least half a dozen proteins in addition to dynein and tubulin have been identified in extracts derived from isolated axonemes. Among these are the proteins that make up the interdoublet link (nexin), the spokes, and the central sheath.

Basal Bodies, Centrioles, and the Biogenesis of Cilia and Flagella   Cilia and flagella are anchored to the cell by a cylindrical structure, termed the **basal body,** that contains nine sets of *triplet tubules* (Figures 9-24 and 9-25). Of the 3 tubules that comprise each triplet only one, the A-tubule, has a complete circular wall of 13 protofilaments. The B-tubule is a horseshoe-shaped structure that shares part of the wall of the A-tubule, while the C-tubule is a horseshoe-shaped structure sharing part of the wall of the B-tubule. The A- and B-tubules are continuous with the analogous tubules of the cilium or flagellum, but the C-tubules are restricted to the basal body. Another difference between the basal body and the cilium or flagellum to which it attaches is that the basal body lacks the two central microtubules. A **transition zone** of variable size occurs between the termination of the C-tubules of the basal body and the beginning of the two central microtubules of the axoneme. Just below the origin of the two central microtubules a dense partition, the **basal plate** or **axosome,** crosses the lumen of the transition zone. **Transitional fibers** help to anchor the basal body to the plasma membrane, and larger, striated **rootlet fibers** pass from the basal body to the cell interior. These stabilizing fibers

help the basal body to anchor the cilium or flagellum to the cell interior.

In addition to functioning as an anchoring mechanism, basal bodies play a crucial role in the development of cilia and flagella. The formation of cilia and flagella is triggered by small cylindrical organelles, termed **centrioles,** that closely resemble basal bodies in ultrastructure. But unlike basal bodies, centrioles lie adjacent to the nucleus rather than attached to cilia or flagella (Figure 9-26). The first stage in the formation of a cilium or flagellum involves migration of a centriole to the cell surface, where it makes contact with the plasma membrane. The A- and B-tubules of the centriole then act as nucleation sites for microtubule assembly, initiating polymerization of tubules that will form the nine outer doublets of the ciliary axoneme. After this process of tubule assembly has begun, the centriole is referred to as a basal body.

The role of centrioles in triggering formation of new cilia and flagella raises the question of how the cell's supply of centrioles is replenished. At least three different mechanisms exist for reproducing centrioles. The first, typical of cells requiring relatively few new centrioles, involves the development of new centrioles at right angles to existing centrioles or basal bodies. This type of reproduction begins with the appearance of a centriole precursor, or **procentriole,** next to an existing centriole. In the newly forming procentriole the A-tubules are the first element to form. The B-tubules then grow from the A-tubule walls, followed by growth of the C-tubules from the B-tubule walls (Figure 9-27). The newly formed triplet tubules then elongate, generating a mature centriole lying at right angles to the original centriole.

FIGURE 9-25 Diagram of the basal body and its relationship to the cilium or flagellum to which it attaches.

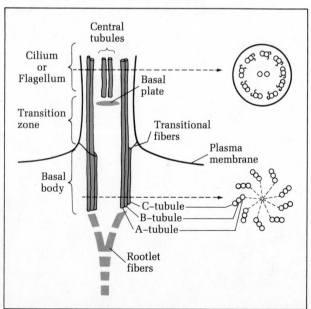

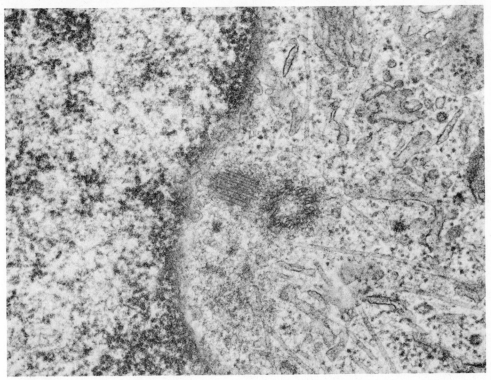

FIGURE 9-26 Electron micrograph showing two centrioles lying near the nuclear envelope. One centriole is seen in cross section and the other in longitudinal section. Courtesy of M. McGill.

In contrast to the above process, alternative modes of centriole assembly do not require an intimate association between newly forming and previously existing centrioles. One such mechanism occurs when cells that contain a small number of centrioles (usually a single pair) are called upon to produce large numbers of cilia. In such cases dozens of centrioles form adjacent to elongated granular masses of material, called **deuterosomes,** which function as organizing centers for the microtubules of the newly developing centrioles (Figure 9-28). As the centrioles are assembled, the deuterosomes diminish in size and eventually disappear. The centrioles then migrate to the cell surface and induce the assembly of new cilia.

The remaining mechanism of centriole formation is characteristic of cells that initially lack centrioles. This situation is most commonly encountered in plant cells, which typically do not have centrioles. Plant sperm cells are an exception to this general rule because they require centrioles for initiating development of the sperm flagella. During the differentiation of plant sperm, centrioles arise *de novo* adjacent to dense accu-

FIGURE 9-27 Development of centrioles adjacent to existing centrioles. (A) Cross section of an early stage of procentriole development, showing the formation of the A-tubules. (B) In the next stage the B-tubules are seen growing from the A-tubule walls. (C) Formation of the C-tubules is then completed. (D) A newly developing basal body is seen in longitudinal section adjacent to an existing basal body. The newly forming basal body is about one-third grown.

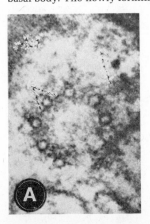
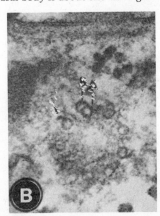
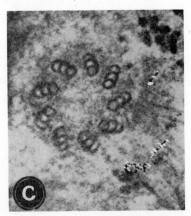

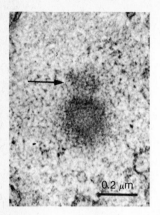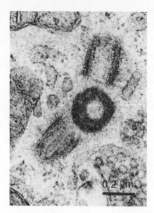

FIGURE 9-28 Development of centrioles adjacent to deutero-somes. *(Left)* A procentriole (arrow) is seen forming adjacent to a deuterosome. *(Middle)* Six newly forming procentrioles arranged symmetrically around a deutero-some that is partly hidden by an overlying procentriole. *(Right)* A deuterosome and two well-developed procentrioles. Courtesy of S. P. Sorokin.

mulations of granular material that presumably function as microtubule-organizing centers. In some plants this mechanism leads to the formation of thousands of centrioles, generating a large mass called a **blepharoplast** that is readily visible with the light microscope.

Once a centriole has started the development of a new cilium or flagellum, the question arises as to how the newly forming axoneme elongates. Studies carried out in the laboratory of Joel Rosenbaum on *Chlamydomonas*, a green alga containing two flagella, have provided some interesting insights into this process. The dynamics of flagellar growth in this organism were investigated by briefly shearing live cells in a homogenizer to break off the existing flagella, and then reincubating the cells in the presence of radioactive amino acids under conditions that permit flagellar regrowth. Electron microscopic autoradiographs of the newly forming flagella reveal that radioactivity is most concentrated over the flagellar tips, indicating that flagellar growth occurs mainly by addition of tubulin to the tip, rather than to the base or middle, of the growing axoneme. One of the more interesting findings to emerge from these studies involved cells in which only one of the two flagella had been amputated during homogenization. In such cases initial growth of the missing flagellum is accompanied by a shortening of the existing flagellum; when the two flagella reach the same size, both elongate until the mature size is reached (Figure 9-29). Presumably the initial shortening of the existing flagellum provides tubulin subunits to the newly forming flagellum, but the mechanism that coordinates the lengths of the two flagella is not understood.

Although the above experiments indicate that the bulk of flagellar growth occurs at the tip, a small fraction of the tubulin incorporated into newly forming flagella appears to be selectively incorporated at the flagellar base. This conclusion derives mainly from experiments in which isolated flagella were incubated in the presence of unpolymerized tubulin and then examined microscopically to determine where microtubule growth occurs. Under these conditions the outer doublets, but not the central pair of microtubules, are found to contain newly added tubulin subunits at the tip of the flagellum. Addition of tubulin to the tip of the central pair of microtubules will only occur if the cap structure is first removed artificially. Since the cap is normally present in elongating flagella, this observation suggests that the central pair of microtubules is assembled from the base rather than from the tip of the growing flagellum.

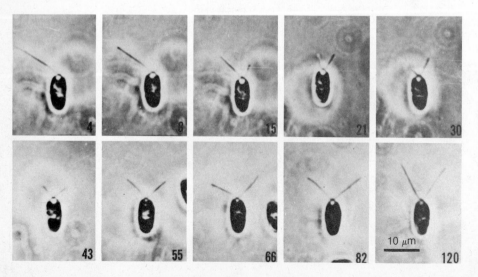

FIGURE 9-29 Experiment showing that after amputation of one of the two flagella of a single algal cell (*Chlamydomonas*), regeneration of the amputated flagellum is accompanied by a dramatic shortening of the intact flagellum. This sequence is shown beginning 4 min after amputation of the right flagellum. Courtesy of J. L. Rosenbaum.

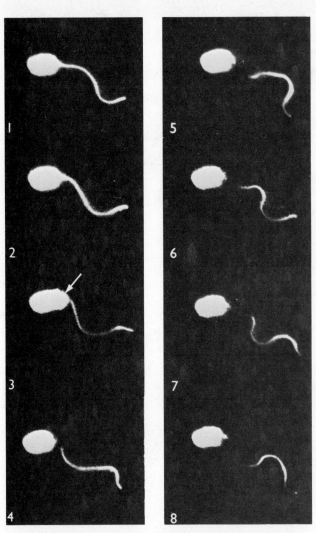 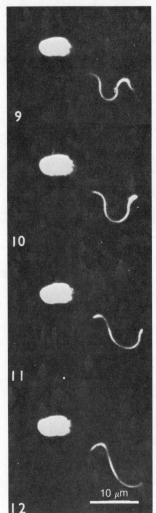

FIGURE 9-30 Micrographs showing the continued movement of a flagellum after detachment from the cell. (1,2) Intact flagellum prior to detachment. (3,4) A laser beam is directed to the region indicated by the arrow, severing the flagellum from the cell. (5–12) Following detachment, the flagellum continues to beat. Courtesy of M. Holwill.

The Mechanism of Ciliary and Flagellar Motility Several kinds of observations indicate that eukaryotic cilia and flagella are inherently motile. Perhaps the most striking is the finding that flagella removed from living cells by laser microbeam irradiation continue to beat until their supply of stored ATP is exhausted (Figure 9-30). Even axonemes isolated from detached cilia have been found to be motile when provided with ATP as an energy source. It can therefore be concluded that the axoneme itself is capable of generating movement when supplied with an appropriate source of energy.

Though ciliary and flagellar movement involve relatively complex patterns (see Figure 9-17), the basic motion underlying these patterns is a simple bending of the axoneme at appropriate points along its length. One of the earliest theories proposed to explain axonemal bending was that microtubules contract in the presence of ATP, causing the axoneme to bend toward the side of the cilium where the microtubules have contracted. Such a simple mechanism was effectively ruled out, however, by a classic series of studies carried out by Peter Satir in the mid-1960s. Satir reasoned that if microtubules contract, the doublets on one side of a bend-

ing cilium should be shorter and thicker than those on the opposite side. But when Satir examined sections through the tips of bending cilia, this expectation was not confirmed. Instead, the microtubules on one side of the bending cilium were found to project further into the ciliary tip. As illustrated in Figure 9-31, such an observation suggests that microtubules slide past one another during ciliary bending. Further support for this conclusion was obtained by measuring the repeat distance between the radial spokes viewed in longitudinal section. Because these spokes are fixed to the doublet walls, the distance between spokes should decrease if microtubules contract. The distance between spokes was found, however, to be the same in straight and bent regions of the axoneme, indicating that ciliary microtubules do not contract.

A radically different approach taken by Keith Summers and Ian Gibbons has provided independent support for the idea that ciliary microtubules slide past one another rather than contract. These investigators discovered that exposure of sperm tail axonemes to the proteolytic enzyme trypsin causes the radial spokes and interdoublet nexin links to be selectively degraded. In-

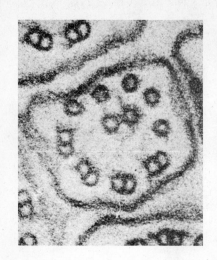
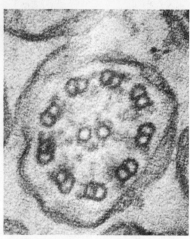
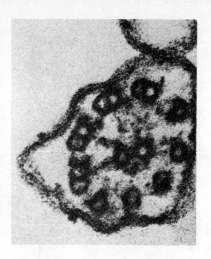

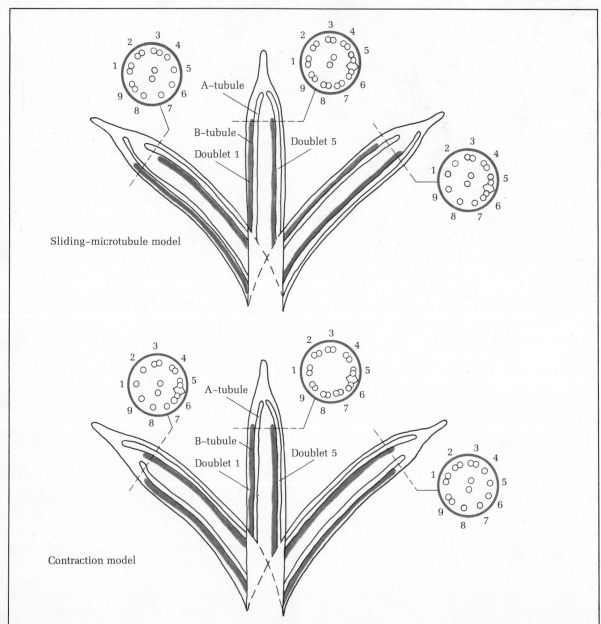

A–tubule

B–tubule

Doublet 1

Doublet 5

Sliding–microtubule model

A–tubule

B–tubule

Doublet 1

Doublet 5

Contraction model

FIGURE 9-31 (*opposite page*) Summary of the experimental approach employed to distinguish between the sliding-microtubule and contraction theories of ciliary motion. The discovery that the A-tubule of each microtubule doublet extends further into the tip of the cilium than the B-tubule provided a convenient way to identify the tip end of each doublet. The sliding-microtubule theory predicts that the B-tubules will be missing on the outer side of the bend, while the contraction theory predicts the opposite result. Electron micrographs (*top*) showing sections through the tips of bending cilia are consistent with the predictions of the sliding-microtubule theory. Micrographs courtesy of P. Satir.

stead of contracting in response to the addition of ATP, such axonemes extend to several times their original length and then fall apart (Figure 9-32). This ATP-dependent increase in length is most compatible with the theory that microtubules slide past one another during ciliary movement, and that the destruction of the radial spokes and nexin links by trypsin simply exaggerates the amount of sliding that can occur by destroying the linkages that normally restrain the sliding process.

If the energy released by ATP hydrolysis is utilized to drive the sliding of microtubules past one another, how do these sliding forces cause the axoneme to bend? The above experiments suggest that the radial spokes and nexin links of intact cilia provide architectural constraints that limit microtubule sliding. This physical limitation is translated into bending of the microtubule. Support for this idea has been obtained from studies on mutant cells whose cilia lack radial spokes and the central-tubule/central-sheath complex. Trypsin-treated axonemes obtained from such mutant cilia undergo elongation and fall apart just as do normal axonemes, confirming that microtubule sliding occurs. These mutant cilia are not capable of normal bending movements, however, indicating that radial spokes and the central-tubule/central-sheath complex are required for converting microtubule sliding into axonemal bending. A role for the nexin links as well has been inferred from experiments in which mutant axonemes that lack radial spokes but contain nexin links were exposed to ATP without trypsin treatment. Under such conditions the axonemes do not fall apart, suggesting

that the nexin links may also function in the restraint of microtubule sliding.

The mechanism by which ATP induces microtubule sliding is not understood in detail, but the ability of isolated dynein preparations to hydrolyze ATP suggests that the dynein-containing arms play a crucial role. It is currently believed that ATP hydrolysis powers a cycle in which the dynein arms of the A-tubule of one doublet bind to specific sites on the B-tubule of the adjacent doublet, causing the two doublets to move relative to one another (Figure 9-33). Because the two central microtubules contain no dynein arms, this model suggests that only the nine outer doublets are involved in ciliary and flagellar motility. This conclusion is consistent with the discovery that certain sperm cells have motile flagella that lack the central pair of microtubules.

Although the details of the above model remain to be verified, there is ample evidence implicating the dynein arms in ciliary motility. For instance, flagellar motility declines in sperm cells from which dynein has been extracted. This decrease can be largely reversed by adding back purified dynein. It is also relevant that nonmotile cilia, such as those occurring in sensory organs such as the inner ear, lack dynein arms.

Several disease syndromes are known to be associated with defects in ciliary activity. A prime example is cystic fibrosis, an inherited disease in which defects in ciliary motility are responsible at least in part for abnormalities in the secretions of the respiratory tract, pancreas, and salivary glands. Another example involves the degenerative changes observed in the lung tissue of chronic smokers that result from the inhibition of ciliary activity by tobacco smoke. Finally, some individuals suffering from inherited respiratory abnormalities and associated sterility have been found to be lacking dynein arms in their cilia and flagella.

Other Axonemal Structures

Not all cilia are motile. A variety of sense organs, including eyes, acoustico-vestibular systems, and olfac-

FIGURE 9-32 Darkfield light micrographs showing the behavior of trypsin-treated axonemes exposed to ATP. The successive photographs in the series were taken at intervals of 10–30 sec. The white lines indicate the initial position of each end of the axoneme. Note the dramatic lengthening of the axoneme under these conditions. Courtesy of I. R. Gibbons.

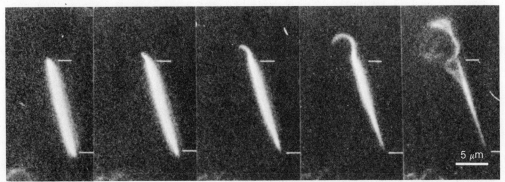

5 μm

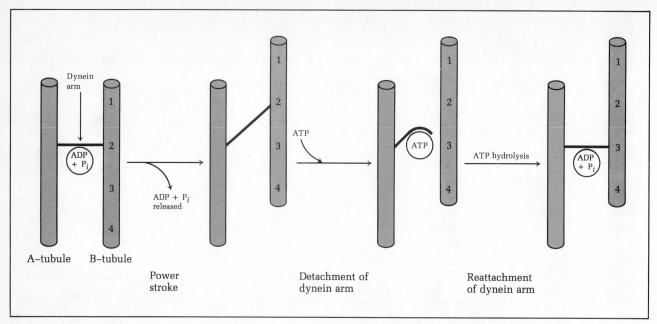

FIGURE 9-33 Schematic diagram illustrating how the sequential binding and release between the dynein arms of one microtubule doublet and binding sites on the adjacent microtubule doublet could induce microtubule sliding. The energy driving this cycle is derived from the dynein-induced hydrolysis of ATP.

tory organs, contain nonmotile cilia. The axonemes of such cilia generally lack central microtubules and dynein arms, and may exhibit other unusual features as well. For example the outer segment of the retinal rod cell, discussed in detail in Chapter 18, is derived from a single cilium whose outer membrane has undergone an elaborate series of invaginations. In this and other sensory cells, cilia serve to increase the plasma membrane surface area available for the reception of incoming stimuli. What role, if any, the axonemal microtubules play in the reception and transmission of sensory signals remains to be ascertained.

Cilia and flagella are not the only organelles comprised of ordered arrays of microtubules (Figure 9-34). Certain protozoa have long thin projections, termed **axopods,** protruding from their cell surface. These structures, which are involved in cellular locomotion and in attaching cells to various surfaces, contain an axonemal backbone constructed from spiral coils of microtubules. In other protozoa movements are produced by intracellular bundles of cross-linked microtubules known as **axostyles.** Other examples of organized microtubular arrays occurring in unicellular eukaryotes are **cortical fibers,** which induce rapid changes in cell shape, and the **cytopharyngeal basket,** a complex microtubular organelle through which some single-cell organisms feed. In contrast to cilia and flagella, where microtubule doublets are a prominent structural feature, the above organelles are comprised of collections of single microtubules. The term *axoneme* is nevertheless employed in referring to such microtubule arrays because, like their counterparts in cilia and flagella,

they consist of an ordered arrangement of microtubules held together by intertubule linkages.

Cytoplasmic Microtubules and the Movement of Subcellular Components

The microtubules of eukaryotic cells can be divided into two general groups. One includes the relatively stable microtubules found in cilia, flagella, and other axonemal structures. The other is a more loosely organized, dynamically changing network of cytoplasmic microtubules. In contrast to axonemal structures, which generally function to move either the cell as a whole or fluids surrounding the cell, cytoplasmic microtubules are involved in the movements and interactions of subcellular components and in determining and maintaining cell shape. The most widely used experimental approach for implicating microtubules in such activities has been to determine whether the activity in question is inhibited by drugs that promote microtubule disassembly, such as *colchicine, vinblastine,* and *vincristine.*

In this way cytoplasmic microtubules have been implicated in the movement of a variety of subcellular constituents. The most obvious example involves the cell's chromosomes, whose movements at the time of cell division are intimately associated with the microtubules of the mitotic spindle. The mechanism by which chromosomes move as cells divide and the role played by the spindle microtubules will be discussed in detail in Chapter 13, which is devoted to the topic of cell division.

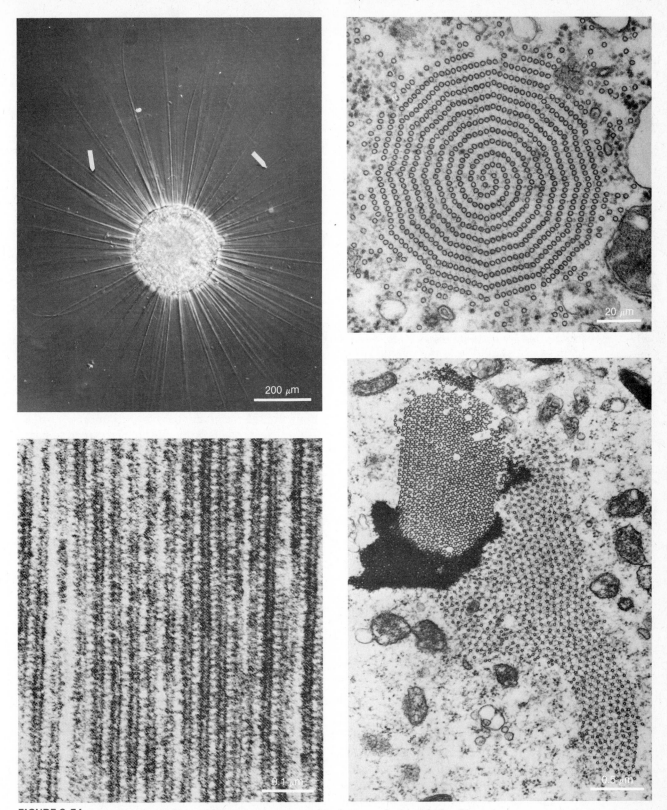

FIGURE 9-34 Examples of nonciliary structures containing ordered arrays of microtubules. *(Top left)* A protozoan exhibiting numerous axopods protruding from the cell surface. *(Top right)* Cross section through an axopod showing the spiral arrangement of microtubules. Courtesy of L. E. Roth. *(Bottom left)* Longitudinal section through an axostyle showing an array of parallel microtubules held together by cross-links between the adjacent microtubule walls. Courtesy of L. G. Tilney. *(Bottom right)* Cross section of a cytopharyngeal basket in a ciliated protozoan showing the organized array of microtubules held together by cross-links between adjacent microtubules. Courtesy of J. B. Tucker.

Several lines of indirect evidence suggest that microtubules also play a role in the movement of intracellular components other than chromosomes. It has been observed, for example, that vesicles, pigment granules, and even organelles as large as mitochondria often move in patterns directed along lines parallel to microtubule arrays (see Chapter 18 also). In cells specialized for secretion, microtubules are usually concentrated near the Golgi complex and in the area through which secretory vesicles pass prior to discharge at the cell surface. Disruption of such microtubules with the drug colchicine has been shown to inhibit secretion of hormones by endocrine cells, plasma proteins by liver cells, and enzymes by salivary gland and pancreatic cells.

Several problems arise in interpreting such experiments, however. The first is that drugs such as colchicine may affect cellular constituents other than microtubules. And even if the studies involving such drugs accurately indicate that microtubules are required for cell secretion, it is difficult to distinguish whether microtubules are actively responsible for the movement of secretory vesicles and the expulsion of their contents from the cell, or whether they play a more passive role, guiding vesicles toward the cell surface and/or serving as an anchoring framework for other elements (e.g., microfilaments) whose contraction provides the propelling force.

Cytoplasmic Microtubules and the Cell Surface

One organelle whose properties appear to be significantly influenced by interaction with cytoplasmic microtubules is the plasma membrane. As discussed in Chapter 4, the plasma membrane consists of a fluid matrix of lipid in which membrane proteins are embedded. Because of the fluid nature of the lipid bilayer, one might expect membrane proteins to randomly diffuse throughout this matrix. Though this is clearly the case for some membrane proteins, others appear to be more restrained in their movements. Independent experiments carried out in the laboratories of Richard Berlin and Gerald Edelman suggest that microtubules are involved in restraining and regulating the movements of such membrane proteins.

The studies of Berlin and his colleagues focused on the plasma membrane of phagocytic white blood cells. During active phagocytosis by these cells, as much as 50 percent of the plasma membrane surface area is involved in forming phagocytic vesicles that become incorporated into the cell interior. In spite of this extensive internalization of plasma membrane, however, the ability of the cell surface to carry out the active transport of nutrients remains unchanged. This finding suggests that the total number of plasma membrane transport carriers is not significantly decreased in the

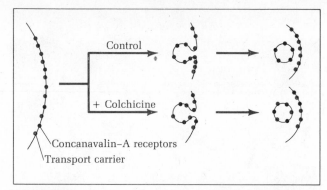

FIGURE 9-35 Summary of the experiments on phagocytosis implicating microtubules in the regulation of membrane protein behavior. During normal phagocytosis, membrane transport carriers (color) are selectively excluded from invaginating membrane vesicles while concanavalin-A receptors are selectively included. The discovery that colchicine abolishes this selectivity suggests that submembranous microtubules are involved.

face of a 50 percent reduction in total plasma membrane area. It can therefore be concluded that membrane transport proteins are selectively excluded from those regions of the plasma membrane that invaginate to form phagocytic vesicles.

Not all membrane proteins behave in this same way, however. When Berlin investigated the behavior of membrane receptors that bind to **concanavalin A,** a lectin that interacts with specific carbohydrate chains of membrane glycoproteins, more than 50 percent of these receptors were found to be lost from the cell surface during phagocytosis. Thus neither transport carriers nor lectin receptors behave as if they are randomly distributed throughout the plasma membrane; the former are selectively excluded from areas of membrane invagination while the latter are preferentially included (Figure 9-35). One of the more interesting findings to emerge in this context was the discovery that disruption of microtubules with the drug colchicine destroys this selectivity without affecting the overall rate of phagocytosis. In other words the presence of colchicine causes plasma membrane transport proteins and lectin receptors to disappear from the cell surface at the same rate. Such observations suggest that microtubules, which are known to lie beneath phagocytic areas of the plasma membrane, are required for the selective inclusion or exclusion of membrane components from phagocytic vesicles.

Experiments carried out in the laboratory of Gerald Edelman have provided additional support for the conclusion that microtubules influence the behavior of specific membrane components. These studies were based on the initial observation that low concentrations of concanavalin A cause membrane receptors that bind this lectin to become cross-linked into small patches, which in turn coalesce with one another to form a larger aggregate of receptors referred to as a **cap.** High concentrations of concanavalin A, on the other hand, cause

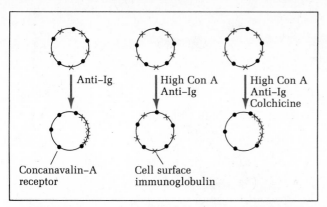

FIGURE 9-36 Summary of experiments on receptor capping suggesting that microtubules are involved in the regulation of membrane protein mobility. In this experiment immobilization of concanavalin-A receptors by high concentrations of concanavalin A is found to inhibit the capping of cell surface immunoglobulin (Ig) molecules by antibodies to Ig. When microtubules are disrupted by colchicine treatment, however, this inhibition of capping is no longer observed.

lectin receptors to become immobilized rather than to diffuse together into patches and caps. Especially interesting is the discovery that high concentrations of concanavalin A also influence the mobility of other membrane components. For example cell-surface immunoglobulin, a membrane protein that normally diffuses into patches and caps when exposed to antibodies that bind to it, no longer acts in this manner if the membrane's concanavalin-A receptors have first been immobilized by exposure to high concentrations of concanavalin A. This ability of immobilized concanavalin-A receptors to hinder the mobility of other plasma membrane receptors can be abolished by disrupting microtubules with colchicine (Figure 9-36), suggesting that the mobility of membrane proteins is influenced by interactions with submembranous microtubules.

The above observations have led to the proposal that membrane proteins exist in two alternative states, a free state in which they are capable of lateral diffusion and hence can be induced to form caps when exposed to an appropriate cross-linking reagent, and a bound state in which they are immobilized by linkage to a submembranous network of microtubules and so cannot diffuse into caps. It has been further postulated that high concentrations of concanavalin A cause its receptors to become attached to these submembranous microtubules, propagating the assembly of a lattice that restricts not only the mobility of concanavalin-A receptors but the mobility of other membrane proteins as well. When the submembranous microtubule network is disrupted by colchicine treatment, the general inhibition of membrane protein mobility is reversed.

The existence of a submembranous network of microtubules that modulates the behavior of cell surface proteins has implications for a variety of cellular processes. One possibility that has been suggested is that

this network is involved in the propagation of signals to and from the cell surface. One phenomenon of this type that has been extensively studied is the ability of the lectin, concanavalin A, to induce certain types of cells to divide. In such cells interaction of concanavalin A with its cell surface receptor triggers a series of intracellular changes that eventually culminate in the process of cell division. In the presence of colchicine, however, these intracellular changes do not occur, providing support for the notion that microtubules are intermediates in the transmission of signals from the cell surface to the cell interior.

Submembranous microtubules have also been implicated in cell wall formation in plants. In this case the microtubule network is thought to dictate the organized pattern of cellulose microfibrils characteristic of cell walls. Support for this idea has come from the observation that submembranous microtubules are oriented in the same direction as the cellulose microfibrils of the overlying wall. If such a microtubular network is disrupted by exposing dividing cells to colchicine, the cell walls that are subsequently produced contain disordered arrays of microfibrils rather than the parallel bundles normally observed. Though it is not known for certain how a microtubule network separated from the cell wall by the plasma membrane can influence the orientation of newly forming cellulose microfibrils, it has been suggested that contact between microtubules and the overlying plasma membrane causes the cellulose-synthesizing enzymes localized within the membrane to become oriented parallel to the underlying microtubules.

Because many of our ideas concerning the functional significance of submembranous microtubules are based upon experiments involving the drug colchicine, the possibility of artifacts arising from actions of this substance on structures other than microtubules should always be kept in mind. One test that can be carried out to help validate such experiments is to compare the effects produced by colchicine treatment with those induced by *lumicolchicine,* a closely related analog that resembles colchicine in chemical structure but is unable to disrupt microtubules. The usefulness of this approach is illustrated by experiments in which treating cells with colchicine was found to inhibit both phagocytosis and active transport, raising the possibility that microtubules play a role in these two activities. It was subsequently shown, however, that lumicolchicine inhibits these same events without breaking down microtubules. Such observations point to the need for caution in interpreting experiments involving the use of colchicine when appropriate controls have not been included.

Cytoplasmic Microtubules and Cell Shape

When the ubiquitous presence of microtubule networks in the cytoplasm of eukaryotic cells first became appar-

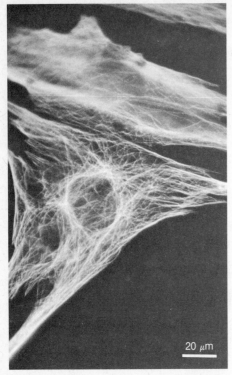

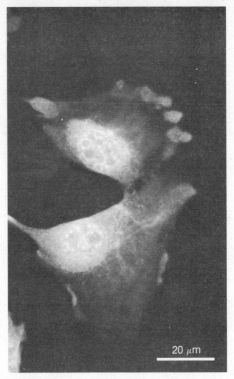

20 μm

20 μm

FIGURE 9-37 Immunofluorescent light micrographs showing the changes in cell shape that occur subsequent to drug-induced microtubule disassembly. *(Left)* Mouse fibroblasts stained with fluorescein-labeled anti-tubulin. *(Right)* The same cell type after treatment with the drug Colcemid, which breaks down microtubules. Note that the disappearance of cytoplasmic microtubules is accompanied by a rounding up of the cell. Courtesy of B. R. Brinkley.

ent in the early 1960s, one of the initial functions to be ascribed to this network was a skeletal role in determining and maintaining the various shapes that animal cells can assume. Subsequent support for this notion emerged from reports that indicated that disruption of microtubules with drugs, low temperature, or high pressure causes ellipsoid or elongated cells to assume a more spherical shape (Figure 9-37). When such treatments are stopped and microtubules allowed to re-form, the original asymmetric shapes are again established.

Microtubules influence the shape of animal cells in at least two different ways. First, a properly aligned network of microtubules appears to be a necessary prerequisite for the initial establishment of cellular asymmetry. Second, the maintenance of such asymmetry, once established, requires the continued presence of microtubules. In plant cells microtubule networks play a less important role in the maintenance of cell shape than in animal cells because this function is served by the rigid plant cell wall.

MICROFILAMENTS

In addition to microtubules, the cytoskeleton of most eukaryotic cells contains a second type of element, known as the **microfilament,** which is morphologically, chemically, and functionally distinct from the microtubule. Like microtubules, microfilaments have been implicated in a wide variety of cellular activities. Among these are cytoplasmic streaming, amoeboid locomotion, cell division, cell shape changes, and cell surface events such as endocytosis and secretion. The unifying feature

that appears to link these apparently diverse events is that they all exploit the ability of microfilaments to produce movement.

FIGURE 9-38 Electron microscopic appearance of microfilaments. *(Top)* Longitudinal section through a mouse fibroblast showing bundles of microfilaments (arrow). A single microtubule can be seen between the microfilament bundles. *(Bottom)* Cross section of a cell exhibiting three distinct cytoskeletal structures: microtubules (mt), microfilaments (mf), and intermediate filaments (f). Courtesy of R. D. Goldman.

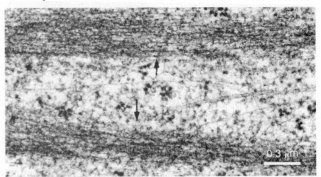

0.3 μm

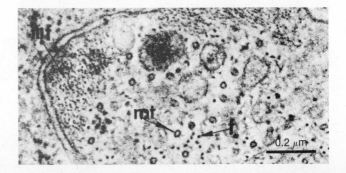

mf

mt

f

0.2 μm

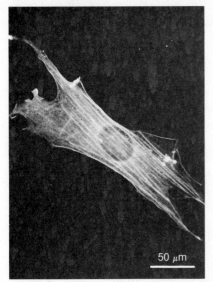
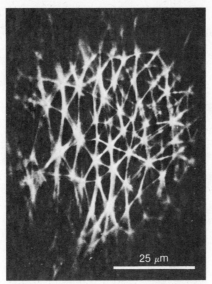

FIGURE 9-39 Fluorescence micrographs of nonmuscle cells stained with antibodies against actin and myosin. *(Left)* Fluorescence micrograph employing antibodies against actin showing microfilament bundles (stress fibers) spanning the length of the cell. Courtesy of E. Lazarides. *(Middle)* Example of a cell in which microfilament bundles stained with anti-actin are seen to converge on several focal points. Courtesy of E. Lazarides. *(Right)* Cell stained with antibody against myosin. The myosin appears to be arranged like the microfilament bundles observed in the left micrograph. Courtesy of T. D. Pollard.

Microfilaments are considerably smaller than microtubules, measuring roughly 6 nm in diameter (Figure 9-38). In some cells microfilaments are packed into parallel bundles that form structures large enough to be seen with the light microscope. These microfilament bundles, or **stress fibers,** either span the length of the cell or converge on several focal points (Figure 9-39 *left, middle*). Microfilaments also occur in organized arrays within microvilli and in looser networks localized beneath the plasma membrane, adjacent to the nucleus, and dispersed through the cytoplasm.

Microfilaments are constructed from the protein **actin,** a molecule whose biochemical properties and role in muscle contraction are described in detail in Chapter 19. Many observations indicate that the actin of muscle cells and the actin that comprises the microfilaments of nonmuscle cells are closely related molecules. Antibodies directed against muscle actin, for example, bind to the microfilaments of nonmuscle cells as well. In spite of this underlying similarity, several distinct forms of actin can be distinguished. Nonmuscle cells contain two major types of muscle actin, designated β- and γ-*actin*, whose amino acid sequences differ slightly from that of muscle, or α, *actin*.

When isolated under conditions of low ionic strength, actin molecules behave as monomeric proteins of roughly 43,000 daltons. Increasing the ionic strength induces actin molecules to polymerize into long filaments consisting of a double-helical array of actin monomers (see Figure 19-19). Several experimental approaches have been utilized to demonstrate that these actin filaments exhibit an intrinsic polarity. Per-

haps the most striking involves the ability of actin filaments to bind **heavy meromyosin,** a fragment of the muscle cell protein, myosin. When actin fibers (from muscle or nonmuscle cells) are incubated with heavy meromyosin, the meromyosin molecules bind to the fibers in a distinctive "arrowhead" pattern in which all the bound meromyosin molecules point in the same direction (see Figures 9-40 and 19-20). This polarity reflects functional differences in the behavior of the two ends of the filament; addition of new actin monomers to growing microfilaments occurs mainly at one end of the fiber, while disassembly occurs at the other.

Within living cells, actin occurs in a variety of structural forms including free monomers, individual organized filaments, regularly cross-linked filament bundles, and less regularly cross-linked filament networks that behave like gels (Figure 9-41). Although the factors regulating the structural state of actin are not completely understood, more than 60 different **actin-binding proteins** isolated from cells have been implicated in the process. In general terms this group of proteins can be divided into three classes: length-regulating proteins, depolymerizing proteins, and cross-linking proteins (Table 9-1). *Length-regulating proteins* bind preferentially to one end of actin microfilaments, inhibiting the addition of monomers to the growing filament and thereby slowing the overall rate of polymerization. *Depolymerizing proteins* bind to monomeric actin, shifting the equilibrium toward filament depolymerization and therefore decreasing the overall number of filaments that can be produced. Finally, *cross-linking proteins* promote the formation of fila-

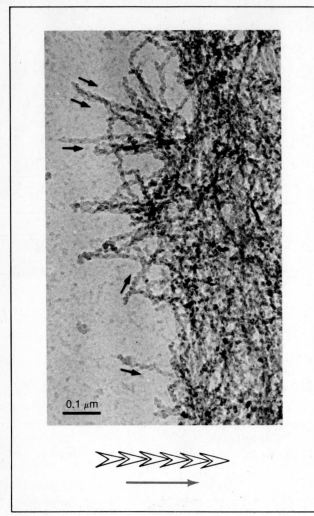

FIGURE 9-40 *(Top)* Thin-section through the advancing edge of a moving cell. The actin filaments have been stained with heavy meromyosin, creating "arrowhead patterns" (arrows) that point toward the cell interior. Courtesy of K. B. Pryzwansky. *(Bottom)* Diagram of arrowhead pattern showing its polarity.

TABLE 9-1
Three classes of actin-binding proteins

Class (With Examples)	Source	Ca²⁺ Sensitivity
Cross-linking Proteins		
α-Actinin	Muscle	No
α-Actinin	Fibroblasts	Yes
Actinogelin	Ascites cells	Yes
Fascin	Sea urchin eggs	No
Filamin	Fibroblasts, muscle	No
Fimbrin	Intestinal epithelium	?
Gelactin	Amoeba	No
Spectrin	Erythrocyte	No
Villin	Intestinal epithelium	Yes
Vinculin	Fibroblasts, muscle	No
Length-regulating Proteins		
β-Actinin	Muscle	No
Brevin	Serum	Yes
Fragmin	Cellular slime mold	Yes
Gelsolin	Macrophage	Yes
Villin	Intestinal epithelium	Yes
Depolymerizing Proteins		
γ-Actinin	Muscle	?
Profilin	Mammalian tissues	No

ment bundles and actin gels by cross-linking individual actin filaments to one another. The interaction of many of these proteins with actin can be regulated by Ca²⁺, suggesting a role for this ion in the regulation of actin behavior.

Because the actin present in the cytoplasmic microfilaments of nonmuscle cells closely resembles muscle actin, the question arises as to whether the behavior of actin in the two systems is based on similar principles. Since the mechanism of muscle contraction will be described in detail in Chapter 19, a brief summary of this

FIGURE 9-41 Diagrammatic summary of the various interconversions that actin undergoes in intact cells, indicating the roles played by the major classes of actin-binding proteins.

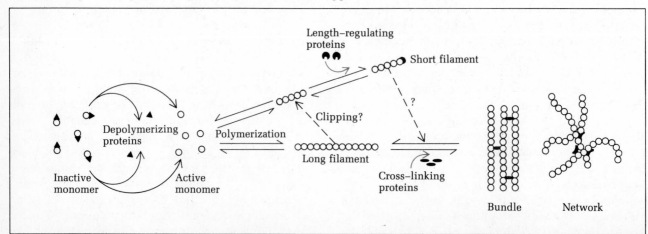

process will suffice in the present context. In essence, muscle contraction is mediated by interactions between two types of filaments: thin filaments of actin and thicker filaments of an ATP-hydrolyzing protein, called **myosin.** The energy released during hydrolysis of ATP allows myosin to bind to actin and to slide along its surface, just as dynein arm attachments mediate the movement of ciliary microtubules past one another during ciliary or flagellar bending. The interaction between actin and myosin is regulated by the proteins **troponin** and **tropomyosin,** whose properties are in turn influenced by the presence of calcium ions. Normally the troponin-tropomyosin complex inhibits muscle contraction by interfering with the interaction between actin and myosin. When Ca^{2+} concentrations are elevated, however, calcium ions bind to the troponin molecule and trigger a conformational change in the troponin-tropomyosin complex that permits actin and myosin to interact with each other and contraction to ensue.

The close association between actin and myosin in muscle cells raises the question of whether myosin is also present in nonmuscle cells. Staining of nonmuscle cells with fluorescent antibodies specific for myosin has revealed that myosin is associated with actin-containing stress fibers (Figure 9-39, *right*), as well as with the band of actin microfilaments that underlies the cleavage furrow of dividing cells (see Figure 13-23). Tropomyosin, which modulates actin-myosin interactions in muscle cells, has also been detected in stress fibers and other microfilament structures. Taken together, the preceding observations have fostered the notion that actin-myosin interactions similar to those occurring in muscle cells are involved in at least some kinds of nonmuscle cell movement.

Although the actin microfilaments of nonmuscle cells may, in certain situations, mediate contractile movements analogous to those occurring in muscle, the diversity of microfilament organization, composition, and interactions suggests the existence of other mechanisms as well. In attempting to determine the diverse roles played by microfilaments, cell biologists have been greatly aided by use of the drugs **cytochalasin B,** which hinders the polymerization of actin filaments, and **phalloidin,** which blocks the depolymerization of actin. By monitoring the effects of such drugs on various aspects of cell architecture and physiology, it has been possible to at least tentatively identify those events in which microfilaments might play a role. In the following sections we shall review some of the major activities in which microfilaments have been implicated.

Microfilaments and Cytoplasmic Streaming

When observed by light microscopy, the cytoplasm of most living cells appears to be in constant motion. Some

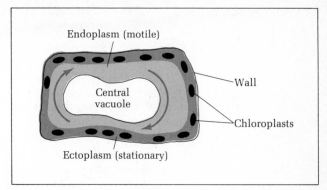

FIGURE 9-42 A diagram of cyclosis, the pattern of cytoplasmic streaming typical of plant cells.

of this activity is random Brownian movement generated by the kinetic energy of the water molecules in which the cytoplasmic constituents are dissolved or suspended. But certain movements occur in patterns that are too organized to be random. For example, many plant and animal cells manifest a regular pattern of cytoplasmic flow known as **cytoplasmic streaming.** Cytoplasmic streaming in plant cells follows a circular path around a central vacuole, and is therefore referred to as **cyclosis** (circling). In such cells the outermost region of cytoplasm, or **ectoplasm,** is relatively nonmotile and gelatinous, while the inner moving region of cytoplasm, or **endoplasm,** is more fluid (Figure 9-42). Parallel rows of microfilaments are situated at the interface between ectoplasm and endoplasm, anchored to the stationary chloroplasts of the ectoplasm (Figure 9-43). An active involvement of these microfilaments in cytoplasmic streaming is suggested by the discovery that disrupting

FIGURE 9-43 Scanning electron micrograph of chloroplasts and aligned actin filaments in a green alga viewed from the side facing the inner moving region of cytoplasm (endoplasm). Courtesy of N. K. Wessells.

the cell's microfilaments with the drug cytochalasin B causes streaming to stop. Although the mechanism by which microfilaments induce streaming is not precisely understood, it has been proposed that these anchored actin filaments might propel a myosinlike protein along their surfaces in the direction of streaming, moving this protein by an ATP-dependent cycle of binding and release analogous to the one occurring in muscle.

Cytoplasmic streaming in animal cells, which lack the rigid cell wall characteristic of plants, differs from cyclosis in that the pressure produced by streaming can induce changes in cell shape that in turn lead to movements of the cell as a whole. Streaming of this type is referred to as **shuttle streaming** because of its tendency to change direction. When focused in a particular direction, shuttle streaming leads to the formation of cytoplasmic projections (pseudopodia) and an associated type of cell locomotion, called amoeboid movement, which will be described later in the chapter.

In addition to cytoplasmic streaming, many cells exhibit a more erratic type of cytoplasmic movement in which cytoplasmic particles or organelles suddenly move several micrometers in a particular direction at a much greater velocity than they were previously moving. Although these **saltatory movements** often occur in areas enriched in microfilaments, the role played by microfilaments in this type of activity is unclear. In order to trigger such movements the microfilaments would have to be anchored to some fixed structure, since it is difficult to envision how microfilaments floating freely in the cytoplasm could produce directed motion. In some cells saltatory movements occur in regions enriched in microtubules and can be inhibited by colchicine, suggesting that microtubules may also be involved. It is thought in such cases that microtubules function as the anchoring structure or as a passive skeleton that guides the direction of particle movement, while the motive force for particle movement emanates from the microfilaments.

Microfilaments and Amoeboid Movement

Cell locomotion can be accomplished by several different mechanisms, depending largely on the cell type involved. We saw earlier in the chapter that some cells are propelled by external motile appendages such as cilia and flagella. The other principal type of locomotion is mediated by bulk movements of the cytoplasm and is referred to as **amoeboid movement** because it was first studied in amoebae and related unicellular eukaryotes. A similar kind of locomotion occurs in some blood cells, embryonic cells, cancer cells, and cells grown in tissue culture. Though differing in detail, amoeboid locomotion in all cell types appears to involve microfilament-mediated movements of the cytoplasm. Because this type of locomotion has been most extensively studied in

amoebae, we shall describe the process in this organism first.

Locomotion in Amoebae It was first proposed more than a hundred years ago that amoeboid movement is driven by contraction of the cytoplasm. This idea was later strengthened by the discovery that amoebae exposed to glycerin, which makes the plasma membrane permeable to solutes present in the external medium, contract when ATP is added. Although such experiments clearly revealed that the mechanical properties of the cytoplasm can be altered by energy sources such as ATP, the question of how such changes lead to an organized movement of the cell has been very difficult to answer.

Some important clues have been provided by studies on the behavior of the cytoplasm in actively migrating amoebae and related protozoa. Locomotion of such organisms has been found to be associated with the formation and retraction of cytoplasmic projections, termed **pseudopodia** (Figure 9-44), which can be long and thin (filopodia), thick and cylindrical (lobopodia), or broad and flat (lamellipodia). Movement occurs in the direction of the advancing pseudopodium, with the rear portion or "tail" of the cell being pulled forward as the cell advances. The cytoplasm located in the cell interior, termed the **endoplasm,** flows from the tail toward the advancing pseudopodium. As it reaches the pseudopodium, the stream of flowing endoplasm is diverted toward the sides of the cell, where it becomes transformed into the more rigid **ectoplasm.** Meanwhile, the

FIGURE 9-44 Light micrograph showing one large and several smaller pseudopodia in the giant amoeba *Chao*. The pseudopod on the right is advancing while the one on the left is retracting. Courtesy of R. D. Allen.

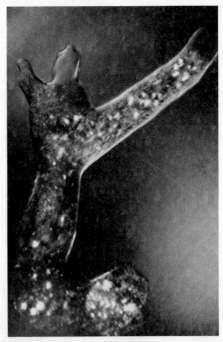

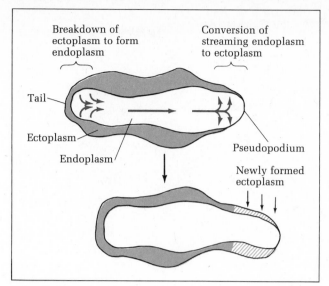

FIGURE 9-45 Pattern of cytoplasmic streaming accompanying locomotion in amoebae.

ectoplasmic wall is being broken down at the rear of the cell to provide new endoplasm for the forward flow. Thus a cyclic interconversion of ectoplasm and endoplasm is established whose net effect is to propel the cell forward (Figure 9-45). This general process of converting cytoplasm from a rigid, gelatinous consistency to a more fluid, relaxed state is referred to as a *gel-sol transition.*

In trying to understand how such a gel-sol mechanism might operate, several questions immediately arise. First of all, where within the cell does the crucial event driving this cycle take place? The first idea to gain prominence was that the driving force occurs at the rear of the cell, generating a positive pressure gradient that propels the endoplasm toward areas of lower pressure at the front of the cell. This hypothesis was generally accepted until the early 1960s, when Robert Allen was led to question this idea by an accident that occurred while he was examining an amoeba contained within a glass capillary tube. Using an unfamiliar microscope, he mistakenly moved the lens the wrong way, breaking the glass tube. To his surprise the amoeba cytoplasm continued to stream, even though the plasma membrane of the cell had been ruptured. According to the positive-pressure theory, the streaming should have stopped because any pressure within the cell would have been released upon disruption of the plasma membrane.

Allen cast further doubt on the validity of the positive-pressure theory when he later demonstrated that application of suction to the rear end of an amoeba with a small micropipette does not prevent the forward flow of cytoplasm toward the advancing pseudopodium. If the flow of endoplasm toward the pseudopodium is caused by a positive-pressure gradient generated at the rear of the cell, then the application of suction to reduce

the pressure should have slowed, stopped, or even reversed the normal direction of streaming. On the basis of these and other observations, Allen concluded that amoeboid movement does not involve propulsion of the endoplasmic stream by pressure generated in the tail.

The most obvious alternative is that the endoplasmic stream is pulled forward by a force generated at the front of the cell. Insight into the way in which such a frontal pull mechanism might operate has emerged from electron microscopic studies carried out on isolated amoeba cytoplasm. Isolated cytoplasm is an attractive model for study because its movements can be regulated by the direct addition of various substances to the incubation medium. In the absence of any added source of energy, the movement of such isolated cytoplasm preparations quickly comes to a halt. If ATP and Ca^{2+} are then added, however, organized cytoplasmic streaming and the formation of pseudopodiumlike structures occur again, even though a plasma membrane is not present. When preparations treated in this way are examined with the electron microscope, the most striking morphological change observed after ATP and Ca^{2+} have been added is the formation of large arrays of actin microfilaments. Such observations have led to the proposal that ATP and Ca^{2+} promote the assembly of microfilament arrays at the front of the cell, triggering a sol-gel transition that pulls the endoplasm forward and converts it to the more rigid ectoplasm. At the same time, depolymerization of the actin network at the rear of the cell has been postulated to trigger a gel-sol transition that converts the ectoplasm back to endoplasm (Figure 9-46).

It is not certain how the preceding changes in the state of actin are regulated, but one possibility is suggested by the existence of the large family of actin-binding proteins that modify the polymerization state and cross-linking of actin molecules (see Table 9-1). Many of these actin-binding proteins are regulated by Ca^{2+}, and

FIGURE 9-46 Model depicting how changes in the polymerization state of actin might account for locomotion in amoebae.

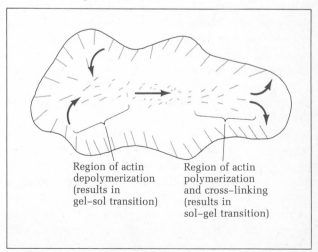

Region of actin depolymerization (results in gel–sol transition)

Region of actin polymerization and cross–linking (results in sol–gel transition)

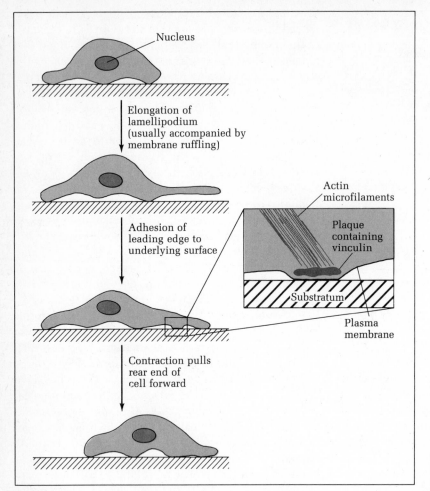

Nucleus

Elongation of
lamellipodium
(usually accompanied by
membrane ruffling)

Adhesion of
leading edge to
underlying surface

Actin
microfilaments

Plaque
containing
vinculin

Substratum

Plasma
membrane

Contraction pulls
rear end of
cell forward

FIGURE 9-47 Schematic representation of the
steps involved in the locomotion of mammalian
cells in tissue culture.

some of them have been shown to have binding sites for ATP as well. Hence the formation of large actin networks at the proposed site of sol-gel transition may be triggered by the appropriate actin-binding proteins in the presence of the proper concentration of ATP and Ca^{2+}.

Locomotion in Other Cell Types Cell locomotion in multicellular organisms differs in several ways from the comparable events in amoebae. In fibroblasts grown in tissue culture, where locomotion has been extensively studied, the overall process can be subdivided into a series of discrete stages (Figure 9-47). First, a number of broad flattened **lamellipodia** begin to protrude from one side of the cell. When one of these extensions adheres to the underlying substratum, it becomes dominant and the cell moves in its direction. The leading edge of this lamellipodium is usually characterized by fluttering or undulating movements of the cell surface referred to as *ruffling*. Finally, the trailing end of the cell is rapidly pulled forward as its attachment to the substratum is broken. When viewed by time-lapse photography, this sequence of events appears as a series of jerky crawling movements oriented in the direction in which the cell is moving.

Two aspects of this overall sequence have received special attention in studies investigating the mechanisms of cell locomotion. The first is the process by which the leading edge of the cell adheres to its underlying substratum. This adhesion occurs in discrete regions, termed **focal contacts,** measuring $1-2\ \mu m$ long and roughly $0.5\ \mu m$ wide. In regions of focal contact, a plaque of electron-dense material enriched in the protein **vinculin** occurs directly beneath the plasma membrane (see Figure 9-47). Emerging from such plaques are parallel bundles of actin-containing microfilaments (stress fibers) that pass toward the nuclear region, where they mesh with a more diffuse microfilament network. The evidence indicating that focal contacts represent regions of firm adhesion between a cell and its underlying substratum is quite good; for example, when the tail end of a cell is pulled forward during cell locomotion, it actually ruptures, leaving behind a small tip of cytoplasm bound to the substratum by focal contacts.

The preceding arrangement has fostered the notion that microfilament bundles attached to regions of focal contact play an active role in propelling cell locomotion, drawing the nuclear region forward toward the leading edge and the tail forward toward the nuclear region. In spite of the simple attractiveness of this theory, several experimental observations suggest it may not be valid. It has been shown, for example, that retraction of the trailing edge is triggered by a process that is not inhibited by lowering the temperature or blocking ATP synthesis. Such observations suggest that instead of being caused by an active contraction, retraction of the tail is more likely due to the passive recoil of a tautly stretched system. The notion that microfilament bundles do not function as little "muscles" propelling the cell forward is further supported by the observation that some of the fastest moving cells contain relatively few such bundles, while immobilized, stretched-out cells often exhibit the most spectacular microfilament arrays.

The second aspect of cell locomotion to be intensively studied in recent years, and the one considered more likely to provide the main driving force for movement, is the protrusion of the lamellipodium at the leading edge of the cell. An involvement of microfilaments in this process of cytoplasmic protrusion is suggested by at least two observations: (1) The cytoplasm at the front of the advancing edge contains a loosely organized network of microfilaments, and (2) disrupting these filaments with the drug cytochalasin B inhibits both ruffling and protrusion of the lamellipodium. Although the microfilament network contains most of the major proteins involved in muscle contraction, there is no compelling evidence that a musclelike contraction actually occurs, nor is it straightforward to envision how a contractile event could cause the forward protrusion of the lamellipodium. For this reason alternative theories need to be considered. We have already discussed the possibility that cytoplasmic movements in amoebae involve local alterations in the physical properties of the cytoplasm (sol-gel transitions) triggered by changes in the polymerization state and cross-linking of actin filaments. A change in the polymerization state of actin has likewise been implicated in the formation of the specialized protrusions that emerge from the head of certain sperm cells when they touch the egg plasma membrane (Chapter 15). Such examples suggest that changes in actin cross-linking and/or polymerization may be as viable a mechanism for inducing cytoplasmic protrusion as the more classical notion of contraction per se.

Although the evidence implicating actin microfilaments in cell locomotion is fairly convincing, other factors may be involved as well. The increase in plasma membrane surface area that accompanies the formation of lamellipodia has led to the suggestion that membrane proliferation also plays an important role. According to this view, development of lamellipodia may be induced at least in part by insertion of new membrane components into the plasma membrane, followed by the flow of cytoplasm into this area. It is worth noting in this context that the previously mentioned ability of cytochalasin B to inhibit cell locomotion is not incompatible with the idea that changes in the plasma membrane play an important role because cytochalasin B is known to bind directly to plasma membrane components as well as to microfilaments. In fact, concentrations of cytochalasin B that are too low to visibly disrupt microfilaments still bind to the plasma membrane and inhibit locomotion, so a direct involvement of the plasma membrane in the locomotion process remains a distinct possibility.

Regulation of Cell Locomotion The direction and velocity of cell locomotion are subject to regulation by both chemical and physical factors. The general term **chemotaxis** is employed to refer to movements guided by chemical substances present in the environment. Migration may be directed either toward (positive chemotaxis) or away from (negative chemotaxis) a particular substance.

One of the best understood examples of chemotaxis occurs in the cellular slime mold, an organism that exists as both free-living individual cells and as multicellular aggregates. In the presence of adequate nutrients, slime mold cells live and function independently. If the food supply becomes depleted, however, some of the cells begin releasing pulsatile waves of a chemical attractant. In several species of cellular slime mold this attractant has been identified as **cyclic AMP,** a molecule that also functions as a chemical messenger in higher organisms. Cyclic AMP binds to the surface of neighboring cells and directs their movement toward the cells that are releasing the cyclic AMP. The net result is the gradual coalescence of cells into a large multicellular aggregate. The mechanism by which cyclic AMP bound to the cell surface determines the direction in which cells migrate is yet to be clearly elucidated.

Chemotaxis also takes place in higher organisms. One of the more striking examples involves the attraction of phagocytic white blood cells (leukocytes) to invading bacteria. Leukocyte chemotaxis is thought to be mediated at least in part by small peptides terminating in *N-formylmethionine*, an amino acid derivative synthesized by bacteria but rarely occurring in animal cells. Analogous to the action of cyclic AMP on slime mold cells, peptides ending in *N*-formylmethionine have been found to bind to leukocyte cell surface receptors. Again, however, is it not clear how the interaction of such a compound with its cell surface receptor ultimately influences the direction of cell locomotion.

In addition to chemical signals, physical factors also influence amoeboid locomotion. In 1954 Michael Abercrombie noted that the movement of cells in tissue culture is inhibited when physical contact is made with adjacent cells. This **contact inhibition of movement** prevents cells from overlapping and piling up on one another, though in cancer cells this control mechanism may be absent (Chapter 20). Physical contact between cells and their underlying substratum likewise influences movement. In general, cells move toward areas in

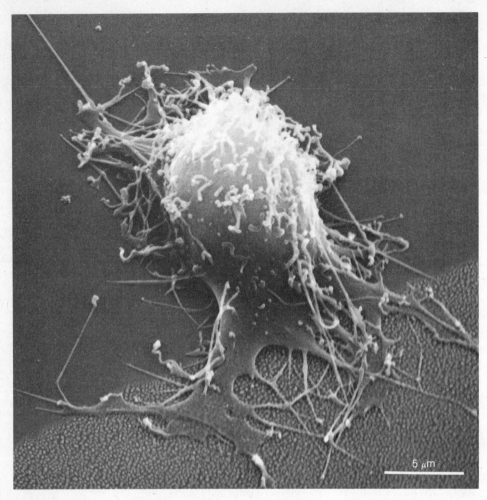

5 µm

FIGURE 9-48 Scanning electron micrograph of a mouse fibroblast showing numerous filopodia extending from the cell. Courtesy of G. Albrecht-Buehler.

which the adhesive attractiveness of the substratum is high. This selective movement is mediated by the formation of long slender filopodia that explore the substratum in the areas immediately surrounding the cell (Figure 9-48). If the filopodia make contact with a favorable surface, lamellipodia are extended in the same direction and locomotion proceeds.

Microfilaments and Cell Division

During cell division, two major types of movement occur: cleavage of the cell in half (cytokinesis) and migration of the two sets of chromosomes to the newly forming cells. Since the topic of cell division will be discussed in Chapter 13, we shall delay a detailed consideration of the role of actin microfilaments in cell cleavage and chromosome movement until that time. To briefly summarize, however, it can be mentioned that actin microfilaments have been implicated in the process of cell cleavage by the discovery that a band of circumferentially arranged microfilaments lies directly beneath the plasma membrane in the region where the cell is being pinched in two. Disruption of this ring of microfilaments with cytochalasin B prevents cell cleavage. The role of actin microfilaments in chromosome movement, if any, has not been as clearly established,

though actin as well as myosin have been identified in the mitotic spindle.

Microfilaments, Cell Shape, and the Cell Surface

Like microtubules, microfilaments influence the development and maintenance of asymmetrical cell shapes. In elongated cells bundles of microfilaments are often observed to be organized in parallel to the direction of cell asymmetry. If such microfilament bundles are disrupted by exposing cells to cytochalasin B, the cells lose their asymmetrical shape and become rounded (Figure 9-49). In addition to influencing cell shape as a whole, microfilaments provide the structural core for cell surface projections such as microvilli and filopodia. Such cell surface projections function both in increasing the surface area of the cell and in mediating overall changes in cell shape and behavior. A vivid illustration of this latter phenomenon occurs when freshly isolated animal cells, which usually assume a spherical shape, are placed in tissue culture. Upon making contact with the surface of the culture vessel, a dramatic outgrowth of microfilament-containing microvilli occurs. These microvilli then elongate into filopodia that gradually spread out, causing the cell to become thinly flattened over the vessel surface (Figure 9-50).

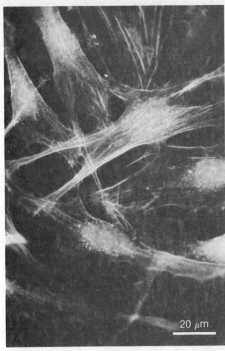

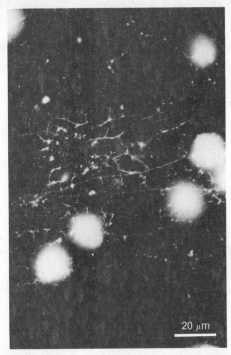

FIGURE 9-49 *(left)* Effects of cytochalasin B on cell morphology. *(Left)* Control fibroblasts spread on glass and stained with fluorescein-labeled anti-actin. *(Right)* Appearance of cells after treatment with cytochalasin B. The bundles of distinct actin filaments have disappeared and the cells have rounded up. Courtesy of R. Norberg.

FIGURE 9-50 *(below)* Scanning electron micrographs showing the flattening and spreading that occurs when fibroblasts are placed in tissue culture. *(Top left)* Spherical cell showing early stages of filopodial outgrowth. *(Top right)* A slightly later stage in which the filopodia are beginning to advance (arrows). *(Bottom left)* Extensive blebbing of the cell surface occurs as flattening begins. *(Bottom right)* In the final stages of flattening, cytoplasm is seen flowing along the filopodia in droplets (arrows). Courtesy of R. Rajaraman.

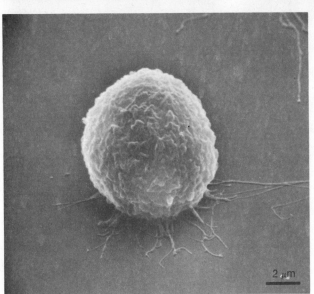

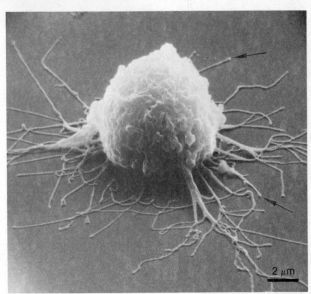

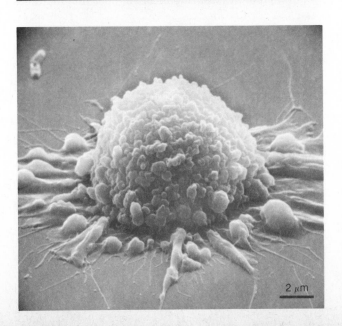

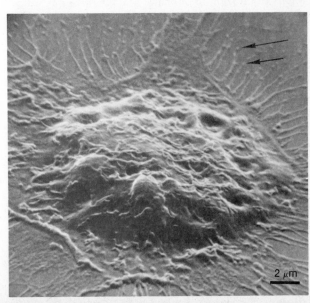

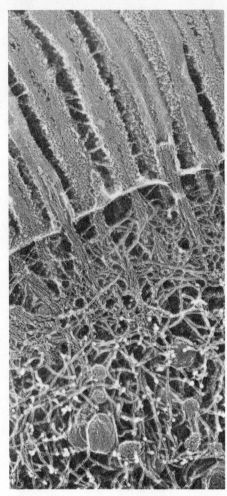

0.2 μm

FIGURE 9-51 Two views of the microvilli that comprise the brush border of intestinal lining cells. *(Left)* Thin-section electron micrograph of the brush border. Each microvillus contains a core of actin filaments that extend below the microvillus into the terminal web region, which contains a meshwork of cytoplasmic microfilaments. Courtesy of M. S. Mooseker. *(Right)* Appearance of the brush border prepared by the quick-freeze, deep-etch technique. After freeze drying, a thin metal replica of the sample is made, which is viewed with the transmission electron microscope. The filament network making up the terminal web shows clearly directly beneath the cell surface. Courtesy of J. Heuser.

The relationship between microfilaments and cell surface projections has been extensively investigated in intestinal lining cells, where the luminal cell surface is covered with a dense accumulation of microvilli, known as a **brush border,** that functions to increase the plasma membrane surface area available for the absorption of nutrients. Each microvillus measures 1–2 μm in length and 0.1 μm in diameter and contains a bundle of several dozen microfilaments oriented parallel to its long axis (Figure 9-51). Staining with heavy meromyosin has revealed that the microfilaments are all oriented with the same polarity. At least two kinds of attachments occur between the microfilament bundles and the plasma membrane: At the tip of each microvillus, the ends of the microfilaments are attached to the overlying membrane through an electron-dense plaque, and along the length of the microvilli, periodic cross-bridges extend laterally from the microfilament surfaces to the plasma membrane (Figure 9-52).

Biochemical analyses of isolated microvilli have revealed that in addition to actin, four major cytoskeletal (detergent-insoluble) proteins are present: **fimbrin, villin, calmodulin,** and a protein of 110,000 molecular weight termed the **110-kDa protein.** Staining cells with fluorescent antibodies specific for these proteins has led to the conclusion that a complex between calmodulin and the 110-kDa protein is responsible for forming the lateral bridges that connect actin to the plasma membrane, while the proteins fimbrin and villin are associated with the core of the microfilament bundle, where they function at least in part to connect the microfilaments to one another.

At the base of each microvillus, the microfilament core extends into a network of perpendicularly arranged nonactin filaments, called the **terminal web** (see Figure 9-51), which consists predominantly of myosin and nonerythroid spectrin. The presence of myosin within the terminal web suggests the possible existence of some type of contractile interaction between the terminal web and the actin microfilaments that emerge from the base of the microvilli. In support of this idea, the terminal web of isolated brush border preparations has been found to contract upon the addition of ATP, causing the associated microvilli to fan out. Although the relevance of such observations to the behavior of intact cells is not clear, it is conceivable that contraction of the terminal web normally functions to elicit movements of the microvilli that increase the overall effec-

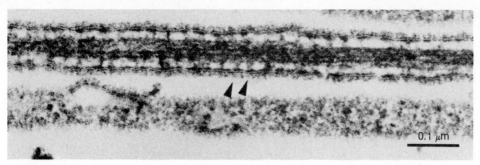

FIGURE 9-52 Longitudinal section through a microvillus of an isolated chicken intestinal cell brush border. The core bundle of microfilaments is attached at numerous points to electron-dense patches on the inner surface of the plasma membrane (arrowheads). Courtesy of D. R. Burgess.

tiveness of the brush border as an absorptive surface.

Regions containing microvilli are not the only area of the cell surface to be closely associated with microfilaments. Earlier in the chapter we discussed the existence of a submembranous network of microtubules and its relationship to the plasma membrane. Microfilaments also constitute part of this submembranous assembly, though the exact relationship between the microtubules and microfilaments is unclear. The notion that microfilaments are components of this network first emerged from studies on the "capping" of lectin receptors caused by exposing cells to low concentrations of concanavalin A. Under such conditions membrane proteins that are initially free to diffuse laterally in the membrane become clustered and immobilized into a structure called a *cap*; these caps, in turn, are closely associated with submembranous microfilament networks. These microfilament networks have been implicated in the capping process by the discovery that cytochalasin B inhibits the formation of membrane caps, although the interpretation of such experiments is complicated by several factors. First, cytochalasin B is known to bind directly to the plasma membrane as well as to actin microfilaments; hence the inhibition of receptor capping observed in the presence of this drug might result from a direct effect on the plasma membrane rather than disruption of microfilaments. The situation is further complicated by the discovery that in some cell types, cytochalasin B induces rather than inhibits the coalesence of membrane receptors. In fibroblasts treated with cytochalasin B, for example, membrane receptors that had previously been distributed randomly over the cell surface become concentrated in one area, and submembranous microfilaments become localized in this same area. Removal of the cytochalasin B reverses the clustering of both the membrane receptors and the actin microfilaments.

The function of membrane-associated actin has perhaps been most clearly elucidated in red blood cells, which contain an extensive submembranous network of spectrin filaments cross-linked by actin molecules (Chapter 4). This filament network is in turn linked to the plasma membrane by another protein, ankyrin. Several lines of evidence suggest that the submembranous actin-spectrin network functions in regulating the movement of membrane proteins. First, treatments that cause spectrin to aggregate into clumps, such as exposing membranes to antibodies against spectrin, have been found to induce aggregation of other membrane proteins, such as glycophorin. In other words, conditions that hinder the mobility of spectrin lead to a secondary restriction of the mobility of other membrane proteins. The notion that the actin-spectrin network influences the mobility of plasma membrane constituents is further supported by experiments performed on membrane vesicles reconstituted from purified phospholipids and band-3 protein, a molecule normally visible as intramembrane particles in freeze-fracture micrographs (see Figure 4-18). In normal red cell membranes these intramembrane particles aggregate when exposed to low pH; in reconstituted membrane vesicles, on the other hand, such aggregation is only observed if spectrin and actin are added along with the band-3 protein. When considered in total, the preceding observations clearly suggest that the actin-spectrin complex functions in the red blood cell to regulate the mobility of other membrane proteins.

Microfilaments and Endocytosis/Exocytosis

Another category of cell surface event in which microfilaments have been implicated consists of the related processes of endocytosis and exocytosis. Of the various types of endocytosis known to occur, both the uptake of particulate matter (phagocytosis) and the uptake of fluid in large vesicles (micropinocytosis) have been found to be inhibited by cytochalasin B. The possibility that these inhibitory effects of cytochalasin B result from a direct action of the drug on the plasma membrane rather than from the disruption of microfilaments has been investigated in experiments employing an alternative form of the drug, called **cytochalasin D.** Unlike cytochalasin B, cytochalasin D disrupts microfilaments without binding to the plasma membrane. The discovery that cytochalasin D also inhibits endocytosis has reinforced the conclusion that microfilaments are in some way involved. The most obvious possibility is that the contraction of submembranous microfilaments during endocytosis plays a role in the invagination and pinching off of plasma membrane vesicles. It has been suggested that this contraction of microfilaments may in turn be induced by calcium ions released from intracellular stores during endocytosis.

Exocytosis (cell secretion), like endocytosis, has been found to be inhibited by cytochalasin B in a variety of cell types. Since calcium ions are known to influence the behavior of actin filaments, the conclusion that microfilament contraction is involved in cell secretion is consistent with the known dependence of secretion on the presence of calcium ions (Chapter 6). This relatively straightforward view is complicated, however, by the discovery that cytochalasin B stimulates exocytosis in certain cases, for example, secretion of lysosomal enzymes by phagocytic cells and histamine by leukocytes. A possible explanation for this apparent discrepancy has emerged from studies of secretory granule discharge from exocrine cells of the pancreas, where low concentrations of cytochalasin B are stimulatory and high concentrations are inhibitory. Low concentrations of cytochalasin B do not disrupt microfilaments in such cells, but they do inhibit active transport by binding directly to the plasma membrane. High concentrations of cytochalasin B, in contrast, do disrupt microfilaments. Hence cytochalasin B appears to *inhibit* exocytosis when microfilaments are affected, and *stimulate* exocytosis when the plasma membrane is its primary target.

Microfilament/Microtubule Interaction

Many of the cellular properties and activities discussed in the present chapter, such as cell shape, movement of intracellular and cell surface components, endocytosis/exocytosis, and cell locomotion involve an interplay between microfilaments and microtubules. In cases where such interactions occur, microtubules tend to play a structural or architectural role while microfilaments provide the motile force. The advantage of this arrangement is that the presence of long, rigid microtubules permits the forces generated by microfilaments to acquire a directionality and long-range organization that would not be possible with microfilaments acting alone.

A classic example of cooperative interactions between microfilaments and microtubules occurs in embryonic nerve cells at the time when they develop long cytoplasmic projections known as *axons* (Chapter 18). At the growing tip of newly forming axons, membrane ruffling and cell locomotion are prominent. Exposing such cells to cytochalasin B inhibits membrane ruffling

and locomotion, which in turn blocks further axon elongation, but the existing axon remains intact. Treatment with colchicine, in contrast, causes the axon to retract and disappear, even though membrane ruffling is unaffected. This suggests that axon elongation is dependent upon the presence of both microfilaments and microtubules. The microfilaments mediate membrane ruffling and cell locomotion, which are essential ingredients of the elongation process. But as elongation proceeds, microtubules assemble into a skeletal framework that supports the growing axon and whose integrity is essential for maintaining the elongated state.

INTERMEDIATE FILAMENTS

The third major class of cytoskeletal element occurring in eukaryotes consists of a group of related structures referred to as **intermediate filaments** because their diameter of 10–15 nm is intermediate between that of the 6-nm microfilaments and the 25-nm microtubules. In contrast to microtubules and microfilaments, whose main protein subunits (tubulin and actin) are virtually identical in all cell types, the intermediate filaments of different cells have distinctly different protein compositions. In recent years it has become apparent that this heterogeneous group of filaments can be subdivided into five major types: (1) the keratin filaments of epithelial cells (cells covering organ and body surfaces); (2) the neurofilaments of nerve cells; (3) the desmin filaments of muscle cells; (4) the glial filaments of glial cells (supporting cells of the nervous system); and (5) the vimentin filaments of mesenchymal tissues (connective tissue, blood cells, etc). In spite of their inherent differences in protein composition (Table 9-2), the various types of intermediate filaments are similar to one another in size and morphology and tend to be the most stable component of the cytoskeleton and the most difficult to solubilize.

The function of intermediate filaments has been most clearly established for the keratin filaments (tonofilaments), which are restricted to the cell layers that

TABLE 9-2

Major types of intermediate filaments

Cell Type	Protein Makeup of Intermediate Filament	Protein Molecular Weight (Daltons)
Epithelial	Keratin (several forms)	45,000–60,000
Neuronal	Neurofilament (3 proteins)	68,000, 145,000, and 220,000
Mesenchymal	Vimentin	57,000
Glial	Glial fibrillary acidic protein	55,000
	Vimentin[a]	57,000
Muscle	Desmin	53,000
	Vimentin[a]	57,000

[a] Sometimes present.

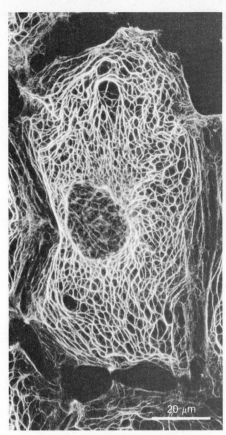

20 μm

FIGURE 9-53 Immunofluorescence microscopy of cultured cells stained with antibodies to bovine epidermal prekeratin, showing the extensive network of keratin filaments. Courtesy of W. W. Franke.

Such observations, in combination with the marked variability in the protein composition of intermediate filaments, suggest that these filaments may function in other unknown ways. A major obstacle to investigating such potential functions is the absence of a well-defined inhibitor (such as colchicine for microtubules and cytochalasin B for microfilaments) that specifically disrupts intermediate filaments.

CYTOSKELETAL INTERCONNECTIONS

Although traditional electron microscopy has provided a vast amount of information concerning subcellular architecture, the general requirement for slicing specimens into thin sections has precluded an in-depth view of the three-dimensional organization of the cell's contents. One way around this obstacle is to utilize the high-voltage electron microscope, an instrument capable of accelerating electrons across a potential of a million volts (an order of magnitude greater than that of the conventional electron microscope). The greatly increased penetration power of the resulting electron beam makes it possible to examine intact cells without the need for prior sectioning.

In the early 1970s Keith Porter and his associates utilized this approach to investigate the interrelationships between the various elements of the cytoskeleton. By growing cells on grids designed for electron microscopic examination and then examining the intact cells directly by high-voltage electron microscopy, these investigators were able to expose a delicate meshwork of interconnected fibers dispersed throughout the cytoplasm and attached to the inner surface of the plasma membrane (Figure 9-54). This latticework, which appeared to contain tapering filaments of varying thickness, was referred to as the *microtrabecular network*. More recent studies by Hans Ris have suggested, however, that the variable thickness of these interconnected fibers is a distortion generated by the fixation conditions employed, and that the network is simply composed of microtubules, microfilaments, and intermediate filaments interconnected by small thin filaments (2–3 nm in diameter) of unknown composition.

Even this more conservative conclusion suggests that the various elements of the cytoskeleton are connected to one another, forming an organizing framework for the cell. Although the functional significance of such an interconnected scaffolding is yet to be ascertained, it clearly has the potential for organizing and facilitating interactions between various cytoplasmic components. It has long been contended, for example, that metabolic pathways occurring in the cytoplasm, such as glycolysis, are too efficient to be based upon a random interaction between enzymes and substrates floating freely in solution. Recent evidence suggests that some metabolic enzymes may be loosely bound to the

cover organ and body surfaces. It was noted in Chapter 5 that the cells that comprise such layers are held together by mechanical junctions referred to as desmosomes. Keratin filaments radiate from the inner surfaces of these desmosomes into the underlying cytoplasm, forming an interconnected filament network that reinforces the mechanical strength of the epithelial cell layer. Large numbers of keratin filaments accumulate in the outer cell layers of the skin (Figure 9-53), eventually becoming cross-linked into a massive network of great strength and rigidity. Even after the cells die, this keratin network persists as a protective covering at the outermost surface of the skin. Similar keratin networks are involved in the formation of hair and nails.

Although the functional significance of keratin filaments is well established, the same cannot be said for the other types of intermediate filaments. Because these filaments often occur in areas of the cell that are subject to mechanical stress, it has been proposed that they too play a tension-bearing role. The intermediate filaments of muscle cells, for example, are thought to maintain the proper alignment of the contractile fibrils. But in some cells intermediate filaments are arranged as irregular networks whose mechanical role, if any, is not obvious.

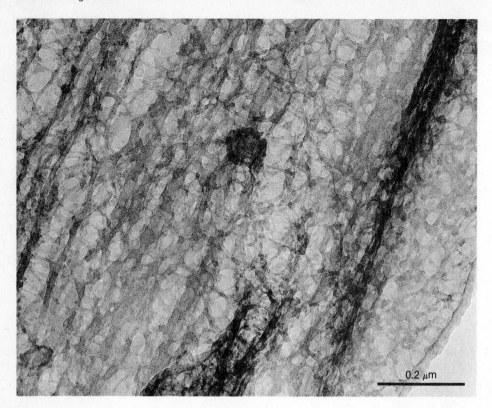

0.2 μm

FIGURE 9-54 High-voltage electron micrograph of a rat cell grown in culture showing an interconnected meshwork of cytoskeletal fibers. Courtesy of K. L. Andersen and K. R. Porter.

cytoskeletal network, perhaps oriented in ways that facilitate the operation of metabolic pathways.

Metabolic enzymes are not the only "free" components of the cell cytoplasm that may in reality be associated with the cytoskeleton. It was stated earlier in the text that eukaryotic ribosomes occur in two forms: attached to the endoplasmic reticulum and "free" in the cytoplasm. Careful examination of high-voltage electron micrographs suggests, however, that many of the cell's free ribosomes may actually be bound to the cytoskeleton. The functional significance of these bound ribosomes has been investigated in the laboratory of Sheldon Penman, where gentle detergent extraction techniques were developed for removing membranes and soluble proteins without destroying the cytoskeleton. Analysis of the resulting fractions has revealed that the ribosomes present in the soluble fraction represent single ribosomes that are not active in protein synthesis. The ribosomes associated with the cytoskeleton, on the other hand, occur in the form of multiple-ribosome aggregates, or polysomes, which are associated with messenger RNA and are active in the process of protein synthesis. Such observations suggest that the free cytoplasmic ribosomes are in reality attached to the cytoskeleton when actively involved in protein synthesis. Significant support for this conclusion has come from studies carried out on cells infected with poliovirus, an event that causes the cell to stop synthesizing its own proteins and to begin synthesizing viral protein. In such infected cells the host cell messenger RNA normally

attached to the polysomes associated with the cytoskeleton is replaced by poliovirus messenger RNA.

Another striking discovery that has emanated from studies on detergent-extracted cells concerns the relationship between the cytoskeleton and the plasma membrane. Since detergent treatment solubilizes membrane lipids, it might be expected to cause the release of membrane proteins as well. However, electron microscopic analysis has revealed the presence of a **surface lamina** at the outer boundary of detergent-extracted cells that, according to biochemical analyses, retains many plasma membrane proteins (Figure 9-55). Moreover, scanning electron micrographs of normal and detergent-extracted cells reveal a strikingly similar appearance (Figure 9-56), indicating the extent to which the surface lamina and the underlying cytoskeleton form an interconnected network that holds the cell together.

BACTERIAL FLAGELLA

In contrast to eukaryotes, bacteria do not contain a cytoskeletal network of microtubules, microfilaments, and related elements. Bacteria do share with eukaryotes, however, the presence of cell surface appendages whose movements are capable of propelling the cell as a whole. Though referred to as **flagella,** these motile appendages do not resemble eukaryotic flagella in chemical composition, structure, or mechanism of action. The confusion caused by using the same term *flagella* to refer to two such different organelles is unfortunate.

Bacterial flagella are more than an order of magnitude smaller than those of eukaryotic cells, measuring 12–20 nm in diameter versus 500 nm for typical eukaryotic flagella. In spite of its small size, however, the bacterial flagellum is a complex structure; at least 35 different genes are required for its proper assembly and functioning, and another 20 or more are involved in its modulation by sensory stimuli. Three distinct regions of the bacterial flagellum can be distinguished: a long helical **filament,** a short **hook** located between the filament and the cell surface, and a **basal body** embedded within the cell envelope (Figure 9-57). The plasma membrane stops at the basal body, rather than enveloping the flagellum as it does in eukaryotic cells. The basal body, which bears no resemblance to the similarly named structure of eukaryotes, consists of a central **rod** surrounded by a series of **rings.** In gram-negative bacteria four rings are present, while in gram-positive organisms there are only two (Figure 9-58). This difference occurs because the two outer rings are embedded within the outer lipopolysaccharide of the envelope, which is present in gram-negative but not gram-positive bacteria. The two inner rings, on the other hand, are embedded within the plasma membrane itself and are therefore present in both gram-negative and gram-positive organisms.

The major building block of the bacterial flagellum is **flagellin,** a protein whose monomers are packed to-

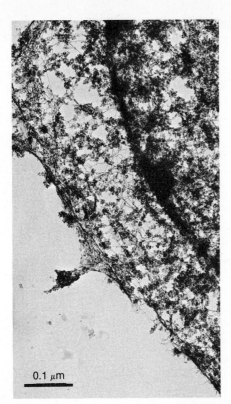

FIGURE 9-55 Thin-section electron micrograph of a detergent-extracted HeLa cell. Note that in spite of the removal of the plasma membrane by the detergent treatment, a distinct outer boundary termed the surface lamina remains at the cell surface. Courtesy of S. Penman.

FIGURE 9-56 Scanning electron micrograph of normal and detergent-extracted myotube cells. *(Left)* Normal cell. *(Right)* Detergent-extracted cell. Note that in spite of the removal of the plasma membrane by the detergent treatment, the surface lamina and underlying cytoskeleton maintain the cell's normal shape and general appearance. Courtesy of S. Penman.

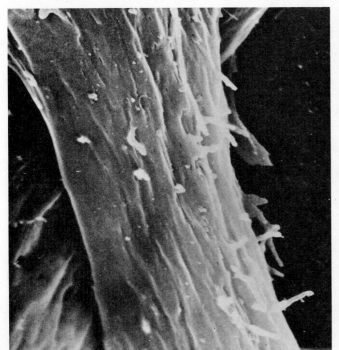
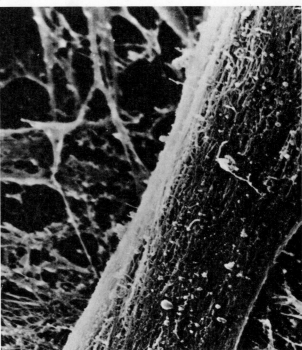

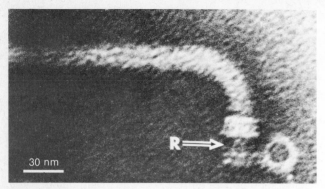

FIGURE 9-57 High-power electron micrograph of a negatively stained bacterial flagellum showing the filament, the curved hook, and the basal body. The rod connecting the inner and outer rings of the basal body is labeled R. Courtesy of J. Adler.

gether with helical geometry to form the flagellar filament (Figure 9-59). When isolated from the cell, flagellar filaments can be induced to undergo reversible cycles of breakdown and repolymerization that are analogous in a superficial sense to the depolymerization/polymerization cycles observed with eukaryotic microtubules and microfilaments. In the case of bacterial flagella, warming promotes depolymerization of filaments into flagellin monomers while cooling triggers their reassembly into filaments, provided small fragments of intact flagella are present to serve as nucleation sites for assembly.

Although a similar polymerization mechanism is believed to underlie the assembly of the flagellar filament by intact bacterial cells, the question arises as to whether flagellin monomers are added to the growing flagellum at the base, at its tip, or throughout its length. In order to distinguish between these alternatives, Tetsuo Iino performed a series of experiments in which bacterial cells were grown in the presence of *fluorophenylalanine*, an amino acid analog that causes the formation of abnormally curly flagella. When cells with partially formed flagella were switched to a growth medium containing fluorophenylalanine, the flagella developed curly tips, indicating that growth must be occurring at the tip (Figure 9-60). Subsequent autoradiographic experiments employing radioactive amino acids confirmed that flagellar growth occurs by addition of newly synthesized flagellin monomers to the tip of the organelle.

A perplexing issue raised by the above conclusion concerns the question of how flagellin monomers reach the flagellar tip prior to incorporation into the elongating end. It should be recalled in this context that the plasma membrane does not envelop the bacterial flagellum, leaving the flagellar tip directly exposed to the external medium. The possibility that flagellin is secreted into the external environment prior to incorporation into the flagellar tip has been tested by growing two strains of bacteria, characterized by different types of flagella, in the same culture flask. If flagellin were secreted into the external medium where it could freely exchange between the two bacterial strains, one might expect the flagella of each strain to acquire characteristics of the other. Such an exchange does not occur, however, indicating that each cell transports flagellin

FIGURE 9-58 Organization of the basal region of the flagellum in gram-negative and gram-positive bacteria. The arrows indicate the direction in which flagella rotate to produce straight swimming. Rotation in the opposite direction produces tumbling.

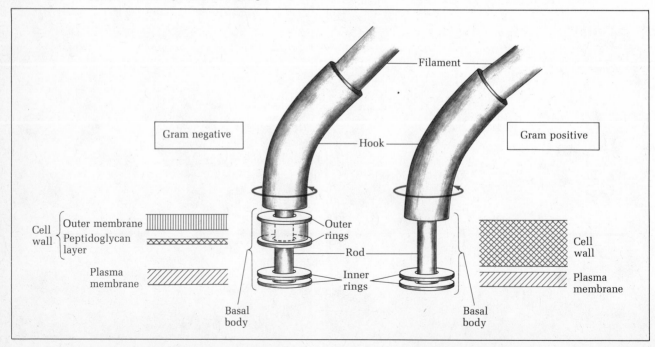

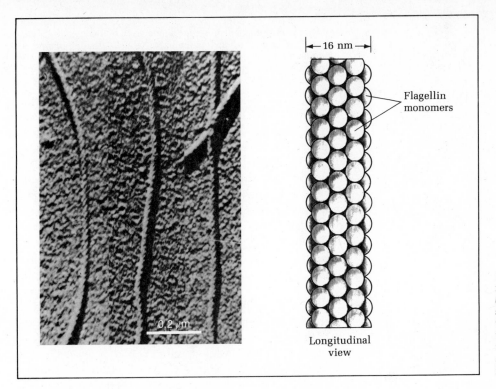

← 16 nm →

Flagellin
monomers

Longitudinal
view

FIGURE 9-59 Substructural organization of the bacterial flagellum. *(Left)* Electron micrograph of a shadowed flagellum isolated from the bacterium *Bordetella,* showing the helical packing of flagellin subunits. *(Right)* Model depicting the arrangement of flagellin subunits in this type of flagellum.

monomers directly to the tip of its own flagella. The mechanism by which such transport occurs, however, remains to be elucidated.

The movement of bacterial flagella is quite different from that of eukaryotic flagella. When viewed by high-speed cinematography, bacterial flagella appear to be rotating in a screwlike fashion. There are two possible explanations for such an observation. One possibility is that the flagellum is a rigid helix rotated by a circular movement of its base. Alternatively, the propagation of helical waves along the length of the filament might cause it to bend in such a fashion that it appears to be rotating, when in fact it is only bending. In order to distinguish between these possibilities, Michael Silverman and Melvin Simon designed an experiment in which live bacteria were placed on a glass slide to which antibodies against flagellin had previously been bound.

FIGURE 9-60 Summary of the results obtained from the experiments in which bacteria were grown in the presence of fluorophenylalanine, which induces the formation of "curly" flagella. The results are compatible with the interpretation that flagellar growth occurs at the tip rather than the base.

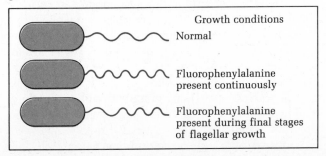

Growth conditions

Normal

Fluorophenylalanine
present continuously

Fluorophenylalanine
present during final stages
of flagellar growth

The presence of the antibodies caused the bacteria to become attached to the slide via their flagella. When observed under the light microscope, such bacteria could be seen to be spinning in circles (Figure 9-61), indicating that the flagellar motion is based upon rotation of the flagella relative to the cell body rather than waves of helical bending.

The realization that the bacterial flagellum behaves like a rigid rod rotated by a circular motion at its base suggests that the forces that drive the movement of this appendage are exerted at the basal region of the flagellum where it lies buried within the plasma membrane. This arrangement is quite different from that occurring in eukaryotic cells, where movement is driven by the interaction of ATP with dynein arms located along the entire length of the flagellum. This difference may explain why eukaryotic flagella are membrane enclosed and bacterial flagella are not. In eukaryotic cells the membrane enclosure prevents ATP from escaping into the environment, while in bacterial cells such an enclosure is not required because movement is driven at the flagellar base rather than along its length.

The most logical candidate for the "motor" that drives the rotation of bacterial flagella is the set of inner rings buried within the plasma membrane of both gram-negative and gram-positive bacteria. The outer rings, which are restricted to gram-negative bacteria, appear to function only as a nonrotating seal through which the central rod of the rotating flagellum passes (see Figure 9-58). Investigations into the mechanism that drives the flagellar motor have revealed that ATP, the general

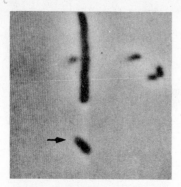

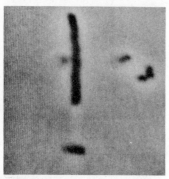

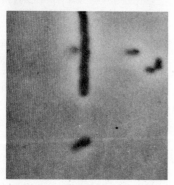

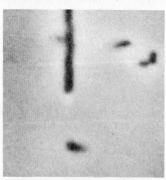

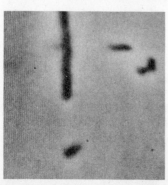

FIGURE 9-61 Results of an experiment in which a bacterial cell (arrow) was attached to a microscope slide by its flagellum. The sequence of pictures shows the bacterial cell spinning in circles, indicating that flagellar motion is based upon rotation of the flagellum relative to the cell body. Courtesy of M. Simon.

energy source for motility in eukaryotes, is not directly involved. Such a conclusion was first suggested by the discovery that bacterial cells in which oxidative phosphorylation is blocked, either by chemical inhibitors or the presence of a specific mutation, can still make ATP by glycolysis but cannot move their flagella. Such observations suggest that an intermediate created during oxidative phosphorylation is the direct energy source for flagellar motility. It was subsequently shown that flagellar movement can be induced by changing the proton concentration of the external medium, pointing to the proton gradient as the most likely candidate. In further studies it was demonstrated that the velocity of flagellar rotation is directly related to the magnitude of the electrochemical gradient, indicating that each turn of the motor requires the flow of a fixed number of protons across the plasma membrane.

The behavior of bacterial flagella is complicated by the fact that the flagellar motor can rotate in either the clockwise or counterclockwise direction, and that the direction of rotation profoundly influences the overall movement of the cell. In the absence of attracting or repelling substances in the environment, bacteria move about in a random pattern in which short periods of swimming in smooth straight lines are interrupted by brief episodes of thrashing around ("tumbling") that lead to random changes in direction. Careful observation of mutant bacteria that either never tumble or always tumble has led to the discovery that swimming in a straight line is associated with counterclockwise rotation of flagella, while tumbling is produced by clockwise rotation. This difference occurs because counterclockwise rotation facilitates the organization of individual flagella into coordinated bundles whose individual members move in unison to propel the cell. Clockwise rotation, on the other hand, causes these organized bundles to fly apart into individual flagella whose uncoordinated movements give rise to random tumbling.

The direction in which bacteria move has long been known to be influenced by chemical substances present in the external environment, a phenomenon termed **chemotaxis.** Recent observations suggest that chemical attractants function by suppressing clockwise rotation and its associated tumbling movements. Hence when bacteria move toward an area where the concentration of a chemical attractant is higher, their swimming is interrupted less and less by random tumbling and they tend to continue moving toward the attractant. The opposite occurs with repellants, which promote clockwise rotation of flagella and therefore enhance tumbling. The presence of chemical substances in the environment is detected by several dozen types of cell surface receptors, each specific for a particular substance. Although it might seem that a cell's survival should require the ability to sense the presence of more than a few dozen compounds, it would be superfluous to have receptors for every compound that might be useful or toxic to the cell. For example, environments that contain decaying foodstuffs would contain all the amino acids normally generated by the degradation of protein molecules, so the presence of receptors for only a few different amino acids would be adequate for sensing the presence of such an area.

Once a chemical substance present in the medium has bound to its cell surface receptor, this binding must in turn influence flagellar rotation. Although the steps in transferring this signal are yet to be completely elucidated, methylation of membrane proteins appears to be an essential stage in the process. This conclusion is based upon the discovery that bacteria in which protein methylation has been blocked by an appropriate inhibitor or genetic mutation can only swim in straight lines. In contrast, bacteria defective in the enzyme that removes protein methyl groups are only capable of tumbling. It has therefore been suggested that binding of chemical attractants to the cell surface promotes methylation of membrane proteins, while binding of chemical repellants promotes removal of protein methyl groups. At least three different membrane proteins have been found to undergo such methylation reactions, but the mechanism by which these methylated proteins in turn regulate the direction of flagellar rotation is not well understood. It is interesting to note, however, that the direction of flagellar rotation can be changed experimentally by altering the magnitude of the proton gradient, suggesting that the ultimate effect of protein methylation may be exerted at this level.

SUMMARY

The cytoplasm of eukaryotic cells contains an interconnected network of microtubules, microfilaments, and intermediate filaments that together constitute the cytoskeleton. This cytoskeletal framework functions both as a scaffolding that supports and organizes the cell interior and as a source of motile elements that permit movement of intracellular components as well as the cell as a whole.

Microtubules are long rigid cylinders measuring 25 nm in diameter that occur both dispersed through the cytoplasm and as organized arrays localized within cilia, flagella, centrioles, basal bodies, and the mitotic spindle. The diverse activities in which microtubules are involved are based upon two underlying properties: (1) their long rigid shape, which facilitates structural roles such as anchoring, guiding, orienting, and supporting, and (2) their ability to generate movement. The major component of the microtubule is tubulin, a protein dimer that reversibly assembles into microtubules under cell-free conditions. Assembly is stimulated by microtubule-associated proteins, while depolymerization into free tubulin is promoted by cooling, high pressure, calcium ions (in the presence of calmodulin), and drugs such as colchicine and vinblastine. Within the intact cell the locations at which microtubules form is determined by the presence of microtubule-organizing centers such as basal bodies, chromosomal kinetochores, and centrosomes.

Among the various microtubule-containing structures that occur within eukaryotic cells, cilia and flagella are perhaps the best understood. Though differing in size, number, and overall pattern of movement, cilia and flagella are closely related organelles whose movements function to propel fluids across the cell surface or to move the cell as a whole. Both of these membrane-enclosed appendages contain a highly organized backbone of microtubules, termed an axoneme, in which nine doublet microtubules surround a pair of central microtubules. This orderly $9 + 2$ arrangement of the axoneme is maintained by an intricate system of connections among its various elements. Formation of cilia and flagella is initiated by contact between centrioles and the plasma membrane. The centriole, which contains nine sets of triplet microtubules, acts as a nucleation site for the assembly of the axoneme and comes to be known as a basal body. Centrioles are produced in three different ways: by assembly of individual centrioles at right angles to existing centrioles, by simultaneous formation of multiple centrioles adjacent to masses of granular material called deuterosomes, and by *de novo* formation in cells completely lacking centrioles.

Cilia and flagella are inherently motile structures that even continue to beat after being isolated from cells, provided they are supplied with ATP as an energy source. Electron microscopic observations have revealed that the microtubule doublets slide past one another during ciliary or flagellar bending. Because of the constraints placed on the sliding process by the presence of the radial spokes and the nexin links, the forces generated during sliding are converted to bending. Microtubule sliding is driven by interactions between the dynein arms of one microtubule doublet and the B-tubule of the adjacent doublet, powered by the energy released during the dynein-catalyzed hydrolysis of ATP.

A variety of sense organs contain nonmotile cilia whose axonemes lack dynein arms and central microtubules. Axonemes with structures different from those typically encountered in cilia and flagella also occur in the axopods, axostyles, cortical fibers, and cytopharyngeal basket of unicellular eukaryotes.

In contrast to axonemal structures, which generally function to move the cell as a whole or fluids surrounding the cell, cytoplasmic microtubules are involved in intracellular movements and interactions, and in determining cell shape. One of the most widely used experimental approaches for implicating microtubules in such activities is to determine whether disruption of microtubules with a drug such as colchicine results in an inhibition of the activity in question. In this way it has been discovered that microtubules play a role in chromosome movement (to be discussed in Chapter 13), the migration of vesicles and granules during cell secretion, mobility of membrane proteins during phagocytosis, propagating signals to and from the cell surface, cell wall formation in plants, and in the establishment and maintenance of asymmetrical cell shapes.

Microfilaments are actin-containing fibrils 6 nm in diameter that occur in massive parallel bundles called stress fibers, in organized arrays within microvilli, and in looser networks localized beneath the plasma membrane, adjacent to the nucleus, and dispersed through the cytoplasm. Isolated actin molecules reversibly assemble into microfilaments under cell-free conditions. These microfilaments exhibit an inherent polarity, with assembly occurring preferentially at one end and disassembly at the other. The actin of intact cells occurs in a variety of states, including free monomers, individual microfilaments, regularly cross-linked microfilament bundles, and less regularly cross-linked filament networks that behave like gels. The interconversion between these various states is controlled by three classes of actin-binding proteins: length-regulating proteins, depolymerizing proteins, and cross-linking proteins. The interaction of many of these proteins with actin is subject to regulation by calcium ions.

Experiments in which microfilament polymerization is inhibited by exposing cells to drugs such as cytochalasin B have implicated microfilaments in various types of cellular movements, including cytoplasmic streaming, amoeboid locomotion, membrane ruffling, cytokinesis (to be discussed in Chapter 13), diffusion of cell surface receptors, and endocytosis/exocytosis. Microfilaments also play a structural role in the maintenance of cell shape and in supporting cell surface projections such as microvilli and filopodia. Some of these functions involve an interplay between microfilaments and microtubules, with microtubules playing a structural or architectural role and microfilaments generating the motile force. The advantage of this arrangement is that the long, rigid microtubules permit the motile forces generated by microfilaments to acquire a directionality and long-range organization that would not be possible with microfilaments acting alone.

The third class of cytoskeletal element occurring in eukaryotic cells is a group of related structures termed intermediate filaments because their 10- to 15-nm diameter is intermediate in size between the 6-nm microfilaments and the 25-nm microtubules. Intermediate filaments can be subdivided into five classes of differing cell origin and protein composition: (1) the keratin filaments of epithelial cells, (2) the neurofilaments of nerve cells, (3) the desmin filaments of muscle cells, (4) the glial filaments of glial cells, and (5) the vimentin filaments of mesenchymal tissues such as connective tissue and blood cells. The function of intermediate filaments has been most clearly established for keratin filaments, which reinforce the mechanical strength of epithelial cell layers. Although it has been proposed that the other types of intermediate filaments play a tension-bearing role as well, the existence of alternative functions has not been ruled out.

Recent studies utilizing high-voltage electron microscopy to investigate the three-dimensional organization of the cell have suggested that microtubules, microfilaments, and intermediate filaments are connected to one another by small thin filaments, forming an interconnected cytoskeletal framework for the cell. In addition to providing a mechanism for facilitating interactions between the various cytoskeletal components, this interconnected network has been proposed to aid in the organization of metabolic pathways and to provide a site for the synthesis of protein molecules by ribosomes that are not associated with the endoplasmic reticulum.

Although bacteria do not contain a cytoskeletal network of microtubules, microfilaments, and related elements, they do share with eukaryotes the presence of flagella. The bacterial flagellum differs from eukaryotic cilia and flagella in several important ways: It is more than an order of magnitude smaller in diameter; it is constructed from flagellin rather than tubulin; it is not membrane enclosed; its movement is based upon a circular motion at its base rather than bending along its length; and the energy source that drives its movement is a proton gradient rather than ATP. Counterclockwise rotation of bacterial flagella facilitates the organization of individual flagella into coordinated bundles whose movement produces smooth, straight swimming. Clockwise rotation, on the other hand, causes these bundles to fly apart into individual flagella whose uncoordinated movements give rise to random tumbling. The ability of environmental chemicals to influence the direction of cell movement (chemotaxis) appears to be mediated by effects on protein methylation. The binding of chemical attractants to cell surface receptors triggers methylation of membrane proteins and a subsequent increase in counterclockwise flagellar rotation and smooth swimming. Binding of repellants to the cell surface, on the other hand, is believed to promote removal of methyl groups and a subsequent increase in clockwise flagellar rotation and tumbling.

SUGGESTED READINGS

Books

Amos, W. B., and J. G. Duckett (1982). *Prokaryotic and Eukaryotic Flagella*, Cambridge University Press, New York.

Borisy, G. G., D. W. Cleveland, and D. B. Murphy (1984). *Molecular Biology of the Cytoskeleton*, Cold Spring Harbor Laboratory, Cold Spring Harbor, NY.

Dustin, P. (1984). *Microtubules*, 2nd Ed., Springer-Verlag, New York.

Fulton, A. B. (1984). *The Cytoskeleton. Cellular Architecture and Choreography*, Chapman and Hall, London.

Ishikawa, S., S. Hatano, and H. Sato, eds. (1986). *Cell Motility, Mechanism, and Regulation*, Alan R. Liss, New York.

Lackie, J. M. (1986). *Cell Movement and Cell Behavior*, Allen & Unwin, Winchester, MA.

Lloyd, C. W., ed. (1982). *The Cytoskeleton and Plant Growth*, Academic Press, New York.

Organization of the Cytoplasm (1982). Proceedings of a symposium published as Cold Spring Harbor Symp. Quant. Biol., Vol. 46, Cold Spring Harbor Laboratory, Cold Spring Harbor, NY.

Schliwa, M. (1986). *The Cytoskeleton: An Introductory Survey*, Cell Biology Monographs, Vol. 13, Springer-Verlag, New York.

Shay, J. W., ed. (1986). *Cell and Molecular Biology of the Cytoskeleton*, Plenum Press, New York.

Sheterline, P. (1983). *Mechanisms of Cell Motility*, Academic Press, New York.

Soifer, D., ed. (1986). *Dynamic Aspects of Microtubule Biology*, Vol. 466, Ann. N.Y. Acad. Sci., New York.

Traub, P. (1985). *Intermediate Filaments*, Springer-Verlag, New York.

Trinkaus, J. P. (1984). *Cells into Organs. The Forces that Shape the Embryo*, Prentice-Hall, Englewood Cliffs, NJ.

Wang, E., D. Fischman, R. K. H. Liem, and T.-T. Sun (1985). *Intermediate Filaments*, Vol. 455, Ann. N.Y. Acad. Sci., New York.

Wheatley, D. N. (1982). *The Centriole: A Central Enigma of Cell Biology*, Elsevier, New York.

Articles

Allen, R. D. (1981). Motility, *J. Cell Biol.* 91:148s–155s.

———(1987). The microtubule as an intracellular engine, *Sci. Amer.* 256(Feb.):42–49.

Boyd, A., and M. Simon (1982). Bacterial chemotaxis, *Ann. Rev. Physiol.* 44:501–517.

Brinkley, B. R. (1985). Microtubule organizing centers, *Ann. Rev. Cell Biol.* 1:145–172.

Geiger, B. (1985). Microfilament–membrane interactions, *Trends Biochem. Sci.* 10:456–461.

Gibbons, I. R. (1981). Cilia and flagella of eukaryotes, *J. Cell Biol.* 91:107s–124s.

Goodenough, U. W., and J. E. Heuser (1982). Substructure of the outer dynein arm, *J. Cell Biol.* 95:798–815.

Haimo, L. T., and J. L. Rosenbaum (1981). Cilia, flagella and microtubules, *J. Cell Biol.* 91:125s–130s.

Macnab, R. M. (1984). The bacterial flagellar motor, *Trends Biochem. Sci.* 9:185–188.

Margolis, R. L., and L. Wilson (1981). Microtubule treadmills—possible molecular machinery, *Nature* 293:705–711.

Mitchison, T., and M. Kirschner (1984). Microtubule assembly mediated by isolated centrosomes, *and* Dynamic instability of microtubule growth, *Nature* 312:232–236 and 237–242.

Mooseker, M. S. (1985). Organization, chemistry, and assembly of the cytoskeletal apparatus of the intestinal brush border, *Ann. Rev. Cell Biol.* 1:209–242.

Oster, G. F. (1984). On the crawling of cells, *J. Embryol. Exp. Morphol.* 83 (suppl):329–364.

Payne, M. R., and S. E. Rudnick (1984). Tropomyosin as a modulator of microfilaments, *Trends Biochem. Sci.* 9:361–363.

Pollard, T. D., and S. W. Craig (1982). Mechanism of actin polymerization, *Trends Biochem. Sci.* 7:55–58.

Porter, K. R., and J. B. Tucker (1981). The ground substance of the living cell, *Sci. Amer.* 244 (March):57–67.

Ris, H. (1985). The cytoplasmic filament system in critical point-dried whole mounts and plastic-embedded sections, *J. Cell Biol.* 100:1474–1487.

Weber, K., and M. Osborn (1985). The molecules of the cell matrix, *Sci. Amer.* 253 (Oct.):110–120.

Wiche, G. (1985). High-molecular weight microtubule associated proteins (MAPS): a ubiquitous family of cytoskeletal connecting links, *Trends Biochem. Sci.* 10:67–70.

CHAPTER

10

Information Flow in the Cell: General Principles

Virtually every activity that takes place in living cells is made possible in one way or another by information encoded within molecules of DNA. Although our current understanding of the role played by DNA is taken largely for granted, achieving this understanding required almost a hundred years of experimentation and the independent contributions of hundreds of different investigators. In this chapter we shall summarize both the basic principles that govern DNA action and the major lines of experimentation and theory that led to the formulation of these general principles. In the process it will become apparent that scientific progress is not always straightforward. Biologists' views concerning the significance and organization of DNA molecules have undergone many shifts over the years and for long periods of time many biologists even argued that DNA could not possibly store and transmit genetic information.

One of the most important lessons to be learned from examining the experimental history underlying our current ideas concerning DNA is that scientific data can be easily misinterpreted, and that once incorrect ideas have become established, they can be very difficult to dislodge. Scientists must therefore always be on guard against the view that a particular concept, no matter how widely accepted, is a "fact." Science is not a collection of facts, but an ongoing intellectual process involving hypotheses, experimentation, formulation of models, and modification of these hypotheses and models as new information becomes available. In this and the following chapter, we shall see several instances in which apparently well-established concepts underwent complete reversal or major revision as the result of an unexpected new experimental finding. It therefore follows that some of our present ideas, no matter how firmly entrenched, may be subject to a similar upheaval in the future.

IDENTIFICATION OF DNA AS THE GENETIC MATERIAL

Discovery of DNA and Early Views Concerning Its Significance

Every time an organism reproduces, the instructions needed to guide the development of a new organism must be transmitted from parent to offspring. Prior to the late nineteenth century most biologists believed that transmission of these instructions involves a complete "blending" of the traits of the two parents, and that the characteristics of the individual parents disappear in the process. Such a theory made experimentation on the chemical basis of inheritance virtually impossible since it is difficult to carry out a systematic study of traits that are continually disappearing. This viewpoint changed dramatically, however, as a result of experiments carried out in the 1860s by the Austrian monk, Gregor Mendel. These experiments, in which Mendel quantitated the frequency of inheritance of several traits in garden peas, are thoroughly covered in textbooks of introductory biology and genetics and so will not be described in any detail here. For our purposes it is simply necessary to point out that Mendel's data led him to conclude that hereditary information is transmitted in the form of distinct entities, later termed *genes*, which do not blend together and disappear but maintain their unique identities over many generations.

Although the chemical makeup of genes was not investigated by Mendel himself, the essential molecular component of these hereditary units was unknowingly identified a few years later by a Swiss physician, Johann Friedrich Miescher. Miescher was one of the first biologists to use the tools of biochemistry, rather than microscopy, to study the molecular organization of cells. Since Miescher was especially interested in the chemical composition of the nucleus, his first major objective was to develop a procedure for isolating and purifying nuclei from intact cells. The procedure he developed, which involved incubating human white blood cells with proteolytic enzymes to destroy the cytoplasm, represented an important milestone in the field of cell biology because it was the first time any subcellular organelle had been successfully isolated in pure form. Miescher subsequently discovered that exposing these isolated nuclei to dilute sodium hydroxide results in extraction of a chemical substance, enriched in phosphorus, that had never before been identified. Because of its localization in the nucleus it was termed *nuclein*, but when later analysis proved this material to be highly acidic, it was renamed **nucleic acid.**

Miescher initially believed that nucleic acids function in the transmission of hereditary information, a conclusion based largely on the discovery that nucleic acid accounts for the bulk of the mass of sperm cells. This notion was later rejected, however, because of data that suggested that the amount of nucleic acid in a hen's egg is hundreds of times greater than that in the sperm cell that fertilizes it. Since both parents must contribute roughly equal amounts of hereditary information to the offspring, it seemed to Miescher that nucleic acid could not be the responsible agent. This erroneous conclusion occurred because of an extremely unfortunate coincidence. The criterion Miescher utilized to identify and quantitate nucleic acid was its high content of phosphorus. Unfortunately hen's eggs contain a protein, called *phosvitin*, that is exceptional in that it too contains large amounts of phosphorus. Of the thousands of proteins that have been identified and characterized in the years since Miescher's studies, phosvitin is the only one containing so much phosphorus that it could be readily mistaken for a nucleic acid. The presence of phosvitin caused Miescher to inadvertently overestimate the nucleic acid content of the hen's egg, and hence to conclude that nucleic acid cannot be involved in the transmission of hereditary information.

Although Miescher was led astray concerning the significance of the substance he had discovered, other biologists were soon back on the right track. In the early 1880s a botanist named Eduard Zacharias reported that extracting nucleic acids from cells also causes the staining of the chromosomes to disappear. Zacharias concluded on this basis that chromosomes contain nucleic acids. It was known at the time that chromosomes are duplicated prior to cell division, and that each of the two duplicate sets of chromosomes is passed on to one of the newly forming cells. This observation suggested that chromosomes transmit hereditary information at the time of cell division. When Zacharias discovered that chromosomes contain nucleic acids, it seemed logical to infer that nucleic acids are involved in the transfer of hereditary information. Within a few years this viewpoint came to be widely accepted, as is evidenced by the following statement written in the most prominent cell biology textbook of the era:

> It is the remarkable substance, nuclein [nucleic acid], which is almost certainly identical with chromatin. The most essential material handed on by the mother cell to its progeny is the chromatin, and this substance therefore has a special significance in inheritance. (From E. B. Wilson, The Cell in Development and Inheritance, 1896, pp. 332, 352.)

This statement makes it clear that by the year 1896, biologists had reached a conclusion that is in essence identical to our current views concerning the role of DNA in heredity.

Unfortunately, this view was discredited a few years later when it was discovered that the staining intensity of nuclear DNA fluctuates according to the physiological state of the cell. Such results were in direct contrast to the expected constancy in the amount of genetic information and soon led to a repudiation of the idea that nucleic acids carry genetic information. In the following decades attention turned away from the study of nucleic acids, and most biologists gradually came to believe that genes were made of protein molecules.

Fortunately, the prominent chemist Phoebus Levene and a small number of other investigators retained an interest in nucleic acids and proceeded to investigate their structural organization. As a result of the combined efforts of these individuals it was known by the mid-1930s that the fundamental building block of nucleic acids is the **nucleotide,** a chemical unit containing a five-carbon sugar joined to a phosphate group and a nitrogenous base. Adjacent nucleotides are linked together by a *phosphodiester bond* that joins the phosphate group on the 5′ carbon of one sugar to the hydroxyl group on the 3′ carbon of another sugar (see Figure 1-23). The nucleic acid originally discovered by Miescher contains the sugar *deoxyribose* and was therefore named **deoxyribonucleic acid** or **DNA.** A second type of nucleic acid was discovered to contain the sugar *ribose* and was therefore called **ribonucleic acid** or **RNA.** Analytical measurements revealed that DNA contains the bases adenine (A), guanine (G), cytosine (C), and thymine (T), while in RNA the base thymine is replaced by uracil (U).

Although the above information was crucial in furthering our understanding of nucleic acids, the lack of sensitivity of the analytical methods employed by Levene also had some unfortunate consequences. Levene's data suggested that DNA contains roughly equivalent amounts of the four bases A, G, C, and T, leading him to propose that DNA is a short chain containing one each of the four bases. When improved isolation techniques later revealed that DNA molecules are much longer than four nucleotides, Levene modified his so-called *tetranucleotide theory* and claimed that DNA is a repetitious polymer containing the same sequence of four bases repeated again and again (e.g., (AGCT)(AGCT)(AGCT) . . .). Since it is obvious that such a monotonous pattern cannot store genetic information, biologists were further reinforced in their belief that DNA, in spite of its localization within chromosomes, cannot be the genetic material.

By this time it had been established that chromosomes contain protein molecules as well as DNA. Since DNA did not appear to exhibit the attributes expected of a molecule encoding genetic information, most biologists concluded that genes must be made of protein. It was argued that proteins contain 20 different amino acids that can be assembled into a vast number of different sequences, thereby generating the sequence diversity and complexity expected of a molecule that stores and transmits genetic information. The conclusion that genes must therefore consist of protein rather than DNA seemed so logical and inescapable that it was widely taught for many years as an established fact.

Genetic Transformation and the Role of DNA

A great surprise was in store, however, for biologists who were busy studying protein molecules to determine how genetic information is stored and transmitted. The background for this surprise was laid in England in the late 1920s by Fred Griffith, a microbiologist who was studying the organism that causes bacterial pneumonia. Griffith discovered that this bacterium, *Pneumococcus pneumoniae*, exists in two genetic forms. One was designated the **S strain** because it forms colonies with smooth edges when grown on agar plates, while the other was named the **R strain** because its colonies have rough edges. When injected into mice, S-strain but not R-strain bacteria were found to trigger a fatal pneumonia. One of the most intriguing discoveries,

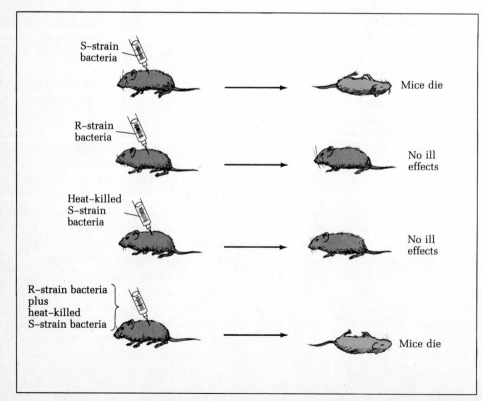

FIGURE 10-1 Summary of Griffith's genetic transformation experiments on S- and R-strain cells of the bacterium *Pneumococcus pneumoniae*.

TABLE 10-1
Base composition of DNA obtained from various sources

Source of DNA	A	T	G	C	A:T	G:C	(A + T)/(G + C)
Bovine thymus	28.4	28.4	21.1	22.1	1.00	0.95	1.31
Bovine liver	28.1	28.4	22.5	21.0	0.99	1.07	1.30
Bovine kidney	28.3	28.2	22.6	20.9	1.00	1.08	1.30
Bovine brain	28.0	28.1	22.3	21.6	1.00	1.03	1.28
Human liver	30.3	30.3	19.5	19.9	1.00	0.98	1.53
Locust	29.3	29.3	20.5	20.7	1.00	1.00	1.41
Sea urchin	32.8	32.1	17.7	17.3	1.02	1.02	1.85
Wheat germ	27.3	27.1	22.7	22.8	1.01	1.00	1.19
Marine crab	47.3	47.3	2.7	2.7	1.00	1.00	17.5
Aspergillus (mold)	25.0	24.9	25.1	25.0	1.00	1.00	1.00
Yeast	31.3	32.9	18.7	17.1	0.95	1.09	1.79
Clostridium (bacterium)	36.9	36.3	14.0	12.8	1.02	1.09	2.73

however, was that pneumonia can also be induced by injecting animals with a mixture of live R-strain bacteria and dead S-strain bacteria. This finding was particularly surprising because neither the live R-strain nor the dead S-strain bacteria induce pneumonia when injected alone. When Griffith autopsied the animals that had been injected with the mixture of live R-strain and dead S-strain bacteria, he found them teeming with live S-strain organisms. Since the animals had not been injected with any such live S-strain bacteria, it was concluded that the dead S-strain bacteria had induced the live R-strain cells to undergo a **genetic transformation** that converted them to S-strain cells (Figure 10-1).

Griffith's pioneering discovery opened the door for studies on the biochemical nature of the substance responsible for genetic transformation. The first important observation emerged a few years later, when James Alloway reported that simple extracts of dead S-strain cells are sufficient for inducing R-strain bacteria to undergo genetic transformation. Then for the next decade, Oswald Avery and his associates devoted themselves to the task of identifying the component in such bacterial extracts that is responsible for inducing genetic transformation. His first important discovery was that treatment of bacterial extracts with nucleases, but not proteases, abolishes their ability to cause genetic transformation, suggesting that the transforming agent is a nucleic acid rather than a protein. After many years of work, Avery succeeded in isolating the active nucleic acid and identifying it as DNA. Microgram quantities of this DNA were found to be capable of transforming bacterial cells from one type to another. In 1944, after a decade of painstaking work, Avery announced his inescapable conclusion to the world: The chemical substance responsible for genetic transformation is DNA!

Verification of the Genetic Role of DNA

At first, Avery's conclusions fell on deaf ears. Since scientists had known for the previous 40 years that genes are made of protein, not DNA, a variety of arguments were formulated to discredit Avery's work. It was claimed, among other things, that Avery's DNA preparations were contaminated with protein, that his nucleases and proteases were impure, and that genetic transformation in bacteria is a special case with no applicability to inheritance in other organisms. Within a few years, however, evidence began to accumulate that supported Avery's contentions. In 1949 several investigators reported the results of a series of careful chemical analyses on the amount of DNA in different cells. These data revealed that the amount of DNA is the same in all cells of an organism other than sperm and eggs, which contain half the normal amount. This DNA constancy, and even the presence of half the normal amount of DNA in the two cell types that join together to form a new organism, is exactly what would be expected of the genetic material. These conclusions were in direct opposition to reports early in the century that, it will be recalled, suggested that nucleic acid is present in variable amounts and therefore cannot be the genetic material. In retrospect, it is clear that these early reports were in error because the staining process used to measure nuclear DNA concentration is influenced by the amount of protein associated with DNA; variations in nuclear proteins, rather than DNA, were therefore responsible for the observed variations in staining intensity.

Shortly after these observations were published, Erwin Chargaff provided further support for the idea that DNA is the genetic material. Using sensitive analytical techniques to investigate the base composition of DNA samples obtained from various organisms, Chargaff discovered that Levene's generalization about the tetranucleotide nature of DNA was incorrect. Instead of equal amounts of the four bases, Chargaff found that the DNAs of different species are characterized by differing base compositions (Table 10-1). Thus instead of being constructed from a simple repeating sequence of four bases, DNA must have a more complex sequence that varies from organism to organism. This is exactly what would be expected, of course, of the genetic material. It was also found that DNA molecules isolated from different cell types within the same organism have similar base compositions, which is likewise compatible with a genetic role because all cells of an individual organism would be expected to inherit the same genetic information.

The final evidence that convinced the remaining skeptics that DNA is the genetic material was obtained in 1952 by Alfred Hershey and Martha Chase, who studied the infection of bacterial cells by a virus called T2-bacteriophage. During infection, the T2 virus attaches to the surface of bacterial cells and injects something into the cell (Figure 10-2). Shortly thereafter the cells burst, releasing thousands of new virus particles into the medium. Since this scenario suggests that the material injected into the bacterial cells contains information encoding the production of new virus particles, Hershey and Chase set about to ascertain the chemical nature of the injected material. The search was simpli-

fied by the discovery that the T2 virus contains only two kinds of molecules, DNA and protein. DNA molecules were known to contain phosphorus but no sulfur, while viral proteins contain sulfur but no phosphorus. Hershey and Chase therefore infected bacterial cells with virus particles that had previously been labeled with either radioactive phosphorus (^{32}P) or radioactive sulfur (^{35}S), and then determined which type of radioactivity entered the bacterial cells. The results revealed that ^{32}P, but not ^{35}S, enters bacterial cells during infection, indicating that viral DNA rather than protein contains the genetic information coding for new virus particles (Figure 10-3).

Thus by the early 1950s, the role of DNA as the carrier of genetic information came to be widely accepted. Unfortunately, Oswald Avery, the visionary most responsible for this complete turnabout in our views concerning the function of DNA, never received the credit he so richly deserved. The Nobel Prize Com-

FIGURE 10-2 Electron micrograph showing bacteriophage particles attached to the surface of a bacterial cell and injecting material into the cell (arrow). Courtesy of L. D. Simon and Photo Researchers Inc.

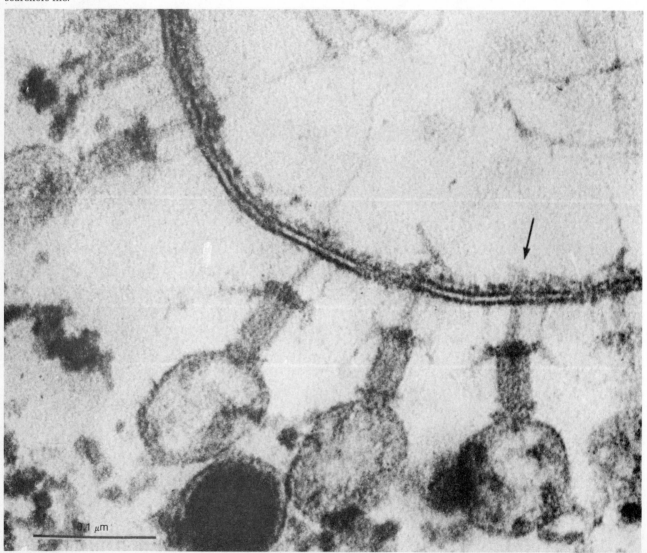

0.1 µm

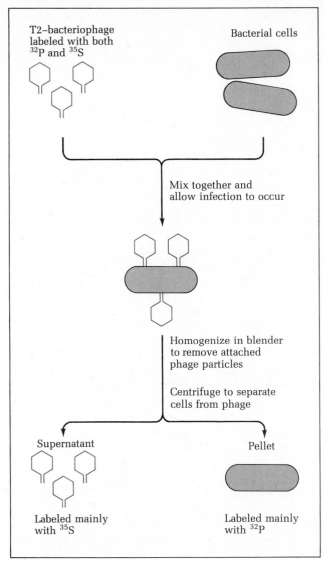

FIGURE 10-3 Summary of the Hershey–Chase experiment that demonstrated that DNA enters bacterial cells during infection by T2-bacteriophage.

mittee discussed Avery's work but decided he had not done enough. Perhaps Avery's modest and unassuming nature was responsible for this lack of recognition. After Avery died in 1955, Erwin Chargaff wrote in tribute: "He was a quiet man; and it would have honored the world more, had it honored him more" (1971:639).

As the scientific community gradually came to the realization that DNA encodes genetic information, an entire new set of questions began to emerge. The most fundamental issues concerned three basic problems: How is DNA accurately replicated so that duplicate copies of the genetic information can be passed on from cell to cell and from parent to offspring? How is the information contained within the DNA molecule utilized to determine the properties of cells and organisms? How does the information encoded within DNA undergo the changes, or *mutations*, required for evolution?

In the remainder of this chapter, we shall provide an overview of the major mechanisms involved in each of these three areas. Subsequent chapters devoted to the nucleus, ribosomes, and cell division will then examine these mechanisms in more detail, emphasizing their relationship to the subcellular organelles involved.

HOW IS DNA REPLICATED?

One of the first insights into the mechanism by which DNA molecules are replicated emerged from the previously mentioned studies of Erwin Chargaff on DNA base composition. In the process of analyzing DNA molecules isolated from a wide variety of organisms, Chargaff discovered a remarkable pattern: The bases adenine and thymine were always found to be present in equal amounts, and the bases guanine and cytosine were likewise found to occur in equivalent amounts. This pattern, which came to be known as *Chargaff's rule*, appeared to hold regardless of the source of DNA, and no matter what its overall base composition (see Table 10-1).

Although Chargaff felt that this regularity must reflect some fundamental property of DNA, its exact significance eluded him. The importance of the adenine-thymine and guanine-cytosine relationships became apparent in 1953, however, when James Watson and Francis Crick published a three-dimensional model of the structure of DNA that was to have far-reaching implications. This model was based on X-ray diffraction pictures of DNA, taken by Rosalind Franklin and Maurice Wilkins, which suggested to Watson and Crick that DNA is coiled into a spiral or helix. When Watson and Crick utilized molecular models to examine what type of helical structure might be involved, they discovered that two intertwined helical chains of DNA could be held together by hydrogen bonds between the nitrogenous bases of the two opposing chains. In such a DNA double helix, the sugar-phosphate backbones of the two DNA strands exhibit opposite polarity, one running in the $5' \rightarrow 3'$ direction and the other running in the $3' \rightarrow 5'$ direction (Figure 10-4). The nitrogenous bases of the two strands project toward the interior of the DNA molecule, where they can interact with one another. The most remarkable feature of this model was the discovery that the hydrogen bonds holding together the two strands of the helix can form only between the base adenine in one chain and thymine in the other, or between the base guanine in one chain and cytosine in the other (Figure 10-5). Because the base sequence of one chain therefore determines the base sequence of the opposing chain, the two chains are said to be **complementary** to one another. It was immediately obvious, of course, that such a model would explain why Chargaff had observed an equivalence between adenine and thymine, and between guanine and cytosine.

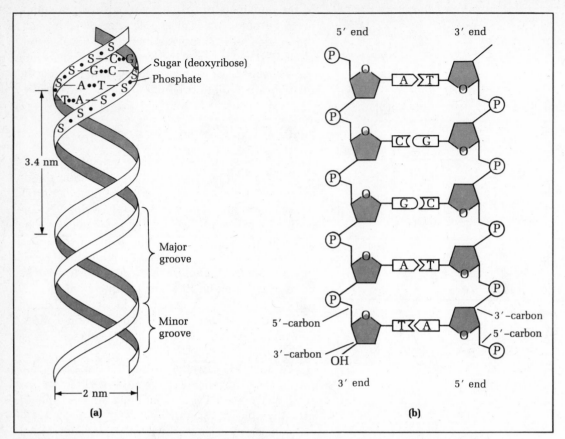

FIGURE 10-4 The DNA double helix. (a) A three-dimensional representation of the DNA double-helix model proposed by Watson and Crick. The two chains of the helix are held together by hydrogen bonds that join the base A to the base T, and the base G to the base C. (b) A linear representation of a short stretch of the DNA double helix that facilitates visualization of the relationships between the various chemical groups. Note that the two chains run in opposite directions.

FIGURE 10-5 Location of the hydrogen bonds linking guanine to cytosine, and adenine to thymine.

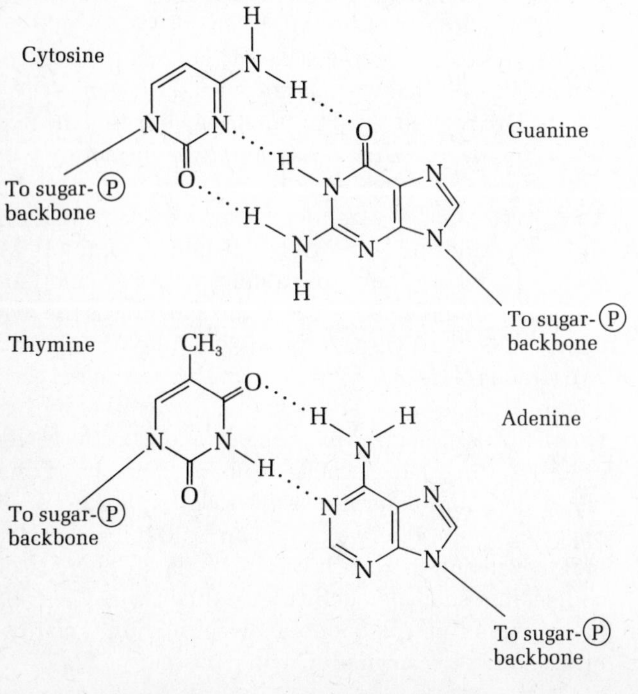

The most profound implication of the Watson–Crick model was that it provided a relatively simple molecular explanation for the ability of cells to duplicate their genetic information. It was simply proposed that the two chains of the DNA double helix unwind from each other, and that each chain serves as a template upon which a new opposing chain is copied by complementary base pairing (Figure 10-6). Because each new DNA molecule synthesized by such a mechanism would contain one old DNA strand and one newly synthesized strand, this process is referred to as **semiconservative replication.** A few years after this model was first proposed, an elegant experimental test of its validity was carried out by Matthew Meselson and Franklin Stahl. In these experiments two isotopic forms of nitrogen, ^{14}N and ^{15}N, were utilized to distinguish newly synthesized strands of DNA from old strands. Because nitrogen accounts for almost 20 percent of the total mass of DNA, it was reasoned that the overall density of the DNA molecule could be appreciably altered by allowing cells to grow in the presence of ^{15}N, an isotope that is heavier than the normally prevailing form of nitrogen, ^{14}N.

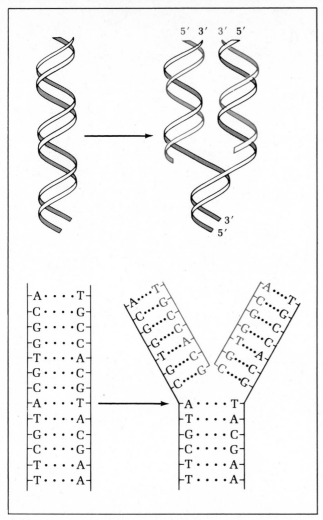

FIGURE 10-6 A schematic model of DNA replication, illustrating how the two strands of the double helix might unwind and serve as templates for the synthesis of new complementary chains.

found to exhibit the density of DNA containing only ^{14}N. As is illustrated in Figure 10-7, such a result is again consistent with the predictions of the Watson–Crick model. As a final test of the idea that the two strands of the DNA molecule separate during replication, Meselson and Stahl isolated DNA from cells that had undergone one round of division in the presence of ^{14}N, and heated the DNA to separate the two strands of the double helix from each other. When the individual strands were analyzed by isodensity centrifugation, half were found to exhibit the density characteristic of normal ^{14}N-containing strands, and half were found to exhibit the heavier density of ^{15}N-containing strands. This provided final confirmation for the conclusion that newly synthesized DNA molecules consist of one old strand of DNA and one new strand.

The DNA Polymerase Reaction

At about the same time that the pioneering studies of Meselson and Stahl were being carried out, biochemists began to explore the enzymatic reactions involved in the synthesis of DNA. At first DNA synthesis was thought to be inseparable from living cells, but in the mid-1950s Arthur Kornberg succeeded in isolating an enzyme from bacteria that is capable of synthesizing DNA under cell-free conditions. This enzyme, which he named **DNA polymerase,** catalyzes the polymerization of deoxynucleoside triphosphates into DNA and is greatly stimulated by the presence of small amounts of DNA in the initial reaction mixture:

$$\begin{vmatrix} dATP \\ dGTP \\ dCTP \\ dTTP \end{vmatrix}_n \xrightarrow[\text{DNA (template)}]{\text{DNA polymerase}} DNA + nPP_i$$

The four deoxynucleoside triphosphates utilized as substrates for this reaction are energy-rich compounds containing the four bases present in DNA: A, G, C, and T. Though differing from ATP in that they contain the sugar deoxyribose instead of ribose, each deoxynucleoside triphosphate contains the energy equivalent of one ATP molecule. Hence the equivalent of one molecule of ATP is consumed for every base incorporated into DNA. This energy input drives what would otherwise be a thermodynamically unfavorable polymerization reaction.

The most important feature of the DNA polymerase reaction became apparent when Kornberg compared the properties of the newly synthesized DNA molecules with the DNA initially added to stimulate the reaction. Not only were the starting and final DNA molecules found to resemble each other in base composition, but measurements of the frequencies with which the various bases occurred adjacent to one another indicated that the sequential arrangement of the bases was the

Bacterial cells were therefore inoculated into a growth medium containing ^{15}N-labeled ammonium chloride and were allowed to grow until their DNA molecules were extensively labeled with this heavy isotope of nitrogen. The ^{15}N-labeled cells were then transferred to a medium containing the normal light isotope of nitrogen (^{14}N), and the effect of this change on the density of their DNA molecules was monitored by isodensity centrifugation. After one round of cell division in the ^{14}N-containing medium, the bacterial DNA was found to exhibit a density halfway between that of normal DNA and DNA obtained from organisms grown entirely in the presence of ^{15}N. This is exactly what would be expected if the DNA double helix unwinds during replication and each of the original, ^{15}N-containing strands serves as the template for the synthesis of a new, ^{14}N-containing strand. After the second round of cell division, 50 percent of the DNA molecules were found to exhibit this intermediate density characteristic of ^{15}N/^{14}N hybrids, and 50 percent of the molecules were

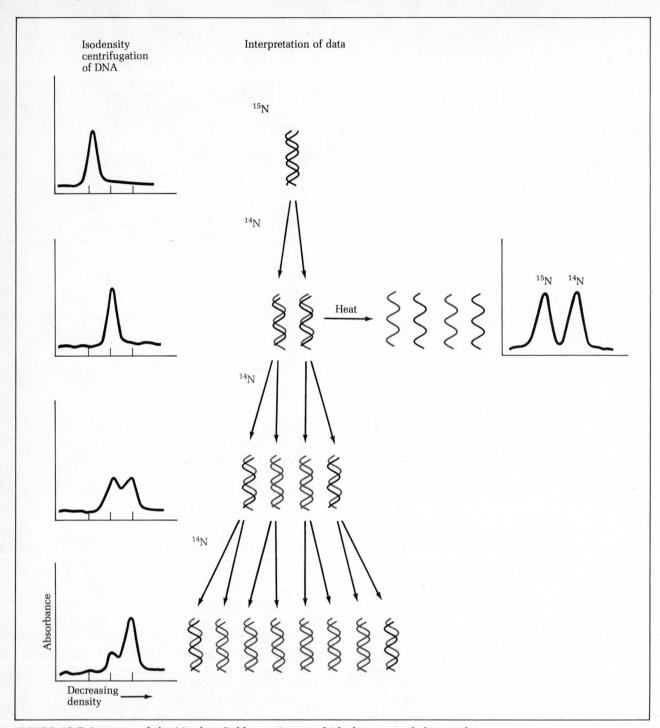

FIGURE 10-7 Summary of the Meselson-Stahl experiment, which demonstrated that newly synthesized DNA molecules consist of one old strand and one new strand.

same. Kornberg therefore concluded that the starting DNA served as a **template** whose base sequence was accurately copied by the enzyme DNA polymerase. Definitive proof for this idea was ultimately obtained in 1967, when it was shown that DNA polymerase can utilize purified viral DNA as a template for the synthesis of new DNA molecules that are capable of infecting cells just like the natural virus. This milestone experiment was the first direct demonstration that an accurate and

biologically active copy of a naturally occurring DNA molecule can be synthesized by DNA polymerase under cell-free conditions.

In spite of the importance of the above discoveries, the enzyme isolated by Kornberg does not appear to be responsible for DNA replication in intact cells. This fact became apparent in 1969 when John Cairns and his associates reported that mutant strains of bacteria defective in the Kornberg enzyme still replicate their DNA and divide in a normal manner. Subsequent studies revealed that bacteria contain at least two other enzymes capable of synthesizing DNA. These enzymes, which appear to be associated with the plasma membrane, are referred to as *DNA polymerase II* and *DNA polymerase III* to distinguish them from the Kornberg enzyme, or *DNA polymerase I*. Though mutant organisms defective in enzymes I or II are capable of replicating their DNA normally, this is not the case for mutants defective in enzyme III. DNA polymerase III is therefore thought to mediate chromosomal DNA replication, while DNA polymerases I and II are involved in special types of DNA synthesis. It has been found, for example, that mutant strains of bacteria deficient in DNA polymerase I recover poorly from radiation-induced DNA damage, suggesting that this enzyme functions in DNA repair. DNA polymerase I has also been implicated in the replication of *plasmids*, which are small molecules of DNA that replicate independently from the chromosomal DNA.

Although it is clear that at least some forms of DNA

polymerase play a role in normal DNA replication, the molecular mechanisms involved are quite complex and require the participation of several additional proteins. One reason for the need for additional proteins is that DNA polymerases can only catalyze the attachment of nucleotides to the 3′-hydroxyl end of existing DNA chains; hence DNA chains can only be synthesized in the 5′ → 3′ direction (Figure 10-8). The two chains of the DNA double helix, however, run in opposite directions. Hence one chain running in the 5′ → 3′ direction and one chain running in the 3′ → 5′ direction must be replicated by an enzyme that appears to be capable of synthesizing DNA only in the 5′ → 3′ direction.

A solution to this problem was first proposed in 1968 by Reiji Okazaki, whose experiments suggested that DNA is synthesized in the form of small pieces that are later joined to form the growing DNA molecule. In these studies DNA was first isolated from cells that had been briefly exposed to a radioactive precursor of DNA. Subsequent fractionation of this DNA on the basis of size revealed that the majority of the radioactivity was

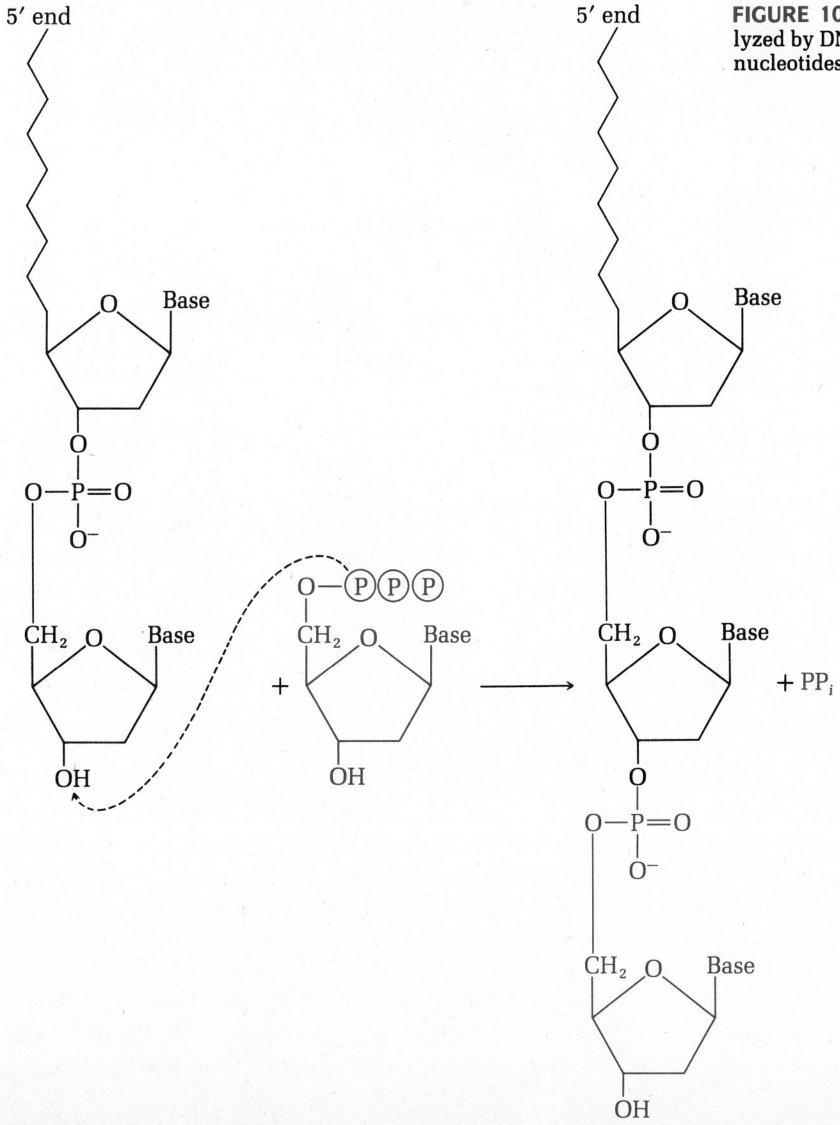

FIGURE 10-8 The polymerization reaction catalyzed by DNA polymerase involves the addition of nucleotides to the 3′ end of DNA chains.

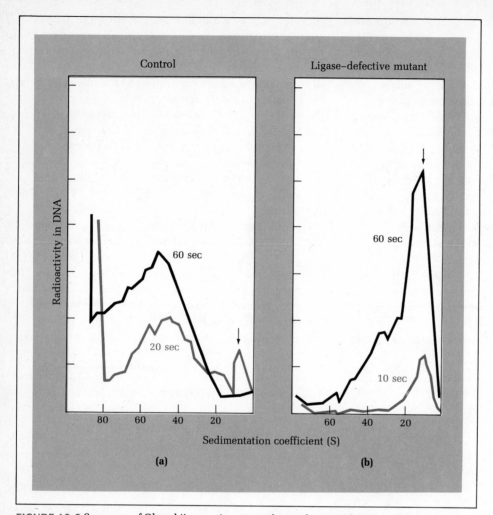

FIGURE 10-9 Summary of Okazaki's experiments on the mechanism of DNA replication in bacterial cells. After incubation of bacteria for brief periods with ³H-thymidine, DNA was isolated and analyzed by moving-zone centrifugation under conditions where the DNA is dissociated into its individual strands. (a) In normal bacteria, the ³H-thymidine is initially incorporated into small fragments (arrow), and later appears in larger molecules of DNA. (b) In bacterial mutants deficient in the enzyme DNA ligase, the radioactivity remains in the small DNA fragments. DNA ligase is therefore thought to catalyze the joining together of the small DNA fragments in normal cells.

associated with small fragments of DNA measuring about a thousand nucleotides in length (Figure 10-9). After longer labeling periods, on the other hand, the radioactivity was found to be associated with larger molecules of DNA. This finding suggested to Okazaki that the smaller DNA pieces, often referred to as **Okazaki fragments,** are precursors of the newly forming DNA molecule. The incorporation of Okazaki fragments into larger DNA molecules does not occur in mutant organisms known to be deficient in **DNA ligase,** an enzyme that joins DNA fragments together by catalyzing the formation of a covalent phosphodiester bond between the 3' end of one nucleotide chain and the 5' end of another (Figure 10-10).

When considered in total, the above observations suggest a model of DNA replication which overcomes the problem that DNA polymerase only functions in the 5' → 3' direction. According to this model (Figure 10-11), DNA polymerase copies both strands of the double helix by synthesizing DNA in the 5' → 3' direction. Copying one strand in this direction could generate a continuous 5' → 3' DNA chain, but copying the opposite strand would produce a series of small, discontinuous fragments that must later be joined together by the action of DNA ligase.

Since DNA polymerase can only join nucleotides to the 3'-hydroxyl end of existing nucleotide chains, another fundamental issue that must be addressed concerns the mechanism by which the synthesis of a DNA chain is initiated. Several findings reported shortly after the discovery of Okazaki fragments suggested that RNA might be involved. Among the more important of these observations were that (1) Okazaki fragments often have short stretches of RNA at their 5' ends; (2) RNA polymerase, unlike DNA polymerase, can initiate the formation of new RNA chains without the need for an

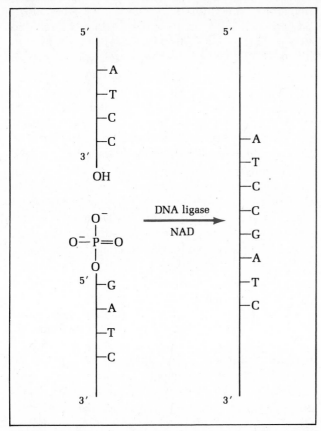

FIGURE 10-10 The DNA ligase reaction. This reaction, which joins the 3'-end of one nucleic acid molecule to the 5'-end of another, often utilizes NAD as a cofactor.

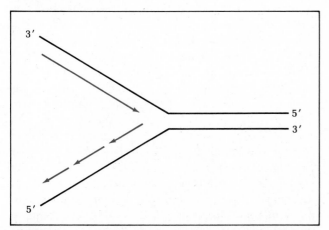

FIGURE 10-11 A general model of DNA replication involving the discontinuous synthesis of small DNA fragments on at least one strand of the double helix. It is possible that DNA synthesis may be discontinuous on the other strand as well.

existing chain to add to; and (3) DNA polymerase can catalyze the addition of deoxyribonucleotides to the 3' end of RNA chains. Based on such information, it is now widely believed that DNA replication is initiated by an enzyme, termed a **primase,** that catalyzes the formation of an RNA primer that is later removed and replaced by DNA (Figure 10-12).

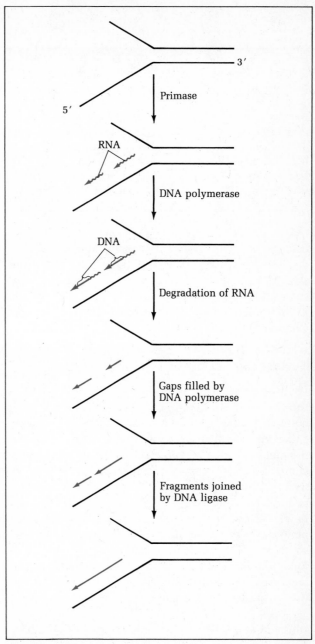

FIGURE 10-12 Illustration of the individual steps that may be involved in synthesizing small DNA fragments and joining them together to form longer strands.

Besides the enzymes already described, DNA replication requires several additional proteins (Table 10-2). Some of these proteins promote unwinding of the double helix, exposing the individual strands so that they can serve as templates for the synthesis of new complementary strands. This unwinding process is also facilitated by enzymes that introduce single-stranded breaks or "nicks" into the individual strands, a process that allows localized regions of the DNA molecule to unwind. Yet other proteins are thought to be required for initiating and terminating the replication process. Within intact cells, DNA replication is further compli-

TABLE 10-2
Partial list of proteins involved in bacterial DNA replication

Protein	Function
Helicase (rep protein)	Unwinds DNA double helix
Helix destabilizing protein	Stabilizes single-stranded DNA
DNA gyrase	Relaxes supercoiled double helix
Primase	Synthesizes primer
dnaB protein	Interacts with primase
dnaC protein	Interacts with dnaB protein
DNA polymerase III	DNA elongation
DNA polymerase I	Gap filling and repair
DNA ligase	Joining together of DNA fragments

cated by the fact that the DNA is associated with various types of protein molecules and folded into complicated structures known as chromatin fibers and chromosomes. The relationship of the DNA replication process to these more complex structures will be discussed in Chapter 13.

Reverse Transcriptase

While most forms of DNA polymerase utilize DNA as the template for directing the synthesis of new molecules of DNA, some tumor viruses contain an enzyme, called **reverse transcriptase,** that employs RNA as a template instead. Reverse transcriptase utilizes the complementary base-pairing mechanism to copy the base sequence information from a single-stranded molecule of RNA into a complementary strand of DNA. This newly formed DNA strand in turn functions as a template for the synthesis of a second complementary chain of DNA, yielding a double-stranded DNA molecule as final product. The only physiological function that has been clearly established for reverse transcriptase is in the life cycle of RNA tumor viruses, where it allows the viral RNA to be copied into DNA molecules that are subsequently integrated into the host cell DNA (Chapter 20). Reverse transcriptase has been an exceptionally important experimental tool because it allows one to artificially synthesize DNA molecules that are complementary in sequence to a given messenger RNA. As we shall see later, such complementary DNA molecules, or **cDNAs,** can be used in various ways to detect the presence of DNA or RNA sequences that are related to a particular messenger RNA.

HOW IS DNA UTILIZED?

The One Gene–One Polypeptide Chain Theory

The second major question regarding the biological role of DNA concerns the issue of how the genetic information encoded within DNA molecules influences the properties of cells and organisms. Although the notion that genes function as discrete inherited units that govern the individual traits of organisms was set forth by Gregor Mendel in 1866, nearly a century passed before the mechanism underlying gene action was clearly elucidated. The first significant contribution to our understanding of how genes function was made in the early twentieth century by an English physician, Sir Archibald Garrod, who was interested in the relationship between inherited diseases and cellular metabolism. Garrod was especially intrigued by *alkaptonuria*, a disease that causes the urine of afflicted individuals to blacken upon standing. When Garrod analyzed urine specimens obtained from patients suffering from alkaptonuria, he discovered large quantities of an amino acid breakdown product known as *homogentisic acid*. In normal individuals homogentisic acid is degraded to carbon dioxide and water, and hence does not appear in the urine. Garrod therefore concluded that patients suffering from alkaptonuria are deficient in one or more of the enzymes involved in catalyzing the breakdown of homogentisic acid. Because alkaptonuria is an inherited condition, Garrod's conclusion implied that hereditary information controls the production of enzyme molecules. Although many present-day textbooks credit Garrod with being ahead of his time in seeing this relationship, the published reports of these experiments, which appeared in 1909, suggest that he did not clearly perceive it. As a physician, Garrod was more interested in disease than in the question of how genes function, and so the profound implications of his discovery went largely unappreciated.

Thirty years passed before the next significant opportunity arose for discovering the relationship between genes and enzymes. In the early 1940s, George Beadle and Edward Tatum embarked upon a study of the inheritance of metabolic traits in *Neurospora crassa*, the common bread mold. This eukaryotic microorganism was chosen for study because its genetic characteristics had been well worked out, its nutritional requirements were precisely known, and it grows and divides readily in culture. Given a "minimal" growth medium containing sugar, salts, and a single vitamin (biotin), *Neurospora* can synthesize all the chemical building blocks needed for its survival and proliferation. In order to investigate the genetic basis of the metabolic pathways employed by *Neurospora* for the synthesis of these various building blocks, Beadle and Tatum utilized X-irradiation to induce genetic mutations. They reasoned that if genes regulate metabolism, X-ray-induced genetic changes might result in altered nutritional requirements. This expectation was confirmed by the discovery that after X-irradiation, mutant strains of *Neurospora* can be isolated that are no longer capable of growing in the minimal medium. Beadle and

Tatum discovered, however, that addition of specific amino acids or vitamins to the minimal medium permitted the mutant strains of *Neurospora* to grow and divide normally. It was therefore concluded that the mutant cells had lost the ability to synthesize certain amino acids or vitamins, and hence could grow only in media in which these substances were provided.

The identity of the specific amino acid or vitamin required by a particular mutant strain of *Neurospora* was investigated by transferring the affected cells to a variety of different growth media, each containing a single amino acid or vitamin added to the normal minimal medium. In this way it was ascertained that one mutant strain would grow only on a medium supplemented with the amino acid arginine, another would grow only on a medium supplemented with vitamin B_6, and so forth (Figure 10-13). This general approach eventually led to the identification of a large number of *Neurospora* mutants, each impaired in its ability to synthesize a particular amino acid or vitamin.

Because the synthesis of amino acids and vitamins involves multistep pathways, Beadle and Tatum next attempted to identify the particular step in each pathway that had been affected. This question was investigated by supplementing the minimal medium with metabolic precursors of a given amino acid or vitamin, rather than with the amino acid or vitamin itself. By determining which precursors support the growth of a particular mutant organism, the defective step in each metabolic pathway could be pinpointed (Figure 10-14). The results of these experiments indicated that most mutant strains of *Neurospora* are deficient in a single metabolic reaction. Since individual metabolic reactions were known to be catalyzed by enzymes, Beadle

FIGURE 10-13 Experimental approach utilized by Beadle and Tatum to identify the nature of the metabolic defects occurring in *Neurospora* cells subjected to X-irradiation to induce genetic mutations.

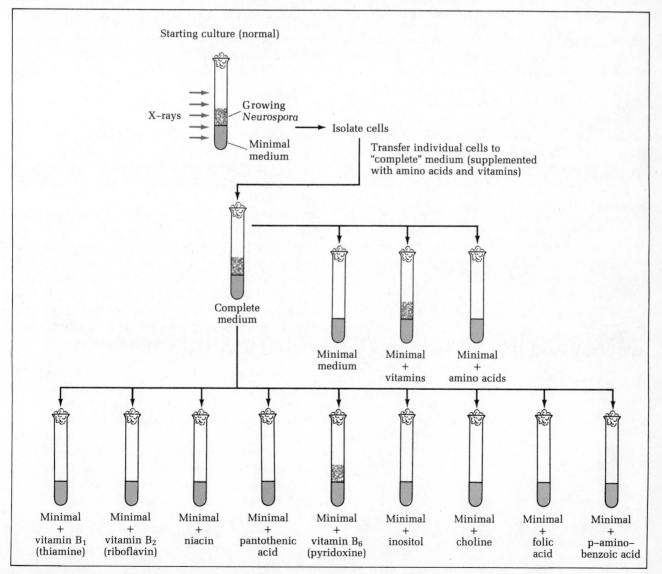

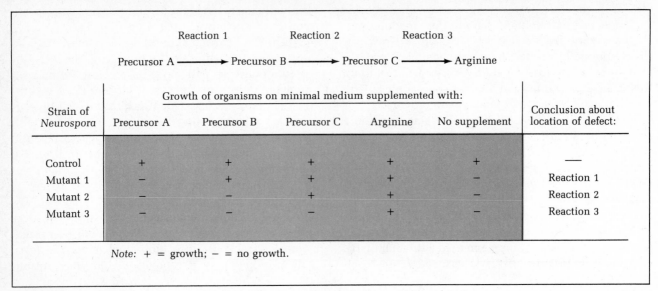

	Reaction 1	Reaction 2	Reaction 3		
Precursor A ⟶	Precursor B ⟶	Precursor C ⟶	Arginine		

| Strain of *Neurospora* | Growth of organisms on minimal medium supplemented with: | | | | | Conclusion about location of defect: |
	Precursor A	Precursor B	Precursor C	Arginine	No supplement	
Control	+	+	+	+	+	—
Mutant 1	–	+	+	+	–	Reaction 1
Mutant 2	–	–	+	+	–	Reaction 2
Mutant 3	–	–	–	+	–	Reaction 3

Note: + = growth; – = no growth.

FIGURE 10-14 Outline of the experimental approach employed to identify the particular metabolic step that is defective in a given *Neurospora* mutant. In this example several mutants requiring arginine for survival were differentiated from one another by comparing their abilities to grow on various precursors in the metabolic pathway for synthesizing arginine.

and Tatum concluded that each mutant strain is defective in a single enzyme. Moreover, genetic studies of the inheritance pattern of these defects revealed that each defect is associated with a mutation in a single gene. Beadle and Tatum therefore combined these observations into a unifying concept, termed the **one gene–one enzyme theory,** which stated that the function of a gene is to control the production of a single enzyme.

Although this theory represented a crucial mile-

FIGURE 10-15 Comparison of the electrophoretic behavior of normal and sickle-cell hemoglobin. Normal hemoglobin behaves as if it is more negatively charged than sickle-cell hemoglobin.

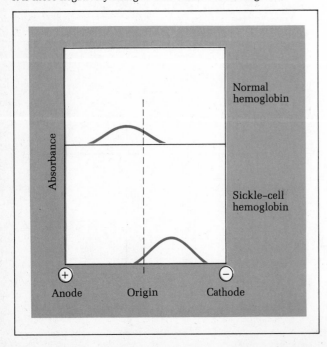

stone in the development of our understanding of gene action, it did not provide much insight into the question of how genes control the production of enzyme molecules. The first clue to the underlying mechanism emerged a few years later in the laboratory of Linus Pauling, where the inherited disease *sickle-cell anemia* was under investigation. Because individuals suffering from sickle-cell anemia have abnormal red blood cells, Pauling decided to analyze the properties of the major red cell protein, hemoglobin. When normal and sickle-cell hemoglobin were analyzed by electrophoresis, it was discovered that they exhibit different migration rates (Figure 10-15). Since some amino acids have electrically charged side chains, and electrical charge influences electrophoretic migration, Pauling proposed that sickle-cell hemoglobin has an abnormal amino acid makeup. In order to determine the exact nature of the amino acid abnormality, it seemed necessary to determine the entire amino acid sequence of hemoglobin. Though this task has been successfully accomplished with present-day techniques, the largest protein to have been sequenced at the time of Pauling's discovery was less than one-tenth the size of hemoglobin. Determining the complete amino acid sequence of hemoglobin would therefore have been a monumental undertaking in the early 1950s.

Fortunately, an ingenious short-cut devised by Vernon Ingram made it possible to detect the amino acid abnormality in sickle-cell hemoglobin without the necessity for determining its complete amino acid sequence. In the approach designed by Ingram, the proteolytic enzyme trypsin was utilized to degrade hemoglobin into a series of peptide fragments. These fragments were then separated from one another by a combination of paper electrophoresis and chromatography, a process that generates a two-dimensional pattern of spots referred to as a *fingerprint* (Figure 10-16). When the fingerprint patterns of sickle-cell and normal hemoglobin were compared, Ingram discovered that only a single fragment was altered. Isolation of this peptide and analysis of its amino acid composition revealed that a glutamic acid residue present in the fragment derived from normal hemoglobin was replaced by a valine residue in the fragment obtained from sickle-cell hemoglobin. Since glutamic acid is negatively charged and valine is neutral, this substitution explains the difference in electrophoretic behavior between normal and sickle-cell hemoglobin originally observed by Pauling.

Ingram's studies, which were published in the mid-1950s, confirmed the idea that genes specify the amino acid sequence of protein molecules. In the case of hemoglobin, a genetically induced alteration in a single amino acid changes the properties of the protein sufficiently to induce a debilitating, and often fatal, disease. Subsequent studies have revealed that other mutant forms of hemoglobin occur, and that some of these involve changes in a gene that maps to a different region of chromosomal DNA than the gene altered in sickle-cell anemia. The discovery that two different genes affect the amino acid sequence of the same protein is easily explained, however, by the fact that hemoglobin contains two different kinds of polypeptide chains (α- and β-chains). The amino acid sequences of these two chains are specified by separate genes.

The preceding discoveries necessitated several refinements in the view of gene action originally encapsulated by the one gene – one enzyme theory of Beadle and Tatum. First, the fact that hemoglobin is not an enzyme suggests that genes encode the amino acid sequences of all proteins, not just enzymes. In addition, the discovery that different genes encode the amino acid sequences of the α- and β-chains of hemoglobin implies that each

FIGURE 10-16 Fingerprints of normal *(left)* and sickle-cell *(right)* hemoglobins. The colored spots in the accompanying diagrams highlight the region where the two patterns differ.

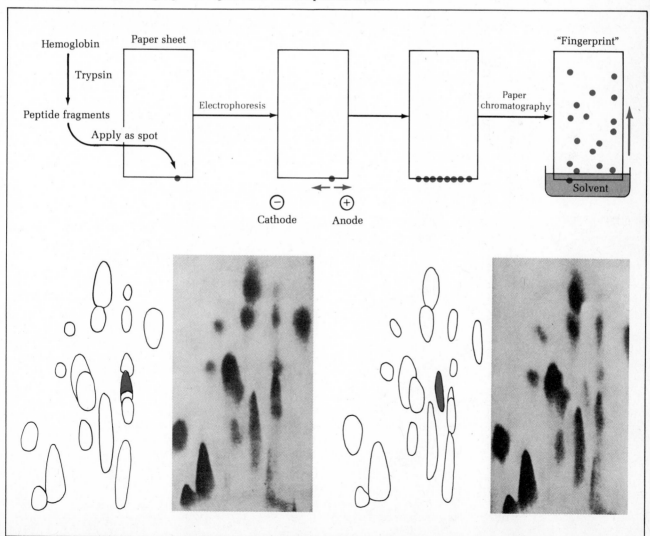

gene encodes the amino acid sequence of a specific polypeptide chain. To accommodate these new developments, the one gene–one enzyme concept eventually evolved into the notion of **one gene–one polypeptide chain.**

The most straightforward interpretation of the relationship between a gene and the polypeptide chain it encodes is that the sequence of bases contained within the gene determines the linear order of amino acids in the polypeptide chain. This prediction was confirmed in the mid-1960s in the laboratory of Charles Yanofsky, where the precise locations of dozens of mutations in the bacterial gene coding for the enzyme tryptophan synthetase were determined. The linear sequence of these mutations was found to correlate precisely with the sequence of the resulting amino acid alterations in the tryptophan synthetase molecule (Figure 10-17).

These crucial experiments provided independent proof for the theory that the nucleotide base sequence of DNA encodes the amino acid sequence of polypeptide chains. Although the simple linear relationship between DNA base sequence and polypeptide amino acid sequence discovered by Yanofsky appears to be typical for bacterial genes, recent observations have revealed that the organization of many eukaryotic genes is significantly more complex. This additional complexity will be described in Chapter 11, which deals with the details of nuclear organization and gene expression.

The Role of Messenger RNA

Once biologists realized that genes determine the amino acid sequence of polypeptide chains, the question arose as to how this transfer of information occurs. Several

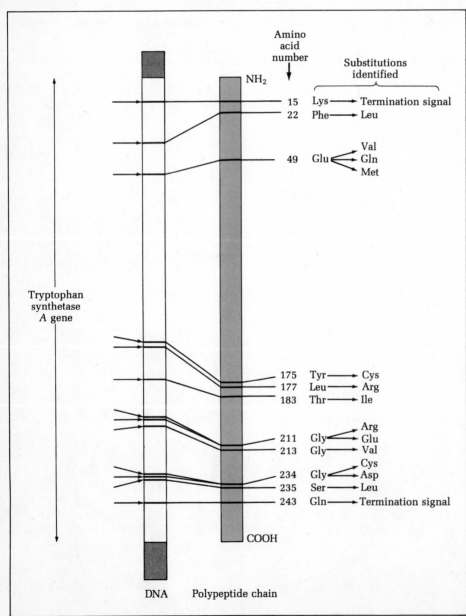

FIGURE 10-17 Relationship between the positions of several mutation sites *(arrows)* in the bacterial gene coding for the A subunit of tryptophan synthetase and the resulting amino acid substitutions in the polypeptide chain encoded by this gene. Note that the linear order of mutations corresponds to the linear order of the amino acid alterations.

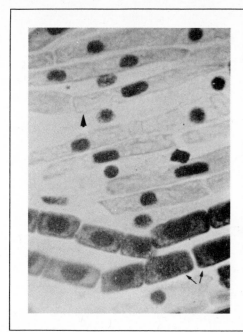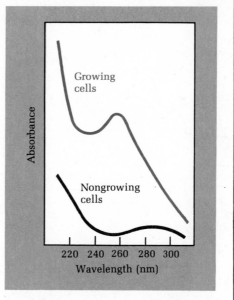

FIGURE 10-18 Identification of cytoplasmic RNA in cells of the growing onion root tip. *(Left)* Cytochemical staining of RNA in root cells reveals darkly stained cytoplasm in cells located near the growing tip (arrows), while cells in the nongrowing region of the root are unstained (arrowhead). Courtesy of T. O. Caspersson. *(Right)* Absorption spectra of the cytoplasm of root tip cells analyzed by quantitative ultraviolet light microscopy. The cytoplasm of the cells from the root tip exhibits an absorption peak at 260 nm, which is characteristic of nucleic acids. The curve for nongrowing cells shows less overall absorption and a peak at 280 nm, which is characteristic of proteins. Such data indicate that cytoplasmic nucleic acid is present in higher concentration in the growing cells.

observations suggested that DNA cannot directly guide the assembly of amino acids into polypeptide chains. First, autoradiographic studies carried out on cells incubated with radioactive amino acids revealed that incorporation of radioactivity into protein occurs in the cytoplasm, while DNA was known to be located predominantly in the nucleus. Moreover, cells that have had their nuclei artificially removed continue synthesizing proteins for considerable periods of time, in spite of the absence of the nuclear DNA. Such observations suggested that some intermediary molecule must exist that carries information from the nuclear DNA to the site of cytoplasmic protein synthesis.

Experiments carried out in the early 1940s in the laboratories of Jean Brachet and Torbjöern Caspersson provided some early clues as to the nature of this intermediate even before the role of DNA in genetic information transfer had been elucidated. Both laboratories were attempting to measure the quantity of RNA present in the cytoplasm, either by staining cells with basic dyes or with ultraviolet microscopy (Figure 10-18). The results of these studies revealed that cells active in protein synthesis tend to contain large amounts of cytoplasmic RNA.

The significance of this cytoplasmic RNA did not become apparent until a decade later, however, when subcellular fractionation techniques were first applied to the study of protein synthesis. In these pioneering

experiments animals were injected with radioactive amino acids to allow the most recently formed protein molecules to become radioactive. Various subcellular fractions were then isolated and analyzed for radioactivity incorporated into newly synthesized protein. In all cells examined the bulk of the radioactivity was detected in the microsomal fraction. When this microsomal fraction was further separated into its ribosomal and membrane constituents, the newly synthesized proteins were found to be associated predominantly with the ribosomes (Figure 10-19).

Since RNA accounts for roughly half the mass of the ribosome, the discovery that proteins are synthesized on ribosomes explained why Brachet and Caspersson had detected large quantities of RNA in cells that are active in protein synthesis. But the mechanism by which the genetic information guiding protein assembly is transmitted from the nuclear DNA to cytoplasmic ribosomes remained to be determined. The notion that RNA molecules might serve as intermediates first surfaced in the late 1950s, primarily as the result of autoradiographic studies carried out on cells incubated with radioactive precursors to label newly synthesized RNA molecules. In cells briefly exposed to such precursors, the newly formed radioactive RNA was found predominantly within the nucleus. During subsequent incubation in a nonradioactive medium, however, the radioactivity was seen to migrate to the cytoplasm

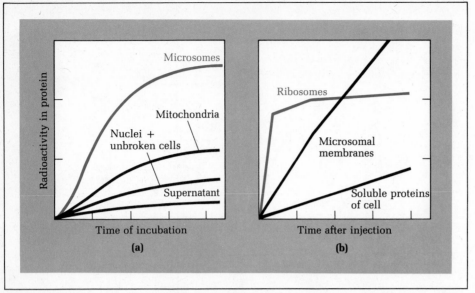

FIGURE 10-19 Data from two early experiments designed to determine the subcellular location of the process of protein synthesis. (a) When homogenates of rat liver are incubated with radioactive amino acids and then fractionated, the highest concentration of newly synthesized radioactive protein is found in the microsomal fraction. (b) When the microsomal fraction is further divided into its membrane and ribosomal components, newly synthesized protein appears first in the ribosomal fraction and then later appears in other fractions. This result suggests that proteins are synthesized on ribosomes and subsequently shipped to other parts of the cell. In this particular experiment proteins were labeled by injecting radioactive amino acids into intact animals, and then isolating liver tissue for subcellular fractionation.

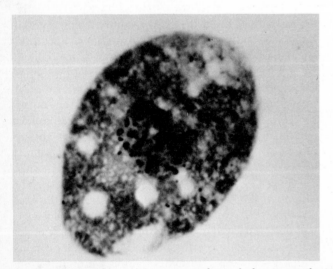
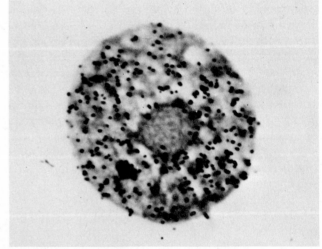

FIGURE 10-20 Light microscopic autoradiograph showing synthesis of RNA in the nucleus and its transfer to the cytoplasm. *(Left)* Autoradiograph of a single *Tetrahymena* cell incubated in the presence of radioactive cytidine for 15 min. The newly formed radioactive RNA is localized predominantly in the nucleus. *(Right)* Autoradiograph of a cell incubated with radioactive cytidine for 12 min to label the newly synthesized RNA, and then transferred to a nonradioactive medium for an additional 88 min of incubation. Under such conditions most of the original radioactive RNA is seen to migrate to the cytoplasm. Courtesy of D. M. Prescott.

(Figure 10-20). It was therefore concluded that RNA is synthesized in the nucleus and later travels to the cytoplasm, a behavior expected of a molecule that transfers information from the nuclear DNA to the cytoplasmic protein-synthesizing machinery.

At first it was thought that **ribosomal RNA,** which accounts for the bulk of the RNA present in the ribosome, is the molecule that migrates to the cytoplasm to guide the synthesis of specific proteins. It was soon discovered, however, that ribosomal RNA molecules obtained from different cells and organisms tend to resemble each other in size and base composition, while the proteins synthesized by different kinds of ribosomes vary widely in size and amino acid sequence. Hence ribosomal RNA did not appear to be a likely candidate for the molecule that specifies the type of protein being synthesized. A clever study carried out by Francois Jacob and Jacques Monod reinforced the view that ribosomal RNA is not the intermediate that carries protein-coding information from DNA to the ribosome. In this particular experiment bacteria were grown in media containing abnormally high concentrations of radioactive phosphate. Upon incorporation into newly forming DNA molecules, this excessive amount of radioactivity caused the DNA to become degraded. Shortly after this degradation occurred, the synthesis of proteins was found to stop, even though the cell's ribosomal RNA remained intact. It was therefore concluded that the molecules that guide protein synthesis are relatively unstable and must be continually transmitted from DNA to the ribosomes. Ribosomal RNA, which is a stable component of the ribosome, clearly could not be the molecule involved.

In view of the difficulties engendered by the idea that ribosomal RNA guides the process of protein synthesis, Jacob and Monod proposed in 1961 that the molecule that carries information from DNA to ribosomes is a less stable type of RNA, which they termed **messenger RNA** or **mRNA.** According to this theory, each gene produces its own particular type of messenger RNA that becomes subsequently associated with ribosomes to guide the synthesis of a particular type of polypeptide chain. Ribosomes were thus viewed as nonspecific machines that are in theory capable of producing any type of protein, depending upon the particular messenger RNA to which they happen to be attached at the moment.

As Jacob and Monod were quick to point out, chemical evidence for the existence of an RNA fraction exhibiting some of the properties predicted for messenger RNA molecules already existed. A few years earlier Elliot Volkin and Lazarus Astrachan had discovered that infection of bacterial cells with T2-bacteriophage triggers the production of a new type of RNA whose base composition resembles that of the phage DNA. However, in order to verify that this RNA functions as a

messenger that transmits genetic information from the viral genes, the following three predictions of the messenger RNA model had to be confirmed: (1) that the base sequence of RNA is copied from DNA; (2) that such RNA becomes transiently associated with ribosomes; and (3) that the base sequence of these transiently associated RNA molecules directs the process of protein synthesis, determining the linear order of amino acids in newly forming polypeptide chains. As shall be described below, these three predictions were all confirmed shortly after the existence of messenger RNA was first postulated.

1. The prediction that the base sequence of RNA molecules is copied from the base sequence of DNA was confirmed in several ways. The first major breakthrough in this area occurred when Samuel Weiss, Jerard Hurwitz, and Audrey Stevens independently isolated the enzyme responsible for catalyzing the synthesis of RNA. This enzyme, termed **RNA polymerase,** catalyzes the polymerization of nucleoside triphosphates into RNA, utilizing DNA as template:

$$\begin{vmatrix} \text{ATP} \\ \text{GTP} \\ \text{CTP} \\ \text{UTP} \end{vmatrix}_n \xrightarrow[\text{DNA (template)}]{\text{RNA polymerase}} \text{RNA} + n\text{PP}_i$$

The above reaction, which in several ways resembles the DNA polymerase reaction, utilizes the complementary base-pairing mechanism to copy base sequence information from one strand of the DNA double helix into a complementary chain of RNA. The same base-pairing rules apply as in DNA synthesis, with the single exception that the base uracil (U) substitutes in RNA for the base thymine (T) in DNA. This substitution is permitted because uracil and thymine can both form hydrogen bonds with adenine; hence adenine pairs with the base thymine in a complementary chain of DNA and with the base uracil in a complementary chain of RNA.

The most convincing support for the conclusion that the base sequence of the RNA molecule being synthesized is copied from a DNA template has emerged from experiments utilizing nucleic acid hybridization, a technique to be described in detail later in the chapter. In essence, nucleic acid hybridization permits one to determine the extent to which the base sequences of two nucleic acid chains are complementary to one another. Using this approach, it has been shown that RNA molecules synthesized in the reaction catalyzed by RNA polymerase are complementary in sequence to the DNA molecules employed as templates. Similar results have been obtained in experiments carried out on RNA isolated from intact cells. RNA molecules synthesized after viral infection of bacteria, for example, have been shown to be complementary in sequence to viral, but not bacterial, DNA.

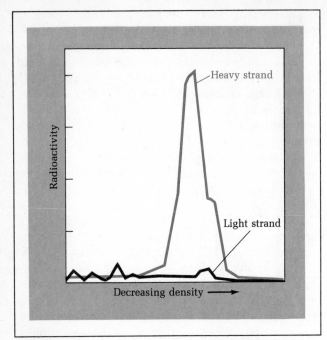

FIGURE 10-21 Experimental evidence for the asymmetry of RNA synthesis. The two strands of SP8-bacteriophage DNA were separated from each other by isodensity centrifugation and individually mixed with radioactive phage RNA to determine which strand is complementary to the RNA. Isolation of the resulting DNA-RNA hybrids by isodensity centrifugation revealed that the radioactive RNA hybridized to only one of the two DNA strands.

Because DNA molecules contain two strands while most RNA is typically single stranded, the question arises as to whether one or both of the DNA strands is copied into complementary molecules of RNA. Investigation of this issue has been facilitated by the existence of certain viruses, such as *SP8-bacteriophage*, whose two DNA strands differ enough in base composition to permit them to be separated from each other by isodensity centrifugation. Exploiting this difference, Julius Marmur and his colleagues separated the two strands of the SP8 viral DNA from one another and tested their ability to form complementary hybrids with RNA molecules synthesized in virally infected cells. Only one of the isolated DNA strands was found to be complementary to the RNA (Figure 10-21), indicating that RNA polymerase selectively copies one of the two DNA strands. In longer DNA molecules the strand being copied may not necessarily be the same for every gene, but for each gene it is still generally true that transcription is *asymmetric*, that is, only one strand of the DNA double helix is copied.

2. The messenger RNA theory postulated that once the base sequence of a DNA gene has been copied into a molecule of complementary RNA, the RNA in turn becomes associated with ribosomes. This prediction was first supported by experiments in which bacteria that had been growing in a ^{15}N-containing medium were switched to ^{14}N-containing media and infected with T4-phage. It was reasoned that any new ribosomes made

after infection by this virus would contain ^{14}N, and hence could be distinguished by isodensity centrifugation from the heavier, ^{15}N-labeled ribosomes formed prior to infection. When the ribosomes from infected cells were analyzed by isodensity centrifugation, it was found that the old, ^{15}N-containing ribosomes contained the bulk of the newly synthesized, viral RNA (Figure 10-22). In addition, newly synthesized viral proteins were also found to be associated with these same ribosomes. Such data are compatible with the idea that the viral RNA has become associated with preexisting bacterial ribosomes, where it directs the synthesis of viral proteins.

3. The final prediction of the messenger RNA concept is that the base sequence of messenger RNA molecules contains information capable of determining the linear order in which amino acids are joined together during protein synthesis. This crucial prediction was initially verified in 1961 by Marshall Nirenberg, who pioneered the development of *cell-free* systems for studying the process of protein synthesis. Nirenberg discovered that protein synthesis will take place in a test tube containing a mixture of isolated ribosomes, amino acids, an energy source, and a cellular extract containing soluble components of the cell sap. More-

FIGURE 10-22 Data obtained in one of the early experiments supporting the existence of messenger RNA. Bacteria that had been growing in ^{15}N-containing media were switched to a medium containing ^{14}N, and were then infected with T4-bacteriophage in the presence of radioactive precursors of RNA and protein. When bacterial homogenates were subsequently analyzed by isodensity centrifugation, the newly synthesized viral RNA (labeled with ^{32}P) and viral-specific proteins were both found to be associated with ^{15}N-containing ribosomes. The colored lines indicate the positions in which "old" (^{15}N-containing) and "new" (^{14}N-containing) ribosomes would appear in such gradients.

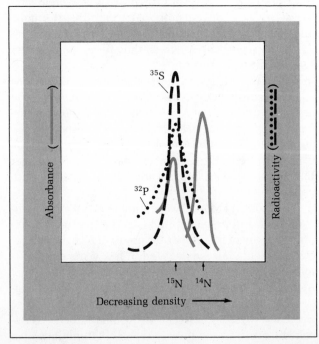

TABLE 10-3
Selected data illustrating the effects of synthetic RNAs on a cell-free protein-synthesizing system

Synthetic RNA Added	Radioactive Amino Acid(s) Added	Amount of Radioactivity Incorporated into Polypeptides (Counts/Min)
None	Phenylalanine	68
Poly(U)	Phenylalanine	38,300
Poly(U)	Glycine + alanine + serine + aspartic acid + glutamic acid	33
Poly(A)	Phenylalanine	50
Poly(C)	Phenylalanine	38

Source: Data from M. W. Nirenberg and J. H. Matthaei, *Proc. Natl. Acad. Sci. USA* 47(1961):1588–1602.

over, the rate of protein synthesis by such cell-free systems was found to be significantly enhanced by the addition of RNA.

To investigate the question of whether the added RNA molecules function as messengers that determine the amino acid sequences of each protein being manufactured, Nirenberg decided to add synthetic RNA molecules of known base sequence. One such RNA is polyuridylic acid, or **poly(U),** an RNA molecule containing the same base, uracil, repeated again and again. When poly(U) was added to the cell-free protein-synthesizing system, a marked stimulation was observed in the incorporation of one particular amino acid, phenylalanine, into polypeptide chains. Synthetic RNA molecules containing bases other than uracil did not stimulate phenylalanine incorporation, while poly(U) enhanced the incorporation of only phenylalanine (Table 10-3). It was therefore concluded that poly(U) directs the synthesis of polypeptide chains consisting solely of phenylalanine. This conclusion represented a crucial milestone in the development of the messenger RNA concept, for it was the first time that it had been demonstrated that the nucleotide base sequence of an RNA molecule influences the sequence of amino acids incorporated during protein synthesis.

As a result of the three lines of evidence outlined in the preceding paragraphs, it became obvious that the expression of genetic information encoded in DNA involves a two-stage process. During the first stage, termed **transcription,** the information encoded in the base sequence of DNA is utilized to direct the synthesis of complementary molecules of messenger RNA. In the second stage, referred to as **translation,** the base sequence of messenger RNA molecules is used to guide the sequential incorporation of amino acids into polypeptide chains. In spite of the apparent simplicity of this two-stage model, the individual steps involved in the transcription/translation sequence are exceedingly complex. In the remainder of this chapter, and in Chapters 11 and 12, the mechanisms by which these individual steps are carried out will be discussed in detail.

The Genetic Code

Once it had been established that the base composition of messenger RNA molecules determines the amino acid sequence of newly synthesized proteins, the question arose as to the nature of the coding process involved. Nirenberg's discovery that poly(U) stimulates the synthesis of polypeptide chains containing only phenylalanine does not in itself define the coding relationship between the base U in RNA and the phenylalanine residues that make up the polypeptide chain. It is clear that a single base in RNA cannot code for an amino acid because 20 different amino acids must be encoded and there are only 4 different kinds of bases in RNA. If the code for each amino acid were a doublet involving 2 adjacent bases (i.e., UU = phenylalanine), there would still only be $4 \times 4 = 16$ possible combinations of bases. To accommodate 20 different amino acids, the coding system must be based at the very least on nucleotide triplets (i.e., UUU = phenylalanine). In such a **triplet code** a total of $4 \times 4 \times 4 = 64$ different combinations of 4 bases is possible, which is more than sufficient to encode 20 different amino acids.

Although the above mathematical argument first led biologists to suspect the existence of a triplet code in the early 1950s, experimental evidence for such a code did not appear until 1961. In that year Francis Crick, Sidney Brenner, and their associates reported the results of a series of studies in which the dye *acridine orange* was utilized to induce mutations in T4-bacteriophage. Acridine dyes insert between adjacent bases in DNA, distorting the double helix in such a way that extra bases tend to be incorporated during DNA replication. Treatment of phage-infected bacteria with acridine orange therefore induces the production of virus particles with varying numbers of extra bases inserted in their DNA. When one or two extra bases were inserted in this way, the resulting virus particles were found to be unable to carry out subsequent infections. The presence of three extra bases, on the other hand, altered the properties of the virus only slightly.

On the basis of these observations, it was concluded that the insertion of extra bases causes a shift in the reading frame of DNA. To illustrate this concept of a *frameshift* mutation, let us consider a hypothetical gene in which the triplet CAT is repeated many times:

. . . (CAT)(CAT)(CAT)(CAT)(CAT)(CAT) . . .

A mutation involving the insertion a single extra base (X) will shift the reading frame beyond the point of insertion, creating a garbled sequence:

. . . (CAT)(XCA)(TCA)(TCA)(TCA)(TCA)(T . . .

The insertion of a second base (Y) shifts the reading frame one base further, still leaving a distorted sequence:

. . . (CAT)(XYC)(ATC)(ATC)(ATC)(ATC)(AT . . .

If a third base (Z) is inserted, however, all triplets beyond the point of insertion become properly aligned again:

. . . (CAT)(XYZ)(CAT)(CAT)(CAT)(CAT)(CAT) . . .

Even though this DNA sequence contains a small segment of misinformation in the region of the three inserted bases (XYZ), the vast majority of the sequence again codes for the proper amino acid sequence. This phenomenon, which would only be expected if the genetic code were read in units of three bases, explains why phage mutants containing three insertions behave relatively normally, while those with one or two do not.

Once the existence of a triplet code had been established, the next step involved the identification of the triplets that specify each of the 20 amino acids. Nirenberg's experiments with poly(U) clearly indicated that the sequence UUU in messenger RNA must code for the amino acid phenylalanine. The coding properties of other homogeneous polymers, such as poly(A), poly(C), and poly(G), were also studied in cell-free protein-synthesizing systems to determine the amino acids encoded by the triplets AAA, CCC, and GGG. Determining the coding assignments of triplets containing more than one type of base, however, required more complex experimental approaches. One clever idea, devised by Nirenberg and his associate Philip Leder, involved the use of individual triplets, rather than long RNA molecules, to identify specific amino acid codes. Since single triplets cannot code for the synthesis of polypeptide chains, monitoring the incorporation of radioactive amino acids into protein could not be employed as an assay for determining which amino acid is encoded by a given triplet. However, amino acids are normally carried to the ribosome by small RNA molecules termed *transfer RNAs*, and Nirenberg and Leder reasoned that they could measure the ability of synthetic triplets to induce binding of these amino acid–transfer RNA complexes to the ribosome. Adding the synthetic triplet UUU to a mixture of ribosomes and amino acid–transfer RNA complexes, for example, was found to cause the transfer RNA containing phenylalanine to bind to the ribosome. Utilizing this same approach with other RNA triplets, Nirenberg and Leder were able to assign amino acids to most of the triplets.

Though the above assay allowed unequivocal coding assignments to be made for many of the triplets, some triplets did not seem to induce the clear-cut binding of one specific type of amino acid–transfer RNA complex to the ribosome. These questionable cases were soon resolved by Har Gobind Khorana, who developed techniques for synthesizing short messenger RNAs of defined sequence. Examination of the amino acid composition of the polypeptides synthesized in the presence of these synthetic RNAs allowed independent ver-

ification of the assignments generated by the triplet binding assay. Khorana's approach also revealed the presence of several triplets that do not code for amino acids. The existence of such noncoding triplets was first suspected when Khorana studied the coding properties of synthetic RNAs containing certain repeated patterns of three bases such as

AGUAGUAGUAGUAGU . . .

Depending on where the ribosome begins reading such a message, one might expect three possible products:

(AGU)(AGU)(AGU)(AGU)(AGU) . . . ⟶ product 1

A(GUA)(GUA)(GUA)(GUA)GU . . . ⟶ product 2

AG(UAG)(UAG)(UAG)(UAG)U . . . ⟶ product 3

When this particular RNA was added as a messenger RNA to a cell-free protein-synthesizing system, however, only two products were detected: a polypeptide containing only serine (the amino acid encoded by AGU), and a polypeptide containing only valine (the amino acid encoded by GUA). The lack of a third product suggested one of two possibilities. Either the third triplet, UAG, also codes for serine or valine, or it does not code for any amino acid.

To distinguish between these alternatives, Khorana synthesized an RNA molecule containing the repeating sequence –AGAU–:

AGAUAGAUAGAUAGAUAGAU . . .

Examination of this pattern reveals that no matter where protein synthesis begins, the triplet UAG will soon be encountered. When this particular RNA is added to a cell-free protein synthesizing system, short chains of two or three amino acids are synthesized instead of the usual long polypeptide chains. The most reasonable explanation for this phenomenon is that every time the triplet UAG is encountered, it causes the growing peptide chain to be terminated:

Start

A G A U A G A U A G A U A G A U A G A U A G A U A

Arg   Ile   Asp   Arg   Ile   Asp

Stop   Stop

Start

A G A U A G A U A G A U A G A U A G A U A G A U A

Asp   Arg   Ile   Asp   Arg   Ile

Stop   Stop

Start

A G A U A G A U A G A U A G A U A G A U A G A U A

Ile   Asp   Arg   Ile   Asp   Arg

Stop

TABLE 10-4
The genetic code

First Base	Second Base				Third Base
	U	C	A	G	
U	UUU = phe	UCU = ser	UAU = tyr	UGU = cys	U
	UUC = phe	UCC = ser	UAC = tyr	UGC = cys	C
	UUA = leu	UCA = ser	(UAA = STOP)	(UGA = STOP)	A
	UUG = leu	UCG = ser	(UAG = STOP)	UGG = trp	G
C	CUU = leu	CCU = pro	CAU = his	CGU = arg	U
	CUC = leu	CCC = pro	CAC = his	CGC = arg	C
	CUA = leu	CCA = pro	CAA = gln	CGA = arg	A
	CUG = leu	CCG = pro	CAG = gln	CGG = arg	G
A	AUU = ile	ACU = thr	AAU = asn	AGU = ser	U
	AUC = ile	ACC = thr	AAC = asn	AGC = ser	C
	AUA = ile	ACA = thr	AAA = lys	AGA = arg	A
	AUG = met	ACG = thr	AAG = lys	AGG = arg	G
G	GUU = val	GCU = ala	GAU = asp	GGU = gly	U
	GUC = val	GCC = ala	GAC = asp	GGC = gly	C
	GUA = val	GCA = ala	GAA = glu	GGA = gly	A
	GUG = val	GCG = ala	GAG = glu	GGG = gly	G

Notes: Triplets represent codons in messenger RNA.
Amino acid abbreviations:

ala = alanine	gln = glutamine	leu = leucine	ser = serine
arg = arginine	glu = glutamic acid	lys = lysine	thr = threonine
asn = asparagine	gly = glycine	met = methionine	trp = tryptophan
asp = aspartic acid	his = histidine	phe = phenylalanine	tyr = tryrosine
cys = cysteine	ile = isoleucine	pro = proline	val = valine

The preceding data suggests that rather than coding for an amino acid, the triplet UAG functions as a punctuation mark that signals the end of the polypeptide chain. Two other triplets, UAA and UGA, have subsequently been found to play a similar role. Although these three **termination** or **stop signals** are sometimes referred to as *nonsense codes* because they do not code for amino acids, it should be emphasized that they play an important biological function in signaling when the synthesis of a particular polypeptide chain is complete.

As a result of the pioneering work carried out in the laboratories of Nirenberg and Khorana, the coding assignments for all 64 triplets were established by the mid-1960s. A summary of this information is provided in Table 10-4, which lists the amino acids specified by each of the 64 possible triplets occurring in messenger RNA. By convention such messenger RNA triplets, or **codons,** are always written in the 5' → 3' direction. Examination of the genetic code assignments reveals several important features. Perhaps the most striking is that 61 of the 64 triplets code for amino acids, even though there are only 20 different amino acids. Hence most amino acids are encoded by more than one codon. When more than one codon specifies the same amino acid, the first two bases are often the same. For example, the four codons for the amino acid proline all begin with CC. An analogous pattern holds for threonine (AC–), alanine (GC–), and glycine (GG–). For amino acids encoded by only two triplets, the first two bases are again

always the same. These general observations indicate that the first two bases of a triplet are more important than the third in determining codon specificity.

Because most amino acids are specified by more than one codon, the genetic code is said to be **degenerate.** In spite of the somewhat derogatory connotation of this word, degeneracy in no way implies malfunction. Each codon specifies a particular amino acid, and hence there is no ambiguity during protein synthesis as to which amino acid is to be incorporated for each triplet. It should be pointed out in this context that degeneracy even serves a useful function in enhancing the adaptability of the coding system. If only 20 codons existed, one for each of the 20 amino acids, then any mutation in DNA leading to formation of one of the other 44 possible triplets in messenger RNA would stop the synthesis of the growing polypeptide chain at that point. Since incomplete polypeptide chains are almost always nonfunctional, the susceptibility of such a coding system to disruption would be very great. With a degenerate code, on the other hand, DNA base substitutions produce codon changes in messenger RNA that usually result only in a change in the amino acid being encoded by a particular triplet. The change in protein function caused by such a single amino acid substitution is often minimal, and in some cases may even be advantageous. In addition, mutations in the third base of a codon frequently do not even change the amino acid being encoded.

Another noteworthy feature of the genetic code is the existence of special punctuation signals. In addition to the three terminator codons mentioned earlier, special codons are also utilized to signal the beginning of a polypeptide chain. The principal codon utilized for this purpose is AUG, which codes for the amino acid methionine. Methionine is therefore the first amino acid incorporated into almost all polypeptide chains, though it is frequently removed in the process of generating the final protein product. Unlike terminator codons, which function only at the end of polypeptide chains, AUG serves as both an initiator codon and as the codon that specifies the incorporation of methionine elsewhere in the polypeptide chain.

Although the codon assignments summarized in Table 10-4 were originally derived from experiments involving synthetic nucleotides and cell-free protein-synthesizing systems, their validity has since been confirmed by amino acid sequence analysis of mutant forms of naturally occurring proteins. It was seen earlier in the chapter, for example, that sickle-cell hemoglobin differs from normal hemoglobin in a single amino acid: Valine is present at one position instead of glutamic acid. Examination of the genetic code reveals that glutamic acid may be encoded by either GAA or GAG. Whichever triplet is used, a single base change could lead to the formation of a codon for valine. For example, GAA could have become mutated to GUA, or GAG might have been changed to GUG. Numerous other mutant proteins have been examined in a similar way, and

TABLE 10-5
Examples of amino acid substitutions observed in mutant proteins

Amino Acid Substitution	Corresponding Alteration in Triplet Code
Hemoglobin	
Glu → Val	GAA → GUA
	or GAG → GUG
His → Tyr	CAU → UAU
	or CAC → UAC
Asn → Lys	AAU → AAA
	or AAC → AAG
Tobacco mosaic virus coat protein	
Glu → Gly	GAA → GGA
	or GAG → GGG
Ile → Val	AUU → GUU
	or AUC → GUC
	or AUA → GUA
Tryptophan synthetase	
Tyr → Cys	UAU → UGU
	or UAC → UGC
Gly → Arg	GGA → CGA
	or GGG → CGG
Lys → Stop	AAA → UAA
	or AAG → UAG

in nearly all cases the amino acid substitutions can be explained by a single base substitution (Table 10-5).

In recent years sophisticated new sequencing techniques have provided a wealth of information concerning the nucleotide base sequences of genes and the amino acid sequences of their corresponding protein products. In addition to providing further verification for the codon assignments summarized in the genetic code table, such observations have revealed that the same set of codons, with a few minor exceptions, is employed by all organisms and cell types. The reason that the genetic code has remained virtually unchanged throughout evolution is not difficult to comprehend. Any mutation that alters the identity of the amino acid specified by a given triplet would cause amino acid substitutions to occur in every protein whose messenger RNA contains that particular triplet. Although an occasional mutation may be advantageous, the simultaneous occurrence of amino acid substitutions in virtually every protein would certainly be deleterious, and organisms exhibiting such widespread mutations would not be expected to survive.

The Role of Transfer RNA

Since the linear sequence of codons in messenger RNA ultimately determines the amino acid sequence of polypeptide chains, a mechanism must exist that enables codons to arrange amino acids in the proper order. The general nature of this mechanism was first proposed in 1957 by Francis Crick, before the triplet code or the existence of messenger RNA had even been demonstrated. With remarkable foresight Crick postulated that amino acids are not capable of directly recognizing RNA base sequences, and that "adaptor" molecules must therefore exist to mediate the interaction between amino acids and RNA. He further proposed that each adaptor molecule contains two sites, one specific for the binding of a particular amino acid and the other capable of recognizing the RNA base sequence coding for that amino acid.

In the year following Crick's proposal, the existence of such an adaptor molecule was confirmed by Mahlon Hoagland. While studying the process of protein synthesis in cell-free systems, Hoagland discovered that radioactive amino acids become attached to small molecules of RNA prior to incorporation into newly forming protein molecules. These small RNA molecules are localized predominantly in the soluble fraction of the cytoplasm (cytosol or cell sap fraction). Incubation of the cytosol fraction with radioactive amino acids and ATP leads to the covalent attachment of the amino acids to these small RNA molecules, even in the absence of ribosomes. When the resulting amino acid–RNA complexes are added to isolated ribosomes containing bound messenger RNA, protein synthesis begins, and the radioac-

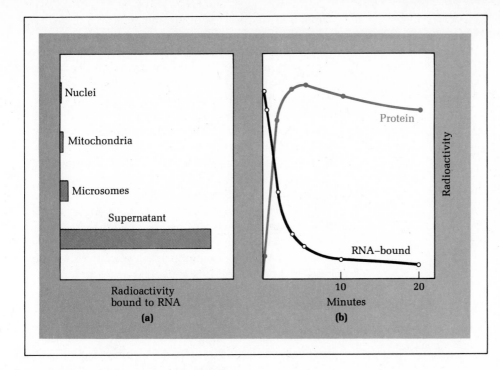

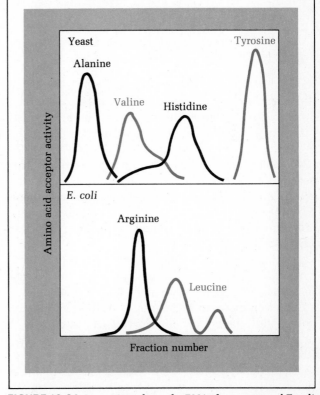

FIGURE 10-23 Experimental data obtained by Hoagland that led to the discovery of transfer RNA. (a) When various subcellular fractions were tested for their ability to bind radioactive amino acids to their RNA, only the supernatant (cytosol) fraction was found to exhibit appreciable activity. (b) When the supernatant fraction containing RNA-bound radioactive amino acids was added to microsomes, the labeled amino acids were found to be transferred from the RNA into newly synthesized proteins.

tive amino acids become incorporated into newly forming polypeptide chains (Figure 10-23). These observations led Hoagland to conclude that amino acids are initially bound to small RNA molecules, which then bring the amino acids to the ribosome for subsequent insertion into newly forming polypeptide chains.

The RNA discovered by Hoagland, termed **transfer RNA or tRNA,** consists of a heterogeneous family of RNA molecules with differing amino acid specificities. Evidence for the existence of multiple forms of tRNA was initially obtained from experiments in which the transfer RNA fraction was incubated with a single radioactive amino acid until no additional incorporation of radioactivity occurred. When a second radioactive amino acid was then added, more radioactivity was found to be incorporated. A similar increase can be observed with all 20 amino acids, suggesting that a separate population of transfer RNA molecules exists for each amino acid. The ultimate verification of this conclusion was obtained by separating the transfer RNA fraction into its individual components using biochemical fractionation techniques. This approach led to the discovery that at least one transfer RNA exists for each amino acid, and that for many amino acids, more than one occurs (Figure 10-24). Different transfer RNAs for the same amino acid are referred to as **isoaccepting tRNAs** and are designated by the amino acid abbreviation in superscript (e.g., tRNA^{ser}). Members of an isoacceptor family are indicated by subscript numerals (e.g., tRNA$_1^{leu}$, tRNA$_2^{leu}$, etc.).

The existence of specific transfer RNAs for each of the 20 amino acids raises the question of how the proper amino acid becomes joined to each transfer RNA mole-

FIGURE 10-24 Separation of transfer RNAs from yeast and *E. coli* into several distinct fractions specific for different amino acids. The separation technique employed in these particular experiments was countercurrent distribution, a special type of liquid–liquid chromatography in which molecules are successively partitioned between two immiscible solvents. It is clear from the bottom graph that more than one transfer RNA may exist for the same amino acid. Using this and other physical separation techniques, transfer RNAs specific for all twenty amino acids have been identified.

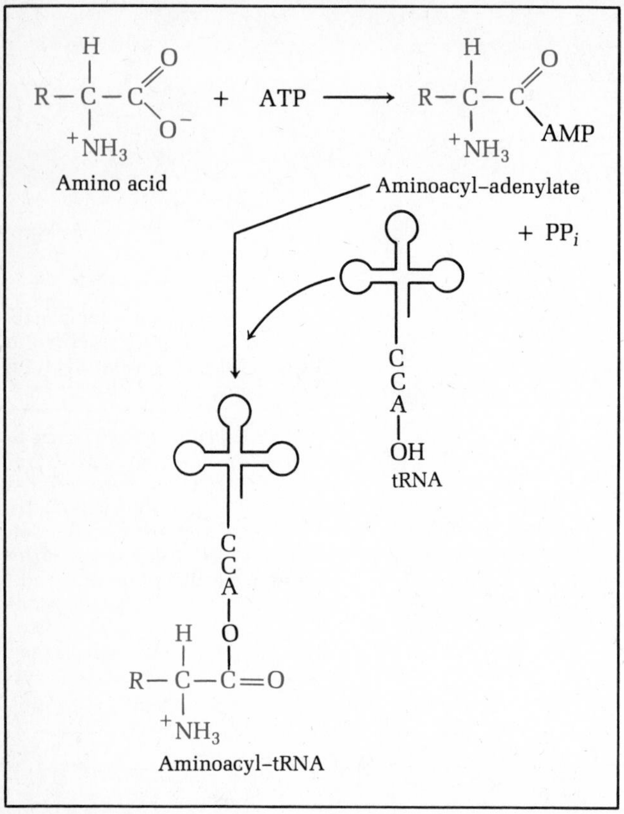

FIGURE 10-25 Summary of the net reaction catalyzed by aminoacyl-tRNA synthetases.

cule. This selectivity is made possible by a family of enzymes, termed **aminoacyl-tRNA synthetases,** which catalyze a two-step reaction that results in the covalent attachment of the carboxyl group of an amino acid to the 2′-hydroxyl group of the 3′-terminal ribose of a transfer RNA molecule (Figure 10-25). The final product of this reaction is referred to as an **aminoacyl-tRNA.** At least 20 different aminoacyl-tRNA synthetases have been identified in the cell sap, each catalyzing the covalent attachment of a particular amino acid to the appropriate isoacceptor tRNAs.

Once a particular aminoacyl-tRNA has been formed, the amino acid group must be transferred to a newly forming polypeptide chain. In order for such incorporation to occur at the proper location within the polypeptide chain, the aminoacyl-tRNA complex must be able to recognize the proper codon in messenger RNA. If transfer RNA molecules act as adaptors in the sense originally suggested by Crick, it follows that it is the transfer RNA molecule rather than the attached amino acid that recognizes the appropriate codon in messenger RNA. The first critical test of this idea was carried out by Francois Chapeville and Fritz Lipmann, who designed an elegant experiment involving the transfer RNA for the amino acid cysteine. It was known at the time that treatment of cysteine with a nickel catalyst converts it to the amino acid alanine. Chapeville and Lipmann therefore set out to determine what would

happen to the process of protein synthesis if cysteine were converted to alanine by nickel treatment *after* it had been attached to its appropriate tRNA. The immediate result, of course, is an alanine residue covalently attached to a transfer RNA molecule that normally carries cysteine. When this abnormal aminoacyl-tRNA complex was added to a cell-free protein-synthesizing system, alanine was found to be inserted into polypeptide chains in locations normally occupied by cysteine. It was therefore concluded that it is the transfer RNA molecule, rather than its attached amino acid, that is recognized by codons in messenger RNA.

The simplest mechanism that can be envisioned for this recognition process is that each codon in messenger RNA recognizes and binds to a complementary three-base sequence in an appropriate transfer RNA molecule. In order to test this view, it is necessary to know the nucleotide sequences of individual transfer RNA molecules. The first person to successfully determine the complete nucleotide base sequence of a transfer RNA molecule was Robert Holley, who reported the sequence of an alanine transfer RNA from yeast in 1965. In the ensuing years, sequence information has been obtained for more than a hundred other transfer RNA molecules. One of the most important generalizations to emerge from this data is that all tRNA molecules contain a triplet sequence that is complementary to a codon that specifies the amino acid carried by the tRNA in question. Moreover, this triplet is always situated in a similar location within the overall structure of the tRNA molecule. This three-base sequence is referred to as an **anticodon** because its identical location in all tRNAs suggests that it recognizes and binds to mRNA codons during the process of protein synthesis. As in all situations where complementary base pairing is involved, the messenger and transfer RNA molecules are oriented in opposite directions in regard to 5′ → 3′ polarity. Since base sequences are always written in the 5′ → 3′ direction, one must take care to write anticodon triplets properly. For example, the anticodon that recognizes the codon CUU (leucine) is properly written as AAG, not GAA:

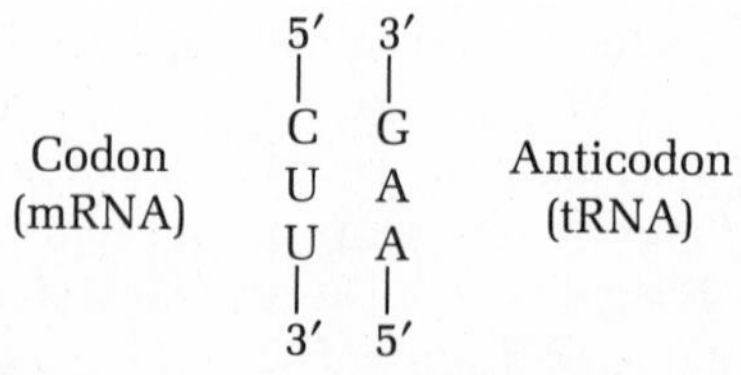

$$
\begin{array}{ccc}
 & 5' & 3' \\
 & | & | \\
\text{Codon} & \text{C} & \text{G} & \text{Anticodon} \\
\text{(mRNA)} & \text{U} & \text{A} & \text{(tRNA)} \\
 & \text{U} & \text{A} & \\
 & | & | & \\
 & 3' & 5' &
\end{array}
$$

In addition to the presence of an anticodon, transfer RNA molecules exhibit several other characteristic features. All possess the sequence CCA at their 3′ ends; the terminal A residue of this sequence is the site to which amino acids are covalently bound during the aminoacyl-tRNA synthetase reaction. Of the roughly 75–100 bases that comprise a typical tRNA molecule, 10–15

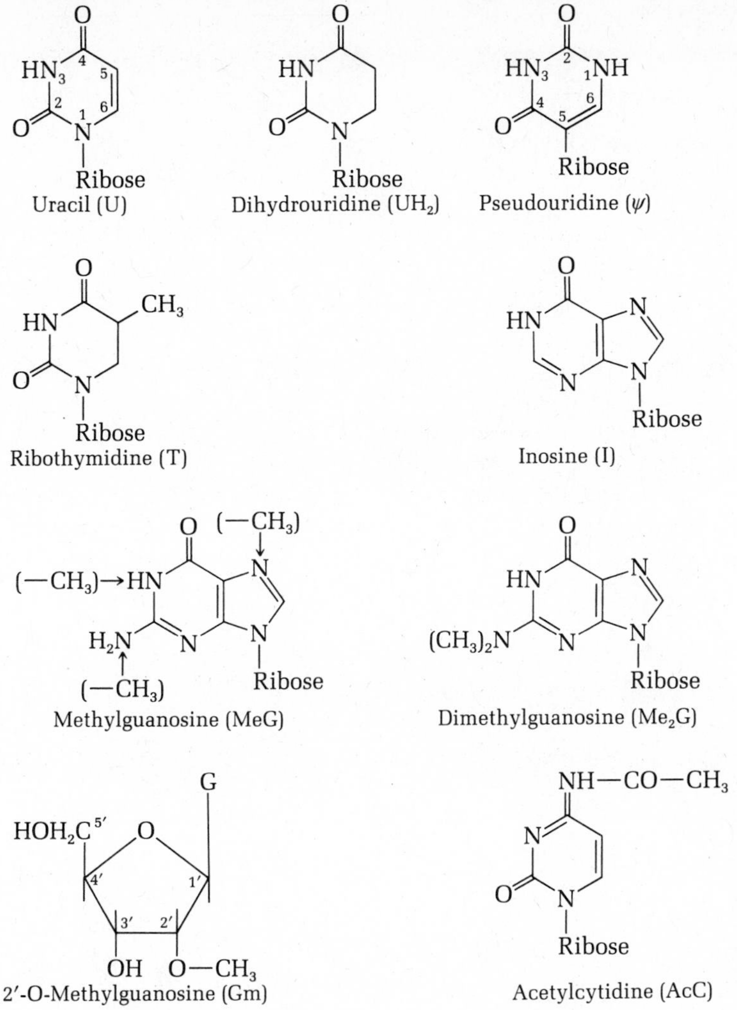

Uracil (U)

Dihydrouridine (UH₂)

Pseudouridine (ψ)

Ribothymidine (T)

Inosine (I)

Methylguanosine (MeG)

Dimethylguanosine (Me₂G)

2'-O-Methylguanosine (Gm)

Acetylcytidine (AcC)

FIGURE 10-26 Structures of some of the unusual bases occurring in transfer RNA molecules. The structure of uracil is presented for comparative purposes. Arrows in methylguanosine point to the three places where methyl groups can be attached.

percent are usually modified in some way. The principal modifications include methylation of bases and sugars, as well as creation of unusual bases like pseudouridine (ψ), inosine, dihydrouridine, and ribothymidine (Figure 10-26). An additional generalization to emerge from the sequencing data is that all transfer RNA molecules contain base sequences that are capable of hydrogen bonding with each other in such a way that the molecule is induced to fold into a "cloverleaf" pattern containing three loops and four base-paired stems (Figure 10-27). The anticodon always occurs in the middle loop, while the other two loops are characterized by a particular type of base composition. One of these loops is termed the **TψC-loop** because it always contains the sequence ribothymidine-pseudouridine-cytidine, while the other loop tends to be enriched in dihydrouridine and is therefore referred to as the dihydrouridine, or **D-loop.**

Our general view of transfer RNA structure has recently been refined by X-ray crystallographic studies carried out independently in the laboratories of Alexander Rich and Aaron Klug. This approach has permit-

ted the three-dimensional structure of the transfer RNA molecule to be elucidated to a resolution of 0.3 nm. In addition to confirming the presence of the base-paired regions predicted by the cloverleaf model, these studies have revealed that the overall shape of the molecule resembles a twisted "L." The amino acid acceptor site and the anticodon loop are situated at the opposite ends of the L, while the TψC- and D-loops occur at the bend (Figure 10-28).

HOW IS GENETIC INFORMATION ALTERED?

We have now seen that the complementary base-pairing mechanism mediates both the replication of DNA molecules and the transcription of DNA into messenger RNA molecules that direct the synthesis of specific proteins. Before proceeding to the chapters in which these events will be described in detail, the remaining property of DNA that needs to be described in general terms concerns its ability to undergo alterations in base sequence, or **mutations.** Though the short-term survival

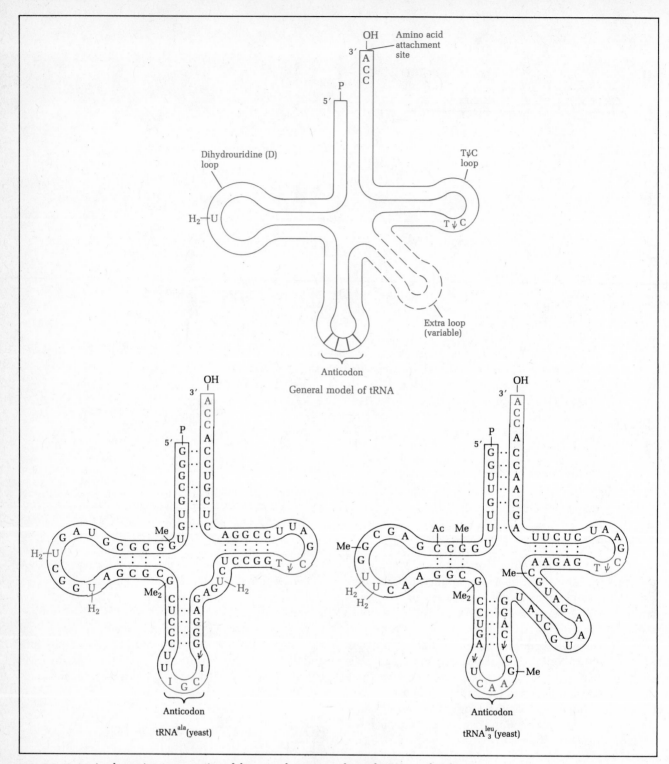

FIGURE 10-27 A schematic representation of the general structure of transfer RNA molecules. In spite of major differences in their overal base sequences, all transfer RNAs are organized according to the same cloverleaf pattern. Abbreviations for the unusual bases are defined in Figure 10-26.

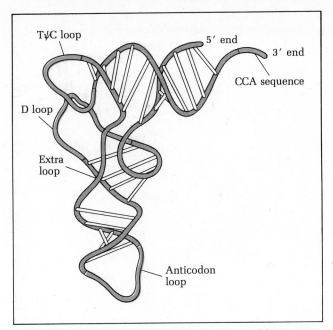

FIGURE 10-28 Three-dimensional representation of the structure of yeast phenylalanine transfer RNA.

and reproduction of a given individual require that DNA replication be extremely accurate, long-term evolution is dependent upon changes in the information encoded within DNA. It has therefore been interesting to discover that the DNA replication process, in spite of its high degree of accuracy, has built-in mechanisms that ensure the occurrence of a finite number of errors.

One reason for the spontaneous occurrence of

FIGURE 10-29 Structures of the two tautomeric forms of thymine, showing how the rare "enol" configuration base-pairs with guanine instead of with adenine.

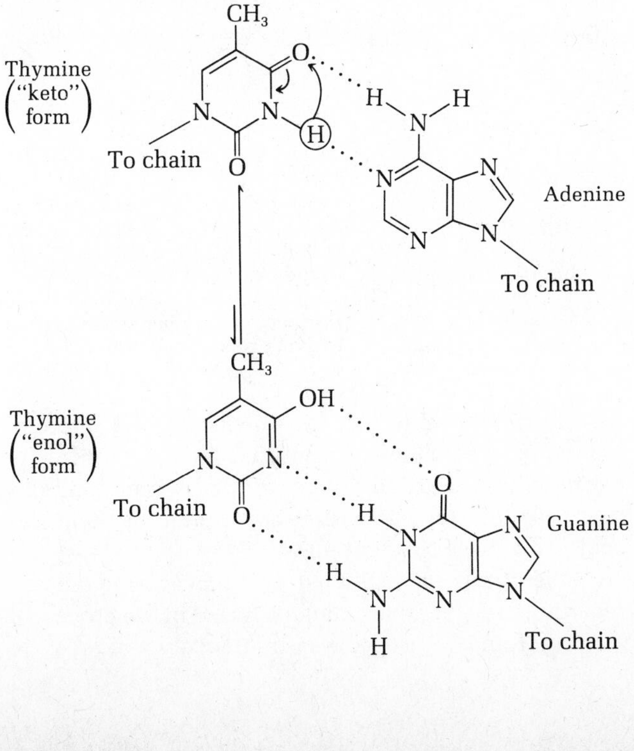

errors in the DNA replication process is that nitrogenous bases exist in multiple structural forms, called **tautomers,** which are in equilibrium with one another. The base thymine, for example, exists in two forms that differ in the arrangement of the chemical groups involved in hydrogen bonding. In the predominant form of thymine, called the **keto** form, these groups are arranged in a way that facilitates hydrogen bonding to the base adenine (Figure 10-29). In the less prevalent **enol** form of thymine, these same chemical groups are oriented in a way that promotes hydrogen bonding to the base guanine. Since a small amount of the enol form of thymine is always in equilibrium with the more predominant keto form of thymine, thymine-guanine base pairs will always be generated in small but finite numbers during normal DNA replication. When a DNA molecule containing such an abnormal T-G base pair is subsequently replicated, the guanine residue will serve as a template for the incorporation of its complement, cytosine; the final result is therefore a G-C base pair where a T-A pair had originally existed. Since DNA repair mechanisms would not recognize anything inherently wrong with a G-C base pair, such a mutation would be replicated indefinitely.

The existence of tautomeric forms of the nitrogenous bases is not the only factor that promotes spontaneous mutation. Within a given strand of the DNA double helix, regions often exist that are complementary to one another. The existence of such self-complementarity within a given DNA chain fosters the formation of "hairpin" loops that can cause either insertion or deletion mutations to occur as DNA is replicated (Figure 10-30).

In addition to such localized changes in DNA base sequence, rearrangements also occur over longer distances. In the early 1940s analyses of the inheritance patterns of various traits in corn carried out by Barbara McClintock led her to postulate the existence of movable genetic elements that permit genes to migrate from one location to another. Although these ideas did not fit into the main body of genetic knowledge available at the time, the reality of such mobile DNA sequences has recently been confirmed using modern DNA isolation and sequencing techniques. Such mobile DNA sequences, now referred to as **transposable elements,** have been identified in a variety of eukaryotic and prokaryotic organisms. Transposable elements can be subdivided into two general categories: **insertion sequences,** which are less than a few thousand bases in length and contain only genes involved in the movement of the transposable element itself, and **transposons,** which tend to be much larger and may contain numerous genes in addition to those involved in moving the sequence.

Transposable elements are generally flanked on both sides by an inverted repeat sequence of nucleotides that functions to facilitate recognition and removal

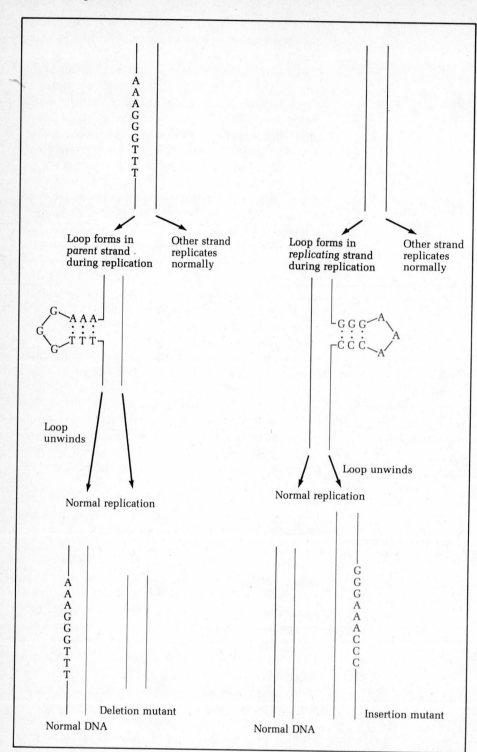

FIGURE 10-30 This diagram illustrates how deletion and insertion mutations can be generated by the formation of loops within DNA strands during the process of replication. Such loops are caused by the existence of complementary regions within a given DNA strand.

of the transposable DNA segment (Figure 10-31). Some transposable elements are selectively inserted into specific DNA sites, while others move about more randomly. The movement of transposable elements is controlled at least in part by two proteins encoded by genes located within the elements themselves. One of these proteins, termed a **transposase**, initiates excision of the element from one region of a DNA molecule and

its insertion somewhere else. The second protein, known as a **resolvase**, is required for completing the excision and insertion process, and also functions as a repressor that regulates the transcription of the genes coding for these two proteins.

One of the more interesting discoveries to emerge regarding transposable elements is that in the process of moving from site to site, they influence the expression

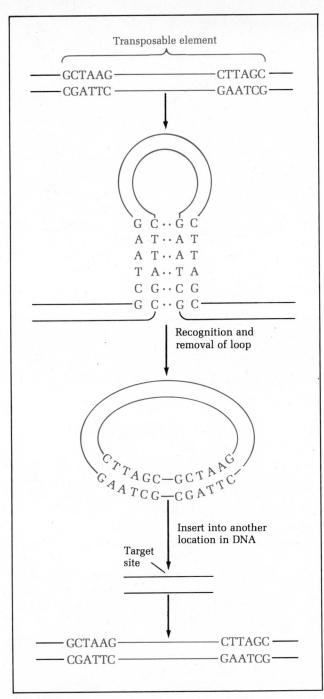

FIGURE 10-31 A schematic model summarizing one way in which inverted repeat sequences might facilitate the movement of transposable elements.

portant role in generating the genetic diversity required for evolution, the immediate effects of most mutations are deleterious and excessive changes in DNA base sequence are therefore undesirable. In recent years there has been an alarming increase in the concentration of mutation-inducing chemicals in the environment. Some of these mutagenic chemicals interact directly with the nitrogenous bases of DNA, causing their structures and hence their base-pairing properties to be altered. Other mutagens, known as *intercalating agents*, insert between adjacent bases of the DNA double helix, distorting the helical structure and increasing the chances for the insertion or deletion of bases. In addition to chemicals, various types of physical radiation also induce mutations. Ultraviolet light, for example, causes adjacent thymine residues to become covalently joined to each other, forming **thymine dimers** that interfere with proper DNA replication. X-rays, on the other hand, remove electrons from DNA, generating highly reactive intermediates that induce various kinds of chemical alterations.

Once a DNA molecule that has been exposed to a chemical or physical mutagen undergoes a subsequent round of replication, the mutation becomes very difficult, if not impossible, to repair. To illustrate this point, let us consider what happens to a DNA molecule that has been briefly exposed to the chemical mutagen, nitrous acid. Under acid conditions this substance removes amino groups from nitrogenous bases. In the case of the base cytosine, removal of an amino group converts it to the base uracil (Figure 10-32). Since the cytosine was originally hydrogen-bonded to guanine, the immediate result is the conversion of a G-C base pair to a G-U base pair. Since a G-U base pair is obviously mismatched, it could conceivably be recognized and repaired by enzymes that normally function in DNA repair. However, if a DNA molecule containing such an abnormal base pair undergoes replication prior to being repaired, the "incorrect" base (uracil) will serve as a template for its complement (adenine), yielding a new DNA double helix with an A-U base pair. Since an A-U base pair is properly hydrogen bonded, it can be faithfully replicated by the normal base-pairing mechanism. During the next round of replication, the adenine will serve as a template for the incorporation of its normal

and organization of neighboring genes. Since transposable elements can migrate with considerable frequency, this process has the potential for causing rather rapid and dramatic genetic changes. Such changes are now believed to be a major contributing factor to an organism's ability to undergo evolutionary change.

Although a certain amount of spontaneous mutation and rearrangement of DNA sequences plays an im-

FIGURE 10-32 Exposure of DNA to nitrous acid causes bases to undergo deamination (loss of amino groups), producing new bases with differing base-pairing properties. In the case of cytosine, for example, nitrous-acid-induced deamination produces the base uracil, which forms hydrogen bonds with adenine instead of guanine.

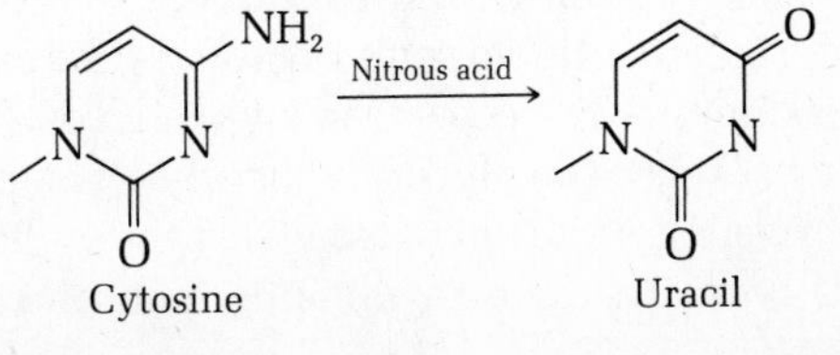

Cytosine → Nitrous acid → Uracil

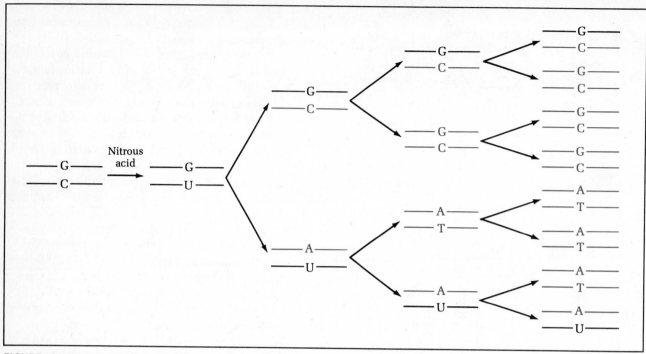

FIGURE 10-33 Schematic diagram illustrating how the nitrous-acid-induced conversion of the base C to the base U leads to the substitution of A-T base pairs for G-C base pairs during subsequent rounds of replication.

complement, thymine. Hence the final result is the creation of an A-T base pair where a G-C base pair had originally been present (Figure 10-33). Since DNA repair mechanisms will not recognize anything abnormal about an A-T base pair, this mutation will now be indefinitely replicated. Hence the process of DNA replication eventually stabilizes mutations of this type.

Fortunately, molecular mechanisms exist for repairing many types of DNA damage before they have become stabilized by DNA replication. These DNA repair pathways involve several types of enzymes, including nucleases, DNA polymerases, and DNA ligases. For example, in one widely employed method for repairing thymine dimers (Figure 10-34), an endonuclease initially cleaves the DNA strand containing the dimer, allowing the two strands to separate locally. DNA polymerase then catalyzes the synthesis of a new stretch of complementary DNA and degrades the old, abnormal region of DNA. Finally, DNA ligase joins the repaired segment to the rest of the DNA molecule. This overall process is referred to as **excision-repair** because it involves the removal and replacement of the mutated DNA segment. Although excision-repair mechanisms are quite common, the initial mechanism for recognizing and cleaving the strand containing the mutation may vary significantly. In some organisms, for example, thymine dimers are recognized by a *glycosylase* enzyme that cleaves the glycosidic bond between one of the dimerized thymines and its sugar. The resulting site, which is now missing a thymine, is then nicked by a special nuclease, followed by excision-repair mediated by DNA polymerase and DNA ligase.

Mutant bacteria deficient in DNA polymerase I have been found to exhibit a reduced ability to repair thymine dimers, indicating that this particular form of DNA polymerase is crucial to the normal repair process. Cells defective in their ability to repair DNA have also been identified in humans suffering from an inherited disease known as *xeroderma pigmentosum*. Such individuals are extremely sensitive to sunlight and tend to develop multiple skin cancers, suggesting that DNA repair plays an important role in protecting people from the mutagenic effects of ultraviolet light.

TECHNICAL ADVANCES IN THE STUDY OF INFORMATION FLOW

The intent of the present chapter has been to describe the general principles that govern the flow of genetic information within cells. Given this general background, the following chapters shall investigate the relationship of these processes to the organelles and subcellular structures involved. Much of our current understanding in this area has been made possible by the development of sophisticated biochemical techniques for isolating and analyzing the properties of nucleic acid molecules. We shall conclude the present chapter, therefore, with a description of the techniques that have been most useful in this regard.

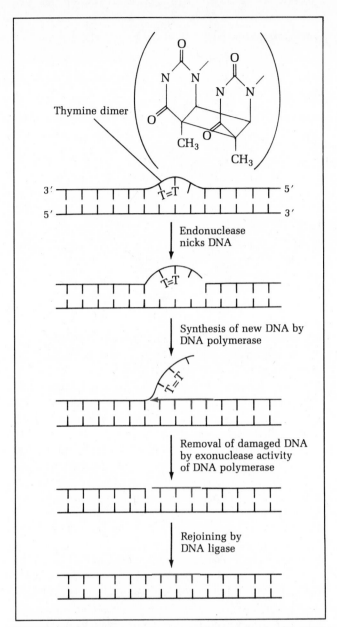

FIGURE 10-34 Summary of the enzymatic steps involved in DNA repair. In this particular example the repair process is employed to remove a thymine dimer, a type of defect triggered by exposure to ultraviolet light.

Nucleic Acid Hybridization

In 1960, Julius Marmur and Paul Doty discovered that the two strands of a DNA double helix can be dissociated from each other and then put back together in proper alignment under appropriate experimental conditions. This pioneering discovery has formed the basis for the development of a diverse family of techniques, collectively referred to as *nucleic acid hybridization,* in which the base sequences of two nucleic acid chains are compared by testing their ability to form hybrids with each other. Such hybrids can form only when the two chains have complementary sequences that are capable of hydrogen bonding to one another. This approach, which has been utilized to study DNA-DNA, DNA-RNA, and even RNA-RNA interactions, has had a profound impact on our understanding of gene organization and expression. Because the following chapters describe many experiments involving nucleic acid hybridization, it is important to understand how such experiments are carried out and interpreted.

When DNA is exposed to high temperature or alkaline pH, the hydrogen bonds holding the two strands together are disrupted and the molecule dissociates into two separate polynucleotide chains. DNA molecules that have been dissociated in this way are said to be **denatured.** Denaturation is accompanied by several changes in the physical properties of the DNA solution, including a decrease in viscosity, an increase in density, and an increase in the ability to absorb ultraviolet light in the 260-nm region of the electromagnetic spectrum.

Of the three physical changes mentioned above, the increase in absorption at 260 nm is most easily used to monitor the progress of denaturation. If the absorbance of a DNA solution is monitored as the sample is gradually heated, a temperature is eventually reached at which the absorbance begins to increase. The absorbance will then continue to increase until denaturation is complete (Figure 10-35). The total increase in absorbance, referred to as the **hyperchromic effect,** varies among DNA samples, but generally ranges from 20 to 60 percent of the initial value. For DNA samples that are properly base paired, the temperature range over which denaturation occurs is relatively small. Since denatura-

FIGURE 10-35 A graph showing the increase in the absorption of ultraviolet light that accompanies the heat-induced conversion of DNA from the double-stranded to the single-stranded state.

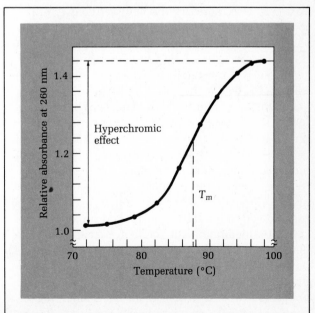

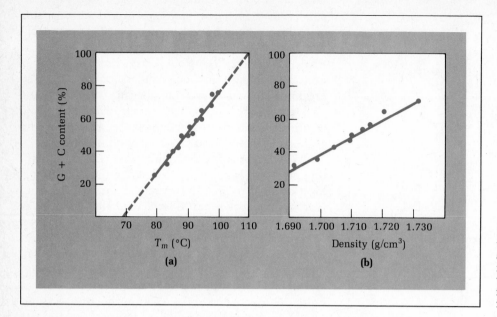

FIGURE 10-36 Effects of GC content on the melting temperature (a) and buoyant density (b) of DNA molecules. Data obtained from various types of DNA with differing GC contents have been pooled for each graph.

tion occurs over a narrow temperature range, just like the melting of a simple crystal, thermal denaturation of DNA is often referred to as "melting." The **melting temperature,** or T_m, is defined as the temperature at which the transition from double-stranded to single-stranded DNA has proceeded halfway to completion.

One prominent factor influencing the denaturation of double-stranded DNA molecules is their relative content of A-T and G-C base pairs. The G-C base pair, held together by three hydrogen bonds, is tighter and more compact than the A-T base pair, which is held together by only two hydrogen bonds (see Figure 10-5). Since more energy is therefore required to disrupt G-C base pairs, DNA samples with high G-C contents will melt at higher temperatures than samples with lower G-C contents (Figure 10-36). The greater compactness of the G-C base pair also increases the buoyant density of DNA, making it possible to employ isodensity centrifugation to separate DNA molecules with differing G-C contents.

In nucleic acid hybridization experiments, double-stranded DNA must first be dissociated into its individual strands before the ability of these strands to form complementary hybrids with other DNA or RNA molecules can be tested. Hybridization of dissociated DNA chains to other single-stranded DNA or RNA molecules is carried out by mixing the two nucleic acids together and incubating them under appropriate conditions of temperature and ionic strength. Hybridization is generally carried out either in solution or with one of the nucleic acids immobilized on an inert support such as a synthetic membrane surface. It has been empirically determined that a complementary region as short as a dozen bases is sufficient to cause two nucleic acids to hybridize to one another. Since nucleic acid molecules are generally much longer than this, a small region of complementarity could conceivably serve to hold together two molecules whose sequences are largely

noncomplementary. For this reason nucleic acid hybridization is usually estimated after the hybridized sample has been treated with an enzyme, such as **S1 nuclease,** that selectively degrades single-stranded nucleic acids. Only base sequences that have successfully hybridized to one another will survive treatment with such nucleases (Figure 10-37).

The principles of nucleic acid hybridization have been applied to biological problems in a wide variety of ways. These various approaches can be subdivided, however, into the major categories summarized below.

Saturation Hybridization   One common application of nucleic acid hybridization is in determining the relative abundance of a particular type of sequence in a DNA sample. To cite an example, this approach has been used to determine the percentage of an organism's nuclear DNA that is complementary to ribosomal RNA. In this case, denatured DNA is hybridized to increasing amounts of radioactive ribosomal RNA. When the RNA concentration reaches a level at which no further increase in hybridization occurs, the DNA is said to be saturated (Figure 10-38). At this point, all the DNA sequences complementary to ribosomal RNA have become hybridized. By measuring the initial amount of DNA and the quantity of radioactive RNA that has become hybridized, the percentage of the total DNA that is complementary to ribosomal RNA can be calculated. Since the length of the ribosomal RNA molecule is known, such data can be used to estimate the total number of ribosomal RNA genes present in DNA. Similar approaches have been employed to determine the relative abundance of many other kinds of gene sequences.

Competition Hybridization   In contrast to saturation hybridization, competition hybridization is employed to determine the extent to which two nucleic acid sam-

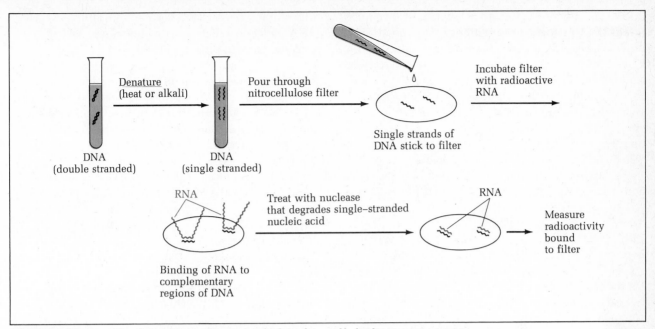

FIGURE 10-37 An outline of one of the many ways in which nucleic acid hybridization experiments can be carried out. In this example, denatured DNA is immobilized on a membrane filter to prevent its individual strands from hybridizing back to one another. After RNA has been hybridized to the immobilized DNA, nonhybridized nucleic acid is destroyed with a nuclease that degrades single-stranded nucleic acid. Similar procedures can be employed for DNA–DNA hybridization.

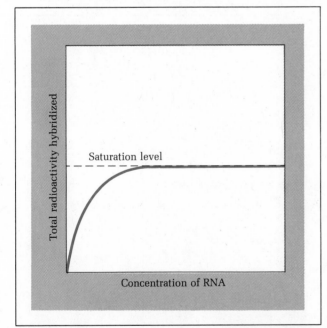

FIGURE 10-38 Kinetics of a saturation hybridization experiment in which radioactive DNA is hybridized to increasing concentrations of unlabeled RNA. When further increases in RNA concentration cause no additional hybridization, saturation is said to be achieved. By calculating the percentage of the total radioactivity that has become hybridized at this point, one can estimate how much of the DNA is complementary to the RNA sample in question.

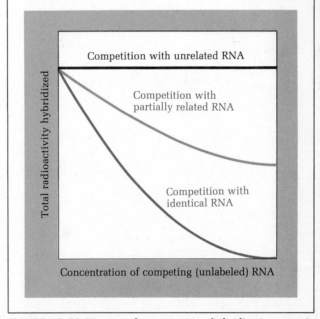

FIGURE 10-39 Kinetics of a competition hybridization experiment in which a sample of radioactive RNA is hybridized to DNA in the presence of varying concentrations of another type of unlabeled RNA. The degree to which hybridization of the radioactive RNA is decreased by the presence of the unlabeled RNA indicates how closely related the two RNA samples are to each other.

ples resemble each other in sequence. This approach is most commonly used to compare two different samples of RNA, one of which has been radioactively labeled. If the two RNA samples resemble each other in sequence, they would be expected to hybridize to the same regions

of a denatured DNA molecule. Because the two RNAs are competing for hybridization to the same DNA sites, increasing the amount of unlabeled RNA in the hybridization mixture causes a decrease in the amount of radioactive RNA that can hybridize. RNA samples that differ in sequence, on the other hand, hybridize to different DNA regions and so do not compete with each other. The extent to which increasing amounts of unlabeled RNA interfere with the hybridization of a labeled RNA is therefore a reflection of the degree to which the base sequences of the two RNA populations resemble each other (Figure 10-39).

Southern and Northern Blotting Hybridization can also be used to localize specific nucleic acid sequences that have been separated by electrophoresis. In the Southern blotting procedure, DNA that has been fractionated by gel electrophoresis is transferred to a membrane filter by "blotting" the gel with the membrane. This process yields a replica of the gel that can then be hybridized with a radioactive nucleic acid. Autoradiography of the hybridized membrane allows one to localize the DNA fragments that are complementary to the radioactive nucleic acid employed as probe (Figure 10-40). An analogous procedure, termed Northern blotting, utilizes a similar approach to analyze electropho-

retically fractionated RNA samples for the presence of RNA molecules that are complementary to a radioactive nucleic acid probe.

In Situ and Colony Hybridization In addition to being used to characterize isolated nucleic acid molecules, hybridization techniques can also be applied to situations in which nucleic acids reside in their original location (*in situ*) within cells. This approach of ***in situ hybridization*** has been used, for example, to determine the subcellular localization of the DNA sequences coding for ribosomal RNA. To accomplish such an objective, cells are first fixed and sectioned for microscopic examination. The tissue section is then incubated at high pH to denature the DNA, and radioactive ribosomal RNA is added. After incubation under suitable hybridization conditions, unhybridized radioactivity is washed away and the section is examined by autoradiography. As we shall see in a later chapter, such experiments have revealed that the DNA that codes for ribosomal RNA is localized within the nucleolus (see

FIGURE 10-40 A schematic representation of the blotting techniques utilized to localize specific nucleic acid fragments that have been fractionated by gel electrophoresis. In "Southern blotting" the starting sample is DNA, while in "Northern blotting" the sample is RNA.

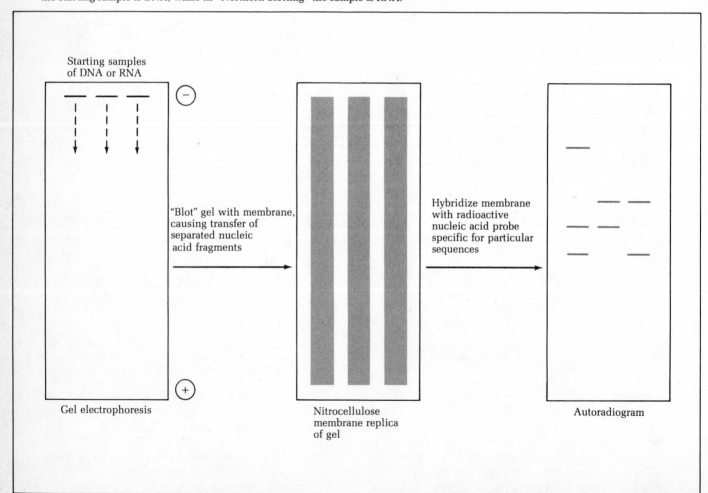

Starting samples of DNA or RNA

$\ominus$

$\oplus$

Gel electrophoresis

"Blot" gel with membrane, causing transfer of separated nucleic acid fragments

Nitrocellulose membrane replica of gel

Hybridize membrane with radioactive nucleic acid probe specific for particular sequences

Autoradiogram

Figure 12-7). By varying the type of radioactive nucleic acid utilized as a probe, this approach can be used for determining the subcellular localization of other types of DNA and RNA.

The principle of nucleic acid hybridization has also been utilized to determine whether or not a given population of cells contains a particular nucleic acid sequence. In one widely used version of this approach, termed **colony hybridization,** an agar plate containing bacterial colonies is blotted with a piece of nitrocellulose membrane or filter paper to make a replica of the pattern of colonies. The replica is then hybridized with a specific radioactive nucleic acid and analyzed by autoradiography. In this way it is possible to identify those colonies that contain nucleic acid molecules complementary to the radioactive probe. This technique has been widely used in conjunction with the recombinant DNA cloning techniques described below.

Recombinant DNA Technology and Gene Cloning

The nuclear DNA of a typical eukaryotic cell contains more than a billion base pairs, while an average gene measures only a few hundred to several thousand base pairs in length. It is therefore obvious that the organization and behavior of an individual gene can only be studied after it has been separated from the large mass of DNA sequences with which it is normally associated. An extremely powerful approach for accomplishing this objective, termed **recombinant DNA technology,** has evolved in recent years as a result of several experimental developments.

The first of these was the discovery, pioneered in the laboratories of Werner Arber, Hamilton Smith, and Daniel Nathans, that prokaryotic cells produce enzymes that cleave double-stranded DNA molecules at specific locations. Such enzymes, termed **restriction endonucleases,** recognize specific DNA sequences four to six

base pairs in length. The enzyme EcoRI, for example, cleaves DNA every place the sequence -GAATTC- occurs, the enzyme HaeIII cuts DNA at -GGCC- sequences, BamHI cleaves at -GGATCC- sequences, and so forth. Over a hundred restriction enzymes, most of which recognize different cleavage sites, have been isolated from bacteria. Each restriction enzyme is named using a three-letter abbreviation for the host organism, followed by a single-letter strain designation (if applicable) and a Roman numeral. Hence "EcoRI" refers to the first restriction enzyme isolated from Escherichia coli strain R, "HaeIII" refers to the third restriction enzyme to be isolated from the bacterium Haemophilus aegyptius, and so forth. Restriction enzymes normally function to degrade foreign DNA that has entered bacterial cells, thereby providing a defense against infection by DNA viruses. The DNA of the bacterium itself is protected from degradation by methylating some of the bases that are normally recognized by the restriction enzymes involved.

Because restriction endonucleases cleave DNA at precisely defined *restriction sites,* these enzymes have become an invaluable experimental tool for cutting large DNA molecules into reproducible sets of smaller fragments. By comparing the fragment sizes obtained after DNA samples have been digested with various restriction enzymes, it is possible to create a **restriction map** that identifies the locations of the various restriction sites relative to one another. Such maps have become extremely useful in analyzing and comparing DNA sequences. Another use of restriction nucleases has emerged from the discovery that many of these enzymes cut the two strands of the DNA double helix in a staggered fashion, generating short single-stranded regions that are complementary to each other (Figure 10-41). Since these complementary sequences tend to hybridize to one another, they are referred to as *sticky* or *cohesive* ends.

The cohesive ends generated by cutting DNA with

FIGURE 10-41 Mechanism of action of two kinds of restriction enzymes. BamHI is an example of a restriction enzyme that cuts DNA in a staggered fashion, generating single-stranded cohesive ends. HaeIII is an example of an enzyme that cuts both strands in the same location, generating "blunt" ends. Both kinds of restriction sites have two-fold rotational symmetry, that is, the sequence reads the same on each strand when examined in the 5′ → 3′ direction.

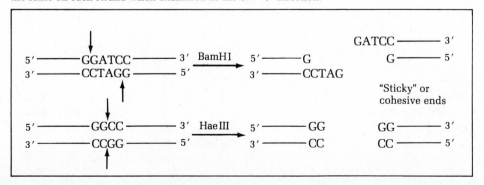

restriction enzymes have turned out to be an extremely important tool because they provide a relatively simple mechanism for joining together DNA fragments obtained from different sources. In essence, any two DNA molecules that have been cut with the same restriction enzyme can be joined together in this way. Pioneering experiments carried out in the early 1970s in the laboratories of Herbert Boyer, Stanley Cohen, and Paul Berg utilized this approach to join together DNA fragments obtained from different sources, generating **recombinant DNA** molecules that were subsequently introduced into bacterial cells for replication. By identifying bacterial colonies whose recombinant DNA molecules contain particular sequences and then growing up such colonies in mass, it is possible to obtain large quantities of specific types of DNA. This general approach of DNA **cloning** has permitted hundreds of different types of gene sequences to be isolated and characterized. Although the specific details of such cloning experiments vary among experiments, the same four steps are always involved: (1) insertion of foreign DNA into a cloning vector, (2) introduction of the resulting recombinant DNA molecule into a host cell for amplification, (3) selection of those host cells containing the recombinant DNA, and (4) identification of cells containing the foreign DNA sequences of interest.

1. *Insertion of foreign DNA into cloning vectors.* DNA preparations containing gene sequences to be cloned can be obtained in two general ways. One approach is to start with total nuclear DNA, breaking it into smaller fragments by physical shearing or treatment with restriction enzymes. An alternative possibility is to generate DNA molecules by copying messenger RNA with the enzyme reverse transcriptase. This reaction generates a population of molecules referred to as **complementary DNA** or **cDNA** molecules because they are complementary in sequence to the messenger RNA employed as template.

In order to clone DNA molecules of either of the above types, it is first necessary to attach them to a second kind of DNA, termed a **cloning vector,** that is capable of rapid replication in an appropriate host cell (usually a bacterium or yeast, though certain cloning vectors can be incorporated directly into mammalian cells as well). Some cloning vectors are related to viruses, while others are small circular DNA molecules, known as **plasmids,** that replicate in the cytoplasm of bacterial cells independent of the chromosomal DNA. The usefulness of restriction enzymes for facilitating the insertion of foreign DNA sequences into cloning vectors is nicely illustrated by experiments involving the plasmid **pBR322,** a cloning vector that is widely used in recombinant DNA experiments. This plasmid contains single cleavage sites for several restriction enzymes that generate cohesive ends, including PstI, BamHI, and HindIII. When pBR322 is cleaved with one

of these enzymes, the circular plasmid is converted to a linear molecule of DNA. If the foreign DNA to be inserted into the vector is also cut with the same restriction enzyme, it will contain the same kinds of sticky ends and will thus hybridize to the cut ends of the plasmid. The enzyme DNA ligase, which catalyzes the end-to-end joining of DNA molecules, can then be used to link the inserted DNA to the plasmid (Figure 10-42).

Several other approaches also exist for inserting foreign DNA sequences into plasmids or viruses. One alternative, illustrated in Figure 10-43, is to utilize the enzyme **terminal transferase** to add complementary nucleotides to the 3′ ends of the two DNA molecules being joined. This enzyme is often employed, for example, to add G residues to the 3′ ends of a cut plasmid and C residues to the 3′ ends of the foreign DNA to be cloned. Hybridization of the plasmid to the foreign DNA is then mediated by complementary base pairing between the G and C residues. After such hybridization has occurred, the two DNAs can be covalently joined by the action of DNA ligase. Terminal transferase has been used in an analogous fashion to add complementary A and T tails to DNA samples being joined.

The other alternative for recombining DNA molecules involves the use of short DNA fragments of defined sequence known as **linkers.** The main application of linkers is in the cloning of DNA fragments that do not possess a particular restriction site needed for the cloning process. In such cases DNA linkers containing the appropriate restriction site are joined to the ends of the DNA fragment to be cloned. Because synthetic linkers containing a wide variety of restriction sites are now commercially available, this tactic can be utilized to join together many different kinds of DNA fragments.

2. *Transformation of host cells with recombinant DNA.* Once foreign DNA has been inserted into a cloning vector, the resulting population of recombinant DNA molecules is *amplified* by introducing it into a host cell in which the vector can replicate. The simplest way of accomplishing this uptake is to introduce recombinant DNA molecules into the medium surrounding the target cells. Although prokaryotic and eukaryotic cells will both take up DNA from the external medium under appropriate conditions, the efficiency of this process is quite low and some selective step is generally necessary to screen out cells that have not taken up the foreign DNA (step 3 below). After entering into the host cell the recombinant DNA molecules undergo replication and, in some cases, their incorporated gene sequences may even be transcribed and translated into protein products.

In cases where viral cloning vectors are utilized instead of plasmids, virus particles containing recombinant DNA sequences are employed to infect an appropriate host cell. Although the efficiency of viral infection is higher than the efficiency with which plasmid DNA molecules are taken up from the medium, the

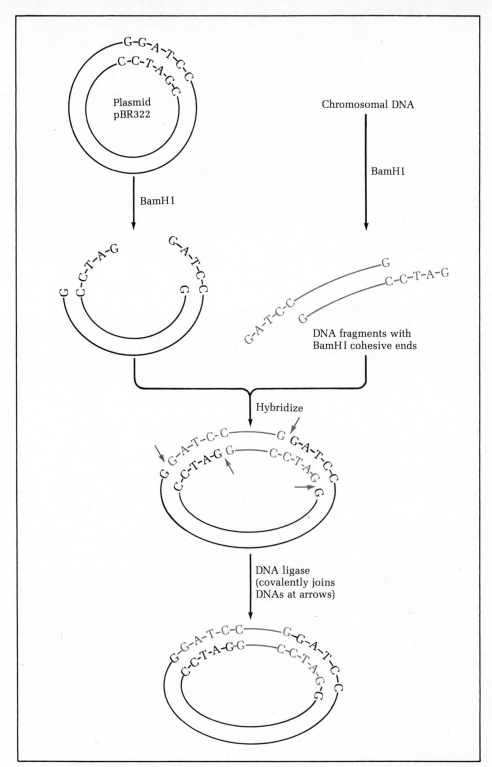

FIGURE 10-42 An example of one way in which restriction enzymes can be used to generate recombinant DNA molecules. This general approach can be used with restriction enzymes that cut the cloning vector in a single location, and that also generate suitable fragments of the DNA sample being cloned.

net result is still the same. In either case the host cell is genetically transformed by the introduction of recombinant DNA sequences.

3. *Selection of cells containing recombinant DNA.* No matter which type of cloning vector is utilized, it is eventually necessary to identify those cells that have taken up the recombinant DNA. Plasmid cloning vectors generally contain genes coding for *antibiotic resistance*, making it easy to use antibiotics as a selection tool.

The plasmid pBR322, for example, contains genes coding for resistance to *tetracycline* and *ampicillin* (a type of penicillin). The most effective way of utilizing these genes is to insert the DNA being cloned into a restriction site that causes one of the two genes to be inactivated. If the Pst I site is used, for example, the ampicillin-resistance gene will no longer be functional because foreign DNA has been inserted in the middle of the gene. Hence bacteria that have been transformed with such a recom-

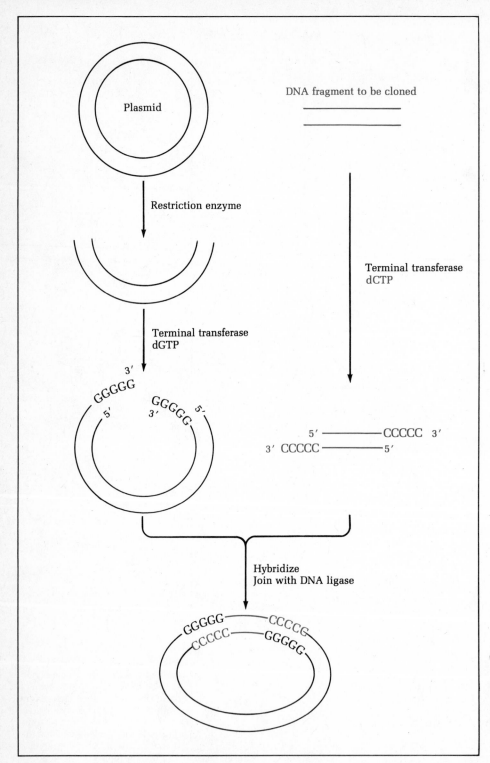

FIGURE 10-43 The use of terminal transferase is an alternative approach for generating recombinant DNA molecules. This enzyme catalyzes the addition of homogenous sequences, such as poly(G) or poly(C), to the 3' ends of DNA molecules. By adding a poly(G) tail to one type of DNA molecule and a poly(C) tail to another, joining of the two molecules can be facilitated.

binant plasmid will be resistant to tetracycline but not to ampicillin (Figure 10-44).

A different type of selection approach has been utilized with lambda phage, a bacterial virus that is also widely used as a cloning vector. A lambda phage derivative designed for use in cloning experiments has been artificially created by using the restriction enzyme EcoRI to remove a segment from the middle of the lambda phage DNA molecule. The remaining two frag-

ments, when joined back together, create a DNA molecule that is only about 70 percent of the length of normal lambda DNA. As a result, it is too small to be packaged into a functional virus particle. If a fragment of foreign DNA is inserted between the two pieces, however, a larger piece of DNA is created that is capable of being incorporated into functional lambda phage particles. Hence when cloning experiments are carried out using this mutant phage, virtually all phage particles that

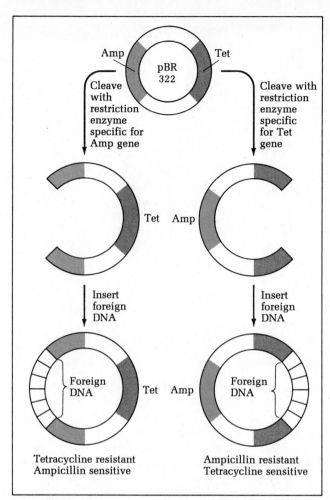

FIGURE 10-44 Diagram showing how insertion of DNA into the ampicillin or tetracycline genes of the plasmid pBR322 generates recombinant DNA molecules that confer different types of antibiotic resistance. Bacteria containing such recombinant DNA molecules can be selected for by growing them in media containing the appropriate combination of antibiotics.

successfully infect the host bacteria can be expected to contain recombinant sequences (Figure 10-45).

4. *Identification of specific cloned genes.* The final step in any recombinant DNA experiment is to identify those transformed cells that contain the cloned DNA sequences of interest. The particular approach chosen for this screening step will depend upon what is known about the gene being cloned. If the messenger RNA for the gene in question has been purified, reverse transcriptase can be used to make a radioactive cDNA copy of the message. When employed in a colony hybridization assay, such a cDNA permits colonies containing complementary sequences to be identified (Figure 10-46). Recent advances in the colony hybridization technique permit thousands of colonies to be screened simultaneously in this fashion.

An alternative tactic is available in situations in which a portion of the amino acid sequence of the gene's protein product is already known. In such cases knowledge of the genetic code is utilized to predict the nucleotide base sequences that might code for such an amino

acid sequence. Short DNA fragments containing these sequences are then chemically synthesized and used as probes for colony hybridization.

A third alternative for screening recombinant DNA clones involves a direct assay for the protein product encoded by a particular gene. When the insulin gene was cloned in bacteria, for example, some laboratories utilized antibodies directed against insulin to detect the appropriate colonies containing insulin gene sequences. Although this approach has the advantage of identifying recombinant colonies that produce a functional product from the cloned gene, it is a relatively insensitive screening technique because many recombinant organisms contain cloned gene sequences that for various reasons are not transcribed and translated into a detectable protein product.

In instances where the ultimate purpose of a cloning experiment is to produce large amounts of the protein product encoded by a particular gene, steps must be

FIGURE 10-45 The use of bacteriophage lambda as a cloning vector.

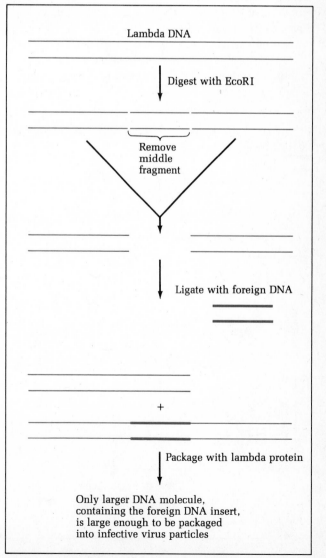

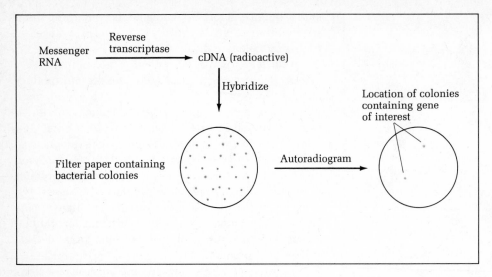

FIGURE 10-46 The use of the colony hybridization technique for locating bacterial colonies that contain cloned DNA sequences coding for a particular messenger RNA.

taken to increase the efficiency of transcription and translation of the cloned gene in its foreign environment. This is especially true when the cloning of eukaryotic genes is carried out in bacteria, for the transcription/translation mechanisms of bacterial cells are slightly different than those of higher organisms. Significant advances have been made, however, in modifying cloning vectors so as to optimize the efficiency with which eukaryotic genes are expressed as functional protein products in bacterial cells. In addition, cloning vectors have been developed that replicate

FIGURE 10-47 The Maxam-Gilbert procedure for DNA sequencing.

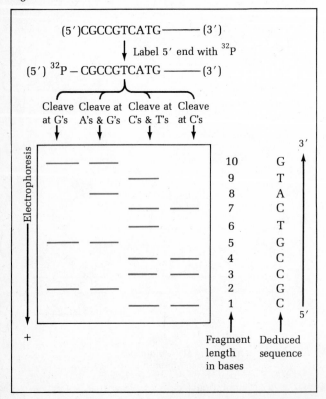

in yeast, a eukaryotic microorganism whose transcription and translation mechanisms more closely resemble those of higher organisms.

DNA Sequencing

Shortly after recombinant DNA techniques were first introduced, their usefulness was significantly enhanced by the development of two rapid procedures for determining the linear sequence of bases in DNA. The more commonly used procedure for DNA sequencing, developed by Allan Maxam and Walter Gilbert, utilizes four specific chemical reagents to cleave single-stranded DNA chains adjacent to particular bases. These reactions cleave DNA adjacent to either (1) guanine, (2) cytosine, (3) adenine and guanine, or (4) cytosine and thymine. If such reactions are carried out on a single-stranded DNA molecule that has first been labeled at its 5′ end with ^{32}P, radioactive fragments are produced whose relative sizes are determined by the location of the base for which the reaction is specific. For example, if a DNA molecule contains guanine as its second, fifth, and tenth bases, the guanine-specific cleavage reactions will yield radioactive fragments 1, 4, and 9 bases in length. Analysis of the pattern of radioactive fragments by acrylamide gel electrophoresis therefore allows one to elucidate the sequential arrangement of bases in the original DNA molecule (Figure 10-47).

An alternative sequencing procedure, developed by Frederick Sanger, utilizes *dideoxynucleotides* (nucleotides lacking a 3′-hydroxyl group) to interfere with the normal enzymatic synthesis of DNA. When a dideoxynucleotide is incorporated into a growing DNA chain instead of the normal deoxynucleotide, subsequent DNA synthesis stops because the absence of the 3′-hydroxyl group makes the formation of a bond with the phosphate group of the next nucleotide impossible. In the Sanger procedure DNA synthesis is carried out in the presence of the four normal deoxynucleotide sub-

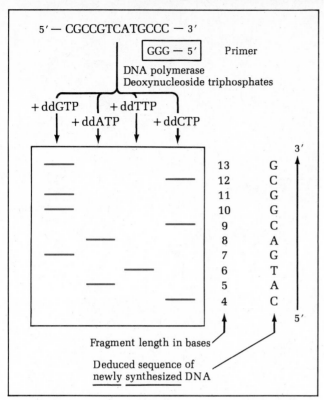

5′ — CGCCGTCATGCCC — 3′

GGG — 5′   Primer

DNA polymerase
Deoxynucleoside triphosphates

+ ddGTP   + ddTTP
  + ddATP   + ddCTP

13	G	3′
12	C	
11	G	
10	G	
9	C	
8	A	
7	G	
6	T	
5	A	
4	C	5′

Fragment length in bases

Deduced sequence of
newly synthesized DNA

FIGURE 10-48 The Sanger dideoxy method for DNA sequencing. A short primer must be provided in order to initiate synthesis of the DNA strand to be sequenced. Abbreviation: dd = dideoxy.

strates plus low concentrations of one of the four dideoxynucleotides. Every time a dideoxynucleotide is incorporated, the synthesis of the DNA chain will be prematurely terminated. Hence a series of incomplete DNA fragments of varying sizes is produced whose sizes provide information concerning the sequential arrangement of the bases (Figure 10-48).

The relative ease with which DNA sequences can now be determined has also provided us with a new way of elucidating the amino acid sequences of protein molecules. Until quite recently, the most straightforward way of determining the amino acid sequence of protein molecules involved the use of proteolytic enzymes to cleave proteins into smaller fragments, followed by chemical reactions that sequentially remove amino acids from the ends of these fragments. Although this approach has been automated and successfully applied to a large number of proteins, it is not nearly as efficient and rapid as DNA sequencing techniques, especially for large proteins. For this reason, an increasingly common alternative is to clone and sequence the gene coding for a protein, and to utilize the genetic code to predict the amino acid sequence encoded by the gene. Although there are six possible reading frames for any DNA sequence (three in each direction), the one that codes for the protein product can be recognized because it is the only one to proceed for long stretches without interrup-

tion by termination codons. In several instances this approach has made it possible to predict the amino acid sequence of a protein molecule before the protein itself has even been isolated and purified.

Chemical Synthesis of DNA Sequences

In addition to the important contributions of DNA sequencing techniques, the power of recombinant DNA technology has also been enhanced by the development of rapid procedures for the chemical synthesis of DNA fragments of defined sequence. Synthetic procedures for linking together nucleotide bases in a defined sequence were initially pioneered by Har Gobind Khorana, the first to create an entirely synthetic gene in this fashion. Machines that automate the procedure for joining together bases in a defined sequence are now available, and many gene fragments and small DNA pieces needed for recombinant DNA experiments have been made in this way. DNA sequences generated by chemical synthesis can be incorporated directly into recombinant DNA cloning vectors, opening the possibility for creating new genes and modifying existing ones in predetermined ways.

Recombinant RNA

Although nucleic acid biochemists have focused most of their attention in recent years upon the development of techniques for isolating and manipulating DNA molecules, a recently reported procedure has opened the possibility for carrying out similar kinds of manipulations on RNA molecules. This new approach, pioneered by Fred Kramer, Donald Mills, and Eleanor Miele, utilizes an enzyme called **Qβ replicase** to synthesize RNA under cell-free conditions. Qβ replicase normally functions in the replication of the RNA contained within a small bacterial virus known as Qβ. Test-tube conditions have been worked out, however, under which this enzyme will replicate RNA molecules that contain artificially inserted RNA sequences. Using this approach, it may eventually be possible to generate large amounts of specific types of RNA molecules that would be difficult to obtain in adequate quantities by alternative procedures.

Implications of Recent Advances in Nucleic Acid Technology

The rapid advances that have recently occurred in our ability to isolate and manipulate nucleic acid molecules have begun to exert profound and far-reaching effects on biological research. One of the more widely publicized applications of this type of technology is its potential for cloning genes that code for medically useful proteins, such as insulin, growth hormone, and interferon. Classical methods for isolating and purifying these proteins from natural sources are quite cumbersome and yield only tiny amounts of material; hence prior to the advent of recombinant DNA technology, inadequate supplies of such substances were available and their cost was extremely high. By cloning the appropriate genes in bacteria and yeast, however, it has been possible to create cell lines that produce these proteins in large quantities.

Another potential application of recombinant DNA technology involves the use of cloned genes for inducing desired genetic alterations in organisms that do not normally contain the gene in question. This approach is being most actively pursued in the agricultural research area, where there would be obvious practical benefits to developing new plant strains containing genes whose expression leads to better photosynthetic efficiency, nitrogen-fixing abilities, disease resistance, ability to grow under adverse weather conditions, and nutritional value. The most successful vector for introducing foreign DNA molecules into plant cells thus far has been the **Ti plasmid,** a naturally occurring DNA molecule that can be introduced into a broad spectrum of plants by the bacterium *Agrobacterium tumefaciens*. In nature, infection of plant cells by this bacterium leads to an insertion of a small part of the Ti plasmid into the plant cell chromosomes; expression of a portion of this inserted DNA then leads to the formation of an uncontrolled growth referred to as a *crown gall* tumor. It has recently been shown that the DNA sequences involved in triggering crown gall tumor formation can be removed from the Ti plasmid without preventing the transfer of DNA from the plasmid to the host cell chromosome. By inserting genes of interest into such modified plasmids, vectors can thus be created that are capable of transferring foreign genes into plant cells.

Genes isolated by recombinant DNA technology have also been successfully introduced into animals. In one of the first experiments of this type, Richard Palmitter and his associates injected a recombinant DNA plasmid, containing the gene coding for human growth hormone, into fertilized mouse eggs. The eggs were then reimplanted into mice and allowed to develop normally. In order to increase the likelihood that the human gene would be expressed in mouse cells, the injected plasmid also contained a segment of a normal mouse gene coding for the protein *metallothionein*. This particular DNA segment contains a sequence that normally facilitates transcription. By inserting this mouse sequence adjacent to the human growth hormone gene, it was hoped that it would facilitate transcription of the human gene. When the organisms that developed from these injected eggs were born and allowed to develop into mature mice, many were found to grow significantly larger than littermates that had not been injected with the growth hormone gene (Figure 10-49). The con-

FIGURE 10-49 The mouse on the right developed from an egg into which a plasmid containing the human growth hormone gene had been injected. Due to expression of this human gene, the affected mouse has grown to more than twice the size of its control littermate on the left. Courtesy of R. L. Brinster and R. E. Hammer, School of Veterinary Medicine, University of Pennsylvania.

clusion that this increased growth rate was actually caused by the production of human growth hormone was confirmed by assays that detected the presence of this hormone in the serum of the affected mice.

Given the success of the above experiments, the question has arisen as to whether genetic engineering techniques should be applied to human beings. Humans certainly suffer from many genetic defects, such as the lysosomal storage diseases and sickle-cell anemia, that might conceivably be cured by introduction of the appropriate normal gene. Although certain viruses capable of infecting human cells might be usable as vectors for introducing cloned gene sequences into such individuals, many practical barriers remain before experiments of this type will become feasible. Applications of this sort also raise serious ethical issues that need to be thoroughly discussed before any experiments involving humans proceed.

Although the practical benefits of recombinant DNA technology have received the most attention in the public media, the impact of this technology has actually been more impressive at the basic research level. As we shall see in the following chapters, many new insights into the organization, regulation, and functioning of eukaryotic genes have emerged whose discovery would have been virtually inconceivable without this powerful new way of isolating and characterizing DNA sequences. It is through this increased basic knowledge that the long-term benefits of recombinant DNA technology are most likely to be achieved.

SUMMARY

Although our understanding of the role played by DNA in the transmission of hereditary information is now taken for granted, achieving this understanding required almost a century of experimental investigation. Shortly after DNA was discovered by Miescher in the late 1860s, staining experiments at the light microscopic level revealed that DNA is localized predominantly in chromosomes. Though this observation suggested to many biologists that DNA functions in heredity, such conclusions were undermined by the subsequent discovery that DNA staining fluctuates with different physiological states of the cell. The idea that genes are made of DNA was further discredited by Levene's analytical data, which suggested that DNA consists of a simple repeating tetranucleotide.

For the above reasons it was widely believed during the first half of the twentieth century that genes consist of protein rather than DNA. During the early 1940s, however, Oswald Avery discovered that bacteria can be transformed from one genetic type to another by purified preparations of DNA. Though Avery's conclusion that DNA must be the genetic material at first fell on largely deaf ears, other data soon supported this idea. Most important were the discoveries that an organism's DNA content is constant from cell to cell, that the base composition of DNA differs among different types of organisms, and that DNA is injected into bacterial cells by infecting virus particles.

The realization that DNA is the genetic material raised a series of fundamental questions concerning the mechanisms by which information contained within DNA molecules is replicated, utilized, and mutated. The first clue to the mechanism underlying DNA replication emerged from the discovery that the base composition of DNA molecules exhibits a remarkable pattern: The amount of adenine always equals the amount of thymine, and the amount of guanine always equals the amount of cytosine. The importance of this relationship became apparent in 1953, when Watson and Crick formulated the double-helix model of DNA structure. According to this model, the two intertwined chains of the DNA double helix are held together by hydrogen bonds between the base adenine in one chain and thymine in the other, and between guanine in one chain and cytosine in the other. This model immediately suggested a mechanism for DNA replication in which the two chains of the double helix first separate and are then employed as templates upon which new opposing chains are copied by complementary base pairing.

The first experimental support for this view of DNA replication was provided by the experiments of Meselson and Stahl, in which DNA strands isolated from bacteria incubated with heavy or light isotopes of nitrogen were found to behave exactly as predicted by the Watson–Crick model. Around the same time Arthur Kornberg discovered DNA polymerase, an enzyme capable of replicating DNA molecules under cell-free conditions. Subsequent investigations have revealed that several types of DNA polymerase exist, some of which are involved in DNA replication and others functioning in more specialized processes such as DNA repair. Because DNA polymerase can only catalyze the polymerization of DNA chains in the $5' \rightarrow 3'$ direction, the synthesis of DNA on at least one of the two chains is discontinuous, generating small DNA fragments that are later joined together by the enzyme DNA ligase. DNA replication is initiated by an enzyme, termed a primase, that catalyzes the formation of an RNA primer that is later removed and replaced by DNA. In addition to DNA polymerase, DNA ligase, and primase, DNA replication involves several proteins involved in unwinding and nicking the double helix.

While most forms of DNA polymerase utilize DNA as a template, some tumor viruses contain an enzyme, termed reverse transcriptase, that employs RNA instead. Reverse transcriptase permits RNA viruses to copy their genetic information into DNA molecules that can subsequently be inserted into the host cell DNA.

The first major breakthrough in our understanding of gene function evolved from Beadle and Tatum's studies of X-ray induced mutations in *Neurospora*, which led them to conclude that genes control the production of enzyme molecules. Subsequent investigations revealed that a direct relationship exists between the sequence of bases in DNA and the sequence of amino acids in polypeptide chains. The transfer of information from DNA to polypeptide chains is mediated by messenger RNA molecules whose base sequences are copied from DNA by the enzyme RNA polymerase. The information encoded within messenger RNA is organized into units of three bases, called codons, which specify particular amino acids. In addition to coding for amino acids, some codons serve special functions as punctuation marks signaling the beginning or end of an amino acid sequence. The coding assignments for the various triplets were established by experiments that involved the use of synthetic RNA triplets, as well as longer RNA molecules of defined sequence. These studies revealed that of the 64 possible codons, 61 code for amino acids. The interaction between messenger RNA codons and the amino acids they encode is mediated by transfer RNA molecules. Each transfer RNA molecule carries one particular amino acid and contains a triplet sequence, termed an anticodon, that is complementary to messenger RNA codons specific for that particular amino acid.

The ability of DNA to undergo changes in base sequence (mutations) is necessary for the long-term evolution of organisms. One source of mutation is the built-in mechanisms that ensure the occurrence of a finite number of errors during DNA replication. The existence of tautomeric forms of the nitrogenous bases, for example, causes a certain number of incorrect base pairs to be generated. Inverted repeats, which foster the formation of hairpin loops, may cause insertion or deletion mutations to occur during replication. Finally, transposable elements facilitate DNA rearrangements over longer distances.

In addition to spontaneous and physiological mechanisms for generating changes in DNA base sequence, many mutation-inducing chemicals and physical agents occur in our environment. Once DNA molecules altered by exposure to chemical or physical mutagens have undergone subsequent rounds of replication, the mutations become difficult to recognize and repair. DNA repair pathways involving the interaction of nucleases, DNA polymerase, and DNA ligase do exist, however, that can recognize and repair DNA defects before they have become stabilized by the act of replication.

Much of our current understanding of the principles that govern the flow of genetic information has been made possible by technical advances for studying nucleic acids. In nucleic acid hybridization techniques, isolated nucleic acid molecules are tested for their ability to form complementary hybrids under cell-free conditions. Saturation hybridization is employed to determine the quantity of a particular type of sequence in a population of DNA molecules, while competition hybridization is used to determine how closely two nucleic acid samples resemble each other. Southern and Northern blotting techniques employ hybridization to localize specific types of nucleic acid sequences separated by electrophoresis; *in situ* hybridization is used to localize specific types of nucleic acids within cells; and colony hybridization is employed to determine which bacterial colonies contain a particular type of sequence.

Recombinant DNA techniques permit individual gene sequences to be isolated from large masses of DNA and produced in virtually unlimited quantities. Such technology depends heavily on the use of restriction endonucleases, which cleave DNA molecules at specific sequences four to six bases long. The general approach for cloning gene sequences by recombinant DNA technology involves four steps: (1) insertion of foreign DNA into cloning vectors such as plasmids and viruses, (2) transformation of appropriate host cells with this recombinant DNA, (3) selection of cells containing the foreign DNA, and (4) identification of those cells containing the particular gene sequences of interest.

The utility of DNA cloning techniques has been sig-

nificantly enhanced by the development of rapid procedures for DNA sequencing, as well as procedures for the chemical synthesis of DNA fragments of defined sequence. The technology now available for isolating, purifying, characterizing, and manipulating nucleic acid sequences offers the hope for both immediate practical benefits and for advancing our understanding of the organization, regulation, and functioning of genetic information.

SUGGESTED READINGS

Books

Genetics, Readings from *Scientific American*, 1981, Freeman, San Francisco.

Kornberg, A. (1980 and 1982). *DNA Replication* and *Supplement to DNA Replication*, Freeman, San Francisco.

Portugal, F. H., and J. S. Cohen (1977). *The Century of DNA: A History of the Discovery of the Structure and Function of the Genetic Substance*, MIT Press, Cambridge, MA.

Scientific American (1981) 245 (Sept.). This entire issue is devoted to recombinant DNA and genetic engineering.

Stent, G. S., and R. Calendar (1978). *Molecular Genetics. An Introductory Narrative*, 2nd Ed., Freeman, San Francisco.

Strickberger, M. (1985). *Genetics*, 3rd Ed., Macmillan, New York.

Watson, J. D., and J. Tooze (1981). *The DNA Story*, Freeman, San Francisco.

Watson, J. D., J. Tooze, and D. T. Kurtz (1983). *Recombinant DNA A Short Course*, Scientific American Books, Freeman, San Francisco.

Wilson, E. B. (1896). *The Cell in Development and Inheritance*, Macmillan, New York.

Articles

Chargaff, E. (1971). Preface to a grammar of biology, *Science* 172:637–642.

Chilton, M.-D. (1983). A vector for introducing new genes into plants, *Sci. Amer.* 248 (June):51–59.

Cohen, S. N., A. C. Y. Chang, H. W. Boyer, and R. B. Helling (1973). Construction of biologically functional bacterial plasmids *in vitro*, *Proc. Natl. Acad. Sci. USA* 70:3240–3244.

Cohen, S. N., and J. A. Shapiro (1980). Transposable genetic elements, *Sci. Amer.* 242 (Feb.):40–49.

Dickerson, R. E. (1983). The DNA helix and how it is read, *Sci. Amer.* 249 (Dec.):94–111.

CHAPTER

11

The Nucleus and Transcription of Genetic Information

Although genetic information is stored and transmitted in the form of DNA base sequences, the protein molecules encoded by DNA are actually responsible for most of the properties and activities associated with living cells. As outlined in Chapter 10, the process by which information is transferred from DNA to the amino acid sequences of protein molecules involves two steps: transcription and translation. These two stages in the flow of genetic information will be considered in detail in this and the following chapter.

To understand the process of gene transcription, one must first be acquainted with the structural organization of the genetic material within the cell. In eukaryotic organisms the vast bulk of the DNA is localized within the nucleus, a large membrane-enclosed organelle that has long been known to play a key role in directing cellular activities. Although prokaryotic cells lack a true, membrane-enclosed nucleus, they too contain a nuclear region in which the genetic material is localized. In this chapter we shall begin by examining the morphological organization of these two kinds of cell structures, and shall then describe the mechanisms by which the genetic information they contain is transcribed and processed for subsequent use by the protein-synthesizing machinery of the cytoplasm.

ULTRASTRUCTURAL ORGANIZATION OF THE NUCLEUS

The importance of the nucleus in cell heredity and the role it plays in directing the activities of the cell was appreciated by cell biologists even before the discovery that the DNA contained in this organelle stores genetic information. One of the earliest experimental demonstrations of the control exerted by the nucleus over the rest of the cell emerged from the studies of Joachim Hämmerling and Jean Brachet on the large unicellular alga, *Acetabularia*. The large size of this single-celled organism, and the localization of the nucleus at one end of the cell, make it relatively simple to remove the nucleus from one cell and transplant it to another. Moreover, different species of *Acetabularia* have cytoplasmic "caps" of different shapes, making it possible to investigate the question of whether the shape of the cap is determined by the nucleus. To study this issue, Hämmerling and Brachet removed the nucleus from one type of *Acetabularia* and transplanted it to another cell whose nucleus and cap had been removed. Such hybrid

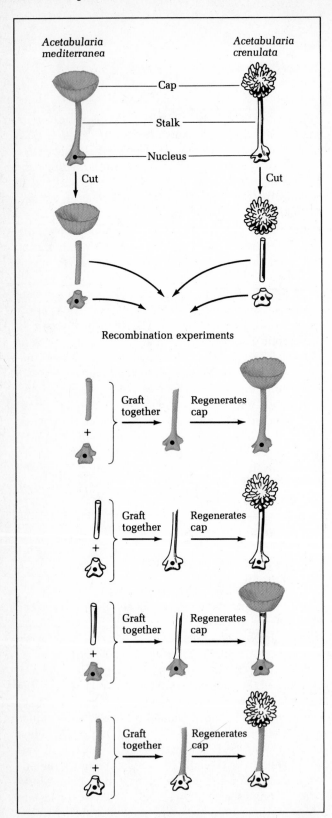

FIGURE 11-1 Summary of experiments carried out on two species of the unicellular alga *Acetabularia* that demonstrated the existence of nuclear control over cytoplasmic behavior.

cells were found to regenerate new caps whose shape is determined by the type of nucleus implanted (Figure 11-1).

Such observations stimulated a great deal of interest in investigating the morphological and functional properties of the nucleus. Although the nucleus is typically pictured as a large, spherical, membrane-enclosed structure located near the center of the cell, considerable variations exist in nuclear size, shape, and position. In terms of size the nucleus generally accounts for about 5–10 percent of the total cell volume, but in some cases this value may reach 80 percent or more. Such variations in size are determined both by the amount of DNA present and the physiological state of the cell. Nuclei range from one to several hundred micrometers in diameter, with typical mammalian nuclei measuring about 5–10 μm. The shape of the nucleus is generally spherical, but can also be ellipsoid, flattened, lobed, or irregular, depending on the particular cell type. The position of the nucleus within the cell is also variable, and often characteristic for a given cell type. Although most cells contain only one nucleus, some specialized cell types such as skeletal muscle are *multinucleate*, containing from a few to hundreds of separate nuclei.

At the ultrastructural level, at least five distinct nuclear constituents can be distinguished: (1) a double-membrane **nuclear envelope** that separates the contents of the nucleus from the cytoplasm, (2) a fluid **nucleoplasm** (or nuclear sap) that contains the soluble material of the nucleus, (3) small spherical **nucleoli** involved in the formation of ribosomes, (4) DNA-containing fibers, known as **chromatin** when dispersed throughout the nucleus, and **chromosomes** when they become organized into discrete structures, and (5) a **nuclear matrix** that provides a skeletal framework for the nucleus. Before proceeding to describe each of these five components in detail, it should be noted that the nuclear region of prokaryotic cells is considerably less complex in organization. This region, often referred to as the **nuclear body** or **nucleoid,** is not membrane enclosed nor are nucleoli or a nuclear matrix present. Although there is no membrane to define a separate compartment, the nuclear body consists of a mass of DNA-containing fibers that are tightly packed together in a way that maintains a fairly sharp border between this region and the cytoplasm. These DNA-containing fibers carry out functions comparable to those performed by the chromatin and nucleoli of eukaryotic nuclei.

The Nuclear Envelope

The presence of a nuclear envelope in eukaryotic cells establishes a membrane barrier between the nuclear components and the cytoplasm. Since prokaryotic cells function perfectly well without such a barrier, it is clear that the presence of a nuclear membrane is not an absolute requirement for living cells. Yet if the nuclear envelope of a eukaryotic cell is disrupted with a tiny needle, the cell generally dies. Although it is not clear why the nuclear envelope has become essential in eukaryotes, this need for compartmentalizing the nuclear contents is presumably caused by the larger size and DNA content of eukaryotic cells.

Structure and Composition of the Nuclear Envelope

Light microscopic examination reveals only a narrow, fuzzy border at the edge of the nucleus, so little was known about the structure of the nuclear envelope prior to the advent of electron microscopy. In thin-section electron micrographs, the nucleus can be seen to be enveloped by two concentric membranes: an **outer membrane** abutting the cytoplasmic compartment and an **inner membrane** adjacent to the nuclear interior (Figure 11-2). Between the two membranes of this **nuclear envelope** is a **perinuclear space** typically measuring from 15 to 30 nm across. The outer nuclear membrane usually contains cytoplasmic ribosomes on its outer surface, and may exhibit points of continuity with membranes of the endoplasmic reticulum. This arrangement suggests that the perinuclear space is continuous with the cisternae of the endoplasmic reticulum. Cytoskeletal elements such as microtubules, microfilaments, and intermediate filaments are often associated with the outer surface of the nuclear envelope, helping to anchor the nucleus and to maintain its proper shape. Closely adhering to the nucleoplasmic surface of the inner membrane is a thin layer of tough fibrous material, termed the **nuclear lamina,** which acts as a structural support for the nuclear envelope and as a scaffolding for the nuclear pores. The nuclear lamina is constructed from proteins, called **lamins,** which closely resemble the proteins present in intermediate filaments.

One of the most striking morphological features of the nuclear envelope is the presence of **nuclear pores.** These structures, which pass through both membranes of the envelope, measure approximately 60 nm in diameter and exhibit an eight-sided, or octagonal, symmetry (Figure 11-3). Although the central cavity of the pore looks like an open space in some electron micrographs, it is usually filled with a granular or fibrous **annular material** that surrounds the pore and extends into it (Figures 11-4 and 11-5). The failure to see this material after certain types of fixation and sectioning is believed to be an artifact resulting from the method of sample preparation. The space occupied by the nuclear pores accounts for a significant portion of the nuclear envelope, generally ranging from 10 to 25 percent of the total surface area. Cells with transcriptionally active nuclei tend to contain large numbers of pores while metabolically inert nuclei possess few, if any.

The first direct evidence that nuclear pores mediate the transfer of material between nucleus and cytoplasm was obtained in 1965 by Carl Feldherr, who injected gold particles of various sizes into the cytoplasm of amoebae and then examined the cells by thin-section electron microscopy. Shortly after injection, gold particles were seen first within the pores themselves, and then inside the nucleus (Figure 11-6). The rate of entry of these particles into the nucleus was found to be inversely related to particle diameter, with particles larger than 10–15 nm in diameter excluded entirely. Since the nuclear pore is significantly larger than 10–15 nm in diameter, such observations suggest that the pores contain a narrow central channel through which small particles and molecules can freely pass. Subsequent studies monitoring the entry rate of radioactive

FIGURE 11-2 Thin-section electron micrograph of the nuclear envelope in the giant amoeba *(Pelomyxa).* The double membrane of the envelope separates the nucleus *(bottom)* from the cytoplasm *(top).* Several nuclear pores (single arrows) are evident. In some regions fuzzy material representing the nuclear lamina is seen adhering to the nucleoplasmic surface of the inner membrane (double arrow). Courtesy of E. W. Daniels.

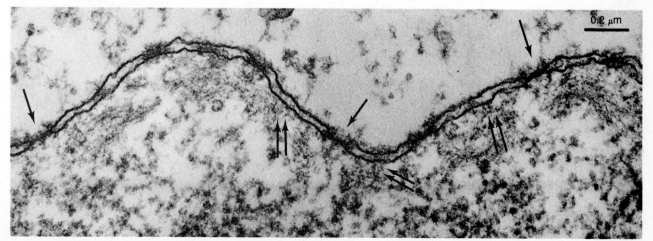

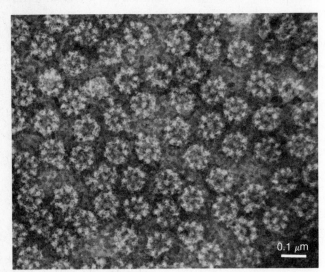

FIGURE 11-3 Structure of the nuclear pore as seen by scanning electron microscopy, negative staining, and freeze-fracture analysis. *(Top left)* Surface view of a detergent-treated nucleus viewed with the scanning electron microscope. Nuclear pore complexes are prominent and in some (arrow) eight subunits can be discerned. Courtesy of R. H. Kirschner. *(Bottom left)* The nuclear envelope of a newt oocyte viewed after negative staining. The eight subunits of the pore complex are clearly visible. Courtesy of A. C. Fabergé. *(Bottom right)* Surface view of the inner membrane of the nuclear envelope of an onion root tip after freeze-etching, showing several nuclear pores. Courtesy of D. Branton.

FIGURE 11-4 Ultra-thin section of a nuclear envelope from an amphibian oocyte showing nuclear pore complexes containing annular material (arrows) within them. Courtesy of U. Scheer.

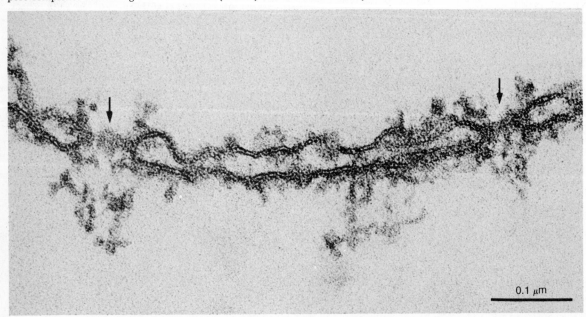

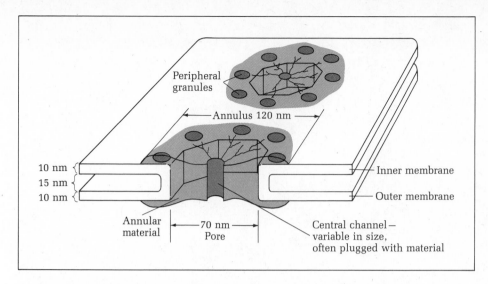

FIGURE 11-5 Schematic diagram of the main elements present in a "typical" nuclear pore complex. Significant variations occur in the arrangement of the annular material.

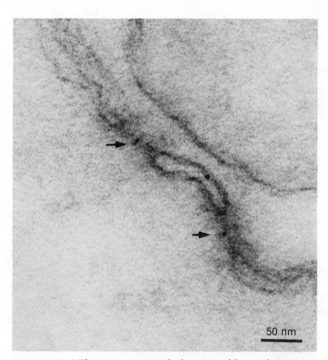

FIGURE 11-6 Electron micrograph showing gold particles passing through the nuclear pore. The giant amoeba *(Chaos)* was injected with colloidal gold particles one minute prior to fixation for electron microscopy. In this section two nuclear pores are seen, each containing a centrally located gold particle (arrows). Courtesy of C. M. Feldherr.

molecules of various sizes into the nucleus revealed that the functional diameter of this channel is about 10 nm, which is sufficient to permit the unhindered passive diffusion of globular proteins up to about 50,000 in molecular weight.

Since passive diffusion is a relatively nonselective mechanism, the question arises as to how the movement of specific molecules between cytoplasm and nucleus is regulated. It is known, for example, that certain proteins synthesized on cytoplasmic ribosomes are se-

lectively accumulated within the nucleus. In the case of small proteins capable of diffusing unhindered through the nuclear pores, this selective accumulation could be explained by the presence of substances within the nucleus that selectively bind to particular proteins after they have diffused through the pores. This theory does not explain, however, how the movement of larger proteins and RNAs into and out of the nucleus is controlled. Since the size of such molecules would severely hinder their passive diffusion through the pores, a more complex transport process is required. Electron microscopic observations have suggested that the nuclear pores, in spite of their relatively small size, are nonetheless involved in the transport of much larger molecules. Perhaps the most graphic evidence is the pictures that have been obtained of large ribonucleoprotein granules, 50 nm or more in diameter, in the process of moving through the nuclear pores (Figure 11-7). Since this size is much greater than the functional diameter of the pore, the granules become compressed and elongated during passage, reducing their diameter to about 15 nm.

The preceding observation raises the question of how nuclear pores recognize those particular molecules that need to be transported into the nucleus but are too large to diffuse passively through the pores. One substance that has served as a model for investigating this issue is **nucleoplasmin,** a protein of 165,000 molecular weight that is normally accumulated within the nucleus. In order to investigate the mechanism of nucleoplasmin uptake, experiments have been carried out in which gold particles too large to normally pass unhindered through the nuclear pore were coated with nucleoplasmin and then injected into the cytoplasm of frog oocytes. Such gold particles were found to rapidly pass through the nuclear pores and into the nucleus. It was further discovered that cleavage of a small peptide fragment from the nucleoplasmin molecule prior to coating the gold particles prevents this rapid uptake from occur-

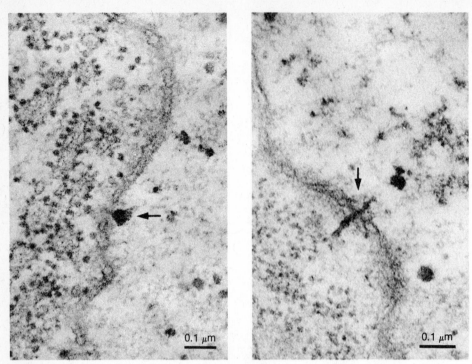

FIGURE 11-7 Electron micrographs showing ribonucleoprotein (RNP) granules passing through the nuclear pore. *(Left)* A nuclear RNP granule (arrow) is drawn out slightly in the direction of the nuclear pore. *(Right)* An RNP granule in the actual process of passing through the nuclear pore. The originally spherical particle is seen to be stretched out into an elongate shape during passage. The nucleus is on the right. Courtesy of B. J. Stevens.

ring, indicating that a specific peptide sequence targets the molecule for uptake through nuclear pores.

Although it is now known that the nuclear pore functions in both the passive diffusion of smaller molecules and the mediated transport of larger molecules, nuclear pores are not the only route available for the movement of materials between the nucleus and cytoplasm. Like other cellular membranes, the nuclear envelope is selectively permeable and is capable of carrying out active transport and facilitated diffusion. In addition, small membrane vesicles occasionally form within the nucleus and then fuse with the inner nuclear membrane, releasing their contents either into the perinuclear space or directly into the cytoplasm. It has also been proposed that proteins synthesized on ribosomes attached to the outer nuclear membrane can be inserted directly into the perinuclear space, facilitating their entry into the nucleus. Thus a variety of different mechanisms are potentially available for transporting materials through the nuclear envelope.

In addition to its involvement in regulating the flow of materials between nucleus and cytoplasm, the nuclear envelope performs several other functions. First, it facilitates the spatial organization of the cell by serving as an anchoring point for other structures. The DNA-containing chromatin fibers of the nucleus, for example, are attached at multiple sites to the nuclear lamina, which is associated with the nucleoplasmic surface of the inner nuclear membrane. These attachments,

which often occur at the edges of the nuclear pores, are important in organizing the chromatin fibers, especially prior to their condensation into chromosomes at the time of cell division. Connections have also been observed between the outer nuclear membrane and a variety of cytoplasmic fibers, membranes, and organelles, including the endoplasmic reticulum, Golgi complex, and mitochondria.

Another cytoplasmic membrane system associated with the nuclear envelope is the **annulate lamellae.** This unusual structure occurs primarily in developing oocytes and spermatocytes, and in certain embryonic and tumor cells. Annulate lamellae consist of several dozen stacks of cytoplasmic membranes containing pore complexes similar to those present in the nuclear envelope (Figure 11-8). These structures, which form by budding off from the outer nuclear membrane, are characterized by a high RNA content and are thought to function in the transport, storage, and/or protection of certain types of RNA. This viewpoint is indirectly supported by the observation that developing oocytes, which tend to have extensive annulate lamellae, are known to store large amounts of messenger RNA in their cytoplasm (Chapter 15).

Another function that appears to be carried out by the nuclear envelope is *oxidative metabolism.* Many enzymes and proteins involved in electron transfer and oxidative phosphorylation have been detected in isolated nuclear envelope preparations, and in certain types of isolated nuclei, oxidative phosphorylation has even been observed. Unfortunately, the difficulty in isolating nuclear envelope fractions that are completely free of contamination by mitochondrial membranes makes such observations difficult to interpret. It has been demonstrated, however, that certain inhibitors have different effects on the oxidation pathways of mitochondrial and nuclear membranes, suggesting that the oxidative components detected in isolated nuclear envelope preparations are not derived from mitochondrial contamination. Hence it is possible that some type of electron transfer, accompanied in certain nuclei by the synthesis of ATP, occurs within the nuclear envelope.

The Nucleoplasm

The **nucleoplasm** (or nuclear sap) is an amorphous, fluidlike material that contains the soluble material of the nucleus. By extracting isolated nuclei with dilute buffers or salt solutions, it is possible to isolate and identify the various components present in this fraction. The substances isolated under such conditions fall into several categories: proteins, RNA and ribonucleoprotein complexes, and small molecules.

1. *Proteins.* Various kinds of enzymatic proteins have been identified in the nucleoplasmic fraction, including some involved in metabolic path-

ways and others functioning in the replication and transcription of DNA. In addition to enzymes, some of the proteins present in the nucleoplasm are thought to be involved in the regulation of chromatin structure and function.

2. *RNA and ribonucleoprotein complexes.* Most of the RNA molecules present in the nucleoplasm are complexed with proteins, forming ribonucleoprotein granules of varying sizes and types. Some of these RNAs belong to a class of molecules, called **small nuclear RNAs,** which are never exported to the cytoplasm. As we shall see later, RNAs in this category have been implicated in the processing of messenger RNA. Other ribonucleoprotein granules present in the nucleoplasmic fraction contain larger RNA molecules and have been observed passing through the nuclear pores (see Figure 11-7), indicating that they contain RNA destined to carry out a cytoplasmic function. Some particles in this category are precursors of cytoplasmic ribosomes, while others contain messenger RNA or messenger RNA precursors eventually destined for translation in the cytoplasm.

3. *Small molecules.* A variety of low-molecular-weight substances, including coenzymes, metabolites, and ions, are dissolved in the nucleoplasm. Especially important members of this category are the nucleoside and nucleotide precursors utilized in the synthesis of DNA and RNA.

FIGURE 11-8 Electron micrograph of an array of annulate lamellae consisting of a stack of several dozen membranes in the frog oocyte. Courtesy of R. G. Kessel.

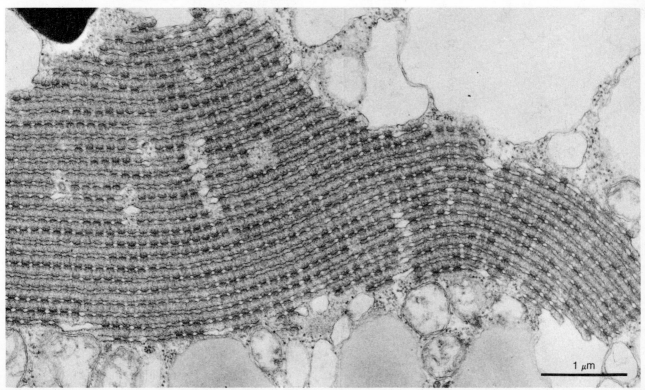

1 μm

The Nucleolus

One of the most striking morphological features of the eukaryotic nucleus is the presence of a large spherical structure known as a **nucleolus** (Figure 11-9). Though most cells contain only one or two nucleoli, the occurrence of several more is not uncommon and in certain situations hundreds or even thousands may be present. Nucleoli are formed under the direction of special chromosomal sites called *nucleolar organizer regions* (NOR). Although the maximum number of nucleoli that can be formed per cell is determined by the number of nucleolar organizers present, the number of nucleoli present is often less than the total number of organizers. In such cases, an individual nucleolus may be formed under the direction of several different nucleolar organizers.

A typical nucleolus is a roughly spherical structure measuring several micrometers in diameter, but a wide range in size and shapes occurs. Because of their relatively large size, nucleoli are readily visible with the light microscope (see Figure 11-9) and hence were first visualized over 200 years ago. Although silver staining techniques gave early light microscopists an especially striking view of nucleolar structure and even led to the description of a "threadlike" substructure within the

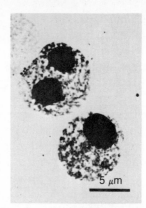

FIGURE 11-9 Light micrograph of two isolated nuclei stained with silver. Two prominent nucleoli are visible in one nucleus, and one nucleolus can be seen in the other. Courtesy of J. Olert.

nucleolus, it was not until the advent of electron microscopy that the structural components that make up the nucleolus became clearly evident. In thin-section electron micrographs, several distinct components can be distinguished. Thin **fibrils** approximately 5 nm in diameter predominate in the core of the nucleolus. Surrounding this fibrillar core is a region termed the **nucleolar cortex** that is enriched in 15-nm diameter **ribonucleoprotein granules** (Figure 11-10). These

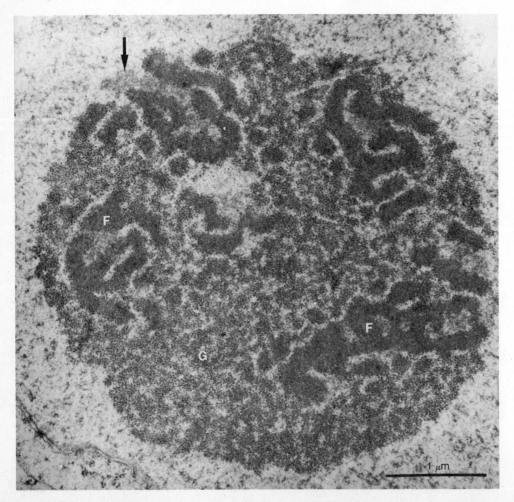

FIGURE 11-10 Thin-section electron micrograph of the nucleolus in the alga *Spirogyra*. The most lightly stained area is the nucleolar chromatin (arrow), which follows a meandering course throughout the nucleolus. The fibrillar (F) and granular (G) components are readily distinguished from one another. Courtesy of E. G. Jordan.

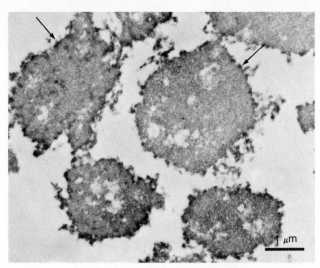

FIGURE 11-11 Electron microscopic appearance of isolated nucleoli. Chromatin fibers (arrow) are seen adhering to the nucleolar surface. Courtesy of T. S. Ro-Choi, K. Smetana, and H. Busch.

fibrils and granules are embedded in a homogeneous ground substance called the **nucleolar matrix.** In addition, chromatin fibers representing extensions from the nucleolar organizer can be seen projecting into the nucleolus from the surrounding chromatin.

The biochemical composition of the nucleolus has been investigated both by cytochemical staining and by direct chemical analysis of purified nucleoli. Procedures for obtaining nucleoli generally begin with isolated nuclei, which are then disrupted by physical means and subjected to differential centrifugation to sediment the nucleoli. After employing electron microscopy to confirm that the pellet contains purified nucleoli (Figure 11-11), biochemical analyses are carried out. Such studies have revealed that about 80 percent of the nucleolar mass is accounted for by protein, with the remainder consisting of a mixture of RNA and DNA.

As we shall see in Chapter 12, the nucleolus is the site within the cell where ribosomes are made. When

the process of ribosome formation is discussed in detail at that time, the functional significance of the various nucleolar components summarized in this section will be elaborated upon.

Chromatin Fibers

The DNA contained within the nucleus of eukaryotic cells is tightly complexed with a specific group of proteins, forming a series of intertwined fibers that collectively account for the bulk of the nuclear mass. Although these **chromatin fibers** are generally dispersed throughout the nucleus, under certain circumstances (most notably at the time of cell division) they become condensed into larger, more discrete structures known as **chromosomes.** Our current understanding of the chemical composition of chromatin fibers has been made possible by the development of techniques for extracting and purifying chromatin from isolated nuclei. One commonly employed approach is to break isolated nuclei by sonication or by exposure to hypotonic solutions, and to then sediment the chromatin fibers by differential centrifugation. It is also possible to solubilize chromatin by extracting nuclei with high ionic strength solutions, such as 1.0M NaCl. In either case, experimental protocols can be designed that will minimize contamination with other nuclear components (Figure 11-12). Although a certain amount of variability exists in the chemical composition of chromatin fractions isolated from different sources and by different procedures, protein is generally found to account for 60–70 percent of the total mass, DNA for 30–40 percent, and RNA a few percent. Because protein is present in such high concentration, considerable effort has been expended in characterizing the protein components of chromatin and in investigating the functions in which they are involved.

Histone and Nonhistone Proteins   Although chromatin proteins are a heterogeneous group of molecules,

FIGURE 11-12 Outline of two nuclear subfractionation approaches for isolating chromatin. The purified nuclear fraction employed as starting material for these two procedures is obtained by centrifuging a tissue homogenate first at low speed, and then recentrifuging the crude nuclear pellet at high speed through a concentrated sucrose solution. *Abbreviations:* P = pellet, S = supernatant.

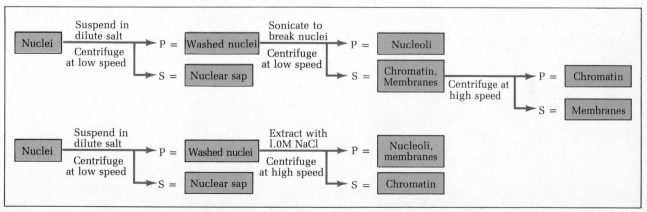

TABLE 11-1
Five major types of histones

		Mole %		Molecular Weight	Acetylation	Methylation	Phosphorylation
		Lysine	Arginine				
H1	Very lysine rich	20–26	2–3	21,000–24,000	±	±	+++
H2A	Moderately lysine rich	10–11	8–12	13,000–16,000	±	+	±
H2B		14–18	4–8	14,000–16,000	±	+	+++
H3	Arginine rich	9–10	13	14,000–15,000	+++	+++	+
H4		10–11	14	11,300	+++	+++	+

they can be divided into two general categories: histones and nonhistones. **Histones** are a group of basic proteins whose high content of lysine and arginine give them a strong positive charge at physiological pH. Years of painstaking work on the isolation and characterization of these proteins has led to the conclusion that there are five major types of histones: **H1, H2A, H2B, H3, and H4** (Table 11-1). Chromatin contains roughly equal numbers of H2A, H2B, H3, and H4 molecules, and about half that number of histone H1 molecules. These proportions are remarkably constant among all eukaryotes, regardless of the cell type, organism, or physiological state (Figure 11-13). In addition to the five major types of histones, certain specialized histones occur in particular cell types. In avian red blood cells, for example, an unusual histone called H5 is formed during the developmental stage when the chromatin begins to undergo condensation and inactivation.

Amino acid sequencing studies have revealed that the primary structures of histone molecules have been highly conserved during evolution. Histone H3 isolated from pea seedlings and from calf thymus, for example, exhibit only two differences in amino acid sequence, in spite of the great evolutionary distance between these four organisms. This resistance to evolutionary change suggests that even minor changes in these histone molecules would have harmful repercussions on nuclear function. Although histone heterogeneity is limited by the relatively small number of histone classes, some diversity in histone structure is generated by other mechanisms. First, multiple forms of histone H1 have been found to exist, each characterized by a slightly different amino acid sequence. In addition histones are subject to enzymatic modification reactions, such as *acetylation*, *methylation*, and *phosphorylation*, which alter amino acid side chains after the protein has been synthesized (Figure 11-14). Such alterations, which are readily reversible, are believed to influence both the structural and functional properties of histone molecules. We shall see later that acetylated histones are preferentially associated with chromatin fibers that are being actively transcribed, while phosphorylated histones have been implicated in the condensation of chromatin fibers into chromosomes.

Although histones are a virtually universal constituent of eukaryotic nuclei, some exceptions do occur. In certain eukaryotes the formation of sperm cells is accompanied by the replacement of histones with another class of basic proteins known as **protamines.** Roughly two-thirds of the amino acids present in these small basic proteins are arginines, and the remaining amino acids are neutral. Ionic bonds between these positively charged arginines and the negatively charged phosphate groups present in DNA allow protamine molecules to wrap tightly around the DNA double helix, lying in the minor groove. This arrangement, which has been deduced from X-ray diffraction analyses of intact sperm heads, creates an extremely stable crystalline structure. Such a compact structure facilitates the packaging and protection of the DNA of sperm cells, whose main function is to deliver an intact set of DNA molecules to the egg.

In contrast to the histones and protamines, the **nonhistone** protein fraction consists of a diverse and variable group of macromolecules (Figure 11-15). Unlike the histones, which are present in virtually the same concentration (relative to the total nuclear DNA content) in all eukaryotic cells, the nonhistone proteins of different cell types tend to vary in both quantity and type. Among the best characterized components of the nonhistone protein fraction are enzymes involved in the metabolism of DNA and in the modification of chromatin proteins (Table 11-2). Aside from such enzymes, the best characterized nonhistone proteins are the **high-mobility group** or **HMG proteins.** These proteins, so named because of their high electrophoretic mobility, have been identified in chromatin isolated from a

FIGURE 11-13 Polyacrylamide gel electrophoresis of histones isolated from various tissues of the calf. Courtesy of R. Chalkley.

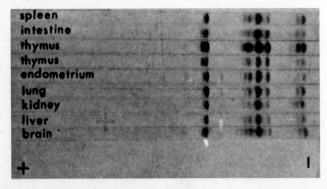

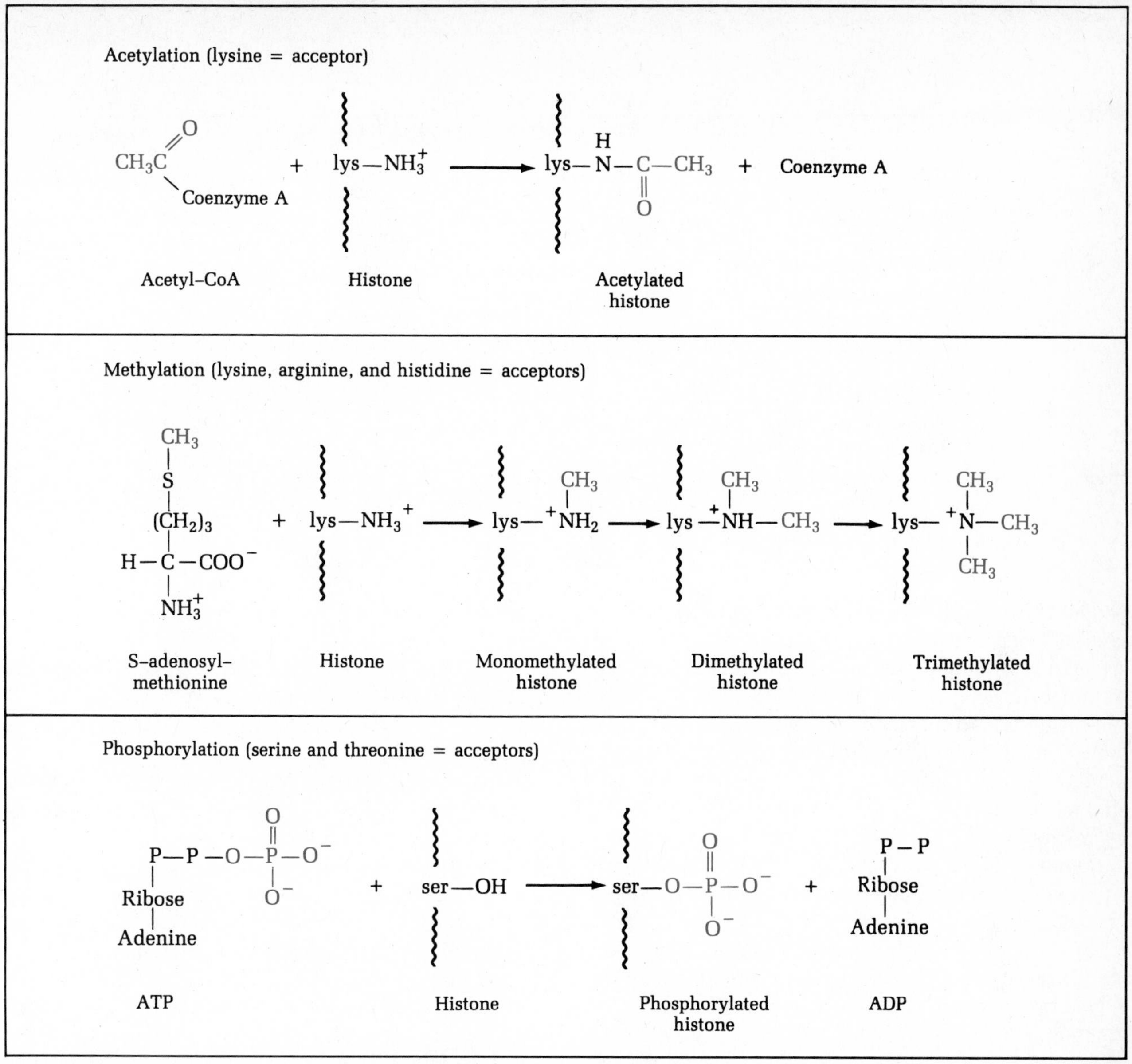

Acetylation (lysine = acceptor)

$$CH_3C \overset{O}{\diagdown} \text{—Coenzyme A} \quad + \quad lys-NH_3^+ \quad \longrightarrow \quad lys-\overset{H}{\underset{}{N}}-\underset{O}{\overset{}{C}}-CH_3 \quad + \quad \text{Coenzyme A}$$

Acetyl–CoA   Histone   Acetylated histone

Methylation (lysine, arginine, and histidine = acceptors)

$$\begin{array}{c} CH_3 \\ | \\ S \\ | \\ (CH_2)_3 \\ | \\ H-C-COO^- \\ | \\ NH_3^+ \end{array} \quad + \quad lys-NH_3^+ \quad \longrightarrow \quad lys-{}^+NH_2 \quad \longrightarrow \quad lys-{}^+\overset{CH_3}{\underset{}{NH}}-CH_3 \quad \longrightarrow \quad lys-{}^+\overset{CH_3}{\underset{CH_3}{N}}-CH_3$$

S–adenosyl–methionine   Histone   Monomethylated histone   Dimethylated histone   Trimethylated histone

Phosphorylation (serine and threonine = acceptors)

$$\begin{array}{c} P-P-O-\overset{O}{\underset{O^-}{P}}-O^- \\ | \\ \text{Ribose} \\ | \\ \text{Adenine} \end{array} \quad + \quad ser-OH \quad \longrightarrow \quad ser-O-\overset{O}{\underset{O^-}{P}}-O^- \quad + \quad \begin{array}{c} P-P \\ | \\ \text{Ribose} \\ | \\ \text{Adenine} \end{array}$$

ATP   Histone   Phosphorylated histone   ADP

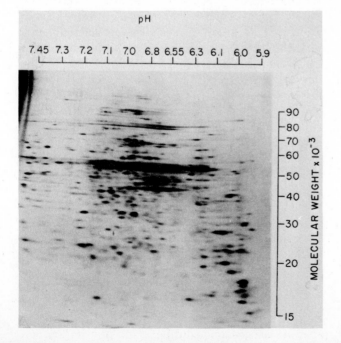

FIGURE 11-14 *(Above)* Examples of enzymatic modification reactions involving histone molecules. Acetyl, methyl, and phosphoryl groups attached to histone molecules by the above reactions are subject to removal by specific hydrolytic enzymes (histone deacetylase, demethylase, and phosphatase, respectively).

FIGURE 11-15 *(Left)* Two-dimensional polyacrylamide gel electrophoresis of HeLa cell nonhistone proteins illustrating the diverse spectrum of components present. Courtesy of E. H. McConkey.

TABLE 11-2
Partial list of enzymes identified in the nonhistone protein fraction

Enzyme Category	Function
I. Nucleic acids as substrates	
DNA polymerase	Polymerization of deoxyribonucleotides into DNA
RNA polymerase	Polymerization of ribonucleotides into RNA
Nucleases	Cleavage of DNA and/or RNA
DNA ligase	Joining DNA segments during DNA replication and repair
Poly(A) polymerase	Addition of poly(A) tails to RNA
DNA methylase	Methylation of DNA
Helix-destabilizing enzymes	Unwind DNA double helix and stabilize resulting single-stranded DNA
II. Chromatin proteins as substrates	
Proteases	Protein cleavage
Histone acetylases and deacetylases	Acetylation and deacetylation of histones
Histone kinases	Phosphorylation of histones
Nonhistone kinases	Phosphorylation of nonhistones
Histone methylases	Histone methylation

wide variety of cell types and organisms. As we shall see later, at least some HMG proteins are preferentially associated with chromatin fibers that are being actively transcribed.

In addition to enzymes and HMG proteins, the nonhistone fraction contains a diverse group of other proteins. Although these molecules have not been well characterized, many appear to be extensively modified by reversible cycles of phosphorylation and dephosphorylation. Since phosphorylation and dephosphorylation reactions are known to regulate the functional properties of other, better characterized proteins (Chapter 16), the extensive occurrence of nonhistone protein phosphorylation suggests that these proteins may be involved in the regulation of chromatin structure and/or function.

The Nucleosome and Chromatin Structure The DNA contained within a typical eukaryotic nucleus would measure a meter or more in overall length if it were completely unfolded. Since an individual nucleus is roughly a million times smaller than this in diameter, a significant topological problem is encountered in packing such an enormous length of DNA into such a small space. This packaging is made possible by the presence of chromatin proteins that associate with DNA and trigger the formation of chromatin fibers. Great strides have been made in recent years in unraveling the way in which the DNA and protein components are arranged in such fibers.

One of the earliest insights into the organization of chromatin fibers emerged in the late 1960s, when X-ray diffraction analyses of isolated chromatin by Maurice Wilkins and his colleagues revealed the presence of a repeating structure with a periodicity of 10 nm. Such a repeat structure is not observed with isolated DNA or histones, but does appear when histones and DNA are mixed together. It was therefore concluded that his-

tones impose an organized repeat structure on DNA. One of the first speculations to arise concerning the nature of this interaction was that histones cause the DNA molecule to coil into a twisted fiber in which the distance between adjacent turns of the twisted fiber is 10 nm.

Although this model was consistent with the X-ray diffraction data, there was a scarcity of other supporting information. Intact chromatin fibers had been visualized in electron micrographs of ruptured, unsectioned nuclei, but little substructure could be seen in fibers prepared in this way. The only clear-cut conclusion to emerge from such observations was that intact chromatin fibers measure about 30 nm in diameter, which is an order of magnitude greater than the diameter of a single DNA double helix. No signs of a supercoiled substructure were apparent, nor were there any other clues as to the arrangement of the DNA molecule within the fiber. Then in 1974 a new perspective was suddenly shed upon the subject when Ada and Donald Olins published electron micrographs of chromatin fibers prepared in a new way. In this procedure nuclei swollen by exposure to hypotonic conditions were gently fixed with formalin and then stained, a sequence that avoids the harsh solvents employed in earlier procedures for examining chromatin fibers. Examination of material prepared in this way revealed that chromatin fibers consist of a series of tiny particles attached to one another by thin filaments (Figure 11-16). This "beads on a string" arrangement immediately triggered the proposal that the spherical particles or beads are made of histones, while the thin filaments contain DNA molecules passing from particle to particle.

On the basis of electron microscopic evidence alone, it would have been difficult to determine whether the particles observed in the preceding experiments are a normal component of chromatin structure or an artifact generated during sample preparation. For-

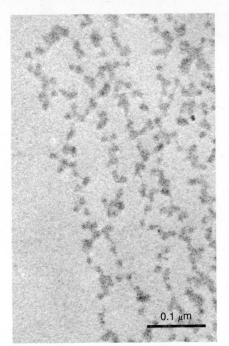
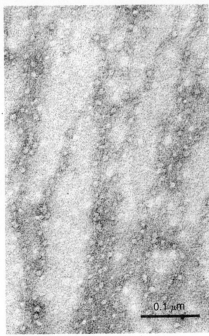

0.1 μm   0.1 μm

FIGURE 11-16 Electron micrographs showing beads-on-a-string appearance of isolated chromatin fibers prepared using gentle fixation procedures and viewed after positive staining *(left)* or negative staining *(right)*. Courtesy of D. E. Olins and A. L. Olins.

tunately, independent evidence for the existence of a repeating protein substructure in chromatin emerged at about the same time from the studies of Dean Hewish and Leigh Burgoyne, who discovered that rat liver nuclei contain a nuclease capable of degrading the DNA molecules present in chromatin fibers. When the partially degraded DNA generated by the action of this nuclease on rat liver chromatin was isolated and analyzed by gel electrophoresis, a distinct pattern of fragments was revealed in which the size of each DNA fragment corresponds to an exact multiple of the size of the smallest fragment (Figure 11-17). Subsequent measurements revealed that this smallest fragment measures about 200

FIGURE 11-17 Results of experiments in which the DNA fragments generated by partial nuclease digestion of intact chromatin are analyzed by gel electrophoresis. The discovery that the DNA fragments are all multiples of 200 base pairs in length suggests that chromatin proteins are clustered at 200 base-pair intervals along the DNA molecule, thereby conferring a regular pattern of protection against nuclease digestion.

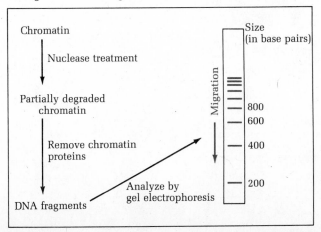

base pairs in length. Since nuclease digestion of naked DNA does not generate any such pattern of fragments, it can be concluded that (1) chromatin proteins are clustered along the DNA molecule in a regular pattern that repeats at roughly 200-base-pair intervals, and (2) DNA is more susceptible to nuclease digestion between these protein clusters.

Such observations immediately raised the question of whether the particles observed in electron micrographs of chromatin fibers are responsible for the 200-base-pair periodicity detected in the nuclease digestion experiments. To investigate this issue, a combination of the nuclease digestion and electron microscopic approaches was required. Experiments were therefore carried out in which chromatin was first exposed to micrococcal nuclease, a bacterial enzyme that, like the rat liver nuclease, cleaves chromatin DNA at 200-base-pair intervals. After brief exposure to this enzyme, the fragmented chromatin was separated into fractions of varying sizes by moving-zone centrifugation. Examination by electron microscopy revealed that the smallest fraction contained single particles, the next fraction contained dimers, the next trimers, and so forth (Figure 11-18). When DNA was isolated from these various fractions and analyzed by gel electrophoresis, the DNA obtained from the fraction containing single particles was found to measure 200 base pairs in length, the DNA from the dimer fraction measured 400 base pairs in length, and so on. It was therefore clear that the spherical particles observed in electron micrographs are each associated with 200 base pairs of DNA. This basic repeat unit, consisting of 200 base pairs of DNA associated with a protein particle, is called a **nucleosome.**

The relationship between histone molecules and

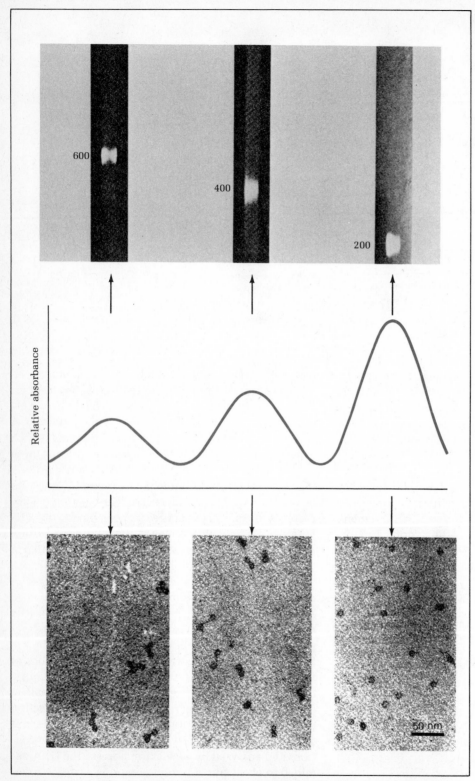

FIGURE 11-18 Results of experiments in which chromatin partially degraded by treatment with micrococcal nuclease was size-fractionated by moving-zone centrifugation. The individual fractions were then analyzed either by electron microscopy or by gel electrophoresis after removal of chromatin proteins. The results indicate that the spherical particles observed in electron micrographs are each associated with 200 base pairs of DNA. Photographs courtesy of R. D. Kornberg.

the nucleosomal particle was initially investigated by Roger Kornberg and his associates, who approached this problem by attempting to reconstitute chromatin from a mixture of DNA and various types of histones. Although successful reconstitution of the chromatin repeat structure was obtained with DNA and total unfractionated histones, successful results were not obtained with mixtures of individually purified histones (i.e., H1, H2A, H2B, H3, and H4). It soon became apparent that this failure occurs because histone purification procedures that disrupt normal histone-histone interactions inhibit the ability of histones to participate in proper chromatin reconstitution. If histones are purified by gentler techniques, histones H3 and H4 tend to remain bound to each other, as do histones H2A and H2B. When these H3-H4 and H2A-H2B aggregates are mixed with DNA, chromatin exhibiting normal nucleosomal structure can be reconstituted. Kornberg therefore concluded that histone H3-H4 and H2A-H2B complexes are an integral part of nucleosome structure.

To investigate the nature of these histone-histone interactions in more depth, Kornberg and his colleague, Jean Thomas, treated isolated chromatin with a chemical reagent that forms covalent cross-links between protein molecules situated adjacent to one another. After treatment with this reagent, the chemically cross-linked proteins were isolated and analyzed by acrylamide gel electrophoresis. Protein complexes the size of eight histone molecules were found to be a prominent component in such gels, suggesting that the nucleosomal particle contains an **octomer** of histones. Given the knowledge that histones H3-H4 and histones H2A-H2B each form tight complexes and that these four histones are present in roughly equivalent amounts, it was proposed that the histone octomer contains two H2A-H2B dimers and two H3-H4 dimers.

The proposal that nucleosomes contain a histone octomer represented a significant advance in our understanding of chromatin structure, but still left unanswered the question of how the DNA and histone octomer are bound to one another. Although the original phrase "beads on a string" suggested that the histone particles cover the DNA, subsequent experimentation has revealed that the DNA actually surrounds the histones. This point first emerged from studies involving **pancreatic DNAase-I**, an enzyme that, unlike micrococcal nuclease, is capable of cleaving DNA molecules that are bound to a solid surface. Treatment of chromatin with DNAase-I was found to generate DNA fragments that are multiples of about 10 base pairs in length. Since 10 base pairs is the approximate length of one complete turn of the DNA double helix, the most straightforward interpretation of these data is that the DNA double helix is tightly wound over the surface of the histone octomer, and that a small region of each turn

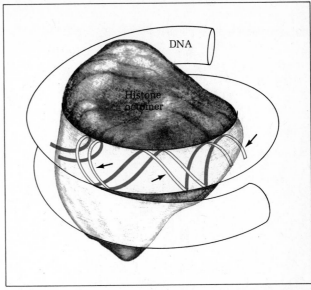

FIGURE 11-19 Model illustrating why a DNA double helix wound tightly over the surface of a protein particle is susceptible to DNAase-I cleavage at approximately 10 base-pair intervals *(arrow).*

of the double helix is exposed and potentially subject to cleavage by DNAase-I (Figure 11-19).

One question not addressed by the above model of the nucleosome concerns the localization of histone H1, the one histone that is not part of the octomer. In nucleosomes prepared by brief digestion of chromatin with micrococcal nuclease, histone H1 is still present (along with the other four histones and 200 base pairs of DNA). If nuclease digestion is carried out for longer periods, however, the DNA fragment is further degraded until it reaches a length of about 146 base pairs; during the final stages of this digestion process, histone H1 is released. The remaining particle, consisting of a histone octomer associated with 146 base pairs of DNA, is referred to as a **core particle.** The DNA degraded during digestion of the DNA from 200 to 146 base pairs in length is referred to as the **linker DNA** because it joins one nucleosome to the next (Figure 11-20). The length of the linker DNA varies somewhat among organisms, but the DNA associated with the core particle always measures close to 146 base pairs. Due to their uniform size, core particles can be readily crystallized for analysis by X-ray diffraction and electron microscopy. Results from such analyses have revealed that the core is a disk-shaped particle, measuring approximately 11 nm in diameter and 5.7 nm in height. The 146 base pairs of DNA associated with the core particle is sufficient to wrap around the particle roughly 1.8 times.

Since histone H1 is released from the nucleosome during digestion of the linker DNA, it has been concluded that H1 is associated with the linker rather than with the core nucleosomal particle. In this peripheral location histone H1 is thought to facilitate the packing of

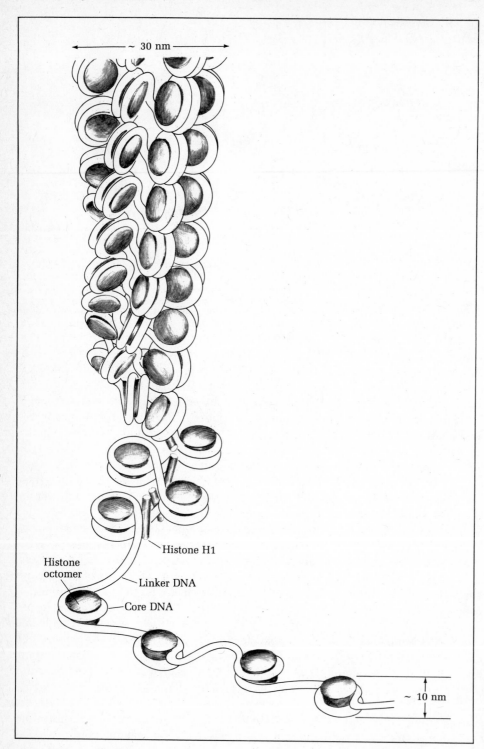

~ 30 nm

Histone H1

Histone octomer

Linker DNA

Core DNA

~ 10 nm

FIGURE 11-20 One of several models that has been proposed to explain how the 10-nm chromatin fiber consisting of an extended chain of nucleosomes might be folded to form the 30-nm chromatin fiber.

nucleosomes into higher levels of organization. Such higher-order packing is known to be important under physiological conditions because intact chromatin fibers often exhibit a diameter of 30 nm in electron micrographs, while isolated nucleosomes and nucleosome chains measure closer to 10 nm in diameter. The relationship between an extended 10-nm diameter chain of nucleosomes and a 30-nm diameter chromatin fiber has been investigated under cell-free conditions, where the two kinds of fibers can be interconverted by changing the salt concentration of the medium. In electron micrographs of chromatin preparations undergoing transition from the 10-nm diameter to the 30-nm diameter state, the extended chain of nucleosomes appears as if it is coiling into a helical chain or *solenoid*. Such observations have led to the formulation of a variety of models and theories as to how the individual nucleosomes might be packed in such chains (see Figure 11-20). This

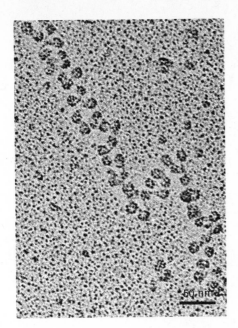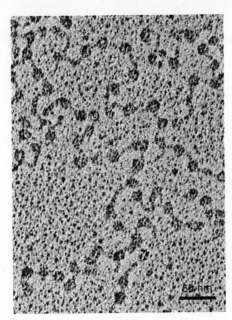

FIGURE 11-21 Electron micrographs illustrating the requirement for histone H1 in the higher-order packing of nucleosomes. *(Left)* Normal appearance of nucleosomes in isolated rat chromatin fibers. *(Right)* Appearance of chromatin after removal of histone H1. Note the disruption of the normal packing arrangement. Courtesy of F. Thoma.

packing process does not occur in chromatin preparations whose histone H1 molecules have been extracted, suggesting that histone H1 mediates the folding of the 10-nm chain of nucleosomes into the tightly packed 30-nm fiber (Figure 11-21). X-ray diffraction analyses have revealed that the characteristic 10-nm periodicity of chromatin appears during conversion of the 10-nm chain to the 30-nm solenoid, implying that the spacing between turns of the solenoid, rather than between adjacent nucleosomes, is responsible for the 10-nm repeat.

It is clear that the existence of nucleosomes makes a major contribution to solving the topological problem of packing several meters of DNA into the space the size of a cell nucleus. At the most fundamental level of organization, a 200-base-pair stretch of DNA, measuring about 70 nm in extended length, is wrapped around a nucleosome roughly 10 nm in diameter. Hence at this level of organization the **packing ratio** (i.e., the ratio of extended length to packed length of DNA) is about 7. When nucleosomes are packed into helical arrays to form 30-nm chromatin fibers, the calculated packing ratio increases to about 40.

Under certain conditions the 30-nm chromatin fiber is subjected to further folding, forming structures with even greater packing ratios. The most common situation of this type occurs during cell division, when chromatin fibers are packaged into highly condensed chromosomes with packing ratios approaching 10^4. When the structure of such chromosomes is described in detail in Chapter 13, we shall see that a highly organized folding of the chromatin fiber is involved. Although the molecular basis for this folding is not clearly understood, chromosomes from which histones have been extracted retain a protein backbone whose basic shape resembles that of the intact chromosome. This backbone or **scaffold,** which consists of a mixture of several dozen nonhistone proteins, serves as an anchoring point to which long loops of DNA are attached (Figure 11-22). Although it is more difficult to detect, the chromatin fibers of nondividing cells are also thought to be organized into long loops anchored at periodic intervals to the nuclear matrix.

In addition to the chromosomes that form at the time of mitotic cell division, chromatin fibers can be packaged into several other types of chromosomes. Classic examples are the giant polytene chromosomes of insects (page 404) and the lampbrush chromosomes formed during meiotic division of certain types of oocytes (page 526). In spite of the striking differences between the morphological appearance of these various types of chromosomes, they all appear to be constructed by folding and condensation of the same underlying type of 30-nm chromatin fiber. The overall picture of eukaryotic chromatin structure that therefore emerges is of a DNA backbone that undergoes several levels of coiling and folding as a result of its interactions with histones and other proteins. The existence of these varying levels of organization makes chromatin an inherently dynamic entity whose physical and functional properties are subject to constant alterations in response to the changing needs of the cell.

DNA Packaging in Prokaryotic Cells Because prokaryotic cells contain several orders of magnitude less DNA than do typical eukaryotic cells, the topological problem of DNA packaging is not as severe. This may explain why neither histones nor nucleosomes can be detected in prokaryotes. Although bacterial DNA is sometimes said to be "naked," several observations indicate that this is not the case. For example, electron

FIGURE 11-22 Electron micrograph showing the protein scaffold that remains after the removal of histones from HeLa cell metaphase chromosomes. The chromosomal DNA remains attached to this scaffold. The arrow points to a region at the edge of the DNA mass where the DNA molecule can be clearly seen. Courtesy of U. K. Laemmli.

microscopic studies utilizing disrupted bacterial cells have revealed the presence of DNA-containing fibrils averaging 12 nm in diameter (Figure 11-23). Since a naked DNA double helix measures about 2 nm in diameter, these findings suggest that other substances are associated with bacterial DNA.

Chemical analyses of DNA-containing fibrils isolated from bacterial cells have revealed the presence of a significant amount of protein, further reinforcing the idea that bacterial DNA is not naked. Two particular proteins present in this fraction account for a large proportion of the DNA-associated protein and are present in roughly equimolar amounts, suggesting that they may play a structural role in DNA packaging. Partial nuclease digestion of bacterial chromatin has been reported to generate 120-base-pair fragments of DNA in association with these two proteins. Although such results bear a superficial resemblance to those obtained with nucleosomes, a direct analogy to the nucleosomal structure of eukaryotic chromatin appears unjustified because a repeating pattern of multimers of the 120-base-pair fragment is not observed.

The Nuclear Matrix

It is clear from both electron microscopic and biochemical analyses that chromatin is the dominant component of the nucleus, accounting for 80–90 percent of the total mass. A nucleus that has had its chromatin fibers removed might therefore be expected to collapse into a relatively structureless mass. It was therefore somewhat surprising when Ronald Berezney and Donald Coffey reported in 1974 that an insoluble substructure, which retains the overall shape of the nucleus, can be observed in electron micrographs of nuclei that have had more than 95 percent of their chromatin removed

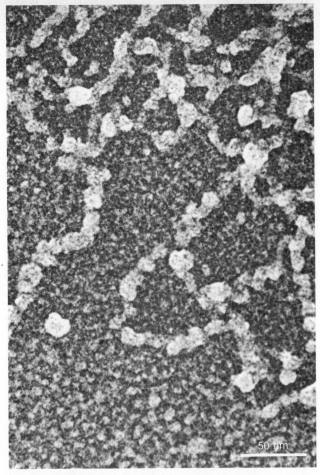

FIGURE 11-23 Electron micrograph of DNA-containing fibrils released from a disrupted bacterial cell *(E. coli)*. The fibrils exhibit a knobby appearance and are considerably thicker than a naked DNA double helix. Courtesy of J. D. Griffith.

by a combination of nuclease and detergent treatments (Figure 11-24). This substructure, termed the **nuclear matrix,** is thought to be analogous to the cytoskeletal filament network of the cytoplasm, though the nuclear matrix differs significantly in both the morphological appearance and protein composition of its constituent fibers. Nonetheless several proteins, including actin, are found in both the nuclear matrix and the cytoskeleton.

Although the functional role of the nuclear matrix is not clearly understood, a close association of this structure with chromatin fibers is suggested by the discovery that isolated nuclear matrix preparations always contain a small amount of tightly bound DNA and RNA. Using cloned genes as hybridization probes, it has even been shown that DNA sequences that are being actively transcribed remain preferentially associated with the nuclear matrix when the bulk of the DNA is removed by enzymatic digestion. Labeling experiments employing ^{3}H-thymidine have also revealed that newly synthesized DNA is preferentially associated with the matrix. Although the preceding observations raise the possibility that the nuclear matrix plays a role in organizing chromatin fibers involved in transcription and replication, alternative interpretations have not been ruled out. It is possible, for example, that the preferential association of certain types of DNA and RNA with the nuclear matrix is an artifact generated by the harsh isolation conditions employed.

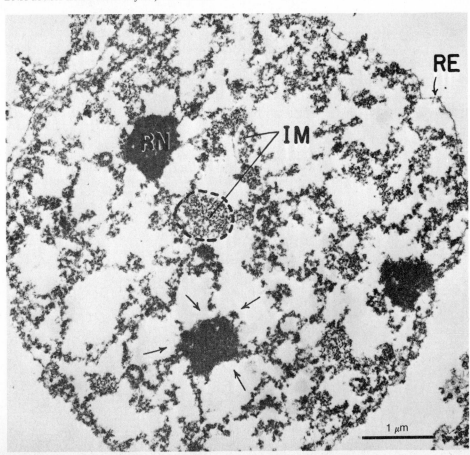

FIGURE 11-24 Electron micrograph of an isolated nuclear matrix preparation showing the internal framework of the nuclear matrix (IM), the residual nuclear envelope (RE), and the residual nucleolus (RN) surrounded by empty spaces (arrows). Courtesy of R. Berezney.

GENE TRANSCRIPTION

Now that the general features of nuclear organization have been introduced, we can begin to investigate the functional activities in which this organelle is involved. The most crucial activities of the nucleus relate to its roles in the transcription and replication of genetic information. In the remainder of this chapter we shall focus upon the transcription process, while the events associated with DNA replication will be deferred until the chapter devoted to the topic of cell division (Chapter 13).

Microscopic Visualization of Gene Activity

Because the DNA of most eukaryotic cells is dispersed throughout the nucleus as a mass of intertwined chromatin fibers, the organization and activity of individual genes is usually difficult to visualize. However, an exceptional situation occurring in certain insect cells has provided a way around this obstacle. In flies belonging to the order Diptera, metabolically active tissues such as the salivary glands, intestines, and excretory organs grow by an increase in the size of, rather than the number of, their constituent cells. This growth process eventually produces "giant" cells whose volumes may be thousands of times greater than normal. The development of these giant cells is accompanied by the formation of enormous chromosomes that measure several hundred micrometers in length and several micrometers in width (Figure 11-25). Such chromosomes are roughly ten times greater in length and a hundred times greater in width than the chromosomes associated with normal cell division.

In the early 1960s Hewson Swift utilized cytochemical staining methods to monitor the increase in DNA content that accompanies the formation of giant chromosomes in the fruit fly *Drosophila*. He discovered that the increase in DNA occurs as a series of about ten discrete doublings, at which point the DNA content has increased roughly a thousandfold ($2^{10} = 1024$). This observation suggests that giant chromosomes are multistranded or **polytene,** that is, composed of a large number of identical DNA molecules generated by successive rounds of replication. Because this replication occurs in a cell that is not dividing, the newly synthesized DNA molecules accumulate in the nucleus and line up in parallel to form a polytene chromosome.

One of the more striking morphological features of polytene chromosomes is the presence of a distinct pattern of darkly staining regions, termed **bands,** which are separated from one another by lighter staining areas known as **interbands** (see Figure 11-25). It has long been known that mutations involving the deletion of specific genes are accompanied by the disappearance of particular bands, an observation that led Calvin Bridges to propose in the early 1930s that each band corresponds to a gene. More recent experimentation has reinforced this basic view, though there are indications that essential gene functions occasionally map in the interband regions as well. As a general rule, no more than one protein-coding gene can be identified within a given band-interband region. Since the polytene chromosomes of an organism such as *Drosophila* contain a total of about 5000 bands, it has been concluded that a maximum of roughly 5000 genes is present.

Electron microscopy has provided several clues as to the way in which polytene chromosomes are constructed. In thin-section electron micrographs, 30-nm fibers resembling the chromatin fibers of typical eukaryotic cells can be observed in both the band and interband regions. The major difference in the morphological appearance of the two regions is that the chromatin fibers are densely packed together in the bands, while in the interbands they are arranged as loosely

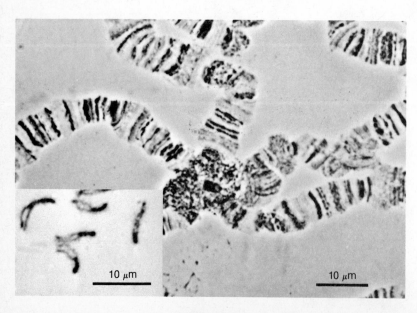

FIGURE 11-25 Light micrograph of a polytene chromosome of *Drosophila virilis* along with a photograph of a typical metaphase chromosome (inset) for size comparison. The bands of the polytene chromosomes appear as dark regions separated from one another by the interbands. Courtesy of J. Gall.

10 μm

10 μm

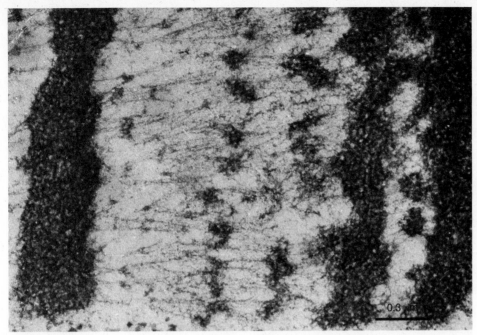

0.3

FIGURE 11-26 Electron micrograph of a small region of a polytene chromosome showing the fine structure of the band and interband regions. The light staining region, which represents an interband, contains a loosely packed array of chromatin fibers. In the darker staining regions, which correspond to bands, the chromatin fibers are more densely packed. Courtesy of V. Sorsa.

packed parallel bundles (Figure 11-26). On the basis of the preceding information, it has been proposed that polytene chromosomes are constructed from chromatin fibers that are lined up in register and pass continuously from one end of the chromosome to the other (Figure 11-27a). According to this model the individual chromatin fibers, each containing a single replicated DNA molecule, are tightly folded and packed together in the band regions and more loosely assembled in the interbands.

FIGURE 11-27 Schematic model of a polytene chromosome illustrating the normal organization of bands and interbands (a), and the uncoiling of chromatin fibers that occurs during puff formation (b).

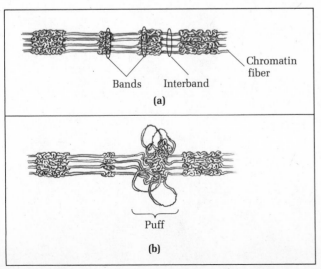

Chromatin fiber

Bands   Interband

(a)

Puff

(b)

Although the sequential pattern of bands is relatively constant for each type of polytene chromosome, certain bands undergo changes in appearance that are specific for a particular tissue or stage of development. Such bands uncoil into a looser, puffed-out configuration known as a chromosome **puff** or, when it becomes exceptionally large, a **Balbiani ring** (Figure 11-28). In 1952, Wolfgang Beerman postulated that puffing is a reflection of the activity of individual genes. This idea was originally based on the observation that each tissue type exhibits its own characteristic pattern of puffs, and that this puffing pattern changes in a reproducible way as development proceeds (Figure 11-29).

Beerman supported this hypothesis with a series of experiments in which the expression of a specific genetic trait was correlated with the development of a particular Balbiani ring. In these studies Beerman focused his attention on the salivary secretions of the midge *Chironomus*. In one species of midge, *Ch. pallidivittatus*, the salivary gland produces a granular fluid; comparable cells of a closely related organism, *Ch. tentans*, produce a clear fluid. Upon comparing the salivary gland chromosomes of these two organisms, Beerman noted that a large Balbiani ring present in *Ch. pallidivitattus* is missing in *Ch. tentans*. In order to determine whether this Balbiani ring is responsible for the granular secretion, hybrids were formed between the two species of midge. The salivary gland secretion of the hybrids was found to be about half as granular as that normally occurring in *Ch. pallidivitattus*, and the Balbiani ring in question was half as large. Beerman there-

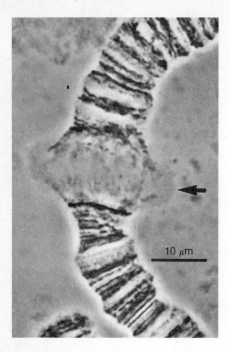

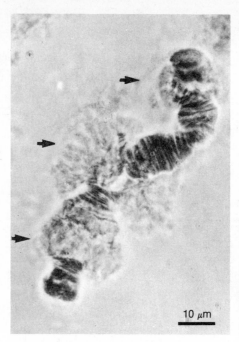

FIGURE 11-28 Structural appearance of puffs and Balbiani rings in light micrographs of polytene chromosomes. *(Left)* An insect salivary gland polytene chromosome exhibiting a puff (arrow). Courtesy of J. M. Amabis. *(Right)* A polytene chromosome from *Chironomus* exhibiting three Balbiani rings (arrows). Courtesy of B. Daneholt.

fore concluded that the formation of this Balbiani ring represents activation of the gene responsible for the granularity of the salivary gland secretion.

The correlation observed between the puffed state of a specific chromosomal region and gene activity implies that puffs and Balbiani rings are the sites of RNA synthesis. This idea was first confirmed experimentally in the early 1960s by Claus Pelling, who employed microscopic autoradiography to monitor the incorporation of ^{3}H-uridine into RNA. His studies revealed that newly

synthesized RNA is localized almost exclusively in the puffed regions of polytene chromosomes, and that inhibitors of RNA synthesis (such as actinomycin D) prevent this labeling from occurring (Figure 11-30). Efforts to characterize the RNA synthesized in the puffed regions have been facilitated by the discovery that the Balbiani rings of *Chironomus* salivary gland cells are large enough to be isolated by microsurgery. This approach has been elegantly exploited by Bo Lambert, who extracted radioactive RNA from several isolated

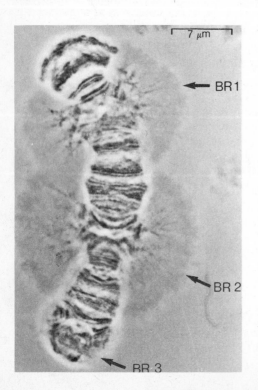

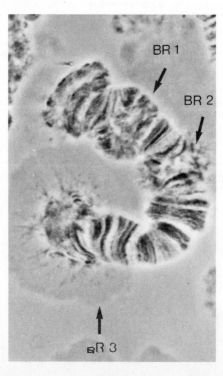

FIGURE 11-29 Illustration of the way in which chromosomal puffing patterns can be altered in response to changing conditions. *(Left)* A normal *Chironomus* salivary gland chromosome exhibiting three Balbiani rings (BR1, BR2, BR3). Note that BR2 is normally the largest and BR3 the smallest. *(Right)* When cells are exposed to dimethylsulfide, BR3 is stimulated and BR2 is inhibited. Similar changes in puffing patterns occur during normal development. Courtesy of H. Sass.

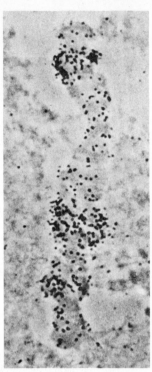

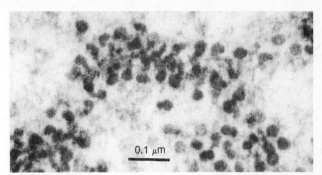

FIGURE 11-32 Electron micrograph of a looped-out region of a large Balbiani ring showing the presence of ribonucleoprotein granules. Courtesy of B. Daneholt.

FIGURE 11-30 Autoradiograph of a polytene chromosome from cells incubated with ³H-uridine in the absence and presence of actinomycin D. *(Left)* Extensive labeling of chromosomal puffs occurs in control cells. *(Right)* In cells treated with actinomycin D to inhibit RNA synthesis, the puffs decrease in size and their incorporation of ³H-uridine is markedly reduced. Courtesy of W. Beerman.

Balbiani rings and hybridized it back to unlabeled chromosomes. When RNA isolated from Balbiani ring II was analyzed in this way, it was found to preferentially hybridize back to Balbiani ring II (Figure 11-31). Likewise, RNA extracted from Balbiani ring I preferentially hybridized back to Balbiani ring I. It was therefore concluded that each Balbiani ring synthesizes a specific type of RNA unique to that particular segment of DNA.

FIGURE 11-31 Hybridization properties of labeled RNA isolated from specific Balbiani rings in *Chironomus*. The two autoradiographs show the appearance of chromosome IV after hybridization with RNA isolated from either Balbiani ring 2 *(left)* or Balbiani ring 1 *(right)*. Note that the two RNA preparations hybridize to different regions of the chromosome. Courtesy of B. Lambert.

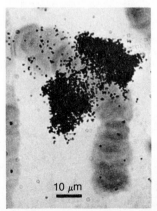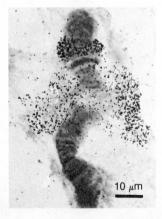

Subsequent experiments have revealed that RNAs that hybridize to specific Balbiani rings can also be isolated from the nucleoplasm, where they are present in ribonucleoprotein granules, and in the cytoplasm, where they are associated with ribosomes. Taken together, the preceding observations suggest that each chromosomal puff synthesizes a specific type of RNA that is immediately complexed with protein to form a ribonucleoprotein granule (Figure 11-32). After being released into the nucleoplasm, such granules are transported through the nuclear pores to the cytoplasm (see Figure 11-7). The RNA component of the ribonucleoprotein complex then becomes associated with ribosomes, where it presumably functions as a messenger RNA that directs the synthesis of a particular polypeptide chain. Later in the chapter we shall see that biochemical approaches to the study of gene transcription have yielded similar conclusions, though they have also revealed that RNA molecules may undergo significant changes in structure during this process.

Although our understanding of the factors that regulate puff formation is by no means complete, one of the first changes that can be detected at newly forming puffs is the accumulation of nonhistone proteins (Figure 11-33). Some nonhistone proteins are common to multiple puffs, while others are specific to certain ones. As nonhistone proteins accumulate, the puff begins to swell and the chromatin fibers uncoil, becoming thinner and less condensed than the typical 30-nm fibers characteristic of a normal band (see Figure 11-27b). Shortly after the uncoiling process begins, the synthesis of RNA is initiated.

One situation in which the molecular events that trigger the puffing process are well established is the normal molting process in insects. During molting, a highly reproducible and characteristic sequence of puffs occurs. In the early 1960s Ulrich Clever reported that this same sequence of puffs can be induced prematurely by injecting nonmolting insects with the steroid hormone **ecdysone.** The conclusion that ecdysone stimulates the formation of these puffs was further reinforced by the discovery that identical puffing patterns

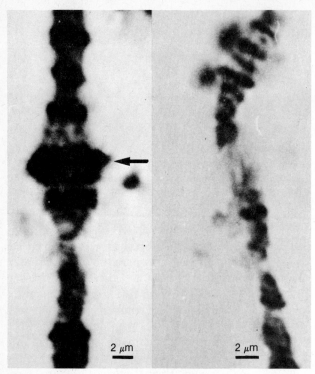

2 μm   2 μm

FIGURE 11-33 Light micrograph illustrating the accumulation of protein in *Drosophila* salivary gland puffs. The intense staining of the puff observed in the control polytene chromosome *(left)* is abolished when the chromosome is pretreated with an enzyme that breaks down protein molecules *(right)*, indicating that the puff is enriched in protein. Courtesy of Th. K. H. Holt.

are induced when isolated salivary glands are incubated with ecdysone (Figure 11-34). The puffing induced by ecdysone occurs in two phases: "early puffs," which develop within a few minutes of hormone administration, and "late puffs," which form several hours later and are accompanied by regression of the early puffs. Inhibitors of protein synthesis block formation of the late, but not early, puffs, suggesting that the RNAs synthesized by the early puffs code for the synthesis of proteins required for development of the later puffs. As in the case of other steroid hormones (Chapter 16), the action of ecdysone is mediated by a specific protein receptor to which this steroid binds. As we shall see later in the chapter, steroid hormone-receptor complexes bind to specific sites on the DNA molecule and activate gene transcription. The ecdysone receptor is thus an example of a nonhistone protein that selectively regulates the transcriptional activity of a specific group of puffs.

Although chromosome puffs clearly represent sites of active gene transcription, the absence of visible puff formation does not necessarily indicate the absence of transcription. Autoradiographic analyses of cells incubated with ^{3}H-uridine has revealed the presence of radioactivity over bands where obvious puffs are not present, suggesting that these bands also represent sites of active gene transcription. In the case of the histone

genes of *Drosophila*, it has been clearly demonstrated that transcription occurs in such banded regions lacking puffs.

Puffing in polytene chromosomes is not the only situation in which the activity of individual eukaryotic genes has been studied at the microscopic level. During the meiotic division that precedes the development of certain kinds of egg cells, large "lampbrush" chromosomes are formed. Although these chromosomes are not polytene, gene transcription occurs in uncoiled, looped-out areas that may be analogous to the puffs of polytene chromosomes. Since the formation of lampbrush chromosomes occurs during meiotic cell division, a detailed discussion of their morphology and behavior will be delayed until Chapter 13, which is devoted to the topic of cell division.

In contrast to the situation in cells containing polytene or lampbrush chromosomes, it is relatively difficult to identify specific active genes within the intermingled chromatin fibers of a typical eukaryotic nucleus. However, if nuclei are first ruptured in hypo-

FIGURE 11-34 A series of four light micrographs illustrating the sequence of puffs induced in the third chromosome of *Drosophila* by exposure to the hormone ecdysone for 1, 2, 4, and 6 hours *(top to bottom)*. Note the changing appearance of the numbered bands. Courtesy of M. Ashburner.

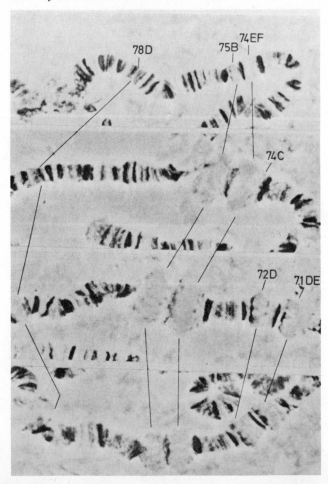

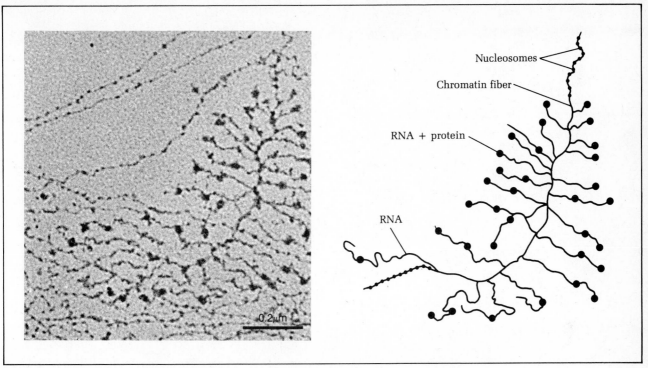

FIGURE 11-35 Electron microscopic appearance of chromatin fibers in the process of being transcribed into RNA. The newly forming RNA molecules, which can be seen protruding from the central chromatin fiber, are already complexed with protein and exhibit a beaded appearance. In a few regions the nucleosomal organization of the chromatin fiber can be detected. Courtesy of A. L. Beyer.

tonic media and the chromatin fibers are then spread out for electron microscopic examination, individual chromatin fibers in the process of synthesizing RNA can be identified (Figure 11-35). It is interesting to note that nucleosomes are usually observed in such fibers, indicating that the nucleosomal organization is largely retained even during the process of gene transcription. The thickness and staining properties of the newly forming RNA molecules indicate that they are becoming complexed with protein while the process of gene transcription is still occurring.

The RNA Polymerase Reaction

Although microscopy has provided some important insights into the process of gene transcription, our understanding of the molecular events involved has emerged largely from biochemical studies. The first major step in unraveling the biochemical basis of gene transcription came with the isolation and purification of the enzymes, called **RNA polymerases,** which catalyze the transcription of DNA base sequences into RNA. In prokaryotic cells, a single RNA polymerase catalyzes the synthesis of all the major types of RNA. Purified preparations of bacterial RNA polymerase can be fractionated by column chromatography into two components: a **core enzyme** containing four subunits and an accessory protein termed the **sigma factor.** In the absence of sigma factor,

the core enzyme tends to transcribe both strands of template DNA molecules randomly. When sigma factor is added, however, RNA synthesis is restricted to sites known to be transcribed within intact cells. Once RNA synthesis has begun, the sigma factor is released from the core enzyme and can be utilized by a second RNA polymerase molecule to initiate another transcription event.

In contrast to prokaryotes, eukaryotes contain several distinct kinds of RNA polymerase. The existence of these multiple RNA polymerases first became apparent in the mid-1960s when it was reported that the base composition of the RNA synthesized by isolated nuclei can be varied by altering the ionic conditions of the incubation medium. When incubated in low-ionic-strength solutions containing Mg^{2+}, isolated nuclei synthesize ribosomal RNA; in higher-ionic-strength solutions containing Mn^{2+}, on the other hand, the product resembles a population of messenger RNA molecules. Autoradiographic studies revealed that RNA synthesis occurs predominantly in the nucleolus when low-ionic-strength solutions containing Mg^{2+} are employed, and in the remainder of the chromatin when high-ionic-strength solutions containing Mn^{2+} are used (Figure 11-36).

The most straightforward interpretation of the preceding observations is that eukaryotic cells contain multiple forms of RNA polymerase localized in different

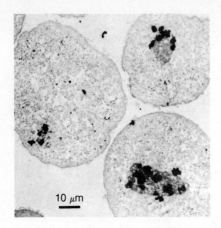

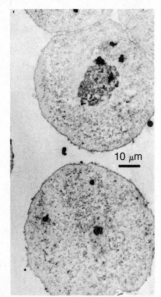

FIGURE 11-36 Autoradiogram showing subnuclear localization of RNA synthesized by isolated nuclei under different ionic conditions. *(Top)* In low-ionic-strength solutions containing Mg^{2+}, silver grains accumulate over the nucleolus. *(Bottom)* In high-ionic-strength solutions containing Mn^{2+}, more silver grains appear over the chromatin. Courtesy of V. G. Allfrey.

regions of the nucleus and activated by different ionic conditions. This interpretation was verified in the late 1960s by Robert Roeder and William Rutter, who successfully solubilized the RNA polymerases from eukaryotic nuclei and fractionated them by ion exchange chromatography into distinct enzymes designated **RNA polymerases I, II,** and **III.** When nuclei are subfractionated into nucleolar and chromatin fractions prior to purification of these enzymes, RNA polymerase I activity is detected mainly in the nucleolar fraction, and polymerases II and III in the chromatin fraction (Figure 11-37). In addition to this difference in subnuclear localization, the three RNA polymerases have been found to differ in ionic requirements, sensitivity to inhibitors, and the type of RNA products they synthesize (Table 11-3). RNA polymerase I preferentially transcribes genes coding for large ribosomal RNA molecules, RNA polymerase II transcribes genes coding for messenger

RNAs, and RNA polymerase III transcribes genes coding for several small RNAs, including transfer RNAs and the small 5S RNA molecule present in ribosomes. These three enzymes can be readily distinguished from one another by the inhibitor **α-amanitin,** a highly toxic peptide isolated from the poisonous mushroom, *Amanita phalloides.* RNA polymerase II is highly susceptible to inhibition by this toxin; enzyme III is inhibited only at high concentrations; and enzyme I is insensitive (Figure 11-38).

In spite of the differences between the RNA polymerases of prokaryotic and eukaryotic cells, the transcription reactions catalyzed by these various enzymes share several basic features. As we shall now describe, in each case RNA synthesis is a multistep process in which four major stages can be distinguished: binding, initiation, elongation, and termination.

FIGURE 11-37 Fractionation by ion exchange column chromatography of RNA polymerases extracted from isolated rat liver nuclei. Of the three forms of RNA polymerase detected in this way, RNA polymerase I is preferentially associated with nucleoli and polymerases II and III are preferentially associated with the remainder of the nuclear chromatin.

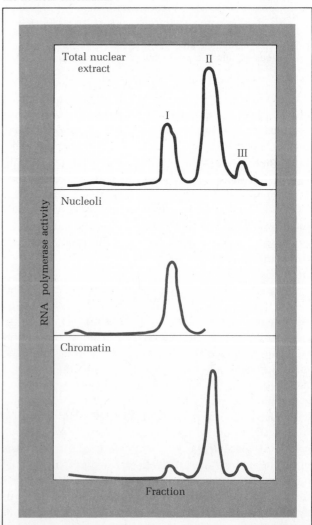

TABLE 11-3
Distinguishing features of the three eukaryotic RNA polymerases

RNA Polymerase	Location	Optimal Activity When Assayed in the Presence of	Sensitivity to Inhibition by α-Amanitin	Types of Genes Transcribed
I	Nucleolus	Mg^{2+}, low ionic strength	Insensitive	Genes coding for large ribosomal RNAs
II	Chromatin	Mn^{2+}, high ionic strength	Very sensitive	Genes coding for proteins
III	Chromatin	Mn^{2+}, high ionic strength	Moderately sensitive	Genes coding for transfer RNAs, 5S ribosomal RNA

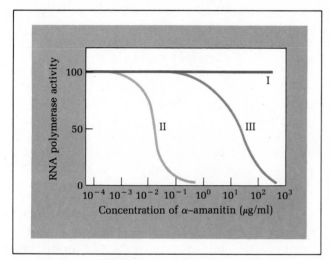

FIGURE 11-38 Differential sensitivity of RNA polymerases I, II, and III to inhibition by the mushroom toxin, α-amanitin.

1. *Binding of RNA polymerase to DNA.* The first step in gene transcription is the binding of RNA polymerase to an appropriate location, or **promoter site,** on the DNA molecule being transcribed. Promoter sequences have been extensively investigated using recombinant DNA techniques, which make it possible to purify specific genes and their surrounding regions for experimental manipulation. By deleting or adding specific base sequences to such recombinant DNA molecules and then testing their ability to be transcribed by RNA polymerase, several kinds of essential sequences have been identified. To facilitate discussion of the location of such sequences, the terms **upstream** and **downstream** are commonly used by molecular biologists to refer to sequences located toward the 5′ or 3′ end of the DNA strand being transcribed (Figure 11-39).

One widely occurring type of promoter sequence identified using the above approach is the **TATA box,** a 6–8 base region beginning with the sequence TATA- and consisting solely of the bases adenine and thymine. TATA boxes have been identified about 20–30 bases upstream from the transcriptional start sites of prokaryotic genes, as well as eukaryotic genes transcribed by RNA polymerase II. Removal of the TATA box from cloned genes results in a marked decrease in their ability to be transcribed by RNA polymerase, while removal of sequences adjacent to the TATA box generally has little effect on transcription rates. In addition to the TATA box, additional sequences that enhance RNA polymerase binding are often located further upstream. In bacterial genes a second promoter sequence is often detectable a few dozen bases upstream from the TATA box, while in eukaryotic genes such additional sequences may be located anywhere from a few dozen to hundreds of bases upstream.

Although promoter sequences are generally situated upstream from the starting site for transcription, a rather striking exception has been observed in the gene coding for 5S ribosomal RNA, a small RNA molecule that forms part of the structure of the ribosome (Chapter 12). A series of elegant experiments carried out by Donald Brown and his colleagues have led to the rather surprising discovery that the DNA base sequences crucial for the binding of RNA polymerase III to the 5S gene are located in the middle of the gene itself, rather than upstream from the transcription start site. In these experiments the 5S gene and its surrounding sequences were first cloned using recombinant DNA techniques. Various portions of the gene and/or its surrounding se-

FIGURE 11-39 Definition of the terms *upstream* and *downstream* when used in relation to DNA sequence organization.

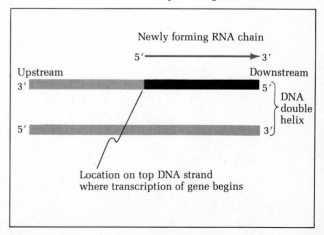

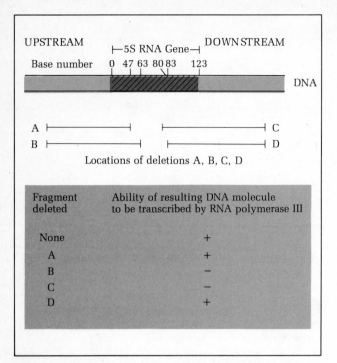

FIGURE 11-40 Experimental approach employed to identify the location of the DNA base sequence that promotes transcription of the 5S RNA gene by RNA polymerase III. Note that transcription is only inhibited when sequences situated within the middle of the 5S RNA gene are deleted.

FIGURE 11-41 Comparison of the organization of promoter sequences for RNA polymerases II *(top)* and III *(bottom)*. RNA polymerase II binds directly to DNA promoter sequences located upstream from the transcription start site, while RNA polymerase III interacts with a protein factor that binds to promoter sequences situated downstream from the transcription start site.

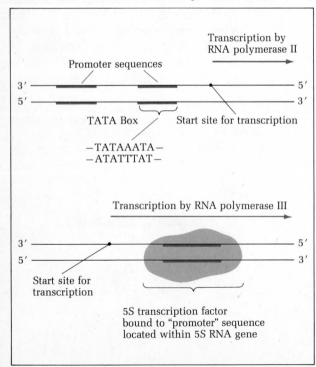

quences were then removed by nuclease digestion, and the ability of the resulting DNA molecule to be transcribed was tested. In this way it was discovered that removal of sequences located upstream from the transcription start site does not inhibit the rate of RNA synthesis, nor does removal of sequences from the beginning of the 5S gene itself. Removal of sequences beyond base 47, however, abolishes the ability of the 5S gene to be transcribed, suggesting that sequences in this middle region of the 5S gene are required for transcription. By removing DNA sequences from the other end of the 5S gene and again monitoring the resulting effects on RNA synthesis, the crucial area was pinpointed to a segment about 30 base pairs in length situated approximately 50 bases into the gene (Figure 11-40).

To confirm the importance of this 30-base-pair segment for promoting RNA synthesis, it was removed from the center of the 5S RNA gene and cloned into the middle of other DNA molecules. The recombinant DNA molecules formed in this way were found to be transcribed by RNA polymerase III starting at a point located about 50 bases *upstream* from the site at which the fragment was inserted. This behavior is in striking contrast to the promoters for other RNA polymerases, which cause RNA synthesis to be initiated downstream from the promoter signal.

Another important discovery to emerge from studying the promoter for RNA polymerase III is that a particular nonhistone protein, termed the **5S transcription factor,** binds specifically to this DNA sequence. Efficient transcription of recombinant DNA molecules containing the 5S RNA gene only occurs when this protein factor is present, suggesting that the binding of RNA polymerase to the 5S gene is mediated by the transcription factor rather than by the promoter sequence itself (Figure 11-41). Because of this arrangement, the rate of transcription of the 5S genes can be regulated within the cell by altering the availability of the 5S transcription factor. It has subsequently been discovered that the existence of such factors is a widespread phenomenon that applies to the binding of other forms of RNA polymerase to their respective promoters as well.

2. *Initiation of RNA synthesis.* Once an RNA polymerase molecule has bound to an appropriate DNA site, the process of transcription can be initiated. In bacterial genes and in eukaryotic genes transcribed by RNA polymerase II, transcription usually begins about 20–30 bases downstream from the TATA box. If DNA sequences located between the TATA box and the normal start site for transcription are experimentally removed, transcription is still initiated about 30 bases downstream from the TATA box, rather than at the normal start site (Figure 11-42). Such experiments indicate that the site at which transcription is initiated is largely determined by the location of the promoter sequences. Similar studies carried out on the 5S gene have demon-

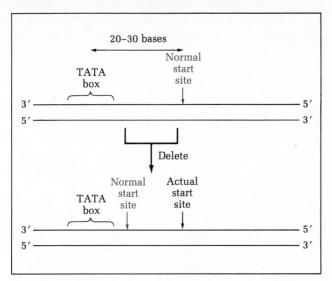

FIGURE 11-42 Design of a typical experiment demonstrating the role of promoter sequences in determining the location of the initiation site for transcription. After sequences adjacent to the TATA box have been removed, transcription is still initiated about 30 bases downstream, even though this no longer corresponds to the normal start site.

strated that the promoter sequences situated in the middle of this gene always cause transcriptional initiation to occur about 50 bases *upstream*.

The nucleotide utilized to begin the synthesis of a new RNA chain is always an ATP or GTP. Since RNA polymerase can utilize only these two nucleotides to start an RNA chain, there can be no ambiguity as to which DNA strand is the template strand; that is, if the DNA base pair situated at the initiation site is a G-C pair,

the strand containing the C must be the template strand, while in the case of an A-T base pair, the strand containing the T must be the template strand. Unlike the nucleotides incorporated during the elongation phase of RNA synthesis, the initiating ATP or GTP retains all three phosphate groups.

The discovery of agents that selectively interfere with the binding or initiation stages of RNA synthesis has made it clear that these two steps are distinct processes. The most thoroughly studied initiation inhibitors are the **rifamycins,** a family of antibiotics that prevent initiation of RNA chains in prokaryotic systems without affecting the binding of RNA polymerase to DNA. In contrast to the rifamycins, polyanions such as **heparin** inhibit the reaction in which RNA polymerase binds to DNA, but do not interfere with initiation by RNA polymerase molecules already bound to DNA.

3. *Elongation step of RNA synthesis.* As soon as the initial ATP or GTP has been incorporated, the RNA chain is elongated by the formation of phosphodiester bonds between the first (or α) phosphate group of each new incoming ribonucleotide and the 3'-hydroxyl of the preceding one (Figure 11-43). During this process, the terminal two phosphates of each incoming nucleotide are released as pyrophosphate. Elongation of the nucleotide chain therefore takes place in the $5' \rightarrow 3'$ direction, just as it does during DNA synthesis. Because the base sequence of the newly forming RNA chain is determined by complementary base pairing between each new base and the corresponding base on the template strand of the DNA molecule, localized unwinding of the DNA double helix must occur. The region of un-

FIGURE 11-43 Mechanism of the elongation step of RNA synthesis.

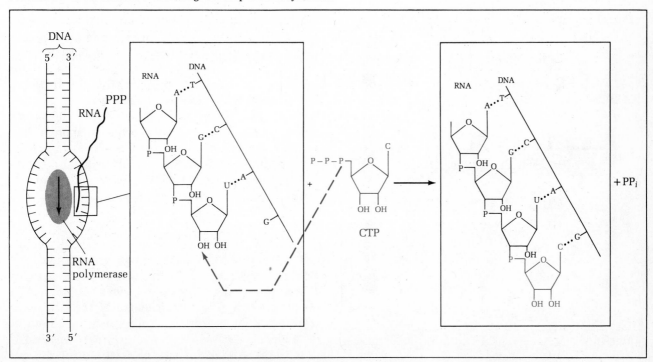

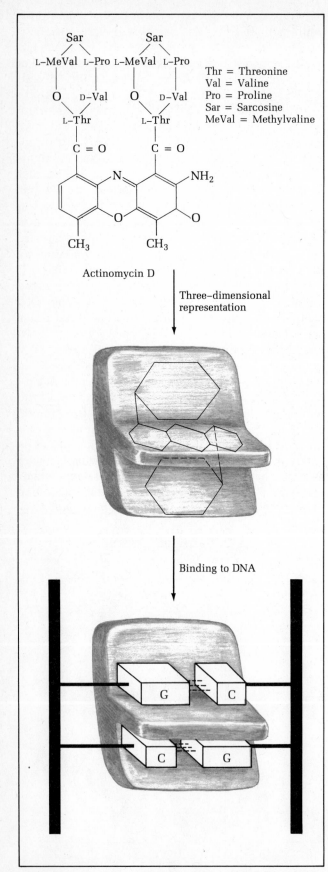

FIGURE 11-44 Structure of actinomycin D and its mode of binding to DNA.

winding need not be very large, however, because the growing RNA chain is continually peeled off the DNA template as the RNA polymerase molecule moves along. Once elongation has proceeded far enough into the gene to free up the promoter and initiation sites, another RNA polymerase molecule can bind and begin the synthesis of a second molecule of RNA. In this way the transcription rate of a single gene can be maximized by having many RNA polymerase molecules transcribing it simultaneously (see Figure 12-39).

One of the most widely employed inhibitors of RNA synthesis is **actinomycin D,** an antibiotic that acts primarily at the elongation stage. Actinomycin D selectively binds between adjacent G-C and C-G base pairs, becoming inserted or *intercalated* into the DNA double helix in a way that prevents passage of RNA polymerase (Figure 11-44). Because of the way in which actinomycin binds to DNA, the transcription of genes enriched in G-C base pairs is more susceptible to inhibition by this antibiotic.

4. *Termination of transcription.* Elongation of newly forming RNA molecules proceeds along the DNA template until a special region is reached whose base sequence signals that transcription should be terminated. In many genes this signal has been found to consist of a short self-complementary sequence of bases directly preceding a stretch of contiguous adenine residues. The self-complementary sequence generated in the new RNA molecule can form a hairpin loop that is thought to be an important part of the signal for ending transcription (Figure 11-45). Upon recognition of the termination signal, the completed RNA transcript, and then the RNA polymerase molecule, are released from the DNA.

Termination of gene transcription is also subject to regulation by specific protein factors. In the case of bacteriophage lambda DNA, cell-free transcription studies have revealed that certain newly transcribed RNA molecules are longer than normal unless a protein factor known as **rho** is added, suggesting that rho facilitates the recognition of transcriptional termination sites. Lambda DNA also codes for the production of specific antiterminator proteins that selectively suppress transcriptional termination of certain genes. This suppression permits transcription to proceed through to adjacent genes whose transcription would otherwise not be promoted (Figure 11-46).

Processing of the Primary Transcript

The process of DNA transcription usually generates RNA molecules that are longer than would be expected for the transcripts of individual genes. Several factors contribute to the increased length of these **primary transcripts.** In prokaryotic cells primary transcripts generally contain sequences derived from several adjacent genes. Although these **polycistronic** messenger

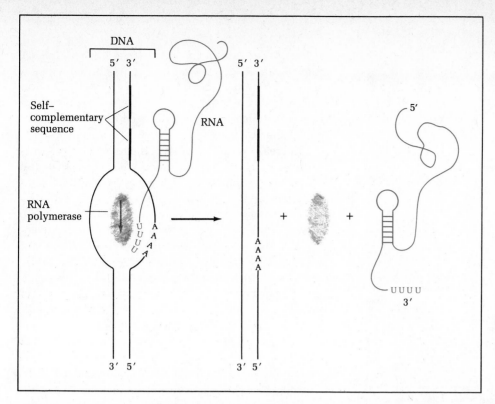

FIGURE 11-45 The termination of transcription. A short self-complementary sequence that allows the newly formed RNA chain to form a hairpin loop, and a stretch of contiguous adenine residues in the DNA template strand, have been identified at the ends of a number of different genes.

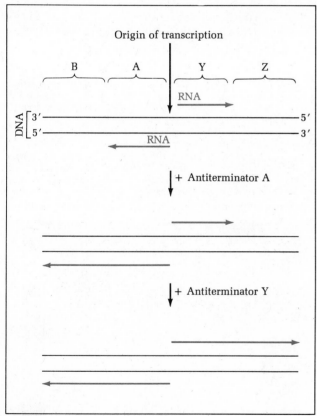

FIGURE 11-46 Diagram illustrating how specific antiterminator proteins can regulate the transcription of adjacent genes. In the above case, antiterminator A activates the transcription of gene B, and antiterminator Y activates the transcription of gene Z. Note that in this example, which is based on the behavior of lambda bacteriophage DNA, transcription proceeds in opposite directions from a common origin. Since transcription always proceeds in the 5' → 3' direction, this means that the opposite DNA strands are transcribed on the two sides of the origin.

RNAs often remain intact during translation, directing the synthesis of several different polypeptides in succession, they may also be cleaved into individual messenger RNAs prior to translation. Such conversion of a primary transcript into smaller RNA products is referred to as **RNA processing,** and is observed during formation of transfer and ribosomal as well as messenger RNAs. The processing steps associated with the production of messenger RNA in eukaryotic cells will be described below, while the processing of ribosomal RNA is discussed in the following chapter.

Properties of Heterogeneous Nuclear RNA (HnRNA)
In prokaryotic cells most of the RNA sequences transcribed from DNA function as messages that guide the synthesis of polypeptide chains. In eukaryotes, on the other hand, the situation is significantly more complex. The first indication of this complexity was reported in 1962 by Henry Harris, who utilized microscopic autoradiography to study the behavior of newly synthesized RNA molecules that had been labeled by exposing cultured cells for brief periods to ^{3}H-uridine. By counting the number of autoradiographic grains present over the nucleus and cytoplasm at varying time intervals after exposure to this radioisotope, Harris was able to obtain a quantitative estimate of where these newly synthesized RNA molecules go within the cell. Surprisingly, his data revealed that the quantity of radioactive RNA appearing in the cytoplasm is less than 10 percent of that which had initially become labeled in the nucleus (Figure 11-47). This unexpected result led Harris to conclude that the bulk of the nuclear RNA is degraded in the nucleus without ever entering the cytoplasm.

This remarkable conclusion initially met with con-

415

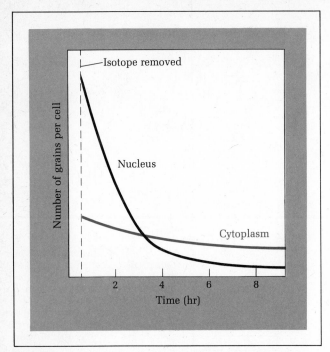

FIGURE 11-47 Evidence for the rapid degradation of newly synthesized nuclear RNA. Cultured cells were incubated for 30 min in the presence of ³H-uridine, after which the isotope was removed. Incubation was continued for varying amounts of time in fresh medium, and microscopic autoradiography was then employed to quantitate the amount of radioactive RNA present in the nucleus and cytoplasm. The data reveal that the rapid disappearance of radioactive RNA from the nucleus is not accompanied by a comparable increase in radioactivity in the cytoplasm. It was therefore concluded that most of the labeled nuclear RNA has been degraded to nucleotides (which cannot be seen in autoradiographic experiments because they are washed out of the cell during preparation of the specimen).

siderable skepticism, for it did not seem compatible with the idea that DNA is transcribed into messenger RNA molecules that function in the process of cytoplasmic protein synthesis. But experiments on the behavior of isolated nuclear RNA carried out soon thereafter by James Darnell and his associates provided further support for Harris's contention. In these experiments HeLa cells were first incubated with ³H-uridine for varying periods of time, and the nuclear RNA was then extracted for analysis by moving-zone centrifugation. When labeling periods of only a few minutes were utilized, the radioactive RNA was found to sediment as a heterogeneous array of molecules whose sedimentation coefficients ranged from 20S to 100S. Because of the heterogeneity of the molecules contained within this RNA fraction, they were referred to as **heterogeneous nuclear RNAs (HnRNAs).**

To investigate the fate of these molecules, Darnell performed an experiment in which cells were first incubated briefly with ³H-uridine to label the HnRNA fraction. Actinomycin D was then added to inhibit the formation of any additional radioactive HnRNA, and the cells were reincubated for varying periods of time to determine the fate of the previously labeled HnRNA molecules. Within a few hours the bulk of the radioactivity was found to disappear from the RNA fraction (Figure 11-48), a finding which supported Harris's contention that most newly synthesized RNA is subject to rapid degradation.

The realization that most of the radioactivity incorporated into HnRNA is degraded without entering the cytoplasm is compatible with two possible interpretations: HnRNA is either selectively degraded, with certain sequences preferentially preserved for ultimate export from the nucleus, or the degradation is nonselective, and all sequences present in the HnRNA fraction are represented in the small number of HnRNA molecules that escape degradation and enter the cytoplasm. These alternatives were first investigated in the late 1960s by Ruth Shearer and Brian McCarthy, who employed competitive DNA-RNA hybridization to compare the RNA sequences present in the nucleus and cytoplasm. Results from such experiments revealed that unlabeled cytoplasmic RNA is less effective than unlabeled HnRNA in preventing the binding of radioactive HnRNA to DNA; hence many of the base sequences present in HnRNA molecules are not represented in cytoplasmic RNA. It was therefore concluded that a selective degradation of HnRNA occurs, yielding only a

FIGURE 11-48 Experiment demonstrating the rapid degradation of HnRNA. HeLa cells were incubated for 5 min in the presence of radioactive uridine, at which point actinomycin D was added to prevent the synthesis of any additional radioactive RNA, and incubation was then continued for 2.5 hr. Total RNA was isolated and analyzed by moving-zone centrifugation. Note that in cells labeled for 5 min *(black curve)*, the newly synthesized RNA sediments as a heterogeneous family of molecules. Two and a half hours after the addition of actinomycin D *(colored curve)*, most of the radioactive RNA has disappeared.

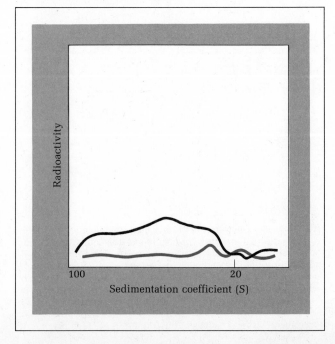

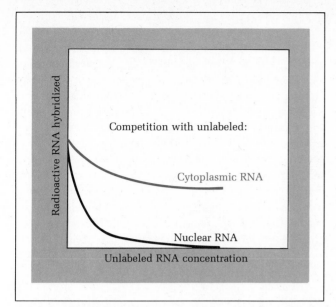

FIGURE 11-49 Typical results obtained from competition hybridization experiments that compare the abilities of unlabeled nuclear and cytoplasmic RNA to compete with radioactive HnRNA for hybridization to DNA. Such experiments have revealed that cytoplasmic RNA does not contain a large fraction of the sequences present in nuclear RNA.

specific subset of RNAs for transport to the cytoplasm. This conclusion has subsequently been verified by nucleic acid hybridization data from a variety of cell types, which indicate that only 10–20 percent of the sequences present in HnRNA are typically represented in cytoplasmic messenger RNA (Figure 11-49).

The preceding observations imply that HnRNA is processed in such a way that certain sequences are preferentially conserved for export to the cytoplasm as messenger RNAs. One of the first eukaryotic gene products for which this processing was directly demonstrated is the messenger RNA coding for β-globin, a protein produced in large quantities in developing red blood cells. Because of the high concentration of β-globin messenger RNA in such cells, it can be easily purified and utilized as a template for the synthesis of cDNA using reverse transcriptase. In the late 1970s several laboratories utilized globin cDNA molecules made in this way as hybridization probes for determining whether globin-specific sequences are present in HnRNA. In these experiments RNA isolated from cells that had been incubated with ³H-uridine was size fractionated by either moving-zone centrifugation or polyacrylamide gel electrophoresis. When the RNA fractions obtained in this manner were hybridized to globin cDNA, it was discovered that globin-specific sequences are present in two classes of RNA molecules measuring 15S and 9S in size (Figure 11-50). To determine the relationship between these two kinds of globin-specific RNA, experiments were carried out in which cells exposed for a few minutes to ³H-uridine were subsequently reincubated in the absence of ³H-uridine. Under these conditions all the radioactive globin-specific sequences are eventually converted to the 9S form, which corresponds to the size of mature globin messenger RNA. Such results suggest that the 9S globin messenger RNA is derived from a larger, 15S precursor. Subsequent investigations involving a variety of other eukaryotic genes have provided further support for the conclusion that messenger RNAs are often derived from larger precursor molecules. HnRNAs are therefore often referred to now as messenger-RNA precursors, or "pre-mRNAs."

The conclusion that eukaryotic messenger RNAs are generally derived from larger precursors raises a series of fundamental questions concerning the nature of the conversion process. In the following sections we shall see that several kinds of chemical changes are involved, including packaging of the initial RNA transcript with protein, addition of special "cap" and "tail" structures to its two ends, and the removal of certain types of base sequences from its interior.

Packaging of HnRNA with Protein One of the first steps that occurs during HnRNA processing is packaging the RNA molecule with protein to form a ribonucleoprotein complex. An important insight into the organi-

FIGURE 11-50 An experiment demonstrating the existence of a 15S RNA precursor to 9S β-globin messenger RNA. In RNA isolated from cells labeled for 10 min with ³H-uridine and fractionated by moving-zone centrifugation, radioactive molecules sedimenting at 15S and 9S are found to be capable of hybridizing to β-globin cDNA. During subsequent incubation of cells in the absence of ³H-uridine, radioactivity accumulates entirely in the 9S form.

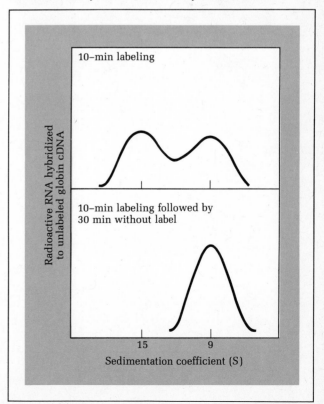

zation of these complexes was provided in the mid-1960s by Georgii Georgiev and collaborators, who extracted HnRNA molecules along with their tightly bound proteins from purified nuclei and analyzed the behavior of this material by moving-zone centrifugation. Under conditions designed to prevent nuclease activity, Georgiev found that most of the HnRNA-protein fraction sediments as a series of discrete peaks ranging from 30S to 200S. Electron micrographs revealed the presence of single particles in the first peak, clusters of two particles in the second peak, and so on. HnRNA-protein preparations exposed to ribonuclease, on the other hand, yielded only the 30S peak containing single particles (Figure 11-51).

On the basis of the preceding observations, Geor-

giev proposed that 30S protein particles are distributed along the length of each HnRNA molecule (Figure 11-52). Such an organization is directly compatible, of course, with the finding that ribonuclease cleavage of the RNA chain releases a homogeneous population of single particles. Several years after this model was first proposed by Georgiev, electron microscopy of transcribing chromatin fibers revealed that newly forming RNA transcripts exhibit a beaded appearance (see Figure 11-35), just as would be expected of molecules with protein particles bound along their length. This beads-on-a-string arrangement is somewhat reminiscent of the nucleosomal organization of chromatin fibers, although the nature of the protein particle present in the HnRNA-protein complex has not been clearly eluci-

FIGURE 11-51 Sedimentation profile and electron microscopic appearance of HnRNA-protein complexes isolated in the presence or absence of an inhibitor of ribonuclease (RNAase) activity. Under conditions that prevent ribonuclease activity *(left)*, HnRNA-protein complexes sediment as a series of peaks containing clusters of protein particles. When ribonuclease is active *(right)*, these clusters are degraded to single protein particles that sediment at approximately 30S. Micrographs courtesy of G. P. Georgiev.

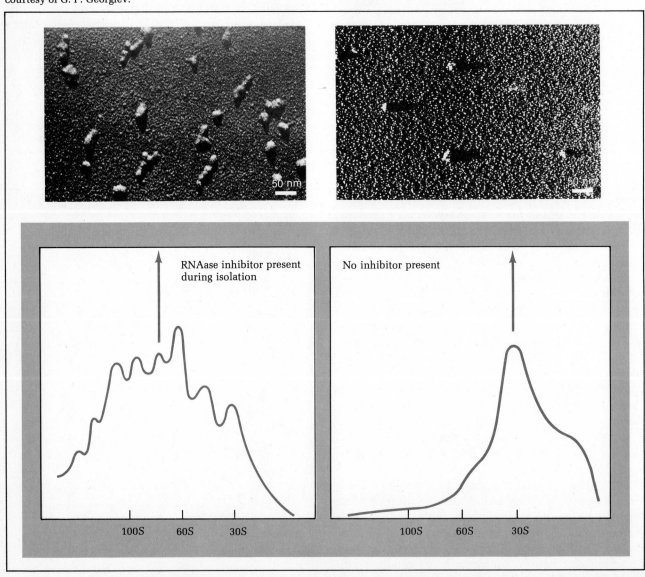

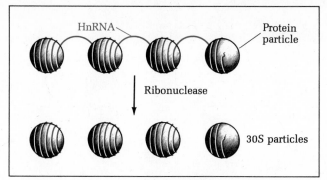

FIGURE 11-52 A model that has been proposed to explain the organization of HnRNA-protein complexes.

FIGURE 11-53 Organization of the cap structure located at the 5′ end of HnRNA and messenger RNA molecules. Both methyl groups on the two riboses are not always present.

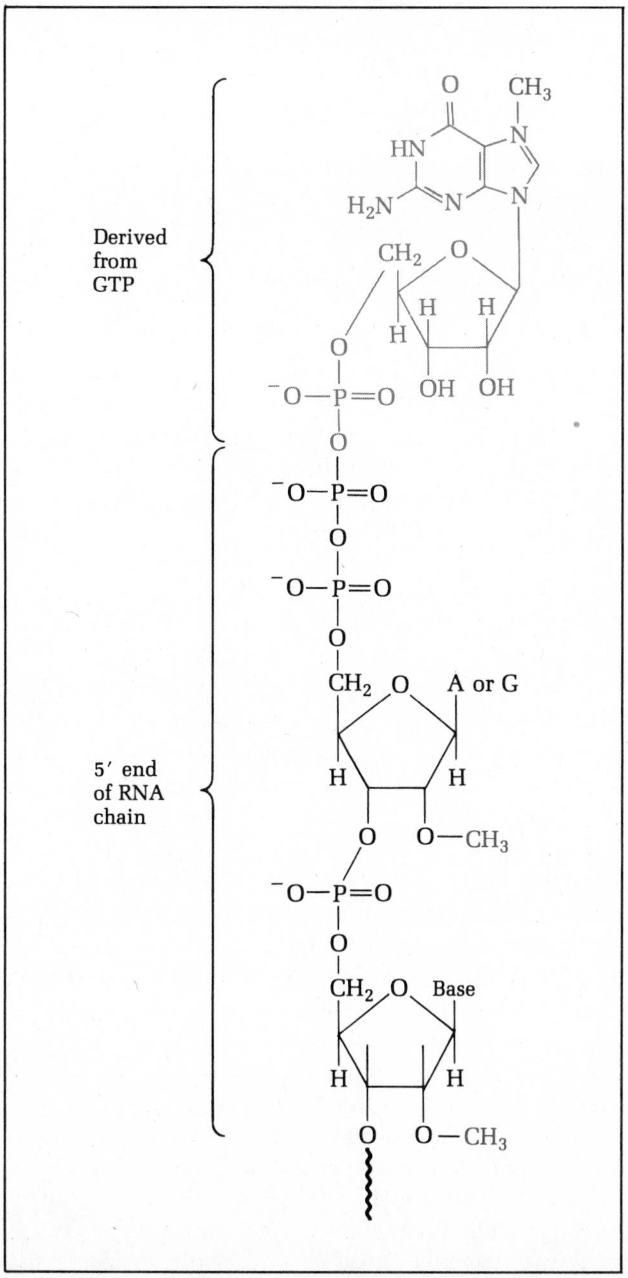

dated. A protein of about 40,000 molecular weight is thought to be a major constituent, but other proteins have been detected as well. Many of these proteins vary among different cell types, suggesting that they may play specific roles in the processing and transport of particular types of RNA sequences.

Capping and Addition of Poly(A) Tails It was pointed out earlier in the chapter that RNA synthesis is initiated by the incorporation of an ATP or GTP residue that, unlike other nucleotides in the RNA chain, retains its three phosphate groups. In eukaryotic RNA transcripts destined to be processed into messenger RNA, this triphosphate group undergoes a subsequent modification reaction whose net result is the creation of a 7-methylguanosine residue at the 5′ end of the RNA chain (Figure 11-53). During this modification reaction the ribose groups of the first, and often the second, bases of the RNA chain become methylated. Together these alterations create a special structure, known as a **cap,** which is characteristic of eukaryotic messenger RNAs and which appears to play an important role in the initiation of protein synthesis (Chapter 12).

The 3′ end of HnRNA molecules is also characterized by an unusual region, termed a **poly(A) tail,** which consists of roughly 200 adenine residues. The presence of poly(A) tails is easily detected because their unique base sequence renders them resistant to degradation by ribonuclease. Hence digestion of HnRNA or messenger RNA preparations with ribonuclease results in the degradation of everything except the poly(A) tail (Figure 11-54). Since genes lack the long stretches of thymine that would be required to encode such homogeneous poly(A) sequences, it has been concluded that poly(A) sequences are added to the 3′ end of HnRNA molecules after transcription has been terminated. Independent support for this conclusion has come from the isolation of an enzyme, known as **poly(A) polymerase,** which catalyzes the synthesis of poly(A) sequences without the requirement for a DNA template.

Although poly(A) tails are present on almost all eukaryotic messenger RNAs, the exact role played by this sequence is not clear. Because histone messenger RNAs have been shown to lack poly(A), it is obvious that the presence of a poly(A) tail is not an absolute requirement for the production of functional messenger RNAs. Conversely, the presence of a poly(A) tail does not in itself determine that a given RNA molecule is destined to become a functional message. For example, in tissues that do not produce hemoglobin, globin-coding sequences have been detected in HnRNA molecules containing poly(A) tails. Yet in spite of the presence of this poly(A), such HnRNA molecules are not processed into functional messenger RNAs. It is therefore clear that the addition of poly(A) to an HnRNA molecule is not in itself the determining factor that governs whether a given

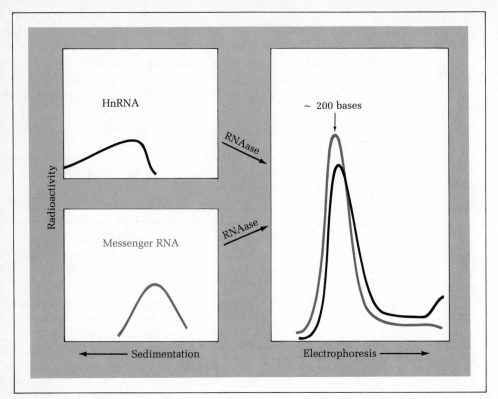

FIGURE 11-54 Experimental demonstration of the presence of similarly sized poly(A) sequences in HnRNA and messenger RNA. Radioactive HnRNA and messenger RNA preparations were isolated separately and each subjected to digestion with ribonuclease (RNAase). In both cases the ribonuclease-resistant material is found to consist of poly(A) sequences about 200 bases long.

messenger RNA sequence is processed and transported to the cytoplasm.

An alternative possibility is that poly(A) tails serve a cytoplasmic function associated with the process of protein synthesis. In regard to this alternative it has been shown that the poly(A) tails of messenger RNA molecules undergo a gradual decrease in size after transport to the cytoplasm. In the following chapter we shall describe some experiments that suggest the possibility that such changes in poly(A) length may be involved in regulating the stability of messenger RNA.

The Discovery of Intervening Sequences One of the more striking changes that occurs in HnRNA molecules during their conversion to messenger RNA is a reduction in size. In extreme cases the length of the final messenger RNA may be less than 10 percent of the length of the HnRNA molecule from which it is derived. One of the earliest and most straightforward models proposed to account for this size reduction was that HnRNA molecules contain extraneous sequences on the two ends of the messenger RNA sequence, and that this extra material must be removed prior to transport of the message to the cytoplasm. This relatively simple view was shown to be incorrect in the late 1970s when a series of surprising discoveries led to a complete reevaluation of our concept of the eukaryotic gene.

One of the first of these discoveries involved the properties of RNA molecules produced by *adenovirus*, a DNA virus that infects a variety of eukaryotic cells. Upon entry into the host cell, the viral DNA is transcribed into a large RNA molecule that is subsequently cleaved into smaller messenger RNAs. In 1977 Susan Berget, Claire Moore, and Phillip Sharp reported the results of experiments in which these messenger RNAs were hybridized to adenovirus DNA, and the resulting hybrids then examined by electron microscopy. The pictures obtained in this way revealed that a single molecule of adenovirus messenger RNA hybridizes to several regions of the adenovirus DNA molecule that are not contiguous with each other, causing the nonhybridizing regions of the DNA to form characteristic loops (Figure 11-55). This rather surprising observation indicated that the DNA sequences that code for a given molecule of messenger RNA are not continuous with each other, but are instead separated by sequences that do not appear in the final message. Sequences disrupting the linear continuity of the message-encoding regions of a gene are known as **intervening sequences** or **introns,** while sequences destined for retention in the final messenger RNA are termed **exons** (Figure 11-56).

Shortly after the initial discovery of intervening sequences in adenovirus DNA, a similar type of organization was detected in other eukaryotic genes (Table

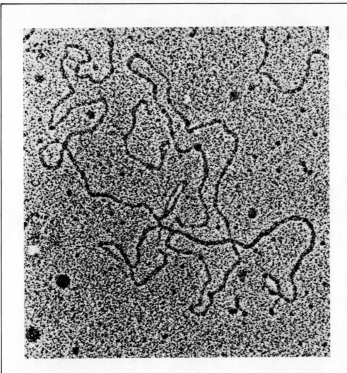

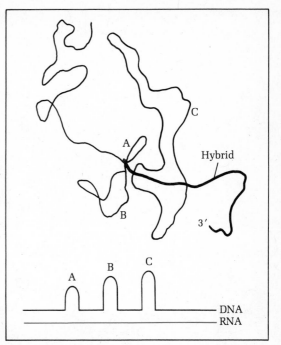

FIGURE 11-55 Experimental evidence indicating that adenovirus messenger RNA is derived from noncontiguous regions of adenovirus DNA. The electron micrograph pictures a single-stranded adenovirus DNA molecule that has been hybridized to messenger RNA. In the line drawing on the right, the region in the micrograph that corresponds to DNA hybridized to RNA is indicated by a heavy line. Loops A, B, and C represent single-stranded DNA that has not hybridized to the RNA. Below the line drawing is a schematic diagram that illustrates why the existence of such loops indicates that the messenger RNA molecule is encoded by nonadjacent DNA sequences. Courtesy of P. A. Sharp.

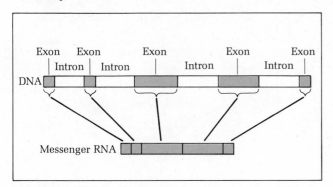

FIGURE 11-56 Organization of eukaryotic genes in which the DNA sequences coding for a given messenger RNA are interrupted by intervening sequences (introns).

TABLE 11-4
Examples of genes containing introns

Gene	Organism	No. of Introns	No. of Exons
Actin	Drosophila	1	2
β-Globin	Mammals	2	3
Insulin	Chicken, human	2	3
Lysozyme	Chicken	3	4
Actin	Chicken	3	4
Ovalbumin	Chicken	7	8
Collagen	Chicken	50	51

11-4). Although electron microscopy of DNA-RNA hybrids has been a major tool in such studies, digestion of DNA with restriction enzymes has also played an important experimental role. In this latter approach to the study of gene organization, DNA samples are digested with various combinations of restriction enzymes. By comparing the sizes of the DNA fragments generated by different enzymes, a map can be created that summarizes the linear order of restriction sites of various types. When the restriction map of a cloned gene is compared to that of its corresponding cDNA (made by transcribing the gene's messenger RNA with reverse transcriptase), significant differences are often observed because of the presence of intervening sequences in the gene that are not represented in the final messenger RNA (Figure 11-57).

The discovery that eukaryotic genes are disrupted by intervening sequences that do not appear in the final messenger RNA raises the question of whether these sequences are transcribed into the initial HnRNA transcripts. This question has been cleverly addressed by experiments in which hybrids formed by incubating together HnRNA and DNA are subsequently examined under the electron microscope. In contrast to the appearance of hybrids between messenger RNA and DNA,

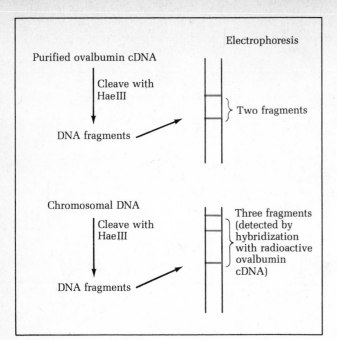

Purified ovalbumin cDNA

Electrophoresis

Cleave with
HaeIII

↓

DNA fragments → } Two fragments

Chromosomal DNA

Cleave with
HaeIII

↓

DNA fragments → } Three fragments (detected by hybridization with radioactive ovalbumin cDNA)

FIGURE 11-57 An example showing how information obtained from restriction enzyme digestion studies may suggest the existence of introns. Purified ovalbumin cDNA cleaved with the restriction enzyme HaeIII yields two fragments, indicating the presence of one HaeIII site in the cDNA molecule. In contrast, treatment of chromosomal DNA with HaeIII generates three fragments containing ovalbumin sequences. The extra HaeIII site responsible for generating the additional fragment is situated within an intron, and therefore does not appear in the final ovalbumin messenger RNA molecule (from which ovalbumin cDNA is derived).

FIGURE 11-58 Comparison of the hybridization of 15S and 9S globin RNAs to cloned β-globin gene DNA. *(Right)* Hybridization of the 9S RNA to the DNA is interrupted by a large loop corresponding to an intervening sequence. *(Left)* The 15S RNA, in contrast, hybridizes along the entire length of the DNA template strand, indicating that the intervening sequence must be present in the RNA. Abbreviations: dsDNA = double-stranded DNA, ssDNA = single-stranded DNA. Photographs courtesy of P. Leder.

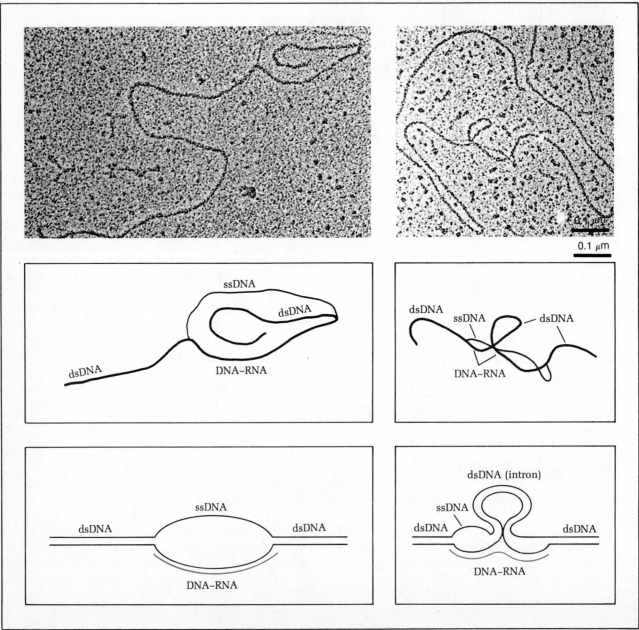

which exhibit looped-out regions where the DNA molecule contains sequences not present in the messenger RNA, HnRNA hybridizes to the entire length of the DNA molecule (Figure 11-58). It has therefore been concluded that HnRNA molecules represent complete transcripts of their corresponding genes, containing introns as well as sequences destined to be part of the final messenger RNA. The realization that introns must be removed from HnRNA during its conversion to messenger RNA explains why many of the base sequences represented in HnRNA never reach the cytoplasm.

Mechanism of RNA Splicing Since the base sequences located on the two sides of an intron present within an HnRNA molecule must be joined together after the intron itself has been removed, the conversion of HnRNA to messenger RNA is often referred to as RNA **splicing.** An important insight into the mechanism underlying RNA splicing has emerged from an examination of the base sequences of a variety of introns. Such studies have

revealed that the base sequences located at each end of the intron tend to be similar among different introns. If these short base sequences are altered, proper removal of the intron during RNA processing does not occur. Because of the importance of these sequences for proper intron removal and RNA splicing, they are referred to as **splice junctions.** In the early 1980s it was discovered that the base sequence of these splice junctions is complementary to sequences present in RNA molecules contained within small nuclear ribonucleoprotein particles known as **snurps.** This sequence complementarity prompted the suggestion that the RNA molecules present in snurps play a role in the process of intron removal. In the case of one particular snurp, which contains a small RNA molecule known as *U1*, much supporting evidence for this idea has been obtained. It has been shown, for example, that RNA splicing in nuclear extracts can be inhibited by removing a few nucleotides from the end of the U1 RNA, or by treating the extract with antibodies that react with purified snurps. An ad-

FIGURE 11-59 Schematic representation of several mechanisms of RNA processing. (a) The removal of introns mediated by small nuclear ribonucleoprotein granules (snurps). Base pairing between the RNA component of the snurp and splice junction sequences is an important step in this process. Note that the intron is released in a "lariat" configuration. (b) Snurps have also been implicated in promoting cleavage reactions that generate the proper 3′ ends of messenger RNAs. Base pairing involving the snurp RNA again plays an important role. (c) Certain self-splicing RNAs employ a nonlariat mechanism for intron removal. Self-splicing RNAs can also utilize a lariat mechanism. Abbreviation: G-OH = guanosine.

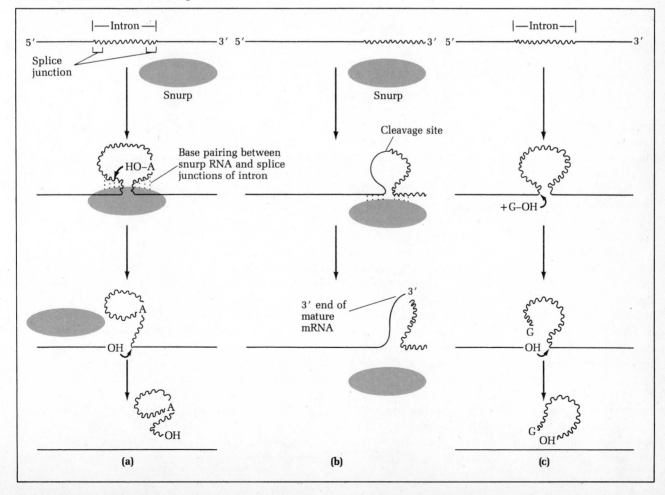

ditional insight into the mechanism of RNA splicing has emerged from studies that have revealed that during intron removal, the RNA sequence being removed is released in the form of a "lariat" molecule in which one of the ends of the RNA fragment is joined by a 2',5'-phosphodiester bond to an adenine residue located internally within the fragment. Taken together, the preceding information has led to the formulation of an RNA splicing model in which snurp RNA molecules bind to the splice sequences situated at the two ends of an intron, triggering a looping out of the intron and its subsequent removal in a lariat configuration (Figure 11-59a).

In addition to a role in intron removal, snurps have also been implicated in producing the proper 3' ends of mature messenger RNA molecules. Our best understanding of this role comes from studies on the production of histone messenger RNA molecules, where the U7 snurp triggers the formation of an RNA loop that is in turn required for the cleavage reaction that produces the proper 3' end of the messenger RNA (see Figure 11-59b). Although snurps have thus been implicated in both intron removal and in the production of correct 3' ends, they are not required for all RNA processing reactions. Several classes of genes, including ribosomal RNA genes in ciliated protozoa and several kinds of mitochondrial genes, have been found to be **self-splicing,** that is, intron removal appears to proceed in the absence of proteins. Some self-splicing RNAs release the excised intron in a lariat configuration, as is typical of nuclear messenger RNA splicing, while others utilize a nonlariat mechanism in which a guanosine is added to the 5' end of the intron prior to its removal as a linear fragment (see Figure 11-59c).

Although many eukaryotic genes contain intervening sequences destined for removal during the formation of messenger RNA, not all genes are organized in this fashion. Introns are extremely rare in the genes of prokaryotic cells, nor do the genes coding for the five histone molecules of eukaryotic cells contain introns. The realization that many, but not all, eukaryotic genes contain intervening sequences raises some profound questions concerning the nature of the gene. At first glance it may seem inefficient and potentially hazardous to interrupt genes with sequences that do not appear to serve any useful function and that are simply destined for removal. One possible explanation for this type of arrangement has emerged from the discovery that exons often code for different functional regions of polypeptide chains, each of which can independently fold into a separate *domain* (Figure 11-60). The presence of intervening sequences between such exons may facilitate their evolutionary rearrangement into new combinations, creating new proteins more efficiently than would be possible by the random mutation of DNA base sequences. In an analogous fashion, changes in the way

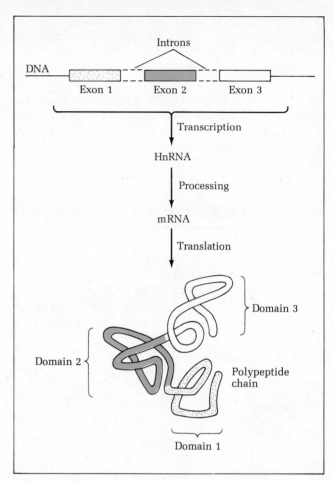

FIGURE 11-60 The proposed role of exons in coding for protein domains. Each domain represents a separate region of the polypeptide chain that is capable of independently folding into a functional unit.

in which exons are spliced together during RNA processing would allow a given gene sequence to code for more than one protein. As we shall see below, such a mechanism has been found to be one of the many ways in which the process of gene expression is regulated.

REGULATION OF GENE EXPRESSION

The DNA of any given cell contains the genetic information coding for all the proteins the cell can potentially synthesize. But the need for these various kinds of proteins differs considerably, depending upon the cell type, its physiological state, the availability of external nutrients, and so forth. To cite a few examples, bacterial cells contain genes coding for enzymes involved in the synthesis of all 20 essential amino acids. Such enzymes are clearly important when amino acids are not present in the external medium, but they may become superfluous when adequate amounts of amino acids are available. Likewise bacteria contain genes coding for enzymes involved in the metabolism of a wide variety of sugars, but these enzymes are useful only when the appropriate sugars are present in the cell's external environment. In the case of multicellular organisms, individual cell types become specialized to perform different kinds of functions. As a result, each cell requires a somewhat different spectrum of proteins. Thus red blood cells must produce the protein hemoglobin, nerve cells need to synthesize enzymes involved in the formation of neurotransmitters, and so forth.

A variety of mechanisms are available in both prokaryotic and eukaryotic cells for regulating gene expression so that each gene product is produced at appropriate times and in appropriate quantities. In the remainder of this chapter we shall focus upon the regulation of gene expression at the DNA and nuclear level, while in the following chapter we shall examine some of the additional controls that can be exerted after messenger RNA has entered the cytoplasm.

Cytoplasmic Influence over Nuclear Activity

Because DNA contains the genetic information that ultimately directs the synthesis of all cellular proteins, it is clear that the nucleus (or nuclear region in prokaryotes) exerts a profound influence over events occurring in the cytoplasm. The extent to which the cytoplasm influences the activity of the nucleus, on the other hand, is not so intuitively obvious. One experimental approach that has been especially useful in investigating this issue is nuclear transplantation. If the nucleus removed from a cell that is synthesizing neither DNA nor RNA is transplanted into an egg cell whose nucleus has been removed, the previously inactive nucleus generally

begins to synthesize RNA and, in some cases, DNA. In the most dramatic experiments of this type, carried out on the African clawed toad *Xenopus laevis*, John Gurdon removed the nuclei from cells lining the intestine of swimming tadpoles and transplanted them to enucleated eggs. Under these conditions a small percentage of the transplanted nuclei were found to direct the development of the recipient egg cells into complete new organisms. Clearly the egg cytoplasm must exert profound effects on nuclear activity if it can induce the nucleus of an intestinal cell to behave in this way.

The technique of cell-cell fusion has also provided important insights into the factors involved in the cytoplasmic control of gene activity. Using this approach, Henry Harris and his associates have investigated the biochemical and structural changes that occur after cells containing active nuclei are fused with cells in which the nuclei are not active. One cell type extensively studied in this way is the chicken red blood cell, which contains a nucleus that is inactive in both DNA and RNA synthesis. Although the chromatin of the red cell nucleus is normally extremely condensed, after fusion with a more active cell type the red cell nucleus begins to swell and its intensely staining, compacted chromatin gradually uncoils (Figure 11-61). These structural changes are accompanied by the accumulation within the red cell nucleus of nonhistone proteins derived from the other cell type, and by the gradual onset of red cell nuclear RNA and DNA synthesis.

In certain cell-cell fusion experiments it has been clearly demonstrated that the new cytoplasmic environment activates the expression of specific nuclear

FIGURE 11-61 Light micrographs showing reactivation of a red cell nucleus after fusion of chick embryo red blood cells with cultured mouse cells. *(Left)* Twelve hours after fusion the red cell nucleus (arrow) still contains condensed, darkly staining chromatin. The double arrow points to the larger mouse cell nucleus. *(Middle)* Two days after fusion the red cell nucleus has swelled and its chromatin has uncoiled. *(Right)* Four days after fusion a prominent nucleolus has appeared within the red cell nucleus. Courtesy of E. Sidebottom.

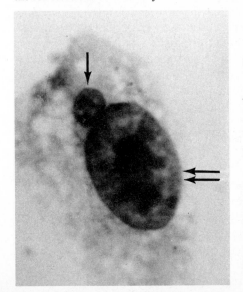
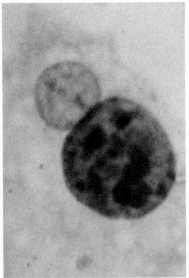
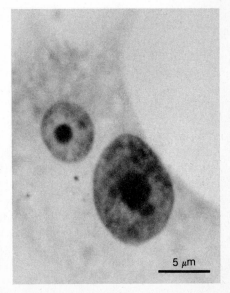

genes that otherwise would not be active. During the fusion of chick red blood cells with mouse cells, for example, the enzyme *inosinic acid pyrophosphorylase* (IMPase) begins to be synthesized. Since the mouse cells used in these studies lack the gene for IMPase and red cells do not normally synthesize it, one can conclude that some factor present in the mouse cell cytoplasm has caused the previously inactive chick IMPase gene to be expressed.

The preceding observations clearly indicate that the pattern of genes expressed in a given cell type is not rigidly or permanently determined within the nucleus itself, but is instead subject to influence by external conditions. In the remainder of the chapter we shall investigate some of the molecular mechanisms by which changes in cytoplasmic and external conditions can trigger such alterations in gene expression.

DNA Organization in Prokaryotic and Eukaryotic Cells

Any discussion of the mechanisms involved in the control of gene expression requires a basic familiarity with the way in which DNA sequences are organized in prokaryotic and eukaryotic cells. We have already seen that eukaryotic genes are often interrupted by intervening sequences that are not represented in the mature RNA product, while intervening sequences are extremely rare in prokaryotes. Another important difference between prokaryotic and eukaryotic cells is that eukaryotes are characterized by a disproportionate excess of DNA. Higher eukaryotes typically contain about a thousand times more DNA than do bacterial cells, but nowhere near a thousandfold increase in the number of genes. The bacterium *E. coli*, for example, has about 4000 genes, most of which code for proteins. While the exact number of genes present in the cells of higher eukaryotes is not known, two-dimensional gel electrophoresis of cellular extracts has led to the identification of several thousand distinct polypeptide chains, while the hybridization properties of the messenger RNA fraction indicate the possible existence of messages for an additional 10,000 polypeptide chains that may be synthesized in quantities too small to be detected by electrophoresis. Although the number of genes present in human cells may therefore be several times greater than in bacterial cells, it does not approach the thousandfold excess that would be expected on the basis of the thousandfold difference in DNA content. The presence of extra DNA that does not perform an obvious genetic function is not restricted to higher animals. Certain plants and amphibians contain an order of magnitude more DNA than do human cells, yet these organisms do not have a significantly greater number of genes.

The preceding considerations indicate that a large fraction of the DNA sequences present in eukaryotic cells does not code for polypeptides. A significant advance in our understanding of the nature of some of this excess DNA occurred in the late 1960s when DNA hybridization experiments carried out by Roy Britten and David Kohne led to the discovery of repetitive DNA. In these experiments DNA was broken into small fragments by physical shearing forces and dissociated into single strands by heating. The temperature was then lowered to permit the single-stranded fragments to hybridize back to one another. The rate of this **DNA reassociation** reaction depends on the concentration of each individual DNA sequence; the higher the concentration of a specific DNA sequence in the initial DNA solution, the greater the probability that it will collide with a complementary strand to which it can hybridize.

Given these considerations, how would the reassociation properties of different kinds of DNA be expected to compare? As an example, let us consider DNA derived from a bacterial cell and from a typical mammalian cell containing a thousandfold more DNA. If this difference in DNA content actually reflects a thousandfold difference in the total number of different kinds of DNA sequences present, then bacterial DNA should reassociate a thousand times faster than does mammalian DNA. The rationale for this statement is that in samples of mammalian and bacterial DNA of *equal concentration*, any particular DNA sequence should be present in a thousandfold lower concentration in the mammalian DNA sample because there are a thousand times more kinds of sequences present.

When experiments comparing the reassociation kinetics of mammalian and bacterial DNA were actually performed by Britten and Kohne, however, the results were not as expected. Figure 11-62 summarizes the data obtained when the reassociation kinetics of typical mammalian and bacterial DNAs were compared. In this graph the percentage of the total DNA that has become reassociated is plotted as a function of the starting concentration of DNA multiplied by the length of time. This parameter of DNA concentration × time, or **Cot,** is employed in place of time alone because it allows the direct comparison of data obtained from reactions run at different DNA concentrations. When plotted in this way, the data obtained by Britten and Kohne reveal that a significant portion of the mammalian DNA molecules reassociate more rapidly (i.e., at a lower Cot value) than bacterial DNA. The most straightforward explanation for this unexpected result is that some DNA sequences present in mammalian DNA are repeated many times. Such repetition would increase the relative concentration of such sequences, generating more collisions and a faster rate of reassociation than would be expected if each sequence were present in only a single copy. In contrast to the mammalian DNA sequences that reasso-

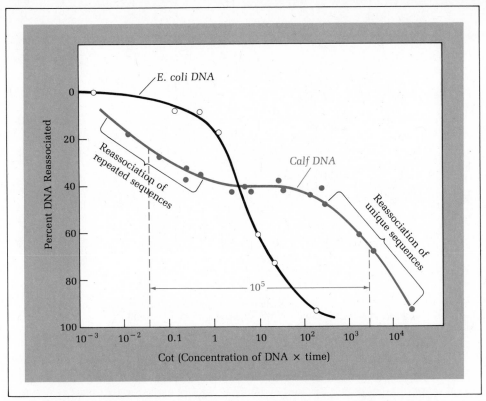

FIGURE 11-62 A Cot curve comparing the reassociation kinetics of calf and *E. coli* DNAs. The calf DNA that reassociates more rapidly than the bacterial DNA consists of repeated sequences.

ciate more rapidly than bacterial DNA, a major fraction also reassociates more slowly. Close examination of Figure 11-62 reveals that this fraction reassociates about a thousand times more slowly than the bacterial DNA, which is the behavior expected of DNA sequences that are not repeated. This latter fraction is therefore referred to as **unique** or **single-copy** DNA to distinguish it from the **repetitive** DNA that reassociates more quickly.

Examination of the reassociation or Cot curve generated by a given sample of DNA provides information concerning both the prevalence of repeated sequences and their relative degree of repetition. In the case of the calf DNA whose behavior is illustrated in Figure 11-62, for example, examination of the Cot curve reveals that about 40 percent of the sequences reassociate more quickly than expected for unique sequences, and are hence repetitive. Examination of this graph also reveals that the bulk of repetitive sequences reassociate between a Cot value of 10^{-2} and 10^{-1}, while the unique sequences reassociate at Cot values that are 100,000 times higher, that is, between 10^3 and 10^4. Since the repetitive DNA fraction reassociates about 100,000 times faster than the unique fraction, it follows that the individual sequences that comprise the repetitive fraction are present on average in about 100,000 copies each.

Repetitive and unique DNA sequences can be easily separated from each other by carrying out the DNA

reassociation reaction at a Cot value that is high enough to promote the reassociation of repetitive, but not unique, DNA sequences. Under such conditions repetitive DNA will reassociate into double-stranded molecules, while unique sequences remain single stranded. The single- and double-stranded DNAs can then be separated from each other by physical means, such as chromatography on hydroxylapatite columns. Repetitive DNA sequences purified in this way have been found to exhibit melting temperatures that are lower than that expected from normal DNA molecules of comparable base composition. From studies involving DNA molecules of defined sequence, it is known that a decreased melting temperature occurs when two DNA strands are not properly matched throughout their sequence; for every 1 percent of the bases that are not properly paired with a complementary base, the melting temperature decreases by about 1 °C. On the basis of this information, it has been determined that reassociated repetitive DNA is usually about 10 percent mismatched, although in some cases the mismatching may be considerably more extensive. The existence of mismatching in the reassociated molecules indicates that the multiple copies of repetitive sequences are not exact replicas of each other, and that during reassociation, molecules of slightly different sequence can reassociate with one another (Figure 11-63).

Though it is common to speak of the average repeti-

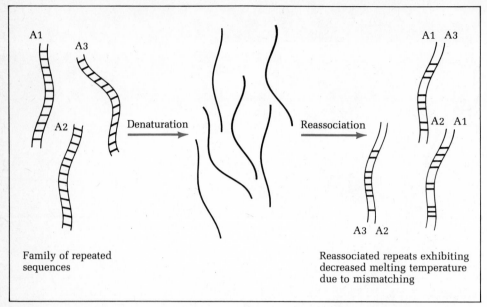

FIGURE 11-63 Schematic diagram illustrating why reassociated repeated DNA sequences tend to exhibit a decreased melting temperature. The magnitude of the decrease in melting temperature indicates the extent of mismatching between the various members of the repeat family.

tion frequency and extent of mismatching of an organism's repetitive DNA, it is important to emphasize that the repetitive DNA fraction contains a heterogeneous group of sequences repeated to varying extents and with varying degrees of precision. A small contribution to the repetitive DNA fraction is made by a few genes, such as those coding for histones and ribosomal RNAs, which are present in hundreds or thousands of copies per cell. The functional significance of the remainder of the re-

FIGURE 11-64 Fractionation of mouse DNA by isodensity centrifugation. About 10 percent of the total DNA has a base composition different enough from the rest of the DNA to cause it to sediment as a separate satellite peak.

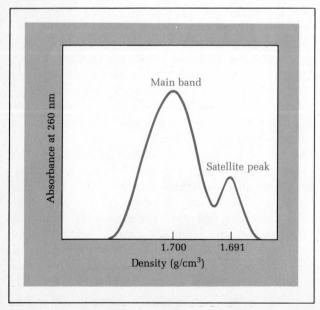

petitive DNA fraction, which does not appear to be comprised of genes, is largely a mystery. At least two major kinds of sequences can be distinguished among this noncoding repetitive DNA: satellite DNAs and interspersed repetitive DNAs. The term **satellite DNA** was originally coined to refer to repeated sequences whose unique base composition allows them to be detected during isodensity centrifugation as a shoulder or satellite peak distinct from the main band of DNA (Figure 11-64). Base sequence analyses of the material present in such satellite peaks have revealed the presence of relatively short sequences that are extensively repeated. In the case of mouse satellite DNA, for example, a simple sequence about a dozen bases long is repeated with minor variations more than ten million times, accounting for roughly 10 percent of the organism's total DNA content. Satellite DNAs of varying lengths, sequence arrangements, and chromosomal locations have been detected in a variety of animal and plant cells, indicating that such sequences tend to change rapidly during evolution. In spite of this variation most satellite DNA sequences share three basic features: (1) They consist of multiple copies of the same sequence repeated in tandem; (2) they are not transcribed into RNA; and (3) they are preferentially associated with constitutive heterochromatin (page 429) and chromosomal centromeres (page 506). Because such chromosome regions tend to be permanently coiled up and transcriptionally inactive, it has been proposed that satellite DNA sequences play a structural role in creating or maintaining this inactive state. There is little direct evidence supporting this notion, however.

In addition to satellite DNAs, the repetitive DNA

fraction contains a large number of sequences several hundred base pairs in length that are interspersed throughout the chromosomal DNA rather than being localized in tandemly repeated clusters. Unlike satellite DNA, these **interspersed repetitive sequences** are transcribed into RNA and are relatively stable in terms of sequence and location. It has been proposed that interspersed repetitive sequences function in the regulation of RNA synthesis and processing, as initiation sites for DNA replication, and/or in the structural organization of chromatin. None of these functions has been clearly established, however, and the possibility has even been raised that some of this repeated DNA represents "selfish" sequences that perform no function at all. According to this view, repetitive DNA sequences may have arisen during evolution from viruses or transposable elements that are capable of multiplying and inserting into various regions of the chromosome. Such inserted sequences may then behave like parasites that are replicated along with the rest of the chromosomal DNA but perform no function for the host cell.

The realization that a major portion of the DNA present in eukaryotes, but not prokaryotes, consists of repetitive sequences whose functional significance is unknown should be kept in mind as we proceed in our discussion of gene regulation. As we shall see, there are a number of differences between the genetic regulatory mechanisms of eukaryotic and prokaryotic cells, and it is possible that the existence of at least some types of repetitive DNA may ultimately be found to be related to genetic regulatory processes that are specific to eukaryotes.

Gene Regulation: The Role of Chromatin Structure

It has been known for many years that alterations in eukaryotic gene expression are often associated with changes in chromatin structure. One of the earliest realizations of this point dates back to the 1920s when it was first noted that certain regions of eukaryotic chromosomes do not uncoil after cell division is completed, but instead remain tightly condensed throughout the cell cycle. Subsequent genetic and biochemical studies revealed that this condensed form of chromatin is not transcribed into RNA. Such condensed, genetically inactive chromatin is referred to as **heterochromatin,** while the more loosely packed, transcriptionally active form of chromatin is known as **euchromatin.** Within the general category of heterochromatin, two subtypes can be distinguished: constitutive heterochromatin and facultative heterochromatin. **Constitutive** heterochromatin contains permanently inactivated DNA sequences never destined to be transcribed, such as the satellite DNAs present in chromosomal centromeres. **Facultative** heterochromatin, on the other hand, contains gene

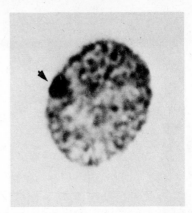

FIGURE 11-65 Light micrograph showing the dark staining mass of chromatin (arrow), called the Barr body, which occurs in the cells of female mammals. Courtesy of G. R. Wilson.

sequences that are potentially capable of being transcribed into functional products but that become inactivated in certain cell types as part of the normal developmental process.

One of the most dramatic examples of facultative heterochromatin occurs in mammalian sex chromosomes. In the early 1950s Murray Barr and his collaborators observed that the cells of female mammals often exhibit a small, darkly staining mass of chromatin that is not detectable in the comparable cells of males (Figure 11-65). Subsequent studies revealed that this structure, termed a **Barr body,** represents an X chromosome that has been completely converted to heterochromatin. Mary Lyon was the first to suggest that Barr bodies are produced by the random inactivation of one of the two X chromosomes normally present in females, leaving only one X chromosome active. Hence every cell in an adult female has only one active X chromosome, just as in males where only one X chromosome is present in the first place. Lyon further proposed that once a given X chromosome has been inactivated in a particular cell, all future descendants of that cell will have the same X chromosome inactivated. Since the initial inactivation process was thought to be random, some cells in the adult would be expected to derive from a cell lineage in which one X chromosome is inactive, and the rest would be expected to derive from cells in which the other X chromosome is inactive. In support of this hypothesis, Lyon pointed out that female animals containing genes for different coat colors on their two X chromosomes always have mottled coats. A classic example of this phenomenon occurs in the tortoise-shell cat, whose coat consists of patches of yellow and black fur. The tortoise-shell pattern is generally restricted to female cats and can be readily explained by expression of one X chromosome in the yellow patches of fur and expression of the other X chromosome in the black patches.

The Lyon hypothesis has derived subsequent support from the direct analysis of protein products known

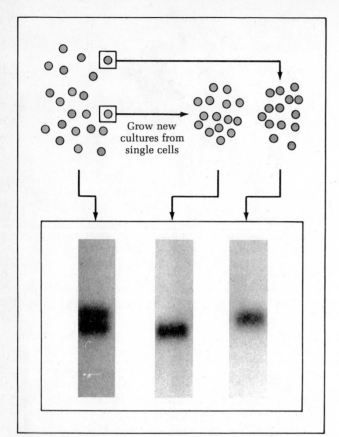

FIGURE 11-66 Summary of experiment showing that cultured cells from human females are a mixture of two populations, each expressing a different form of the gene for the enzyme glucose-6-phosphate dehydrogenase. Since the gene for this enzyme is localized on the X chromosome, these studies indicate that only one of the two X chromosomes is expressed in any given cell. Photograph courtesy of R. G. Davidson and H. M. Nitowsky.

to be encoded by genes localized on the X chromosome. The behavior of one such product, the enzyme *glucose-6-phosphate dehydrogenase*, has been studied in cultured cells obtained from an individual known to synthesize two different forms of this enzyme. Although extracts from the original cell culture were found to contain both forms of the enzyme, clones obtained by growing up new cultures from isolated single cells produce only one form or the other (Figure 11-66). Hence the original culture must have consisted of a mixed population of two cell types, each expressing the genes contained on only one of the two X chromosomes.

Although the preceding example shows that heterochromatin formation can inactivate large blocks of DNA, such a process obviously represents a relatively coarse type of control over gene activity. More subtle changes in chromatin structure appear to be associated with the selective activation and inactivation of specific genes. In the mid-1970s Harold Weintraub and his associates demonstrated that actively transcribed genes are more susceptible than inactive genes to digestion with pancreatic DNAase-I, an enzyme that cleaves DNA bound to the nucleosome surface. It was found, for example, that globin-specific DNA sequences present in red blood cell chromatin, where the globin gene is expressed, are more sensitive to digestion by DNAase-I than are the globin sequences of brain cell chromatin, where the globin gene is not expressed. Equal sensitivity to DNAase-I digestion appears to be exhibited by genes that are rapidly transcribed, by genes that are less frequently transcribed, by genes that have recently been transcribed but are no longer active, and by DNA sequences adjacent to genes of the preceding types. Such observations indicate that the increased sensitivity to DNAase-I digestion is not caused by the process of gene transcription per se, but instead reflects an altered nucleosomal configuration in regions associated with active or potentially active genes.

Some sequences located adjacent to active genes are so sensitive to digestion by trace amounts of DNAase-I that they are referred to as DNAase-I *hypersensitive* sites. Such sites, which tend to occur a few hundred bases upstream from the transcriptional start sites of active genes, are thought to represent short areas in which nucleosomes are missing. This altered DNA configuration presumably facilitates the binding of RNA polymerase and/or regulatory proteins involved in gene transcription.

Although the structural differences between the nucleosomal organization of inactive and active genes are not generally detectable at the electron microscopic level, several biochemical differences between the two classes of nucleosomes have been detected. It has been discovered, for example, that incubation of chromatin with DNAase-I under conditions that foster the selective degradation of active genes leads to the preferential release of the acetylated form of histones H3 and H4, suggesting that acetylated histones are preferentially associated with the nucleosomes of active genes. Further support for this conclusion has emerged from studies involving sodium butyrate, a fatty acid that inhibits the normal process of histone deacetylation and hence causes histones to accumulate an excessive number of acetyl groups. Chromatin isolated from cells treated with sodium butyrate has been found to be more susceptible than normal to digestion with DNAase-I, suggesting that histone acetylation alters nucleosomal structure in a way that enhances the susceptibility of the nucleosome-associated DNA to enzymatic cleavage.

Another type of histone modification preferentially associated with active nucleosomes is the covalent attachment of a polypeptide known as **ubiquitin** to histone 2A. Though the significance of this type of histone modification is unknown, ubiquitin is a highly conserved protein present in virtually all prokaryotic and eukaryotic cells. The occurrence of ubiquitin in prokaryotic cells lacking histones suggests that this polypeptide serves a general role in modifying proteins for special purposes. One general function in which ubiquitin has been implicated is in earmarking proteins for degradation, presumably by changing the conformation of the protein to which it attaches to make it more susceptible to proteolytic degradation. Although there is no indication that the association of ubiquitin to histone 2A promotes histone degradation, this association may alter histone conformation in such a way that promotes availability of the DNA for transcription.

Yet another unique biochemical feature of transcriptionally active nucleosomes is their high content of HMG nonhistone proteins. An important role for this class of proteins in the structure of active chromatin is suggested by the discovery that the active genes of chromatin preparations whose HMG proteins have been removed lose their preferential sensitivity to degradation by DNAase-I. This sensitivity can be restored by adding two particular HMG proteins, HMG14 and HMG17, to HMG-depleted chromatin, indicating that these two proteins are directly responsible for conferring DNAase-I sensitivity upon active genes. It is not clear, however, how HMG14 and HMG17 recognize and bind specifically to the nucleosomes of active genes.

A final structural feature that appears to be characteristic of actively transcribed genes is a close relationship with the nuclear matrix. This property of active genes was first clearly established for the chicken gene coding for ovalbumin, a protein formed only in cells lining the oviduct. In studies carried out by Bert Vogelstein and his collaborators, the nuclear matrix was isolated from chick oviduct cells, which actively transcribe the ovalbumin gene, and from chick liver cells, which do not. When the small fraction of the total cellular DNA that remains associated with the isolated nuclear matrix was hybridized to cloned ovalbumin gene sequences, it was found that ovalbumin sequences are significantly enriched in nuclear matrix fractions isolated from oviduct, but not liver. Such data suggest that DNA sequences undergoing transcription become more closely associated with the nuclear matrix.

It is clear from the preceding discussion that a variety of structural alterations occur in chromatin regions containing active genes. Such findings do not address the underlying question, however, of how specific DNA segments are initially selected for activation. As we shall see in the following sections, several kinds of mechanisms have been implicated in this process.

Gene Regulation: The Role of Changes in DNA Structure

One way in which the transcription of individual gene sequences can be regulated is by altering the DNA molecule itself. Among the various kinds of changes in DNA structure that have been implicated in gene regulation are gene amplification, gene deletion, DNA methylation, DNA rearrangements, enhancers, and changes in

DNA conformation. These six types of regulation will be discussed in the following sections.

Gene Amplification   The maximum rate at which any particular gene transcript can be synthesized is directly related to the number of copies of the gene that are present. We have already seen that thousands of identical copies of the total chromosomal DNA are generated in polytene chromosomes, thereby facilitating the synthesis of the large quantities of RNA needed by giant cells. Although this represents an extreme case in which multiple copies of all gene sequences are produced, a more selective type of gene amplification also occurs. For example, extra copies of the DNA sequence coding for ribosomal RNA are formed in many types of egg cells to facilitate the production of the large number of ribosomes needed during early development. In this case the DNA sequence coding for ribosomal RNA is utilized as a template for the formation of independent copies of new genes coding for ribosomal RNA. In amphibians this process can generate a thousandfold increase in the number of ribosomal genes, and the concomitant formation of a thousand or more nucleoli.

Another example of selective amplification involves the genes coding for the chorion proteins that make up the tough outer coat of insect eggs. Like ribosomal RNA genes, chorion genes are preferentially replicated just prior to the time at which their gene product is required in large amounts. In contrast to ribosomal gene amplification, however, the amplified chorion genes remain attached to the chromosomal DNA, generating a localized area of polyteny.

Despite the clear-cut evidence that selective amplification of the genes coding for specific proteins can occur, this does not appear to be a particularly widespread phenomenon. DNA hybridization experiments using DNA probes specific for various kinds of protein-coding genes have revealed that most cellular genes are present in only one or a few functional copies. A notable exception to this generalization involves the genes coding for the various types of histones. In contrast to true gene amplification, however, these multiple histone gene copies are a permanent component of the nuclear DNA rather than the product of an amplification process that occurs in a particular cell type at a specific stage of development. It is interesting to note, however, that the number of histone gene copies present in different organisms appears to vary according to the maximum need for histone synthesis. In the sea urchin, for example, where embryonic cell division is extremely rapid and histone synthesis must proceed rapidly to package the newly synthesized DNA into chromatin fibers, the histone genes are repeated up to a thousandfold. Organisms whose embryos do not divide so rapidly, on the other hand, have histone genes that are less extensively repeated.

Gene Deletion In contrast to the selective amplification of gene sequences whose products are in great demand, it is also possible to delete gene sequences whose transcripts are not required. Perhaps the most extreme example of DNA deletion (also called DNA diminution) occurs in mammalian red blood cells, which discard their nuclei entirely after adequate amounts of hemoglobin have been made. A less extreme example of gene deletion occurs in a group of tiny crustaceans known as *copepods*. During the early development of these organisms, the heterochromatic regions of the chromosomes are excised and discarded from all cells except the ones destined to give rise to the germ cells. In this way up to half of the total DNA of the organism is removed from the cells that comprise the bulk of the adult organism.

DNA Methylation The DNA of most organisms contains small amounts of a modified base, 5-methylcytosine, which is generated by enzymatic transfer of a methyl group from S-adenosylmethionine to cytosine residues present within the DNA chain. Most methylated cytosines are located immediately adjacent to a guanine residue, forming the dinucleotide sequence –CG–. The enzyme *DNA methylase*, which catalyzes the methylation reaction, is most efficient at methylating –CG– sequences that are base paired to opposing –GC– sequences that are already methylated. Hence once an opposing pair of –CG– sequences has become methylated, this methylated state tends to be reproduced in the opposing DNA strands that are formed during the process of DNA replication (Figure 11-67). In this way DNA methylation patterns can be passed on from generation to generation.

A relationship between DNA methylation and gene activity is suggested by several kinds of observations. First, the pattern of DNA methylation has been found to differ among cells expressing different genes. For example, certain cytosine residues located near the 5′ end of the globin gene are methylated in tissues that do not produce hemoglobin but are unmethylated in red blood cells. A similar correlation between gene activity and decreased levels of DNA methylation has been observed for several other genes, though exceptions to this generalization have also been noted.

Independent support for the conclusion that DNA methylation influences gene activity has emerged from experiments employing 5-azacytidine, a cytosine analog that cannot be methylated because it contains a nitrogen atom in place of carbon at the site where methylation normally occurs. Because of the tendency of methylation patterns to be inherited, incorporation of 5-azacytidine into DNA triggers an undermethylated state that is maintained for many cell generations after the drug has been removed. Exposure of cells to 5-azacytidine has been found to trigger activation of several kinds of gene sequences, including some present on heterochromatically inactivated X chromosomes.

Although the preceding observations suggest that DNA methylation is associated with the suppression of transcriptional activity, certain observations make it difficult to arrive at any simple generalizations concerning the nature of this relationship. It has been reported, for example, that methyl groups are removed from DNA *after* gene transcription has been activated rather than before, suggesting that methylation might be an effect rather than a cause of gene activation. Moreover, certain genes are found to be actively expressed in spite of extensive methylation, while the DNA of the fruit fly *Drosophila* has been reported to contain no methylated cytosine at all. Hence instead of representing a general requirement for gene regulation in eukaryotes, DNA methylation might simply be one of a multitude of factors contributing to the overall control of gene expression.

DNA Rearrangements Another category of DNA alteration known to influence gene activity includes various kinds of rearrangements in DNA base sequence organization. It has been known for many years that inherited rearrangements in the chromosomal locations of partic-

FIGURE 11-67 The inheritance of DNA methylation patterns. This inheritance is based on the higher efficiency of the methylation reaction for CG sequences that are base paired to CG sequences that are already methylated.

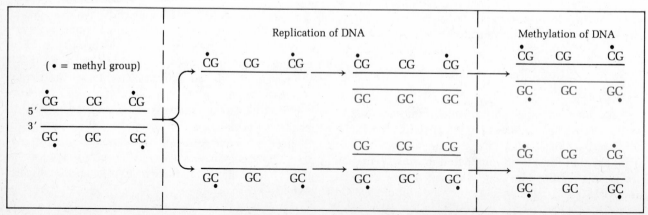

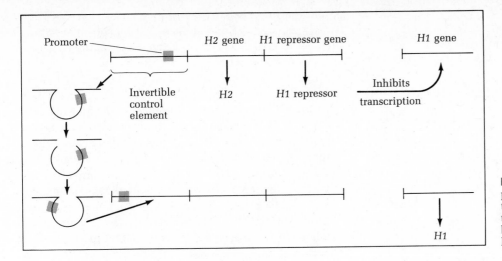

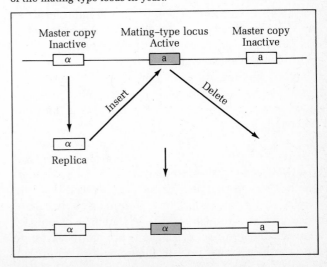

FIGURE 11-68 The mechanism of phase variation in *Salmonella*. Alternation between expression of the *H1* and *H2* genes is mediated by inversion of a control element located adjacent to the *H2* gene.

ular genes can exert dramatic effects on gene activity. Only recently has it become apparent, however, that DNA rearrangements also occur as part of normal developmental processes for regulating gene expression. Such rearrangements, which have been detected in both prokaryotic and eukaryotic cells, occur in several different ways.

One of the simplest rearrangement mechanisms is observed in the bacterium *Salmonella* in association with **phase variation,** a phenomenon that involves the alternate expression of two genes coding for differing forms of the flagellar protein, flagellin. This switching between the expression of two alternate genes helps the bacterium evade the immune response of the host organism being infected. The activities of the two flagellin genes, **H1** and **H2**, are regulated by an invertible control sequence located immediately upstream from the *H2* gene. This 970-base-pair control region contains both a promoter sequence and sequences resembling those present in transposable elements. In cells in which the *H2* gene is active, this control element is oriented so that its promoter sequence is situated at the appropriate position upstream from the *H2* gene. RNA polymerase can therefore transcribe both the *H2* gene and an adjacent gene that codes for a repressor protein that inhibits transcription of the *H1* gene (Figure 11-68). However, in cells in which the *H1* gene is active, the control element is inverted so that its promoter sequence is disconnected from the *H2* gene. Under these conditions neither the *H2* gene nor the repressor gene can be transcribed; moreover, the resulting absence of repressor production permits transcription of the *H1* gene to be activated. Hence the event that triggers alternate transcription of the *H1* and *H2* genes is the excision of the 970-base-pair segment and its reinsertion in the opposite orientation. The resemblance of this invertible control sequence to a transposable element suggests that the mechanism of the excision and reinsertion process is analogous to the mechanisms by which transposable elements move from site to site within DNA.

A somewhat different mechanism for switching expression back and forth between alternate genes has been discovered in the yeast *Saccharomyces cerevisiae*. This organism contains two mating-type genes, termed α and a, which determine whether or not given cells can mate with each other. Cells of the α type readily mate with cells of the a type, but cells of the same type cannot mate with one another. Since mating type is inherited, yeast colonies descended from the same cell might be expected to be of the same type and therefore incapable of mating. This obstacle is overcome, however, by a mechanism that allows yeast cells to occasionally convert from one mating type to the other. During the switching process, the gene present at the **mating-type locus** is excised and replaced with a gene of the opposite type. This replacement is made possible by the existence of "master copies" of the α and a genes that are not normally capable of being transcribed. However, replica copies of these inactive master genes can be synthesized and inserted in place of the currently existing gene at the mating-type locus (Figure 11-69). In this new

FIGURE 11-69 The cassette mechanism for regulating expression of the mating-type locus in yeast.

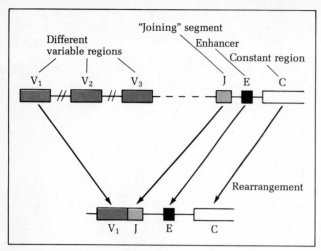

FIGURE 11-70 Schematic diagram summarizing the organization of the DNA sequences coding for immunoglobulin light chains. Different kinds of light chains can be created by joining different variable regions (V1, V2, V3, etc.) with the same constant region. Transcriptional promoter sequences are located upstream from the variable regions, but are not efficient enough to promote transcription until rearrangement brings them closer to the enhancer sequences located adjacent to the constant region.

location the replica of the master gene is activated, presumably because the DNA sequences and/or chromatin structure surrounding the mating-type locus facilitate transcription. This type of regulation is referred to as the "cassette" mechanism because the insertion of different gene sequences into the same location determines which gene will be expressed, just as putting different cassettes into a tape player determines what will be heard.

Yet another type of DNA rearrangement is associated with the production of antibody or **immunoglobulin** molecules in the white blood cells of higher organisms. As will be described in detail in Chapter 17, the polypeptide chains that make up immunoglobulin molecules contain two regions: a *variable* region whose amino acid sequence varies among different kinds of antibodies and a *constant* region whose sequence is always the same. The DNA sequences that code for the variable and constant regions of a given immunoglobulin chain are not normally adjacent to one another in the DNA molecule, but are brought together by a specific type of rearrangement that only occurs in antibody-producing cells. As a result of this rearrangement, multiple kinds of antibodies can be produced by bringing together differing variable regions with the same constant region (Figure 11-70).

Enhancers Although the preceding model explains how DNA sequences coding for a large number of different kinds of antibodies can be generated, it does not explain why transcription of immunoglobulin genes does not occur until after the DNA sequences coding for the variable and constant regions are brought close to one another. Recent investigations suggest that this ac-

tivation process is triggered by a special type of DNA sequence termed an **enhancer.** Enhancers are a family of DNA sequences, first discovered in animal viruses, that are capable of increasing the transcription rate of neighboring sequences. In the case of immunoglobulin genes, an enhancer sequence is located immediately upstream from the DNA sequences coding for the constant region. However, a promoter sequence is not present in this same area, so transcription by RNA polymerase cannot occur. The promoter for immunoglobulin gene transcription is located upstream from the variable region, but it is not efficient enough to promote transcription in the absence of an enhancer sequence. Thus prior to DNA rearrangement, the promoter and enhancer sequences of an immunoglobulin gene are far apart and transcription does not occur; only after rearrangement are they close enough for transcription to be activated (see Figure 11-70).

Although it is not yet known how widely enhancers are employed in the regulation of eukaryotic gene expression, such sequences exhibit some unusual attributes that make them quite flexible as potential regulatory elements. It has been discovered, for example, that enhancers activate transcription when placed either upstream or downstream from the start site of transcription, and are even active when inserted in reversed orientation. Enhancers are also active when inserted adjacent to genes other than the ones with which they are normally associated, and are capable of activating transcription when situated several thousand bases away from the start site of transcription. This set of unusual properties suggests that enhancer sequences may exert a generalized effect on chromatin conformation that makes the DNA molecule more accessible for transcription.

Changes in DNA Conformation In addition to alterations in DNA base sequence, changes in the three-dimensional conformation of DNA also influence gene activity. One type of conformational change that has been clearly implicated in the regulation of gene activity involves changes in the coiling of the DNA double helix. During transcription by RNA polymerase, localized unwinding of the DNA double helix must take place in order to allow base pairing between the DNA template strand and the newly forming RNA chain. If the DNA double helix were free to rotate at one end, this localized unwinding would simply cause the DNA molecule to rotate. However, the DNA of both prokaryotic and eukaryotic cells is organized into long loops whose ends are relatively fixed and unable to rotate. Localized unwinding within such a fixed structure would create tension in the double helix and the resulting formation of **positive supercoils** (Figure 11-71). This supercoiled state is energetically unfavorable and therefore tends to slow the rate of transcription.

In bacterial cells this obstacle is overcome by the

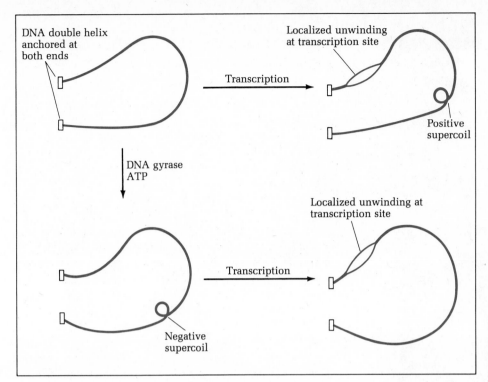

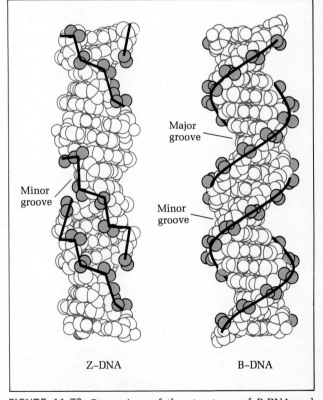

FIGURE 11-71 The role of DNA gyrase in facilitating transcription. When the action of this enzyme creates negative supercoils in DNA, the tension normally generated during transcription by unwinding of the DNA double helix is used to remove these negative supercoils instead.

FIGURE 11-72 Comparison of the structures of B-DNA and Z-DNA. In B-DNA the sugar-phosphate backbone forms a smooth right-handed double helix, while in Z-DNA the backbone forms a zig-zag left-handed helix.

enzyme **DNA gyrase,** which utilizes energy released from the breakdown of ATP to generate **negative supercoils** whose handedness is the opposite of the positive supercoils generated by localized unwinding of the DNA double helix. When transcription takes place in a negatively supercoiled molecule of DNA, the tension normally generated by localized unwinding of the double helix is utilized to remove such negative supercoils rather than to generate positive supercoils. The possibility that this type of phenomenon is involved in regulating the expression of specific genes is suggested by the discovery that inhibitors of DNA gyrase depress the transcription rate of some bacterial genes but not others. It has therefore been proposed that certain regions of the bacterial chromosome have specific sites for interaction with DNA gyrase, and that the negative supercoils generated in these regions facilitate gene transcription.

Another type of alteration in the three-dimensional structure of the DNA double helix has been observed in eukaryotic cells. This alternative DNA structure was initially discovered in the laboratory of Alexander Rich, where X-ray diffraction studies of synthetic DNA molecules containing an alternating sequence of guanine and cytosine led to the discovery of a new form of DNA referred to as **Z-DNA.** In the normal or *B-form* of the DNA molecule, the sugar-phosphate backbone is arranged as a smooth right-handed double helix; Z-DNA, in contrast, is a left-handed helix with a zigzag backbone (Figure 11-72). Subsequent investigations have revealed that this Z-DNA structure tends to occur in DNA regions that are enriched in alternating sequences of purines and pyrimidines. Such alternating purine-

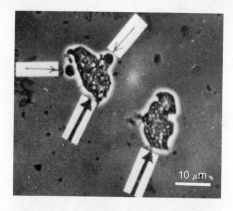

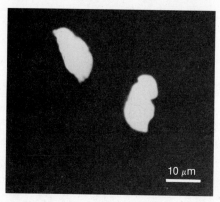

FIGURE 11-73 Light micrographs showing the binding of fluorescein-labeled antibodies against Z-DNA to the macro- and micronuclei of the ciliated protozoan *Stylonychia*. (*Top*) Phase contrast micrograph showing the micronuclei (light arrow) and macronuclei (heavy arrow). (*Bottom*) Immunofluorescence microscopy of the same field showing that the antibodies to Z-DNA bind only to the macronucleus. Courtesy of A. Rich.

pyrimidine stretches are preferentially localized in enhancer and other control sequences of viral DNA molecules, raising the possibility that the Z-DNA structure plays a role in regulating gene activity.

Although Z-DNA was first detected in synthetic DNA molecules, antibodies against Z-DNA also react with the DNA present in normal cells. In order to study the relationship between Z-DNA and gene activity, Rich and his colleagues have investigated the localization of Z-DNA in the nuclei of ciliated protozoa. These organisms contain two kinds of nuclei, a large transcriptionally active *macronucleus* and a smaller inactive *micronucleus*. When such cells are stained with antibodies against Z-DNA, only the transcriptionally active macronucleus reacts (Figure 11-73). Since the DNA of the macronucleus is derived from replication of the micronuclear DNA, the question arises as to why the Z-DNA structure forms in the macronucleus when it was not originally present in the micronucleus. Though the answer to this question is not known at present, several factors have been shown to facilitate the formation of the Z-DNA structure in DNA regions where it is potentially capable of forming. Among these factors are methylation of –CG– sequences, the presence of special

proteins that bind to and stabilize Z-DNA, and negative supercoiling of the double helix.

Gene Regulation: The Role of DNA-associated Proteins

We have now seen that regulation of gene expression is associated with several kinds of changes in the DNA molecule. Another general way in which gene expression can be modulated is by proteins that interact with DNA. The most obvious example of a protein capable of influencing gene activity is RNA polymerase, the enzyme responsible for catalyzing the transcription process. Because RNA polymerase exists in multiple forms, alterations in this enzyme can be utilized to change the spectrum of genes being transcribed. Some excellent examples of this type of regulation occur in association with bacteriophage infection of bacterial cells. During infection of *E. coli* by T4-phage, for example, a small number of phage genes are first transcribed by the bacterial RNA polymerase into messenger RNAs. These messages in turn code for enzymes and proteins that interact with the bacterial RNA polymerase, causing chemical modification of its subunits and the addition of several new subunits. This altered RNA polymerase now selectively transcribes the remaining phage genes. A somewhat different mechanism is employed by T7-phage, whose DNA is transcribed into a messenger RNA that codes for a new, phage-specific RNA polymerase and for factors that inactivate the previously existing bacterial RNA polymerase.

Regulation of RNA polymerase activity is also observed in eukaryotic cells. During embryonic development, for example, gradual changes occur in the rates at which RNA polymerases I, II, and III are formed. Since these three enzymes transcribe genes coding for ribosomal RNA, HnRNA, and small RNAs, respectively, changes in the relative rates of synthesis of these three RNA classes can be accomplished in this way. The existence of protein factors that influence the activity of eukaryotic RNA polymerases has also been reported, but the physiological role played by such molecules is yet to be clearly elucidated.

Although alterations in RNA polymerase activity are capable of exerting a relatively coarse level of control over gene activity, a more powerful mechanism for regulating the transcription of individual genes involves DNA-associated proteins that selectively interact with specific genes. Since the information currently available concerning the existence and properties of such genetic regulatory proteins differs significantly for prokaryotic and eukaryotic cells, we shall discuss these two classes of cells separately.

Genetic Regulatory Proteins of Prokaryotic Cells Our current understanding of the mechanisms utilized by bacterial cells for regulating the activity of individual

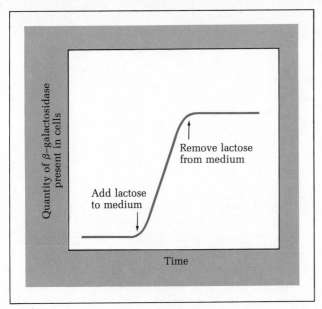

FIGURE 11-74 Induction of β-galactosidase by lactose in *E. coli.*

genes owes a great debt to the pioneering work of Fran-cois Jacob and Jacques Monod on the control of lactose metabolism in *E. coli.* The key enzyme in this pathway, **β-galactosidase,** cleaves the disaccharide lactose into the monosaccharides glucose and galactose. Since bac-teria have no use for this enzyme when lactose is not available as a nutrient, a control mechanism has evolved that ensures that bacteria only synthesize β-ga-lactosidase when lactose or a closely related sugar is present (Figure 11-74). This type of regulation, in which a small molecule such as lactose triggers the synthesis of a specific enzyme, is referred to as **enzyme induction,** and the molecules that trigger the process of enzyme induction are referred to as **inducers.**

Through a combination of biochemical and genetic studies on the genes involved in lactose metabolism, Jacob and Monod developed a general theory of gene regulation that has had far-reaching implications. The cornerstone of this theory rested on the discovery of two distinct classes of genes, designated **structural** and **regulatory genes.** The structural genes code for the pro-duction of enzymes involved in lactose uptake and metabolism, and are actively transcribed only when an inducer such as lactose is present. There are three of these so-called *lac* genes: a **z gene** coding for the enzyme β-galactosidase, a **y gene** coding for a permease involved in lactose transport, and an **a gene** coding for an acety-lating enzyme. These three genes are situated adjacent to one another in the bacterial DNA.

Genetic studies revealed that the activity of the three *lac* genes is controlled by a regulatory gene, termed the *i* **gene,** which is located elsewhere in the bacterial chromosome. Bacteria in which the *i* gene has been deleted synthesize all three *lac* gene products, re-gardless of whether an inducer such as lactose is present; thus the *i* gene must normally generate a prod-uct that inhibits expression of the *lac* structural genes. This type of inhibitory regulation is known as **negative control,** and the inhibitory product of the regulatory gene is termed a **repressor.** Subsequent experiments by Jacob and Monod revealed that mutations situated near the beginning of the z gene also generate a condition in which the *lac* gene products are synthesized regardless of the presence of inducer. Since mutations in this area, designated the **operator,** affect the transcription of all three structural genes, it was concluded that the three *lac* genes are transcribed as an integrated unit. This in-tegrated transcriptional unit is termed the *lac* **operon,** and contains the following five elements in sequence: a promoter (p site) for RNA polymerase, the operator (o site), and the three structural genes z, y, and a.

Based on the above information Jacob and Monod proposed that the *lac* repressor binds specifically to DNA sequences present within the *lac* operator, and that this binding of repressor prevents RNA polymerase from initiating transcription of the adjacent structural genes. It was further postulated that in the presence of an inducer such as lactose, the *lac* repressor becomes inactive and is no longer capable of binding to the *lac* operator. Under such conditions, RNA polymerase would be free to initiate transcription of the *lac* genes. When this model was first proposed in 1961, the only molecular mechanism known to be capable of recogniz-ing specific DNA base sequences was base pairing with a complementary nucleic acid. Since the above model re-quired that the *lac* repressor be able to specifically rec-ognize and bind to the *lac* operator, it was initially believed that the repressor would turn out to be a nucleic acid.

However, in 1966 Walter Gilbert and Benno Müller-Hill successfully isolated the *lac* repressor and demonstrated that it is a protein molecule. Because a typical bacterial cell contains only 10–20 molecules of this particular protein, a sensitive assay was required for its identification and isolation. For this purpose Gil-bert and Müller-Hill utilized *isopropyl-thio-galactoside (IPTG),* an extremely potent inducer of the *lac* genes. They reasoned that if inducer molecules inactivate the *lac* repressor by physically binding to it, then it might be possible to identify the *lac* repressor by looking for mol-ecules that bind IPTG. To measure IPTG binding, the technique of *equilibrium dialysis* was employed. In this procedure molecules to be tested for their ability to bind IPTG were placed in a membrane sac and dialyzed against radioactive IPTG. When IPTG-binding macro-molecules are absent, the radioactive IPTG should dif-fuse back and forth across the membrane until its concentration is equal on both sides. If the sac contains macromolecules that bind IPTG, however, the radio-active IPTG will become more concentrated within the sac.

When a crude extract of bacterial cells was tested in the above manner, a small amount of IPTG-binding activity was detected. By fractionating the crude extract in various ways and testing the individual fractions by equilibrium dialysis against radioactive IPTG, Gilbert and Müller-Hill were eventually able to purify a single protein molecule that binds IPTG with high affinity. This protein could not be detected in mutant bacteria defective in *lac* repressor activity; mutants exhibiting increased sensitivity to induction of the *lac* genes, on the other hand, were found to contain a protein that bound IPTG with even higher affinity than normal. Although these observations strongly suggested that the isolated IPTG-binding protein was the *lac* repressor, the most direct support for this conclusion came from studying the ability of this protein to bind to purified DNA. When DNA samples containing the *lac* operon were mixed with the IPTG-binding protein, strong binding between the protein and DNA was observed. The protein did not bind, however, to other kinds of DNA or to *lac* operon DNA containing a mutated operator sequence (Figure 11-75). Final support for the con-

clusion that the IPTG-binding protein is the *lac* repressor came from the discovery that the ability of this protein to bind to *lac* operon DNA is inhibited in the presence of IPTG. Such an observation is directly compatible with the known ability of inducers of the *lac* operon to inhibit *lac* repressor activity.

The preceding data clearly demonstrated that the *lac* repressor is a protein molecule whose binding to the *lac* operon is regulated by inducer molecules such as IPTG and lactose. From this information a relatively complete model of *lac* operon regulation could be constructed. According to this model (Figure 11-76), the *lac* repressor normally binds to the *lac* operon and inhibits its transcription by RNA polymerase. When an inducer of the *lac* operon binds to an allosteric site on the *lac* repressor, however, the conformation of the repressor is changed so that it can no longer bind to DNA. Under these conditions the genes of the *lac* operon can be transcribed by RNA polymerase.

Soon after the formulation of the above model, the development of mutant strains of bacteria that produce excessive amounts of *lac* repressor made it possible to

FIGURE 11-75 Experiments demonstrating the binding of the *lac* repressor to DNA. Radioactive repressor was mixed with various types of DNA, and the mixtures were then subjected to moving-zone centrifugation to separate free repressor from the repressor-DNA complex. (a) Repressor only binds to DNA samples containing the *lac* genes. (b) Binding of the repressor to DNA is abolished in the presence of the inducer, IPTG.

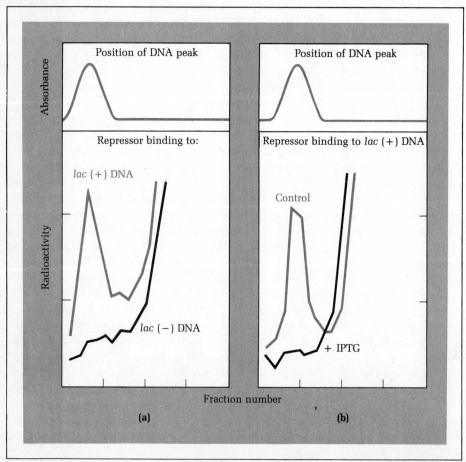

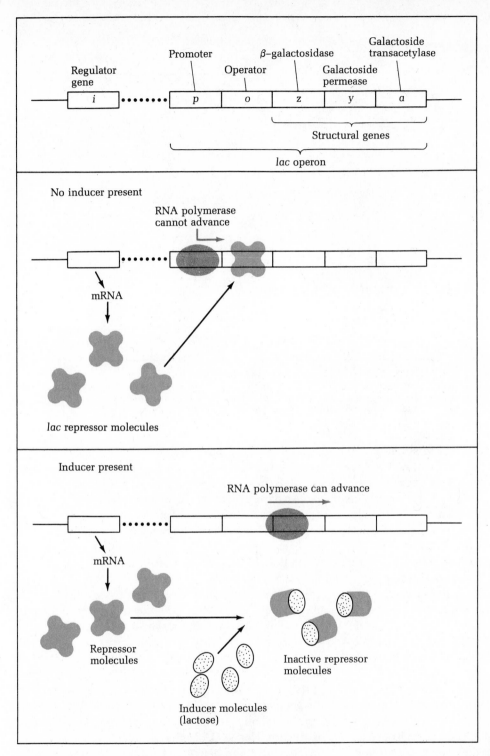

FIGURE 11-76 A model summarizing the regulation of the *lac* operon by the *lac* repressor. The sizes of the various genes are not drawn to scale.

isolate and purify large quantities of this repressor protein. Biochemical studies on this purified repressor revealed that it is a tetramer consisting of four identical subunits, each containing a binding site for an inducer molecule. The availability of large quantities of purified repressor also facilitated the subsequent isolation of the *lac* operator. In these experiments, purified *lac* repressor molecules were mixed with DNA samples containing *lac* operon sequences, and the mixture was then digested with deoxyribonuclease. Under such conditions the DNA sequences containing bound repressor are protected from digestion, and can be isolated and characterized. Experiments of this type revealed that the *lac* operator is roughly two-dozen bases in length, and exhibits a twofold axis of symmetry, that is, the sequence of bases in each strand reads roughly the same in the 5' $\longrightarrow$ 3' direction (Figure 11-77).

The type of regulation utilized to control the activ-

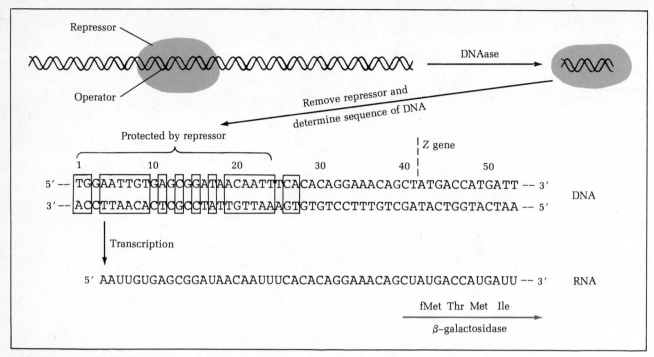

FIGURE 11-77 Isolation of the *lac* operon. Purified *lac* repressor was first mixed with a DNA sample containing the *lac* operon, and the mixture was then digested with deoxyribonuclease. The DNA sequences containing bound repressor were protected from digestion, allowing them to be purified for subsequent sequence analysis. The bases of the *lac* operator exhibiting twofold symmetry are enclosed in boxes.

ity of the *lac* operon is ideally suited to situations in which the presence of a particular nutrient, such as lactose, signals the need for the enzymes encoded by a certain set of genes. In some situations, however, the presence of a particular substance indicates that the products encoded by a certain set of genes are *not* needed. For example, bacteria placed in an environment containing adequate amounts of the amino acid tryptophan no longer require the enzymes involved in the biosynthesis of this particular amino acid. Transcriptional regulation of the genes involved in tryptophan synthesis is also mediated by a repressor, but in this case the repressor binds to DNA only when tryptophan is bound to the repressor. Tryptophan is therefore said to act as a **co-repressor.** In spite of their different modes of action, inducers and co-repressors both trigger conformational changes that influence the ability of a repressor molecule to bind to DNA.

Although inducers and co-repressors exert opposite effects on repressor molecules, they both function in negative control systems in which a DNA-binding protein inhibits transcription when bound to its appropriate operator. Bacterial cells also have **positive control** systems that utilize **activator** proteins to stimulate the transcription of particular gene sequences. One of the more thoroughly investigated examples of this type of control is associated with the ability of glucose to depress the formation of degradative or **catabolic** enzymes such as α-galactosidase, arabinose isomerase, and tryp-

tophanase. This phenomenon, known as **catabolite repression,** allows cells to turn off the synthesis of enzymes that are not needed when glucose is abundant.

An early clue to the mechanism underlying catabolite repression was provided in the late 1960s by Ira Pastan and Robert Perlman, who discovered that exposing bacterial cells to glucose lowers the intracellular concentration of **3′,5′-cyclic AMP,** and that adding cyclic AMP to the extracellular medium inhibits catabolite repression. Although these two observations implicated cyclic AMP in the mechanism of catabolite repression, the exact role played by this nucleotide did not become clear until Geoffrey Zubay and his associates isolated a DNA-binding protein that interacts with both cyclic AMP and catabolite-repressible genes. Mutant bacteria lacking this protein, termed the **catabolite gene activator protein** or **CAP,** are unable to transcribe catabolite-repressible genes, even in the absence of glucose. Hence CAP must be required for the transcription of these genes. Subsequent studies revealed the basis for this requirement; CAP, when complexed with cyclic AMP, binds upstream from certain promoter sites and facilitates the binding of RNA polymerase to such promoters. It is interesting to note that like operator sequences, the DNA sequence recognized by the CAP protein exhibits a twofold axis of symmetry.

Among the various genes subject to control by catabolite repression are the structural genes of the *lac* operon. The *lac* operon is thus subject to independent regulation by both positive and negative control systems, the former based upon an activator protein that binds at the upstream end of the promoter, and the latter employing a separate repressor protein that binds downstream from the promoter (Figure 11-78). In some cases negative and positive control can even be exerted by the same protein molecule. An example of this type of regulation occurs in the arabinose operon, where transcription is normally inhibited by a specific repressor protein. When bound to arabinose, however, this protein no longer binds to the operator and inhibits transcription, but instead binds adjacent to the promoter and activates transcription.

Genetic Regulatory Proteins of Eukaryotic Cells   The above examples clearly indicate that the use of small molecules to regulate the DNA-binding properties of genetic regulatory proteins provides bacterial cells with a powerful mechanism for tailoring the process of gene transcription to the changing needs of individual cells. Control over the transcription of individual genes also occurs in eukaryotes, although the larger DNA content of eukaryotic cells has made it significantly more difficult to investigate the mechanisms involved. In a typical eukaryotic cell only 5–10 percent of the total DNA is transcribed into RNA. The first insight into the mechanism responsible for keeping the vast majority of the DNA transcriptionally inactive was provided in the early 1960s, when it was reported that the addition of histones to purified DNA inhibits the ability of the DNA to be transcribed by RNA polymerase. These observa-

tions caused a great stir in the biological community because it at first appeared that the specific repressors of eukaryotic gene transcription had been discovered. However, the subsequent discovery that there are only five kinds of histones put to rest the idea that these proteins might be serving as specific repressors for individual genes. It is now clear, of course, that the primary function of histones is packaging DNA into nucleosomes, but this does not rule out the possibility that such packaging is in turn responsible for a generalized restriction in the availability of DNA for transcription. According to this view, histones maintain the bulk of the DNA in a transcriptionally inactive form, while the small fraction of the nuclear DNA that needs to be transcribed is selected by specific activator proteins.

Although the existence of specific gene-activator proteins in eukaryotic cells therefore seems likely, progress in isolating and characterizing such proteins has been relatively slow. One activator protein that has been clearly identified is the 5S transcription factor, a protein described earlier in the chapter that is required for the efficient transcription of the 5S ribosomal RNA genes by RNA polymerase III. The enzyme responsible for messenger RNA synthesis, RNA polymerase II, also requires the presence of additional protein factors in order to transcribe DNA properly. Some of these protein factors appear to facilitate the initiation of transcription at all promoter sites, while others are specific for the promoters of specific genes. Several of these promoter-specific factors have been isolated and characterized, including one that activates transcription of SV40 viral

FIGURE 11-78 Summary of the positive and negative control systems operating on the *lac* operon.

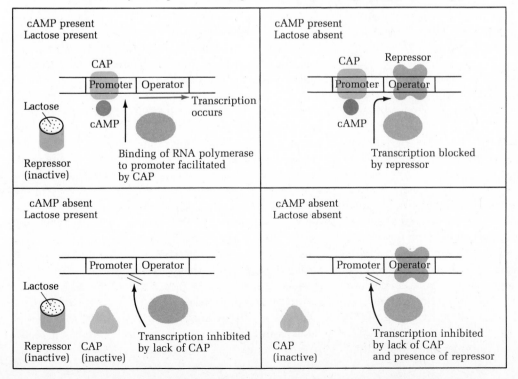

DNA sequences and another that activates the transcription of a set of genes that becomes active when cells are exposed to abnormally high temperatures ("heat-shock" genes).

Another class of activator proteins whose interactions with DNA are beginning to be clarified are the protein receptors that mediate the actions of steroid hormones (Chapter 16). Steroid hormones are known to stimulate the synthesis of selected proteins in the cells with which they interact, and at least some of this enhanced synthesis is due to the activation of specific genes. The transcriptional activation induced by a steroid hormone requires the presence of a specific receptor protein to which the steroid binds. In recent years it has been shown that steroid hormone-receptor complexes bind selectively to DNA sequences located upstream from the transcriptional start sites of the genes being activated, suggesting that this binding is required for the initiation of transcription. It is believed that all the genes whose transcription is activated by a particular steroid hormone contain similar receptor-binding sequences, explaining how a single regulatory agent such as a steroid hormone can activate the transcription of a particular family of genes. If this type of organization is widespread in eukaryotic DNA, it might also explain why certain kinds of DNA sequences are repeated in various locations.

Gene Regulation: The Role of RNA Processing and Export

In eukaryotic cells gene transcription is only the first step in the production of a functional molecule of messenger RNA. As we saw earlier in the chapter, the initial product of gene transcription is an HnRNA transcript that undergoes extensive processing during its conversion to messenger RNA. In recent years evidence has begun to accumulate that suggests that a significant amount of control over gene expression can be exerted at the level of RNA processing. The existence of such control was first discovered in animal viruses, such as adenovirus and SV40, whose DNA codes for primary transcripts that can be spliced in different ways to produce several distinct messenger RNAs. Examples of multiple splicing pathways have also been demonstrated for several cellular genes. In the case of the gene coding for the lens protein αA_2 crystalline, for example, the same primary transcript has been found to code for two different polypeptide chains, one of which contains an extra 22-amino-acid segment within its interior. This extra segment is encoded by an exon that is spliced out of the primary transcript during formation of the messenger RNA that codes for the alternative form of the protein (Figure 11-79).

In addition to its role in normal gene regulation,

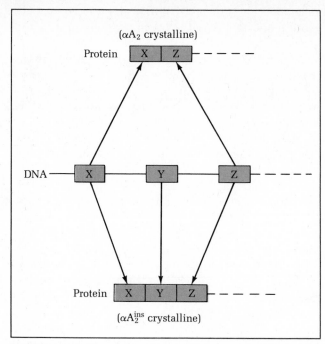

FIGURE 11-79 Alternative splicing of the gene coding for the lens protein, αA_2 crystalline, generates two different forms of the protein.

alternative splicing patterns can also be induced by mutations. For example some forms of *thalassemia*, a group of diseases characterized by abnormalities in globin gene expression, involve mutations localized near the ends of intervening sequences. Such mutations disrupt the normal splice sites, leading to an abnormal splicing pattern and the resulting production of a defective form of hemoglobin.

Once a primary transcript has been converted to a mature messenger RNA, this message must still become associated with ribosomes before it can be translated into a protein product. In prokaryotic cells, which lack a nuclear envelope, ribosomes have direct access to newly synthesized RNA molecules and translation of newly forming messenger RNA molecules usually begins before transcription has been completed. In eukaryotic cells, on the other hand, messenger RNA must pass through the nuclear envelope and into the cytoplasm before protein synthesis can be initiated. Relatively little is known about the factors that control the export of messenger RNA from the nucleus, but some evidence indicates that RNA processing and export may be closely coupled to each other. The most direct support for this statement comes from experiments in which introns have been artificially removed from purified genes, and the genes then inserted back into eukaryotic cells. Although such genes are actively transcribed, their transcripts are not exported from the nucleus unless at least one intron is initially present. Although such findings suggest the existence of a close relationship between RNA processing and export, this is clearly not an absolute requirement because not all genes contain introns.

Although we are just beginning to understand the ways in which the processing and export of RNA might be regulated in eukaryotic cells, there are good reasons for believing that such regulatory mechanisms exist. It has been shown by DNA-RNA hybridization, for example, that the HnRNA molecules present in nuclei obtained from different cell types resemble one another more than do the messenger RNA molecules present in the cytoplasm. This means that cells producing different kinds of messenger RNAs may transcribe a similar spectrum of genes into HnRNA, and that regulation of RNA processing and/or export determines which of these sequences end up in the cytoplasm as mature messenger RNAs. The existence of control at this level has been directly demonstrated in immature red blood cells transformed by a virus that causes them to stop producing hemoglobin. By comparing the hybridization of nuclear and cytoplasmic RNAs to globin cDNA, it was shown that globin-specific sequences are present in both the nucleus and cytoplasm before viral infection, but only in the nucleus after viral infection. Such results suggest that the cessation of hemoglobin synthesis was due to an alteration at the level of RNA processing and/or export rather than control at the level of gene transcription.

We have thus seen in this chapter that a wide variety of control mechanisms are available for regulating which gene sequences will be utilized for the production of functional molecules of messenger RNA. Once a molecule of messenger RNA appears in the cytoplasm, the next stage in the overall process of gene expression is its translation into a polypeptide chain. The role of the ribosome in this translation process, and the ways in which the translation step can itself be regulated, are discussed in the following chapter.

SUMMARY

The bulk of the DNA present in eukaryotic cells is localized within the nucleus. Though prokaryotic cells lack a true, membrane-enclosed nucleus, they too contain a defined nuclear region (nucleoid) in which the cell's DNA-containing fibrils are localized. In contrast to the relatively simple organization of the prokaryotic nucleoid, the eukaryotic nucleus consists of at least five distinct components: the nuclear envelope, nucleoplasm, nucleoli, chromatin fibers or chromosomes, and the nuclear matrix.

The nuclear envelope is constructed from two concentric membranes known as the outer and inner nuclear membranes. The outer nuclear membrane is characterized by the presence of cytoplasmic ribosomes on its outer surface, and may exhibit points of continuity with the endoplasmic reticulum. Beneath the inner membrane is a thin layer of tough fibrous material, the nuclear lamina, which provides a structural support for the nuclear envelope. One of the most striking morphological features of the nuclear envelope is the presence of octagonal-shaped nuclear pores that measure about 60 nm in diameter and function in both the passive diffusion of smaller molecules and the mediated transport of larger molecules. In addition to its role in the transport of material between the nucleus and cytoplasm, the nuclear envelope serves as an anchoring point for chromatin fibers, and has been implicated in certain types of oxidation reactions.

The nucleoplasm is an amorphous, fluidlike material that, like the cell sap in the cytoplasm, contains soluble molecules and small particles. Among the constituents occurring in the nucleoplasm are enzymatic and regulatory proteins, small RNA molecules, ribonucleoprotein granules, and low-molecular-weight metabolites, coenzymes, and ions.

The nucleolus is a distinctive spherical structure consisting of a mixture of fibrils, granules, and chromatin fibers embedded in a homogeneous ground substance. This organelle functions in the formation of ribosomes by a process whose details will be discussed in the following chapter.

Chromatin fibers, which consist of a tightly bound complex of DNA and protein, account for the bulk of the nuclear mass. Although chromatin fibers are generally dispersed throughout the nucleus, at certain times they become organized into larger, more discrete structures known as chromosomes. The protein components of chromatin can be divided into two major types: histones and nonhistones. Nonhistone proteins are a heterogeneous collection of mildly acidic molecules that have been implicated in a variety of enzymatic, structural, and regulatory roles. The histones, in contrast, are highly basic proteins that occur in only five major types: H1, H2A, H2B, H3, and H4. Histone molecules are responsible for packaging DNA into particles known as nucleosomes. Each nucleosome contains an octomer of histones (two molecules each of histones H2A, H2B, H3, and H4) tightly associated with about 146 base pairs of DNA. Another 50 or so base pairs of DNA serve as a linker that, in association with histone H1, joins one nucleosome to the next. While extended chains of nucleosomes measure about 10 nm in diameter, intact chromatin fibers are generally closer to 30 nm in diameter. Formation of the 30-nm fiber from an extended chain of nucleosomes involves packing the 10-nm diameter nucleosomes into a helical coil termed a solenoid. The 30-nm chromatin fiber can be further folded to form highly compact chromosomes, such as the polytene chromosomes present in the giant cells of certain insects or the metaphase chromosomes that form at the time of cell division. The organization of chromatin fibers into metaphase chromosomes is mediated in part by a nonhistone protein scaffold or backbone to which DNA is attached at multiple points, forming a series of loops.

The nuclear matrix is a fibrous substructure that provides an underlying framework for the nucleus. Although it is basically analogous to the cytoskeletal network of the cytoplasm, the nuclear matrix differs from the cytoskeleton in both the morphological appearance and protein composition of its constituent fibers. The nuclear matrix appears to be preferentially associated with DNA that is undergoing transcription or replication, though the role played by the matrix in such activities is unclear.

Because the chromatin fibers of most eukaryotic cells are dispersed and intertwined throughout the nucleus, the activity of individual genes is generally difficult to visualize. An exception occurs, however, in the giant cells of certain insects, where the DNA is extensively replicated and organized into polytene chromosomes containing multiple copies of each DNA molecule. Such chromosomes exhibit distinctive banding patterns that make it possible to map the locations of individual genes. Gene transcription in such chromosomes is often associated with the formation of regions of localized uncoiling known as chromosome puffs or Balbiani rings. Each puff makes a specific type of RNA that becomes associated with protein as it is being formed. The resulting ribonucleoprotein complex is transported through the nuclear pores to the cytoplasm, where its RNA component becomes associated with ribosomes and directs the synthesis of a particular protein. One agent that has been shown to play a pivotal role in inducing the formation of puffs is the steroid hormone ecdysone, which triggers the normal molting process in insects. Ecdysone acts by associating with a protein receptor that binds to specific DNA sites and induces gene transcription.

The transcription process is catalyzed by the enzyme RNA polymerase. Eukaryotic cells contain three major forms of this enzyme: RNA polymerases I, II, and III. RNA polymerase I is localized in the nucleolus, where it catalyzes the formation of the large ribosomal RNAs. RNA polymerases II and III are associated with the rest of the nuclear chromatin, where RNA polymerase II transcribes genes coding for messenger RNAs and RNA polymerase III transcribes genes coding for small RNA molecules such as transfer RNA and 5S ribosomal RNA. The reaction catalyzed by RNA polymerase involves four main stages: (1) the binding of RNA polymerase to promoter sequences, (2) initiation of RNA synthesis by the incorporation of a single molecule of ATP or GTP, (3) elongation of the RNA chain in the

$5' \rightarrow 3'$ direction, and (4) termination of RNA synthesis and release of the completed RNA chain. Commonly employed inhibitors of RNA synthesis may act at the binding step (heparin), initiation step (rifamycins), or elongation step (actinomycin).

The process of DNA transcription generally produces RNA molecules that are longer than would be expected for the transcripts of individual genes. In prokaryotic cells these primary transcripts are often polycistronic messenger RNA molecules that direct the synthesis of several different proteins. In eukaryotic cells, on the other hand, most of the RNA molecules generated by the process of transcription are HnRNA molecules that are extensively degraded within the nucleus; only a small fraction of this RNA finds its way to the cytoplasm to function as messenger RNA. The conversion of HnRNA into messenger RNA involves a complex series of steps, collectively known as RNA processing, which includes packaging of the RNA with protein, addition of a methylated cap structure to the 5' end, joining of a poly(A) tail to the 3' end, and removal of intervening sequences (introns) by a process that is generally mediated by ribonucleoprotein particles called snurps, but is in some cases self-splicing. Although it is not clear why the coding regions (exons) of eukaryotic genes are interrupted by introns, one possible explanation is that exons code for separate protein domains; this type of organization may facilitate the rearrangement of exons into new combinations, resulting in the creation of new types of proteins.

Gene expression is subject to regulation by a variety of cytoplasmic and environmental factors. An understanding of gene regulation in eukaryotes is complicated by the existence of a large number of repetitive DNA sequences of unknown functional significance. These repetitive sequences fall into two major categories: satellite DNAs, which are tandemly repeated, nontranscribed sequences that are preferentially associated with heterochromatin and centromeres; and interspersed repetitive sequences, which are dispersed throughout the DNA and often transcribed.

A relatively coarse level of gene control is exerted in eukaryotic cells by condensing some of the chromatin fibers into a tightly packaged, transcriptionally inactive form known as heterochromatin. Constitutive heterochromatin contains sequences never destined for transcription, while facultative heterochromatin involves dynamically regulated sequences that are potentially capable of being transcribed into functional products. A dramatic example of heterochromatin occurs in female mammals, where one of the two X chromosomes is randomly inactivated by conversion to the heterochromatic state. More subtle changes in chromatin structure are associated with the selective activation and inactivation of individual genes. The existence of an altered nucleosomal configuration in the regions of active genes has been revealed by the increased susceptibility of the DNA in these areas to digestion by DNAase-I. Nucleosomes associated with active genes are enriched in acetylated histones, the protein ubiquitin (which is attached to histone 2A), and HMG proteins. The DNA of such nucleosomes also appears to be closely associated with the nuclear matrix.

One way in which the transcription of individual genes is selectively controlled is via alterations in the DNA molecule. Among the changes in DNA structure that have been implicated in the regulation of gene expression are gene amplification, gene deletion, DNA methylation, DNA sequence rearrangements, enhancers, and changes in DNA conformation such as supercoiling and the formation of Z-DNA.

The transcription of individual genes is also subject to regulation by DNA-associated proteins, such as RNA polymerase and DNA-binding regulatory proteins. Our current understanding of the mechanism of action of DNA-binding regulatory proteins owes a great debt to the pioneering experiments carried out on the *lac* operon of *E. coli*. The transcription of these genes is regulated by a specific protein molecule, the *lac* repressor, which inhibits transcription of the *lac* operon by binding to the *lac* operator DNA. Inducers of the *lac* operon, such as lactose, bind to the *lac* repressor and trigger a change in conformation that prevents it from binding to the *lac* operon. Subsequent to the discovery of the *lac* repressor, many other kinds of DNA-binding regulatory proteins have been identified in bacteria. Some repressors, such as the one that regulates transcription of the genes involved in tryptophan metabolism, can only bind to DNA and inhibit transcription when they are associated with an appropriate corepressor. Bacterial cells also contain activator proteins, such as CAP, that stimulate the transcription of specific gene sequences. When complexed with cyclic AMP, the CAP protein binds upstream from the promoter sites of catabolite-repressible genes and facilitates the binding of RNA polymerase.

In eukaryotic cells the role of DNA-associated proteins in regulating gene transcription is somewhat more complex. Since only 5–10 percent of the DNA of a typical eukaryotic cell is ever transcribed into RNA, the bulk of the DNA is maintained in a transcriptionally inactive state by the presence of histones. Promoter-specific activator proteins can then bind to particular DNA sequences and facilitate transcription. Control of gene expression in eukaryotic cells is also exerted at the level of RNA processing and/or transport. One way in which RNA processing can be regulated is by changes in the pattern of RNA splicing that allow the same primary transcript to be converted into different messenger RNAs.

SUGGESTED READINGS

Books

Bradbury, E. M., N. Maclean, and H. R. Matthews (1981). *DNA, Chromatin and Chromosomes*, Wiley, New York.

Busch, H., ed. (1974–1981). *The Cell Nucleus*, Vols. 1–9, Academic Press, New York.

Cold Spring Harbor Laboratory (1978). *Chromatin*. Proceedings of a symposium published as *Cold Spring Harbor Symp. Quant. Biol.*, Vol. 42.

Cold Spring Harbor Laboratory (1983). *DNA Structures*. Proceedings of a symposium published as *Cold Spring Harbor Symp. Quant. Biol.*, Vol. 47.

Gluzman, Y., and T. Shenk, eds. (1983). *Enhancers and Eukaryotic Gene Expression*, Cold Spring Harbor Laboratory, Cold Spring Harbor, NY.

Lewin, B. (1987). *Genes III*, 3rd Ed., Wiley, New York.

Risley, M. S., ed. (1986). *Chromosome Structure and Function*, Van Nostrand Reinhold, New York.

Watson, J. D., N. H. Hopkins, J. W. Roberts, J. A. Steitz, and A. M. Weiner (1987). *Molecular Biology of the Gene*, 4th Ed., Benjamin/Cummings, Menlo Park, CA.

Articles

Berezney, R. (1984). Organization and functions of the nuclear matrix, In: *Chromosomal Nonhistone Proteins*, Vol. IV (L. S. Hnilica, ed.), CRC Press, Boca Raton, FL.

Borst, P., and D. R. Greaves (1987). Programmed gene rearrangements altering gene expression, *Science* 235:658–667.

Britten, R. J., and D. E. Kohne (1968). Repeated sequences in DNA, *Science* 161:529–540.

Cech, T. R. (1986). The generality of self-splicing RNA: relationship to nuclear mRNA splicing, *Cell* 44:207–210.

Chambon, P. (1981). Split genes, *Sci. Amer.* 244(May): 60–71.

Darnell, J. E. (1985). RNA, *Sci. Amer.* 253(Oct.):68–78.

Dreyfus, G. (1986). Structure and function of nuclear and cytoplasmic ribonucleoprotein particles, *Ann. Rev. Cell Biol.* 2:459–498.

Dynan, W. S., and R. Tjian (1985). Control of eukaryotic messenger RNA synthesis by sequence-specific DNA-binding proteins, *Nature* 316:774–778.

Feldherr, C. M., E. Kallenbach, and N. Schultz (1984). Movement of a karyophilic protein through the nuclear pores of oocytes, *J. Cell Biol.* 99:2216–2222.

Felsenfeld, G. (1985). DNA, *Sci. Amer.* 253(Oct.):58–67.

Felsenfeld, G., and J. D. McGhee (1986). Structure of the 30 nm chromatin fiber, *Cell* 44:375–377.

Green, M. R. (1986). Pre-mRNA splicing, *Ann. Rev. Genet.* 20:36–72.

Jackson, D. A. (1986). Organization beyond the gene, *Trends Biochem. Sci.* 11:249–252.

Kornberg, R. D., and A. Klug (1981). The nucleosome, *Sci. Amer.* 244(Feb):52–64.

Miller, O. L. (1981). The nucleolus, chromosomes, and the visualization of genetic activity, *J. Cell Biol.* 91:15s–27s.

Mount, S. M., and J. A. Steitz (1984). RNA splicing and the involvement of small ribonucleoproteins, *Modern Cell Biol.* 3:249–297.

Pelham, H. R. B., and D. D. Brown (1980). A specific transcription factor that can bind either the 5S RNA gene or 5S RNA, *Proc. Natl. Acad. Sci. USA* 77:4170–4174.

Richmond, T. J., J. T. Finch, B. Rushton, D. Rhodes, and A. Klug (1984). Structure of the nucleosome core particle at 7Å resolution, *Nature* 311:532–537.

CHAPTER
12

The Ribosome: Translation of Genetic Information

Once the nucleotide base sequence of a protein-coding gene has been transcribed into RNA and the resulting transcript processed into a mature molecule of messenger RNA, the next step in the pathway of information flow is the utilization of this messenger RNA to guide the synthesis of polypeptide chains. In both prokaryotic and eukaryotic cells, this latter process occurs on specialized cytoplasmic particles known as ribosomes. Although ribosomes were once viewed as passive structures whose sole purpose was to provide a physical surface for the interaction between messenger RNA and the other components involved in its translation, it is now known that ribosomes are dynamic organelles that actively contribute to the process of protein synthesis. In essence, the behavior of the ribosome therefore resembles that of a large, complicated enzyme.

The ribosome is an unusual "enzyme," however, for it is composed of more than 50 different proteins and several kinds of RNA. The presence of so many molecules within a single particle has made it difficult to unravel the three-dimensional ultrastructure of the ribosome and to determine the roles played by its individual constituents. Yet an understanding of this ultrastructural organization is clearly essential if we are to ultimately comprehend the mechanisms involved in the synthesis of polypeptide chains. In this chapter we shall therefore begin by exploring what is known about the structural organization of the ribosome and shall then discuss the function of this particle in the process of protein synthesis.

INTRODUCTION TO THE RIBOSOME

Typical ribosomes measure about 25 nm in diameter, which is an order of magnitude smaller than the wavelength of visible light and hence beneath the resolving power of the light microscope. For this reason ribosomes were not discovered until procedures for examining cells by electron microscopy were developed in the early 1950s. Several electron microscopists of this period noted the presence of small, electron-dense granules in the cytoplasm of thin-sectioned cells, but it was the extensive observations of George Palade and his associates that revealed that similar granules are present in the cytoplasm of widely diverse cell types, and are especially numerous in cells that are actively engaged in protein synthesis. When Palade and his colleague, Philip Siekevitz, isolated these granules by subcellular fractionation, they were found to sediment along with

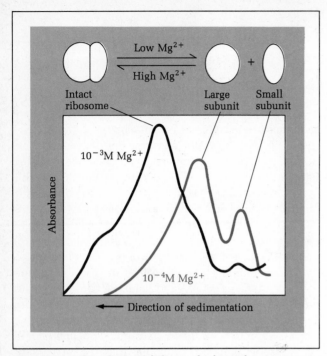

FIGURE 12-1 Identification of ribosomal subunits by moving-zone centrifugation. At low Mg^{2+} concentration, intact ribosomes dissociate into a large and a small subunit.

membranes of the endoplasmic reticulum in a fraction referred to as the microsomal fraction. When biochemical analyses revealed that the microsomal fraction is enriched in RNA, the term **ribosome** was coined to refer to the granules.

The realization that ribosomes are involved in protein synthesis first emerged from experiments in which cells exposed for brief periods to radioactive amino acids were homogenized and subjected to subcellular fractionation. Under such conditions the highest concentration of radioactive protein was detected in the microsomal fraction. Separation of this fraction into its ribosome and membrane components subsequently revealed that the radioactive protein is associated predominantly with ribosomes (see Figure 10-19), implicating the ribosome as the site of protein synthesis.

Soon after the ribosome was identified as the site of protein synthesis, centrifugational analyses revealed that isolated ribosomes dissociate into two subunits when the magnesium ion concentration is lowered (Figure 12-1). In prokaryotic cells these two subunits sediment at approximately 30S and 50S, and combine to form a ribosome of about 70S. Cytoplasmic ribosomes of eukaryotes, on the other hand, consist of 40S and 60S subunits that combine to form an 80S ribosome. These differences in sedimentation coefficient reflect a major difference in the overall size of typical eukaryotic ribosomes (about 4.5 million molecular weight) and prokaryotic ribosomes (roughly 2.5 million molecular weight). Ribosomes of other sizes occur in chloroplasts and mitochondria, but a discussion of their properties

will be delayed until the discussion of mitochondrial and chloroplast biogenesis (Chapter 14).

Although the notion that ribosomes are composed of two separable subunits originally emerged from studies on the behavior of isolated ribosomes during moving-zone centrifugation, the existence of these subunits was subsequently confirmed by high-power electron microscopy (Figure 12-2). With negative staining techniques, it has even been possible to detect the presence of characteristic protuberances and ridges on the two types of subunits. On the basis of such observations a three-dimensional model of the *E. coli* ribosome has been constructed (Figure 12-3). According to this model, the small subunit consists of three regions termed the **head,** the **base,** and the **platform,** while the large subunit contains three regions referred to as the **central protuberance,** the **stalk,** and the **ridge.** When the two subunits are bound together to form an intact ribosome, the central protuberance of the large subunit and the head of the small subunit face one another.

Two major approaches have been utilized for isolating and characterizing the RNA and protein molecules from which ribosomes are constructed. Ribosomal RNA can be isolated by homogenizing ribosomes in a mixture of water and the organic solvent phenol, which denatures and precipitates proteins and releases the RNA into the aqueous phase. Ribosomal proteins, in contrast,

FIGURE 12-2 Electron micrograph of a negatively stained preparation of ribosomes and ribosomal subunits isolated from the bacterium *E. coli*. The large arrowheads point to intact ribosomes, the medium-sized arrowheads point to the large ribosomal subunit, and the small arrowheads point to the small subunit. Courtesy of J. A. Lake.

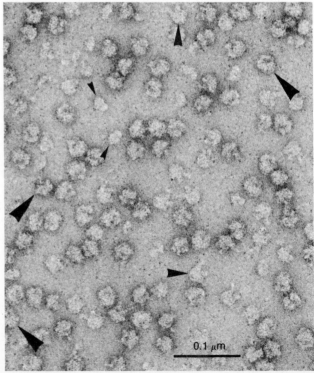

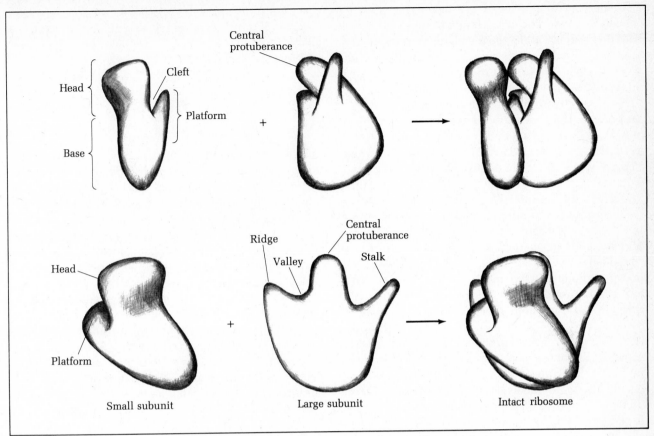

FIGURE 12-3 Three-dimensional model of the *E. coli* ribosome illustrated from two different angles.

FIGURE 12-4 Chemical composition of typical prokaryotic (*E. coli*) and eukaryotic (rat liver) cytoplasmic ribosomes. The numbers in parentheses represent average molecular weights.

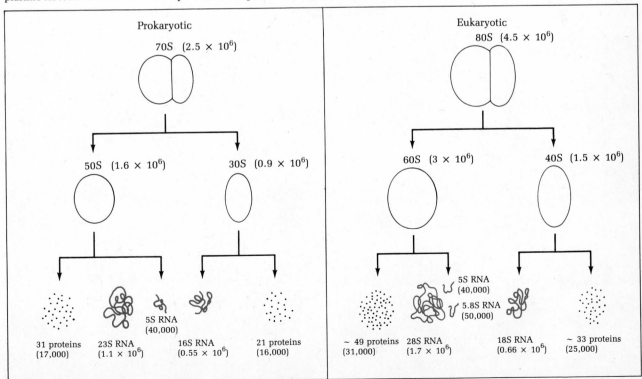

are generally isolated by extracting ribosomes with a concentrated solution of urea and salt. The urea solubilizes the proteins, while the high ionic strength renders the RNA insoluble. Using such approaches, the various kinds of molecules present in prokaryotic and eukaryotic ribosomes have been isolated and identified (Figure 12-4). In both kinds of ribosomes several forms of RNA are present. Prokaryotic ribosomes contain a single type of RNA in the small subunit (16S RNA), and two RNAs in the large subunit (23S and 5S RNAs). In eukaryotic ribosomes the small subunit also contains a single type of RNA (18S RNA), but the large subunit contains three RNAs (28S, 5.8S, and 5S).

Although the number of different proteins present in ribosomes has been more difficult to assess, electrophoretic and chromatographic fractionation of *E. coli* ribosomal protein preparations has led to the identification of 21 proteins in the small subunit, designated **S1 to S21**, and 31 proteins in the large subunit, designated **L1 through L34** (the numbering system extends to 34 because of mistakes in early work on ribosomal protein identification). All these proteins are present in a single copy per ribosome except one, protein L7/L12, which is present in four copies. This particular protein is designated with two numbers because it exists in two forms that differ from one another in their state of acetylation. In contrast to the precise information available for the *E. coli* ribosome, the number of proteins present in the ribosomes of eukaryotic cells is yet to be precisely determined. It is clear, however, that there is a somewhat larger number of different proteins than in bacterial ribosomes, and that their average size is slightly greater.

RIBOSOME BIOGENESIS

Role of the Nucleolus

A great deal has been learned in recent years concerning the mechanisms by which cells manufacture ribosomes. In eukaryotic cells it has been well established that the nucleolus plays a central role. The first definitive evidence for an involvement of the nucleolus in the process of ribosome formation was provided in the early 1960s by Robert Perry, who employed a microbeam of ultraviolet light to destroy the nucleoli of living cells. Such treatment was found to selectively inhibit the synthesis of ribosomal RNA, suggesting that the nucleolus plays a role in the formation of ribosomes. In later experiments Perry found that low concentrations of actinomycin D inhibit the incorporation of ^{3}H-uridine into ribosomal RNA without affecting the formation of other RNAs. Autoradiographic examination of cells treated in this manner revealed that RNA synthesis is selectively depressed in the nucleolus (Figure 12-5). Further support for the conclusion that the nucleolus is involved in synthesizing ribosomal RNA came from studies carried

FIGURE 12-5 Autoradiographic evidence demonstrating that ribosomal RNA synthesis occurs in the nucleolus. *(Left)* Autoradiograph of control HeLa cells incubated with ^{3}H-uridine to label newly synthesized RNA. Note the presence of diffuse labeling over the chromatin, as well as the dense labeling of the nucleoli (arrows). *(Right)* Autoradiograph of cells preincubated with actinomycin D at low concentration (10^{-8} M) prior to exposure to ^{3}H-uridine. Note the selective reduction in the amount of radioactivity incorporated into the nucleoli. Courtesy of R. P. Perry.

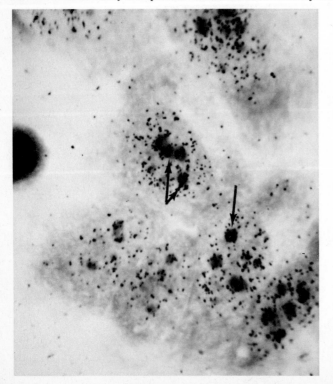
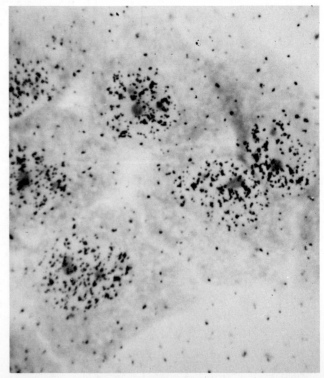

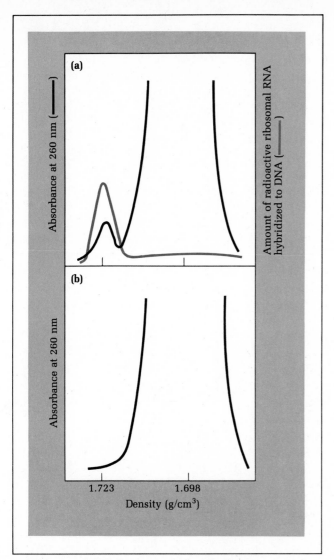

FIGURE 12-6 Behavior of DNA coding for 18S and 28S ribosomal RNAs in *Xenopus laevis*. (a) Isodensity centrifugation of total nuclear DNA reveals the presence of a small satellite peak that contains the DNA sequences that are complementary to 18S and 28S ribosomal RNAs. (b) In DNA obtained from anucleolar mutants, this satellite peak is absent.

high content of the bases G and C, the density of this DNA is significantly greater than that of the remainder of the nuclear DNA. The DNA coding for ribosomal RNA therefore forms a separate band, called a **satellite** peak, when nuclear DNA is analyzed by isodensity centrifugation (Figure 12-6). This satellite peak is missing in DNA samples obtained from anucleolar mutant embryos, indicating that organisms lacking nucleoli also lack the DNA sequences coding for ribosomal RNA. The most straightforward interpretation of this finding is that the genes coding for ribosomal RNA are localized within the nucleolus. This conclusion has been directly verified by *in situ* hybridization experiments in which radioactive ribosomal RNA has been hybridized directly to tissue sections. Autoradiographs of such sections clearly reveal that radioactive 18S and 28S ribosomal RNAs hybridize preferentially to DNA sequences localized within the nucleolus (Figure 12-7). Radioactive 5S ribosomal RNA, on the other hand, hybridizes to chromatin fibers situated outside the nucleolus, indicating that the 5S genes are not localized within this organelle.

FIGURE 12-7 Light microscopic autoradiography of an insect salivary gland cell that has been hybridized with radioactive 18S and 28S ribosomal RNAs. The localization of radioactivity within the nucleolus (arrow) supports the conclusion that the nucleolus contains the genes coding for these ribosomal RNAs. Courtesy of M. L. Pardue.

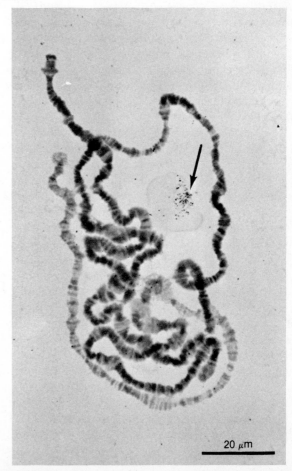

out by Donald Brown and John Gurdon on the African clawed toad, *Xenopus laevis*. Through appropriate genetic crosses, it is possible to produce *Xenopus* embryos whose cells lack nucleoli, as well as the chromosome sites called **nucleolar organizer regions (NOR)** that give rise to nucleoli. Brown and Gurdon discovered that such embryos, termed *anucleolar mutants*, are incapable of synthesizing new ribosomal RNA and therefore die during early development.

The idea that ribosomal RNA is synthesized in the nucleolus has received additional support from the discovery that the DNA sequences coding for 18S and 28S ribosomal RNAs are localized within the nucleolus. This association has been elegantly demonstrated in *Xenopus* embryos, where the genes coding for ribosomal RNA are extensively amplified. Because DNA coding for ribosomal RNA is characterized by an exceptionally

Amplification of Ribosomal Genes

Cells active in protein synthesis generally contain hundreds of thousands of ribosomes, creating a need for large amounts of ribosomal RNA. As indicated in Chapter 11, one way to increase the rate of formation of any particular gene transcript is to increase the number of genes coding for it. In the case of the genes coding for ribosomal RNAs, an increase in gene frequency is accomplished in two different ways. First, the ribosomal RNA genes are normally present in multiple copies. This repetition is much more extensive in eukaryotic cells, which typically contain anywhere from a few hundred to many thousand copies of the 18S and 28S genes, and up to 50,000 copies of the 5S genes.

In situations where the demand for new ribosomes is excessive, however, this normal level of repetition of the ribosomal genes may be insufficient to meet the needs for ribosomal RNA. In such cases additional copies of the ribosomal genes are produced by the process of **gene amplification.** Amplification of the ribosomal RNA genes has been most thoroughly studied in developing oocytes of *X. laevis*, where the ribosomal RNA genes are amplified more than a thousandfold during the early growth phase of oogenesis. This amplification of ribosomal RNA genes is associated with the formation of several thousand nucleoli, each of which receives several hundred copies of the genes coding for 28S and 18S ribosomal RNA. Hence each oocyte ends up with over a hundred thousand copies of these gene sequences, while other cells of this organism contain only a few hundred copies. This extensive amplification facilitates the synthesis of the large number of ribosomes that need to be produced and stored in the developing egg cell for utilization by the embryo after fertilization takes place.

Investigation of the mechanism underlying the amplification of DNA sequences coding for ribosomal RNA has led to the discovery that oocytes contain large numbers of circular DNA molecules of a size sufficient to encompass one or more copies of the ribosomal RNA genes. Although it is not known how such circular DNA molecules are initially generated from ribosomal genes that are part of the linear chromosomal DNA molecule, a special type of replication process is thought to be responsible for increasing the number of these circular molecules once a few have been generated. This special mechanism, termed **rolling-circle replication,** is related to the way in which the DNA of certain single-stranded DNA viruses is replicated. In the rolling-circle mechanism the 3′ end of one of the two strands of a circular DNA molecule is extended by a replication process that displaces the 5′ end of the same strand from the circle (Figure 12-8). The newly formed single-stranded DNA is then cleaved and utilized as a template for the synthesis of the appropriate complementary strand.

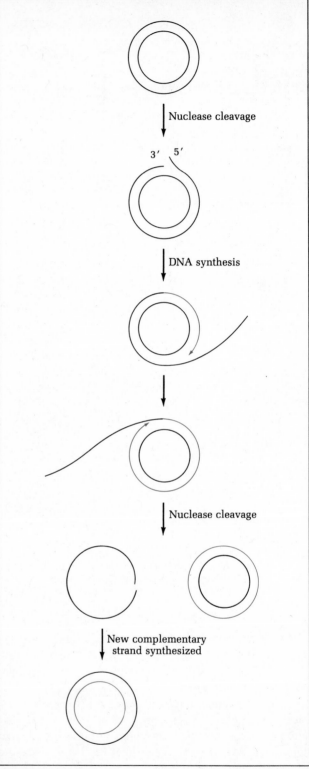

FIGURE 12-8 The rolling-circle mechanism of DNA replication.

Synthesis and Processing of Ribosomal RNA

It was pointed out in Chapter 11 that eukaryotic cells contain a specific form of RNA polymerase, designated RNA polymerase I, which is localized within the nucleolus and is responsible for transcribing the genes coding for 18S and 28S ribosomal RNAs. As in the case of genes coding for messenger RNAs, transcription of ribosomal RNA genes by RNA polymerase yields a primary transcript that is subsequently processed into smaller molecules of RNA. The existence of this processing pathway for ribosomal RNA was first discovered in the early 1960s by Klaus Scherrer and James Darnell, who incubated cells with ³H-uridine for varying periods of time and then analyzed the newly synthesized radioactive RNA by moving-zone centrifugation. When labeling was carried out for relatively short periods of time, a peak of radioactive RNA sedimenting at roughly 45S was found to appear. With somewhat longer labeling periods, a radioactive RNA sedimenting at 32S was also detected; finally, after about 30 minutes radioactivity was detected in 18S and 28S ribosomal RNAs as well. In theory there are two possible explanations of such data: Either the 45S and 32S RNAs are synthesized more rapidly than 28S and 18S RNAs and can therefore be detected earlier, or the 45S and 32S RNAs are detected first because they are precursors of the 28S and 18S RNAs.

To distinguish between these two alternatives, experiments were carried out in which cells were incubated for a short period of time with ³H-uridine to allow the 45S RNA, but not the 32S, 28S, or 18S RNAs, to become radioactive. Actinomycin D was then added to the culture medium to prevent any additional synthesis of radioactive RNA, and incubation was continued. Under such conditions the radioactive 45S RNA was found to gradually disappear, accompanied by the appearance of radioactive RNAs of 32S, 28S, and 18S (Figure 12-9). Since new radioactive RNA cannot be synthesized in the presence of actinomycin D, it was concluded that the 45S RNA is a precursor for these smaller RNAs.

By employing polyacrylamide gel electrophoresis instead of moving-zone centrifugation, a more complete picture of the intermediates produced during the conversion of 45S to mature ribosomal RNA has been obtained (Figure 12-10). Such data have led to the formulation of the ribosomal RNA processing pathway summarized in Figure 12-11. Although the exact sizes of the RNA intermediates illustrated in this pathway may differ somewhat among organisms, a similar kind of ribosomal RNA processing scheme has been observed in a large number of different animal and plant cell types. In such pathways, the initial product generated by transcription of the ribosomal RNA genes is a 45S ribosomal RNA precursor containing both 28S and 18S sequences within it. This 45S RNA, whose total length is about twice that of the combined lengths of a 28S and an 18S RNA molecule, is subsequently cleaved and trimmed by ribonucleases in such a way that roughly half the sequences present in the original 45S RNA molecule are discarded. The first step in this processing pathway involves the conversion of the 45S RNA to a slightly smaller molecule of 41S RNA, which is then cleaved into molecules of 32S and 20S RNA. Subsequent trimming of these molecules generates the 28S and 18S RNAs, respectively. Detailed examination of the prop-

FIGURE 12-9 Experiment demonstrating that 45S RNA is a precursor of 32S, 28S, and 18S RNAs. The curve on the left was obtained by labeling cells for 25 minutes with ³H-uridine and then analyzing total cellular RNA by moving-zone centrifugation. Under such conditions, the bulk of the labeled RNA sediments at 45S. The subsequent fate of this 45S RNA was determined by adding actinomycin D to a culture of cells that had been first labeled with ³H-uridine for 25 minutes. During subsequent incubation, the radioactivity is converted to 32S, and then 28S and 18S RNAs.

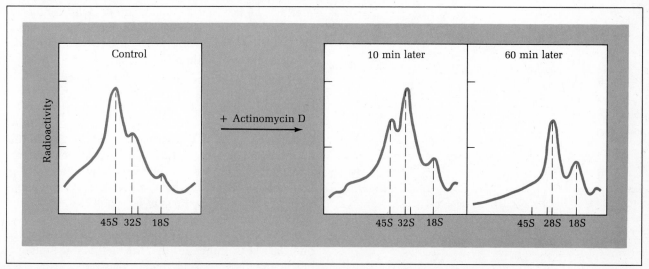

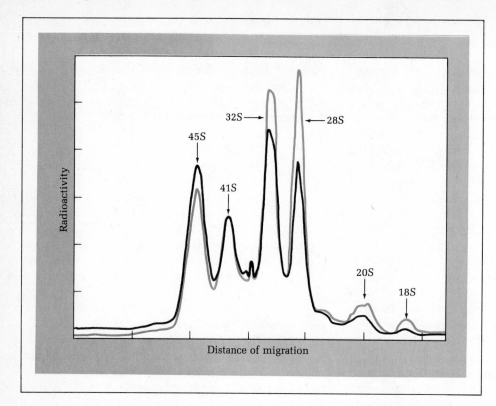

FIGURE 12-10 Polyacrylamide gel electrophoresis of radioactive RNA isolated from the nucleoli of human cells, showing the various intermediates involved in the synthesis of ribosomal RNA. The black line represents ^{3}H-uridine incorporated into the RNA chain, while the colored line indicates the labeling of methyl groups obtained when ^{14}C-S-adenosyl methionine is utilized as a precursor.

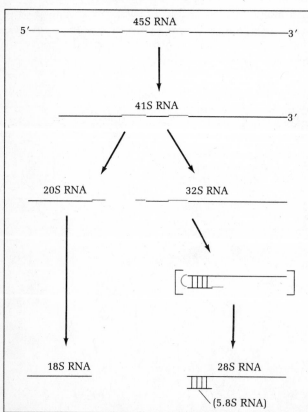

FIGURE 12-11 General pathway of ribosomal RNA processing in human cells. The colored segments represent regions discarded during the formation of the mature ribosomal RNAs. Although the sizes of the intermediates vary among different eukaryotic organisms, a similar kind of processing is involved.

erties of the final 28S RNA has revealed that a small piece of RNA, about 150 nucleotides in length, is released when the 28S RNA is exposed to agents that disrupt hydrogen bonds. This fragment, referred to as **5.8S RNA,** represents part of the original 45S precursor whose covalent attachment to the rest of the RNA chain is broken during RNA processing.

In addition to being cleaved and trimmed during ribosomal RNA processing, the primary RNA transcript is also enzymatically modified by the addition of methyl groups at specific sites in the RNA chain. The main site of methylation is the 2′-hydroxyl group of the sugar ribose, although some methylation of the bases also occurs. Since the amino acid derivative S-adenosylmethionine serves as the methyl group donor for biological methylation reactions, methylation of ribosomal RNA has been studied by incubating cells with radioactive S-adenosylmethionine. The pattern of RNA labeling obtained under these conditions resembles that observed when cells are labeled with ^{3}H-uridine: The 45S RNA becomes radioactive first, followed by the 41S, 32S, and 20S RNAs, and finally the 28S and 18S RNAs (see Figure 12-10). If transcription is blocked by adding actinomycin D, the labeling of 45S RNA with radioactive methyl groups ceases within a few minutes, suggesting that RNA methylation occurs simultaneous with or shortly after the synthesis of 45S RNA. Essentially all the radioactive methyl groups incorporated into 45S RNA eventually become associated with 28S and 18S RNAs, indicating that the methylated segments of the 45S RNA are selectively conserved during ribosomal RNA processing. Although the biological significance of this type of RNA methylation is not known with certainty, it has been suggested that it helps to guide RNA processing by protecting specific regions of the primary transcript from nuclease attack.

In addition to 28S, 18S, and 5.8S RNAs, the assembly of ribosomes also requires 5S RNA. Although eukaryotic ribosomes contain one molecule each of 28S, 18S, 5.8S, and 5S RNA, it appears that these RNAs are not all synthesized in equivalent amounts. The 28S, 18S, and 5.8S RNAs are automatically produced in equal quantities because they are derived by cleavage of the same 45S precursor. Molecules of 5S RNA, on the other hand, are transcribed from an independent set of genes that are localized outside the nucleolus and are present in larger numbers than the genes coding for the 45S ribosomal RNA precursor. The 5S genes are also transcribed at a rate that is severalfold greater than that of the

45S RNA genes. The excess 5S RNA is then degraded. The discovery of the 5S transcription factor (page 412) has given us a glimpse of transcriptional control of the 5S RNA genes, but the rationale for the synthesis of extra copies of 5S RNA remains a mystery.

Joining of RNA with Ribosomal Proteins

To be complete, any model of ribosome formation must address the question of when and how the newly forming ribosomal RNAs become associated with ribosomal proteins. One experimental approach that has provided important information regarding this issue involves the extraction of isolated nucleoli with solvents that permit RNA to be isolated along with any proteins that may be bound to it. The analysis of such extracts by moving-zone centrifugation has revealed the presence of ribo-

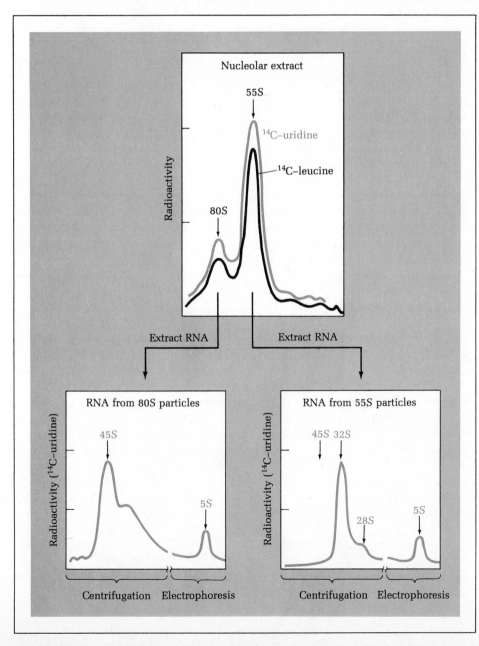

FIGURE 12-12 Moving-zone centrifugation of a nucleolar extract obtained from cells that had been incubated with radioactive leucine or uridine. The two major peaks, sedimenting at 80S and 55S, both contain a mixture of protein and RNA. The lower two graphs show the types of RNA that can be isolated from these two peaks by phenol extraction. For illustrative purposes the results obtained from centrifugational and electrophoretic analyses of these RNAs are combined.

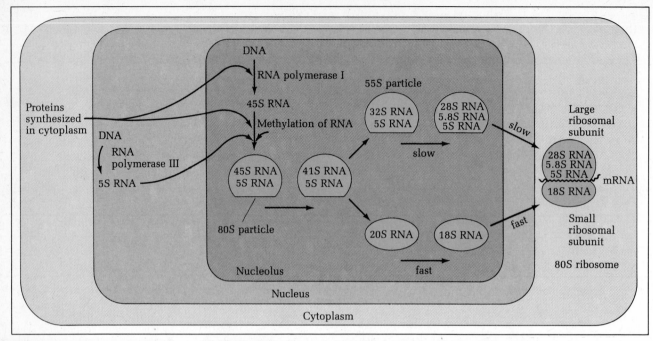

FIGURE 12-13 Schematic model summarizing the main steps in ribosome formation in human cells.

nucleoprotein particles sedimenting at 80S and 55S (Figure 12-12). When the RNA molecules present in these particles are isolated by phenol extraction and characterized by moving-zone centrifugation or polyacrylamide gel electrophoresis, 45S and 5S RNAs are found in the 80S particles and 32S and 5S RNAs are detected in the 55S particles. Since the 45S RNA molecule present in the 80S ribonucleoprotein particle is a precursor for RNAs present in the large subunit (28S and 5.8S RNAs) as well as the small subunit (18S RNA), it can be concluded that the 80S particle is a precursor of both the small and large ribosomal subunits. The 55S particle, on the other hand, must function as a precursor of the large ribosomal subunit because it contains the 32S RNA. The preceding conclusions have been supported by electrophoretic and peptide fingerprinting analyses of the protein constituents of the ribonucleoprotein particles; the proteins of the 80S particle are found to resemble those of both the large and small ribosomal subunits, while the proteins of the 55S subunit resemble those of the large subunit only.

On the basis of the above information and a variety of other experimental observations, the model of ribosome biogenesis summarized in Figure 12-13 has been formulated. As indicated in this model, the newly forming 45S transcript becomes associated with protein even before its transcription is complete; hence processing takes place on RNA molecules that are already complexed with protein. Two types of proteins are associated with the RNA during these processing steps; some are ribosomal proteins, while others function during ribosomal RNA processing and are released prior to

formation of the final ribosome. After processing has cleaved the 80S particle into precursor particles for the large and small subunits, the small-subunit precursor is quickly processed and transported to the cytoplasm, while the large-subunit precursor is processed more slowly in the nucleolus prior to export. This difference in transport rates explains why large ribosomal subunit precursor particles (55S particles) are more readily detected in nucleolar extracts than are small subunit precursor particles.

Visualization of Ribosome Formation

Microscopic observations have provided some significant insights into the relationship between nucleolar substructure and the events associated with ribosome formation. In cells that have been briefly exposed to ^{3}H-uridine to label the newly forming 45S RNA molecules, light microscopic autoradiography reveals the radioactivity to be localized predominantly over the nucleolar core, which is enriched in the fibrillar component of the nucleolus (see Figure 11-10). Such observations suggest that the fibrillar core is the region where the ribosomal RNA genes are transcribed into 45S RNA. If the fate of this newly formed 45S RNA is followed by briefly labeling cells with ^{3}H-uridine and then adding actinomycin D to block further RNA synthesis, the radioactivity is found to migrate to the granular cortex of the nucleolus as the 45S RNA is cleaved and processed (Figure 12-14). The overall picture that therefore emerges is that the 45S ribosomal precursor RNA is transcribed in the fibrillar core of the nucleolus, where

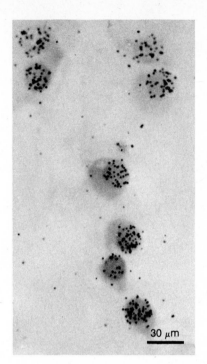
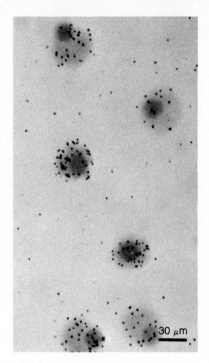

30 μm
30 μm

FIGURE 12-14 Light microscopic autoradiographs of nucleoli isolated from the marine worm *Urechis* and incubated with radioactive uridine to label the newly synthesized ribosomal RNA. *(Left)* After a short exposure to ³H-uridine, the radioactivity is seen to be localized predominantly within the nucleolar core. *(Right)* Nucleoli briefly labeled with ³H-uridine followed by several hours of incubation in the presence of actinomycin D to prevent further synthesis of radioactive RNA. During this subsequent incubation, it can be observed that the radioactivity has shifted toward the outer cortex region of the nucleolus. Courtesy of M. Alfert.

it becomes associated with protein prior to being cleaved and transported to the nucleolar cortex. According to this view, the granules that comprise the nucleolar cortex consist of partially processed ribosomal RNA with its associated proteins.

A detailed examination of the ultrastructure of the fibrils that make up the nucleolar core has been made possible by electron microscopic procedures pioneered in the laboratory of Oscar Miller. Miller discovered that exposure of isolated oocyte nuclei to distilled water causes the nuclei to rupture and the granular outer layers of the nucleoli to disperse, leaving the fibrous cores of the nucleoli readily visible. Staining of these isolated cores with heavy metals followed by electron microscopic examination reveals that they are comprised of a long **central fiber** from which clusters of shorter **lateral fibrils** project (Figure 12-15). In each cluster of lateral fibrils the length of the projecting fibrils increases progressively from one end of the cluster to the other, forming an "arrowhead" pattern. Between these clusters the central fibril is bare.

By treating such preparations with a variety of stains and enzymes, Miller was able to determine that the central fibril consists of a complex of DNA and protein, while the lateral fibrils represent a complex of RNA and protein. It was therefore proposed that the central fiber represents an extended chromatin fiber in a nonnucleosomal configuration, and that each region of the chromatin fiber containing a cluster of lateral fibers corresponds to a gene coding for 45S ribosomal RNA. According to this view, the arrowhead pattern of lateral fibrils observed in each such region represents

45S ribosomal RNA molecules in varying stages of transcription. The short fibrils located at one end of each cluster represent RNA molecules whose transcription has just been initiated, while the progressively longer fibrils observed as one proceeds toward the other end of the cluster represent RNA molecules whose transcription is further along. In support of these interpretations it has been calculated that the length of DNA required to code for one complete 45S RNA molecule is the same as the length of the central DNA-containing fiber within each cluster. The presence of protein in the lateral fibrils indicates that the 45S ribosomal RNA precursor becomes associated with protein even before its transcription is complete. The DNA regions lacking lateral fibers, which are located in between the repeated 45S genes, represent nontranscribed **spacer** DNA.

Ribosome Formation in Prokaryotic Cells

Although prokaryotic cells do not contain nucleoli specialized to carry out the task of ribosome formation, the way ribosomes are manufactured in prokaryotes exhibits a number of underlying similarities to the analogous process in eukaryotes. In both cases the genes coding for ribosomal RNA are present in multiple copies and are transcribed into precursor molecules destined for processing into mature ribosomal RNAs. The organization of the ribosomal genes and the processing of the resulting transcripts exhibit some significant differences in prokaryotes and eukaryotes, however. One notable difference involves the organization and extent of repetition of the genes coding for the ribosomal RNAs.

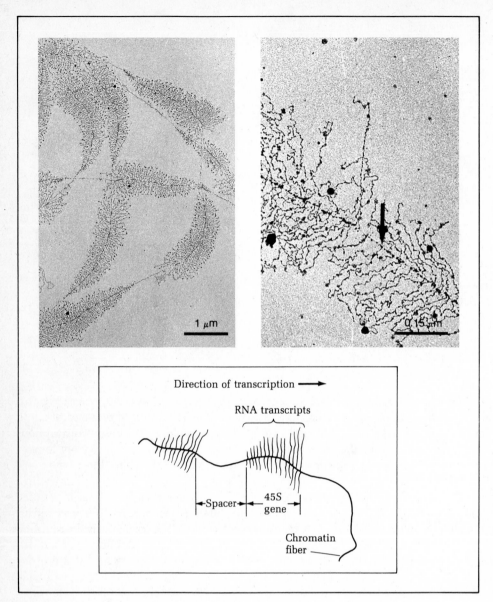

Direction of transcription ⟶

RNA transcripts

←Spacer→ | 45S gene

Chromatin fiber

1 µm

0.15 µm

FIGURE 12-15 Electron micrographs and accompanying diagram of a dispersed nucleolar core showing a series of newly forming ribosomal RNA transcripts. In the higher magnification view *(right)*, RNA polymerase molecules (arrow) can be identified. Micrographs courtesy of O. L. Miller, Jr.

In contrast to the extensive repetition of the ribosomal RNA genes observed in eukaryotes, the comparable genes of bacteria are typically present in about half a dozen copies. Moreover in eukaryotic cells the 5S ribosomal RNA genes are not associated with the genes coding for the 18S and 28S RNA, while in prokaryotes the genes coding for the three ribosomal RNAs (23S, 16S, and 5S) are closely linked to one another. Within these ribosomal RNA gene clusters, the order of the three genes is 16S–23S–5S.

Because these three ribosomal RNA genes lie adjacent to one another in prokaryotic DNA, the question arises as to whether they are transcribed into a single precursor RNA, which is then processed into the individual ribosomal RNAs. Support for this idea has been obtained from studies of mutant bacteria defective in the enzyme **ribonuclease III;** in such organisms, RNA molecules containing 16S, 23S, and 5S ribosomal RNA

sequences joined together in a single precursor can be detected. Since a precursor of this type is not observed in normal bacteria, it has been concluded that ribonuclease III normally functions to cleave the growing ribosomal RNA transcript while transcription is still occurring. Such an interpretation is supported by electron micrographs of prokaryotic ribosomal genes undergoing transcription, which reveal the presence of separate clusters of RNA transcripts emerging from the 16S and 23S RNA genes (Figure 12-16). This type of arrangement is exactly what would be expected if the completed transcript of the 16S gene is cleaved by the action of ribonuclease III and released from the DNA template before the RNA polymerase molecule proceeds from the 16S gene to the adjacent 23S gene.

Ribonuclease III is not the only enzyme that has been implicated in the processing of prokaryotic ribosomal RNAs. Bacterial cells deficient in other types of

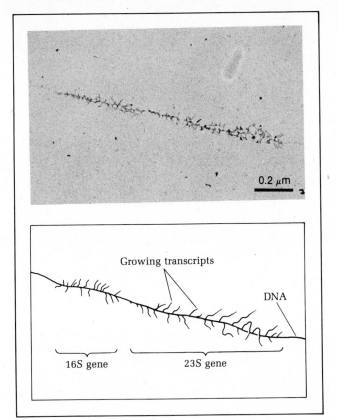

FIGURE 12-16 Electron micrograph and accompanying diagram of bacterial ribosomal RNA genes in the process of being transcribed. Separate clusters of transcripts emerging from the 16S and 23S RNA genes can be detected. Courtesy of O. L. Miller, Jr.

FIGURE 12-17 Ribosomal RNA processing in the bacterium *E. coli*. The small arrows represent sites at which the RNA chain is cleaved by various nucleases. Cleavage of the primary transcript begins before transcription of this set of genes is completed.

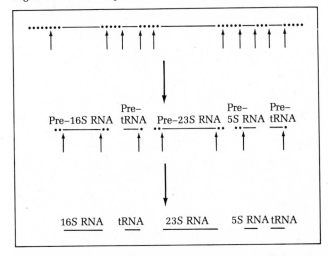

TABLE 12-1
Evolutionary changes in the sizes of the precursors for the two large ribosomal RNAs

	Sedimentation Coefficient of Precursor(s)	Approximate Percentage of Precursor Conserved During Processing
Bacterium (*E. coli*)	17S	90
	23S	90
Fish (trout)	38S	81
Amphibian (frog)	38S	79
Insect (*Drosophila*)	38S	72
Plant (tobacco)	38S	71
Bird (chicken)	45S	57
Mammal (human)	45S	50

esting discovery to emerge from such studies is that in addition to generating the three ribosomal RNAs, processing of the primary transcript of the ribosomal RNA genes yields several transfer RNA molecules, one derived from sequences located between the 16S and 23S genes, and one or more derived from sequences downstream from the 5S RNA gene.

In contrast to the situation in eukaryotes, bacterial ribosomal RNA precursors contain relatively few methyl groups compared to the number present in the final mature ribosomal RNAs. It is therefore believed that most of the ribosomal RNA methylation in prokaryotic cells occurs after ribosomal proteins have become associated with the ribosomal RNA precursors and processing and assembly of the ribosomal subunits has begun. Binding of ribosomal proteins to the ribosomal RNA precursors generates two classes of ribonucleoprotein particles, one containing the 23S and 5S precursor RNAs and functioning as an intermediate in the formation of the large ribosomal subunit, and the other containing the 16S precursor and serving as an intermediate in the formation of the small subunit. Only after becoming bound to these proteins are the precursor RNAs trimmed to form mature ribosomal RNAs.

A major difference exists between eukaryotic and prokaryotic cells in terms of the relative amounts of RNA that are degraded during processing. In prokaryotic cells roughly 10 percent of the total RNA is degraded during conversion of the precursors of the 16S and 23S RNAs to their final forms. In mammalian cells, on the other hand, about half of the 45S RNA molecule is degraded during processing. An insight into the evolutionary history of this difference can be obtained by comparing the sizes of the ribosomal RNA precursors in various organisms (Table 12-1). Such data suggest that two major events must have occurred during evolution. The first event involved a change from the production of separate precursors subject to minimal trimming, as occurs in prokaryotes, to the formation of a single large precursor of about 38S coding for the two larger ribosomal RNAs. This 38S precursor, which is trimmed

ribonuclease have been found to accumulate a variety of intermediates slightly larger in size than mature 16S, 23S, and 5S RNAs. On the basis of such observations, a model has been proposed that involves the participation of a combination of different nucleases in the processing of the three ribosomal RNAs (Figure 12-17). One inter-

about 20–30 percent during processing, appears to have remained relatively uniform through long periods of animal and plant evolution. At some point during the evolution of birds and mammals, however, an abrupt increase in the size of the precursor to 45S took place, with a resulting increase in the extent of trimming to about 50 percent. What is not clear, however, is why an increase in precursor size and therefore the amount of RNA degraded during processing has accompanied advancing evolution.

Regulation of Ribosome Formation

Since protein synthesis by living systems requires the participation of ribosomes, factors that influence the rate at which ribosomes are produced can exert a profound influence on a cell's metabolic state. Several distinct mechanisms exist for regulating the rate of ribosome formation. We have already seen, for example, that oocytes increase their capacity for manufacturing ribosomes by amplifying the number of genes coding for ribosomal RNA. Hormones that stimulate cell growth, on the other hand, stimulate ribosome production by increasing the rate at which ribosomal RNA genes are transcribed. Posttranscriptional regulation of ribosome formation appears to occur in lymphocytes, where more than 50 percent of the 45S RNA molecules produced by transcription of the ribosomal RNA genes are normally destroyed rather than being processed into 18S and 28S RNAs; when these cells are stimulated to grow and divide, however, a higher percentage of the 45S RNA molecules are converted to mature ribosomal RNAs.

In prokaryotic cells the rate of ribosome formation is regulated by the concentration of free ribosomes, that is, ribosomes that are not actively synthesizing protein. If an excessive number of free ribosomes accumulates within the cell, the transcription rate of the ribosomal RNA genes is decreased, limiting the quantity of new ribosomes that can be produced. Since ribosomes consist of protein molecules as well as ribosomal RNA, efficient control requires that the rate of ribosomal protein formation also be regulated under such conditions. This control is exerted in a rather unique fashion; in essence, the ribosomal proteins serve as feedback inhibitors of their own synthesis. The mechanism of this process, which involves binding of the ribosomal proteins to the messenger RNAs from which they were translated, will be discussed later in the chapter under the general topic of translational control mechanisms.

Although the overall scheme for controlling ribosome formation in prokaryotic cells is fairly well understood, some major issues remain unresolved. We have just seen, for example, that a key element in this scheme is the ability of an increase in free ribosomes to trigger a decrease in ribosomal RNA transcription. It is not known, however, whether this decrease in ribosomal RNA transcription is caused by the free ribosomes themselves or by intermediary molecules generated in response to the increase in free ribosomes. Some evidence for the existence of intermediaries has emerged from studies on cells grown in the presence of inadequate amounts of essential amino acids. Under such conditions the rate of protein synthesis is sharply curtailed, increasing the number of free ribosomes. As a result, ribosomal RNA formation slows down. In searching for the mechanism that triggers this decrease in ribosomal RNA synthesis, it has been discovered that cells starved for amino acids accumulate two unusual nucleotides, **guanosine tetraphosphate** (ppGpp) and **guanosine pentaphosphate** (pppGpp). The conclusion that these nucleotides mediate the inhibition of ribosomal RNA synthesis is supported by the discovery that mutant bacteria unable to regulate the rate of ribosomal RNA formation in response to amino acid starvation do not accumulate either ppGpp or pppGpp.

Although a tiny amount of ribosomal RNA transcription still occurs under conditions of amino acid starvation, the RNA produced accumulates in the form of precursor molecules rather than mature ribosomal RNAs. Thus in addition to inhibiting transcription of the ribosomal RNA genes, amino acid starvation leads to a block in ribosomal RNA processing. Although the exact mechanism underlying this block in processing is yet to be determined, the fact that protein synthesis slows down during amino acid starvation suggests that newly synthesized proteins must normally become associated with the precursor RNAs in order for successful processing to ensue. The inhibition of protein synthesis that occurs during amino acid starvation is also thought to hinder the formation of proteins that normally function to promote the transcription of ribosomal RNA. One protein that might be involved in this type of control is the so-called **psi factor,** a protein complex consisting of a combination of protein synthesis elongation factors Tu and Ts (page 472). When added to cell-free transcription systems, this protein complex preferentially stimulates the formation of ribosomal RNA. Hence the overall picture that emerges is that control of ribosome formation in prokaryotic cells is exerted at multiple levels and invokes a variety of molecular mechanisms.

THREE-DIMENSIONAL MODELS OF RIBOSOMAL SUBSTRUCTURE

To fully understand the role played by the ribosome in mediating the process of protein synthesis, it is necessary to know the location within the ribosome and the functional roles of the protein and RNA molecules that comprise this organelle. Though this goal is yet to be fully attained, many experimental approaches have been brought to bear on the problem of ribosomal substructure in recent years and a number of important findings have begun to emerge. These approaches and the types of information they yield are summarized below.

Ribosome Reconstitution

One way in which the molecular organization of the ribosome has been investigated is by reassembling ribosomes from mixtures of ribosomal proteins and RNA. This reconstitution approach was pioneered in the late 1960s in the laboratory of Masayasu Nomura, where it was first discovered that 16S RNA and the 21 isolated proteins of the small ribosomal subunit of *E. coli* will spontaneously reassemble into 30S ribosomal subunits when mixed together and briefly warmed. Not only do subunits reconstituted under these conditions resemble native subunits in morphology, but they exhibit normal function when mixed with large subunits and assayed for their ability to participate in protein synthesis.

By carrying out reconstitution experiments on mixtures of RNA and proteins added in differing sequences, it has been possible to investigate the roles played by individual components in ribosome assembly and function. It has been discovered in this way, for example, that an intact molecule of 16S RNA is required for the assembly of the 30S ribosomal subunit, and that the 18S ribosomal RNA of eukaryotic ribosomes will not substitute for this 16S RNA. Experiments in which selected ribosomal proteins are added in a defined sequence have revealed that assembly occurs in a stepwise fashion in which the initial binding of specific ribosomal proteins to the 16S ribosomal RNA induces conformational changes that then facilitate the binding of other proteins. During this assembly process particles of intermediate size can be detected that resemble ribosomal precursor particles present in living cells. For example, when cell-free reconstitution is carried out at low temperatures, a 21S particle accumulates that contains approximately two-thirds of the proteins normally found in the mature 30S subunit. In intact cells a 21S particle with a similar protein composition can be detected as an intermediate in the process of ribosome formation. When the 21S particles formed at low temperature during cell-free reconstitution are incubated at higher temperatures in the presence of the remaining ribosomal proteins, these particles undergo a conformational change that permits them to associate with the missing proteins and to form complete 30S subunits.

Compared to the 30S subunit, cell-free assembly of the 50S subunit is more difficult to achieve, and has only been accomplished by incubating mixtures of ribosomal RNA and proteins at relatively high temperatures (40°C to 50°C). Since proteins generally become denatured at such temperatures, successful reconstitution of the large subunit was not achieved until ribosomal proteins

and RNA were obtained from an unusual type of bacterium, *Bacillus stearothermophilus*, which is capable of living at high temperatures. The ribosomal proteins of this organism can therefore be heated during reconstitution without undergoing denaturation. Although successful reconstitution of the 50S subunit of this organism can be obtained by utilizing temperatures of 50°C, it should be realized that the same subunit is assembled within the intact organism at lower temperatures. Hence reconstitution under cell-free conditions cannot be an exact mimic of the assembly process occurring in intact cells.

In contrast to the behavior of bacterial ribosomes, neither the large nor the small subunit of eukaryotic ribosomes has been successfully reconstituted from purified proteins and RNA. This failure is no doubt related to the fact that the 18S and 28S RNAs present in mature eukaryotic ribosomes differ substantially in structure from the 45S RNA precursor involved in ribosome assembly under physiological conditions. Hence some of the RNA sequences crucial to the early stages of ribosome assembly must be removed from the 45S RNA during processing, making it impossible for the final 18S and 28S RNAs to substitute for 45S RNA in mediating ribosome assembly.

In addition to illuminating the process of ribosome assembly, cell-free reconstitution has been used to investigate the functions of individual ribosomal proteins. This objective is generally pursued by omitting or substituting specific proteins during the reconstitution process, and then assessing the effects of such alterations on the properties of the reconstituted ribosome. This approach has been utilized, for example, to study the ribosomal proteins involved in conferring resistance to antibiotics that inhibit protein synthesis. One such antibiotic, **streptomycin,** causes ribosomes to incorporate many incorrect amino acids when translating messenger RNAs into polypeptide chains. Mutations occur, however, that make ribosomes resistant to this effect of streptomycin. The possibility that the streptomycin-resistant state is caused by an alteration in a particular ribosomal protein has been investigated by the technique of ribosome reconstitution. Such experiments have revealed that when the ribosomal proteins of streptomycin-resistant bacteria are purified and substituted for normal proteins during ribosome reconstitution, inclusion of a single protein (S12) from the streptomycin-resistant ribosomes is sufficient for conferring the property of streptomycin resistance upon the reconstituted ribosomes.

Biochemical and Microscopic Probes of Ribosome Substructure

The technique of ribosome reconstitution has provided a number of important insights into the relationship between the various ribosomal constituents, but this approach is inadequate in itself for determining the three-dimensional organization of molecules within the ribosome. Since the large size and complex shape of the ribosome make it difficult to apply the technique of X-ray diffraction to this problem, indirect approaches for probing ribosomal substructure have had to be devised. Though these newly developed techniques have not yet provided us with a complete three-dimensional picture of ribosomal substructure, their use has generated a considerable amount of information concerning the localization of the various ribosomal constituents. Among the main experimental approaches in this category are: (1) chemical cross-linking, (2) enzymatic digestion, (3) radioactive labeling, (4) deuterium labeling, and (5) immunological labeling.

1. Chemical cross-linking. One approach for determining the location of individual proteins within the ribosome involves the use of reagents that chemically link neighboring proteins to one another. Reagents of this type are referred to as **bifunctional cross-linking agents** because they contain two reactive chemical groups, each capable of forming a covalent bond with an amino acid side chain. The distance between the two functional groups, which generally measures about one nanometer, defines the maximum distance between any two ribosomal proteins that can be cross-linked.

In a typical cross-linking experiment ribosomes are first incubated in the presence of a bifunctional reagent to link adjacent ribosomal proteins. The ribosomal proteins are then extracted and analyzed by polyacrylamide gel electrophoresis. Linked proteins appear as bands that migrate more slowly than normal because their molecular weights are higher than those of typical ribosomal proteins. If the material in these bands is isolated and treated with reagents that disrupt protein cross-linking, the involved proteins can be identified (Figure 12-18). Using this general approach, a variety of neighboring protein pairs have been identified in the ribosome.

2. Enzymatic digestion. A second approach for investigating the location of specific molecules within the ribosome involves enzymatic digestion with proteases and/or ribonuclease. Protease digestion is useful for identifying proteins that are located near the ribosome surface because such exposed proteins tend to be more susceptible to degradation. Ribonuclease digestion, on the other hand, has been employed to cleave ribosomal subunits into subparticles consisting of RNA and small numbers of bound proteins. Determining the protein composition of such subparticles has permitted the identification of groups of proteins that are located close to one another in the ribosome.

Ribonuclease digestion has also been useful in identifying those regions of the ribosomal RNA mole-

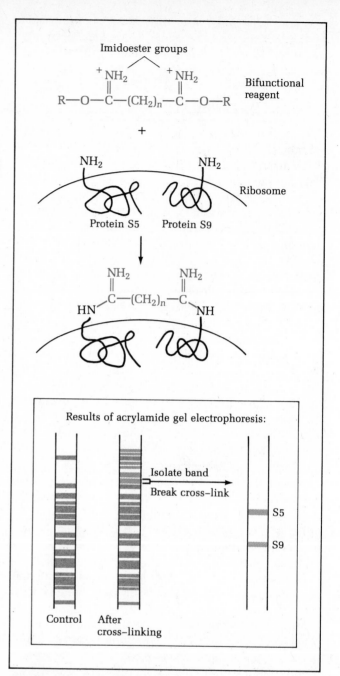

FIGURE 12-18 Diagram illustrating how bifunctional cross-linking reagents can be employed to detect neighboring ribosomal proteins.

cule to which proteins attach. One way of obtaining this type of information is to allow isolated ribosomal proteins to bind to purified ribosomal RNA, and to then digest the resulting ribonucleoprotein complex with ribonuclease. The portion of the ribosomal RNA molecule containing bound protein will be protected from digestion and can subsequently be isolated and sequenced. Alternatively the RNA can be first degraded to fragments with ribonuclease, and the individual RNA fragments then tested for their ability to bind ribosomal proteins. Data obtained from both these approaches indicate that in the small ribosomal subunit, the proteins that bind to RNA during the early stages of the assembly process tend to interact with the 5′ end of the 16S RNA molecule: Such observations suggest that within the intact cell, ribosomal proteins may begin binding to the 5′ end of the ribosomal RNA precursor before transcription at the 3′ end has been completed.

3. *Radioactive labeling.* Many radioactive reagents capable of forming covalent bonds with protein side chains have been utilized for labeling ribosomal proteins. Among the more commonly used reagents in this category are those that attach radioactive iodine to tyrosine residues (iodination), radioactive aldehydes to lysine residues (alkylation), and radioactive sulfhydryl-containing groups to cysteine residues (thiolation). Incubation of intact ribosomes with such reagents permits the identification of those proteins that are exposed on the ribosome surface. A more specific type of labeling can be obtained with **affinity labels,** which are analogs of substances that normally interact with the ribosome. Affinity labels differ from normal substrates, however, in that they contain highly reactive side chains that cause them to form covalent linkages with the sites to which they bind. Affinity labels have been designed to mimic several of the normal components involved in protein synthesis, including messenger RNA and transfer RNA. Ribosomal proteins that become radioactive after exposure to such affinity labels are presumed to be located at or near the active site with which the affinity label interacts.

4. *Deuterium labeling.* In an alternative type of labeling procedure, pioneered by Peter Moore and Donald Engelman, cells are grown in the presence of heavy water (D_2O) to introduce atoms of deuterium into ribosomal proteins. When ribosomes are reconstituted from a mixture of ribosomal proteins in which two of the proteins have been obtained from cells grown in D_2O and the remaining proteins are normal, the distance between the two deuterium-labeled proteins in the reconstituted particle can be investigated by placing it in a neutron beam and analyzing the resulting interference pattern. Such interference patterns are dominated by the diffraction of neutrons by deuterium, and thus provide important information regarding the distance between the deuterated proteins.

5. *Immunological labeling.* In the early 1970s, Georg Stöffler and his associates purified the various proteins present in the *E. coli* ribosome and injected them into rabbits to induce the formation of antibodies. These pioneering studies opened the way for the use of specific antibodies for mapping both the topographical and functional organization of the ribosome. One unique feature of this approach not shared by the techniques described above is that antibody molecules are large enough to be seen with the electron microscope. It is therefore possible to visualize the location of particu-

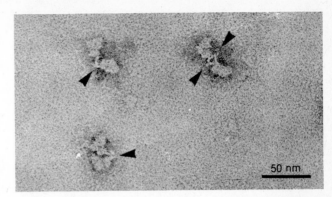

FIGURE 12-19 Electron micrograph of *E. coli* small ribosomal subunits following incubation with antibody that binds to a specific ribosomal protein (S14), thereby linking the subunits into pairs. The arrowheads point to the antibody molecules. The site where the antibody molecules join the subunits together corresponds to the location of protein S14. Courtesy of J. A. Lake.

lar proteins on the ribosome surface by noting where their respective antibodies bind. This approach is facilitated by the fact that each antibody molecule contains two binding sites for the antigen with which it interacts; hence antibodies to ribosomal proteins can link ribosomal subunits into pairs. The orientation of the subunits within such antibody-joined pairs provides crucial information regarding the location of the protein to which the antibody is binding (Figure 12-19).

Since proteins buried within the ribosome interior are not accessible for antibody binding, differences in the abilities of antibodies directed against various ribosomal proteins to bind to intact ribosomes indicates which proteins are most exposed at the ribosome surface. By combining this approach with ribosome reconstitution, antibodies can be employed to identify proteins that lie over one another. For example, it has been found that antibodies against the small subunit protein S4 only bind to reconstituted small subunits when proteins S5 and/or S12 have been deleted during the reconstitution process. It has therefore been concluded that proteins S5 and S12 normally cover protein S4.

General Conclusions Concerning Ribosome Substructure

Although the goal of obtaining a complete three-dimensional picture of ribosome substructure is yet to be achieved, data obtained from the approaches described in the previous sections have allowed some rather detailed models of the bacterial ribosome to be constructed (Figure 12-20). In such models the locations of certain proteins are relatively well-established because relevant data have been obtained from several independent sources. For example, the proposed existence of a cluster of proteins S4, S5, S8, and S12 on the side of the small subunit facing away from the large

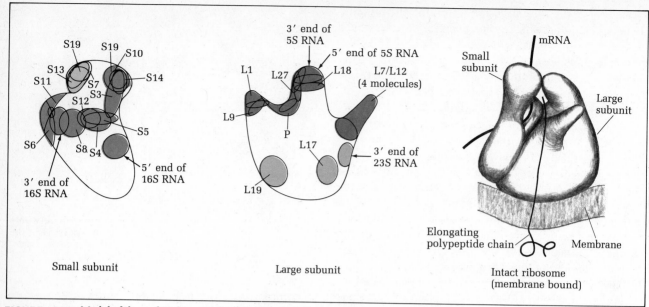

FIGURE 12-20 Model of the molecular organization of the *E. coli* ribosome, indicating the location of several kinds of ribosomal proteins and RNAs. Individual ribosomal proteins are indicated with the letter L or S (for large or small subunit) followed by a number. P designates the site where peptide bond formation occurs. The diagram on the right shows the relationship between the ribosomal subunits, messenger RNA, and the growing polypeptide chain on a membrane-bound ribosome.

subunit is compatible with data obtained from immune-labeling, deuterium-labeling, reconstitution, and cross-linking studies. In terms of the RNA constituents of the ribosome, the 5S RNA molecule has been localized to the central protuberance, but the arrangement of the 16S and 23S RNA molecules has been more difficult to ascertain because of their large sizes. The data obtained thus far indicate that these RNA molecules are partly exposed on the ribosome surface and partly buried within the interior. Among the regions located on the ribosome surface are the 3′ end of the 23S ribosomal RNA and the 3′ and 5′ ends of the 16S ribosomal RNA.

The most important generalization to emerge concerning the localization of specific ribosomal functions is that the small subunit is involved in the binding of messenger and transfer RNAs to the ribosome, while the large subunit catalyzes peptide bond formation. Current evidence also suggests that factors required for the initiation of protein synthesis bind near the cleft of the small subunit, GTP hydrolysis is associated with proteins L7/L12 situated in the stalk of the large subunit, and peptide bond formation takes place at the central protuberance of the large subunit.

THE RIBOSOME AND PROTEIN SYNTHESIS

It was pointed out in Chapter 10 that ribosomes utilize the genetic information encoded in the base sequence of messenger RNA molecules to direct the sequence of amino acids incorporated into polypeptide chains, and that transfer RNA molecules serve as the intermediaries

that recognize the triplet coding system inherent in messenger RNA and insert the appropriate amino acids into the growing polypeptide chain. Little was said, however, about the molecular mechanisms underlying this complex series of events. Now that the structural organization of the ribosome has been described, we can discuss the biochemistry of protein synthesis in more detail.

Ingredients Required for Protein Synthesis

Our current understanding of the biochemistry of protein synthesis is derived largely from studies on cell-free systems, whose development has greatly facilitated the identification of the large number of components involved. The first studies on the cell-free synthesis of proteins were initiated shortly after World War II, when biological compounds labeled with radioisotopes first became generally available to the scientific community. The availability of radioactive isotopes for experimental use was a crucial development because only with radioactive amino acids is it possible to detect the tiny amount of protein synthesis that occurs under cell-free conditions.

In the early 1950s Paul Zamecnik and his collaborators established the basic conditions required for the incorporation of radioactive amino acids into polypeptide chains by cell-free systems. One of the more extensive studies carried out in Zamecnik's laboratory was undertaken by Philip Siekevitz, who showed that radioactive amino acids are incorporated into newly synthesized proteins in homogenates of liver tissue. When homogenate preparations that had been briefly incubated with radioactive amino acids were subjected to subcellular fractionation, the newly synthesized radioactive protein was localized predominantly in the microsomal fraction, which consists of a mixture of endoplasmic reticulum and ribosomes. Although this finding implicated the microsomes as the site of protein synthesis, isolated microsomes by themselves were found to synthesize protein very slowly. Faster rates of protein synthesis could be obtained, however, by adding the mitochondrial and supernatant (cell sap) fractions back to isolated microsomes. It was later discovered that ATP and GTP can be substituted for the mitochondrial fraction, indicating that the mitochondria simply function as a source of energy. The role played by the cell sap was clarified when the presence of transfer RNAs and aminoacyl-tRNA synthetases in this fraction was discovered a few years later.

After the conditions required for protein synthesis by isolated microsomes had been worked out, investigations were carried out to determine whether the ribosomes or the membranes of the microsomal fraction are the actual site of protein synthesis. This issue was studied by solubilizing the microsomal membranes with detergent, and then testing the ability of the resulting membrane-free ribosomes to carry out protein synthesis. Ribosomes prepared in this way were found to retain their capacity to synthesize proteins, indicating that the ribosome and not the membrane is the essential component. It has therefore been concluded that in spite of the attachment of many ribosomes to membranes (mainly the endoplasmic reticulum in eukaryotic cells and the plasma membrane in prokaryotic cells), membrane attachment is not an absolute requirement. As indicated in Chapter 6, the main function of the membranes associated with cytoplasmic ribosomes is in the intracellular transport and packaging of newly synthesized protein molecules, especially those proteins destined for secretion or insertion into membranes.

The Direction of Messenger RNA Translation

The synthesis of polypeptide chains is an ordered, stepwise process that begins at the amino-terminal end of the polypeptide chain and involves the sequential addition of individual amino acids to the growing chain until the carboxyl end of the molecule is reached. The first experimental evidence supporting this view was provided in 1961 by Howard Dintzis, who investigated the pattern of incorporation of radioactivity into hemoglobin chains in immature red blood cells that had been incubated briefly with radioactive amino acids. He reasoned that if the time of incubation with radioisotope were relatively short, then the radioactivity present in completed hemoglobin chains should be most highly concentrated at one end of the molecule, that is, the end

representing the amino acids most recently incorporated prior to completion of the chain. The amino acids present elsewhere in the chain would be expected to be nonradioactive because they would have been incorporated prior to the addition of radioisotope to the cells. This idea was tested by isolating hemoglobin from cells that had been briefly incubated with radioactive leucine, digesting the hemoglobin into peptide fragments with the protease trypsin, and then analyzing the individual peptides for radioactivity. The results of these experiments indicated that the highest concentration of radioactivity is present in the carboxyl-terminal end of the hemoglobin chains, and that a gradually decreasing gradient of radioactivity occurs as one progresses toward the amino-terminal end (Figure 12-21). Dintzis therefore concluded that during protein synthesis, amino acids are sequentially added to the growing polypeptide chain in the amino to carboxyl direction.

The first experiments indicating the direction in which messenger RNA is read during this process involved the use of relatively simple synthetic RNAs. One example is the synthetic RNA polymer that can be made by adding a cytosine residue to the 3′-terminal end of poly(A):

$$(5') \ A \ A \ A \ A \ A \ A \ A \ A \ A \ A \ldots A \ A \ C \ (3')$$

When this particular RNA is added to a cell-free protein-synthesizing system, a polypeptide consisting of a long chain of lysine residues ending with an asparagine at the carboxyl-terminal end is synthesized. Since AAA codes for lysine and AAC codes for asparagine, it follows that the translation of messenger RNA proceeds in the 5′ → 3′ direction. The conclusion that translation proceeds in the 5′ → 3′ direction has subsequently been verified by direct comparison of the base sequences of natural messenger RNAs and the amino acid sequences of the polypeptide chains they encode.

The Role of Ribosome Dissociation

Early workers in the field of protein synthesis believed that the translation of messenger RNA into polypeptide chains occurs entirely on intact ribosomes. However, the presence of two, easily dissociable subunits within the ribosome suggests the possibility that these subunits might dissociate and reassociate as part of the process of protein synthesis. The first evidence for the existence of such a dissociation-reassociation cycle was provided in 1968 by Raymond Kaempfer, who utilized heavy and light isotopes of nitrogen (^{15}N and ^{14}N) to monitor the behavior of the small and large subunits of bacterial ribosomes during the process of protein synthesis. In these experiments ribosomes isolated from cells that had been grown in the presence of both a radioisotopic precursor and heavy nitrogen were mixed with an excess of nonradioactive ribosomes containing light nitro-

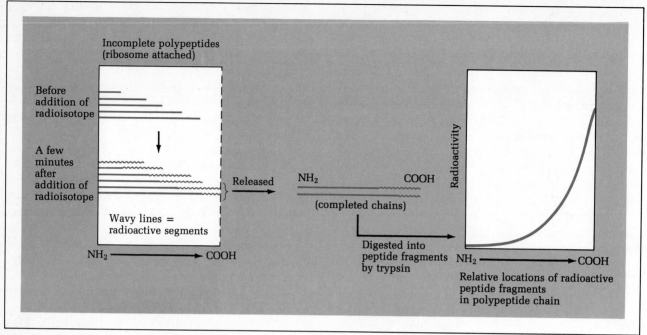

FIGURE 12-21 Diagrammatic summary of the experiment demonstrating that polypeptide chains are synthesized in the amino-terminal to carboxyl-terminal direction. After exposing reticulocytes to radioactive leucine for 4 min, completed chains of radioactive hemoglobin were isolated and digested into peptide fragments with trypsin. The highest concentration of radioactivity was found in those peptide fragments located near the carboxyl-terminal end of hemoglobin (see graph on right).

gen. When this mixture of ribosomes was incubated under conditions that allow protein synthesis to occur, the radioactive heavy ribosomes were found to be replaced by radioactive ribosomes intermediate in density between heavy and light ribosomes (Figure 12-22). Such findings indicate that the subunits of the heavy ribosomes have first dissociated from one another, and then reassociated with subunits derived from the nonradioactive light ribosomes. This behavior does not occur if an inhibitor of protein synthesis is added to the ribosome mixture prior to incubation, indicating that the dissociation and reassociation of ribosomal subunits is occurring as part of the process of protein synthesis.

The above experiments provided one of the first clear indications that ribosomal subunits come apart and reassociate during the protein synthetic cycle. Many subsequent biochemical investigations have addressed the problem of defining the individual steps associated with this overall cycle. From these various studies it is now clear that protein synthesis involves three distinct phases: (1) an **initiation** phase in which messenger RNA is bound to the ribosome and positioned for proper translation, (2) an **elongation** phase in which amino acids are sequentially joined to one another in an order specified by the arrangement of codons in the messenger RNA bound to the ribosome, and (3) a **termination** phase in which the messenger RNA and newly formed polypeptide chain are released from the ribosome.

Initiation of Protein Synthesis

The first step in protein synthesis involves the binding of messenger RNA to the ribosome. This event is of special importance because it is usually the rate-limiting step in protein synthesis, and therefore determines whether or not a given messenger RNA will be translated into a protein product. Experiments involving dissociated ribosomal subunits have revealed that messenger RNA binds to the dissociated small subunit rather than to intact ribosomes, and only afterwards does the large subunit become associated with the complex. In addition to messenger RNA and ribosomal subunits, this interaction requires GTP, several ribosome-binding proteins known as **initiation factors,** and a special type of transfer RNA called an **initiator tRNA.** The sequence in which these components interact is outlined in Figure 12-23, and is discussed in more detail in the following sections.

Binding of Messenger RNA to the Small Subunit Our current understanding of the mechanisms underlying the interaction of messenger RNA with the ribosome owes a great debt to the development of techniques for isolating and characterizing individual messenger RNAs. The purification of messenger RNAs from eukaryotic cells has been facilitated by the presence of the poly(A) tail at the 3' end of such messages. Since this stretch of adenine residues can form complementary

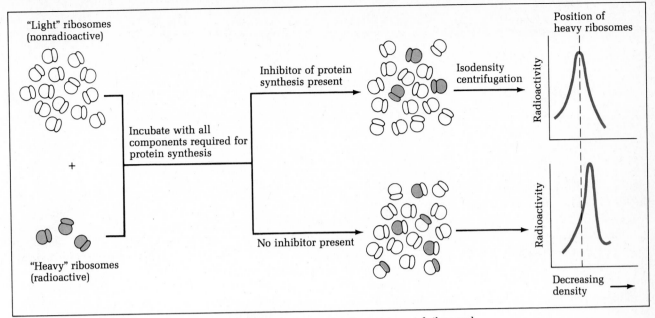

FIGURE 12-22 Diagrammatic summary of experiment demonstrating the existence of ribosomal subunit dissociation/reassociation during protein synthesis.

hydrogen bonds with stretches of uracil (U) or thymine (T), purification of eukaryotic messenger RNAs has been routinely carried out by column chromatography on matrices containing synthetic poly(U) or oligo(dT). Poly(A)-containing RNAs bind to such columns while other RNAs do not, resulting in a considerable purification of messenger RNA. The subsequent isolation of individual messages from such partially purified preparations is accomplished by size fractionation using standard centrifugation and electrophoresis techniques. Since typical cells contain thousands of different messenger RNAs of overlapping sizes, purification of individual messenger RNAs requires careful selection of a cell type or experimental conditions under which a particular type of messenger RNA predominates.

Careful size measurements have revealed that purified messenger RNA molecules are always larger than would be expected on the basis of the polypeptide chains whose amino acid sequences they encode. In prokaryotic messages, where more than one protein-coding sequence is often present, these additional non-coding sequences are present at the 5' and 3' ends as well as between the individual coding sequences. Eukaryotic messenger RNA molecules, which contain only a single protein-coding sequence, are generally characterized by the presence of several dozen non-translated bases at the 5' end and from several dozen to

FIGURE 12-23 The main steps involved in the initiation of protein synthesis in prokaryotic cells. The minor differences that characterize initiation in eukaryotic cells are discussed in the text.

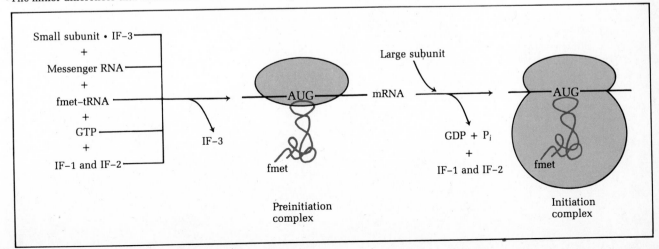

TABLE 12-2
Length of the coding and untranslated regions of several eukaryotic messenger RNAs

	Length in Bases of			
	Complete mRNA	Coding Region	5'-Untranslated Region	3'-Untranslated Region
α-Globin (rabbit)	551	429	36	86
α-Globin (human)	575	429	37	109
β-Globin (rabbit)	589	444	53	92
β-Globin (human)	628	446	50	132
Ovalbumin (chick)	1859	1164	64	631
Insulin (rat)	443	333	57	53

several hundred untranslated bases situated between the termination signal for translation and the beginning of the poly(A) sequence (Table 12-2).

The untranslated bases located at the 5' end of messenger RNAs are thought to play a role in the initiation of protein synthesis. This idea was originally proposed in 1974 by J. Shine and L. Dalgarno, whose sequencing studies on the 3'-terminal end of 16S ribosomal RNA revealed the presence of a short pyrimidine-rich sequence (-CCUCCUU-) that is complementary to an -AG-rich segment located 5–10 bases upstream from the initiation codon of many bacterial messenger RNAs. The existence of this complementary relationship between messenger and ribosomal RNA raised the possibility that the initial binding of messenger RNA to the ribosome is mediated by base pairing between a short stretch of 5' untranslated sequences in messenger RNA and the pyrimidine-rich **Shine–Dalgarno sequence** located at the 3' end of 16S ribosomal RNA (Figure 12-24). More direct support for this notion emerged from cell-free protein synthesis experiments in which the addition of short synthetic RNAs, complementary to the Shine–Dalgarno sequence, were found to inhibit protein synthesis, presumably because they bind to the Shine–Dalgarno sequence and thereby make it unavailable for binding to the 5' untranslated region of messenger RNA. Additional evidence for the physiological importance of the 3' end of the 16S ribosomal RNA has been derived from studies on the mechanism of action

of **colicin E3,** a toxic protein produced by certain strains of *E. coli.* Upon entrance into the cytoplasm of susceptible strains of bacteria, colicin E3 catalyzes the removal of a 49-base fragment from the 3' end of 16S ribosomal RNA. Ribosomes altered in this way are unable to initiate protein synthesis.

The 18S ribosomal RNA of eukaryotic cells also contains a sequence near its 3' end that has been implicated in the binding of messenger RNA. In contrast to the Shine–Dalgarno sequence, this eukaryotic sequence is enriched in the bases A and G. Though many eukaryotic messenger RNAs have been found to exhibit 5' untranslated sequences that are to some extent complementary to this 3' sequence present in 18S ribosomal RNA, the exact location, length, and base composition of these 5' untranslated sequences differ significantly among messages, and some viral messenger RNAs lack them entirely. Thus if complementary base pairing between messenger RNA and 18S ribosomal RNA is involved in the initiation of translation in eukaryotic cells, it does not appear to be an absolute requirement.

An alternative structural element implicated in the binding of eukaryotic messenger RNA molecules to the small ribosomal subunit is the methylated "cap" present at the 5' end of eukaryotic messages. Two kinds of experiments suggest that this cap structure is required for efficient translation. One is that messenger RNAs whose caps have been removed are poorly bound to the small ribosomal subunit and are therefore poorly

FIGURE 12-24 Diagram illustrating how complementary base-pairing between the 3'-end of 16S ribosomal RNA and the 5'-untranslated end of messenger RNA mediates the binding of messenger RNA to the ribosome in prokaryotic cells. The exact sequence present in the messenger RNA varies somewhat among messages. This particular example illustrates the 5'-end of the messenger RNA transcribed from the z-gene of the *lac* operon.

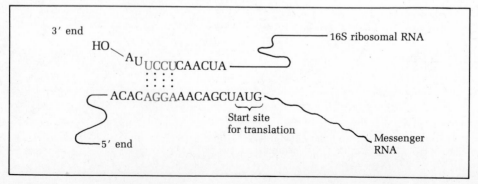

translated in cell-free protein-synthesizing systems. The other is that addition of free 7-methyl-GMP to such systems has been found to inhibit the translation of capped messages, presumably because the 7-methyl-GMP molecule mimics the cap structure and binds to a site on the small ribosomal subunit that normally interacts with the capped end of the messenger RNA. It has therefore been concluded that during normal protein synthesis, the capped end of the messenger RNA first binds to the small ribosomal subunit; the messenger RNA molecule then moves over the surface of the ribosomal subunit until the first AUG codon is encountered. At this point translation of the coding sequence begins. The notion that translation is simply initiated at the first AUG codon is supported by the discovery that with few exceptions, no extra AUG triplets can be detected between the 5′ capped end of eukaryotic messenger RNAs and the AUG triplet at which translation begins.

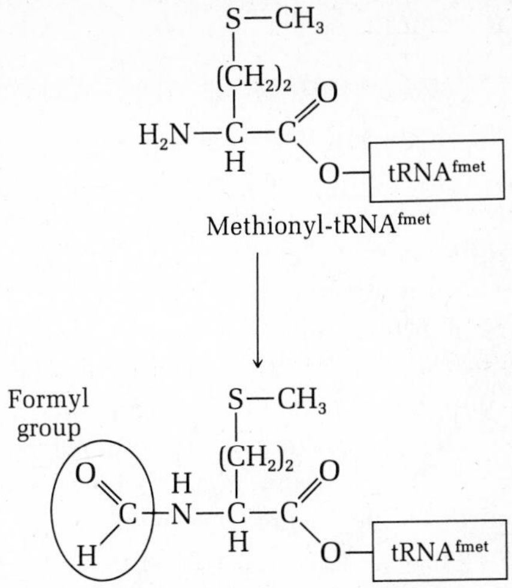

Initiator Transfer RNAs In addition to the obvious need for messenger RNA, the initiation of protein synthesis also requires an appropriate transfer RNA. The general role of transfer RNA molecules in protein synthesis, as well as the reaction that catalyzes the attachment of amino acids to transfer RNAs, has already been discussed in Chapter 10. Like messenger and ribosomal RNAs, transfer RNA molecules are initially transcribed from their corresponding genes in the form of precursor RNAs that must be cleaved and trimmed to generate the mature transfer RNAs.

The first clue that a special kind of transfer RNA might be associated with the initiation of protein synthesis surfaced in the early 1960s when it was discovered that roughly half the proteins in the bacterium *E. coli* contain methionine at their amino-terminal end. This observation was surprising because methionine is a relatively rare amino acid, accounting for no more than a few percent of the total amino acids present in bacterial proteins. The importance of this discovery became apparent in 1964, when K. Marcker and F. Sanger succeeded in isolating two different methionine-specific transfer RNAs from *E. coli*. One, designated *tRNA^{met}*, carries a normal methionine destined for insertion into the internal regions of polypeptide chains. The other, *tRNA^{fmet}*, carries a methionine that is converted to formylmethionine after it has become bound to the transfer RNA (Figure 12-25). It is evident from this figure that the amino group of methionine is blocked by the addition of the formyl group and is therefore not able to form a peptide bond with another amino acid. Hence formylmethionine can only be situated at the amino-terminal end of a polypeptide chain. This realization led to the proposal that tRNA^{fmet} is involved in the initiation of protein synthesis, an idea soon verified by the discovery that polypeptide chains in the process of being synthesized in bacterial cells contain formyl-

FIGURE 12-25 Reaction by which methionyl-tRNA^{fmet} is converted to formylmethionyl-tRNA^{fmet} by the addition of a formyl group (circled).

methionine at the amino-terminal position. Following completion of the polypeptide chain (and in some cases while it is still in the process of being synthesized), the formyl group, and often the methionine itself, is enzymatically removed.

The above observations led to the conclusion that bacterial cells contain two forms of methionyl tRNA that play distinctly different roles. Transfer RNA^{met} recognizes AUG codons that call for methionine to be inserted into internal positions in the polypeptide chain, while tRNA^{fmet} functions as an **initiator tRNA** that binds to AUG (and to a much lesser extent, GUG) codons located at the start site for translation. Eukaryotic cells, like prokaryotes, also contain two forms of methionyl tRNA, one of which is specialized for initiation. In this case, however, formylation does not come into play; both forms of eukaryotic methionyl tRNA carry unmodified methionine.

The existence of two types of transfer RNA recognizing the same codon raises the question of how the correct transfer RNA is selected by initiating and non-initiating AUG sequences. The answer to this question is that the initiation of protein synthesis involves conditions that are quite different from those prevailing during the elongation phase. Initiation occurs on the small subunit and involves the participation of several initiation factors; the elongation phase, in contrast, occurs on intact ribosomes in the absence of initiation factors. The notion that initiation factors play a significant role in selecting tRNA^{fmet} during the initiation of protein synthesis is supported by the discovery that purified initiation factor IF-2 selectively binds to formylmethionyl-tRNA^{fmet} under cell-free conditions. Moreover in eukaryotic protein-synthesizing systems, a complex between initiation factors and the initiator transfer RNA has been shown to attach to the small ribosomal subunit *before* the binding of messenger RNA.

Comparing Initiation in Prokaryotes and Eukaryotes

Two significant differences between the initiation of protein synthesis in prokaryotic and eukaryotic cells have already been mentioned: The role played by the cap structure in messenger RNA binding in eukaryotic but not prokaryotic cells, and the formylation of the methionine associated with the initiator tRNA of prokaryotes but not eukaryotes. Another notable difference is that three well-defined initiation factors (IF-1, IF-2, and IF-3) have been identified in prokaryotes, while a more complex series of initiation factors appears to be involved in eukaryotes. In addition the initiation phase of protein synthesis in eukaryotes has been found to require the hydrolysis of ATP as well as GTP, while in prokaryotes only GTP hydrolysis is involved. Finally, in eukaryotic cells the initiator transfer RNA binds to the small ribosomal subunit before it binds to messenger RNA, while this is not a requirement in prokaryotes.

More is known about the relationship of specific ribosomal constituents to the initiation process in prokaryotic cells than in eukaryotes. Experiments in which small ribosomal subunits and initiation factors were incubated in the presence of chemical cross-linking reagents have led to the identification of a small group of proteins (S11, S12, S13, and S19) thought to be closely associated with the binding of initiation factors to the ribosome. These proteins are situated at the head and platform end of the small subunit. A similar approach has revealed that the 3'-terminal end of the 16S RNA is closely associated with some of these same proteins, and with the initiation factors as well. In addition, proteins S12 and S21 have both been shown to be necessary for the binding of formylmethionyl tRNA^{fmet} to the initiator codon. Although the level of resolution of such studies is still relatively crude, a general picture is beginning to emerge in which the 3' end of the ribosomal RNA, the initiation factors, the initiator tRNA, and the 5' end of the messenger RNA are closely associated with one another and with a small group of proteins localized in the head and platform region of the small subunit.

In spite of the differences that have been noted about the details of the initiation process in prokaryotic and eukaryotic cells, the net results are similar. During the first phase of initiation, messenger RNA, the initiator tRNA (containing either methionine or formylmethionine), GTP, and a series of initiation factors become associated with the ribosomal small subunit. Once this **preinitiation complex** has been formed, the large subunit joins the small subunit in a reaction accompanied by the hydrolysis of GTP and the release of the initiation factors. The resulting **initiation complex,** consisting of messenger RNA and an aminoacyl-initiator tRNA bound to an intact ribosome, is now ready for the elongation phase of protein synthesis. In this complex the

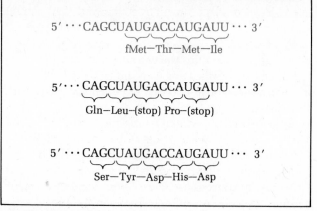

FIGURE 12-26 Diagram illustrating how the initiation codon (AUG) sets the proper reading frame for messenger RNA translation, using the gene coding for β-galactosidase as an example. If the initiation codon were not recognized, the two incorrect reading frames shown in the lower portion of the diagram might be selected instead of the proper one.

initiator transfer RNA is positioned with its anticodon base paired to the initiator codon (AUG or GUG) of the messenger RNA. Hence the process of initiation ensures that the message will be translated in the proper reading frame. Without such a mechanism an individual messenger RNA might be translated into three different polypeptides, each encoded by a one-base shift of the reading frame (Figure 12-26). Such improper initiation can be induced experimentally by certain conditions, such as elevated Mg^{2+} concentrations. This effect of Mg^{2+} is occasionally exploited for a deliberate purpose, such as in forcing ribosomes to initiate the translation of artificial messenger RNAs, like poly(U), which would not otherwise be translated because they lack a proper initiation codon.

Elongation Phase of Protein Synthesis

Once an initiation complex has been formed, the newly initiated polypeptide chain is elongated by joining together amino acids one at a time in an order specified by the codons in the messenger RNA. This elongation process is mediated by a cyclic series of reactions involving the participation of aminoacyl-tRNAs, several proteins known as **elongation factors,** and **GTP.** Each iteration of the cycle can be subdivided into three major steps: (1) the binding of an aminoacyl-tRNA to its appropriate codon in messenger RNA, (2) the formation of a peptide bond between the amino acid carried by this tRNA and the previously incorporated amino acid, and (3) translocation (movement) of the messenger RNA along the ribosome so that the next codon is brought into position for translation.

1. *Binding of aminoacyl-tRNA to the ribosome.* The first elongation cycle begins with the binding to the ribosome of the aminoacyl-tRNA whose anticodon is complementary to the codon situated immediately downstream from the initiation codon in messenger RNA. The ribosomal site to which this incoming aminoacyl-tRNA binds is referred to as the **aminoacyl or A site,** while the previously bound initiator tRNA (formylmethionyl-tRNA or methionyl-tRNA) is said to occupy the **peptidyl or P site.** The latter site is referred to as the peptidyl site because at the comparable stage of subsequent elongation cycles, it will contain the partially synthesized, or *nascent,* polypeptide chain.

The main evidence for the existence of separate A and P sites on the ribosome has come from experiments involving the antibiotic **puromycin,** an inhibitor of protein synthesis whose structure mimics a portion of an aminoacyl-tRNA molecule. Because of this resemblance, puromycin can bind to ribosomes in place of an aminoacyl-tRNA and become incorporated into the growing polypeptide chain. Once puromycin is incorporated, however, it cannot participate in any subsequent steps; polypeptide synthesis therefore ceases and the incomplete polypeptide containing puromycin joined to its carboxyl-terminal amino acid is released from the ribosome. The idea that ribosomes contain two tRNA binding sites is supported by the observation that incubating ribosomes containing bound formylmethionyl-tRNA^{fmet} in the presence of puromycin leads to the formation and release of formylmethionyl-puromycin; if ribosomes containing formylmethionyl-tRNA^{fmet} are first reacted with the appropriate aminoacyl-tRNA, however, a dipeptide is formed that will not react with puromycin (Figure 12-27). Hence this dipeptide must be

FIGURE 12-27 Mode of action of puromycin. *(a)* Puromycin resembles an aminoacyl-tRNA in structure, and therefore is capable of reacting with peptidyl-tRNA. The resulting peptidyl-puromycin cannot undergo further elongation, and therefore falls off the ribosome. *(b)* Puromycin will react with ribosome-bound initiator tRNA but not with other aminoacyl-tRNAs. This observation led to the proposed existence of separate A- and P-sites on the ribosome.

Peptidyl–tRNA

Puromycin
(note similarity to
aminoacyl–tRNA)

Aminoacyl–tRNA

(a)

Reacts ← Puromycin → No reaction

fmet

P   A

Initiator–tRNA
bound to ribosome

P   A

Noninitiating
aminoacyl–tRNA
bound to ribosome

(b)

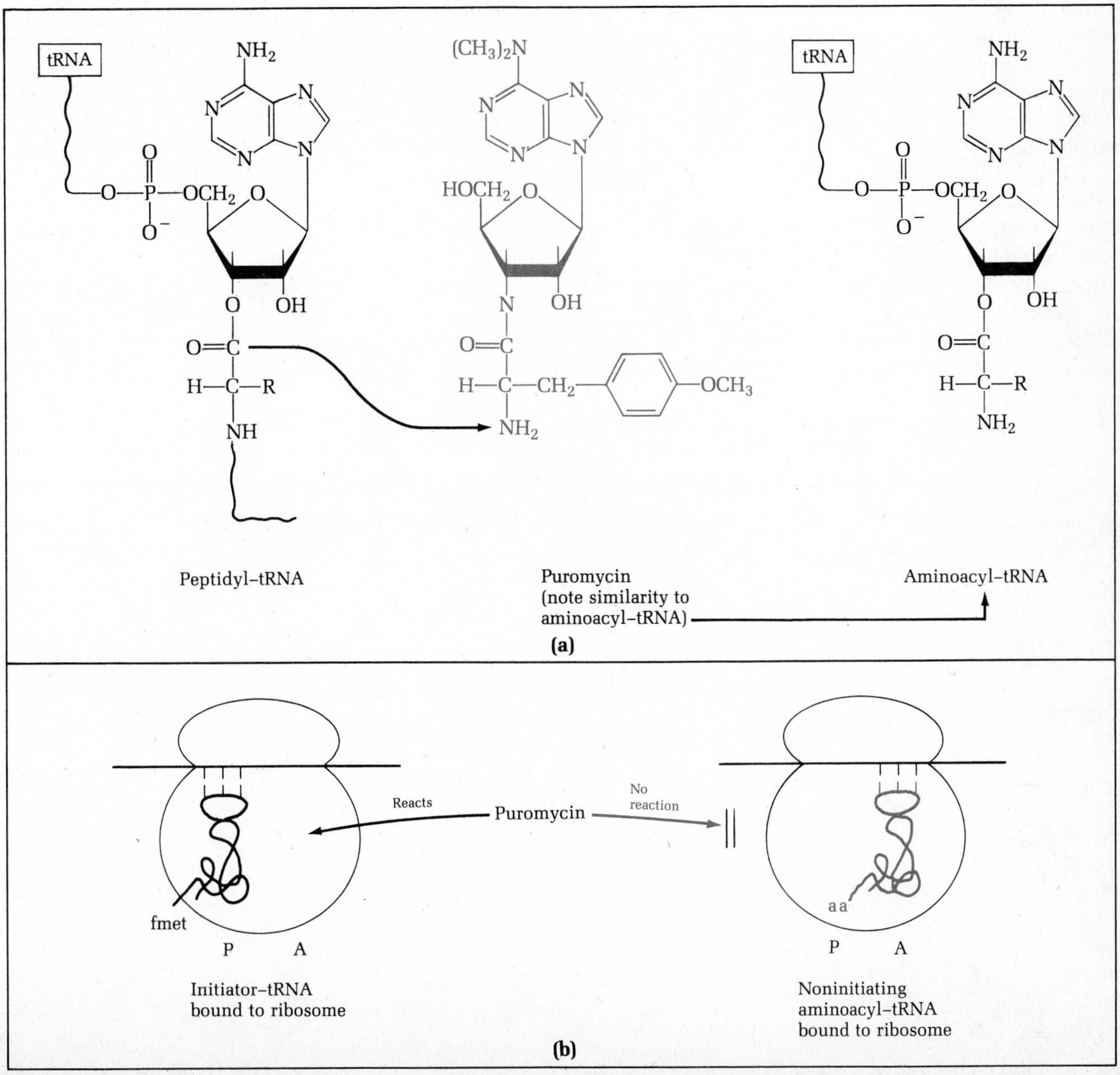

attached to a tRNA that is bound to a different ribosome binding site (the A site) than was the original formylmethionyl-tRNA (the P site). Only after movement of this dipeptidyl-tRNA back to the P site, a process that requires GTP and a specific elongation factor, can puromycin again bind to the A site and become incorporated into the nascent polypeptide chain.

Although the precise physical locations of the P and A sites on the ribosome are not known, the small and large subunits both appear to be involved. The small subunit has been implicated by the discovery that initiator tRNA molecules will bind to the small subunit in the absence of the large subunit; an involvement of the large subunit has been demonstrated by affinity labeling experiments in which the radioactive analog bromoacetylmethionyl-tRNA[fmet] was reacted with ribosomes in place of formylmethionyl-tRNA[fmet]. The bromoacetyl analog group reacts with the free amino groups of ribosomal proteins, labeling those ribosomal proteins that are in close proximity to the tRNA binding site. Under such conditions ribosomal proteins L2 and L27 become radioactive, suggesting that these particular large subunit proteins are situated in close proximity to the aminoacyl end of transfer RNA molecules bound to the ribosome.

The binding of incoming aminoacyl-tRNA molecules to the A site of prokaryotic ribosomes requires both GTP and elongation factor **EF-Tu,** a protein that promotes the ribosomal binding of all aminoacyl-tRNAs other than the initiator tRNA. During the binding of an aminoacyl-tRNA to the ribosome, GTP is hydrolyzed to GDP and EF-Tu is released bound to this GDP. A second elongation factor, **EF-Ts,** promotes conversion of the re-

sulting EF-Tu/GDP complex back to EF-Tu-GTP (Figure 12-28). In eukaryotic cells a similar cycle occurs, although a single elongation factor, **EF-1,** appears to carry out the functions of bacterial factors EF-Tu and EF-Ts.

The elongation factor–GTP complex responsible for promoting the binding of aminoacyl-tRNAs to the ribosome does not recognize specific anticodons, so aminoacyl-tRNAs of all varieties are indiscriminately brought to the vicinity of the A site by this reaction. Some mechanism must therefore exist for ensuring that only the correct aminoacyl-tRNA is retained at the A site for participation in the subsequent peptide bond forming step. If there is a poor match between the anticodon of the incoming aminoacyl-tRNA and the messenger RNA codon exposed at the A site, the aminoacyl-tRNA will not remain bound to the ribosome at all. If the match is close, however, transient binding may take place but the mismatched aminoacyl-tRNA is eventually rejected. The basis of this error-checking mechanism is not completely understood, but it does cost the cell energy because GTP hydrolysis accompanies the transient binding of even incorrectly paired aminoacyl-tRNAs.

Although the rules of complementary base pairing suggest that each transfer RNA should recognize only one codon in messenger RNA, many transfer RNA molecules have been found to bind to more than one codon. This phenomenon does not lead to the insertion of incorrect amino acids, however, because the multiple codons recognized by a single transfer RNA differ only in the last base. Examination of the genetic code (see Table 10-4) reveals that codons differing in the third

FIGURE 12-28 The role of elongation factors Tu and Ts in the binding of aminoacyl-tRNA to bacterial ribosomes.

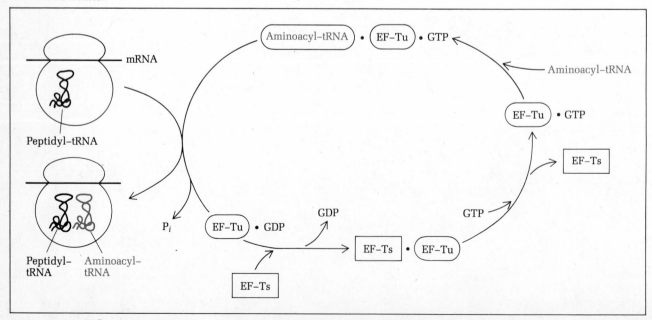

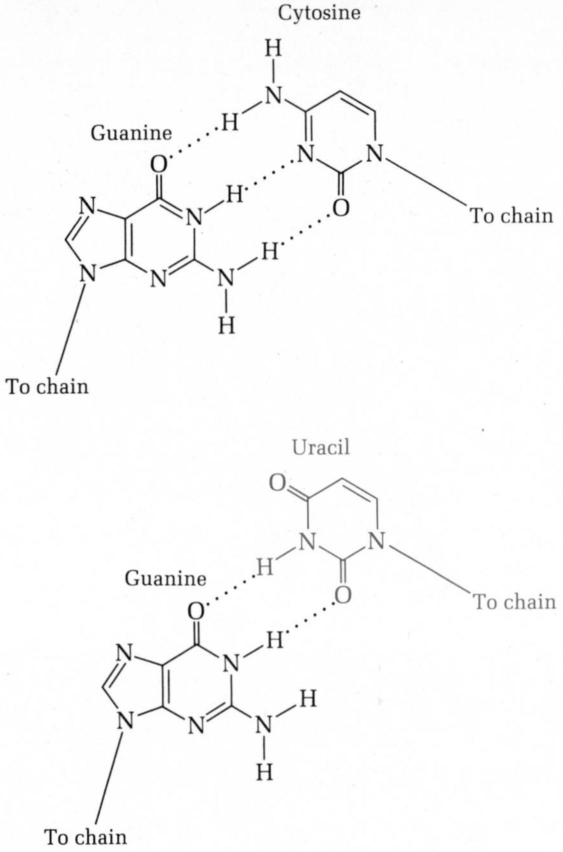

FIGURE 12-29 Diagram illustrating how the existence of "wobble" permits guanine to pair with uracil instead of its normal complementary base, cytosine.

TABLE 12-3
Base pairs permitted at the third position of a codon according to the wobble hypothesis

Base in Anticodon	Bases Recognized in Codon (Third Position Only)
A	U
C	G
U	A or G
G	U or C
I[a]	U, C, or A

[a] The unusual base inosine (I) appears in several anticodons.

base often code for the same amino acid. For example, UUU and UUC both code for phenylalanine, UCU, UCC, UCA, and UCG all code for serine, and so forth. In such cases the same transfer RNA is often capable of binding to more than one codon. For example, a single transfer RNA molecule recognizes both the UUU and UUC codons for phenylalanine.

The above considerations led Francis Crick to propose that messenger RNA and transfer RNA line up on the ribosome in a way that permits flexibility or "wobble" in the pairing between the third base of the codon and the corresponding base in the anticodon. According to this **wobble hypothesis,** such flexibility allows some unusual base pairs, such as G-U, to form (Figure 12-29). Table 12-3 summarizes the base pairs permitted by the wobble hypothesis. It should be noted that the base inosine (I), which is extremely rare in other RNA molecules, often occurs in the wobble position of the anticodon of transfer RNAs. Though the existence of wobble means that the same transfer RNA molecule can recognize more than one codon (and vice versa), in no case does this alter the identity of the amino acid being incorporated.

2. *Peptide bond formation.* Once an appropriate aminoacyl-tRNA binds to the A site of the ribosome, the

next step in the elongation cycle is the formation of a peptide bond between the free amino group of the amino acid attached to the tRNA at the A site and the carboxyl group by which the initiating amino acid (or growing peptide chain) is attached to the tRNA at the P site. In the process the newly formed peptide is transferred to the A site of the ribosome (Figure 12-30). The peptide bond forming reaction is catalyzed by a group of ribosomal proteins, collectively known as **peptidyl transferase,** which are constituents of the large subunit. The peptidyl transferase reaction is the only step in protein synthesis that requires neither nonribosomal protein factors nor an outside source of energy such as GTP or ATP. The energy required for driving peptide bond formation is provided entirely by the cleavage of the high-energy bond that joins the amino acid or peptide chain to the transfer RNA located at the P site. (Recall from Chapter 10, however, that the original formation of this bond during the aminoacyl-synthetase reaction is accompanied by the hydrolysis of ATP to AMP.)

3. *Translocation.* At the conclusion of the peptide bond forming step the nascent polypeptide is attached to the tRNA located at the A site, while the P site contains an empty tRNA whose amino acid has just been joined to the polypeptide chain. The messenger RNA now advances by one triplet unit, bringing the next codon into proper position for translation. Simultaneously the peptidyl-tRNA moves from the A site to the P site, displacing the empty tRNA from the ribosome. This movement of messenger RNA and peptidyl-tRNA relative to the ribosome, termed **translocation,** requires both GTP and elongation factor **EF-G** (**EF-2** in eukaryotes). The binding of EF-G to the ribosome can be experimentally inhibited by treating ribosomes with antibodies to proteins L7 and L12, implicating these two related proteins as components of the binding site for EF-G. The large ribosomal subunit has also been implicated in the mechanism of translocation by the discovery that the isolated large subunit catalyzes GTP hydrolysis when EF-G is added. Since the messenger RNA molecule undergoing translocation is associated with the small subunit, the effects of GTP hydrolysis must somehow be conveyed from the GTP-hydrolyzing

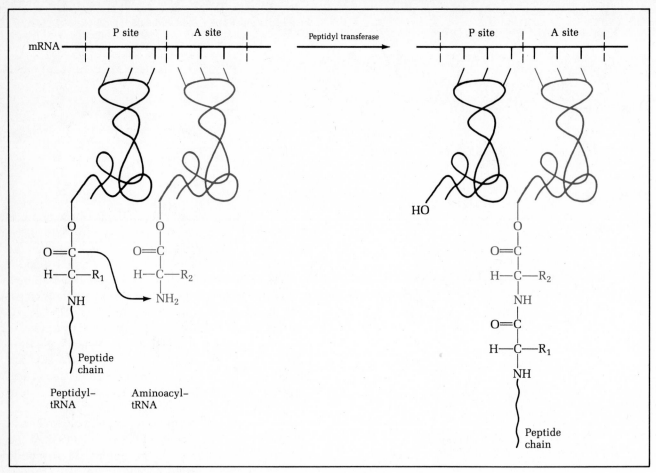

FIGURE 12-30 The peptidyl transferase reaction.

site on the large subunit to the messenger-RNA binding site on the small subunit.

The net effect of translocation is to vacate the A site and to bring the next codon into proper position. Binding of a new aminoacyl-tRNA to the A site then starts the next elongation cycle (Figure 12-31). The only difference between succeeding elongation cycles and the first one is that peptidyl-tRNA now occupies the P site at the beginning of each cycle, while initiator formylmethionyl-tRNA (methionyl-tRNA in eukaryotes) occupies the P site at the beginning of the first elongation cycle. As elongation proceeds, the newly forming polypeptide chain gradually gets longer and eventually protrudes from the ribosome. Digestion of active ribosomes with proteolytic enzymes has revealed that a segment of the growing polypeptide chain about 30–35 amino acids long is protected from degradation, suggesting that a portion of the newly forming polypeptide is buried within a cavity on the ribosome in such a way as to make it inaccessible to enzymatic degradation. Similar studies employing ribonuclease digestion have revealed that a messenger RNA fragment about 25 bases long resists digestion, indicating that a small region of the messenger RNA molecule is also buried within the ribosome.

Termination of Protein Synthesis

The elongation cycle is repeated again and again, adding one amino acid at a time to the growing polypeptide chain, until a termination codon (UAA, UAG, or UGA) is reached. Amino acid incorporation ceases at this point because no normal transfer RNAs contain an anticodon capable of binding to these particular triplets. In prokaryotic cells three proteins called **release factors** have been implicated in the recognition of the termination codons. Factor **RF-1** binds to UAA and UAG, **RF-2** binds to UAA and UGA, and **RF-3** stimulates the activity of both RF-1 and RF-2. In eukaryotes a single release factor appears to recognize all three termination codons.

The binding of an appropriate release factor to a termination codon located at the A site triggers the activation of peptidyl transferase, just as the normal binding of an aminoacyl-tRNA to the A site triggers this enzyme. With a release factor bound to the A site, however, there is no amino acid available with which the polypeptide chain in the P site can form a peptide bond. Instead a bond is formed with water, that is, the linkage between the polypeptide chain and the tRNA is hydrolyzed, releasing the finished polypeptide into solution (Figure 12-32). At this point the messenger RNA and

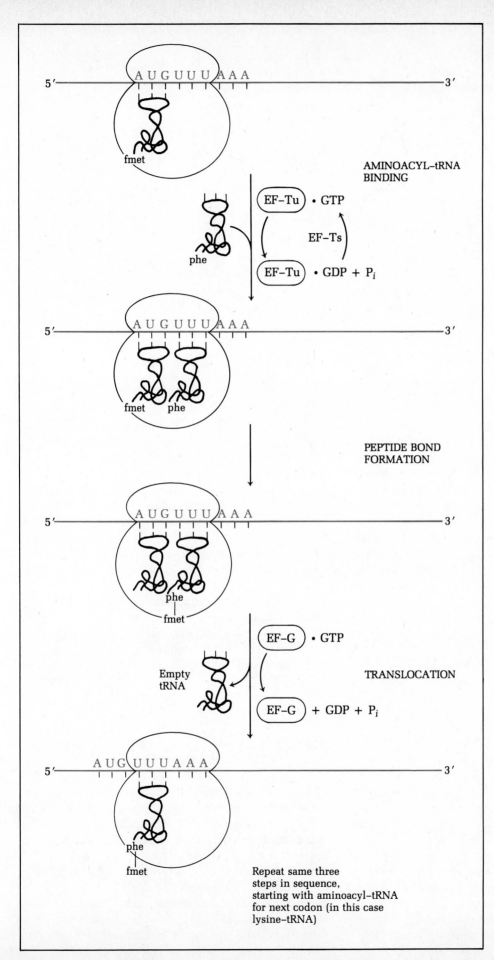

FIGURE 12-31 Summary of the three steps involved in the elongation phase of protein synthesis: aminoacyl-tRNA binding, peptide bond formation, and translocation.

AMINOACYL–tRNA BINDING

EF–Tu · GTP

EF–Ts

EF–Tu · GDP + P$_i$

PEPTIDE BOND FORMATION

EF–G · GTP

Empty tRNA

TRANSLOCATION

EF–G + GDP + P$_i$

Repeat same three steps in sequence, starting with aminoacyl–tRNA for next codon (in this case lysine–tRNA)

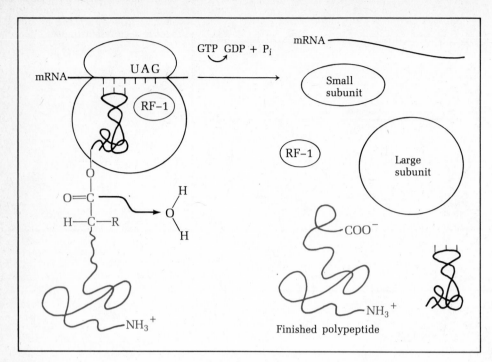

FIGURE 12-32 The termination step of protein synthesis in prokaryotic cells.

empty tRNA are also released from the ribosome, which then dissociates into its constituent subunits. The details of this overall process are not well understood, but the hydrolysis of GTP appears to be involved at some point. Once the individual ribosomal subunits have dissociated, their reassociation is inhibited by the binding of initiation factor IF-3 to the small subunit. This small subunit/IF-3 complex is then ready for the initiation of a new round of protein synthesis.

Some important insights into the mechanism of termination have emerged from the study of naturally occurring mutations in which a codon for a normal amino acid is mutated to a termination codon. For example, a single-base change in DNA can result in the production of a messenger RNA in which a CAG triplet, which codes for glutamine, is instead converted to the termination codon UAG. If such a mutation occurs in a protein-coding gene, the introduction of a termination codon in a position normally coding for an amino acid will cause the polypeptide chain encoded by this gene to be released prior to completion. The deleterious effects of this type of mutation are sometimes overcome, however, by the independent occurrence of **suppressor mutations** involving alterations in the anticodon sequences of transfer RNA molecules. An example of a suppressor mutation would be a mutation in the anticodon of a tyrosine tRNA from 3'--AUG--5' to 3'--AUC--5'. Instead of pairing with the normal tyrosine codons UAC and UAU, this mutated transfer RNA anticodon would base pair with the termination codon UAG. In the presence of such a transfer RNA, the codon UAG will specify the incorporation of tyrosine into the growing polypeptide chain where premature termination would otherwise have occurred. Though the resulting polypeptide

will contain a tyrosine residue where a glutamine would normally have resided, such a polypeptide has a better chance of being functionally active than a prematurely terminated peptide fragment.

Since the above type of suppressor mutation permits a normal terminator codon to be read as if it coded for an amino acid, the question arises as to how proper termination is signaled at the end of polypeptide chains in cells containing suppressor tRNAs. Most polypeptides are in fact terminated normally in such cells, indicating that a termination codon located in its proper place at the end of a coding sequence still triggers termination, while the same codon in an abnormal location does not. It can therefore be concluded that normal termination involves the recognition of a special sequence or three-dimensional configuration at the end of messenger RNA in addition to the termination codon itself.

Polysomes

Although the major steps involved in the synthesis of polypeptide chains have now been described, we have only taken into account the interaction of a single ribosome with a single molecule of messenger RNA. In the early 1960s it was independently discovered in the laboratories of Hans Noll and Alexander Rich that ribosomes function in groups during the translation of messenger RNA. The experiments carried out in Rich's laboratory focused on the synthesis of hemoglobin in immature red blood cells that had been briefly incubated with radioactive amino acids. After incubation, the cells were gently broken open and centrifuged at moderate speed to remove nuclei and mitochondria. The resulting supernatant was placed on top of a sucrose

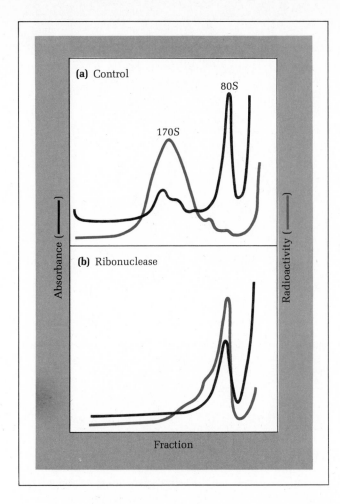

FIGURE 12-33 Experiment demonstrating that protein synthesis occurs on polysomes. (a) Moving-zone centrifugation of ribosomes from reticulocytes incubated with radioactive amino acids. The newly synthesized radioactive protein is associated with ribosome clusters (170S peak) rather than with individual ribosomes (80S) peak. (b) Moving-zone centrifugation of ribosomal fraction after degradation of messenger RNA with ribonuclease. The ribosome clusters that comprise the 170S peak are converted to individual 80S ribosomes by such treatment.

gradient and further fractionated by moving-zone centrifugation. Analysis of the resulting fractions revealed the presence of two major peaks of material, a sharp peak sedimenting at 80S and a broader peak sedimenting at 170S. Since eukaryotic ribosomes were known to sediment at 80S, it was expected that the newly synthesized radioactive protein would be associated with the 80S peak. Surprisingly, the radioactivity was found to sediment in the 170S region instead (Figure 12-33).

These results suggested the possibility that the 170S peak contained clusters of active ribosomes held together by messenger RNA. If this were the case, one would predict that exposing the 170S material to ribonuclease digestion under conditions known to degrade messenger RNA should result in the conversion of the 170S peak to individual 80S ribosomes. When such experiments were carried out, the 170S peak disappeared and the 80S peak increased in size as expected. Additional support for the preceding interpretations emerged from electron microscopic studies, which revealed the presence of single ribosomes in the 80S peak and clusters of four to six ribosomes in the 170S peak (Figure 12-34). In high-power electron micrographs these ribosome clusters were found to be held together

FIGURE 12-34 Electron micrographs of reticulocyte ribosome fractions isolated as described in the preceding figure. *(Left)* Material present in the 80S peak is seen to consist of single ribosomes. *(Right)* Material from the 170S peak consists of ribosome clusters (polysomes). Courtesy of A. Rich.

by a thin strand whose length was close to that predicted for an RNA molecule containing the base sequence information coding for a hemoglobin chain.

From the above experiments it was postulated that individual messenger RNA molecules are simultaneously translated by several ribosomes, creating complex structures known as polyribosomes, or **polysomes.** According to the polysome model of protein synthesis (Figure 12-35), translation begins when a small ribosomal subunit forms a preinitiation complex with the 5′ end of a messenger RNA molecule. After the large ribosomal subunit has joined the complex and several cycles of elongation have ensued, the 5′ end of the messenger RNA gradually begins to move away from the surface of the ribosome. The initiation site at the 5′ end of the messenger RNA thus becomes free to bind to another small ribosomal subunit and begin a new initiation cycle. In this way multiple ribosomes eventually become associated with the same messenger RNA, each synthesizing the same polypeptide chain but at a different stage of completion. This general mechanism increases the overall efficiency of messenger RNA utilization by permitting many ribosomes to translate the same message simultaneously.

One obvious prediction of the polysome model is that the number of ribosomes present within a given polysome should be directly related to the length of the messenger RNA sequence being translated. The immature red blood cells studied in Rich's laboratory are highly specialized for the production of a single type of

FIGURE 12-35 Model illustrating how the simultaneous translation of a messenger RNA molecule by several ribosomes creates a polysome.

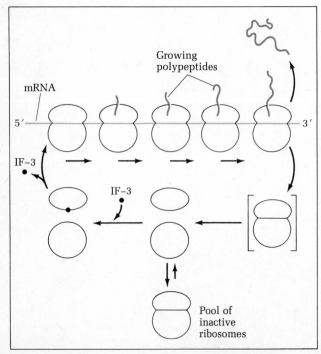

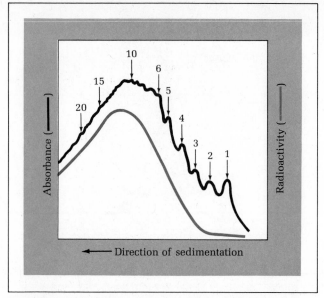

FIGURE 12-36 Moving-zone centrifugation of the ribosome fraction from rat liver following administration of radioactive amino acids. The multiple peaks represent ribosome clusters (polysomes) of varying sizes. The number above each peak represents the number of ribosomes present per polysome. Note that most of the protein synthetic activity occurs in polysomes consisting of ten or more ribosomes.

protein (hemoglobin), so the simple polysome profile exhibited by these cells reflects a situation in which a relatively small number of different messenger RNAs is present. In the more typical case of cells synthesizing a variety of proteins of differing sizes, the messenger RNA population will be heterogeneous in length and a wider variety of polysome sizes is therefore expected. In such cases, like the liver cells in which Hans Noll independently discovered polysomes, a more complex polysome profile consisting of varying numbers of ribosomes has been observed (Figure 12-36). In cells where very large proteins are being synthesized, polysomes containing as many as 50 or more ribosomes have even been detected (Figure 12-37).

Since the polysome concept was initially based on the behavior of ribosomes in isolated subcellular fractions, it has been reassuring to find structures resembling polysomes in electron micrographs of thin-sectioned tissues that have not been subjected to fractionation procedures. Polysomes visualized in this way often exhibit a considerable degree of order, and may even assume highly organized configurations such as spirals or helices (Figure 12-38). In prokaryotic cells polysomes are formed on messenger RNA molecules while the latter are still in the process of being transcribed from the chromosomal DNA. As transcription proceeds so does translation, continually freeing the 5′ end of the newly forming messenger RNA for the binding of additional ribosomes (Figure 12-39). Such coupling of transcription and translation does not take place in eukaryotic cells because the site of transcription (the

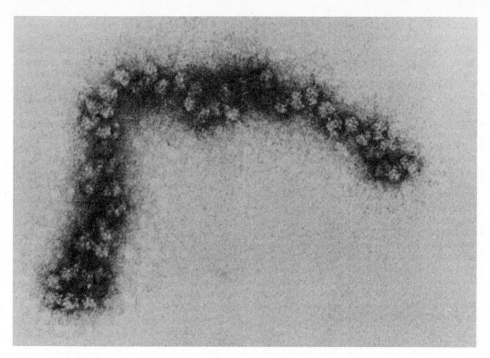

FIGURE 12-37 Electron micrograph of a giant polysome isolated from chick embryo cells. Courtesy of A. Rich.

nucleus) is separated from the site of translation (the cytoplasm) by the membranes of the nuclear envelope.

In the case of messenger RNA molecules that code for polypeptides destined for secretion from the cell or for incorporation into membranes, translation occurs on polysomes that are bound to membranes (the endoplasmic reticulum in eukaryotes or the plasma membrane in prokaryotes). As translation takes place in such polysomes, the elongating polypeptide chain is directed into or through the associated membrane regions. The mechanisms that guide this association of ribosomes with cellular membranes and the subsequent transport of newly forming polypeptides were described in detail in Chapter 6.

Inhibitors of Protein Synthesis

Much has been learned about the mechanism of protein synthesis through the use of inhibitors that act at specific steps in the overall process (Table 12-4). It was already pointed out, for example, that the ability of puromycin to compete with aminoacyl-tRNA for binding to the ribosome led to the realization that ribosomes contain at least two tRNA binding sites. **Streptomycin** is another antibiotic whose use has shed light on the events associated with protein synthesis. This oligosaccharide interferes with the binding of both initiating and noninitiating transfer RNAs to the ribosome, thereby inhibiting initiation as well as elongation. One

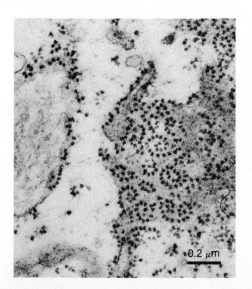

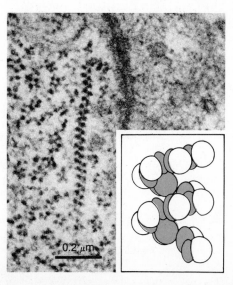

FIGURE 12-38 Examples of highly organized polysome configurations. (*Left*) Electron micrograph of radish root cell showing polysomes arranged in spiral patterns. Courtesy of E. H. Newcomb. (*Right*) A rat intestinal epithelium cell containing polysomes constructed from helically packed ribosomes. (*Inset*) A model illustrating how helically packed ribosomes are thought to be arranged. Courtesy of O. Behnke.

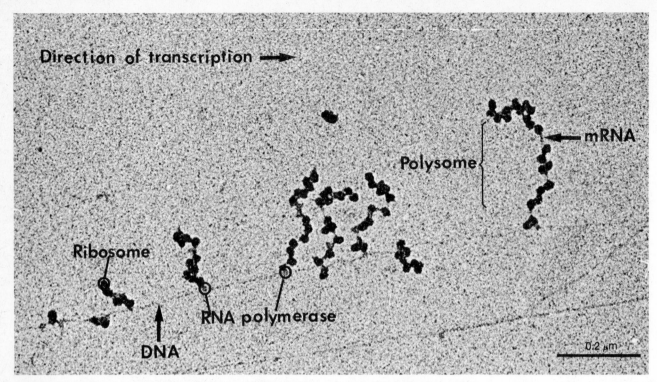

FIGURE 12-39 Electron micrograph illustrating the coupling of transcription and translation that occurs in bacterial cells. In this particular DNA molecule, isolated from *E. coli*, transcription is proceeding from left to right. Before transcription is completed, ribosomes become associated with the newly forming messenger RNA molecule and begin translation. Courtesy of O. L. Miller, Jr.

of the most interesting aspects of streptomycin action is that its presence also causes misreading of the genetic code. For example, the artificial message poly(U), which normally codes only for phenylalanine, also triggers the incorporation of isoleucine, leucine, serine, and tyrosine when streptomycin is present. It was mentioned earlier that mutations in protein S12 render ribosomes resistant to these misreading effects of strep-

tomycin. It has therefore been proposed that protein S12 participates in the error-checking mechanism that normally ensures that the proper aminoacyl-tRNA is bound to the A site of the ribosome.

Many of the commonly employed inhibitors of protein synthesis, such as streptomycin, tetracycline, and chloramphenicol, are only effective against prokaryotic cells. For this reason they are commonly employed to

TABLE 12-4
Commonly used inhibitors of protein synthesis

Inhibitor	Subunit or Factor Affected	Step(s) Blocked	Reaction Affected	Cell Type Affected
Streptomycin	30S	Initiation	Binding of initiator tRNA	Prokaryotes
		Elongation	Binding of aminoacyl-tRNA (induces misreading)	
Tetracycline	30S	Elongation	Binding of aminoacyl-tRNA	Prokaryotes
		Termination	Binding of RF-1 and RF-2	
Spectinomycin	30S	Elongation	Translocation	Prokaryotes
Kasugamycin	30S	Initiation	Binding of initiator-tRNA	Prokaryotes
Colicin E3	30S	Initiation	Binding of messenger RNA	Prokaryotes
		Elongation	Binding of aminoacyl-tRNA	
Chloramphenicol	50S	Elongation	Peptide-bond formation	Prokaryotes
Aurintricarboxylic acid	30S/40S	Initiation	Binding of messenger RNA	Prokaryotes and eukaryotes
Puromycin	50S/60S	Elongation	Peptide-bond formation (triggers chain release)	Prokaryotes and eukaryotes
Fusidic acid	EF-G/EF-2	Elongation	Translocation	Prokaryotes and eukaryotes
Cycloheximide	60S	Initiation	Binding of initiator-tRNA	Eukaryotes
		Elongation	Translocation (tRNA release from P-site)	
Diphtheria toxin	EF-2	Elongation	Translocation	Eukaryotes

combat bacterial infections in humans. The specificity of such antibiotics for prokaryotic systems presumably reflects the fact that the ribosomal constituents with which they interact are significantly different in prokaryotes and eukaryotes. The one major antibiotic that effectively inhibits both prokaryotic and eukaryotic protein synthesis, namely puromycin, acts by mimicking an aminoacyl-tRNA rather than by interfering with the function of specific ribosomal constituents.

A few inhibitors, such as cycloheximide and diphtheria toxin, are selective for eukaryotic protein synthesis. Diphtheria toxin is of special interest because of its association with the infectious disease, diphtheria. Produced by the bacterium *Corynebacterium diphtheriae,* diphtheria toxin is a protein of 65,000 daltons that is lethal to unimmunized humans in doses as small as a few micrograms. This toxin inhibits the translocation step of protein synthesis by catalyzing a reaction in which the adenosine diphosphate ribose (ADPR) portion of an NAD molecule is covalently joined to elongation factor EF-2:

$$\text{NAD} + \text{EF-2} \xrightarrow{\text{Diphtheria toxin}} \text{ADPR-EF-2} + \text{Nicotinamide}$$

The resulting ADP-ribosylated EF-2 is incapable of participating in the translocation step of protein synthesis. In prokaryotic cells the comparable elongation factor, EF-G, is sufficiently different in structure to render it inactive as a substrate for diphtheria-toxin-catalyzed ADP-ribosylation.

TRANSLATIONAL AND POSTTRANSLATIONAL CONTROL MECHANISMS

In Chapter 11 the various kinds of transcriptional and posttranscriptional control mechanisms utilized by cells for regulating the formation of messenger RNA were described. Since the ultimate expression of the information encoded within protein-coding genes requires that these messenger RNAs ultimately be translated into polypeptide chains, the translational level represents another potential site for the control of gene expression. Although translational control mechanisms do not appear to be as numerous and widely utilized as those at the transcriptional level, several kinds of translational regulation are known to occur. These control mechanisms can be grouped into four general categories, depending on whether they involve messenger RNA-binding proteins, alterations in protein synthesis factors, effects on messenger RNA stability, or changes in transfer RNAs.

Messenger RNA-Binding Proteins

The binding of specific proteins to messenger RNA has been shown to influence translational efficiency in a number of different situations. One striking example involves the expression of the bacterial genes coding for ribosomal proteins. The genes coding for these proteins in *E. coli* are distributed among at least ten different operons, each exhibiting its own unique promoter. Since the various ribosomal proteins are needed in roughly equal amounts during the formation of ribosomes, some mechanism must exist to coordinate their synthesis. A shared regulatory sequence has not been detected in the promoter region of the ribosomal protein operons, suggesting that the expression of these genes is not coordinated at the transcriptional level. Recent investigations in the laboratory of Masayasu Nomura suggest that this coordination is achieved at the translational level instead. Each operon examined thus far has been found to code for the synthesis of at least one ribosomal protein that is capable of binding to and inhibiting the translation of the messenger RNA transcribed from that operon. Normally such translational suppression does not occur because these particular ribosomal proteins have an even greater binding affinity for ribosomal RNA. Hence when ribosomal RNA is available, these ribosomal proteins bind to it and initiate the process of ribosome assembly. When the pool of ribosomal RNA is depleted, however, these particular ribosomal proteins bind to and inhibit the translation of their respective messenger RNAs. In this way a feedback loop is established that ensures that the rate of translation of ribosomal proteins is coordinated with the availability of ribosomal RNA (Figure 12-40).

FIGURE 12-40 Feedback loop utilized to control the production of ribosomal proteins in *E. coli*. One protein produced by each ribosomal protein operon appears to be capable of binding either to ribosomal RNA or the messenger RNA from which it was translated. When ribosomal RNA supplies are low, binding of such proteins to messenger RNA inhibits translation and thereby slows the production of ribosomal proteins.

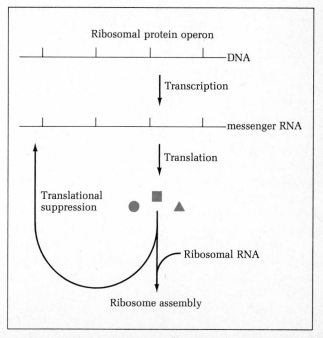

Evidence also exists in eukaryotic systems for the occurrence of messenger RNA-binding proteins. In many eukaryotic cell types a significant proportion of the cytoplasmic messenger RNA is complexed with protein to form **messenger ribonucleoprotein (RNP) particles.** Messenger RNA in this form is not associated with polysomes, nor is it readily translated by cell-free protein-synthesizing systems. Hence it is believed to represent messenger RNA whose ability to undergo translation is inhibited by the presence of RNA-binding proteins. Such an interpretation is supported by the discovery that removal of the protein constituent of messenger RNP particles by digestion with proteases releases messenger RNA that is capable of being translated. A dramatic example of this type of regulation occurs in unfertilized egg cells, which contain large numbers of messenger RNP complexes but are relatively inactive in protein synthesis. Upon fertilization the rate of protein synthesis increases markedly, even when actinomycin D has been added to prevent the formation of new messenger RNA. It has therefore been concluded that the unfertilized egg contains messenger RNA molecules "masked" by the presence of bound protein, and that the release or degradation of this protein at the time of fertilization frees the messenger RNA for translation.

Alterations in Protein Synthesis Factors

An alternative mechanism for regulating translation involves modifications in the activity of the specific factors required for protein synthesis. An example of this type of control occurs in immature red blood cells, where hemoglobin is the major product of translation. Protein synthesis in such cells has been found to be dependent upon **hemin,** an iron-containing prosthetic group that becomes associated with globin chains to form hemoglobin. In cell-free protein-synthesizing systems, a lack of hemin causes the rate of protein synthesis to fall; this inhibitory effect can be overcome by adding an excess of initiation factors, suggesting that the absence of hemin depresses the initiation step of protein synthesis (Figure 12-41). Recent investigations have revealed that hemin functions by inhibiting the activity of a cyclic-AMP-dependent protein kinase that catalyzes the phosphorylation of eukaryotic initiation factor eIF-2. In its phosphorylated state, eIF-2 is inactive. Hence in the presence of hemin the phosphorylation of eIF-2 is prevented and initiation proceeds as normal, while in the absence of hemin, eIF-2 undergoes phosphorylation and hence inactivation (Figure 12-42).

Phosphorylation of eIF-2 has also been implicated in the mechanism of action of **interferon,** a protein produced by white blood cells and fibroblasts in response to viral infection. Interferon induces the formation of messenger RNAs that code for proteins that help to combat viral infection. Among the proteins induced by interferon is a protein kinase that catalyzes the phosphorylation of eIF-2, leading to an inhibition of protein synthesis. Though this effect of interferon is believed to prevent viral replication by inhibiting the translation of viral messenger RNAs, the phosphorylation of eIF-2 inhibits the translation of host messenger RNAs as well.

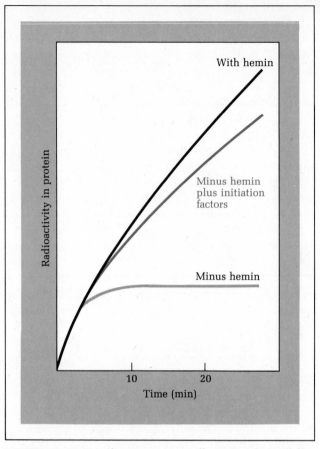

FIGURE 12-41 Data from experiment illustrating that cell-free protein synthesis in reticulocyte systems is stimulated by hemin. The hemin requirement can be largely overcome by adding excess initiation factors, suggesting that hemin functions by stimulating the initiation step of protein synthesis.

Messenger RNA Stability

The average lifespan of messenger RNA molecules varies widely. Most bacterial messenger RNAs are degraded rapidly, exhibiting half-lives of only a few minutes or so; the half-lives of typical eukaryotic messages, on the other hand, range from several hours to a few days. Since the rate at which a given messenger RNA is degraded determines the length of time it is available for translation into polypeptide chains, alterations in messenger RNA degradation rate influence the amount of protein product that can be formed.

If messenger RNA stability is to be considered an important factor in the regulation of gene expression,

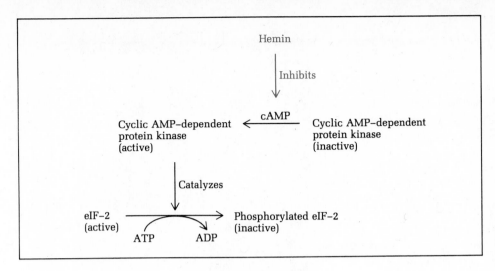

FIGURE 12-42 Mechanism by which hemin regulates protein synthesis in eukaryotic red blood cells. *Abbreviation:* cAMP = cyclic AMP.

some mechanism must exist for controlling the rate at which specific messenger RNAs are degraded. The existence of such mechanisms is suggested by the finding that the half-lives of individual messenger RNAs are subject to alteration in response to changing physiological conditions. For example in the chick oviduct, the half-life of the messenger RNA coding for ovalbumin has been found to be regulated at least in part by estrogen levels. When the estrogen concentration falls, this serves as a signal to the oviduct that ovalbumin is no longer needed. In response the degradation rate of ovalbumin messenger RNA increases by an order of magnitude, accompanied by a corresponding drop in the rate of ovalbumin synthesis.

FIGURE 12-43 Comparison of the relative rates of globin synthesis in *Xenopus* oocytes injected with normal globin messenger RNA or with globin messenger RNA from which the poly(A) tails had been removed.

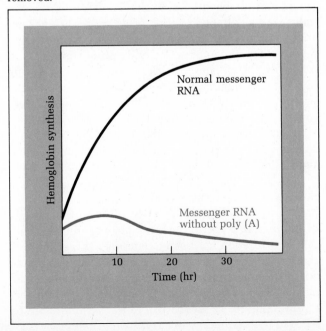

The molecular mechanisms that control the rate of messenger RNA degradation have not been elucidated in detail, but several alternatives can be envisioned. One possibility is that alterations occur in the nucleases responsible for messenger RNA degradation. One enzyme capable of degrading messenger RNA is an endonuclease known to be activated by a short oligomer of adenosine; this adenosine derivative is termed **2-5A** because it is linked together by $2' \rightarrow 5'$ phosphodiester bonds instead of the usual $3' \rightarrow 5'$ bonds characteristic of nucleic acids. Synthesis of 2-5A is catalyzed by **oligoadenylate synthase,** an enzyme whose formation is in turn subject to induction by interferon. Hence one of the ways in which interferon combats viral infection is through the oligoadenylate-stimulated degradation of viral messenger RNA.

Another element that may influence the lifespan of messenger RNA molecules is the poly(A) sequence present at their 3' ends. The most direct evidence implicating poly(A) in such regulation has been obtained from experiments involving the study of messenger RNA molecules whose poly(A) tails have been artificially removed. In one of the more dramatic experiments of this kind, G. Huez and associates monitored the translation of globin messenger RNA injected into the cytoplasm of *Xenopus* oocytes. Shortly after injection the globin messenger RNA lacking poly(A) tails was found to be translated at the same rate as normal globin message. Several days later, however, the translation rate of the messenger RNA molecules lacking poly(A) sequences was markedly depressed relative to that of normal messenger RNA (Figure 12-43). Such results indicate that at least for some messages, the length of the poly(A) tail may influence messenger RNA stability. Since the poly(A) tails of many messenger RNAs are known to undergo shortening after entry into the cytoplasm, it is possible that such alterations play a role in regulating messenger RNA lifespan.

Regulatory Role of Transfer RNA

Another factor capable of influencing the rate of messenger RNA translation is the availability of aminoacyl-tRNAs. We have already seen that mutant messenger RNAs containing premature chain termination codons require the presence of a suppressor tRNA in order to be translated into a complete polypeptide product. Even in the absence of such mutations, it is possible that the rate of messenger RNA translation is affected by the availability of rate-limiting transfer RNAs. The extent to which such control is a significant factor in normal cells is unclear at present, but it is known that the spectrum of transfer RNAs present in the cytoplasm differs between cell types and is subject to changes during development. Alterations have also been observed in aminoacyl-tRNA synthetases and tRNA methylases, both of which affect the production of functional aminoacyl-tRNAs.

Although it is clear that the intracellular concentrations of the various aminoacyl-tRNAs are subject to dynamic alteration, such observations do not prove that these changes play a role in regulating protein synthesis. More compelling evidence has emerged from studies utilizing cell-free protein-synthesizing systems, where the effects of aminoacyl-tRNA availability on messenger RNA translation can be directly tested. One set of transfer RNAs whose effects have been tested in this way are those encoded by T4-phage. When these phage-encoded tRNAs are added to a bacterial cell-free protein-synthesizing system, the translation of phage messenger RNA is stimulated to a greater extent than the translation of bacterial messenger RNA. Such observations suggest that the formation of these phage-encoded transfer RNAs, which only occurs in virus-infected cells, leads to a preferential stimulation in the translation of viral messenger RNA.

Selective effects of tRNA on messenger RNA translation have also been demonstrated during the cell-free translation of globin messenger RNA. In the presence of high concentrations of tRNA, roughly equal amounts of the α- and β-chains of hemoglobin are synthesized. When the tRNA concentration is lowered to rate-limiting levels, however, the synthesis of α-chains is preferentially inhibited (Figure 12-44). This discovery suggests that one or more particular types of tRNA is utilized to a greater extent (or even exclusively) by the messenger RNA coding for α-chains, and that a reduction in the concentration of this tRNA therefore inhibits the translation of α-chain messenger RNA preferentially.

In certain situations aminoacyl-tRNA availability has been found to trigger changes in transcription as well as translation. An example of this type of regulation occurs in bacteria starved for essential amino acids. We saw earlier in the chapter that amino acid starvation stimulates the formation of ppGpp and pppGpp, which

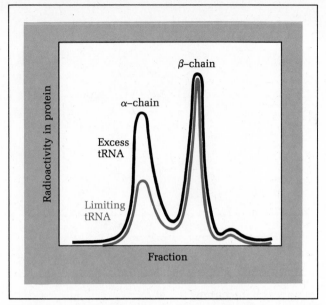

FIGURE 12-44 Effects of transfer RNA on the relative rates of synthesis of the α- and β-chains of hemoglobin in a cell-free protein-synthesizing system. Protein synthesis was carried out in the presence of excess or rate-limiting quantities of transfer RNA, and the newly synthesizing hemoglobin chains were then purified by ion exchange chromatography. Note that the synthesis of α-chains is preferentially inhibited in the presence of low concentrations of transfer RNA.

in turn triggers a decrease in the transcription rate of the ribosomal RNA genes. The synthesis of ppGpp and pppGpp is mediated by a reaction in which the terminal pyrophosphate of ATP is transferred to the 3'-hydroxyl group of either GDP or GTP. This reaction is catalyzed by a protein, known as the **stringency factor,** which is activated by ribosomes containing bound messenger RNA and an A site occupied by a transfer RNA molecule lacking an attached amino acid. Transfer RNA molecules lacking attached amino acids are prevalent only when amino acids are scarce; hence the conditions that favor the formation of ppGpp and ppGppp only prevail when insufficient amino acids are available for carrying out protein synthesis. Under these conditions the ppGpp and ppGppp formed inhibit the synthesis of ribosomal RNA, providing cells with a feedback mechanism for regulating the rate of ribosome production in response to changes in the ability to carry out protein synthesis.

Another type of tRNA-mediated regulation, termed **attenuation,** also combines effects at the transcriptional and translational levels. The best understood example of attenuation involves the tryptophan operon of *E. coli,* whose regulation has been extensively investigated in the laboratory of Charles Yanofsky. The tryptophan operon, which codes for five enzymes involved in the biosynthesis of the amino acid tryptophan, contains a typical operator sequence where transcription can be blocked by the binding of a tryptophan-repressor com-

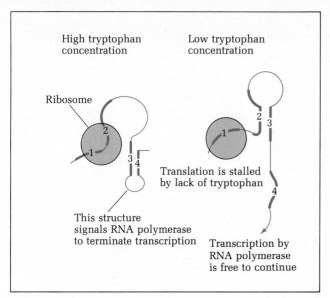

FIGURE 12-45 Mechanism of attenuation in the trytophan operon of *E. coli*. When the intracellular concentration of tryptophan is high, translation of the messenger RNA proceeds through the leader sequence and permits the formation of stem-loop structure (3-4) that signals RNA polymerase to terminate transcription prematurely. When the concentration of tryptophan is low, translation of the messenger RNA stalls because of the lack of tryptophanyl-tRNA. Under these conditions the stem-loop structure that signals transcriptional termination does not form, and hence transcription of the tryptophan operon is free to continue.

plex. In addition to the operator sequence, a second type of regulatory element called the **attenuator site** is present in this operon. Located near the beginning of the gene coding for the first messenger RNA, the attenuator site represents a region at which transcription is subject to premature termination. Premature termination only occurs, however, when tryptophan is present and the enzymes produced by the tryptophan operon are therefore unnecessary.

The mechanism by which the attenuator site senses the presence of tryptophan and triggers premature termination is based on the coupling between transcription and translation that occurs in bacterial cells. As the 5′ end of the messenger RNA encoded by the tryptophan operon is transcribed from the DNA template, it binds to the ribosome and begins to be translated. Within this region of the message, two adjacent UGG triplets coding for tryptophan are present. If the concentration of tryptophan in the cell is low, tryptophanyl-tRNA is not synthesized in adequate amounts and translation stalls when these triplets are encountered. RNA molecules stalled at this point assume a particular three-dimensional configuration that permits the transcription process to continue (Figure 12-45). When tryptophan is present in adequate amounts, however, stalling does not occur; under these conditions translation proceeds past the two UGG codons, allowing the RNA molecule to assume a different configuration that causes transcription to terminate at the attenuator site. Hence

the lack of tryptophanyl-tRNA functions both to stall translation and, through the resulting effect on messenger RNA structure, to facilitate transcription of the corresponding operon by inhibiting premature termination of the transcript.

Posttranslational Modifications

Although the translation of messenger RNA into polypeptide chains represents the final stage in the flow of information from DNA to protein, the polypeptides generated by this process must often undergo structural modifications before they can carry out their normal functions. We have already seen, for example, that the initiating formylmethionine or methionine residue is usually removed from the amino-terminal end of newly forming polypeptide chains. Peptide bond cleavage reactions are also employed to remove additional amino acids from the ends of polypeptide chains and to break polypeptides into smaller fragments. Cleavages of both types occur during the production of insulin, whose messenger RNA encodes the synthesis of a precursor polypeptide known as pre-proinsulin. The amino-terminal end of pre-proinsulin contains a signal sequence that directs insertion of the growing polypeptide through the endoplasmic reticulum membrane, and is then removed by a protease after the growing polypeptide chain emerges into the cisterna. After removal of the signal sequence, the completed polypeptide product, called proinsulin, is converted to insulin by a second cleavage that removes a 22-amino-acid long fragment from the internal region of the molecule, leaving two separate chains held together by disulfide bonds (Figure 12-46).

The covalent structure of polypeptide chains is also subject to alteration through modification of amino acid side chains. Among the more commonly encountered modifications are the formation of disulfide bridges, hydroxylation, phosphorylation, acetylation, methylation, ADP-ribosylation, amidation, adenosylation, and glycosylation. Many of these reactions are reversible, giving cells a mechanism for dynamically altering the structural and functional properties of its protein molecules. A classic example of regulation of this type involves the enzyme phosphorylase, whose control by reversible cycles of phosphorylation and dephosphorylation was described earlier (page 50).

The most extreme way in which a protein can be modified is by degradation to free amino acids. Since the quantity of any particular protein present in the cell is a function of its rate of degradation as well as its rate of synthesis, control of protein degradation rates is another level at which the expression of gene products can potentially be controlled. The occurrence of regulation at this level is suggested by the discovery that the half-lives of typical protein molecules vary from a few

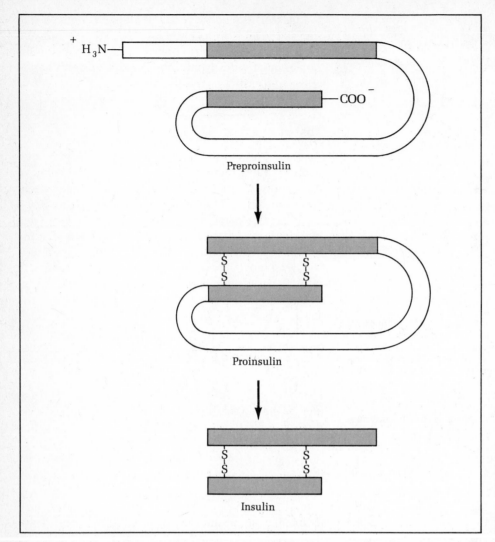

$H_3\overset{+}{N}$ — Preproinsulin

COO$^-$

Proinsulin

Insulin

FIGURE 12-46 The formation of insulin illustrates how proteolytic cleavage reactions can be used to convert a primary product of messenger RNA translation into a final, functional protein.

minutes to several weeks, and that the lifespans of individual proteins change under different physiological conditions (see Chapter 15 for a further discussion of the importance of protein degradation rates in cellular regulation). Although the factors that control protein degradation are not well understood, a small polypeptide called **ubiquitin** has been found to be covalently attached to at least some proteins destined for degradation. Such findings suggest that ubiquitin functions to mark polypeptides for destruction, but the factors that determine which proteins will have ubiquitin attached to them are yet to be clarified.

We have thus seen that posttranslational alterations in protein structure, including the ultimate degradation of proteins to amino acids, provide the final level at which cells can regulate the expression of specific gene products. When considered in combination with the previously discussed transcriptional, posttranscriptional, and translational controls, it becomes apparent that cells have available a powerful and diverse battery of weapons for regulating the expression and properties of the various products encoded within their DNA.

SUMMARY

Protein synthesis occurs on small ribonucleoprotein granules, called ribosomes, which measure roughly 25 nm in diameter. Eukaryotic cytoplasmic ribosomes typically sediment at about 80S and consist of 60S and 40S subunits, while prokaryotic ribosomes sediment at 70S and are comprised of 50S and 30S subunits. Small ribosomal subunits contain a single molecule of RNA (16S in prokaryotes and 18S in eukaryotes), while the larger subunit has two molecules of RNA in prokaryotes (23S and 5S) and three molecules of RNA in eukaryotes (28S, 5.8S, and 5S). Fifty-two different proteins are present in prokaryotic ribosomes, accounting for roughly half the total ribosomal mass; in eukaryotic cells the total mass and number of different ribosomal proteins is slightly larger. Close examination of negatively stained ribosomes has revealed the presence of several protuberances and ridges. These demarcations partition the small subunit into three distinct regions termed the head, the base, and the platform, and create three regions in the large subunit known as the central protuberance, the stalk, and the ridge.

Eukaryotic ribosomes are manufactured within a specialized nuclear substructure, the nucleolus, which contains from several hundred to several thousand copies of the genes coding for 18S, 28S, and 5.8S (but not 5S) ribosomal RNAs. During oogenesis these repeated genes are amplified up to a thousandfold more, giving rise to the formation of large numbers of nucleoli. The DNA segments coding for the 18S, 28S, and 5.8S ribosomal RNAs lie adjacent to one another and are transcribed into a single RNA molecule that sediments at about 45S in higher eukaryotes. As this RNA molecule is being transcribed, it becomes associated with protein; shortly thereafter the RNA undergoes methylation, predominantly at the 2′-OH groups of the ribose moieties. During the subsequent processing of the primary transcript into 18S, 28S, and 5.8S ribosomal RNAs, the methylated regions of the RNA molecule are preferentially retained. Transcription of the 45S ribosomal RNA precursor occurs in the fibrillar core of the nucleolus, while the granular cortex contains partially processed ribosomal RNA with its associated proteins.

Ribosome formation in prokaryotes, which lack nucleoli, differs in several ways from the comparable process in eukaryotic cells. Prokaryotic ribosomal RNA genes are less extensively repeated, and occur in clusters in which all three ribosomal genes (16S-23S-5S) are closely linked. Transcription begins at the 16S end of this cluster, and the primary transcript is cleaved into separate precursors for the three ribosomal RNAs as transcription is occurring. In prokaryotes most ribosomal RNA methylation occurs after processing and assembly of the ribosomal subunits has begun. Only about 10 percent of the total RNA is degraded during ribosomal RNA processing, while in higher eukaryotes this value approaches 50 percent.

A variety of experimental techniques have been utilized to investigate the three-dimensional organization and functional roles of the individual protein and RNA molecules that comprise the ribosome. One approach involves the reconstitution of functional ribosomes from isolated RNA and protein components. By carrying out reconstitution on varying mixtures of RNA and protein, it has been possible to investigate the roles played by individual components in ribosome assembly and function. Ribosomal organization has also been explored using: (1) chemical cross-linking reagents to join neighboring proteins, (2) enzymatic digestion with proteases and ribonucleases to investigate the topographical relationships between proteins and RNA, (3) radioactive and affinity labeling to determine which proteins are associated with particular active sites, (4) deuterium labeling to investigate the proximity of neighboring proteins, and (5) immunological labeling to investigate the location and function of ribosomal proteins to which specific antibodies have been developed. Although a complete three-dimensional model of ribosomal substructure is yet to emerge, the data obtained from the various approaches have permitted some preliminary models of ribosomal organization to be constructed. Among the more important generalizations to emerge are that the small subunit is involved in the binding of messenger RNA and transfer RNAs to the ribosome, while the large subunit catalyzes peptide bond formation.

Our current understanding of the biochemistry of protein synthesis has been derived largely from studies on cell-free systems, the use of which has facilitated the identification of the large number of components involved. Cell-free protein synthesis requires the presence of ribosomes, transfer RNAs, aminoacyl-tRNA synthetases, amino acids, several kinds of protein synthesis factors, and an energy source capable of generating GTP and ATP. Messenger RNA is translated in the $5′ \rightarrow 3′$ direction, specifying the incorporation of amino acids starting at the amino-terminal end of the polypeptide chain and proceeding one amino acid at a time until the carboxyl-end is reached. The overall process of translation can be divided into three phases: initiation, elongation, and termination. During the initiation phase, messenger RNA, an initiator aminoacyl-tRNA (containing formylmethionine in prokaryotes and methionine in eukaryotes), GTP (and ATP in eukaryotes), and a series of initiation factors become associated with the small ribosomal subunit to form a preinitiation complex. The binding of messenger RNA to the ribosome is facilitated by the cap structure present in eukaryotic messenger RNAs, and by an interaction between the 3′ end of the 16S ribosomal RNA and a complementary untranslated 5′ sequence present in prokaryotic messenger RNAs. Once the preinitiation complex has been formed, the large subunit joins the small subunit in a reaction accompanied by the hydrolysis of GTP and the release of the initiation factors.

During the elongation phase, amino acids are joined together in a linear order specified by the codons in messenger RNA. Elongation proceeds by a cyclic series of reactions requiring the participation of aminoacyl-tRNAs, elongation factors, and GTP. Each iteration of the cycle can be subdivided into three steps. In the first step an incoming aminoacyl-tRNA binds to the A site of the ribosome in a reaction involving GTP and elongation factors EF-Tu and EF-Ts in prokaryotes, or EF-1 in eukaryotes. Although the rules of complementary base pairing suggest that the anticodon of each incoming aminoacyl-tRNA should recognize only one specific codon in messenger RNA, flexibility or "wobble" in the pairing between the third base of the codon and the corresponding base in the anticodon permits some unusual base pairs to form. As a result many transfer RNAs can bind to more than one type of codon; coding accuracy is not affected by this phenomenon, however, because the multiple codons recognized by a given transfer RNA always specify the same amino acid. The second step in the elongation cycle involves the formation of a peptide bond between the free amino group of the amino acid attached to the tRNA bound to the A site, and the carboxyl group by which the initiating amino acid (or growing peptide chain) is attached to the tRNA bound to the P site. The peptide bond-forming reaction, which requires neither nonribosomal protein factors nor an energy source such as GTP or ATP, is catalyzed by a group of proteins from the large ribosomal subunit collectively known as peptidyl transferase. During the third step of the elongation cycle, called translocation, the messenger RNA advances by one triplet unit, accompanied by displacement of the empty tRNA from the ribosome and movement of the peptidyl-tRNA from the A site to the P site. Translocation is accompanied by the hydrolysis of GTP and the binding of elongation factor EF-G (EF-2 in eukaryotes) to the large ribosomal subunit. The net effect of translocation is to vacate the A site and to bring the next codon into proper position to begin the subsequent elongation cycle.

The final phase of protein synthesis, called termination, occurs when a termination codon (UAA, UAG, or UGA) is reached. Elongation ceases at this point because no normal transfer RNAs contain an anticodon capable of binding to these triplets. Termination codons are instead recognized by release factors that bind to the A site and activate peptidyl transferase. Since no amino acid is present at the A site to participate in peptide bond formation, the bond joining the polypeptide chain to the transfer RNA located at the P site is hydrolyzed and the finished polypeptide chain is released into solution. At this point messenger RNA and the empty tRNA are also released from the ribosome, which then dissociates into its constituent subunits. Although the preceding summary has only taken into account the interaction between messenger RNA and a single ribosome, each molecule of messenger RNA is normally translated by several ribosomes simultaneously, producing a complex structure known as a polysome.

Much has been learned about protein synthesis through the use of inhibitors that act at specific stages in the translation process. For example, the ability of puromycin to compete with aminoacyl-tRNAs for binding to the ribosome led to the realization that ribosomes contain separate A and P binding sites for binding tRNAs. Many inhibitors of protein synthesis, such as streptomycin, tetracycline, and chloramphenicol, only interact with the constituents of prokaryotic protein-synthesizing systems, making them useful agents for combatting bacterial infections in humans.

Translational control mechanisms can be divided into four major categories: (1) messenger RNA-binding proteins that inhibit translation, (2) alterations in protein synthesis factors such as the phosphorylation of eIF-2, (3) regulation of messenger RNA stability me-

diated by alterations in specific ribonucleases or changes in the length of the poly(A) tail, and (4) changes in the availability of transfer RNAs, such as occurs when the lack of aminoacyl-tRNAs triggers the production of ppGpp/pppGpp or when the absence of a specific aminoacyl-tRNA prevents premature transcriptional termination in genes subject to attenuation.

Regulation can also be achieved at the posttranslational level by modifying polypeptide chains after they have been synthesized. Among the more commonly encountered modifications are the formation of disulfide bridges, hydroxylation, phosphorylation, acetylation, methylation, ADP-ribosylation, amidation, adenosylation, and glycosylation. The most extreme type of modification is the degradation of a protein into its constituent amino acids. Although the factors that regulate protein degradation are not well understood, the existence of regulation at the level of protein degradation is suggested by the discovery that the half-lives of proteins vary with changing physiological conditions.

SUGGESTED READINGS

Books

Bielka, H., ed. (1982). *The Eukaryotic Ribosome*, Springer-Verlag, New York.

Chambliss, G., G. R. Craven, J. Davies, K. Davis, L. Kahan, and M. Nomura, eds. (1980). *Ribosomes: Structure, Function, and Genetics*, University Park Press, Baltimore.

Clark, B. F. C., and H. U. Petersen, eds. (1984). *Gene Expression: The Translational Step and Its Control*, Raven Press, New York.

Hadjiolov, A. A. (1985). *The Nucleolus and Ribosome Biogenesis*, Springer-Verlag, New York.

Hardesty, B., and G. Kramer, eds. (1986). *Structure, Function, and Genetics of Ribosomes*, Springer-Verlag, New York.

Lewin, B. (1987). *Genes III*, Wiley, New York.

Matthews, M. B., ed. (1986). *Translational Control*, Cold Spring Harbor Laboratory, Cold Spring Harbor, NY.

Articles

Dreyfus, G. (1986). Structure and function of nuclear and cytoplasmic ribonucleoprotein particles, *Ann. Rev. Cell Biol.* 2:459–498.

Lake, J. A. (1981). The ribosome, *Sci. Amer.* 245 (Aug.):84–97.

Lake, J. A. (1985). Evolving ribosome structure: domains in archaebacteria, eubacteria, eocytes and eukaryotes, *Ann. Rev. Biochem.* 54:507–530.

Nomura, M. (1984). The control of ribosome synthesis, *Sci. Amer.* 250 (Jan):102–114.

Nomura, M., R. Gourse, and G. Baughman (1984). Regulation of the synthesis of ribosomes and ribosomal components, *Ann. Rev. Biochem.* 53:75–117.

Subramanian, A. R. (1985). The ribosome: Its evolutionary diversity and the functional role of one of its components, *Essays in Biochemistry* 21:45–85.

Wittman, H. G. (1983). Architecture of prokaryotic ribosomes, *Ann. Rev. Biochem.* 52:35–65.

Wool, I. G. (1979). The structure and function of eukaryotic ribosomes, *Ann. Rev. Biochem.* 48:719–754.

Zamecnik, P. (1984). The machinery of protein synthesis, *Trends Biochem. Sci.* 9:464–466.

CHAPTER
13
The Cell Division Cycle

In the preceding three chapters, we described the process by which the genetic information encoded within DNA molecules is ultimately utilized to guide the synthesis of specific proteins. Because of the essential role played by DNA in determining the kinds of proteins that can be made, the base sequence information contained within a cell's DNA must be faithfully replicated before the cell divides. Only in this way can it be ensured that the two new cells formed during cell division each inherit a complete set of genetic instructions. The process of cell division therefore has two distinct aspects: replication of the genetic information encoded within DNA, and physical separation of the parent cell into two new cells, each containing a complete copy of this duplicated DNA. The mechanisms by which these objectives are accomplished in prokaryotic and eukaryotic organisms are described in this chapter.

INTRODUCTION TO THE CELL CYCLE

Although cells in the process of undergoing division were first observed with the light microscope over a century ago, it is only in the past few decades that the mechanisms underlying this crucial event have begun to be unraveled. It is now clear that dividing cells pass through a reproducible series of stages, collectively referred to as the **cell cycle,** which begins when two new cells are formed by the division of a single parental cell and ends when one of these cells then divides to form two new cells. Much of our current knowledge concerning the biochemical events associated with the cell cycle has emerged from studies on **synchronous** cell cultures, which consist of a large number of cells whose division cycles have been brought into phase with one another. Two general approaches have been employed for generating synchronized cell populations. One approach, referred to as **induction synchrony,** utilizes conditions that temporarily halt cell division at a particular stage, thereby causing all cells to eventually stop at the same point in the cycle. If the inhibitory conditions are then removed, the cells proceed through the rest of the cell cycle in phase with one another. The most common adaptation of this approach is the **thymidine block** technique, which is based on the effects of adding large amounts of thymidine to the culture medium. This high thymidine concentration triggers a feedback inhibition of the synthesis of other essential precursors of DNA synthesis, such as deoxycytidine, leading to a cessation of DNA synthesis. Hence in the presence of high concentrations of thymidine, cells proceed through the cell cycle until they reach the point at which DNA synthesis would normally occur. After all cells have been blocked at this point the excess thymidine can be removed, allowing the entire cell population to initiate DNA synthesis simultaneously.

Though the induction approach for generating synchronous cell populations is widely utilized, a major disadvantage inherent to such procedures is that the conditions employed for temporarily halting the cell cycle may exert toxic side effects. This potential problem is avoided in the other approach to cell synchronization, called **selection synchrony,** in which cells at a particular stage in the cell division cycle are selectively removed from a culture and grown as a separate population. Selection synchrony techniques usually exploit the fact that cells in the process of dividing tend to round up and detach from the surface of culture vessels. Cells at this particular stage of the cell cycle can therefore be selectively removed from a culture flask by gently shaking the flask and removing the medium containing the detached cells, leaving behind those cells at other phases of the cycle.

The study of synchronized cultures has led to the discovery that the synthesis of certain cell constituents is restricted to particular phases of the cell cycle. In eukaryotic cells this kind of pattern has been most clearly established for the synthesis of DNA. Classical cell biologists originally believed that chromosome replication takes place at the time of **mitosis,** the period that begins when the chromosomes condense and ends when the cell physically divides into two new cells. Although mitosis accounts for only a short period of time relative to the total time between cell divisions, the remaining portion of the cell cycle, termed **interphase,** was not viewed as a likely time for chromosome replication because no dramatic changes in chromosome appearance occur during this period. In the early 1950s Hewson Swift investigated this issue by using the Feulgen staining technique to measure the quantity of DNA present at various stages of the cell cycle. His data revealed that the amount of DNA present in dividing cell populations doubles during interphase, rather than at the time of mitosis.

This conclusion was later verified by experiments in which the rate of DNA synthesis was monitored directly by measuring the incorporation of ^{3}H-thymidine. Such experiments revealed that DNA synthesis is restricted to a defined portion of interphase referred to as **S-phase** (Figure 13-1). S-phase is separated from the previous mitosis by a time gap known as the **G_1-phase;** another gap, termed the **G_2-phase,** occurs between the end of S-phase and the beginning of the next mitosis. Mitosis is usually the shortest phase of the cell cycle, lasting only an hour or two. In typical eukaryotic cells DNA synthesis occupies 6–10 hr and G_2 lasts an average

of 3–5 hr. The most variable phase of the cell cycle is G_1, which is entirely absent in the most rapidly dividing cells and lasts for many hours or days in slowly dividing cells. The variable length of G_1 is thus one of the major factors determining the overall time required for the cell cycle. Once a cell leaves G_1 and initiates DNA synthesis, the progression through S, G_2, and mitosis proceeds at a relatively fixed rate.

Although the terms G_1 and G_2 were originally coined to refer to "gaps" in the cell cycle situated before and after DNA replication, this terminology is potentially misleading because many important events occur during G_1 and G_2, some of which play essential roles in the process of cell growth and division. The initiation of DNA synthesis, for example, appears to depend on events occurring in G_1, while certain events required for chromosome condensation and formation of the mitotic spindle take place during G_2.

The cell cycle of prokaryotic cells differs in several ways from that of eukaryotic cells. Unlike the situation in eukaryotes, where DNA synthesis always occupies a restricted period of the cell cycle, the pattern of DNA synthesis within the bacterial cell cycle has been found to vary significantly, depending upon the overall growth rate of the cells involved. One theory designed to account for such differing patterns of DNA synthesis is the **Cooper–Helmstetter model,** which proposes the existence of two time constants within the bacterial cell cycle; one constant, designated **C,** corresponds to the amount of time required for replication of the chromosomal DNA, while the other constant, **D,** refers to the minimum amount of time necessary between the end of DNA replication and the splitting of the cell in two.

Experiments utilizing radioactive precursors to monitor the rate of DNA synthesis in bacterial cells have revealed that in cell populations that divide once per hour, DNA replication is restricted to the first 40 min (Figure 13-2a). It therefore follows that $C = 40$ min and $D = 20$ min. In bacteria that divide more rapidly than once per hour, the C and D phases begin to overlap one another. For example, in cells that divide every 40 min, a new round of DNA replication is started each time a previous round ends. Although division of the cell still does not occur until 20 min after completion of the first round of DNA synthesis, this overlapping of the C and D phases allows a net cell cycle of 40 min to be generated (Figure 13-2b). It is even possible for a new round of DNA replication to be initiated before the previous round has been completed. The multiple replication sites generated in this way permit cell division cycles that are even shorter than 40 min to be created (Figure 13-2c).

It should be apparent from the preceding description of the cell division cycles of eukaryotic and prokaryotic organisms that two events are crucial in both instances. One is replication of the chromosomal DNA,

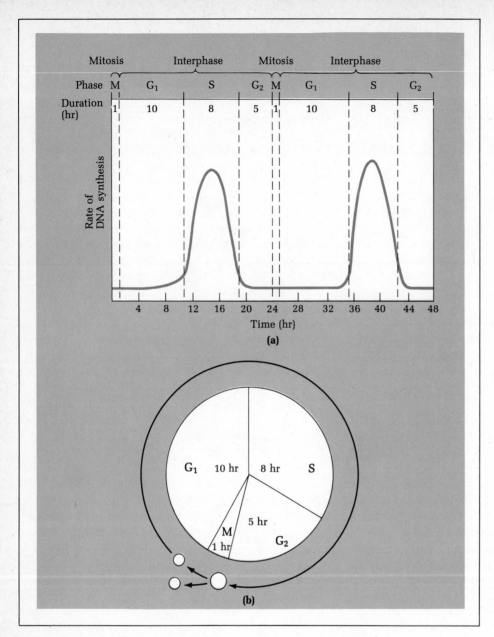

FIGURE 13-1 The eukaryotic cell cycle. (a) Example of the type of data obtained when a synchronous population of eukaryotic cells is incubated in the presence of ^{3}H-thymidine to monitor DNA incorporation. (b) Diagrammatic representation of the cell cycle for such a cell population.

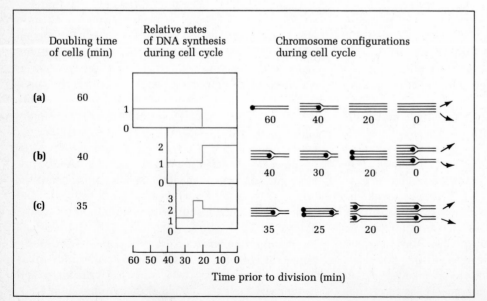

FIGURE 13-2 Model of the cell cycle in the bacterium, *E. coli*. (a) In cultures with a doubling time of 60 min, DNA replication occupies the first 40 min of the cell cycle. (b) In cultures with a doubling time of 40 min, a new round of DNA replication begins as soon as the preceding one ends. (c) In cultures with doubling times shorter than 40 min, a new round of DNA replication begins *before* the preceding round has been completed.

and the other is the physical division of the cell into two separate entities, each containing a complete set of chromosomal DNA sequences. The remainder of the chapter will focus on the mechanisms by which these objectives are accomplished.

DNA REPLICATION AND THE CELL CYCLE

The molecular mechanisms involved in the enzymatic synthesis of DNA have already been described in Chapter 10. However, at that time we did not consider the broader question of how entire chromosomes, containing vast amounts of DNA, are replicated as a whole. Because of differences in the size and molecular organization of prokaryotic and eukaryotic chromosomes, the details of chromosome replication in these two classes of cells will be discussed separately.

Prokaryotic Chromosome Replication

The replication of the bacterial chromosome was first studied in the early 1960s by John Cairns, who incubated cultures of *E. coli* with ³H-thymidine to label the chromosomal DNA. When these cells were gently disrupted and the radioactive DNA visualized by autoradiography, the bacterial chromosome was found to consist of a circular molecule of DNA measuring approximately 1100 µm in circumference. In replicating chromosomes a portion of this circular DNA molecule could be seen to be split into two strands, presumably representing the region where the DNA was unwinding into two chains and undergoing replication (Figure

13-3). This picture raised the question as to whether replication proceeds in one or both directions around the circular DNA molecule. To investigate this issue, experiments were carried out in which bacterial cells grown for brief periods in the presence of a low concentration of ³H-thymidine were subsequently grown in a medium containing a high concentration of ³H-thymidine. If replication proceeds in both directions around the circular chromosome, autoradiographs of DNA isolated from such cells should reveal DNA molecules in which a region containing a low concentration of silver grains (representing replication in the presence of a low concentration of ³H-thymidine) is surrounded on both sides by regions containing a higher concentration of silver grains (representing replication in the presence of a high concentration of ³H-thymidine). This predicted pattern is exactly what was observed, indicating that replication is bidirectional, proceeding in both directions around the circle from a single point of origin (Figure 13-4). Bacterial DNAs thus typically contain a single replicating unit, or **replicon,** a situation very much simpler than that observed in eukaryotic cells.

The discovery that the *E. coli* chromosome contains about 1100 µm of DNA and can be replicated in roughly 40 min led to the conclusion that DNA is replicated at a rate approaching 30 µm/min. To accommodate this replication rate, the DNA double helix must unwind at about 10,000 turns per minute. Such a rapid unwinding rate would generate considerable physical stress on a double-helical molecule existing as a completely closed circle. It has therefore been proposed that nucleases introduce temporary single-strand breaks into the DNA molecule, permitting a more localized unwinding of the

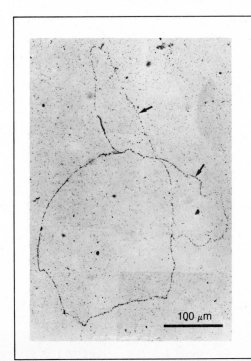
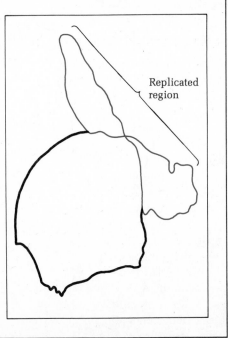

100 µm

Replicated region

FIGURE 13-3 Autoradiograph and accompanying diagram of a replicating *E. coli* chromosome isolated from cells incubated with ³H-thymidine. The two strands marked with arrows represent the region where the chromosomal DNA has split and replicated. Micrograph courtesy of J. Cairns.

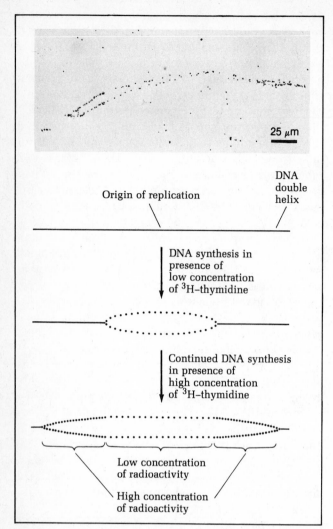

FIGURE 13-4 Autoradiographic evidence for the existence of bidirectional replication of *E. coli* chromosomal DNA. Cells were briefly incubated in the presence of a low concentration of ³H-thymidine, followed by growth in a medium containing a high concentration of ³H-thymidine. In this autoradiograph two newly replicated DNA strands can be seen. Note that the labeling is heavier at the two ends than in the middle. As is shown in the diagram, bidirectional replication from a single origin would be expected to produce such a pattern. Micrograph courtesy of D. M. Prescott.

double helix (Figure 13-5). Such breaks could then be repaired by the action of DNA ligase.

Within the intact cell the replicating chromosome is thought to be intimately associated with the plasma membrane. This notion was originally prompted by electron microscopic studies showing that the nuclear region (nucleoid) of bacterial cells is often associated with invaginations of the plasma membrane known as **mesosomes.** Biochemical analyses carried out on bacterial cells grown in the presence of radioactive DNA precursors subsequently revealed that the newly synthesized DNA, as well as DNA polymerase activity, is enriched in the plasma membrane fraction. Although the preceding observations suggest the existence of a close relationship between bacterial DNA replication and cellular membranes, the chemical nature of this

association is yet to be defined, nor is there any clear indication that replication actually requires membrane attachment. As we shall see later, it is possible that the observed chromosome-membrane association is related more to the process of parceling out the replicated DNA molecules to the two newly forming cells than it is to the actual process of DNA replication per se.

Since DNA synthesis is a crucial prerequisite for the process of cell division, the factors that trigger the initiation of DNA synthesis have been the focus of considerable study. In bacterial cells a specific DNA sequence, about 250 base pairs in length, has been identified as the start site for chromosomal DNA replication. Genetic analyses of mutant organisms defective in initiating DNA synthesis at this site have implicated several kinds of proteins in the initiation process. A requirement for the synthesis of specific initiator proteins is also suggested by the discovery that the initiation of DNA replication can be blocked by exposing cells to inhibitors of protein synthesis. Such inhibitors do not block a round of DNA replication that is already in progress, however, indicating that these proteins are specifically required for the initiation of replication rather than for the process of DNA synthesis per se. Taken together, the preceding observations suggest that specific initiator

FIGURE 13-5 Schematic diagram showing how a nick in one strand of the DNA double helix could permit localized unwinding during chromosomal DNA replication.

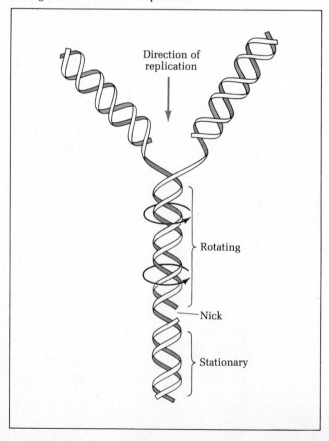

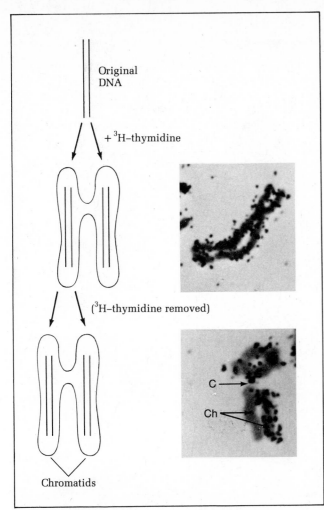

Original
DNA

+ ³H–thymidine

(³H–thymidine removed)

C

Ch

Chromatids

FIGURE 13-6 Experiment demonstrating semiconservative distribution of newly synthesized DNA into chromatids of eukaryotic cells. Cells of the broad bean *Vicia* were incubated for one round of cell division in the presence of ³H-thymidine, followed by one round in the absence of ³H-thymidine. Note that the two chromatids of each chromosome are radioactive after the first round of DNA replication, while the radioactivity is predominantly localized in one of the two chromatids after the second round of DNA replication. As shown in the diagram, this pattern is compatible with the existence of semiconservative replication. Micrographs courtesy of J. H. Taylor. Abbreviations: C = centromere, Ch = chromatid.

proteins trigger the onset of bidirectional replication at a particular site in the bacterial DNA molecule, and that DNA synthesis then proceeds in both directions around the circular DNA without the need for any new initiator proteins or initiation events.

Eukaryotic Chromosome Replication

Several factors make the replication of eukaryotic chromosomes more complex than that of prokaryotic chromosomes. First, the DNA content of the typical eukaryotic nucleus is roughly three orders of magnitude greater than that of the prokaryotic nucleoid. Though this increased DNA content is distributed among multi-ple chromosomes, these individual chromosomes still contain from ten to a hundred times more DNA than a typical bacterial chromosome. An additional level of complexity is introduced by the nucleosomal organization of eukaryotic chromatin that, as we shall see shortly, must be disassembled at the site of DNA replication and reassembled on the two newly forming DNA molecules.

In spite of these additional complexities, DNA synthesis in eukaryotic cells follows a semiconservative replication pattern similar to that occurring in prokaryotes. This replication mechanism was first investigated in the late 1950s by J. Herbert Taylor, who briefly exposed proliferating plant cells to ³H-thymidine and then examined the chromosomal distribution of the radioactivity using light microscopic autoradiography. In eukaryotic organisms the two new chromosomes generated when each chromosome is replicated remain attached to one another until just prior to cell division. As long as they remain attached to each other these newly replicated chromosomes are referred to as **chromatids,** while the region holding each chromatid pair together is referred to as the **centromere.** Taylor discovered that during the first cell division after exposure of cells to ³H-thymidine, both chromatids are radioactive. During the next round of cell division, however, most replicated chromosomes contain one radioactive chromatid and one nonradioactive chromatid. The most straightforward interpretation of such an organization is that each chromosome contains a single molecule of DNA that replicates by a semiconservative mechanism. The two DNA molecules generated by such a process, each containing one old strand and one new strand, are then parceled into the two paired chromatids (Figure 13-6).

Although eukaryotic chromosome replication resembles prokaryotic chromosome replication in that both are based on a semiconservative mode of DNA synthesis, a significant difference in mechanism still distinguishes the two processes from one another. As we saw earlier, replication of the bacterial chromosome is initiated at a single site on the DNA molecule. Autoradiographic analysis of DNA isolated from eukaryotic cells briefly exposed to ³H-thymidine, on the other hand, has revealed the presence of multiple replication sites within each DNA molecule (Figure 13-7). These individual replication units, or **replicons,** vary from a few micrometers to several hundred micrometers in length. In order to determine the direction of replication within each replicon, experiments have been carried out in which cells were first exposed to low concentrations of ³H-thymidine and then to high concentrations of ³H-thymidine. Under such conditions the first DNA to be synthesized should have less radioactivity than the DNA synthesized later. When DNA molecules isolated from such cells are examined by autoradiography, the DNA in the center of each replicon is found to be less

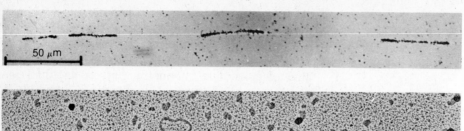

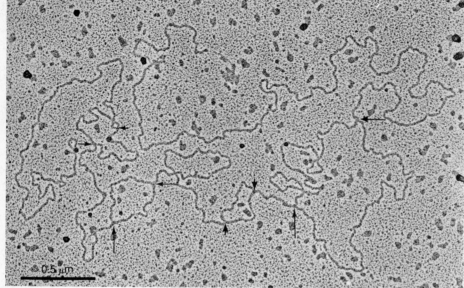

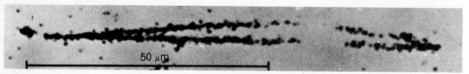

FIGURE 13-7 Autoradiographic and direct electron microscopic visualization of eukaryotic repli-cons. *(Top)* Autoradiograph of a replicating eukaryotic DNA molecule isolated from cells briefly incubated in the presence of ³H-thymidine. Note the presence of several distinct regions of radioac-tive DNA, each of which represents a separate area of replication. *(Middle)* Electron micrograph of a eukaryotic DNA molecule showing four separate replicons undergoing replication. The two ends of each replicon are indicated with arrows. *(Bottom)* Autoradiographic evidence for the occurrence of bidirectional DNA replication in eukaryotic replicons. Chinese hamster cells were briefly exposed to a low concentration of ³H-thymidine, followed by growth in the presence of a high concentration of ³H-thymidine. Note that the labeling is heavier at the two ends of this replicon than in the middle, as would be predicted if replication were bidirectional (compare Figure 13-4). Top and bottom micrographs courtesy of J. A. Huberman. Middle micrograph courtesy of D. R. Wolstenholme.

radioactive than the DNA at either end (see Figure 13-7, *bottom*). This pattern is consistent with the theory that DNA replication is initiated at a central site in each replicon, forming two replication forks that move away from one another. When the replication forks of adja-cent replicons reach one another, the newly synthe-sized DNA from the two replicons is joined together by the action of DNA ligase. In this way DNA synthesized at multiple replication sites is ultimately spliced to-gether to form a complete new chromosomal DNA mol-ecule (Figure 13-8).

At least three reasons can be proposed to explain why DNA replication involves multiple replication sites in eukaryotic, but not prokaryotic, cells. The first is that eukaryotes possess more than one chromosome, a situa-tion demanding more than one replicon. Eukaryotic chromosomes also contain several orders of magnitude

more DNA than do prokaryotic chromosomes, which would take too long to replicate if initiation occurred at only a single origin. Finally, the rate of DNA synthesis is inherently slower per replication site in eukaryotic DNA than it is in prokaryotic DNA. Direct measure-ments of the length of radioactive DNA synthesized in eukaryotic cells exposed to ³H-thymidine for varying periods of time have indicated that replicons synthesize DNA at a rate of about 0.5 μm/min, which is consider-ably slower than the 30 μm/min typical of bacterial cells. Because of the slower pace of eukaryotic DNA synthesis, it would take many weeks to replicate a typi-cal eukaryotic chromosome if multiple replication sites were not invoked.

DNA replication is thought to be inherently slower in eukaryotic cells because the nucleosomal organiza-tion of the chromatin fibers must be disassembled before the two strands of the DNA double helix can unwind and replicate. Although the details of this disassembly process are yet to be completely elucidated, the behavior of the histone octomers during DNA replication has been investigated by Harold Weintraub and his associates. In these experiments cells actively synthesizing DNA were grown in the presence of density-labeled amino acids to label newly formed histones. Because they are denser than normal, these newly synthesized histones could then be separated from previously existing histones by isodensity centrifugation. Analysis of the resulting centrifugation patterns revealed the presence of two classes of histone octomers: octomers consisting entirely of newly synthesized dense histones, and octomers consisting of old, less dense histones. Since old and new histones did not become mixed in the octomers, it was concluded that the histone octomers remain as intact particles when they dissociate from DNA during the replication process.

As DNA replication proceeds, nucleosomes quickly appear again on each of the two newly forming DNA molecules. Since the total amount of DNA is doubled during this process, the total number of nucleosomes must also be doubled. Half these nucleosomes are generated from "old" histone octomers that randomly reassociate with the two newly forming DNA molecules as replication progresses. The remaining nucleosomes are assembled from newly synthesized histones. Although the details of the assembly process are not completely understood, a nonhistone protein called **nucleoplasmin** appears to be required for generating new nucleosomes from individual histones.

At one time it was thought that eukaryotic DNA replication occurs in association with the nuclear envelope, just as replication in prokaryotic cells appears to be membrane associated. Rigorous autoradiographic studies have revealed, however, that incorporation of ^{3}H-thymidine into newly synthesized DNA occurs throughout the nucleus in eukaryotic cells, and not just adjacent to the membranes of the nuclear envelope. It has recently been discovered that newly synthesized DNA molecules are present in high concentration in

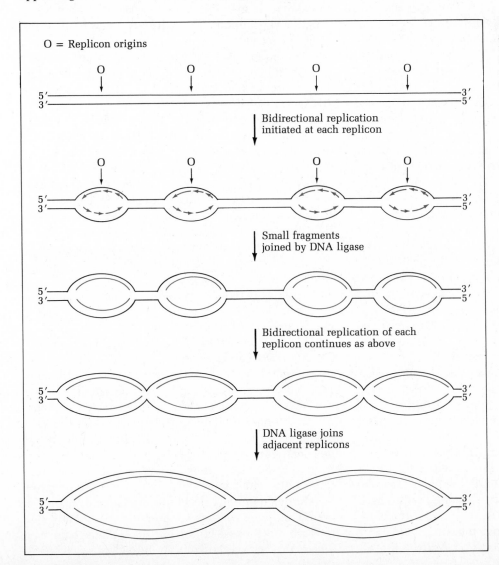

FIGURE 13-8 Schematic model summarizing the relationship between the bidirectional replication of individual replicons and replication of a eukaryotic chromosomal DNA molecule as a whole.

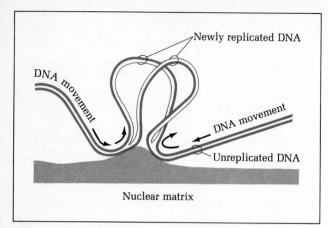

FIGURE 13-9 Hypothetical model illustrating how bidirectional DNA replication might be achieved by fixed replication sites associated with the nuclear matrix. The newly replicated DNA emerges as a loop from the center of the replication site as the unreplicated DNA enters from both sides.

isolated nuclear matrix preparations, raising the possibility that initiation sites for DNA replication are anchored to the nuclear matrix. This information has led to the suggestion that DNA is synthesized as the chromatin fibers pass through fixed replication sites associated with the nuclear matrix (Figure 13-9). Although this model is compatible with electron microscopic observations showing that chromatin fibers loop out from attachment sites on the nuclear matrix, the significance of this arrangement for DNA replication remains to be experimentally verified.

Since the average DNA replication rate for higher eukaryotes is 0.5 μm per replicon per minute, and a typical S-phase lasts about 8 hr (480 min), it can be calculated that a single replicon would achieve a maximum length of about $0.5 \times 480 = 240$ μm *if* it were active during the entire S-phase. Direct autoradiographic analyses of replicating DNA molecules have revealed, however, that the average replicon is significantly smaller than this. Typical replicon sizes fall in the range of 15–60 μm in length, which corresponds to roughly 45,000–180,000 base pairs of DNA. Since it would take less than an hour or two to synthesize this much DNA, it can be concluded that each individual replicon is active for only a small portion of the S-phase.

As in the case of prokaryotic cells, specific factors appear to be required for the initiation of DNA synthesis in eukaryotes. Experimental support for this idea has emerged from nuclear transplantation studies in which nuclei removed from cells in G_1 or G_2 have been found to initiate DNA synthesis when transplanted into the cytoplasm of cells in S-phase. Conversely, nuclei from S-phase cells stop synthesizing DNA when they are transplanted into the cytoplasm of G_1 or G_2 cells. These two kinds of experiments suggest that either the cytoplasm of S-phase cells contains molecules that trigger the initiation of DNA synthesis or the cytoplasm of G_1

and G_2 cells contains an inhibitor of DNA synthesis. To distinguish between these alternatives, studies have been carried out in which entire G_1-phase cells were fused with S-phase cells. Such fusion triggers the initiation of DNA synthesis in the G_1 nucleus rather than cessation of DNA replication in the S-phase nucleus, confirming the presence of initiator substances in the S-phase cytoplasm.

Although these initiator molecules have not yet been clearly identified, they are thought to be proteins because treatment of late G_1-phase cells with inhibitors of protein synthesis has been found to prevent the subsequent initiation of DNA synthesis. Unlike the situation in prokaryotic cells, however, protein synthesis inhibitors also depress eukaryotic DNA synthesis that is already in progress. This difference is thought to reflect the fact that prokaryotic chromosomes have only one site at which replication is initiated, while eukaryotic DNA replication involves multiple replicons that initiate DNA replication at various times throughout S-phase. The requirement for continued protein synthesis in eukaryotes may therefore be related to the need to progressively initiate DNA synthesis at new replicons as S-phase proceeds.

The notion that DNA replication is not simultaneously initiated in all replicons has received further support from experiments in which newly synthesized DNA was labeled by exposing synchronized cell populations to ^{3}H-thymidine at varying times during S-phase. Such studies have revealed that DNA sequences that replicate early in S-phase are enriched in the bases G and C, while DNA synthesized later in S-phase is enriched in the bases A and T. If cells labeled at different times during S-phase are later examined during mitosis, when individual chromosomes can be visualized, light microscopic autoradiography reveals that different regions of the chromosomes are radioactive, depending on when during S-phase the ^{3}H-thymidine is administered. The most striking observation to emerge from such studies is that heterochromatic chromosome regions tend to be replicated relatively late in S-phase in comparison to euchromatic regions.

The length of time required to complete the overall process of DNA replication in eukaryotic cells is determined by the frequency with which DNA synthesis is initiated at the multiple replicons dispersed throughout the DNA. This principle is nicely illustrated by a comparison of the overall rates of DNA synthesis in embryonic and adult cells of the fruit fly, *Drosophila*. In the developing embryo S-phase is reduced to a few minutes to allow rapid cell division to occur; under these conditions, active replicons are numerous and closely spaced together. In dividing cells of the adult organism, on the other hand, replicons appear much less frequently and the resulting S-phase lasts about 10 hr. Since the rate at which DNA is synthesized by individual replicons is

FIGURE 13-10 Chromosome separation during bacterial cell division. The schematic diagram illustrates how attachment of bacterial chromosomes to the plasma membrane is thought to facilitate the parceling out of the chromosomes to the two newly forming cells during bacterial cell division. The electron micrograph of a dividing *Streptococcus* shows the transverse septum in the process of dividing the cell in two. Courtesy of M. L. Higgins and L. Daneo-Moore.

about the same in both cases, it is clear that the frequency at which new replicons are being initiated, rather than the rate at which each one synthesizes DNA, is the major factor determining the length of S-phase.

Once replication of the chromosomal DNA has been carried out, preparation can be made for the process of cell division. In addition to physically dividing the cytoplasm in half, this division must ensure that a complete set of replicated DNA molecules is passed to each of the two newly formed cells. In the following sections we shall see that the mechanism by which this separation is accomplished differs between prokaryotic and eukaryotic cells, and even among different eukaryotic cell types.

PROKARYOTIC CELL DIVISION

In bacterial cells separation of the duplicated chromosomes into the two newly forming cells is facilitated by the fact that the chromosomes are attached to the plasma membrane. As illustrated in Figure 13-10, the gradual addition of new membrane material to the plasma membrane in the region located between the attachment sites of the two chromosomes causes the chromosomes to gradually separate from one another. After adequate chromosome separation has occurred the cytoplasm is divided, creating two cells containing complete copies of the chromosomal DNA. During the process of cytoplasmic division, or **cytokinesis,** the plasma membrane first grows across the center of the cell, forming a doubled-layered **transverse septum.** The cell wall then extends into the space between these newly forming membranes, producing a **cross-wall** that is considerably thicker than a normal cell wall. This thickened wall is then cleaved down the middle, generating two separate cells. The importance of cell wall formation to the process of cell division has been demonstrated by experiments in which inhibitors of cell wall formation, such as penicillin, have been found to trigger the formation of wall-less cells (protoplasts) that exhibit great difficulty in dividing. Although such observations suggest that the cell wall plays an important role in the normal process of cell division, it should be pointed out that some bacteria, such as *Mycoplasma*, lack cell walls and yet are still capable of dividing.

A special type of cell division occurs in certain bacteria when they are exposed to harsh environmental

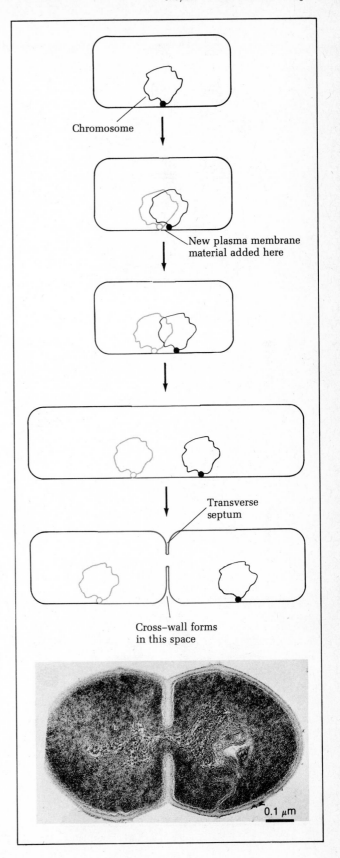

Chromosome

New plasma membrane material added here

Transverse septum

Cross-wall forms in this space

0.1 μm

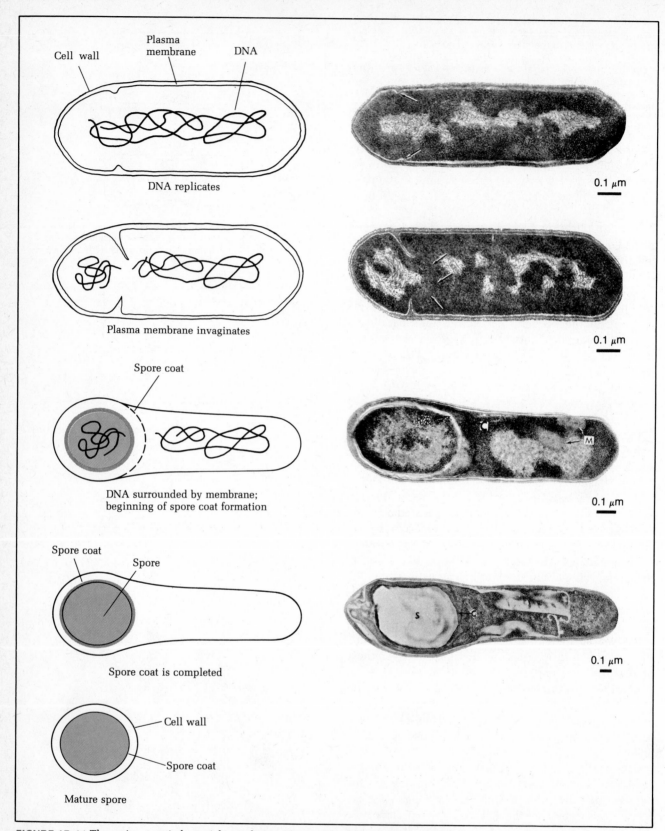

FIGURE 13-11 The major steps in bacterial sporulation. Micrographs courtesy of H. R. Hohl.

conditions that threaten cell survival. Under such conditions these bacteria are converted into dormant, inert structures known as **spores** (Figure 13-11). The process of spore formation begins with a round of chromosome replication, followed by migration of one of the two newly formed chromosomes to one end of the cell. A plasma membrane partition then forms across this end of the cell, sequestering the chromosome in a small cytoplasmic compartment. This region of the cell is next enveloped by a series of walls that together form a thick protective **spore coat** around the chromosome. Most of the water is then removed, generating a dehydrated spore capable of withstanding extremely harsh conditions and existing in a dormant state for many years. When environmental conditions become more hospitable, the spore germinates and becomes a normal cell. During germination the cell water is replenished, the spore coat is discarded, and the cell resumes the characteristics of a normal cell, all within a few hours.

EUKARYOTIC CELL DIVISION: MITOSIS

In eukaryotic cells the prokaryotic system of chromosome attachment to the plasma membrane is replaced by an association of the chromosomes with the nuclear envelope and microtubules. In some primitive eukaryotes, such as the unicellular algae known as dinoflagellates, the nuclear envelope appears to pull the chromosomes apart in a manner reminiscent of prokaryotic cell division. An added feature in even these most primitive eukaryotes, however, is the use of microtubules to aid chromosome movement. In most higher eukaryotes the nuclear envelope disintegrates at the time of cell division and the microtubules assume prime responsibility for chromosome separation, forming an elaborate structure known as the **spindle apparatus** or **mitotic spindle.** In such organisms the process of nuclear division is referred to as **mitosis** or **meiosis,** depending upon the behavior of the duplicated chromosomes. Since mitosis is the more general type of cell division, it will be described first.

Summary of Mitotic Stages

Mitosis consists of four stages known as **prophase, metaphase, anaphase,** and **telophase** (Figure 13-12). Though these four phases can be readily distinguished from one another by differences in chromosome appearance and behavior, in reality mitosis is a continuous process that has been arbitrarily subdivided into these stages to facilitate discussion. Before proceeding with a detailed discussion of the mechanisms underlying the overall process of mitosis, we shall summarize the main events associated with each of these four stages.

Prophase. The chromosomal DNA of most interphase cells is present in the form of a diffuse network of intertwined chromatin fibers, making it impossible to discern individual chromosomes. Prophase marks the point in the cell cycle in which these chromatin fibers first condense into discrete structures recognizable as separate chromosomes. As each chromosome becomes visible, it can be seen to be composed of a pair of chromatids joined together by a centromere. As the chromatin fibers condense into chromosomes, transcription of the chromosomal DNA slows down and eventually stops. In animal cells the nucleoli generally disperse during prophase, while plant cell nucleoli either remain as discrete entities, undergo partial disruption, or disappear entirely as in animal cells.

In addition to these changes in chromosome and nucleolar morphology, assembly of the mitotic spindle is also initiated during prophase. The onset of mitotic spindle formation in animal cells is associated with the migration of two pairs of cytoplasmic **centrioles** to opposite sides of the nucleus (centrioles are microtubule-containing structures whose structural organization and mode of replication were described in Chapter 9). After the two pairs of centrioles have become situated on opposite sides of the nucleus adjacent to the nuclear envelope, short microtubules assemble close to each centriole pair. These short microtubules radiate in all directions from the centrioles, forming structures known as **asters.** At about the same time, longer microtubules destined to form the bulk of the spindle apparatus begin to assemble between the opposite pairs of centrioles. The membranes of the nuclear envelope then begin to fragment, reducing the envelope to a series of small vesicles dispersed throughout the cytoplasm. Soon thereafter the nuclear pore complexes disappear from these vesicles, making them indistinguishable from membranes of the endoplasmic reticulum. This breakdown of the nuclear envelope marks the end of prophase.

We shall return for a more detailed discussion of the assembly and functioning of the mitotic spindle after summarizing the events that occur during the remaining stages of mitosis. It should be noted at this point, however, that most plant cells lack both centrioles and asters, and yet are capable of forming functional mitotic spindles.

Metaphase. As the nuclear envelope disappears, the developing spindle apparatus penetrates into the region originally occupied by the nucleus. A small number of spindle microtubules become attached to the centromere region of each chromosome, followed by migration of the chromosomes toward the equator of the spindle. Upon reaching the equator, the chromosomes line up adjacent to one another. At this point the centromere region of each chromosome is attached to two sets of microtubules, one set emerging from each pole of the spindle.

Metaphase is the easiest time to count and identify

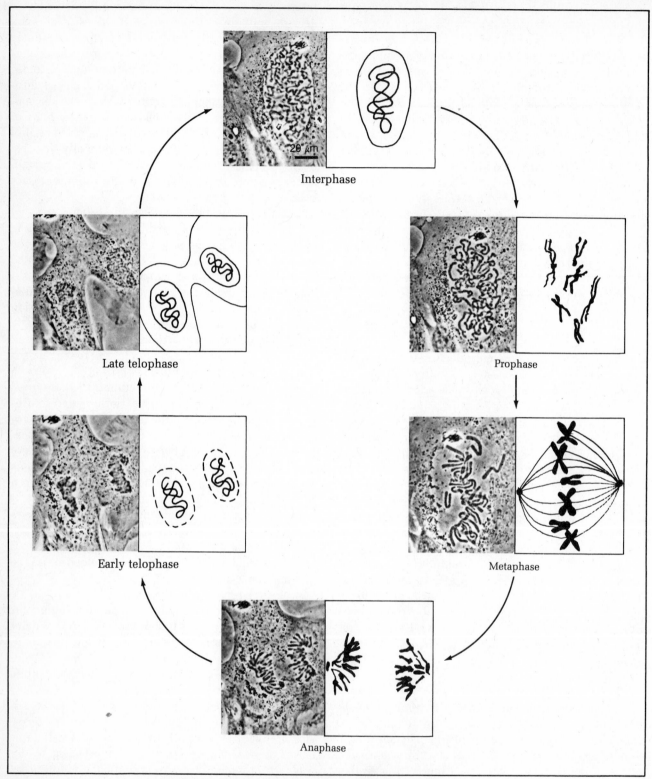

FIGURE 13-12 The stages of mitosis as illustrated by the appearance of lung cells of the newt. As the cell enters prophase, the nuclear envelope disintegrates and the chromosomes condense. At metaphase the chromosomes become aligned at the equator of the newly formed spindle apparatus. At this stage each chromosome can be seen to consist of two chromatids held together by a centromere. At anaphase the two chromatids comprising each metaphase chromosome separate from one another and these newly formed chromosomes move toward the spindle poles. During early telophase the spindle apparatus breaks down, a new nuclear envelope begins to form (dotted line), and the chromosomes begin to uncoil. By late telophase the nuclear envelope is intact and cytokinesis is almost finished. Micrographs courtesy of C. L. Rieder.

posite poles of the spindle. During this migration the centromere region of the chromosome leads the way, suggesting that the chromosomes are being pulled by the microtubules attached to these centromeres. While this type of chromosome movement is taking place, the two poles of the mitotic spindle also move away from one another. Thus chromosome-to-pole migration and pole-pole separation are both involved in separating the two sets of chromosomes from each other (Figure 13-14). The theories that have been proposed to explain these two types of movements will be discussed later in the chapter when the properties of the mitotic spindle are described in more detail.

Telophase. The beginning of telophase is marked by the arrival of the two sets of chromosomes at the opposite poles of the spindle. Upon arrival at the poles the individual chromosomes unfold, dispersing into

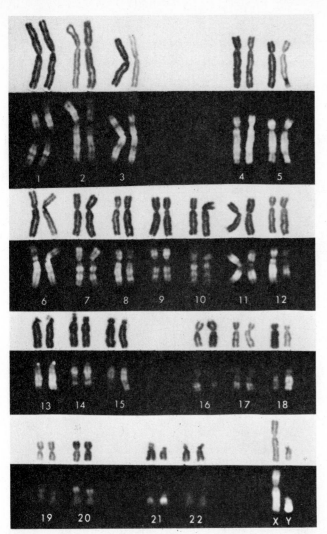

FIGURE 13-13 A mitotic karyotype of human chromosomes obtained from metaphase-arrested cells. On the black background are chromosomes stained with the dye quinicrine. On the white background are the same chromosomes stained with orcein after quinicrine treatment. Although orcein is a good stain for giving defined chromosome outlines, quinicrine or some other banding stain is necessary for matching up many homologs. Courtesy of I. Uchida.

individual chromosomes. Agents that interfere with the assembly of the mitotic spindle, such as colchicine, halt the normal mitotic process at metaphase and can therefore be used to generate cell populations enriched in metaphase-arrested cells. Microscopic examination of the chromosomes of metaphase-arrested cells allows individual chromosomes to be identified and classified on the basis of differences in size and shape, generating a chromosomal analysis known as a **karyotype** (Figure 13-13).

Anaphase. At the beginning of anaphase the two chromatids comprising each metaphase chromosome separate from one another, thereby generating two independent chromosomes. The two newly formed chromosomes derived from each chromatid pair then migrate in opposite directions until they reach the op-

FIGURE 13-14 Diagram illustrating the two types of movement that contribute to chromosome separation during anaphase: chromosome-to-pole migration (reflected by a gradual decrease in the distance "CP"), and pole-pole separation (reflected by an increase in the distance "PP").

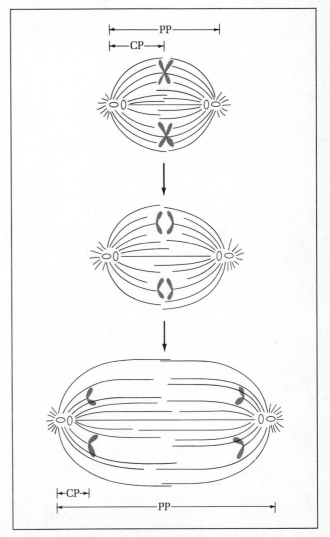

typical interphase chromatin fibers. These changes are accompanied by the reappearance of nucleoli (if they had disappeared), the breakdown of the mitotic spindle, and the resumption of DNA transcription. At the same time vesicles that are indistinguishable from membranes of the endoplasmic reticulum begin to condense around the two sets of uncoiling chromosomes. These membrane vesicles gradually coalesce to form two double-layered membrane envelopes, each enclosing the chromatin fibers derived from one of the two chromosome sets. Nuclear pore complexes develop close to points where chromatin fibers become attached to the inner membranes of the newly forming envelopes. Division of the cytoplasm, or **cytokinesis,** occurs in synchrony with these nuclear changes, leading to the physical separation of the original cell into two new cells.

Now that the general features of the four stages of mitosis have been summarized, we can proceed with a more detailed discussion of the crucial steps involved. The topics of central importance in this regard are: (1) the structure of mitotic chromosomes, (2) the structure and assembly of the mitotic spindle, (3) the mechanism of chromosome movement, and (4) the mechanism of cytokinesis.

Structure of Mitotic Chromosomes

During the early stages of mitosis, the dispersed mass of chromatin fibers present in the interphase nucleus condenses into a series of discrete chromosomes. By the time metaphase has arrived, each individual chromatin fiber has been packaged into a mitotic chromosome whose overall length is about 7000-fold shorter than the extended length of the DNA molecule it contains. Electron microscopic examination of intact metaphase chromosomes has revealed that they are composed of an elaborately folded mass of 25-nm fibers that are morphologically indistinguishable from interphase chromatin fibers (Figure 13-15). Some fibers appear to be looped out from the longitudinal axis of the chromatid, while others are oriented parallel to one another longitudinally. A variety of experimental evidence suggests that each chromosome consists of a single, intricately folded chromatin fiber containing a single continuous molecule of DNA. During S-phase the DNA molecule contained within such a chromatin fiber is replicated, forming two chromatin fibers. Since the two chromatids derived from each round of replication remain attached to one another until the beginning of anaphase, which occurs several hours later, it has been proposed that a small region of DNA may remain unreplicated during S-phase. According to this theory, the two DNA molecules generated during the replication of each chromosome remain joined to each other at this unreplicated region, holding the two chromatids together until the

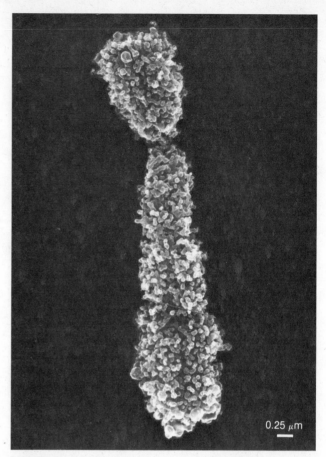

FIGURE 13-15 Scanning electron micrograph of an intact metaphase chromosome isolated from Chinese hamster cells. Courtesy of S. M. Gollin and W. Wray.

appropriate stage of mitosis (Figure 13-16). Replication of this segment then permits the two chromatids to separate, forming two independent chromosomes.

Although the preceding description provides an overview of the relationship between chromatin fibers and mitotic chromosomes, it does not explain the nature of the forces that guide the folding of the chromatin fiber. As mentioned in Chapter 11, electron microscopic observations of mitotic chromosomes from which histones and most nonhistone proteins have been removed reveal the presence of a **scaffold** of nonhistone proteins

FIGURE 13-16 Schematic model illustrating how the two chromatids comprising each metaphase chromosome may be held together by an unreplicated region of the original chromatin fiber.

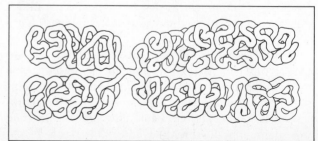

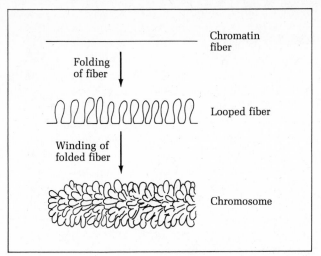

FIGURE 13-17 The radial loop model of chromosome organization.

to which loops of DNA are attached (see Figure 11-22). Isolated scaffolds retain the same overall shape as the chromosome from which they are derived, suggesting that this protein complex is the structural backbone of the chromosome. Such observations have led to the formulation of the **radial loop model** of chromosome structure, which proposes that the conversion of interphase chromatin fibers into mitotic chromosomes involves the folding of these fibers into a series of successive loops that become attached to a central nonhistone-protein scaffold (Figure 13-17).

The pattern in which a given chromatin fiber is folded in its respective chromosome appears to be highly reproducible. Some of the most striking evidence for the existence of reproducible patterns of chromatin folding has emerged from studies on chromosome staining. This approach received a great impetus in the late 1960s when Torbjöern Caspersson discovered that staining of chromosomes with certain dyes generates a reproducible pattern of stained or fluorescent bands. Staining of metaphase chromosomes with quinacrine mustard, for example, gives rise to a reproducible series of fluorescent bands referred to as **Q-bands.** A similar banding pattern is observed when chromosomes are stained with a dye known as Giemsa stain. By altering the conditions under which the chromosomes are prepared and stained, other types of bands, such as **C-, R-,** and **G-bands,** have also been observed. The banding patterns exhibited by a given type of chromosome are so characteristic for that particular type of chromosome that they can be used to identify chromosomes that are difficult to distinguish from one another on the basis of size and centromere position alone (Figure 13-18). The discovery that each type of staining technique always produces the same pattern of bands on a given type of chromosome clearly indicates that each chromosome is always folded in the same precise manner.

Although the molecular basis for specific chromo-

some-banding patterns is not completely understood, the staining of some bands appears to be caused by the presence of heterochromatin. As discussed in Chapter 11, the term *heterochromatin* refers to chromosomal material that remains condensed and transcriptionally inactive during interphase. The banding patterns observed in metaphase chromosomes appear to be related to at least two distinct types of constitutive, or permanently inactivated, heterochromatin. Constitutive heterochromatin located near the centromere, termed **centromeric heterochromatin,** is enriched in A-T residues located in highly repeated satellite DNA sequences and is intensely stained by techniques that visualize C-bands. In contrast, constitutive heterochromatin located in the chromosome arms, known as **intercalary heterochromatin,** is selectively stained by techniques that visualize Q- and G-bands. Unlike centromeric heterochromatin, intercalary heterochromatin is not enriched in repetitive DNA. The reason for distinguishing between these two subclasses of constitutive heterochromatin is that even though they share certain properties such as late replication, transcriptional inactivity, and permanent condensation, the observed differences in their biochemical properties suggest that they perform different functions.

The region of the metaphase chromosome whose

FIGURE 13-18 Staining of human chromosomes 3, 7, and 12 for R-bands *(left)* and G-bands *(right).* The broader lines connect the centromere regions. The narrow lines show that the R-bands on the left chromosome correspond to nonstaining regions between the G-bands on the right chromosome. Courtesy of O. J. Miller.

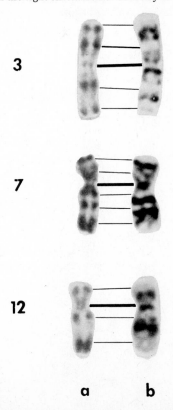

3

7

12

a   b

structural and functional properties are best understood is the **centromere.** Under the light microscope the centromere appears as a narrowing of the chromosome where the two paired chromatids join one another. The distance between the centromere and the two ends of the chromosome varies among different chromosomes, but is always the same for a given type of chromosome. During cell division two important functions are carried out by the centromere: (1) It provides the site of attachment for the spindle microtubules, and (2) it holds the two chromatids together until the beginning of anaphase. Electron micrographs of metaphase chromosomes have revealed that these two functions are related to separate structural entities. Attachment of the chromatids to the spindle microtubules is mediated by two specialized structures, known as **kinetochores,** which are attached to opposite sides of the centromere. The kinetochores are disk- or ball-shaped structures that measure about 200 nm across and consist of several layers of finely granular or fibrillar material to which a dozen or so microtubules are attached (Figure 13-19). Since the two kinetochores lie on opposite sides of the centromere and therefore face in opposite directions, they attach to microtubules emanating from opposite poles of the spindle. This arrangement ensures that the two chromosomes are pulled to opposite poles of the spindle during anaphase.

As was mentioned earlier, the mechanism by which the centromere functions to hold the two chromatids together may be related to the presence of a DNA segment whose replication was not completed during the previous S-phase. An important corollary of this theory is that a small amount of DNA synthesis must occur before the chromatids can separate. Support for this notion has been obtained from studies in which it was shown that chromatid separation during anaphase can be blocked by adding inhibitors of DNA synthesis to the culture medium. Such observations suggest that replication of the centromeric DNA may be responsible for triggering the onset of chromosome separation. The factors responsible for delaying the replication of this critical DNA region from S-phase until anaphase have not been identified, though it is known that the centromere is enriched in an unusual type of highly repetitive, satellite DNA.

In addition to the centromere, other kinds of narrowed regions called **secondary constrictions** have also been observed in metaphase chromosomes. Some secondary constrictions, called *nucleolar organizer regions* (page 451), contain DNA sequences coding for ribosomal RNA. The significance of other types of secondary constrictions is unknown, though their reproducible occurrence in specific chromosome locations has been useful in distinguishing individual chromosomes from one another.

The discovery that centromeres, secondary constrictions, and intercalary heterochromatin bands (Q- and G-bands) always appear in reproducible locations within a given type of chromosome suggests the existence of an orderly and reproducible folding of the chromatin fiber within metaphase chromosomes. Although the role of the nonhistone-protein scaffold in anchoring the chromosomal fiber has already been pointed out, it is not clear whether this scaffolding plays an active role

FIGURE 13-19 Relationship between kinetochores and chromosome organization. *(Left)* Diagrammatic representation of a metaphase chromosome showing the relationship between the centromere and the kinetochores. *(Right)* Thin-section electron micrograph of a metaphase chromosome showing the attachment of spindle microtubules to the kinetochore (arrowhead). Courtesy of C. L. Rieder.

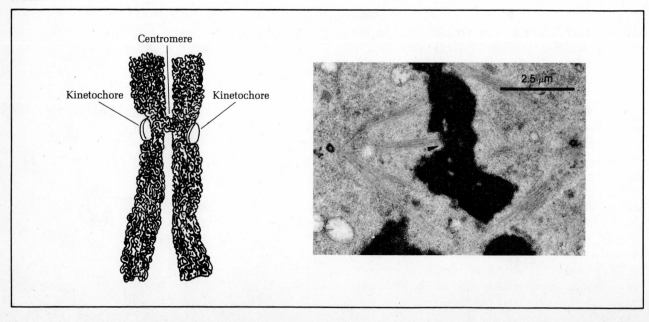

in organizing the pattern in which the chromatin fiber condenses, or is simply an inert framework upon which condensation occurs. Another factor that may help to guide chromosome folding is attachment of the interphase chromatin fibers to the nuclear envelope or nuclear matrix. If these types of chromatin attachment sites are stable and specific for certain regions of the chromatin fibers, they could serve as fixed structural points that organize the condensation process.

At the molecular level, several lines of evidence suggest that protein phosphorylation plays a crucial role in triggering the onset of chromosome condensation and folding. The initial stimulus for inducing protein phosphorylation is provided by **maturation promoting factor (MPF),** a substance that has been isolated from cells that are undergoing the transition from interphase to mitosis or meiosis. When injected into cells that have already passed through S-phase, MPF is found to trigger nuclear envelope breakdown, chromosome condensation, and spindle formation. One of the most dramatic biochemical responses to MPF administration is an increased phosphorylation of the major structural proteins of the nuclear lamina, **lamins A and C.** Phosphorylation of the nuclear lamina is followed by depolymerization of the lamina, condensation of the chromatin, and breakdown of the nuclear envelope, though it is not clear that these are all causally related to lamin phosphorylation.

In addition to lamin phosphorylation, a significant increase in the phosphorylation of histone H1 also occurs just prior to chromosome condensation. In order to investigate the question of whether this change in histone structure plays a critical role in chromosome condensation and the initiation of mitosis, experiments have been carried out in which the enzyme responsible for catalyzing this phosphorylation reaction, **histone kinase,** was injected into cells emerging from S-phase. Under such conditions chromosome condensation and mitosis were found to be initiated promptly, rather than after the normal G_2 delay of several hours. Such experiments suggest that the onset of chromosome condensation and the initiation of mitosis is induced by the phosphorylation of histone H1. The conclusion that phosphorylation of this particular histone is involved in the mechanism of chromosome condensation is consistent with the localization of histone H1 between adjacent nucleosomes and other evidence implicating this histone in regulating the higher-order folding of the chromatin fiber (page 401).

Structure and Assembly of the Mitotic Spindle

In most eukaryotic cells, chromosome movement is dependent on the presence of the mitotic spindle. This spindle apparatus contains at least three different categories of microtubules: (1) **astral microtubules,** which

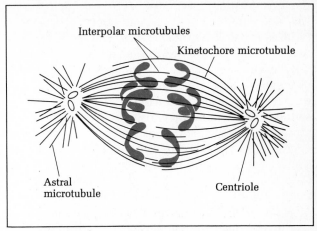

FIGURE 13-20 Diagrammatic model of the spindle apparatus, illustrating the three types of spindle microtubules.

radiate from the centrioles (in animal cells only), (2) **kinetochore microtubules,** which attach to the chromosomal kinetochores, and (3) **interpolar microtubules,** which span the area between the two spindle poles and account for the bulk of the spindle's structure (Figure 13-20). Although microtubules are the main structural component of the spindle apparatus, other macromolecules have also been detected in close association with the spindle. Included in this category are proteins such as calmodulin, dynein, actin, and myosin. As we shall see shortly, the presence of these proteins has been used to support various theories of chromosome movement and spindle function.

Pioneering experiments carried out in the laboratory of Sinya Inoué provided one of the first indications that assembly of the spindle is regulated by changes in the equilibrium between the polymerized and unpolymerized states of tubulin. In these experiments, which were mentioned in Chapter 9, it was found that exposing cells to low temperature, high pressure, or the drug colchicine causes breakdown of the mitotic spindle and the cessation of mitosis. The spindle spontaneously reforms, however, as soon as the temperature is raised, the pressure lowered, or the colchicine removed. This rapid reappearance of the spindle occurs even when inhibitors of protein synthesis are present, indicating that assembly of the spindle is not dependent on the synthesis of new spindle proteins. Such observations led to the conclusion that assembly of the mitotic spindle is based on the polymerization of preexisting tubulin subunits, while spindle breakdown stems from tubulin depolymerization.

The realization that spindle formation and breakdown is controlled by the assembly and disassembly of microtubules rather than by the synthesis and degradation of tubulin has stimulated interest in identifying the factors that regulate microtubule assembly/disassembly during normal mitosis. Several lines of evidence suggest that calcium ions may play an important role in this process. Microinjection of Ca^{2+} into metaphase cells has been found to hasten the onset of anaphase, while conditions that lower the cytoplasmic Ca^{2+} concentration slow the progression into anaphase. The mitotic spindle has also been found to contain large amounts of **calmodulin,** a calcium-binding protein that mediates the regulatory effects of calcium ions upon a variety of cellular processes (Chapter 16). Finally, calcium ions are known to influence the assembly and disassembly of microtubules under cell-free conditions. Taken together, the preceding observations suggest a role for Ca^{2+} in regulating the assembly/disassembly of the mitotic spindle.

In addition to the three types of microtubules already mentioned, the spindle apparatus of animal cells also contains a pair of centrioles associated with each spindle pole. Centrioles are small, microtubule-containing structures whose morphology, reproduction, and role in the formation of cilia and flagella were discussed in Chapter 9. Most cells contain a single pair

of centrioles that lie at right angles to each other adjacent to the nuclear envelope. Under the light microscope these centrioles appear to be surrounded by a clear zone or halo, but electron microscopy has revealed that the clear zone is occupied by a finely granular, amorphous material. The term **centrosome** is generally employed to refer to the complex between the centrioles and this closely associated granular material. Centrioles usually replicate during the S-phase of the cell cycle, yielding two pairs of centrioles destined to migrate to opposite sides of the nuclear envelope at the time of prophase. Although centriole replication looks like "division" under the light microscope, careful electron microscopic examination has revealed that this process is not a true division, but rather a sequence of events in which each preexisting centriole induces the formation of a new centriole (see Figure 9-27).

During interphase, centrioles are surrounded by a scattered cluster of short microtubules. At the beginning of mitosis the number and length of these asso-

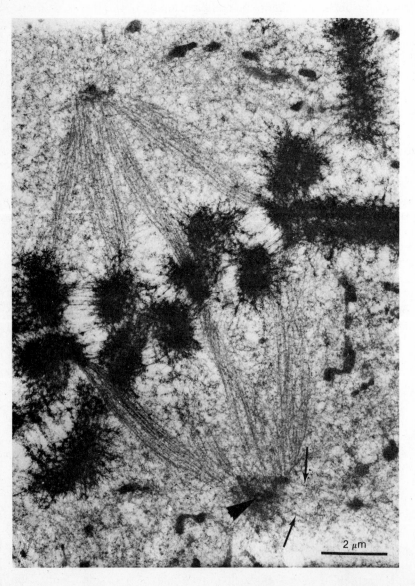

FIGURE 13-21 Electron micrograph of a mitotic spindle showing the aster microtubules (arrows) radiating from the region containing the centrioles (arrowhead). Courtesy of C. L. Rieder.

2 μm

ciated microtubules increase dramatically, generating a structure that consists of microtubules radiating in all directions from a central centriole pair. The microtubules comprising this structure, termed an **aster** (Figure 13-21), are too small to be seen with the light microscope, but their precise geometric arrangement causes the surrounding cytoplasmic material to become organized into a starlike pattern that is readily visible. Although the microtubules of the aster appear to be emerging directly from the centrioles, careful electron microscopic examination has revealed that there is no direct physical contact between the microtubules and the centrioles. The microtubules emerge instead from the granular material surrounding the centrioles, which is believed to function as a nucleation site for microtubule assembly.

At the beginning of prophase, the two pairs of centrioles and their associated asters begin to separate and move toward opposite ends of the cell. At the same time, spindle microtubules start forming between the centriole pairs. By the time the centrioles have reached opposite sides of the cell, the spindle microtubules are almost completely assembled. It was pointed out in Chapter 9 that microtubules exhibit an intrinsic polarity, one end functioning in tubulin assembly and the other functioning in disassembly. In order to investigate the polarity of the microtubules contained within the mitotic spindle, experiments have been carried out on mitotic cells pretreated with appropriate detergents to destroy the plasma membrane. Large molecules added to the external medium of such *permeabilized* cells pass directly into the cell instead of being stopped by the plasma membrane, making it possible to study the interaction between externally added macromolecules and the mitotic spindle. The polarity of the spindle microtubules has been investigated in permeabilized cells by adding an excess of free tubulin to the external medium and observing the direction in which this tubulin is added to the spindle microtubules. This approach has revealed that the microtubules situated in the two halves of the spindle are oriented in opposite directions, with each group of tubules growing outward from its respective spindle pole.

It might seem logical to conclude from these studies that the centriole pairs located at the two poles of the spindle are responsible for nucleating the assembly of the spindle microtubules. This conclusion is inconsistent, however, with several kinds of observations. One of the most glaring problems is that plant cells assemble functional spindles in the absence of centrioles and asters. Even in animal cells, where centrioles are normally present, it has been found that the centrioles are not a prerequisite for mitosis. They can be experimentally removed without preventing the subsequent formation of a functional mitotic spindle. There is even a mutant *Drosophila* cell line that lacks centrioles, but

proliferates normally by mitosis. The preceding observations indicate that centrioles themselves are not directly involved in generating the mitotic spindle. It is possible, however, that the granular material surrounding the centrioles nucleates the assembly of the spindle microtubules, and that an analogous type of microtubule-organizing material is present in cells that do not contain centrioles.

If centrioles are not required for nucleating formation of the mitotic spindle, one might wonder why the two pairs of centrioles move to opposite ends of the cell during mitosis. One intriguing possibility is that the centriole pairs are simply being pushed farther and farther away from each other by the growing spindle. By ensuring that the centrioles are situated at opposite poles of the cell when cytokinesis occurs, this process guarantees that each of the two newly formed cells will receive a pair of centrioles. Such a model has significant merit when one considers the fact that centrioles carry out important functions that have nothing to do with the process of cell division, and that it is therefore important for newly formed cells to inherit centrioles. Some support for this viewpoint can be found in the fact that the only kinds of plant cells known to contain centrioles are those cells destined to produce cilia and flagella, a process that requires centrioles. The presence of centrioles in the spindle apparatus of such cells may therefore be related more to the role of centrioles in the biogenesis of cilia and flagella than with the mechanism of cell division per se.

The Mechanism of Chromosome Movement

Several theories have been proposed to explain the mechanism underlying the movement of the two sets of chromosomes to opposite poles of the spindle during anaphase. As was mentioned earlier in the chapter, two distinct types of movements occur at this time: migration of the chromosomes toward the poles and movement of the two poles away from one another. Several observations suggest that the spindle microtubules play a central role in these two types of movements. First, the kinetochore microtubules are the only external structures that appear to be attached to the moving chromosomes. Moreover the kinetochore region of the chromosome leads the way during chromosome movement, suggesting that the chromosome is being pulled toward the pole by the microtubules attached to the kinetochore. Finally, chromosome movements can be inhibited by exposing cells to microtubule-disrupting agents such as colchicine.

Although the preceding observations suggest that the spindle microtubules play an important role in chromosome movement, the exact nature of their role is yet to be clearly elucidated. The various theories that have been proposed fall into three distinct categories depending on whether the motive force is postulated to emanate from microtubule-microtubule interactions, microtubule polymerization/depolymerization, or an interaction of microtubules with some other component. A general description of these three kinds of theories is presented below.

Microtubule-Microtubule Interactions In Chapter 9 we saw how ciliary and flagellar movements are based on the active sliding of microtubules past one another. This sliding movement is driven by the transient formation and breakage of bonds between adjacent microtubules in a reaction that involves the ATP-hydrolyzing protein, **dynein.** The discovery that dyneinlike proteins are also present within the mitotic spindle raises the possibility that a similar type of microtubule-sliding mechanism is involved in chromosomal movements. In order to test this idea, permeabilized mitotic cells have been exposed to antibodies against dynein to determine whether the resulting inhibition of dynein function affects chromosome movement. Under such conditions pole-pole separation is severely inhibited, but chromosome-to-pole movement is not. Such observations are further reinforced by the discovery that pole-pole separation, but not chromosome-to-pole movement, requires ATP. Taken together the preceding observations suggest that chromosome separation during mitosis is driven by two sets of forces: a dynein-based, ATP-requiring process of microtubule sliding that pushes the spindle poles apart, and a separate ATP-independent mechanism for pulling the chromosomes toward the poles.

Microtubule Polymerization/Depolymerization Although the preceding observations suggest that microtubule sliding plays a role in chromosome movement, at least two aspects of spindle behavior are left unexplained by this mechanism. One is that the overall length of the interpolar microtubules increases as the two poles of the spindle separate from one another, and the other is that the individual kinetochore microtubules shorten as the chromosomes move toward the poles. These changes suggest that interpolar microtubules are being lengthened by the addition of tubulin subunits to their polymerizing ends, while the kinetochore microtubules are being shortened by the loss of tubulin subunits from their depolymerizing ends.

Although there is little doubt that such alterations in microtubule length occur during anaphase, the significance of the observed changes is more difficult to ascertain. Shinya Inoué has proposed that the depolymerization process responsible for shortening the kinetochore microtubules generates sufficient force to actively pull the chromosomes toward the spindle poles. Though there is little precedent for the idea that microtubule depolymerization can act this way as a mo-

tile force, it should be noted that chromosome movement is a relatively slow process that requires tiny amounts of energy when compared to more vigorous cellular movements like ciliary motion or muscle contraction. It is therefore conceivable that a fundamentally different kind of mechanism is involved. Support for the idea that microtubule depolymerization plays an active role in chromosome movement has emerged from studies in which increased atmospheric pressure was used to increase the rate of microtubule depolymerization in anaphase cells. Under such conditions, the rate of movement of the chromosomes toward the spindle poles is found to increase. Although such observations are compatible with the idea that microtubule depolymerization provides the driving force for chromosome-to-pole movement, alternative interpretations are possible. It could be postulated, for example, that other forces are responsible for pulling chromosomes toward the spindle poles, but that such movement cannot occur unless the kinetochore microtubules shorten and allow the chromosomes to approach closer to the poles. According to this viewpoint microtubule depolymerization would be the rate-limiting step in chromosome movement, but not the source of the motile force.

Interaction of Microtubules with Other Components In recent years a number of proteins other than tubulin have been detected within the mitotic spindle, raising the possibility that components other than microtubules provide the driving force for chromosome movement. Such theories do not deny the need for spindle microtubules, but suggest that the motile force is provided by some external agent acting on these microtubules. Among the more obvious candidates for this role are the proteins actin and myosin, both of which have been detected in close association with the spindle apparatus. Although the role of actin and myosin in other types of motility is well established, experiments designed to determine whether they play a role in chromosome movement have been largely negative. It has been shown, for example, that chromosome movement is not adversely affected in permeabilized mitotic cells that have been treated with agents that inhibit actin function, such as the drug cytochalasin B or antibodies against actin. Such observations do not, of course, rule out the possibility that other nonmicrotubular components of the cytoskeletal network play a role in the movement of chromosomes.

The Mechanism of Cytokinesis

Division of the cell as a whole, or **cytokinesis,** normally begins in conjunction with the final stages of chromosome separation. Although the mechanism that triggers the onset of cytokinesis is unknown, there are several

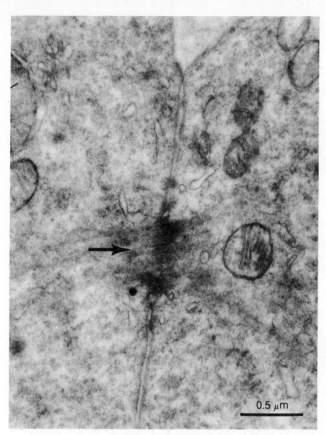

FIGURE 13-22 Formation of the midbody during the anaphase of mitosis in a developing red blood cell of the rat. This thin-section electron micrograph shows the beginning of the cleavage furrow and the appearance of electron-dense material (arrow) that represents an early stage of midbody formation. Courtesy of R. C. Buck.

reasons for believing that it is not inextricably coupled to the process of chromosome movement and nuclear division. First of all, it is known that chromosome replication and nuclear division proceed repeatedly in certain cell types in the complete absence of cytokinesis, generating multinucleated cells in the process. In other situations cytokinesis has been observed to occur in the absence of nuclear division. Even in normal cells a significant time lag can occur between nuclear division and cytokinesis, indicating that these two processes are not tightly coupled to one another.

The cleavage plane along which the cell divides usually passes through the equatorial region of the spindle, suggesting that the location of the spindle determines where the cytoplasm is to be divided. This idea has been verified by experiments in which the position of the mitotic spindle was artificially altered either by micromanipulation or by exposing cells to large gravitational forces. When the location of the mitotic spindle is altered in these ways, the position of the cleavage plane is found to change accordingly. If the position of the mitotic spindle is not altered until anaphase, however, the cleavage plane passes through an area correspond-

ing to the original equatorial region of the spindle. It can therefore be concluded that the site of cytoplasmic division has been programmed by the end of metaphase. The role of the spindle in determining the site of cytokinesis applies to animal and plant cells alike, though cytokinesis itself proceeds by somewhat different mechanisms.

In animal cells the first signs of cytokinesis become visible at mid-anaphase, when electron-dense material condenses in the equatorial region of the spindle to form a layer known as the **midbody** (Figure 13-22). As the spindle begins to disintegrate during telophase, the microtubules embedded in the midbody remain intact and may even increase in number. At this same time a slight infolding of the plasma membrane starts to appear, generating a slight depression in the cell surface that passes around the circumference of the cell. Shortly thereafter a band of actin microfilaments appears in the cytoplasm directly beneath this depression or **cleavage furrow** (Figure 13-23). In order to determine whether these microfilaments generate the force that pinches the cell

FIGURE 13-23 Electron microscopic appearance of the cleavage furrow of the sea urchin egg *(Arbacia). (Top)* The inset shows a low-power cross section through the dividing cell after cytokinesis has begun. The enlarged view of the boxed region shows that a dense layer of material (arrow), termed the contractile ring, is present beneath the cleavage furrow. *(Bottom)* In a higher-power micrograph of the cleavage furrow, individual microfilaments can be seen within the contractile ring (arrow). Courtesy of T. E. Schroeder.

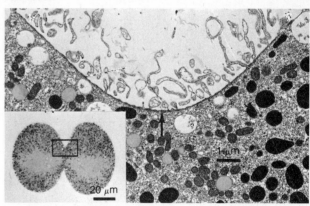

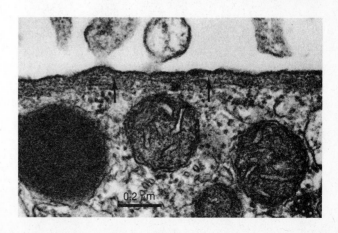

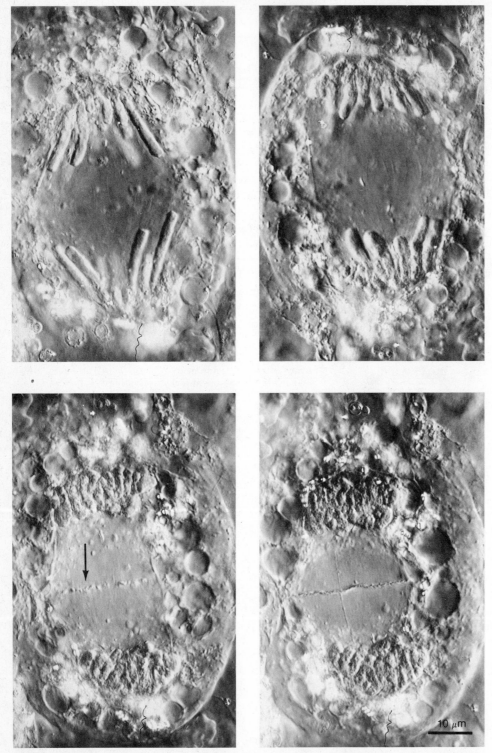

FIGURE 13-24 Formation of the phragmoplast and cell plate in plant endosperm cells as viewed by light microscopy employing Nomarski optics. As the chromosomes condense during late anaphase and early telophase, phragmoplast vesicles appear in the center of the cell (arrow). At late telophase these vesicles coalesce to form a cell plate. Courtesy of A. S. Bajer.

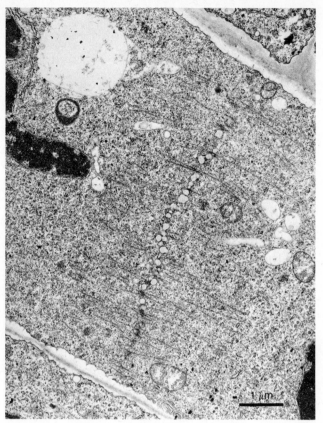

FIGURE 13-25 Thin-section electron micrograph through the center of a dividing plant cell showing phragmoplast vesicles in the process of fusing to form the cell plate. Courtesy of J. D. Pickett-Heaps.

cell in two, this band of filaments is referred to as the **contractile ring.** Although myosin has been found to be associated with the actin microfilaments of this contractile ring, the nature of the force-generating mechanism remains to be elucidated.

Cytokinesis in plant cells proceeds by a somewhat different mechanism. The first signs again occur in late anaphase, when a layer of membrane vesicles begins to appear across the equatorial region of the spindle. These vesicles form a structure called a **phragmoplast** that is in many ways analogous to the midbody (Figure 13-24). Rather than following the formation of the phragmoplast with an infolding of the plasma membrane, however, the phragmoplast continues to grow outward until it extends across the entire cell. The vesicles of the phragmoplast then fuse with one another, generating a double-layered membrane sheet, or **cell plate,** which divides the cell in half (Figure 13-25). Soon thereafter cell wall material begins to accumulate between the two membranes of the cell plate. After the cell wall reaches sufficient thickness it splits down the middle, producing two independent cells with intact plasma membranes and walls. Thus in plant cells cytokinesis progresses from the center of the cell toward the periphery, while in animal cells it is mediated by an infolding of the plasma membrane that proceeds from the outside of the cell toward the cell interior (Figure 13-26).

Because the two sets of chromosomes are initially situated at opposite poles of the spindle and the cleavage plane passes through the equatorial region, cytokinesis usually divides the chromosomes equally between the two newly formed cells. This process does not necessarily ensure that each cell receives an equal distribution of other organelles, however. But as long as the two newly formed cells receive at least a minimum representation of the various organelles (chloroplasts, mitochondria, microtubules, endoplasmic reticulum, ribosomes, etc.), a shortage of any particular component can be overcome by the formation of additional organelles during the next cell cycle.

in half, studies have been carried out employing the drug cytochalasin B to interfere with microfilament function. When cytochalasin B is applied to dividing cells with shallow cleavage furrows, the band of microfilaments disappears, cytokinesis stops, and the furrow disappears. Removal of the drug leads to the reappearance of the microfilaments and the resumption of cytokinesis. Because microfilaments therefore appear to be responsible for generating the tension that cleaves the

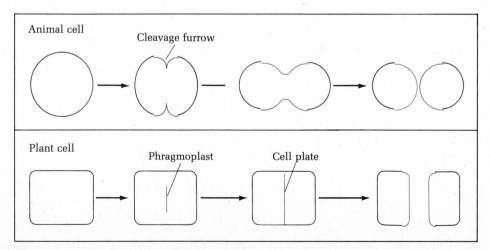

FIGURE 13-26 Diagram summarizing the difference between the mechanism of cytokinesis in animal and plant cells. In animal cells invagination of the plasma membrane proceeds from the periphery of the cell toward the interior; in plant cells the phragmoplast begins to form in the center of the cell and then expands toward the periphery.

EUKARYOTIC CELL DIVISION: MEIOSIS

Although most eukaryotic cells divide by mitosis, a special type of division termed **meiosis** occurs during the production of reproductive cells such as sperm and eggs. The need for this special type of division arises because the formation of reproductive cells is accompanied by a reduction in the chromosome number from the normal **diploid** state (two sets of chromosomes) to the **haploid** state (one set of chromosomes). In addition to reducing the chromosome number in half, meiosis is also associated with an exchange of genetic information between the two sets of chromosomes. This process of genetic **recombination** creates individual chromosomes that have a new combination of paternal and maternal genetic information.

The overall process of meiosis involves two separate cell divisions occurring in sequence. The first meiotic division reduces the chromosome number from diploid to haploid, while the second meiotic division separates the two paired chromatids that comprise each chromosome (Figure 13-27). Prior to the first meiotic division DNA is replicated as in any normal cell cycle, though this replication takes somewhat longer than in a typical mitotic cell. The delay is caused by a decrease in the frequency at which replication is initiated at different replicons rather than by a decrease in the overall rate of DNA elongation per replicon. DNA replication is

FIGURE 13-28 *(Opposite page)* Appearance of the chromosomes during the various stages of meiosis in grasshopper cells viewed by light microscopy. The homologous chromosomes become paired during zygotene, and shorten and thicken at the pachytene stage. During diplotene the four chromatids comprising each bivalent chromosome can be distinguished; at this stage chiasmata hold the homologs together. During metaphase I the paired homologs come together, followed by their separation at anaphase I so that each newly forming cell receives one homolog of each pair. At metaphase II each chromosome is seen to consist of a pair of chromatids held together by a centromere. During anaphase II these chromatids separate and migrate in opposite directions, ensuring that each new cell receives one chromatid from each pair. Courtesy of B. John.

followed by a normal G_2-phase and subsequent entry into the first meiotic division. Although meiosis can be subdivided into phases analogous to the prophase, metaphase, anaphase, and telophase of mitosis (Figure 13-28), we shall see below that a number of significant differences distinguish the comparable phases of mitosis and meiosis from one another.

Summary of Meiotic Stages

Prophase I. In diploid cells each chromosome inherited from the maternal parent is matched by a similar, or **homologous,** chromosome inherited from the paternal parent. During prophase of the first meiotic division (prophase I), each chromosome pairs with its homolog

FIGURE 13-27 Summary of the two stages of meiosis.

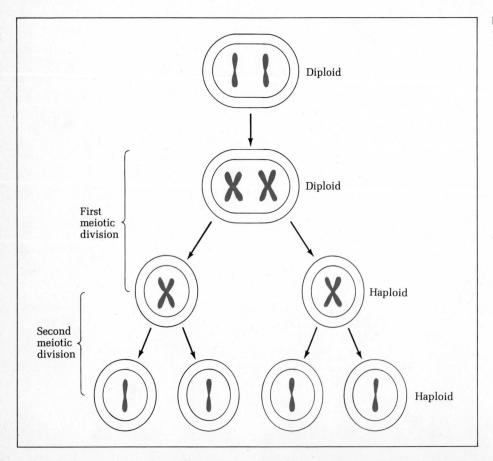

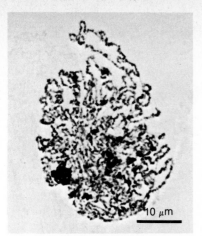

Prophase I—Leptotene

Prophase I—Zygotene

Prophase I—Pachytene

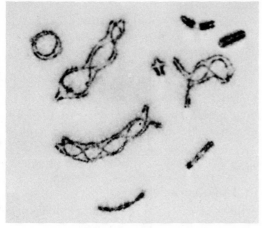

Prophase I—Diplotene

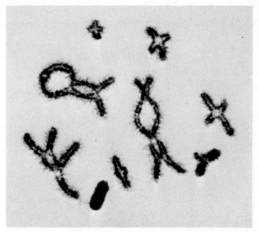

Prophase I—Diakinesis

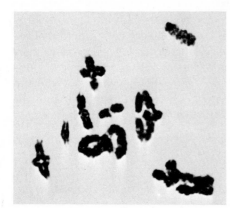

Metaphase I

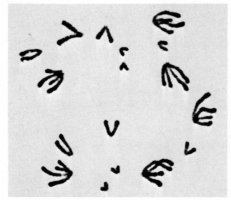

Anaphase I

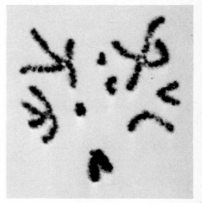

Prophase II

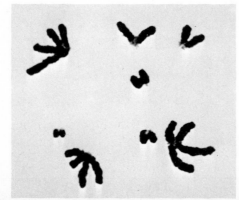

Metaphase II

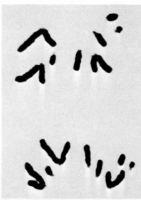

Anaphase II

10 μm

so that interchange of genetic information can occur. Because this phase of meiosis requires a considerable amount of time and involves a complex series of events, prophase I is commonly subdivided into five subphases designated *leptotene, zygotene, pachytene, diplotene,* and *diakinesis.*

The **leptotene** (lepto = thread) stage of prophase I begins with the condensation of the chromatin into long, thin threads. This condensation superficially resembles the process that occurs at the beginning of mitotic prophase, except that the chromatin threads are more extended and contain a series of granulelike swellings called **chromomeres.** Separate chromosomes cannot be distinguished from one another at this point, nor can the two chromatids that comprise each chromosome.

The **zygotene** (zygo = yolked, paired) phase of prophase I is characterized by the onset of close pairing, or **synapsis,** between homologous chromosomes. Although the mechanism of attraction between homologous chromosomes is unknown, it is usually initiated by physical association of the two homologs at a few points, followed by the progressive alignment of the rest of the chromosome until all the chromomeres are lined up adjacent to one another. The only exception to this behavior occurs with the chromosomes involved in sex determination. In most vertebrates these sex chromosomes are referred to as X and Y chromosomes. Normal pairing can occur between two X chromosomes, but there may or may not be sufficient homology between the X and Y chromosomes to permit close pairing when one of each is present.

In electron micrographs a small space, measuring about 20 nm across, can be seen to separate the two homologous chromosomes of each pair. This space contains a specialized component, termed the **synaptonemal complex,** whose structural organization and role in genetic recombination will be discussed later in the chapter. During zygotene the paired chromosomes often remain attached to the nuclear envelope at their two ends.

The **pachytene** (pachy = thick) stage of prophase I begins at the completion of homolog pairing, and is marked by a dramatic thickening and shortening of the chromosomes. This process reduces each chromosome to less than a quarter of its previous length. The synaptonemal complex becomes fully developed and the two chromosomes come to be so tightly joined to one another that they behave as a single unit. This overall unit is referred to as either a **tetrad** or a **bivalent** chromosome (because it contains a total of *four* chromatids, *two* derived from each chromosome). At this stage genetic information is exchanged between chromatids by a recombination process whose detailed mechanism will be described later in the chapter.

During the **diplotene** stage of prophase I, the synaptonemal complex disappears and the homologs of each bivalent chromosome become partially separated. The homologs still remain attached to each other, however, by connections known as **chiasmata** (chiasma = singular). These connections are situated in regions where segments of two chromatids have been exchanged with one another, and hence represent visual evidence that genetic recombination has occurred (Figure 13-29). As diplotene progresses, the individual chromatids that comprise each homolog become visible for the first time. It is common for the chromosomes to uncoil at this point and for many nucleoli to form. In certain types of developing egg cells a period of extensive growth ensues, increasing the cell mass by many orders of magnitude. During this growth phase, which may last for several years, the chromosomal and nucleolar DNA is very actively transcribed and the chromosomes assume a characteristic morphology

FIGURE 13-29 Chromosomal events occurring during prophase I of meiosis.

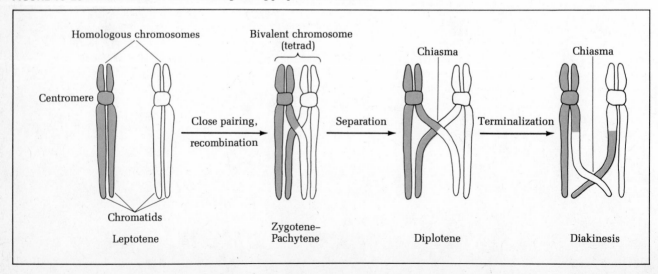

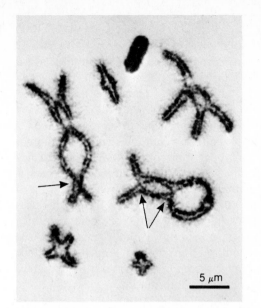

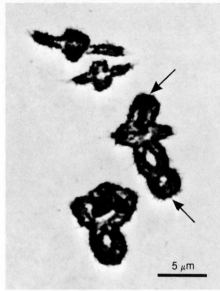

FIGURE 13-30 The behavior of chiasmata (arrows) during prophase I of meiosis. *(Left)* The chiasmata, which represent regions of genetic recombination, are seen to hold together the homologous chromosomes during the diplotene stage of prophase I. *(Right)* At the end of prophase I, the chiasmata move to the ends of the chromosomes by a process known as terminalization. Courtesy of B. John.

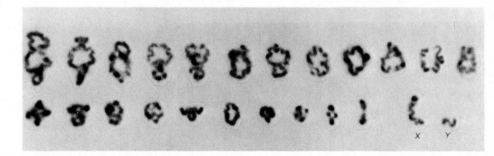

FIGURE 13-31 A meiotic karyotype of human sperm chromosomes isolated from metaphase-I-arrested cells. Note the difference in appearance between these chromosomes and typical mitotic chromosomes (see Figure 13-13).

consisting of a darkly staining central axis from which numerous loops project. One of the most thoroughly studied examples of this kind of chromosome is the "lampbrush" chromosome of amphibian oocytes discussed later in the chapter.

During the **diakinesis** stage of prophase I the chromosomes return to a highly condensed state. At the same time a process known as **terminalization** causes the chiasmata to migrate toward the ends of the chromosomes (Figure 13-30). The nucleoli disappear as in the comparable stage of mitosis, and the centrioles and asters (if present) move to opposite sides of the nuclear envelope. The spindle apparatus then begins to form and the nuclear envelope breaks down and disappears, marking the end of prophase I.

Metaphase I. After the first meiotic prophase has been completed, the paired chromosomes become attached to spindle microtubules and migrate to the equatorial region of the spindle. At this stage the chromosomes look quite different from typical mitotic chromosomes (compare Figures 13-13 and 13-31). Typical mitotic chromosomes consist of two chromatids joined at the centromere, while the bivalent chromosomes of metaphase I consist of a pair of homologous chromosomes held together by chiasmata (the two chromatids comprising each homolog are usually not

distinguishable). In spite of these obvious differences in the organization of meiotic and mitotic metaphase chromosomes, both appear to be comprised of extensively folded 25-nm chromatin fibers.

Because bivalent chromosomes are composed of four chromatids, a total of four kinetochores are present. The kinetochores present on the two chromatids of each homolog face in the same general direction (rather than in opposite directions as in mitosis), and therefore attach to microtubules emanating from the same pole of the spindle. The kinetochores on the chromatids of the opposite homolog face in the opposite direction, and are therefore attached to microtubules emerging from the other spindle pole. In this way the stage is set for separation of the homologs during anaphase (Figure 13-32).

Anaphase I. During anaphase I, the homologous chromosomes separate from each other and migrate toward opposite poles of the spindle. This phenomenon is fundamentally different from the events of mitotic anaphase, when the two chromatids that comprise each chromosome separate from one another and migrate to opposite spindle poles. In meiosis the paired chromatids of each chromosome remain attached to each other as the homologous chromosomes separate and move in opposite directions.

Telophase I. The onset of telophase I is marked by

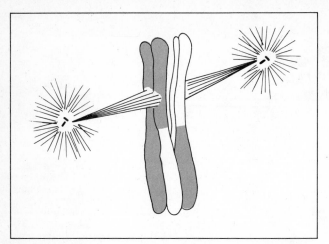

FIGURE 13-32 Schematic diagram illustrating the attachment of the homologous chromosomes to opposite spindle poles during metaphase I of meiosis. The colored shading is used to highlight the region where crossing over has occurred.

the arrival of the chromosomes at the spindle poles. Since the homologous chromosomes have become separated from one another, the poles contain half the normal number of chromosomes, that is, the haploid number. After arriving at the polar region, the chromosomes uncoil and disperse into chromatin fibers as the spindle disappears. In some cell types two new nuclear envelopes form around the chromosomes and cytokinesis ensues, generating two haploid cells. In other cases the telophase chromosomes remain condensed and simply proceed directly into the second meiotic division.

Interphase. The interphase between the two meiotic divisions is generally quite short, and may even be absent entirely. If a recognizable interphase does occur at this stage, it is not accompanied by DNA replication.

Prophase II. In cases where a distinct interphase intervenes between the first and second meiotic divisions, entry into prophase II requires condensation of the dispersed chromatin fibers into visible chromosomes and breakdown of the nuclear envelope. When an interphase does not occur, the chromosomes bypass prophase II and move directly from telophase I to metaphase II. If a centriole-aster complex is present, the two centrioles separate and migrate toward the poles as the new spindle forms. This generates a spindle apparatus having only one centriole at each pole. Replication of these unpaired centrioles may occur at a later time.

Metaphase II. During metaphase II the chromosomes become associated with spindle microtubules and migrate to the equatorial region of the spindle. The structure of the chromosomes at this point resembles that of typical mitotic chromosomes. Each chromosome is composed of two paired chromatids joined at a common centromere, with two sets of microtubules attached to kinetochores located on opposite sides of the centromere. The two chromatids comprising each chromosome are thus attached to opposite poles of the spindle.

Anaphase II. At the beginning of anaphase II, the two chromatids present within each metaphase chromosome separate from one another. As in the case of mitosis, it has been proposed that this separation may be triggered by replication of DNA situated within the centromere. The newly formed chromosomes are then free to migrate to opposite poles of the spindle.

Telophase II. The onset of telophase II is marked by the arrival of a haploid set of chromosomes at each pole of the spindle. These chromosomes then uncoil and disperse into chromatin fibers, the spindle disintegrates, a new nuclear envelope and nucleoli appear, and cytokinesis divides the cell. Although the timing of these final processes varies among cell types, the final product is a cell containing a haploid set of chromosomes. As a result of the recombination events of prophase I, each of these chromosomes is comprised of a mixture of maternal and paternal genetic information.

The Molecular Basis of Genetic Recombination

Two unique events are associated with the process of meiotic division: reduction of the chromosome number from diploid to haploid and recombination of genetic information. We have already seen how the behavior of the homologous chromosomes during meiosis leads to reduction of the chromosome number, but the process of genetic recombination is considerably more complex and requires a more detailed examination. Although this discussion of genetic recombination is prompted by our consideration of meiotic cell division, it should be noted that genetic recombination also occurs in bacterial cells where meiosis does not take place. At least three distinct types of genetic recombination have been observed in bacteria. In the first, called **transduction,** viruses incorporate genetic information from the chromosomal DNA of one bacterium and transfer it to another bacterium by recombination. In the second process, termed genetic **transformation,** naked molecules of DNA are taken up from the external environment and integrated into the host cell's chromosomes by the process of recombination. And third, some bacteria undergo a modified type of sexual reproduction termed **conjugation** in which the chromosomal DNA from one bacterium is injected into another bacterium. Genetic recombination then takes place in the recipient cell, generating a new type of chromosomal DNA containing a mixture of information from both parents. Each of the three preceding types of recombination generates DNA molecules comprised of sequences derived from two different sources, just as in the process of meiosis. Although the following discussion of genetic recombination will focus to a large extent on the events associated with meiosis, many of the underlying molecular phenomena apply to genetic recombination in prokaryotic systems as well.

The notion that chromosomal material is exchanged during meiosis was initially proposed in the early 1900s when it was first discovered that homologous chromosomes become tightly paired during prophase I. It was even suggested that chiasmata represent sites at which material is being exchanged between chromosomes, though there was no direct evidence at the time supporting the existence of such an exchange. It was not until several decades later that studies on the inheritance patterns of specific genes revealed that homologous chromosomes really do exchange genetic information during meiosis. It was also discovered that the number of such exchanges correlates with the number of chiasmata, providing support for the earlier suggestion that chiasmata represent sites of genetic recombination.

Shortly after the discovery of genetic recombination, two alternative theories were proposed to explain its underlying mechanism. One, termed the **breakage-and-exchange model,** postulated that genetic recombination involves physical breakage of the chromatids followed by exchange and rejoining of the broken segments. In contrast, the **copy-choice model** postulated that genetic recombination occurs during chromosome replication. According to this latter view, chromosome replication begins by copying the genetic information contained within one chromosome and then switches at some point to copying the other chromosome, generating a new chromosome containing information derived from both of the original homologs. This latter model led to several predictions that could be easily tested, the most straightforward being that chromosome replication and genetic recombination should occur at the same time. Subsequent studies revealed, of course, that DNA replication takes place during S-phase while recombination does not occur until prophase of the first meiotic division. This and other incompatible observations eventually led to the rejection of the copy-choice model.

The breakage-and-exchange model was also viewed with skepticism for many years because geneticists felt that it was potentially too inaccurate to account for the precise interchange that characterizes genetic recombination. Unless the site of breakage is precisely the same in the two chromatids, exchange of material would be expected to create chromosomes with fewer genes in one chromatid and additional genes in the other. Yet such errors are known to occur very infrequently. Only after isotopic tracers became widely available for research purposes did it become possible to experimentally demonstrate that in spite of this apparent limitation, breakage and exchange does occur during genetic recombination.

Some of the first experiments of this kind were performed in 1961 by M. Meselson and J. Weigle, who employed bacterial viruses labeled with either the heavy or light isotope of nitrogen. Infection of bacterial cells with a mixture of two strains of virus, one labeled with ^{15}N and the other with ^{14}N, was found to result in the production of new virus particles whose DNA contained a mixture of ^{15}N and ^{14}N. Since these experiments were performed under conditions in which new DNA synthesis was prevented, it seemed reasonable to conclude that the hybrid viruses were generated by a breakage-and-exchange process involving the two original kinds of viral DNA. Subsequent experiments carried out on bacterial cells whose chromosomal DNA had been labeled with either ^{15}N or ^{14}N similarly revealed that DNA containing a mixture of heavy and light nitrogen is produced during genetic recombination. In these latter experiments denaturation of the DNA into single strands revealed the presence of both ^{15}N and ^{14}N within the same strand, indicating that individual DNA strands are being broken and rejoined.

A similar phenomenon was also demonstrated in eukaryotic cells by J. Herbert Taylor, who employed radioactive tracers to monitor the behavior of individual molecules of DNA. In these experiments cells were briefly exposed to ^{3}H-thymidine during the S-phase preceding the last mitosis before meiosis. In this way chromatids containing one radioactive DNA strand per double helix were generated. The radioactive isotope was then removed from the medium so that DNA replication during the interphase prior to meiosis would generate chromosomes containing one labeled chromatid and one unlabeled chromatid (Figure 13-33). If breakage and exchange occurs during the subsequent prophase I, it would be expected to generate DNA molecules containing a mixture of radioactive and nonradioactive segments. Unfortunately the chromatids are too intimately associated with one another during prophase I to detect such a pattern autoradiographically, but the individual chromatids can be readily distinguished during metaphase II. When the chromatids were examined at this stage, some were observed to contain a mixture of labeled and unlabeled segments as predicted by the breakage-and-exchange model. Moreover the frequency of such exchanges was found to be directly proportional to the frequency with which the genes located in these regions undergo genetic recombination. Such observations provided strong support for the notion that genetic recombination in eukaryotic cells is based upon a breakage-and-exchange mechanism.

In addition to providing evidence for the existence of breakage and exchange, the preceding experiments also revealed that most DNA exchanges involve the chromatids of opposing homologous chromosomes rather than the two chromatids of a given homolog. This selectivity in the exchange process is important for an obvious reason: Exchange between the two chromatids of a single homolog would be of little functional significance since, discounting the possibility of rare mutations, two such chromatids would be expected to be

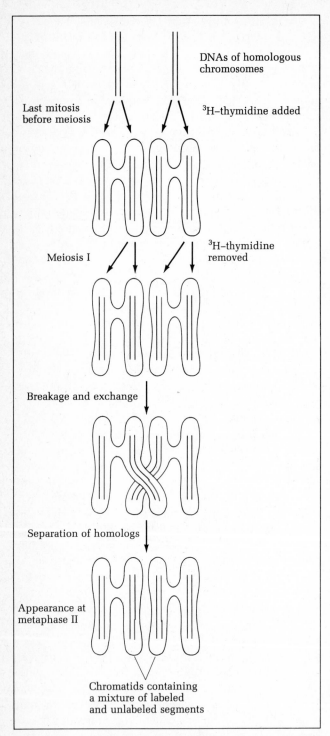

DNAs of homologous chromosomes

Last mitosis before meiosis

³H–thymidine added

Meiosis I

³H–thymidine removed

Breakage and exchange

Separation of homologs

Appearance at metaphase II

Chromatids containing a mixture of labeled and unlabeled segments

FIGURE 13-33 Diagram summarizing the experimental approach utilized to demonstrate the existence of breakage and exchange during genetic recombination in eukaryotic cells.

genetically identical. Although the molecular mechanism responsible for the selectivity of breakage and exchange is yet to be ascertained, it has been suggested that the synaptonemal complex may play a role.

The conclusion that genetic recombination involves some type of breakage-and-exchange mechanism does not in itself provide much information concerning the molecular events involved. One of the most straightforward mechanisms that might be envisioned would involve nuclease cleavage of the two DNA molecules at the same location followed by physical exchange of the cut ends and rejoining of the two molecules by DNA ligase (Figure 13-34). In spite of its attractive simplicity, such a model does not explain how the cut ends of the DNA molecules are held in close proximity to one another so that ligation can take place. An even more serious limitation of this simple model is that it makes some predictions that are not borne out by experimental evidence. For example, such a symmetrical process of breakage and exchange would ensure that genetic recombination would be completely reciprocal, that is, any genes exchanged from one chromatid should appear on the other chromatid, and vice versa. However, in some situations genetic recombination has been found to be nonreciprocal.

To illustrate what is meant by nonreciprocal recombination, let us consider a hypothetical situation involving two genes designated A and B. If one chromatid were to contain forms of these genes designated A1 and B1, while the other chromatid contained alternate forms of the same genes designated A2 and B2, genetic recombination would be expected to generate one chromosome containing genes A1 and B2 and one chromosome containing genes A2 and B1. Although this is the pattern normally observed, nonreciprocal combinations occasionally occur, for example, one chromosome containing genes A1 and B2 and the other chromosome containing genes A2 and B2. In this particular example, the B1 gene expected on the latter chromosome appears to have been "converted" to a B2 gene. For this reason nonreciprocal recombination is generally referred to as **gene conversion.** Gene conversion is most commonly observed with recombining genes that are located very close to one another on the chromosome. Because the frequency with which two genes recombine is directly related to the distance separating them, recombination between such closely spaced genes is a rare event. These rare recombinational events, and hence gene conversion, are therefore most readily detectable in organisms that reproduce rapidly and generate large numbers of offspring, such as yeast and *Neurospora*.

In addition to providing evidence for the existence of gene conversion, studies on organisms such as *Neurospora* have yielded another set of observations difficult to reconcile with the occurrence of a simple, reciprocal type of breakage-and-exchange mechanism. *Neurospora* is a convenient organism in which to study meiosis because its meiotic cells are enclosed in a small sac, called an **ascus,** which prevents the cells from moving around and hence allows the lineage of each cell to be easily followed (Figure 13-35). Initially each ascus contains a single diploid cell. Meiotic division of this cell produces four haploid cells that subsequently divide by

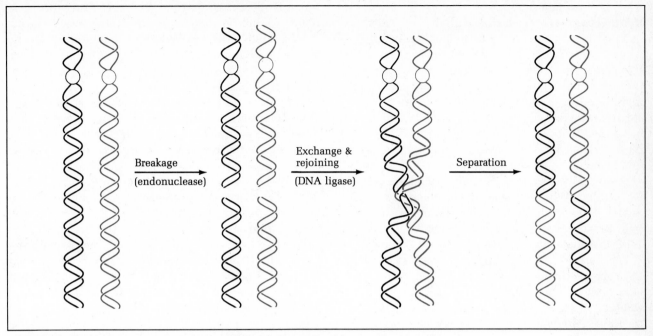

FIGURE 13-34 One of the most straightforward molecular models of DNA breakage and exchange. Although endonucleases capable of cleaving double-stranded DNA, and DNA ligases capable of rejoining DNA fragments, are both known to exist, certain events associated with genetic recombination are not readily explained by such a simple mechanism (see text for details).

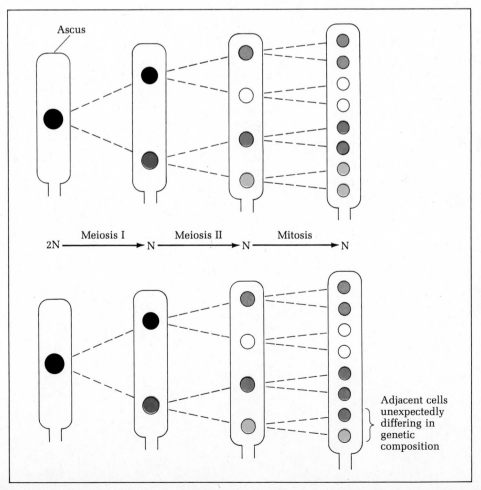

Adjacent cells unexpectedly differing in genetic composition

FIGURE 13-35 In *Neurospora*, cells undergoing meiotic division are contained within a special structure known as an ascus. This structure keeps the individual cells lined up in a row, making it easy to trace each cell's lineage. The separation of homologous chromosomes during the first meiotic division generates two cells with different genetic compositions. The cells produced by the second meiotic division may also differ from each other genetically, in this case because of the crossing over that occurred during meiosis I. The third division is a simple mitosis, and therefore normally produces a pair of genetically identical cells. Hence each pair of adjacent cells in the final, eight-celled ascus normally contains the same set of genes (*top*). Occasionally, however, this terminal mitosis generates a pair of nonidentical cells (*bottom*). This unexpected phenomenon has led to the formulation of new models of genetic recombination.

mitosis, yielding a final total of eight cells. Since this final division is mitotic, it should produce two identical progeny cells for each cell that divides. Yet in a significant number of cases, this final division has been found to generate a pair of cells that differ from each other with respect to a given gene. Such unexpected results are most often observed with genes that are close to a site of genetic recombination. One possible explanation for the fact that two cells exhibiting different forms of the same gene can arise during mitotic division is that the cell undergoing mitosis contains DNA molecules whose two strands are not entirely complementary to one another. When these noncomplementary regions separate from each other during the process of DNA replication, they will serve as templates for the synthesis of DNA molecules of slightly different sequence.

The preceding observations, as well as the existence of gene conversion, suggest that genetic recombination may be more complicated than envisioned in the simple breakage-and-exchange model illustrated in Figure 13-34. Although several alternative theories have been proposed in response, a common feature of most of them is the notion that the recombination process involves an intermediate in which a *single strand* of each DNA double helix has been broken and joined to the opposing DNA molecule. Such **single-strand exchange intermediates,** whose existence was first proposed in the early 1960s by Robin Holliday, are subsequently cleaved to yield two separate molecules of DNA, each containing regions in which the two opposing strands were derived from different DNA molecules (Figure 13-36). If the two original DNAs differed slightly in base sequences, the result will be partially noncomplementary, or **heteroduplex,** regions within the recombined DNA. There are two possible fates for such noncomplementary regions. If they are left intact, an ensuing mitotic division will separate the mismatched DNA strands and each can then serve as a template for the synthesis of a new complementary strand. The net result would be two new DNA molecules with different base sequences, and hence two cells containing slightly different gene sequences in the affected region. We have just seen that this type of behavior is occasionally observed in *Neurospora,* where the formation of two

FIGURE 13-36 Genetic recombination models based on the initial formation of a single-strand exchange intermediate in which one strand in each of two adjacent DNA double helices has been broken and joined to the opposing DNA molecule *(left).* This single-strand exchange intermediate is subsequently cleaved to yield two separate DNA molecules that contain heteroduplex regions. If these heteroduplex regions remain intact *(top),* cells with slightly different DNA sequences will be produced during the next mitosis (as occurs in *Neurospora).* Alternatively one or both of these heteroduplex regions may be repaired *(bottom),* resulting in gene conversion. See text for further details.

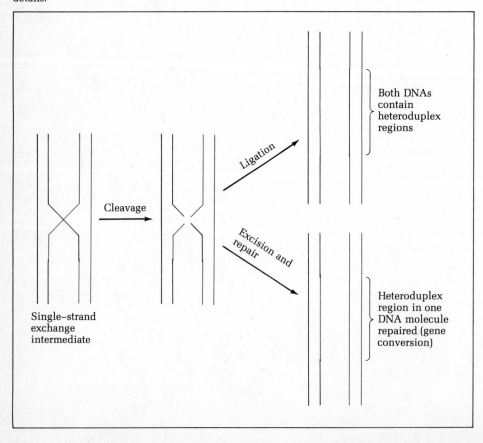

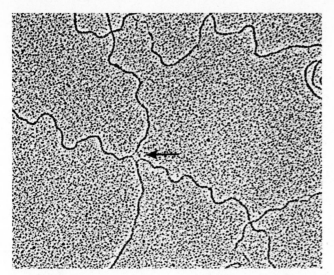

FIGURE 13-37 Electron micrograph of two bacteriophage lambda DNA molecules joined by a single-strand crossover (arrow). This structure is thought to represent an intermediate stage in the process of genetic recombination. Courtesy of R. B. Inman.

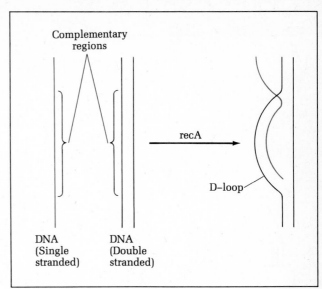

FIGURE 13-38 Reaction catalyzed by the recA protein.

genetically different cell types can occur during the mitosis following meiosis.

The alternative fate for the mismatched DNA regions is for one or both of them to be corrected by excision and repair. As illustrated in Figure 13-36, the net effect of such repair would be to convert genes from one form to another, that is, it would lead to gene conversion. It is worth noting in this context that a small amount of DNA synthesis can be detected during the pachytene phase of meiosis, which is the time when recombination occurs. The DNA synthesized at this time resembles DNA that has already been replicated during the preceding interphase, suggesting that it represents repair synthesis.

Independent support for the idea that single-strand exchange intermediates play a role in genetic recombination has emerged from electron microscopic examinations of viral DNA molecules presumed to be in the process of undergoing recombination. These micrographs reveal the presence of DNA molecules joined by single-strand crossovers (Figure 13-37). Intermediates exhibiting this type of structure can also be created by incubating DNA in the presence of bacterial cell extracts. Such extracts contain a protein, called the **recA protein,** whose presence is crucial to the recombination process. Mutant bacteria with a defective recA gene cannot carry out genetic recombination, nor are extracts prepared from such cells capable of generating DNA molecules that have undergone strand exchange. Studies on the properties of the isolated recA protein have revealed that it catalyzes a reaction in which a single-stranded DNA molecule displaces one of the two strands of a DNA double helix, generating a structure referred to as a displacement loop or **D-loop** (Figure 13-38).

The properties of the recA protein have some interesting implications for the way in which the single-strand exchange intermediate might originally arise during genetic recombination. When Holliday first proposed the existence of a single-strand exchange intermediate, it was thought to be generated by single-strand cleavage of two DNA molecules in the same location followed by joining of each broken strand to the opposite DNA molecule (Figure 13-39a). Although such a mechanism has not been completely ruled out, the requirement that the two DNA molecules be simultaneously cleaved at exactly the same location is a severe limitation. The recently discovered properties of the recA protein suggest alternative ways for inducing single-strand exchange that do not require simultaneous cleavage of two DNA molecules in the same location. One possibility, which requires only a single initial cleavage in one DNA strand, is illustrated in Figure 13-39b. In this diagram the single-stranded DNA region created by cleavage of one DNA molecule is seen to displace one strand of the opposing DNA molecule in a reaction catalyzed by the recA protein. Subsequent cleavage of the second DNA molecule *anywhere* within the displaced strand, followed by repair and ligation, would ultimately generate a single-strand exchange intermediate in which a single strand of each DNA double helix has been broken and joined to the opposing strand.

Role of the Synaptonemal Complex

Although significant progress has been made in unraveling some of the molecular events associated with genetic recombination, such studies have not revealed much about the relationship between the structural organization of meiotic chromosomes and the molecular

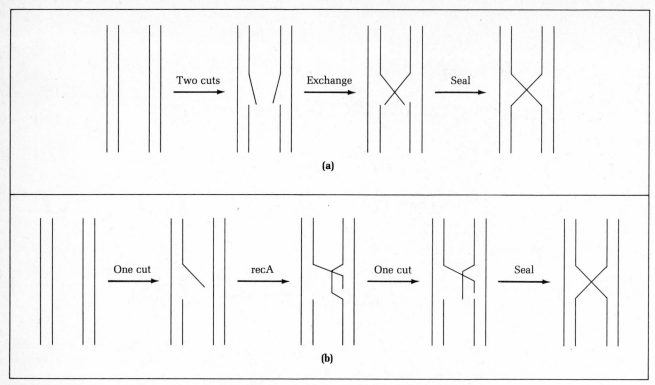

FIGURE 13-39 Two possible ways in which single-strand exchange intermediates might be formed during genetic recombination.

events of recombination. Little is known, for example, about chromatid organization during the recombination process or the mechanism responsible for the close pairing of the homologous chromosomes. One structure thought to be intimately involved in these events is the **synaptonemal complex.** This complex, which occupies the space between paired homologous chromosomes, was discovered independently in 1956 by Montrose Moses and Don Fawcett. In electron micrographs the synaptonemal complex can be seen to contain an electron-dense **central element** that runs lengthwise between the two chromosomes, plus two parallel **lateral elements,** one associated with each of the two homologs (Figure 13-40). The central and lateral elements are connected to each other by a series of **transverse fibers.**

Synaptonemal complexes have been detected in almost all eukaryotic cells undergoing meiotic division, although variations in size and organization do occur. The overall width of the complex, for example, is about 160 nm in tomatoes, 180 nm in mammals, and 200 nm in crickets. The central element generally ranges from 12–20 nm in diameter, and is separated from the lateral elements by spaces of approximately 40 nm. Cross sections through the central element reveal the presence of fibers about 5–7.5 nm in diameter, while the lateral elements consist of granules and fibers that are slightly larger. In some cell types the termini of the synaptonemal complex appear to abut upon the inner membrane of the nuclear envelope. The relationships among these

various elements of a "typical" synaptonemal complex are summarized in Figure 13-41.

Cytochemical studies have revealed that the bulk of the synaptonemal complex is accounted for by protein, though small amounts of DNA can be detected in the lateral elements. This DNA is thought to be derived from the chromatids with which the lateral elements are closely associated. The transverse fibers crossing the space between the central and lateral elements may also contain some DNA. This overall arrangement suggests that the synaptonemal complex is composed of a framework of protein within which DNA-containing chromosomal fibers in various stages of uncoiling are interspersed.

The first step in the development of the synaptonemal complex occurs during the leptotene stage of prophase I, at which time a lateral element is formed adjacent to each homologous chromosome. At zygotene the two chromosomes of each homologous pair come together, leaving a space of about 100 nm between their respective lateral elements. The central element then forms in this space, generating a fully developed synaptonemal complex. This structure is retained until the beginning of diplotene, at which point it disassembles as the homologous chromosomes separate. In some cells remnants of the synaptonemal complexes persist through the second meiotic division, forming aggregates known as **polycomplexes** (Figure 13-42).

A number of indirect arguments suggest that the

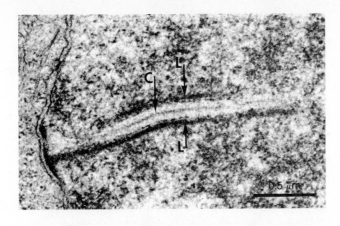

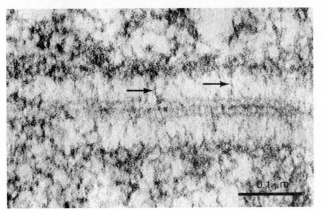

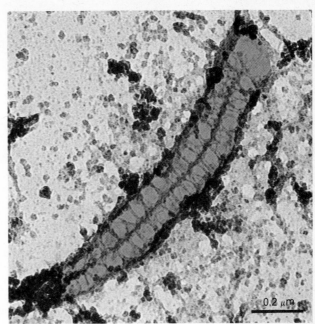

FIGURE 13-40 The synaptonemal complex. *(Upper left)* Thin-section electron micrograph of a synaptonemal complex at relatively low magnification. The central (C) and lateral (L) elements are clearly visible, but the transverse fibers are not distinguishable at this magnification. Courtesy of D. E. Comings. *(Lower left)* Higher-power view of a synaptonemal complex showing the transverse fibers (arrows) passing between the central and lateral elements. Courtesy of R. Wettstein. *(Upper right)* An unsectioned synaptonemal complex (color) viewed after removal of adjacent chromosomal fibers by digestion with DNAse. The major elements of the synaptonemal complex are clearly discernible. Courtesy of D. E. Comings.

FIGURE 13-41 Schematic model of the synaptonemal complex.

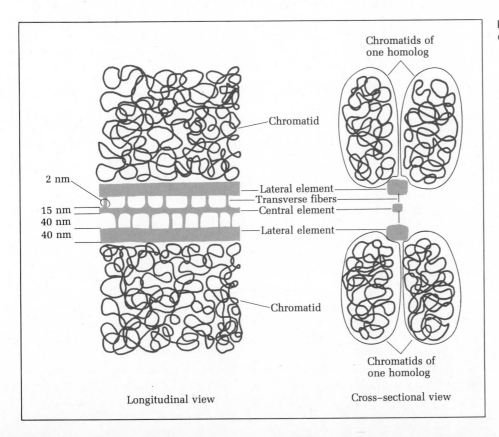

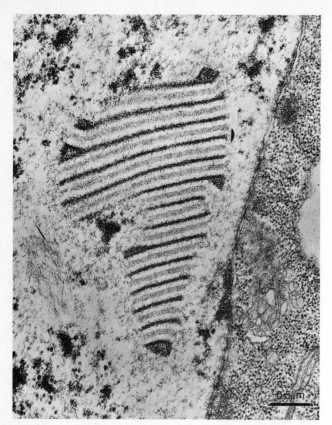

FIGURE 13-42 Thin-section electron micrograph showing a large polycomplex in the nucleus of a cell from the mosquito *Aedes aegypti*. This structure consists of a series of synaptonemal complexes stacked upon one another. The nuclear envelope lies to the right of the polycomplex. Courtesy of T. F. Roth.

synaptonemal complex plays an important role in genetic recombination. First of all, the synaptonemal complex is formed at the same time that recombination takes place, and its physical location between the opposed homologous chromosomes corresponds to the region where crossing over occurs. More direct support for the notion that recombination involves the synaptonemal complex has emerged from experiments in which genetic recombination was prevented by exposing cells to high temperatures or drugs that block DNA synthesis. Under such conditions synaptonemal complexes are not formed. Likewise synaptonemal complexes have been found to be absent from the meiotic cells of male fruit flies, which are unusual in that they do not carry out genetic recombination.

Although a correlation between the presence of a synaptonemal complex and the occurrence of genetic recombination clearly exists, the exact role played by this structure is not known. Since synaptonemal complexes occasionally form between nonhomologous chromosomes, it is thought that they cannot be responsible for precise alignment of the chromosomes. An alternative possibility is that the synaptonemal complex mediates a relatively crude alignment between two homologous chromosomes, bringing comparable regions

of their respective DNA molecules into the general vicinity of one another. It has been suggested that each chromatid then undergoes localized uncoiling, generating a series of looped-out DNA regions that project into the area of the central element. The transverse fibers observed in the space between the central and lateral elements may represent such loops. The molecular events associated with DNA strand exchange would then take place between these looped-out regions. Such exchange can only occur, of course, when the looped-out regions of DNA contain homologous base sequences, since base pairing between complementary DNA sequences is a crucial ingredient in the molecular models of DNA strand exchange discussed in the previous section. If the simultaneous looping out of comparable regions of opposing chromatids is a relatively rare event, this might explain why recombination occurs so infrequently.

In spite of the evidence implicating the synaptonemal complex in genetic recombination, some observations are difficult to reconcile with this general notion. It has been reported, for example, that certain strains of *Drosophila* carry out genetic recombination in the absence of a synaptonemal complex. Although this may represent a special case in which the close chromosomal pairing required for recombination is accomplished by another mechanism, an alternative possibility is that the synaptonemal complex normally performs a subsidiary, nonessential role in genetic recombination. It might, for example, be responsible for imparting a structural organization that ensures that genetic exchange occurs between chromatids of opposing homologs, but not between chromatids of the same homolog.

Lampbrush Chromosomes

The events associated with genetic recombination in meiotic cells occur during the zygotene-pachytene stage of prophase I. After recombination has been completed, the synaptonemal complex disappears and the chromosomes uncoil. These changes mark the beginning of the diplotene phase. In certain cell types, such as the oocytes of amphibians and birds, diplotene lasts for several years and is associated with an extensive increase in cellular mass. This growth is dependent on the formation of large amounts of protein whose synthesis in turn depends on the transcription of new messenger RNAs. As a result of this intense transcriptional activity, the diplotene chromosomes acquire a unique appearance that is characteristic for this stage of meiosis. In the late nineteenth century the name **lampbrush chromosome** was coined to refer to such chromosomes because their shape is reminiscent of the brushes then used for cleaning the chimneys of oil-burning lamps.

Each lampbrush chromosome consists of a pair of homologous chromosomes that, though well separated

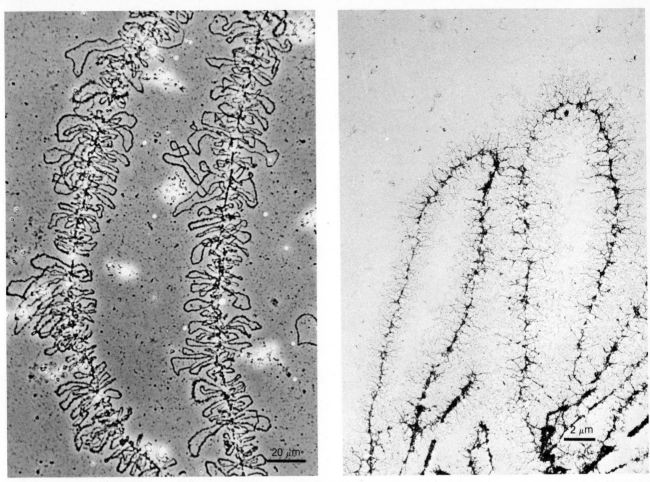

FIGURE 13-43 Micrographs of lampbrush chromosomes. *(Left)* Phase contrast microscopy of two homologous lampbrush chromosomes isolated from the newt. Note the numerous loops extending from the central axis of each chromosome. Courtesy of J. G. Gall. *(Right)* Electron micrograph of individual lampbrush chromosome loops, showing the tiny fibrils that emerge from each loop. Courtesy of O. L. Miller, Jr.

from each other, remain attached through one or more chiasmata. The individual homologs each contain a long central fiber, measuring up to a millimeter in length, and numerous thinner fibers that project from this central axis (Figure 13-43). Close examination of the projecting fibers reveals that they actually represent loops emerging from the central axis. These loops, which tend to occur in pairs projecting from opposite sides of the chromosome, originate from beadlike swellings in the central axis referred to as **chromomeres.** About 5 percent of the total mass of the chromosome is present in the loops at any given time.

The relationship between the loops and the central axis of the lampbrush chromosome has been extensively investigated in the laboratories of H. Callan and J. Gall. In one set of experiments, isolated lampbrush chromosomes were stretched until a gap was created in the central axis at the region of one of the chromomeres. Under these conditions the loops that normally emerge from this chromomere could be seen spanning the break (Figure 13-44). Such an arrangement suggests that the

two loops are continuous with the central axis of the chromosome, and that the central axis must therefore consist of two closely adjoined fibers. Since lampbrush chromosomes occur during the phase of meiosis when the individual homologs are still comprised of two chromatids, it is thought that each of these fibers corresponds to one chromatid. It logically follows that each pair of loops contains one loop derived from each of the two chromatid fibers present within the central axis.

Most of the loops emerging from the lampbrush chromosome are associated with ribonucleoprotein granules and filaments that occur in gradually increasing concentration as one proceeds around the loop, giving the loop a "thin" end and a "thick" end (see Figure 13-44). Autoradiographic examination of lampbrush chromosomes obtained from cells incubated in the presence of ^{3}H-uridine has revealed that the loops represent active sites of RNA synthesis, suggesting that these ribonucleoprotein complexes contain newly forming RNA transcripts. The most straightforward interpretation of the asymmetrical arrangement of ribonucleopro-

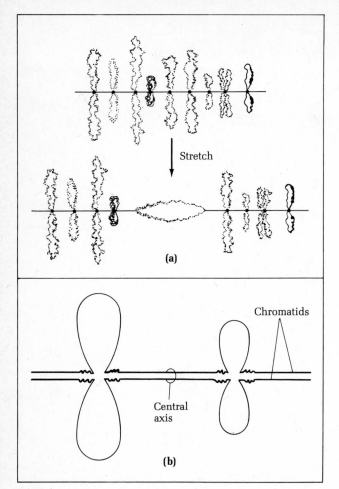

FIGURE 13-44 The organization of lampbrush chromosomes. (a) A diagram summarizing the behavior of lampbrush chromosomes when they are stretched. (b) A model of the lampbrush chromosome suggested by the above behavior. According to this model the central axis is composed of the two chromatids that comprise a given chromosome, while each pair of loops contains one loop derived from each chromatid. Normally the two chromatids present within the central axis are too closely paired to be distinguished from one another.

tein along the length of the loop is that the thin end represents the site at which transcription is being initiated. The RNA transcripts are therefore small in this region, but as one proceeds around the loop the transcripts get longer and the amount of RNA therefore becomes greater.

The intense transcriptional activity of the lampbrush chromosome loops generates the large amounts of RNA needed during the diplotene phase of meiosis, when extensive growth of the developing egg cell takes place. Although this RNA is needed in part to support the extensive amount of protein synthesis that occurs during this growth phase, a majority of the RNA manufactured at this time is actually stored in the egg for use after fertilization.

REGULATION OF CELL DIVISION

We have now described the major events associated with the cell division cycles of both prokaryotic and eukaryotic cells. One fundamental question that still remains, however, concerns the nature of the mechanisms by which cell division is controlled. This question of how cell division is regulated encompasses at least two separate issues. One involves the mechanisms by which the cell cycle is controlled in cells that are already dividing, while the other concerns the mechanism by which external agents or conditions trigger the onset of cell division in cells that are not actively proliferating.

Control of the Cell Cycle in Dividing Cells

It was pointed out earlier in the chapter that cells generally shorten or lengthen the time between successive cell divisions by shortening or lengthening the time spent in G_1. As a result, regulation of the eukaryotic cell cycle is thought to be intimately associated with events that occur during the G_1-phase. Direct evidence for the importance of G_1 in regulating the cell cycle has emerged from studies involving cells exposed to conditions that limit cell growth, such as lack of nutrients or appropriate growth factors. Under such conditions, cells halt at a specific point in the cell cycle situated near the end of G_1. Cells stopped at this stage, referred to as the **restriction point,** then enter into a special phase called G_0, where they may exist for long periods outside the normal cycle of cell proliferation.

Many experimental approaches have been utilized in attempting to identify the molecular events associated with the passage of cells from G_1 into S-phase. The cell fusion experiments described earlier in the chapter, for example, revealed that cytoplasmic factors present in S-phase cells are capable of triggering the premature entrance of G_1 nuclei into S-phase. Although such observations suggest that the production of specific substances during G_1 is ultimately responsible for triggering the entry into S-phase, the relevant molecules are yet to be clearly identified. One possibility is that they are simply the enzymes required for DNA replication (e.g., DNA polymerase) or enzymes involved in the metabolism of the bases (e.g., thymidine kinase, thymidylate kinase, or deoxycytidine monophosphate deaminase). However, this viewpoint is difficult to reconcile with the discovery that such enzymes are present in high concentrations in certain cells that do not replicate their DNA and that, conversely, some of these same enzymes can be absent from cells that are actively synthesizing DNA.

Alternative candidates for the molecular event that

triggers the transition from G_1 to S-phase have emerged from studies on yeast cells. Yeasts are an exceptionally powerful tool for investigating the cell cycle because of the ease with which genetic mutants can be obtained and characterized. Such approaches have led to the discovery of a small group of genes that control an event occurring in G_1, called *start*, which represents the point when cells become committed to entering the mitotic cell cycle. After this commitment has been made cells enter into a program leading to S-phase, mitosis, and cell division. Among the genes found to be essential for the *start* function, at least one appears to code for a protein resembling a *protein kinase* (an enzyme catalyzing the phosphorylation of other proteins). The possibility that protein phosphorylation may play a role in starting the cell division cycle during G_1 is especially interesting in view of our previous discussion concerning the role of protein phosphorylation in triggering the transition from G_2 to mitosis (page 507).

In bacterial cells commitment to the cell division cycle has been found to require the production of DNA-binding proteins that destabilize the double helix, thereby allowing the enzymes involved in replication to have direct access to the DNA. Proteins capable of destabilizing the DNA double helix are also thought to be involved in the initiation of DNA replication in eukaryotic cells, though the situation is more complicated in eukaryotes because of the need to disassemble the nucleosomal organization of the DNA. The mechanism responsible for unpacking this nucleosomal organization has not been elucidated in detail, but it is worth noting that phosphorylation of a specific class of histone molecules occurs during S-phase. If this histone modification reaction influences nucleosomal packing, then the protein kinase that catalyzes histone phosphorylation might be an essential requirement for the entrance into S-phase.

If the onset of S-phase is dependent on the synthesis of specific protein molecules, then the question arises as to how the synthesis of these particular proteins is regulated. The most straightforward assumption would be that such molecules are synthesized at a rate directly proportional to the rate of protein synthesis as a whole; if this were true, then the rate at which cells enter into S-phase would be directly related to the increase in cell mass caused by the synthesis of new proteins during G_1. In support of this simple notion, it has been shown by David Prescott that cell division in *Amoebae* only occurs after the mass of the cell has doubled, and that division can be prevented by experimentally removing small portions of cytoplasm. Unfortunately this simple relationship does not always hold. Yeast grown in a culture medium depleted of nutrients divide without doubling their mass, yielding cells that are smaller than normal.

Likewise embryonic development in higher organisms involves a long series of mitotic divisions in the absence of cell growth, thereby generating smaller and smaller cells. Since an increase in mass does not appear to be an absolute requirement for cell division, the synthesis of proteins required for the commitment to cell division is thought to be controlled by mechanisms that are distinct from those that determine the overall rate of protein synthesis.

Although G_1 appears to be the main site of regulation of the cell cycle in eukaryotes, some cells stop their cycle in G_2. Such G_2-arrested cells are unique in that they are capable of quickly initiating mitosis in response to an appropriate stimulus, and therefore occur in places such as skin where rapid cell division may be required in response to an injury. In contrast to G_2-arrested cells, cells in either G_0 or G_1 require many hours to enter mitosis because of the necessity of first passing through the remainder of G_1, S, and G_2.

External Factors That Stimulate Cell Division

In addition to the importance of metabolic events occurring within the cell, external agents and conditions also exert controls over the process of cell division. In prokaryotic cells the major controlling factor appears to be the presence of sufficient nutrients in the external environment. As long as adequate nutrients are available, cell division generally continues. Eukaryotic cells, on the other hand, typically occur in multicellular organisms in which the division of each cell must be carefully regulated to suit the needs of the organism as a whole. A variety of external factors are involved in coordinating the division of such cells. Certain steroid hormones, for example, have been found to stimulate the growth and division of specific types of animal cells, while hormones called auxins, cytokinins, and gibberellins stimulate the division of plant cells. Although the preceding types of hormones act by entering directly into their target cells (Chapter 16), other growth-promoting factors have been identified that bind to specific receptors in the plasma membrane. Well-studied examples of such growth factors include *epidermal growth factor, nerve growth factor, platelet-derived growth factor, thymosin,* and *erythropoietin.* Each of these proteins stimulates the proliferation of a specific group of target cells.

Although the mechanism by which such molecules stimulate cell division is not completely understood, considerable progress has been made in unraveling some of the early steps involved. One of the more thoroughly investigated examples involves the response of cultured cells to the addition of **epidermal growth factor (EGF),** a polypeptide of approximately 6400 molecular weight. Experiments with radioactive EGF have revealed that this polypeptide first binds to specific membrane receptors located on the surface of the cell being stimulated to divide. These receptors, which are initially spread randomly over the cell surface, become clustered into patches after binding to EGF and are subsequently taken into the cell by endocytosis. One of the first biochemical changes observed in cells treated with EGF is an increased rate of phosphorylation of tyrosine residues located in certain intracellular proteins as well as the EGF receptor itself. The enzymatic activity responsible for catalyzing this tyrosine-specific phosphorylation reaction resides in a portion of the EGF receptor that faces the cell interior. In other words the EGF receptor is a transmembrane protein that binds to EGF molecules at the outer surface of the plasma membrane and functions as a **tyrosine-specific protein kinase** at the inner surface of the plasma membrane (Figure 13-45). Binding of EGF to the membrane receptor causes its protein kinase activity to become activated. Although it has been suggested that the resulting phosphorylation of intracellular proteins is the immediate stimulus that triggers the cell to divide, there is little direct evidence for this idea because the functions of such phosphorylated proteins are poorly understood. It is interesting to note, however, that the ability of certain tumor viruses to transform normal cells into cancer cells also involves the production of a tyrosine-specific protein kinase (Chapter 20).

The preceding information indicates that one way in which the interaction of external factors with the plasma membrane can trigger intracellular changes is by altering the activity of an enzyme localized within the plasma membrane. In addition to the protein kinase activated by EGF, several other membrane-localized enzymes are known to be regulated by agents that bind to plasma membrane receptors. We shall see in Chapter 16 that the binding of many protein hormones to the plasma membrane triggers activation of the membrane-bound enzyme **adenylate cyclase,** which catalyzes the formation of **3′,5′-cyclic adenosine monophosphate (cyclic AMP).** Intracellular concentrations of cyclic AMP have been found to change both during the normal cell cycle and in response to agents that stimulate or inhibit cell division. Moreover the addition of cyclic AMP to the external medium has been reported to influence the division rate of certain kinds of cultured cells. However, the interpretation of such effects is complicated by the fact that increased levels of cyclic AMP appear to be associated with an increase in cell division in some situations and a decrease in others. Hence the role of cyclic AMP in the cell cycle remains to be clarified.

In addition to stimulating or inhibiting membrane-bound enzymes such as protein kinase and adenylate cyclase, agents that bind to the cell surface can also influence intracellular events by inducing alterations in the permeability and transport properties of the plasma

FIGURE 13-45 Summary of the mechanism by which epidermal growth factor (EGF) influences protein phosphorylation. Binding of EGF to the exterior surface of its plasma membrane receptor causes a conformational change in the receptor that permits it to function as a tyrosine-specific protein kinase at the cytoplasmic surface of the plasma membrane. This enzymatic activity in turn catalyzes the phosphorylation of the amino acid tyrosine (tyr) in specific cytoplasmic proteins.

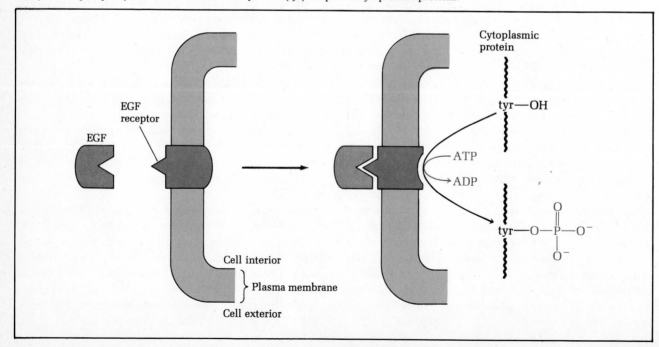

membrane. Such changes can be detected shortly after the binding of various kinds of growth factors to the plasma membrane, raising the possibility that the resulting alteration in the concentration of intracellular solutes is involved in signaling the cell to divide. Independent support for the idea that changes in membrane permeability and transport properties influence the cell cycle has emerged from the discovery that the membrane potential at the cell surface varies in a reproducible manner during progression through the normal cell cycle, indicating that the concentration of charged ions within the cell is changing. Moreover, cell division can be artificially stimulated or inhibited by manipulating the ionic composition of the medium in which the cells are growing. The addition of potassium ions, which increase the membrane potential, has been found to block cell division while sodium ions, which lower the membrane potential, stimulate cell division.

Another solute whose intracellular concentration changes during the cell cycle is the calcium ion. Some of the most striking evidence linking calcium ions to the regulation of cell division has come from studies involving sea urchin eggs. When such eggs are fertilized, the intracellular concentration of calcium ions quickly rises and the egg begins to divide. In order to determine whether this increase in intracellular Ca^{2+} is directly involved in triggering the onset of cell division, studies have been performed in which the intracellular Ca^{2+} concentration was artificially increased in unfertilized eggs by exposing them to a calcium ionophore. In response to such treatment the egg cells begin to divide, even though fertilization has not taken place. The preceding observations suggest that an elevation in the intracellular concentration of free Ca^{2+} is capable of triggering the onset of cell division. Although the mechanism by which calcium ions act is not clearly understood, many of the effects of Ca^{2+} are known to be mediated by the calcium-binding protein, *calmodulin*. At least two lines of evidence support the idea that calmodulin mediates the effects of calcium ions on the cell cycle: Calmodulin levels have been observed to increase during the transition from G_1 to S, and treatment of cells with drugs that inhibit calmodulin function causes a delay in the entry into and progression through S-phase.

An involvement of calcium ions in triggering cell division is further supported by the discovery that the interaction of growth factors with the cell surface often triggers an increase in the intracellular concentration of free Ca^{2+}. In the case of **platelet-derived growth factor (PDGF),** the pathway leading to the increase in free Ca^{2+} is fairly well understood. Interaction of PDGF with its plasma membrane receptor leads to the activation of **phospholipase C,** a membrane-bound enzyme that in turn catalyzes breakdown of the lipid phosphatidylinositol 4,5-bisphosphate to *diacylglycerol* and *inositol*

trisphosphate. The diacylglycerol stimulates protein phosphorylation by activating a plasma membrane-bound enzyme called **protein kinase C,** while the inositol trisphosphate triggers the release of calcium ions stored within vesicles of the endoplasmic reticulum (see Figure 16-11).

The discovery that many extracellular growth regulators trigger cell division by interacting with the cell surface indicates that the plasma membrane may be a crucial element in the normal regulation of cell proliferation. It is worth pointing out in this context that the morphological appearance of the cell surface changes during the normal cell cycle (Figure 13-46), suggesting that altered structural interactions with other cells may be occurring. The idea that cell-cell interactions exert an influence on cell proliferation was first proposed several decades ago on the basis of observations made on the growth of normal eukaryotic cells in tissue culture. Such cells generally proliferate until a confluent monolayer of cells completely covers the surface of the culture vessel. At this point, cell division ceases. Since proliferation appears to stop at about the same time that the cells are beginning to make close contact with one another, this phenomenon was first termed **contact inhibition.** However, the evidence that physical contact between cells is actually responsible for inhibiting cell division is not very compelling. It has been shown, for example, that cultures in which cell division has been halted by contact inhibition can be induced to start dividing again by adding large quantities of serum to the medium. Such observations suggest that cell division is inhibited in crowded cultures not by physical contact per se, but because the availability of essential nutrients or growth factors has become rate limiting.

Since the role of cell-cell contact in regulating cell division is unclear, the more general term **density-dependent inhibition** is now commonly used in place of the term *contact inhibition* to indicate that cell division slows or stops in crowded cultures when the concentration of cells reaches a certain density. The changeover to this newer term does not necessarily preclude, however, a role for cell-cell interactions in the regulation of cell growth and division. Cell surface alterations are frequently observed in cancer cells, which differ from normal cells in that their growth in culture is much less susceptible to density-dependent inhibition. One of the more intriguing findings to emerge from the study of cancer cells is that they do not establish gap junctions as frequently as do normal cells, suggesting that the control of normal cell growth may require the ability of adjacent cells to communicate through such junctions.

It should be clear from the foregoing discussion that much remains to be learned about the mechanisms involved in the regulation of cell growth and division. Because of the obvious relevance of this problem to the question of why cancer cells exhibit uncontrolled

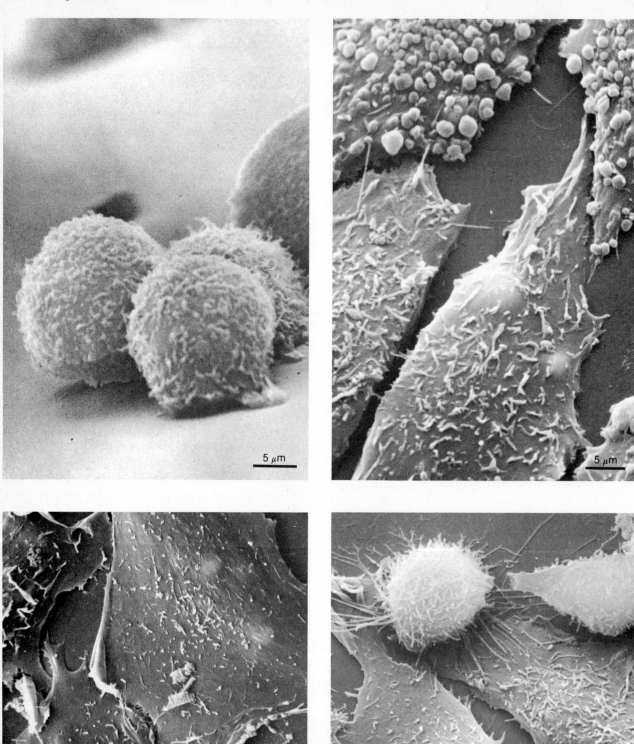

FIGURE 13-46 Scanning electron micrographs of cultured Chinese hamster ovary cells at various stages during the cell cycle. *(Top left)* Cells in early G_1 are settling down on the surface of the culture vessel. At this stage the cell surface is covered with small microvilli. *(Top right)* By late G_1, the cells have spread out and are covered with blebs and microvilli. A little ruffling appears at the cell margins. *(Bottom left)* Cells in S-phase are completely flattened, with microvilli and ruffles on their surfaces. *(Bottom right)* The flattened cells containing numerous microvilli are in late G_2. The rounded cell is in the early stages of mitosis. Courtesy of K. R. Porter.

growth, research in this area is currently very active. Cancer cells in fact present the cell biologist with such complex and intriguing problems that we will devote an entire chapter to their study later in the book.

SUMMARY

Although cells in the process of undergoing division were first observed microscopically over a century ago, we have only recently begun to understand the mechanisms involved. In eukaryotic cells DNA synthesis is restricted to a defined portion of the cell cycle referred to as S-phase. S-phase is separated from the previous mitosis by a time gap known as the G_1-phase, while a second gap, G_2, intervenes between S-phase and the beginning of the next mitosis. In the bacterial cell cycle the time period occupied by DNA synthesis and the time period that intervenes between DNA synthesis and the onset of cell division can overlap one another, allowing the creation of extremely short cell division cycles.

In bacterial cells semiconservative DNA replication is initiated at a single site on the chromosome and proceeds in both directions around the circular DNA molecule. Although this replicating chromosome is closely associated with the plasma membrane, it is not clear whether such an association is required for replication or is simply related to the mechanism by which the newly formed chromosomes are parceled out to the progeny cells. The DNA of eukaryotic cells is replicated by a semiconservative mechanism similar to that of prokaryotic cells, but the increased DNA content and the need to disassemble the nucleosomal organization make the replication process slower and more complex. To speed up replication of the large amounts of DNA involved, bidirectional replication is initiated at many separate sites termed replicons. The DNA synthesized by individual replicons is later joined together to form complete new chromosomal DNA molecules. The total time required to complete replication of an individual chromosomal DNA molecule is determined by the frequency with which DNA synthesis is initiated at the multiple replicons, rather than by the rate at which DNA is synthesized at a given replicon. Isolated nuclear matrix preparations have been found to be enriched in recently synthesized DNA, raising the possibility that DNA is replicated as the chromatin fibers pass through replication sites associated with the nuclear matrix.

At the time of cell division, a complete set of replicated DNA molecules is transmitted to each of the newly formed cells. In bacterial cells this transmission is accomplished by adding new membrane material to the plasma membrane in the region between the attachment sites of the two chromosomes, causing the two chromosomes to gradually separate from one another. After separation has occurred, the cytoplasm is divided by involution of the plasma membrane and formation of a cross-wall. Cleavage of this wall then generates two separate cells, each containing one of the original chromosomes.

In eukaryotic cells the events associated with chromosome separation and division of the cytoplasm (cytokinesis) are more elaborate. The most common type of eukaryotic cell division, termed mitosis, consists of four stages termed prophase, metaphase, anaphase, and telophase. During prophase the chromatin fibers condense into recognizable chromosomes, each consisting of two paired chromatids joined by a centromere. Nucleoli often disperse at this time, and formation of the mitotic spindle is initiated. In animal cells this latter process is associated with migration of two pairs of centrioles to opposite sides of the nucleus and formation of a group of small microtubules, termed an aster, around each centriole pair. The end of prophase is marked by breakdown of the nuclear envelope. During metaphase, microtubules become attached to the chromosomes and the chromosomes migrate toward the equator of the spindle, where they line up adjacent to one another. At the beginning of anaphase the two chromatids comprising each chromosome separate, generating two independent chromosomes. These separated chromosomes then migrate to opposite poles of the spindle, accompanied by movement of the two poles of the spindle away from one another. The beginning of telophase is marked by the arrival of the two sets of chromosomes at opposite poles of the spindle. These chromosomes then disperse into chromatin fibers, nucleoli reappear (if they had disappeared), the spindle breaks down, and nuclear envelopes form around the two sets of chromosomes. Division of the cytoplasm takes place in synchrony with these nuclear changes.

The two chromatids comprising each metaphase chromosome are composed of an elaborately folded mass of fibers that are morphologically indistinguishable from interphase chromatin fibers. It has been suggested that the two chromatids are held together by an unreplicated region of DNA whose replication during anaphase permits the chromosomes to separate. Chromosome-staining procedures have revealed the presence of a reproducible pattern of stained bands in each type of metaphase chromosome, suggesting that the way in which chromatin fibers are folded is highly reproducible for a given type of chromosome. Although the molecular basis underlying this precise folding pattern is not well understood, linkage of the chromatin fibers to an underlying nonhistone protein scaffolding is thought to be involved.

In most eukaryotic cells chromosome movement is dependent on the presence of the mitotic spindle. The spindle apparatus contains three types of microtubules: astral microtubules radiating from the centrioles (in animal cells only), interpolar microtubules spanning the area between the spindle poles, and kinetochore microtubules attached to the kinetochores located on opposite sides of the chromosomal centromeres. In animal cells centriole pairs surrounded by a finely granular, amorphous material are present at each spindle pole. Because the mitotic spindles of plant cells develop in the absence of centrioles, it is thought that the granular material surrounding the centrioles, rather than the centrioles themselves, nucleates the assembly of the spindle microtubules. Such observations suggest that the migration of centrioles to opposite sides of the spindle may be related more to the need to guarantee that each new cell receives a pair of centrioles than to a direct involvement of centrioles in the process of spindle formation per se.

Although several lines of evidence indicate that the presence of spindle microtubules is required for chromosome movement, the exact nature of their role is yet to be clearly elucidated. Microtubule-sliding mechanisms analogous to those involved in ciliary and flagellar motility have been implicated in the process of pole-pole separation, but there is little evidence that such mechanisms are responsible for chromosome-to-pole movements. As an alternative, it has been proposed that gradual depolymerization of the kinetochore microtubules generates sufficient force to actively pull the chromosomes toward the spindle poles. Yet another suggestion is that the motile force for chromosome movement may be provided by some external agent acting on these microtubules. Although several candidates for such an external agent have been proposed (e.g., actin, myosin, cytoskeletal elements), there is no experimental support for a direct involvement of such structures in chromosome movement.

Cytokinesis normally begins in association with the final stages of chromosome separation, though these two processes are not inextricably coupled to one another. The location of the cleavage plane is determined by the position of the mitotic spindle, passing through its equatorial region. In animal cells cytokinesis proceeds from the outside of the cell toward the interior, powered by the contraction of a band of microfilaments that passes around the circumference of the cell directly beneath the plasma membrane. In plant cells cytokinesis is mediated by a coalesence of membrane vesicles and deposition of cell wall material that begins in the center of the cell and moves toward the periphery.

During the production of reproductive cells such as sperm and eggs, a special type of division termed meiosis occurs. This process, which involves two cell divisions occurring in sequence, reduces the chromosome number from diploid to haploid and is associated with the process of genetic recombination. The prophase of the first meiotic division is divided into five stages known as leptotene, zygotene, pachytene, diplotene, and diakinesis. During leptotene the chromatin condenses into long thin threads that become recogniz-

able as individual chromosomes at the beginning of zygotene. At this point the homologous chromosomes become tightly paired with one another. Next comes pachytene, which is characterized by a thickening and shortening of the chromosomes and the onset of genetic recombination. During diplotene the homologous chromosomes become partially separated, but remain connected by chiasmata that represent areas where recombination has occurred. In certain kinds of egg cells the chromosomes uncoil at this point and an extensive period of cell growth ensues. During diakinesis the chromosomes return to a highly condensed state, the spindle apparatus begins to appear, and the nuclear envelope disintegrates. The subsequent metaphase, anaphase, and telophase exhibit several important differences from the comparable stages of mitosis. First, the chromosomes of meiotic metaphase I consist of a pair of homologous chromosomes held together by chiasmata, while mitotic chromosomes are comprised of two chromatids joined together by a centromere. The second major difference is that during meiotic anaphase I, the homologous chromosomes separate and move to opposite spindle poles; in mitosis, it is the two chromatids comprising each chromosome that separate and move to opposite poles. The second meiotic division resembles mitosis more closely, except that only one homolog from each pair of chromosomes is present. Since each homolog is comprised of two chromatids, it is these chromatids that separate from one another during meiotic anaphase II.

Genetic recombination, which occurs during the first meiotic division, as well as during transduction, transformation, and conjugation in bacterial cells, is mediated by the physical breakage and rejoining of DNA molecules. Genetic recombination is sometimes accompanied by gene conversion (i.e., nonreciprocal recombination) as well as the formation of individual DNA molecules whose two strands are not completely complementary to one another. These unusual phenomena have led to the formulation of a model of genetic recombination based on the formation of a single-strand exchange intermediate in which one strand of each DNA double helix has been broken and joined to the opposing DNA molecule. In bacterial cells the recA protein, which catalyzes a reaction in which a single-stranded DNA molecule displaces one of the two strands of a DNA double helix, has been implicated in this process of genetic recombination. Based on the ability of the recA protein to catalyze single-strand displacement, detailed models of genetic recombination have been proposed that require the initial cleavage of only one DNA strand.

Several lines of evidence suggest that the synaptonemal complex plays an important role in the process of genetic recombination in eukaryotic cells, most likely by mediating a relatively crude alignment between homologous chromosomes. After such alignment has occurred the chromatid fibers appear to undergo localized uncoiling, generating a series of DNA loops that project into the central region of the synaptonemal complex. The molecular events associated with strand exchange presumably take place between these looped-out regions.

In oocytes that undergo extensive growth during the diplotene stage of the first meiotic division, the chromosomes often assume a lampbrush configuration. Each lampbrush chromosome consists of a pair of homologous chromosomes that, though well separated from each other, remain attached through one or more chiasmata. The individual homologs each contain a long central fiber from which a series of paired loops project. Each pair of loops contains one loop derived from each of the chromatids present within the central axis. The loops represent regions where the chromosomal DNA is being transcribed into the large amounts of RNA needed for the growth phase of meiosis and the early events of embryonic development.

Regulation of the cell cycle of actively dividing eukaryotic cells appears to be intimately associated with events that occur during G_1. When actively proliferating cells are exposed to conditions that are not conducive to cell division, the cells halt at a specific point in the cell cycle situated near the end of G_1. Cells stopped at this stage, called the restriction point, are said to be in G_0. Although G_1 appears to be the main site of regulation of the cell cycle, tissues in which rapid cell division may be required in response to injury contain some cells whose cycles have been arrested during G_2.

In addition to internal controls, external factors also influence the process of cell division. In prokaryotic cells a crucial external factor influencing the rate of cell division is the presence of sufficient nutrients. Although the proliferation of eukaryotic cells is likewise dependent on the availability of adequate nutrients, higher organisms must also have mechanisms for ensuring that the division of each cell is regulated to suit the needs of the organism as a whole. Such regulation is mediated at least in part by circulating proteins, called growth factors, that stimulate the division of specific target cells by binding to plasma membrane receptors. Interaction of these agents with the plasma membrane triggers alterations in membrane permeability and

transport as well as changes in the activity of membrane-localized enzymes such as protein kinase, adenylate cyclase, and phospholipase C. Such events trigger alterations in intracellular molecules and solute concentrations that in turn signal the cell to begin dividing.

SUGGESTED READINGS

Books

Baserga, R. (1981). *The Biology of Cell Reproduction,* Harvard University Press, Cambridge, MA.

Callan, H. G. (1986). *Lampbrush Chromosomes,* Springer-Verlag, Berlin.

Cold Spring Harbor Laboratory (1984). *Recombination at the DNA Level,* proceedings of a symposium published as *Cold Spring Harbor Symp. Quant. Biol.,* Vol. 49, Cold Spring Harbor, NY.

John, P. C. L., ed. (1981). *The Cell Cycle,* Cambridge University Press, Cambridge, England.

Lloyd, O. D., S. W. Edwards, and R. K. Poole (1982). *The Cell Division Cycle: Temporal Organization and Control of Cellular Growth and Reproduction,* Academic Press, New York.

Moens, P. V., ed. (1986). *Meiosis,* Academic Press, Orlando.

Articles

Berridge, M. J., and R. F. Irvine (1984). Inositol trisphosphate, a novel second messenger in cellular signal transduction, *Nature* 312:315–321.

Das, M. (1982). Epidermal growth factor: mechanisms of action, *Int. Rev. Cytol.* 78:233–256.

Inoué, S. (1981). Cell division and the mitotic spindle, *J. Cell Biol.* 91:132s–147s.

McIntosh, J. R. (1984). Mechanisms of mitosis, *Trends Biochem. Sci.* 10:195–198.

Murray, A. W., and J. W. Szostak (1985). Chromosome segregation in mitosis and meiosis, *Ann. Rev. Cell Biol.* 1:289–315.

Nurse, P. (1985). Cell cycle control genes in yeast, *Trends Genet.* 1:51–55.

Rozengurt, E. (1986). Early signals in the mitogenic response, *Science* 234:161–166.

Stahl, F. W. (1987). Genetic recombination, *Sci. Amer.* 256(Feb.):90–101.

CHAPTER

14

Biogenesis of Mitochondria and Chloroplasts

The last several chapters have focused on the replication and expression of genetic information stored within a cell's chromosomal DNA molecules. Although the vast majority of the DNA present in typical eukaryotic cells is localized within the nucleus, some DNA is also present in mitochondria and chloroplasts. In recent years the genetic information contained within this DNA has been extensively characterized, leading to a new understanding of the extent to which mitochondria and chloroplasts exert control over their own formation. The preceding developments have in turn raised some intriguing questions concerning the evolutionary origin of these two organelles. This chapter will therefore concentrate on the following interrelated questions: How are mitochondria and chloroplasts generated in present-day cells? How did mitochondria and chloroplasts arise?

GENETIC EVIDENCE FOR CYTOPLASMIC INHERITANCE

Chloroplast Genetics

The initial realization that genes may be located in the cytoplasm as well as in the nucleus can be traced to two reports appearing in 1909, one by Carl Correns and the other by Erwin Baur. Both men were studying the inheritance of genetic mutations that result in the failure of chloroplasts to acquire their normal green pigmentation. The leaves of affected plants are a mixture of green, colorless, and variegated (green and colorless) regions. Because some flowers occur in the regions of the plant that are entirely green, while others grow from areas that are colorless, it is possible to selectively fertilize normal flowers with pollen derived from colorless tissue, and vice versa.

When Correns carried out such experiments on the four-o'clock flower *Mirabilis jalapa*, he found that the source of the pollen had no effect on the result obtained. Flowers from normal green areas yielded only green plants, even if the fertilizing pollen was taken from the mutant colorless segments. Conversely, flowers from the colorless areas always gave rise to colorless plants, even when normal pollen was employed. Because the properties of the offspring were always determined by the female parent, this phenomenon was referred to as **maternal inheritance.** Such a pattern is different from the behavior of nuclear genes, which had been previously shown by Mendel to be contributed equally by the male and female parents. It was quickly realized that the maternal pattern of inheritance might be accounted for by genes in the cytoplasm, since the cytoplasm of the fertilized egg is contributed almost solely by the female parent.

Slightly different results were obtained by Baur, who chose the geranium *Pelargonium* as his experimental material. Starting with one green and one colorless parent, he found some of the offspring to be variegated, others pure green, and the rest colorless. The green and colorless offspring both bred true, meaning that each had received genes from only one of the two parents. This observation is also inconsistent with classical Mendelian inheritance patterns in which genes are contributed by both parents to their offspring. This type of non-Mendelian information transfer, in which genes are contributed by either the male or the female parent, is referred to as **biparental inheritance** (Figure 14-1). Genes that are inherited by either maternal or biparental means do not behave as if they are located in the nucleus, and so it was concluded that these genes must be located in the cytoplasm. Thus it was logical for both Correns and Baur to propose that the genes for chloroplast pigmentation are located within the chloroplasts themselves. Nevertheless, these early genetic experiments could not rule out the possibility of an alternate cytoplasmic site for such genes.

In the years since these early discoveries, many other examples of non-Mendelian inheritance of plastid pigmentation have been described. Most, but not all, involve maternal inheritance like that observed by Correns. One unusual and illuminating case, however, is the loss of pigmentation produced by the *iojap* mutation in maize. Some *iojap* plants lack green pigment entirely, while others are green with white striping. Though the *iojap* mutation maps to a nuclear chromosome, Marcus Rhoades has demonstrated that once chloroplasts have lost their pigmentation as a result of the influence of this mutation, the nuclear *iojap* gene is no longer required to maintain the unpigmented state. Even after prolonged breeding with normal individuals to remove the *iojap* mutation from the nucleus, the plants continue to inherit unpigmented plastids. The most straightforward interpretation of this result is that the nuclear *iojap* gene induces a cytoplasmic mutation that, once it has been established, no longer requires the *iojap* gene for its inheritance. As Rhoades graphically put it, "although induced by a nuclear . . . gene, the mutated plastid, like a Frankenstein monster, is no longer under the control of its maker" (1943:328–329).

Electron microscopic examination of *iojap* maize has further substantiated the partial autonomy of the mutated chloroplasts from nuclear control. In areas of these plants where normal green tissue borders on mutant colorless tissue, individual cells contain a mixture of two types of chloroplasts: normal pigmented ones and abnormal chloroplasts lacking chlorophyll-containing

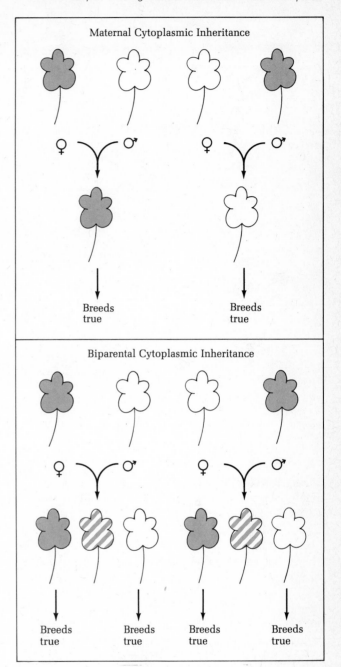

FIGURE 14-1 Patterns of maternal and biparental cytoplasmic inheritance. Colored shading represents green pigmentation.

thylakoids (Figure 14-2). The coexistence of these two kinds of chloroplasts *within the same cell* (and therefore under the influence of the same nuclear genes) is dramatic proof that major properties of the chloroplast are not under nuclear control.

Most cytoplasmic mutations leading to abnormalities in chloroplast development occur spontaneously, but the rate of mutation can be artificially enhanced with physical or chemical mutagens. Ultraviolet light in the 260-nm region of the spectrum is especially effective, suggesting that nucleic acids, which absorb

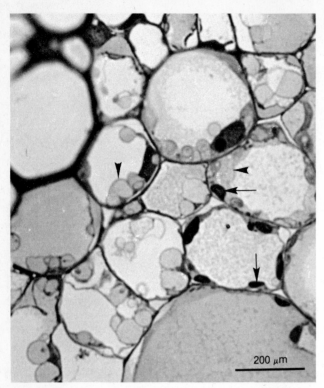

200 μm

FIGURE 14-2 Light micrograph of *iojap* maize seedling showing mesophyll cells containing both normal, chlorophyll-containing plastids (arrows) and colorless *iojap*-affected plastids (arrowheads). Courtesy of E. H. Coe.

strongly in this region, are the target molecules. Using a finely focused microbeam of ultraviolet light, Aharon Gibor and Sam Granick have demonstrated that a stable loss of chloroplast pigmentation and photosynthetic ability can be produced in the protozoan *Euglena* when the cytoplasm alone is irradiated. No comparable effect occurs when the nucleus is treated in this same way, clearly indicating that the site of the altered gene is cytoplasmic.

The usefulness of chemical mutagens in studying chloroplast genes is limited by the fact that both nuclear and cytoplasmic mutations are induced. Due to the vast preponderance of nuclear DNA relative to that in chloroplasts, most mutations resulting from such agents will be nuclear. But one important exception to this generalization is the antibiotic **streptomycin,** whose effects on the green alga *Chlamydomonas* have been extensively studied by Ruth Sager. Streptomycin is not a mutagen for nuclear genes, nor is it an effective mutagen for bacteria or viruses. But by some unknown mechanism it causes a wide variety of chloroplast mutations, including loss of photosynthetic activity, inability to grow at temperatures above 25°C, low growth rates, and resistance to the inhibitory effects of a variety of antibiotics, including streptomycin itself.

The availability of this vast array of mutants has permitted Sager to map the location of many cytoplas-

mic genes affecting the chloroplast. The creation of a genetic map generally requires that genes undergo recombination, for it is the frequency with which genes recombine that allows their relative distances from each other to be calculated. Since cytoplasmic inheritance in most organisms, including *Chlamydomonas*, is entirely maternal, it is not possible to measure the rates at which maternal and paternal cytoplasmic genes recombine with one another. This obstacle has been overcome with the discovery that in *Chlamydomonas*, irradiation of the female parent prior to mating results in progeny that carry cytoplasmic genes of both parents rather than just the mother. Analysis of the relative rates of recombination of cytoplasmic genes in such matings has led to the construction of a genetic map showing that these genes are arranged on a circular rather than a linear DNA molecule (Figure 14-3).

Although these data clearly demonstrate that genes governing certain chloroplast properties are linked together to form an extranuclear circular chromosome, they do not tell us where this chromosome is located. The most obvious site is within the chloroplast, but such a localization cannot be unequivocally established from genetic data alone. What is needed is the direct demonstration that an alteration in chloroplast DNA is associated with changes in genes that affect chloroplast behavior. Evidence of this type, which requires a combination of genetic and biochemical techniques, will be discussed later in the chapter.

Mitochondrial Genetics

The first cytoplasmically inherited alteration in mitochondrial behavior to be discovered was the **petite** or ρ^- (rho$^-$) mutation of yeast, described by Boris Ephrussi and his colleagues in France in 1949. Yeast bearing this mutation are incapable of carrying out electron transfer and oxidative phosphorylation because, among other defects, they lack cytochromes a, c_1, and aa_3. Associated with these biochemical deficiencies is a drastic reduction in the number of mitochondrial cristae (Figure 14-4). This defect in respiration is not lethal because the cells are able to convert foodstuffs into useful chemical energy by the glycolytic pathway (see Figure 7-5). But the small amount of energy derived from this metabolic pathway as compared to mitochondrial respiration leads to slow growth rates, and therefore yeast colonies that are smaller than usual. It is for this reason that such mutant yeast are called *petites*.

Analysis of the inheritance pattern of the *petite* condition has shown that a variety of different mutations can produce the same effect. A small number of them, designated **nuclear petites,** exhibit classical Mendelian inheritance and are linked to other known nuclear genes. In most cases, however, the *petite* state is cytoplasmically inherited. There are two basic classes of

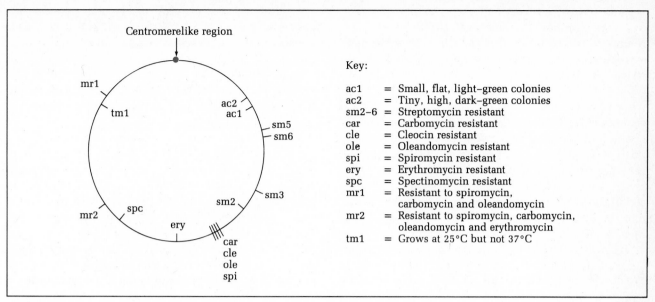

FIGURE 14-3 Map positions of 15 genes on the chloroplast chromosome of *Chlamydomonas*.

cytoplasmic petites: neutral and suppressive. When yeast bearing the neutral *petite* mutation are crossed with a normal strain, the resulting progeny are always normal, no matter how long they are inbred. In other words the *petite* character disappears altogether. But with suppressive *petites* the opposite occurs, namely, all offspring acquire the *petite* state. Neither inheritance pattern is like that observed with nuclear genes, in which the traits of both parents are ultimately expressed.

Several features distinguish cytoplasmic *petite* mutations from typical nuclear mutations. First, the cytoplasmically inherited *petite* condition occurs spontaneously in roughly 1 percent of the cells of a typical yeast colony, while the spontaneous mutation rate of most nuclear genes is closer to 0.01 percent. Second,

certain agents that are not particularly effective as mutagens for nuclear genes, such as *ethidium bromide*, *euflavine*, and *5-fluorouracil*, can convert virtually an entire population of yeast to the *petite* state within one generation of growth. Mutagens with close to 100 percent efficiency are unheard of when dealing with nuclear genes. Finally, the cytoplasmic *petite* mutation is also unusual in that it never reverts to the normal state, while nuclear mutations usually revert at low but finite frequencies. This apparent irreversibility of the *petite* state led Ephrussi to suggest that it involves the loss of a cytoplasmic hereditary factor. As will be seen later, the recent demonstration that mitochondrial DNA from *petite* yeast is grossly altered in structure lends support to this view.

In 1952 Mary and Herschel Mitchell discovered a

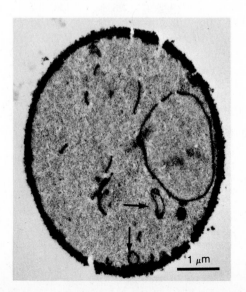

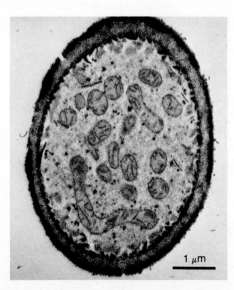

FIGURE 14-4 Electron micrographs showing mitochondria from normal and *petite* yeast. *(Left) Petite* yeast have few mitochondria with poorly developed internal structure (arrows). *(Right)* Normal yeast have mitochondria with well-developed cristae. Courtesy of Y. Yotsuyanagi.

mutant strain of the bread mold *Neurospora* that is analogous in many ways to *petite* yeast. These slow-growing *Neurospora*, called the **poky** strain, lack cytochromes *b* and *aa₃*, have an excess of cytochrome *c*, and contain mitochondria exhibiting abnormal cristae. Crosses between *poky* "males" and normal "females" yield all wild-type offspring, while the offspring of the reciprocal cross express only the *poky* trait. Because the "female" parent contributes most of the cytoplasm during mating, this pattern of maternal inheritance suggests that the gene governing the *poky* trait is located in the cytoplasm. However, some nuclear genes that produce a condition similar to *poky* also exist, indicating that mitochondrial behavior is under the influence of both nuclear and cytoplasmic genes.

A clue to the nature of the biochemical lesion underlying the *poky* condition was provided in 1956 by Thad Pittenger, who induced wild-type and *poky* strains of *Neurospora* to fuse with each other, forming cells containing two types of nuclei and a mixture of two cytoplasms. At first the cells grew normally, but eventually the *poky* condition prevailed. Pittenger therefore concluded that the *poky* condition is not caused by the *absence* of some normal cytoplasmic entity, but is instead due to the *presence* of an abnormal component that converts normal mitochondria into abnormal ones. Consistent with this notion, subsequent biochemical studies have shown that *poky* mitochondria contain an enzymatic activity, not present in normal mitochondria, that rapidly degrades cytochromes.

In addition to *poky*, several other slow-growing strains of *Neurospora* exhibiting maternally inherited mitochondrial defects have been isolated. One such strain, identified in Edward Tatum's laboratory, has been used to provide the first direct evidence that cytoplasmic genes affecting mitochondrial traits are actually located within the mitochondrion. In these studies Elaine Diacumakos injected purified nuclei or mitochondria from slow-growing mutants into cells of a normal strain. Injection of nuclei produced no detectable change in the recipient cells, but introduction of mitochondria derived from the mutant strain caused defects in growth rate and cytochrome content similar to those present in the mutant. These results strongly suggest that the mitochondrion contains the genes responsible for the mutant condition.

Though *petite* yeast and *poky Neurospora* have been extremely useful in establishing the existence of cytoplasmic genes that influence mitochondrial structure and function, such gross alterations are of limited utility in studying individual genes. More recently, yeast strains exhibiting mutations in single cytoplasmic genes have been isolated by exposing cells to antibiotics that interfere with specific mitochondrial functions (e.g., *erythromycin*, *chloramphenicol*, and *oligomycin*). An ingenious approach has been devised to select for such mutants. Yeast are grown on lactate instead of glucose, thereby bypassing the glycolytic pathway. Under these conditions only yeast with a functional respiratory pathway can grow. Hence in the presence of an antibiotic inhibiting a crucial mitochondrial function, only mutant cells resistant to that particular antibiotic will be able to grow. The numerous antibiotic-resistant mutants isolated in this way have provided both a way to map a large number of mitochondrial genes and an approach for identifying the products specified by these genes (such products would be expected to be altered in the antibiotic-resistant mutants). In addition to antibiotic-resistant mutants, mutations involving deficiencies in particular mitochondrial proteins (e.g., cytochrome oxidase, F_1-ATPase) have also been identified. Genetic analyses of the mutants obtained via these various approaches have permitted a rather extensive genetic map of the yeast mitochondrial genome to be constructed.

MOLECULAR BASIS OF CYTOPLASMIC INHERITANCE

The genetic evidence supporting the view that mitochondria and chloroplasts contain genes distinct from those located in the nucleus was overshadowed for many years by the classical notion that DNA resides only in the nucleus. At best biologists underrated the significance of cytoplasmic inheritance, and at worst, they disregarded it entirely. But in the early 1960s the situation was abruptly changed by the almost simultaneous discovery of DNA in mitochondria and in chloroplasts. The study of cytoplasmic genes suddenly became respectable because of the demonstrated existence of DNA molecules with which these genes could be associated. As will be outlined in the following sections, the discovery of mitochondrial and chloroplast DNA opened this field up to the modern tools of molecular biology, leading to the discovery that these organelles contain their own protein-synthesizing systems based on a unique set of messenger RNAs, transfer RNAs, and ribosomes.

The Discovery of DNA in Mitochondria and Chloroplasts

The idea that DNA might exist outside the nucleus dates back to the early 1920s when Ernst Bresslau and Luigi Scremin used the newly developed Feulgen stain to localize DNA in a parasitic protozoan known as a *trypanosome*. This organism, which is responsible for African sleeping sickness and several other diseases, has a large cytoplasmic organelle called the **kinetoplast,** which stains intensely with Feulgen dye. Because the connection between DNA and genes was not recognized at the time, the significance of DNA in the

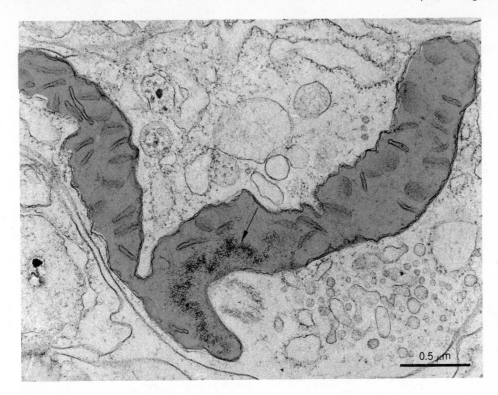

FIGURE 14-5 Electron micrograph of kinetoplast DNA of a trypanosome. The V-shaped kinetoplast (color) can be seen to contain a granular material that is Feulgen-positive (arrow). Courtesy of P. R. Burton.

kinetoplast remained unappreciated. When electron microscopic thin-sectioning techniques were developed in the 1950s, it was soon discovered that the kinetoplast has the structural characteristics of a typical mitochondrion, including separate outer and inner membranes, cristae invaginated from the inner membrane, and a finely granular matrix (Figure 14-5). The Feulgen-positive reaction observed 30 years earlier by Bresslau and Scremin was found to be associated with a mass of fine fibrils contained within the matrix of this organelle.

The presence of DNA in trypanosome kinetoplasts, thought by some to be a biological oddity, was soon reaffirmed by the discovery of DNA-containing fibrils in the chloroplasts and mitochondria of a wide variety of organisms. The first such report was that of Hans Ris and Walter Plaut, who showed in 1962 that *Chlamydomonas* chloroplasts react with the Feulgen reagent as well as with another nucleic-acid stain, acridine orange. It was further shown that the staining reaction can be abolished by prior treatment of the cells with deoxyribonuclease, verifying that DNA is responsible for the observed staining. Electron microscopic examination of these chloroplasts revealed 3-nm diameter fibrils scattered throughout the stroma (Figure 14-6). These fibrils are destroyed by deoxyribonuclease and resemble the DNA-containing fibrils observed in the nucleoids of prokaryotic cells.

Shortly thereafter Margit and Sylvan Nass reported that deoxyribonuclease-sensitive fibrils are present in the mitochondria of a number of different organisms.

With osmium fixation these DNA-containing fibrils appear as large aggregates 20–40 nm in diameter, but following a rinse in uranyl acetate after osmium treatment, the clumps disperse to reveal individual 3-nm fibrils (Figure 14-7). Interestingly, uranyl-acetate rinsing procedures have also been found to be effective in preventing the clumping of the DNA fibrils of prokaryotic cells.

In spite of extensive microscopic evidence supporting the view that mitochondria and chloroplasts contain

FIGURE 14-6 Electron micrograph showing Feulgen-positive fibrils (arrows) in the chloroplast stroma of *Chlamydomonas*.

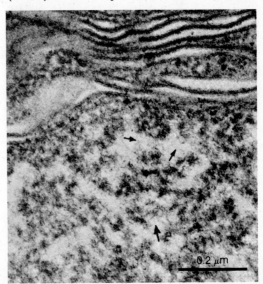

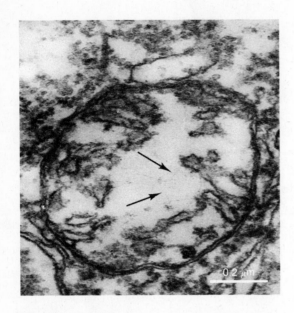

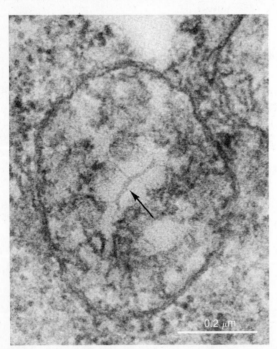

FIGURE 14-7 Electron micrographs showing DNA fibrils in the mitochondria of the sea urchin. *(Top)* DNA fibrils in a dispersed state, revealing a fine fibrillar network in the matrix (arrows). *(Bottom)* DNA fibrils in a clumped array (arrows). Courtesy of M. M. K. Nass.

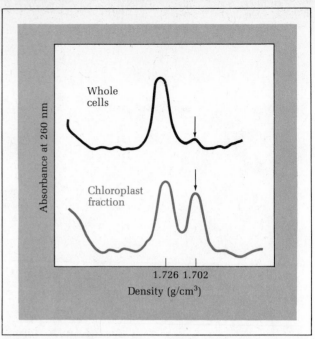

FIGURE 14-8 Identification of chloroplast DNA via cesium chloride isodensity centrifugation. DNA was obtained from whole cells and from the purified chloroplast fraction of *Chlamydomonas*. The fact that the minor DNA peak (arrow) is selectively enriched in the chloroplast fraction identifies it as chloroplast DNA.

FIGURE 14-9 Identification of mitochondrial DNA via cesium chloride isodensity centrifugation. DNA was obtained from nuclear and mitochondrial fractions of *Neurospora*. The band marked "S" is a viral DNA marker included so that the two gradients can be aligned precisely with one another.

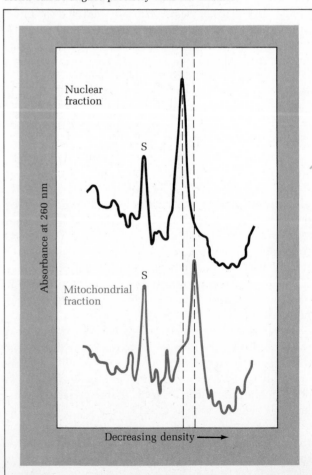

their own DNA, this idea was slow to be accepted. Part of the problem can be traced to conflicting data as to whether or not mitochondria and chloroplasts stain with the Feulgen reaction. Though several laboratories reported positive results, the situation was confused by the presence of many negative reports published in the older literature. It is now known that the lack of staining reported in these early studies was due to an inability of the reaction to detect minute amounts of DNA. Though a logical approach to resolving this dilemma might conceivably involve isolation of mitochondria and chloroplasts followed by analysis to detect the presence of DNA, a serious problem with this tact is the difficulty in ruling out contamination from nuclear or bacterial sources.

Fortunately, a solution to this problem was provided by the newly developed technique of cesium chloride isodensity centrifugation, which permitted mitochondrial and chloroplast DNAs to be identified on the basis of their unique buoyant densities. In one of the first studies of this type, R. Sager and M. R. Ishida used cesium chloride centrifugation to compare the DNA obtained from whole cell extracts and purified subcellular fractions of *Chlamydomonas*. The total cellular DNA was found to separate into two peaks, a major one (density = 1.726 g/cm³) and a minor one (density = 1.702 g/cm³). When the DNA obtained from the purified chloroplast fraction was analyzed, it was seen to be enriched in the minor DNA (Figure 14-8). This observation suggested that the minor, lower density DNA, is specifically localized within the chloroplast. Shortly thereafter David Luck and Edward Reich used the same approach to identify *Neurospora* mitochondrial DNA, which also is less dense than the corresponding nuclear DNA (Figure 14-9).

Cesium chloride isodensity centrifugation has turned out to be such a powerful tool that in some cases nuclear, mitochondrial, and chloroplast DNAs can all be resolved from one another in the same gradient. Because isodensity centrifugation provides a way to separate mitochondrial and chloroplast DNAs from nuclear DNA, it has opened the way to the chemical, physical, and genetic characterization of these organellar DNAs. The remainder of this section will focus on the characterization of these mitochondrial and chloroplast DNA molecules and their interaction with genes located within the nucleus.

Properties of DNA from Mitochondria and Chloroplasts

Mitochondrial DNA The relative amount of mitochondrial DNA present in eukaryotic cells varies considerably. In most animal cells the mitochondria account for 0.1–1.0 percent of the total DNA content, but in some eukaryotic microorganisms, such as yeast, this value may approach 15 percent. In amphibian oocytes, which have a massive cytoplasmic volume and a large number of mitochondria, 90 percent or more of the cell's DNA may be mitochondrial in origin.

Mitochondrial DNA exhibits several features that distinguish it from nuclear DNA. One difference is that mitochondrial DNA lacks histones and hence is not packaged into nucleosomes. Mitochondrial DNA is also characterized by a unique base composition that generally permits separation of nuclear and mitochondrial DNAs by isodensity centrifugation (Figure 14-10). A comparison of the reassociation kinetics of mitochondrial and nuclear DNAs reveals that mitochondrial DNA reassociates much more rapidly than does nuclear DNA, indicating that it contains a smaller number of different sequences (Figure 14-11). The conclusion that mitochondrial DNA has a smaller number of different genes than does nuclear DNA is not surprising given the relatively small size of the mitochondrial DNA molecule.

Electron microscopic examination of purified DNA preparations has revealed another difference between mitochondrial and nuclear DNAs: Mitochondrial DNA is generally circular (Figure 14-12). In animal cells this circular DNA molecule generally contains a total length of about 5 μm of DNA (~16,000 base pairs). The mitochondrial DNA of protozoans ranges from 10 to 15 μm in length, while in fungi the total length ranges from 8 to 30 μm. In higher plants the length diversity of mitochondrial DNA is even more extensive, with estimated genome sizes extending from 30 to 150 μm (Table 14-1). Due to the tendency of these larger circular DNAs to break during isolation, there are many reports in the early literature of DNA from plants and protozoa that is linear, rather than circular. Today we know that most of these reports can be attributed to breakage of circular DNA molecules. However, in a few cases, such as the protozoans *Paramecium* and *Tetrahymena*, and the common bean, *Phaseolus*, genuine linear molecules have been identified, apparently free of isolation artifacts.

Mitochondrial DNA preparations typically consist of a mixture of **open** circles and twisted or **supercoiled** circles (see Figure 14-12). The supercoiled structure is a result of twists in the DNA helix that cannot be unwound once the DNA molecule assumes the form of a closed circle. The untwisted or open form occurs when one or more single-stranded breaks are introduced in the DNA. The actual extent of supercoiling of mitochondrial DNAs *in vivo* is not known, since nucleases released during isolation of mitochondrial DNA may nick the DNA, converting it from the supercoiled to the open configuration.

Mitochondrial DNA preparations generally contain a small percentage of oligomers. Some of these appear as **circular dimers**, which are circles of DNA twice the

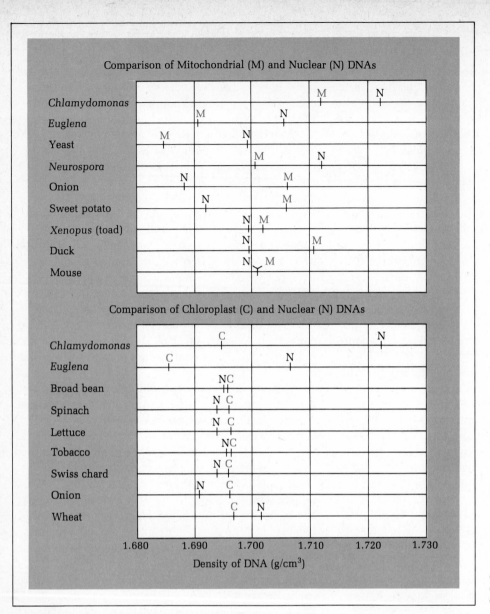

FIGURE 14-10 Buoyant densities of nuclear, mitochondrial, and chloroplast DNAs in selected species.

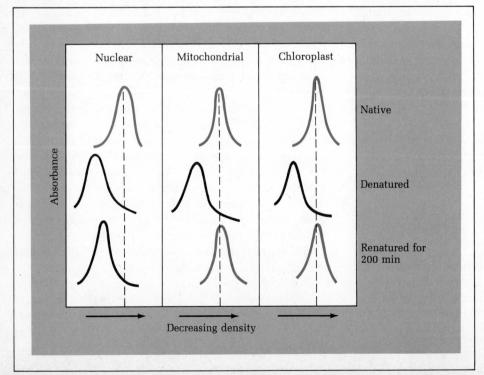

FIGURE 14-11 Reassociation of nuclear, mitochondrial, and chloroplast DNAs. Because single-stranded DNA has a greater buoyant density than double-stranded DNA, the two can be distinguished from each other via isodensity centrifugation. The top panels show the density of the native double-stranded DNAs, the middle panels show the increase in density that occurs after heat denaturation, and the botton panels show the degree to which the density characteristic of the double-stranded state is regained after 200 minutes under renaturing conditions. Note that mitochondrial and chloroplast DNAs renature much more readily than does nuclear DNA.

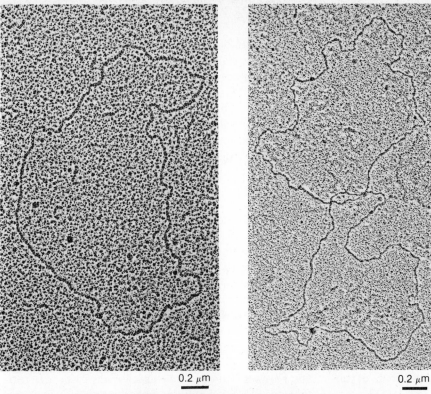

0.2 μm   0.2 μm

0.2 μm

FIGURE 14-12 Electron micrographs of mouse fibroblast mitochondrial DNA in three different configurations. *(Left)* Circular DNA in the open configuration. *(Right)* Catenated DNA with two interlocked monomers. *(Bottom)* Supercoiled DNA. Courtesy of M. M. K. Nass.

TABLE 14-1
Properties of mitochondrial and chloroplast DNAs

Species	Length (μm)	Shape
MITOCHONDRIA		
Animals		
Chicken	5.4	Circular
Fruitfly *(Drosophila)*	6.0	Circular
Human	5.0	Circular
Salamander	4.8	Circular
Turtle	5.3	Circular
Higher Plants		
Corn *(Zea)*	150.0	Circular
Pea *(Pisum)*	30.0	Circular
Wheat *(Triticum)*	70.0	Circular
Fungi		
Mold *(Neurospora)*	20.0	Circular
Yeast *(Saccharomyces)*	25.0	Circular
Protozoans		
Acanthamoeba	12.8	Circular
Paramecium	14.0	Linear
Tetrahymena	14.0	Linear
CHLOROPLASTS		
Unicellular Plants		
Chlamydomonas	62.0	Circular
Euglena	44.0	Circular
Higher Plants		
Corn	43.0	Circular
Lettuce	44.0	Circular
Oats	43.0	Circular
Pea	45.0	Circular

normal circumference, while other **catenated dimers** consist of two normal-size circles interlocked like the links of a chain [see Figure 14-12 (right)]. How such oligomers arise has not been clearly established, though a reasonable assumption is that they originate from breakage and exchange of DNA during recombination. Although circular and catenated oligomers generally account for no more than a few percent of the total mitochondrial DNA, values in the range of 10–30 percent have been reported in some human tumor cells.

Electron microscopy has revealed the presence of membrane fragments bound to mitochondrial DNA. It is not certain whether this association reflects a preparatory artifact or a real DNA-membrane linkage. However, the fact that the addition of purified DNA to isolated mitochondria generates little artificial binding of membranes to DNA argues that the observed connection between mitochondrial DNA and mitochondrial membranes is not an artifact.

The circularity of mitochondrial DNA is responsible for some unique physical properties that have aided in its isolation. For example, Jerome Vinograd and his co-workers have shown that circular DNA molecules bind *less* of the dye ethidium bromide than do linear DNA molecules. The reason for this difference is that intercalation of ethidium bromide into the helix causes the DNA to partially unwind, and the degree to which this takes place is greatly restricted in a closed-circular molecule. Because the binding of ethidium bromide normally causes a *decrease* in the density of DNA, the density of the closed-circular DNA-dye complex, containing less dye than normal, is considerably *greater* than that of the linear or nicked circular DNA-dye complex. Hence in the presence of ethidium bromide, differences in density between nuclear DNA and the various forms of mitochondrial DNA are enhanced, thereby facilitating their separation.

The kinetics of reassociation of denatured mitochondrial DNA reveal that there is no extensive repetition of sequences within a given mitochondrial DNA molecule, and that all the mitochondrial DNA molecules within a cell are the same. An understanding of the average size of these DNAs and the total mass of mitochondrial DNA per cell has led to the calculation that a typical mitochondrion contains anywhere from five to ten molecules of DNA. This conclusion should be viewed cautiously, however, because of the previously described data indicating that mitochondria may form dynamically interconnected networks (Chapter 7). Such an arrangement could conceivably lead to mitochondrial fragmentation and a spurious estimate of the number of DNA molecules per organelle.

Although the existence of mitochondrial DNA is now an accepted fact, this in itself does not constitute proof that the genes governing cytoplasmically inherited mitochondrial traits are contained within this DNA. The first direct evidence for an association be-

tween a cytoplasmically inherited trait and mitochondrial DNA was provided in 1966 by Piotr Slonimski and co-workers, who demonstrated that mitochondrial DNA obtained from yeast bearing the *petite* mutation has a different buoyant density than the mitochondrial DNA of normal yeast (Figure 14-13). The molecular basis for this density shift was studied in several other laboratories, employing ethidium-bromide-induced *petites* as a model system. It was found that ethidium bromide both inhibits incorporation of radioactive precursors into mitochondrial DNA and induces the breakdown of preexisting mitochondrial DNA (Figure 14-14). Long exposure to ethidium bromide results in the complete destruction of mitochondrial DNA, but if treatment is stopped before all the DNA has been degraded, the remaining DNA sequences resume replication until the total content of mitochondrial DNA again reaches normal levels. This new mitochondrial DNA is grossly abnormal, however, containing very few of the sequences present in normal mitochondrial DNA. The presence of this defective mitochondrial DNA in *petite* yeast argues strongly that this DNA governs the inheritance of the abnormalities in mitochondrial structure and respiratory activity characteristic of this condition.

Chloroplast DNA Early studies of chloroplast DNA utilized isodensity centrifugation as the major tool for its isolation. Though this approach works well with unicellular plants (see Figures 14-8 and 14-10), in higher plants most of the DNA species initially claimed to be chloroplast DNA (and on which the belief in the existence of chloroplast DNA was largely based) turned out not to be chloroplast DNA at all. These early experiments, in which DNA prepared from isolated chloroplasts was analyzed by cesium chloride isodensity centrifugation, revealed the presence of three DNA components: a major band thought to represent contaminating nuclear DNA, and two minor DNA components denser than nuclear DNA (Figure 14-15). For several years similar reports confirming the existence of this denser chloroplast DNA appeared. Then around 1968 these reports abruptly ceased. Higher plant chloroplasts were found to contain only one type of DNA, with a buoyant density differing slightly from the major band earlier attributed to contaminating nuclear DNA. Careful measurements revealed that the density of this chloroplast DNA is generally close to 1.696 g/cm^3, a value that is slightly greater than the density of the nuclear DNA of most higher plants (see Figure 14-10). In green algae this distinction is even more apparent, since the nuclear DNA has a density considerably greater than that of chloroplast DNA. Thus in the original reports the major band actually represents the real chloroplast DNA, while the denser components were probably due to contamination by mitochondrial and/or bacterial DNA species. This example illustrates that

FIGURE 14-13 Cesium chloride isodensity centrifugation of mitochondrial DNAs from normal and *petite* strains of yeast.

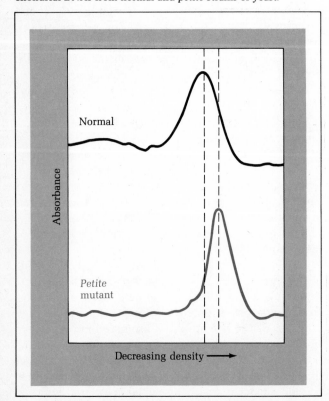

in spite of its great usefulness, cesium chloride isodensity centrifugation is not without its artifacts. (It is worth noting that biologists were reluctant to accept the electron microscopic evidence for the existence of mitochondrial and chloroplast DNA because of the possibility of artifacts, and were only convinced after the isolation of these DNAs by cesium chloride centrifugation; yet a substantial part of the centrifugation data that convinced them was itself an artifact!)

Because the density of chloroplast DNA is often close to that of nuclear DNA, other properties are needed to distinguish these two types of DNA. One useful indicator is the content of **5-methylcytosine,** a methylated derivative of cytosine present in as much as 10 percent of the nuclear DNA of higher plants, but absent from chloroplast DNA. A second property employed to distinguish chloroplast DNA from nuclear DNA is that after denaturation, chloroplast DNA reassociates much more rapidly than does nuclear DNA (see Figure 14-11). As in the case of mitochondrial DNA, such a rapid reassociation rate indicates that chloroplast DNA contains a smaller number of different sequences.

In addition to the absence of 5-methylcytosine and rapid renaturation kinetics, the structural organization of chloroplast DNA is also distinctive. In most early

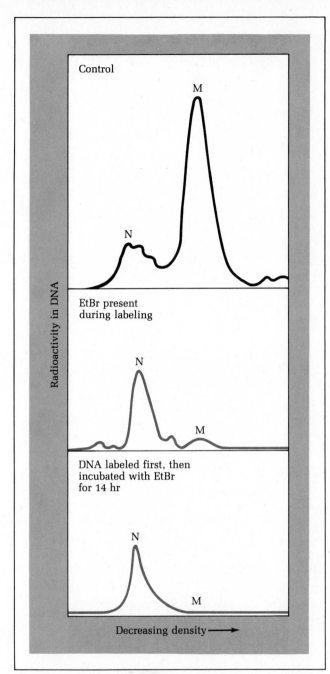

FIGURE 14-14 Effects of ethidium bromide (EtBr) on replication and stability of yeast DNA. All graphs represent cesium chloride isodensity centrifugations of total yeast DNA, which is readily separated by this technique into its nuclear (N) and mitochondrial (M) fractions. The top curve is the labeling pattern obtained when control cultures are exposed to a radioactive DNA precursor. The middle curve shows that inclusion of ethidium bromide during the labeling period blocks the synthesis of mitochondrial but not nuclear DNA. The bottom curve reveals what happens when yeast are first labeled to obtain the pattern shown in the top curve, the isotope removed, and ethidium bromide then added. A 14-hr exposure to ethidium bromide under these radioisotope-free conditions causes the prelabeled mitochondrial DNA to undergo almost complete destruction, but the nuclear DNA is unaffected.

FIGURE 14-15 Cesium chloride isodensity centrifugation of spinach DNA fractions. In 1963 the chloroplast fraction was reported to be enriched in two DNAs (β and γ) with a density greater than nuclear DNA. The main band (α) occurring in the chloroplast fraction was interpreted as contamination by nuclear DNA. By 1968 reports of the β- and γ-components disappeared, and the major band (α) was found to be the true chloroplast DNA, differing slightly in density from the nuclear DNA. The β- and γ-bands are now thought to represent mitochondrial and bacterial contamination. See text for additional details.

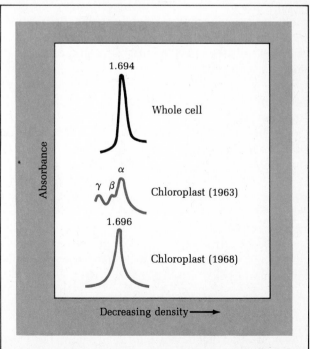

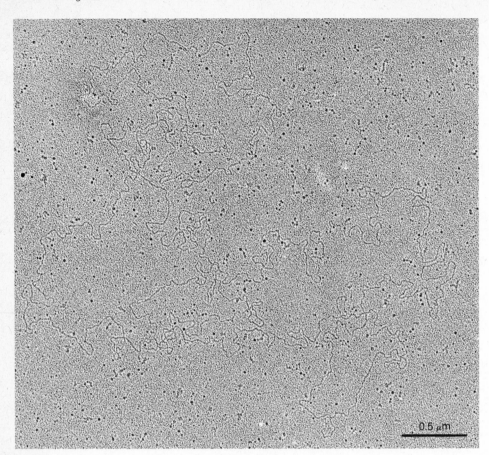

0.5 μm

FIGURE 14-16 Electron micrograph of a circular DNA molecule isolated from the chloroplast of *Euglena* and prepared by rotary shadowing. Courtesy of J. E. Manning, D. R. Wolstenholme, R. S. Ryan, J. A. Hunter, and O. C. Richards.

electron micrographs, chloroplast DNA appeared as a mass of linear fragments, but in 1971 it was shown that the chloroplast DNA of *Euglena* can be isolated as a single circular molecule with a contour length of about 20 μm (Figure 14-16). The notion that chloroplast DNA is generally circular is consistent with the previously mentioned genetic evidence for a circular chromosome in *Chlamydomonas*, and suggests that the linear fragments usually seen in chloroplast DNA preparations reflect breakage of the large circles during isolation. The number of copies of the circular DNA molecule present per chloroplast varies from a few dozen to near a hundred.

As in the case of mitochondrial DNA, the presence of DNA in chloroplasts is not indisputable evidence that this DNA contains the genes governing chloroplast traits. But direct evidence for an association between cytoplasmic inheritance and chloroplast DNA does exist. For example, in *Euglena* cells that have been exposed to ultraviolet light to induce cytoplasmically inherited defects in chloroplast development, nuclear and mitochondrial DNA are found to be unaffected while chloroplast DNA disappears.

Mechanism of Organellar DNA Replication

The realization that mitochondria and chloroplasts contain their own DNA raises the question of how the DNA in these organelles is replicated. Autoradiography of cells exposed to ^{3}H-thymidine has provided morphological evidence that DNA is synthesized directly within these organelles (Figure 14-17). The conclusion that DNA replication occurs within mitochondria and chloroplasts is further supported by: (a) the demonstration that isolated mitochondria and chloroplasts incorporate radioactive deoxynucleoside triphosphates into DNA, and (b) the identification of DNA polymerase and DNA ligase activities in extracts prepared from these organelles.

Experiments employing radioisotope incorporation are of limited use in studying DNA replication, however, because DNA repair reactions can result in the incorporation of radioactive label into molecules that actually are not replicating. A more reliable approach is to use a density label, which must be extensively incorporated throughout the DNA helix before its effect on DNA density is large enough to be detected. In 1966 Edward Reich and David Luck used this density-labeling approach to determine whether the replication of mitochondrial DNA, like that of nuclear DNA, follows a semiconservative pattern. Modeling their experiments after the original studies of Meselson and Stahl (see Figure 10-7), *Neurospora* were initially maintained on media containing ^{15}N and were then switched to ^{14}N. If mitochondrial DNA replication is semiconservative, then after one round of replication the DNA molecules

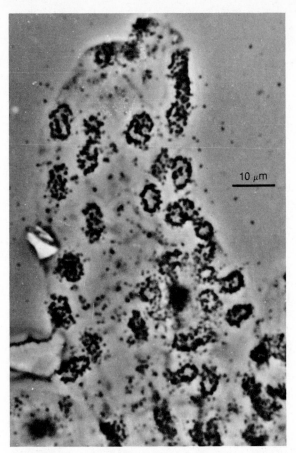

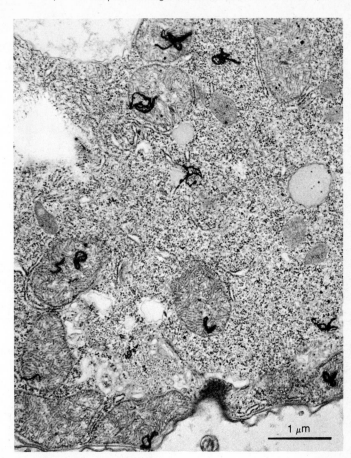

FIGURE 14-17 Autoradiographs showing that DNA synthesis takes place in both chloroplasts and mitochondria. *(Left)* Light-microscope autoradiograph of a spinach leaf cell following incubation with radioactive thymidine. The silver grains are localized primarily over the chloroplasts. Courtesy of J. V. Possingham. *(Right)* Incubation of the ciliated protozoan *Tetrahymena* with radioactive thymidine. This electron micrograph reveals that the majority of the silver grains occur over mitochondria. Courtesy of R. Charret and J. André.

should consist of one ^{15}N strand and one ^{14}N strand; however, Reich and Luck found that most of the DNA was of a greater density than expected for such a ^{14}N/^{15}N hybrid. The most likely explanation for this unexpected finding is that *Neurospora* contain pools of deoxynucleotides that are prelabeled with ^{15}N and require several generations to wash out in the presence of the ^{14}N-containing medium. The continued incorporation of the stored ^{15}N-labeled precursors therefore accounts for the delay in the appearance of the expected ^{15}N^{14}N hybrids. After the second round of replication, however, DNA with an intermediate density close to that of a ^{15}N^{14}N hybrid did appear, and after the third round of replication, "light" DNA, composed mainly of ^{14}N, was found. Therefore, in spite of the complication of the ^{15}N-labeled pools, these results are generally compatible with the pattern of semiconservative replication of mitochondrial DNA.

Shortly thereafter Kwen-Sheng Chiang and Noboru Sueoka employed the same approach to study chloroplast DNA replication, obtaining even clearer results. These investigators observed that after transfer of *Chla-*

mydomonas cultures from ^{15}N- to ^{14}N-containing media, DNA of intermediate (^{15}N^{14}N) density appears after one round of replication, and equal amounts of intermediate and light density DNA appear after two rounds (Figure 14-18). This is the same pattern as Meselson and Stahl observed in their original experiments on bacterial DNA replication, and is thus strong evidence that chloroplast DNA is replicated semiconservatively.

Another tool that has been useful in probing the mechanism of DNA replication is **deoxybromouridine**, a thymidine analog that cells will incorporate into DNA in place of thymidine. Because deoxybromouridine is denser than thymidine, it causes a shift in DNA density. When the incorporation of radioactive deoxybromouridine into mitochondrial DNA is studied, the newly synthesized DNA can be recognized both by its radioactivity and by its increased density. Denaturation of the newly synthesized DNA has revealed that only one strand contains the density label, an observation clearly in support of the semiconservative replication scheme.

Visualization of replicating DNA molecules with the electron microscope has also been of value in

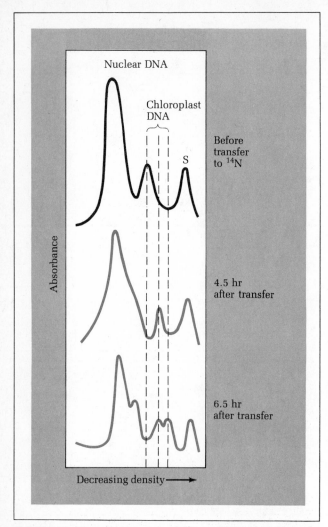

Nuclear DNA

Chloroplast DNA

S

Before transfer to ¹⁴N

4.5 hr after transfer

6.5 hr after transfer

Absorbance

Decreasing density ——→

FIGURE 14-18 Semiconservative DNA replication in *Chlamydomonas* chloroplasts demonstrated via a ¹⁵N to ¹⁴N density-shift experiment. Graphs represent cesium chloride isodensity gradient centrifugation of total cell DNA before and after transfer to ¹⁴N-containing media. The stepwise shift that occurs in the density of the chloroplast DNA is consistent with semiconservative replication. See text for additional details.

formulating models of DNA synthesis. In 1971 it was independently discovered in the laboratories of Jerome Vinograd and Piet Borst that mitochondrial DNA often exhibits a small loop displaced from the main circle [Figure 14-19 *(left)*]. Denaturation of the DNA releases a small fragment of single-stranded DNA about 450 bases long, suggesting that the loop is caused by the displacement of one of the two strands of the double helix by this small fragment. The loop is therefore referred to as a **displacement** or **D-loop.**

In addition to the relatively short D-loop, longer loops have also been observed [see Figure 14-19 *(right)*]. In some cases these loops are single stranded, while in others they are partially or completely double stranded. A model of mitochondrial DNA replication consistent with the existence of these various forms is presented in Figure 14-20. There are five essential features of this model: (1) Replication is initiated at a single origin by synthesis of one new complementary DNA strand approximately 450 bases in length, forming a D-loop. (2) Synthesis of this new strand continues in a 5′ → 3′ direction. (3) After synthesis of the first strand has proceeded for a certain amount of time, synthesis of DNA complementary to the opposing strand is initiated, also in the 5′ → 3′ direction. Up to 80 percent of the first strand may be replicated before synthesis of the second strand begins. (4) The two newly forming DNA circles may separate before replication is complete, resulting in circles containing single-stranded gaps. (5) The overall pattern of DNA replication proceeds in one direction away from a single fixed origin.

Chloroplast DNA appears to replicate by a slightly different mechanism than that utilized by mitochondrial DNA (Figure 14-21). First, electron micrographs reveal two D-loops in each molecule, with replication proceeding from the two loops toward each other until they merge to form a single loop. Replication then continues in both directions, making the net process bidir-

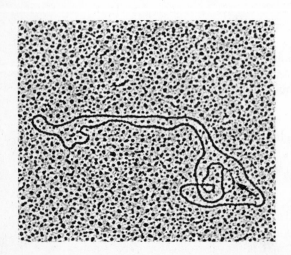

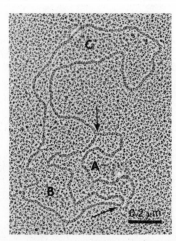

FIGURE 14-19 Electron microscopic analysis of replicating mitochondrial DNA. *(Left)* Closed-circular DNA showing a displacement loop (arrow). Courtesy of D. L. Robberson. *(Right)* Circular DNA molecule showing two, equal-length daughter DNA segments, labeled A and B (arrows indicate the replication forks). The unreplicated DNA segment (labeled C) is longer than the other two. Courtesy of K. Koike and D. R. Wolstenholme.

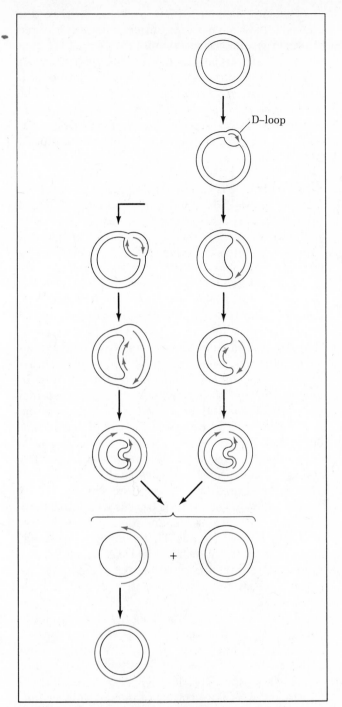

FIGURE 14-20 A general model for mitochondrial DNA replication based on the various forms of mitochondrial DNA observed in electron micrographs. See text for details.

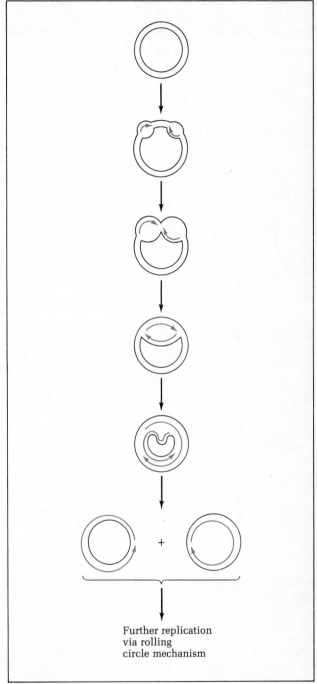

FIGURE 14-21 A general model of chloroplast DNA replication based on the various forms of chloroplast DNA observed in electron micrographs.

ectional rather than unidirectional as in mitochondria.

The mechanism responsible for initiating the synthesis of mitochondrial and chloroplast DNAs has not been identified, but the fact that such replication is independent of nuclear DNA synthesis indicates that separate signaling systems are employed.

Mechanism of Maternal Inheritance

The development of biochemical techniques for studying the DNA of mitochondria and chloroplasts has provided new insights into the question of why cytoplasmic inheritance is usually maternal. The classical explanation is that the male gamete (sperm, pollen, etc.) contains little cytoplasm, and so the cytoplasm present in the zygote is mainly that of the egg. Whether or not this explanation is correct for organisms in which the male gamete contains little cytoplasm, it is clearly not valid for the green alga *Chlamydomonas*, where the male and female gametes are of equal size and contribute equal amounts of cytoplasm when they fuse. Yet in spite of the mixing of equal amounts of cytoplasm, inheritance of cytoplasmically governed chloroplast traits still follows the maternal pattern.

An explanation for this paradox has been proposed by Ruth Sager and her associates, who have followed the fate of the chloroplast DNA derived from each parent in *Chlamydomonas*. By labeling the chloroplast DNA of one parent with ^{14}N and that of the other parent with ^{15}N, they were able to show that only one type of chloroplast DNA survives after fusion of the male and female gametes, and that its density always corresponds to that of the female parent. Apparently the paternal chloroplast DNA is selectively destroyed after fusion of the two gametes. The mechanism by which the cell's degradative machinery marks the paternal DNA for destruction is apparently related to lack of methylation of the DNA. The DNA of maternal origin is methylated first in gametes and again in the zygote. However, the paternal DNA is not methylated prior to fusion of the male and female gametes. This DNA is degraded just after zygote formation but before the two parental chloroplasts have fused. Thus maternal inheritance in *Chlamydomonas* does not result from the absence of paternally derived cytoplasm, but rather is explained by the lack of methylation of nucleotides in the mitochondrial DNA molecule from the male parent. Because the DNA is not methylated, it is subject to destruction by endonucleases.

Transcription in Mitochondria and Chloroplasts

Expression of genetic information encoded in DNA requires that it first be transcribed into RNA. Several lines of evidence support the conclusion that mitochondrial and chloroplast DNA act as templates for RNA synthesis. First, isolated mitochondria and chloroplasts incorporate radioactive precursors into RNA in a reaction whose dependence on a DNA template is revealed by its sensitivity to inhibition by actinomycin D and deoxyribonuclease. Second, mitochondria have been shown to contain an RNA polymerase that differs from nuclear RNA polymerases in being sensitive to inhibition by rifampicin, an antibiotic that is generally effective against bacterial but not eukaryotic RNA polymerases. Third, ethidium bromide inhibits the synthesis of mitochondrial but not nuclear RNA, further supporting the conclusion that independent RNA synthetic pathways exist in nuclei and mitochondria.

Though the above evidence strongly suggests that RNA synthesis takes place inside mitochondria and chloroplasts, unequivocal proof that it is the DNA contained within these organelles that codes for this RNA requires the demonstration that the RNA formed will hybridize to the organellar DNA. As we shall see in the following section, the use of DNA/RNA hybridization techniques, coupled with sequencing of DNA as well as its RNA products, have provided a great deal of support for this idea.

Mitochondrial RNA   In the last several years a variety of mitochondrial DNAs have been completely or partially sequenced. This approach is feasible because of the relatively small size of the mitochondrial genome, and has led to important advances in our understanding of how the genes present within mitochondrial DNA are arranged. The earliest study of mitochondrial DNA was conceived by Guiseppe Attardi and his collaborators, and is an exquisite example of the power of nucleic acid hybridization techniques in characterizing the types of RNA transcribed from organellar DNA. These studies on HeLa (human) DNA were facilitated by the fact that the two strands of HeLa mitochondrial DNA differ in buoyant density, and therefore can be separated into **heavy** and **light** strands by isodensity centrifugation under denaturing conditions. When Attardi labeled mitochondrial RNA for a short time and then tested its ability to hybridize to these heavy and light DNA strands, he unexpectedly discovered that the RNA hybridized to virtually all the sequences in both DNA strands. When longer labeling periods were used, however, the radioactive RNA bound primarily to the heavier of the two DNA strands. These findings suggest that both strands of the mitochondrial DNA are completely transcribed, but that most of the RNA copied from the lighter DNA strand is quickly destroyed in the mitochondrion.

Localizing the genes coding for specific mitochondrial RNAs has been made possible by several adaptations of the basic DNA/RNA hybridization methodology. In one, the two strands of mitochondrial DNA are separately hybridized with a mitochondrial RNA tagged with the electron-dense molecule *ferritin*. The position of the ferritin on the DNA molecule can then be visualized with the electron microscope, permitting the localization of the gene coding for the RNA to which the ferritin is attached (Figure 14-22). Another approach involves hybridization of mitochondrial RNAs to DNA, followed by digestion of all single-stranded (nonhybridized) molecules by the enzyme **S1 nuclease.** The double-stranded hybrid is protected from enzyme attack, and can therefore be recovered and analyzed.

In 1981 the complete sequence of human mitochondrial DNA was determined. Subsequently, complete or

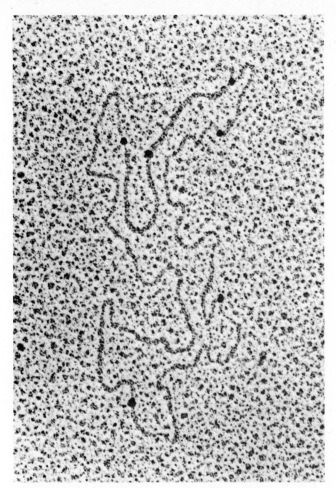

FIGURE 14-22 Electron micrograph of mitochondrial DNA hybridized to ferritin-labeled mitochondrial transfer RNAs, which can be seen as large, black granules. Courtesy of N. Davidson.

Mitochondrial DNAs vary in size among organisms (see Table 14-1), leading to some differences in organization of the mitochondrial genome. In the relatively small mammalian mitochondrial DNAs, there are few noncoding sequences present. However, in yeast the coding sequences are spread out over a circular DNA molecule that is nearly five times larger than mammalian mitochondrial DNA. In this case long stretches of AT-rich sequences separate the known coding sequences from each other. Another striking aspect of yeast mitochondrial DNA is that several genes are split by introns.

In mammalian mitochondria, transcription of each strand of DNA is initiated at a single site, located in the D-loop region of the heavy strand, and at another site on the light strand. This produces two giant primary transcripts that must be cleaved to generate individual ribosomal, transfer, and messenger RNA molecules. Attardi and his colleagues have shown that the bulk of the RNA

FIGURE 14-23 Map of HeLa mitochondrial DNA showing heavy (H) and light (L) strands. Colored segments represent genes transcribed on the opposite strand. Ribosomal RNA genes are labeled 16S and 12S. Open circles reflect the positions of transfer RNA genes. Protein coding genes are indicated by URF (unassigned reading frames) or by the abbreviation of the encoded polypeptide: cyt *b*, cytochrome *b*; CO I, II, III, cytochrome oxidase subunits; ATPase 6, subunit 6 of ATPase. See text for additional details.

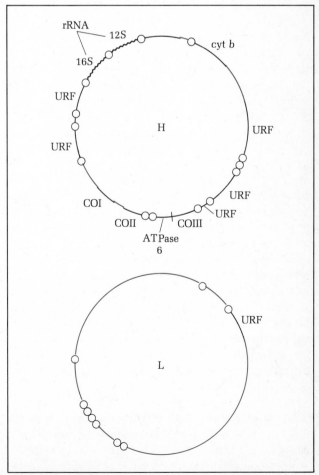

nearly complete sequences of many other types of mitochondrial DNA have been ascertained, including those from other mammals, yeast, and *Xenopus*. The map of HeLa mitochondrial DNA illustrates the general principles of organization shared by all mammalian mitochondrial DNAs examined thus far (Figure 14-23). Two ribosomal RNA genes, coding for RNAs of 16S and 12S, are present, in addition to genes coding for 22 different transfer RNA molecules. These genes are distributed throughout the entire DNA molecule. Thirteen genes encode protein chains, several of which have had their products identified; the remaining are referred to as **unassigned reading frames (URF)** because they contain sequences that code for amino acids of as yet unidentified polypeptides. The genes coding for proteins include those specifying subunits of cytochrome oxidase (aa_3), subunits of ATPase, and cytochrome *b*. Because only some of the subunits of these mitochondrial proteins are encoded by mitochondrial DNA, it follows that the nuclear DNA must also play a role in the synthesis of mitochondrial proteins. This subject will be discussed in more detail later in the chapter.

transcript encoded by the light strand of DNA is destroyed almost immediately upon synthesis. Mapping of each strand has revealed that both 16S and 12S ribosomal RNAs, as well as 14 transfer RNAs and 12 of 13 protein coding regions are localized to the heavy strand of DNA, while the light strand encodes 8 transfer RNA genes and a single unassigned reading frame (see Figure 14-23). Although mapping the mitochondrial genome has revealed important structural information, our understanding of the processing of the resultant transcripts is severely limited. One of the facets of mammalian mitochondrial RNA processing that is not yet understood involves the mechanism by which the enormous RNA transcript from the heavy strand of mammalian DNA is cleaved into its component RNAs. Inspection of the genetic map in Figure 14-23 shows that the genes for ribosomal RNAs, as well as structural genes encoding polypeptide chains, are bounded by transfer RNA genes. It has been proposed that these sequences flanking ribosomal and messenger RNAs fold into a configuration that can be recognized by site-specific restriction endonucleases, allowing cleavage of the RNA transcript to occur. The messenger RNAs formed after this cleavage lack a cap at their 5′ ends, but they do acquire poly(A) tails.

A comparison of the transcriptional machinery in mammalian mitochondria with that of yeast reveals several striking differences. In yeast, all genes except for that of a single transfer RNA are sequestered on one of the two DNA strands, and transcription is initiated at multiple sites throughout the genome. Another unique feature is the presence of introns within certain protein-coding genes. Such split genes are common in the chromosomal DNA of eukaryotic cells (see Table 11-4), but in such cases the introns represent noncoding sequences that must be excised in order to generate a mature messenger RNA. In the case of at least some yeast mitochondrial introns, however, the intron sequences appear to contain information needed for the proper processing of the RNA transcript into functional messenger RNA. In work carried out by Piotr Slonimski and his colleagues, it was demonstrated that mutation of a specific intron within the cytochrome *b* gene leads to incorrect splicing of the message. The factor responsible for this activity has been termed an **RNA maturase,** since its effect is to ensure the removal of appropriate introns so that the exon sequences can be spliced together to form the mature messenger RNA molecule. The general applicability of this pathway to processing of other yeast mitochondrial messenger RNAs is not yet known.

Chloroplast RNA The circular chloroplast genome ranges from 85 to about 190 kilobases, with most higher plants having about 150,000 base pairs (see Table 14-1). Using the same basic tools that have been applied to the

study of mitochondrial RNAs, it has been found that chloroplast DNA also contains genes coding for ribosomal and transfer RNAs as well as for specific polypeptides. However, unlike the situation in mitochondrial DNA, the ribosomal RNAs have an operonlike arrangement reminiscent of prokaryotic cells, with 16S, 23S, and 5S ribosomal RNAs separated by spacer DNA sequences. Depending on the source of the DNA, up to five sets of these operons may be present within a single chloroplast DNA molecule. In addition, hybridization of total chloroplast transfer RNA to chloroplast DNA indicates that there are between 25 and 45 transfer RNA genes; some are clustered together, while others are dispersed throughout the circular DNA molecule. The map of spinach chloroplast DNA illustrates the basic features of chloroplast DNAs from higher plants so far examined (Figure 14-24). Genes coding for specific chloroplast proteins, including the large subunit of RuBP carboxylase, certain subunits of the ATPase complex, and specific components of photosystems I and II have been identified thus far. As was the case for mitochondrial DNA, the analysis of transcriptional initiation and termination in chloroplasts is still in its early stages, although preliminary evidence suggests that multiple promoters similar in sequence to prokaryotic promoters are present in chloroplast DNA.

Protein Synthesis Within Mitochondria and Chloroplasts

Reports that isolated mitochondria incorporate radioactive amino acids into protein date as far back as the 1950s, but several problems have hampered interpretation of such observations. The most intractable issue is that mitochondrial protein synthesis usually represents but a tiny fraction of total cellular protein synthesis, and

FIGURE 14-24 Map of spinach chloroplast DNA showing the location of genes for transfer RNAs (open circles), the two ribosomal DNA operons containing the sequences for 16S, 23S, 5S, and 4.5S ribosomal RNAs (color), as well as five protein coding regions identified to date: ATPase subunits α, β, and ϵ; rbc-L, ribulose bisphosphate carboxylase large subunit; PS II, thylakoid membrane protein in the PS II complex. See text for additional details.

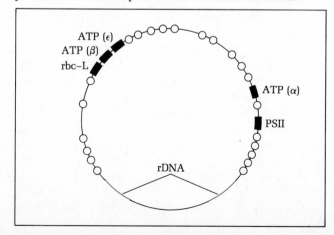

thus even a small contamination of isolated mitochondria by cytoplasmic ribosomes or bacteria might account for the observed activity. Fortunately, the discovery that mitochondria and chloroplasts contain ribosomal and transfer RNAs differing from those in the cytoplasm has helped to establish the legitimacy and uniqueness of the protein synthetic events occurring within these organelles. In addition, the finding that mitochondrial and chloroplast protein synthesis differs from cytoplasmic protein synthesis in its sensitivity to certain inhibitors has permitted the proteins synthesized by these organelles to be identified under conditions where cytoplasmic protein synthesis is selectively blocked. In the following sections we shall see that mitochondrial and chloroplast protein synthesis must be coordinated both temporally and spatially with translation in the cytoplasm so that proper formation of mitochondrial and chloroplast proteins occurs.

Ribosomes Mitochondria and chloroplasts contain ribosomes that are distinctly different from those of the cell sap in size, sedimentation coefficient, RNA and protein composition, and sensitivity to inhibitors. Because of the many similarities between mitochondrial, chloroplast, and bacterial ribosomes, it has been common practice to categorize mitochondrial and chloroplast ribosomes as being of the "70S" prokaryotic type rather than the typical "80S" eukaryotic type; it is now becoming clear, however, that this generalization is an oversimplification.

Generalizations are especially difficult to formulate in the case of mitochondrial ribosomes because they vary significantly among organisms (Table 14-2). Mammalian mitochondria contain ribosomes sedimenting at 55–60S that dissociate into large and small subunits containing RNAs of approximately 16S and 12S. The mitochondrial ribosomes of unicellular eukaryotes and higher plants are somewhat larger, sedimenting in the range of 70–80S; the large subunit contains RNAs that range from 21S to 25S in size, while the smaller subunit contains a 16S RNA. In no case have mitochondrial ribosomes been found to contain 5.8S RNA hydrogen bonded to the larger ribosomal RNA, a configuration characteristic of eukaryotic cell sap ribosomes but absent in prokaryotic ribosomes.

One striking feature that distinguishes mitochondrial ribosomes of animals and fungi from all other types of ribosomes, eukaryotic as well as prokaryotic, is the apparent absence of 5S RNA. Since 5S RNA is an essential functional component of all other ribosomes, it is quite possible that animal and fungal mitochondria contain an analogous but slightly smaller RNA that is lost from the ribosome during isolation and escapes detection because its smaller size causes it to be obscured by the 4S transfer RNAs. The likelihood that animal and fungal mitochondria contain such an undetected RNA

TABLE 14-2
Characteristics of mitochondrial, chloroplast, and cell sap ribosomes

Source	Sedimentation Coefficients (S)		
	Ribosomes	Subunits	RNAs
Mitochondria			
Human (HeLa)	60	45	16
		35	12
Pig, rat	55	39	16
		28	13
Yeast	75	53	21
		35	15
Neurospora	73	50	25
		37	19
Euglena	71	50	21
		32	16
Mung bean	77	60	24, 5
		44	18
Chloroplasts			
Euglena	68	50	23, 5
		29	16
Bean, spinach	70	50	28, 5
		30	16
Cell sap			
Prokaryotic (*E. coli*)	70	50	23
		30	16, 5
Eukaryotic animal	80	60	28
		40	18, 5.8, 5
Eukaryotic plant	80	60	25
		40	18, 5.8, 5

that is at least functionally equivalent to 5S RNA is supported by the fact that 5S RNA has been detected in the mitochondria of higher plants.

The ribosomes of chloroplasts exhibit less variability in size than those of mitochondria, sedimenting near 70S and consisting of subunits of 50S and 30S (see Table 14-2). Like prokaryotic ribosomes, the large subunit of most chloroplast ribosomes contains molecules of 23S and 5S RNA while the small subunit contains a single molecule of 16S RNA. However, in some cases additional small RNA species have been detected in the large ribosomal subunit. For example, in *Chlamydomonas*, RNAs sedimenting at 3S and 7S are found in addition to the normally occurring 5S and 23S RNAs.

The process by which mitochondrial and chloroplast ribosomes are formed involves the initial synthesis of large precursor ribosomal RNAs that are trimmed to their final sizes as they are assembled with ribosomal proteins, much like the process that occurs during the formation of cytoplasmic ribosomes (Chapter 12). The mature mitochondrial and chloroplast ribosomes then become associated with messenger RNAs to form polysomes. In yeast the binding of messenger RNA to the small ribosomal subunit involves an association between the 5' leader of the messenger RNA and the 3' end of the small ribosomal RNA. A similar pairing occurs in prokaryotic cells. In mammalian mitochondria the messenger RNAs do not contain leader sequences, nor has

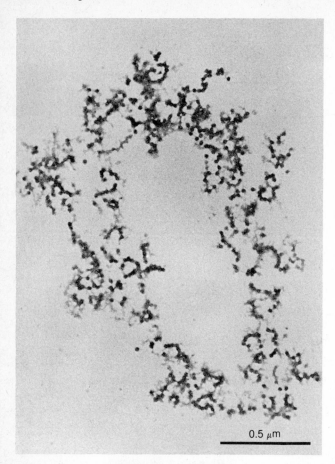

0.5 µm

0.1 µm

FIGURE 14-25 Electron micrographs of mitochondrial and chloroplast ribosomes. *(Top)* Mitochondria in the eggs of the fruit fly contain circular clusters of polyribosomes associated with mitochondrial DNA. Courtesy of C. D. Laird. *(Bottom)* The membranes of chloroplast grana have ribosomes attached to their outer surfaces. Courtesy of H. Falk.

the mechanism of association of these molecules with the ribosome been clarified. As in bacterial cells, both mitochondrial and chloroplast ribosomes become associated with messenger RNA still attached to the circular DNA template [Figure 14-25 *(top)*]. The majority of the active polysomes in mitochondria and chloroplasts sediment with the membrane fraction, suggesting that the polysomes are attached to these membranes just as some cytoplasmic polysomes are attached to the endoplasmic reticulum. This conclusion is supported by thin-section electron micrographs that reveal the presence of ribosomes attached to the internal mitochondrial and chloroplast membranes [see Figure 14-25 *(bottom)*]. As will be seen later, this attachment may be related to the fact that some of the proteins synthesized

by mitochondrial and chloroplast ribosomes are destined for incorporation into these membranes.

Transfer RNAs and Protein Synthesis Factors Mitochondria and chloroplasts contain a set of transfer RNAs and aminoacyl-tRNA synthetases that are at least in part unique to these organelles. Though the aminoacyl-tRNA synthetases are encoded by the nuclear DNA, all of the transfer RNAs are transcribed from organellar DNA and differ in sequence from those found in the cell sap. Taking into account the existence of wobble, a minimum of 32 transfer RNA species is required for translation of cytoplasmic RNAs. Although 32 different transfer RNAs have been identified in chloroplasts, this is not the case in mitochondria. The mitochondrial DNA of mammalian cells has been found to encode only 22 transfer RNAs, and there is no evidence that the remaining 10 transfer RNAs enter the organelle from the cytoplasm. A similar situation exists in yeast and *Neurospora* mitochondria. It appears that in mitochondria the wobble rules have been relaxed so that a single transfer RNA can recognize four, rather than three, codons. Thus in mitochondria, only the first two bases of the codon appear to be required for codon-anticodon recognition. These alterations, coupled with the introduction of several codon substitutions, has led to the realization that the universal genetic code is not scrupulously adhered to in the mitochondrial genetic apparatus. For example, mammalian mitochondria translate the codons AGA and AGG into stop signals, rather than into the amino acid arginine. In addition, the codons AUA and AUU can act as starting points for translation in this system, instead of the AUG codon that is typically used as the start codon (see Table 10-4). There are other substitutions that have been found, such as the translation of AUA as methionine instead of isoleucine in both yeast and mammalian mitochondria, and the reading of the four codons beginning with CU as threonine, rather than leucine, in yeast mitochondria. Why mitochondria have evolved in this way while chloroplasts continue to use the normal genetic code is bewildering, but it may be related to the different evolutionary histories of these two semiautonomous organelles.

One novel property of protein synthesis in mitochondria and chloroplasts is the existence of a special transfer RNA for formylmethionine. At one time this **tRNA^fmet**, as well as the enzyme **transformylase** that formylates its methionine, were thought to be unique to bacterial cells, where they function in the initiation of protein synthesis. It is now clear that these components also occur in mitochondria and chloroplasts, where the machinery of protein synthesis appears to resemble that of prokaryotic cells more closely than it does cytoplasmic protein synthesis in eukaryotic cells.

In addition to the components already described (ribosomes, messenger RNA, transfer RNA, aminoacyl-tRNA synthetases, and transformylase), protein synthesis in mitochondria and chloroplasts requires an appropriate set of initiation, elongation, and termination factors. Most of these appear to be structurally distinct from their counterparts in the cell sap. Again there appears to be a close resemblance to the analogous factors of prokaryotic cells; several mitochondrial factors have been found to be functionally interchangeable with those obtained from bacterial cells, but not interchangeable with those present in the eukaryotic cytoplasm.

Inhibitors of Protein Synthesis In the early 1960s several investigators observed that protein synthesis by isolated chloroplasts is inhibited by **chloramphenicol.** Since chloramphenicol is an inhibitor of bacterial but not eukaryotic cytoplasmic protein synthesis, it was concluded that mitochondrial and chloroplast ribosomes are of the prokaryotic type rather than the eukaryotic type. This notion was soon reinforced by the finding that other antibiotics selective for bacterial protein synthesis inhibit mitochondria and chloroplasts as well, while inhibitors selective for eukaryotic cell sap ribosomes, such as **cycloheximide,** are inactive against mitochondrial and chloroplast ribosomes (Table 14-3).

Thus the pattern of susceptibility to antibiotics, the similarity in protein synthesis factors, and the utilization of formylmethionine all indicate a close resemblance between the protein synthetic systems of mitochondria, chloroplasts, and bacteria. It is somewhat of an oversimplification, however, to refer to mitochondrial and chloroplast ribosomes as the 70S prokaryotic type, both because some mitochondrial ribosomes are significantly smaller or larger than 70S and because there are significant differences between the RNAs and proteins that make up mitochondrial, chloroplast, and bacterial ribosomes.

Polypeptides Manufactured in Mitochondria and Chloroplasts The question of which proteins are synthesized by mitochondria and chloroplasts has been resolved using two basic experimental tools. The most obvious one is to incubate isolated mitochondria or chloroplasts with radioactive amino acids, followed by identification of the labeled products. The only draw-

back to this approach is that due to the vast excess of 80S cytoplasmic ribosomes in most cells, even a small carryover of these ribosomes into isolated mitochondrial or chloroplast preparations might confuse the results. The alternative approach is to exploit the above-mentioned selectivity of antibiotics, searching for proteins whose synthesis is inhibited by chloramphenicol but unaffected by cycloheximide.

Mitochondria. The first experiments designed to explore polypeptide synthesis within mitochondria involved incubating isolated mitochondria with radioactive amino acids and isolating the labeled polypeptides. It was found that the radioactive polypeptides are localized primarily in the mitochondrial inner membrane and can be readily extracted with organic solvents, suggesting that they are hydrophobic membrane-associated proteins. Electrophoresis of this labeled material in SDS-polyacrylamide gels reveals the presence of eight to ten major polypeptides. Five lines of evidence suggest that the information coding for these polypeptides is contained within mitochondrial DNA: (1) Selective inhibition of mitochondrial RNA synthesis by ethidium bromide blocks the synthesis of these polypeptides, even though cytoplasmic protein synthesis remains unaffected. (2) The number and size of the polypeptides roughly corresponds to what would be expected from the translation of mitochondrial poly(A)-containing RNAs known to be transcribed from mitochondrial DNA. (3) Isolated mitochondrial DNA has been transcribed into RNA in a cell-free system, and this RNA then translated into polypeptides similar to those made in intact mitochondria. (4) Cytoplasmically inherited mutations in mitochondrial DNA lead to alterations in the structure of two mitochondrially synthesized polypeptides, namely subunits of cytochrome oxidase and oligomycin-sensitive ATPase. (5) Recent advances in mapping of mitochondrial DNA have revealed not only the number but also the location of the genes coding for mitochondrial polypeptides (see Figure 14-23). Taken together, this evidence strongly argues that proteins synthesized in mitochondria are translated from messenger RNAs transcribed within the mitochondrion.

The first mitochondrially synthesized polypeptides to be identified turned out to be components of **oligomycin-sensitive ATPase,** an enzyme complex consisting of ten polypeptide subunits. Five of these subunits comprise the F_1-ATPase, one is the **oligomycin-sensitivity conferring protein (OCSP),** and the other four are hydrophobic membrane polypeptides. Alexander Tzagoloff and his associates identified the components of this complex that are synthesized in mitochondria by incubating yeast with cycloheximide to block cytoplasmic protein synthesis; under these conditions the four hydrophobic polypeptides continue to be synthesized, indicating that they are manufac-

TABLE 14-3
Differential effects of the major protein synthesis inhibitors

Ribosome Source	Inhibitor		
	Puromycin	Chloramphenicol	Cycloheximide
Prokaryotic	+	+	−
Eukaryotic			
Cytoplasmic	+	−	+
Mitochondrial	+	+	−
Chloroplast	+	+	−

Note: + = inhibited; − = no effect.

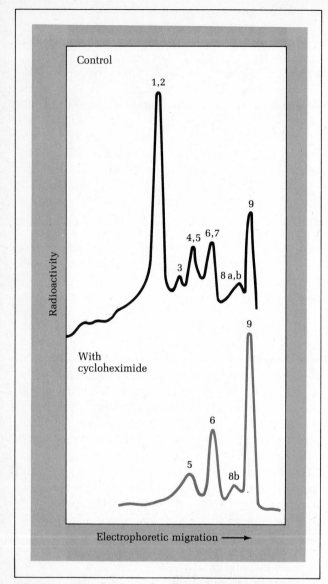

FIGURE 14-26 SDS-polyacrylamide gel electrophoresis of components of the oligomycin-sensitive ATPase complex synthesized by yeast in the presence or absence of cycloheximide. In the presence of cycloheximide only four of the ten polypeptides are synthesized.

tured within the mitochondrion (Figure 14-26). Conversely, in the presence of chloramphenicol only the F_1-ATPase subunits and OSCP are made, confirming that they are synthesized in the cytoplasm. It is interesting to note, however, that the F_1-ATPase subunits formed in the cytoplasm in the presence of chloramphenicol do not become attached to the inner mitochondrial membrane as normally occurs, suggesting that the hydrophobic polypeptides synthesized in the mitochondrion are required to anchor the F_1-ATPase to the inner membrane.

Another mitochondrial enzyme complex composed of a mixture of subunits synthesized in the cytoplasmic and mitochondrial compartments is **cytochrome**

oxidase (cytochrome aa_3). Of the seven polypeptides that make up this enzyme, the three largest continue to be synthesized in the presence of cycloheximide, indicating that they are made within the mitochondrion. In *petite* yeast these three large subunits are missing, providing direct evidence that they are encoded in mitochondrial DNA. Though three of the four cytochrome oxidase subunits made in the cytoplasm are still present in the mitochondria of *petite* yeast, their binding to the inner membrane is abnormally loose and they are easily detached. Thus, analogous to the situation with oligomycin-sensitive ATPase, the portions of cytochrome oxidase made in the mitochondrion appear to be required for the tight binding of the remaining subunits to the inner membrane.

In addition to the ATPase and cytochrome oxidase complexes, antibiotic studies utilizing cycloheximide and chloramphenicol have also revealed that cytochrome b (one of the eight subunits making up the coenzyme Q–cytochrome c reductase complex) is encoded by the mitochondrial genome. Genes coding for nine mitochondrially synthesized polypeptides have thus been identified so far: four of the ten subunits comprising the ATPase complex, three of seven subunits of cytochrome oxidase, cytochrome b, and, in yeast, a single ribosomal protein (Table 14-4). The presence of unassigned reading frames suggests that several more polypeptides may also be translated on mitochondrial ribosomes, though it is clear that the bulk of the mitochondrial proteins must be made in the cytoplasm and subsequently transported into this organelle.

TABLE 14-4

Polypeptides manufactured within mitochondria and chloroplasts

Polypeptide	Location in Organelle
MITOCHONDRION	
Cytochrome oxidase (aa_3)	
Subunits I, II, III	Inner membrane
ATPase complex	
F_0: subunits 6, 8, 9	Inner membrane
F_1: subunit α	Inner membrane
Cytochrome b	Inner membrane
Ribosomal protein (yeast)	Ribosome
CHLOROPLAST	
RuBP carboxylase	
Large subunit	Stroma
ATPase complex	
CF_0: 2 subunits	Thylakoid membrane
CF_1: 3 subunits	Thylakoid membrane
Cytochrome b/f complex	
3 subunits	Thylakoid membrane
32,000 dalton protein	Thylakoid membrane
Photosystems I and II proteins	Thylakoid membrane
Ribosomal proteins (2?)	Ribosome

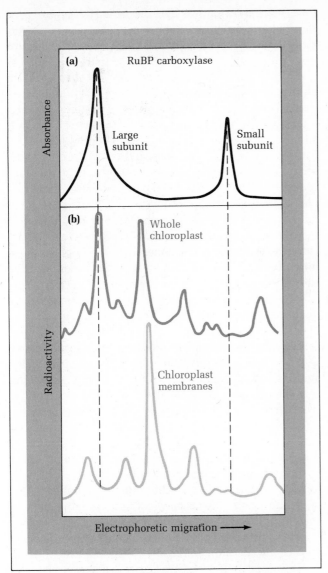

FIGURE 14-27 SDS-polyacrylamide gel electrophoresis of chloroplast proteins. (a) Absorbance profile of purified RuBP carboxylase. (b) After incubation of isolated chloroplasts with radioactive amino acids, labeled proteins present in the whole chloroplast as well as in the chloroplast membranes were analyzed. The results show that: (1) chloroplasts synthesize the large but not the small subunit of RuBP carboxylase, and (2) most polypeptides synthesized by the chloroplast (with the major exception of the large subunit of RuBP carboxylase) are membrane components.

Chloroplasts. Like mitochondria, chloroplasts synthesize a relatively small number of proteins. A second similarity is that the majority of the proteins synthesized on chloroplast ribosomes are hydrophobic membrane polypeptides, though a few soluble and ribosomal proteins are manufactured as well. The approaches used to identify the products of chloroplast protein synthesis resemble those that have been employed with mitochondria, though studies with inhibitors such as chloramphenicol and cycloheximide often yield ambiguous results when employed with intact cells and

therefore require independent verification by direct examination of the proteins synthesized by isolated chloroplasts. In addition, chloroplast DNA fragments can be isolated by restriction enzyme digestion and purified, thereby providing a source of DNA for direct investigation of transcriptional and translational events.

The first protein to be identified as a product of chloroplast protein synthesis was **RuBP carboxylase,** the carbon-fixing enzyme of the Calvin cycle that consists of a large catalytic subunit and a smaller regulatory subunit. The discovery that synthesis of the small subunit is selectively inhibited by cycloheximide while formation of the large subunit is blocked by chloramphenicol suggests that the small subunit is made in the cytoplasm and the large subunit in the chloroplast. This conclusion has been directly verified by the demonstration that isolated chloroplasts synthesize the large but not the small subunit (Figure 14-27). Moreover, messenger RNA isolated from chloroplasts has been translated in a cell-free protein-synthesizing system and the large subunit identified as a polypeptide product. Finally, the position of the gene for the RuBP carboxylase large subunit has been localized in a number of chloroplast DNAs (see Figure 14-24), and alterations in the amino acid sequence of the large subunit have been shown to be governed by maternal inheritance.

The second major polypeptide synthesized in cell-free systems containing chloroplast messenger RNAs is a 32,000 dalton protein component of thylakoid membranes. During light-induced plastid development the messenger RNA for this polypeptide is rapidly synthesized, suggesting that its translation is coupled to the need for this polypeptide as thylakoid membranes are assembled. The gene coding for this product is termed a **photogene,** since it is expressed only when plants are exposed to light.

Studies on antibiotic-resistant mutants of *Chlamydomonas* have provided indirect evidence that at least two ribosomal proteins may be synthesized within the chloroplast. An erythromycin-resistant mutation discovered by Lawrence Bogorad has been shown to be maternally inherited and is characterized by the presence of a single altered protein in the 50S subunit of the chloroplast ribosome. A cytoplasmically inherited mutation inducing streptomycin resistance, on the other hand, has been found by Ruth Sager to involve alterations in a protein present in the chloroplast 30S ribosomal subunit.

In addition to the large subunit of RuBP carboxylase, the 32,000 dalton thylakoid membrane protein, and the ribosomal proteins described above, chloroplasts also synthesize several other polypeptides. Among these are 5 subunits of the thylakoid membrane ATPase complex, 3 subunits of the cytochrome *b/f* complex, and thylakoid membrane polypeptides associated with photosystems I and II (see Table 14-4).

Translocation of Polypeptides into Mitochondria and Chloroplasts

Since mitochondria and chloroplasts synthesize a relatively small number of proteins, the vast majority of mitochondrial and chloroplast polypeptides must be encoded by nuclear genes, synthesized in the cytoplasm, and then transported into the appropriate organelle. Even those polypeptides that are manufactured within mitochondria or chloroplasts require interaction with other polypeptides synthesized on cytoplasmic ribosomes in order to form biologically active proteins. This required association between polypeptides encoded in nuclear and organelle compartments raises many questions concerning the coordination of protein synthesis on ribosomes in the organelles and in the cytoplasm. For example, how are the appropriate products of cytoplasmic protein synthesis selectively transported into mitochondria and chloroplasts? And what mechanism coordinates the rate of synthesis of these products with the synthetic rate of polypeptides synthesized within these two organelles?

Because polypeptides do not freely diffuse through membranes, the answer to the first question must involve some special mechanism for selectively transporting polypeptides manufactured on cytoplasmic ribosomes into mitochondria and chloroplasts. The discovery that secretory proteins are synthesized by ribosomes bound to the rough endoplasmic reticulum led to the suggestion that a similar mechanism involving ribosomes bound to the outer mitochondrial and chloroplast membranes might exist. Some support for this model was obtained with the discovery that, at least in yeast, the cytoplasmic ribosomes bound to the outer membrane of mitochondria are enriched in nascent mitochondrial preproteins.

In 1977 Walter Neupert and his colleagues reported the results of pulse-chase studies with *Neurospora* that strongly argued against translation-dependent import of proteins into mitochondria. These studies, as well as others carried out in yeast by Gottfried Schatz, revealed that the transport of newly synthesized polypeptides into mitochondria occurs after a lag, and that this transport process is insensitive to inhibitors of protein synthesis. This suggests that mitochondrial proteins are transported into the organelle *following* synthesis. Based on studies of the transfer of several different polypeptides into mitochondria and chloroplasts, a general description of the translocation process for many imported proteins is beginning to emerge. The sequence of events can be broken down into several steps. (1) A precursor polypeptide, longer than the mature protein, is synthesized on cytoplasmic ribosomes. This synthesis does not occur in association with the outer membrane of the mitochondrion or chloroplast. (2) The precursor polypeptide often, but not always, contains an amino-terminal extension. (3) Specific **import-receptor proteins** in the outer membrane serve to recognize and bind those polypeptides that are destined to enter the organelle. (4) After binding, the protein is either inserted into or transferred across the mitochondrial or chloroplast membranes, depending on its ultimate location in the organelle. This uptake of proteins into the matrix, inner membrane, or into the intermembrane space requires an electrochemical potential across the inner membrane. Those polypeptides containing extra amino acids are then cleaved by proteases to generate the mature protein molecule. It is now evident that import of at least some chloroplast polypeptides may also follow the same general pathway. For example, the large subunit of RuBP carboxylase has been found to be synthesized on free ribosomes in the cytoplasm as a much larger molecule than is actually present within the chloroplast.

The transport of proteins into double-membrane enclosed organelles such as mitochondria and chloroplasts is a more complicated event than the simple transfer that occurs across the single endoplasmic reticulum membrane. Insertion of some cytoplasmically synthesized polypeptides into the outer membrane or into the intermembrane space is relatively straightforward, involving import-receptor proteins localized in the outer membrane. Other proteins (such as cytochrome *c* oxidase) destined to be incorporated into the inner membrane (cristae or thylakoids) or inner compartment (matrix or stroma) may be transferred via *contact sites* joining the outer and inner membranes of the organelle. These regions may facilitate the movement of specific proteins into the mitochondrial and chloroplast interior. The existence of such contact sites has been verified by electron microscopy, lending credence to the idea that such a transport pathway may exist.

The coordination of synthesis of polypeptides made in separate locations but destined for incorporation into the same organelle is not well understood. Zoltan Barath and Hans Kuentzel have observed that when mitochondrial protein synthesis is selectively inhibited, the rates of formation of many cytoplasmically synthesized mitochondrial proteins increase, leading to the speculation that mitochondria normally synthesize a "repressor" that regulates transcription of mitochondria-specific genes in the nucleus. In the case of chloroplasts an analysis of the coordination of synthesis of the large and small subunits of RuBP carboxylase has provided some insight into this process. R. John Ellis has found that an inhibitor that blocks the synthesis of the small subunit by cytoplasmic ribosomes ultimately inhibits formation of the large subunit by chloroplasts, even though the inhibitor has no direct effect on chloroplast protein synthesis. It has therefore been proposed that the small subunit of RuBP carboxylase, made on cytoplasmic ribosomes, acts as a positive control factor required for either the transcription or translation of chloroplast messenger RNA coding for the large subunit of this enzyme. These two examples may provide important clues concerning the mechanism that coordinates cytoplasmic, mitochondrial, and chloroplast protein synthesis.

Why some mitochondrial and chloroplast polypeptides are synthesized within these organelles while others are made in the cytoplasm is not yet fully understood. In part the answer may lie in the evolutionary history of mitochondria and chloroplasts, a subject to be discussed later in the chapter. A related question is whether copies of genes coding for organelle polypeptides are present in both the organelle and the nucleus. This arrangement does apply to certain mitochondrial genes. One of the first techniques employed to investigate this question was that of DNA-DNA hybridization. The technical problems inherent in this approach are formidable, since one copy of the mitochondrial gene within a typical vertebrate nucleus would represent about 0.001 percent of the total nuclear DNA. Such a gene would be difficult to detect, particularly if small amounts of mitochondria contaminated the nuclear DNA preparation. The DNA-DNA hybridization experiments revealed no homology between mitochondrial and nuclear DNAs, suggesting that the nucleus lacks mitochondrial DNA sequences. However, using the more sensitive Southern blot technique, mitochondrial DNA sequences have been identified in the nuclei of a variety of organisms, including insects, mammals, sea urchins, and yeast. Frances Farrelly and Ronald Butow have shown that in yeast these nuclear genes follow the Mendelian pattern of inheritance and behave as if they are integrated into the nuclear DNA. However, these integrated mitochondrial sequences have been found to contain segments of different mitochondrial genes that have become fused together, suggesting that they are genetically inactive. Conversely, the gene coding for a mitochondrial ATPase subunit in *Neurospora* has been detected in both nuclear and mitochondrial DNAs, and though it is genetically silent in the organelle, it is transcribed into the appropriate messenger RNA in the nucleus. These results indicate that when similar sequences are present in both nuclear and mitochondrial DNA, in some cases the nuclear sequences may be active and in other cases the mitochondrial sequences will be active. Further study is required to elucidate the significance of these results and their relevance to the origins of both chloroplasts and mitochondria in eukaryotic cells.

DEVELOPMENTAL ORIGINS OF MITOCHONDRIA AND CHLOROPLASTS

The occurrence of DNA, RNA, and protein synthesis within mitochondria and chloroplasts raises the question of the degree to which these two organelles regulate their own activities and development. It is clear that products specified by nuclear DNA and synthesized within the cytoplasm are essential for the long-term growth and reproduction of mitochondria and chloroplasts, but this does not rule out the possibility that the genetic systems of these organelles permit them at least partial autonomy in governing their own behavior. Several examples of such partial autonomy will be discussed in the following sections on the development and evolutionary origins of mitochondria and chloroplasts.

Mitochondrial Development

Three mechanisms have been proposed to explain the way in which new mitochondria originate: (1) growth and division of existing mitochondria; (2) complete *de novo* (from the beginning) formation of new mitochondria from a mixture of the required proteins, nucleic acids, lipids, and so forth; and (3) assembly of new mitochondria from other membrane systems, such as the plasma membrane, nuclear envelope, or endoplasmic reticulum.

Morphological evidence from light and electron microscopy has tended to support the first theory, namely, that mitochondria arise by growth and division. Mitochondria in the process of dividing in half have been observed in living tissues viewed under the phase microscope, and membranous partitions that appear to be in the process of separating mitochondria in half have been observed with the electron microscope (Figure 14-28). In the case of the small green alga *Micromonas*, which contains only one mitochondrion and one chloroplast, each of these organelles can be clearly seen to divide as the cell divides. But the above observations indicating that mitochondria *can* divide do not rule out the possibility that mitochondria also arise by *de novo* synthesis or development from other membrane systems, especially in more complex cells containing large numbers of mitochondria.

In the early 1960s David Luck designed a series of experiments on *Neurospora* aimed at determining whether such alternative mechanisms of mitochondrial formation occur. First, *Neurospora* cultures were incubated with ^{3}H-choline, a phospholipid precursor that becomes incorporated into mitochondrial membranes. The labeled cells were then transferred to nonradioactive medium and allowed to grow for varying periods of time. Finally, the cells were analyzed by autoradiography to determine the distribution of radioactivity in the mitochondria. It was reasoned that, depending on the mode of mitochondrial development, one of the following three possible patterns of autoradiographic grains would emerge: (1) If mitochondria form by growth and division, each doubling of the mitochondrial population

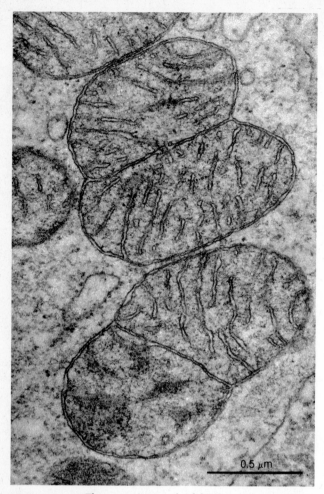

0.5 μm

FIGURE 14-28 Electron micrograph of dividing mitochondria in the fat body cells of a developing larva. Each mitochondrion is constricted into two compartments by a transecting crista membrane. Courtesy of W. J. Larsen.

after the ^{3}H-choline is removed should be accompanied by a 50 percent reduction in the number of autoradiographic grains per mitochondrion. (2) If *de novo* synthesis occurs, then new mitochondria formed after removal of ^{3}H-choline should be unlabeled, while the radioactivity in the preexisting mitochondria would remain essentially constant. The average grain count per *labeled* mitochondrion would therefore not change. (3) If mitochondria develop from other membranes, then the preexisting labeled mitochondria should maintain the

same level of radioactivity, and new mitochondria should at first also be labeled (because the membranes from which they develop would be labeled). As time passes, however, the absence of ^{3}H-choline in the medium would result in a progressive decrease in the amount of radioactivity in the cellular membranes from which the new mitochondria develop. The average grain count per labeled mitochondrion would therefore gradually decrease, but not as fast as 50 percent per generation as expected from prediction 1. The data Luck obtained (Table 14-5) clearly follows the first pattern of distribution of silver grains, and is consistent with the conclusion that mitochondria arise by growth and division.

In a second set of experiments addressing the same basic question, Luck exploited the fact that the concentration of choline in the medium influences the density of the mitochondria formed: incubation with high choline concentrations results in the production of mitochondria of low density, while incubation in the presence of low choline concentrations results in mitochondria of high density. Luck therefore grew cells in the presence of low choline, and then switched them to high choline. He reasoned that if mitochondria are formed either *de novo* or from other preformed membranes, the high-density mitochondria originally manufactured in the presence of low choline should remain undisturbed, while a new population of low-density mitochondria would appear as a result of subsequent growth in high choline. If mitochondria arise by growth and division, however, the material added to the preexisting high-density mitochondria as they grow would be of low density, and so the mitochondria should gradually get lighter. When Luck analyzed the density of the mitochondria formed after shifting cells from low to high choline, this is exactly what occurred, with no persistence of low-density mitochondria (Figure 14-29). In concert with the widespread microscopic evidence that mitochondria divide, Luck's two sets of experiments strongly argue that new mitochondria arise solely by growth and division of previously existing mitochondria.

Growth and division of mitochondria are affected by the characteristics of the cell in which the mitochondria reside as well as by the external environment. In

TABLE 14-5
Average number of silver grains per labeled mitochondrion following exposure of *Neurospora* to ^{3}H-choline

Number of Generations After Labeling	Actual Data	Pattern Predicted By		
		Growth and Division	*De Novo* Synthesis	Assembly from Nonmitochondrial Membranes
0	2.0	2.0	2	Intermediate between growth and division and *de novo* synthesis
1	1.1	1.0	2	
2	0.5	0.5	2	
3	0.25	0.25	2	

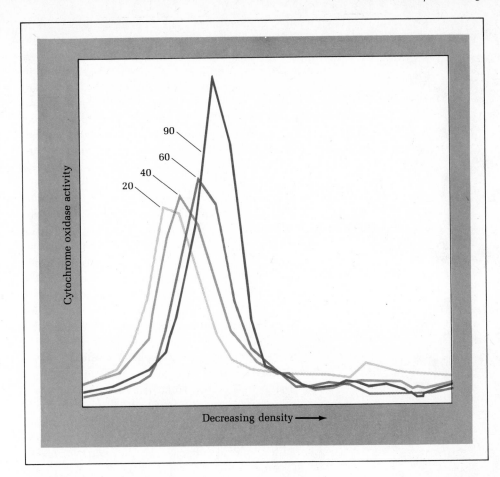

FIGURE 14-29 Isodensity centrifugation of mitochondria isolated 20, 40, 60, and 90 min after shifting *Neurospora* cultures from a low to a high choline growth medium. The gradual shift in density of the mitochondria (detected by the presence of cytochrome oxidase activity) is not consistent with formation of mitochondria *de novo* or from nonmitochondrial membranes, but is compatible with the theory that mitochondria are formed via growth and division. See text for details.

cells that are rapidly dividing, for example, the mitochondria must also rapidly grow and divide to ensure that there are sufficient numbers of mitochondria to populate the newly forming cells. In rapidly dividing HeLa cells, mitochondrial DNA replication has been

FIGURE 14-30 Long-term retention of radioactivity in nuclear and mitochondrial DNA of rat liver after a single injection of ³H-thymidine. Note that in contrast to nuclear DNA, mitochondrial DNA is continually being degraded.

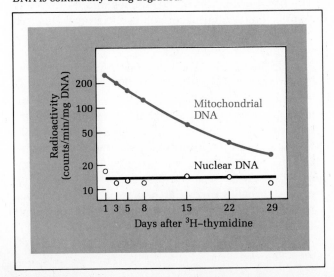

found to occur during late S and G₂, while in *Tetrahymena* the actual division of mitochondria takes place in late S-phase. Though it is risky to generalize these results to other cell types, it seems safe to conclude that DNA replication and division of mitochondria are restricted to a particular portion of the cell cycle rather than occurring randomly. As described earlier in the chapter, mitochondrial DNA replicates by the classical semiconservative mechanism, but the means by which the resulting DNA molecules are equally distributed to the two newly forming mitochondria is unknown. The fact that mitochondrial DNA appears to be associated with the inner mitochondrial membrane suggests the possibility that membrane growth accomplishes separation of the new DNAs in a fashion similar to the process that occurs in dividing bacterial cells.

Formation of new mitochondria is not restricted to proliferating cells. Even in cells that never divide or replicate their nuclear DNA, mitochondrial DNA synthesis has been observed. This DNA synthesis is needed to replace mitochondrial DNA that is continually being destroyed. In rat liver, for example, half the total mitochondrial DNA is degraded every nine days (Figure 14-30). At least some of this destruction is due to the breakdown and recycling of mitochondrial components, necessitating continual formation of new mitochondria by growth and division of preexisting ones.

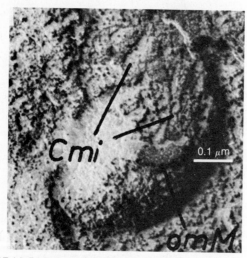

FIGURE 14-31 Freeze-etch electron micrographs of anaerobically grown yeast showing the presence of promitochondria. *(Top)* Promitochondrion seen partly in cross section and partly in surface relief. Two distinct membranes are visible. *(Bottom)* Promitochondrion in surface relief, with outer membrane visible. Several cristae are also present. Abbreviations: omM = outer membrane, Cmi = cristae. Courtesy of H. Plattner.

In addition to being influenced by cellular growth rates, mitochondrial biogenesis has been shown to be affected by environmental factors such as oxygen tension. An extreme example of oxygen-induced modulation of mitochondrial development occurs in yeast, which can survive both in the presence and absence of oxygen. Anerobically grown yeast, whose energy requirements are met solely by glycolysis, gradually lose the capacity to carry out the Krebs cycle and electron transfer. Electron micrographs of such yeast reveal a lack of any clearly recognizable mitochondria. Subsequent exposure to oxygen, however, results in the rapid reappearance of functional mitochondria, a phenomenon at one time considered to be evidence for the *de novo* formation of these organelles. However, when anaerobically grown yeast are properly fixed prior to elec-

tron microscopic observations, small double-membrane enclosed vesicles about 1 μm in diameter can be seen (Figure 14-31). Though lacking cristae and cytochromes, these vesicles have been found to contain oligomycin-sensitive ATPase and mitochondrial DNA. In the presence of oxygen the inner membrane of these vesicles invaginates to form cristae, which then acquire newly synthesized cytochromes and other components of the electron transfer chain. Because of this ability to differentiate into mitochondria, the double-membrane enclosed vesicles of anaerobically grown yeast are referred to as **promitochondria.**

Inhibitors of mitochondrial and cytoplasmic protein synthesis have been utilized to explore the relative contributions of mitochondrial and nuclear genes to mitochondrial biogenesis. In the presence of chloramphenicol, mitochondria continue to grow and divide for several days. Mitochondrial DNA replication and RNA synthesis proceed as usual, and the mitochondria appear normal in number and appearance. However, the lack of mitochondrial protein synthesis eventually leads to functional defects, such as gradual loss of cytochrome oxidase activity. This deficiency causes a depression in electron transfer and oxidative phosphorylation and eventual arrest of cell growth. Such observations indicate that the mitochondrial genetic system is not required for the gross assembly and division of mitochondria, but is clearly necessary for maintaining normal mitochondrial function.

Chloroplast Development

As in the case of mitochondria, considerable microscopic evidence suggests that chloroplasts arise by division from preexisting chloroplasts. In algae, which contain only one or two large chloroplasts per cell, the fate of individual chloroplasts is especially easy to watch. As long ago as the late nineteenth century, light microscopic observations of the green alga *Spirogyra* revealed that each chloroplast divides into two prior to cell division. Similar observations have been reported for other algae, and films of the process of chloroplast division have even been made.

In higher plant cells the large number of chloroplasts makes it difficult to follow the fate of individual chloroplasts with the light microscope. However, micrographs occasionally reveal the presence of membrane partitions dividing chloroplasts in two (Figure 14-32). Moreover, chloroplasts isolated from higher plant tissues undergo spontaneous division when cultured for a few days in an artificial growth medium (Figure 14-33), confirming that higher plant chloroplasts can divide in two, and revealing that the division mechanism must be inherent to the chloroplast itself.

Mature chloroplasts are not present in all plant cell types. The various kinds of cells that comprise a typical

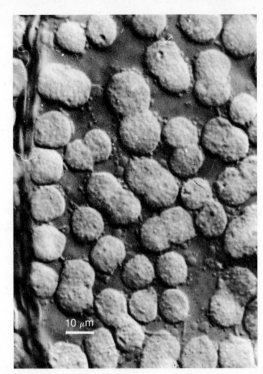

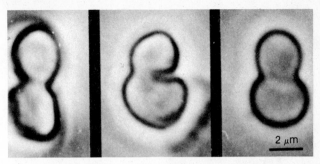

FIGURE 14-33 Phase contrast micrographs of dividing spinach chloroplasts. Courtesy of R. M. Leech.

FIGURE 14-32 Micrograph of a living spinach leaf viewed by Nomarski optics. Dividing chloroplasts can be seen in various stages from dumbbell shaped to fully separated. Courtesy of J. V. Possingham.

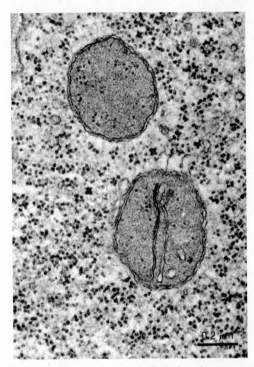

FIGURE 14-34 Electron micrograph of proplastids in a root tip cell of the bean *(Phaseolus)*. A few scattered ribosomes are evident and in one proplastid a few internal membranes can be seen. Micrograph by W. P. Wergin, courtesy of E. H. Newcomb.

FIGURE 14-35 Major pathways of plastid differentiation. All forms in the smaller box are leukoplasts (colorless).

higher plant arise from a rapidly dividing, undifferentiated type of tissue called **meristem.** Meristem cells do not have chloroplasts, but they do contain small vesicles, 1 μm or less in diameter, composed of an undifferentiated stroma surrounded by a double membrane (Figure 14-34). These **proplastids** resemble promitochondria in appearance, but their relationship to chloroplasts is revealed by the presence of chloroplast DNA rather than mitochondrial DNA. The developmental fate of proplastids depends on both the region of the plant in which they occur and the amount of light they receive (Figure 14-35). Some proplastids differentiate into **amyloplasts,** which are starch-filled particles that predominate in starchy vegetables such as potatoes. Others evolve into organelles specialized for the storage of protein **(proteinoplasts)** or lipids **(lipoplasts).** Since proplastids, amyloplasts, proteinoplasts, and lipoplasts are all colorless, they are collectively referred to by the general term **leukoplast** (leuko = white).

In spite of the variety of organelles into which proplastids can develop, their major fate in photosynthetic tissues is to evolve into chloroplasts. This process, which is triggered by light, begins with an enlargement of the proplastid and an invagination of the inner membrane into tubules that project into the stroma. The tubules then spread into flat sheets that line up in parallel to form grana. These morphological changes in the inner membrane are accompanied by the acquisition of

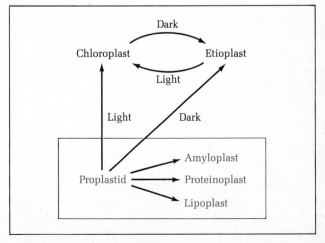

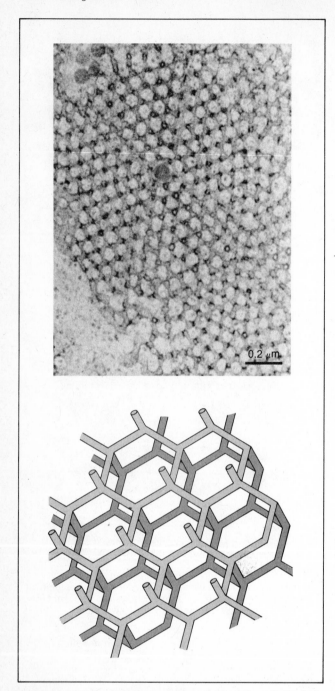

FIGURE 14-36 *(Top)* Electron micrograph of a prolamellar body in an etioplast of a dark-grown seedling. The hexagonal arrangement of tubules is clearly evident. Courtesy of T. E. Weier. *(Bottom)* Model of the arrangement of tubules in the prolamellar body.

chlorophyll and other components of the photosynthetic electron transfer chain.

If light is not present, a somewhat different sequence of events takes place. In seedlings kept in the dark, the proplastids still enlarge and membranous sheets invaginate from the inner membrane, but these sheets then contract into an ordered set of interconnected tubules known as a **prolamellar body.** The tubules of the prolamellar body are generally arranged in

a regular crystalline pattern, the basic unit of which is a hexagonal array of tubules (Figure 14-36). At this stage the plastid is referred to as an **etioplast.** In addition to developing from proplastids in dark-grown seedlings, etioplasts are also formed when green plants containing mature chloroplasts are moved from light to darkness. In this case the thylakoid membranes become disrupted and contract into tubules that fuse together to produce a typical prolamellar body.

Regardless of whether etioplasts develop from proplastids in dark-grown seedlings or from mature chloroplasts in plants switched from light to darkness, illumination triggers the conversion of etioplasts to chloroplasts. Within a few hours in the presence of light the highly ordered prolamellar body is transformed into an irregular mass of tubules. Membranous sheets then grow out from this mass, and grana develop by overlapping or budding of these extending membranes (Figure 14-37). As the grana increase in size and number, the prolamellar body gradually disappears, and within a day the transformation into a mature chloroplast is usually complete.

This differentiation of the etioplast into a chloroplast provides an excellent model system for exploring the biochemistry of chloroplast biogenesis. The main pigment present in etioplasts is **protochlorophyllide,** which differs from chlorophyll in two ways: It lacks two hydrogens in the porphyrin ring, and the phytol chain is not attached (see Figure 8-23). When etioplasts are illuminated, the absorption of light by protochlorophyllide stimulates its conversion to chlorophyllide *a*; phytol is then added in a light-independent reaction to produce chlorophyll *a* (Figure 14-38).

Since the quantity of protochlorophyllide present in etioplasts has been found to be relatively small, this light-induced conversion of protochlorophyllide to chlorophyll accounts for only a tiny fraction of the chlorophyll actually needed by the developing chloroplast. The bulk of the chlorophyll forms *de novo*, that is, from small building blocks rather than by conversion of a preexisting pigment such as protochlorophyllide. Such *de novo* chlorophyll synthesis does not occur until several hours after illumination, during which time enzymes required for chlorophyll formation are being synthesized. When etiolated leaves are experimentally provided with preformed **5-aminolevulinate,** the first major product in the biosynthetic pathway for chlorophyll (see Figure 14-38), this lag period in *de novo* chlorophyll synthesis is abolished. Such findings suggest that the enzyme whose synthesis is being induced during the lag period is **5-aminolevulinate synthetase (ALA synthetase),** the enzyme that catalyzes the formation of 5-aminolevulinate.

Induction of ALA synthetase does not require the continuous presence of light. If etiolated leaves are

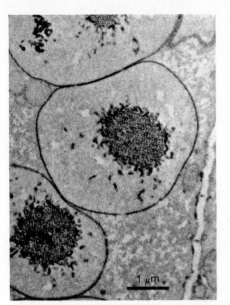
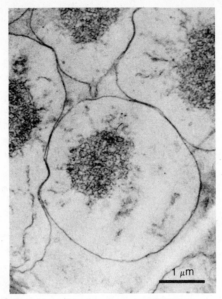
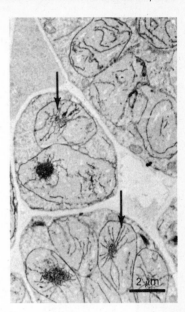

FIGURE 14-37 Electron micrographs of etioplast to chloroplast conversion. *(Left)* Etioplast in the leaf of a dark-grown bean plant showing the regular arrangement of the prolamellar body. *(Middle)* Plastids exposed to red light (655 nm) for 10 sec. Note the irregular appearance of the prolamellar body. *(Right)* Plastids exposed to red light for 5 hr. The prolamellar body has expanded and the beginnings of chloroplast lamellae are evident (arrows). Courtesy of L. Bogorad.

FIGURE 14-38 Major steps in chlorophyll biosynthesis.

Glycine + Succinyl-CoA $\xrightarrow{\text{ALA synthetase}}$ 5-Aminolevulinate + CO_2

5-Aminolevulinate (two molecules) → Porphobilinogen

Porphobilinogen (four molecules) → Protochlorophyllide

Protochlorophyllide $\xrightarrow[\text{(light dependent)}]{2H}$ Chlorophyllide *a* $\xrightarrow{\text{Phytol}}$ Chlorophyll *a* ⤏ Other chlorophylls

exposed to red light (660 nm) for a few minutes and returned to darkness for several hours, chlorophyll synthesis will occur without lag upon subsequent illumination. This means that once the induction of ALA synthetase and the resulting formation of 5-aminolevulinate are triggered by the brief exposure to red light, these processes continue to occur in the ensuing dark period. The triggering of the induction process by red light can be reversed by a subsequent exposure to far-red light (>700 nm). This pattern of activation by red light and inhibition by far-red light is not unique to the induction of *de novo* chlorophyll synthesis, but is shared by a variety of light-inducible metabolic activities in plants, including Calvin cycle and Hatch–Slack enzymes. The light-receptor pigment mediating these effects is **phytochrome,** a large protein attached to a linear pyrrole pigment. Phytochrome exists in two interconvertible forms, an active form absorbing maximally at 725 nm and an inactive form absorbing maximally at 665 nm:

$$\underset{\text{(Inactive)}}{\text{Phytochrome}_{665}} \xrightleftharpoons[\text{Light at 725 nm}]{\text{Light at 665 nm}} \underset{\text{(Active)}}{\text{Phytochrome}_{725}}$$

This interconversion between active and inactive forms of phytochrome explains why pretreatment with red light (near 665 nm) triggers the induction of chlorophyll synthesis, while subsequent irradiation with far-red light (>700 nm) halts the process. It is thought that the active form of phytochrome exerts its effects by stimulating the transcription of specific messenger RNAs, since treatment of cells with actinomycin D blocks the light-induced activation of *de novo* chlorophyll synthesis.

Though a brief exposure of etioplasts to red light is sufficient to trigger the formation of ALA synthetase as well as Calvin cycle and Hatch-Slack enzymes, continuous light is needed to sustain the conversion of the prolamellar body into thylakoids, indicating the involvement of a light-activated pathway other than the one mediated by phytochrome. The efficiency of various wavelengths of light in inducing the first stages of prolamellar body breakdown is similar to the absorption spectrum of protochlorophyllide, implicating this pigment as the photoreceptor triggering the initial membrane transformations. The later events of thylakoid outgrowth and grana formation appear to depend on other as yet unidentified photoreceptors. Whatever the identity of these substances, a chain of events is triggered that seems to involve cytoplasmic as well as plastid protein synthesis. For example, the development of both photosystem I activity, which appears as the thylakoids begin to grow out from the prolamellar body, and photosystem II activity, which appears shortly before the onset of grana stacking, can be blocked by either cycloheximide or chloramphenicol. This finding emphasizes that the assembly of a functional chloroplast depends on cooperation between plastid and cytoplasmic protein-synthesizing systems.

A summary of the major events we have been discussing is presented in Figure 14-39. It should be clear from examining this diagram that chloroplast biogenesis proceeds by a stepwise process. Some of the steps, such as breakdown of the prolamellar body, formation of thylakoids, and conversion of protochlorophyllide to chlorophyll require the continuous presence of light. Others, like the induction of enzymes involved in chlorophyll synthesis and the Calvin and Hatch-Slack pathways, are mediated by phytochrome activation, which requires only a brief exposure to red light. In the final analysis a considerable degree of cooperation between different photoreceptors and protein-synthesizing systems is involved in the formation of the mature, functionally competent chloroplast.

EVOLUTIONARY HISTORY OF MITOCHONDRIA AND CHLOROPLASTS

The evolutionary origin of eukaryotic cells has been a topic of vigorous debate ever since biologists became aware of the basic differences between prokaryotes and

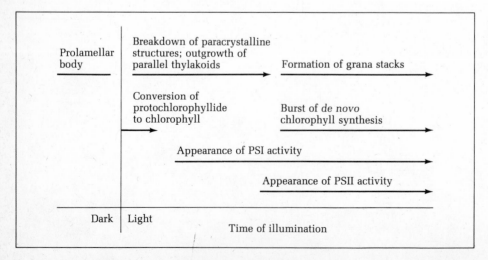

FIGURE 14-39 Diagrammatic representation of the stepwise changes that occur in etioplasts after exposure to light. The time scale is not calibrated in hours because the rapidity with which this sequence of events occurs varies with cell type and experimental conditions.

eukaryotes. This topic was briefly discussed in Chapter 1, but in the absence of detailed knowledge of the biochemistry of eukaryotic cell organelles, it could only be discussed in general terms. Now that the prerequisite background material has been provided, current theories concerning the evolutionary origin of mitochondria and chloroplasts can be elaborated upon in detail.

The Endosymbiont Theory

The debate on the evolutionary origins of mitochondria and chloroplasts has a long history. As early as 1890, R. Altmann referred to mitochondria as primitive "organisms" analogous to bacteria, and suggested that they can exist as free-living forms outside the host cytoplasm. In 1910 C. Mereschovsky extended this idea, proposing that chloroplasts originally arose from a symbiotic association between primitive blue-green algae and nonphotosynthetic nucleated cells. A few years later J. F. Wallin claimed to support these theories with the demonstration that mitochondria can grow outside of living cells. However, his data were questionable and his case overstated, and the whole notion of a symbiotic origin of mitochondria and chloroplasts soon fell into disrepute.

After several decades of ridicule and neglect, these ideas regained prominence when it was discovered in the early 1960s that mitochondria and chloroplasts contain their own DNA. Especially influential was the gradual realization that nucleic acid and protein synthesis in mitochondria and chloroplasts resemble the comparable events of prokaryotic cells in many ways. Among the similarities shared by mitochondria, chloroplasts, and prokaryotic cells are the following:

1. The DNA of chloroplasts and mitochondria, like that of prokaryotic cells, is generally circular and is not complexed with histones, as is eukaryotic nuclear DNA.
2. Like prokaryotic ribosomes, chloroplast and mitochondrial ribosomes are smaller than 80S eukaryotic ribosomes. The chloroplast 70S ribosome closely resembles bacterial ribosomes in size, and active ribosomes have been artificially assembled containing the small subunit of *Euglena* chloroplast ribosomes and the large subunit of *E. coli* ribosomes.
3. The base sequence of the 16S ribosomal RNA of chloroplasts closely resembles that of the 16S ribosomal RNA of bacteria and blue-green algae, but not that of the 18S ribosomal RNA of plant cell sap ribosomes.
4. The poly(A) tails of mitochondrial messenger RNA are considerably shorter than those of nucleus-derived messages, but are similar in size to those reported for some bacterial mRNAs.
5. Protein synthesis is initiated by formylmethionine in mitochondria, chloroplasts, and prokaryotes, but not in the cytoplasm of eukaryotes.
6. Antibiotics that block protein synthesis on prokaryotic but not eukaryotic cytoplasmic ribosomes (e.g., chloramphenicol) also inhibit chloroplast and mitochondrial protein synthesis. Conversely, inhibitors selective for eukaryotic cytoplasmic protein synthesis (e.g., cycloheximide) exert no effect on chloroplast or mitochondrial protein synthesis.
7. Protein synthesis initiation and elongation factors are interchangeable between mitochondria and bacteria, but not between mitochondria and the eukaryotic cell sap.
8. Certain viral messenger RNAs are efficiently translated by protein-synthesizing systems prepared from bacteria and mitochondria, but not by the protein-synthesizing apparatus of eukaryotic cells.
9. Bacterial and mitochondrial RNA polymerases are inhibited by rifampicin, while nuclear RNA polymerase is not.
10. Eukaryotic cells contain distinct mitochondrial and cytoplasmic forms of superoxide dismutase, the enzyme responsible for breaking down the superoxide anion. The form occurring in mitochondria resembles bacterial superoxide dismutase more closely than it does the eukaryotic cytoplasmic enzyme.

The above list of similarities between mitochondria, chloroplasts, and prokaryotic cells is not intended to imply the absence of differences. The 55–60S ribosomes of animal mitochondria are significantly smaller than 70S prokaryotic ribosomes, and the subunits of animal mitochondrial and bacterial ribosomes are not functionally interchangeable. Yet, in spite of such qualifications, the general process of information flow in mitochondria and chloroplasts appears to exhibit more similarities to the processes occurring in prokaryotic cells than to the analogous events taking place in the nucleus and cytoplasm of the eukaryotic cell.

In addition to these biochemical parallels, mitochondria and chloroplasts bear a certain structural resemblance to prokaryotic cells. Mitochondria are similar to bacteria in size and shape, while the somewhat larger chloroplasts are closer to blue-green algae in appearance. The plasma membrane of prokaryotic cells and the inner membrane of mitochondria are both characterized by the presence of protruding spheres that contain the ATPase involved in oxidative phosphorylation. And finally, blue-green algae have pigment-containing thylakoid membranes that resemble the internal thylakoid membranes of chloroplasts.

The resemblance of mitochondria and chloroplasts to prokaryotic cells has led to the formulation of the **endosymbiont theory,** which views mitochondria and chloroplasts as having evolved from aerobic bacteria and photosynthetic prokaryotes that were ingested by primitive cells a billion or more years ago. A detailed accounting of the events that might have occurred has been put together by Lynn Margulis (Figure 14-40). This accounting begins with the assumption that the absence of oxygen in the primitive atmosphere caused the first cells to be anaerobic. Some of these primitive anaerobic cells then synthesized porphyrin pigments capable of absorbing light energy, and evolved the capability to use this trapped energy to drive photosynthesis. These first photosynthetic cells, which employed hydrogen sulfide or molecular hydrogen as electron donors, became the ancestors of today's photosynthetic bacteria. But later photosynthetic cells, the ancient blue-green algae, began to utilize water as an electron donor, resulting in the release of oxygen into the atmosphere.

As oxygen accumulated in the primitive atmosphere, some of these primitive anaerobes may have adapted by moving to oxygen-free environments, such as muddy flats, where they exist today as anaerobic bacteria. Others are thought to have evolved into aerobic bacteria by developing the oxygen-utilizing pathways of electron transfer and oxidative phosphorylation. Finally, a third type of primitive cell may have become the progenitor of contemporary eukaryotic cells. Two features developed in this hypothetical **protoeukaryote** that distinguish it from other primitive prokaryotic cells: large size and the ability to import nutrients into the cell from the environment by phagocytosis. According to the endosymbiont theory, one result of phagocytosis was the uptake of primitive aerobic bacteria by the larger protoeukaryotic cell. Margulis feels that prior to this event the protoeukaryote was anaerobic, dependent solely on glycolysis for energy; it follows that because the respiratory pathway present in aerobic bacteria generates more usable energy than glycolysis, the introduction of these aerobes into the cytoplasm of the protoeukaryote would be beneficial. As the protoeukaryotic cell and the bacteria residing in its cytoplasm mutually adapted to living with one another, the bacteria gradually lost unnecessary functions and eventually evolved into mitochondria.

Critics of the preceding scenario claim that it is not accurate to view the eukaryotic cell as a primitive anaerobic cytoplasm containing aerobic mitochondria. They instead contend that the nonmitochondrial portion of the eukaryotic cell must be considered fundamentally aerobic because it contains enzymes such as superoxide dismutase and cytochrome P-450 that require oxygen. The presence of these aerobic features in eukaryotic cytoplasm suggests that the ancestral protoeukaryote was already aerobic, rather than anaerobic as the Margulis model requires. But if the protoeukaryote was already capable of aerobic metabolism, what would have been the advantage of acquiring an aerobic symbiont? Christian de Duve has proposed that the aerobic metabolism of the primitive protoeukaryote might

have depended on peroxisomal respiration, which was described in Chapter 6 as being less efficient than electron transfer because it is not coupled to ATP formation. Hence the acquisition of an aerobic symbiont capable of coupling electron transfer to oxidative phosphorylation would have been advantageous even if the protoeukaryote were already aerobic.

Introduction of mitochondria is only one of several steps that must have taken place during the transition from prokaryotic to eukaryotic cells. Among the other required events are the elaboration of an extensive set of intracellular membranes (including the nuclear envelope), and the formation of cytoskeletal structures (including the mitotic apparatus). Subsequent to these steps a second major symbiotic event, leading to the development of chloroplasts, is thought to have taken place. Since chloroplasts are not present in all eukaryotic cells, this symbiosis would have had to involve a subgroup of ancestral eukaryotic cells destined to give rise to the plant kingdom. The organisms thought to have been ingested in this case are ancient photosynthetic prokaryotes, which would have conferred upon their host eukaryotes the ability to photosynthesize. With time these photosynthetic prokaryotes would have lost those functions found to be unnecessary in their new cytoplasmic environment, until they ultimately evolved into chloroplasts as we know them today.

Unlike mitochondria, which are similar in all eukaryotic cells, the chloroplasts of different cell types exhibit significant biochemical differences. The major chloroplast pigments are chlorophylls a and b in higher plants and green algae, chlorophylls a and c in brown algae, dinoflagellates, and diatoms, and chlorophyll a and phycobilins in red algae. Peter Raven has suggested that the simplest explanation for such pigment differences is that the chloroplast in each of these three plant groups arose from an independent symbiont event (see Figure 14-40). The three symbiotic events would have occurred as follows: (1) one line of ancestral protoeukaryote, destined to become red algae, ingested ancient blue-green algae containing chlorophyll a and phycobilins; (2) a second line of protoeukaryote, destined to evolve into brown algae, dinoflagellates, and diatoms, ingested a hypothetical photosynthetic prokaryote ("yellow prokaryote") containing chlorophylls a and c; and (3) a third line of protoeukaryote, destined to evolve into green algae and higher plants, ingested a hypothetical photosynthetic prokaryote ("green prokaryote") containing chlorophylls a and b. Since no prokaryotes living today contain either chlorophylls b or c, it must be further postulated that these hypothetical green and yellow photosynthetic prokaryotes later became extinct.

The endosymbiont theory is based predominantly on the biochemical and structural similarities between mitochondria, chloroplasts, and prokaryotic cells.

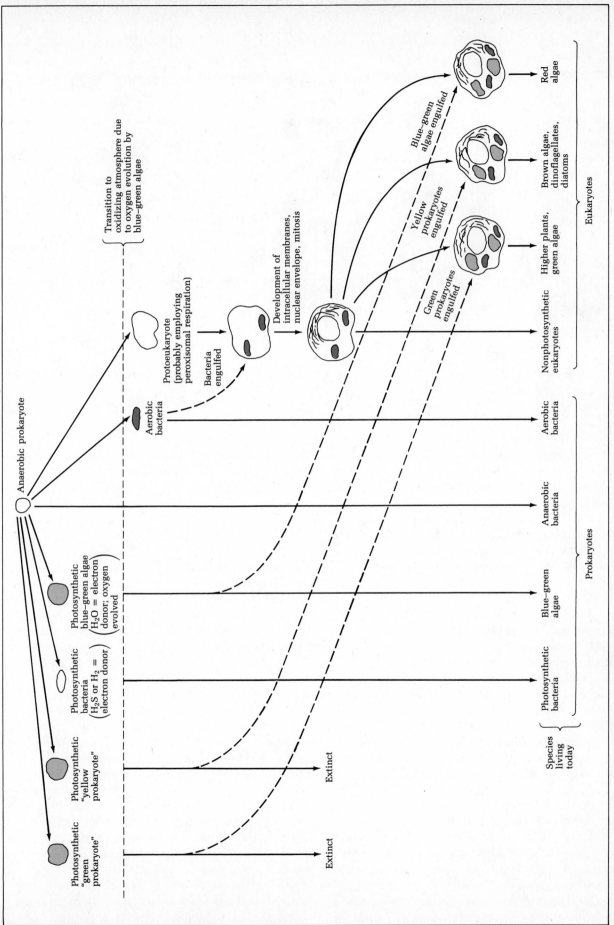

FIGURE 14-40 The endosymbiont theory of the evolutionary origins of mitochondria and chloroplasts. See text for details.

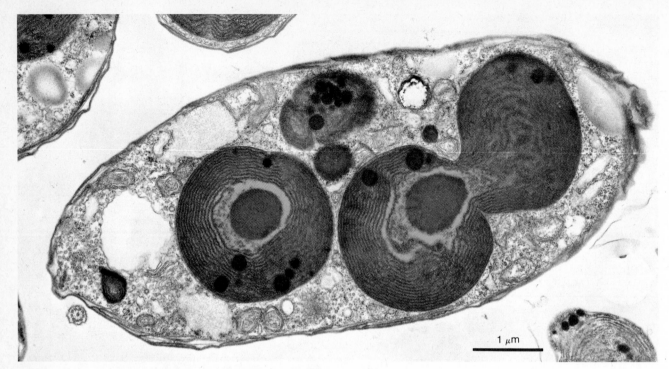

FIGURE 14-41 Electron micrograph of a blue-green photosynthetic alga (color) inhabiting the cytoplasm of a flagellated protozoan *(Cyanophora).* Courtesy of W. T. Hall.

Support of another kind is provided by numerous examples of symbiotic events occurring today that resemble what may have occurred a billion or more years ago. Because all present-day eukaryotes already have mitochondria, new symbiotic events involving the uptake of a respiratory prokaryote would be of little value. But not all eukaryotes have chloroplasts, so the acquisition of a photosynthetic symbiont might be of value to some animal cells. The search for examples of such symbiotic events has led to the discovery that algae, dinoflagellates, and diatoms occur as symbionts in the cytoplasm of cells occurring in over 150 kinds of invertebrates (Figure 14-41). In many cases the cell wall of the symbiont is no longer present, leaving a "naked" algal cell within the host cytoplasm. In a few instances the reduction in structure goes even further, with only the chloroplasts of the symbiont remaining.

A striking example of this latter type of symbiosis occurs in certain mollusks, where the cells lining the digestive tract contain clearly identifiable chloroplasts. These chloroplasts, which originate from the green algae these mollusks feed upon, continue to carry out photosynthesis after being incorporated into the mollusk's cells, and the reduced carbon generated in the process is distributed as a source of nutrients to other tissues not containing chloroplasts. Though the chloroplasts do not reproduce, they continue to function in the animal cell cytoplasm for at least several months. That chloroplasts from ingested algae are taken up by animal cells and assume a stable symbiotic relationship with their host is certainly consistent with the notion that

chloroplasts may have had their evolutionary origins in endosymbiotic associations between ancient photosynthetic prokaryotes and ancestral eukaryotes.

Direct Filiation Theories

Though a variety of arguments supporting the endosymbiont theory have been described, none provide unequivocal evidence for the evolutionary origins of mitochondria and chloroplasts. Critics of the theory rightly point out that: (1) the fact that chloroplasts *can* occasionally act as symbionts does not prove that this is their evolutionary origin, and (2) the impressive resemblance of mitochondria and chloroplasts to prokaryotic cells cannot be taken as direct support for the endosymbiont hypothesis because it is equally possible that these similarities represent retained primitive characteristics. In other words, if mitochondria and chloroplasts had arisen *de novo* in a primitive prokaryote rather than by symbiosis, these two newly formed organelles would resemble the prokaryotic cell in which they were formed. During subsequent evolution the remainder of the cell might have acquired the characteristic traits of the eukaryote while the mitochondria and chloroplasts, isolated by their membranes from the rest of the cell, retained the primitive characteristics of the prokaryote in which they were originally formed.

One of the major shortcomings of the endosymbiont theory is that it does not easily explain why most mitochondrial and chloroplast components are encoded in nuclear DNA. If mitochondria and chloroplasts arose as symbionts, then they initially contained all their own genetic information. This means that either this same information was also present in the host chromosomes and could therefore be dispensed with in the endosymbionts, or else it was in some way transferred from the endosymbionts to the host cell chromosomes as evolution of the cell proceeded. But if the information were originally present in the host cell chromosomes, endosymbionts would be unnecessary because they would not be providing any new functions; on the other hand, if the information was not already present in the host chromosomes, why and how was it transferred there from the endosymbionts?

In response to these objections to the endosymbiont hypothesis, several alternative theories have been proposed based on the idea of **direct filiation**, that is, that mitochondria and chloroplasts gradually arose *de novo* rather than through sudden incorporation by symbiosis. The first detailed theory based on this idea was that of

Rudolf Raff and Henry Mahler. According to their view the ancestral protoeukaryote was an enlarged aerobic prokaryote in which the respiratory pathways of electron transfer and oxidative phosphorylation were localized in the plasma membrane. As this protoeukaryote gradually became larger, cell volume increased faster than the area of the cell surface. In order to provide for the cell's increasing energy needs, the plasma membrane with its energy-yielding respiratory pathway had to increase in size faster than the surface area of the cell. This was accomplished by invaginations of the plasma membrane, which eventually budded off to form cytoplasmic vesicles containing the respiratory pathway (Figure 14-42a). Such vesicles, which later acquired a second outer membrane for protection, provided an enclosed environment whose composition could be regulated for maximal efficiency in carrying out electron transfer and oxidative phosphorylation.

However, the compartmentalization of the respira-

FIGURE 14-42 Three versions of the direct filiation theory of mitochondrial and chloroplast evolution. (a) Model of Raff and Mahler. (b) Model of Cavalier–Smith. (c) Model of Reijnders. See text for explanations of each model.

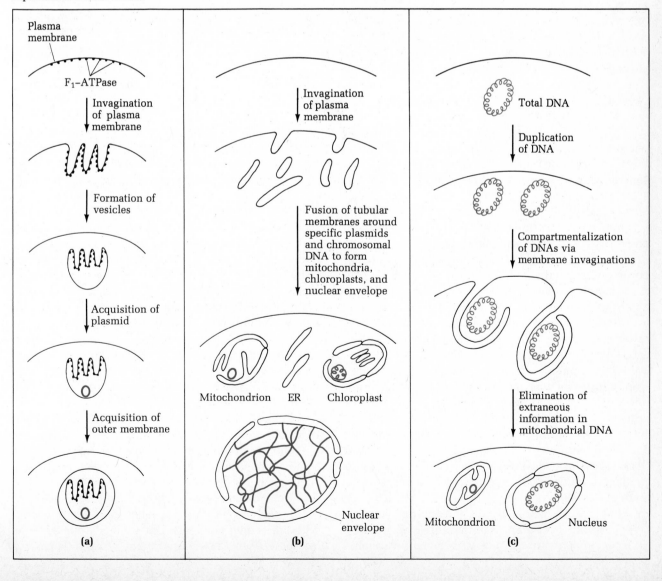

tory pathway within an enclosed mitochondria-like vesicle provided a problem for certain hydrophobic constituents of the respiratory chain, whose hydrophobic nature would prevent them from readily diffusing into the vesicle if they were synthesized in the cell sap. This limitation required that at least some respiratory proteins be synthesized within the vesicle itself. While the establishment of a genetic system within these newly forming vesicles might seem to be a formidable problem, Raff and Mahler propose that it simply involved the incorporation into the vesicles of an extrachromosomal piece of DNA, or *plasmid*, containing genes for transfer RNA, ribosomal RNA, and several messenger RNAs coding for hydrophobic proteins of the respiratory membrane.

A slightly different view of direct filiation has been proposed by T. Cavalier-Smith (see Figure 14-42b), who suggests that the critical event leading to the evolution of the eukaryotic cell was the acquisition of the ability to carry out endocytosis. Instead of viewing endocytosis as a means for ingesting prokaryotic symbionts destined to evolve into mitochondria and chloroplasts, this model considers endocytosis as part of a more general mechanism for the budding and fusing of membranes. Among the membranous organelles formed in this way could have been the endoplasmic reticulum, Golgi complex, peroxisomes, lysosomes, and the nuclear envelope. The formation of mitochondria and chloroplasts is viewed as occurring in a similar fashion, with the important addition that these particular membranous vesicles entrapped plasmids coding for respiratory or photosynthetic functions. This model is thus similar to the one of Raff and Mahler, except that by stressing the general role of membrane fusion and budding it can account for the origin of all the membranous organelles of the eukaryotic cell, not just mitochondria.

A third type of direct filiation model has been proposed by L. Reijnders (see Figure 14-42c), who claims that it is very unlikely that a single plasmid would happen to contain all the transfer RNA, ribosomal RNA, and messenger RNA genes needed for a mitochondrion or chloroplast. It is instead proposed that the first step leading to the evolution of these organelles was a duplication of the entire DNA genome of the protoeukaryotic cell, followed by a trapping of the duplicated sets of DNA in separate membrane-enclosed compartments. Eventually selection pressures eliminated redundant information, with those DNA sequences required for mitochondrial functions retained in the membrane compartment destined to become the mitochondrion, and the remainder of the DNA sequences retained in the compartment destined to become the nucleus. A similar mechanism could be envisioned for the development of chloroplasts.

One feature shared by all three direct filiation models is the formation of intracellular vesicles by primitive prokaryotic cells. This is not an unreasonable

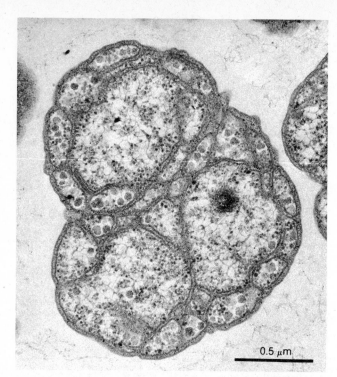

FIGURE 14-43 Electron micrograph of the bacterium *Nitrosolobus* showing an internal membrane system that compartmentalizes the cytoplasm. Courtesy of S. W. Watson.

assumption, for even today some bacteria are capable of producing complex internal membrane systems (Figure 14-43). It is also worth emphasizing that once the ancestral mitochondria and chloroplasts acquired their own membrane-enclosed DNA, these genetic systems would be subject to different selection pressures than the genetic system of the rest of the cell. The DNA, messenger RNA, and protein-synthesizing machinery of mitochondria and chloroplasts are all present within the same compartment, analogous to the situation in prokaryotic cells. Nuclear DNA and its messenger RNA, on the other hand, are separated from the protein synthetic machinery of the cytoplasm by the nuclear envelope. Perhaps it is not surprising, therefore, that the genetic systems of mitochondria and chloroplasts have continued to resemble those of prokaryotic cells, while the nucleocytoplasmic informational system has undergone considerable change in adapting to the separation created by the presence of the nuclear envelope. Hence the direct filiation theories are as compatible with the observed similarities between mitochondria, chloroplasts, and prokaryotic cells as is the endosymbiont theory.

Though the notion of direct filiation overcomes some of the difficulties with the endosymbiont theory, such as the problem of how genes coding for mitochondrial and chloroplast functions have come to reside within the nucleus, the concept of direct filiation is not without its own problems. One difficulty concerns the question of why mitochondria and eukaryotic cell sap contain different forms of the enzyme superoxide dismutase, with the mitochondrial enzyme resembling the one occurring in prokaryotic cells. The endosymbiont theory is readily compatible with such findings, but if mitochondria did not arise by symbiosis, it is necessary to postulate that the protoeukaryote sequestered its prokaryotic superoxide dismutase within vesicles destined to become mitochondria, and then evolved a structurally unrelated type of superoxide dismutase to take its place in the cell sap.

Hence none of the theories of mitochondrial and chloroplast evolution is without its drawbacks. It may even be erroneous to assume that the same theory of evolution applies to both mitochondria and chloroplasts. The evidence for the present-day occurrence of symbiotic relationships between photosynthetic algae and animal cells suggests the possibility that chloroplasts originated by symbiosis, while mitochondria could have evolved by direct filiation. Even such a qualified conclusion needs to be viewed with extreme caution, however, for we are contemplating events that occurred a billion or more years ago. All evidence relating to such events must of necessity be circumstantial, and so it is quite likely that it will never be known with certainty exactly what happened.

Cellular Evolution and Organelle Biogenesis

The commonly held view that all cells can be classified into two major groups, prokaryotes and eukaryotes, has been a fundamental tenant of biology for many years. This dichotomy has fostered the assumption that the relatively "simple" prokaryotes served as progenitors for the development of eukaryotic cells, which are structurally and genetically more complex. In recent years this view has been challenged. The cell retains evidence of its evolutionary history in the amino acid sequence of its proteins and the nucleotide sequences of its DNA and RNA. The ability to "read" the evolutionary details of the cells' history has awaited appropriate methodological developments, and now we are beginning to gain a fuller understanding of the origins of the cells that make up our world today.

The work of Carl Woese, C. Fred Fox, and their collaborators has been instrumental in modifying our views of cellular evolution. Based on an analysis of the nucleotide sequences of 16S (prokaryotic) and 18S (eukaryotic) cytoplasmic ribosomal RNAs, they suggest that living things comprise three distinct evolutionary lineages that diverged at some point from a common ancestor. These three classes are termed the **archaebacterial 16S, eubacterial 16S,** and **nuclear-cytoplasmic 18S.** It has now been clearly documented that contemporary cells falling into these three categories are also distinguishable by a variety of other molecular properties, including distinctive transfer RNA modification patterns, RNA polymerases, and antibiotic sensitivities.

The second conclusion derived from the data of Woese and Fox is that the nuclear-cytoplasmic 18S class of cells, which corresponds to present-day nucleated eukaryotic cells, originally diverged from the eubacterial lineage during an early stage in cellular evolution. Divergence of this sort may have occurred at a time when genetic organization and the mechanisms and machinery of gene expression were undergoing rapid evolution toward increased efficiency and accuracy. If this be true, then it would support the remarkable differences in genomic organization seen in prokaryotic and eukaryotic cells. This information, coupled with the many other differences in the transcriptional and translational machinery, can be most easily fit into a model in which the nuclear-cytoplasmic components of eukaryotic cells diverged from eubacteria before final solutions to the problem of effective genetic organization and expression had been found.

We need to know much more about the relationship between the three cell lineages because this information will impinge upon our views of the evolutionary development of mitochondria and chloroplasts. Nonetheless our perception of evolutionary matters may always be based on a lack of hard facts, so it is likely that we will never really know the exact path that led to the appearance of these essential organelles in eukaryotic cells.

SUMMARY

Mitochondria and chloroplasts contain genetic systems that play a significant role in regulating the development of these two organelles. The initial realization that genes may be located in organelles other than the nucleus can be traced back to the turn of the century, when several disorders in chloroplast pigmentation were discovered to be inherited solely from the female parent. Because the female parent usually contributes most of the cytoplasm to the offspring, it was concluded that the genes determining these traits must be located in the cytoplasm, presumably in the chloroplast itself. Many years later defects in mitochondrial structure and function, exemplified by the *petite* condition in yeast and the *poky* trait in *Neurospora*, were also found to be cytoplasmically inherited.

In spite of this genetic evidence for the existence of genes in mitochondria and chloroplasts, cytoplasmic inheritance was not given serious attention until the presence of mitochondrial and chloroplast DNA was demonstrated by electron microscopy and isodensity centrifugation. Mitochondrial and chloroplast DNA were found to be distinguishable from nuclear DNA by their small size, base composition, circularity, absence of histones, rapid renaturation kinetics, and (in the case of plants) the absence of 5-methylcytosine. Mitochondrial DNA varies in length from 5 μm (about 16.5 kilobases) in animal cells to 10–15 μm in unicellular eukaryotes and 30–150 μm in higher plants. Chloroplast DNA is even larger, averaging 40–60 μm in length. Genetic proof that mitochondrial and chloroplast DNA contain genes governing cytoplasmically inherited traits has been provided in several instances, including *petite* yeast, in which the occurrence of mitochondrial defects is correlated with an alteration in the buoyant density of mitochondrial DNA, and in ultraviolet-irradiated *Euglena*, where cytoplasmically inherited defects in chloroplast development are associated with the disappearance of chloroplast DNA. The behavior of organellar DNA may provide at least a partial explanation for the phenomenon of maternal inheritance, since in *Chlamydomonas* the paternal chloroplast DNA is selectively destroyed in the zygote, leaving only the maternal chloroplast DNA in the offspring.

Density-labeling experiments indicate that mitochondrial and chloroplast DNA are replicated by a semiconservative mechanism. Electron micrographs have revealed that replication is accompanied by the formation of displacement loops that expand around the DNA circle. DNA sequencing and DNA/RNA hybridization experiments have shown that organellar DNA is transcribed into ribosomal, transfer, and messenger RNAs. The ribosomes constructed from the ribosomal RNAs are generally smaller than typical 80S eukaryotic ribosomes; mitochondrial ribosomes of animal cells sediment at about 55–60S, mitochondrial ribosomes of unicellular eukaryotes and plants at 70–80S, and chloroplast ribosomes at 70S. The transfer RNAs produced by mitochondria and chloroplasts are also distinct from those of the eukaryotic cell sap. Chloroplasts appear to manufacture a complete set of transfer RNAs, but in mitochondria only 22–24 different transfer RNAs have been detected. The ability of certain transfer RNAs to recognize four codons rather than three reduces the need for the full set of 32 transfer RNAs required for cytoplasmic protein synthesis. One unique property of organellar transfer RNAs is the presence of a transfer RNA for formylmethionine, which plays a role in the initiation of protein synthesis just as it does in prokaryotic cells.

In addition to ribosomes, transfer RNA, and messenger RNA, mitochondria and chloroplasts contain all the other components required for protein synthesis, including aminoacyl-tRNA synthetases, transformylase, and initiation, elongation, and termination factors. Identification of the products of organellar protein synthesis has been aided by the fact that chloramphenicol selectively inhibits protein synthesis by mitochondrial and chloroplast ribosomes, while cycloheximide selectively inhibits protein synthesis by eukaryotic cytoplasmic ribosomes. The major products of mitochondrial protein synthesis include four subunits of the oligomycin-sensitive ATPase complex, three subunits of the cytochrome oxidase complex, and cytochrome *b*. These mitochondrially synthesized polypeptides are

generally hydrophobic, and may function to anchor their respective complexes to the inner membrane. The major identified products of chloroplast protein synthesis include the large subunit of RuBP carboxylase, five subunits of the thylakoid membrane ATPase complex, three subunits of the cytochrome *b-f* complex, a 32,000 dalton protein of the thylakoid membrane, thylakoid membrane polypeptides associated with photosystems I and II, and possibly two ribosomal proteins. It is interesting that animal mitochondria, yeast mitochondria, and chloroplasts all appear to synthesize about the same number of polypeptides (8–12), in spite of the fact that yeast mitochondria and plant chloroplasts contain much more DNA. At least in the case of yeast mitochondria, this extra DNA is accounted for by the presence of AT-rich sequences interspersed among the genes coding for transfer, ribosomal, and messenger RNA molecules.

The vast majority of mitochondrial and chloroplast proteins are encoded in the nucleus and synthesized on cytoplasmic ribosomes. Many, but not all, of these polypeptides contain an amino-terminal extension that is removed after entry into the mitochondrion or chloroplast. Receptor proteins on the outer membrane are thought to recognize polypeptides to be translocated, and the appearance in electron micrographs of contact points connecting the outer and inner membranes of mitochondria and chloroplasts suggests that entry into organelle membranes or compartments may be facilitated by such sites. Products of cytoplasmic, mitochondrial and chloroplast protein synthesis control each other's rates of synthesis, thereby helping to ensure that the relative amounts of the various components are properly balanced.

Of the three theories proposed to account for the biogenesis of mitochondria (growth and division, *de novo* formation, and assembly from nonmitochondrial membranes), microscopic and biochemical evidence provides the strongest support for the notion of growth and division. Not only have dividing mitochondria been observed microscopically, but experiments monitoring changes in mitochondrial labeling and density argue that all mitochondria arise by growth and division of preexisting mitochondria. Mitochondrial development is influenced by both the rate of cell growth and by ambient oxygen levels. In the absence of oxygen, for example, yeast mitochondria are replaced by small double-membrane vesicles, called promitochondria, which lack cristae and are incapable of carrying out respiration. When oxygen is reintroduced, these promitochondria develop cristae and the components required for carrying out the Krebs cycle, electron transfer, and oxidative phosphorylation.

Chloroplasts also appear to arise by growth and division. Undifferentiated meristematic cells contain small double-membrane vesicles called proplastids that are capable of developing into storage vesicles for starch (amyloplasts), protein (proteinoplasts), and lipids (lipoplasts), as well as into chloroplasts. In seedlings exposed to light, the proplastids enlarge and their inner membranes invaginate into membranous sheets that project into the stroma and eventually differentiate into thylakoids. If light is not present, the invaginated membranous sheets instead contract into an ordered crystalline array of interconnected tubules, the prolamellar body, to form a nonphotosynthetic plastid termed an etioplast. Exposure of etioplasts to light results in the conversion of the prolamellar body to thylakoids, accompanied by the synthesis of chlorophyll and other components of the photosynthetic pathway. Activation of *de novo* chlorophyll synthesis is mediated by the absorption of red light by phytochrome; in its light-activated form phytochrome induces the synthesis of the enzyme ALA synthetase, which catalyzes the formation of 5-aminolevulinate, the first major product in the chlorophyll biosynthetic pathway.

Two major theories have been proposed to explain the evolutionary origins of mitochondria and chloroplasts. The endosymbiont theory views mitochondria and chloroplasts as having arisen from aerobic bacteria and photosynthetic prokaryotes that were ingested by primitive cells a billion or more years ago. This theory is supported by the observation that mitochondria and chloroplasts resemble prokaryotic cells in a large number of ways, including the presence of circular DNA, absence of histones, presence of ribosomes smaller than 80S eukaryotic ribosomes, similar ribosomal RNAs, messenger RNAs with short or nonexistent poly(A) tails, initiation of protein synthesis with formylmethionine, sensitivity of protein synthesis to chloramphenicol but not cycloheximide, sensitivity of RNA synthesis to rifampicin, similar initiation and elongation factors, and similar forms of superoxide dismutase. Further support for the endosymbiont theory is provided by the existence of many present-day examples of symbiotic relationships between unicellular photosynthetic organisms and animal cells.

Opponents of the endosymbiont theory argue that the impressive resemblance of mitochondria and chloroplasts to prokaryotic cells does not provide unambiguous support for this theory because it is equally possible that such similarities represent no more than retained primitive characteristics. They instead propose that mitochondria and chloroplasts arose directly in a primitive prokaryote, and have continued to resemble the primitive cell in which they first arose. This idea of direct filiation has been introduced to overcome one of the major shortcomings of the endosymbiont theory, namely, the issue of how and why mitochondrial and chloroplast genes have come to be located in the nucleus. Though there are three variations of the direct filiation theory, all have in common the idea that mitochondria and chloroplasts arose by invaginations of the plasma membrane of a primitive prokaryote that en-

trapped within them DNA coding for mitochondrial and chloroplast functions. At present it appears as if the endosymbiont and direct filiation theories both have advantages and shortcomings, and since we are discussing events that occurred a billion or more years ago, it is likely that it will never be known with certainty which, if either, of these theories is correct.

SUGGESTED READINGS

Books

Attardi, G., P. Borst, and P. P. Slonimski (1982). *Mitochondrial Genes,* Cold Spring Harbor Laboratory, Cold Spring Harbor, NY.
Ellis, R. J., ed. (1984). *Chloroplast Biogenesis,* Cambridge University Press, New York.
Margulis, L. (1981). *Symbiosis and Cell Evolution,* Freeman, San Francisco.
Schiff, J. A. (1982). *On the Origins of Chloroplasts,* Elsevier Science Publishing, New York.
Tzagoloff, A. (1982). *Mitochondria,* Plenum Press, New York.

Articles

Attardi, G. (1986). The elucidation of the human mitochondrial genome: a historical perspective, *BioEssays* 5:34–39.
Bogorad, L. (1981). Chloroplasts, *J. Cell Biol.* 91:256s–270s.
Borst, P., and L. A. Grivell (1984). Organelle DNA, *Trends Biochem. Sci.* 9:128–130.
Clayton, D. A. (1984). Transcription of the mammalian mitochondrial genome, *Ann. Rev. Biochem.* 53:573–594.
Ellis, R. J. (1981). Chloroplast proteins: synthesis, transport and assembly, *Ann. Rev. Plant Physiol.* 32:111–137.
Ernster, L., and G. Schatz (1981). Mitochondria: a historical review, *J. Cell Biol.* 91:227s–255s.
Gray, M. W., and W. F. Doolittle (1982). Has the endosymbiont hypothesis been proven? *Microbiol. Rev.* 46:1–42.
Grivell, L. A. (1983). Mitochondrial DNA, *Sci. Amer.* 248 (Mar.):78–89.
Hurt, E. C., and A. P. G. M. van Loon (1986). How proteins find mitochondria and intramitochondrial compartments, *Trends Biochem. Sci.* 11:204–207.
Palmer, J. D. (1985). Comparative organization of chloroplast genomes, *Ann. Rev. Genet.* 19:325–354.
Rhoades, M. M. (1943). Genetic induction of an inherited cytoplasmic difference, *Proc. Natl. Acad. Sci. USA* 29:327–329.
Wickner, W. T., and H. F. Lodish (1985). Multiple mechanisms of protein insertion into and across membranes, *Science* 230:400–407.
Woese, C. R. (1981). Archaebacteria, *Sci. Amer.* 244 (June):96–122.
Yang, D., Y. Oyaizu, H. Oyaizu, G. J. Olsen, and C. R. Woese (1985). Mitochondrial origins, *Proc. Natl. Acad. Sci. USA* 82:4443–4447.

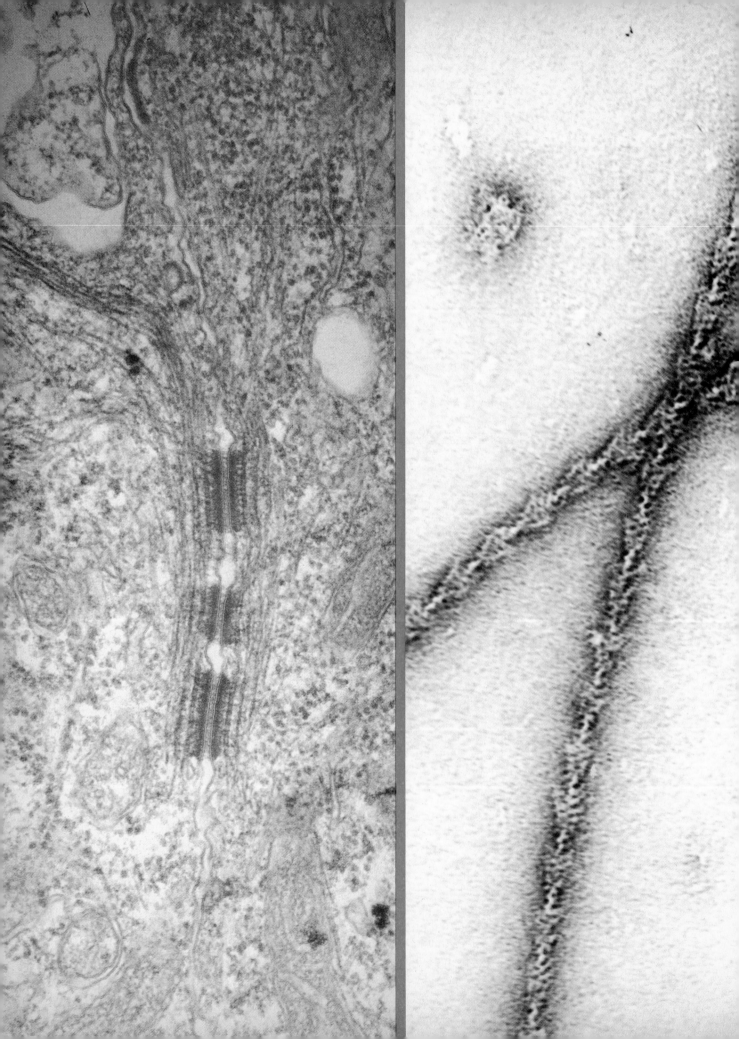

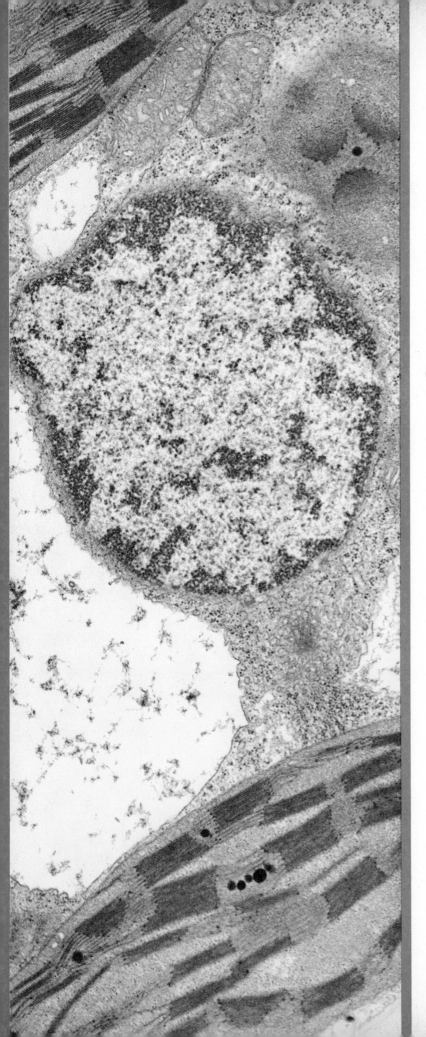

3
Formation of Specialized Cells

Germ Cells, Early Development, and Cell Differentiation

One of the most awe inspiring of all biological phenomena is triggered every time an egg is fertilized. Through a complex series of divisions, differentiations, and interactions, this single cell gives rise to billions of new cells of differing types, organized in precise patterns to form the tissues and organs of the adult organism. The great mystery of life is how each egg cell, be it a frog egg, a sea urchin egg, a human egg, or any other type of egg, always manages to generate an organism of the proper type. Long ago it was believed that every fertilized egg contains within it a complete miniature version of the adult organism, and that this miniature organism simply increases in size as maturation proceeds. But embryonic development is now known to be considerably more complex than is implied by this naive picture. There is no visible sign of the mature organism within the egg; instead, each of the myriad of intricate steps of development is programmed within the egg cell in an invisible and as yet largely undeciphered form.

The key to understanding development lies in unraveling the mechanism of cell differentiation, the process by which cells gradually acquire new properties and become specialized for particular tasks. Not only is cell differentiation involved in the gradual transition from a fertilized egg cell to specialized cell types such as nerve, muscle, liver, blood, and bone, but it also plays a role in the formation of the egg and sperm cells whose union initiates the entire developmental sequence in the first place. The great question posed by the process of cell differentiation concerns the issue of how cells inheriting the same genetic information become so different from one another. How are significant differences in biochemical activities, morphological appearance, and intercellular interactions established in cells possessing the same information encoded in their DNA? Though a comprehensive answer to this question is still far from our grasp, recent discoveries are beginning to shed some light on the molecular mechanisms involved. It is in the context of this general question that this chapter will discuss the major features of eukaryotic development: gametogenesis, fertilization, cell differentiation, cell-cell interactions, and aging.

OVERVIEW OF DEVELOPMENT

Before proceeding to a detailed consideration of the events involved, it will be useful to summarize the major stages associated with the development of multicellular organisms. It is beyond the scope of this chapter to describe the numerous variations in developmental patterns that occur within the animal and plant kingdoms, so we shall instead direct our attention to the general principles of development that appear to be widely applicable. Although much of this beginning overview will be focused on animal development, many of the same principles apply to plants as well; a few of the more unique features of plant development will be described later in the chapter.

Sexual reproduction involves the fusion of two haploid cells, the male and female **gametes,** to form a diploid **zygote.** The male gamete, or **sperm,** is a small cell generally propelled by one or more flagella. The female gamete (**egg** or **ovum**), on the other hand, is non-motile and much larger than the sperm. In animals where embryonic development occurs outside the reproductive tract, this larger size permits the egg to store all the nutrients required for sustaining embryonic development. Despite their differing sizes, the sperm and egg transmit equal amounts of nuclear DNA to the zygote; however, the larger cytoplasmic volume of the egg allows it to contribute more cytoplasmic DNA (mainly from mitochondria, but also from chloroplasts in the case of plant ova).

During **fertilization** the egg cell is penetrated by a sperm, thereby reestablishing the diploid chromosome number. Two rapid changes occur in response to fertilization: A **fertilization membrane** develops around the

FIGURE 15-1 Summary of the major stages in the development of a frog embryo. Gastrulation (stages F through I) is illustrated in both surface view and cross section to illustrate how cell movement produces a three-layered embryo. Those cells on the surface of the blastula destined to migrate through the blastopore are darkly shaded to facilitate the tracing of their movements.

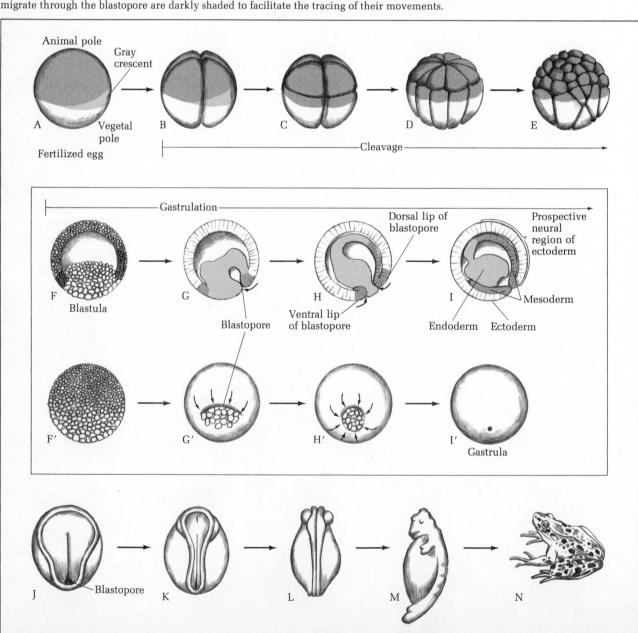

egg to prevent the entry of additional sperm, and the biosynthetic machinery required for DNA synthesis is activated. Soon thereafter the fertilized egg enters a phase of rapid cell division known as **cleavage.** Because no growth occurs during this stage, cleavage causes the massive egg cell to be partitioned into smaller and smaller cells, ultimately producing a spherical mass of small cells termed a **blastula.**

As soon as cell division ceases, the cells of the blastula embark upon a series of elaborate movements designed to alter their spatial relationships with each other (Figures 15-1 and 15-2). Though the details of this

FIGURE 15-2 Stages in the development of a frog egg as viewed by scanning electron microscopy. (1) An unfertilized egg. (2) Two-cell stage. (3) Four-cell stage. (4) Eight-cell stage. (5) Sixteen-cell stage. (6) The 32- to 64-cell stage. (7) Late blastula. (8) Gastrula with blastopore. (9) The yolk-plug stage showing the yolk plug surrounded by the blastopore. (10) An early neurula stage. (11) Late neurula stage showing the folding that gives rise to the neural tube. (12) Tailbud stage. Courtesy of R. G. Kessel.

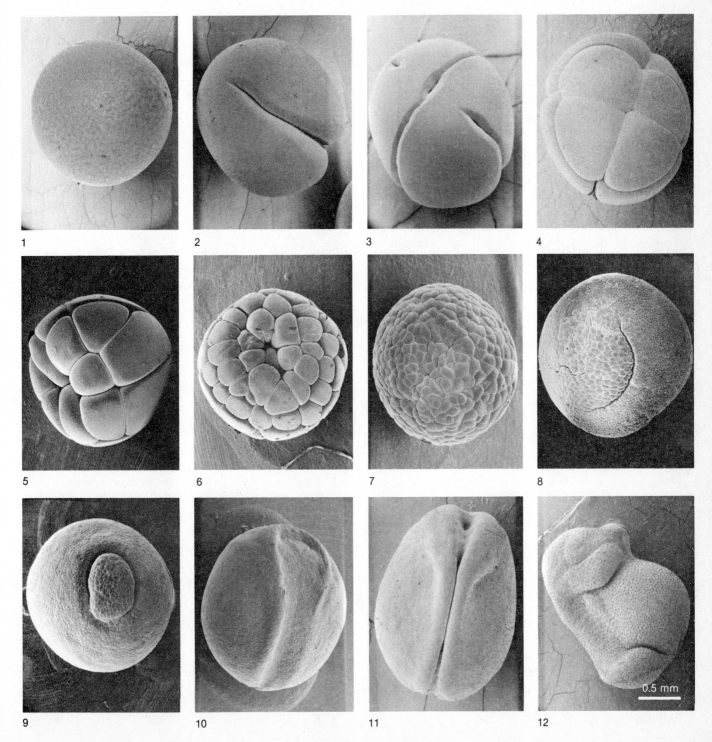

process vary among species, a universal feature is the migration of cells through a cleft or opening in the blastula surface, known as the **blastopore,** into the blastula interior. The final result is a **gastrula** composed of three embryonic cell layers termed the **ectoderm** (outer layer), **mesoderm** (middle layer), and **endoderm** (inner layer). Ectoderm and endoderm are destined to form sheets of cells called **epithelia** that cover the exposed surfaces of the organism; ectoderm produces epithelia covering external surfaces (skin and related structures), while endoderm differentiates into epithelia covering internal surfaces (gastrointestinal tract and related structures). In addition, ectoderm gives rise to neural tissue. Mesoderm, on the other hand, gives rise to a diffuse spongework of **mesenchyme** cells that ultimately differentiate into supportive tissues such as muscle, cartilage, bone, blood, and connective tissue.

The new interrelationships established between cells during formation of the gastrula trigger a combination of cell divisions, cell differentiations, and cell movements that ultimately lead to the formation of tissues and organs. In many regions of the developing embryo the presence of one cell type causes neighboring cells to differentiate in a particular way. As a result of this **induction** process, cells that at early stages of development are potentially capable of developing into almost any kind of cell have their developmental fates **determined.**

Cell differentiation does not cease at the end of embryonic development. Throughout an organism's life certain cells, such as those present in the blood, lining the gastrointestinal tract, and covering the body, must continually be replenished by division and differentiation of appropriate precursor or **stem** cells. As the organism grows older, the more subtle and gradual process of aging occurs. Though the mechanisms involved are poorly understood, it appears as if the gradual loss of function that occurs during aging may be to some extent programmed into cells as part of the normal process of development.

GAMETOGENESIS

With the preceding overview in mind, we can now begin to discuss the individual stages of development in more detail. The process of gamete formation, or **gametogenesis,** takes place in the ovary and testis, where meiotic division and differentiation of initial precursor cells known as **primordial germ cells** lead to the formation of egg and sperm cells, respectively (Figure 15-3). The details of meiosis were described in Chapter 13, and will therefore not be elaborated upon again. Instead we shall focus our attention on the morphological and biochemical changes that accompany the transformation of primordial germ cells into functionally specialized egg and sperm cells.

Oogenesis

The process of female gamete formation, or **oogenesis,** begins with a **proliferation** phase in which the primordial germ cells (**oogonia**) increase their numbers by mitotic division. In some species, such as humans, this proliferation phase is completed by the end of embryonic development, meaning that all germ cells capable of differentiating into ova for the organism's entire lifetime are present by the time of birth. Amphibians, on the other hand, continue to produce new germ cells by mitotic division of oogonia throughout most of their adult lives.

After mitotic proliferation the oogonia enter into meiosis, at which point they are renamed **oocytes.** During meiosis, genetic recombination occurs and the chromosome number is reduced to the haploid level. In nonmammalian oocytes, a dramatic increase in cell size also occurs during this period. To provide time for this growth to occur, meiosis is delayed at prophase I (late pachytene or early diplotene). The time required for oocyte growth varies widely among species, from a few days or weeks to several years. The extent of the growth process also varies significantly. In amphibian oocytes the growth phase produces a thousandfold increase in cell mass, accompanied by an increase in cell diameter from 50 to 100 μm to a millimeter or more. The increase in cell size is even greater in birds, creating cells as large as the chicken egg. Mammalian oocytes, on the other hand, grow relatively little during oogenesis. In oocytes that do undergo an extensive growth phase, a series of unique metabolic and morphological changes occurs. These changes, which have been best characterized for amphibian oocytes, are summarized below.

1. RNA synthesis increases dramatically during the growth phase of amphibian oogenesis. Production of ribosomal RNA and the formation of ribosomes is especially conspicuous, though the majority of the newly made ribosomes do not engage in protein synthesis at this time. Messenger RNA is also produced, but most of it is stored in an inactive state for use after fertilization. The increase in RNA synthesis that occurs at this stage is associated with an uncoiling of the prophase chromosomes, producing the characteristic lampbrush chromosome configuration (page 526).

2. The large demand for ribosomal RNA places a severe strain on ribosomal gene transcription because the nucleus is being asked to supply ribosomal RNA to a cytoplasm that has a greater volume than normal. To accommodate this increased demand the ribosomal RNA genes are selectively **amplified,** causing an increase of a thousandfold or more in the number of ribosomal genes available to serve as templates for transcription. Amplification of the ribosomal RNA genes leads to the formation of additional nucleoli, which may number in the thousands in some amphibia.

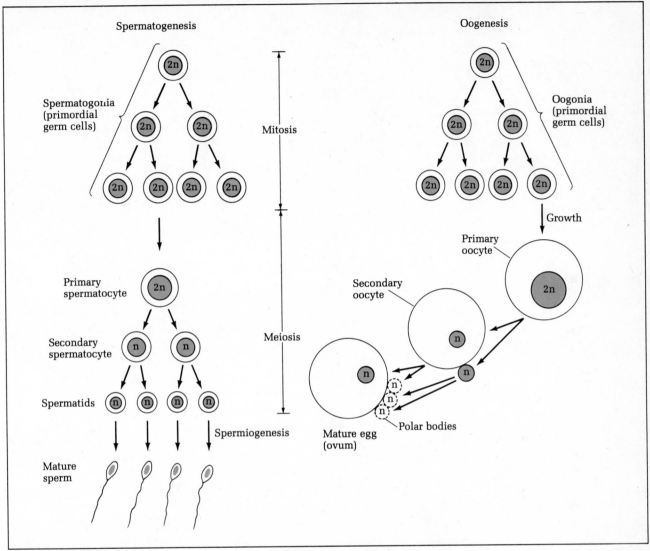

FIGURE 15-3 Comparison of male and female gametogenesis in animals. Note that in oogenesis, only one of the four haploid products of meiosis yields a functional gamete.

3. The above metabolic and morphological changes are accompanied by a dramatic increase in the size of the nucleus. The enlarged nucleus produced by this process is often referred to as a **germinal vesicle.**

4. A massive enlargement of the amphibian oocyte occurs, caused primarily by the accumulation of vast amounts of protein, lipid, and carbohydrate destined for later use by the embryo. Though some of this material is manufactured by the oocyte itself, much of it is derived from external sources. In organisms where the egg provides virtually all the nutrients for early embryonic development, such as amphibians and birds, the liver plays a critical role in manufacturing these nutrients. In response to stimulation by female sex hormones the liver manufactures large amounts of **vitellogenin,** a complex composed of the phosphoprotein **phosvitin** and the lipoprotein **lipovitellin.** Vitellogenin is secreted into the bloodstream and passes to the ovary, where it

enters primary oocytes by the process of pinocytosis. The individual pinocytic vesicles formed by this process eventually fuse together, forming nutrient-containing vesicles known as **yolk platelets** (Figure 15-4). It should be emphasized that the term *yolk* does not refer to a single molecular species, but to the varying mixtures of proteins, lipids, and polysaccharides stored in such vesicles. In addition to stored nutrients, yolk platelets contain hydrolytic enzymes such as phosphatases and proteases. These enzymes are activated upon fertilization, digesting the yolk into nutrients that can be used by the developing embryo.

The liver is not the only source of nutrients stored in the developing oocyte; additional metabolites are provided by accessory cells called **follicle cells** that surround most animal oocytes (Figure 15-5). Protruding from the plasma membranes of both oocyte and follicle cells are numerous microvilli that function to increase

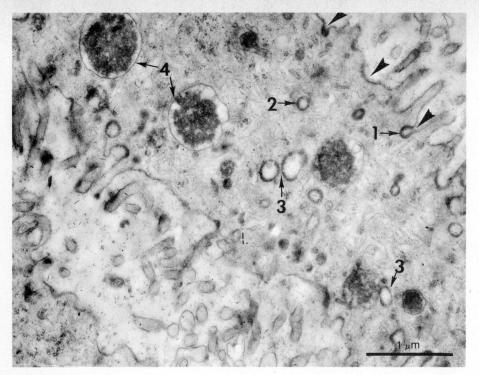

FIGURE 15-4 Electron micrograph of the periphery of a small oocyte from the milkweed bug. The arrowheads in the upper right-hand corner point to regions where pinocytosis is occurring at the cell surface. The numbered arrows point to sequential stages in which the pinocytic vesicles invaginate, bud off from the cell surface, and fuse together to form yolk platelets. Courtesy of R. G. Kessel.

the area of contact between the two cell types, thereby facilitating the transport of nutrient materials from the follicle cells to the oocyte. The nutritional role of follicle cells is not limited to oocytes that produce yolk proteins. Mammalian oocytes, for example, do not produce and accumulate yolk proteins because the developing embryo receives its nourishment directly from the mother's bloodstream and hence has little need for food reserves of its own. Nonetheless most of the metabolites utilized by the developing oocyte are first taken up and metabolized in follicle cells before introduction into the oocyte through gap junctions. Oocyte growth depends on this nutritional role of follicle cells.

Another type of accessory cell, termed the **nurse cell,** supports oocyte growth in certain invertebrates such as insects, annelids, and mollusks. Unlike follicle cells, these nurse cells are derived from the same germ cell lineage as the oocyte itself. For example, in the fruit fly a sequential series of four mitotic divisions of an oogonium yields 1 oocyte and 15 nurse cells. Nurse cells are connected to the developing oocyte by cytoplasmic bridges through which a variety of macromolecules are passed. The nutritional contribution of nurse cells to oocyte growth is generally greater than that made by follicle cells; in some cases, the entire nurse cell may even be engulfed by the oocyte.

Not all oocytes depend on other tissues to manufacture their stored nutrients. In some amphibian and bird oocytes, yolk proteins are synthesized within the rough endoplasmic reticulum and are either packaged directly into yolk platelets or are transported to the Golgi

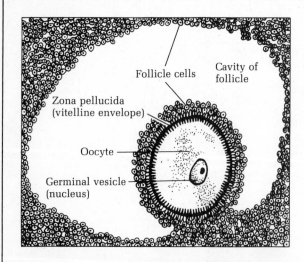

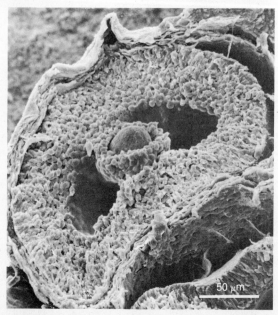

FIGURE 15-5 Diagram and accompanying scanning electron micrograph illustrating the relationship between an oocyte and the surrounding follicle cells in a typical mammalian ovary. Micrograph courtesy of P. Bagavandoss.

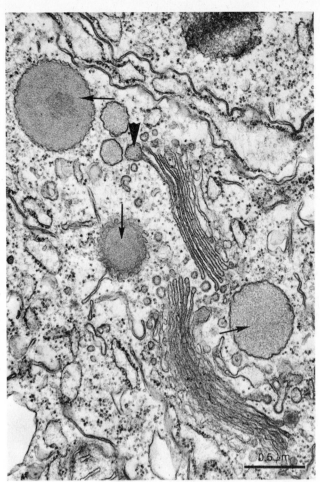

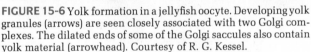

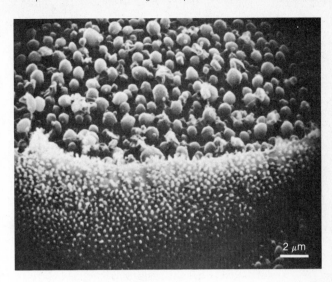

FIGURE 15-6 Yolk formation in a jellyfish oocyte. Developing yolk granules (arrows) are seen closely associated with two Golgi complexes. The dilated ends of some of the Golgi saccules also contain yolk material (arrowhead). Courtesy of R. G. Kessel.

region for complexing with carbohydrates prior to assembly into platelets (Figure 15-6). Yolk can also be produced between the inner and outer membranes of mitochondria, followed by disintegration of the mitochondrial inner membrane and transformation of the mitochondrion into a yolk platelet. It is not clear whether the mitochondrion is actively involved in the synthesis of yolk materials or is simply accumulating substances manufactured elsewhere.

In addition to yolk platelets, the cytoplasm of amphibian oocytes typically contains **lipid droplets** with no enclosing membrane, as well as membrane-enclosed **pigment granules** and **cortical granules** (Figure 15-7).

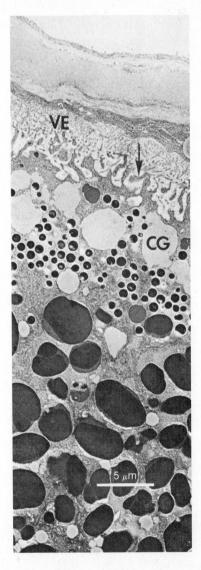

FIGURE 15-7 Structural features of the cell surface of oocytes. *(Top)* The surface of an unfertilized sea urchin egg observed by scanning electron microscopy. The vitelline envelope and plasma membrane are broken away from the top half of the egg, revealing the cortical granules that lie just beneath the cell surface. Courtesy of G. Schatten. *(Bottom)* Thin-section electron micrograph of the surface region of a nearly mature amphibian oocyte, showing the vitelline envelope (VE) and cortical granules (CG). The darker granules are pigment granules. The undulating plasma membrane (arrow) separates the cytoplasm from the vitelline envelope. Courtesy of J. N. Dumont.

The cortical granules, which are localized just beneath the plasma membrane, contain proteolytic enzymes destined to play an important role in the early stages of fertilization.

5. Dramatic changes in the cell surface occur during the growth phase of oogenesis. Most striking is the formation of an external coat of polysaccharide-rich material known as the **vitelline envelope** in nonmammalian oocytes or the **zona pellucida** in mammalian oocytes (see Figure 15-5). This external coat, which protects the egg from chemical and physical injury, may remain intact well past the time of fertilization. When the external envelope is produced by the oocyte itself, it is referred to as a **primary coat.** When formed by the follicle cells it is called a **secondary coat,** and when produced by the oviduct or other maternal tissue, it is termed a **tertiary coat.** The appearance of these various layers differs significantly among species. Especially variable is the tertiary layer, which ranges from the jelly coat of amphibian eggs to the egg white and shell of hens' eggs. Impervious shells of this latter type must of course be added to the egg *after* fertilization, lest they prevent entry of sperm.

6. In nonmammalian oocytes the biochemical and morphological changes that occur during the growth phase of oogenesis impart an obvious asymmetry or *polarity* to the developing oocyte (Figure 15-8). One end of the cell, designated the **vegetal pole,** contains most of the yolk platelets and stored nutrients. The opposite or **animal pole** of the cell contains little in the way of stored nutrients, but is enriched in ribosomes, mitochondria, and pigment granules (when present). We shall see later that this polarity plays an important role in early embryonic development.

To summarize the events described above, the growth phase of oogenesis is a time when the developing

FIGURE 15-8 Schematic diagram of an amphibian oocyte illustrating the polarity imparted to this cell by the asymmetric distribution of yolk platelets, mitochondria, and ribosomes.

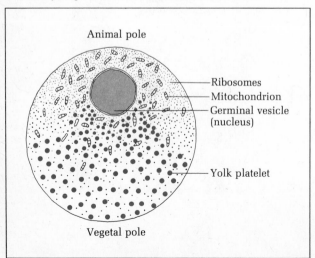

Animal pole

Ribosomes
Mitochondrion
Germinal vesicle (nucleus)

Yolk platelet

Vegetal pole

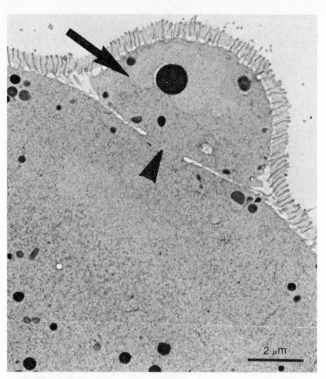

FIGURE 15-9 Electron micrograph of a polar body (arrow) forming at the surface of an egg cell. A small region of cytoplasmic continuity (arrowhead) still exists between the developing polar body and the egg cell itself. Courtesy of E. Anderson.

oocyte is preparing for the future. In nonmammalian oocytes this requires the building up of large reserves of nutrients, organelles, and macromolecules, since in such organisms the embryo develops outside the mother and hence must be self-sufficient.

After the growth phase is completed, the oocyte is ready to complete meiosis. Rather than occurring spontaneously, this so-called **maturation phase** is usually triggered by appropriate hormonal signals. In response, the germinal vesicle breaks down, a spindle forms near the plasma membrane, and the first meiotic division, which had been delayed at prophase to permit the growth phase to occur, is completed. The cytoplasmic cleavage associated with this first meiotic division is highly asymmetrical; instead of dividing the oocyte into two equal cells, the cleavage furrow pinches off one set of chromosomes into a small cytoplasmic protrusion, the **polar body,** which forms near the animal pole (Figure 15-9). The resulting cell is called a **secondary oocyte** to distinguish it from the **primary oocytes** that have not yet completed the first meiotic division (see Figure 15-3). The secondary oocyte then goes through a second meiotic division, producing another polar body and a mature **egg;** at this time the first polar body may either divide, remain quiescent, or disintegrate. In some species this maturation sequence proceeds rapidly to completion, while in others it halts at an intermediate stage (e.g., after the first polar body has been extruded) and is not completed until after fertilization. The polar bodies appear to play no future role in development and eventually disappear. The advantage of having only one of the four haploid products of meiosis produce a functional gamete is that the cytoplasm that would otherwise have been distributed among four cells is concentrated into one, maximizing the content of stored nutrients in each egg.

The mature egg produced by meiosis is a highly differentiated cell specialized for the task of producing a new organism in much the same sense as a muscle cell is specialized to contract, or a red blood cell is specialized to transport oxygen. This inherent specialization of the egg is vividly demonstrated by the observation that even in the absence of fertilization by a sperm cell, nonmammalian egg cells can be artificially stimulated to develop into a complete embryo (a phenomenon known as **parthenogenesis**). Hence everything needed for programming the early stages of development must already be present in the egg.

Spermatogenesis

The process by which male gametes are formed, called **spermatogenesis,** differs in several ways from oogenesis. To begin with the male primordial germ cells, or

FIGURE 15-10 Relationship between Sertoli cells and developing sperm cells in the seminiferous tubule of the mammalian testis. Invaginations of the Sertoli cell cytoplasm (color) create compartments in which the differentiating sperm cells develop. Mature sperm are released from the Sertoli cell into the lumen of the tubule.

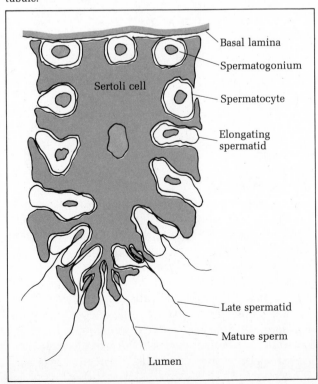

spermatogonia, continually proliferate by mitotic division because of the need for the constant production of large numbers of new sperm cells. Some of the cells resulting from these mitotic divisions enter into meiosis. Though a small amount of growth may be associated with meiosis, no striking morphological changes occur at this stage as they do in the comparable phase of oogenesis. In mammals the developing sperm cells are intimately associated with nongerminal **Sertoli cells** that provide physical and perhaps nutritional support for the developing sperm cells (Figure 15-10). The recent discovery of gap junctions connecting Sertoli cells with developing sperm suggests that direct communication between these two cell types may be involved in the regulation of spermatogenesis.

Meiosis yields four haploid cells of equal size, termed **spermatids,** which remain connected to each other by cytoplasmic bridges. Unlike oogenesis, where the ova produced by meiosis are functionally competent gametes, the spermatids generated by meiosis must undergo further differentiation before functional **sperm** are formed (Figure 15-11). This differentiation process, termed **spermiogenesis,** involves loss of most of the spermatid's cytoplasm and differentiation of the remainder of the cell into two major structures, the head and the tail (Figure 15-12). The **head** contains a highly compacted nucleus and a membranous vesicle, termed the **acrosome,** which is designed to aid in the penetration of the egg coats. The tail consists of a long flagellum enveloped for part of its length by a sheath of mitochondria that provide the ATP needed for sperm motility.

Development of the Sperm Head During the early stages of spermiogenesis the chromatin fibers, whose initial diffuse appearance resembles that of typical interphase cells, become thickened and tightly packed together, forming a compact amorphous mass (Figure 15-13). As a result the nuclear volume decreases markedly and nucleoli disappear, leaving the final sperm nucleus with no discernible substructure. As chromatin condensation proceeds, the spermatid nucleus gradually acquires a shape characteristic for the organism, which may be elongate, spheroidal, or a flattened ovoid. It is not clear whether these characteristic shapes are produced by external guiding forces, such as microtubules or microfilaments, or whether nuclear shape results from some property of the chromatin fibers that dictates their pattern of aggregation.

Chromatin condensation and the associated reduction in nuclear volume is accompanied by removal of nonhistone proteins and replacement of the normal histones with either special arginine-rich histones or protamines (page 394). Nucleosomes occur in the sperm nuclei of some organisms, although they do not appear to be universally present. X-ray diffraction of intact sperm heads has revealed that the chromatin is packed

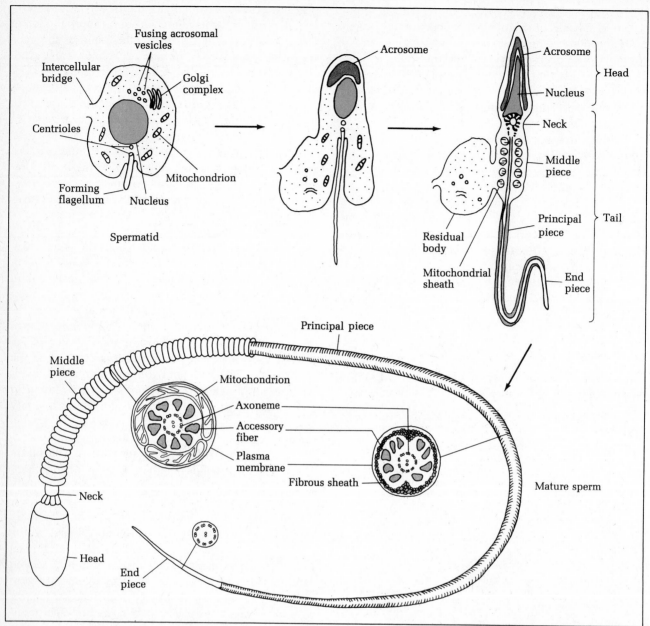

FIGURE 15-11 The major stages of spermiogenesis.

in a highly ordered, almost crystalline configuration. In this form the DNA is metabolically inert, unable to act as a template for either RNA or DNA synthesis. Packaging of the sperm cell DNA in such a compact, inert form serves at least two functions: It protects the paternal genes from physical or chemical damage and, by reducing nuclear volume, it makes the cell easier to propel.

At the tip of the sperm head, lying between the nucleus and the overlying plasma membrane, is a variably shaped vesicle known as an **acrosome.** The acrosome is formed by the fusion of small vesicles derived from the Golgi complex (Figure 15-14), yielding a large membrane-enclosed organelle of spherical, conical, or elongate shape. After the acrosome develops, the Golgi com-

plex moves from its location adjacent to the nucleus toward the tail region of the spermatid, where it will be expelled from the cell. The acrosome contains a variety of hydrolytic enzymes, including acid phosphatase, proteases, and hyaluronidase, fostering the idea that it represents a specialized lysosome. The function of these hydrolytic enzymes in aiding penetration of the egg coats during fertilization will be discussed later.

Development of the Sperm Tail Directly beneath the spermatid nucleus one usually finds two centrioles, a **proximal centriole** adjacent to the nucleus and a **distal centriole** lying at right angles to the proximal centriole. During spermiogenesis, the distal centriole acts as a

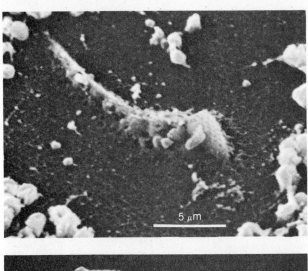

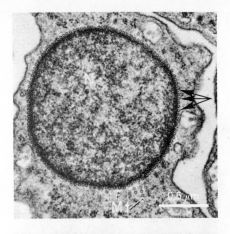

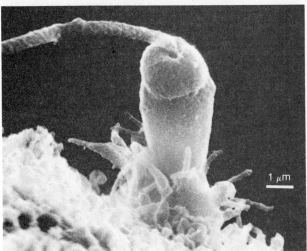

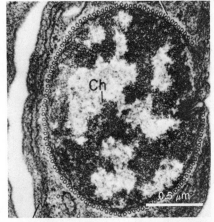

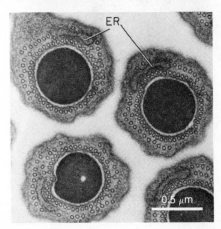

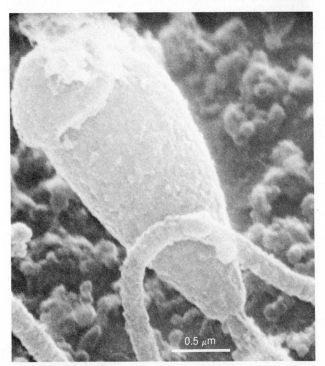

FIGURE 15-13 Electron micrographs illustrating the process of chromatin condensation that occurs during formation of the sperm head in the earthworm. *(Top)* During early spermatid development the chromatin fibers are evenly dispersed throughout the nucleus. Note the single layer of microtubules (arrows) surrounding the nucleus. *(Middle)* In a slightly later stage of development the chromatin fibers (Ch) are beginning to condense into clumps. *(Bottom)* In the final stages of maturation the chromatin has become highly condensed and the cytoplasmic volume is reduced. Although small amounts of endoplasmic reticulum (ER) are retained, microtubules are the main cytoplasmic component. Courtesy of W. A. Anderson.

FIGURE 15-12 Scanning electron micrographs of sea urchin sperm lying on the surface of an egg cell. Courtesy of G. Schatten.

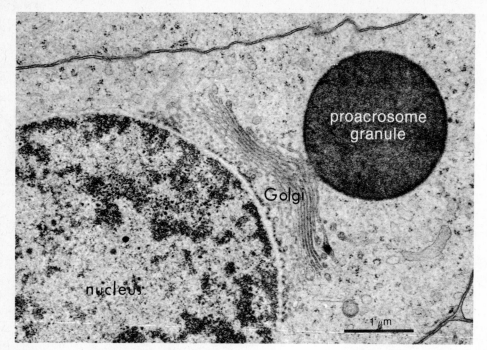

FIGURE 15-14 Electron micrograph illustrating the relationship between the Golgi complex and acrosome formation in a developing grasshopper sperm cell. Vesicles derived from the convex side of the Golgi complex fuse with one another, forming a structure called a proacrosome granule. This granule subsequently differentiates into the mature acrosome. Courtesy of D. M. Phillips.

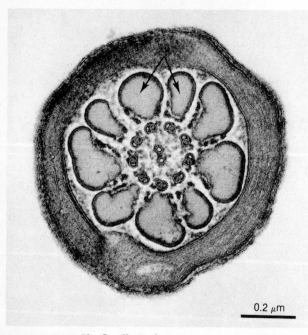

FIGURE 15-15 The flagellum of a guinea pig sperm cell observed in cross section. In this electron micrograph nine large accessory fibers (arrows) are seen surrounding a central group of microtubules arranged in the classical 9 + 2 pattern. Courtesy of D. S. Friend.

nucleation site for assembly of a flagellar axoneme that forms the core of the developing sperm tail. In addition to the nine outer doublets and central pair of microtubules generally characteristic of flagellar axonemes, some sperm flagella contain nine thick **accessory fibers,** one adjacent to each of the outer doublets (Figure 15-15). The chemical nature or function of these fibers is unknown, but they occur almost solely in species where fertilization takes place within the organism's reproductive tract rather than externally. It is therefore possible that they provide extra strength or motile power needed for propelling sperm through the viscous environment of the reproductive tract.

In mammalian sperm, assembly of the flagellum is accompanied by the development of dense strands of material that form a **connecting piece** attaching the flagellum to the proximal centriole and the nuclear envelope. The distal centriole, which functioned earlier to nucleate the assembly of the flagellar axoneme, disintegrates at this time. Formation of the flagellum is also accompanied by a gradual migration of mitochondria to a position directly beneath the nucleus. Here they form a **middle piece** that surrounds the base of the flagellar axoneme. In many organisms the mitochondria of the middle piece line up end to end, forming a helical sheath around the flagellar axoneme. In insects and certain invertebrates the individual mitochondria actually fuse together to form a composite mass, called the **nebenkern,** which separates into two bodies of equal size that come to lie on either side of the developing flagellum (Figure 15-16).

As the flagellum elongates and the mitochondria take up their position in the middle piece, the remaining cytoplasm of the spermatid becomes localized in a large cellular protrusion called the **residual body.** This entire mass of cytoplasm is eventually pinched off and discarded, yielding a mature sperm cell that lacks many of the organelles of the original spermatid. The highly specialized cell produced by the process of spermiogenesis is uniquely constructed for the task of transporting paternal genes to the egg. The sperm head is designed for protecting these genes and for penetrating the egg, while

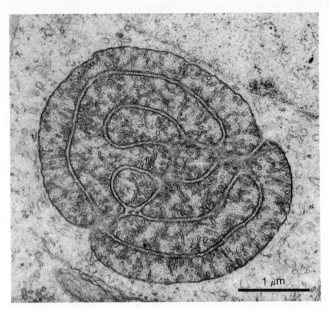

FIGURE 15-16 Electron micrograph of a four-layered nebenkern lying next to the nucleus of an insect spermatid. Such structures, which consist of a mass of aggregated mitochondria, occur in the spermatids of many insects. Courtesy of S. A. Pratt.

the tail, with its flagellum and associated mitochondria, provides motility. Most other cellular functions, which would be extraneous to this task, have been eliminated.

FERTILIZATION AND EARLY DEVELOPMENT

The union of sperm and egg, or **fertilization,** may occur either inside or outside the female reproductive tract. In mammals, birds, and reptiles fertilization is internal, while in amphibians, fish, and invertebrates it takes place after the eggs have been released into the external environment. Most sperm cells are propelled to the vicinity of the egg by their flagella, although external forces, such as contraction of the oviduct or movement of cilia lining its walls, also appear to facilitate sperm movement. Sperm that do not have flagella either travel by amoeboid motion or are passively carried by movements of the fluid in which they are suspended.

The movement of sperm cells toward the egg may be either random or guided by chemical attractants produced by the egg cell. The latter phenomenon, termed **chemotaxis,** is utilized by a variety of plants as well as some invertebrates. In higher animals, however, there is no evidence that eggs produce chemotactic substances that attract sperm; in such organisms the meeting of sperm and egg seems to occur by chance, making it necessary for animals to produce a vast excess of sperm to ensure that sufficient numbers reach the egg. In the case of mammals, for example, millions of sperm are deposited in the female reproductive tract during copulation. Yet no more than a few hundred manage to reach even the general vicinity of the egg.

As soon as a sperm cell collides with an egg, the two cells become tightly bound to one another. In the case of organisms that employ external fertilization, it is important that this binding be selective since sperm cells produced by a variety of different organisms may coexist in the same seawater. The molecular basis for this specificity is thought to involve an interaction between sperm cell proteins and specific receptors present on the egg cell surface. For example, a protein isolated from sea urchin sperm called **bindin** has been found to preferentially bind to eggs obtained from the same species of sea urchin. Moreover, sperm from different types of sea urchins contain slightly different forms of bindin. Hence the preferential binding of a sea urchin sperm cell to the appropriate type of egg is thought to be facilitated by an interaction between specific types of bindin and corresponding receptors on the egg cell surface. The egg cell receptor with which bindin interacts appears to be a cell surface glycoprotein, since treatment of egg cells with lectins that bind to glycoproteins prevents fertilization from occurring. Observations of this type may one day yield practical benefits in the development of new methods of human birth control. For example, if women could be immunized against either the sperm protein or the egg receptor involved in sperm-egg binding, the resulting antibodies might prevent fertilization without interfering with normal hormone function as existing contraceptive pills do.

Interaction of sperm cells with the exterior surface of the egg (or with the zona pellucida in mammals) triggers the **acrosomal reaction,** a process involving fusion of the acrosomal membrane with the sperm plasma membrane and the subsequent liberation of the acrosome's hydrolytic enzymes into the external environment (Figures 15-17 and 15-18). The discharged enzymes digest a pathway for the sperm through the outer coatings of the egg. In many marine invertebrates and in a few vertebrates, rupture of the acrosome is followed by the emergence of a long thin cytoplasmic protrusion from the region directly in front of the sperm head. Covered by an extension of the plasma membrane, this **acrosomal process** is produced by polymerization of actin molecules stored in the region between the acrosome and the nucleus. The acrosomal process quickly traverses the outer coats and makes contact with the egg plasma membrane. In organisms where it is present, the acrosomal process is the major site of the egg-binding protein, bindin.

Early Responses of Fertilization

After the plasma membrane of the sperm makes contact with the plasma membrane of the egg, the two membranes fuse together. This fusion, which marks the beginning of fertilization, triggers a dramatic sequence of events within the egg. The earliest of these events,

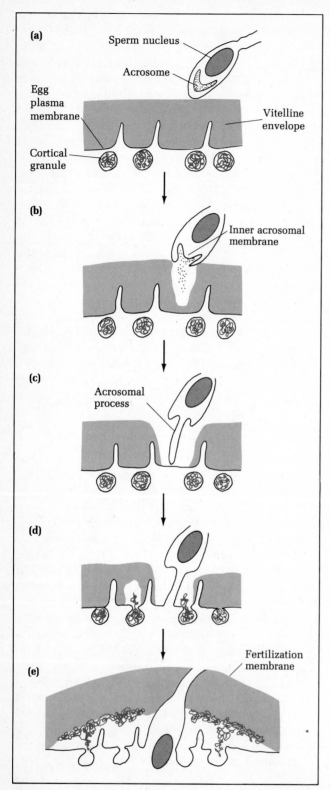

FIGURE 15-17 Outline of the main steps in sperm-egg fusion. Physical contact between sperm cell and vitelline envelope (a) triggers exocytosis of acrosomal contents (b) and formation of the acrosomal process (c). The subsequent fusion of sperm and egg plasma membranes (d) leads to cortical granule discharge and formation of the fertilization membrane (e). In organisms whose sperm do not form an acrosomal process, digestion of the vitelline envelope by hydrolytic enzymes released from the acrosome permits the entire sperm head to penetrate the vitelline envelope and fuse with the egg plasma membrane.

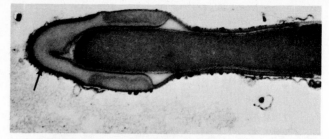

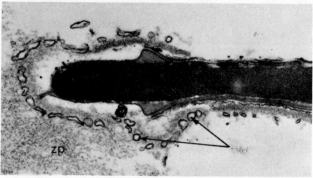

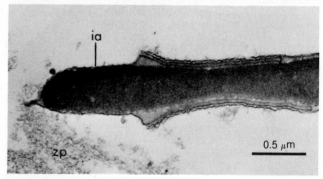

FIGURE 15-18 A series of electron micrographs illustrating the acrosome reaction in the sperm cell of the golden hamster. *(Top)* A sperm cell with an intact acrosome (arrow). *(Middle)* Upon contact with the egg surface the acrosomal membrane fuses with the sperm cell plasma membrane, creating a series of membrane vesicles (arrows). This loss of membrane continuity leads to the release of the acrosomal contents. *(Bottom)* The inner acrosomal membrane (ia) is now in direct contact with the zona pellucida (zp). Courtesy of R. Yanagemachi.

which have been especially well analyzed in sea urchins, are designed to prevent **polyspermy,** the fertilization of the egg by more than one sperm. Establishment of a rapid block to polyspermy is important because hundreds or even thousands of sperm may be present at the egg surface when the first sperm fuses with the egg plasma membrane. Experiments carried out by Larinda Jaffe suggest that changes in membrane potential are important in preventing subsequent fertilization by additional sperm. Jaffe discovered that the electrical potential across the membrane of the sea urchin egg increases from its resting value of -60 mV to a value between -3 and $+20$ mV within a few seconds after fertilization. This change in membrane potential, or **depolarization,** is caused by a transient influx of sodium ions into the egg. In order to test the hypothesis that this transient depolarization inhibits polyspermy,

Jaffe employed an electric current to artificially raise the membrane potential of unfertilized eggs to +5 mV. Under these conditions eggs remained unfertilized, even in the presence of a vast excess of sperm. Only after the electric current was turned off did fertilization occur, indicating that fertilization is prevented by a positive membrane potential.

The rapid block to polyspermy caused by membrane depolarization is transitory in nature, generally lasting for no more than a few minutes. By the time membrane depolarization subsides, however, a more permanent block to polyspermy has been established through changes in the structural organization of the egg cell surface. This set of changes, referred to as the **cortical reaction,** is initiated in response to fusion of the sperm and egg plasma membranes, and is completed within about a minute in the sea urchin. The first step of the cortical reaction is fusion of the membranes of the cortical granules with the egg's plasma membrane, resulting in discharge of the granules' contents into the space between the plasma membrane and the overlying vitelline envelope (Figure 15-19). This discharge event originates at the site of sperm-egg binding and quickly passes over the entire egg surface, causing a fluid-filled space to form between the plasma membrane and the vitelline envelope.

Among the components discharged from the cortical granules is an enzyme that destroys the sperm receptors of the vitelline envelope, causing release of bound sperm and preventing the binding of any new sperm (Figure 15-20). Other components discharged from the cortical granules cause the vitelline envelope to change from a soft elastic structure into a tough rigid envelope, or **fertilization membrane,** which serves as a highly effective barrier to further sperm penetration (Figure 15-21). In spite of the name *fertilization membrane,* this structure is not a membrane in the classical sense of a lipid bilayer containing embedded proteins. Instead it is created by a hardening of the vitelline envelope caused by the formation of covalent bonds between tyrosine residues present in the envelope's glycoprotein constituents. These bonds are formed by **ovoperoxidase,** a hydrogen-peroxide-requiring enzyme released from cortical granules at the time of fertilization.

The discovery that the cortical reaction can be artificially induced in unfertilized eggs by treatment with calcium ionophores has led to the conclusion that calcium ions play an important role in triggering the early

FIGURE 15-19 Cortical granule discharge in the sea urchin egg immediately after fertilization. *(Left)* In this scanning electron micrograph of the egg surface, the region where cortical granule discharge has occurred is clearly evident (arrow). This area corresponds to the region where a sperm cell has just entered the egg. Courtesy of G. Schatten. *(Right)* A thin-section electron micrograph of a comparable region in the sea urchin egg reveals the remnants of two cortical granules that have just fused with the plasma membrane and discharged their contents. Courtesy of E. Anderson.

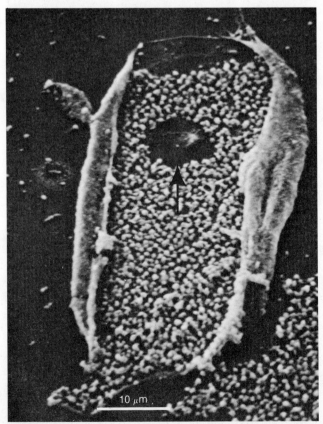
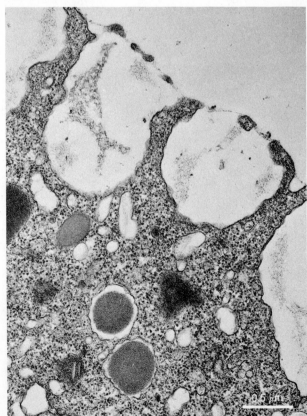

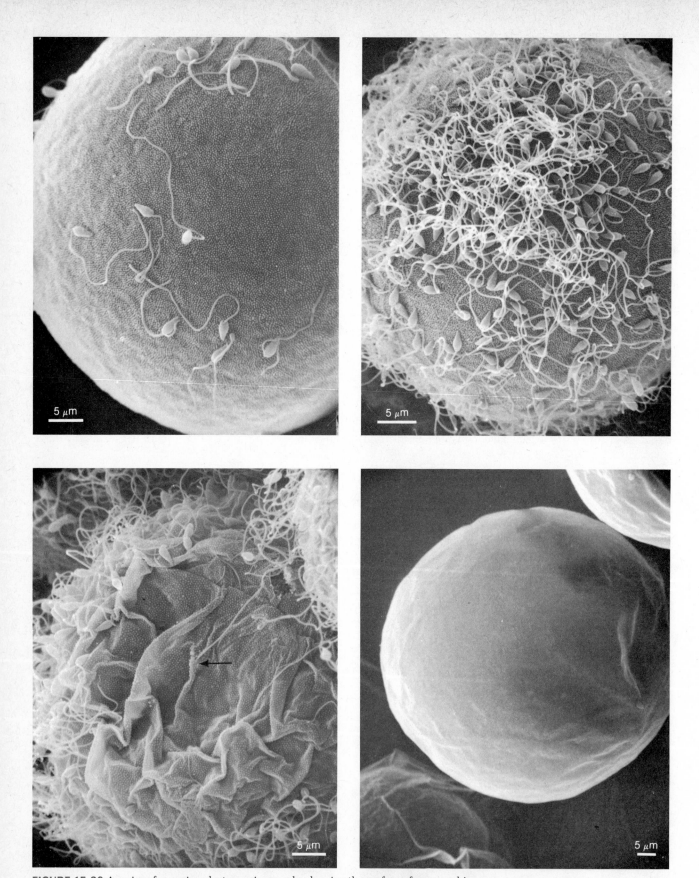

FIGURE 15-20 A series of scanning electron micrographs showing the surface of a sea urchin egg during the first few minutes after fertilization. *(Top left)* At the time of fertilization, several sperm cells are seen scattered across the egg surface. *(Top right)* Fifteen seconds after fertilization the number of sperm on the egg surface is still increasing. *(Bottom left)* At 30 sec a circular zone lacking bound sperm has developed around the point of sperm entry (the arrow points to the tail of the entering sperm). This sperm-free zone corresponds to the area of cortical-granule breakdown. *(Bottom right)* By 3 min, the hardened fertilization membrane has formed, and no sperm remain bound to the egg surface. Courtesy of M. Tegner.

FIGURE 15-21 Light micrograph of a recently fertilized egg cell revealing the thick fertilization membrane. Note that the sperm cells (arrows) surrounding the egg are unable to penetrate this barrier. Courtesy of R. D. Grey.

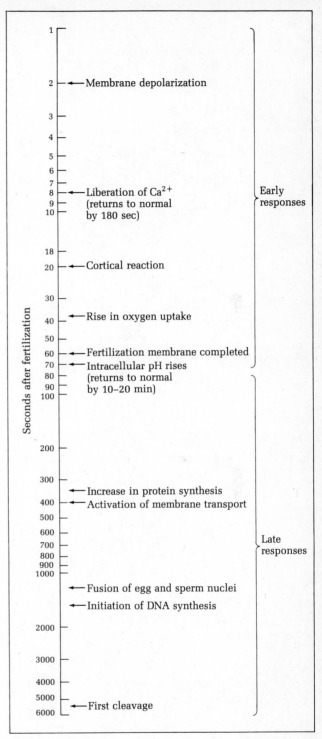

FIGURE 15-22 Sequence of events following fertilization in sea urchin eggs.

events associated with fertilization. Further support for this conclusion has been obtained from experiments utilizing the dye **aequorin,** which luminesces in the presence of calcium ions. Aequorin injected into unfertilized eggs is barely luminescent, indicating a low internal concentration of free calcium ions. A few seconds after fertilization, however, the luminescence of aequorin increases over four orders of magnitude, indicating a dramatic increase in free calcium ions within the cell. These calcium ions are thought to be derived from intracellular stores of bound calcium, rather than from the external environment, because calcium ionophores can trigger the cortical reaction in the absence of calcium ions in the external medium.

Depolarization of the egg plasma membrane, an increase in cytoplasmic Ca^{2+}, and the cortical reaction all occur within the first minute after fertilization, and are therefore referred to as "early" responses (Figure 15-22). As we shall see below the "late" responses, which occur shortly thereafter, involve major changes in the biosynthetic activity of the egg.

Late Responses of Fertilization

Depending on the organism involved, the egg may be at any stage of maturation when fertilization occurs (e.g., primary oocyte, metaphase of first meiotic division, secondary oocyte, metaphase of second meiotic division, or at the end of meiosis). In eggs where maturation is not yet complete, contact of the sperm cell with the egg plasma membrane triggers the remaining stages of meiosis. Meanwhile, the cytoplasmic continuity between sperm and egg produced by fusion of their plasma membranes permits the egg cytoplasm to creep up between the nucleus and the plasma membrane of the sperm, producing a **fertilization cone** [Figure 15-23 (*top left*)]. As this process draws the sperm nucleus into

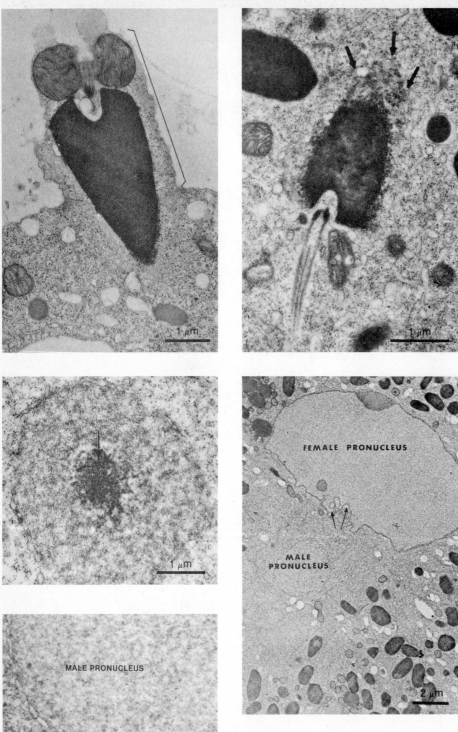

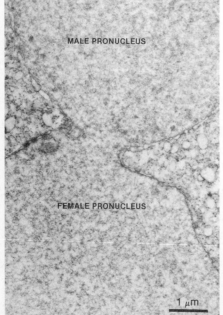

FIGURE 15-23 Changes in the appearance of the sperm cell nucleus after fertilization. *(Top left)* Shortly after fertilization the egg cytoplasm creeps up around the entering sperm nucleus, creating a fertilization cone (bracket). *(Top right)* As the sperm head moves through the egg cytoplasm, its chromatin begins to disperse (arrows). Note the absence of a surrounding nuclear envelope at this stage. *(Middle left)* Uncoiling of the sperm chromatin continues as a new nuclear envelope begins to form. In this micrograph only a small region of the sperm chromatin remains condensed (arrow). *(Bottom right)* After a new nuclear envelope has formed around the sperm chromatin, it is called the male pronucleus. As the male pronucleus and egg cell nucleus (female pronucleus) approach one another, projections appear on the surface of the female pronucleus (arrows). *(Bottom left)* Fusion of the female and male pronuclei leads to a mixing of their chromatin contents. Courtesy of E. Anderson.

the egg cytoplasm the compact chromatin of the sperm begins to uncoil, gradually dispersing to the diffuse fibrillar state typical of interphase cells [Figure 15-23 (*top right, middle, bottom right*)]. This structural transformation appears to be caused by replacement of the highly basic chromatin proteins of the sperm nucleus with less basic proteins present in the egg cytoplasm.

As the sperm head enters the egg cytoplasm its nuclear envelope disintegrates into small vesicles, momentarily leaving the sperm chromatin without an enclosing membrane. But a new nuclear envelope quickly forms, derived either from the original sperm envelope, from newly synthesized membranous material, or from preexisting membranes present in the egg cytoplasm. The newly enveloped sperm chromatin is referred to as the **male pronucleus.** The male pronucleus migrates toward the egg nucleus (often termed the **female pronucleus** at this stage), where their envelopes may or may not fuse, depending on the species [Figure 15-23 (*bottom left*)]. In either case the envelope(s) soon breaks down, releasing the paternal and maternal chromosomes into the cytoplasm where they are free to associate with the mitotic spindle that is forming in preparation for the first cleavage.

While the preceding events are taking place, significant alterations in the metabolic condition of the egg are also occurring. In lower species such as the sea urchin, one of the first detectable changes is an enhanced uptake of small molecules like amino acids and nucleosides. But the most striking alteration in such organisms involves the synthesis of proteins and nucleic acids. Changes in protein synthesis are especially prominent in sea urchins, where less than 1 percent of the ribosomes of the unfertilized egg are active in protein synthesis. Soon after fertilization, however, the rate of protein synthesis increases 10-fold or more (Figure 15-24). Though at first glance it might appear as if this sudden burst of protein synthesis is due to the synthesis of new messenger RNAs, two observations argue against this interpretation. First, artificial activation of sea urchin eggs that have had their nuclei removed still leads to a stimulation of protein synthesis, even though new messenger RNAs can no longer be synthesized. And second, the increase in protein synthesis that accompanies fertilization of sea urchin eggs is not blocked by the transcription-inhibitor actinomycin D, again suggesting the absence of a requirement for new RNA synthesis.

Such experiments have led to the conclusion that the messenger RNAs employed during the burst of protein synthesis are formed prior to fertilization and are stored in the unfertilized egg in an untranslated state. There is a difference of opinion, however, as to whether the mechanism limiting messenger RNA translation in unfertilized eggs involves direct inhibition or "masking" of the messenger RNA, or inhibition of one or more components of the protein-synthesizing system, such

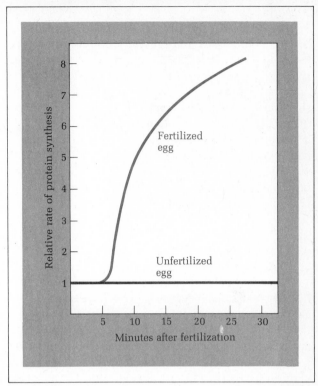

FIGURE 15-24 Relative rates of protein synthesis in sea urchin eggs before and after fertilization.

as ribosomes, transfer RNAs, and initiation factors. Though it has been discovered that fertilization stimulates the addition of poly(A) tails to the 3′ ends of RNA molecules already present in the egg cytoplasm, this reaction does not seem to be responsible for messenger RNA activation because inhibitors of poly(A) formation do not prevent the increase in protein synthesis from occurring. An alternative possibility is that the activation mechanism involves proteolytic degradation of inhibitory proteins associated with messenger RNAs and/or ribosomes. The discovery that protein synthesis in cell-free systems prepared from unfertilized eggs can be stimulated by the addition of proteases is consistent with this idea, though the physiological significance of such experiments has been questioned.

The marked stimulation of protein synthesis seen in the fertilized eggs of lower organisms is not a general response in the eggs of higher organisms. One metabolic response to fertilization that is universally observed, however, is a stimulation of DNA synthesis. It is the onset of DNA synthesis and the cell division cycle that is of course the biologically important response to fertilization. Penetration of the egg by the sperm triggers DNA synthesis in both male and female pronuclei, preparing them for the ensuing mitotic division. Some RNA synthesis may also occur at this point, though it differs in amount and type in different organisms.

The mechanism responsible for triggering the late responses of fertilization is known to differ from that triggering the early responses because certain treatments, such as exposure to ammonia, activate late responses like DNA and protein synthesis without increasing the internal Ca^{2+} concentration or triggering the cortical reaction. The fact that ammonia readily penetrates into the egg and raises the cytoplasmic pH suggests that pH changes may induce the late responses of fertilization. Such a notion has found support in the discovery that protons normally move out of the egg cell (accompanied by an uptake of sodium ions) about 90 seconds after fertilization. The extent of the observed proton efflux is sufficient to cause the pH of the egg to increase several tenths of a pH unit.

Cleavage

Following activation of DNA synthesis, the egg undergoes a series of cell divisions collectively referred to as the **cleavage** stage. Because there is little or no detectable increase in mass during this period, the successive mitotic divisions gradually partition the egg into smaller and smaller cells. Cleavage proceeds normally even when the synthesis of new RNA molecules is artificially blocked by treatment with actinomycin D, indicating that the events that take place during cleavage are dependent solely on maternal messenger RNAs stored in the egg.

The length of the cell cycle during cleavage is significantly shorter than in the dividing cells of the adult organism. This reduction in cell-cycle length is accomplished mainly through elimination of the G_1-phase of the cell cycle. Presumably this elimination of G_1 is made possible by the fact that cell growth does not accompany cell division at this time. As the cleavage stage progresses, a G_1-phase is gradually introduced, so that by the end of cleavage a typical cell cycle is present and cell growth accompanies cell division.

Because the cytoplasm of the egg has an inherent polarity to its organization (see Figure 15-8), division of this asymmetric cytoplasm by successive cleavages eventually yields cells inheriting different types of cytoplasm. Additional differences in the properties of the cells generated by cleavage result from the fact that eggs with large amounts of yolk tend to divide unequally, generating larger and more slowly dividing cells in the vegetal region. Hence the beginnings of cell differentiation occur in the cleavage state as a result of the asymmetry originally inherent in the egg.

Unique Features of Plant Development

Although the developmental principles described thus far are applicable to a wide variety of organisms, plant development entails a number of unique features requiring special comment. In comparison to animal eggs the female gamete of plants is relatively simple in structure, lacking elaborate outer coats and stores of nutrients. In lower plants such as algae, mosses, and ferns the male gamete resembles the motile sperm of animals, though it may contain more than one flagellum. Flowering plants, on the other hand, produce nonmotile male gametes that differ considerably from animal sperm. As will be briefly outlined below, development in such higher plants is a unique and complex process entailing two distinct fertilization events, one producing the embryo and the other producing a specialized nutritive tissue known as **endosperm.**

The flowers of higher plants contain both female and male reproductive organs, termed the **ovary** and **anther,** respectively (Figure 15-25). In the ovary, meiotic division of the diploid female gamete or **megasporocyte** generates four haploid **megaspores,** three of which disintegrate. The remaining haploid megaspore undergoes three mitotic divisions, yielding an **embryo sac** that contains eight haploid nuclei distributed as follows: one **egg** cell, two **polar nuclei** suspended in a common cytoplasm, two **synergid** cells, and three **antipodal** cells (see Figure 15-25). The egg cell is small and undifferentiated, containing few organelles and no significant store of nutrients. The antipodal cells usually die without contributing to further development; in contrast, the synergids function in transporting nutrients to the egg and in receiving the pollen tube of the male gamete during fertilization, while the polar nuclei eventually fuse with a sperm nucleus to form the triploid (3n) endosperm.

Development of the male gamete begins with meiotic division of **microsporocytes** located in the anther. Each of the four haploid **microspores** produced by meiosis becomes enveloped in a thick spore coat, forming **pollen grains.** At about the same time the haploid nucleus present in each pollen grain undergoes a mitotic division that yields two nuclei: a **generative nucleus** destined to produce the sperm nuclei used in fertilization, and a **vegetative nucleus** destined to control germination of the pollen grain (Figure 15-26). The pollen grain is a metabolically inert structure, protected from the environment by its hard impervious coat.

Mature pollen grains are easily dislodged from the anther by the action of wind, rain, or insects. In this way they may be brought in contact with the **stigma,** a sticky projection of the female reproductive organ specialized for trapping pollen. Contact with the stigma activates the pollen grain, triggering a burst in respiration, synthesis of a small amount of RNA, rupture of the thick spore coat, and the gradual outgrowth of a cytoplasmic projection termed the **pollen tube.** The pollen tube, still surrounded by the plasma membrane and a primary cell wall, penetrates through the tissues of the stigma and grows toward the ovary. As the tip of the pollen tube advances, carrying the generative and vegetative nuclei with it, the older areas of the pollen tube gradually

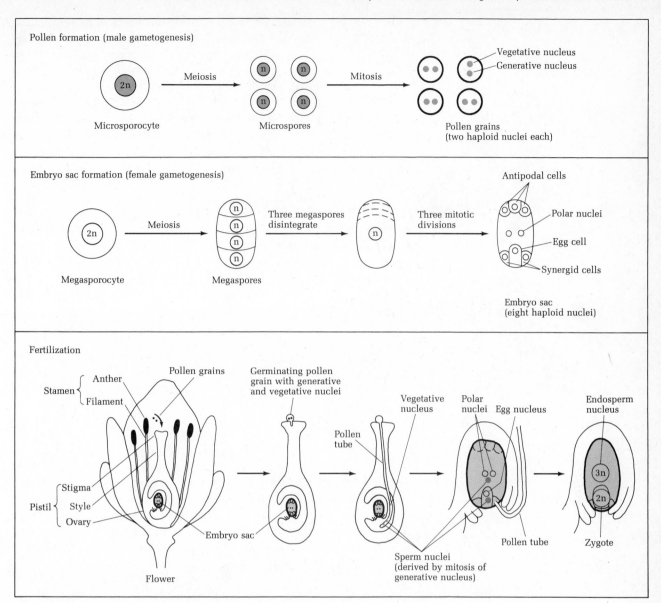

FIGURE 15-25 Summary of gametogenesis and fertilization in flowering plants.

lose their cytoplasm and become sealed off by partitions. Pollen tubes that are produced by pollen grains germinated in the absence of external nutrients grow to normal length, indicating that all the materials required for growth of the pollen tube are present in the initial pollen grain.

During the period of pollen tube elongation the generative nucleus undergoes another mitosis, yielding two haploid **sperm nuclei** that become enclosed within their own plasma membranes. When the embryo sac is finally reached, the pollen tube penetrates one of the two synergids and releases the two nuclei. One nucleus fuses with the egg, forming a diploid zygote as in animal fertilization. Because no fertilization membrane is formed, however, it is not clear how polyspermy is prevented. The other sperm nucleus fuses with the two polar nuclei, which may themselves have fused together by this time. The net result is the triploid nucleus (three sets of chromosomes) of the endosperm.

Once the diploid zygote has been formed, it divides mitotically to generate the plant embryo. Cell migration does not play a significant role in plant embryogenesis as it does in animals, presumably because the presence of cell walls hinders such movements. Thus only growth, division, and differentiation of cells derived from the fertilized egg are involved in shaping the developing plant embryo. The endosperm also undergoes extensive growth and division at this time, accumulating vast stores of starch and protein from other regions of the parent plant. The result is a rich nutritive tissue designed to nourish the embryo. Development is usually arrested at this stage, yielding a **seed** composed of the diploid embryo surrounded by the triploid endosperm. The seed may remain in an arrested state for long

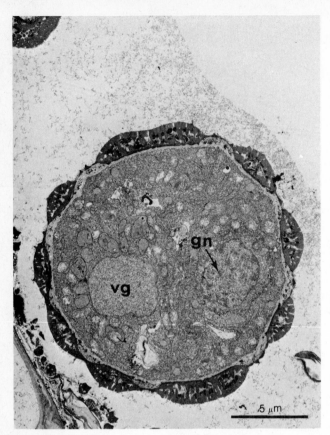

FIGURE 15-26 Electron micrograph of a pollen grain from sugar beets, showing the vegetative nucleus (vg) and generative nucleus (gn). Courtesy of L. L. Hoefert.

periods of time, until appropriate conditions trigger its germination and subsequent development into a new plant.

CELL DIFFERENTIATION

The major unanswered question concerning the mechanism of both animal and plant development can be stated quite simply: How do cells produced by successive mitotic divisions of a single fertilized egg and there-fore inheriting the same genes come to be so different from one another in appearance and function? In 1934 Thomas Hunt Morgan suggested that cell differentiation can occur because gene expression is influenced by the cytoplasmic environment in which a nucleus resides. Since the organelles and other constituents initially present in the egg cytoplasm are asymmetrically distrib-uted, cells derived from successive mitotic divisions of the egg contain dissimilar cytoplasms. If the nuclei in these cells express different genes because of their dif-ferent cytoplasms, the resulting gene products will in turn make the cytoplasms even more dissimilar, pro-ducing a cascading effect. We shall now describe a few of the many experimental observations that support Morgan's idea that the key to cell differentiation lies in this ability of identical nuclei to express different genes, depending on the particular cytoplasmic environment in which they reside.

Cytoplasmic Influence on Cell Differentiation

The general influence of the cytoplasm on the early pro-cess of cell differentiation has been demonstrated in a variety of different kinds of egg cells. In snail eggs, for example, cell differentiation has been shown to be pro-foundly affected by the **polar lobe,** a large cytoplasmic protrusion temporarily formed at the vegetal pole dur-ing the first few cleavages (Figure 15-27). If this cyto-plasmic lobe is surgically removed, an embryo with a defective foot, eye, and shell will form. In amphibian eggs, cell differentiation is influenced by the **gray cres-cent,** a pigmented area of the cell surface that develops shortly after fertilization (see Figure 15-1). If the gray crescent is injured by pricking, an embryo with a grossly abnormal nervous system will result. The criti-cal area of the gray crescent appears to be the **cortex,** a region encompassing the plasma membrane and the thin layer of cytoplasm lying directly beneath it.

Although the preceding observations indicate that the cytoplasm can exert a significant influence on the

FIGURE 15-27 Polar lobe formation in mollusk eggs. The first polar lobe is formed prior to the first cleavage, and is subsequently resorbed into the cytoplasm of one of the two cells produced by this division. The second polar lobe is formed by this same cell prior to the second cleavage, and is again resorbed into one of the two cells produced by this division. The net result is that the cytoplasmic material segregated into the polar lobe is selectively passed to one of the four cells present at the end of the second cleavage.

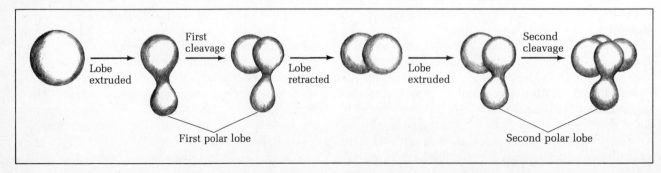

pattern of cell differentiation, they do not directly demonstrate that such effects involve changes at the level of the nuclear genes. Direct evidence that the cytoplasm does influence the behavior of nuclear genes during cell differentiation was provided many years ago, however, by the classic studies of Theodor Boveri on embryonic development in the roundworm, *Ascaris*. As is typical of nonmammalian oocytes, the cytoplasm of the *Ascaris* egg exhibits a pronounced cytoplasmic polarity that creates distinct animal and vegetal poles. *Ascaris* is unusual, however, in that the first cleavage after fertilization occurs perpendicular to the animal-vegetal axis, dividing the egg into one cell that contains cytoplasm derived from the vegetal pole and a second cell containing cytoplasm derived from the animal pole. As the cell containing the animal-pole cytoplasm prepares for its next division, the heterochromatic regions of its chromosomes are shed into the cytoplasm and degenerate, while the remaining chromosomal regions break up into numerous small chromosomes. This process, referred to as **chromosome diminution,** yields cells that have lost a portion of the normal chromosomal material.

In order to determine whether the type of cytoplasm a cell receives is responsible for triggering this unusual chromosomal behavior, Boveri carried out a series of experiments in which *Ascaris* eggs were centrifuged prior to cleavage in order to shift the orientation of the mitotic spindle relative to the cytoplasm. In this way eggs were created in which the first cleavage plane ran in the same direction as the animal-vegetal axis rather than perpendicular to it. When cleavage occurs in this direction, the resulting two cells each contain a mixture of both animal and vegetal cytoplasm. Under such conditions neither cell was found to undergo chromosome diminution during the next division, indicating that the cytoplasmic environment is normally responsible for inducing this type of chromosomal behavior.

The effects of the cytoplasm on cell differentiation are also vividly demonstrated by the behavior of developing germ cells in *Drosophila* embryos. In insect embryos the cells destined to become primordial germ cells initially arise from the posterior end of the egg. In order to determine whether it is the cytoplasm in this region of the egg that is responsible for triggering germ cell differentiation, Karl Illmensee and Anthony Mahowald removed a tiny bit of cytoplasm from the posterior end of one *Drosophila* egg and injected it into the opposite (anterior) end of another egg. Under such conditions germ cells were found to develop in the anterior region of the injected egg where they normally would not have done so. Though the cytoplasmic component responsible for inducing this differentiation is yet to be clearly identified, the injected cytoplasm derived from the posterior end of the egg is known to contain distinctive **polar granules** that may be involved. Polar granules are comprised of a mixture of RNA and protein, but the role

played by these components in inducing germ cell differentiation, if any, remains to be determined.

Although the examples of cytoplasmic control of cell differentiation described thus far involve embryonic cells, cytoplasmic control of nuclear activity has also been observed in adult cell types. We saw in Chapter 11, for example, that inactive red blood cell nuclei can be induced to become transcriptionally active again by introducing them into the cytoplasm of a more active cell type. Not only is RNA synthesis in the red cell nucleus stimulated in the new cytoplasmic environment, but specific genes that were previously inactive begin to be transcribed. The fact that such results can be obtained using cells derived from adult tissues raises the general question of whether the cytoplasm continues to control the state of cell differentiation in the mature organism. Cell differentiation certainly continues throughout life in tissues such as skin, intestine, and blood, and there is good reason for believing that cytoplasmic factors are intimately associated with this type of differentiation. One extensively studied example of postembryonic differentiation involves the grasshopper neuroblast, a cell whose every division leads to the formation of one differentiated ganglion cell and a new neuroblast destined to divide again and repeat the process. Dividing neuroblasts are always oriented in a particular direction, making it possible to predict which regions of the neuroblast cytoplasm will give rise to the new neuroblast and to the ganglion cell, respectively. In an elegant microdissection study J. Gordon Carlson inserted a fine needle into a grasshopper neuroblast to rotate the mitotic spindle in such a way that those chromosomes normally destined for the ganglion cell would enter the cell normally destined to become a neuroblast, and vice versa. In spite of this manipulation of the chromosomes, the cells destined to form either ganglion cell or neuroblast did so as expected. This result clearly reveals the existence of regional cytoplasmic differences in the dividing neuroblast responsible for determining the developmental fate of the two progeny cells. A similar phenomenon may be responsible for differentiation in other adult cell types, such as skin, intestine, and blood.

The Concept of Nuclear Totipotency

The discovery that the cytoplasm exerts a strong influence in regulating cell differentiation does not in itself reveal much about the molecular events involved. Differentiated cells differ from one another primarily in the proteins they produce; proteins, in turn, are encoded by genes, so a crucial question concerning the molecular basis of cell differentiation concerns the issue of how the cytoplasm influences the pattern of genes being expressed as functional proteins.

The most drastic possibility, initially proposed by

August Weismann in 1897, is that genes are progressively destroyed during differentiation, with each specialized cell retaining only those genes needed for the particular functions it will carry out. Alternatively, every cell may retain all the organism's genes, but only express certain ones. A critical test of these alternatives was not possible until the early 1950s, when Robert Briggs and Thomas King developed a technique for transplanting nuclei from differentiated cells into fertilized eggs whose own nuclei had been surgically removed. The rationale underlying these studies was simple: If a nucleus derived from a differentiated cell can, upon transplantation into egg cytoplasm, support the development of a normal embryo with all its tissue types, then this transplanted nucleus cannot have lost any genes during differentiation. When Briggs and King injected nuclei obtained from late frog gastrulae into enucleated frog eggs, a few normal tadpoles developed, though many of the eggs differentiated abnormally. They interpreted the preponderance of abnormal development as indicating that the nuclei used for transplantation had already undergone permanent changes that limited their capacity to promote differentiation. But the inability of these nuclei to promote normal development could just as well have been due to damage inflicted by the transplantation procedure itself.

The nuclear transplantation approach was subsequently extended to other cell types by John Gurdon, whose experiments led him to a conclusion different from that of Briggs and King. Instead of embryonic nuclei, Gurdon utilized nuclei from differentiated cells lining the intestine of feeding tadpoles. When transplanted into enucleated eggs, about 10 percent of these nuclei programmed the development of normal tadpoles; some of these tadpoles even matured into adult frogs capable of sexual reproduction (Figure 15-28). Those eggs that failed to develop normally were found to exhibit chromosomal abnormalities not present in the original donor nuclei, suggesting that chromosomal damage had occurred during the transplantation procedure. Gurdon therefore concluded that nuclei from intestinal cells are normally **totipotent,** that is, they possess all the genes required for programming the development of an entire new organism.

Genetic totipotency has also been demonstrated for differentiated plant cells. In this case, however, the nuclei do not need to be removed from their normal cytoplasmic environment and transplanted into an egg cell in order to express totipotency. During the late 1950s Frederick Steward discovered that individual cells isolated from carrot roots and placed in tissue culture will eventually grow into complete carrot plants with normal roots, shoots, and leaves. Subsequent to these pioneering observations, differentiated cells isolated from many other kinds of plants have also been shown to be totipotent. In such cases the expression of totipotency appears to be triggered simply by freeing cells from their normal contacts with neighboring cells.

In 1971 Itaru Takebe and his colleagues in Japan introduced an important refinement to the technique for growing plants from single cells. By treating plant tissues with the enzymes pectinase and cellulase, they were able to generate a suspension of wall-less cells termed **protoplasts** (Figure 15-29). When such protoplasts are placed in an appropriate culture medium they grow, synthesize new walls, and begin to divide. Each protoplast gives rise in this way to a small clump of undifferentiated cells, termed a **callus,** which eventually begins to develop shoots and roots. At this stage the embryonic plant can be placed in soil, where it will develop into a mature plant identical to the original parent from which the protoplast was derived. One of the major advantages of working with protoplasts is that the absence of the cell wall allows one to carry out genetic

FIGURE 15-28 Photograph of a mature frog that developed from an egg cell in which the nucleus had been replaced with a nucleus obtained from a highly differentiated intestinal epithelial cell. The frog is normal in all respects, indicating that the intestinal nucleus contains all the genes required for directing normal development. Courtesy of J. B. Gurdon.

FIGURE 15-29 Light micrograph of protoplasts prepared from mesophyll tissue of wild tobacco plants. The removal of the cell wall allows the plant cells to assume a spherical shape. Under appropriate conditions, individual protoplasts can be induced to develop into normal plants. Courtesy of D. A. Evans.

manipulations by either fusing together protoplasts of different genetic strains or by introducing purified genes isolated by recombinant DNA techniques. Such approaches open the possibility of creating new plant strains characterized by increased crop yields, enhanced disease resistance, or an enhanced ability to fix nitrogen.

The Role of Gene Expression in Development

The realization that the nuclei of differentiated animal and plant cells retain all the genes required for programming the development of a complete new organism implies that some mechanism other than a permanent change in the genes themselves must be responsible for cell differentiation. Since differentiated cells contain different kinds of proteins but the same set of genes, it can be concluded that it is the pattern of gene expression that is changing, that is, a somewhat different spectrum of genes is being expressed in each cell type. Since a large number of steps intervene between a DNA sequence and the ultimate formation of a functional protein product, a large number of potential mechanisms for controlling gene expression can be envisioned.

One of the most obvious ways of changing the pattern of gene expression is to alter the spectrum of genes

being transcribed into messenger RNA. Once a particular gene has been transcribed, many subsequent steps intervene prior to the formation of the final protein product encoded by the gene. As discussed in Chapter 11, the initial RNA transcript must undergo posttranscriptional processing steps such as cleavage, addition of poly(A), and methylation before the final RNA product is produced. After transport to the cytoplasm, translation of messenger RNA may be influenced by the availability of transfer RNAs, ribosomes, and initiation factors, as well as by the presence of enzymes that degrade messenger RNA. Even after the messenger RNA has been translated into a polypeptide chain, the functional activity of the protein may require posttranscriptional events such as covalent modification of amino acid side chains, cleavage of the polypeptide chain, or interaction with cofactors, allosteric regulators, or other protein molecules. For a description of the molecular mechanisms involved in these various levels of control, the reader is referred to appropriate chapters earlier in the book.

Although the preceding discussion indicates that a multitude of mechanisms are capable of altering the expression of individual genes during cell differentiation, the question still exists as to which particular genes play the crucial roles in guiding this developmental process. In recent years two organisms have become the favorite models of study in attempts to identify genes involved in specific developmental processes. One of these is the roundworm *Caenorhabditis elegans*, an organism that in spite of its small size contains many of the diverse cell types present in higher animals. During the early 1980s every cell division, cell migration, and cell death leading from the fertilized egg to the adult roundworm was traced out, making *C. elegans* the most complex organism for which a complete cell lineage is available. Such an accomplishment allows the identification of the pathways followed by all the organisms' cells as they become committed to their particular fates, although it provides no direct information concerning the mechanisms involved. The next major step, therefore, has been to identify mutant organisms exhibiting developmental abnormalities and to attempt to identify and characterize the specific genes involved. Although this approach is still in its infancy, some interesting findings are beginning to emerge. For example, a so-called *binary switch* gene has been identified that causes one cell to follow the developmental pathway normally followed by another type of cell. Unraveling the molecular mechanism of action of this type of gene might provide some major insights into the factors that cause a cell to differentiate in a particular way.

The other organism being intensively studied in attempts to identify the genes responsible for guiding cell differentiation and development is the fruit fly, *Drosophila*. One of the more interesting findings to emerge from studies of this organism involves an unusual class of genes known as **homeotic genes.** When mutations occur in one of these genes, a strange thing happens during embryonic development; one part of the body is replaced by a structure that normally occurs somewhere else. For example, mutations in the homeotic gene called *Antp* cause the antennae of the fly to be replaced by a pair of middle legs. Although such a phenomenon might at first glance appear to be no more than an oddity of nature, the fact that legs are formed in the wrong place suggests that the *Antp* gene normally functions in controlling the proper spatial organization of the developing fly. One of the most interesting discoveries to emerge from the study of the *Antp* gene is that it contains a ~180 base-pair sequence near its 3′ end that is homologous to similar sequences located in other homeotic genes. Termed the **homeo box,** this sequence has been used as a hybridization probe to identify additional genes that might play a controlling role in development. Such studies have also led to the identification of homeo box sequences in organisms as diverse as sea squirts, frogs, mice, and humans, which represents an evolutionary distance of more than 500 million years. This widespread occurrence suggests that homeo-box containing genes serve an important developmental function that has been highly conserved during evolution. It is therefore hoped that the study of such genes may ultimately provide a key to the understanding of development not only in *Drosophila,* but in higher organisms, including humans, as well.

Gene Injection Techniques

Another new tool for investigating the behavior of specific genes during development involves the injection of purified genes directly into embryos. Beatrice Mintz and her associates were the first to show that injected foreign genes can become stably integrated into the chromosomes of recipient embryos. In these experiments a cloned DNA molecule containing both a viral gene coding for the enzyme thymidine kinase and the human *β*-globin gene was injected into fertilized mouse eggs (Figure 15-30). The eggs were then implanted into a mouse oviduct and allowed to develop into normal embryos. Some of the mice born in these experiments were found to contain the human and viral genes incorporated into their DNA. In a few cases these animals even passed the foreign DNA sequences on to their offspring, indicating that these genes had become incorporated into their germ cells.

Although foreign genes have been successfully introduced into mouse embryos, expressing these genes is a bigger problem. Some eggs injected with the human growth hormone gene have been found to yield a functional protein product (page 382), but detection of globin gene expression has thus far been less successful. It

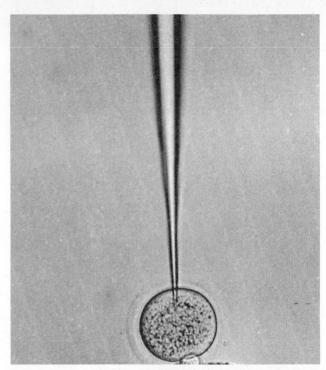

FIGURE 15-30 Photograph showing a microneedle in the process of injecting DNA into the pronucleus of a fertilized mouse egg. DNA injected in this way can become permanently incorporated into the mouse cell chromosomes. Courtesy of J. W. Gordon and F. H. Ruddle.

is not yet clear whether the problem lies in the cloned globin genes themselves, or something that happens to these cloned genes after they are injected into the mouse eggs (e.g., methylation or integration into an inappropriate location in the host cell chromosome). In either case, work is progressing rapidly in this area and the conditions necessary for optimizing foreign gene expression in mice are being worked out.

A slightly different approach has been employed for introducing foreign genes into the fruit fly *Drosophila.* Gerald Rubin and Allan Spradling have facilitated the entry of cloned genes into *Drosophila* embryos by joining them to a **transposable element,** which is a segment of DNA that readily moves from place to place in the genome (page 367). In one such experiment the *rosy* gene, which codes for the enzyme xanthine dehydrogenase that is involved in the production of normal eye color, was linked to a transposable element called a **P-element.** This hybrid gene was then cloned in a bacterial plasmid, and the cloned plasmid injected into *Drosophila* embryos lacking xanthine dehydrogenase. A large proportion of the injected embryos developed into flies with normal eye color, indicating that the injected xanthine dehydrogenase gene is functional. Furthermore, these flies give rise to offspring with normal eye color, indicating that the injected gene is passed in a functional state from generation to generation through the germ cell line. The use of transposable elements seems to significantly increase the efficiency of the gene transfer process, and may eventually be adapted to genetic manipulations in higher animals and plants.

The development of reliable techniques for injecting foreign genes into embryos provides great promise for future investigations on the mechanisms underlying the control of gene expression during development. By introducing defined changes into a cloned gene prior to injection and then watching the fate of this DNA as it is passed on to various cell types and cytoplasmic environments, it should be possible to unravel the exact contributions of gene structure and cytoplasmic environment in the developmental control of gene expression.

Protein Turnover and Cell Differentiation

Although our discussion of cell differentiation has been focused largely at the level of gene expression, regulation does not stop once the appropriate set of proteins is produced by a cell undergoing differentiation. Even after gene expression has led to the formation of the proper set of proteins, these proteins are subject to *turnover*, that is, breakdown and resynthesis. In this way a dynamic state is established in which the spectrum of proteins present in a given cell is subject to future change.

It follows that at any given moment, the quantity of a particular protein present in a cell is a function of both its rate of formation and its rate of breakdown. Mathematically, this relationship can be expressed by the equation

$$P = \frac{K_s}{K_d} \qquad (15\text{-}1)$$

where P is the amount of a particular protein present within the cell, K_s is the rate order constant for its synthesis, and K_d is the rate order constant for its degradation. Rate constants for degradation are often expressed in the form of a **half-life** ($t_{1/2}$), which corresponds to the amount of time required for half the molecules existing at any moment to be subsequently degraded. Thus, $t_{1/2}$ can be calculated from K_d by the equation

$$t_{1/2} = \frac{\ln 2}{K_d} \qquad (15\text{-}2)$$

The half-lives of cellular proteins range from as rapid as a few minutes to as long as several weeks. Such differences in half-life can exert rather dramatic and surprising effects on the protein contents of differentiating cells. A striking example of this phenomenon involving the liver enzymes *tryptophan pyrrolase* and *arginase* was uncovered in the laboratory of Robert Schimke in the mid-1960s. When liver cells are exposed to the steroid hormone cortisone, the quantity of tryptophan pyrrolase present in the liver increases almost tenfold, while arginase increases only slightly (Figure

15-31). At first glance such data would seem to indicate that cortisone selectively stimulates the synthesis of tryptophan pyrrolase. In fact, careful analysis of the data has revealed that cortisone stimulates the synthesis of both tryptophan pyrrolase and arginase to about the same extent, in spite of the fact that there is a much larger increase in the amount of tryptophan pyrrolase. The explanation for this apparent paradox is to be found in the discovery that tryptophan pyrrolase has a much shorter half-life than arginase. Because of the relation-

FIGURE 15-31 Results of experiment revealing the influence of protein half-lives on protein behavior. (a) Graph summarizing the relative amounts of tryptophan pyrrolase and arginase present in the livers of rats before (black bars) and 4 hr after (colored bars) injection of hydrocortisone. (b) Graph summarizing the relative rates of synthesis of these two enzymes under the same conditions. Note that the rate of synthesis of these two enzymes is equally stimulated by hydrocortisone, even though the total amount of tryptophan pyrrolase present in the liver increases much more dramatically than does the total amount of arginase. This difference in behavior is caused by the fact that tryptophan pyrrolase has a shorter half-life than arginase (2.5 hr versus 96 hr).

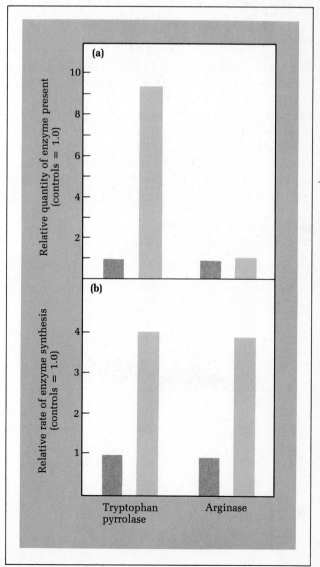

ship summarized in Equation 15-1, it can be shown mathematically that enzymes with short half-lives (large values of K_d) are more sensitive to changes in synthetic rate than proteins with long half-lives. For this reason increases or decreases in the overall rate of protein synthesis will cause the levels of some proteins to be more dramatically affected than others.

The factors determining the rates at which various proteins are normally degraded are not well understood, but it is clear that a certain amount of control can be exerted over this process. In rats deprived of food, for example, the amount of arginase present in the liver doubles within a few days in spite of the fact that the rate of arginase synthesis does not change from its normal value. Arginase degradation, however, virtually comes to a halt in starving rats, and it is this change in degradation rate that is responsible for the observed increase in arginase. The decrease in arginase degradation is a highly selective effect; the degradation rates for most proteins in the starving rat have been found to increase, not decrease (Figure 15-32).

Differentiation and Cell Division

During the development of multicellular organisms cell differentiation is inevitably associated with cell division because of the need for transforming a single cell, the fertilized egg, into an organism composed of a vast number of different kinds of cells. This intimate relationship between cell division and cell differentiation raises the question of whether there is any obligatory relationship between the two. According to a theory first proposed by Howard Holtzer, passage through a

FIGURE 15-32 Experiment demonstrating selective regulation of the degradation rate (half-life) of the liver protein, arginase. Degradation rate is measured by first injecting rats with a radioactive amino acid to allow radioactive protein to be synthesized, and then measuring the rate at which the radioactivity is lost from the protein fraction due to breakdown of the radioactive protein molecules. The data reveal that in starved rats arginase degradation is inhibited, while the degradation rate of total protein is slightly enhanced.

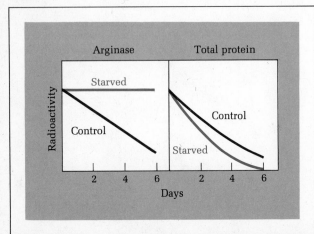

specific stage of the cell cycle is required for producing the changes associated with cell differentiation. Some proponents of this theory believe that the critical stage of the cell cycle involves unraveling the DNA in such a way as to make new regions available for transcription. Cell cycles producing changes in differentiation are referred to as *quantal* cycles to distinguish them from normal cell divisions. Some support for the above theory is to be found in the discovery that inhibitors of DNA replication and cell division block differentiation of certain cell types. Unfortunately, such results may also be due to toxic effects of the drugs employed, so definitive evidence for the existence of quantal cell cycles is yet to be obtained.

CELL-CELL INTERACTIONS AND MORPHOGENESIS

In multicellular animals and plants the differentiation of individual cells does not occur in isolation, but as a coordinated part of the development of the organism as a whole. This coordination is made possible by cell-cell interactions through which cells influence each others' movements, metabolism, and states of differentiation. The net result of such interactions is the patterning of cells into tissues and organs. In the following sections we shall consider some of the cellular activities that underlie this process of **morphogenesis** (development of form).

Cell Movements and Cell Shape Changes

In animal morphogenesis, cell movements and changes in cell shape play important roles in early development. The first major contribution of cell movement to embryonic development occurs during formation of the gastrula, when a specific group of cells migrates from the surface of the blastula to its interior cavity, converting a hollow ball of cells into a multilayered structure. Later in development cells migrate over longer distances. For example, the **neural crest,** a structure located above the developing spinal cord, contains cells that migrate to diverse regions of the embryo, ultimately differentiating into cartilage, nerve cells, glial cells, pigment cells, and adrenal cells. Long-distance migration also occurs with primordial germ cells, which originate near the developing gastrointestinal tract and only later migrate to their final position in the ovary or testis.

The crucial question regarding the preceding kinds of movements concerns the issue of how cells are induced to migrate in the proper directions. One important factor in guiding cell migration appears to be the composition and organization of the extracellular matrix. It has been shown in amphibian embryos, for example, that cells located in the roof of the blastula are surrounded by an extracellular matrix enriched in

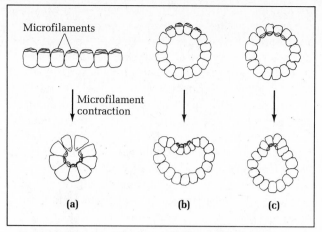

FIGURE 15-33 Schematic diagram illustrating how changes in cell shape caused by microfilament contraction can convert a layer of cells into a spherical mass (a) or can cause a specific group of cells to invaginate (b) or evaginate (c).

fibronectin-containing fibrils. If antibodies against fibronectin are injected into such a blastula, the cell migrations responsible for changing the blastula into a gastrula are eliminated. The organization of the basal lamina also apparently plays a role in cell migration. For example, migration of neural crest cells has been found to be preceded by the breakdown of the surrounding basal lamina.

Directional movements can also be caused by changes in cell shape, even when cells remain in the same positions relative to their neighbors. This point is illustrated in Figure 15-33, which shows that if each cell in a flat sheet of cells were to contract on the same side, the sheet would be converted into a rounded mass. Likewise, constriction of a selected group of cells within a larger cell layer would cause the involved portion of the tissue to invaginate or evaginate, depending on where the constriction were to occur. An example of this latter phenomenon occurs in the developing chick oviduct, where glands form by evagination of cells from the oviduct wall. Joan Warren and Norman Wessels have shown that development of these evaginations is preceded by the formation of bands of microfilaments across the inner ends of the cells destined to evaginate. These microfilaments are thought to be involved in a contractile event that narrows the inner ends of the cells, causing them to evaginate. In support of this interpretation, it has been found that the drug cytochalasin B causes both the microfilaments to disappear and the evaginating cell mass to recede back into the oviduct.

Cell Adhesion

In addition to the mechanisms described in the preceding section, cell-cell adhesion also plays an important role in guiding the cellular interactions and movements that occur during development. The importance of se-

lective cell-cell adhesion for the construction of tissues and organs was first suggested by the pioneering studies of Aaron Moscona on embryonic cells that had been artificially dissociated from each other and then recombined. These studies revealed that when cells of two different types are dissociated and then intermixed, they first bind to one another randomly; eventually, however, they become arranged into two groups, a central clump composed of cells of one type and an outer layer of cells of the second type.

In analyzing this sorting-out behavior in a variety of cell types, Malcolm Steinberg has uncovered the existence of an interesting pattern. If a mixture of cell types A and B forms aggregates in which the A cells surround the B cells, and a mixture of cell types B and C forms aggregates in which the B cells surround the C cells, then it can be predicted that a mixture of cell types A and C will yield aggregates in which the A cells surround the C cells. As an example of this phenomenon, let us consider the behavior of heart, cartilage, and liver cells. In a mixture of heart and cartilage cells, the heart cells take up the external position; in a mixture of liver and heart cells, the liver cells take up the external position. It can therefore be predicted that liver cells will aggregate external to cartilage cells (Figure 15-34). Using this approach Steinberg has arranged a large group of cells into a hierarchical list in which each cell type reaggregates external to those cell types preceding it, and internal to those cell types listed below it (Table 15-1).

What mechanism underlies this selective sorting out that takes place when cells of different types are mixed together? According to the simplest interpretation, the first event that occurs upon the mixing together of two cell populations (A and B) is a random, reversible binding between cells. If the bond between two A cells or two B cells is stronger than the bond between an A cell and a B cell, however, a gradual selection for A-A and B-B interactions and a resulting sorting out of pure A and B populations will occur. According to Steinberg, the reason that one cell population always ends up on the inside and the other on the outside is related to the existence of differences in adhesive strength. If the A-A bond is stronger than the B-B bond, the more cohesive A cells will maximize their cell-cell

TABLE 15-1
Several tissue types organized according to pattern of aggregation[a]

Epidermis
Cartilage
Pigmented epithelium (eye)
Myocardium (heart)
Nervous tissue (neural tube)
Liver

[a] The cells of each listed tissue tend to aggregate internal to the cells of the tissues situated lower in the list.

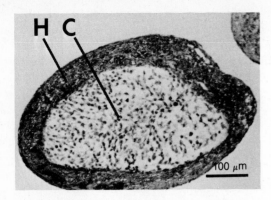

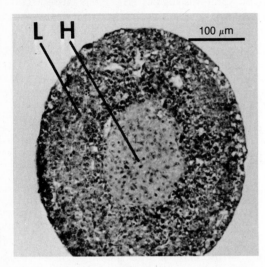

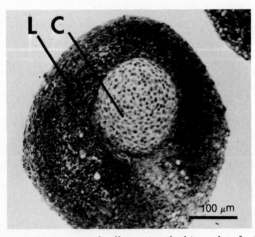

FIGURE 15-34 Micrographs illustrating the hierarchy of cell-cell aggregation manifested by liver (L), heart (H), and cartilage (C) cells. *(Top)* When cartilage and heart cells are mixed together and allowed to reaggregate, the heart cells end up on the exterior. *(Middle)* When liver and heart cells are mixed together, the heart cells now end up in the interior. *(Bottom)* In a mixture of cartilage and liver cells, the liver cells end up on the exterior. Data from such experiments has provided the information summarized in Table 15-1. Courtesy of M. S. Steinberg.

interactions by forming a spherical mass on the inside of the aggregate; the less cohesive B cells will be forced to accumulate on the outer free surface of the mass, where there is relatively less cell-cell contact and more exposure of cell surfaces to the external medium.

The obvious importance of cell-cell adhesion in guiding tissue and organ formation has led to a considerable interest in identifying the molecules responsible for mediating these effects. One way of identifying such molecules is to develop antibodies to various embryonic tissues and to then test the ability of the resulting antibodies to block cell-cell adhesion. In this way Gerald Edelman and his associates have been able to identify several cell-surface glycoproteins that appear to function as **cell adhesion molecules (CAMs).** One of these, designated **N-CAM** because of its association with neural development, is detectable on all neurons in the peripheral as well as central nervous systems. N-CAM has been implicated in neural development by the discovery that exposure of embryonic cells to antibodies against N-CAM disrupts the orderly formation of neural tissues. Moreover during early development, N-CAM as well as other cell adhesion molecules have been found to appear and disappear in various regions of the embryo, again suggesting a role for these molecules in guiding tissue and organ formation.

In addition to its role in guiding the formation of tissues and organs, the establishment of proper cell-cell contacts also appears to be an important prerequisite for the acquisition of certain types of cell functions. In embryonic retina, for example, it has been shown that hormonal regulation of enzyme synthesis requires the existence of cell-cell contacts. The enzyme glutamine synthetase, which is normally synthesized in embryonic retina in response to stimulation by the hormone hydrocortisone, is not manufactured in isolated retina cells exposed to this hormone. But if the isolated retinal cells are first permitted to reaggregate into multicellular clusters, glutamine synthetase is once again formed in response to hormonal stimulation (Figure 15-35).

Several possible mechanisms may underlie the ability of cell-cell contacts to influence tissue function. One is that contact between adjacent cell surfaces induces rearrangements of plasma membrane macromolecules that in turn trigger metabolic alterations within the cells themselves. Another possibility is that cell-cell contacts stimulate the formation of communicating junctions that permit direct passage of molecules or electrical signals between adjacent cells. Intimate coupling of this type may facilitate the coordinated development of tissues and organs by permitting chemical or electrical signals to pass directly from cell to cell.

Embryonic Induction

Another role of cell-cell interactions during development is to coordinate the differentiation of adjacent

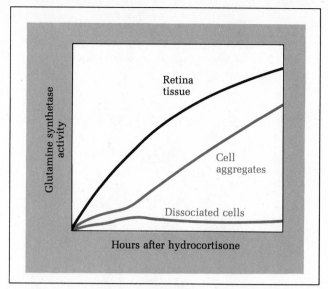

FIGURE 15-35 Data from experiment revealing the existence of a relationship between cell-cell contacts and tissue function. In intact retina the enzyme glutamine synthetase is formed in response to treatment with the hormone hydrocortisone (black curve). This response to hormone treatment does not occur in isolated retina cells (lower colored curve) unless they are allowed to reaggregate into clusters (upper colored curve).

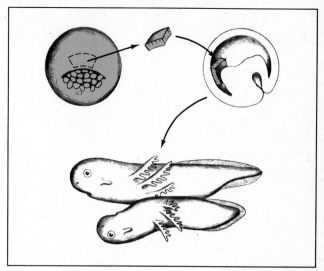

FIGURE 15-36 Outline of experiment demonstrating the existence of primary embryonic induction. When cells obtained from the dorsal lip of the blastopore of a darkly pigmented newt gastrula were transplanted to a nonpigmented newt gastrula, a second embryo developed that was joined to the host embryo like a Siamese twin. Since the Siamese twins are both nonpigmented, the pigmented dorsal lip cells of the graft must have induced the host's nonpigmented cells to differentiate into the second embryo.

groups of cells. Experiments in which cells from one embryo are grafted to a new location on another embryo have dramatically demonstrated the degree to which one group of cells may influence the pathway of differentiation pursued by neighboring cells. For example, if cells normally destined to become nervous tissue are removed from an amphibian embryo during early gastrula stage and transplanted to another embryo in a region destined to become epidermis, the grafted cells will evolve into epidermis instead of nervous tissue. Likewise, prospective epidermal cells grafted to an area fated to become neural tissue will differentiate into neural tissue.

But this ability or **competence** of cells to differentiate in a variety of different ways, depending upon their location, is a transient property. If the same experiment is performed on a late gastrula (about two days later), the results are quite different. Prospective neural tissue placed in the region of prospective epidermal tissue no longer differentiates into epidermis like its neighbors, but rather forms neural tissue as it would normally do. Conversely, prospective epidermis placed in the region of prospective neural tissue develops into epidermis. The developmental potential of the prospective epidermal and neural cells has thus become restricted or **determined.**

What occurs between early and late gastrula to account for such changes in developmental potential? During this period cells lying directly above the blastopore invaginate into the interior of the embryo to produce a multilayered structure in which some of the invaginated cells lie directly under the prospective

neural cells (see Figure 15-1). In order to ascertain whether it is the presence of these newly invaginated cells that causes the overlying cells of the gastrula to differentiate into neural tissue, Hans Spemann and Hilde Mangold performed a classic series of transplantation experiments on newt embryos in the early 1920s in which the **dorsal (upper) lip** of the blastopore from one gastrula was grafted to a different position in another gastrula. In order to distinguish cells of the donor from those of the host, two strains of newt with different pigmentation were employed.

The remarkable result obtained from these experiments was that in its new location in the host gastrula, the grafted blastopore material triggered formation of an entire new embryo joined to the host embryo like a Siamese twin (Figure 15-36). The pigmentation of the second embryo was similar to that of the host rather than the graft donor, indicating that the grafted blastopore cells had *induced* the host cells to form a wide variety of new structures they would not normally have formed. This type of induction, in which an entire new embryo is formed, is referred to as **primary embryonic induction.**

In addition to primary embryonic induction, various kinds of **secondary inductions** exist in which one group of cells induces a neighboring group of cells to differentiate in a particular way. A classic example of secondary induction occurs in the developing eye, where formation of the lens is induced by contact between the prospective lens cells and the overlying ectoderm. Likewise, differentiation of epithelial cells into glandular structures such as the pancreas, thyroid, kidney, thymus, and salivary glands involves an interaction between the epithelial cells and underlying mesenchymal cells.

Considerable effort has been expended in attempting to ascertain whether induction is mediated by diffusion of soluble chemical signals from inducing to target cells, or whether direct cell-cell contact is required. To investigate this issue, Clifford Grobstein devised a method for studying inductive interactions between cells separated from each other by a porous membrane filter designed to prevent the migration of whole cells, but not the passage of diffusible molecules. The first experiments of this type, reported in the early 1950s, revealed that salivary gland epithelium will continue to differentiate into glandular structures even when it is separated from its inducing mesenchyme by a thin membrane filter. Similar results were soon obtained for several other tissue inductions.

Though originally interpreted as support for the conclusion that diffusible molecules trigger induction, the filters employed in these pioneering studies contained an irregular network of pores, making it difficult to determine by electron microscopy whether or not cell-cell contacts are actually prevented by the filter. When filters with straight uniform pores subsequently became available, it was quickly discovered that secondary embryonic inductions are associated with the establishment of cell-cell contacts through the pores of the membrane (Figure 15-37). If membranes with extremely small pores (0.1 μm) are employed, cell-cell contacts can be completely prevented, but under such conditions induction is also blocked. It therefore appears as if cell-cell contact is required for secondary induction.

Such a generalization does not appear to hold for primary induction, however; dorsal blastopore cells can induce ectoderm to differentiate into neural tissue even across membranes with pores small enough to prevent cell-cell contact, indicating the existence of a diffusible signal. Unfortunately, the search for the diffusible substance responsible for primary induction has had a long and disappointing history. The fact that primary induction does not require living cells was first discovered in 1930, when boiled dorsal lip material was shown to trigger differentiation of ectoderm into neural tissue. Though this observation stimulated attempts to purify the inducing substance from dorsal lip extracts, a vast array of materials was soon found to trigger primary induction. The list of inducers includes not only natural substances such as steroids, fatty acids, glycogen, and nucleotides, but also artificial materials such as the dye methylene blue. The final blow came when Johannes Holtfreter reported that even a simple change in the pH of the medium in which fragments of ectoderm are suspended causes them to differentiate into nerve cells. It thus appears as if ectoderm is in an arrested state poised for differentiation into neural tissue, and that even non-

FIGURE 15-37 Cross section through a membrane filter being utilized in embryonic induction experiments. The membrane is being used to separate mesenchymal cells (M) lying above the filter from spinal cord tissue located beneath the filter. This electron micrograph shows that cells from opposite sides of the filter can make close contacts (C) with each other through pores in the filter. Induction across such membrane filters may therefore be mediated by cell-cell contacts rather than by diffusible molecules. Courtesy of J. Wartiovaara.

specific stimuli are capable of releasing it from this arrested state.

Though this turn of events generally discouraged the search for specific inducers of differentiation, some evidence does support the idea that certain macromolecules can direct differentiation in specific ways. For example, Tuneo Yamada and Heinz Tiedemann have isolated a "mesodermalizing" protein fraction from amphibian embryos that will cause cultured newt ectoderm to differentiate into muscle, notochord, and kidney tubules, even though this ectoderm is normally destined to differentiate into epidermis. This *same* ectoderm can be made to differentiate into neural tissue by exposing it to a "neuralizing" protein fraction that these investigators prepared from chick brain.

The above observations suggest that in spite of the ability of nonspecific stimuli to trigger primary induction under experimental conditions, normal development may involve control of differentiation by specific proteins or other molecules diffusing between cells. Even in the case of secondary induction, where cell-cell contacts appear to be required, the participation of inducing molecules is not ruled out because it is possible that close contact is needed for the passage of such molecules between cells.

Hormone-mediated Interactions

Both primary and secondary induction are based on interactions between cells located relatively close to each other in the embryo. Cells separated by larger distances may also interact with one another, in this case utilizing chemical signals such as hormones or growth factors to facilitate communication. Most, if not all, hormones regulate some phase of development in their target organs. For example, the gonadotropic hormones stimulate growth of the sex organs, thyrotropin fosters growth of the thyroid, adrenocorticotropin triggers growth of the adrenals, and so forth.

In addition to hormones, related substances termed **growth factors** play important roles during early development. Two well-characterized agents in this category are *nerve growth factor*, which is required for the normal development of nerve fibers, and *epidermal growth factor*, which is involved in regulating the maturation of epithelial tissues. As we will see later the distinction between growth factors and hormones is somewhat arbitrary because these two classes of regulatory molecules closely resemble each other in both structure and mechanism of action. In fact, nerve growth factor and epidermal growth factor bear a resemblance in some of their properties and behavior to the protein hormone, insulin. The mechanisms underlying the specificity and mode of action of hormones and growth factors will be elaborated upon in the following chapter.

AGING

As organisms grow older, they gradually deteriorate and eventually die. Though disease and natural disasters may accelerate the aging process, they are not responsible for it. In spite of the common belief to the contrary, modern medicine has not significantly increased the maximum lifespan of the human species. By preventing early deaths, the conquest of infectious diseases has improved our *average* life expectancy, but the *maximum* age attainable by humans has not changed significantly since the beginning of recorded history. Medical progress has simply allowed more of us to attain the limits of what appears to be a relatively fixed maximum lifespan of about a hundred years.

For many years the aging process was thought to be restricted to intact organisms; individual cells were considered to be potentially immortal when cultured under appropriate conditions. This idea evolved primarily from the work of Alexis Carrel, who claimed to have kept chick heart cells growing in culture for over 30 years, which is well past the normal lifespan of the organism itself. But in 1961 Leonard Hayflick undermined this view by showing that embryonic human fibroblasts placed in culture divide vigorously at first, but then slow down, deteriorate, cease to multiply at about the 50th division, and subsequently die. The relationship of this phenomenon to aging of the organism as a whole became evident when the behavior of fibroblasts obtained from humans of different ages was compared. In contrast to embryonic fibroblasts, which are capable of dividing about 50 times in culture, fibroblasts from adults were found to multiply only 15–30 times before dying. Furthermore, fibroblasts obtained from young children suffering from *progeria* or *Werner's syndrome*, two rare diseases that cause youngsters to age prematurely, divide only 2–10 times in culture. The maximum number of divisions normal cells undergo in culture is also related to the average lifespan of the species. For example cells of the Galapagos tortoise, whose maximum lifespan is about 175 years, survive for over 100 doublings in culture; in contrast cells obtained from

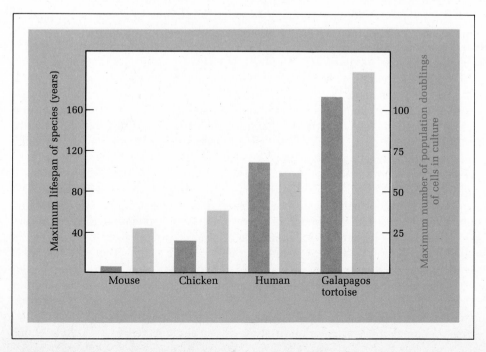

FIGURE 15-38 Histogram illustrating the general relationship between the maximum lifespan of an organism (black bars) and the maximum number of times its cells will divide in culture before dying (colored bars).

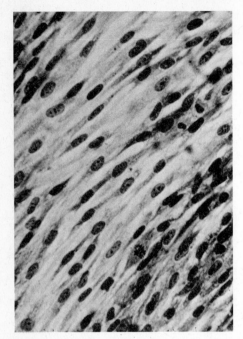 

FIGURE 15-39 Microscopic appearance of young and old fibroblasts maintained in tissue culture. *(Left)* Young fibroblasts that have divided a relatively small number of times in culture exhibit a thin, elongated shape. *(Right)* These aging fibroblasts, which have divided roughly 50 times in culture, are undergoing a variety of degenerative changes and will soon die. Note the striking difference in cell morphology. Courtesy of L. Hayflick.

mice, whose life expectancy is only a few years, divide less than 30 times in culture before dying (Figure 15-38).

The idea that normal cells can divide only a limited number of times, and that this number is correlated with the age and lifespan of the organism from which the cells are obtained, has opened the way for employing tissue culture as a model system for studying the process of aging at the cellular and molecular levels. Many cellular properties have been found to change as cultured cells age and approach their maximal lifespan. Among these are a decreased activity of certain metabolic pathways, an increase in hydrolytic enzymes, a decrease in the ability to repair DNA and to synthesize RNA, and alterations in cell morphology (Figure 15-39). Such findings have led Hayflick to attribute senescence to gradual losses in cell function that occur as the maximum number of doublings is approached. The problem in using this approach for unraveling the mechanism of aging, however, is determining which of the changes observed in aging cells are causally related to the aging process, and which alterations are simply by-products of aging. Because each species exhibits its own characteristic lifespan, it is hard to escape the conclusion that the cellular events associated with aging are at least in part genetically determined. Some biologists believe that a genetic program for aging and death is a normal part of development, an idea readily compatible with the fact that the successive replacement of each generation of organisms with a new generation is an integral part of evolution. Despite this argument for the existence of a normal genetic program for aging, random damage of

cells has also been implicated in the aging process.

A good illustration of the contribution of both genetic factors and random damage to aging can be found in the work of Lester Packer and James Smith on vitamin E, an antioxidant known to interfere with chemical reactions involving highly reactive intermediates termed *free radicals*. It has been proposed that free radicals, which are produced by normal cell metabolism as well as by the interaction of noxious chemicals or radiation with cells, react with and chemically modify other cellular macromolecules, thereby inducing the gradual deteriorations in molecular function that accompany aging. Packer and Smith have claimed that culturing human fibroblasts in the presence of vitamin E, which inhibits free-radical reactions, extends the lifespan of these cells to over 100 cell divisions from its normal value of around 50 divisions. Such results suggest that the normal lifespan of cells in culture does not reflect a cell's full genetic potential for longevity, but is a value restricted by free-radical induced damage.

Free-radical reactions are not the only potential source of damage in aging cells. Because the nucleic acid base-pairing mechanism is inherently imperfect, spontaneous errors in DNA replication, RNA synthesis, and protein synthesis continually occur. Such errors contribute to the formation of defective proteins, some of which may be abnormal nucleic acid polymerases or ribosomal proteins that foster the accumulation of other errors. If such errors contribute to the aging mechanism, then chemicals that interfere with normal transcription or translation might be expected to accelerate aging, while mechanisms for repairing or destroying abnormal nucleic acids or proteins would be expected to retard aging.

Though the general applicability of this view of cellular aging is yet to be firmly established, it has been discovered that errors in protein synthesis *can* contribute to cell death under certain conditions. The most spectacular observations involve the *leu-5* mutant of *Neurospora*, a strain containing an abnormal leucyl-tRNA synthetase that permits the joining of amino acids other than leucine to the transfer RNA normally designed to carry leucine. The *leu-5* strain of *Neurospora* has been found to age much more rapidly than normal, presumably due to the multiple errors in protein synthesis caused by the defective synthetase.

The above experiments do not, however, clarify the issue of whether or not such errors are involved in the normal aging process. In an attempt to determine whether accumulation of defective cytoplasmic proteins plays a significant role in determining the fixed proliferative capacity of normal cells, Hayflick has fused nuclei obtained from young cells with old cells whose nuclei have been removed. Such cells, with their "young" nuclei and "old" cytoplasm, continue to divide like young cells; on the other hand, cells constructed of "young" cytoplasm and old "nuclei" exhibit the limited proliferative capacity characteristic of old cells. Hence the control of aging in cultured cells appears to reside in the nucleus rather than in the cytoplasm. This finding tends to support the conclusion that senescence is not caused by accumulation of defective cytoplasmic macromolecules but is rather a programmed genetic event, analogous to cell differentiation, in which the cell is ultimately directed by its genes to die. However, the possible accumulation of errors in nuclear DNA and/or proteins in old nuclei cannot be ruled out as an alternative interpretation of these experiments.

In spite of a vast amount of data and theorizing, we still do not know to what extent aging is genetically programmed, and to what extent it results from accumulated errors. Perhaps new insight will eventually be obtained by studying the properties of cells that do not appear to age at all. The major cell type in this category is the cancer cell. As we shall see in Chapter 20, cancer cells are immortal when placed in culture, continuing to divide for as long as they are provided with adequate nutrients. Though transformation of cells to the malignant state is certainly detrimental to the organism as a whole, something learned from studying these immortal cells may one day help us to prevent the deleterious losses of function that accompany the normal aging process.

SUMMARY

This chapter focuses on the question of how a single cell, the fertilized egg, gives rise to billions of new cells of differing types organized in precise patterns to form the tissues and organs of the adult organism. The first step in this process is the formation of the haploid gametes destined to unite during fertilization. Female gametogenesis, or oogenesis, involves mitotic proliferation of oogonia followed by meiotic division to reduce the chromosome number to the haploid level. In organisms where the egg must provide most of the nutrients for early embryonic development, oogenesis is associated with a dramatic burst in cell growth that increases the cell mass as much as several orders of magnitude. In such oocytes, like those of amphibians and birds, this growth phase is associated with an increased synthesis of ribosomal and messenger RNAs, development of lampbrush chromosomes, amplification of ribosomal RNA genes, formation of multiple nucleoli, nuclear enlargement, accumulation of stored nutrients in the form of yolk platelets, formation of cortical granules, elaboration of microvilli and a vitelline envelope, and establishment of a cytoplasmic asymmetry in which distinct animal and vegetal poles are evident. Three of the four haploid products of meiosis are the small polar bodies that disintegrate, leaving the mature egg with the vast majority of the nutrients accumulated by the oocyte.

In male gametogenesis (spermatogenesis), spermatogonia increase their numbers by mitotic proliferation and then enter into meiosis. Meiosis yields four haploid spermatids that subsequently differentiate into mature sperm by the process of spermiogenesis. During spermiogenesis most of the spermatid's cytoplasm is discarded and the remainder of the cell differentiates into two major structures, the head and the tail. The head contains the highly compacted nucleus and the acrosome, while the tail consists of a long flagellum partially enveloped by a sheath of mitochondria that provide the ATP needed for sperm motility.

Interaction of sperm cells with the exterior region of the egg triggers fusion of the acrosomal membrane with the sperm plasma membrane and the subsequent discharge of the acrosome's hydrolytic enzymes into the external environment, where they digest a pathway through the egg coats. Contact of the sperm membrane with the egg plasma membrane triggers fusion of the two membranes and marks the beginning of fertilization. The early responses of fertilization, which occur within the first minute, include membrane depolarization, an increase in cytoplasmic calcium ion concentration, and the cortical reaction with its associated formation of the fertilization membrane. Membrane depolarization and the fertilization membrane both function to prevent polyspermy. The late responses of fertilization, which appear to be triggered by an increase in cytoplasmic pH, include activation of DNA synthesis and, in some organisms, protein and RNA synthesis.

The stimulation of DNA synthesis prepares the fertilized egg for the first mitotic division. The subsequent series of mitotic cleavages causes the massive egg to be partitioned into smaller and smaller cells, ultimately yielding a spherical mass of cells termed a blastula. When cleavage ends, cells located above the blastopore migrate into the blastula interior, producing a gastrula composed of three cell layers (ectoderm, mesoderm, and endoderm). The new interrelationships established among cells by this process trigger a combination of divisions, differentiations, and movements that ultimately lead to the formation of tissues and organs.

Development in higher plants differs in several respects from the pattern described above. The haploid egg cell is produced by meiotic division of a megasporocyte, followed by three mitotic divisions of one of the resulting megaspores. These three mitoses yield eight haploid nuclei distributed as follows: one egg cell, two polar nuclei in a common cytoplasm, two synergid cells (which function in providing nutrients to the egg), and three antipodal cells (which usually die). During male gamete formation, meiotic division of microsporocytes yields haploid microspores that develop into pollen grains by forming thick spore coats. The haploid nucleus of each pollen grain then divides mitotically, forming a generative and a vegetative nucleus. When a pollen grain comes in contact with the stigma (a projection of the female reproductive organ specialized for trapping pollen), the pollen grain develops a pollen tube that grows toward the ovary. During this time the generative nucleus divides again, yielding two nuclei. Upon reaching the embryo sac, one nucleus fuses with the egg to form a diploid zygote, while the other nucleus fuses with the two polar nuclei, yielding the triploid endosperm. Mitotic division of the fertilized egg produces a plant embryo, while the endosperm divides and accumulates nutrients designed for later nourishment of the embryo. The final result is a seed composed of a diploid embryo surrounded by the triploid endosperm; further development of the seed is arrested until appropriate conditions trigger its germination.

The major unanswered question concerning the mechanism of both animal and plant development can be stated quite simply: How do cells produced by successive divisions of a single fertilized egg and therefore inheriting the same genes come to be so different from each other? Considerable evidence suggests that cell differentiation is made possible by the fact that gene expression is influenced by the cytoplasmic environment in which a nucleus resides. Because the components present in the initial egg cytoplasm are asymmetrically organized, cells derived from successive mitotic divisions of the egg contain dissimilar cytoplasms, and thus express different genes. Experimental examples supporting the existence of cytoplasmic influences on cell differentiation include the behavior of the gray crescent of amphibian eggs and the polar lobe of molluscan eggs, the effects of animal and vegetal cyto-

plasm on chromosome diminution in *Ascaris,* and the effects of posterior egg cytoplasm on germ cell differentiation in *Drosophila.* Nuclear transplantation experiments on amphibian eggs have further revealed that nuclei removed from differentiated cells and transplanted into the cytoplasm of enucleated eggs can program the development of entire new organisms, indicating that cell differentiation does not involve the loss or irreversible inactivation of specific genes. Such genetic totipotency has also been demonstrated in plants, where removing cells from their normal contacts with neighboring cells is sufficient stimulus for triggering individual cells to grow into complete new plants. Isolated plant cells from which the cell walls have been removed (protoplasts) can be genetically manipulated by cell fusion or genetic engineering techniques prior to developing into complete plants, providing the possibility of experimentally creating new plant strains with improved genetic characteristics.

The discovery that the nuclei of differentiated cells are genetically totipotent suggests that differentiation proceeds by changes in gene expression rather than by irreversible gene loss or inactivation. These changes in gene expression may involve regulation at the transcriptional, posttranscriptional, translational, and posttranslational levels. The nature of some of the genes that control the early stages of embryonic development is beginning to emerge from studies on both *C. elegans* and *Drosophila* embryos. Among the findings to emerge is the discovery of the homeo box, a DNA sequence shared by a variety of developmental regulatory genes in *Drosophila* and present in many other organisms as well. Another new tool for investigating the mechanisms by which gene expression is controlled during development involves injection of cloned genes into embryos. By introducing defined changes into these genes prior to injection and then monitoring their expression as they are passed on to various cells, it should eventually be possible to unravel the relative contributions of gene structure and cytoplasmic environment in the developmental control of gene expression.

Inherent differences in protein turnover rates can also influence the protein composition of differentiating cells. For example, a nonselective increase or decrease in the overall rate of protein synthesis will cause a more dramatic change in protein content for proteins with short half-lives than for those with long half-lives. The factors determining protein half-lives are not well understood, but it is known that the half-lives of individual proteins can be selectively altered in response to changing environmental conditions.

Formation of tissues and organs is made possible by cell-cell interactions through which cells influence each others' movements, metabolism, and differentiation. Cell movements play their first critical role during gastrula formation, when a particular group of cells migrates from the surface of the blastula to its interior. Later in development migration occurs over longer distances; for example, the primordial germ cells originate near the gastrointestinal tract and only later migrate to the ovary or testis. Interactions with the extracellular matrix and localized cell shape changes appear to be among the factors that help to create movements that are guided in the proper direction.

Selective cell-cell adhesion, which permits cells of differing types to arrange themselves into organized patterns, is also crucial for tissue and organ formation, and may even influence tissue function by permitting direct passage of molecules or electrical signals between cells. Another role of cell-cell interaction is to coordinate the differentiation of neighboring groups of cells. In primary induction, for example, ectoderm is induced to differentiate into neural tissue by cells that have invaginated into the gastrula from the dorsal lip of the blastopore. Dorsal lip tissue grafted from one gastrula to another triggers the formation of an entire new embryo. Various secondary inductions, such as the induction of lens tissue by overlying ectoderm or induction of epithelial gland formation by underlying mesenchyme, also occur during embryogenesis. While primary induction appears to be mediated by diffusible substances, secondary induction requires direct cell-cell contact.

As multicellular organisms grow old, they gradually deteriorate and eventually die. This process of aging is reflected at the cellular level by the fact that cells placed in culture divide for only a limited number of times before they too deteriorate and die. During the aging of cells in culture, gradual losses in cell function occur, but it is difficult to distinguish changes that are primary to the aging process from those that are secondary by-products. Cellular aging appears to be influenced by a built-in genetic program as well as by the gradual accumulation of random damage. The relative importance of these two factors may become clearer one day through the study of cancer cells, for such cells are capable of dividing indefinitely in culture and thus appear to have somehow overcome the aging process.

SUGGESTED READINGS

Books

Browder, L. W. (1984). *Developmental Biology*, 2nd Ed., CBS College Publishing, New York.

Davidson, E. H. (1986). *Gene Activity in Early Development*, 3rd Ed., Academic Press, Orlando.

Graham, C. F., and P. F. Wareing, eds. (1984). *Developmental Control in Animals and Plants*, Blackwell Scientific, London.

Karp, G., and N. J. Berrill (1981). *Development*, 2nd Ed., McGraw-Hill, New York.

Malacinski, G. M., and S. V. Bryant, eds. (1984). *Pattern Formation: A Primer in Developmental Biology*, Macmillan, New York.

Metz, C. B., and A. Monroy (1985). *Biology of Fertilization*, Vols. 1–3, Academic Press, Orlando.

Sang, J. H. (1984). *Genetics and Development*, Longman, New York.

Slack, J. M. W. (1986). *From Egg to Embryo*, Cambridge University Press, Cambridge, England.

Trinkhaus, J. P. (1984). *Cells into Organs: The Forces That Shape the Embryo*, 2nd Ed., Saunders, Philadelphia.

Yamada, K. M., ed. (1983). *Cell Interactions and Development: Molecular Mechanisms*, Wiley, New York.

Articles

Blau, H. M., G. K. Pavlath, E. C. Hardeman, C.-P. Chiu, L. Silberstein, S. G. Webster, S. C. Miller, and C. Webster (1985). Plasticity of the differentiated state, *Science* 224:758–766.

Carlson, J. G. (1952). Microdissection studies of the dividing neuroblast of the grasshopper, *Chortophaga viridifasciata* (De Geer), *Chromosoma* 5:199–220.

Cunningham, B. A. (1986). Cell adhesion molecules: a new perspective on molecular embryology, *Trends Biochem. Sci.* 11:423–426.

DiBerardino, M. A., N. J. Hoffner, and L. D. Etkin (1984). Activation of dormant genes in specialized cells, *Science* 224:946–952.

Edelman, G. M. (1984). Cell-adhesion molecules: a molecular basis for animal form, *Sci. Amer.* 250(April):118–129.

Epel, D. (1978). Mechanisms of activation of sperm and egg during fertilization of sea urchin gametes, *Current Topics Develop. Biol.* 12:185–246.

Gehring, W. J. (1985). The homeo box: a key to the understanding of development, *Cell* 40:3–5.

———. (1985). The molecular basis of development, *Sci. Amer.* 253 (Oct.):152–162.

Greenwald, I. S., P. W. Sternberg, and H. R. Horvitz (1983). The *lin-12* locus specifies cell fates in *Caenorhabditis elegans*, *Cell* 34:435–444.

Gurdon, J. B. (1962). Adult frogs derived from the nuclei of single somatic cells, *Develop. Biol.* 4:256–273.

Hayflick, L. (1980). The cell biology of human aging, *Sci. Amer.* 242 (Jan):58–65.

Illmensee, K., and A. P. Mahowald (1974). Transplantation of posterior polar plasm in *Drosophila*. Induction of germ cells at the anterior pole of the egg, *Proc. Natl. Acad. Sci. USA* 71:1016–1020.

Rubin, G. M., and A. C. Spradling (1982). Genetic transformation of *Drosophila* with transposable element vectors, *Science* 218:348–353.

Schimke, R. T. (1973). Control of enzyme levels in mammalian tissues, *Advan. Enzymol.* 37:135–187.

Shepard, J. F. (1982). The regeneration of potato plants from leaf-cell protoplasts, *Sci. Amer.* 246 (May): 154–166.

Steinberg, M. S. (1970). Does differential adhesion govern self-assembly processes in histogenesis? Equilibrium configurations and the emergence of a hierarchy among populations of embryonic cells, *J. Exp. Zool.* 173:395–434.

Stewart, T. A., E. F. Wagner, and B. Mintz (1982). Human β-globin gene sequences injected into mouse eggs, retained in adults, and transmitted to progeny, *Science* 217:1046–1048.

Sulston, J. E., E. Schierenberg, J. G. White, and J. N. Thomson (1983). The embryonic cell lineage of the nematode *Caenorhabditis elegans, Develop. Biol.* 100:64–119.

Thiery, J. P., J. L. Duband, and G. C. Tucker (1985). Cell migration in the vertebrate embryo: role of cell adhesion and tissue environment in pattern formation, *Ann. Rev. Cell Biol.* 1:91–113.

Wassarman, P. M. (1987). The biology and chemistry of fertilization, *Science* 235:553–560.

16

Hormones and Their Role in Intercellular Communication

As discussed in Chapter 15, the cells of a multicellular organism become structurally and functionally specialized as development proceeds. This division of labor ultimately leads to the formation of specific tissues and organs, each of which is designed to fulfill a particular function. With the evolution of multicellularity and consequent cellular specialization, a system was needed to coordinate and regulate the activities of these various cell types. Specialized regions of the plasma membrane, such as tight junctions and plasmodesmata, expedite the transfer of information between adjacent cells. However, communication between noncontiguous cells involves a more elaborate signaling system based on diffusible chemical messengers. The molecules that make up this long-distance intercellular communication system are collectively known as **hormones.**

Though all cells that comprise a multicellular organism are exposed to circulating hormones, only certain cells recognize each type of hormone and respond to it. Our understanding of the molecular basis of this specificity has been enhanced by the discovery that each target cell contains specific **receptors** designed to bind a particular hormone. Thus the specificity of hormone action is the result of a highly selective hormone-receptor interaction. Though the association of a hormone with its receptor may ultimately affect events occurring throughout the cell, a hormone's primary action is usually exerted on one of two subcellular targets: the plasma membrane or the nucleus. Since it is beyond the scope of a cell biology textbook to describe the actions of all hormones in detail, attention will be focused on the general types of events triggered by hormone-receptor interactions.

INTRODUCTION TO THE VERTEBRATE ENDOCRINE SYSTEM

The Discovery of Hormones

The initial discovery that diffusible, blood-borne substances influence cellular behavior and development occurred in the mid-1800s when biologists were just beginning to move from observational techniques to animal experimentation as a means of investigating biological phenomena. The first experiment to suggest the existence of blood-borne messenger molecules was carried out in 1848 by the German physiologist, Arnold Berthold, who was studying the effects of castration on the subsequent development of immature roosters. He found that castration inhibits the normal development of secondary sexual characteristics, such as wattles and combs, and behavior patterns, including normal crowing, belligerence toward other males, and attraction to hens. Transplantation of a single testis back into a castrated animal resulted in normal physical and behavioral development, so that these animals could not be distinguished from adult control roosters. Later, when Berthold killed the young roosters, he found the transplanted testis had a well-developed blood supply, though the organ lacked a connection to the nervous system. The ability of the implanted testis to induce normal development in the absence of neural input suggested to Berthold that a substance produced by the testis was carried by the blood to all cells of the body, influencing those cells involved in sexual development.

Although Berthold's results provided the impetus for subsequent studies of such blood-borne substances and their role in regulating physiological processes, nearly 90 years was to pass before the identity of the testicular substance, *testosterone*, was established. In the years following the publication of Berthold's results, the analysis of blood-borne substances and their role in regulating various physiological processes became a major area of biological research. Despite the growing belief in the importance of this chemical signaling system, it was difficult to absolutely rule out the possibility of direct nervous involvement. However, in the early 1900s two Canadian physiologists, William Bayliss and Ernest Starling, clearly demonstrated that chemical messengers can trigger a physiological response in a target tissue even in the absence of neural input. In these studies Bayliss and Starling investigated the factors involved in stimulating the release of pancreatic juice *in vivo* when the upper small intestine is exposed to acid released from the stomach. It had previously been shown that in intact animals, injection of acid into the lumen of the small intestine increases the flow of pancreatic juice. Utilizing a denervated segment of small intestine that retained its blood supply, they injected acid into the lumen of the intestine and measured the flow of pancreatic juice. The intestine reacted to the presence of acid with an increased flow of juice from the pancreatic duct. This stimulation could only be induced if the blood supply between the intestine and pancreas remained intact (Figure 16-1). This result was the first unequivocal demonstration that chemical messengers released from one organ (in this case the intestine) are transported by the circulation to a target organ (the pancreas) where a specific physiological effect is elicited (increase in the flow of pancreatic juice).

This conclusion was buttressed by a second experiment, in which mucosal cells lining the lumen of the small intestine were scraped off, treated with acid, and then injected into the blood vessels leading to the pancreas. The resulting increase in flow of pancreatic fluid

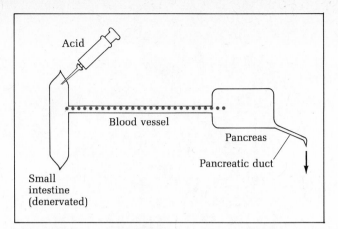

FIGURE 16-1 The Bayliss and Starling experiment showing that a blood-borne substance can stimulate flow of pancreatic juice. Injection of acid into the lumen of a denervated section of small intestine induces the release of a chemical substance (color) into the circulation. When the messenger molecule reaches the target cells in the pancreas, the flow from the pancreatic duct increases.

was nearly twice that observed in the first experiment. Taken together, these data clearly support the view that a chemical transported by the bloodstream can regulate intracellular events in target cells that are spatially removed from the cell secreting the messenger molecule. Some time later the molecule responsible for control of pancreatic juice secretion was identified and named *secretin*.

In 1905 Starling introduced the term **hormone** (derived from a Greek word *horman*, meaning "to arouse to activity") to describe the factor responsible for the control of pancreatic enzyme release. The gradual realization that other physiological processes are also controlled by blood-borne hormone molecules eventually opened an entire new field of biological research. The number of molecules known to be part of this intercellular communication system is steadily growing, and the early simplistic view that hormones are a specialized class of molecules is slowly being modified as new and different chemical messengers are identified.

Endocrine Glands and Their Hormone Products

Hormones are produced by the cells of **endocrine glands** such as the testis, ovary, kidney, pancreas, adrenal, thyroid, parathyroid, and pituitary. The term *endocrine* refers to cells that synthesize and release hormones directly into the circulation. In contrast, **exocrine glands** produce nonhormonal substances secreted directly into ducts leading away from the gland (Figure 16-2). Each endocrine gland is composed of hormone-secreting endocrine cells, plus surrounding tissue including muscle, blood vessels, and nerves; in some cases exocrine cells may also be present. The pancreas is an example of a gland containing both endocrine and exocrine tissue.

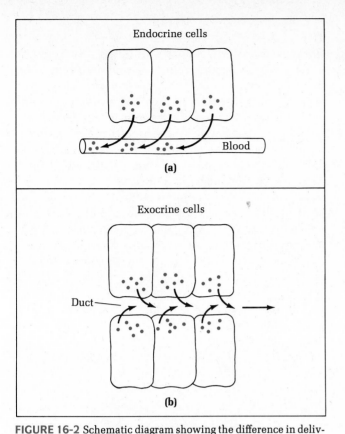

FIGURE 16-2 Schematic diagram showing the difference in delivery of secreted substances (color) from cells of endocrine and exocrine tissue. (a) The hormone made in the endocrine cell is released directly into the bloodstream. (b) Exocrine cells secrete material into ducts leading away from the gland.

The endocrine tissue is composed of *alpha cells* that secrete the hormone *glucagon* and *beta cells* that secrete *insulin* into the bloodstream. As we shall see later, these two hormones work as antagonists in the regulation of blood glucose levels. The exocrine cells synthesize digestive enzymes that are delivered into the intestine through the pancreatic duct.

The many vertebrate hormones that have so far been identified can be classified into three major groups: peptide hormones, steroid hormones, and amino acid derivatives (Table 16-1).

1. The **peptide hormones** represent 80 percent or more of all the vertebrate hormones. They are synthesized in a variety of endocrine glands and are composed of amino acid chains of varying lengths, from as few as 3 to as many as 200. Despite the fact that some of these molecules are in fact polypeptides, they are classified as peptide hormones. A newly discovered class of peptide hormones is the **neuropeptides,** small molecules localized within the nervous system and thought to exert their effects directly on other cells of the nervous system. Peptide hormones are water-soluble molecules stored in membrane-enclosed vesicles until the signal for hormone release is received.

2. **Steroid hormones** are synthesized by enzymes of the smooth endoplasmic reticulum, utilizing cholesterol as the parent molecule. The adrenal gland and gonads (testes and ovaries) are the principal sites of steroid hormone synthesis.

TABLE 16-1
Examples of the three classes of vertebrate hormones

Hormone Class	Structure	Site of Action
PEPTIDE HORMONES		Plasma membrane
Oxytocin	9 amino acids	
Insulin	51 amino acids	
Somatotropin	191 amino acids	
STEROID HORMONES		Nucleus
Cortisol		
AMINO ACID DERIVATIVES		
Epinephrine		Plasma membrane
Thyroxine		Nucleus

Since steroid hormones are hydrophobic molecules, they cannot be stored in membrane-enclosed vesicles, but are instead synthesized just before release into the bloodstream. Because they are not water-soluble, steroid hormones are usually bound to proteins in the blood for efficient transport to their target cells.

3. **Catecholamines** and **thyroid hormones** are amino acid derivatives that comprise the third structurally distinct class of hormones. These small molecules are derivatives of the amino acid tyrosine. Catecholamines are synthesized in the adrenal gland as well as in certain cells of the nervous system and so can be classified as both hormones and neurotransmitters. In both cases these hormones are stored within vesicles in the cell and then released when needed. The thyroid hormones are synthesized first as a larger precursor molecule called **thyroglobulin** in the cells lining the follicles of the thyroid gland. Thyroglobulin is iodinated and in this form is stored in the lumen of the follicle. On the appropriate signal it reenters the cell and two hormones, triiodothyronine (T_3) and thyroxine (T_4), are cleaved from the larger molecule by intracellular enzymes.

The Neuroendocrine System and Control of Hormone Release

Hormone secretion is generally triggered either by other hormones [e.g., thyrotropin (TSH) stimulates thyroid hormone release] or by the nervous system. The ability of an organism to detect changes in its external environment is due to the presence of sensory neurons that are sensitive to external stimuli such as temperature, sound, and smell. Such stimuli may trigger hormone secretion by three possible neural pathways (Figure 16-3). Nerves connected to the adrenal medulla trigger release of the catecholamines *epinephrine* and *norepinephrine* into the bloodstream. A second pathway is mediated by nerves that lead from the hypothalamus to the posterior pituitary in the brain. These neurosecretory cells secrete the hormones *oxytocin* (acting on the uterus and mammary glands) and *vasopressin* (acting on the kidney). Finally, certain neurons of the hypothalamus secrete small neuropeptides, called *releasing factors* or *inhibitory factors*, into blood vessels leading to the anterior pituitary, where they in turn control the release of another set of hormones: *adrenocorticotropin (ACTH)* acting on the adrenal cortex, *thyrotropin* or *thyroid-stimulating hormone (TSH)* acting on the thy-

FIGURE 16-3 The three major pathways by which the nervous system influences hormone production.

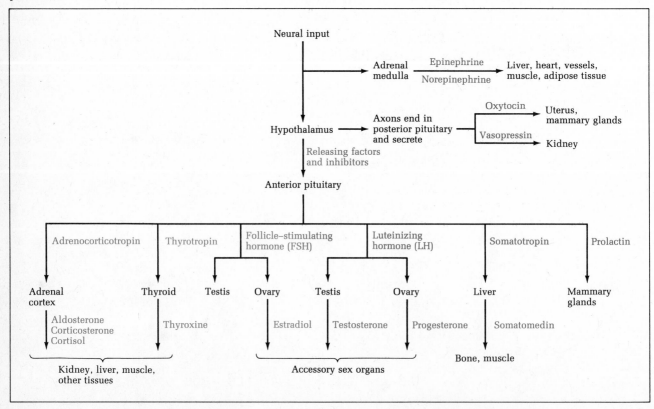

TABLE 16-2
Hormones acting on the plasma membrane

Source and Hormone	Major Target
PEPTIDE HORMONES	
Anterior pituitary	
Adrenocorticotropin (ACTH)	Adrenal cortex
Somatotropin (growth hormone)	Liver
Prolactin	Mammary gland
Thyrotropin (TSH)	Thyroid
Follicle-stimulating hormone (FSH)	Ovary, testis
Luteinizing hormone (LH)	Ovary, testis
Hypothalamus	
Thyrotropin releasing factor (TRF)	Anterior pituitary
Gonadotropin releasing factor (GnRF)	Anterior pituitary
Somatostatin	Anterior pituitary
Posterior pituitary	
Oxytocin	Uterus, mammary gland
Vasopressin (antidiuretic hormone)	Kidney
Small intestine	
Secretin	Pancreas
Pancreas	
Insulin	Adipose tissue, liver, muscle
Glucagon	Adipose tissue, liver, muscle
Parathyroid	
Parathormone	Bone, kidney, intestine
Parathyroid, thyroid, thymus	
Calcitonin	Bone
CATECHOLAMINES	
Adrenal medulla	
Epinephrine	Liver, heart, muscle, blood vessels, adipose tissue
Norepinephrine	Small arteries

roid, *follicle-stimulating hormone (FSH)* and *luteinizing hormone (LH)* acting on the ovary and testes, *somatotropin (growth hormone)* acting on the liver, and *prolactin* acting on the mammary glands. In response to hormonal stimulation, these target tissues synthesize other hormones as well. Thus a cascade of events interconnects sensory information received by the nervous system with the manufacture and release of hormones by specific endocrine cells.

In addition to nervous system control, hormone synthesis and secretion is also subject to **feedback** regulation. For instance, steroid hormones produced by the adrenal cortex inhibit the secretion of the pituitary hormone ACTH, which is initially responsible for stimulating steroid hormone synthesis in the adrenal cortex. In the case of the hormone insulin, which promotes the uptake and metabolism of glucose, high blood levels of glucose stimulate release of insulin while low levels of glucose are inhibitory. Positive and negative feedback mechanisms such as these ensure that circulating hormone levels are continually adjusted to meet the needs of the organism as a whole.

The number of different types of hormones and tar-get cells present in vertebrates is so great that virtually every cell either produces or responds to one hormone or another. Due to this great multiplicity, there is no "typical" target cell whose morphology or function can be studied as a general model. However, most hormone-responsive cells can be classified into one of two general categories depending on whether the hormone first acts on the plasma membrane or the nucleus. This distinction forms the basis of the general description of hormone–target cell interactions that follows.

HORMONES ACTING ON THE PLASMA MEMBRANE

Hormones whose primary action is exerted on the plasma membrane are generally constructed from amino acids. This family includes both the **catecholamines** (epinephrine and norepinephrine), and the **peptide** (or protein) hormones (Table 16-2). Two major discoveries occurring around 1960 were fundamental to our understanding of the mode of action of this group of hormones. Development of the *radioimmunoassay (RIA)* technique by Rosalyn Yalow and Solomon Berson

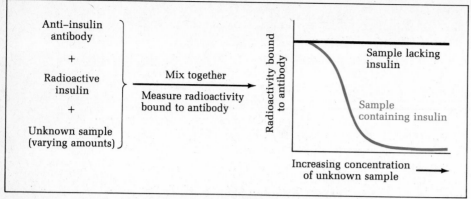

FIGURE 16-4 Diagram summarizing the technique of radioimmunoassay using insulin as an example. Antibodies directed against insulin are mixed with radioactive insulin and varying amounts of a sample containing unknown amounts of insulin. If insulin is present in the sample, it will compete with the radioactive insulin for the antibody binding site. Hence the amount of radioactivity bound to the antibody will decrease with increasing insulin concentration in the unknown sample. This same general approach can be used to quantify trace amounts of any hormone for which specific antibodies are available.

permitted detection and quantification of the tiny amounts of hormone normally present under physiological conditions (Figure 16-4). The second important development was Earl Sutherland's discovery of 3',5'-cyclic adenosine monophosphate (cyclic AMP), a key intermediate in the action of this family of hormones. As we shall see shortly, this latter discovery was instrumental in demonstrating that a common mechanism underlies a broad spectrum of seemingly diverse hormonal effects.

Plasma Membrane Hormone Receptors

The selectivity of hormone–target cell interactions is made possible by the presence of specific **receptor** molecules that bind selectively to particular hormones. The presence of hormone receptors within the plasma membrane can be demonstrated experimentally by incubating cells or subcellular fractions in the presence of a radioactive hormone. Because some nonspecific binding of hormone may occur, it is essential to critically examine the binding data before concluding that a specific receptor is present. The following five criteria are employed in typical hormone-receptor studies to ensure that the binding observed actually reflects an association between the hormone and a specific receptor:

1. Increasing the hormone concentration must eventually lead to **saturation** of the binding sites, indicating that the quantity of receptors per cell is limited. The actual number of receptors may range from a few to several hundred thousand per cell.

2. Saturation of binding sites occurs with small amounts of hormone, indicating that hormone binds to receptor with **high affinity**. The **dissociation constant** (K_d) for hormone-receptor binding, defined as the concentration of hormone at

which binding is half-maximal, is usually in the range of 10^{-8} to $10^{-10}M$.

3. The recognition between hormone and receptor must be **specific**. This can be demonstrated by showing that the association of radioactive hormone with its receptor is susceptible to competition by nonradioactive hormones that are identical to or structurally similar to the radioactive hormone. Unrelated hormones should not reduce the amount of radioactive hormone bound.

4. The binding of a hormone to its receptor is generally **rapid** and **reversible**.

5. Interaction of a hormone with its receptor induces an appropriate **physiological response** in the target cell, the magnitude of which is directly related to the quantity of hormone bound.

There are several reasons for believing that peptide hormone and catecholamine receptors are localized within the plasma membrane of appropriate target cells. First of all, peptide hormones such as adrenocorticotropin (ACTH), insulin, and glucagon retain the ability to stimulate their respective target cells even when artificially linked to large inert beads that block the movement of hormone into the cell. In addition, subcellular fractionation studies utilizing a wide variety of cell types have revealed that peptide hormone receptors are most highly concentrated in the plasma membrane fraction. Moreover, right-side out plasma membrane vesicles isolated from fat cells bind insulin, while inside-out vesicles from the same source do not. This indicates that insulin receptors are localized on the exterior surface of the plasma membrane. Some hormone receptors have even been isolated from the plasma membrane by detergent treatment and shown to retain the ability to bind hormone with high affinity.

Most plasma membrane hormone receptors appear

to be membrane proteins made up of two or more sub-unit polypeptide chains, with carbohydrate or lipid groups often an integral part of the functional protein molecule. The actual number of receptors present in the plasma membrane varies, depending upon the target cell as well as the physiological or developmental state of the organism. In some instances the amount of circulating hormone has been shown to regulate the number of receptors for that hormone, causing either an increase or a decrease in the concentration of its specific receptor. Such effects illustrate the dynamic control mechanisms involved in the regulation of hormone-receptor interactions.

Cyclic Nucleotides and the Second Messenger Concept

Cyclic AMP as a Mediator of Hormone Action Since many kinds of hormone receptors are localized on the outer surface of the plasma membrane, one can ask how the binding of a hormone molecule to such a receptor can influence events occurring inside the cell. Pioneering studies leading to an unraveling of this process were initiated in the late 1950s by Earl Sutherland and Theodore Rall, who were interested in the mechanism by which epinephrine and glucagon stimulate glycogen breakdown in the liver. They began by demonstrating that incubation of liver homogenates with either of these hormones stimulates the activity of **glycogen phosphorylase,** an enzyme catalyzing the breakdown of glycogen into glucose-1-phosphate. However, if the membrane fraction was first removed from the homogenate by centrifugation, the glycogen phosphorylase present in the supernatant could no longer be activated by the addition of epinephrine or glucagon. Addition of hormone directly to the membrane fraction resulted in the appearance of an unidentified substance that could activate glycogen phosphorylase present in the membrane-free supernatant.

Sutherland isolated tiny amounts of this activating substance and tentatively identified it as a nucleotide. By coincidence David Lipkin had just discovered a new, unusual nucleotide, formed by treating ATP with barium hydroxide. This substance turned out to be the same one identified by Sutherland, so Lipkin's procedure for synthesizing the compound in large quantities facilitated its identification as a cyclic derivative of AMP called **3′,5′-cyclic adenosine monophosphate,** or **cyclic AMP,** and provided Sutherland with an abundant supply of this molecule for subsequent studies on its physiological effects.

Sutherland and his associates soon provided four lines of evidence compatible with the notion that cyclic AMP mediates the action of glucagon and epinephrine on liver cells. First, they showed that addition of these hormones to liver elevates intracellular levels of cyclic AMP. Second, they found that adding these hormones to isolated membrane preparations stimulates a plasma-membrane-bound enzyme, called **adenylate cyclase,** which catalyzes the formation of cyclic AMP from ATP (Figure 16-5). Third, they showed that the ability of glucagon and epinephrine to activate glycogen breakdown in liver cells can be mimicked by exposing cells to purified cyclic AMP in the absence of hormone. And finally, they established that the ability of these two hormones to stimulate glycogen breakdown can be amplified by inhibitors of the enzyme **phosphodiesterase,** such as caffeine and theophylline. Phosphodiesterase normally breaks down cyclic AMP into AMP, so inhibiting this enzyme with caffeine or theophylline potentiates hormone action by maintaining abnormally high intracellular levels of cyclic AMP.

FIGURE 16-5 Formation of 3′,5′-cyclic adenosine monophosphate (cyclic AMP) from ATP by action of the enzyme adenylate cyclase. The removal of pyrophosphate results in the formation of a ring (color) in which the remaining phosphate group is linked from the 5′ carbon to the 3′ carbon of the ribose moiety.

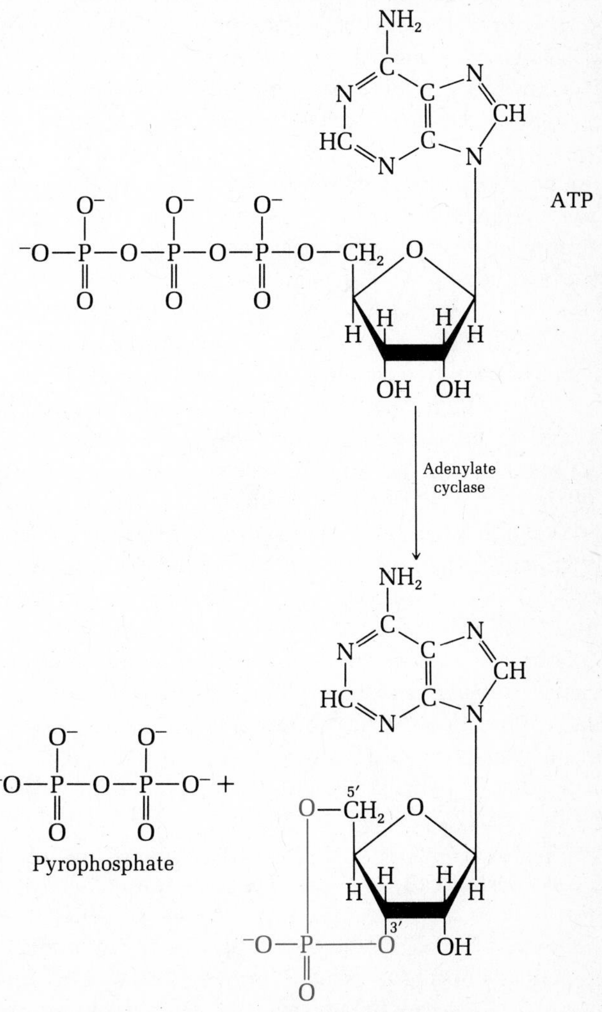

TABLE 16-3
Some hormonal responses associated with elevations in
intracellular cyclic AMP levels

Hormone	Tissue	Response
ACTH	Adrenal	Hydrocortisone formation
TSH	Thyroid	Thyroxine formation
LH	Ovary	Progesterone formation
Vasopressin	Kidney	Water resorption
Glucagon	Adipose tissue	Lipid breakdown
Glucagon	Liver	Glycogen breakdown
Parathormone	Kidney	Phosphate excretion
Parathormone	Bone	Calcium resorption
Secretin	Intestine	Pancreatic enzyme release
Epinephrine	Cardiac muscle	Increased contractility
Epinephrine	Liver	Glycogen breakdown
Epinephrine	Erythrocyte	Increased Na^+ permeability

These four properties have been adopted as general criteria for determining whether the effects of hormones other than glucagon and epinephrine are also mediated by cyclic AMP. In this way cyclic AMP has been implicated in a remarkably diverse spectrum of hormonally induced responses (Table 16-3). Its widespread involvement in hormone action has led to the designation of cyclic AMP as a **second messenger,** since it transmits to the inside of cells signals originally brought to the outer surface by one of a variety of first messengers (i.e., circulating hormones). How the binding of a hormone to its receptor stimulates cyclic AMP production, and how this nucleotide in turn triggers the appropriate cellular response, are questions to which we now have partial answers.

Interaction Between Hormone Receptors and Adenylate Cyclase One can envision several possible mechanisms by which hormone receptors might stimulate the activity of adenylate cyclase. One of the earliest models suggested that the hormone-receptor complex directly binds to adenylate cyclase, forming a multisubunit protein complex. It is easy to see how the binding of a hormone to the receptor component of such a complex could alter the conformation and hence catalytic activity of the cyclase unit. However, most available evidence supports a two-step model, whereby receptor and cyclase normally exist as independent entities in the membrane. In this case binding of hormone to its appropriate receptor could cause the receptor to in turn activate adenylate cyclase by either binding to it directly or by indirectly causing some other molecule to activate adenylate cyclase (Figure 16-6). Such an arrangement would permit cells to respond to more than one type of hormone using the same adenylate cyclase molecule.

Several lines of evidence support this notion of independent hormone receptors interacting with a common pool of adenylate cyclase molecules in the membrane. One example is to be found in fat cells, where adenylate cyclase can be stimulated by at least

six different hormones (ACTH, LH, TSH, epinephrine, glucagon, and secretin). If each of these hormones activates a separate adenylate cyclase, then the effect of each hormone should be additive, a prediction that is not borne out by the experimental data. Instead, once adenylate cyclase has been maximally stimulated by one hormone, addition of another hormone has no further effect, indicating that the same adenylate cyclase system is being activated by each hormone.

A more direct demonstration of the independence of hormone receptors and adenylate cyclase was provided by Joseph Orly and Michael Schramm, who showed that hormone receptors derived from one cell can influence the catalytic activity of the adenylate cyclase of another cell. In these experiments the technique of cell fusion was employed to join together cells of two types: erythrocytes containing catecholamine receptors but no adenylate cyclase, and erythroleukemia cells with adenylate cyclase but no catecholamine receptors. A few minutes after these two cell types were fused together, the adenylate cyclase became susceptible to stimulation by catecholamines, indicating that receptors derived from one cell can quickly establish communication with the adenylate cyclase of another cell.

In addition to the hormone receptor and adenylate cyclase, the cyclic AMP generating system also contains a guanine nucleotide binding-protein called the **G-protein,** which makes cyclic AMP production sensitive to the presence of GTP. In addition to requiring GTP for an optimal response, other factors such as calcium ions, prostaglandins, certain proteins, and the lipid composition of the membrane may also modulate the response of adenylate cyclase to hormones. Though the relationship of these factors to adenylate cyclase activity has not been fully elucidated, the fundamental events involving interaction of membrane components are now partially understood. In the absence of hormone the G-protein is complexed with GDP and cannot activate adenylate cyclase. However, when an appropriate hormone binds to the receptor, the GDP bound to the G-protein is replaced by GTP. The binding of GTP to the G-protein allows it to activate adenylate cyclase and hence stimulate the production of cyclic AMP. A second feature of the G-protein that is important in the overall modulation of adenylate cyclase activity is its ability to hydrolyze GTP to GDP. When hormone is present, the majority of the G-protein is bound to GTP, even though there is a slow hydrolysis of GTP to GDP by the GTPase activity inherent to the G-protein. When the hormone dissociates from its receptor, however, the exchange of GTP for GDP ceases and all of the GTP bound to the G-protein is eventually hydrolyzed to GDP. In this configuration the G-protein can no longer stimulate adenylate cyclase activity (Figure 16-7).

One model system that has provided some support

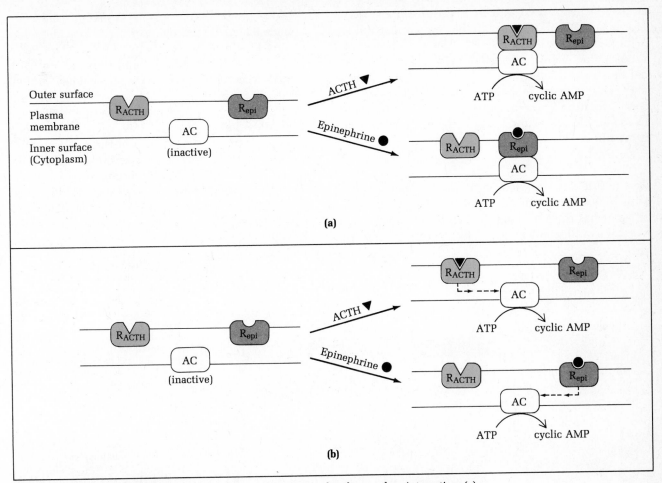

FIGURE 16-6 Two hypothetical models of hormone receptor–adenylate cyclase interaction. (a) Binding of hormone to its receptor induces the receptor to make physical contact with adenylate cyclase, thereby stimulating its catalytic activity. (b) Binding of hormone to its receptor activates adenylate cyclase *indirectly*, without requiring contact between receptor and enzyme. Abbreviations: R = receptor; AC = adenylate cyclase; ACTH = adrenocorticotropin; epi = epinephrine.

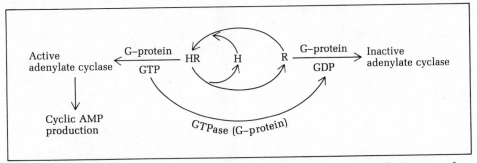

FIGURE 16-7 Control of activation of adenylate cyclase by the G-protein. In the presence of an empty receptor (R), the G-protein is associated with GDP and in this form cannot activate adenylate cyclase. When hormone binds to the receptor (HR), GTP exchanges with GDP and the G-protein can transform adenylate cyclase into an active enzyme capable of converting ATP to cyclic AMP. When the hormone dissociates from its receptor, the GTPase activity in the G-protein transforms the GTP to GDP, converting the G-protein back to its inactive form.

for this proposal involves **choleragen,** the bacterial toxin responsible for the disease cholera. Choleragen binds to gangliosides in the plasma membrane of intestinal cells, leading to an excessive activation of adenylate cyclase. The abnormally large amount of cyclic AMP produced in turn causes a profuse diarrhea. Like some of the large peptide hormones (e.g., TSH, FSH, and LH), choleragen is comprised of two polypeptide chains designated subunits A and B. Subunit B is responsible for binding choleragen to the plasma membrane, while subunit A catalyzes the cleavage of NAD into nicotinamide and an ADP-ribose moiety that becomes covalently attached to the G-protein in the membrane. This process of ADP ribosylation inhibits the normal ability of the G-protein to hydrolyze GTP. The resulting stabilization of the GTP form of the G-protein leads to a continual activation of adenylate cyclase. Although NAD has not been implicated in the stimulation of adenylate cyclase by naturally occurring hormones, the preceding example clearly shows that modification of the G-protein can have marked consequences on adenylate cyclase activity.

Cyclic AMP and the Phosphorylation of Cellular Proteins Once the binding of a hormone to its appropriate receptor has triggered the activation of adenylate cyclase, the cyclic AMP produced by this enzyme in turn influences other cellular events. A major insight into the mode of action of cyclic AMP was made in the late 1960s by Edwin Krebs and Donal Walsh, who discovered that cyclic AMP binds to and activates a class of enzymes known as **cyclic AMP-dependent protein kinases.** These enzymes, which are widespread in eukaryotic animal cells, catalyze the phosphorylation of a variety of proteins using ATP as a phosphate group donor. In liver and muscle one of the main proteins phosphorylated this way is the enzyme **phosphorylase kinase.** This phosphorylation procedure enhances the activity of phosphorylase kinase, which in turn catalyzes the phosphorylation and resulting activation of **glycogen phosphorylase,** the enzyme responsible for promoting glycogen breakdown. This multistep cascade

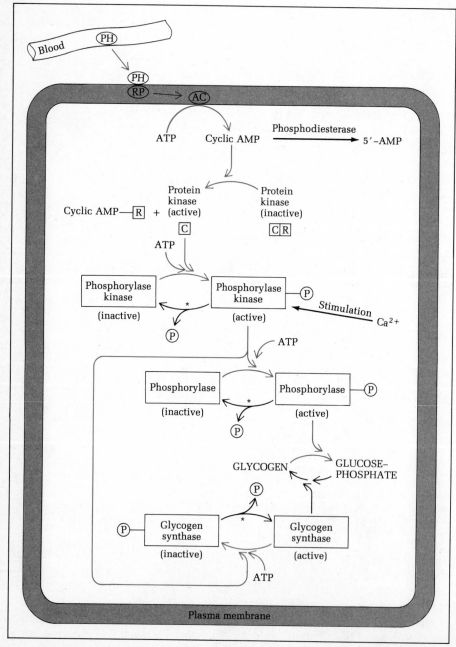

FIGURE 16-8 Schematic diagram of the mechanism of action of peptide hormones such as epinephrine and glucagon on glycogen breakdown in liver and muscle cells. The peptide hormone moves from the circulation to the plasma membrane of the target cell, where it binds to a specific protein receptor. The hormone-receptor complex activates adenylate cyclase via the G-protein to increase intracellular levels of cyclic AMP. Binding of cyclic AMP to the regulatory subunit of the protein kinase causes its dissociation from the catalytic subunit of the enzyme. The free catalytic subunit is the active form of the enzyme. Colored arrows indicate the metabolic pathway responding to hormone stimulation. Black arrows indicate reactions involved in terminating the physiological response. Reactions involving ATP result in phosphorylation (addition of a P)); dephosphorylation reactions involve a protein phosphatase (*) and loss of a phosphate group. Abbreviations: PH = peptide hormone; RP = receptor protein; AC = adenylate cyclase; C = catalytic subunit; R = regulatory subunit.

TABLE 16-4

Examples of proteins subject to phosphorylation by cyclic AMP-dependent protein kinases

Nucleus
   Histones
   Nonhistones
Cytosol enzymes
   Phosphorylase kinase
   Glycogen synthase
   Triglyceride lipase
   Pyruvate kinase
   Cholesterol esterase
Membrane proteins
   Plasma membrane
   Endoplasmic reticulum
   Myelin basic protein
Microtubules
   Tubulin and associated proteins
Ribosomes
   Selected proteins and initiation factors

of enzyme activations therefore explains how hormones elevating cyclic AMP levels in muscle and liver ultimately promote glycogen breakdown (Figure 16-8).

Phosphorylation of enzymes does not always enhance their catalytic activity. The enzyme **glycogen synthase,** for example, becomes less active rather than more active after phosphorylation. Since glycogen synthase is responsible for catalyzing glycogen formation, the net effect of this cyclic AMP-induced protein phosphorylation is to inhibit glycogen phosphorylation (see Figure 16-8).

The widespread occurrence of cyclic AMP-dependent protein kinases in eukaryotic cells and their ability to promote phosphorylation of various proteins (Table 16-4) has led to the idea that most, if not all, of the biological effects of cyclic AMP are mediated by changes in protein phosphorylation. Support for this idea has been provided by the investigations of Philip Coffino and his associates on mouse lymphoid cells. In this cell type cyclic AMP has been shown to trigger a variety of physiological responses including inhibition of growth, induction of the enzyme phosphodiesterase, and cell lysis. By growing cells in the presence of high concentrations of cyclic AMP, mutants resistant to the growth-inhibiting effects of cyclic AMP were selected for. Of several hundred mutants obtained, all were subsequently found to be defective in cyclic AMP-dependent protein kinase activity. In addition, all three biological effects of cyclic AMP (growth inhibition, phosphodiesterase induction, and cell lysis) were altered in every mutant, supporting the conclusion that cyclic AMP-dependent protein kinase mediates each of these actions of cyclic AMP.

Additional support for the involvement of protein kinase in mediating the biological effects of cyclic AMP has come from the discovery that a close correlation exists between the degree of activation of protein kinase

by a given dose of hormone and the degree to which the final target tissue responds to that hormone. For example, the effectiveness of varying concentrations of luteinizing hormone (LH) in stimulating testis cells to make testosterone or ovarian cells to make progesterone is closely related to the degree to which protein kinase is activated by that particular hormone concentration (Figure 16-9).

If cyclic AMP-dependent protein kinases are responsible for mediating the effects of all cyclic AMP-elevating hormones, then how can each target tissue exhibit its own special response to hormonal stimulation? This apparent paradox is illustrated by several hormones. Glucagon stimulates glycogen breakdown in liver but lipid breakdown in fat cells. LH enhances steroid synthesis in ovary and testis, while epinephrine increases sodium permeability in erythrocytes. In each of these cases the triggering hormone increases cyclic AMP levels. Such differing responses to cyclic AMP are thought to occur because each tissue contains a different set of proteins subject to phosphorylation by cyclic AMP-dependent protein kinase (Figure 16-10). Hence in liver, **phosphorylase kinase** and **glycogen synthase** are among the proteins phosphorylated, leading to an increase in glycogen breakdown to glucose-1-phosphate. In fat cells, on the other hand, the enzyme **triglyceride lipase** becomes phosphorylated, enhancing its ability to catalyze lipid breakdown. Though the evidence is more tentative, hormonal stimulation of steroid-producing

FIGURE 16-9 Effects of varying concentrations of luteinizing hormone (LH) on cyclic AMP levels, protein kinase activity, and progesterone synthesis in isolated ovarian tissue. Note that the extent of the final response (progesterone formation) in the target tissue correlates closely with cyclic AMP levels and protein kinase activity.

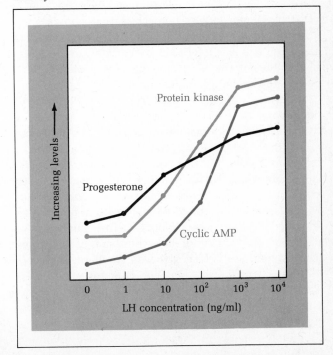

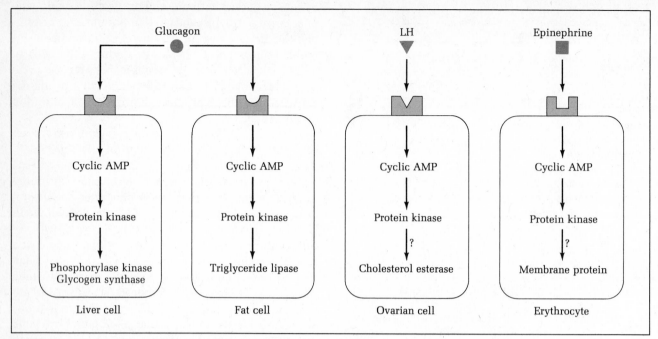

FIGURE 16-10 Schematic representation of the major elements contributing to hormonal specificity. Each hormone affects only those cells containing the appropriate receptor. The effects triggered by cyclic AMP depend on the particular proteins that are phosphorylated by cyclic-AMP-dependent protein kinase. These vary with the cell type.

cells in the ovary, testis, and adrenal cortex is believed to induce phosphorylation of **cholesterol esterase,** a key enzyme of steroid metabolism, while the changes in membrane permeability that occur in hormonally stimulated erythrocytes may involve alterations in the phosphorylation of membrane protein(s).

At first glance it might appear as if the cyclic AMP-mediated pathway of hormone action is unnecessarily complex: The binding of a hormone to its receptor activates an enzyme (adenylate cyclase) that catalyzes the formation of a compound (cyclic AMP) that activates another enzyme (protein kinase) to catalyze the phosphorylation of yet other enzymes and proteins. Why doesn't the hormone simply interact with the final target enzyme or protein directly? The answer appears to be that at least two purposes are served by such a multistep cascade: amplification and regulation. Amplification of the incoming signal occurs because each step in the pathway is catalytic, that is, each molecule of activated adenylate cyclase produces hundreds or thousands of molecules of cyclic AMP per second, each of which can activate a molecule of protein kinase. Every activated protein kinase molecule in turn catalyzes the phosphorylation of hundreds or thousands of molecules of phosphorylase kinase, glycogen synthase, triglyceride lipase, or whatever enzyme is being acted upon. These enzymes may in turn convert hundreds or thousands of substrate molecules into products. In this way a single hormone molecule acting on the cell surface is ultimately capable of triggering a response involving millions of molecules.

The other advantage to a multistep cascade is that its individual steps may be subject to alternative modes of regulation. For instance, in addition to being activated by cyclic AMP-dependent protein kinase, phosphorylase kinase is also stimulated by calcium ions. Hence any stimulus altering the intracellular concentration of calcium ions will influence glycogen breakdown by a mechanism that is independent of changes in cyclic AMP levels.

Cyclic GMP and Hormone Action Several hormones known to bind to plasma membrane receptors, including insulin, growth hormone, prolactin, and oxytocin, do not appear to stimulate cyclic AMP formation in their target cells. In response to insulin administration cyclic AMP levels actually fall, which is perhaps not surprising because insulin acts to promote glycogen formation, an effect opposite to that of the cyclic AMP-elevating hormones glucagon and epinephrine. Though it would be simple to propose that this decrease in cyclic AMP is responsible for the biological effects of insulin, many responses to insulin occur before such a decline in cyclic AMP levels is detectable. This discrepancy has prompted the search for mediators of hormone action other than cyclic AMP.

One candidate for such a mediator is **cyclic GMP.** Though present in much smaller quantities than cyclic AMP, cyclic GMP also occurs widely in eukaryotic cells, and its concentration can be elevated by several hormones, including insulin and oxytocin. One of the most intriguing facets of cyclic GMP action is that its effects on cells are often counter to those of cyclic AMP. Moreover, an increase in the intracellular concentration of cyclic GMP is often (but not always) accompanied by a decline in the concentration of cyclic AMP, and vice versa. These relationships have led Nelson Goldberg to propose that cyclic AMP and cyclic GMP exert opposing regulatory influences on the cell. This notion has been dubbed the *yin-yang hypothesis* because of its similarity to the ancient Oriental principle of opposing, complementary forces.

The involvement of cyclic GMP in hormone action is not as well understood as that of cyclic AMP. Cyclic GMP formation is catalyzed by **guanylate cyclase,** an enzyme present in both the plasma membrane and the cell sap. It cannot be readily activated by adding hormones to isolated membrane preparations, suggesting that it may be regulated indirectly. Cyclic GMP-dependent protein kinase activity has been detected in several cell types, but the normal protein substrates for this enzyme are yet to be identified. Hence at present there are no specific hormone-induced responses where the role of cyclic GMP is well understood at the molecular level. However, it has been shown that cyclic GMP is localized predominantly in the nucleus, while cyclic AMP occurs mainly in the cytoplasm, suggesting that cyclic GMP acts upon substrates that differ from those influenced by cyclic AMP.

Calcium Ion as an Intracellular Messenger

Not all biological responses induced by hormones acting on the plasma membrane can be correlated with changes in either cyclic AMP or cyclic GMP. Several lines of evidence have implicated calcium ion as another intracellular messenger of hormonal signals. Support for this concept begins with the observation that binding of a hormone to its plasma membrane receptor is often followed by an increase in membrane permeability to calcium ions. Since the extracellular calcium ion concentration ($10^{-3}M$) is orders of magnitude higher than that inside the cell ($10^{-7}M$), calcium ions flow inward. If Ca^{2+} is deliberately removed from the external medium, many of the biological responses normally associated with hormone action do not occur. On the other hand, calcium ionophores, which artificially promote Ca^{2+} influx into the cell, can mimic hormones in triggering certain target cell responses.

In most situations where cyclic AMP or cyclic GMP has been implicated in mediating hormone action, the presence of Ca^{2+} is still required for a proper target tissue response. One role played by calcium ions is to activate phosphodiesterase, an enzyme that degrades cyclic AMP, thereby maintaining the intracellular concentration of this nucleotide within prescribed limits. Calcium ions are also involved in implementing certain actions of cyclic nucleotides. For example, the activation of phosphorylase kinase by cyclic AMP-dependent protein kinase is dependent on Ca^{2+}. In addition to influencing the activity of metabolic enzymes, Ca^{2+} has also been implicated in cellular responses such as secretion, motility, muscle contraction, and cell division. In at least some of these instances Ca^{2+} itself, rather than cyclic AMP or cyclic GMP, appears to be the principal controlling factor.

The mechanism by which hormones initiate changes in cytoplasmic levels of calcium ions is not fully understood, but at least two pathways have been implicated: (1) enhanced uptake of calcium across the plasma membrane and (2) release of calcium ions into the cytoplasm from intracellular storage sites. Control of this second pathway involves turnover of inositol phospholipids in the plasma membrane (Figure 16-11). The inner leaflet of the plasma membrane contains several phosphorylated forms of phosphatidylinositol, among

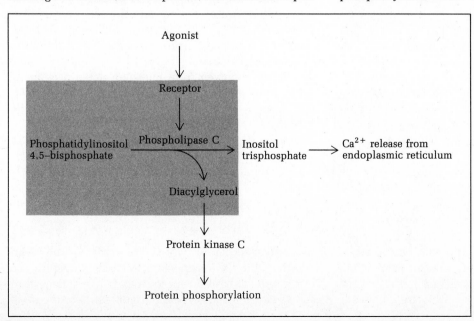

FIGURE 16-11 Diagram illustrating the activation of inositide metabolism by binding of an agonist to its receptor in the plasma membrane (gray shading). The membrane-bound phospholipase C converts phosphatidylinositol bisphosphate into inositol trisphosphate and diacylglycerol. Inositol trisphosphate in turn triggers the release of Ca^{2+} from the endoplasmic reticulum, while diacylglycerol stimulates protein kinase C and hence influences protein phosphorylation.

them **phosphatidylinositol 4,5-bisphosphate.** Binding of a specific agonist (hormone, growth factor, or neurotransmitter) to its receptor in the membrane stimulates the breakdown of this inositol derivative by the membrane-bound enzyme **phospholipase C.** The products of this enzymatic reaction, **inositol trisphosphate** and **diacylglycerol,** are important mediators of cellular processes. They too may be classified as cellular second messengers because they are not only rapidly produced and degraded, but they also act at extremely low concentrations. Although the functions of these molecules are just beginning to be understood, some aspects of their behavior are apparent. Extensive experimental evidence points to inositol trisphosphate as the intracellular linker connecting events at the cell surface with the release of calcium ions from the lumen of the endoplasmic reticulum. Diacylglycerol, on the other hand, may exert its effects by enhancing the phosphorylation of specific cellular proteins. In 1977 Yoskima Takai and Akira Kishimoto reported the discovery of a new protein kinase they called **protein kinase C** to distinguish it from those protein kinases involved in mediating the actions of cyclic AMP and cyclic GMP. This enzyme has been identified in a wide variety of phyla, suggesting it may be involved in some fundamental cellular process. The enzyme is activated specifically by diacylglycerol, and also requires calcium ions and the membrane lipid phosphatidylserine for its activity. In unstimulated cells protein kinase C is inactive; however, on stimulation by an agonist, diacylglycerol is produced and protein kinase C is activated (see Figure 16-11). This is also accompanied by an increase in the intracellular concentration of calcium ions. Protein kinase C acts to phosphorylate specific cellular proteins that are not phosphorylated by other protein kinases in the cell. This type of pathway is utilized in the control of serotonin release from stimulated platelets, and it has been suggested that a similar method of signal transduction may be employed by other hormone-sensitive cell types.

One of the ways in which calcium ions exert their effects within the cell is by associating with a specific calcium-binding protein known as **calmodulin.** The ubiquitous occurrence of this protein in eukaryotic cells suggests it may be involved in a variety of calcium-regulated processes. Calmodulin is a single polypeptide chain folded to expose four calcium binding sites (Figure 16-12). Binding of calcium ions to this protein causes it to assume a new conformation in which it can activate a number of different cellular enzymes, including phosphodiesterase, guanylate cyclase, adenylate cyclase, NAD kinase, Ca^{2+}-dependent ATPase, myosin light-chain kinase, phosphorylase kinase, and glycogen synthase. This activation of calmodulin occurs when the calcium concentration within the cell reaches micromolar levels. The return of calcium ion to normal lower concentrations may occur either by increasing

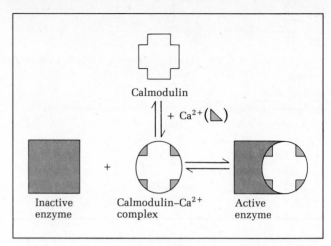

FIGURE 16-12 The structure and action of calmodulin. Calmodulin contains four calcium binding sites. When calcium ions (color) bind, this changes the three-dimensional shape of the calmodulin molecule. The Ca^{2+}-calmodulin complex is then able to activate certain calcium-dependent enzymes (gray). Neither calmodulin alone nor calcium alone can activate the Ca^{2+}-dependent enzyme.

the uptake of Ca^{2+} into intracellular storage sites such as the mitochondrion or by expulsion of calcium ions from the cell via the action of a calmodulin-dependent Ca^{2+}-ATPase. When the Ca^{2+} concentration inside the cell returns to normal levels, calcium ions dissociate from calmodulin, returning the molecule to its inactive form (see Figure 16-12).

Termination of Hormone Action

After a particular cell has responded to a hormone signal, the response must be terminated when circulating hormone levels fall. This may be accomplished in one of two ways. For those responses involving cyclic AMP activation of an enzymatic pathway, degradation of cyclic AMP by the enzyme phosphodiesterase shuts down the activated system and the cell returns to its preinduced state. This is accomplished primarily by activation of phosphodiesterase via the Ca^{2+}-calmodulin complex as discussed above. In addition, phosphate groups are removed from phosphorylated proteins by the enzyme **protein phosphatase.** Even when circulating hormone levels remain elevated, a decline in the capacity of target cells to respond often occurs. This **desensitization** is accounted for by either receptor inactivation or removal of hormone receptors from the plasma membrane by the process of endocytosis.

HORMONES ACTING DIRECTLY ON THE NUCLEUS

In contrast to peptide hormones and catecholamines, where binding of hormone to the plasma membrane of the target cell is sufficient to trigger a physiological response, **steroid** and **thyroid** hormones enter their target

TABLE 16-5
Properties of selected hormones acting directly on the nucleus

Hormone	Structure[a]	Source	Target Tissue
	STEROIDS		
Estrogens: (estradiol)		Ovary (follicle)	Uterus, vagina, breast, bone, brain
Androgens: (testosterone)		Testis	Prostate, bone, brain, seminal vesicle
Progestins: (progesterone)		Ovary (corpus luteum)	Uterus
Glucocorticoids: (cortisol)		Adrenal cortex	General
Mineralocorticoids: (aldosterone)		Adrenal cortex	Kidney, intestines, sweat and salivary glands
	AMINO ACID DERIVATIVES		
Thyroid hormones: (triiodothyronine)		Thyroid	General

[a] Of sample hormone in parenthesis.

cells and interact directly with the nucleus (Table 16-5). This is not to imply that peptide hormones or catecholamines never influence events occurring in the nucleus; however, any nuclear effects produced by these agents appear to be triggered indirectly by a second messenger rather than directly by the hormone.

Cytosol and Nuclear Receptors

Like peptide hormones and catecholamines, steroid hormones elicit physiological responses only in their target cells. The existence of specific hormone receptors in steroid target tissues first came to light around 1960 when Elwood Jensen and his colleagues showed that radioactive estradiol injected into rats is concentrated and retained by estrogen target tissues, such as uterus and vagina, but is rapidly lost from nontarget tissues (Figure 16-13). Such observations suggested the presence in target tissues of receptors designed to bind estradiol and prevent its escape back into the circulation. A few years later the receptor for estradiol was identified

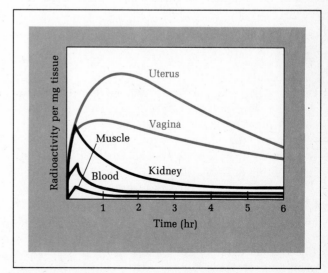

FIGURE 16-13 Uptake of injected radioactive estradiol by various tissues of the rat. Note the concentration and retention of radioactivity in the estrogen target tissues (uterus and vagina).

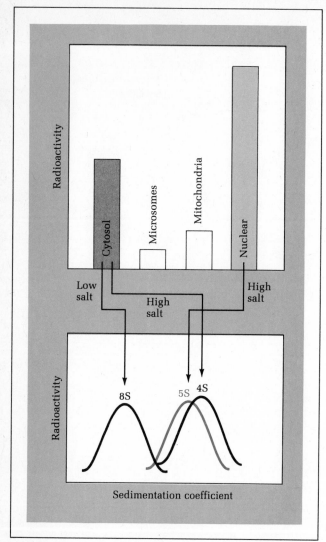

FIGURE 16-14 Radioactive estradiol injected into rats is found predominantly in the cytosol and nuclear fractions of rat uterus. When analyzed by moving-zone centrifugation, the nuclear radioactivity is found associated with a receptor sedimenting at 5S, while the cytosol receptor sediments at 8S in low salt or 4S in high salt.

by David Toft and Jack Gorski, who employed moving-zone centrifugation to analyze the state of the radioactive estradiol present in the cytosol fraction of uterine cells. Rather than existing as free estradiol, the radioactive steroid was found to be associated with a protein sedimenting at approximately 8S. This estradiol-binding protein was identified as a receptor based on the following criteria: (1) it exists only in target cells; (2) its affinity for estradiol is very high (dissociation constant $= 10^{-10}M$); (3) it is present in small amounts (about 10,000 molecules per cell); and (4) it is very selective in binding estrogens. If the ionic strength of the medium containing isolated estrogen receptor is raised enough to discourage nonspecific protein-protein interactions, the 8S receptor is converted to a smaller 4S form thought to more closely resemble its state in the intact cell.

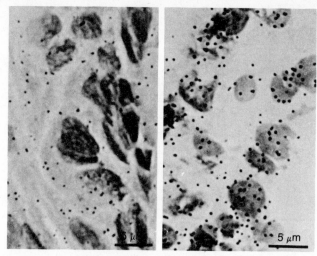

FIGURE 16-15 Light microscopic autoradiograph of toad bladder after exposure to two radioactive steroid hormones. (Left) In the presence of radioactive progesterone, the silver grains are evenly divided between the nuclei and the cytoplasm. (Right) In the presence of radioactive aldosterone, the hormone is preferentially associated with the nuclei. Of the two hormones, only aldosterone acts on the bladder. Courtesy of I. S. Edelman.

In addition to its presence in the cytosol, subcellular fractionation studies have revealed that large amounts of radioactive estradiol are concentrated in the nuclear fraction of target cells (Figure 16-14). The implication that steroid hormones act within the nucleus has been supported by autoradiographic studies in which labeled steroid hormones have been directly visualized inside nuclei in target, but not nontarget, tissues (Figure 16-15). When radioactive steroid hormones are extracted from nuclei, they are found to be associated with a receptor whose sedimentation coefficient differs from

FIGURE 16-16 Graph illustrating the movement of radioactivity from cytosol to nuclear fraction that occurs during incubation of uteri initially exposed to a pulse of radioactive estradiol.

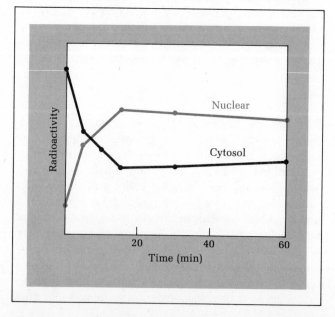

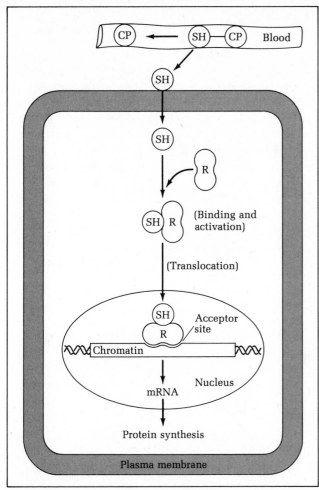

FIGURE 16-17 A general model for the mechanism of action of steroid hormones. The steroid hormone (SH) is transported in the blood bound to a carrier protein (CP). This complex dissociates and the hormone moves into the cytoplasm, where it binds to a receptor (R). This results in activation and translocation of the steroid hormone-receptor complex into the nucleus, where it binds to a specific acceptor site in chromatin, leading to transcription of a specific gene. The resulting mRNA is translated in the cytoplasm into a polypeptide molecule. This model applies to the action of estradiol, as well as to that of other steroid hormones.

action is summarized in Figure 16-17. Though initially based on studies with estradiol, the action of all steroid hormones seems to involve the following common features defined in this model:

1. Steroid hormones enter cells by *diffusing* through the plasma membrane.
2. In target cells the hormone then *binds* to its receptor, preventing the steroid from diffusing back out of the cell. This process allows steroid hormones to be selectively concentrated by cells containing the appropriate receptor.
3. Upon binding to hormone, the cytosol receptor undergoes *transformation* to an activated state.
4. Following receptor transformation the receptor-steroid hormone complex is *translocated* into the nucleus.
5. Within the nucleus the hormone-receptor complex binds to specific sites on the chromatin.
6. The binding of the hormone-receptor complex to chromatin ultimately leads to enhanced transcription of particular genes.

Although this basic model describing the path taken by steroid hormones in reaching the nucleus has been widely accepted since its inception nearly 20 years ago, some new studies have come to light that may lead to a slight modification of this view of steroid hormone action. Using methods unrelated to the subcellular fractionation approach originally taken in the 1960s, it has been demonstrated that, at least in the case of estradiol, the initial binding of hormone and receptor actually may occur inside the nucleus. The most compelling evidence in support of this notion comes from studies by Jack Gorski and his colleagues using enucleated cells. Treatment of cells with the drug cytochalasin B leads to extrusion of the cell nucleus, leaving an enucleated cell termed a **cytoplast.** When Gorski examined the subcellular distribution of the estrogen receptor in the cytoplast and nuclear fractions, he found that the receptor was almost entirely present in the nuclear fraction, while the cytoplast fraction contained little hormone-binding activity. These data suggest that the steroid hormone actually enters the nuclear compartment before it is bound to a receptor protein. Since the original idea that the steroid hormone binds to a cytosol receptor is based on the results of cell fractionation studies, it is not unreasonable to suggest that the presence of receptor in the cytoplasm may be an artifact of the experimental method used. The receptor could be extracted into the cytosol fraction as a result of the low ionic strength conditions employed. Even if this turns out to be the case, this does little to change the substance of the model of steroid hormone action in target cells, since the concept of hormone-receptor binding inside the cell still holds true.

that of the cytosol receptor; in the case of estradiol the nuclear receptor sediments at 5S, while under the same ionic strength conditions the cytosol receptor sediments at 4S. A clue to the relationship between nuclear and cytoplasmic receptors is found in the fact that during prolonged incubation of uteri exposed to an initial dose of radioactive estradiol, the quantity of radioactive nuclear receptor increases as the quantity of cytoplasmic receptor decreases (Figure 16-16). This shift of steroid from cytosol to nucleus, accompanied by a change in the sedimentation properties of the receptor from 4S to 5S, led Gorski and Jensen to independently propose that binding of estrogen to the cytosol receptor transforms the receptor into a new physical state in which it is capable of interacting with the nucleus.

A current version of this model of steroid hormone

Our understanding of steroid hormone receptors

has been useful in the elucidation of the molecular basis of certain diseases. For example, the inherited condition **testicular feminizing syndrome,** in which males develop many of the external characteristics of females, involves a deficiency in receptors for the male sex steroids (androgens). Afflicted individuals have testes that make testosterone, but the absence of normal receptors in the cells of the target tissues blocks any response, and so the male secondary sex characteristics fail to develop. Steroid receptors also play a role in the growth of malignant breast tumors, some of which require estrogen for growth and can therefore be treated by removing the ovaries responsible for estrogen synthesis. A quick assessment of whether or not such treatment is likely to be effective is to assay the tumor cells for the presence of estrogen receptors; the absence of receptors almost always indicates that this treatment will be ineffectual in slowing the disease.

It is not clear to what extent the general model of steroid hormone-receptor interactions is applicable to the thyroid hormones, which also interact with the nucleus. To date, specific receptors for triiodothyronine and tetraiodothyronine (thyroxine) have been detected only in nuclei of target cells. This evidence may be indirectly supportive of the view that cytosol receptors in general are artifacts of the experimental procedure used in studying hormone-cell interactions.

Activation of Transcription of Specific Genes

The discovery that steroid and thyroid hormones become localized in the nuclei of their respective target cells clearly suggests that these agents influence some nuclear process. The first clue to the nature of these events was provided in the late 1950s, when injections of the steroid hormone hydrocortisone were shown to cause an increase in activity of the rat liver enzyme, tyrosine transaminase. Because this rise in enzyme activity can be prevented by inhibitors of protein synthesis, it must involve synthesis of new enzyme molecules rather than activation of previously existing ones. Subsequent to this pioneering observation, many other steroid hormones have been found to stimulate the production of specific enzymes or proteins in their particular target tissues.

Many observations support the hypothesis that the increased production of such proteins is caused by an enhanced transcription of their respective genes. It has been widely observed, for example, that RNA synthesis in steroid hormone target tissues increases within a few minutes after hormone administration. Furthermore, many of the normal target tissue responses to hormonal stimulation can be experimentally prevented by blocking transcription with the drug actinomycin D. It is also known that the steroid hormone ecdysone induces puff-

ing and associated RNA synthesis in specific regions of giant polytene chromosomes in insects.

Some of the most direct evidence for the specific activation of gene transcription by steroid hormones has emerged from the experiments of Bert O'Malley and his collaborators on the estrogen-stimulated chick oviduct. In mature egg-laying hens, the egg-white protein **ovalbumin** accounts for up to 60 percent of the total soluble protein of oviduct cells. Young chicks, on the other hand, do not synthesize ovalbumin unless injected with estrogen. When estrogen administration is discontinued, ovalbumin synthesis ceases. This model system is ideal for exploring the mode of action of steroid hormones in regulating the synthesis of specific proteins.

By extracting messenger RNA from oviduct tissue and translating it in a cell-free system, O'Malley has shown that the intracellular level of ovalbumin messenger RNA rises in response to estrogen administration, and falls upon estrogen removal. These changes in ovalbumin mRNA activity closely parallel the rise and fall in the rate of ovalbumin synthesis that occurs in the intact oviduct under these conditions (Figure 16-18a). In order to distinguish whether this increase in ovalbumin mRNA activity is due to the production of new mRNA molecules or the activation of previously existing ones, hybridization experiments were carried out employing a complementary DNA (cDNA) probe that had been generated by copying purified ovalbumin mRNA with the enzyme reverse transcriptase. As described in Chapter 10, such cDNAs can be used as tools to measure the exact amount of a particular type of mRNA within a given RNA sample. When RNA isolated from chick oviduct was hybridized to the ovalbumin cDNA, it was discovered that the number of ovalbumin mRNA molecules per cell increases from less than ten to several thousand or more within a few hours after estrogen treatment (see Figure 16-18b). Using the same hybridization assay, it was also shown that RNA transcribed under cell-free conditions using chromatin isolated from estrogen-stimulated oviducts contains sequences coding for ovalbumin, while RNA transcribed from chromatin isolated from unstimulated oviducts has no detectable ovalbumin sequences. Taken together, the above findings provide strong support for the conclusion that estrogen activates transcription of the ovalbumin gene. Enhanced production of specific messenger RNAs has also been documented in several other steroid hormone-induced systems, including progesterone stimulation of ovalbumin, conalbumin, lysozyme, and avidin formation in the chick oviduct; vitellogenin synthesis in amphibian liver; hydrocortisone stimulation of tyrosine transaminase and tryptophan oxygenase synthesis in mammalian liver; androgen stimulation of alpha-2-microglobulin synthesis in mammalian liver;

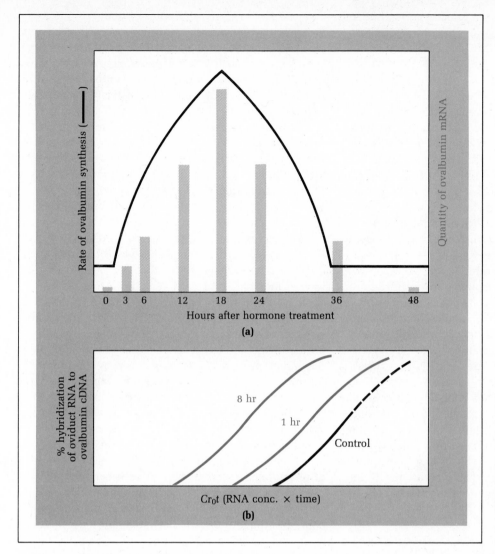

FIGURE 16-18 Data demonstrating induction of ovalbumin mRNA in chick oviduct by estrogen. (a) Effect of one injection of estrogen on ovalbumin synthesis and ovalbumin mRNA activity of extracted RNA. (b) RNA extracted from chick oviduct at varying times after estrogen treatment is tested for its ability to hybridize to ovalbumin cDNA. Displacement of such hybridization curves to the left indicates an increase in the relative proportion of ovalbumin sequences.

and glucocorticoid stimulation of mouse mammary tumor virus RNA in cultured rat hepatoma cells. In addition, virtually all steroid hormones cause an increase in the transcription of the ribosomal RNA genes.

Some insights into the mechanism that permits each hormone to activate the transcription of a specific set of genes have emerged from studies on the effects of glucocorticoids on the transcription of genes contained within the mouse mammary tumor virus (MMTV). The glucocorticoid receptor has been found to bind specifically to a DNA sequence situated near the promoter of the gene being activated. Cloning of this DNA region and sequence analysis has shown that it is similar to enhancer elements identified as transcription activators in other systems (page 434). Thus, the glucocorticoid receptor appears to be able to recognize and bind to a particular DNA sequence located upstream from the initiation point of transcription, and then to activate transcription of the gene involved. Other steroid hormone receptors are thought to function in a similar fashion.

The widespread evidence implicating steroid hormones in the regulation of gene transcription does not necessarily imply that all biological responses induced by steroids are mediated at the level of gene transcription. For example, in the maturation of frog oocytes, direct effects of steroid hormones on the plasma membrane are known to occur. Though the biological significance of such findings remains to be determined, the possibility that steroid hormones have more than one mode of action should not be discounted.

THE EVOLUTION OF INTERCELLULAR COMMUNICATION

Although the models of hormone action developed in this chapter are widely applicable to vertebrates, their relevance to nonvertebrates and to plants is less certain. The discovery of molecules closely resembling vertebrate hormones in unicellular organisms as well as in fungi and invertebrates suggests that the vertebrate hormones may have a long evolutionary history.

Plant Growth Substances

Plants clearly lack specialized endocrine glands designed to secrete hormones into a circulatory system, but some plant cells do synthesize *growth substances* that influence the growth and differentiation of other regions of the plant. These growth substances can be divided into five major categories (Table 16-6): **auxins** and **gibberellins,** which stimulate cell enlargement; **cytokinins,** which promote cell division; and **abscisic acid** and the gas **ethylene,** which inhibit growth, especially during winter or the dry season when environmental conditions are unfavorable.

The mechanisms by which plant growth substances act are not as well understood as in the case of animal hormones, but at least some of the same principles seem to apply. The major effects induced by plant growth substances include alterations in membrane permeability and transport, enzyme activities, and RNA synthesis, all of which have precedent in the case of animal hormones. However, in plants it has often been difficult to distinguish the primary site of action of a growth substance from its secondary effects. For example, the responses induced by auxin treatment include an increased proton efflux through the plasma membrane, enhanced synthesis of RNA, and an increase in the activity of the enzymes glucan synthetase and cellulase.

The increase in glucan synthetase, but not cellulase, can be prevented by blocking transcription with actinomycin D, indicating that the increase in cellulase activity is based upon the activation of previously existing molecules, while the increase in glucan synthetase requires gene transcription. One way to investigate the nature of the primary target of auxins is to search for the specific receptor molecules with which they interact, an approach that has been widely successful with animal hormones. A clear picture is yet to emerge, however, for auxin-binding proteins have been identified in the plasma membrane, cytosol, and nuclear fractions; which, if any, represent the true auxin receptor is not clear.

The confusion that has surrounded the investigation of plant growth substances is well illustrated by the question of whether or not cyclic AMP is employed as a second messenger in plant cells. Both the presence of cyclic AMP in higher plants and the ability of added cyclic AMP to mimic certain effects of plant growth substances have been reported. Nonetheless, the general failure to detect adenylate cyclase and cyclic AMP-dependent protein kinases in plants provides reason for questioning the significance of such reports. The assay procedures employed for measuring cyclic AMP in plants have often failed to discriminate against the possibility of contamination by interfering substances, and experiments involving artificially added cyclic AMP do not necessarily demonstrate that the observed effect is of physiological significance. Even if cyclic AMP does turn out to be a genuine constituent of higher plant cells, its concentration is much lower than in animal cells, and the four criteria used to define a cyclic AMP-mediated hormonal response are still to be satisfied in a plant system. Since plant growth substances appear to enter the cells they act upon, there is most likely no need for a second messenger such as cyclic AMP to transmit signals from the plasma membrane to the cell interior. Like our ideas concerning many other aspects of plant growth regulation, however, this conclusion must remain tentative until more definitive evidence fills in the hazy picture we now have concerning hormone action in higher plant cells.

Hormones Produced by Nonendocrine Cells

For many years it was believed that hormones comprise a distinct and highly specialized class of organic molecules, unique with respect to their site of synthesis in specialized endocrine glands, and their physiological effects on a limited array of target cells. However, this rather narrow view has been challenged of late and it is now believed that endocrine glands do not hold a monopoly as sites of hormone synthesis. In the 1970s, evidence began to appear suggesting that cells of the nervous system can synthesize hormones. These

TABLE 16-6
Structure of plant growth substances

Substance	Derivative of	Structure
Auxins	Indole	
Gibberellins	Gibbane ring system	
Cytokinins	Adenine	
Abscisic acid	—	
Ethylene	—	$CH_2 = CH_2$

neurohormones are manufactured and secreted by a neuron and act on a specialized target cell, either another nerve cell, a muscle cell, or a secretory cell. In addition, it became apparent that some neurohormones are synthesized in nonneural tissue (e.g., epinephrine is synthesized by nerve cells as well as by the adrenal gland). Finally, neurohormones have been found in the brain and have been shown to exert their effects in localized areas. The **endorphins,** which serve as analgesic agents to modulate the pain response, exemplify this rapidly expanding class of messenger molecules.

In addition to the cells of the nervous system, other nonendocrine cells are also known to manufacture and release certain hormones. For instance, glucagon-producing cells have been found in the intestine, and ACTH production has been localized to the placenta. Certain cancers have also been shown to produce and release hormones. The realization that hormone production is not limited to endocrine glands raises the possibility that such molecules might have a more widespread distribution throughout the animal kingdom as well. In order to explore this possibility, Jesse Roth and his collaborators have investigated the question of whether nonvertebrate organisms contain molecules analogous to vertebrate hormones. Using radioimmunoassay as a tool to probe for molecules that are structurally similar to vertebrate hormones, Roth identified substances structurally similar to insulin in extracts obtained from the bacterium *E. coli*, from several different protozoa (unicellular eukaryotes), and from certain fungi. When this approach was extended to other hormones, including ACTH and somatostatin, similar results were obtained. Molecules resembling catecholamines and steroid hormones have also been identified in nonvertebrates. Although the presence of materials structurally resembling vertebrate hormones has been confirmed in many of these "lower organisms," in no case has a clear function for these molecules been identified. Thus, it is not clear whether these counterparts of vertebrate hormones are simply vestiges of an ancient intercellular communication system, or whether they actually serve a hormonelike function. Regardless, the data suggest that the vertebrate endocrine system has its origins deep in evolutionary history.

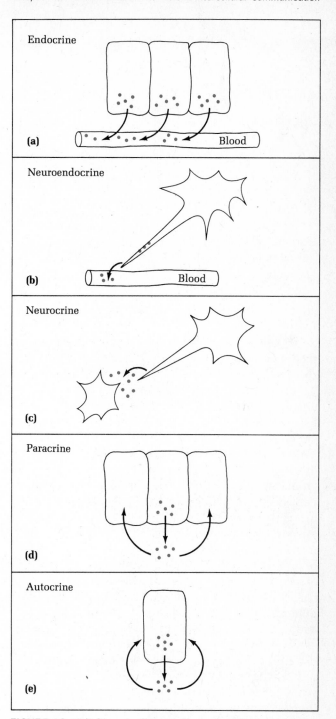

FIGURE 16-19 Schematic diagram illustrating the five principal delivery routes of vertebrate hormones (color): (a) endocrine, (b) neuroendocrine, (c) neurocrine, (d) paracrine, (e) and autocrine. See text for details.

Other Mediators of Intercellular Communication

Analysis of the various kinds of hormonal communication observed in vertebrates has led to the realization that at least five different routes of communication exist: (1) **endocrine,** the traditional pathway of hormone transport via the circulatory system (Figure 16-19a); (2) **neuroendocrine,** whereby the hormonal agent manufactured by a nerve cell is released into the bloodstream (Figure 16-19b); (3) **neurocrine,** typified by the release of a neurotransmitter into the intercellular space adjacent to a target cell (Figure 16-19c); (4) **paracrine,** whereby hormone molecules synthesized by nonneural cells diffuse to neighboring target cells through the intercellular milieu (Figure 16-19d); and (5) **autocrine,** in which the hormone molecule may reenter the cell of

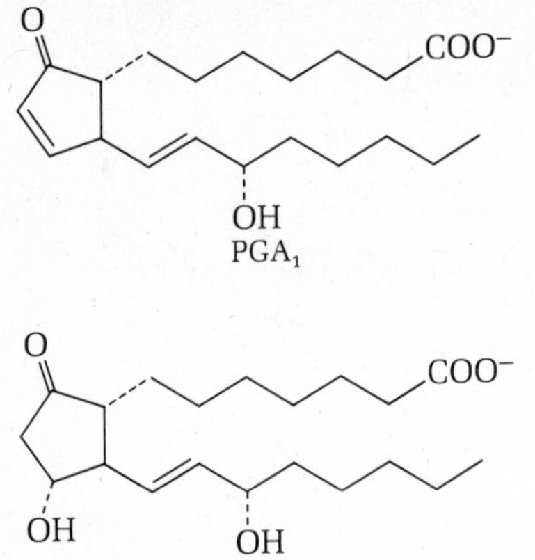

FIGURE 16-20 Structure of two prostaglandins, PGA_1 and PGE_1. Other prostaglandins have the same general form.

origin to effect some physiological change (Figure 16-19e).

When intercellular communication is viewed in this broad way, it becomes apparent that many substances that are not formally labeled as "hormones" or "neurohormones" are also involved. For example, growth factors follow either the paracrine or autocrine pathway to reach their target cells. Another class of closely related intercellular mediators are the **prostaglandins,** first discovered in the early 1930s by Swedish investigators examining the effects of semen on uterine motility. The active factors in semen were named prostaglandins because it was incorrectly believed that they arose in the prostate gland. Prostaglandins are lipid molecules synthesized from fatty acids (principally arachidonic acid) within cellular membranes. They are composed of a 20-membered fatty acid carbon skeleton folded into a loop by incorporating a ring at the bend in the molecule (Figure 16-20). Many different kinds of prostaglandins have been identified, and their ubiquitous distribution suggests that they are synthesized in a wide variety of cell types. Their effects on target cells are equally diverse, ranging from inducing platelet aggregation, to stimulating contraction of smooth muscle in the uterus and bronchi of the lung, to involvement in the control of inflammation. In some cases they may even modify the action of certain hormones. An example of such an effect occurs with the prostaglandin called PGE_1, which is known to inhibit the breakdown of fat in adipose tissue after stimulation of this tissue by the hormones epinephrine or glucagon. PGE_1 acts by inhibiting the activity of adenylate cyclase, thereby leading to an arrest in the elevation of cyclic AMP levels that normally accompanies hormone action. Prostaglandins exert control over processes in nearby cells as well as serving as feedback modifiers of physiological processes occurring within the cells in which they themselves are synthesized.

Cell Specialization and Hormone Evolution

It is not unreasonable to consider that as cellular specialization and division of labor became a common feature of multicellular organisms, a means of coordination of increasingly diverse cell types was needed. Perhaps the localized tissue factors and prostaglandins that are now emerging as important controlling elements in vertebrate intercellular relationships are in fact the evolutionary end products of ancestral molecules that served similar functions as cells made the transition from the unicellular to multicellular level of organization. Such chemical mediators may at first have acted in a highly localized environment. However, with advances in structural and functional differentiation, and the development of organs and organ systems, a subset of these tissue factors may have evolved into specialized intercellular messengers. This view would help to explain why some vertebrate hormones apparently have structural, if not functional, counterparts in unicellular and invertebrate organisms. Thus the intimate relationship between the endocrine and nervous systems that is so clearly important in coordinating various physiological processes in vertebrates may be the end result of a long evolutionary process in which cell specialization and diversification required a continual fine-tuning of the intercellular communication network. Just how plants fit into this scheme is not clear, though the close functional similarity of plant growth substances to the vertebrate growth factors may reflect a common ancestry.

SUMMARY

Three major classes of hormones occur in vertebrates: peptide hormones, steroid hormones, and amino acid derivatives (catecholamines and thyroid hormones). Hormones are synthesized in endocrine glands and are secreted directly into the bloodstream. Hormones are the communicating link between cells located in various tissues in the organism, and their presence facilitates the coordination of metabolism, growth, and differentiation. Each hormone acts on a selected group of target cells possessing specific receptors that recognize the hormone. The response of target cells to hormonal stimulation depends both on the cell type and hormone involved, with the principal effects including alterations in membrane permeability, the metabolism of carbohydrates and lipids, and the biosynthesis of nucleic acids and proteins required for cell growth, division, and differentiation. In spite of the diverse effects of hormones on their target cells, the primary site of hormone action is generally localized to either the plasma membrane or the nucleus.

The plasma membrane is the main site of action of the peptide hormones and catecholamines. The target cell of each hormone contains binding sites on the plasma membrane that satisfy the major criteria of hormone receptors; that is, binding of hormone to such sites is saturable, of high affinity, specific for a particular hormone, rapid, reversible, and leads to a specific physiological response inside the cell. The binding of many peptide hormones or catecholamines to plasma membrane receptors leads to activation of adenylate cyclase, a membrane-localized enzyme responsible for catalyzing the formation of cyclic AMP. This small molecule, in turn, functions as a second messenger, carrying the signal from the cell surface to the cell interior. The main action of cyclic AMP is to stimulate cyclic AMP-dependent protein kinase, an enzyme catalyzing the phosphorylation of a wide variety of proteins. Because each differentiated cell type has its own unique family of proteins subject to phosphorylation by this enzyme, the responses induced by cyclic AMP will vary from cell to cell. For example in liver, cyclic AMP-dependent protein kinase catalyzes the phosphorylation of phosphorylase kinase and glycogen synthase, stimulating the former enzyme and inhibiting the latter one; the net result is that glycogen breakdown is stimulated while glycogen formation is inhibited. In fat cells, on the other hand, cyclic AMP-dependent protein kinase catalyzes phosphorylation of triglyceride lipase, stimulating its ability to catalyze lipid breakdown. Though the multistep cyclic AMP-mediated pathway of hormone action might appear to be unnecessarily complex, such a cascade serves both to amplify the incoming signal and to provide a series of steps subject to alternative modes of regulation.

Some hormones acting on the plasma membrane, such as insulin, growth hormone, prolactin, and oxytocin, do not stimulate cyclic AMP synthesis. Insulin appears to lower intracellular cyclic AMP levels, though this effect has been ruled out as the explanation for most of the responses induced by insulin. In the search for other possible mediators of hormone action, two major candidates have emerged: cyclic GMP and calcium ions. Cyclic GMP is widely distributed in eukaryotic cells and its intracellular concentration increases in response to several hormones, including insulin. Interestingly, the cellular effects associated with increased cyclic GMP levels are often opposite to those associated with increased cyclic AMP; moreover, increases in cyclic AMP levels are often accompanied by a decline in cyclic GMP, and vice versa. These relationships have led to the speculation that cyclic AMP and cyclic GMP exert opposing regulatory influences on the cell, a notion dubbed the yin-yang hypothesis. The molecular mechanisms underlying cyclic GMP involvement in hormone action have not been well worked out. Cyclic GMP-dependent protein kinase activity has been detected in several cell types, but the proteins normally phosphorylated by this enzyme in response to cyclic GMP-elevating hormones are yet to be clearly identified.

Several lines of evidence implicate calcium ions as another mediator of hormone action. The binding of hormones to plasma membrane receptors often leads to an influx of calcium ions into the cytoplasm, either by transport across the plasma membrane or by release from the endoplasmic reticulum. This latter pathway is activated by inositol trisphosphate, which is generated when the membrane lipid phosphatidylinositol 4,5-bisphosphate is broken down by phospholipase C. Many effects of calcium ions on the cell are mediated by the calcium-binding protein, calmodulin, which is activated when calcium ions bind to it. Diacylglycerol, the second reaction product of phospholipase C action, stimulates protein kinase C and thereby enhances the phosphorylation of certain proteins.

In contrast to peptide hormones and catecholamines, steroid and thyroid hormones enter their target cells and interact with the nucleus in order to induce a physiological response. In the case of steroids, the specific steps involved have been fairly well worked out. Steroid hormones initially enter cells by diffusing through the plasma membrane. Upon reaching the cell interior, they bind to specific receptor molecules, preventing the hormone from diffusing back out of the cell. The receptor-hormone complex then undergoes transformation to an activated state in which it is capable of interacting with acceptor sites on the chromatin. This binding of hormone-receptor complexes to chromatin ultimately leads to enhanced transcription of selected genes.

Some of the earliest direct evidence for the ability of steroid hormones to activate specific gene transcription has come from studies on the estrogen-stimulated chick oviduct. Young chicks produce none of the egg-white protein ovalbumin, but its synthesis can be experimentally induced by estrogen treatment. Ovalbumin synthesis triggered in this fashion has been shown to be accompanied by an increase in ovalbumin messenger RNA. Furthermore, chromatin isolated from oviducts of estrogen-stimulated chicks and transcribed under cell-free conditions synthesizes ovalbumin mRNA, while chromatin isolated from nonstimulated oviducts does not, indicating that estrogen enhances transcription of the ovalbumin gene. Steroid receptors are thought to induce transcription of specific genes by binding to DNA sequences located upstream from the particular genes being activated.

In addition to hormones and neurotransmitters, vertebrates possess other molecules that act in intercellular communication to trigger physiological responses. Polypeptide growth factors and prostaglandins are examples of such molecules. The discovery of substances resembling vertebrate hormones in unicellular organisms, fungi, and invertebrates suggests the possibility that these molecules may have served as chemical mediators, when cells were undergoing the transition from the single-celled to multicellular level of organization. Plants possess growth substances as well, though their mode of action is not as well understood as in the case of vertebrate hormones. It is possible that the first tissue regulatory factors were needed for coordinating the activity of groups of cells and that these organic molecules served as precursors for the development of growth factors in plants and animals. A subset of these chemical messengers could then have evolved in parallel with vertebrate organ systems to reach the level of complexity seen in today's vertebrate hormones.

SUGGESTED READINGS

Books

Goldberger, R. F., and K. R. Yamamoto, eds. (1984). *Biological Regulation and Development*, Vol. 3B, *Hormone Action*, Plenum Press, New York.

Hadley, M. E. (1984). *Endocrinology*, Prentice-Hall, Englewood Cliffs, NJ.

Moore, T. C. (1979). *Biochemistry and Physiology of Plant Hormones*, Springer-Verlag, New York.

Wallis, M., S. L. Howell, and K. W. Taylor (1986). *The Biochemistry of Polypeptide Hormones*, Wiley, New York.

Articles

Bell, R. M. (1986). Protein kinase C activation by diacylglycerol second messengers, *Cell* 45:631–632.

Berridge, M. J. (1984). Inositol trisphosphate and diacylglycerol as second messengers, *Biochem. J.* 220:345–360.

Cheung, W. Y. (1982). Calmodulin, *Sci. Amer.* 246 (June):62–70.

Gilman, A. G. (1986). Receptor-regulated G proteins, *Trends Neurosci.* 9:460–463.

Hokin, L. E. (1985). Receptors and phosphoinositide-generated second messengers, *Ann. Rev. Biochem.* 54:205–235.

Nishizuka, Y. (1984). Protein kinases in signal transduction, *Trends Biochem. Sci.* 9:163–166.

Rasmussen, H., and P. Q. Barrett (1984). Calcium messenger systems: an integrated view, *Physiol. Rev.* 64:938–984.

Ringold, G. M., D. E. Dobson, J. R. Grove, C. V. Hall, F. Lee, and J. L. Vannice (1983). Glucocorticoid regulation of gene expression: mouse mammary tumor virus as a model system, *Rec. Prog. Hormone Res.* 39:387–424.

Strosberg, A. D. (1984). Receptors and recognition: from ligand binding to gene structure, *Trends Biochem. Sci.* 9:166–169.

Vaughan, M. (1982). Cholera and cell regulation, *Hosp. Prac.* 12:145–152.

Yalow, R. S. (1978). Radioimmunoassay: a probe for the fine structure of biologic systems, *Science* 200:1236–1245.

CHAPTER

17

The Cellular Basis of the Immune Response

The chief function of the immune system is to protect an organism against potentially harmful foreign substances. This system of defensive strategies has reached its zenith in the higher vertebrates, where the evolution of acquired immunity has made it possible to rapidly inactivate foreign material and microorganisms. Although lower animals possess a relatively simple system for removing invading microorganisms or viruses, these nonspecific immune responses lack the three critical features of an acquired immune response: (1) the ability to distinguish "self" from "nonself," that is, to identify a substance as foreign; (2) the ability to discriminate among different foreign substances (antigens), that is, to exhibit specificity; and (3) the ability to retain a memory of a particular antigen so that a second, rapid response to that antigen can be mounted. These preceding properties are made possible by the functioning of white blood cells called lymphocytes.

In this chapter we shall examine the fundamental mechanisms by which lymphocytes make such an immune response possible. We shall see that although an immune response can be mounted against an almost infinite array of antigens, only a limited set of cell types is required. The way in which these cells perceive material as foreign, and respond to remove or inactivate the offending substance, provides a fascinating insight into the mechanisms that underlie the immune system.

GENERAL PROPERTIES OF THE IMMUNE SYSTEM

The ability to ensure protection against entry of foreign substances into the tissues is to a large extent dependent on a particular type of cell called the **lymphocyte.** These white blood cells enter most tissues of the body from the bloodstream by penetrating capillary walls. They then make their way to a return vascular system called the **lymphatic system,** which is made up of blind capillaries that collect the intercellular fluid and cells and ultimately deposit this material into the major lymph vessel that empties into the venous side of the circulatory system (Figure 17-1).

Lymphocytes can be separated into two major types of cells based on differences in developmental origin, tissue source, and method of recognizing foreign material. These two classes of cells, called **B-** and **T-lymphocytes,** arise from a common, undifferentiated *stem cell* that is formed very early in the development of both birds and mammals. These progenitor cells

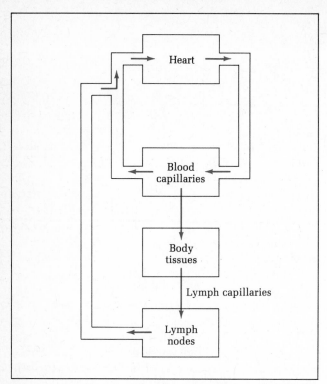

FIGURE 17-1 Schematic diagram showing how lymphocytes move through three compartments, the circulatory system, body tissues, and lymphatic system. The route of lymphocytes is indicated by arrows (color).

The second type of response, which involves antigen recognition by the T-lymphocyte population, is called **cell-mediated immunity.** In this case T-lymphocytes containing antibodylike molecules on their surfaces interact directly with foreign materials rather than indirectly by secreted antibodies. This type of immune reaction is responsible for rejection of skin grafts and organ transplants. T-cells involved in cell-mediated immunity look very different from antibody-synthesizing B-lymphocytes, having little rough endoplasmic reticulum but possessing a large number of free ribosomes in the cytoplasm (Figure 17-4). T-lymphocytes also release soluble factors that are important in coordinating the immune response.

Perhaps the most remarkable feature of the immune response, whether it is cell mediated or humoral, is its diversity and specificity. Immune reactions can be generated against a virtually endless array of foreign antigens, yet the antibodies or mature T-cells formed in each instance are extraordinarily specific in reacting with the particular antigen involved. In trying to elucidate how this diversity and specificity are generated, biologists have focused their attention on two major questions: What are the properties of cells involved in the immune response? What mechanisms give rise to the astounding diversity of antibodies? In the following discussions we shall see how the investigation of these two questions has led to a greater understanding of the fundamental principles underlying the immune response in higher vertebrates.

FIGURE 17-2 Pathway of lymphocyte development from a progenitor stem cell in the yolk sac of birds and mammals. Note that maturation of T- and B-cells into immunologically functional entities requires the presence of antigen.

traverse a specific route through embryonic organs and become deposited in the primary lymphoid organs, the **thymus** and the **bone marrow** (or in birds, the bursa of Fabricius). At these sites the stem cells begin to differentiate, the thymus cells giving rise to the T-lymphocytes and the bone marrow (or its equivalent in birds) giving rise to B-lymphocytes. Full biological function of these cells is not attained until they migrate to the secondary lymphoid organs, the **lymph nodes** and **spleen** (Figure 17-2).

The B- and T-cells differ in terms of their response to foreign substances, or **antigens.** In some cases proteins called **antibodies** are synthesized by B-lymphocytes and secreted into the bloodstream. In such instances the B-lymphocytes develop into highly differentiated **plasma cells** containing a well-developed rough endoplasmic reticulum (Figure 17-3). As we will see later in the chapter, antibodies are designed to bind specifically to a particular antigen. If the foreign material is toxic, binding of antibody generally inactivates the toxin. If the foreign substance is a microorganism, antibody binding leads to destruction of the bacterial cell. This release of antibodies into the bloodstream in response to a particular antigenic stimulus is termed **humoral** or **antibody-mediated immunity.** (The term *humoral* comes from the time when the blood and other body fluids were referred to as "humors.")

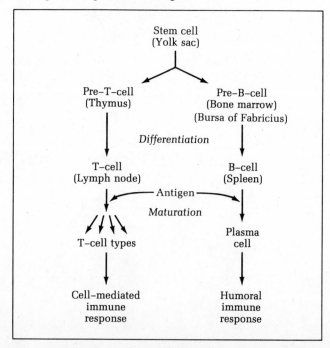

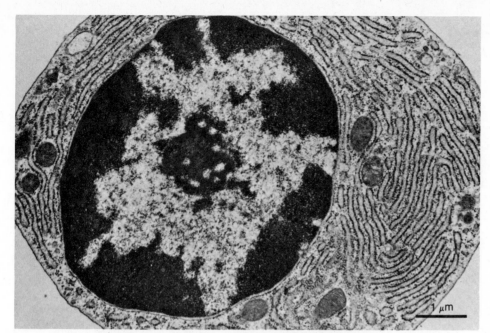

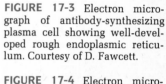

FIGURE 17-3 Electron micrograph of antibody-synthesizing plasma cell showing well-developed rough endoplasmic reticulum. Courtesy of D. Fawcett.

FIGURE 17-4 Electron micrograph of a T-cell. Note the absence of an extensive rough endoplasmic reticulum. Courtesy of D. Fawcett.

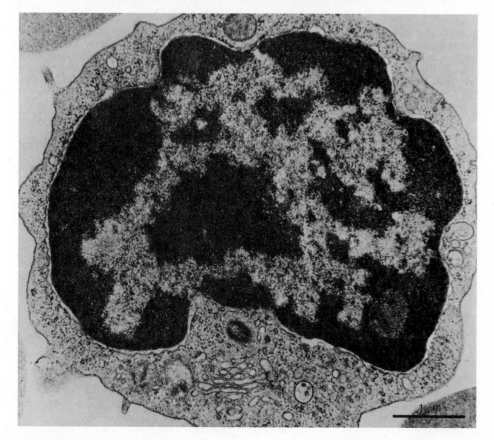

THE B-LYMPHOCYTE AND ANTIBODY SYNTHESIS

Evidence That B-Cells Synthesize Antibodies

Two pioneering discoveries made in the mid-1950s laid the foundation for our current understanding of the role of cellular interactions in antibody production. First, Robert Good, Jacques F. A. P. Miller, and Byron Waksman independently found that surgical removal of the thymus from newborn mice and rabbits prevents these animals from generating antibodies or rejecting skin grafts. This result is observed only in very young animals, indicating that the impact of the thymus gland on development of the immune system is critical at a specific time early in the life of the animal.

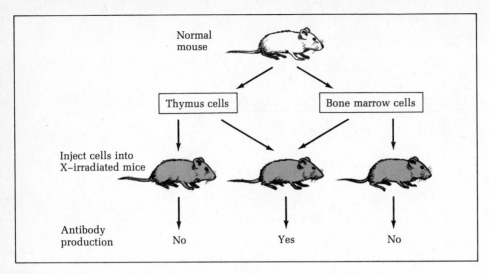

FIGURE 17-5 Reconstitution experiment showing that cooperation between thymus- and bone-marrow-derived lymphocytes is required for the production of circulating antibodies.

The second discovery was made in 1954 by Bruce Glick, who showed that the development of humoral immunity in chickens was markedly restricted by the removal, just after hatching, of a lymphoid organ called the bursa of Fabricius. Loss of this organ had no effect on cell-mediated immunity. Though a bursa of Fabricius is not present in mammals, lymphocytes developing in the bone marrow are now believed to carry out the same functions as those derived from the bursa.

The role of lymphocytes derived from the thymus and bone marrow (or bursa) in the production of circulating antibodies was subsequently clarified by Henry Claman and his co-workers in studies utilizing X-irradiated mice. Such animals no longer produce antibodies, even if they are injected with either thymus or bone marrow cells obtained from healthy animals. If both thymus *and* bone marrow cells are injected, however, the ability to generate circulating antibodies is restored (Figure 17-5). This result fosters the view that antibody production requires the participation of both T-cells (thymus derived) and B-cells (bone marrow derived).

The fact that both T- and B-lymphocytes are necessary for antibody production raises the question of which cell type is actually responsible for antibody synthesis. A definitive answer to this question was provided by Miller and his colleague Graham Mitchell. These investigators cleverly took advantage of the fact that all cells contain on their surfaces a set of identifying proteins called **histocompatibility** (or **transplantation) antigens.** These molecules are encoded by a closely linked group of genes called the **major histocompatibility complex,** or **MHC.** In the mouse this group of genes is referred to as the **H-2 complex.** Animals of an inbred strain exhibit the same **H-2 antigens** on the surfaces of all cells. These antigens differ from those of other inbred strains.

In the Mitchell–Miller experiment the thymus was

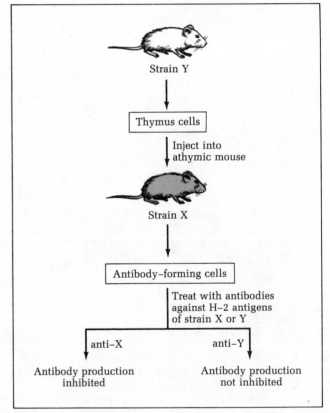

FIGURE 17-6 Experiment showing that a cell type other than the T-cell is responsible for antibody production.

first removed from mice of one strain (X) to obliterate all antibody production. Thymus cells obtained from mice of another strain (Y) were then used to repopulate the strain-X animals with thymus cells. As expected, the ability of the strain-X animals to generate antibodies was restored. In order to determine whether the antibodies were being synthesized by cells derived from the host (strain X) or donor (strain Y) animals, lymphocytes were removed from the host's spleen and tested for their ability to synthesize antibody *after* exposure to antibodies directed against the H-2 antigens of strain X or strain Y. This latter treatment would be expected to specifically destroy either strain-X or strain-Y cells, respectively. Under these conditions it was found that only antibodies against strain-X cells caused an inhibition of antibody synthesis by the spleen lymphocytes. This result clearly indicates that the antibody-producing cells are of host (strain X) rather than donor (strain Y) origin (Figure 17-6). Since the host animal lacked a thymus and was therefore devoid of strain-X T-cells, this result signified that another cell type must be responsible for antibody synthesis after injection of the foreign T-cells. The T-lymphocyte was therefore judged to be acting as a **helper cell,** in some way aiding the synthesis of antibody by another cell type.

To ascertain whether this other cell type responsible for antibody synthesis is the B-cell, reconstitution experiments have been carried out in which normal T-cells are mixed with B-cells derived from animals bearing an abnormal chromosome. This mixture is then injected into irradiated animals incapable of producing antibodies. When the antibody-producing cells are removed from the host, they are found to contain the marker chromosome. However, the reciprocal experiment using T-cells possessing the marker chromosome and B-cells lacking this marker generates antibody-forming cells in the reconstituted host that exhibit a normal chromosome complement. Taken together, these experiments verify that it is the B-cell that actually is responsible for antibody synthesis, even though this requires stimulation by helper T-cells.

Cellular Cooperation in Antibody Synthesis

The realization that B-cell:T-cell interaction is required for normal antibody synthesis led to a variety of experiments designed to investigate this phenomenon. Marc Feldmann and Anthony Basten tested the hypothesis that physical contact between T- and B-cells is necessary for antibody synthesis. They placed these two cell types on different sides of an artificial membrane that prevents direct cell-to-cell contact but permits diffusion of soluble molecules. In such a situation B-cells can synthesize antibody in response to antigenic stimulation, as long as the T-cells are present on the other side of the membrane. This result suggests that cooperation between B- and T-lymphocytes does not require direct cell-to-cell contact but is instead mediated by a diffusible substance.

This diffusible factor appears to be antigen specific since the effect appears only with T-cells that have been previously exposed to the same antigen used to stimulate the B-cells in the chamber. Such results support a model in which antigenic stimulation of T-lymphocytes induces them to release a soluble factor that in turn activates B-lymphocytes to synthesize antibodies to that same antigen. The existence of such soluble factors has been verified using extracts obtained from T-cells. Such preparations are equivalent to the intact T-cell in being able to promote antibody production by B-cells. The soluble factors released by the T-lymphocytes are called **lymphokines;** as we shall see later in the chapter, they play a variety of important roles in coordinating the immune response.

In the experiments demonstrating cooperation between B-cells and T-cells separated by an artificial membrane, **macrophages** are generally present as well. These large phagocytic cells are essential for the induction of antibody formation *in vitro.* If macrophages and lymphocytes are separated and foreign antigen added to each individual population, no antibody is produced. However, when the cells are mixed together and exposed to the same antigen, antibody synthesis occurs. This type of experiment demonstrates that production of antibodies requires the cooperation of at least three cell types: B-lymphocytes, T-lymphocytes, and macrophages. The role of macrophages differs from that of the lymphocytes, however, in terms of antigenic specificity. Macrophages obtained from any animal function equally well in promoting antibody formation by lymphocytes *in vitro.* But the source of the B- and T-cells is critical. To obtain significant levels of antibody production, both B-lymphocytes and T-lymphocytes must be derived from animals previously exposed to the antigen employed to induce antibody formation *in vitro.* B- and T-cells are therefore said to be *antigen specific,* while macrophages are not.

To further investigate the role of macrophages in antibody synthesis, experiments have been performed in which T-lymphocytes are present on one side of an artificial membrane and macrophages on the other. When the macrophages are subsequently removed and mixed with B-lymphocytes, specific antibody production by the B-cells is induced. Unlike the T-cells, the macrophages do not need to be obtained from an animal already exposed to the particular antigen being utilized. Hence it appears as if T-cells produce an antigen-specific lymphokine capable of reacting with any macrophage, which in turn becomes capable of stimulating B-cells to produce antibodies to the specific antigen in question.

Although the details are not well understood, interaction between macrophages, T-cells, and B-cells appears to involve a histocompatibility antigen termed **Ia.** Both macrophages and B-cells possess Ia antigens on their surfaces. The binding of a foreign antigen to the surface of the macrophage results in the activation of the helper T-cell by the macrophage. This activated cell can then recognize the combination of Ia antigen and foreign antigen on the B-cell surface and induce B-cell maturation and subsequent antibody synthesis (Figure

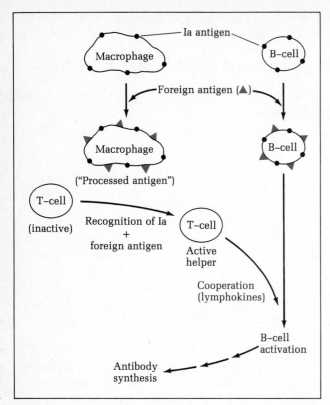

FIGURE 17-7 Summary of interactions of macrophage, T-cell, and B-cell in antibody synthesis. See text for details.

17-7). Thus, antibody synthesis is dependent on input from both the macrophage and helper T-cell, and the newly made antibodies are specific for the particular antigen that induced activation of the B-cell.

Antibody Structure

In order to understand how antibodies recognize specific antigens and thereby facilitate their destruction or removal from the host organism, one must examine the chemical structure of antibodies. Antibodies belong to a family of serum proteins, called **immunoglobulins,** which can be subdivided into five distinct classes of molecules differing in size, amino acid composition, carbohydrate content, and function. These five groups are designated immunoglobulins G, M, A, D, and E, and

are commonly abbreviated **IgG, IgM, IgA, IgD,** and **IgE.** The properties of these varying types of immunoglobulins are summarized in Table 17-1. Of the five, immunoglobulin-G is present in highest concentration in the bloodstream and is the major form of circulating antibody.

The study of antibody structure was first made possible by the development of techniques for splitting these large molecules into fragments of manageable size. In the late 1950s Rodney Porter found that treatment of a crude mixture of antibodies with the proteolytic enzyme **papain** produces three fragments. Two of these, called **Fab** fragments, are of the same size and retain the ability to bind antigen. The remaining fragment, termed **Fc,** does not react with antigen but is capable of binding to **complement,** a group of about 20 enzymes present in serum that functions in both humoral and cell-mediated immunity. The Fc fragment readily forms crystals, an indication that the Fc fragments obtained from a heterogeneous collection of antibodies are all identical; in contrast, the Fab fragments derived from a collection of antibody molecules differ from one another in composition and hence will not form crystals. It therefore has been concluded that each antibody molecule contains two Fab regions with antigen binding sites that vary from antibody to antibody, and a single Fc region that is common to all antibodies and is responsible for binding complement.

Another approach to dissecting antibody structure was taken by Gerald Edelman, who used mercaptoethanol (which cleaves disulfide bonds) and the denaturing agent urea to dissociate immunoglobulins into their constituent polypeptide chains. Such treatment was found to release two **heavy chains** of about 50,000 daltons and two **light chains** of about 20,000 daltons. To ascertain the relationship of these heavy and light chains to Fab and Fc fragments, Porter ran a series of tests utilizing antiserum specifically directed against either the Fab or Fc fragments. Isolated light chains were found to react only with antiserum against Fab, while heavy chains react with both anti-Fab and anti-Fc. From this information, a four-chain model of immunoglobulin-G was constructed (Figure 17-8). Electron micrographs of the IgG molecule show that it is

TABLE 17-1

Properties of the major human immunoglobulin classes

Property	Immunoglobulin Class				
	IgG	IgM	IgA	IgD	IgE
Molecular weight	150,000	900,000	160,000	180,000	200,000
Light chains (κ or λ)	2	10	2	2	2
Heavy chains					
Number	2	10	2	2	2
Type	γ	μ	α	δ	ϵ
Percent of total					
immunoglobulins	80	6	13	0–1	0.001

Y-shaped, with a hinge region that allows the arms to swing out further to an angle of 180°. Subsequent investigations have revealed that the immunoglobulins of each of the four remaining classes are constructed from heavy and light chains, though the actual shape of each antibody is unique. Part of the uniqueness of each immunoglobulin class is imparted by its own special heavy chain, designated γ, μ, α, δ, and ϵ for IgG, IgM, IgA, IgD, and IgE, respectively. Each type of heavy chain is combined with one of two basic types of light chains, **kappa** (κ) or **lambda** (λ), to generate a mature antibody molecule (see Table 17-1).

Because the immunoglobulins present in normal serum consist of a mixture of thousands of antibodies with differing antigenic specificities, it has been relatively difficult to probe the structural details of individual antibody molecules. One of the earliest attempts to circumvent this obstacle emerged from studies of patients with multiple myeloma, a disease characterized by uncontrolled proliferation of antibody-producing

FIGURE 17-8 Four-chain model of the immunoglobulin G molecule, based on treatment with papain or mercaptoethanol-urea. The colored line represents cleavage sites under each condition. Abbreviations: L = light chain; H = heavy chain.

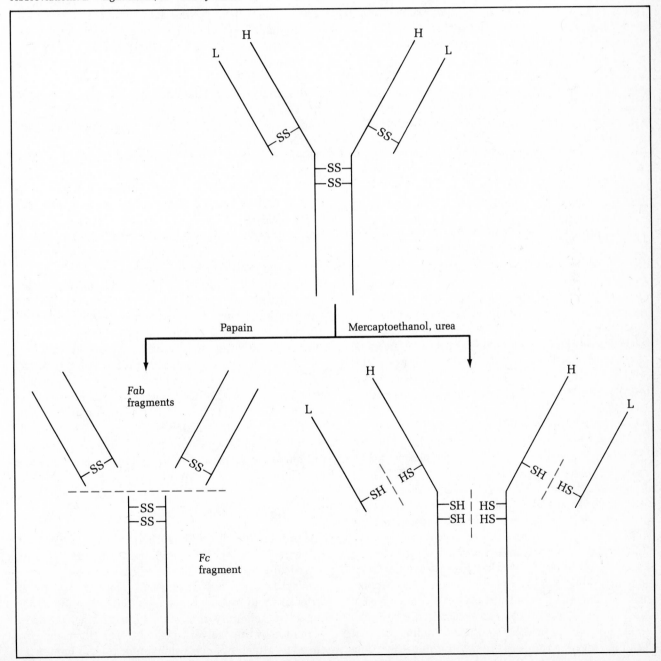

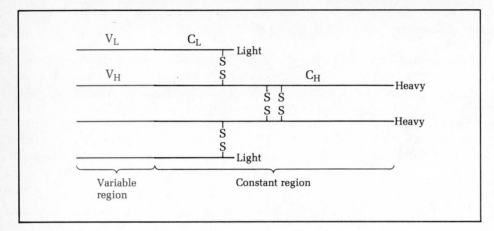

FIGURE 17-9 Schematic diagram showing the location of constant (black) and variable (color) amino acid sequences in immunoglobulin G. Abbreviations: V_L = variable region of light chain, C_L = constant region of light chain, V_H = variable region of heavy chain, C_H = constant region of heavy chain.

plasma cells. Patients with this disorder excrete in their urine large amounts of a homogeneous antibodylike substance known as **Bence–Jones protein.** Edelman has shown that this protein consists of immunoglobulin light chains synthesized in excess by the proliferating plasma cells and present in high concentrations in the urine. Any given patient produces only a single type of Bence–Jones protein, making it ideally suited for structural studies. The first detailed information on the amino acid sequence of such proteins was reported in 1965 by Norbert Hilschmann and Lyman Craig, who compared two Bence–Jones light chains obtained from different individuals. Each light chain was found to exhibit a different amino acid sequence. But an unexpected finding, destined to have a dramatic impact on the field of immunology, also emerged from these studies: The differences between the amino acid sequences of the two light chains all occurred in one end of the molecule. This suggested that light chains contain a **constant region,** whose sequence is the same from molecule to molecule, and a **variable region,** whose sequence varies among antibodies of differing antigenic specificities (Figure 17-9). This variability in amino acid sequence at one end of the light-chain molecule was thought to give rise to the specificity of the antigen binding site.

In addition to the light chains present in the urine of patients with multiple myeloma, the complete IgG molecule produced by the proliferating plasma cells is present in high concentrations in the serum. This *myeloma protein* is the equivalent of a homogeneous population of normal immunoglobulins of a single antigenic specificity, and is therefore a candidate for structural studies. Edelman and his colleagues were the first to determine the complete amino acid sequence of such a myeloma protein. By comparing these data with the sequences of other light and heavy chains, it has become clear that the heavy as well as the light chains consist of a combination of variable and constant regions. The variable regions of the light and heavy chains occur at the amino terminus and are about 110 amino acids long.

The constant region of the light chain is also about 100 amino acids long, while the constant region of the larger heavy chain totals nearly 350 amino acids.

Examination of the amino acid sequence of IgG molecules has provided some interesting insights into antibody structure and function. The constant portion of the heavy chain is composed of three domains, C_{H1}, C_{H2}, and C_{H3}, which resemble one another and the constant region of the light chain in sequence (Figure 17-10). These domains form intrachain globular structures that are stabilized by disulfide bonds. Both the C_{H1} and C_{H2} regions are necessary for binding to complement, while the C_{H3} region is involved in binding of the antibody to the surface of certain cells of the immune system. The antigen binding site, or **idiotype,** is associated with regions of disulfide bond-linked tertiary structure located within the variable regions of both the light and heavy chains. The idiotypes of different antibodies share some similarity in sequence, but sites of extensive amino acid substitution, termed **hypervariable regions,** also occur, giving rise to considerable diversity in the shapes of the antigen binding sites of different antibodies.

Since an organism must manufacture a different type of antibody with a unique antigen binding site every time a new foreign antigen is encountered, the question arises as to how such a vast array of different antibodies is produced. In the following section we shall examine the theories and experiments that underlie our current understanding of how this diversity in antibody structure is generated.

The Origin of Antibody Diversity

The process underlying the formation of antibodies that recognize the shape of specific antigens was first systematically examined early in this century by Karl Landsteiner, who was interested in the antigenic properties of small organic molecules. Though small molecules are generally incapable of triggering antibody formation by themselves, antibody formation can be induced by attaching such a small molecule, or **hapten,** to

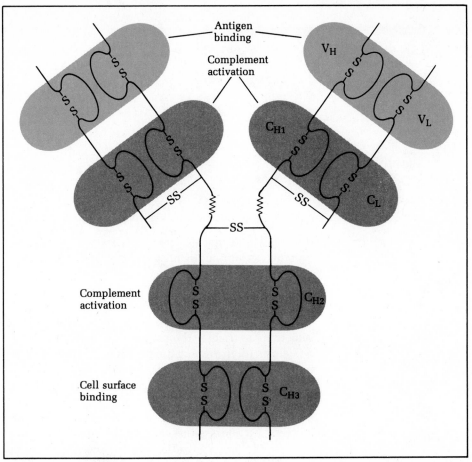

FIGURE 17-10 Arrangement of constant and variable regions of the immunoglobulin G molecule. The hinge region is indicated by the jagged line in the heavy chains. The constant region of the heavy chain is divided into three domains (C_{H1}, C_{H2}, and C_{H3}) that are homologous to each other and to the constant region of the light chain. The variable regions of the light and heavy chains (color) exhibit a weaker homology and possess hypervariable regions. See text for details.

a larger molecule, or **carrier.** Landsteiner discovered that every time he synthesized a new small molecule by subtly changing one of its side chains, a new antibody uniquely specific for that hapten could be produced by injecting animals with the new hapten-carrier complex. This observation raised a very puzzling question: How can an organism synthesize a seemingly endless array of unique antibodies specifically directed against substances that it or its ancestors could never before have encountered in nature?

To attempt to account for this incredible diversity, Linus Pauling proposed that antigens interact with antibodies as the latter are being synthesized, directing the folding of the newly forming antibody chains into a structure specific for the stimulating antigen. According to this theory any given antibody molecule would theoretically be capable of binding to a large number of different antigens, with the final choice being determined by the particular antigen present during folding of the polypeptide chain. Because antigens were believed to instruct the cell to make a certain type of antibody, this

notion came to be called the **instructive theory** of antibody formation (Figure 17-11).

As an alternative to this idea, Niels Jerne and later Macfarlane Burnet postulated that an organism's lymphocytes are inherently capable of producing a vast spectrum of different antibodies independent of any exposure to antigen. According to this view, each lymphocyte is programmed to make one particular type of antibody, and the role of antigen is not to instruct cells to form a proper-fitting antibody, but simply to select those cells that already happen to be making one that fits the antigen. The antigen then stimulates such cells to divide, producing a large population of mature antibody-secreting cells specific for the triggering antigen. The resulting cell population constitutes a **clone,** since it is derived by a series of successive divisions from a single precursor cell (a B-lymphocyte). Because this theory proposes that the antigen functions to *select* rather than to *instruct* formation of the proper antibody, it is termed the **clonal selection theory** (Figure 17-12).

In the early 1960s the first definitive experiments

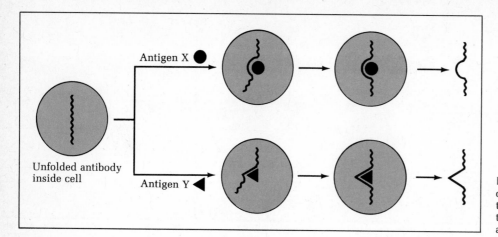

FIGURE 17-11 Instructive theory of antibody formation. According to this theory, antigens act as a template directing the folding of antibody into the proper shape.

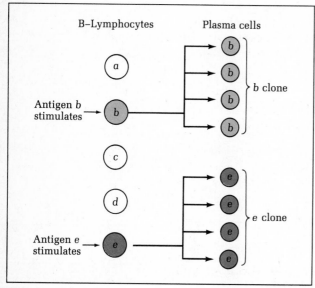

FIGURE 17-12 Clonal selection theory of antibody formation. Each lymphocyte is programmed to make one type of antibody. An antigen selectively stimulates proliferation of B-lymphocytes to produce a clone of cells, which synthesizes a single antibody species capable of binding to a particular antigen.

were carried out to determine whether or not the presence of antigen is essential for proper antibody folding, as is required by the instructive theory. These studies revealed that antibodies whose antigen-binding specificity has been destroyed by exposure to denaturing agents can spontaneously regain the ability to bind to their specific antigens when exposed to nondenaturing conditions. This reappearance of antigen-binding specificity occurs even when the antigen is not present during the renaturation process, indicating that antigen specificity is encoded in an antibody's amino acid sequence rather than instructed by an antigen during folding of the antibody chain. This finding is compatible with the selective, but not the instructive, theory of antibody formation.

Additional support for the clonal selection theory was provided by experiments that showed that B-lymphocyte populations derived from a single cell pro-

duce the same type of antibody. This has been revealed both by examining the offspring of single lymphocytes grown in culture and by studies on multiple myeloma patients. In both cases a single type of antibody is produced by all cells derived from the original parent cell.

As a result of these kinds of observations, clonal selection has become the major working theory of antibody formation. Probing the details of this theory, however, quickly leads to several new questions: How can the organism produce so many different kinds of antibodies? How is the proper antibody-producing plasma cell induced to proliferate in the presence of a particular antigen? To address these issues we must first understand the basic organization of the genes that code for antibodies.

Genetic Control of Antibody Organization

It has been estimated that a given individual can synthesize a million or more different kinds of antibodies. It therefore follows that if every antibody molecule were encoded by a separate gene, virtually all of an organism's DNA would be devoted to antibody production. However, the construction of antibodies from separate heavy and light chains provides a simpler mechanism for creating a million different antibodies. As few as a thousand different light chains and a thousand heavy chains assembled in all possible combinations would generate $1000 \times 1000 = 1,000,000$ distinct antibody molecules. Hence in theory, a total of 2000 genes would be enough to produce a million antibody molecules, and would utilize less than one percent of an organism's DNA. This notion that all the genes needed for antibody diversity are separately encoded in an individual's DNA is called the **germ line theory,** for it assumes that all antibody genes are inherited through the germ cells (sperm and egg).

The assumption that 2000 genes may theoretically encode the required number of antibody molecules does not discount the alternative possibility that individuals inherit an even smaller number of antibody genes, and that additional diversity is generated during the many cell divisions leading to the formation of B-lymphocytes. The mechanisms producing such diversity might involve mutation and/or genetic recombination among the smaller number of inherited genes. The terms **somatic mutation** and **somatic recombination** are used to refer to such theories that postulate that antibody diversity is generated mainly during proliferation and differentiation of the B-lymphocytes. The essential difference between the somatic and germ line theories boils down to the question of whether the information required for antibody diversity is transmitted from parent to offspring by sperm and egg or is created as the B-cells develop from the precursor stem cell (see Figure 17-2).

The pros and cons of the germ-line and somatic theories have been extensively debated over the years, but only recently have the experimental tools become available for directly examining the arrangement and number of antibody genes. The first major breakthrough in this area occurred in 1976 when Nobumichi Hozumi and Suzumu Tonegawa reported that the variable and constant regions of antibody molecules are encoded by separate genes, termed **V** and **C genes,** whose location within the DNA differs in embryonic and antibody-forming cells. This conclusion came from studies in which messenger RNA coding for an antibody light chain was hybridized to DNA fragments produced by cleaving either embryonic or plasma cell DNA with restriction endonucleases. Surprisingly, the light-chain mRNA was found to hybridize to two different fragments of embryonic DNA, but to only one fragment of plasma cell DNA (Figure 17-13). By cleaving the light-

chain mRNA into its constant and variable regions and repeating the experiment, it could be shown that one of the two embryonic DNA fragments contains the variable sequences and the other contains the constant sequences. It was therefore concluded that genes coding for light-chain variable and constant regions are not adjacent to each other in embryonic DNA, and that these genes are brought closer together during the formation of B-lymphocytes. This finding led to the startling realization that **DNA rearrangement** occurs during cell differentiation.

Though the Hozumi–Tonegawa experiments revealed that somatic rearrangement of antibody genes does occur, they did not indicate whether embryonic DNA contains a large number of different V genes, as is required by the germ line theory. One limitation in their experimental approach was that the DNA fragments tested for the presence of V- and C-gene sequences were isolated by one-dimensional electrophoresis, which is not capable of discriminating between pieces of DNA exhibiting small differences in size or base sequence. Philip Leder, Jonathan Seidman, and their colleagues were the first to overcome this difficulty by devising a two-dimensional chromatographic-electrophoretic scheme for separating DNA restriction fragments. Using this method they demonstrated that a kappa light-chain mRNA obtained from one type of plasma cell contains sequences complementary to at least seven different DNA fragments, while a second kappa light-chain mRNA obtained from a different plasma cell line contains sequences that hybridize to at least nine DNA fragments. Furthermore, only one of the DNA fragments recognized by the two light-chain mRNAs is the same, and this particular fragment codes for the constant region of the light chain. The remaining DNA fragments all contain variable sequences. It was therefore initially concluded that the variable region of each light-chain mRNA is related to a family of about seven different V genes. Such families of V genes were found to occur in embryonic as well as plasma cell DNA, indicating that this diversity is not created during somatic cell differentiation.

Since each of the two light-chain mRNAs used as probes in the above studies hybridized to a different family of V genes, the next step is to determine how many different V-gene families one would expect to find if additional light-chain mRNAs were employed as probes. Antibody light chains can be divided into at least 30 *subgroups*, each one made up of a family of molecules of similar, but not identical, amino acid sequences. The two light-chain mRNAs employed by Leder and his colleagues were members of different subgroups, so it is reasonable to assume that each subgroup corresponds to a family of about 7 V genes. If each of the 30 subgroups contains about 7 genes, then a minimum of about 200 different V genes must be present.

FIGURE 17-13 Hybridization of radioactive light-chain mRNA to DNA restriction fragments separated by electrophoresis. Note that light-chain sequences are present in two fragments of embryonic DNA, but in only one fragment of plasma cell DNA.

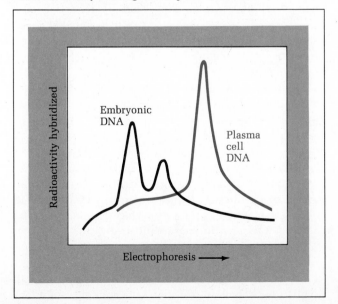

Moreover, new light-chain subgroups are continually being discovered, so it is possible that this number may eventually approach the 1000 or so V genes required by the germ line theory. Though the heavy-chain genes have been less extensively studied, multiple V genes for heavy chains have also been identified, thereby completing the picture of germ line diversity.

Though the number of different V genes inherited through the germ line may ultimately be sufficient to account for much of antibody diversity, the potential for further increasing diversity by somatic mechanisms is not ruled out. This possibility is even supported by base-sequencing studies that have revealed that the DNA sequences surrounding the members of a V-gene family resemble one another. This similarity in the DNA sequences surrounding the V genes of a given subgroup would facilitate recombination, thereby allowing further diversity in variable-region gene structure to be introduced during cell division.

Our current picture of antibody diversity thus involves both multiple germ line V genes and the possibility of somatic recombination between these genes. A crucial question that remains, however, is how a particular V gene is chosen to be expressed. Though a clear answer is yet to emerge, a clue may be provided by the discovery that not all the DNA coding for light chains can be accounted for by the V and C genes. A segment of about 11 or so amino acids located between the variable and constant regions of the immunoglobulin light chain has been found to be encoded in a separate region of DNA situated about 1250 base pairs from the C gene and at least 10 kilobases from the V gene. This DNA segment is termed the **J region** because it codes for amino acids joining the variable and constant regions of light chains. Though the function of the J region is unclear, joining of one of a cell's many V genes to the J region may be a critical step in selecting that particular V gene for expression.

Though the 1250 base-pair sequence separating the J region from the C gene does not code for any portion of the antibody molecule, it remains in the middle of the antibody gene, even in mature antibody-forming cells, and is therefore an example of an intervening sequence or intron. This intervening sequence is transcribed into the precursor RNA molecule, which is subsequently processed to form the final messenger RNA. The light chain contains at least one other intron, located between the V gene and a leader sequence coding for a signal peptide involved in sequestering the newly forming light chain in the endoplasmic reticulum.

The overall complexity of the organization of light-chain genes is well illustrated by what is now known about the structure of mouse IgG genes (Figure 17-14). For the lambda light chain there is a single V gene (V_λ) encoding the variable region, a J region encoding the linker sequence, and a single C gene (C_λ) coding for the

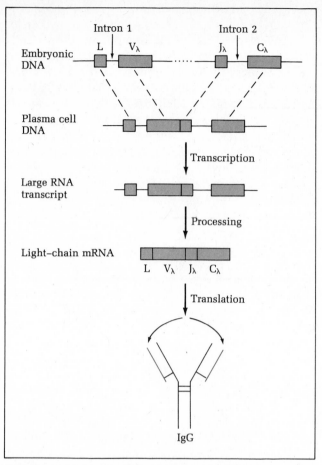

FIGURE 17-14 Organization of mouse lambda light-chain coding sequences in embryonic and plasma cell DNA, and their fates during RNA synthesis and processing. Abbreviations: L = leader sequence; V = variable region; J = joining region; C = constant region.

constant portion of the light chain. The genes coding for the kappa light chains are considerably more complex. To date about 50 V gene (V_κ) clusters have been found, each of which contains 6 to 8 closely related genes. There are 5 different J genes, but a single constant gene. Thus a kappa light chain might result from the linkage of any one of the many V_κ genes with one of the five J genes and one C gene.

The heavy chains of mouse IgG also are put together from a multitude of gene possibilities. Like the mouse kappa light-chain genes, the heavy chain in encoded by many sets of V genes and several J regions. An added feature, however, is the presence of **diversity genes,** or **D genes,** which lie between the V and J sequences. Thus, any combination of V-D-J genes can arise. The constant portion of the heavy chain arises from any of several subclasses of each constant gene class (for example $C_{\gamma 1}$, $C_{\gamma 2}$, etc.). A schematic diagram illustrating these general properties of mouse heavy-chain organization is shown in Figure 17-15.

Thus, the differentiation of a single type of B-cell involves the selection of a single type of light chain (ei-

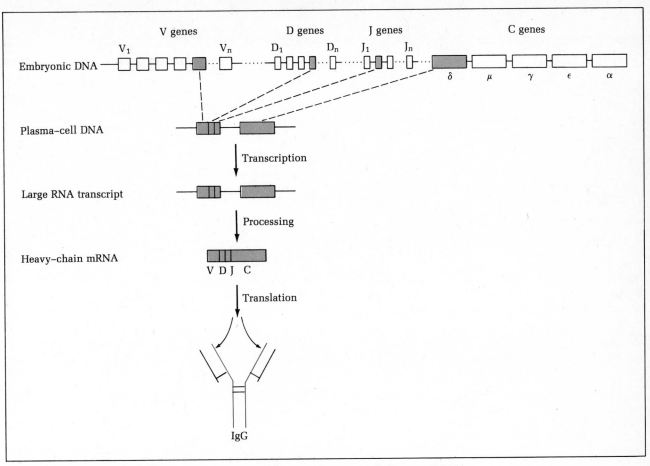

FIGURE 17-15 Organization of mouse heavy-chain genes in embryonic and plasma cell DNA. Abbreviations: V = variable genes; D = diversity genes; J = joining genes; C = constant genes.

ther kappa or lambda) and a single type of heavy chain, each of which arises by genetic rearrangement during differentiation. These two polypeptides come together to form an antibody that, by virtue of the shape of its variable regions, can recognize a specific type of foreign antigen.

The potential for generating a diverse array of different antibodies is thus the result of several factors: (1) the number of different light and heavy coding sequences present in the DNA, (2) the rearrangement of these sequences to produce specific combinations of V-J or V-D-J sequences, and (3) the joining of these resulting variable sequences with particular C genes. The final result is the production of a group of antibodies to fit virtually any conceivable molecular shape, and thus able to protect the organism against a whole host of foreign substances.

Although the above model reveals how different kinds of antibodies can be synthesized, each by a given B-cell, the question next arises as to how an appropriate B-cell is selected to divide and synthesize antibody in response to a foreign antigen. In the next section we shall see how specific B-lymphocytes are induced to proliferate when a foreign antigen enters the scene.

Activation of Antibody Synthesis in B-Cells

At the heart of the clonal selection theory lies the idea that each foreign antigen selectively stimulates division of one or a few B-cells programmed to make antibodies against the antigen in question. This notion has been tested experimentally by exposing lymphocytes to ^{3}H-thymidine in concentrations high enough to kill the cells into which it has become incorporated. If isolated lymphocytes are incubated with ^{3}H-thymidine shortly after being exposed to antigen, antibody formation is inhibited. It has therefore been concluded that the cells responsible for making antibody must be undergoing division, since only cells undergoing DNA synthesis incorporate the ^{3}H-thymidine. To determine whether these dividing cells are specifically programmed to make antibodies against the stimulating antigen, lymphocytes have been exposed to ^{3}H-thymidine during stimulation with one antigen, and then later stimulated

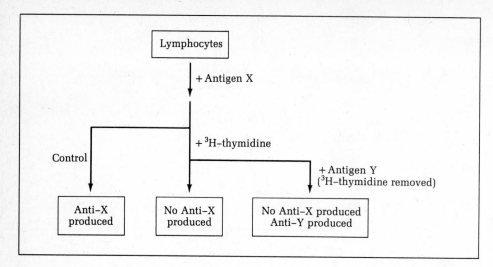

with a second antigen (after the radioactive thymidine has been removed). Under these conditions the response to the first, but not the second, antigen is inhibited, indicating that the cells stimulated to divide by the first antigen are specifically responsible for making antibodies against that antigen (Figure 17-16).

FIGURE 17-17 Experiment with antigen-coated column demonstrating the existence of specific antigen receptors on lymphocytes. See text for details.

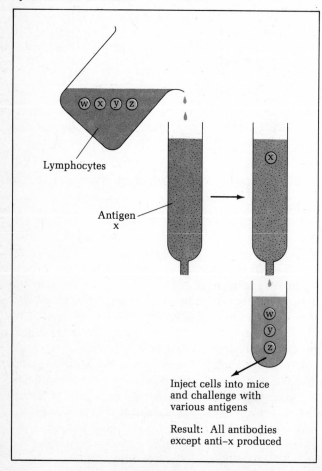

How does an antigen selectively stimulate division of a few lymphocytes out of millions? The idea that lymphocytes contain specific antigen receptors on their surfaces has received strong support from an elegant set of experiments utilizing chromatography columns containing bound antigen. A heterogeneous population of lymphocytes is first passed over a column of antigen x, and the cells that do not stick are then injected into irradiated mice and tested for their ability to generate antibodies. Such animals were found to be capable of making antibodies against all antigens except x, suggesting that those lymphocytes programmed to synthesize antibodies against this antigen were removed by the column because they have surface receptors that bind to antigen x (Figure 17-17).

The above experiments do not, however, indicate whether it is the effector (B-cell) or helper (T-cell), or both, that binds the antigen. To distinguish between these alternatives an "antigen suicide" technique has been devised in which highly radioactive antigen is bound to lymphocytes, killing them by radiation damage. Such experiments have revealed that treatment of *either* B- or T-cells with radioactive antigen prior to injection into an irradiated animal inhibits the ability of the animal to respond to that particular antigen (Figure 17-18). Hence, both B- and T-cells must have specific antigen binding sites.

Having established that antigens bind to surface receptors on B- and T-cells, the question then arises as to how this binding triggers cell division. Studies utilizing fluorescein-labeled antibodies to visualize surface immunoglobulins have implicated membrane reorganization in the process. When lymphocytes are exposed to fluorescent anti-immunoglobulin, the entire cell surface initially fluoresces, indicating that immunoglobulin molecules are present throughout the plasma membrane. Soon, however, the fluorescence pattern becomes patchy, and within a short time all fluorescence coalesces at one end of the cell, forming a struc-

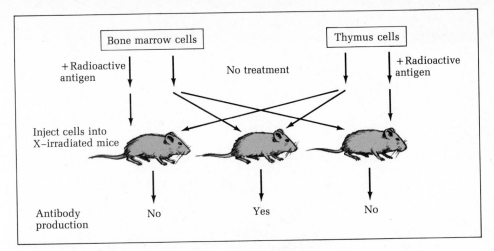

FIGURE 17-18 Antigen suicide experiment showing the existence of specific antigen receptor on both effector and helper lymphocytes.

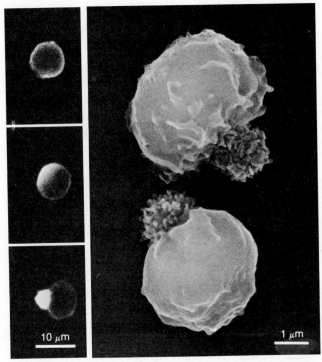

FIGURE 17-19 Micrographs of lymphocyte capping. *(Top left)* Light micrograph of lymphocyte shortly after exposure to fluorescent anti-IgG showing diffuse distribution of cell-surface immunoglobulins. *(Middle left)* Somewhat later a patchy distribution develops. *(Bottom left)* The final stage in cap formation. *(Right)* Scanning electron micrograph of capped cells. Courtesy of I. Yahara.

ture called a **cap** (Figure 17-19). The material in the cap is subsequently removed from the cell surface by pinocytosis. Hence, in response to treatment with anti-immunoglobulin, cell surface immunoglobulins undergo massive reorganization within the fluid plasma membrane, presumably because they are being cross-linked into large aggregates by the antibody molecules. Though this clearly represents an artificial situation in that cell surface immunoglobulins normally react with antigen rather than an antibody, the patching-capping-pinocytosis scenario also occurs in lymphocytes exposed to antigen, provided the antigen has more than one reactive site and can cross-link receptors (Figure 17-20). Though it is therefore appealing to conclude that patching, capping, and internalization of membrane receptors play a role in the mechanism by which antigens stimulate lymphocytes to divide, there is as yet no compelling evidence in support of this theory.

Although our understanding of the molecular basis of B-cell activation is still limited, several features of this process can be described (Figure 17-21). The presence of a foreign antigen within an organism causes the macrophages to bind the antigen and process it. The processed antigen is then presented to B-cells, as well as to helper T-cells. The T-cells are stimulated to release a lymphokine called **interleukin-1.** This protein then stimulates cell division in these B-cells, accompanied by the emergence of cell surface receptors that recog-

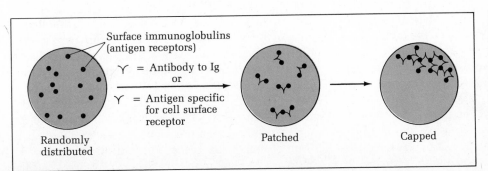

Randomly distributed

Patched

Capped

FIGURE 17-20 Model explaining how either an antibody directed against cell surface immunoglobulin or an antigen specific for the same cell surface immunoglobulin could cause patching and capping. Implicit in this model is the assumption that the surface immunoglobulins are freely mobile in the fluid membrane.

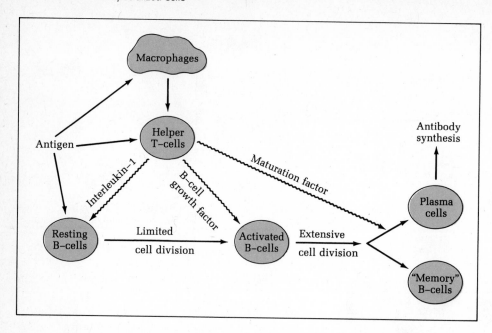

FIGURE 17-21 Summary of interactions among antigen, macrophage, helper T-cell, and B-cells that are required for antibody synthesis. See text for details.

nize a second lymphokine produced by the helper T-cells. This factor is termed **B-cell growth factor,** or **interleukin-2,** and its presence induces rapid proliferation of the B-cell population. The resulting clone of B-cells, which is specific for the particular antigen that began the activation sequence, is now ready to synthesize antibody. A subgroup of this clone is set aside as **memory cells,** while the remainder of the population is stimulated by a **maturation factor** released by the T-cells. This substance induces B-cell differentiation into plasma cells that are specialized for producing large quantities of antibody.

A major difficulty in studying the molecular events associated with antigenic stimulation of lymphocyte proliferation stems from the fact that only a few cells out of a large population of lymphocytes normally respond to any given antigen, making biochemical studies virtually impossible. Because of this problem many investigators have turned to the use of plant lectins, which trigger division of a substantial percentage of lymphocytes regardless of the antigenic specificity of the individual cells. As in the case of stimulation by antigen, lectins bind to the cell surface and cause patching and capping of cell surface macromolecules. Moreover, lectins have been found to elevate cyclic AMP levels, suggesting that alterations in cyclic nucleotide metabolism may also play a role in signaling the cell to divide.

In addition to cell division, the other major activity triggered in antigen-stimulated B-lymphocytes is antibody production. Initially IgM is the predominant form of antibody made, but later IgG takes over. Two possible explanations for this phenomenon can be envisioned: Either a clone of lymphocytes making IgM appears first, followed by a second clone making IgG, or else the same cell clone makes both antibodies, switching from one to

the other. To investigate this problem, single antibody-producing cells have been isolated and analyzed for the type of immunoglobulin being formed. Most cells are found to be making either IgM or IgG, but some lymphocytes produce both, suggesting they may be in the process of switching from one to the other. Certain myelomas that also derive from a single cell likewise make both IgM and IgG, further supporting the idea that a single cell can make these two forms of antibody. In such myelomas not only are the IgG and IgM light chains identical, but the variable regions of the heavy chains are also identical; hence, the cell must have a mechanism for linking the same V gene to two different C genes, one C gene (C_μ) coding for the IgM heavy chain and one C gene (C_γ) coding for the IgG heavy chain.

Monoclonal Antibodies

In 1975 Georges Köhler and César Milstein reported a new method for producing large amounts of homogeneous antibody specific to a particular antigen. The availability of these **monoclonal antibodies** has led to rapid advances in the study of the immune system and has proved to have broad application for other areas of research as well. The technique, for which Köhler and Milstein received the 1984 Nobel Prize in Medicine, relies on the production of hybrid cells that grow well in culture and also synthesize a single type of antibody. These hybrid cells, or **hybridomas,** are formed by mixing together mutant myeloma tumor cells with a population of antibody-producing B-cells isolated from the spleen of a mouse injected with the antigen of interest (Figure 17-22). In the presence of killed Sendai virus, fusion of cells takes place. Three fusion products are possible: myeloma-myeloma fusions, B-cell–B-cell fusions, and hybrid cells derived from fusion of a myeloma cell and a B-cell. The hybrid cell retains the best qualities of each parent cell: ability to grow well in tissue culture (myeloma cell) and ability to synthesize antibody (B-cell). A critical feature of the monoclonal antibody technique is to select the hybrid cells while removing the parental cells from the culture medium. This is accomplished using a selective medium that permits only the hybrid cells to survive. The parent myeloma cell employed in this process is a mutant lacking **hypoxanthine-guanine phosphoribosyltransferase (HGPRT),** an enzyme used in an alternate pathway for synthesis of purines and pyrimidines from the precursors xanthine and thymine. These cells cannot grow in a selective medium containing hypoxanthine, aminopterin, and thymine **(HAT medium)** because aminopterin blocks the remaining metabolic pathway for

FIGURE 17-22 The production of monoclonal antibodies. B-cells and myeloma cells are fused to create a hybrid cell that makes a single type of antibody and can be maintained in tissue culture indefinitely. See text for details.

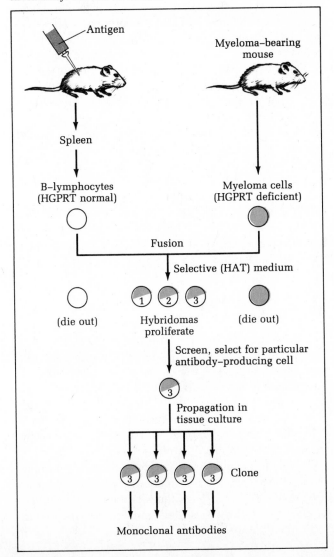

synthesis of purines and pyrimidines. B-cells cannot survive in tissue culture at all. Thus in the presence of HAT medium, only the hybrid cells grow (since they contain normal HGPRT derived from the parental B-cell), while the parental cell types die out.

Many hybridomas are made in this process, since the parental B-cells are derived from a population of cells in the spleen of the immunized mouse. Each hybridoma is secreting a specific antibody, which recognizes a particular antigen. The key is to identify and isolate the hybridoma that is making the antibody against the original injected antigen. This is achieved by cloning individual hybrid cells, creating populations of cells that are tested for the antibody of interest. Because each population is derived from a single cell, only a single type of antibody, a **monoclonal antibody,** is made. Identification of the monoclonal antibody specific for the immunizing antigen is made either by radioimmunoassay or by a newer method called the **enzyme-linked immunosorbent assay (ELISA).** In this assay the immunizing antigen is coated on the surface of wells in a tissue culture plate and then the surface is saturated with an unrelated protein, such as bovine serum albumin. The culture medium containing monoclonal antibody secreted from a particular cell clone is then added to the well and after a period of time to allow for binding of antigen to monoclonal antibody, the medium is removed. In order to detect the antigen-antibody complex, a conjugate made up of an enzyme linked to the Fab portion of a mouse antibody is used. The Fab region binds to the antibody portion of the antigen-antibody complex in the well. The enzyme is chosen because it generates a particular colored reaction product in the presence of its substrate. The conjugate is added to the well and after a period of time the unbound conjugate is removed by washing. When substrate is added, the appearance of a colored reaction product signals the presence of the antigen-monoclonal antibody complex in the well. If the complex is absent from the well, then the enzyme-Fab conjugate does not bind at all and no color is visible. By placing a single clone in each well of the plate, the clone synthesizing the monoclonal antibody specific for the immunizing antigen can be identified. This clone can be maintained in culture indefinitely, providing an inexhaustible source of monoclonal antibody of a given specificity.

How Do Antibodies Recognize and Inactivate Foreign Antigens?

The initiation of the humoral immune response depends on recognition of an antigen by specific immunoglobulin molecules that serve as receptors on the surface of B-cells. It has been estimated that the B-cell surface contains 100,000 randomly distributed immunoglobulin molecules. These molecules differ slightly

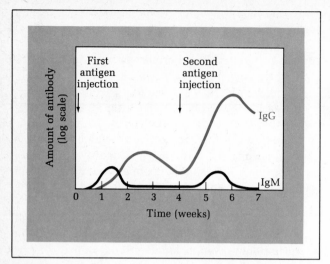

FIGURE 17-23 The synthesis of IgM and IgG antibodies in response to an initial and second injection of a foreign antigen.

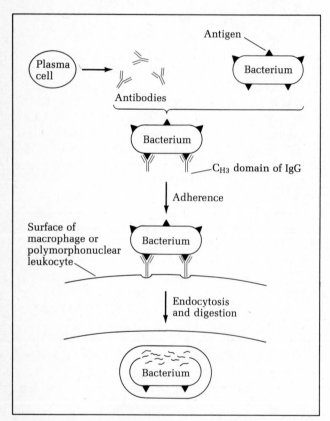

FIGURE 17-24 Schematic diagram of indirect removal of antigens via antibody-mediated phagocytosis.

from secreted immunoglobulins in their C_H region, but apparently are similar to secreted immunoglobulins in the V_L and V_H domains, since both the surface and secreted versions of the molecule recognize the same idiotype.

The initial entry of a foreign antigen into tissues triggers the synthesis of two types of antibody: **IgM** and **IgG.** IgM antibodies are the third most abundant class of

immunoglobulins and are the largest in terms of size (see Table 17-1). These molecules act as the first line of defense against bacterial infection, and so comprise the first of the two antibody species made. The amount of circulating IgM begins to decline as the synthesis of IgG antibodies begins (Figure 17-23). If an organism is exposed to the same antigen a second time, the response is quicker and magnified. In this case the amount of IgM produced is similar to that observed during initial exposure to antigen, but the amount of circulating IgG rises to a level greater than that seen initially. This **secondary response** to the same antigen occurs more rapidly than the primary response, and is dependent on an adequate supply of B-cells previously exposed to the antigen.

The removal of antigen by antibody may be achieved either directly or indirectly, depending on the type of antigen involved. If the antigen is a **toxin,** antibody binding leads to neutralization of the toxin's effect, either by binding to its active site or by changing its conformation. If the antigen is localized on the surface of an invading bacterium, antibody acts indirectly to mediate the removal of the bacterium from the tissues. This is accomplished in a two-step process. First, circulating IgG antibodies bind to the antigen on the bacterial cell surface. This sets the stage for binding of the bacterium to phagocytic cells that have receptors for the C_{H3} domain of the bound antibody (see Figure 17-10). The bound bacterium is then engulfed by phagocytic cells, which are most often either macrophages or polymorphonuclear leukocytes. The resulting vesicle then fuses with lysosomes, which disgorge their hydrolytic enzymes and digest the bacterium (Figure 17-24).

These examples illustrate two of the simplest mechanisms utilizing antibodies to remove foreign substances from the tissues. However, some infectious agents are adapted to live *within* the cells of the host. How the immune system is able to deal with such invaders is the subject of the following section.

THE CELL-MEDIATED IMMUNE RESPONSE

The existence of foreign organisms that live within the cells of the host presents special problems for the immune system. Viruses, the bacteria that cause tuberculosis and leprosy, and fungi living in the cytoplasm of host cells are examples of agents that cannot be attacked by circulating antibodies. The evolution of the cell-mediated immune response, which is orchestrated by specialized T-lymphocytes and lymphokines, has made it possible to defend the organism against such foreign agents. In this section of the chapter we shall see that complex interactions between intact cells and soluble factors produced by these cells must be coordinated in this type of immune response.

The Discovery of T-Cell Involvement in Cell-mediated Immunity

The first definitive evidence that T-cells are involved in cell-mediated immunity came from studies of the **graft-versus-host** reaction, a cell-mediated immune response that occurs when tissue containing lymphocytes is grafted into an immunologically incompetent host. The host can be either a newborn animal that has not yet developed its immune system, a mature animal whose immune system has been inactivated by irradiation, or a special strain of mice, called "nude" mice, that lack a thymus and hence are immunologically deficient. As the term *graft-versus-host* implies, the *donor* lymphocytes recognize host antigens as foreign and react against them. This response is the opposite of the usual grafting situation where the host lymphocytes react against the graft. The advantage of studying the graft-versus-host response as a model of cell-mediated immunity is that the host's spleen becomes enlarged during the process, an effect that can be readily measured. Utilizing this experimental approach, Harvey Cantor and Richard Asofsky were able to show that injection of a mixture of thymus and peripheral blood cells into an immunologically deficient host induces a graft-versus-host reaction that is greater than the sum of the responses obtained when each type of cell is administered separately. This synergistic effect suggests that at least two different cell types, one present in the thymus and one in peripheral blood, cooperate with each other during cell-mediated reactions.

In order to identify the type of cells involved, Cantor and Asofsky took advantage of the fact that mouse T-cells contain a unique surface marker known as the **Thy 1 antigen** (formerly called the theta antigen). The presence of T-lymphocytes in a given population can therefore be monitored using antibodies directed against Thy 1. Cells are simply incubated with anti-Thy 1 in the presence of *complement*, a group of enzymes in normal serum that bind to antigen-antibody complexes. Under these conditions lysis of cells possessing such surface complexes takes place. When either thymus or blood cells are pretreated with antibodies to Thy 1 in the presence of complement, the cooperative effect between the two cell populations is abolished in a subsequent graft-versus-host reaction. Hence the two cell types involved in this cell-mediated response must both be T-cells, one present in the thymus and the other in peripheral blood.

How Do T-Cells Recognize Antigen and Confer Immunity?

When an antigen binds to the T-cell surface, the cell is transformed into a lymphoblast that begins dividing to generate a population of **sensitized lymphocytes** spe-

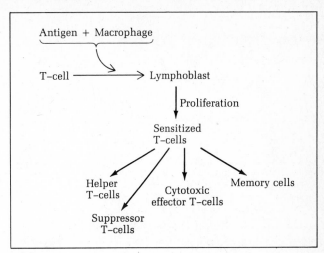

FIGURE 17-25 Major classes of T-lymphocytes produced in response to antigen stimulation.

cific for the initiating antigen. Some of the sensitized lymphocytes are set aside as **memory cells,** while the remaining cells differentiate to perform a variety of functions necessary for cell-mediated immunity. The discovery of a variety of cell surface protein markers has been important in developing a classification system for these different kinds of T-lymphocytes. More than a dozen such markers have now been found, and based on this information, several major classes of T-cells have been identified, each of which performs a specific function in regulating the immune response (Figure 17-25). The **helper T-cell** stimulates the immune response by activating other T-cells and B-cells; the **suppressor T-cell** acts to suppress the activity of helper T-cells, thereby modulating the action of both T-cells and B-cells in an indirect way; and the **cytotoxic effector T-cell** kills virus-infected cells that bear the virus antigen.

Since the cytotoxic T-cells are directly responsible for the destruction of foreign organisms, tissues, and viruses, considerable efforts have been made to understand how they act. The mechanism by which cytotoxic effector cells recognize cells as foreign was first investigated in the 1970s by P. C. Doherty and R. M. Zinkernagel, who were studying the interaction of virus-infected cells and cytotoxic T-lymphocytes. Virus-infected cells possess certain antigens that can be recognized by the T-cells, leading to the killing of the infected cell by the T-cells. However, Doherty and Zinkernagel observed that the T-cells could kill only those virus-infected cells that came from the same inbred strain of mouse as the T-cells themselves (Figure 17-26). This suggested that recognition of the virus antigen might also involve proteins encoded by the major histocompatibility complex (MHC), which would be specific to the mouse strain employed.

This proposal was subsequently confirmed and has

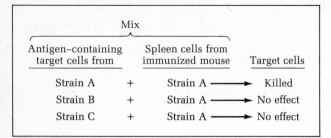

Mix		
Antigen–containing target cells from	Spleen cells from immunized mouse	Target cells
Strain A	+ Strain A ⟶	Killed
Strain B	+ Strain A ⟶	No effect
Strain C	+ Strain A ⟶	No effect

FIGURE 17-26 Experiment showing that killing of target cells by spleen-derived lymphocytes only occurs when the target cells and spleen cells are from the same strain.

led to the realization that a major distinction in the method of antigen recognition by T-cells and B-cells involves the ability of the T-cell to "see" the antigen only when it is associated on the cell surface with a protein product of the major histocompatibility complex. (Receptors of B-cells can recognize an antigen without the help of other cell surface molecules). Each **T-cell receptor** is made up of a heterodimer of α- and β-glycoprotein subunits, which are like the immunoglobulin light and heavy chains in that they contain constant and variable regions. This suggests that the genes encoding the α- and β-subunits belong to the immunoglobulin gene family. Another important feature of the T-cell receptor is that different classes of T-cells apparently recognize different classes of molecules encoded by the major histocompatibility complex. For example, helper T-cells recognize antigen when it is associated with class II-MHC proteins that are expressed selectively on antigen-presenting cells such as B-cells or macrophages. Cytotoxic T-cells respond to antigens of cellular origin, such as virus or tumor antigens, in concert with class I-MHC proteins. We do not yet understand how these different recognition devices are reflected in the T-cell receptor structure. Finally, at least in mice and humans, the T-cell receptor is associated with another multiprotein cell surface complex called **T3.** Whether T3 will turn out to be a common feature of T-cell receptors remains to be determined. It has been proposed that as T-cells differentiate in the thymus, they are "imprinted" with the ability to recognize foreign antigens that coexist on the membrane with MHC polypeptides. Since these surface antigens are unique to the individual, this provides a mechanism to evaluate whether a cell is an alien or an accepted member of the host-cell community.

One of the most deadly diseases to strike the human race in recent memory involves a dismantling of the immune system. **Acquired immune deficiency syndrome,** commonly known as **AIDS,** is a disease that is transmitted by a virus lying concealed inside helper T-cells. When a virus-laden T-cell is transmitted through the blood or semen to an uninfected individual, the helper T-cells of that individual immediately recognize the virus-infected cell as foreign. However, before the

individual's helper T-cells can alert the remaining cells of the immune system to the danger, the invading virus particles move into the T-cells and inactivate them. As the virus moves from helper cell to helper cell in the victim, more and more of the immune system is disabled, since the very cells that are required to mount an immune response are rendered nonfunctional. The AIDS virus may lie dormant inside the new inactive helper T-cells for some time, only to be stimulated by a subsequent infection. Division of virus-infected helper cells simply overwhelms the few remaining normal helper T-cells, and in a relatively short period of time the body is left defenseless. Thus the inactivation of the helper T-cell—which serves as the communicating center for coordinating the immune response of both B-cells and other types of T-cells—wreaks havoc on the immune system. Individuals usually die of persistent and varied infections. The challenge we face today is to understand the AIDS virus and its effects more completely so that we may soon have an effective approach to the treatment of this disease.

Lymphokines Some of the sensitized T-cells that arise during lymphoblast proliferation play an indirect role in antigen destruction by secreting lymphokines into the surrounding interstitial fluid. For the most part, these factors act on other T-cells, B-cells, macrophages, and polymorphonuclear leukocytes to modify their behavior and enhance the host response to foreign antigens. We have already discussed several lymphokines that are important in B-cell development (see Figure 17-21). Another well-studied group of lymphokines are those that influence the movement and activity of macrophages. **Macrophage chemotactic factor** causes the accumulation of macrophages at the site of lymphokine release. Once the macrophages are collected together, they are kept in place by a **macrophage migration inhibition factor.** Another cell product, **macrophage activating factor,** triggers changes in morphology of these cells. This includes ruffling of the surface membrane and an increase in lysosomal enzyme content, both of which enhance the phagocytic capabilities of these cells. Another lymphokine that has been identified is **lymphocyte inhibitory factor,** which blocks division of T-cells. Thus the lymphokines act along with the various subgroups of T-lymphocytes to regulate the destruction of foreign substances that may be intracellular parasites (bacteria, fungi, and viruses), aberrant cells like cancer cells that are no longer seen as "self," or foreign cells such as those present in grafted tissues.

IMMUNOLOGICAL TOLERANCE

The ability of the immune system to react against a vast array of foreign substances raises the paradoxical question of how an organism is prevented from making antibodies against its own cells. In the few instances where such antibodies are produced, severe **autoimmune diseases** result, so that a mechanism for preventing such responses is clearly needed. The first important observation relevant to this problem was made in the 1940s, when Ray D. Owen noted that some nonidentical twin cattle exchange blood cells during embryonic development and retain these "foreign" cells even as adults. If such foreign cells had been injected directly into an adult, an immune response would have been initiated. This observation led Macfarlane Burnet to propose that antigens present during early development are recognized as "self," and induce a permanent state of **immunological tolerance.** Experimental verification of this proposal was later provided by Peter Medawar, who injected embryos of one mouse strain with tissues from another strain. At maturity these animals were found to be incapable of rejecting skin grafts from mice whose tissues they had been exposed to during embryonic life, though response to other antigens was normal.

Subsequently it has been shown that immunological tolerance can be induced in a variety of other ways, even in adults. If, for example, animals are treated with drugs that inhibit cell division at the same time an antigen is administered, not only is an immediate immune response prevented, but a permanent state of tolerance to that specific antigen is induced, even after the drug is removed. Very high concentrations of antigen, or very low ones, can also induce tolerance.

The simplest theory put forth to explain immune tolerance is **clonal deletion,** which states that lymphocytes specific for a particular antigen are destroyed or incapacitated if they are exposed to antigen under conditions where they cannot respond (e.g., during early development when the lymphocytes are still immature, or in the presence of drugs inhibiting cell division). An alternative explanation of tolerance is that some lymphocytes act as **suppressor** cells, specifically inhibiting the ability of other lymphocytes to respond to antigenic stimulation. Though there is evidence that certain T-cells do act to suppress the immune response to particular antigens, their role in immune tolerance is yet to be clearly demonstrated. Hence the question of whether clonal deletion, or suppressor cells, or some other unknown mechanism, is responsible for immune tolerance remains to be clarified.

INNATE VERSUS ACQUIRED IMMUNITY

The ability to respond to a particular foreign substance and to generate a memory bank of lymphocytes that become activated if the same antigen reappears is characteristic of a type of immunity known as an acquired immune response. It is this type of immunity, involving lymphocytes and macrophages, that has been discussed so far in this chapter. Another line of defense against infection is termed **nonspecific** or **innate** immunity, a strategy that involves a more general response to infectious agents and has no memory component.

For example, entry of bacteria into an organism results in an immediate nonspecific response that involves two principal cell types, the macrophage and the polymorphonuclear leukocyte. These phagocytes can destroy invading microorganisms even in the absence of specific immune recognition. In order for this process to occur, the phagocytic cells must be able to "sense" the presence of the invading bacteria. This is achieved by a complex series of events that involves the **complement system,** which consists of a group of about 20 different proteins. Polysaccharides in the bacterial wall cause this complement system to generate two specific molecules, called **C3a** and **C3b** (Figure 17-27). The C3b fragment binds to the surface of the bacterium and renders it recognizable by the phagocyte, which contains receptors for C3b on its surface. The C3a fragment stimulates another cell type, called the **mast cell,** to release two substances. One substance is a chemotactic factor that "attracts" the phagocytes to the vicinity of the bacterial population. The second factor released by the mast cell is the small molecule **histamine,** which increases the permeability of nearby capillaries so that plasma, lymphocytes, and polymorphonuclear leukocytes enter the tissue. The resulting **inflammation** is accompanied by phagocytosis of the bacteria by the

FIGURE 17-27 Mechanism of innate immunity against bacterial infection, accomplished by phagocytosis and requiring the presence of complement.

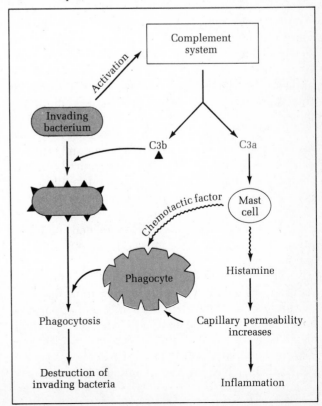

polymorphonuclear leukocytes, which soon die and are removed by scavenger macrophages.

Another type of innate response involves soluble molecules that are capable of inactivating foreign antigens. One such factor is the enzyme **lysozyme,** which degrades the peptidoglycan cell wall of certain bacteria. Protection against virus infection involves a protein called **interferon,** which is released from virus-infected cells. Interferons have now been identified in fish and reptiles as well as birds and mammals, and so are thought to represent an evolutionarily ancient mechanism for preventing the spread of a virus. Rather than exhibiting the specificity of antibodies, interferons act to inhibit the reproduction of a broad spectrum of viruses. Cells that are infected by a virus begin to synthesize interferon and subsequently secrete it into the extracellular fluid. The protein then binds to receptors on neighboring cells, preventing virus replication in those cells. This nonspecific response thus provides an effective protection against virus spread throughout a tissue.

SUMMARY

The immune system functions to protect an organism against potentially harmful foreign substances, or antigens. In higher vertebrates, such as birds and mammals, the immune response may take two forms. Innate, or nonspecific immunity involves a slow-acting response that most often results in inflammation and subsequent removal of the invading microorganism by phagocytosis. This type of response is limited in scope and effectiveness. The most efficient response to a foreign antigen is acquired immunity, in which a highly sophisticated interplay of intact cells or soluble products released by these cells specifically interacts with the antigen identified as foreign. Acquired immunity is thus characterized by three unique features: the ability to distinguish "self" from "nonself," specificity in its response to different antigens, and a memory for particular antigenic configurations so that a second response can be mounted more rapidly. The acquired immune response occurs in two forms, humoral and cell mediated. The humoral response involves the formation of circulating antibodies that are able to recognize a specific antigen and destroy it. Cell-mediated immunity utilizes the direct involvement of cells as well as soluble cell products.

The main cell types involved in the immune response are the B-lymphocyte, arising from the bone marrow (or bursa of Fabricius in birds) and responsible for antibody synthesis, and the T-lymphocyte, which arises from the thymus and is the principal effector of cell-mediated immunity.

Evidence that circulating antibodies are synthesized by B-lymphocytes has come from reconstitution experiments in which immunologically incompetent animals were injected with various lymphocyte populations and then tested for their ability to make antibody. Such experiments demonstrate that antibody synthesis by B-cells requires help from two other cell types: T-lymphocytes and macrophages. Other soluble factors called lymphokines, which are released from helper T-cells, also play a role in activating antibody synthesis by B-cells.

Antibodies are released from B-cells into the bloodstream and become part of the serum protein fraction known as immunoglobulins. Five classes of immunoglobulins can be distinguished on the basis of differences in both physical properties and biological functions. These five fractions are referred to as immunoglobulins G, M, A, D, and E and are abbreviated using the prefix "Ig" before each identifying letter. IgG is the major form of circulating antibody, and its structure has been the most thoroughly investigated. The IgG molecule is composed of four polypeptide chains, two identical light chains and two identical heavy chains. The light chains may be either of two types, called kappa and lambda, but there is only one type of heavy chain. The light and heavy chains are held together by disulfide bonds to form a Y-shaped or T-shaped molecule. Both types of chain contain regions of variable amino acid sequence and regions whose sequence is constant. The variable regions of the light and heavy chains come together to form the antigen binding site, while the constant regions are involved in complement activation and in binding antibody to the surface of other cells of the immune system.

It is now widely believed that each B-lymphocyte is programmed to make one specific type of antibody. According to this clonal selection theory, antigen simply acts to stimulate proliferation of those cells that make antibody that recognizes the antigen in question. This clonal selection theory is supported by experiments showing that: (1) denatured antibodies regain their proper antigenic specificity when allowed to spontaneously refold in the absence of antigen, and (2) only a single type of antibody is produced by a single B-lymphocyte clone.

Several theories have been proposed to account for the genetic origin of the numerous variable sequences required for antibody diversity. According to the germ line theory, all the genes necessary for antibody production are inherited through the germ cells (sperm and egg). It can be calculated that about a thousand different light-chain genes and a thousand heavy-chain genes would be sufficient to account for the million or so antibodies a typical organism is thought to be capable of producing. Alternatively, the somatic mutation and somatic recombination theories envision that organisms inherit only a small number of antibody genes, with additional diversity being introduced by mutation or genetic recombination during the cell divisions leading to the formation of differentiated B-lymphocytes.

The germ line theory has received support from nucleic acid hybridization studies that reveal that embryonic DNA contains at least several hundred different genes coding for the variable (V) regions of kappa light chains; in contrast, only one gene for the constant (C) region is present. The V and C genes are not adjacent in embryonic DNA but during formation of antibody-producing cells, the V gene destined to be expressed in a particular lymphocyte is brought close to the C gene, joining to a special region called the J region. The genes encoding the heavy chain are arranged similarly, except that another set of genes called diversity (D) genes is also involved. The existence of at least hundreds of different V genes in embryonic DNA provides strong support for the germ line theory of antibody diversity. However, the possible occurrence of somatic recombination as an additional means of generating diversity is suggested by the finding that the DNA sequences surrounding the V genes of a given family resemble one another, an arrangement that could enhance crossing-over during somatic cell divisions.

The presence of specific receptors on lymphocyte surfaces permits each antigen to selectively bind to those cells making antibodies specific for the antigen in question. Such receptors, which occur on both helper and effector cells, are believed to consist of immunoglobulins or immunoglobulinlike molecules. Binding of an antigen to its receptor leads to a stimulation of cell division, an event that may be triggered by membrane reorganization (patching, capping, and internalization of the antigen-receptor complex). In addition, macrophages and lymphokines are involved in initiating and modulating the proliferation and maturation of B-cells.

Rapid advances in our understanding of how the immune system functions have been made possible by a new technique in which antibodies of a single type can be made in large quantities. These "monoclonal" antibodies are synthesized by hybrid cells derived from the fusion of a myeloma cell (which can grow indefinitely in tissue culture) and a B-cell (which synthesizes antibody). These hybrids, or hybridomas, can provide an abundant source of individual purified antibodies.

Cell-mediated immune reactions are orchestrated by the T-lymphocytes and their soluble products, the lymphokines. Studies of the graft-versus-host reaction reveal that two different subpopulations of T-cells must interact in order for cell-mediated immune reactions to take place. Analysis of cell surface antigens has led to the classification of these T-cells into three basic types: the helper T-cell aids in activation of other T-cells as

well as B-cells; the suppressor T-cell acts to minimize the activity of the helper T-cells and B-cells; and the cytotoxic effector T-cell (also called the killer cell) interacts directly with cells bearing a foreign antigen, destroying these cells and hence removing the antigen in the process. The release of lymphokines from T-lymphocytes is also important, since these soluble molecules diffuse locally and affect the behavior of not only the lymphocytes but macrophages and polymorphonuclear leukocytes as well.

Organisms normally recognize their own cells as "self" and do not mount an immune response against them. This state of immunological tolerance is automatically induced by antigens that are present during embryonic development. Tolerance can also be produced in adults by exposing them to exceptionally high or low doses of a particular antigen, or to inhibitors of cell division during antigenic stimulation. It is not clear whether tolerance is due to destruction of lymphocyte clones specific for particular antigens (clonal deletion) or to the action of specific suppressor lymphocytes.

Although the acquired immune response is the most specific means of removing foreign antigens from an organism, a more general type of immune response also exists. This more primitive form of immunity, called innate or nonspecific immunity, is a relatively slow and inefficient means of protecting the organism from foreign substances. The acquired immune response provides a rapid and efficient way of destroying invading foreign agents and, together with innate immunity, provides a protective umbrella that defends an organism from foreign organisms and products.

SUGGESTED READINGS

Books

Bellanti, J. A. (1985). *Immunology III,* Saunders, Philadelphia.

Golub, E. S. (1981). *The Cellular Basis of the Immune Response,* Sinauer Associates, Sunderland, MA.

Kindt, T. J., and J. D. Capra (1984). *The Antibody Enigma,* Plenum Press, New York.

Nisonoff, A. (1984). *Introduction to Molecular Immunology,* Sinauer Associates, Sunderland, MA.

Roitt, I. (1984). *Essential Immunology,* Blackwell Scientific Publications, London.

Tizard, I. R. (1984). *Immunology: An Introduction,* Holt Saunders/Saunders College, New York.

Articles

Baltimore, D. (1984). Molecular immunology: growth into adolescence, *Trends Biochem. Sci.* 9:137–138.

Gallo, R. C. (1987). The AIDS virus, *Sci. Amer.* 256 (Jan.):46–56.

Goverman, J., T. Hunkapiller, and L. Hood (1986). A

speculative view of the multicomponent nature of T-cell antigen recognition, *Cell* 45:475–484.

Honjo, T. (1983). Immunoglobulin genes, *Ann. Rev. Immunol.* 1:499–534.

Hood, L., M. Kronenberg, and T. Hunkapiller (1985). T-cell antigen receptors and the immunoglobulin supergene family, *Cell* 40:225–229.

Kennedy, R. C., J. L. Melnick, and G. R. Dreesman (1986). Anti-idiotypes and immunity, *Sci. Amer.* 255 (July):48–56.

Laurence, J. (1985). The immune system in AIDS, *Sci. Amer.* 253(Dec.):84–93.

MacDonald, H. R., and M. Nabholz (1986). T-cell activation, *Ann. Rev. Cell Biol.* 2:231–253.

Manser, T., S.-Y. Huang, and M. L. Gefter (1984). Influence of clonal selection on the expression of immunoglobulin variable region genes, *Science* 226:1283–1288.

Marrack, P., and J. Kappler (1986). The T cell and its receptor, *Sci. Amer.* 254(Feb.):36–45.

Tonegawa, S. (1985). The molecules of the immune system, *Sci. Amer.* 253(Oct.):122–131.

18

The Neuron: Functional Unit of the Nervous System

Because multicellular animals are comprised of a large number of different kinds of cells, mechanisms must exist for coordinating the activities of cells located at considerable distances from one another. In Chapter 16 the role played by hormone molecules in such intercellular signaling was discussed. The other major mechanism for intercellular signaling and coordination, which involves the cells of the nervous system, will be discussed in this chapter.

Nerve cells, or neurons, form a large interconnected network whose properties ensure virtually instantaneous communication among the various regions of the organism. This network receives sensory information from both outside and within the organism, integrates and analyzes this information, and then responds by triggering various kinds of cellular responses. Significant progress has been made in recent years in unraveling the mechanism by which nerve cells communicate with one another. As the chapter unfolds, it will become apparent that this communication is based upon two cellular phenomena: alterations in the electrical properties of the plasma membrane and the release of special chemicals, called neurotransmitters, that transmit this electrical signal from one cell to the next. In order to facilitate this type of signaling, nerve cells have evolved a highly specialized morphology unlike that of any other cell type. The chapter will therefore begin with a discussion of this morphology and its relationship to the mechanism of neural communication.

STRUCTURE AND DEVELOPMENT OF THE NEURON

The Cell Theory and Its Relationship to Nervous Tissue

The enunciation of the cell theory by Schleiden and Schwann in the mid-nineteenth century was followed by widespread experimental support as a growing number of tissues were found to exhibit a cellular basis of organization. The nervous system appeared to be unique, however, in the sense that little evidence of cellular structure could be seen. Because of the difficulty in fixing and staining nervous tissue, the hazy outlines appearing in such preparations suggested a network of connecting channels that were initially believed to transport fluid secreted by the brain and spinal cord to the periphery. Microscopic visualization of nervous tissue suggested that it contained two different structural components: spherical cell bodies and long, fibrous processes intertwined in and among the cells. By the early 1840s, it was recognized that these long processes are actually extensions of the cells themselves. Moreover, these extensions appeared to be joined to one another, suggesting that the nervous system is a network of cell bodies interconnected by cytoplasmic extensions.

A major impediment to understanding the true organization of nervous tissue no doubt lay in the unusual size and shape of typical nerve cells, whose long, branching processes make these cells difficult to visualize in their entirety. In 1873 an impoverished Italian doctor, Camillo Golgi, was experimenting with different staining and fixing methods for nervous tissue. Working by candlelight in his kitchen, Golgi discovered that exposing chromate-hardened brain tissue to silver nitrate revealed the complete outline of nerve cells, even including their branched cytoplasmic extensions (Figure 18-1). The significance of this remarkable advance was not appreciated until several years later, when the technique was used in Spain by Santiago Ramón y Cajal to facilitate the tracing of individual neural processes. This approach led him to the discovery that each nerve cell body has its own set of cytoplasmic extensions that are not continuous with those of other cell bodies. Each cell body plus its associated cytoplasmic extensions was therefore seen to constitute a single nerve cell, or **neuron.**

Ramón y Cajal's conclusion that nervous tissue is comprised of individual cells did not go unchallenged, however, for Golgi and many others still believed nerve cell extensions to be continuous with one another. A heated controversy developed, exacerbated when the two major protagonists, Golgi and Ramón y Cajal, shared the Nobel Prize in 1906. Though the neuron theory eventually gained prominence, definitive evidence in its support had to await the development of electron microscopy some 50 years later. This type of analysis confirmed that distinct membrane boundaries exist between the cytoplasmic extensions of different cell bodies. Thus, over 100 years separated the initial formulation of the cell theory and definitive proof for the idea that nervous tissue is composed of individual cells, each bounded by a single continuous plasma membrane.

Though neurons share many of the morphological and biochemical features exhibited by other kinds of cells, they are also quite unusual. This uniqueness includes a distinctive overall shape, and the presence of a special structure, termed the **synapse,** that is designed for the transfer of signals from one neuron to another. As will become apparent in the following sections, each specialized region of the neuron is morphologically and biochemically differentiated for a particular role in neural communication.

Structure of Neurons

A typical neuron is composed of a **cell body,** which serves as the control center of the cell, and cytoplasmic extensions of the cell body, generally referred to as **neurites,** that fall into one of two categories: **dendrites** and **axons** (Figure 18-2). The cell body contains a centrally placed nucleus, often with a prominent nucleolus, and surrounding cytoplasm in which macromolecules and other materials required for neuron function are synthesized (Figure 18-3). Under the light microscope the cytoplasm is seen to contain large clumps of material that stain intensely with basic dyes [Figure 18-4 *(left)*]. This material, named **Nissl substance** after its discoverer Franz Nissl, was for many years believed to be a precipitation artifact. However, the electron microscope has revealed that this staining pattern is caused by the presence of dense masses of ribosomes and endoplasmic reticulum in this region of the cell [Figure 18-4 *(right)*].

In addition to these organelles, the cell body contains large numbers of mitochondria and lysosomes, and an extensive Golgi complex. The mitochondria provide the ATP needed to drive biosynthetic reactions

FIGURE 18-1 Micrograph of a nerve cell stained by the Golgi method showing branched dendrites (Dn), an axon (A), and the axon hillock (h). Courtesy of A. Peters.

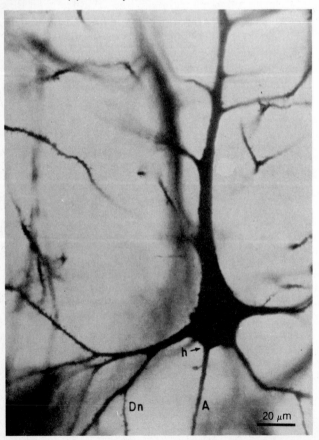

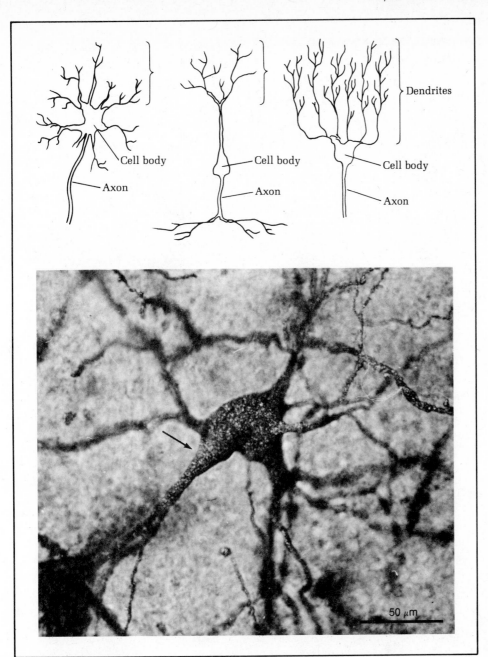

FIGURE 18-2 *(Top)* Organization of axons and dendrites in nerve cells. Schematic diagram illustrating three commonly occurring arrangements of the cell body, axons, and dendrites. *(Bottom)* Photomicrograph of a neuron showing the cell body, dendrites, axon, and axon hillock (arrow). Courtesy of H. Mannen.

occurring in the cell body, while the Golgi complex is thought to function in packaging materials to be transported to other regions of the cell. Lysosomes act as digestive organelles, fusing with endocytic vesicles that form at the ends of axons and dendrites. The formation of these vesicles aids in recirculating components of the plasma membrane. In the neurons of older animals the lysosomes gradually develop into large, pigmented structures, called **lipofucsin granules,** which contain undigested cell debris that accumulates in aging cells (Figure 18-5).

The **dendrites** and **axons** that project from the cell body can be distinguished from one another in a number of ways (Table 18-1). In general, dendrites are

TABLE 18-1

Distinguishing morphological features of typical dendrites and axons

Dendrites	Axons
Usually multiple	Never more than one or two
Relatively short (<1 mm)	May be very long (1 m or more)
Diameter lessens with distance from the cell body	Diameter relatively constant
Contain spines	Smooth surface
Never myelinated	Often myelinated
Contain ribosomes	Do not contain ribosomes
Microtubules usually outnumber neurofilaments	Neurofilaments usually outnumber microtubules

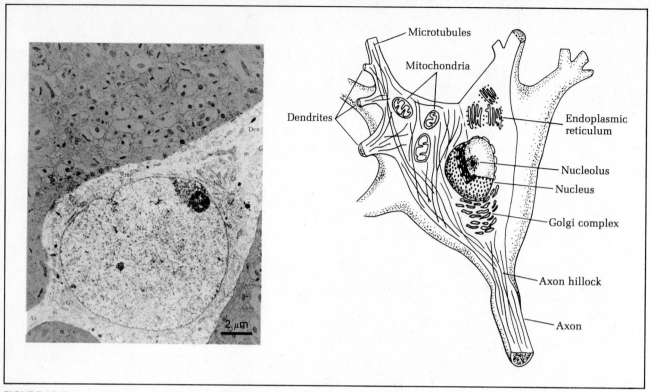

FIGURE 18-3 Internal structure of the neuron. The cytoplasm of surrounding cells is shaded. *(Left)* Electron micrograph of a neuron cell body. Abbreviations: Nuc = nucleus; ncl = nucleolus; NB = Nissl bodies; Den = dendrite; G = Golgi complex; mit = mitochondria; m = microtubules; Ax = axon. Courtesy of A. Peters. *(Right)* Schematic diagram of a typical neuron.

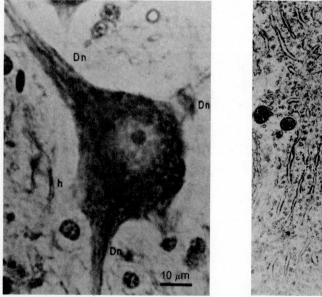
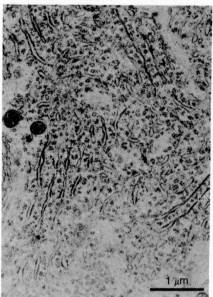

FIGURE 18-4 Light and electron microscopic appearance of Nissl substance. *(Left)* Low-power micrograph showing the appearance of Nissl substance in the dendrites (Dn) and cytoplasm of the neuron. The axon hillock (h) lacks this granular material. *(Right)* Electron micrograph of a region containing Nissl substance showing clusters of free ribosomes and extensive rough endoplasmic reticulum. Courtesy of A. Peters.

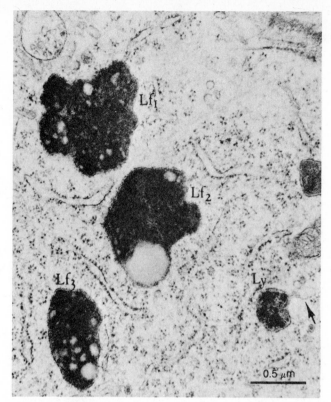

FIGURE 18-5 Electron micrograph of a neuron showing three darkly stained lipofucsin granules (Lf). Nearby is a lysosome (Ly) that appears to be connected to the membranes of the smooth endoplasmic reticulum (arrow). Courtesy of A. Peters.

The cytoplasm within the dendrites contains variable amounts of free and membrane-bound ribosomes, vesicles and flattened cisternae of smooth endoplasmic reticulum, and mitochondria; the most prominent structural feature, however, is the fibrous cytoskeleton. Classical light microscopists long ago noted an interlocking network of slender fibers in the cell bodies and neurites of nervous tissue stained with silver. The electron microscope revealed that these fibers contain two kinds of fibrous elements: 25-nm diameter microtubules resembling those occurring in other cell types, and 10-nm diameter **neurofilaments** that are both morphologically and biochemically distinct from the microfilaments present in other cell types [Figure 18-6 (*left*)]. These neurofilaments are straight, unbranched hollow tubules with walls composed of subunits [Figure 18-6 (*right*)]. They are constructed from fibrous proteins that are distinct from both tubulin and actin, but that are related to the intermediate filament proteins present in other cell types.

Axons are generally unbranched and may range from a few tenths of a millimeter to a meter or more in length. Depending on the type of neuron, there are one or two axons present. Axons emerge from a spherical region of the cell body, called the **axon hillock** (see Figures 18-1 and 18-4), that is characterized by a reduced content of free and membrane-bound ribosomes. Clusters of microtubules extend in parallel from the axon hillock into the axon. The axon contains some smooth endoplasmic reticulum and a few longitudinally oriented mitochondria, but neurofilaments and microtubules predominate (Figure 18-7).

Myelin and Glial Cells

Although neurons are the building blocks of the nervous system, their ability to transmit signals quickly and efficiently depends on the presence of supporting

shorter than axons, rarely extending for more than a millimeter in length, and may occur in large numbers projecting from a single cell body. With increasing distance from the cell body extensive branching may occur, and small protuberances, called dendritic **spines** or **thorns,** emanate from the main stem. These spines are involved in forming **synapses,** the close connections that allow adjacent neurons to communicate with one another.

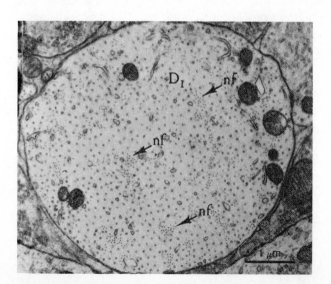

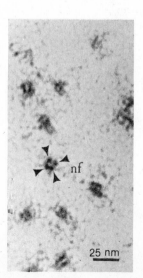

FIGURE 18-6 Electron micrographs showing appearance of neurofilaments. (*Left*) Cross section of a dendrite (D1) showing neurofilaments (nf) throughout the cytoplasm. (*Right*) Higher magnification of a neurofilament in cross section shows four granules (arrowheads) surrounding a hollow core. Courtesy of R. B. Wuerker.

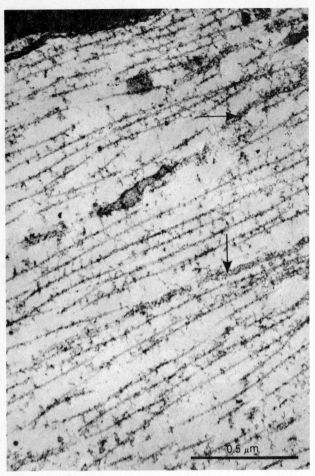

FIGURE 18-7 Electron micrograph of a large axon in longitudinal section showing numerous long, straight neurofilaments running in parallel. A few microtubules are also seen (arrows). Courtesy of Y. J. LeBeux.

FIGURE 18-8 Electron micrograph of a myelinated axon in cross section. The spiral myelin layer starts internally (mes₁) and terminates on the outside of the myelin sheath (mes₂). The interior of the axon is composed of neurofilaments (nf) and larger microtubules (m). The outside is covered by the basal lamina (B). Courtesy of A. Peters.

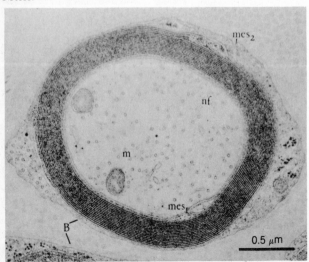

elements called **glial cells.** Gross inspection of the brain and spinal cord of vertebrates reveals the presence of discrete colored areas referred to as the grey and white matter of the central nervous system. In 1864 Rudolph Virchow reported that the white matter contains a closely packed array of cell processes that possess a white covering. He named this substance **myelin,** from the Greek word for *marrow,* because it is present in abundance in the core, or marrow, of the brain. Myelin, which surrounds the axons of many vertebrate neurons, is produced by specialized glial cells called **Schwann cells.** We have already discussed in Chapter 4 how the *myelin sheath* forms by spiral wrapping of the Schwann cell plasma membrane around the axon (Figures 4-5 and 18-8). The thickness of the multilayered myelin sheath is related to axon diameter, with the largest axons having the greatest number of membrane layers making up the myelin sheath.

In 1878 the French pathologist Louis-Antoine Ranvier discovered that the myelin sheath is periodically interrupted, creating regions called **nodes of Ranvier** where one glial cell terminates and the next one begins. As each node is approached, the layers of myelin terminate one by one until the plasma membrane of the neuron is exposed (Figure 18-9). This region, which spans a distance of about one micrometer, may be the site of axonal branching, and is also the place where synapses are most likely to be found. Because both myelinated and unmyelinated axons occur in the nervous system, the term **nerve fiber** is employed to distinguish those axons covered by myelin. Bundles of nerve fibers held together by connective tissue are called **nerves** (Figure 18-10).

Axonal Transport

Because the nucleus and biosynthetic machinery of the neuron are localized in the cell body, far removed from the distant regions of the axon, the question arises as to how the axon's supply of essential metabolites, macromolecules, and organelles is maintained. Simple diffusion is not an adequate explanation, because even small molecules like glucose would take months or years to diffuse the length of a typical axon. Instead materials are actively propelled through the axon cytoplasm, or **axoplasm,** at a rate that varies with the substance transported. This phenomenon was first discovered in 1948 by Paul Weiss and Helen Hiscoe, who found that tying off the axon of a nerve cell causes a swelling to appear proximal to the point of constriction (Figure 18-11a). Removal of the constriction allows the accumulated material to migrate down the axon at a slow and constant rate of 1–4 mm per day. Weiss and Hiscoe named this process **axoplasmic flow.** Weiss was the first to propose that axoplasmic flow is a manifestation of bulk cytoplasmic movement propelled by waves of constric-

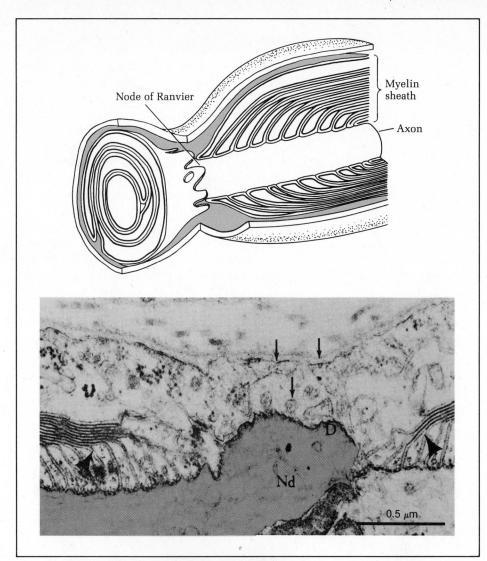

Node of Ranvier

Myelin
sheath

Axon

D

Nd

0.5 μm

FIGURE 18-9 The node of Ranvier. *(Top)* Schematic diagram of the node of Ranvier illustrating the absence of myelin in this region. *(Bottom)* Electron micrograph of a longitudinal section of a node of Ranvier. The Schwann cells extend their processes (arrows) to cover the node (Nd), and the myelin sheath is present on both sides of this region (arrowheads). Color shading is used to highlight the cytoplasm of the axon. Courtesy of A. Peters.

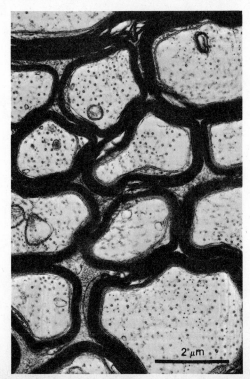

2 μm

FIGURE 18-10 *(Left)* Electron micrograph of a nerve fiber in cross section showing individual axons, each of which is surrounded by a myelin sheath. Courtesy of C. S. Raines.

tion moving down the axon surface. Though constriction waves of this nature have been observed in neurons growing in culture (where bulk cytoplasmic flow may contribute to axon lengthening), comparable evidence for axoplasmic flow in nongrowing axons of mature neurons has not been forthcoming. Indeed it would be hard to explain what happens to the flowing cytoplasm that would continually be arriving at the axon terminus in mature cells. Slow transport still takes place in mature axons, but it involves the movement of substances through the cytoplasm rather than bulk movement of the cytoplasm itself. Among the components transported by this process of **slow axonal transport** are neurofilaments, microtubules, and many proteins including actin, myosinlike proteins, clathrin, and calmodulin. Interestingly, transport of these cellular components occurs in the **orthograde** direction

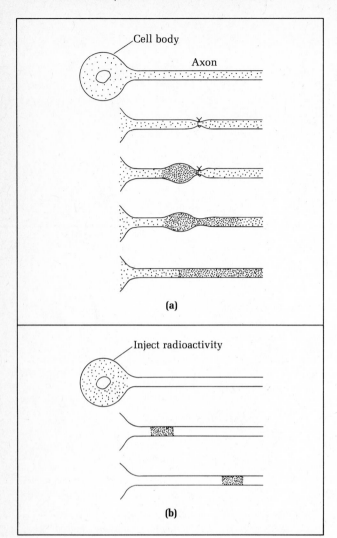

(a)

(b)

FIGURE 18-11 Two ways of detecting axonal transport. (a) Placing a constriction in the axon causes swelling and an accumulation of material proximal to the constriction. After removal of the block, the mass of accumulated material can be seen to migrate down the axon. (b) Radioactive substances injected into the cell body are incorporated into cellular molecules that gradually migrate down the axon. The moving radioactivity can be detected either by autoradiography or by cutting the axon into segments and directly measuring the radioactivity in each segment.

only, that is, moving away from the cell body and toward the axon terminus. Although we are beginning to appreciate the complexity of slow transport, it is still not well understood. We do not know how substances move along the axon, nor do we understand the force-generating mechanism responsible for transport.

Another type of transport that has come to light is of fundamental importance to neurons because it propels neurotransmitter-filled vesicles as well as a variety of membrane-bound organelles. This process of **fast axonal transport** was first detected by putting a radioactive substance into the cell body and then monitoring and timing the appearance of radioactivity at various points along the axon (Figure 18-11b). Under these conditions, rates as fast as 100–400 mm per day have been routinely observed. This type of transport moves mate-

rials up to 100 times faster than does the slower form of axonal transport that is also taking place within the same axon fiber (Figure 18-12).

Structures moved by fast transport are generally membrane bound, and can be moved either in the orthograde direction or in the **retrograde** direction (toward the cell body), depending on the particle being moved. For example, orthograde movement of mitochondria and smaller, neurotransmitter-laden vesicles is commonly observed, while other membrane-bound vesicles, such as lysosomes, move in the retrograde direction. Retrograde transport has been verified by injecting fluorescent or radioactive proteins near axon endings and using fluorescence microscopy or autoradiography to follow their migration toward the cell body. In addition to intact organelles such as lysosomes, pieces of plasma membrane taken into the cell by endocytosis are also moved toward the cell body. This scavenger approach may serve to transport these membrane regions back to the cell body where they may be recy-

FIGURE·18-12 Experiment illustrating the existence of both fast and slow axonal transport in the same axon. After injection of radioactive leucine into the ganglion of a crayfish, the nerve fiber leaving the ganglion was cut into 1-mm sections and the radioactivity in each segment measured. Note that after 24 hr, one peak of radioactivity is found 10 mm down the axon, while a second, slower moving peak of radioactivity has migrated only 1 mm from the cell body.

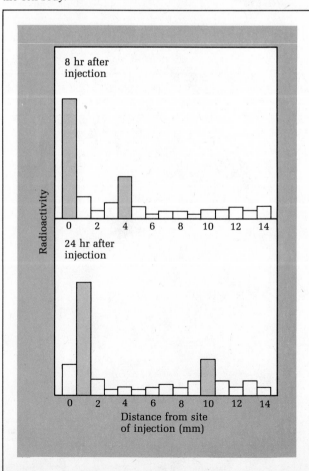

cled. Other substances, such as tetanus toxin and certain viruses including the herpes simplex virus, rabies virus, and poliovirus, reach the central nervous system (brain and spinal cord) by ascending from the axon terminus region toward the cell body via fast axonal transport.

One of the earliest approaches to the study of fast axonal transport involved the use of drugs known to disrupt microtubule organization. Colchicine treatment of microtubules induces their disassembly and blocks fast transport as well. In fact, the relative effectiveness of drugs in inhibiting fast transport correlates well with their abilities to bind to isolated microtubules. This is further supported by the observation that the drug **lumicolchicine,** a colchicine analog that does not bind to microtubules, has no detectable effect on fast transport.

Although direct observation of particle movement in living axons has provided much of the data regarding fast transport, this approach is restricted to observing the movement of larger organelles such as mitochondria. Such observations have suggested that all particle movements are **saltatory;** that is, they occur in short, discontinuous bursts. This view has recently been revised due to the impact of two new experimental approaches. The first involves the use of axoplasm extruded from the giant squid axon as an elegant model system for the study of fast transport. The second approach—a technological breakthrough in the application of video-enhanced microscopy to the study of the axon—was pioneered by Robert D. Allen. Called the *Allen video-enhanced contrast-differential interference contrast microscopy system,* or **AVEC-DIC** microscopy system, this technique has permitted very small (and previously undetected) particles to be observed, circumventing the light-scattering problems encountered with less refined microscopic techniques. The results obtained with isolated axoplasm and AVEC-DIC microscopy have been nothing short of remarkable and have caused major revisions in our interpretation of fast-transport events. For example, it has been clearly shown that single microtubules act as "rails" along which membrane-bound particles travel, not only supporting transport in two directions (orthograde and retrograde), but also allowing two particles moving in opposite directions to pass each other unimpeded. Organelles that move via fast-transport pathways possess multiple binding sites for microtubules, although the chemical nature of the linkage is still unknown. In extruded axoplasm, microtubules transport particles in a continuous (nonsaltatory) fashion and at the same rate (about 2 μm/sec) regardless of the size of the particle. In the intact axon, however, the size of the particle does make a difference in the rate of transport. The general trend suggests that the larger the organelle, the slower and more saltatory is its movement through the axoplasm.

One of the most astonishing results obtained with isolated axoplasm and AVEC-DIC microscopy involves the movement of microtubules that have been sheared to lengths ranging from 1 to 30 μm. These fragments have been observed to *glide* at a constant velocity along glass surfaces, and particles have been seen to attach to these fragments and travel along them in either direction, irrespective of the direction of gliding. These features of fast axonal transport have opened a Pandora's box of questions about particle movement in the axon, not the least of which involves the nature of the driving force guiding this movement. A cell-free system containing purified microtubules (free of microtubule-associated proteins), ATP, and a soluble fraction from squid axoplasm has proved to be a profitable route for the study of some aspects of fast transport. This approach has been used to show that fast axonal transport is ATP-dependent, a conclusion supported by the observation that uncouplers of oxidative phosphorylation, such as 2,4-dinitrophenol, block fast transport. Precisely how this is linked to actual particle movement remains to be determined.

Recently a new protein named **kinesin** has been partially purified from the soluble fraction of the squid axoplasm. Although kinesin does not exhibit detectable ATPase activity, it is able to induce the movement of artificial beads along a microtubule *in vitro,* in a direction corresponding to the orthograde direction in the axon. Because of this property kinesin is believed to act as a **translocator** molecule that may aid in particle transport in the intact cell. A second translocator molecule has been identified that enhances the transport of beads in the opposite direction *in vitro.* It may be possible that these translocator proteins link a particle or organelle to the microtubule, although how movement is actually generated once this linkage takes place is puzzling. One idea put forth some years ago is called the **sliding-vesicle theory.** According to this theory high-energy compounds in the axoplasm drive the cyclic formation and breakage of bonds to microtubules (Figure 18-13), much in the same way that ATP hydrolysis drives microtubule-microtubule sliding in ciliary and flagellar movement. We are not yet in a position to gauge whether this hypothesis applies to the force-generating system of fast axonal transport, nor can we judge how this system is put together and regulated. Nevertheless we now have a powerful model system with which to explore a wide variety of questions concerning the nature of slow and fast axonal transport and the relevance of these processes to other types of cells.

The Synapse

The region where the axon of one neuron makes functional contact with a dendrite, cell body, or axon of a second neuron for the purpose of transmitting an electrical signal is called a **synapse.** Though neurons make

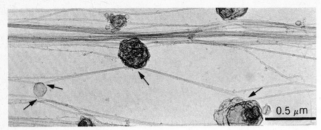

FIGURE 18-13 Electron micrograph of a lysed cell showing attachment of microtubules to vesicles (arrows). Note how the microtubules bend to contact the vesicle surface, suggesting firm associations. Courtesy of M. Schliwa.

frequent contacts with one another, most areas of juxtaposition do not exhibit the characteristic morphological features of synapses, and so do not represent sites of functional interaction. In addition to forming between neurons, synapses also occur between axons and muscle cells, where they are termed *motor endplates* or *neuromuscular junctions* (see Chapter 19).

A single neuron may be involved in thousands of synapses; in addition to synapses between axons and dendrites, these include axon–axon, dendrite–dendrite, and axon–cell body synapses. The most characteristic type of synapse found in the nervous system is

FIGURE 18-14 Thin-section electron micrograph of a chemical synapse. The presynaptic knob containing many synaptic vesicles (ves) is separated from the dendrite (D) by a synaptic cleft (arrows). Courtesy of S. G. Waxman.

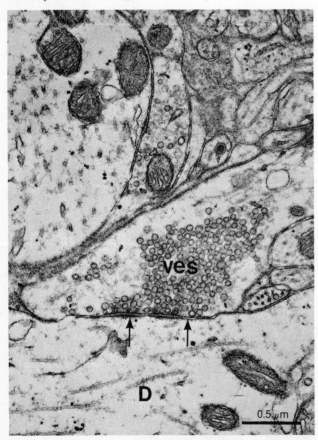

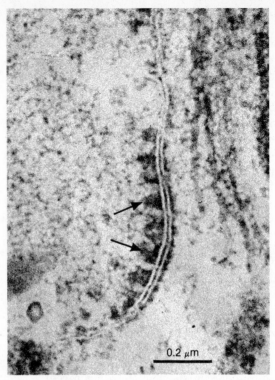

FIGURE 18-15 Electron micrograph showing the dense projections (arrows) that are located beneath the presynaptic plasma membrane. Courtesy of Y. J. LeBeux.

termed the **chemical synapse.** This connection is different from other types of synapses in that it is polarized, meaning that the information travels in one direction. At chemical synapses the terminal portion of the axon is expanded into a bulblike structure, referred to as the **presynaptic knob,** which lacks a myelin sheath. Within this region are numerous **synaptic vesicles** containing chemical neurotransmitters destined to be released from the axon upon arrival of an electrical impulse (Figure 18-14). Mitochondria are numerous in this region, but the microtubules characteristic of the rest of the axon are sparse or absent. Bundles of neurofilaments are common in the presynaptic knobs of the spinal cord, but not in the brain. In addition, networks of actin microfilaments may be present.

The plasma membranes of the presynaptic knob and the **postsynaptic ending** of the adjacent neuron that receives the incoming signal are separated from each other by a space of 20 to 30 nm called the **synaptic cleft.** On the inner surface of the presynaptic membrane lining the synaptic cleft is a series of regularly arranged *dense projections* that protrude into the presynaptic cytoplasm (Figure 18-15). In tangential section these projections appear as a triangular grid whose interstices represent spaces through which synaptic vesicles must pass to make contact with the plasma membrane (Figure 18-16). In freeze-fracture micrographs, 20-nm diameter pits appear on the inner fracture face of the presynaptic membrane, while complementary particles

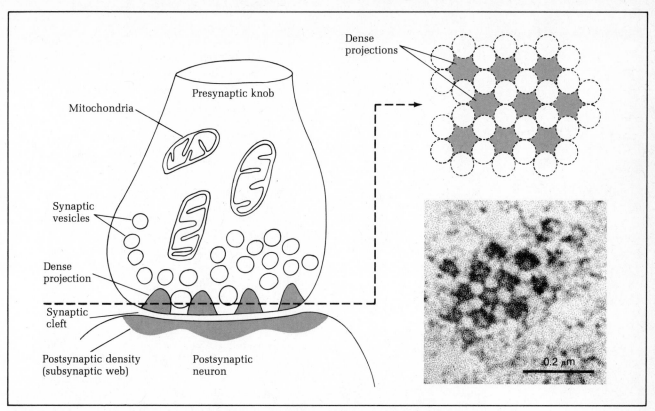

FIGURE 18-16 Schematic diagram of a synapse. The upper right diagram represents a tangential section through the dense projections showing them to be arranged in a triagonal array, with each projection surrounded by a hexagonal pattern of channels through which the synaptic vesicles can pass. The micrograph shows a tangential section through a dense projection revealing the hexagonal array. Micrograph courtesy of Y. J. LeBeux.

FIGURE 18-17 Freeze-etch micrograph of the presynaptic membrane. *(Left)* Inner fracture face showing particles protruding from the surface (arrows). *(Right)* Outer fracture face with many pits (arrows) thought to be synaptic vesicle attachment sites. Courtesy of D. G. Jones.

protrude from the outer fracture face (Figure 18-17). It is thought that these formations represent sites of attachment of synaptic vesicles to the presynaptic membrane. Like the presynaptic membrane, the postsynaptic membrane bordering the synaptic cleft is lined with dense material on its inner surface (see Figure 18-15), though this **postsynaptic density** does not exhibit the regular organization characteristic of the presynaptic dense projections. Microtubules and microfilaments have

been observed emerging from the postsynaptic density, but their functional significance, as well as the significance of the density itself, is not well understood.

The type of synapse described above is termed a chemical synapse because signals are transmitted across it by chemical substances released from the presynaptic knob. Though this is the most commonly encountered kind of synapse some neurons, especially in lower vertebrates and invertebrates, are connected by

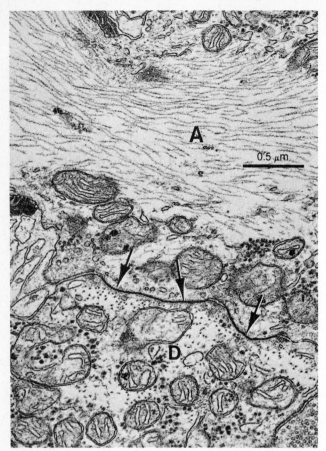

FIGURE 18-18 Thin-section electron micrograph of an electrical synapse showing the closely apposed plasma membranes (arrows) of an axon (A) and a dendrite (D). Courtesy of S. G. Waxman.

FIGURE 18-19 Scanning electron micrograph showing the growth cone of a neuron in culture. Courtesy of S. M. Rothman.

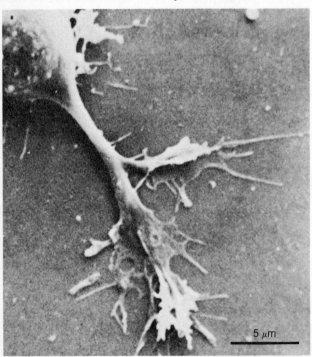

gap junctions, forming **electrical synapses** (Figure 18-18). Cells connected by such electrical synapses are separated from one another by a space of about 2 nm, one-tenth that of the typical synaptic cleft distance in chemical synapses. Since electrical synapses mediate direct electrical coupling between neurons, the impulse is transferred across the synaptic space more rapidly than in chemical synapses, where a significant delay is caused by the need for diffusion of chemical neurotransmitters across the much larger synaptic cleft.

Growth and Development of Neurons

Neurons arise developmentally from embryonic cells that separate from the primitive ectoderm shortly after gastrulation. These cells, called **neuroblasts,** stop dividing and soon differentiate into neurons, a process marked by the outgrowth of neurites destined to become axons and dendrites. Much has been learned about the development of neurites by studying cultures of embryonic neuroblasts induced to form neurites by appropriate incubation conditions. Experiments of this sort were first carried out early in this century by Ross Harrison, one of the pioneers of tissue culture techniques. By carefully observing the development of cultured neuroblasts, he was the first to show that the axon is a direct outgrowth of the neuron rather than an independently formed entity.

In the 1890s Ramón y Cajal had applied the Golgi silver-stain technique to elongating axons and discovered that the tip of each growing axon contains an expanded region, which he named the **growth cone.** Although his material was fixed and stained, and therefore nonliving, Ramón y Cajal suggested that the growth cone represents an area of dynamic change at the growing tip. This hypothesis was soon confirmed by Ross Harrison, who showed that nerve cells in tissue culture exhibit growing tips that are in constant motion. More recently, high-resolution microscopic observations of this region have revealed the presence of slender filopodia, or "microspikes," that protrude and wave about (Figure 18-19). This highly dynamic region of the plasma membrane continually extends and retracts, as though exploring the surrounding environment. In addition to the organelles typically present in axons, such as vesicles, microtubules, and neurofilaments, the growth cone contains a dense network of 6-nm microfilaments found just beneath the plasma membrane; these microfilaments project into the filopodia, forming their major structural component. Norman Wessells and his associates have shown that treatment of elongating axons with cytochalasin B causes the filopodia to become immobile and retract, halting axon elongation within a few minutes. Colchicine, in contrast, does not alter filopodial appearance or motility, nor does it exert an immediate effect on axon elongation. Within a half

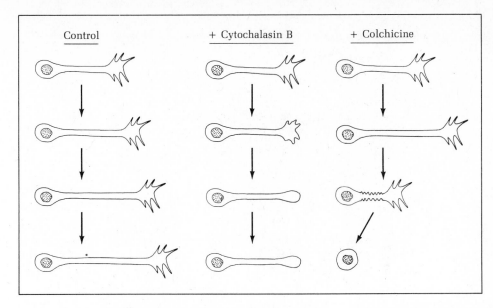

FIGURE 18-20 Diagrammatic summary of the effects of cytochalasin B and colchicine on axon elongation in cultured neurons.

hour of colchicine treatment, however, the axon begins to shorten and, though filopodial activity remains unaltered, the axon eventually collapses back into the nerve cell body. The contrasting effects of these two drugs on axon elongation suggest that cytochalasin-sensitive microfilaments are required for growth cone movement and axon elongation, while colchicine-sensitive axonal microtubules serve as a skeletal framework whose integrity is essential for maintaining the elongated state (Figure 18-20).

The development of at least some types of neurons is known to be under the control of specific growth factors. One of the best studied of these growth-promoting factors was discovered in the early 1950s by Rita Levi-Montalcini and Viktor Hamburger. This substance, called **nerve growth factor (NGF),** selectively maintains the viability and promotes the growth of two cell types: **sensory neurons,** which carry nerve impulses from the periphery to the central nervous system, and **sympathetic neurons,** which are responsible for the regulation of involuntary actions such as breathing and the heart beat. Sensory neurons are most sensitive to NGF during embryonic development, while sympathetic neurons are under the influence of NGF during both embryonic and postnatal development. The first hint of the existence of NGF came from Levi-Montalcini's observation that transplanted tumors stimulate the growth of neurons in the recipient tissue. Working with Stanley Cohen, Levi-Montalcini soon discovered that snake venom and the mouse salivary gland are rich sources of NGF, and today this molecule is known to be synthesized in a wide variety of cell types, including fibroblasts present in virtually all connective tissue. (For their seminal studies on the isolation and characterization of NGF function, Levi-Montalcini and Cohen shared the 1986 Nobel Prize.)

The extreme specificity of the action of NGF has

been dramatically demonstrated by injecting antiserum against NGF into newborn animals; the result is the selective degeneration of the sympathetic nervous system. Likewise in tissue culture both sympathetic and sensory neurons survive poorly in the absence of NGF, while other cells grow well without it. Treatment of sensory or sympathetic neurons with NGF results in a striking and massive outgrowth of neurites, a selective response that forms the basis of the classical **bioassay** for NGF (Figure 18-21). Though the mechanism by which NGF induces its morphological effects is unknown, the discovery that it is composed of several polypeptide chains, one of which resembles insulin in structure, suggests a hormonelike action. Like peptide hormones, NGF binds to plasma membrane receptors of its target cells. However, experiments employing radioactively labeled NGF have shown that the bound molecule is internalized and subsequently delivered by retrograde transport to the cell body, where it appears in membrane-bound cytoplasmic vesicles.

In an attempt to determine whether NGF must be released into the cytoplasm in order to induce neurite outgrowth, John Seeley and his colleagues have microinjected NGF into responsive cells. Under these conditions no physiological response to the presence of the growth factor is detected, suggesting that in the cell, NGF may be effective while still membrane-bound. Although the molecular basis of NGF action is not known, it has been observed that this substance induces proliferation of the Golgi complex, increases in the number of ribosomes, selective stimulation of protein synthesis, and greater activity of the enzyme **tyrosine hydroxylase,** which catalyzes the rate-limiting step in the synthesis of the neurotransmitter norepinephrine.

Once axons begin to elongate under the influence of NGF or any other growth factor, precise pathways must be followed in order to ultimately construct the appro-

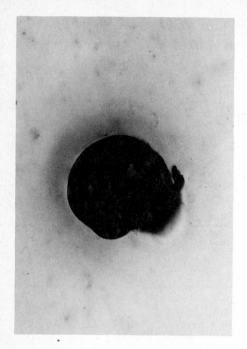
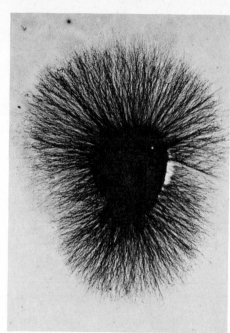

FIGURE 18-21 The classical bioassay for nerve growth factor is based on the effect of this molecule on neurite outgrowth. *(Left)* Untreated ganglion containing many neurons. *(Right)* Ganglion treated with nerve growth factor exhibiting a dense mat of neurities projecting outward. Courtesy of R. Levi-Montalcini.

priate circuits required of coordinated nervous system activity. The axon must not only grow toward a specific region of termination, but must also make contact with specific target cells and, within each cell, a synapse on the cell body, dendrite, or axon must be formed. One of the first attempts to investigate this phenomenon was carried out by Roger Sperry using the amphibian visual system as a model. In the visual system axons emerge from cell bodies located in the retina and migrate to a particular region of the brain called the **optic tectum.** Axons leaving the ventral (lower) portion of the retina migrate specifically to the dorsal (upper) area of the optic tectum, while axons originating in the dorsal retina grow toward the ventral tectum. Sperry's approach was to cut the optic nerve, rotate the eye 180°, and then allow the nerve fibers to regenerate. He found that despite the surgical disruption of the nerves, the damaged axons always sort out appropriately and still migrate to their proper locations. As an explanation of this phenomenon, Sperry proposed that neurons carry chemical identification tags on their surfaces, thereby permitting them to recognize the cells with which they are destined to make synaptic contact. To test this hypothesis, Stephen Roth and his colleagues isolated cells from the dorsal and ventral portions of the retina and tested their abilities to adhere to tissue removed from the dorsal or ventral half of the optic tectum. As predicted by Sperry, cells derived from the dorsal retina adhere preferentially to the ventral tectum, while ventral retinal cells bind selectively to the dorsal tectum. This specific cell-cell recognition presumably helps to define the direction of growth of axons in the nervous system.

The molecular basis of this specific recognition is not yet completely understood, but studies with less complex invertebrate nervous systems are beginning to provide some important clues. One fruitful approach involves the use of monoclonal antibodies directed against specific neurons isolated from the grasshopper nervous system. Working with monoclonal antibodies prepared against neural cells, Corey Goodman and Michael Bastiani have isolated antibodies that bind to surface antigens of subsets of neurons. This suggests that specific axons are chemically distinct from one another, and provides a method for assessing which surface molecules are important in defining the pathways that a growing nerve fiber must follow to establish its proper connections.

In higher vertebrates neurons lose the capacity to divide at about the time they begin to form axons and dendrites. Neurons destroyed after birth can therefore never be replaced, though they can regenerate new axons if their existing axons are cut or damaged. In such an event the old axon degenerates beyond the point of injury because it has become physically separated from the cell body upon which it depends for nutrition. The cell body then synthesizes new axoplasm, allowing the axon stub to begin growing again. The newly forming axon, elongating at a rate of a few millimeters per day, eventually reaches its original destination, where under favorable conditions it may even reestablish its proper synaptic connections.

Though neurons do not normally divide after a certain stage of development, malignant tumors of dividing neuroblasts occasionally arise. These tumors, called **neuroblastomas,** have become an important laboratory research tool because neuroblastoma cells readily grow and divide in culture. They can even be induced to differentiate into cells exhibiting the properties of mature neurons, possessing dendrites, axons, and the ability to synthesize neurotransmitters. Since normal neurons cannot be propagated in culture, the availability of neuroblastoma cells possessing many attributes of mature neurons but capable of growth in culture has been a great aid to neurobiologists.

Though neurons are the cells responsible for the signal-transmitting capabilities of the nervous system, they are not the only cells present. In fact, neurons are outnumbered about ten to one by accessory glial cells, whose role in forming myelin sheaths has already been mentioned. Glial cells occupy essentially all the space of the nervous system not taken up by neurons themselves. Besides their role in myelin formation, glial cells provide mechanical support and metabolic assistance to neurons as well. It has been shown, for example, that certain proteins synthesized by glial cells are ultimately transported to adjacent axons. In addition glial cells synthesize nerve growth factor, indicating a possible role for this cell type in the regulation of neuronal growth and differentiation.

ELECTRICAL PROPERTIES OF THE NEURON

The Action Potential

It was almost 200 years ago when Luigi Galvani first observed that an isolated muscle contracts when its nerves are electrically stimulated. This led Galvani to propose that the signals transmitted by nerve cells are carried by electric currents. To test this possibility, Hermann von Helmholtz measured the time required for a muscle to contract after stimulation of its nerve at varying distances from the neuromuscular junction. His discovery that the nerve impulse travels about 40 m/sec, many orders of magnitude slower than the 186,000 miles/sec traveled by an electric current in a wire, suggested that the nerve impulse could not be thought of as a conventional electric current, but rather involves some active biological process.

Though a direct analogy between nerve impulses and the flow of electric current was ruled out by von Helmholtz, it is now known that the electrical potential across the nerve cell membrane changes during passage of a nerve impulse, indicating that electric currents of some type are involved. Our understanding of this electrical process has been greatly aided by studies on the giant nerve cells of squid. The squid axon, measuring as much as a millimeter in diameter (hundreds of times that of a typical vertebrate axon), is valuable because an electrode can be inserted directly into the axoplasm and the membrane potential measured. In a typical resting axon, the **resting membrane potential** is about −60 mV (the minus sign indicating that the interior of the cell is negative relative to the exterior). As discussed

in Chapter 5, this membrane potential arises because nerve cells actively take up potassium ions and extrude sodium ions, giving rise to a high intracellular concentration of K^+ and a low internal concentration of Na^+ (Figure 18-22a). By itself, this exchange of sodium for potassium would not cause an electrical imbalance across the membrane. However, the membrane is slightly permeable to K^+, allowing some K^+ ions to diffuse out of the cell. This loss of K^+ generates a slight deficit of positive ions within the cell, giving rise to a situation in which the inside of the cell is slightly negative relative to the exterior (Figure 18-22b).

In the late 1930s two teams of scientists, Alan Hodgkin and Andrew Huxley in England, and Howard Curtis

FIGURE 18-22 Illustration of how the resting membrane potential of the neuron is generated. (a) Examination of the relative distribution of ionic species across the neuron plasma membrane shows that the Na^+ concentration is far greater outside the cell than in the cytoplasm, while the opposite is true for K^+. (b) The charge distribution across the membrane is such that the inside of the cell has a net negative charge relative to the outside, with the approximate value of −60 mV.

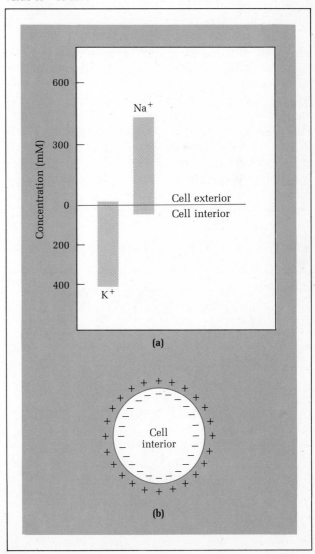

and Kenneth Cole in the United States, measured the changes that occur in the resting membrane potential when a nerve impulse passes down a squid axon. It had been proposed years earlier by Julius Bernstein that during passage of a nerve impulse, the plasma membrane momentarily becomes freely permeable to ions. If this were the case, the membrane potential would fall to zero, since all charged ions would equilibrate across the freely permeable membrane. When the Hodgkin–Huxley and Cole–Curtis groups made their measurements, however, a surprising result was obtained. Instead of merely falling to zero, the potential actually reversed polarity, so that the inside of the cell became positive relative to the outside. This **depolarization** of the membrane lasts for only about a thousandth of a second, after which the membrane potential returns to a negative value. This value, slightly more negative than the initial resting potential, is referred to as the **hyperpolarized** state. It is transient, and is followed by a return to the resting membrane potential. These changes in membrane potential, known collectively as the **action potential** (Figure 18-23), are responsible for the generation of nerve impulses.

Hodgkin and Huxley, in collaboration with Bernhard Katz, soon postulated that a transient increase in the permeability of the axon membrane to sodium ions might explain the unexpected reversal in membrane polarity observed during an action potential. Because the Na^+ concentration is higher outside the cell than inside, such an increase in permeability would cause sodium ions to diffuse into the axon, making the inside of the cell more positive than the exterior. If this interpretation is correct, then action potentials should be dependent on the concentration of sodium ions in the external environment. This prediction was soon con-

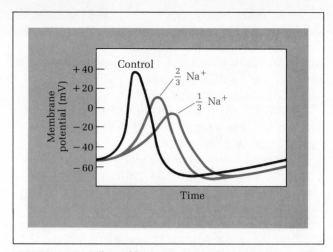

FIGURE 18-24 Effect of lowering the Na^+ concentration of the external medium on action potentials in the squid axon. Decreasing external Na^+ to two-thirds and one-third of its normal concentration in sea water results in successive reductions in the magnitude of the action potential.

firmed; action potentials were shown to decrease in magnitude with decreasing external Na^+ concentration (Figure 18-24), and to disappear entirely in the absence of sodium ions.

But how is the resting membrane potential restored after depolarization by the transient influx of Na^+? An answer to this question was made possible by introduction of the **voltage clamp technique,** which allows one to dissociate membrane potential from ionic movements across the membrane. In this technique two electrodes are inserted into the axon; one monitors the internal potential and the other is used to pass current across the membrane, thereby "clamping" the membrane potential in place at a predetermined level. The movement of Na^+ and K^+ can then be monitored as a function of a given, fixed membrane potential. Hodgkin and Huxley discovered that if a voltage clamp is used to depolarize an axon and then the membrane potential is fixed at zero, electrical current first flows into the axon and then flows in the reverse direction. If the experiment is repeated in a medium lacking Na^+, the inward flow of current is abolished while the outward flow is unchanged. This suggests that the inward current is due to an influx of sodium ions. The simplest explanation for the outward flow of current is that it reflects an efflux of potassium ions from the cell, an idea Hodgkin and Huxley confirmed by following the flow of radioactive K^+ out of the axon.

From the above experiments Hodgkin and Huxley concluded that an action potential is the result of two ionic currents acting sequentially (Figure 18-25). First, a brief increase in membrane permeability to sodium ions allows their entry into the cell and results in depolarization of the membrane. A subsequent increase in membrane permeability to potassium ions results in the

FIGURE 18-23 Graph of the action potential in the squid axon. See text for details.

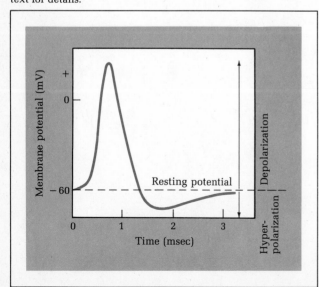

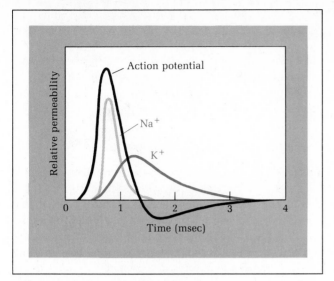

FIGURE 18-25 Graph showing the changes in permeability of the axon plasma membrane to Na$^+$ and K$^+$ with time. The trace of the action potential (black) is superimposed.

diffusion of K$^+$ out of the cell, thereby repolarizing the membrane. According to this theory, nerve impulses can be entirely accounted for by changes in membrane permeability that in turn cause changes in membrane potential. Subsequent support for this model came from the discovery that the cytoplasm can be squeezed out of the squid axon without destroying the axon's ability to transmit nerve impulses. The only requirement is that the inside of the axon be perfused with a solution high in K$^+$ and low in Na$^+$. Not only have such findings on the behavior of perfused axons confirmed the essential roles of K$^+$ and Na$^+$ in generating the action potential, but they have also shown that the nerve impulse is a purely membrane-derived phenomenon; that is, it is not dependent on any aspect of the cytoplasm other than its concentration of sodium and potassium ions.

The model developed thus far describes the changes that occur in membrane permeability during an action potential, but does not explain what causes these changes. The voltage clamp studies described above demonstrate that when the membrane potential is lowered (made less negative), membrane permeability to sodium increases. Based on this information it has been proposed that there are **channels** or **gates** in the plasma membrane of the nerve cell that are specific for the passage of sodium or potassium ions. Control of ion passage through such gates is dependent on the magnitude of the membrane potential. Hence if the potential across the membrane is lowered, either by a physiological stimulus acting on the organism or by an electrical stimulus applied to an isolated nerve, this change in potential increases the permeability of the Na$^+$ gates and hence leads to an influx of Na$^+$ into the cell. This influx of positive ions makes the membrane potential even less negative, causing these channels to open even

more. This results in a further depolarization of the membrane, leading to an additional increase in Na$^+$ permeability. This positive-feedback loop thus amplifies the original depolarization event until the change in membrane potential causes the K$^+$ channels to open. The resulting efflux of potassium ions then reverses the membrane potential until it reaches resting levels. Such **voltage-gated channels** are unique to the membranes of nerve and muscle cells, explaining why they, but not other cells, exhibit action potentials.

Given such a scheme, with its built-in positive-feedback loop, one might expect an action potential to be triggered any time the slightest depolarization of the nerve cell membrane occurs. That this is not the case, however, can be shown by applying an electrical stimulus to the nerve cell membrane sufficient to depolarize it by only a few millivolts. Under such conditions the membrane quickly repolarizes without inducing an action potential. A somewhat greater stimulus may likewise produce only a transient depolarization. But if the initial stimulus is strong enough to depolarize the membrane to a critical value, termed the **threshold potential,** an action potential is triggered (Figure 18-26). An increase in stimulus strength beyond this point has no further effect on the magnitude of the action potential. The reason subthreshold stimuli do not bring about action potentials is that the Na$^+$ influx caused by small depolarizations is less than the K$^+$ efflux, allowing the

FIGURE 18-26 Graph illustrating the concept of threshold. Small depolarizing stimuli do not trigger action potentials. Once the threshold potential is reached, however, an action potential is triggered; further increases in the size of the depolarizing stimulus do not affect the size of the action potential.

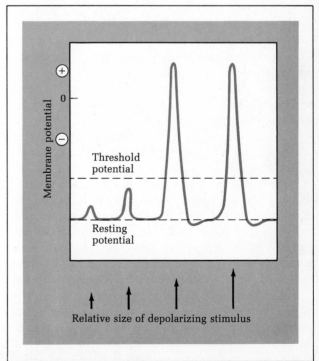

latter to return the membrane potential to its resting value. Only when the initial depolarization is sufficient to produce a net rate of sodium ion influx that is greater than the net rate of potassium ion efflux can the positive-feedback loop leading to an action potential be established.

Although voltage clamp experiments provided many of our early insights into the ionic currents that occur during the action potential, a new method is now contributing more detailed information about the movement of ions across the neuron plasma membrane. This newer **patch clamp technique** permits direct recording of channel events in restricted regions of the membrane. In this procedure a finely polished micropipette tip is placed on the membrane surface and slight suction is applied in order to draw a small portion of the membrane into the pipette tip, forming a tight seal. Current is then applied to this patch of membrane and the flow of current through individual channels is measured. By choosing a membrane possessing few channels per unit area, the opening and closing of single channels can be detected. This information provides a degree of detail not obtainable with the voltage clamp technique, which provides only average measurements for all the channels that are open and closed during the time the current is passed across the membrane.

Many drugs affecting brain and nerve function act by interfering with action potentials. **Tetrodotoxin** is a potent poison produced by the puffer fish that causes paralysis and death in extremely low doses. This molecule blocks action potentials by binding to the outside face of the sodium channel, thereby blocking the entry of Na^+ into the cell. The large size of the tetrodotoxin molecule prevents it from binding to the smaller K^+ channel. Because the binding of this toxin is so selective, measurements of the amount of tetrodotoxin bound to neural membranes can be used to determine the number of Na^+ channels present. Studies of this type have shown that the density of such channels can vary from a few to as many as 12,000 channels per square micrometer of membrane surface.

Certain drugs employed as anesthetics also inhibit action potentials by interfering with sodium channels, though at higher concentrations they may block the K^+ channels as well. Unlike tetrodotoxin, where a small change in toxin structure usually results in a complete loss of activity, a structurally diverse spectrum of compounds are active as anesthetics, their only obvious common property being lipid solubility. Instead of selectively binding to the Na^+ channels, anesthetics of this type are believed to insert into the hydrophobic core of the plasma membrane, causing sodium channels to close as a result of a generalized disruption of membrane structure. Consistent with this theory is the observation that anesthetics bring about an increase in both membrane fluidity (i.e., disorder) and thickness.

Though the action potential is a crucial element in neural communication, it is not the whole story. Once an action potential has been triggered, mechanisms must exist for propagating this impulse from one region of the neuron to another, and eventually to other cells as well. The means by which such propagation is accomplished will be described in the following sections.

Propagation of the Nerve Impulse

To serve as a means of communication within the nervous system, changes in membrane potential must be able to travel from the site of origin of the impulse to the synaptic region of the neuron. Propagation of the nerve impulse in this fashion varies among neurons, depending both on the distances that must be traveled and whether or not the axon is myelinated.

Let us first consider the behavior of neurons with long axons (greater than 1 mm). If any region of the membrane is depolarized to its threshold level, an action potential is triggered. This induces a local flow of current (defined as a flow of positive charges) *away* from the activated area of membrane within the cell, and *toward* the activated area outside the cell (Figure 18-27). These local currents in turn cause the membrane potential in the areas adjacent to the initial site of excitation to fall to their threshold levels, thereby triggering new action potentials. This process is repeated many times, resulting in a series of action potentials traveling across the membrane in all directions away from the original site. Since most action potentials are initiated at the cell body, impulses tend to travel away from this region and toward the end of the axon. Each newly initiated action potential along the way is identical in magnitude to the original one, so that the signal reaching the synaptic region at the end of the axon is virtually identical to that originally initiated at the cell body.

FIGURE 18-27 Mechanism of action potential propagation. See text for description.

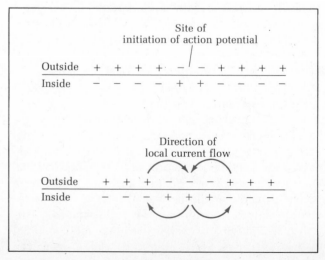

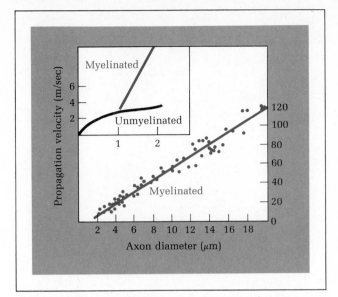

FIGURE 18-28 Relationship between axon diameter and the velocity of nerve impulse propagation. For unmyelinated axons (black), velocity increases with the square root of axon diameter *(inset)*. Myelinated axons (color) propagate impulses much faster because the action potential jumps from node to node. Since the spacing between nodes increases linearly with axon diameter, velocity in myelinated nerves increases linearly with diameter. In very small axons *(inset)* the advantage of myelination is lost because the nodes are so close together.

Two major factors influence the velocity of action potential propagation: *axon diameter* and the extent of *myelination.* Larger diameter axons provide less electrical resistance to the local flow of current because of their larger cross-sectional areas, so the diffusing ions are able to bring adjacent areas of plasma membrane to threshold more rapidly. In quantitative terms this results in a propagation velocity that is roughly proportional to the square root of the fiber diameter (Figure 18-28, *inset*).

In addition, myelination influences propagation velocity by virtue of its insulating effect. In unmyelinated axons some of the ions carrying the local currents that trigger new action potentials leak through the plasma membrane, decreasing the effectiveness of action potential propagation. By reducing this leakage, the myelin sheath increases the speed at which areas adjacent to the initial action potential are brought to threshold. Because action potentials can only occur where the plasma membrane is exposed to the extracellular fluid, myelination also influences propagation velocity by preventing action potentials from occurring anywhere except at the nodes of Ranvier, where the myelin sheath is absent (see Figure 18-9). Action potentials in myelinated axons therefore jump from node to node, a process termed **saltatory conduction.** The time required for the depolarization currents to flow from node to node is small compared to the time needed to generate a new action potential at each node, so propagation velocity increases with increasing distance between nodes.

The distance between nodes, in turn, increases with axon diameter, leading to faster propagation in larger axons (see Figure 18-28). Myelination permits large axons to propagate action potentials at velocities in excess of 100 m/sec, an order of magnitude faster than the velocity of impulse conduction in unmyelinated nerve fibers of similar diameter. The importance of myelination in normal neural function is vividly demonstrated by the disease **multiple sclerosis,** in which symptoms such as speech and vision defects, tremors, and paralysis result from degeneration of myelin sheaths in particular regions of the brain (Figure 18-29).

The triggering of sequential action potentials is certainly an efficient means of propagating impulses in neurons having long axons, but most neurons in the central nervous system have axons measuring less than a millimeter in length and do not utilize (nor can they generate) action potentials for signal transmission. If a neuron of this type is partially depolarized, either by an external stimulus or by a chemical signal released from another neuron, the localized change in membrane potential is passively conducted across the rest of the plasma membrane by local current flow, without ever triggering an action potential. Because of the leakage of ions through the membrane such local currents dissipate rapidly, causing the magnitude of the membrane depolarization to decrease with distance. In spite of this rapid decline in efficiency, local currents can transmit impulses effectively over distances of up to a millimeter, and have the advantage of being equally suitable for the propagation of hyperpolarizations or depolarizations of the plasma membrane. Thus in a general sense, a "nerve impulse" is any change in membrane potential transmitted along nerve cells, including, but not restricted to, action potentials.

Initiation of Electrical Signals by Sensory Stimuli

Though it is common in the laboratory to initiate nerve impulses by stimulating nerve cells electrically, the nervous system must be able to respond to environmental stimuli such as light, sound, odor, taste, and touch. Neurons specialized for responding to such stimuli are called **sensory neurons,** and can be divided into four types: **photoreceptors** sensitive to light, **mechanoreceptors** sensitive to physical pressure (e.g., sound and touch), **chemoreceptors** responding to chemicals (e.g., odor and taste), and **thermoreceptors** sensitive to temperature. These receptors are designed to transduce energy of various kinds to a common form, namely, a change in membrane potential. Such alterations in membrane potential represent the first step in reception of sensory information by the nervous system.

To illustrate how sensory receptors convert natural stimuli into electrical signals, let us consider the photo-

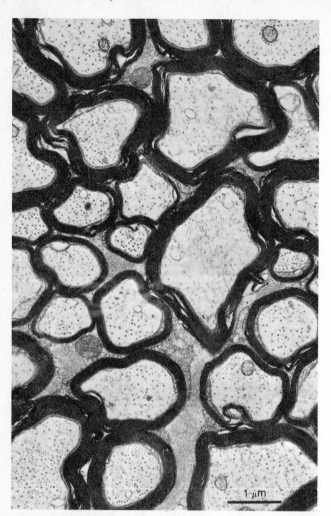

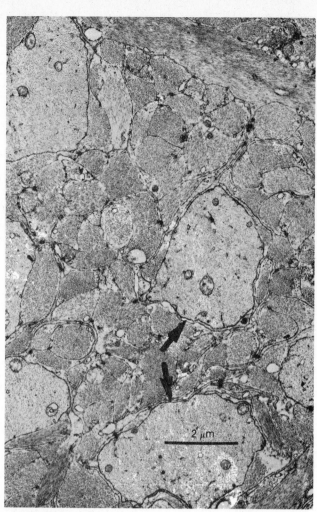

FIGURE 18-29 Electron micrographs showing myelin sheath in normal individuals and those afflicted with multiple sclerosis. *(Left)* Normal myelinated axon. *(Right)* Axons lacking myelin (arrows) found in individuals with multiple sclerosis. Courtesy of C. S. Raines.

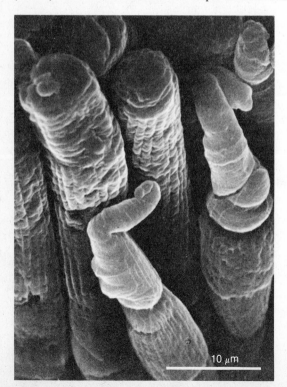

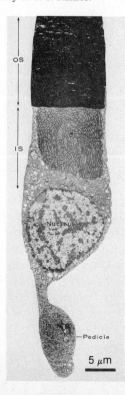

FIGURE 18-30 Electron micrographs of rods and cones. *(Left)* Scanning electron micrograph of rod cells and cone cells in the retina of the mudpuppy. The rod cells have outer segments that are cylindrical in shape, while the outer segments of the cone cells are conical in shape. Courtesy of E. R. Lewis, Y. Y. Zeevi, and F. S. Werblin. *(Right)* Electron micrograph of an isolated rod cell showing the outer segment (OS) made up of tightly packed membrane disks containing the photoreceptor pigment. The inner segment (IS) contains a variety of cell organelles including a large accumulation of mitochondria. The synaptic terminal (pedicle) is filled with synaptic vesicles. Courtesy of E. Townes-Anderson.

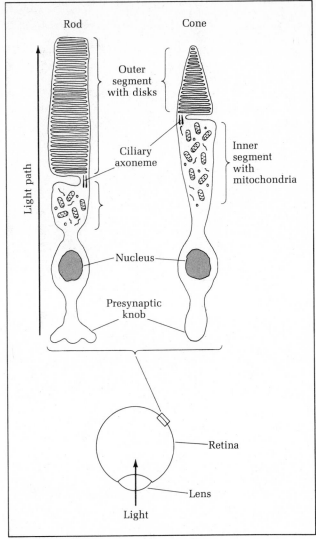

FIGURE 18-31 Schematic diagram of typical rod and cone cells.

FIGURE 18-32 Absorption spectra of the rod pigment, rhodopsin, and the three cone pigments.

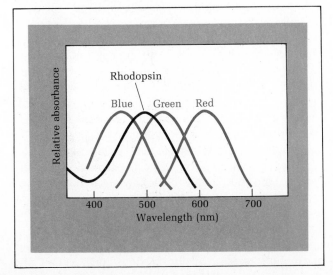

receptor as an example. The vertebrate retina contains two types of photoreceptors, called **rods** and **cones** because of the distinctive shapes of their light-sensitive tips (Figures 18-30 and 18-31). The more numerous rod cells are sensitive to dim light but cannot distinguish colors. Cone cells, on the other hand, only function in bright light, and are responsible for color vision. In both rods and cones photoreception takes place in a specialized region of the cell termed the **outer segment.** Because it is attached to the rest of the cell by a narrow stalk containing a typical ciliary axoneme, the outer segment can be considered to be an elaborately modified cilium.

The outer segment of each photoreceptor cell contains hundreds of flattened **membrane disks** that arise developmentally by invagination of the plasma membrane, though connections to this membrane are not maintained in rod cells. Within the membranes of the disks are embedded light-absorbing pigments consisting of a protein and a light-absorbing derivative of vitamin A called **retinal.** The protein, which occurs in different chemical forms, is given the generic name **opsin.** The retinal molecule also occurs in several configurations, and the complex of retinal and opsin is referred to as **rhodopsin.** All rod cells contain rhodopsin, which absorbs light maximally near 500 nm and whose absorption spectrum resembles the solar output. In contrast there are three types of cone cells, each containing a pigment specifically designed to absorb light in a particular region of the spectrum, either blue, green, or red (Figure 18-32). The same light-absorbing group, 11-*cis*-retinal, is utilized by all the visual pigments, but the linkage of different opsins to the retinal causes the light-absorbing properties to vary. The elaborate membrane specializations characteristic of the outer segments of both rods and cones serve to increase the total surface area for light-catching photopigment molecules, while also permitting the appropriate spatial orientation needed to maximally absorb light energy.

When a photon of light strikes 11-*cis*-retinal, it is converted to its *trans* configuration (Figure 18-33a). This alteration brings about the dissociation of retinal from opsin, whose conformation in turn also changes. But how are such alterations in pigment structure ultimately transduced into changes in membrane potential? Studies using intact retinas have shown that a photon of light induces a transient **hyperpolarization** of the rod outer segment plasma membrane, indicating that a hyperpolarization rather than a depolarization transmits the nerve impulse in this cell type. This change in membrane potential is due to a shift in the permeability of the membrane to sodium ions. In the dark the plasma membrane of the outer segment is highly permeable to Na^+, which flows into the cytoplasm of the outer segment from the external environment where the Na^+ concentration is higher (Figure

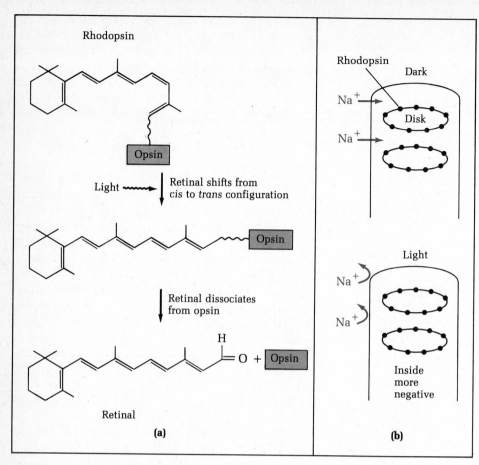

FIGURE 18-33 Effects of light on rod cells. (a) Changes in rhodopsin structure induced by light. (b) Exposure of the rod cell to light results in blockage of sodium ion channels in the membrane, causing membrane hyperpolarization.

18-33b). Upon illumination the entry of Na^+ into the rod cell is blocked, leading to an increased net negative charge inside the cell and hyperpolarization of the membrane. The absorption of a single photon by a rhodopsin molecule can block the transit of millions of sodium ions across the rod outer segment plasma membrane. Just how this amplification of the signal occurs is beginning to be understood. Because rhodopsin molecules are localized to the disk membranes, which are electrically isolated from the sodium channels in the outer segment plasma membrane, it seems reasonable to propose that a diffusible transmitter molecule is responsible for transmitting the signal from rhodopsin to the sodium channels in the plasma membrane. Two molecules, *calcium* and *cyclic GMP*, were at first implicated in this process. The evidence soon showed that the role of Ca^{2+} is secondary, while cyclic GMP actually controls the opening and closing of sodium channels in the plasma membrane. As stated earlier, the absorption of a single photon by a rhodopsin molecule elicits an amplified response, leading to a block in the movement of sodium ions across the rod outer segment plasma membrane. This is achieved by activation of an enzyme cascade, which results in the hydrolysis of thousands of cyclic GMP molecules. In general, cyclic GMP acts to open sodium channels in the dark, while illumination triggers channel closing by de-

creasing the levels of cyclic GMP in the rod outer segment cytoplasm.

These interrelationships between cyclic GMP and membrane polarization are summarized in Figure 18-34. In the presence of light, a photon is absorbed by rhodopsin, causing it to undergo a conformational change. The molecule can now interact with a GTP-binding protein called the **G-protein** located in the disk membrane. One of the three subunits of this protein binds GTP or GDP, while the remaining two subunits are probably important in attaching the G-protein to the membrane. When rhodopsin interacts with the G-protein, bound GDP is replaced by GTP, which in turn leads to the activation of another membrane enzyme, called **cyclic GMP phosphodiesterase.** This enzyme converts cyclic GMP to GMP, thereby lowering the levels of cyclic GMP in this part of the cell. In the dark the cyclic GMP concentration is returned to normal levels (probably by the action of guanylate cyclase on its substrate GTP). An endogenous GTPase activity in the G-protein converts GTP to GDP, which results in inactivation of the phosphodiesterase and aids in the return of cyclic GMP to higher "dark" levels. Amplification occurs at specific points in this photoreception pathway: (1) one activated rhodopsin molecule can activate up to 500 molecules of the G-protein; (2) one activated phosphodiesterase hydrolyzes about 2000 cyclic GMP

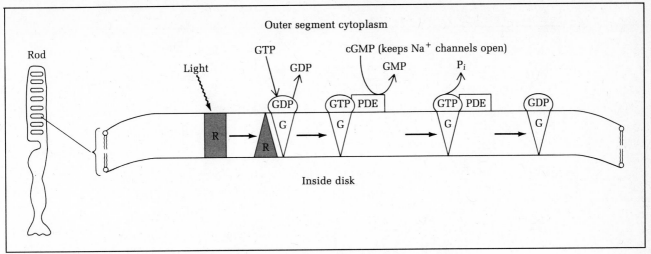

FIGURE 18-34 Stylized illustration of the light-regulated cascade of events leading to cyclic GMP breakdown and closure of Na⁺ channels in rod cells. In the dark the level of cyclic GMP (cGMP) is high and the Na⁺ channels in the rod outer segment plasma membrane are open. Light activates rhodopsin (R), changing its conformation so that it can interact with the G-protein (G). GDP bound to this protein is replaced by GTP. The resulting GTP–G-protein complex activates a cyclic GMP phosphodiesterase (PDE), leading to breakdown of cyclic GMP to GMP and closing of the sodium ion channels. GDP is exchanged for GTP bound to the G-protein and the process is primed for the next cycle. See text for details.

molecules per second. Thus the initial absorption of a photon is amplified $500 \times 2000 = 10^6$-fold by the enzyme cascade.

SYNAPTIC TRANSMISSION OF NERVE IMPULSES

When a nerve impulse (depolarization or hyperpolarization) reaches a synapse, it must be transmitted across the synaptic cleft in order to reach the adjacent neuron. This event occurs so rapidly that for many years it was thought to be purely an electrical phenomenon. In point of fact electrical synapses, which are similar to gap junctions in other cells, do transmit impulses electrically. But most synapses are of the chemical type, in which the signal is carried across the synaptic cleft by the diffusion of chemical substances referred to as **neurotransmitters.**

The first definitive evidence for the existence of chemical neurotransmission was provided by an ingenious experiment carried out by Otto Loewi in 1921. Loewi knew that a heart removed from a frog will continue to beat if it is placed in a container filled with saline solution. Furthermore, if one of the nerves leading to the heart is electrically stimulated, the heart rate will slow down. To investigate this effect Loewi allowed the saline solution bathing a beating heart to flow into a second flask containing another beating heart. Under such conditions stimulation of the nerve of the first heart caused a slowing of the second heart as well, suggesting that the stimulated nerve releases a substance into the saline solution capable of reducing the heart rate. The released substance was subsequently identified as **acetylcholine** by Sir Henry Dale.

It soon became apparent that nerve cells differ from one another in the substances they utilize as neurotransmitters. The following agents are now known to serve as neurotransmitter molecules in specific cells in the nervous system: the amines **acetylcholine, norepinephrine, dopamine, serotonin,** and **histamine;** the amino acids **gamma-aminobutyric acid (GABA), glycine, glutamate,** and **aspartate** (Table 18-2); neuropeptides including **substance P, β-endorphin,** and **enkephalin** (Table 18-3); and the nucleotide **ATP.** Though there are good reasons for believing that each of these substances is utilized as a neurotransmitter by a particular class of neurons, the evidence is not equally convincing for all of them. To be unambiguously identified as a neurotransmitter, a substance must be shown to be released from the nerve endings upon stimulation, to interact with postsynaptic receptors so as to trigger a response, and to be inactivated shortly after release. These steps (neurotransmitter synthesis and release, interaction with postsynaptic receptors, and neurotransmitter activation) are the major events underlying synaptic transmission.

Neurotransmitter Release

Our first important clue to the mechanism underlying neurotransmitter release was provided in the early 1950s by Bernhard Katz and his associates, who discovered that the postsynaptic membrane of resting frog muscle undergoes continual spontaneous depolariza-

TABLE 18-2
Structures of the major nonpeptide neurotransmitters

Substance	Structure	Major Site of Action
Amines		
Acetylcholine	$CH_3\overset{\displaystyle O}{\overset{\|}{C}}-O(CH_2)_2\overset{+}{N}(CH_3)_3$	Neuromuscular junction
Norepinephrine	$HO-\underset{HO}{\bigcirc}-\overset{\displaystyle OH}{\overset{\|}{C}}HCH_2NH_2$	Central nervous system and sympathetic nervous system
Dopamine	$HO-\underset{HO}{\bigcirc}-CH_2CH_2NH_2$	Brain and sympathetic nervous system
Serotonin	$HO-\bigcirc-(CH_2)_2NH_2$ (indole ring, N)	Brain stem
Histamine	$-(CH_2)_2NH_2$ (imidazole ring, N N)	Hypothalamus
Amino acids		
Gamma-aminobutyric acid	$HOOC(CH_2)_3NH_2$	Central nervous system
Glycine	$HOOCCH_2NH_2$	Spinal cord
Glutamate	$HOOC(CH_2)_2\overset{\displaystyle NH_2}{\overset{\|}{C}}HCOOH$	Cerebellum
Aspartate	$HOOCCH_2\overset{\displaystyle NH_2}{\overset{\|}{C}}HCOOH$	?

TABLE 18-3
Selected peptide neurotransmitters

Peptide	Total Amino Acids	Site of Action
Thyrotropin-releasing hormone	3	Hypothalamus
Luteinizing hormone-releasing hormone	10	Hypothalamus
Somatostatin	14	Hypothalamus
Adrenocorticotropin (ACTH)	39	Pituitary
β-Endorphin	31	Pituitary
Substance P	11	Brain
Neurotensin	13	Brain
Leucine enkephalin	5	Brain
Methionine enkephalin	5	Brain

tions measuring about 0.5 mV. These **miniature postsynaptic potentials** (also called *miniature endplate potentials*) are orders of magnitude smaller than the large depolarizations, or **excitatory postsynaptic potentials,** that occur in the postsynaptic membrane when its presynaptic nerve fibers are stimulated. But careful examination of the size of the depolarizations produced by normal nerve stimulation revealed them to be integral multiples of the spontaneous miniature potentials, leading Katz to conclude that neurotransmitter is released from the presynaptic neuron in discrete units or **quanta,** rather than in continuously graded amounts. The quan-

tal nature of synaptic transmission cannot be attributed to a property of the postsynaptic membrane, for depolarizations smaller than the unit size of 0.5 mV can be produced by the administration of tiny amounts of acetylcholine to muscle cells.

At about the same time as the quantal release theory was being formulated, electron microscopists discovered the existence of synaptic vesicles in presynaptic nerve terminals, raising the possibility that each vesicle contains a quantum of neurotransmitter destined to be released from the cell by exocytosis upon nerve stimulation. According to this hypothesis, miniature postsynaptic potentials are due to the spontaneous expulsion of the contents of a few vesicles from the presynaptic nerve ending, while the large excitatory postsynaptic potentials associated with nerve stimulation are caused by the release of hundreds of vesicles triggered by an action potential arriving at the presynaptic nerve ending. Though originally based on little direct evidence, this **vesicle hypothesis** has since garnered considerable experimental support. Subcellular fractionation studies have confirmed the basic premise that neurotransmitters are localized in synaptic vesicles. In some nerve endings the number of vesicles has been found to correspond to the number of quanta the nerve ending is capable of releasing under experimental conditions where formation of new vesicles is prevented.

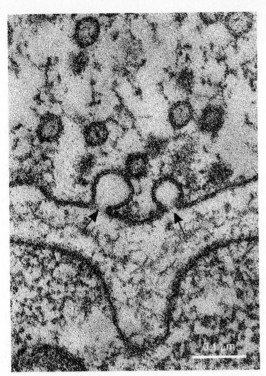

FIGURE 18-35 Electron micrograph of synaptic vesicles (arrows) fusing with the plasma membrane of the presynaptic ending to release their contents into the synaptic cleft. Courtesy of J. E. Heuser.

Consistent with the vesicle hypothesis, extensive stimulation of nerve fibers has been shown to result in a depletion of their presynaptic vesicles. And finally, presynaptic vesicles have been photographed in the process of fusing with the plasma membrane (Figure 18-35).

The mechanism by which an action potential arriving at the presynaptic knob induces fusion of synaptic vesicles with the plasma membrane has been partially elucidated. Like exocytosis in other cell types, this event depends on the presence of calcium ions. In an unstimulated neuron the plasma membrane is not permeable to calcium ions, which are found in a higher concentration outside the cell (about 10^{-3} M) than inside the cell (about 10^{-7} M). Depolarization opens voltage-sensitive calcium channels in the membrane and Ca^{2+} enters the cell, triggering neurotransmitter expulsion from the presynaptic terminal. Evidence for this scenario is derived from studies showing that when *aequorin* (a protein that emits light when exposed to calcium) is added to the presynaptic terminal, light can be detected in this region during normal synaptic transmission; hence Ca^{2+} must have entered the cell. Additional evidence implicating calcium ions in neurotransmitter release includes the observation that microinjection of Ca^{2+} into nerve terminals results in neurotransmitter expulsion, even in the absence of an electrical stimulus. Moreover, removal of calcium ions from the external medium has been found to inhibit

transmission of nerve impulses. Finally, it has been shown that the release of neurotransmitter from isolated nerve endings can be induced by Ca^{2+} ionophores.

Several theories have been proposed to account for how an influx of Ca^{2+} might cause neurotransmitter release. One possibility is that calcium alters the surface charge of vesicle membranes by binding to them or by promoting phosphorylation of their membrane proteins, thereby facilitating vesicle fusion with the plasma membrane. Alternatively it has been suggested that Ca^{2+} activates an actin-myosin type of contractile system, as it does in muscle. Though not providing unequivocal support for this proposal, actin-containing microfilaments have been detected in the presynaptic knob, and the microfilament inhibitor cytochalasin B has been reported to interfere with neurotransmitter release. However, colchicine also inhibits the release of neurotransmitters, raising the possibility of microtubule involvement as well. Unfortunately it is difficult to distinguish whether the results obtained with antimicrotubular drugs such as colchicine indicate an active role of microtubules in vesicle expulsion, a passive role in guiding vesicles to attachment sites on the plasma membrane, or simply reflect side effects of these drugs that have nothing to do with microtubule structure and function.

A growing body of evidence suggests that vesicle fusion is not the only means for releasing neurotransmitters into the synaptic cleft. In the electric fish *Torpedo*, Yves Dunant and Maurice Israël have found that the acetylcholine content of the presynaptic terminus is divided about equally between synaptic vesicles and the cytoplasm of the axon ending. Careful measurements of the loss of neurotransmitter from each of these compartments upon stimulation indicates that initially acetylcholine is released from the cytoplasmic pool. This pool is rapidly regenerated, presumably due to the presence of the enzyme choline acetyltransferase, which synthesizes acetylcholine from the precursors choline and acetate. If the stimulus is maintained, this regeneration of cytoplasmic acetylcholine eventually ceases and the release of acetylcholine from the vesicles begins. Such observations suggest that the vesicle hypothesis may provide only part of the story regarding synaptic transmission.

In cells where the nerve impulse is transmitted by a hyperpolarization rather than a depolarization of the plasma membrane (e.g., photoreceptor cells), a sequence of events opposite to the typical situation is believed to occur. The hyperpolarization arriving at the nerve ending decreases plasma membrane permeability to Ca^{2+}, reducing intracellular levels of calcium and slowing the rate of neurotransmitter release. The nerve ending can therefore be visualized as a finely tuned instrument whose rate of neurotransmitter release at any given moment depends on the prevailing membrane

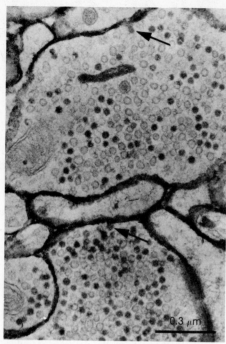

FIGURE 18-36 Micrograph showing the recycling of synaptic vesicles into the presynaptic terminal. The extracellular space is filled with the electron-dense reaction product of horseradish peroxidase and many synaptic vesicles are filled with this tracer. Two vesicles (arrows) are intimately associated with the plasma membrane. Courtesy of A. B. Harris.

brane during synaptic transmission. The possibility that this excess membrane is continually being recycled is suggested by the fact that the number of vesicles present in the presynaptic terminal does not decrease significantly during normal nerve firing, even though neurotransmitter is being released. This recycling process can be directly visualized by incubating neurons in a medium containing *horseradish peroxidase*, a tracer whose uptake into the nerve ending can be easily monitored cytochemically. Nerves stimulated in the presence of horseradish peroxidase collect this tracer in new vesicles that form by invagination of the plasma membrane (Figure 18-36). These vesicles are of a special type, called **coated vesicles** (page 189), which contain a proteinaceous coat or shell constructed of linear rods 6 to 7 nm in diameter arranged in pentagonal and hexagonal arrays. The newly generated coated vesicles coalesce into larger structures, from which new synaptic vesicles eventually arise (Figure 18-37). Though it is not known how neurotransmitter molecules are inserted into these new vesicles, this cycle of membrane flow accompanying neurotransmitter release appears to be analogous to the membrane recycling that occurs in other secretory cells. Studies with horseradish peroxidase have shown, however, that the vesicle membrane is not reutilized indefinitely. During synaptic activity some of this tracer appears in lysosomes that migrate by retrograde transport to cell bodies. Presumably these degraded vesicles are replaced by new vesicles that are delivered to the synaptic region by axonal transport.

Neurotransmitter-Receptor Interaction

Upon release from the presynaptic nerve terminal, neurotransmitter molecules diffuse across the synaptic cleft

potential. A decrease in this potential (depolarization) will enhance the rate of neurotransmitter release, while an increase in the potential (hyperpolarization) will slow neurotransmitter release.

Since neurotransmitter release mediated by exocytosis results in the fusion of vesicle membranes with the plasma membrane, a mechanism must exist for preventing uncontrolled expansion of the plasma mem-

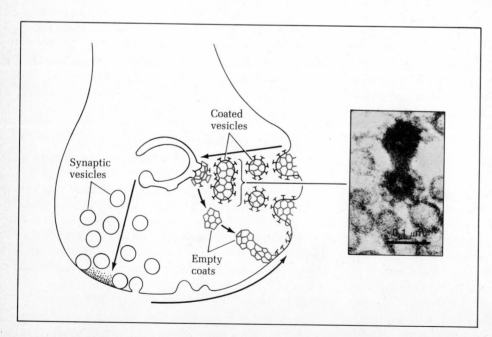

Synaptic vesicles

Coated vesicles

Empty coats

FIGURE 18-37 Schematic diagram of synaptic vesicle recycling. Membrane material added to the plasma membrane during exocytosis of synaptic vesicles is retrieved by budding of coated vesicles that then fuse together, lose their coats, and ultimately subdivide into new synaptic vesicles. The electron micrograph shows the fusion of two coated vesicles with one another. Courtesy of J. E. Heuser.

in a few tenths of a millisecond and bind to the specific **receptors** located in the postsynaptic membrane. Binding of neurotransmitter to receptor brings about a change in permeability and hence electrical potential of the postsynaptic membrane. As will be seen in the following sections, the direction of this change in potential and the mechanism by which it is brought about depends both on the transmitter and receptor in question.

Acetylcholine Receptors Among its many functions, the neurotransmitter acetylcholine is utilized by motor neurons to trigger skeletal muscle contraction, by parasympathetic neurons of the autonomic nervous system to slow the heart beat and constrict smooth muscles, and by **interneurons** to transmit signals between nerve cells in the central nervous system. Acetylcholine is able to exert these varied effects because of the existence of several types of acetylcholine receptors, each of which triggers a different response in the postsynaptic membrane following acetylcholine binding.

The best characterized acetylcholine receptor is the **nicotinic (fast) acetylcholine receptor** found in the

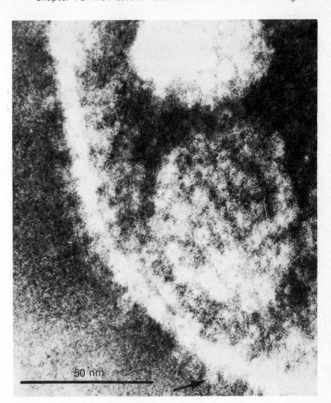

FIGURE 18-39 Negatively stained membrane vesicles from *Torpedo* enriched in acetylcholine receptors. Receptors exhibit a subunit organization and can be seen protruding from the plasma membrane (arrow). Courtesy of F. Hucho.

plasma membrane of vertebrate skeletal muscle cells (Chapter 19) and in the electric organs of certain fish. Binding of acetylcholine to this receptor initiates a rapid increase in permeability of the membrane to both Na^+ and K^+. Since the electrochemical gradient is steeper for Na^+ than for K^+, the influx of sodium exceeds the efflux of potassium, leading to an inward flow of sodium ions. This depolarizes the postsynaptic membrane to its threshold, triggering an action potential. The mechanism by which acetylcholine-receptor binding induces changes in postsynaptic membrane permeability is now being examined by purification and analysis of the nicotinic acetylcholine receptor.

Studies on the acetylcholine receptor have been facilitated by the discovery that certain snake toxins, such as the small protein **α-bungarotoxin,** bind selectively and almost irreversibly to the acetylcholine receptor. Such binding obstructs the interaction of acetylcholine with its receptor and ultimately leads to respiratory paralysis and death. Radioactive α-bungarotoxin has been employed experimentally to localize acetylcholine receptors in tissue sections autoradiographically (Figure 18-38), and to assay for the presence of acetylcholine receptors in subcellular fractions.

A rich supply of receptor is present in the electric organ of the fish *Torpedo* (Figure 18-39). This organ

FIGURE 18-38 Autoradiograph of a neuromuscular junction showing the location of cholinergic receptors (arrows) identified by their binding to radioactive α-bungarotoxin. Courtesy of M. M. Salpeter.

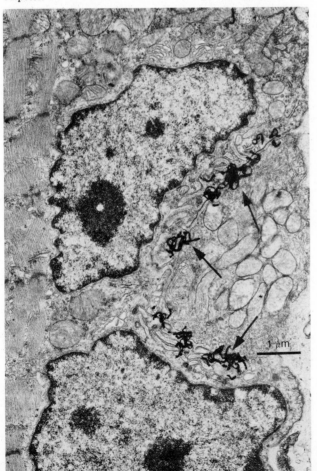

arises from the same embryonic tissue as muscle, but does not accumulate large numbers of contractile filaments. Instead its large flat cells, called **electroplaques,** become extensively infiltrated on one side by acetylcholine-releasing neurons. Upon nerve stimulation the membrane potential on the innervated side of each cell declines from its resting value of -90 mV to $+60$ mV; the opposite side of the cell, however, remains at -90 mV, producing a potential difference of 150 mV across the two outer surfaces. In the electric organ these cells are arranged in series so that their individual voltages are additive. A group of 5000 electroplaques can therefore generate a discharge of 5000×150 mV $= 750$ V, allowing the fish to shock and thereby incapacitate its prey.

Employing the binding of radioactive toxins as an assay, the acetylcholine receptor has been purified by affinity chromatography. The resulting product is a large protein that appears under the electron microscope to consist of five subunits arranged around a central cavity. Electrophoretic analysis of this protein under denaturing conditions generates four different polypeptide chains, (α, β, γ, and δ) in the stoichiometry of $2:1:1:1$. The smallest subunit (α) contains the binding site for acetylcholine. Using cloned cDNAs, the amino acid sequence of each subunit polypeptide has been elucidated and studies are in progress to determine how each subunit inserts into the lipid bilayer in order to create the channel seen in electron micrographs. It has been proposed that the five subunits form an ion channel whose permeability is regulated by interaction of acetylcholine with the subunits. According to this view the acetylcholine receptor behaves like an allosteric enzyme; binding of acetylcholine to the α-subunits induces a conformational change in the polypeptides comprising the channel, resulting in an increase in permeability of the postsynaptic membrane to sodium and potassium ions.

If the purified receptor does possess an ionophorlike component, the reconstitution of receptor with artificial lipid bilayers should regenerate an excitable system. Experiments of this type have been partially successful in that membrane vesicles reconstituted with purified acetylcholine receptor become more permeable to Na^+ in the presence of added neurotransmitter. Such reconstituted systems do not, however, exhibit the specificity, magnitude, and kinetics of ion transport characteristic of the original intact cell. In spite of these problems the reconstitution approach offers considerable promise for elucidating the mode of action of the acetylcholine receptor.

Though binding of acetylcholine to its receptors in skeletal muscle or the electric organ induces depolarization (excitatory postsynaptic potentials) of the postsynaptic membrane, at other synapses, such as those in cardiac muscle, acetylcholine effects hyperpolarization (inhibitory postsynaptic potentials) of the membrane. Thus this neurotransmitter may either induce or inhibit the action potential, depending on the type of receptor to which it binds. This plasticity in receptor behavior is the result of differences in the kind of ion channels opened as a result of acetylcholine binding. We have already seen that interaction of acetylcholine with its skeletal muscle or electric organ receptors causes membrane permeability to both Na^+ and K^+ to increase; the membrane depolarizes because Na^+ influx predominates. Binding of acetylcholine to its receptors in cardiac muscle, on the other hand, leads to a selective increase in permeability of the postsynaptic membrane to K^+. This increase in K^+ efflux causes the membrane to hyperpolarize, inhibiting the rate of contractility of the heart muscle. Thus a given neurotransmitter can be either excitatory or inhibitory, depending on the properties of the receptor to which it binds.

Dopamine Receptors The binding of acetylcholine to its nicotinic receptor, which induces conformational changes in the receptor that lead to an immediate change in membrane permeability, is a mechanism ideally suited for situations placing a great premium on speed, such as the contraction of voluntary muscle. Neural pathways governing events that are slower and more long lasting, on the other hand, tend to utilize more complicated receptor mechanisms. In this latter category are receptors for neurotransmitters like dopamine, serotonin, and norepinephrine, which play regulatory roles such as modulating the activity of other cells (see Table 18-2). Binding of the appropriate neurotransmitter to such a receptor generally activates an enzyme on the inner face of the membrane, causing the synthesis of a second messenger within the cell. This type of mechanism has been postulated for both norepinephrine and dopamine. Several lines of evidence point to cyclic AMP as the second messenger in such situations. In subcellular fractionation experiments on brain tissue the **synaptosome fraction,** which is enriched in pinched-off nerve endings, has been found to contain large amounts of adenylate cyclase, the enzyme catalyzing cyclic AMP formation. Electrical stimulation of nervous tissue causes a substantial elevation in cyclic AMP levels in postsynaptic terminals, as does the application of neurotransmitters such as dopamine and norepinephrine. Finally, direct administration of cyclic AMP to neurons elicits changes in membrane potential similar to those produced by neurotransmitters alone.

The most direct evidence for cyclic AMP involvement, however, has come from the isolation of neurotransmitter-activated adenylate cyclase from neurons. A well-characterized enzyme of this type is the one activated by dopamine, an inhibitory neurotransmitter of the brain, of special interest because it has been implicated both in **Parkinson's disease,** a malady correlated with a deficiency in dopamine-secreting neurons, and in certain types of **schizophrenia,** in which abnormal mental symptoms appear to be related to excess activity of dopamine-utilizing neural pathways. The discovery that the tremors and uncontrollable body movements occurring in Parkinson's disease are accompanied by degeneration of neurons secreting dopamine led to the idea that such patients might improve if the amount of dopamine in the brain could be restored to normal. Since dopamine cannot enter the brain from the circulation, an amino acid precursor to this neurotransmitter, **levo-dihydrophenylalanine (L-DOPA)** has been used instead. This therapy has provided some relief to individuals suffering from Parkinson's disease, though progression of the illness continues. Schizophrenia, on the other hand, has been treated with drugs such as *chlorpromazine (Thorazine),* that block dopamine receptors.

The medical importance of dopamine has led to considerable interest in its receptor. Using brain regions rich in dopamine receptors as starting material, Paul Greengard and his associates have isolated an adenylate cyclase whose activity is stimulated by dopamine. The ability of dopamine to activate this adenylate cyclase is inhibited by antischizophrenic drugs known to block the dopamine receptor, while drugs capable of mimicking the actions of dopamine on its receptor stimulate the dopamine-sensitive adenylate cyclase. These findings suggest that the dopamine receptor is the dopamine-binding portion of dopamine-sensitive adenylate cyclase, and that the actions of dopamine are therefore mediated by cyclic AMP.

The preceding observations raise the question of how neurotransmitter-induced changes in cyclic AMP ultimately trigger the postsynaptic change in membrane potential (in the case of dopamine, the postsynaptic membrane becomes hyperpolarized). Experiments carried out in Greengard's laboratory have demonstrated the presence in synaptic membranes of a cyclic AMP-dependent protein kinase that catalyzes the rapid phosphorylation of a few proteins located in these membranes. It has therefore been suggested that the cyclic AMP-stimulated phosphorylation of one or more of these membane proteins may induce changes in membrane permeability that in turn lead to an alteration in membrane potential. The complex series of steps that must intervene between the initial binding of dopamine to its receptor and the ultimate change in membrane permeability shows why this type of mechanism is associated with synaptic events that are relatively slow and long lasting.

Opiate Receptors The milky exudate obtained from unripe seedpods of the opium poppy has been known from ancient Greek times to relieve pain and induce a sense of well-being. **Morphine,** the major active ingredient in opium, was first isolated in 1803, and was destined to gain widespread medical use before its toxic and addictive properties came to be appreciated. Many attempts have since been made to produce painkillers with the potency, but not the addictive properties, of opiates. For example, in the 1890s two methyl groups were added to morphine in the laboratories of the Bayer company, producing a new opiate, named **heroin,** which was touted as a nonaddictive painkiller. In spite of this failure and others like it, the fact that opiates are still the only drugs effective in the management of severe chronic pain has motivated continued efforts to produce a nonaddictive opiate. Though this goal remains an elusive one, recent research has not only begun to point the way to the design of such drugs, but in the process had led to new insights concerning the physical basis of addiction and to the surprising discovery of a new class of neurotransmitters.

It has long been suspected that opiates such as morphine and heroin produce their effects by binding to specific receptors in nerve cell membranes. Historically this belief arose because of the structural specificity of these molecules. Not only do small changes in the chemical structure of opiates lead to dramatic changes in potency, but only one of the two possible stereoisomers (mirror-image forms) of any given opiate is pharmacologically active. Though these properties point toward the existence of highly specific opiate receptors, early attempts to detect their presence by incubating subcellular fractions with radioactive opiates ended in failure because opiates bind nonspecifically to many substances, overshadowing any specific binding to receptors. In 1971 Avram Goldstein devised a clever method for detecting specific binding based on the premise that the binding of radioactive opiates to their receptors should be displaceable by unlabeled opiate of the active, but not the inactive, stereoisomer form. Though he was only able to detect a small amount of specific binding using this criterion, technical improvements devised independently in the laboratories of Solomon Snyder, Eric Simon, and Lars Terenius soon led to the unequivocal detection of specific opiate binding to membrane fragments of brain tissue. The discovery that the abilities of different opiates to bind to such membrane receptors correlates with their known physiological potencies (Figure 18-40) has led to the use of this simple assay as a screening method for testing the pharmacological potencies of new opiates.

Studies on the distribution of opiate binding sites within the nervous system have provided some significant insights into the mechanism of opiate action. Large numbers of opiate receptors are present in the areas of the spinal cord and the brain that are associated with the conduction and perception of dull chronic pain. The brain's *limbic system,* which mediates emotional behavior, also contains many opiate receptors. It is not

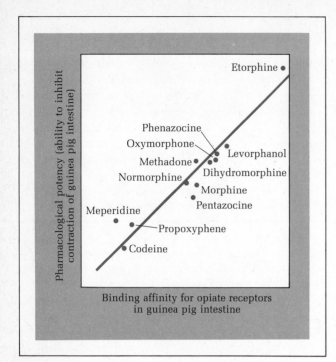

FIGURE 18-40 Graph illustrating the correlation between the affinity of various opiates for the opiate receptor and their pharmacological potencies.

surprising that drugs that selectively interact with such pathways relieve pain and induce euphoria. But why are opiate receptors present in the first place? Since they are clearly not designed to interact with morphine, their presence suggests that the body manufactures its own morphinelike substances. This possibility was decisively confirmed in 1975 when John Hughes and his collaborators reported the isolation and characterization of two pentapeptides from brain tissue that mimic the ability of morphine to inhibit the contraction of intestinal muscle, and whose activities are blocked by morphine antagonists. These short-chain peptides, named **enkephalins** (see Table 18-3), compete with ra-

FIGURE 18-41 Localization of anti-enkephalin antibodies to nerve endings in specific areas of the brain, showing the discrete distribution of the enkephalin molecules. Courtesy of S. H. Snyder.

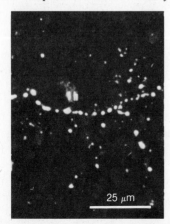

dioactive opiates for binding to opiate receptors in brain membrane fragments. Longer-chain peptides, which bind in a similar fashion to opiate receptors, have also been discovered in nervous tissue; such neuropeptides are termed **endorphins** (endogenous morphinelike substances).

Analysis of the enkephalin content of various regions of the nervous system has revealed that its distribution parallels that of opiate receptors. Moreover the use of fluorescein-labeled antienkephalin antibodies has localized the enkephalins to nerve endings (Figure 18-41). Taken together, these observations suggest that enkephalins may be neurotransmitters designed to interact with opiate receptor sites. Though this idea is still somewhat speculative, enkephalin applied to single neurons has been shown to slow down their normal rate of firing, as would be expected for an inhibitory neurotransmitter.

How do enkephalins and other opiates inhibit neural excitability? Unlike other inhibitory neurotransmitters, enkephalins do not appear to hyperpolarize the postsynaptic membrane. Instead the ability of excitatory neurotransmitters to depolarize enkephalin-treated neurons appears to be diminished, suggesting that enkephalin prevents the increase in Na^+ permeability normally elicited by such neurotransmitters. In addition to this effect on Na^+ channels, cyclic nucleotide alterations have also been implicated in the chronic addiction associated with opiate action. Not only do opiates increase cyclic GMP and decrease cyclic AMP levels in brain slices, but the magnitude of this effect correlates with their relative binding affinities for the opiate receptor. A possible insight into the mechanism of opiate addiction has come from the discovery that continued exposure of neuroblastoma cells to morphine results in a compensatory increase in adenylate cyclase activity. Such cells are then tolerant to the effects of morphine in that opiate concentrations that normally lower cyclic AMP levels no longer do so because of the increased adenylate cyclase activity. In fact such cells are dependent on morphine for the maintenance of normal cyclic AMP levels, since removal of the opiate leads to the formation of abnormally high amounts of cyclic AMP by the excess adenylate cyclase. Though it may be premature to conclude that excess cyclic AMP is related to the symptoms of narcotic withdrawal, the ability of this model system to mimic the hallmarks of addiction, namely tolerance, dependence, and withdrawal, is striking.

The ability of opiates to inhibit neuron firing and neurotransmitter release does not appear to be solely related to cyclic AMP. In certain neuron types it has been suggested that opiates act to modify calcium-sensitive channels in the membrane, leading to an inhibition of neurotransmitter release. Because of the variety of physiological effects induced by opiates in different kinds of neurons, it may be some time before a uniform model of opiate action can be formulated.

Though the evidence supporting a neurotransmitter role for the enkephalins is substantial, the situation is complicated by the existence of related morphinelike substances. The pituitary gland synthesizes a polypeptide 91 amino acids long, called **β-lipotropin,** which contains within it the sequence of one of the enkephalins. Although β-lipotropin itself has very little opiate activity, the pituitary is known to contain a peptide corresponding to amino acids 61–91 of the β-lipotropin molecule. Known as **β-endorphin,** it is 50 times more potent than morphine when injected directly into the brain. The physiological relationship between β-lipotropin, β-endorphin, and the enkephalins is yet to be ascertained. The larger peptides lipotropin and/or endorphin might be biosynthetic precursors of enkephalin, though the lack of a direct vascular connection leading from the pituitary to the brain, and the very low levels of the larger peptides in the brain, raise serious doubts about this possibility. Alternatively the enkephalins and endorphins may function independently, the former in the brain and spinal cord and the latter in the pituitary.

Whatever their relationships to one another, the mere existence of opiatelike substances in the nervous system opens unique new possibilities in the search for nonaddictive painkillers. Although the enkephalins themselves are relatively useless as therapeutic agents because they are susceptible to proteolytic degradation, nondegradable derivatives have already been synthesized in the laboratory. Whether or not such derivatives will turn out to be clinically effective as well as nonaddictive remains to be determined.

Other Receptors In addition to dopamine and the enkephalins, several other neurotransmitters are known to influence cyclic AMP formation. As was the case for dopamine, neurotransmitter-activated adenylate cyclases have been identified for norepinephrine, serotonin, histamine, and octopamine. The serotonin-activated adenylate cyclase is especially interesting because it is inhibited by low doses of LSD, suggesting that the hallucinogenic properties of this substance may be related to an interference with serotonin-stimulated cyclic AMP formation.

Inhibitory neurotransmitters appear to work by several different mechanisms. We have seen how acetylcholine hyperpolarizes cardiac muscle by increasing K^+ permeability, while enkephalins exert their inhibitory effects by preventing the increase in Na^+ permeability normally elicited by excitatory neurotransmitters. Yet another inhibitory mechanism is utilized by the amino acid neurotransmitters **GABA** and glutamate (see Table 18-2). Both molecules increase membrane permeability to chloride ions. Because the negatively charged chloride ion is present in higher concentration outside the cell than inside, an influx of Cl^- results and the membrane becomes hyperpolarized.

Neurotransmitter Inactivation

Once a neurotransmitter has bound to its receptor and triggered an alteration in the permeability of the postsynaptic membrane, its action must be terminated to prevent the postsynaptic cell from remaining in a continual state of excitation (or inhibition). In the case of acetylcholine, the enzyme **acetylcholinesterase** present in the synaptic cleft quickly hydrolyzes the neurotransmitter to choline and acetate. Because this was the first neurotransmitter to be studied in detail, inactivation by enzymatic degradation was thought for many years to terminate the action of other neurotransmitters as well. Recent research suggests, however, that enzymatic inactivation is the exception rather than the rule. A more common mechanism of inactivation involves removal of the neurotransmitter by uptake into the presynaptic nerve ending from which it was released. This transport mechanism is highly selective; for example, norepinephrine-releasing neurons take up only norepinephrine, serotonin-releasing neurons take up only serotonin, and so forth.

TRIGGERING A PHYSIOLOGICAL RESPONSE

We have now seen how nerve impulses are initiated, propagated, and transmitted from neuron to neuron. In order for these impulses to eventually influence the rest of the organism, some neurons must be capable of triggering responses other than the firing of another neuron. Such responses can be divided into three basic types: *glandular secretion, neurosecretion,* and *muscle contraction.*

Nerve cells activate (or inhibit) many types of **glandular secretion,** such as the secretion of saliva from salivary glands or of the hormone epinephrine from the adrenal medulla. The cells of the adrenal medulla are developmentally related to neurons but fail to sprout axons; instead of utilizing their released epinephrine as a neurotransmitter, they secrete it into the bloodstream. The hormone epinephrine circulating throughout the organism exerts effects similar to those produced by the neurotransmitter norepinephrine that is released directly from sympathetic nerves.

The close relationship between neurotransmitters and hormones is underscored by the observation that neurons secrete hormones directly into the bloodstream, a process called **neurosecretion.** For example,

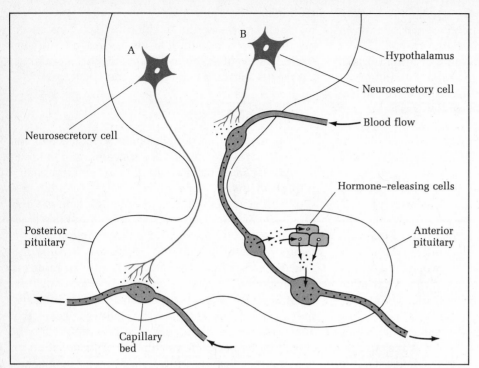

FIGURE 18-42 Schematic diagram illustrating the concept of neurosecretion. In pathway A, molecules (black dots) are released from a neurosecretory cell directly into the bloodstream (gray) in the posterior pituitary. In pathway B, molecules released from a neurosecretory neuron in the hypothalamus enter the bloodstream and travel to the anterior pituitary where they induce the release of hormones from specific cells (light color).

some neurons located in the hypothalamus secrete small peptides into the circulation that travel to the anterior pituitary gland and stimulate the release of other hormones (Figure 18-42). Among the molecules in this category are the peptides *thyrotropin-releasing hormone (TRH)* and *gonadotropin-releasing hormone (GnRH)*, which promote the release of thyrotropin, FSH, and LH, respectively, from the anterior pituitary (see Figure 16-3), and *somatostatin*, which inhibits the release of growth hormone from the anterior pituitary. Other neurons located in the hypothalamus have axons that pass directly into the posterior pituitary, where they secrete the hormones *vasopressin* and *oxytocin* into the bloodstream. Neurosecretion is accomplished by the same basic mechanism as neurotransmitter release. In both cases the arrival of a wave of depolarization at the nerve ending triggers exocytosis of small vesicles. The only difference is that in the case of typical neurons, the product released from the vesicles is a neurotransmitter destined to interact with an adjacent neuron, while in the case of neurosecretory neurons the product released is a peptide hormone that enters the circulation to enable it to interact with more distant cells.

The third type of physiological response triggered by nerve impulses is **muscle contraction.** In addition to initiating contraction, nerve impulses also regulate muscle contraction. For example, cardiac muscle and certain smooth muscles exhibit spontaneous rhythmic contractions that do not require nerve impulses for initiation. The rate of such contractions, however, is under neural control. In the case of the heart, sympathetic nerves releasing norepinephrine increase membrane excitability and therefore speed the rate of spontaneous contraction, while parasympathetic nerves releasing acetylcholine decrease membrane excitability and therefore slow the rate. The next chapter examines the mechanism by which nerve impulses influence muscle contraction. We shall see that as with neural communication, the crucial event is an alteration in the permeability of the muscle cell membrane to specific ions.

SUMMARY

The neuron has a unique morphology unlike that of any other cell type. The cell body contains organelles including the nucleus, ribosomes, Golgi membranes, lysosomes, and mitochondria, and is thus the principal site of biosynthetic activity in the cell. Extending from the cell body are long cytoplasmic projections that are classified as either dendrites or axons. Dendrites are shorter than axons, exhibit an extensively branched pattern, and lack a myelin sheath. Axons are less numerous than dendrites, extend from the cell body for longer distances, and are often closely associated with supporting glial cells (called Schwann cells) that are responsible for forming the myelin sheath. The myelin covering of myelinated axons is interrupted periodically to reveal the neuronal plasma membrane; these sites, called nodes of Ranvier, help to speed propagation of the nerve impulse.

Material synthesized in the cell body must be transported long distances to the tips of dendrites or axons. Two types of transport have been identified in axons: slow axonal transport, which moves substances toward the axon terminus (orthograde direction) at a rate of 1–5 mm per day, and fast axonal transport, which moves material in both directions at much faster rates, often reaching 500 mm per day. The mechanism underlying these modes of transport is not well understood, though microtubules are known to be involved.

The transmission of information through the nervous system depends on functional connections called synapses. At chemical synapses, which are the most commonly encountered type, chemical transmitters are released from the presynaptic ending of one axon, diffuse across the synaptic cleft, and bind to specific receptors located in the plasma membrane of the postsynaptic cell. Electrical synapses, in contrast, are based on the presence of gap junctions that permit electrical coupling of adjacent cells.

The introduction of tissue culture techniques to the study of nervous tissue led to important information regarding the growth and development of nerve cells. One of the first observations to be made was that the growing tips of neurons exhibit an expanded area, termed the growth cone, which is responsible for recognizing the appropriate path the axon must take in establishing connections with other cells. The growth of neurons is also controlled by specific trophic factors such as nerve growth factor (NGF). This protein has been found to control the growth and development of two specific types of nerve cell: the sensory neuron and the sympathetic neuron. Cultured cells treated with nerve growth factor respond by extending hundreds of neurites, giving the cells a fuzzy appearance. Although the mechanism of NGF action is not clearly understood, changes in various biosynthetic activities have been detected in NGF-treated cells, and the movement of this trophic factor from the nerve ending toward the cell body has been recorded.

Formation of the correct pattern of connections between nerve cells is a critical prerequisite for proper functioning of the nervous system. Experimental analysis of the amphibian visual system has suggested that nerve cells are programmed to recognize the specific cells with which they are supposed to make synapses, and that this specificity lies in the types of molecules present in the plasma membrane.

The transmission of signals between cells of the nervous system is mediated by electrical signals that arise from changes in plasma membrane permeability to Na^+ and K^+. The net difference in charged ions distributed across the membrane generates an electrical potential, with the inside of the cell negative relative to the outside. This inherent potential difference is called the resting membrane potential. Initial studies using the squid axon demonstrated that stimulating an axon triggers a transient change in the polarization of the plasma membrane from a negative to a positive potential. This depolarization is followed by hyperpolarization of the membrane to a value more negative than the resting membrane potential, and finally a return of the potential to its resting value. Such a sequence, called an action potential, is caused by transient changes in the permeability of the plasma membrane to sodium and potassium ions. The initial depolarization is caused by the flow of Na^+ into the cell, making the inside of the cell positively charged. This is followed by the exit of K^+ out of the cell, leading to hyperpolarization. The return to a high Na^+ concentration outside the cell and high K^+ concentration inside the cell restores the resting membrane potential. The action potential has a constant amplitude, and is induced only if the stimulus reaches a minimum level termed the threshold potential. Once initiated, the action potential travels in both directions along the axon membrane. If the axon is myelinated, then the action potential jumps from node to node so that the rate of information transfer is considerably enhanced in myelinated fibers.

Reception of stimuli from the external environment is mediated by specialized sensory cells that can detect changes in temperature, pressure, chemicals, and light, and transduce this information into changes in membrane potential. Photoreceptors in the visual system are of two types: rod cells that function in dim light and cone cells that detect colors in bright light. Both types of cells have a unique morphology, characterized by a specialized set of membrane disks containing the light-absorbing molecule rhodopsin. The transformation of light energy into changes in membrane permeability in the rod cell is achieved by blocking Na^+ entry into the cell, leading to a hyperpolarization of the plasma membrane. Although calcium ions may play a secondary role, cyclic GMP appears to be the controlling molecule. A light-regulated cascade of events involving rhodopsin, a GTP-binding protein, and the enzyme cyclic GMP-phosphodiesterase act in concert to lower the levels of cyclic GMP, thereby blocking the sodium ion channels in the plasma membrane.

The action potential is transmitted across the chemical synapse by specialized neurotransmitter molecules that are specific to certain types of neurons. Among the neurotransmitters so far identified are the amines acetylcholine, norepinephrine, dopamine, and serotonin; the amino acids gamma-aminobutyric acid (GABA), glycine, glutamate, and aspartate; the neuropeptides substance P, enkephalins, and endorphins; and the nucleotide ATP. Much of the evidence describing neurotransmitter release from the presynaptic ending comes from studies of acetylcholine. The discovery of miniature postsynaptic potentials first suggested that this neurotransmitter is released in discrete units. Such observations led to the proposal that acetylcholine is released into the synaptic cleft by fusion of acetylcholine-containing vesicles with the presynaptic membrane. However, evidence also exists for the release of acetylcholine into the synaptic cleft directly from a free cytoplasmic pool.

The postsynaptic membrane contains receptors that recognize and bind the neurotransmitters diffusing across the synaptic cleft. The nicotinic (fast) acetylcholine receptor of the electric fish has been extensively studied. This receptor forms a channel in the membrane, and binding of acetylcholine to it leads to an alteration in membrane permeability. Other types of neurotransmitter receptors trigger changes in enzyme activities in the target cell, leading to slower and more long-lasting effects. In the case of the dopamine receptor, for example, alterations in adenylate cyclase activity lead to changes in cyclic AMP levels within the cell. Opiates such as heroin and morphine also bind to specific receptors, a phenomenon that led to the discovery of endogenous opiates such as the enkephalins and endorphins.

Once a neurotransmitter has bound to its receptor and altered the postsynaptic membrane, its action must be terminated. Some neurotransmitters are inactivated by enzymatic hydrolysis within the synaptic cleft, while others are simply taken back up into the presynaptic ending.

The information processed by the nervous system is ultimately used to trigger appropriate physiological responses. These include stimulation of glandular secretion, release of molecules from nerve cells into the bloodstream, and muscle contraction.

SUGGESTED READINGS

Books

Bousfield, D., ed. (1985). *Neurotransmitters in Action*, Elsevier Science Publishing, New York.

Kandel, E. R., and J. H. Schwartz (1985). *Principles of Neural Science*, Elsevier/North Holland, New York.

Kuffler, S. W., A. R. Martin, and J. G. Nicholls (1984). *From Neuron to Brain*, Sinauer Associates, Sunderland, MA, and Blackwell Scientific, Oxford.

Molecular Neurobiology (1983). Proceedings of a symposium published as *Cold Spring Harbor Symp. Quant. Biol.*, Vol. 48, Cold Spring Harbor Laboratory, Cold Spring Harbor, NY.

Purves, D., and J. W. Lichtman (1985). *Principles of Neural Development*, Sinauer Associates, Sunderland, MA.

Shepherd, G. M. (1983). *Neurobiology*, Oxford University Press, New York.

Articles

Bunge, M. B. (1986). The axonal cytoskeleton: its role in generating and maintaining cell form, *Trends Neurosci.* 9:477–482.

Catterall, W. A. (1984). The molecular basis of neuronal excitability, *Science* 223:653–661.

Dunant, Y., and M. Israël (1985). The release of acetylcholine, *Sci. Amer.* 252 (Apr.):58–66.

Goodman, C. S., and M. J. Bastiani (1984). How embryonic nerve cells recognize one another, *Sci. Amer.* 251 (Dec.):58–66.

Kaupp, U. B., and K.-W. Koch (1986). Mechanism of photoreception in vertebrate vision, *Trends Biochem. Sci.* 11:43–47.

Levi-Montalcini, R., and P. Calissano (1986). Nerve growth factor as a paradigm for other polypeptide growth factors, *Trends Neurosci.* 9:473–477.

Miller, R. (1984). How do opiates act? *Trends Neurosci.* 7:184–185.

Reichardt, L. F. (1984). The emergence of molecular neurobiology, *Trends Biochem. Sci.* 9:173–176.

Schnapf, J. L., and D. A. Baylor (1987). How photoreceptor cells respond to light, *Sci. Amer.* 256(Apr.):40–47.

Schnapp, B. J., and T. J. Reese (1986). New developments in understanding rapid axonal transport, *Trends Neurosci.* 9:155–162.

Snyder, S. H. (1984). Drug and neurotransmitter receptors in the brain, *Science* 224:22–31.

CHAPTER 19

The Organization and Function of Vertebrate Muscle Cells

When we discussed the general phenomenon of cell motility in Chapter 9, emphasis was placed on the widespread occurrence of microtubule- and microfilament-related activities such as chromosome migration, constriction of the cleavage furrow, determination of cell shape, cytoplasmic streaming, amoeboid locomotion, modulation of cell surface topography, and ciliary and flagellar motion. Activities of this type involve the movement of only a small portion of a cell or at most a single cell, and so the forces involved are relatively small. But multicellular organisms must also generate forces of great magnitude, not just for the obvious task of moving the organism as a whole, but for propelling blood through the circulatory system, air into and out of the respiratory system, and food through the digestive tract. It is these processes that depend on the function of muscle cells. Whereas in nonmuscle cells the motion-producing microfilaments and/or microtubules occupy a relatively small portion of the cytoplasm, in muscle cells the contractile filaments are the predominant component.

The immediate source of energy driving muscle contraction is ATP. In vertebrates, nearly one-third of the ATP generated by oxidative phosphorylation is devoted to muscle contraction, and this value may reach 90 percent during vigorous physical activity. Hence the major goal underlying the study of muscle cells is to understand how this chemical energy stored in the form of ATP is converted to the mechanical work of contraction. Most of our knowledge concerning this issue has come from the study of vertebrate skeletal muscle, where cell specialization for the purpose of generating movement has reached its greatest degree. It is perhaps surprising that more is known about the highly sophisticated contractile system of skeletal muscle than the simpler motile systems of nonmuscle cells, but this is so because the large forces produced by skeletal muscle require a massive system of contractile filaments whose orderly arrangement facilitates their microscopic and biochemical dissection. As will be seen in this chapter, this ability to correlate microscopic and biochemical information has been largely responsible for our current understanding of how muscle cells contract.

STRUCTURE AND DEVELOPMENT OF SKELETAL MUSCLE CELLS

Structure of Skeletal Muscle Cells

Those muscles attached to the skeleton that are responsible for voluntary movements of the organism are referred to as **skeletal muscles.** A skeletal muscle is composed of long, cylindrical cells, or **muscle fibers,** that are 10 to 100 μm in diameter and as much as several millimeters or centimeters in length. This enormous length means that a single muscle cell may extend over a considerable portion of, and in some cases the entire length of, a given muscle. The muscle fibers are linked together at their ends by collagenous connective tissue that forms the tendons and ligaments connecting muscle tissue to bone or to another muscle. Contraction of individual muscle cells causes the muscle as a whole to shorten, thereby exerting a pulling force on the bones or other structures to which the muscle is attached. If shortening of the muscle is physically prevented during contraction, tension develops instead.

Each muscle fiber is enclosed within a single continuous plasma membrane, or **sarcolemma.** Multiple flattened ovoid nuclei occur at numerous points along the length of the cell just beneath the sarcolemma. Because of the extensive size of a typical muscle fiber, a hundred or more nuclei may be present in a single cell. This unusual multinucleate condition occurs because of the multiple cell fusions that take place during the formation of muscle cells, a topic to be considered in detail shortly.

The cytoplasm of skeletal muscle cells is dominated by three organelles, each with a specialized role in the contraction process. These organelles, the **myofibrils, sarcoplasmic reticulum–transverse tubule system,** and **mitochondria,** are discussed in detail in the following sections.

Myofibrils It has been known since the early nineteenth century that skeletal muscle exhibits a series of alternating light and dark bands when viewed under the light microscope (Figure 19-1). Though there was some initial controversy as to whether these striations are integral to muscle cells, or simply reflect a ringlike surface morphology, William Bowman suggested correctly in 1840 that the striations are due to bands of material of different refractive indices. As light microscopic techniques improved, this striated appearance was soon seen to be caused by the presence of cross-banded structures 1 or 2 μm in diameter, referred to as **myofibrils.** All the myofibrils in a muscle cell are aligned with their cross-bands in register, thereby imparting to the cell its striated appearance.

Under the polarizing light microscope the striations appear as alternating dark and light bands. The dark bands are birefringent or anisotropic, and are therefore termed **anisotropic** or **A-bands.** The light bands are not birefringent, and so are designated **isotropic** or **I-bands.** In addition to the A- and I-bands, light microscopy reveals several other repeating lines within the striation pattern. In the central region of the dark A-band one can see a lighter region, the **H-zone,** which is easiest to ob-

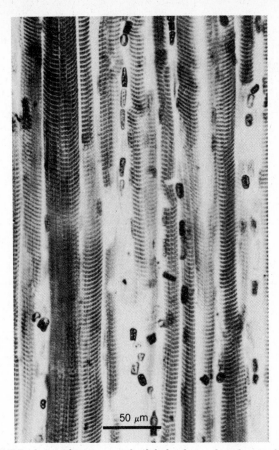

FIGURE 19-1 Light micrograph of skeletal muscle in longitudinal section showing alternating light and dark bands. Courtesy of M. H. Ross.

serve in stretched muscle. The H-zone is bisected by a thin dense band, the **M-line.** The **Z-lines** divide the myofibril into a series of repeating units, each about 2 μm long, called **sarcomeres.**

The underlying substructure responsible for this pattern of repeating striations was revealed in the early 1950s by the pioneering electron microscopic studies of Hugh Huxley and Jean Hanson. Their micrographs showed that myofibrils are composed of two types of longitudinal filaments, **thin filaments** 6 nm in diameter and 1 μm long, and **thick filaments** 15 nm in diameter and 1.6 μm long, arranged in overlapping arrays (Figures 19-2 and 19-3). The I-band contains only thin filaments, accounting for its relatively light appearance, while the A-band contains a mixture of closely interdigitated thick and thin filaments, giving this region its characteristic dense appearance. Though the thick filaments extend throughout the entire length of the A-band, the thin filaments are excluded from the H-zone, making this region somewhat lighter in appearance. Cross sections through those regions of the A-band where the two types of filament interdigitate indicate that the thick and thin filaments are arranged in a hexagonal array, with each thick filament surrounded by six thin filaments. Such an arrangement stipulates that

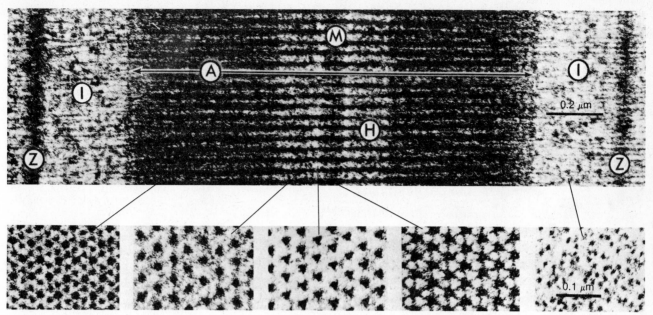

FIGURE 19-2 Structure of the myofibril. *(Top)* Longitudinal section showing the pattern of striations. The various bands in the sarcomere are shown. *(Bottom)* Cross sections through a different region of the sarcomere showing relative positions of thin and thick filaments. Courtesy of F. A. Pepe.

twice as many thin filaments as thick ones are present in each muscle fiber.

In high-resolution electron micrographs **cross-bridges** joining the thick and thin filaments can be seen

FIGURE 19-3 Diagram summarizing the relationship between the striation pattern of skeletal muscle myofibrils and the underlying arrangement of thick and thin filaments.

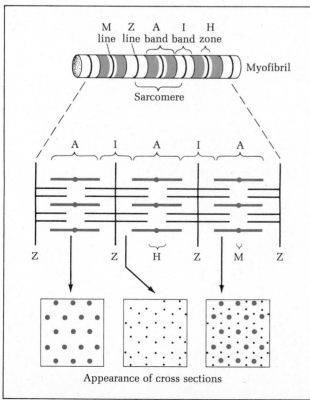

to occur at roughly 43-nm intervals along their length (Figure 19-4). These cross-bridges protrude from the surface of isolated thick filaments, but not thin filaments, suggesting that they are part of the thick filament structure. X-ray diffraction studies of the thick filament suggest that the protruding cross-bridges are arranged in a helical pattern (Figure 19-5). These cross-bridges are the only direct link connecting the thin and thick filaments in the intact myofibril. Because of this the cross-bridges are thought to play a crucial role in muscle contraction. In addition to the cross-bridges between thin and thick filaments, larger bridges connect the thick filaments to one another in the region of the M-line; these bridges are in fact most likely responsible for the dense appearance of the M-line.

Sarcoplasmic Reticulum – Transverse Tubules Skeletal muscle cells contain an elaborate membranous network composed of two distinct membrane systems that come in intimate contact with each other, but whose inner compartments are not interconnected. The more extensive component, the **sarcoplasmic reticulum,** is analogous to the endoplasmic reticulum of other cells and is arranged as a longitudinal system of flattened membranous channels passing over the surfaces of the myofibrils (Figure 19-6). Near each Z-line these channels form enlargements that are known as **terminal cisternae.**

The other component of the cytoplasmic membrane system is a series of thin **transverse tubules** that pass perpendicularly across the myofibrils at the level of the Z-lines. In most vertebrate muscle the tubules of this

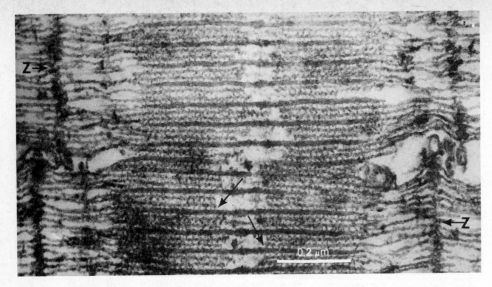

FIGURE 19-4 Electron micrograph showing cross-bridges (arrows) connecting the thick and thin filaments in a sarcomere. The Z-lines are labeled. Courtesy of H. E. Huxley.

FIGURE 19-5 A current model of the organization of the cross-bridges protruding from the thick filament.

FIGURE 19-6 (Below) Diagram illustrating the organization of the sarcoplasmic reticulum and transverse tubules in vertebrate skeletal muscle.

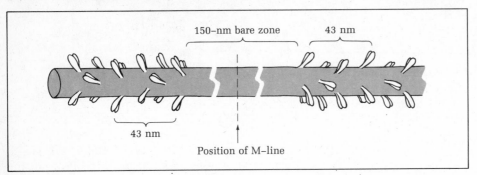

150–nm bare zone

43 nm

43 nm

Position of M–line

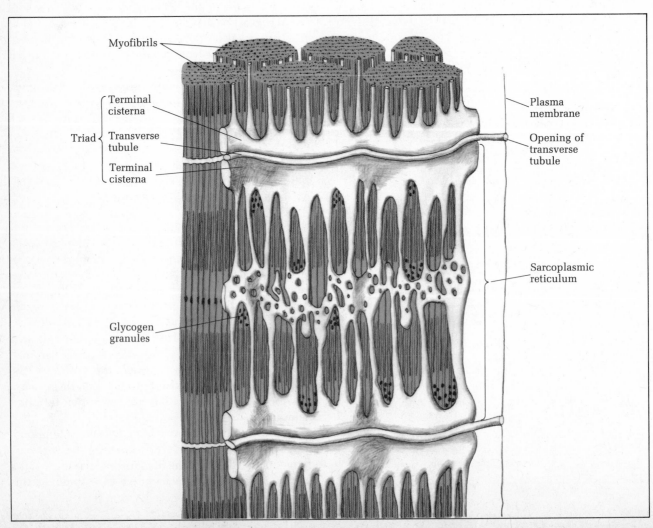

Myofibrils

Terminal cisterna

Triad

Transverse tubule

Terminal cisterna

Glycogen granules

Plasma membrane

Opening of transverse tubule

Sarcoplasmic reticulum

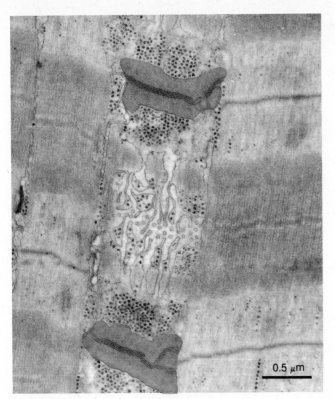

FIGURE 19-7 Thin-section electron micrograph in which the triads are highlighted (color). Courtesy of L. D. Peachey.

FIGURE 19-8 Electron micrograph showing the localization of ferritin (black) within transverse tubules. See text for experimental details. Courtesy of H. E. Huxley.

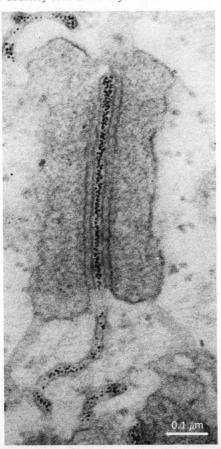

T-system pass between the two terminal cisternae of adjacent sarcomeres, creating an easily recognizable configuration known as a **triad** (Figures 19-6 and 19-7). In some invertebrates the transverse tubules make close contact with only one terminal cisterna in any given region, forming a **dyad.**

Electron micrographs occasionally reveal points of continuity between the plasma membrane and transverse tubules, suggesting that the internal compartment of the T-system is continuous with the outside of the cell. Hugh Huxley has tested this idea by incubating frog muscle in a solution of ferritin molecules that are easily visualized with the electron microscope due to their large size and electron density. Under such conditions ferritin molecules rapidly accumulate within the lumen of the transverse tubules, but do not appear within the sarcoplasmic reticulum or elsewhere in the muscle (Figure 19-8). Since ferritin is too large to pass through the plasma membrane, the simplest explanation is that the transverse tubules, but not the sarcoplasmic reticulum, open directly to the outside of the cell and hence permit the ferritin molecules to diffuse inside. As will be seen later, this structural continuity between the plasma membrane and the transverse tubules is functionally important in transmitting the signal for muscle contraction from the plasma membrane to the cell interior.

Mitochondria Because a replenishable supply of ATP is required to power muscle contraction, it is not surprising to find numerous mitochondria in muscle cells. In highly active muscle, mitochondria are usually arranged in regular patterns adjacent to the myofibrils, where they may occupy as much as 50 percent of the total cytoplasmic volume (Figure 19-9). The flight muscles of insects, for example, exhibit repeating patterns in which one, two, or three mitochondria are lined up adjacent to each sarcomere. Because such muscles are designed for long periods of continuous use, they also contain large amounts of **myoglobin,** an oxygen-binding protein designed to concentrate oxygen from the bloodstream and store it in the muscle cytoplasm for use during periods of intense mitochondrial activity. The high content of myoglobin and mitochondrial cytochromes imparts a reddish color to muscle cells of this type, so they are referred to as **red muscle.** The major source of fuel in red muscle is stored fat, whose fatty acids are oxidized by the mitochondria to produce ATP.

Muscles that are not called upon for prolonged use generally contain fewer, irregularly spaced mitochondria. Because of their reduced content of myoglobin and cytochromes, muscles of this type exhibit little color and so are termed **white muscle.** The major stored fuel in white muscle is glycogen, which is broken down to glucose for oxidation by the glycolytic pathway. It is because of the relative inefficiency of glycolysis com-

FIGURE 19-9 Electron micrograph showing mitochondria (arrows) lined up in a regular pattern adjacent to the sarcomeres of frog skeletal muscle. Courtesy of C. Franzini-Armstrong.

pared to mitochondrial respiration that white muscles are not capable of prolonged use. Common examples of red and white muscle are the "dark" and "white" meat of chickens and turkeys. The red muscle of the legs is designed for continual use in walking and standing, while the white muscle of the breast is used only intermittently.

Developmental Origin of Skeletal Muscle

The formation of skeletal muscle cells during embryonic development generates both the multinucleated condition and the massive accumulation of contractile myofibrils characteristic of mature muscle cells. During muscle formation, or **myogenesis,** muscle cells arise from a group of undifferentiated, spindle-shaped mesenchymal cells called **presumptive myoblasts.** After a period of rapid division some of these cells cease dividing, elongate, and begin synthesizing and accumulating actin and myosin. The resulting immature muscle cells, called **myoblasts,** are gradually transformed into elongated **myotubes** containing hundreds of nuclei lined up in a common cytoplasm (Figure 19-10). In very young embryos striated myofibrils are formed at the myoblast stage, while in older embryos actin and myosin synthesis and the associated formation of striated myofibrils occur predominantly in the myotube. Shortly after the development of myotubes, regions of the plasma membrane invaginate and come to lie next to the Z-lines of the developing myofibrils, producing the transverse tubule system. The longitudinally arranged sarcoplasmic

reticulum, on the other hand, evolves from the endoplasmic reticulum. As these intracellular membranes and the myofibrils both continue to enlarge, the myotube is gradually converted into a typical muscle cell.

The path followed by muscle cells in reaching the multinucleate state was a source of controversy for many years. During this time two major alternatives were considered: either the nuclei of individual myoblasts undergo successive divisions unaccompanied by cell division (the **nuclear-division model**) or myoblasts fuse with one another to create a large, multinucleated cell (the **cell-fusion model**). The first definitive support for the cell-fusion hypothesis was provided in the late 1950s by Howard Holtzer and his colleagues, whose microspectrophotometric measurements revealed that myotube nuclei do not contain more than the diploid amount of DNA; hence these nuclei cannot be preparing to divide. These investigators also reported that ^{3}H-thymidine is not incorporated into the nuclei of myotubes, mitotic figures are not detectable in myotubes, and inhibitors of DNA synthesis or mitosis do not block myotube formation. Such evidence supporting the cell-fusion model was soon reinforced by time-lapse films that revealed myoblasts in the process of fusing with one another. Electron microscopic studies have extended these observations by showing plasma membranes of two cells in the act of fusing (Figure 19-11).

Although the preceding observations support the cell-fusion model, they are to some extent suspect because the results were obtained with cells in culture, and myotube formation under such conditions is

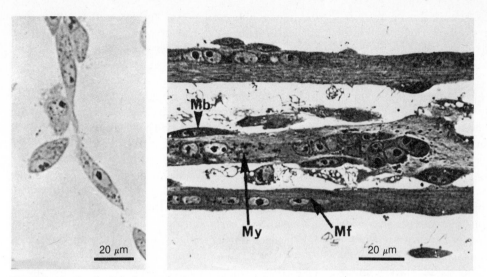

FIGURE 19-10 The development of myotubes in culture. *(Left)* Light micrograph of individual myoblasts after 1 day in culture. *(Right)* After 4 days in culture the myoblasts (Mb) are beginning to form myotubes (My) and striations are visible in some of the myofibrils (Mf). Courtesy of D. A. Fischman.

FIGURE 19-11 Electron micrographs showing the fusion of myoblasts. *(Left)* A dividing myoblast (MB) lies next to a myotube (MT) so that the plasma membranes are in parallel (arrows). *(Middle)* The beginning of fusion is indicated by areas of cytoplasmic continuity (arrows). *(Right)* After fusion the cytoplasmic contents initially remain unmixed, creating a mosaic effect. Courtesy of Y. Shimada.

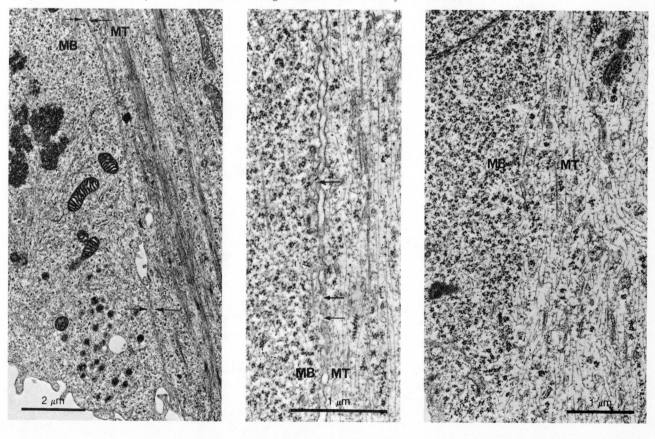

known to differ in several ways from the analogous process in intact organisms. For example, myoblasts are converted into branching myotubes having a thousand or more nuclei after as little as 24 hours in culture. In contrast myotubes formed *in vivo* never branch, and several weeks pass during the development of a myotube containing only a few hundred nuclei.

Because of this discrepancy, it is critical to determine whether cell fusion accounts for myotube formation in intact animals or is simply a peculiarity of cells in culture. To address this issue Beatrice Mintz and Wilber Baker examined the properties of skeletal muscle cells in **allophenic** mice. These animals are composed of a mixture of cells of two different genotypes produced by mixing together cleavage-stage cells from two genetically distinct embryos, and implanting the composite embryo into a mouse uterus so that development can be completed. This approach is ideal for testing the cell-fusion model of myogenesis. As shown in Figure 19-12, if the two initial cell types contain electrophoretically

FIGURE 19-12 Comparison of the predictions made by the nuclear-division model versus the cell-fusion model of myogenesis. If allophenic mice are created from a mixture of two cell types coding for different electrophoretic forms of a multisubunit enzyme, only the fusion model predicts the formation of hybrid enzyme molecules with intermediate electrophoretic mobility. The fact that such hybrid enzymes are found in skeletal muscle provides strong support for the occurrence of cell fusion *in vivo*. See text for details.

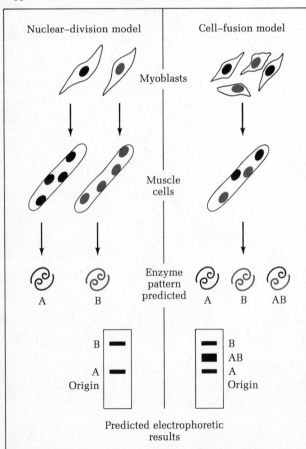

distinguishable forms of a multisubunit enzyme, then the cell-fusion model of myogenesis predicts the appearance of enzymes containing a mixture of both types of subunits in skeletal muscle. Because of this mixture of subunits, such enzyme molecules would be expected to have an intermediate electrophoretic mobility. Utilizing this approach, Mintz and Baker created allophenic mice containing two cell types with differing electrophoretic forms of the multisubunit enzyme, **isocitrate dehydrogenase.** In such mice it was found that skeletal muscle alone contained a new type of isocitrate dehydrogenase whose electrophoretic mobility was intermediate between that of the two original forms. This observation clearly showed that cell fusion must have taken place during skeletal muscle development.

Though the molecular mechanism responsible for triggering myoblast fusion during normal muscle development is not completely understood, it is known to be a highly selective process. If radioactively labeled kidney, cartilage, liver, or fibroblast cells are added to a culture of unlabeled myoblasts in the process of fusing, not a single labeled cell is incorporated into the developing myotubes. Presumably some specific macromolecule present on the myoblast surface is responsible for the specific cell-cell recognition and association that occurs during myotube formation.

BIOCHEMISTRY OF CONTRACTILE PROTEINS

One-half to three-quarters of the total protein contained in a typical vertebrate skeletal muscle cell is present in the myofibrils. Because myofibrillar proteins are relatively insoluble at low ionic strength, the classical approach to their purification has involved an initial extraction of muscle with water to remove the soluble proteins of the cell sap, followed by exposure of the insoluble residue to high ionic strength solutions (e.g., 0.6M KCl) to solubilize the components of the myofibril. For many years the only major protein clearly identified in such extracts was **myosin,** a substance that was thought as early as the middle of the nineteenth century to be the major contractile protein of muscle. But in the early 1940s, the Hungarian biochemist Albert Szent-Györgyi and his colleagues made the intriguing discovery that the properties of isolated "myosin" depend on the conditions employed for its initial extraction. If muscle is extracted for a relatively brief period, the extracted myosin is of low viscosity and is unaffected by the addition of ATP. But when a day-long extraction procedure is used, the myosin obtained is very viscous and this viscosity decreases upon the addition of ATP.

Such observations led to the conclusion that extracting muscle for longer periods of time removes a second protein that interacts with myosin and thereby changes its properties. This second protein, named

actin, was subsequently purified and shown to form complexes with myosin, explaining the higher viscosity of myosin solutions contaminated with actin. The actin-myosin (or actomyosin) complexes formed under these conditions dissociate in the presence of ATP, explaining the viscosity-lowering effect of ATP. The critical importance of the preceding observations for models regarding muscle contraction lay in the realization that not only do actin and myosin bind to each other, but this binding is influenced by ATP, the molecule that furnishes the energy for muscle contraction.

Though myosin and actin are the predominant components of the myofibril, several other proteins are also present. **Tropomyosin** and **troponin** account for about 20 percent of the myofibrillar protein. In addition small quantities of **α- and β-actinin, M-protein, C-protein,** and perhaps other minor proteins yet to be clearly characterized are found in the myofibril. In the following sections the major biochemical features and ultrastructural localization of these various myofibrillar proteins will be discussed.

Myosin: Major Constituent of the Thick Filament

Myosin, a large protein of about 500,000 daltons, accounts for roughly half the protein present in myofibrils. In the early 1930s it was shown that solutions of myosin flowing through a narrow capillary are birefringent, indicating alignment of the molecules in the direction of flow. Since birefringence is also a property exhibited by the A-band, this discovery suggested that myosin might be localized within the A-band. This conclusion was later reinforced by microscopic studies in which extraction of myosin from muscle cells was shown to result in the loss of the thick filaments from

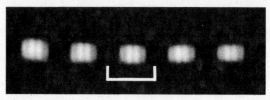

FIGURE 19-13 Localization of muscle myosin in the sarcomere using fluorescent anti-myosin antibodies. The myosin is localized in the A-bands. The bracket indicates the position of two Z-lines. The distance between them is 2.9 μm. Courtesy of F. A. Pepe.

the A-band. The subsequent utilization of fluorescent anti-myosin antibodies to stain myosin within intact myofibrils confirmed the localization of myosin in the A-band (Figure 19-13).

Myosin is composed of six polypeptide subunits: two identical **heavy chains** of about 200,000 daltons each, and four **light chains** of about 20,000 daltons. In electron micrographs the myosin molecule appears as a thin rod, about 140 nm in length, with a pair of **globular heads** protruding at one end (Figure 19-14). The long rodlike portion of the molecule consists of the two intertwined heavy chains, predominantly in the α-helical configuration. Each of the globular heads consists of the terminal end of one of the heavy chains complexed with two light chains. Investigation of the functional properties of the various regions of the myosin molecule has been facilitated by the discovery that the proteolytic enzyme trypsin cleaves the molecule into two specific fragments, **light meromyosin** and **heavy meromyosin** (Figure 19-15). The two globular heads attached to the heavy meromyosin fragment can be removed by treatment with another proteolytic enzyme, **papain.**

Three properties exhibited by myosin are especially important for understanding its biological function; these include the ability to hydrolyze ATP, bind actin, and polymerize into filaments.

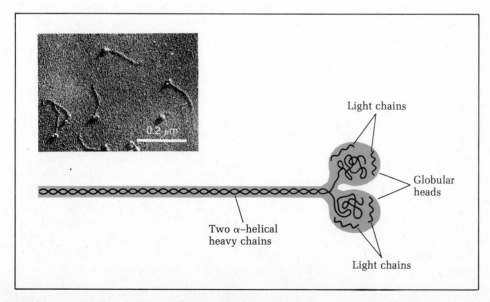

FIGURE 19-14 Electron micrograph and schematic diagram of a single molecule of myosin showing the globular heads and extended tail. Courtesy of A. Elliott.

0.2 μm

Light chains

Globular heads

Two α–helical heavy chains

Light chains

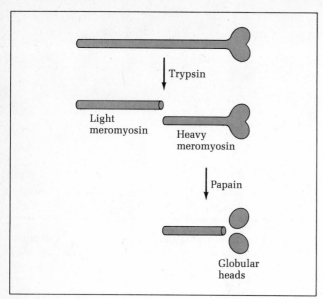

FIGURE 19-15 Fragmentation pattern of myosin resulting from successive treatment by the proteolytic enzymes trypsin and papain.

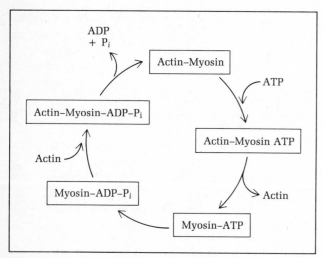

FIGURE 19-16 Diagram of the cyclic association and dissociation of actin and myosin driven by ATP hydrolysis.

1. *ATP hydrolysis.* It was first discovered in 1939 by the Russian biochemists Vladimir Englehardt and Militsa Ljubimowa that myosin hydrolyzes ATP to ADP and inorganic phosphate in the presence of Ca^{2+}. Because myosin had long been recognized as the major structural protein of muscle, this discovery provided the first link between structural studies and the mechanism of contraction. More recent studies on myosin subfragments have revealed that the ATPase activity is localized in the heavy meromyosin fragment; more specifically there are two ATPase sites, one in each of the two globular heads.

2. *Actin binding.* The strong affinity between myosin and actin has already been mentioned. Myosin has two sites for binding actin, one on each of its globular heads. Binding of actin to these sites enhances the ATPase activity of myosin by increasing the rate at which the products of hydrolysis, ADP and P_i, are released from myosin. After the ADP and P_i have been discharged, the myosin head is free to bind another molecule of ATP. This in turn causes the liberation of actin from myosin, for actin has a low affinity for myosin molecules containing bound ATP. After the bound ATP has been hydrolyzed to ADP and P_i, actin is again capable of binding to myosin because actin has a high affinity for the myosin-ADP-P_i complex. The bound actin promotes the release of ADP and P_i, and the cycle repeats (Figure 19-16). We shall see later that this cyclic association and dissociation of actin and myosin, driven by ATP hydrolysis, underlies the mechanism of muscle contraction.

3. *Polymerization into filaments.* When the ionic strength of a solution of myosin is lowered, the individual myosin molecules polymerize into filaments visible under the electron microscope. The shortest filaments are about 300 nm long, which is roughly twice the length of a single myosin molecule. Clusters of globular projections protrude from both ends of such filaments, leaving the middle of the shaft bare. As illustrated in Figure 19-17, the simplest interpretation of this picture is that the short filaments represent bundles of several dozen myosin molecules oriented in opposite directions, with their long tails joined or overlapping in the central bare region and their globular heads forming the projections at the two opposite ends. The longer filaments produced during the cell-free polymerization of myosin may approach 1500 nm in length, and resemble the thick filaments isolated directly from muscle. These extended filaments exhibit the same central bare region of 150 nm, but the regions covered by the globular projections are longer. Such an arrangement suggests that myosin molecules are added to both ends of the growing filament, always orienting in the same direction as the myosin molecules already incorporated at each end. In other words the myosin thick filament is a bipolar structure assembled from myosin molecules that extend in opposite directions from the central bare region.

Other Thick Filament Proteins

At least three types of proteins other than myosin have been identified as minor constituents of the thick filament. One, designated the **C-protein,** is localized within the middle third of each half of the A-band, where it forms nine regular bands across the thick filaments (Figure 19-18). The function of the C-protein is unknown, but its high affinity for myosin and the regularity of its distribution suggest a role in holding the thick filaments together.

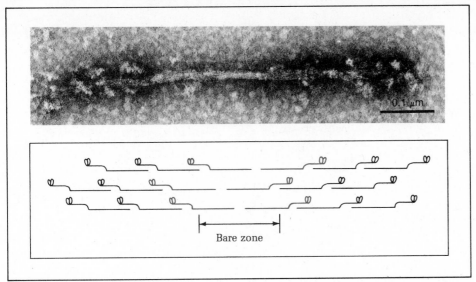

FIGURE 19-17 *(Top)* Negatively stained preparation of thick filaments produced by the cell-free polymerization of myosin. The projections at either end are the globular myosin heads, which form the cross-bridges of the intact myofibril. Courtesy of H. E. Huxley. *(Bottom)* Schematic representation of the arrangement of individual myosin molecules in the thick filament. The smallest filaments assembled *in vitro*, about 300 nm long, consist of two sets of myosin molecules oriented in opposite directions (colored region). Longer filaments are constructed by the addition of myosin molecules to both ends, with each newly added molecule always oriented in the same direction as the molecules already present at the growing end.

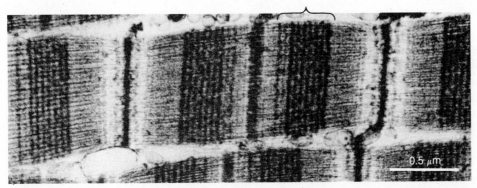

FIGURE 19-18 Electron micrograph showing localization of the C-protein using antibody against this protein. The bracket defines one of several groups of nine bands of C-protein within the A-band. Courtesy of F. A. Pepe.

A second class of minor proteins present in the thick filaments are those localized to the M-line by fluorescent antibody techniques. One polypeptide of 165,000 daltons is firmly attached to this region and so has been claimed to be the true **M-protein** present in the myosin-myosin cross-bridges responsible for the M-line. Another M-protein, more loosely bound to the myofibril, is a dimer of 88,000 molecular weight recently identified as **creatine kinase,** an enzyme whose role in muscle contraction will be described later. A third polypeptide of about 90,000 daltons is also present as a minor constituent in preparations of M-band proteins. This protein has recently been identified as **phosphorylase b,** an enzyme that plays an important role in catalyzing the breakdown of glycogen.

Actin: Major Constituent of the Thin Filament

About 30 percent of the mass of the myofibril is accounted for by α-actin, a slightly different form of actin than the β- and γ-actins in nonmuscle cells. The realization that α-actin is localized within the thin filaments first emerged from the discovery that extraction of this protein from the myofibril causes the thin filaments to disappear. Staining of thin filaments with antibodies directed against actin subsequently confirmed this localization.

Under conditions of low ionic strength, purified actin exists as a single polypeptide chain of 42,000 molecular weight. This monomer, designated **G-actin** because of its globular shape, contains binding sites for one

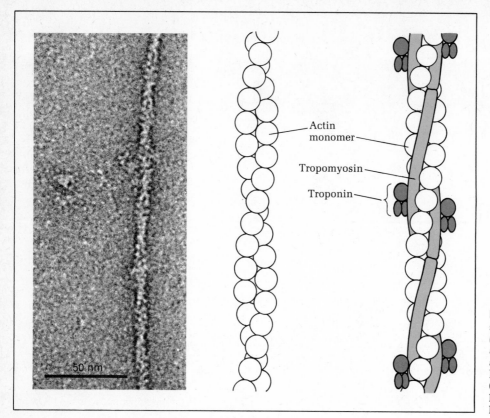

FIGURE 19-19 Molecular substructure of the thin filament. *(Left)* Electron micrograph of a negatively stained preparation of F-actin, showing two intertwined helical strands of actin monomers. Courtesy of H. E. Huxley. *(Middle)* Model of the F-actin helix. *(Right)* Model of the thin filament.

calcium ion and one adenine nucleotide (ATP or ADP). Raising the ionic strength to the physiological range causes G-actin to polymerize into 6-nm diameter filaments of **F-actin** (or filamentous actin) that closely resemble the thin filaments of intact muscle. High-resolution electron micrographs of F-actin filaments reveal two intertwined helical strands of globular actin monomers (Figure 19-19).

Because the formation of F-actin from G-actin is promoted by ATP, the free energy of ATP hydrolysis was at one time believed to drive this polymerization reaction. But nonhydrolyzable analogs of ATP have been shown to stimulate this reaction as effectively as ATP, ruling out any absolute requirement for ATP hydrolysis. The role played by ATP in enhancing actin polymerization therefore remains unclear.

When heavy meromyosin or the myosin head fragment is added to preparations of polymerized actin, the actin filaments exhibit a characteristic *arrowhead pattern* when examined by electron microscopy (Figure

19-20). The arrowheads represent myosin fragments protruding at an angle from the actin filament. That they always point in the same direction over the entire length of a given filament shows the **polarity** inherent in actin filament construction. When thin filaments still joined to the Z-line are examined in this same way, the arrowheads are consistently found to point away from the Z-line, indicating that the thin filaments located on opposite sides of the same Z-line manifest opposite polarity.

Other Thin Filament Proteins

Besides actin, the thin filament contains two major proteins involved in regulating contraction, **tropomyosin** and **troponin,** each of which makes up approximately 10 percent of the total myofibrillar protein. Tropomyosin is a dimer of 70,000 molecular weight comprised of α- and β-subunits of approximately equal size. The two subunits, predominantly in the α-helical configuration,

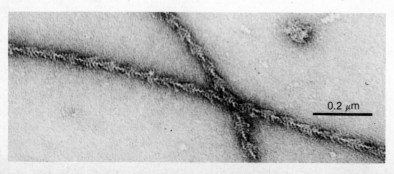

FIGURE 19-20 Electron micrograph of actin filaments decorated with heavy meromyosin. Courtesy of H. E. Huxley.

are intertwined to form a thin, rod-shaped molecule 40 nm long and 2 to 3 nm in diameter. Optical diffraction analyses comparing negatively stained preparations of actin and actin-tropomyosin complexes suggest that the tropomyosin molecules lie extended in the two grooves of the actin filaments (see Figure 19-19). Each tropomyosin molecule is long enough to cover a distance of seven actin monomers, and there is enough tropomyosin present in the thin filaments to completely cover both grooves.

The other major protein of the thin filament, troponin, is a large globular molecule consisting of three polypeptide subunits. One subunit binds selectively to tropomyosin and is therefore called **troponin-T** or **TN-T**. A second subunit, which attaches to actin and *inhibits* its binding to myosin, is designated **troponin-I** or **TN-I**. Finally the remaining subunit, which has two binding sites for Ca^{2+}, is designated **troponin-C** or **TN-C**. Ferritin-labeled antitroponin antibodies have been shown to bind to discrete sites situated every 40 nm along the thin filament, which happens to be the same length as that covered by a single tropomyosin molecule. This observation supports the thin filament model illustrated in Figure 19-19, in which every group of seven actin monomers is associated with one molecule of troponin and one molecule of tropomyosin.

Besides the major proteins actin, tropomyosin, and troponin, small quantities of **α- and β-actinin** are associated with the thin filament. α-Actinin is localized predominantly in the Z-line and is thought to function in cross-linking the thin filaments to one another.

β-Actinin, in contrast, is located at the free ends of the thin filaments where it may function in terminating actin polymerization. Finally, the intermediate filament protein **desmin** is localized to the periphery of the Z-disk and functions to orient and stabilize all the Z-disks so that they remain in register.

MECHANISM OF CONTRACTION IN SKELETAL MUSCLE

The Sliding Filament Hypothesis

The most important breakthrough in our current understanding of the molecular mechanism underlying muscle contraction occurred in the early 1950s when Andrew Huxley and Rolf Niedergerke, and Hugh Huxley and Jean Hanson, independently analyzed and interpreted the banding patterns exhibited by skeletal muscle during various states of contraction. Their measurements revealed that even though individual sarcomeres shorten by as much as 50 percent during muscular contraction, there is no discernible change in the length of the A-bands. The entire decrease in sarcomere length is accounted for by shortening of the I-bands, which disappear entirely in fully contracted muscle. The simplest interpretation of this phenomenon is that during contraction, the thin filaments of the I-band are progressively pulled into the region of the A-band, thereby decreasing the overall length of the I-band (Figure 19-21). The net result is a shortening of

FIGURE 19-21 The sliding filament model of contraction. Diagram at left illustrates the changes in filament overlap that occur as the sarcomere shortens. Electron micrographs at right illustrate the shortening of the I-band produced as a result of this filament sliding. Courtesy of H. E. Huxley.

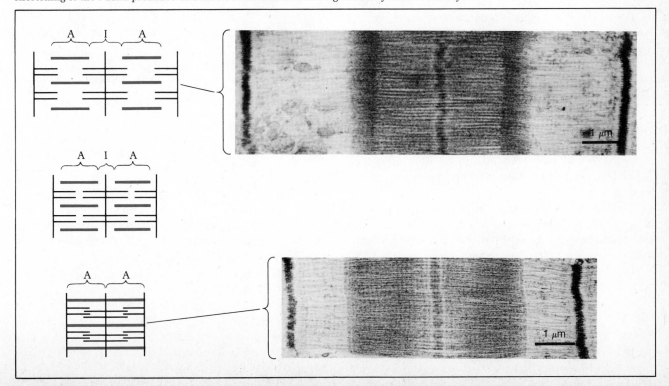

the sarcomeres, and in turn the length of the muscle as a whole.

This **sliding filament hypothesis,** independently formulated by the Huxley–Niedergerke and Huxley–Hanson groups, was first reported in 1954. Substantial support was soon provided by electron microscopic studies revealing that contraction is accompanied not by a change in the length of either the thin or thick filaments, but by an increase in their degree of overlap. Conversely, muscle relaxation is accompanied by a decrease in overlap between the two types of filaments.

If contraction is simply the result of the thick and thin filaments sliding over one another, what causes this sliding to take place? One logical candidate for the driving force is the cross-bridges that extend between the thick and thin filaments. Assuming that each cross-bridge contributes to the contractile force, the total force generated by a given myofibril will be proportional to the total number of cross-bridges. This prediction was rigorously tested in the mid-1960s in the laboratory of Andrew Huxley, where muscles stretched to differing sarcomere lengths were stimulated to contract and the resulting tension measured. The rationale underlying this approach was that the number of cross-bridges that can be formed decreases as the sarcomere is stretched because the degree of overlap between thick and thin filaments decreases. As expected, when muscle fibers were stretched to the point where the thin and thick filaments no longer overlap, and therefore cannot form cross-bridges (a sarcomere length of about 3.7 μm), the fiber did not contract when stimulated. But with the sarcomere length slightly shorter (about 3.5 μm), which is sufficient to permit a small amount of overlap between thin and thick filaments, the muscle fiber is capable of producing contractile tension upon stimulation. As the sarcomere length is progressively set at shorter values, allowing more overlap and hence more cross-bridges between thick and thin filaments, the tension produced upon stimulation gradually increases (Figure 19-22). After reaching an optimum configuration in which all cross-bridges are in use (sarcomere length of about 2.2 μm), further shortening of the sarcomere no longer enhances the tension produced. In fact, at shorter lengths the tension begins to decrease, presumably because the thin filaments extend into the other side of the A-band where they interfere with cross-bridges already formed.

A critical feature of cross-bridge behavior is that it must exhibit polarity. That is, the cross-bridges on one side of the sarcomere must pull all their attached thin filaments in one direction to move them toward the center of the sarcomere, while the cross-bridges in the other half of the sarcomere must pull all their attached filaments in the opposite direction to move them toward the sarcomere center. The physical basis for this polarity resides in the construction of the thin and thick filaments. As discussed earlier, the thick filaments are

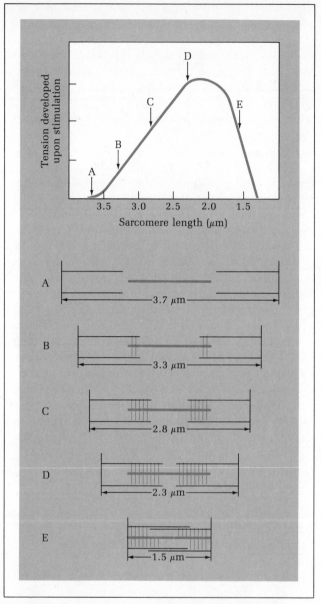

FIGURE 19-22 Tension developed by isolated frog muscle maintained at different sarcomere lengths. (A) When the thick and thin filaments do not overlap, no tension can be produced upon stimulation. (B, C, D) As the degree of overlap, and hence the number of cross-bridges, increases, the tension produced upon stimulation increases. (E) As the sarcomere shortens beyond its optimum configuration, tension decreases because of disruption of cross-bridges by overlapping thin filaments, and collision of the thick filaments with the Z-line. See text for details.

composed of myosin molecules whose orientation is opposite in the two halves of the filament, while thin filaments exhibit a polarity (detectable by the arrowhead pattern of heavy meromyosin binding) that reverses at the Z-line. Hence the polarity of both thick and thin filaments is opposite in the two sides of the sarcomere, explaining why cross-bridge activity moves the two sets of thin filaments in opposing directions (Figure 19-23).

The mechanism by which cross-bridges induce sliding is yet to be elucidated in detail, but several features of the process are generally agreed upon. To begin

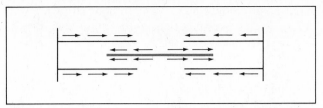

FIGURE 19-23 Schematic representation of the polarity of thin and thick filaments. The reversal in polarity at the middle of the sarcomere is responsible for the two sets of thin filaments moving in opposite directions (toward each other) during contraction.

FIGURE 19-24 A model illustrating how conformational changes in cross-bridge structure induced by ATP hydrolysis might propel thin filament sliding during muscle contraction. See text for details.

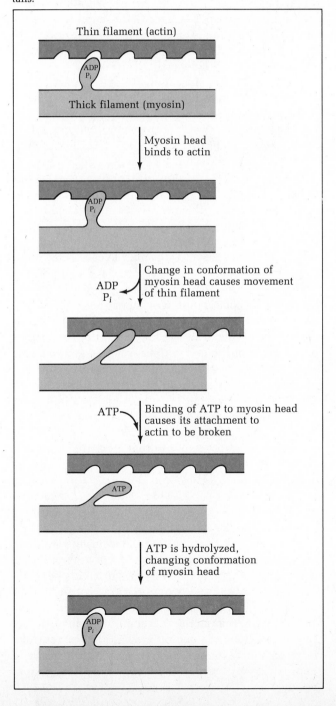

Thin filament (actin)

ADP
Pi

Thick filament (myosin)

Myosin head binds to actin

ADP
Pi

Change in conformation of myosin head causes movement of thin filament

ADP
Pi

Binding of ATP to myosin head causes its attachment to actin to be broken

ATP

ATP

ATP is hydrolyzed, changing conformation of myosin head

ADP
Pi

with, the cross-bridges appear to correspond to the globular heads of the myosin molecules, and so contain both the ATPase and actin-binding sites. In the presence of ATP these actin-binding sites of myosin undergo a repeated cycle of attachment-detachment to actin (see Figure 19-16), which is the same as saying that ATP induces the successive formation and breakage of cross-bridges between the thick and thin filaments. A clue as to how this repeated formation and breakage of bridges between myosin and actin might cause the thin filament to slide relative to the thick filament has come from X-ray diffraction and electron microscopic studies of contracting muscle. These studies indicate that the configuration of myosin cross-bridges changes during contraction. In the relaxed state the bridges are at approximately right angles to the thick filament, while in the contracted state they protrude at an angle of 45°. This change in the spatial orientation of the myosin heads during contraction allows one to postulate that the breakage and reformation of cross-bridges occurs in a progressive fashion, each new bridge assuming a conformation that permits it to bind to an actin site located further along the thin filament than the site from which its attachment was just broken. A successive series of such events, multiplied over thousands of cross-bridges, would cause the thin filaments to slide and the myofibril to shorten.

Figure 19-24 summarizes the preceding model and points out how the energy released during ATP hydrolysis triggers the conformational changes required for the progressive formation of cross-bridges. In the first step of this model the myosin head, retracted from the actin filament and containing bound ADP and P_i generated by a previous ATP hydrolysis, attaches to a specific site on an adjacent actin filament. The second step, designated the **power stroke,** involves the release of ADP and P_i and an accompaning change in the conformation of the myosin head to its nonenergized state. It is this change in the configuration of the cross-bridge that causes the attached actin filament to slide relative to the thick filament. In step three ATP binds to the myosin head, thereby disrupting its attachment to actin. Finally ATP hydrolysis occurs in the fourth step, converting the myosin head back to its energized state ready to initiate another cycle of attachment and sliding. Rapid repetition of this cycle results in a progressive sliding of the thin filament across the thick filament.

Excitation-Contraction Coupling

Though the above model provides a picture of the molecular events underlying the contraction event itself, it fails to shed any light on how muscle contraction is regulated. In fact, considered alone this model makes it appear as if muscles will continue contracting as long as their ATP supplies hold out. Yet this clearly cannot be the case, for exquisite control of muscle activity is required during complex coordinated movements, and such control is exerted in the face of relatively constant intramuscular levels of ATP. In skeletal muscle the presence or absence of incoming nerve impulses is responsible for determining whether or not contraction occurs. The step-by-step process of **excitation-contraction coupling,** through which an incoming nerve impulse ultimately causes the myofibrils to contract, is described below.

Depolarization of Plasma Membrane and Transverse Tubules The first step in excitation-contraction coupling is the arrival of a nerve impulse at the plasma membrane (sarcolemma) of the muscle cell. The termini of the nerve cell include many fine branches that lie in shallow depressions on the muscle cell surface, maintaining a gap of some 50 nm between the nerve and muscle plasma membranes. At these **neuromuscular**

FIGURE 19-25 Electron micrograph of the neuromuscular junction. The axon ending contains many synaptic vesicles (arrows) and mitochondria. The muscle cell membrane is thrown into deep junctional folds (arrowheads). Courtesy of E. G. Gray.

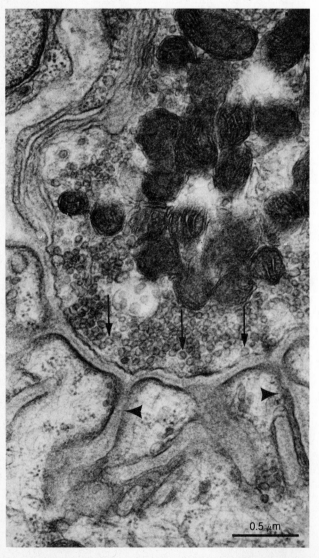

0.5 μm

junctions the muscle cell membrane is thrown into a series of deep invaginations termed **junctional folds** (Figure 19-25). As discussed in Chapter 18, the release of acetylcholine from the presynaptic nerve ending serves as the signal connecting the nerve and muscle cells. The acetylcholine diffuses across the synaptic cleft and ultimately binds to acetylcholine receptors present in the junctional folds of the sarcolemma. Binding of acetylcholine to these receptors causes an increase in the permeability of the sarcolemma to sodium and potassium ions. As in the case of the nerve cells, the muscle cell exhibits a resting membrane potential of -60 to -90 mV. The localized increase in membrane permeability permits an influx of Na^+ and an efflux of K^+, leading to a brief depolarization of the membrane. If this local depolarization exceeds a prescribed threshold value, it triggers an action potential that is propagated across the entire sarcolemma.

For many years muscle physiologists puzzled over the question of how a change in the electrical properties of the surface membrane of a muscle cell can be communicated to the interior of the cell rapidly enough to cause every myofibril to contract virtually simultaneously. Simple diffusion of a chemical signal from the cell surface would be too slow to account for the rapidity and simultaneity of contraction. A clue to the mechanism responsible for this process was provided in 1958 by Andrew Huxley and Robert Taylor, who used a micropipette to apply a small electric current to selected regions on the surface of a frog muscle cell. Though the tiny current induced a local depolarization of the plasma membrane wherever it was applied, the magnitude of the depolarization was below the threshold required for triggering an action potential across the entire muscle cell surface. Therefore, when applied to most regions of the cell surface, such a small current failed to trigger contraction. However, Huxley and Taylor discovered the existence of certain sensitive spots, always located over an I-band, at which stimulation caused the single adjacent I-band to contract (Figure 19-26).

Such observations suggest that the activating effect of depolarization is conducted from the plasma membrane to the interior of the cell along structures located at the level of the I-bands. This location corresponds to the region where the transverse tubules project from the plasma membrane into the cell interior (see Figure 19-6). Support for the idea that the transverse tubules are responsible for the inward spread of the activation signal came when Huxley carried out similar experiments in lizard muscle, where the transverse tubules are located close to the junctions between the A- and I-bands, rather than in the center of the I-bands. In this case contraction occurred only when the tiny current was applied near the A-I junction, and the I-band shortened selectively on its stimulated side. Hence once

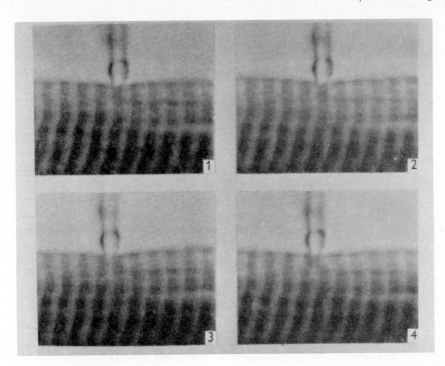

FIGURE 19-26 Light micrographs showing that muscle contraction occurs only when current is applied opposite an I band. Current enters the muscle through a pipet placed above an A band (1 and 2) or an I band (3 and 4). Micrographs 1 and 3 show the muscle before current is applied; micrographs 2 and 4 were taken during the current pulse. Note that a contraction is produced only if the pipet is opposite an I band.

again the site where contraction could be induced corresponded to the location of the transverse tubules. Though the mechanism by which transverse tubules transmit the activation signal was not clear at the time of Huxley's experiments, subsequent electron micrographs revealed regions of continuity between the plasma membrane and the transverse tubule membranes; this physical continuity provides a direct pathway for conducting membrane depolarization from the cell surface to the cell interior.

Calcium Ions and the Sarcoplasmic Reticulum Next we must consider how the wave of depolarization passing from the plasma membrane to the transverse tubular system leads to myofibrillar contraction. The idea that calcium ions initiate the contraction event first received serious consideration in the 1940s when L. V. Heilbrunn and F. J. Wiercinski injected a large number of substances into muscle cells and found Ca^{2+} to be the only biological agent effective in triggering contraction. Further support for the notion of calcium ion involvement appeared a few years later when newly developed cell fractionation techniques led to the isolation of a subcellular fraction from muscle tissue that causes contracted muscle to relax, and whose relaxing effect can be reversed by Ca^{2+}. This **relaxing factor** exhibits a high affinity for binding calcium ions and was therefore thought to promote relaxation by reducing levels of free Ca^{2+}.

When electron microscopic examination of the relaxing factor led to its identification as sarcoplasmic reticulum fragments, a simple theory of muscle contraction could be put together: Contraction is trig-

gered by calcium ions released from the sarcoplasmic reticulum upon electrical stimulation, and relaxation is produced by the subsequent uptake of these calcium ions by the sarcoplasmic reticulum. Two experimental approaches have helped to buttress this interpretation. First of all, direct evidence that changes in free Ca^{2+} accompany normal contraction emerged from studies in which muscle cells were exposed to substances whose color or luminescence varies in direct relationship to changes in free Ca^{2+} concentration. Monitoring the color or luminescence of such indicators in contracting muscle has revealed that free Ca^{2+} levels rise soon after a muscle cell is stimulated, and fall back to normal levels before maximum contractile tension is achieved (Figure 19-27).

The other approach that has been useful in filling in the details of how Ca^{2+} levels are regulated during muscle contraction involves studying the transport properties of the sarcoplasmic reticulum membrane. Homogenization of muscle tissue causes the sarcoplasmic reticulum to become fragmented into small vesicles that can be isolated by differential centrifugation. In the presence of ATP such vesicles actively concentrate calcium ions from the suspending medium, accompanied by the hydrolysis of ATP. The **calcium pump,** or **Ca^{2+}-dependent ATPase** responsible for this transport, is so efficient it can lower the Ca^{2+} concentration of the external medium to less than $10^{-7}M$. Normally, the sarcoplasmic reticulum is relatively impermeable to Ca^{2+} and so the calcium ions taken up by the transport system are effectively trapped. However, depolarization of the transverse tubules induces a transient increase in the permeability of the adjacent sarcoplasmic reticu-

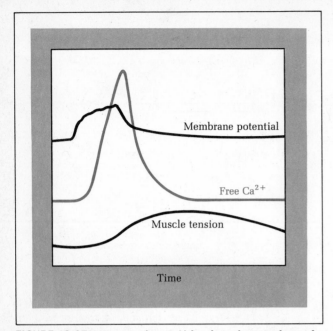

FIGURE 19-27 Change in free Ca^{2+} levels in the cytoplasm of a skeletal muscle cell subsequent to depolarization of the plasma membrane. The rapid fall in free Ca^{2+} is due to the binding of Ca^{2+} to troponin, which is responsible for the activation of cross-bridge formation and filament sliding.

lum, permitting the Ca^{2+} to diffuse out into the surrounding cytoplasm. It is not entirely clear how this depolarization of the transverse tubules triggers the change in permeability of the sarcoplasmic reticulum. Until recently it was believed that the initial action potential triggers the entry of extracellular calcium ions into the cytoplasm, creating an ionic current that changes the permeability properties of the sarcoplasmic reticulum. Evidence against this hypothesis comes from studies showing that muscle contraction can be induced even in the absence of extracellular calcium. A recent discovery that inositol trisphosphate (page 632) controls the release of Ca^{2+} from the lumen of the endoplasmic reticulum in nonmuscle cells has led to an alternative possibility for the mechanism linking depolarization of the transverse tubules to calcium release from the sarcoplasmic reticulum. In experiments using rabbit skeletal muscle, inositol trisphosphate has been found to release calcium ions from the sarcoplasmic reticulum, although the contractile response under such conditions is small, slow, and extremely variable. It remains to be seen whether this molecule will be found to be the missing link connecting excitation to contraction in skeletal muscle cells.

The Ca^{2+}-Receptive Mechanism Our current understanding of how the release of Ca^{2+} from the sarcoplasmic reticulum causes the myofibrils to contract is largely due to studies Setsuro Ebashi and his colleagues initiated in the early 1960s. The original impetus for this work was the observation that mixtures of

actin and myosin prepared in different ways differ in their sensitivities to Ca^{2+}. With "natural" preparations of actomyosin extracted directly from muscle, removal of Ca^{2+} by treatment with a chelating agent causes the actin and myosin to dissociate from each other, while addition of Ca^{2+} enhances actin-myosin binding. In contrast, mixtures of purified actin and myosin are usually unaffected by calcium ions.

Ebashi correctly interpreted this unresponsiveness as an indication that purified actin and myosin lack some component(s) responsible for conferring Ca^{2+} sensitivity upon the actin-myosin interaction. By thoroughly extracting muscle tissue from which actin and myosin had first been removed, he was able to isolate a protein fraction that is capable of restoring Ca^{2+} sensitivity to purified actin and myosin. The general properties of this material closely resembled those of tropomyosin, which had already been identified as a component of muscle some 20 years earlier; however, purified tropomyosin alone was not found to be effective in imparting Ca^{2+} sensitivity to the actin-myosin association. Ebashi discovered that this apparent discrepancy occurred because his tropomyosinlike material was actually a mixture of tropomyosin and a previously unknown protein, which he called troponin. Neither tropomyosin nor troponin alone can confer Ca^{2+} sensitivity on the actin-myosin interaction; only a combination of the two proteins is effective.

From subsequent studies on the tropomyosin-troponin system, a picture of the events following Ca^{2+} release from the sarcoplasmic reticulum has emerged. To begin with, tropomyosin is normally bound to actin in such a way as to block the site at which myosin forms cross-bridges, thereby maintaining a relaxed state. When an increase in the permeability of the sarcoplasmic reticulum causes the Ca^{2+} levels in the area surrounding the myofibrils to rise above $10^{-6}M$, the calcium ions bind to troponin. This binding induces a change in the conformation of troponin that in turn causes tropomyosin to be displaced from its normal position blocking the myosin binding sites of the actin filament (Figure 19-28). The myosin heads can now bind to actin, forming cross-bridges and setting in motion the contraction cycle depicted in Figure 19-24. When stimulation of the muscle ceases and the sarcoplasmic reticulum regains its normal impermeability, the transport of Ca^{2+} back into the cisternae of the sarcoplasmic reticulum lowers the concentration of free Ca^{2+} in the cytoplasm to resting levels of $10^{-7}M$. Under these conditions calcium ions dissociate from their binding sites on troponin, leading to a conformational change back to the state in which tropomyosin blocks the myosin binding sites on actin. Not only does this prevent new cross-bridges from forming, but because old bridges dissociate in the presence of ATP (see Figure 19-24), all the cross-bridges soon disappear and the muscle relaxes.

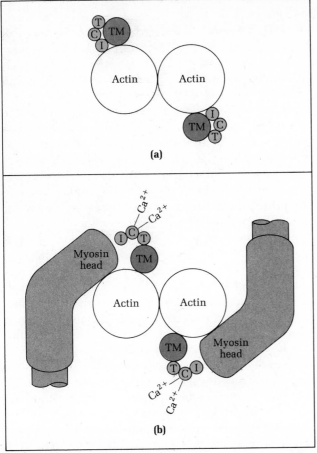

FIGURE 19-28 Model illustrating the role of calcium ions in regulating muscle contraction, shown in cross-sectional view. (a) At low Ca²⁺ concentrations tropomyosin blocks the myosin binding site on actin. (b) At higher concentrations Ca²⁺ binds to troponin, changing the conformation of the troponin-tropomyosin complex so as to free the myosin binding site. Abbreviations: TM = tropomyosin, T = troponin-T, C = troponin-C, I = troponin-I.

Energy Sources for Contraction

Energy provided by ATP hydrolysis plays at least two crucial roles in the model of muscle contraction described above. First, ATP hydrolysis is required for the formation and breakage of cross-bridges underlying the contractile mechanism. And second, transport of Ca²⁺ back into the cisternae of the sarcoplasmic reticulum is dependent on the hydrolysis of ATP by the membrane-bound calcium pump. A mechanism for quickly replenishing ATP supplies is therefore a critical necessity in muscle cells.

It has been known since the early 1930s that the ATP concentration in actively contracting muscle remains virtually constant until exhaustion sets in. Moreover, inhibitors of glycolysis or respiration block neither muscle contraction nor maintenance of ATP levels, at least over the short term, pointing to the existence of some energy source in muscle other than glycolysis, respiration, and ATP. In the early 1930s it was discovered that instead of ATP, the substance whose concentration

does fall in contracting muscle is **phosphocreatine** (Figure 19-29). Though this finding led some investigators to erroneously conclude that phosphocreatine rather than ATP is the immediate energy source driving contraction, phosphocreatine was soon found to donate its phosphate group to ADP in a reaction catalyzed by the enzyme **creatine kinase:**

$$^-OOC-CH_2-\underset{\underset{NH}{\overset{\|}{C}}}{\overset{CH_3}{\underset{|}{N}}}-\overset{H}{\underset{|}{N}}-PO_3^- + ADP \rightleftharpoons$$

Phosphocreatine

$$^-OOC-CH_2-\underset{\underset{NH}{\overset{\|}{C}}}{\overset{CH_3}{\underset{|}{N}}}-NH_2 + ATP$$

Creatine

By replenishing ATP at the expense of phosphocreatine, which is present in high concentration in muscle tissue, the above reaction provided a potential explanation for the relative constancy of ATP levels in contracting muscle.

Though this idea was first proposed in the early 1930s, definitive evidence was not provided until 1962 when D. F. Cain and R. E. Davies reported the effects of a specific inhibitor of creatine kinase on contracting muscle. With the creatine kinase reaction blocked contraction still takes place, but instead of observing constant ATP levels and decreasing phosphocreatine levels, as usually occurs, a decline in ATP levels was observed with little change in the amount of phosphocreatine. This is exactly the predicted result if the ATP were the

FIGURE 19-29 Changes in the concentration of phosphocreatine and ATP in actively contracting skeletal muscle.

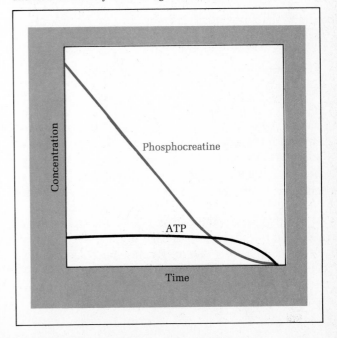

direct energy source for contraction, and phosphocreatine the means for replenishing ATP.

If phosphocreatine and ATP both become depleted, energy can still be salvaged from the ADP accumulating in muscle as a result of ATP hydrolysis. The enzyme **adenylate kinase** (also called **myokinase**) catalyzes the phosphorylation of one ADP molecule at the expense of another:

$$2ADP \rightleftharpoons ATP + AMP$$

This reaction only becomes significant when the ADP concentration rises to very high levels, meaning that all other ATP-generating mechanisms have failed.

Ultimately, of course, the energy used to maintain the phosphocreatine and ATP pools is derived from glycolysis and mitochondrial respiration. The relative contributions of these two pathways varies, depending both on the muscle type and the prevailing conditions. As mentioned earlier red muscle, containing large numbers of mitochondria, generates ATP primarily by mitochondrial oxidation of stored fatty acids. During prolonged contraction, however, the rate of oxygen consumption resulting from mitochondrial respiration may exceed the rate at which oxygen can be delivered from the bloodstream. After depleting the intracellular stores of oxygen bound to myoglobin, such muscles become anaerobic and are forced to shift to anaerobic glycolysis for ATP production, using stored glycogen and blood glucose as sources of fuel. Because of the relative inefficiency of anaerobic glycolysis, glycogen stores are rapidly depleted and there is a massive buildup of lactate, both of which place limits on the length of time contraction can be sustained under these conditions. In white muscle the paucity of mitochondria and the low myoglobin content necessitates the routine utilization of anaerobic glycolysis in the first place, explaining why, unlike red muscle, white muscle is not capable of prolonged activity.

Abnormalities in Skeletal Muscle Function

Several disorders in muscle function have captured the interest of cell biologists because of the potential insights they might provide concerning the workings of skeletal muscle cells. One such malady is **myasthenia gravis,** a disease characterized by progressive weakness and muscle fatigue. Several discoveries made in the early 1970s localized the defect to a decline in the number of functional acetylcholine receptors in the muscle cell plasma membrane. Not only do direct assays reveal a decrease in such receptors, but antibodies capable of binding to the acetylcholine receptor have been identified in the serum of patients suffering from the disease. Hence for some reason these individuals seem to manufacture antibodies that bind to and inactivate the acetyl-

choline receptors needed by their own muscle cells for initiating contraction.

More difficulty has been encountered in attempting to unravel the underlying mechanism of **muscular dystrophy,** a family of diseases characterized by progressive degeneration of skeletal muscle cells. Much controversy has focused on whether the primary defect is in the muscle cells (*myogenic hypothesis*) or in the nerve cells innervating the muscles (*neurogenic hypothesis*). Tissue culture experiments employing various combinations of nerve and muscle from normal and dystrophic mice have shown that normal muscle cells degenerate when cultured in the presence of nerve cells from dystrophic mice, while muscle cells obtained from dystrophic mice cultured in the presence of normal nerve cells grow normally.

While the preceding results support the neurogenic hypothesis for muscular dystrophy in mice, human muscular dystrophy seems to have a different basis. The most common and debilitating form of the human disease, **Duchenne muscular dystrophy,** is inherited by a gene carried on the X chromosome. Because one of the two X chromosomes is inactivated in each cell in human females, one can distinguish between the neurogenic and myogenic hypotheses by examining muscle appearance and function in women containing one normal and one dystrophic X chromosome. Since either the normal or the dystrophic X chromosome will be active in any given nerve cell, the neurogenic hypothesis predicts that muscle cells innervated by dystrophic nerves should be entirely dystrophic, while muscle cells innervated by normal nerves should be entirely normal. But according to the myogenic hypothesis, all muscle cells should be equally affected, for each muscle cell is multinucleate and therefore contains a mixture of nuclei, some with the normal X chromosome active and some with the dystrophic X chromosome active. The evidence indicates that, consistent with the myogenic hypothesis, all muscle cells in women with one dystrophic X chromosome are equally affected, rather than there being distinct populations of normal and dystrophic cells.

A clue to the nature of the defect underlying muscular dystrophy has emerged from the discovery that alterations in membrane fluidity, ion transport, and membrane protein phosphorylation all occur in the red blood cells of patients suffering from various forms of the disease. Though the relationship of such membrane abnormalities to muscle degeneration is yet to be ascertained, the discovery of membrane alterations in cells other than the muscle cell supports the idea that at least some types of muscular dystrophy are reflections of a more widespread underlying defect in cell membranes that, for some unknown reason, is particularly detrimental to muscle cells.

SMOOTH MUSCLE

Structure of Smooth Muscle Cells

In addition to skeletal muscle there are two other types of vertebrate muscle tissue: smooth muscle and cardiac (heart) muscle. **Smooth muscle** occurs in the walls of internal organs such as the intestinal and genital tracts, glands, and arteries. Contraction of this type of muscle is much slower and more prolonged than in the case of skeletal muscle, as exemplified by the contraction of uterine muscle during delivery of the fetus, contraction of blood vessels in redirecting blood flow, contraction of the digestive organs, and contraction of the iris and ciliary body that regulate the entry of light into the eye. The control of smooth muscle contraction can be elicited by hormones, as in the case of the oxytocin-induced uterine contractions, or by input from the autonomic nervous system. Thus smooth muscle contraction is said to be **involuntary,** or unconscious, in contrast to the voluntary control of skeletal muscle contractions.

Smooth muscle cells can be distinguished from skeletal muscle cells in several ways. The first and most obvious difference is that smooth muscle cells lack the striations characteristic of skeletal muscle (Figure 19-30). In fact, it is this "smooth" appearance that gives this muscle type its name. In addition, smooth muscle cells are spindle shaped, and their size is about 100-fold smaller than that of typical skeletal muscle cells.

FIGURE 19-30 Electron micrograph of smooth muscle cells. *(Top)* Longitudinal section of smooth muscle showing absence of striations. *(Bottom)* Micrograph showing that each cell has a single nucleus (arrow). Courtesy of G. Gabella.

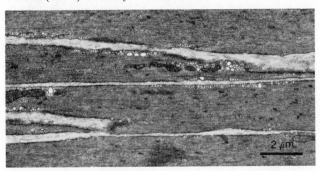

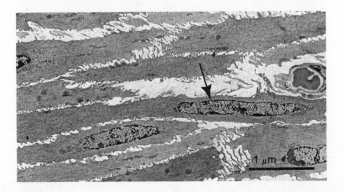

FIGURE 19-31 High-power electron micrograph showing parallel arrays of thin and thick filaments in smooth muscle cells. Courtesy of R. V. Rice.

Smooth muscle cells possess a single nucleus, and the cytoplasm contains both thick and thin filaments, though these are not arranged in a regular, periodic array as in the sarcomere. Finally, the sarcoplasmic reticulum is reduced in extent and T-tubules are completely missing.

The Contractile Machinery

In contrast to the regular arrangement of filaments in skeletal muscle cells, the thin and thick filaments of smooth muscle cells lie in parallel groups that seem to be randomly dispersed in the cytoplasm (Figure 19-31). Analysis of the relative amounts of thin and thick filament proteins has revealed that the ratio of α-actin to myosin is about 15 to 1 in smooth muscle, rather than the 2 to 1 ratio found in the sarcomere of skeletal muscle. The thin filaments are composed of F-actin, but neither troponin nor tropomyosin are present. These actin filaments are linked by α-actinin to electron-dense structures, called **dense bodies,** which are distributed throughout the cytoplasm and beneath the sarcolemma (Figure 19-32). The dense bodies are thought to serve as anchoring sites for the thin filaments, just as the Z-lines function in skeletal muscle.

Although the thin and thick filaments of smooth muscle cells do not appear to be organized into a regular framework, it seems that the contractile mechanism utilized by smooth muscle is similar in principle to that

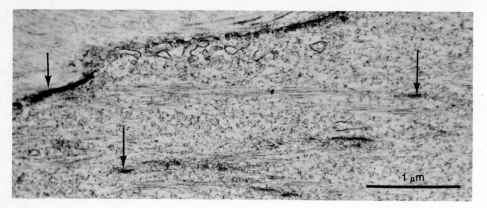

FIGURE 19-32 Electron micrograph showing the dense bodies (arrows) present in the sarcoplasm and at the periphery of smooth muscle cells. To facilitate visualization of these dense bodies, the actin and myosin filaments have been removed prior to electron microscopy. Courtesy of P. Cooke.

FIGURE 19-33 Schematic outline of the principal components that regulate contraction of smooth muscle. See text for details.

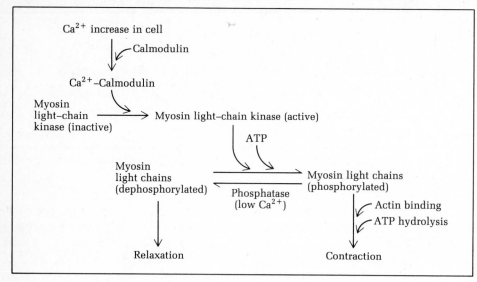

of skeletal muscle cells. In general, smooth muscle contraction is regulated by changes in Ca^{2+} levels. Calcium ions enter the cell as a result of membrane depolarization triggered by an action potential. The calcium may come from the extracellular environment, or it may be derived from the sarcoplasmic reticulum and/or mitochondria. The absence of conducting T-tubules decreases the rate at which the signal can be delivered to the contractile machinery, and so a slower, more prolonged contraction results.

Because smooth muscle lacks troponin and tropomyosin, Ca^{2+} must act by a mechanism different from that in skeletal muscle. This mechanism appears to involve phosphorylation of the myosin molecule. As shown in Figure 19-33, the light chains of myosin are phosphorylated by an enzyme called **myosin light-chain kinase,** which is activated by calcium ions. The effect of Ca^{2+} is mediated by the protein calmodulin, which is inactive unless calcium ions are bound to it. The Ca^{2+}-calmodulin complex activates the kinase that phosphorylates the myosin light chains, permitting the binding of actin to myosin and stimulating the myosin-ATPase activity. The hydrolysis of ATP is then thought to cause movement of actin thin filaments along the

thick filaments in a manner similar to the sarcomere sliding filament mechanism. When intracellular Ca^{2+} levels fall, the kinase is inactivated and a second enzyme, **myosin light-chain phosphatase,** is activated. This enzyme removes phosphate groups from the myosin light chains. The dephosphorylated myosin is unable to bind to actin and the muscle subsequently relaxes. Utilization of calmodulin as a controlling element in the pathway of smooth muscle contraction is similar to the interaction of Ca^{2+} with troponin C in the sarcomere. Calmodulin and troponin C are structurally similar molecules, and it is therefore believed that the latter protein is a specialized molecule that has evolved from the more ubiquitous calmodulin.

In most cases smooth muscle contraction occurs as a result of depolarization of the sarcolemma caused by arrival of action potentials from nerve termini lying on the cell surface. However, in some cells contraction is induced by a specific hormone. For example, the smooth muscle of the uterus contracts upon stimulation by the hormone **oxytocin,** which triggers the entry of Ca^{2+} into the cytoplasm. The calcium ions then trigger myosin phosphorylation just as they do when an action potential is the initiating stimulus.

CARDIAC MUSCLE

Cardiac muscle cells, found in the heart, resemble skeletal muscle in that sarcomeres, T-tubules, and a modified sarcoplasmic reticulum are present. Because heart muscle must contract continually, each cardiac muscle cell contains a large number of mitochondria full of densely packed cristae (Figure 19-34). Many glycogen granules and fat droplets lie nearby to supply the mitochondria with the energy-rich materials needed for ATP synthesis. Instead of the long, multinucleated cells present in skeletal muscle, cardiac muscle is divided by membrane partitions, called **intercalated disks,** into separate cells, each containing a single nucleus (Figure 19-35). These specialized regions of the plasma membrane contain α-**actinin** and **vinculin,** two proteins that are thought to link the plasma membrane to α-actin filaments in the cytoplasm. The intercalated disks contain numerous gap junctions, whose ability to connect cells electrically is important for conducting impulses from cell to cell and thus coordinating the contraction of the heart muscle as a whole. Though the rate of cardiac muscle contraction is regulated by special nerves, cardiac muscle cells are able to contract repetitively in the absence of exogenous stimulation or controls. Because of this unique property, heart muscle contraction is said to be **myogenic.**

The mechanism of cardiac muscle contraction is similar, if not identical, to that occurring in skeletal muscle cells. Cardiac muscle contains actin, myosin, tropomyosin, and troponins I, C, and T. Spontaneous action potentials occurring in specialized nodes of tissue cause an increase in membrane permeability and an influx of calcium ions into the cytoplasm, just as in skel-

FIGURE 19-34 Electron micrograph of cardiac muscle showing the presence of numerous mitochondria (arrows) between the myofibrils. Courtesy of D. J. Scales and T. Yasamura.

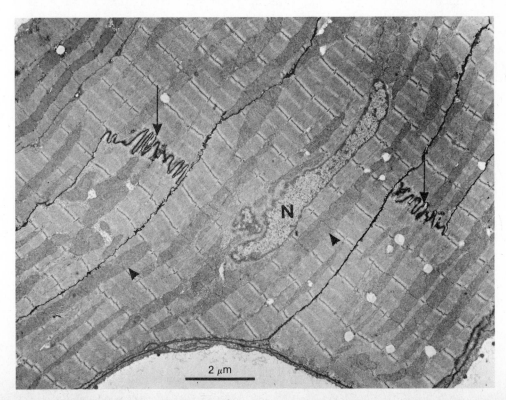

FIGURE 19-35 Electron micrograph of cardiac muscle showing intercalated disks (arrows) between cardiac muscle cells. A single nucleus (N) is centrally located and many mitochondria (arrowheads) are evident. Courtesy of J. R. Sommer.

etal muscle. However, in cardiac muscle Ca^{2+} passes more slowly through the T-tubule system. As a result heart muscle remains depolarized and contracted 20 to 50 times longer than skeletal muscle, a necessary prerequisite to coordinating all the fibers in a given region of the heart. Without this extended period of contraction, the first fibers to receive an impulse would complete their contraction before the last fibers had begun, a situation which would lead to an uncoordinated heart beat that would be ineffective in pumping blood.

Cardiac muscle is thus specialized for a particular type of contraction. The gap junctions permit rapid transit of electrical impulses throughout the heart, the numerous mitochondria provide the needed ATP for continuous contraction, and the prolonged depolarization phase permits the heart to contract as a unit. How this type of tissue is able to spontaneously initiate the contraction process and maintain it in the absence of neural input is not completely understood.

SUMMARY

Most of our understanding of the mechanism that converts chemical energy stored in ATP into the mechanical work of contraction has come from the study of skeletal muscle. This tissue is composed of large cylindrical cells (muscle fibers), each of which measures up to several millimeters or centimeters in length and contains a hundred or more nuclei. The most striking visual feature of skeletal muscle cells is the presence of cross-banded myofibrils lined up in register so as to give the cell a striated appearance. The lightly staining I-bands are composed of thin filaments while the darker A-bands consist predominantly of thick filaments, overlapped to a variable degree by the thin filaments. In the center of the A-band, where the thin filaments have not penetrated, is a light area, termed the H-zone, which is divided in half by a thin partition, the M-line. The I-band is bisected by a thin dense structure, the Z-line, which divides the myofibril into a series of repeating units called sarcomeres. Cross-bridges extend from the thick to the thin filaments, repeating at 43-nm intervals for a given pair of thin and thick filaments.

Also present in the cytoplasm of skeletal muscle cells are two systems of membranes that come in intimate contact with one another, but whose inner compartments are not interconnected. The more extensive of these systems, the sarcoplasmic reticulum, consists of a longitudinal system of flattened membranous channels running over the surfaces of the myofibrils. The other component is comprised of thin transverse tubules that invaginate from the plasma membrane and pass perpendicularly across the myofibrils at the level of the Z-lines.

Red muscles are characterized by the presence of large numbers of regularly arranged mitochondria whose respiratory metabolism provides ATP to the contractile fibrils. Because they are designed for long periods of intense use, such muscle cells contain large quantities of the oxygen-binding protein, myoglobin. White muscles contain few mitochondria and little myoglobin, and are not capable of prolonged activity because of the relative inefficiency of anaerobic glycolysis, their main pathway of ATP formation.

Skeletal muscle cells develop embryonically from spindle-shaped mesenchymal cells, called myoblasts, that fuse with one another to form long myotubes containing hundreds of nuclei lined up in a common cytoplasm. As the actin and myosin synthesized by the myotube are assembled into myofibrils, membranous projections invaginate from the plasma membrane to produce the transverse tubule system, and the endoplasmic reticulum of the myoblasts expands into the longitudinal vesicles of the sarcoplasmic reticulum.

Actin and myosin are the major contractile proteins of the myofibril. Myosin, localized in the thick filaments of the sarcomere, is a large protein that appears under the electron microscope as a long, thin rod containing a pair of globular heads at one end. Three properties exhibited by myosin are especially important for understanding its biological function. First, the globular heads contain catalytic sites for hydrolyzing ATP. Second, the globular heads also contain sites for binding to actin. And third, myosin polymerizes into structures resembling the thick filaments of intact muscle. Clusters of globular projections (myosin heads) protrude from both ends of such filaments, leaving the middle of the shaft bare. The simplest interpretation of this bipolar structure is that it represents a bundle of myosin molecules projecting in opposite directions from the central, bare region.

Actin, the major constituent of the thin filament, can be extracted from muscle as a single polypeptide chain called G-actin. In the presence of ATP and at physiological ionic strength, G-actin polymerizes into filaments (F-actin) composed of two intertwined helical strands of G-actin monomers. If the heavy meromyosin subfragment of myosin is added to actin filaments, a characteristic arrowhead pattern is produced. The arrowheads of heavy meromyosin always point in the same direction on a given filament (away from the Z-line in the myofibril), indicating the existence of polarity in thin filament construction. In addition to actin, the thin filament contains two other major proteins: tropomyosin, lying extended in the two grooves of the actin helix, and troponin, a globular protein situated every 40 nm along the helix.

During muscle contraction, shortening of the sarcomeres is accomplished by sliding of the thin filaments over the thick filaments. The tension produced by contracting muscle is directly related to the number of cross-bridges formed between the thick and thin filaments, supporting the conclusion that the driving force for filament sliding is the formation and breakage of cross-bridges. The most widely accepted model of this progressive formation of cross-bridges involves the following four steps: (1) the energized myosin head, containing bound ADP and P_i, attaches to actin; (2) a change in conformation of the myosin head, accompanied by the release of ADP and P_i, causes the attached actin filament to be moved relative to the thick filament; (3) ATP binds to the myosin head, inducing the separation of myosin from actin; and (4) ATP hydrolysis converts myosin back to its energized state, ready for another cycle of attachment and sliding.

The above cycle is ultimately under the control of an excitation-contraction coupling system in which calcium ions play a crucial role. First, neural (or artificial) stimulation of the muscle cell depolarizes the plasma membrane and the transverse tubule membranes. This depolarization results in an increase in the permeability of the sarcoplasmic reticulum, releasing Ca^{2+} into the cytoplasm bathing the myofibrils. The calcium ions bind to troponin, producing a change in its conformation that leads to a displacement of tropomyosin from its normal position blocking the myosin binding sites of the actin filament. The myosin heads can then form cross-bridges with actin, setting the sliding filament mechanism in motion. Relaxation occurs when the sarcoplasmic reticulum regains its normal impermeability, and is again capable of accumulating Ca^{2+} within its cisternae. Under these conditions Ca^{2+} dissociates from troponin, changing the conformation of the troponin-tropomyosin complex so as to once again block the myosin-binding sites on actin.

Though ATP is the immediate source of chemical energy for both cross-bridge formation and the calcium pump, the ATP levels in actively contracted muscle remain virtually constant until exhaustion sets in. The reason for this constancy is that phosphocreatine serves as a storage reservoir of high-energy phosphate groups, rephosphorylating ADP to ATP as fast as ATP is hydrolyzed.

Considerable effort has been expended in attempting to unravel the defects underlying the muscle diseases myasthenia gravis and muscular dystrophy. In myasthenia gravis afflicted individuals manufacture antibodies against the acetylcholine receptors utilized by their own muscle cells for initiating contraction in response to incoming nerve impulses. In the case of muscular dystrophy the form of the disease occurring in mice is thought to result from a defect in the nerve cells innervating the muscle, while human muscular dystrophy is thought to involve a widespread underlying defect in cell membranes that, for some unknown reason, has an especially deleterious effect on muscle cells.

Two other types of muscle tissue occur in vertebrates. Smooth muscle consists of spindle-shaped cells about 100-fold smaller than those of skeletal muscle. Smooth muscle occurs in the walls of internal organs and its long, slow contraction cycle is controlled either by hormones or by the autonomic nervous system. The thick and thin filaments present contain myosin and actin, respectively, but these filaments are not arranged in regular periodic arrays as in skeletal muscle. In addition the thin filaments lack troponin and tropomyosin, the sarcoplasmic reticulum is reduced, and T-tubules are absent. The actin filaments appear to be linked to α-actinin inserted into dense bodies that serve as attachment sites analogous to the Z-lines of skeletal muscle. Contraction of smooth muscle is initiated by depolarization of the sarcolemma, which increases membrane permeability to Ca^{2+} and thereby allows Ca^{2+} levels inside the cell to rise. Calcium ions bind to calmodulin, and this complex in turn stimulates myosin light-chain kinase to phosphorylate the light chains of the myosin heads. The activated myosin can then form cross-bridges with actin and contraction ensues. The prolonged contraction characteristic of smooth muscle is thought to be at least partly due to the increased length of time needed for Ca^{2+} to diffuse throughout the cell interior in the absence of conducting T-tubules. Dephosphorylation of myosin light chains occurs when a myosin light-chain phosphatase is activated by low Ca^{2+} levels in the cell, and this leads to myosin and actin dissociation and relaxation of the muscle.

Cardiac muscle resembles skeletal muscle in terms of sarcomere structure, but specialized plasma membrane partitions called intercalated disks divide the tissue into individual cells and serve to anchor the actin filaments. In addition, these membrane regions contain gap junctions that connect cells electrically. The mechanism of contraction is similar to that of skeletal muscle cells, except that there is a prolonged depolarization phase to allow the heart muscle to contract as a unit. Large numbers of mitochondria are present, as are extensive supplies of glycogen and lipid droplets that provide the energy stores needed for continual contraction.

SUGGESTED READINGS

Books

Hoyle, G. (1983). *Muscles and Their Neural Control,* Wiley, New York.

Matthews, G. (1985). *Cellular Physiology of Nerve and Muscle,* Blackwell Scientific, Oxford.

Salpeter, M. M., ed. (1987). *The Vertebrate Neuromuscular Junction,* Alan R. Liss, New York.

Shay, J., ed. (1985). *Cell and Muscle Motility,* Vol. 6, Plenum Press, New York.

Squire, J. (1981). *The Structural Basis of Muscular Contraction,* Plenum Press, New York.

Woledge, R. C., N. A. Curtin, and E. Hornsher (1985). *Energetic Aspects of Muscle Contraction,* Academic Press, Orlando.

Articles

Eisenberg, E., and T. L. Hill (1985). Muscle contraction and free energy transduction in biological systems, *Science* 227:199–227.

Franzini-Armstrong, C., and L. D. Peachey (1981). Striated muscle—contractile and control mechanisms, *J. Cell Biol.* 91:166s–186s.

Harrington, W. F., and M. E. Rodgers (1984). Myosin, *Ann. Rev. Biochem.* 53:35–73.

Page, E., and Y. Shibata (1981). Permeable junctions between cardiac cells, *Ann. Rev. Physiol.* 43:431–441.

Pollack, G. R. (1983). The cross-bridge theory, *Physiol. Rev.* 63:1049–1113.

Squire, J. M. (1983). Molecular mechanisms in muscular contraction, *Trends Neurosci.* 6:409–413.

CHAPTER
20

The Cancer Cell

It is difficult to imagine anyone familiar with the details of cell organization and function who does not feel a certain sense of awe at the complexities involved. Given the vast number of intricate events that must be coordinated during the formation, growth, and differentiation of living cells, it is perhaps not surprising to discover that everything does not always work properly. Of the many diseases that directly result from aberrations in cell function, cancer has the most profound effect on human health and mortality. One out of every four people in the United States can now expect to develop cancer, making it the most common cause of death other than cardiovascular disease.

When biologists consider the issue of cancer, three fundamental questions immediately come to mind: What is cancer? What causes cancer? Can cancer be prevented or cured? Although our answers to these questions remain incomplete, there is reason to believe that this dreaded disease can eventually be understood and brought under control. In this chapter we shall explore the experimental advances that provide the basis for this optimism. As we do so, it should quickly become apparent that an understanding of cancer requires an intimate comprehension of the events occurring in normal cells, and that conversely, investigations on cancer cells are helping to broaden our understanding of normal cells.

WHAT IS CANCER?

Cancer is a disease whose existence was first recognized thousands of years ago. The term **cancer,** which means "crab" in Latin, was coined by Hippocrates in the fifth century B.C. to describe diseases in which particular tissues grow and spread unrestrained throughout the body, eventually choking off life. Although cancers can originate in almost any tissue, most of them fall into three general categories: carcinomas, sarcomas, and lymphomas/leukemias. **Carcinomas,** which are the most commonly encountered type of cancer, arise from epithelial cells that cover external and internal body surfaces. Cancer of the lung, breast, and intestine are the most frequently occurring cancers of this type. **Sarcomas** originate in supporting tissues of mesodermal origin, such as bone, cartilage, fat, and muscle. Finally, **lymphomas** and **leukemias** arise from cells of blood and lymphatic origin; the term *lymphoma* is used when the cancer cells are restricted largely to solid tumor masses, while *leukemia* is employed when the

TABLE 20-1
A representative listing of various types of cancer

General Category	Tissue of Origin	Name of Tumor
Carcinomas (~90% of all cancers)	Skin	Basal cell carcinoma
		Squamous cell carcinoma
	Lung	Pulmonary adenocarcinoma
	Breast	Mammary adenocarcinoma
	Stomach	Gastric adenocarcinoma
	Colon	Colon adenocarcinoma
	Uterus	Uterine endometrial carcinoma
	Prostate	Prostatic adenocarcinoma
	Ovary	Ovarian adenocarcinoma
	Pancreas	Pancreatic adenocarcinoma
	Urinary bladder	Urinary bladder adenocarcinoma
	Liver	Hepatocarcinoma
Sarcomas (~5% of all cancers)	Bone	Osteosarcoma
	Cartilage	Chondrosarcoma
	Fat	Liposarcoma
	Smooth muscle	Leiomyosarcoma
	Skeletal muscle	Rhabdomyosarcoma
	Connective tissue	Fibrosarcoma
	Blood vessels	Hemangiosarcoma
	Nerve sheath	Neurogenic sarcoma
	Meninges	Meningiosarcoma
Lymphomas/leukemias (~5% of all cancers)	Red blood cells	Erythrocytic leukemia
	Bone marrow cells	Myeloma or myelocytic leukemia
	White blood cells	Lymphoma or lymphocytic leukemia

cancer cells circulate in large numbers in the bloodstream. Included within each of these three major categories are dozens of different kinds of cancer, each originating from a specific cell type (Table 20-1).

To prevent subsequent confusion several other terms commonly employed by cancer biologists, or **oncologists,** should be defined at this point. The words **tumor** and **neoplasm** are used interchangeably to refer to any growth of new tissue that results from uncontrolled cell division. **Benign** tumors are confined to a circumscribed area and are therefore rarely life threatening. **Malignant** tumors, in contrast, spread throughout the body and are more difficult to treat. The term *cancer* refers to any type of malignant tumor. Specific kinds of tumors are usually named by adding the suffix *-oma* to the name of the cell from which the tumor has arisen. For example, a benign tumor of fibroblast origin is called a fibroma, while a malignant tumor involving this same cell type is referred to as a fibrosarcoma.

Although the various kinds of cancers tend to share certain properties, individual cancers differ from one another in their initiating causes, their clinical behavior, and their susceptibility to treatment. This inherent diversity has made it difficult to create a unifying theory of cancer upon which a rational approach to cancer therapy might be developed. The first step in the creation of such a theory, of course, is to try to determine which properties, if any, are universal characteristics of cancer cells.

Characteristics of Malignant Tumors

Perhaps the most obvious property shared by all malignant tumors is that their constituent cells grow and divide in a way that is not properly controlled. This does not necessarily mean that all tumor cells divide more rapidly than do normal cells. In fact, some rapidly growing normal cells divide more frequently than the cells of slowly growing tumors. The critical issue is not the absolute rate of cell division, but rather the relative balance between cell division and cell loss. In normal tissues the rate of cell division and the rate of cell loss are precisely coordinated with one another. Cells located in the basal layer of the skin, for example, divide at exactly the rate needed to replace the cells that are continually differentiating and eventually being lost from the skin surface. A similar phenomenon occurs in bone marrow, where blood cells are constantly being replaced, and in the intestines, where the epithelial lining cells are continually being replenished. In each of these situations the rate of cell division is carefully balanced with the rate of cell loss so that no net accumulation of new cells occurs. In tumors, on the other hand, the rate of cell division exceeds the rate of cell loss, causing the cell population to gradually become larger and larger. If the rate of cell division is rapid, the tumor will enlarge rapidly; if the rate of cell division is relatively slow, tumor growth will be slower. In either case the tumor continually increases in size. Because new cells are being produced in greater numbers than required, the normal organization of the tissue gradually becomes disrupted as more and more dividing cells accumulate (Figure 20-1).

Uncontrolled cell division is an attribute of malignant and benign tumors alike. The property that distinguishes these two types of tumors from each other is the ability of malignant tumors to spread throughout the body. Such spreading occurs in two ways: direct invasion of tumor cells into surrounding tissues, and penetration of tumor cells into vessels of the circulatory or lymphatic system. This latter process permits cancer cells to move throughout the body, where they establish secondary tumors referred to as **metastases.** Tumors often tend to establish metastases in characteristic locations. Breast cancer, for example, commonly metastasizes to bone and ovary, while lung tumors metastasize principally to the brain.

It is the ability of malignant tumors to invade and metastasize that makes them potentially life threatening, since tumor cells have often spread throughout the body by the time the cancer is diagnosed. Although malignant tumors as a class are therefore more serious than benign tumors, individual exceptions do occur. Basal cell carcinoma of the skin, for example, is a common type of cancer that is rarely lethal because it is very slow to metastasize and the tumors are readily removed surgically. Some benign brain tumors, on the other hand, are surgically inaccessible and are therefore lethal, in spite of the fact that they do not metastasize.

The question of why cancer cells do not behave like their normal counterparts has occupied cancer biologists since the turn of the century. Although some information has been obtained from observations made directly on human cancers, tissue culture and animal model systems have played important roles in helping us to decipher the cellular and molecular alterations underlying the malignant state. In the following section we shall summarize the major conclusions derived from these various kinds of analyses.

Profile of the Cancer Cell

There are several different ways in which the properties of cancer cells can be studied. The most straightforward approach is to analyze tissue taken directly from a naturally occurring tumor. Although such material is a reliable source of malignant cells, intact tumors usually contain a heterogeneous population of cell types, and more variability is encountered when one tries to compare tumors obtained from different individuals. The alternative to the direct examination of tumor tissue is to grow cancer cells in culture. Unfortunately, long periods in culture may cause the behavior and makeup of cell populations to change, and the question therefore arises as to whether the properties exhibited by long-term cultures of cancer cells accurately reflect the behavior of malignant cells *in vivo*. In spite of this potential drawback, the ability to study a defined population of cells under carefully controlled laboratory conditions has made tissue culture a popular approach among cancer biologists.

Cancer cell lines employed for tissue culture studies have been obtained by a variety of means. The most

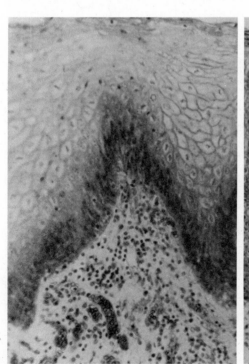
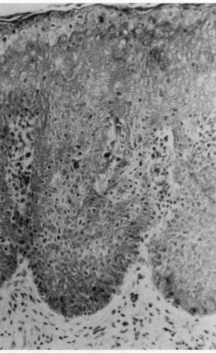

FIGURE 20-1 Light micrographs comparing the arrangement of cells in normal *(left)* and malignant *(right)* skin tissue. Note that in the malignant tumor, the typical layered organization of skin cells has been disrupted.

obvious approach is to remove cells from malignant tumors and culture them directly. Hundreds of human cancer cell lines, encompassing most of the major tissue types, have been established in this way. Alternatively, normal cells can be transformed into malignant ones while in culture. The first indication that malignant transformation can take place under such conditions came from studies carried out by George Gey and his associates in the early 1940s. These investigators made the rather remarkable discovery that rat fibroblasts maintained in culture for prolonged periods of time eventually acquire the ability to form sarcomas when injected back into healthy rats. This tendency of cultured normal cells to undergo spontaneous transformation into malignant cells has subsequently been documented for numerous cell types derived from mouse, hamster, and rat. Cells maintained in culture can also be induced to undergo malignant transformation by exposing them to cancer-causing agents such as chemicals, viruses, and radiation.

A vast amount of data has been acquired concerning the properties of malignant cells obtained in the various ways described above. We shall consider those generalizations that are most useful, keeping in mind that each has its own limitations and exceptions, and that none is completely adequate in itself as a criterion for the malignant state.

1. *Tumorigenicity in experimental animals.* The most obvious property distinguishing malignant cells from normal cells is the ability of malignant cells to produce tumors when injected into appropriate test organisms. For animal tumors this criterion is relatively easy to test, but with human tumor cells the situation is more complicated. Injection of cancer cells into human beings is clearly out of the question, and injecting human tumor cells into laboratory animals is not generally reliable because an animal's immune system may reject the human cells, independent of whether or not they are malignant. One way around this obstacle is to inject tumor cells into **nude mice,** an immunologically deficient strain of animals lacking a thymus gland. Because a thymus gland is required for production of the T-lymphocytes involved in cell-mediated immunity, *nude* mice are incapable of rejecting foreign cells. Hence human tumor cells injected into *nude* mice grow into tumors without immunological rejection. Unfortunately, failure to grow under these conditions does not absolutely rule out the possibility that the same cells might be capable of forming tumors under other conditions.

2. *Indefinite lifespan.* Most normal cells maintained in culture have a limited life expectancy. Human fibroblasts, for example, multiply for about 50–60 generations and then begin to deteriorate and die. In contrast, malignant cells appear to have an indefinite lifespan.

The human **HeLa** cell line, which was derived from a uterine carcinoma in 1953, has been dividing in culture ever since. Cells that have undergone transformation in culture behave in a similar fashion, a fact first revealed by studies carried out on mouse embryo fibroblasts by George Todaro and Howard Green. These investigators noted that within a few weeks after placement in culture the growth rate of these embryonic cells slowed down, suggesting that the cells would eventually stop dividing and die, as occurs with normal human fibroblasts. After two to three months, however, the growth rate of the cells was unexpectedly found to increase again (Figure 20-2). The newly emerging cell population, which had apparently undergone spontaneous transformation, formed permanently dividing cell lines that have now been growing in culture for over 20 years.

Although all malignant cells appear to have an unlimited lifespan when grown in culture, it does not necessarily follow that all nonmalignant cells have a limited lifespan in culture. Some of the "transformed" cell lines isolated by Todaro and Green are capable of indefinite growth in culture, yet do not form tumors when injected back into animals. This observation suggests that acquisition of an indefinite lifespan may be an early stage in the development of malignancy, but that subsequent events must ensue before the malignant state is actually established.

3. *Decreased density-dependent inhibition of growth.* When normal cells are placed in culture, they move about and divide until the surface of the culture vessel is covered by a monolayer of cells. At this point, cell movement and cell division usually cease. In the early 1950s Michael Abercrombie and Joan Heaysman introduced the term **contact inhibition** to refer to the decrease in cell motility that occurs when cells contact

FIGURE 20-2 Growth of normal mouse embryo fibroblasts in culture. After an initial decline in growth rate the cells become transformed and the rate of division increases again, yielding a permanent cell line.

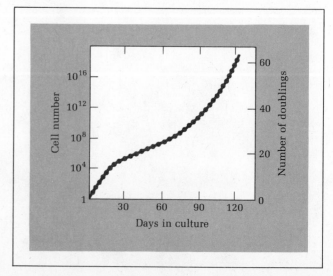

one another in culture. This same term has also been used to refer to the inhibition of cell division that generally takes place when culture conditions become crowded. Because of the confusion that can result from this double meaning of the term contact inhibition, we shall use the phrase **density-dependent inhibition of growth** to refer to the inhibition of cell division that occurs in crowded cultures. This phrase simply describes the fact that cell division is inhibited as the population density increases, without implying anything about the underlying mechanism.

Cells that have undergone malignant transformation typically grow to higher population densities in culture than do their normal counterparts. Instead of growth ceasing when a complete monolayer of cells covers the surface of the culture vessel, malignant cells continue to divide and pile on top of one another, forming multilayered aggregates that contrast markedly with the uniform monolayer typically formed by normal cells. Tumor cell growth in culture is thus less density dependent than growth of normal cells.

The connection between the loss of density-independent growth *in vitro* and the ability to form malignant tumors *in vivo* has been difficult to establish. The first thorough investigation of this issue was carried out by Stuart Aaronson and George Todaro, who examined the properties of several fibroblast cell lines derived from the same batch of mouse embryonic tissue. By manipulating the culture conditions under which these cells were grown, it was found that cell lines differing in their susceptibility to density-dependent inhibition of growth could be established. One type of cell line was produced by growing cells under uncrowded conditions; every time the cell population density increased and crowding was imminent, the cells were diluted and transferred to a new culture flask. The cell line produced under such conditions was found to be very sensitive to density-dependent growth inhibition. Another set of cell lines was established by growing cells under extensively crowded conditions, thereby selecting for cells able to grow under these conditions. The cell lines established by this second procedure were found to be less susceptible to density-dependent inhibition of growth, reaching much higher population densities before growth ceased. When these various cell lines were tested for their ability to produce tumors when injected into mice, tumor-forming ability was found to be inversely correlated with the susceptibility to density-dependent inhibition of growth (Figure 20-3).

The question of why transformed cells are less susceptible to density-dependent inhibition of growth than normal cells requires an understanding of the basic mechanism underlying this phenomenon in normal cells. Although the term *contact inhibition* was used for many years to refer to the tendency of cells to stop dividing in culture when conditions become crowded,

there is no definitive evidence that cell-cell contact per se is responsible. The degree of crowding tolerated before cell division stops in normal cells is subject to considerable variation, depending on the culture conditions employed. For example, if the serum concentration in the growth medium is too low, cells will stop dividing before cell-cell contacts have become established. On the other hand, cell cultures that have stopped dividing because of crowded conditions can be made to start proliferating again if additional serum is added to the medium. Such observations suggest that density-dependent growth inhibition is caused by depletion of nutrients or growth factors, rather than by cell-cell contact.

Transformed cells generally require less serum in the growth medium than do normal cells. An apparent explanation for this difference is suggested by experiments carried out in 1978 by Joseph DeLarco and George Todaro that revealed that transformed mouse fibroblasts manufacture and secrete their own growth factors. The existence of these factors was discovered by taking the culture fluid in which transformed fibroblasts had been growing, and adding it to a flask containing normal rat kidney cells whose growth had been arrested by serum deprivation. Under these conditions, the rat kidney cells began to synthesize DNA and divide. Eventually the culture was converted from a regular monolayer of normal cells to an unorganized mass of cell aggregates more typical of a transformed culture. Such observations suggest that malignant cells are less susceptible to density-dependent inhibition of growth because they produce their own growth factors. The

FIGURE 20-3 Comparison of the tumor-forming ability of several mouse fibroblast cell lines differing in their susceptibility to density-dependent inhibition of growth. Each point represents a different cell line. The data show that cell lines that grow to higher densities in culture require less time to produce tumors when injected back into mice.

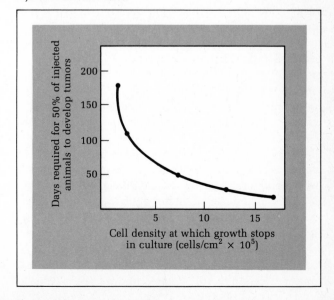

growth factors involved, referred to as **transforming growth factors (TGFs),** have subsequently been isolated from the growth medium of transformed cell cultures. At least two distinct types of TGF have been identified in such media: **TGF-α,** a single polypeptide chain 50 amino acids long that acts by binding to the plasma membrane receptor for epidermal growth factor, and **TGF-β,** a larger protein containing two polypeptide subunits that is present in many normal tissues as well as transformed cells. Stimulation of cell division requires the synergistic action of TGF-β with TGF-α (or epidermal growth factor). The fact that tumor cells can manufacture their own TGF-β and TGF-α may explain why tumors growing within intact organisms manage to grow autonomously without being subject to the normal feedback systems that regulate normal cell growth and division.

4. *Anchorage-independent growth.* Most normal cells derived from solid tissues do not grow well in culture if they are suspended in a liquid medium or a semisolid material such as soft agar or methylcellulose. If provided with an appropriate surface to which they can adhere, however, normal cells will attach, spread out, and commence growth. This type of growth is therefore referred to as **anchorage dependent.** In contrast to the behavior of normal cells, most transformed cells are capable of growing in liquid suspension or a semisolid medium, and are hence said to be **anchorage independent.**

An extensive set of experiments carried out in the mid-1970s in the laboratory of Robert Pollack suggest that the property of anchorage-independent growth is closely correlated with a cell's ability to form malignant tumors *in vivo.* In these studies, a series of transformed fibroblast cell lines were tested for their ability to induce tumors in *nude* mice. The parent cell line exhibited many of the typical characteristics of transformed cells: decreased density-dependent inhibition of growth, low serum requirement for growth, expression of new cell surface antigens, and anchorage-independent growth. By isolating single cells from the parent population and growing a group of clones, Pollack established individual cell lines that had lost one or more of the properties of the parent cell line. Of the four properties studied, only the retention of anchorage-independent growth was consistently associated with the ability to form tumors in animals.

This relationship between anchorage-independent growth and tumorigenicity is not, unfortunately, without exceptions. Some human cell lines exposed to carcinogenic chemicals exhibit anchorage-independent growth, yet are unable to induce tumors when injected into *nude* mice. Conversely mouse *3T3* fibroblasts, which are anchorage-dependent and normally unable to induce tumor formation, gain the capacity to produce tumors when they are attached to glass beads prior to

implantation. Such observations suggest that in spite of its general correlation with tumorigenicity, anchorage-independent growth in culture is not an absolute prerequisite for tumor formation *in vivo.*

5. *Loss of restriction point control.* When normal cells are exposed to suboptimal growth conditions (e.g., inadequate nutrients or growth factors), they stop at a specific point in the cell cycle situated near the end of the G_1-phase. The point in G_1 at which they are blocked is referred to as the **restriction point.** Cells held up at this stage of the cell cycle are said to be in the G_0-phase. Once cells emerge from G_0 or G_1, they are committed to continuing through S, G_2, and M. In contrast to normal cells, transformed cells do not exhibit restriction point control. They continue to cycle and divide under conditions of high cell density, low serum concentration, or suboptimal nutrient concentration that would cause normal cells to stop at the restriction point. Under conditions of severe nutritional deprivation, transformed cells stop at random points in the cell cycle and die, rather than entering into G_0.

Attempts have been made to design therapeutic schemes that might exploit this difference in the behavior of normal and transformed cells. For example, Arthur Pardee and his associates have devised a simple scheme employing a combination of caffeine, which blocks normal cells at the restriction point, and *cytosine arabinoside,* which interferes with DNA synthesis and therefore kills cells situated in the S-phase. When normal cells are treated with a combination of caffeine and cytosine arabinoside, the caffeine blocks cells in G_1 and therefore prevents them from entering S-phase and from being killed by the S-phase specific agent, cytosine arabinoside. If the two drugs are later removed, the cells will begin dividing again. When transformed cells are treated with this drug combination, on the other hand, the caffeine does not stop the cells in G_1 because of the loss of restriction point control. The cells therefore progress into S-phase and are killed by the cytosine arabinoside. The fact that this approach selectively destroys transformed cells in culture suggests that a similar strategy might eventually be adapted to the treatment of human cancer.

6. *Cell surface alterations.* Changes in the behavior and biochemical composition of the plasma membrane are almost universally observed in tumor cells. The significance of such changes is difficult to assess, however, because plasma membrane alterations also occur in normal cells stimulated to divide. For example, active transport systems for the uptake of sugars, amino acids, and nucleosides are frequently activated in tumor cells, but a similar increase in transport activity also occurs in normal cells stimulated to divide by the addition of appropriate nutrients or growth factors. It is therefore important to determine which (if any) membrane changes are restricted to the malignant state and which are char-

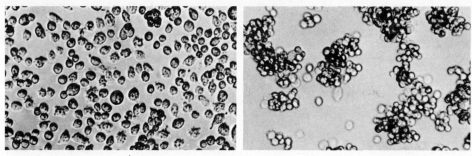

FIGURE 20-4 Light micrographs comparing the behavior of normal *(left)* and malignant *(right)* cells when exposed to the lectin concanavalin A. Note the enhanced tendency of the malignant cells to form clumps under such conditions. Courtesy of L. Sachs.

acteristic of dividing cells in general. Most of the distinctive biochemical changes that occur in the plasma membranes of malignant cells involve alterations in glycoproteins and glycolipids. Among the functional properties of the cell surface that appear to be influenced by these changes are adhesiveness, agglutinability, cell-cell communication, and antigenicity.

Most tumor cells are less adhesive than normal because they are deficient in the cell surface glycoprotein, **fibronectin.** In addition to its role in promoting cell-cell adhesion, fibronectin is known to influence cell shape and motility. Hence the loss of cell surface fibronectin may underlie some of the behavioral properties of tumor cells, such as their tendency to invade neighboring tissues and metastasize. Addition of fibronectin to malignant cells restores cell adhesion, flattened cell morphology, and contact inhibition of movement. It does not, however, restore normal growth control (i.e., density-dependent inhibition of growth), indicating that depletion of fibronectin is not a primary factor in the loss of growth control.

A second cell surface property characteristic of malignant cells is their enhanced tendency to **agglutinate,** or clump together, when exposed to lectins (Figure 20-4). **Lectins** are carbohydrate-binding proteins whose multiple binding sites allow them to link adjacent cells together. A close correlation between loss of growth control and susceptibility to lectin-induced agglutination is suggested by the fact that transformed cells that have spontaneously regained density-dependent growth control simultaneously lose their enhanced agglutinability by lectins. The simplest explanation for the enhanced agglutinability exhibited by tumor cells would be an increased number of lectin receptors, but careful measurements have led to the surprising conclusion that the total number of lectin receptor sites is similar on normal and transformed cells. What does differ, however, is the mobility of the lectin receptors. In tumor cells these receptors diffuse more rapidly within the plane of the membrane, a phenomenon that appears to promote lectin-mediated cell agglutination. Because the plasma membranes of transformed cells do not generally appear to be more fluid than the plasma mem-

branes of normal cells, it has been proposed that a decreased interaction between lectin receptors and the underlying cytoskeletal network is responsible for the enhanced mobility of lectin receptors in tumor cells.

A third cell surface property altered in malignant cells is the ability to establish communicating junctions. This conclusion is based on the observation that malignant cells generally exhibit a decreased number of gap junctions, as well as the discovery that fluorescent dyes injected into normal cells will diffuse rapidly into surrounding cells that are normal, but not into cells that are malignant (Figure 20-5). Experiments carried out in the laboratory of Werner Loewenstein suggest that this inability to communicate through gap junctions is intimately involved with the development of malignancy. In these experiments normal cells were fused with malignant cells that had lost the ability to form gap junctions. In most cases cell fusion resulted in restoration of both normal growth control and the ability to form gap junctions. Eventually, however, some of the fused cells reverted to uncontrolled growth. When this occurred, the ability to form gap junctions was also lost. From these experiments Loewenstein concluded that normal growth control requires the ability to communicate through gap junctions.

The final cell surface property altered in malignant cells is antigenic specificity. Although this area of

FIGURE 20-5 Evidence for a decrease in communicating junctions in malignant cells. *(Left)* Phase contrast micrograph showing four normal liver cells surrounded by numerous malignant cells in culture. Fluorescent dye was injected into one of the four normal cells (arrow). *(Right)* Fluorescence microscopy of the same cells shortly thereafter reveals that the dye has diffused into the three adjacent normal cells, but not into the surrounding malignant cells. Courtesy of W. R. Loewenstein.

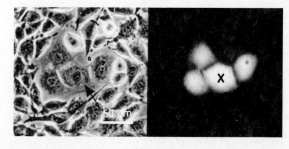

research has had a controversial history, it is now well established that some cancers exhibit **tumor-specific cell surface antigens.** Part of the difficulty in reaching a consensus on this issue has derived from the fact that multiple classes of antigens are involved. In human melanomas, for example, at least three types of antigens have been identified. Antigens of the first type are specific both for melanomas and for the individual from whom a particular melanoma is obtained. Antigens of the second type are specific for melanomas, but not for the individual from which the tumor is obtained. Antigens of the above two kinds are not detectable on normal cells, and can therefore be classified as tumor specific. Antigens of the third type are present on normal cells as well as melanoma cells, but their concentration on melanoma cells is greater.

The existence of tumor-specific antigens (the first two classes described above) raises the question of why individuals afflicted with cancer do not reject their tumors. According to the **immune surveillance theory,** immune rejection of newly forming malignant cells does occur in healthy individuals, and cancer simply reflects the occasional failure of an adequate immune response to be mounted against aberrant cells. If correct, this theory suggests that one possible approach to cancer therapy might involve attempts to stimulate an organism to immunologically reject its own tumor cells. Although this approach has yielded promising results in some animal tumor systems, it is yet to be developed into a practical approach for treating human cancer.

7. *Other molecular changes.* A great deal of effort has recently been expended in searching for specific molecular "markers" unique to malignant cells. Such markers would be useful aids for cancer diagnosis, and might also provide insights into the molecular origins of this disease. Though this search has led to the identification of dozens of molecular properties that appear to be altered in cancer cells, many of these modifications are also exhibited by rapidly dividing normal cells and therefore are not unique to cancer. In this section we shall restrict our discussion to a few molecular changes that have contributed the most to our understanding of malignancy.

It has been known for many years that malignant tumors secrete proteolytic enzymes in high concentration. Although similar enzymes are secreted by some normal tissues, enhanced production of proteases may be responsible for certain properties of cancer cells. For example, the decrease in adhesiveness and loss of anchorage dependence commonly exhibited by tumor cells may be caused by protease-mediated digestion of cell surface components. Secretion of proteases *in vivo* might also provide a partial explanation for the invasive properties of malignant tumors. The main protease secreted by cancer cells is **plasminogen activator,** an enzyme that catalyzes the proteolytic cleavage of an

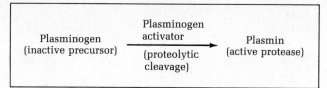

FIGURE 20-6 Mechanism of action of plasminogen activator.

enzyme precursor known as **plasminogen.** This cleavage reaction converts plasminogen to **plasmin,** which is itself a protease (Figure 20-6). Plasminogen is a component of plasma and extracellular fluids, and its conversion to plasmin in the presence of the plasminogen activator may be responsible for many of the proteolytic effects associated with malignant tumors.

Some transformed cells also manufacture proteins typical of embryonic cells. One such molecule is **alpha-fetoprotein,** a protein synthesized by embryonic liver. The amount of alpha-fetoprotein present in the bloodstream of normal adults is quite low, but in patients with liver cancer the concentration increases dramatically. Another embryonic marker occasionally present in tumor cells is **carcinoembryonic antigen,** a glycoprotein normally made in the fetal digestive tract. Fetal hormones, such as chorionic gonadotropin and placental lactogen, are also produced by some human tumors. The production of embryonic components by malignant tumors may relate to the fact that certain properties of cancer cells, such as invasiveness and rapid proliferation, are also manifested by embryonic cells. Testing for the presence of embryonic markers such as alpha-fetoprotein and carcinoembryonic antigen in the bloodstream has been touted as a potential tool for the early diagnosis of cancer, but the fact that these markers are far from universally present in cancer patients limits the usefulness of this approach.

Some tumors also secrete a substance, known as **tumor angiogenesis factor** or **angiogenin,** which stimulates the formation of new blood vessels. Since the presence of an adequate blood supply is a critical factor in determining a tumor's ability to grow and invade surrounding tissues, production of such angiogenesis factors may significantly influence tumor behavior. Certain nonvascular tissues, such as cartilage, have been found to produce inhibitors whose ability to block tumor-induced angiogenesis may eventually be exploited for therapeutic uses. In preliminary studies on animal tumors, such inhibitors have been found to depress the growth rate of tumor cells by interfering with the development of their blood supply.

One of the most striking molecular alterations observed in cancer cells involves changes in the genetic information encoded in their DNA. This alteration has been most dramatically demonstrated by exposing normal cells to DNA extracted from cancer cells. For example, if normal mouse fibroblasts are exposed to DNA extracted from human bladder cancer cells, the fibroblasts undergo malignant transformation. If the fibroblasts are exposed to DNA obtained from normal cells, on the other hand, transformation does not occur. Taken together, such observations suggest that the DNA of the tumor cells is somehow altered.

In order to ascertain whether the same gene sequence is always altered in malignant transformation, Robert Weinberg and his associates have investigated the properties of the DNA isolated from four different lines of transformed fibroblasts. In terms of their susceptibility to inactivation by restriction endonuclease cleavage, the transforming DNAs obtained from all four cell lines were found to behave similarly. It was therefore concluded that the same gene is altered in each of the four transformed cell lines. However, when experiments of the same type were carried out on DNAs obtained from three kinds of human cancer cells (colon cancer, bladder cancer, and leukemia), three different

genes were found to be altered. Thus a different gene is involved when each of these cell types undergoes malignant transformation. As we shall see later, studies on the genes involved in malignant transformation and the proteins they encode are beginning to provide profound insights into the molecular basis of cancer.

8. *Morphological changes.* Although no single morphological criterion is adequate for distinguishing cancer cells from normal cells, alterations in cell structure do occur in malignant tumors. Most notable is the tendency of malignant cells to undergo **anaplasia,** a process involving a loss of differentiation and proper orientation of cells to one another. Cancer cells commonly exhibit large nuclei, prominent nucleoli, and a cell surface covered with microvilli and lamellopodia (Figure 20-7). The number of cells undergoing mitosis is generally elevated in malignant tumors, and abnormal mitoses (Figure 20-8) and multinucleated giant cells are frequently encountered. Irregularities in chromo-

FIGURE 20-7 Scanning electron micrograph of a group of poorly differentiated esophageal cancer cells. Note the numerous radiating lamellipodia and fine microvilli covering the surfaces of these tumor cells. Courtesy of K. M. Robinson.

10 μm

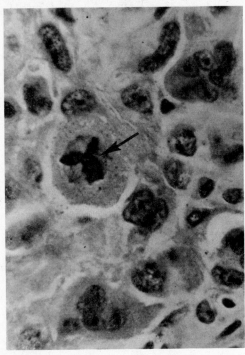

FIGURE 20-8 Light micrograph showing a population of anaplastic tumor cells exhibiting a marked variation in size and shape. The prominent cell in the center has an abnormal tripolar mitotic spindle (arrow).

some number and structure are also common. For example the *Philadelphia chromosome*, an abnormally shaped chromosome caused by translocation of a piece

of chromosome 22 to chromosome 9 (Figure 20-9), occurs in nearly 90 percent of the individuals suffering from chronic granulocytic leukemia.

Although none of these morphological attributes is sufficient in itself for unequivocally identifying cancer cells, in combination they are quite useful for diagnosing the presence of malignancy. It is important to note, however, that many of the above-mentioned characteristics are also manifested by some normal cells, especially in tissues where rapid cell division is occurring. As a result, the final diagnosis of malignancy may depend on criteria other than cell morphology, such as the occurrence of invasion and metastasis, or the presence of embryonic or tumor antigens.

WHAT CAUSES CANCER?

Cancer is often perceived as a disease that strikes randomly and without warning. This common misconception ignores the results of thousands of investigations on the causes of cancer, some of which date back over 200 years. The inescapable conclusion emerging from these investigations is that most human cancers are caused by identifiable environmental factors, most prominently chemicals, radiation, and viruses.

Chemical Carcinogens

The first description of a link between an environmental chemical and human cancer dates back to 1761,

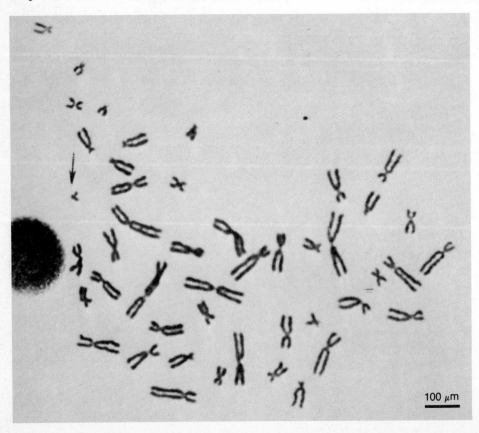

FIGURE 20-9 Metaphase chromosomes from a malignant cell obtained from a patient with chronic granulocytic leukemia. The arrow points to the Philadelphia chromosome, an abnormally small chromosome that represents the remnant of chromosome 22 that remains after the rest of chromosome 22 has been translocated to chromosome 9. Courtesy of P. C. Nowell.

100 μm

when a London doctor named John Hill reported that individuals who use snuff suffer an abnormally high incidence of nasal cancer. A few years later another British physician, Percival Pott, noted an elevated incidence of scrotal cancer among men who had served as chimney sweepers in their youth. It was common practice at the time to employ young boys to clean chimney flues because they fit into these narrow spaces more readily than adults. Pott speculated that the chimney soot gradually became dissolved in the natural oils of the scrotal skin folds, causing constant irritation and eventually triggering the development of cancer. This theory was soon substantiated by the discovery that the scrotal cancer observed among chimney sweeps could be prevented by the judicious use of protective clothing and regular bathing practices.

In the years since these pioneering discoveries, it has become increasingly apparent that cancer is often caused by exposure to environmental chemicals. In many cases the ability of a particular chemical to cause cancer has only become obvious after industrial workers exposed to the agent on a regular basis develop an abnormally high incidence of a specific kind of cancer. At the turn of the century, for example, an elevated incidence of skin cancer was noted among workers in the coal tar industry, and an increased incidence of bladder cancer was observed in workers employed by the newly emerging aniline dye industry. As the industrial revolution moved into the twentieth century, the list of known **carcinogens** (cancer-causing agents) became more and more extensive (Table 20-2). Millions of workers are now known to be exposed to carcinogens on a daily basis, creating the hazard of numerous kinds of occupation-induced cancer. It should be emphasized, however, that exposure to chemical carcinogens is not always work related or involuntary.

TABLE 20-2
Some of the many chemical carcinogens present in our environment (air, food, water, drugs, workplace, etc.)

Agent	Type of Cancer Induced
Asbestos	Lung, esophagus, stomach
α-Naphthylamine	Bladder
4-Aminodiphenyl	Bladder
Carbon tetrachloride	Liver
Acrylonitrile	Colon, lung
Mustard gas	Lung, larynx
Wood and leather dust	Nasal sinuses
Arsenic compounds	Lung, skin
Chromium and chromates	Lung, nasal sinuses
Vinyl chloride	Liver, lung, brain
Benzene	Leukemia
Soot and tars	Skin, scrotum, lung, bladder
Organo-chloride pesticides	Liver
Polychlorinated biphenyls	Liver
Aniline derivatives	Bladder
Lead	Kidney
Nickel	Lung, nose
Cadmium salts	Prostate, lung
Diethylstilbestrol (DES)	Uterus, vagina
Tobacco smoke components	Lung
Aminostilbene, arsenic, benz[*a*]anthracene, benz[*a*]pyrene, benzene, benzo[*b*]fluoranthene, benzo[*c*]phenanthrene, benzo[*j*]fluoranthene, cadmium, chrysene, dibenz[*a,c*]anthracene, dibenzo[*a,e*]fluoranthene, dibenz[*a,h*]acridine, dibenz[*a,j*]acridine, dibenzo[*c,g*]carbazone, N-dibutylnitrosamine, 2,3-dimethylchrysene, indeno[1,2,3-*c,d*]pyrene, 5-methylchrysene, 5-methylfluoranthene, α-naphthylamine, nickel compounds, N-nitrosodimethylamine, N-nitrosomethylethylamine, polonium-210, N-nitrosodiethylamine, N-nitrosonornicotine, N-nitrosoanabasine, N-nitrosopiperidine	

Tobacco smoke, which contains more than 30 known carcinogens, is knowingly inhaled by over a third of the adult population of the United States, despite extensive warnings about the role of smoking in lung cancer, a disease responsible for over 1 in 4 cancer deaths. These statistics, as well as evidence implicating tobacco smoke in a variety of other kinds of cancer, suggest that smoking may be responsible for 5 to 10 times as many cancer deaths as all occupational exposure to carcinogens combined.

Much has been learned about the mechanism of chemical carcinogenesis by studying the effects of known carcinogens on animals or cell cultures. One of the most significant ideas to emerge from such studies is the concept that cancer develops by a multistage process. In the early 1940s Peyton Rous and his associates discovered that tumors induced by painting the skin of rabbits with coal tar regress when application of this carcinogen is stopped. Tumors reappear, however, when the skin is subsequently treated with irritants like turpentine or chloroform, which by themselves do not cause tumors. Based on such observations, Rous proposed that carcinogenesis occurs in two stages, termed **initiation** and **promotion.** In the initiation phase normal cells exposed to a carcinogenic agent are irreversibly altered to a preneoplastic state. These preneoplastic cells remain dormant, however, until they are stimulated by a promoting agent to divide and form tumors.

Independent support for the existence of distinct initiation and promotion phases was obtained at about the same time by Isaac Berenblum, who investigated the induction of skin cancer in mice. Berenblum discovered that if the skin of a mouse is painted a single time with a carcinogen such as 3-methylcholanthrene, tumor formation is quite rare. However, if the same area is later painted with croton oil, which is not in itself carcinogenic, tumors form in most animals. Hence 3-methylcholanthrene is acting as an initiator, and croton oil as a promoting agent. Subsequent experiments involving a variety of other animal tumor and cell culture systems have led to the conclusion that distinct initiation and promotion phases are a general phenomenon in carcinogenesis (Figure 20-10).

The mechanisms underlying initiation and promotion are quite distinct from each other. Initiation requires only brief exposure to a carcinogenic agent and is irreversible. Though the data are largely circumstantial, much evidence suggests that initiators act by mutating DNA. It has been found, for example, that the carcinogenic potency of chemical agents correlates with their ability to bind to DNA (Figure 20-11). In addition, we shall see later in the chapter that most carcinogenic chemicals are mutagenic when tested in bacteria. This capacity to cause mutations provides a simple explanation for the observed ability of carcinogenic agents to trigger a stable, inheritable change in a cell's properties.

In contrast to initiation, promotion is a gradual pro-

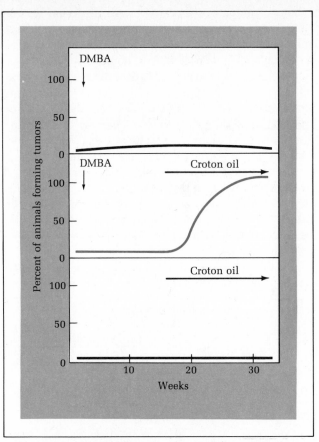

FIGURE 20-10 Experiment illustrating the existence of separate initiation and promotion phases in chemical carcinogenesis. *(Top)* Mice treated with a single dose of the carcinogen dimethylbenzanthracene (DMBA) do not form tumors. *(Middle)* Painting the skin of such animals twice a week with croton oil results in the local appearance of skin tumors. *(Bottom)* Croton oil alone does not produce skin tumors. In this system DMBA is the initiator and croton oil is the promoter.

FIGURE 20-11 Experimental data correlating carcinogenic potency with DNA-binding affinity. Six radioactively labeled carcinogens were applied to mouse skin and allowed to act for 24 hr. Radioactivity bound to DNA and protein was then quantitated, and plotted against the relative carcinogenic potencies of the six substances. Abbreviations: Napth = napthalene, DBA = dibenzanthracene, BP = benzpyrene, MC = 20-methylcholanthrene, DMBA = 9,10-dimethyl-1:2-benzanthracene.

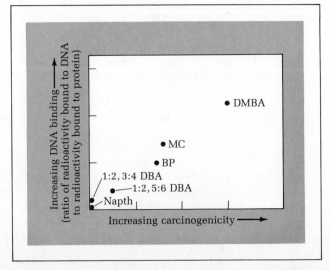

cess that requires prolonged exposure to promoting agents and is at least partially reversible. Exposure to promoting agents generally leads to a stimulation in cell division, thereby increasing the population of damaged cells. As the damaged cells continue to divide, a gradual selection for cells exhibiting enhanced growth rate and invasive properties occurs, eventually leading to the formation of a malignant tumor. The long period that can be occupied by the promotion phase explains why cancer may not develop until many years after exposure to a carcinogenic agent. Some insights into the possible molecular mechanism by which promoting agents stimulate cells has emerged from studies on **phorbol esters,** a family of tumor promotors whose biochemical effects have been extensively investigated. Phorbol esters bind to the cell surface and trigger changes in protein phosphorylation associated with the activation of protein kinase C, a Ca^{2+}-dependent enzyme known to be activated by several other hormones and growth factors. Hence the mechanism by which phorbol esters stimulate cells may not be too different than that employed by naturally occurring growth regulators.

Radiation-induced Cancer

Energy that travels through space is known as **radiation.** Some natural sources of radiation to which humans are normally exposed include ultraviolet rays from the sun, cosmic rays from outer space, and emissions from naturally occurring radioactive elements. Medical, industrial, and military activities have created additional sources of high-energy radiation, mainly in the form of X-rays and radioactivity. Although the various types of radiation to which we are exposed differ in energy and wavelength, many of them are capable of causing cancer.

The role of sunlight in skin cancer was first deduced from the observation that skin cancer is most common in people who spend long hours in the sun, especially in tropical regions where the sunlight is most intense. Only exposed areas of the body, such as the neck, arms, and hands are affected, indicating that protection from this type of radiation-induced cancer can be afforded by covering the skin during prolonged exposure to the sun. Dark-skinned individuals have a much lower incidence of skin cancer than whites, presumably because their skin pigments filter out ultraviolet light. Exposure to sunlight rarely causes any type of malignancy other than skin cancer because ultraviolet radiation is too weak to pass through the skin and into the interior of the body. Fortunately skin cancer rarely metastasizes, and its superficial location makes it relatively easy to treat.

X-rays are a more serious problem because this type of radiation is strong enough to penetrate through the skin and affect internal organs. Shortly after the discovery of X-rays in 1895 by Wilhelm Roentgen, individuals working with this form of radiation were found to ex-

hibit an increased incidence of cancer. The fact that X-rays cause cancer was confirmed shortly thereafter by studies carried out on laboratory animals. Our exposure to X-rays could be minimized if they had no medical uses, but in many situations the health benefits to be gained through the use of X-rays appear to outweigh the risk of inducing cancer. This is not always the case, however. In countries where medical X-rays have been employed for the treatment of superficial skin conditions of the head and neck, such as ringworm and acne, the rate of thyroid cancer is much higher than in countries where this practice is not widespread. Thus the practical benefits to be gained from every X-ray should be prudently weighed against the increased cancer risk.

Many radioactive elements emit radiation whose carcinogenic effects resemble those of X-rays. One of the first scientists to work with radioactivity was Marie Curie, the co-discoverer of the radioactive elements polonium and radium. She later died of a form of leukemia apparently caused by her extensive exposure to radioactivity. One of the more striking examples of radiation-induced cancer occurred during the 1920s in a small group of women employed by a New Jersey factory that produced watch dials that glow in the dark. The luminescent paint used in painting these dials contained radium, and was applied with a fine-tipped brush that the employees frequently wet with their tongues. In the process minute quantities of radium were inadvertently ingested. Several years later this group of women developed an alarmingly high rate of bone cancer caused by the radioactive radium that had gradually become concentrated within their bones.

A more horrifying example of the carcinogenic properties of radiation was provided by the atomic bombs exploded over Hiroshima and Nagasaki in 1945. The massive radioactive fallout produced by these explosions was followed years later by dramatic increases in the incidence of leukemia, lymphomas, and cancers of the thyroid, breast, uterus, and gastrointestinal tract. Individuals exposed to the radioactive fallout produced by the nuclear bombs tested in Nevada in the 1950s also suffered an increased incidence of leukemia years later.

Radiation-induced carcinogenesis resembles chemical carcinogenesis in its basic mode of action. Like most chemical carcinogens, radiation is mutagenic and presumably initiates malignant transformation by inducing changes in DNA. Likewise, a promotion phase is required after exposure to radiation before malignancy is fully expressed. This delay may cause many years to elapse between exposure to radiation and the development of cancer.

Oncogenic Viruses

The idea that viruses can cause cancer was first espoused at the turn of the century. The importance of chemicals and radiation as cancer-inducing agents was already appreciated at the time, but the possibility that certain cancers might be caused by infectious agents was not widely appreciated because cancer does not generally behave like a contagious disease. However in 1908, the Danish scientists V. Ellerman and O. Bang reported that filtered extracts obtained from the blood of leukemic chickens are capable of transmitting this disease when inoculated into other chickens. Several years later Peyton Rous discovered that sarcomas can be transmitted between chickens by injection of tumor extracts that had first been filtered to remove intact cells and bacteria. Rous concluded that the chicken sarcoma is virally transmitted, and he was eventually able to isolate six different viruses from tumors removed from chickens brought to him by local farmers.

Despite the clarity of Rous's data, his work was greeted with skepticism and it was not until years later that the existence of cancer-causing, or **oncogenic** viruses, came to be widely accepted. The eventual acceptance of this idea required the demonstration that other tumors are caused by viruses. In 1933 Richard Shope

demonstrated that a skin cancer known as cutaneous papilloma can be experimentally transmitted between rabbits using filtered cell extracts. In the following year Baldwin Lucké observed that kidney tumors occurring in New England frogs can also be transmitted by a filterable agent. An important milestone occurred a few years later when John Bittner reported that breast cancer in mice is transmitted from mother to offspring by a virus present in the mother's milk. The transmission of a disease from parent to offspring, whether indirectly through the mother's milk or directly through the sperm and egg cells, is known as **vertical transmission.** Since the time of Bittner's pioneering discovery, many oncogenic viruses have been found to be transmitted vertically.

Dozens of different viruses have now been implicated in animal and plant carcinogenesis (Table 20-3). Most of these viruses are relatively selective in the hosts they infect and the types of tumors they cause, but there are exceptions to this rule. **Polyoma** virus, for example, infects a variety of different mammals and causes more

TABLE 20-3
Examples of tumor viruses

Class	Examples	Tumors Induced
DNA VIRUSES		
Herpesviruses	Lucké virus	Kidney adenocarcinoma (frog)
	Epstein–Barr virus	Burkitt's lymphoma and nasopharyngeal carcinoma (humans)
	Marek's disease virus	Lymphoma (chickens)
Papovaviruses	Shope papilloma virus	Papillomas (rabbits)
	SV-40	Subcutaneous, kidney and lung sarcomas (hamsters)
	Polyoma	Cancer of liver, kidney, lung, bone, blood vessels, nervous tissues, and connective tissues (mice)
	Human papillomaviruses (HPVs)	Cervical cancer (humans)
Adenoviruses	Human adenoviruses (many types)	Subcutaneous, intraperitoneal and intracranial tumors (hamsters)
RNA VIRUSES		
B-type viruses	Mouse mammary tumor virus (Bittner)	Mammary carcinoma (mice)
C-type viruses	Rous sarcoma virus	Sarcomas (birds, mammals)
	Murine leukemia viruses (Gross, Moloney, Friend, Rauscher, and others)	Leukemia (mice)
	Feline leukemia virus	Leukemia (cats)
	Murine sarcoma virus	Sarcoma (mice)
	Feline sarcoma virus	Sarcoma (cats)
	Avian leukemia viruses (avian myeloblastosis and others)	Leukemia (chickens)
	Human T-cell lymphotropic viruses (HTLV-I & HTLV-II)[a]	Leukemias/lymphomas (humans)
Plant viruses	Wound tumor virus	Roots and stems

[a] A closely related virus, HTLV-III (also called HIV), causes AIDS.

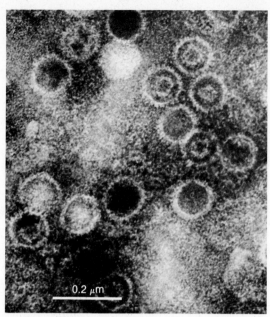

FIGURE 20-12 An electron micrograph of negatively stained Epstein–Barr virus particles. Such particles can be isolated from several different types of human cancer cells. Courtesy of M. A. Epstein.

than 20 different kinds of tumors, including cancer of the liver, kidney, lung, skin, bone, blood vessels, nervous tissues, and connective tissues. Although the overall importance of viruses in human cancer is not clear at present, a viral origin has been demonstrated for at least a few human tumors. The first evidence for the existence of human tumor viruses was provided in the late 1950s by Denis Burkitt, a British surgeon working in Uganda. Burkitt noted that a large number of native children coming to him for medical care suffered from massive tumors of the jaw. This type of cancer, now known as **Burkitt's lymphoma,** is most prevalent in areas where mosquito-transmitted infections are common. The epidemic nature of this lymphoma led Burkitt to propose that it is transmitted by a mosquito-borne infectious agent.

A few years later, electron microscopists examining tumor tissue obtained from patients with Burkitt's lymphoma discovered a virus in these tumor cells (Figure 20-12). This virus, identified as a member of the herpes group, has come to be called the **Epstein–Barr virus** or *EBV* in recognition of the scientists who discovered it. Unlike viruses of animal or plant origin, it is difficult to prove that a virus of human origin causes cancer because the hypothesis cannot be tested directly by injecting the virus back into normal individuals. In spite of this lack of direct proof, the following data are all consistent with the conclusion that the Epstein–Barr virus plays an oncogenic role in Burkitt's lymphoma: (1) viral DNA and proteins can be detected in tumor cells obtained from patients with Burkitt's lymphoma, but not in normal cells; (2) addition of purified Epstein–Barr

virus to normal human lymphocytes in culture causes them to undergo transformation; (3) purified Epstein–Barr virus administered to nonhuman primates induces the formation of lymphomas; and (4) the survival rate of Burkitt's lymphoma patients correlates with the presence of circulating antibodies that react with components of the Epstein–Barr virus.

Besides its role in Burkitt's lymphoma, the Epstein–Barr virus has been implicated in the causation of nasopharyngeal cancer in South China. Perhaps even more interesting is the discovery that in the United States, where more than 90 percent of the population has been exposed to EBV, the major disease induced by this virus is **infectious mononucleosis,** a benign, self-limiting proliferation of white blood cells that causes relatively innocuous flulike symptoms. The reasons why this virus should trigger the formation of deadly forms of cancer in individuals in one country and only a flulike illness in another are not well understood.

Although EBV was the first virus to be clearly associated with human cancer, several other viruses have been implicated in recent years. Among these are the **human T-cell lymphotropic viruses** or **HTLVs,** a family of related RNA-containing viruses. Two members of this group, HTLV-I and HTLV-II, cause certain kinds of human leukemias and lymphomas; a third member, HTLV-III (or HIV), is responsible for **acquired immune deficiency syndrome (AIDS).** Each of these viruses acts by infecting white blood cells; the major difference is that infection by HTLV-I and HTLV-II stimulates abnormal proliferation of the infected lymphocytes, that is, cancer, while infection of lymphocytes by HTLV-III leads to cell death. This lymphocyte destruction triggered by HTLV-III eventually leads to an incapacitated immune system, which is in turn responsible for the various symptoms associated with AIDS. Although patients with AIDS also suffer from an increased incidence of certain kinds of cancer, this appears to be a secondary effect resulting from their depressed immune status rather than a direct effect of HTLV-III itself.

The realization that viruses may be associated with certain forms of human cancer has kindled an intense interest in the question of how oncogenic viruses act. Although important pieces of information are still missing, great strides have been made in recent years in unraveling the mechanism by which oncogenic viruses cause normal cells to become malignant. Because the steps involved seem to differ for oncogenic DNA and RNA viruses, we shall consider these two types of viruses separately.

Malignant Transformation by DNA Viruses Oncogenic DNA viruses can be subdivided into three major groups: herpesviruses, adenoviruses, and papovaviruses. **Herpesviruses,** which include the previously mentioned Epstein–Barr virus, are responsible for a variety of malignant and nonmalignant diseases. Among the malignant tumors caused by herpesviruses are Burkitt's lymphoma and nasopharyngeal carcinoma in humans, kidney cancer in frogs (the Lucké virus), and a contagious type of lymphoma in chickens known as Marek's disease. Members of the herpesvirus group are also the causative agents for several nonmalignant diseases, including chicken pox, cold sores, venereal disease, and, as mentioned earlier, infectious mononucleosis.

Adenoviruses are a group of DNA viruses commonly found in the respiratory tract. Of the more than 30 kinds of adenoviruses isolated from human sources, more than a dozen cause cancer when injected into animals. Although these same viruses are responsible for a variety of respiratory ailments in humans, there is no evidence implicating them in human cancer.

Papovaviruses are the final group of DNA viruses implicated in tumor induction. The most thoroughly studied papovaviruses are polyoma and SV-40, both of which cause cancer in laboratory animals but apparently not in humans. Also in this group are the human papillomaviruses (HPVs), a family of several dozen viruses that have been isolated from a variety of human tissues. While HPVs can cause clearly benign growths such as warts, they have also been implicated in certain forms of human cancer. For example, HPV genes have been detected in the cancer cells of most women suffering from cancer of the uterine cervix, and in many men with cancer of the penis. Such observations raise the alarming possibility that a cancer-inducing virus may be transmitted sexually.

The mode of action of all three groups of DNA tumor viruses appears to be similar. After entering a host cell, the viral DNA is copied into messenger RNA molecules that are then translated into viral proteins. Replication of both cellular and viral DNA ensues, followed by cell division. After several weeks of virus-stimulated cell division, one or more copies of the viral DNA molecule become integrated into the host cell DNA. After this point, the cell continues to replicate the viral DNA as part of its own DNA. The viral genetic information thus becomes a permanent part of the cell's genetic material, and the cell is said to be transformed. During this entire process, no new virus particles need to be produced or released. This explains why the failure to isolate infective virus particles from a human tumor does not rule out a viral origin for the tumor.

Because viruses contain a relatively small number of genes, it should be possible to identify the viral product(s) responsible for conferring the property of malignancy upon infected cells. Oncogenic herpesviruses code for at least 50 polypeptides, so the task of identifying the genes required for malignant transformation has been difficult. With adenoviruses and papovaviruses, on the other hand, the number of genes is much smaller and it has therefore been possible to pinpoint the particular gene required for triggering tumor development. For viruses of this type it has been shown that the gene coding for a protein called the **T antigen** must be intact and functional in order for the virus to cause malignant transformation. Although the exact function carried out by the T-antigen is unclear, it is capable of binding to DNA and stimulating DNA synthesis. The T-antigen has also been reported to exhibit protein kinase activity, although it is not clear whether this activity resides in the T-antigen itself or a contaminating protein. Further work on the nature of the T-antigen is clearly important to our ultimate understanding of how malignant transformation is induced.

Malignant Transformation by RNA Viruses RNA viruses have been linked to a wide variety of animal and plant malignancies, as well as to a few types of human cancer. Among the oncogenic RNA viruses are the Rous sarcoma virus, the Bittner mouse mammary virus, the human T-cell lymphotropic viruses, and several other kinds of viruses that cause leukemias, lymphomas, and sarcomas in rodents, cats, birds, cows, and primates. One of the general characteristics of such viruses is that after infecting a cell, they can remain there for long periods of time in a hidden or "latent" form; such a latent virus does not usually become active until the cell is exposed to appropriate triggering conditions, which may include radiation, chemical carcinogens, hormones, or even other viruses. A common example of this type of behavior is provided by the feline leukemia virus, which can be harbored by otherwise healthy cats for many years. Only upon exposure to a stressful situation, like a relatively mild respiratory infection, does the latent virus become activated to induce the formation of a malignancy. At the same time large numbers of new virus particles are produced and released, triggering a potential cancer epidemic among neighboring cats.

Unlike DNA tumor viruses, RNA tumor viruses cannot insert their genes directly into the host cell DNA because the virus stores its genetic information in the form of RNA rather than DNA. The mechanism by which this limitation is overcome first became apparent in 1970, when Howard Temin and David Baltimore independently discovered that cells infected by RNA tumor viruses produce a virally encoded enzyme, termed **reverse transcriptase,** which catalyzes the synthesis of DNA using the viral RNA as template. The DNA produced by this reaction, termed a **provirus,** is then integrated into the host's chromosome and undergoes replication along with the chromosomal DNA (Figure 20-13). Viruses that employ this reverse-transcriptase-mediated pathway for viral integration and replication are referred to as **retroviruses.**

The portion of the integrated viral DNA responsible for transforming a normal cell into a malignant one is

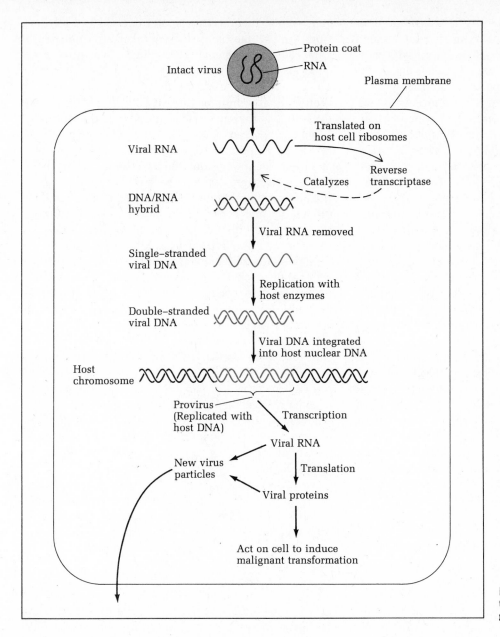

Protein coat

Intact virus — RNA

Plasma membrane

Viral RNA

Translated on
host cell ribosomes

Catalyzes

Reverse
transcriptase

DNA/RNA
hybrid

Viral RNA removed

Single–stranded
viral DNA

Replication with
host enzymes

Double–stranded
viral DNA

Viral DNA integrated
into host nuclear DNA

Host
chromosome

Provirus
(Replicated with
host DNA)

Transcription

Viral RNA

New virus
particles

Translation

Viral proteins

Act on cell to induce
malignant transformation

FIGURE 20-13 Steps involved in transformation of cells by oncogenic RNA viruses.

termed an **oncogene.** Oncogenes may remain masked within the chromosomal DNA for long periods of time, but when they are expressed, the host cell undergoes malignant transformation. As we shall see in the following section, a great deal of progress has been made in recent years in identifying oncogenes and investigating the mechanisms by which they induce malignant transformation.

Identification and Characterization of Oncogenes
One of the first oncogenes to be clearly identified occurs in the Rous sarcoma virus, a relatively small virus containing about half a dozen different genes. Identifying the particular gene responsible for inducing malignancy was facilitated by the discovery of mutant forms of the Rous virus that are incapable of inducing cells to un-

dergo malignant transformation, though they can still infect cells and undergo replication. Since the only missing function in these transformation-defective mutants is the ability to cause malignant transformation, it can be concluded that the gene normally responsible for inducing malignancy, that is, the oncogene, is defective. Subsequent biochemical and genetic analyses of these defective Rous viruses revealed that they are missing a single gene referred to as *src.*

Once the *src* gene had been identified, the next problem was to find out what it codes for. This issue was resolved in the laboratory of Ray Erikson using antibodies isolated from the serum of rabbits bearing Rous sarcomas. When sarcoma extracts were exposed to this antiserum, a 60,000 molecular weight protein called $_{pp}60^{v\text{-}src}$ was found to precipitate. In contrast, extracts of normal cells or cells infected with transformation-defective strains of the Rous virus do not contain $_{pp}60^{v\text{-}src}$, suggesting that this protein represents the product of the *src* oncogene. Biochemical studies on $_{pp}60^{v\text{-}src}$ revealed that it is an enzyme that catalyzes protein phosphorylation reactions, that is, it is a protein kinase. It is a rather unusual protein kinase, however, in that its activity is not regulated by cyclic AMP, and it catalyzes the phosphorylation of tyrosine residues instead of the serine and threonine residues phosphorylated by most other protein kinases. This enzyme is therefore referred to as **tyrosine-specific protein kinase.**

Shortly after the discovery of $_{pp}60^{v\text{-}src}$, a similar protein was discovered in normal cells. To distinguish it from $_{pp}60^{v\text{-}src}$, this normal cellular protein is referred to as $_{pp}60^{c\text{-}src}$ (the "c" stands for cellular, while the "v" stands for viral). Although the normal biological role of $_{pp}60^{c\text{-}src}$ is yet to be elucidated, it has been detected in a wide variety of cell types. The cellular gene coding for $_{pp}60^{c\text{-}src}$ has also been identified in normal cells using purified viral *src* gene sequences as a hybridization probe. In this way DNA sequences homologous to, but not identical with, the Rous *src* gene have been detected in normal cells obtained from salmon, mice, cows, birds, and humans. Since the evolutionary separation of birds and mammals occurred over 400 million years ago, it appears that this gene has been conserved in the cells of higher organisms for hundreds of millions of years and must therefore play an important role in normal cell function. The general term **proto-oncogene** has been introduced to refer to such a normal cellular gene that, in a slightly altered form, is capable of inducing malignancy.

The above considerations have led to the conclusion that malignant transformation induced by the Rous sarcoma virus is mediated by production of a virally encoded protein kinase, $_{pp}60^{v\text{-}src}$, which closely resembles the normal protein kinase, $_{pp}60^{c\text{-}src}$. Because the amount of the $_{pp}60^{v\text{-}src}$ protein kinase produced by virally transformed cells is 30- to 50-fold higher than the normally prevailing levels of $_{pp}60^{c\text{-}src}$ protein kinase, it has been proposed that malignant transformation is caused by an increase in the phosphorylation of key cellular proteins induced by abnormally high levels of protein kinase. Alternatively, it is possible that differences in the spectrum of proteins phosphorylated by $_{pp}60^{v\text{-}src}$ and $_{pp}60^{c\text{-}src}$ is responsible for inducing malignancy.

In either case the close resemblance between the *src* gene of the Rous sarcoma virus and the analogous proto-oncogene of normal cells suggests that the oncogenes of RNA tumor viruses may have originally been derived from normal cellular genes. According to this theory, the first step in the evolutionary development of a retroviral oncogene occurs when a provirus becomes integrated into normal cells adjacent to a proto-onco-gene. When the provirus is later excised and released from the chromosomal DNA, it takes a portion of the adjacent chromosomal DNA, including the proto-oncogene, along with it. During subsequent cycles of viral infection and integration, this proto-oncogene then gradually mutates into an oncogene. This theory thus implies that an ancient ancestor of the Rous sarcoma virus picked up a normal proto-oncogene from an infected cell millions of years ago, and that this proto-oncogene eventually mutated into the *src* oncogene.

The discovery of the *src* gene and the characterization of its protein product has stimulated the search for other viral oncogenes, leading to the identification of more than 20 different kinds of oncogenes among the various oncogenic retroviruses. Moreover, a somewhat different experimental approach has led to the discovery that oncogenes also occur in some cancers that are not caused by oncogenic viruses. This latter realization first emerged from DNA **transfection** experiments in which normal cells were exposed to DNA isolated from cancer cells under conditions that promote the normal cells to take up some of the tumor cell DNA. Such studies revealed that DNA isolated from certain kinds of human cancer cells is capable of causing other cells to become malignant. With the aid of recombinant DNA technology, it has been possible to clone several of the responsible genes. Some of these oncogenes resemble particular oncogenes present in retroviruses, while others do not.

It was soon discovered that oncogenes identified by this approach are similar to genes occurring in normal cells, but differ from their normal counterparts, or proto-oncogenes, in one of several possible ways. The most straightforward type of difference between an oncogene and its normal cellular counterpart is a simple alteration in base sequence. It has been found, for example, that an oncogene isolated from human bladder cancer cells differs from the corresponding proto-oncogene in normal cells by only a single base pair. This single abnormal base pair within the oncogene results in the production of a protein product with one abnormal amino acid. This slightly aberrant protein, in turn, is presumably responsible for triggering malignancy.

A second way in which oncogenes may be generated is through gene amplification, a process leading to an increase in gene number and thus the formation of an excess amount of protein product. In such cases it appears that overproduction of a normal gene product, rather than an abnormal product per se, is responsible for causing cancer. Finally, a third mechanism by which normal genes are converted into oncogenes is through major chromosomal rearrangements, called **translocations,** in which a segment of one chromosome is transferred to another chromosome. A striking example of this phenomenon occurs in patients with Burkitt's lymphoma, a malignancy often associated with a chromosomal abnormality in which a portion of chromosome 8 has been translocated to chromosome 14. The crucial gene transferred by this process is a proto-oncogene located on chromosome 8 known as proto-*myc*. When proto-*myc* is translocated to chromosome 14, it becomes situated adjacent to immunoglobulin genes, which are normally very active in lymphocytes. Being placed in this environment alters the normal rate of transcription of the proto-oncogene, resulting in an abnormal rate of production of its corresponding protein product. In other words, when proto-*myc* is placed in an abnormal chromosomal environment, its normal transcriptional control mechanisms are disrupted, leading to an aberrant pattern of expression that triggers malignancy. Under such conditions this abnormally behaving gene is referred to as a *myc* oncogene.

Classes of Oncogene Products The ability to isolate and characterize specific oncogenes associated with the development of cancer has clearly brought us a step closer to understanding what causes cells to become malignant. Ultimately, however, we need to know exactly how the protein products encoded by oncogenes cause malignancy. We have already seen that the *src* oncogene encodes a tyrosine-specific protein kinase that closely resembles a protein kinase present in normal cells. But what kinds of protein products are encoded by the several dozen other known oncogenes? Although information is still rapidly developing in this area, it appears as if the known oncogene products can be subdivided into at least four major categories (Table 20-4):

1. About a dozen oncogenes code for proteins resembling protein kinases. Some of these are tyrosine-specific protein kinases (like the *src* gene product), while others catalyze the phosphorylation of serine and threonine residues. While many of these oncogene products are cata-

lytically active protein kinases, this category also includes related proteins that may not be enzymatically active. For example, the protein encoded by the **erb-B** oncogene corresponds in amino acid sequence to the portion of the epidermal growth factor receptor that normally catalyzes tyrosine phosphorylation. However, the *erb-B* oncogene codes for only a partial fragment of this protein that, by itself, is not catalytically active.

2. A second type of oncogene codes for proteins related to growth factors. The best characterized example is the *sis* oncogene, which codes for a protein that closely resembles platelet-derived growth factor (PDGF).

3. A small group of oncogenes code for GTP-binding proteins that are closely associated with the inner surface of the plasma membrane. Although the biological role of these proteins is not clearly understood, GTP-binding proteins have been implicated in modulating the activity of other plasma membrane proteins, such as adenylate cyclase. These oncogene-encoded GTP-binding proteins are thought to play a role in controlling the transmission of growth control signals from the plasma membrane to the cell interior.

4. The last category includes those oncogenes that code for protein products that function within the nucleus. Such proteins are thought to be involved in controlling the rate of DNA transcription and/or replication.

In looking over the preceding list, it is clear that oncogene products impinge in various ways upon the normal chain of events by which growth factors regulate cell growth and division. Several oncogene products represent abnormal versions of the protein kinases normally activated by the binding of growth factors to their appropriate cell surface receptors, while at least one oncogene represents an abnormal variant of a

TABLE 20-4
Classes of oncogenes categorized by the nature of their protein products

Class	Name of Oncogene	Nature of Protein Product
Genes coding for protein-kinase-related products	*src, yes, ros, fes, fps, fgr, abl, neu, fms*	Tyrosine-specific protein kinases
	erb-B	Fragment of EGF receptor that normally functions as tyrosine-specific protein kinase
	mil, raf, mos	Serine/threonine-specific protein kinases
Genes coding for growth-factor-related products	*sis*	Analog of PDGF
Genes coding for GTP-binding proteins	Ha-*ras*, Ki-*ras*, N-*ras*	GTP-binding proteins associated with the inner surface of the plasma membrane
Genes coding for proteins localized within the nucleus	*myc*	DNA-binding protein, influences transcription?
	myb, ski, fos	Regulators of DNA transcription or replication?

Note: Abbreviations: PDGF = platelet-derived growth factor; EGF = epidermal growth factor.

growth factor itself. Other proteins encoded by oncogenes reside on the inner surface of the plasma membrane, where they are well situated to influence signals received from the cell exterior. Finally, some oncogene products represent nuclear proteins that may be involved in directing a cell to divide in response to growth factor stimulation.

In addition to providing a unifying view of how cancer might be induced by various alterations in the normal growth control pathway, the discovery of oncogenes also provides a unifying framework for understanding the possible relationships between the various kinds of environmental carcinogens. We have already seen that chemical carcinogens, radiation, and viruses all have in common the ability to alter a cell's DNA. Hence one of the ways in which these diverse agents may act to cause cancer is by converting a proto-oncogene into an oncogene or, in the case of oncogenic viruses, by directly introducing an oncogene. In spite of the attractive simplicity of this model, it should be cautioned that oncogenes have thus far been detected in less than a quarter of all human cancers, and it is therefore premature to assume that they will be able to explain all types of cancer.

Heredity and Cancer

We have now seen that changes in DNA caused by chemicals, radiation, and viruses can all induce cancer. Inherited mutations in DNA have also been implicated in malignancy, although it is the *susceptibility* to cancer

rather than cancer per se that is inherited (Table 20-5). In human breast cancer, for example, the probability of the average woman in the United States developing this condition is 1 in 12. But among women who have a blood relative (mother, sister, aunt, or grandmother) with breast cancer, the chances increase to 1 in 5.

In certain cases the nature of the genetic defect leading to a predisposition to developing cancer is well understood. In patients suffering from **xeroderma pigmentosum,** for example, a specific enzyme involved in DNA repair is deficient. Because the skin cells of such persons are less able to repair DNA damage resulting from exposure to sunlight, an elevated susceptibility to skin cancer results.

Additional insight into the genetic mechanisms underlying susceptibility to cancer has emerged from studies carried out on cells obtained from patients with **familial adenomatosis,** an inherited disease characterized by a predisposition toward cancer of the colon and rectum. Fibroblasts isolated from such individuals will not form tumors when injected into *nude* mice, indicating that these cells have not undergone malignant transformation. However, if the fibroblasts are first placed in culture and then exposed to the promoting agent TPA (12-O-tetradecanoylphorbol-13-acetate, an active component of croton oil), tumors will develop when the cells are injected into *nude* mice. Since TPA treatment does not cause normal cells to undergo malignant transformation under comparable conditions, it has been concluded that cells from adenomatosis patients are already in an initiated state, and that treat-

TABLE 20-5
Genetically inherited conditions causing an increased susceptibility to cancer in humans

Name of Genetic Disorder	Type of Cancer
Familial breast cancer	Breast
Xeroderma pigmentosum	Skin
Familial adenomatosis	Colon, rectum
Hereditary retinoblastoma	Retinoblastoma
Familial hydronephrosis	Kidney
Fibrocystic pulmonary dysplasia	Lung
Bloom's syndrome	Leukemia, intestinal
von Recklinghausen's neurofibromatosis	Fibrosarcoma, neuroma
Ataxia-telangiectasia (Louis–Barr syndrome)	Leukemia, stomach and brain
Paget's disease of bone	Bone
Fanconi's anaplastic anemia	Leukemia, liver and skin
Hereditary polyendocrine adenomatosis	Leukemia
Bruton's agammaglobulinemia	Leukemia
Down's syndrome	Leukemia
Multiple exostosis	Bone and cartilage
Phaeochromocytoma	Adrenal medulla
Multiple polyposis coli (Gardner's syndrome)	Colon, pancreas, thyroid, adrenal
Nevoid basal cell carcinoma syndrome	Skin, ovaries
Tylosis with esophageal cancer	Esophagus
Tuberous sclerosis	Glial tumors, heart muscle, kidney
Chediak–Higashi syndrome	Retina
Multiple endocrine adenomatosis	Pancreas, pituitary, parathyroid and adrenal cortex

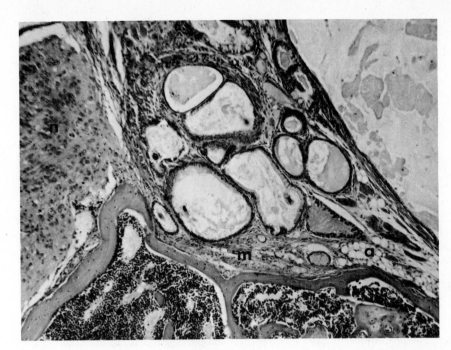

FIGURE 20-14 Light micrograph of a testicular teratocarcinoma of the mouse illustrating the variety of tissue types present. Among the tissues that can be detected within this tumor are nerve tissue (n), adipose tissue (a), and skeletal muscle (m). Each of these differentiated tissues is derived from the tumor itself. Courtesy of L. C. Stevens.

ment with an appropriate promotor causes them to become malignant.

Genetic Versus Epigenetic Views of Cancer

The knowledge that chemical carcinogens are often mutagenic, that radiation causes physical damage to DNA, that oncogenic viruses insert their genetic information into the host DNA, and that susceptibility to certain types of cancer is inherited all support the conclusion that cancer is ultimately produced by changes at the DNA level. Although there is clearly a great deal of truth in this statement, several lines of evidence indicate that genetic mutation is not sufficient in itself for producing cancer. The fact that tumors often do not appear until years after DNA damage has occurred provides a simple basis for concluding that something other than mutation is involved in the ultimate development of malignancy. Even more dramatic evidence has emerged from studies that reveal that under appropriate conditions, malignant cells may spontaneously revert back to nonmalignant behavior.

One well-documented example of such reversion occurs in **crown gall** disease, a malignant tumor of plants induced by infection with a specific strain of bacteria that transmits a cancer-inducing plasmid to the infected cells. An extensive series of investigations carried out by Armin Braun and his colleagues have led to the remarkable conclusion that the crown gall tumor is completely reversible. In these experiments crown gall tumor cells were grown in tissue culture under conditions that promote development of structures resembling plant shoots. These tumor shoots were then grafted to the cut stem tips of healthy plants. In some

cases these grafted shoots responded to the new environment by developing into stems that produced leaves, flowers, and even seeds capable of generating new plants. There was no sign of cancer in the structures emerging from the grafted shoots, and microscopically the cells all appeared normal. However, when cells taken from the leaves growing from these grafted shoots were placed in tissue culture, they again began to grow like tumors. These cells therefore retain the genetic alteration specifying the malignant state, but the expression of this information appears to be controlled by environmental conditions. Such alterations in the expression of genetic information are called **epigenetic** changes, to distinguish them from changes in the genetic information itself.

Reversion from the malignant state is not restricted to plant tumors. One of the most dramatic examples of a reversible animal tumor is the mouse **teratocarcinoma,** a malignant tumor that develops from primordial germ cells. This type of cancer, which also occurs in humans, contains differentiated cell types such as muscle, bone, cartilage, nerve, and skin intermixed with foci of undifferentiated malignant cells termed *embryonal carcinoma* cells (Figure 20-14). Cloning experiments carried out in the early 1960s revealed that single embryonal carcinoma cells injected into mice produce tumors containing all the above cell types, indicating that the embryonal carcinoma cells give rise to the other cell types by differentiation.

An even more dramatic demonstration of the ability of embryonal carcinoma cells to differentiate into apparently normal cells was provided by an elegant study carried out by Beatrice Mintz and Karl Illmensee in 1975. These investigators removed embryonal carci-

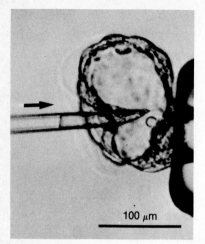

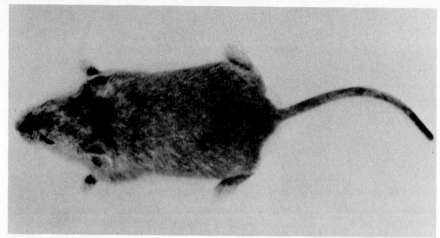

FIGURE 20-15 Fate of malignant teratocarcinoma cells injected into mouse blastocysts. *(Left)* Single teratocarcinoma cells are injected into a mouse blastocyst. *(Right)* Adult female mouse that developed from a blastocyst that had been injected with teratocarcinoma cells. The dark patches of coat color are derived from the teratocarcinoma cells, while the lighter patches are derived from the normal cells of the blastocyst. Courtesy of B. Mintz and K. Illmensee.

noma cells from one strain of mouse and injected them into a blastula obtained from another strain of mice. These blastulae were then reimplanted into a pregnant mouse and allowed to develop to maturity. The mice developing from these injected blastulae grew into normal adults, with no signs of malignancy (Figure 20-15). In addition, genetic markers characteristic of the strain of mouse from which the embryonal carcinoma cells had been obtained were found to be present in many of the normal tissues of these mice. One can therefore conclude that embryonal carcinoma cells retain the ability to differentiate into normal cells and tissues when placed in an appropriate environment.

Although attempts to demonstrate the reversion of malignant cells back to a nonmalignant state have not been generally successful with other kinds of tumor cells, an illuminating experiment carried out by Robert McKinnell and his associates suggests that such reversion may at least be theoretically possible for some other tumor types. In these experiments nuclei transplanted from Lucké adenocarcinoma cells into enucleated frog eggs were found to be capable of programming the development of normal tadpoles. This rather surprising result indicates that even when nuclei contain viral DNA sequences that have already induced malignant transformation, the genetic information specifying malignancy can be overridden by a more normal pattern of gene expression when placed in an appropriate cytoplasmic environment.

The above observations on the behavior of the crown gall tumor, mouse teratocarcinoma, and Lucké adenocarcinoma indicate that epigenetic as well as genetic alterations can play important roles in the development of malignancy. This distinction between genetic and epigenetic changes is of more than academic interest because the potential reversibility of epi-

genetic phenomena opens up new approaches to the problem of cancer chemotherapy. Instead of restricting cancer therapy to methods for killing or surgically removing cancer cells, it is possible, at least in theory, to consider the possibility of converting malignant cells into nonmalignant ones.

CAN CANCER BE PREVENTED OR CURED?

The major force motivating most cancer research is the idea that once this disease is more completely understood, we will be in a better position to prevent and cure it. Until quite recently, more attention was paid to the area of cancer treatment than cancer prevention. But now that we have come to realize that 75 percent or more of all cancer is caused by identifiable environmental factors, strategies for cancer prevention have begun to attract more attention.

Prevention of Cancer

The first step in preventing environmentally induced cancers is to identify the carcinogenic agents involved. To take a simple example, the discovery of the carcinogenic properties of sunlight and X-rays has provided us with a relatively simple way of reducing radiation-induced cancers: Avoid excess and unnecessary exposure to sunlight and medical X-rays. The identification of human tumor viruses, on the other hand, might conceivably lead to the development of preventative vaccines. Such vaccines have already been developed for leukemias occurring in cats and chickens, but the applicability of this approach to human cancer is not clear because few human cancers behave as if they are transmitted by a contagious virus. Another problem

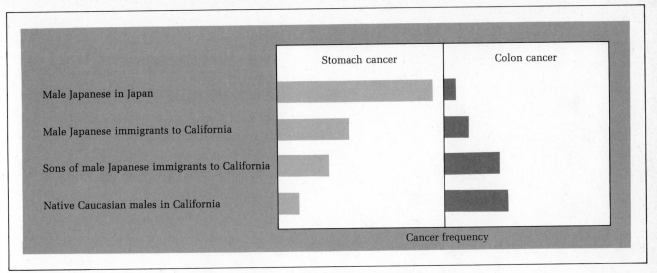

FIGURE 20-16 Comparison of the frequency of stomach and colon cancer in men aged 45 to 64 in Japan, the United States, and in Japanese immigrants to the United States. The fact that the frequency of these cancers changes in the Japanese population when they move to the United States suggests that environmental factors are involved.

with this idea is that any future vaccines made with human tumor viruses would have to be tested in normal human subjects, a prospect posing severe ethical difficulties.

The greatest impact in the area of cancer prevention could be made by eliminating the chemical carcinogens currently present in our food, air, water, clothing, and drugs. As a group, such carcinogens account for a major proportion of all human cancers. In order to remove these carcinogens from the environment, one must first have a reliable way of identifying them. Two approaches have been employed for this purpose: epidemiological analysis and laboratory testing.

In the **epidemiological** approach the incidence of various types of cancer is compared among different populations. This approach has revealed that cancers occur with different frequencies in different parts of the world. For example, stomach cancer is unusually frequent in Japan, breast cancer is especially prominent in the United States, rectal cancer rates are high in Denmark, and esophageal cancer is exceptionally prevalent in Iran. In theory, differences in either hereditary or environmental factors might be responsible for such differences in cancer incidence. Epidemiological studies on individuals who have moved from one country to another suggest that of the two factors, environment is more important. For example, in Japan the incidence of stomach cancer is greater and the incidence of colon cancer is lower than in the United States. In Japanese families immigrating to the United States, however, this difference in cancer incidence gradually disappears (Figure 20-16), indicating that the frequency of these two types of cancer is largely determined by environmental factors.

Epidemiological data have also played an important role in identifying the nature of the environmental factors that cause cancer. One of the most striking examples involves lung cancer, a disease that has increased over tenfold in frequency since 1930 (Figure 20-17). When the environmental factors potentially responsible for this epidemic of lung cancer were investigated, it was discovered that virtually all victims of this disease share one environmental component in common: cigarette smoking. Furthermore, heavy smokers develop lung cancer more frequently than light smokers, and long-term smokers develop lung cancer more frequently than do short-term smokers. Such epidemiological correlations suggest that lung cancer is caused by cigarette smoking, a conclusion reinforced by the discovery that cigarette smoke contains several dozen chemicals known to be carcinogenic (see Table 20-2).

The major drawback to the epidemiological approach is that by definition, it is after the fact. Carcinogenic hazards are only identified after they have caused large numbers of deaths. The epidemiological approach is also limited by the long time span that often intervenes between exposure to a carcinogen and the development of cancer. In the case of cigarette smoking and lung cancer, for example, more than 20 years elapsed before the increase in cigarette smoking was reflected in an increased rate of lung cancer (see Figure 20-17).

The alternative to the epidemiological approach for identifying carcinogenic agents is direct laboratory testing. Until quite recently, such tests were routinely carried out by administering potential carcinogens to laboratory animals. Unfortunately, experiments of this type often take several years to complete, are quite expensive, and the data obtained can be difficult to extrapolate to human populations. This approach is therefore impractical for testing the thousands of chemical pollu-

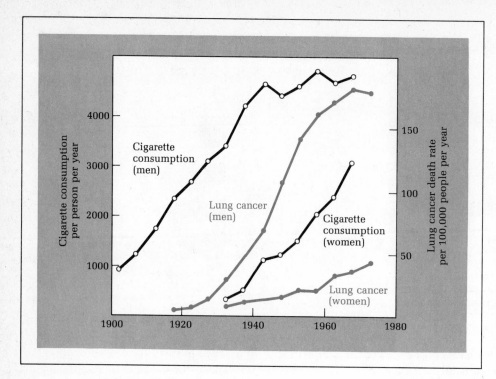

FIGURE 20-17 Epidemiological data illustrating the relationship between increased smoking and increased incidence of lung cancer in Great Britain. A period of 20–30 years intervenes between the increase in cigarette consumption and the increased incidence of lung cancer. Lung cancer is just beginning to increase in the female population, where the increase in cigarette consumption is more recent.

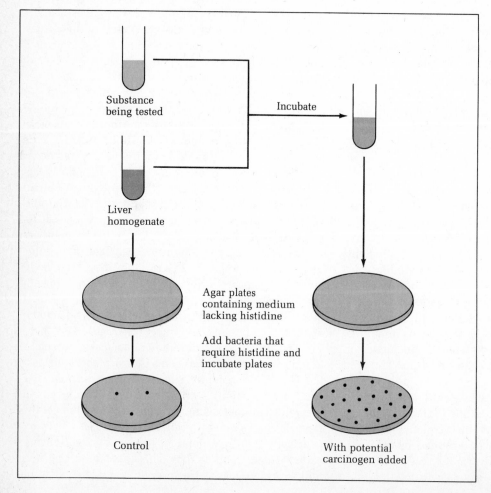

FIGURE 20-18 Summary of the Ames test for identifying potential carcinogens. Each bacterial cell mutating to a form in which it no longer requires histidine in the medium will grow into a colony. The number of bacterial colonies obtained is therefore a function of the mutagenic potency of the substance being tested. See text for details.

tants that have been added to the environment in recent years.

To circumvent this problem, Bruce Ames and his associates have devised a quick laboratory test based on the rationale that many carcinogens are mutagens, and mutagenic activity is relatively easy to measure. In essence, the *Ames procedure* measures the ability of various substances to induce mutations in bacteria (Figure 20-18). A special strain of bacterial cells lacking the ability to synthesize the amino acid histidine is employed in this test. These bacterial cells are placed in culture dishes that contain the chemical substance being tested and a growth medium lacking histidine. Normally, of course, such bacteria will not grow in media lacking histidine. If the substance being tested is a mutagen, however, some of the bacterial cells will undergo mutations that restore their ability to synthesize histidine. Each cell acquiring such a mutation will grow into a visible colony. The total number of bacterial colonies observed on such plates is therefore a direct measure of the mutagenic potency of the substance being tested.

One complication that prevents this strategy from being directly applied to the detection of human carcinogens is the fact that many chemicals are not carcinogenic in the form in which they are present in the environment. Only after they have undergone bio-

chemical modifications in the cells of the host organism are carcinogenic derivatives formed. For this reason chemicals being tested by the Ames procedure are first incubated with a liver homogenate so that such modification reactions can take place before the test for mutagenicity is carried out. Although some exceptions occur, the mutagenic potency of various substances as measured by the Ames assay correlates well their carcinogenic potency (Figure 20-19). Therefore such tests, although not foolproof, provide a rapid and inexpensive method for the preliminary identification of potential carcinogens.

The identification of potential carcinogens by epidemiological analysis or laboratory testing is only the first step in assessing the danger posed by substances present in the environment. Once a particular carcinogen has been pinpointed, the question arises as to how much of this substance can be tolerated in the environment before a clear hazard exists. This is an extremely difficult question to answer experimentally because small doses of carcinogens produce very few tumors, making them difficult to detect statistically. Yet a thorough understanding of the effects of low concentrations of carcinogenic agents is extremely important because of the relevance of such information for public policy decisions. Some investigators believe that a threshold

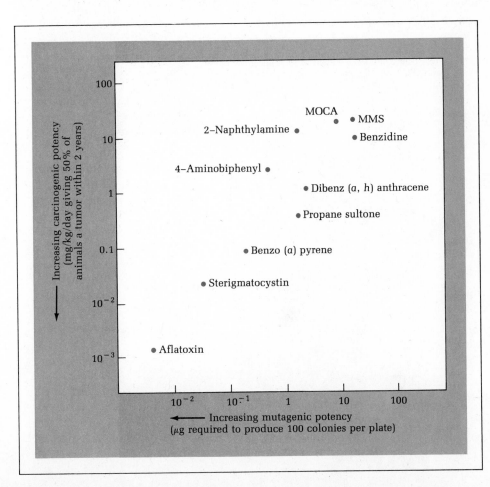

FIGURE 20-19 Correlation between carcinogenic potency in animals and mutagenic potency as measured by the Ames test for ten different substances. *Abbreviations:* MOCA = 4-4′-methylene-*bis*2-chloroanaline, MMS = methyl methanesulfonate.

level exists for each carcinogen, and that doses below the threshold level do not cause cancer. If this is true, it would significantly influence our views about the regulation of environmental carcinogens. Unfortunately, the lack of reliable data on the effects of low doses of carcinogens leaves the point unresolved, and it would therefore be prudent to consider any dose of carcinogen to pose a cancer risk until it can be proven otherwise.

In 1958 the United States Congress passed the Delaney Amendment to the Food, Drug, and Cosmetic Act that instructed the Food and Drug Administration (FDA) to prohibit food additives known to cause cancer in humans or laboratory animals. Under this act the FDA has banned substances such as cyclamate and Red Dyes 2 and 4. However, each banning has been accompanied by a long and controversial debate about the adequacy of the data. Such debates will no doubt continue because enormous pressure is put on Congress by special interest groups whose economic interests would be adversely affected by the banning of particular substances. One need only consider the obvious example of the tobacco industry, whose political and economic power has prevented any significant government regulation of a product implicated in the deaths of over a hundred thousand cancer victims a year. Treatment of these cancer victims alone costs society more than a billion dollars annually, not to mention the billions of dollars in lost earnings and productivity. A society that

continues to condone such a state of affairs is clearly not committed to the idea of preventing cancer.

The tobacco problem also illustrates the point that millions of people are not willing to voluntarily change their habits in order to reduce the risk of cancer. If people were to stop smoking, government regulation of tobacco would become an irrelevant issue. And tobacco is not the only lifestyle factor that influences cancer risk. Extensive meat consumption has been correlated with an increased incidence of colon cancer, and dietary fat intake has been correlated with the frequency of uterine cancer (Figure 20-20). Hence careful attention to personal habits such as smoking and diet, along with avoidance of occupations that involve exposure to known carcinogenic agents, would allow the average person to significantly reduce his or her risk of developing cancer.

Surgery, Chemotherapy, and Radiation

No matter how seriously society pursues avenues for cancer prevention, it is impossible to eliminate every environmental carcinogen. Even if all human-made carcinogens were to be banned from production, which is highly unlikely, natural carcinogens still exist. There are natural sources of radiation, such as cosmic rays and sunlight, as well as naturally occurring carcinogenic chemicals. For example the **aflatoxins,** a group of poi-

FIGURE 20-20 Epidemiological data showing the correlation between diet and two types of cancer in various countries. (a) Data illustrating relationship between meat consumption and the frequency of colon cancer. (b) Data illustrating the relationship between fat consumption and the frequency of uterine cancer.

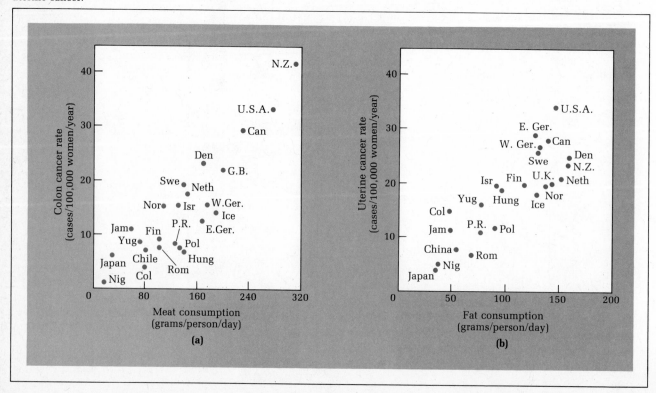

TABLE 20-6
Main types of antitumor agents used in cancer chemotherapy

Class	Examples	Major Mechanism of Action
1. Antimetabolites	Methotrexate 5-Fluorouracil 6-Mercaptopurine	Inhibit enzymatic pathways for biosynthesis of nucleic acids by substituting for normal metabolites
2. Antibiotics (substances produced by microorganisms)	Actinomycin D Adriamycin Daunorubicin	Bind to DNA
3. Alkylating agents	Nitrogen mustard Chlorambucil Cyclophosphamide Imidazole carboximides	Cross-link DNA
4. Mitotic inhibitors	Vincristine Vinblastine	Destroy mitotic spindle
5. Hormones	Estrogen (for prostate cancer) Cortisone Progesterone Androgens	Inhibit growth of hormone-sensitive cells by interacting with hormone receptors
6. Miscellaneous agents	L-Asparaginase Procarbazine	Hydrolyzes asparagine Depolymerizes DNA

sonous substances secreted by the mold *Aspergillus,* are known to be potent inducers of liver cancer. *Aspergillus* grows readily on nuts and grains in humid tropical climates, causing an exceptionally high incidence of liver cancer in such areas. In addition to naturally occurring radiation and chemical carcinogens, inherited mutations in DNA cause a certain amount of cancer that cannot be readily prevented. Thus no matter how successful prevention is as a strategy, better methods for treating cancer will always be needed. The three approaches for treating cancer that have been most successful to date are surgery, chemotherapy, and radiation.

Surgery is the approach that is generally used to remove tumors that are localized to a particular part of the body. This approach is most effective when the cancer is of a nonmetastasizing type, or has been diagnosed early enough to minimize the chances that metastasis has already occurred. Unfortunately cancers involving internal organs often do not produce clinical symptoms until they are well advanced and disseminated throughout the body, reducing the effectiveness of the surgical approach. It is for this reason that early detection of cancer is so important.

Some kinds of cancer are treated by **radiation** therapy because cells engaged in DNA synthesis or mitosis are particularly sensitive to the killing effects of radiation. Although normal dividing cells are also damaged· in this fashion, the ability to focus radiation beams on small areas of the body allows this approach to be somewhat selective. Irradiation with X-rays or radioactive substances such as cobalt-60 has been especially effective in treating Hodgkin's disease (a kind of lymphoma),

skin cancer, and certain forms of testicular and bone cancers. The major disadvantage of radiation therapy is that normal dividing cells, such as bone marrow or intestinal lining cells, are also destroyed by radiation, limiting the doses of radiation that can be safely used. In addition, there is the additional risk posed by the fact that radiation itself is carcinogenic.

The use of antitumor drugs, or **chemotherapy,** is the third widely employed approach for treating cancer. Because drugs are carried by the circulatory system throughout the body, this approach is particularly well suited for the treatment of tumors that have already metastasized. Antitumor drugs can be divided into at least six different classes that reflect their varying mechanisms of action (Table 20-6). The major problem with such drugs is that they inhibit the division of normal cells as well as cancer cells, producing toxic side effects such as diarrhea (caused by destruction of intestinal lining cells), loss of hair (caused by destruction of hair follicle cells), and susceptibility to infections (caused by destruction of dividing blood cells).

For several types of cancer, chemotherapy has been successful in restoring a normal life expectancy to a significant proportion of treated patients. Included in this category are Burkitt's lymphoma, choriocarcinoma, acute lymphocytic leukemia, Hodgkin's disease, non-Hodgkin's lymphomas, mycosis fungoides, Wilms' tumor, Ewing's sarcoma, rhabdomyosarcoma, retinoblastoma, and embryonal testicular tumors. The cure rates for these drug-sensitive tumors range from a low of around 20 percent (metastatic embryonal testicular cancer) to a high of 75 percent (metastatic choriocarcinoma). In some cases drugs alone are responsible for the improved survival rate, while in others chemotherapy is used in conjunction with surgery or X-irradiation. The effectiveness of chemotherapy can occasionally be improved by utilizing several drugs in combination instead of a single therapeutic agent alone. Drug combinations are apparently more effective because the increase in tumor-killing ability obtained when one adds several drugs together tends to exceed the increase in toxic side effects.

Although chemotherapeutic agents have been quite successful in increasing survival rates for certain types of cancer, many of the major cancer killers, such as lung and intestinal cancers, do not respond well to chemotherapy. What cancer biologists continue to search for is a chemical compound that will selectively seek out and destroy cancer cells, without damaging normal cells in the process. Only with such a "magic bullet" will it ever be possible to completely eradicate cancer.

Future Prospects

It is difficult to know at present whether the search for such a magic bullet will be successful. It is clear, however, that two recent technical advances have provided us with powerful new tools for approaching this elusive goal. The first advance is the development of the hybridoma technique for producing monoclonal antibodies (see Figure 17-22). With this technique it should be possible to produce large quantities of antibodies capable of selectively binding to tumor-specific antigens. Such antibodies, if linked to toxic molecules, may one day provide a way of selectively directing toxic agents to tumor cells.

The other technical breakthrough involves the development of recombinant DNA techniques for gene cloning. Hopefully such technology will eventually be used to clone or create genes coding for specific proteins capable of selectively inhibiting the growth of tumor cells. One gene that has already been cloned with this goal in mind is the gene coding for interferon (page 664). This antiviral protein has been reported to retard tumor growth in several clinical studies involving human patients, but conventional protein purification procedures have not been able to provide the quantities of interferon required for extensive testing. The cloning of the interferon gene in bacteria, however, should eventually provide physicians with enough interferon to determine whether or not this is an effective antitumor agent.

In the future the gene cloning approach may allow us to isolate biological growth regulators whose existence is yet to be discovered. It has been long believed, for example, that tissues produce feedback inhibitors of cell division that selectively inhibit cell division in the tissue in which they are produced. These hypothetical substances, termed **chalones,** are yet to be purified and identified, although crude extracts exhibit activities consistent with the existence of chalones. If the failure to purify these agents has been due to the fact that they are present in very low concentration, gene cloning techniques may eventually permit them to be identified and produced in large quantity. The isolation of such tissue-specific inhibitors of cell division, if they exist, would certainly bring us a step closer to the long sought after magic bullet.

SUMMARY

Tumors exhibit uncontrolled growth in which the rate of cell division exceeds the rate of cell loss, causing a progressive increase in cell numbers. Benign tumors are rarely life threatening because they are confined to a localized area, while malignant tumors, or cancers, are more dangerous because they infiltrate neighboring tissues and metastasize via the circulatory system to distant parts of the body. Although no single property appears to be universally present in cancer cells, the following characteristics are commonly observed:

1. Cancer cells produce tumors when injected into animals. Injection of tumor cells into *nude* mice, which lack a thymus gland and are therefore incapable of rejecting foreign cells, is often used to test the tumor-forming capabilities of human cells.
2. Normal cells maintained in culture multiply for a limited number of generations before they die, while malignant cells proliferate indefinitely.
3. Tumor cells are less susceptible to density-dependent inhibition of growth than normal cells. One possible explanation for this difference is that tumor cells produce their own growth factors, rather than being limited by the concentration of endogenous growth factors present in the culture medium.
4. Growth of normal cells in culture requires a surface to which the cells can adhere, while malignant cells can grow in liquid suspension or in a semisolid medium, and are therefore said to be anchorage independent.
5. Normal cells stop near the end of the G_1-phase of the cell cycle when exposed to suboptimal growth conditions, while malignant cells do not exhibit such restriction point control.
6. Changes in the cell surface of cancer cells cause them to be less adhesive than normal, more readily agglutinated by lectins, less capable of forming gap junctions, and more likely to exhibit tumor-specific cell surface antigens.
7. Among the other molecular changes that occur in cancer cells are an enhanced secretion of proteases, production of embryonic markers such as carcinoembryonic antigen and alpha-

fetoprotein, secretion of angiogenesis factors, and alterations in DNA. The latter type of change is especially interesting because it has been shown that DNA extracted from cancer cells is capable of inducing normal cells to undergo malignant transformation.

8. Although no single morphological criterion distinguishes cancer cells from normal cells, alterations in cell structure are often associated with malignancy. Among these are anaplasia, large nuclei and nucleoli, large numbers of microvilli and lamellipodia, increased numbers of mitoses, abnormal mitoses, multinucleated giant cells, and abnormal chromosomes.

Most human cancers are caused by one of three identifiable environmental factors: chemicals, radiation, and viruses. Chemical carcinogens present in our air, food, water, clothing, and drugs account for a major proportion of human cancer. Although exposure to many of these carcinogens is involuntary, tobacco smoke is knowingly inhaled by more than a third of the population of the United States despite warnings about its role in lung cancer, a disease responsible for over one in four cancer deaths. Chemical carcinogenesis generally involves two stages: initiation and promotion. Initiators are thought to act as mutagens, creating preneoplastic cells with irreversible changes in their DNA. For malignancy to develop, however, such cells must be stimulated to divide by promoting agents. Promotion often takes a long period of time, explaining why cancer may not develop until years after initial exposure to a carcinogenic agent.

The major sources of radiation-induced cancer are sunlight, which induces skin cancer, and X-rays and radioactivity, which cause a variety of internal cancers. Like most chemical carcinogens, radiation is mutagenic and hence initiates malignant transformation by inducing changes in DNA. A promotion phase is required after exposure to radiation before malignancy develops.

Viruses are responsible for a variety of animal and plant tumors, including Burkitt's lymphoma, nasopharyngeal carcinoma, and certain forms of leukemia in humans. During infection by oncogenic DNA viruses (members of the herpesvirus, adenovirus, and papovavirus groups), the viral DNA becomes integrated into the host cell DNA. In adenoviruses and papovaviruses the gene responsible for malignant transformation codes for a DNA-binding protein called the T-antigen. Unlike DNA tumor viruses, oncogenic RNA viruses cannot insert their genetic information directly into the host cell DNA. Instead the viral RNA codes for an enzyme, termed reverse transcriptase, which catalyzes the synthesis of a DNA molecule using the viral RNA as template. The DNA provirus produced in this way is then integrated into the host cell DNA.

Genes capable of triggering malignant transformation, called oncogenes, have been identified both in oncogenic viruses and in nonviral human cancers. Oncogenes are closely related to normal cellular genes known as proto-oncogenes. Proto-oncogenes can be converted into oncogenes by simple base changes, gene amplification, or chromosomal translocation. The resulting oncogenes code for protein products that are either abnormal in structure or are produced in inappropriate amounts. Oncogene-encoded proteins fall into four major categories: protein-kinase-related proteins (usually associated with cell surface receptors), growth factor analogs, GTP-binding proteins associated with the inner surface of the plasma membrane, and proteins localized within the nucleus that presumably influence DNA transcription or replication. Each of these protein products appears to impinge in some way upon the normal chain of events by which growth factors control cell growth and division.

In addition to chemicals, radiation, and viruses, heredity also influences susceptibility to cancer. In humans, for example, the chance of developing breast cancer is significantly greater in women having a blood relative with breast cancer. In the inherited disease xeroderma pigmentosum, a defective DNA repair enzyme increases susceptibility to skin cancer.

Although changes in DNA are clearly involved in many types of carcinogenesis, mutation in itself is not sufficient for development of the malignant state. The fact that many types of cancer do not occur until years after exposure to a carcinogenic agent implies that other factors may be involved. More dramatic evidence for the role of epigenetic phenomena in the expression of malignancy comes from the discovery that crown gall tumor cells of plants and embryonal carcinoma cells of mice can revert back to nonmalignant behavior under appropriate environmental conditions.

Because the majority of human cancers are caused by environmental carcinogens, removal of these factors from our surroundings could potentially save hundreds of thousands of lives. Environmental carcinogens can be identified either by analysis of epidemiological data or by testing suspected substances for carcinogenicity in animals or mutagenicity in bacteria. Even after carcinogens have been identified, removing them from the environment is complicated by the political, economic, and behavioral factors involved. Nonetheless careful attention to personal habits such as smoking, diet, and occupation would allow the average person to significantly reduce the risk of developing cancer.

The major approaches for treating cancer involve surgery, radiation, and chemotherapy. Employed singly or in combination, these techniques have been highly successful in the treatment of certain types of cancer. Nonetheless the major cancer killers, such as lung and intestinal cancers, do not respond very well to such treatments. Because it will never be possible to prevent all cancer from occurring, future work on therapeutic agents that selectively act on cancer cells is badly needed. The recent development of the hybridoma technique for producing monoclonal antibodies, and the recombinant DNA approach for gene cloning, should provide us with powerful new weapons for pursuing this goal in the future.

SUGGESTED READINGS

Books

Botchan, M., T. Grodzicket, and P. Sharp, eds. (1986). *Cancer Cells 4/DNA Tumor Viruses: Control of Gene Expression and Replication*, Cold Spring Harbor Laboratory, Cold Spring Harbor, NY.

Braun, A. C. (1977). *The Story of Cancer*, Addison-Wesley, Reading, MA.

Cairns, J. (1978). *Cancer, Science and Society*, Freeman, San Francisco.

Cold Spring Harbor Laboratory (1984). *The Cancer Cell*, Cold Spring Harbor Conference on Cell Proliferation, Vol. II, Cold Spring Harbor, NY.

Farmer, P. B., and J. M. Walker (1985). *The Molecular Basis of Cancer*, Wiley, New York.

Feramisco, J., B. Ozanne, and C. Stiles, eds. (1985). *Cancer Cells 3/Growth Factors and Transformation*, Cold Spring Harbor Laboratory, Cold Spring Harbor, NY.

Franks, L. M., and N. Teich, eds. (1986). *Introduction to the Cellular and Molecular Biology of Cancer*, Oxford University Press, New York.

Kupchella, C. E. (1987). *Dimensions of Cancer*, Wadsworth, Belmont, CA.

Levine, A. J., G. F. Vande Woude, W. C. Topp, and J. D. Watson, eds. (1984). *Cancer Cells 1/The Transformed Phenotype*, Cold Spring Harbor Laboratory, Cold Spring Harbor, NY.

Oppenheimer, S. (1985). *Cancer: A Biological and Clinical Introduction*, 2nd Ed., Jones and Bartlett, Boston.

Prescott, D. M., and A. S. Flexer (1982). *Cancer — The Misguided Cell*, Sinauer Associates, Sunderland, MA.

Ruddon, R. W. (1981). *Cancer Biology*, Oxford University Press, New York.

Vande Woude, G. F., A. J. Levine, W. C. Topp, and J. D. Watson, eds. (1984). *Cancer Cells 2/Oncogenes and Viral Genes*, Cold Spring Harbor Laboratory, Cold Spring Harbor, NY.

Weiss, R. A., N. A. Teich, H. E. Varmus, and J. M. Coffin, eds. (1982). *Molecular Biology of Tumor Viruses, RNA Tumor Viruses*, 2nd Ed., Monograph 10C, Cold Spring Harbor Laboratory, Cold Spring Harbor, NY.

Articles

Ames, B. N., W. E. Durston, E. Yamasaki, and F. D. Lee (1973). Carcinogens as mutagens: a simple test system combining liver homogenates for activation and bacteria for detection, *Proc. Natl. Acad. Sci. USA* 70:2281–2285.

Bishop, J. M. (1987). The molecular genetics of cancer, *Science* 235:305–311.

Braun, A. C., and H. N. Wood (1976). Suppression of the neoplastic state with the acquisition of specialized functions in cells, tissues and organs of crown gall teratomas of tobacco, *Proc. Natl. Acad. Sci. USA* 73:496–500.

Buell, P., and J. E. Dunn, Jr. (1965). Cancer mortality among Japanese Issei and Nisei of California, *Cancer* 18:656–664.

Burkitt, D. (1962). A children's cancer dependent on climatic factors, *Nature* 194:232–234.

Cairns, J. (1985). The treatment of diseases and the war against cancer, *Sci. Amer.* 253 (Nov.):51–59.

Collett, M. S., and R. L. Erikson (1978). Protein kinase activity associated with the avian sarcoma virus *src* gene product, *Proc. Natl. Acad. Sci. USA* 75:2021–2024.

Croce, C. M., and G. Klein (1985). Chromosome translocations and human cancer, *Sci. Amer.* 252 (Mar.):54–60.

Gordon, M. P. (1981). Tumor formation in plants. In: *The Biochemistry of Plants*, (A. Marcus, ed.), Vol. 6, Academic Press, New York, pp. 531–570.

Henle, W., G. Henle, and E. T. Lennette (1979). The Epstein–Barr virus, *Sci. Amer.* 241 (July):48–59.

Hunter, T. (1984). The proteins of oncogenes, *Sci. Amer.* 251 (Aug.):70–79.

_____ (1985). Oncogenes and growth control, *Trends Biochem. Sci.* 10:275–280.

Klein, G., and E. Klein (1985). Evolution of tumors and the impact of molecular oncology, *Nature* 315:190–195.

Massagué, J. (1985). The transforming growth factors, *Trends Biochem. Sci.* 10:237–240.

Mintz, B., and R. A. Fleischman (1981). Teratocarcinomas and other neoplasms as developmental defects in gene expression, *Adv. Cancer Res.* 34:211–278.

Sachs, L. (1986). Growth, differentiation and the reversal of malignancy, *Sci. Amer.* 254 (Jan.):40–47.

Sefton, B. M. (1985). Oncogenes encoding protein kinases, *Trends Genet.* 1:306–308.

Sporn, M. B., and A. B. Roberts (1985). Autocrine growth factors and cancer, *Nature* 313:745–747.

Weinberg, R. A. (1983). A molecular basis of cancer, *Sci. Amer.* 249 (Nov.):126–142.

Appendix: Prefixes, Symbols, and Abbreviations

Prefixes

Prefix	Abbreviation	Equivalent
Mega	M	10^6
Kilo	k	10^3
Hecto	h	10^2
Deca	da	10^1
Deci	d	10^{-1}
Centi	c	10^{-2}
Milli	m	10^{-3}
Micro	μ	10^{-6}
Nano	n	10^{-9}
Pico	p	10^{-12}
Femto	f	10^{-15}

Symbols

~	Approximately
Δ	Change in
<	Less than
>	Greater than
$\leq$	Less than or equal to
$\geq$	Greater than or equal to
log	Common logarithm (base 10)
ln	Natural log (base e)
♂	Male
♀	Female
n	Haploid
2n	Diploid
λ	Wavelength
σ	Sigma
%	Percent (parts per hundred)
π	Pi (3.1416)
ψ	Pseudouridine
$\Delta\psi$	Membrane potential

Abbreviations

A	Adenine or adenosine
Å	Angstrom
A site	Aminoacyl site
Act D	Actinomycin D
ACTH	Adrenocorticotropic hormone
AIDS	Acquired immune deficiency syndrome
Ala	Alanine
AMP, ADP, ATP	Adenosine 5'-mono-, -di-, -triphosphate
Arg	Arginine
Asn	Asparagine
Asp	Aspartic acid
Atm	Atmosphere
ATPase	Adenosine triphosphatase

°C	Degrees Celsius
C	Cytosine or cytidine
Cal	Calorie
CAM	Cell adhesion molecule
cyclic AMP	3',5'-Cyclic adenosine monophosphate
CAP	Catabolite activator protein
CCCP	Carbonyl cyanide *m*-chlorophenylhydrazone
cDNA	Complementary DNA
CF_0, CF_1	Chloroplast coupling factor 0, 1
cyclic GMP	3',5'-Cyclic guanosine monophosphate
Ci	Curie
CMP,CDP,CTP	Cytidine 5'-mono-, -di-, -triphosphate
CoA	Coenzyme A
CoQ	Coenzyme Q (ubiquinone)
Con A	Concanavalin A
C_0t	Concentration of DNA × Time
Cys	Cysteine
Cyt	Cytochrome
D	Dalton ($1.661 × 10^{-24}$ g)
D	Dextro configuration
DAB	Diaminobenzidine
dATP,dCTP,etc.	2'-Deoxyadenosine 5'-triphosphate; 2'-Deoxycytidine 5'-triphosphate; etc.
DCIP (DCPIP)	Dichlorophenolindophenol
DFP (DIPF)	Diisopropylphosphofluoridate
DHAP	Dihydroxyacetone phosphate
DNA	Deoxyribonucleic acid
DNAase	Deoxyribonuclease
DNP	2,4-Dinitrophenol
Dol	Dolichol
DOPA	Dihydroxyphenylalanine
E	Enzyme
E'_0	Redox (oxidation–reduction) potential
EBV	Epstein-Barr virus
EDTA	Ethylenediaminetetraacetic acid
EF	Elongation factor
EGF	Epidermal growth factor
ELISA	Enzyme-linked immunosorbent assay
ESR	Electron-spin resonance
F_0, F_1	Mitochondrial coupling factor 0, 1
$FAD(H_2)$	Flavin adenine dinucleotide (reduced)
Fd	Ferredoxin
FeS	Iron-sulfur protein
fMet	N-Formylmethionine
$FMN(H_2)$	Flavin mononucleotide (reduced)
FP	Flavoprotein
FSH	Follicle-stimulating hormone
Fuc	Fucose
g	Centrifugal force; gram
G	Guanine or Guanosine
ΔG	Free-energy change
$\Delta G\ddagger$	Free-energy of activation
$\Delta G°$	Standard free-energy change
GABA	Gamma-aminobutyric acid
Gal	D-Galactose
GalNAC	N-Acetyl-D-galactosylamine
GH	Growth hormone
Glc	D-Glucose
GLC	Gas-liquid chromatography
GlcNAC	N-Acetyl-D-glucosamine
Gln	Glutamine
Glu	Glutamic acid
Gly	Glycine
GMP,GDP,GTP	Guanosine 5'-mono-, -di-, -triphosphate
GnRH	Gonadotropin-releasing hormone
G3P	Glyceraldehyde-3-phosphate
Hb	Hemoglobin
HbO	Oxyhemoglobin
HEPES	N-2-hydroxyethylpiperazine-N'-2-ethanesulfonic acid
HDL	High-density lipoprotein
His	Histidine
HIV	Human immunodeficiency virus
HMG	High-mobility group
HnRNA	Heterogeneous nuclear RNA
HTLV	Human T-cell lymphotropic virus
I	Inosine
Ig	Immunoglobulin
Ile	Isoleucine
IMP,IDP,ITP	Inosine 5'-mono-, -di-, -triphosphate
K	Degrees Kelvin
K'_{eq}	Equilibrium constant
K_d	Dissociation constant
K_m	Michaelis constant
Kcal	Kilocalorie
l	Liter
L	Levo configuration
LDH	Lactate dehydrogenase
LDL	Low-density lipoprotein
Leu	Leucine
LH	Luteinizing hormone
LHC	Light-harvesting complex
Lys	Lysine
M	Molar concentration
Man	Mannose
MAP	Microtubule-associated protein
Met	Methionine
MHC	Major histocompatibility complex
MMTV	Mouse mammary tumor virus
Mol	Mole
mRNA	Messenger ribonucleic acid
MSH	Melanocyte-stimulating hormone
MtDNA	Mitochondrial deoxyribonucleic acid
MTOC	Microtubule-organizing center
MW	Molecular weight
N	Normal concentration
$NAD^+(H)$	Nicotinamide adenine dinucleotide (reduced)
$NADP^+(H)$	Nicotinamide adenine dinucleotide phosphate (reduced)
NeuNAC	N-Acetylneuraminic acid
NGF	Nerve growth factor

NMR	Nuclear magnetic resonance	Ser	Serine
NOR	Nucleolar organizer region	SER	Smooth endoplasmic reticulum
OAA	Oxaloacetate	SnRNA	Small nuclear RNA
OSCP	Oligomycin-sensitivity conferring protein	Snurp	Small nuclear ribonucleoprotein particle
$\sim$P	High-energy phosphate	SRP	Signal recognition particle
P_i	Inorganic orthophosphate	T	Thymine or thymidine
P site	Peptidyl site	Tm	Melting temperature
PAGE	Polyacrylamide gel electrophoresis	TGF	Transforming growth factor
PC	Plastocyanin	Thr	Threonine
PCMB	p-Chloromercuribenzoate	TLC	Thin-layer chromatography
PDGF	Platelet-derived growth factor	TMP,TDP,TTP	Thymidine 5'-mono-, -di-, -triphosphate
PEP	Phosphoenolpyruvate		
PG	Prostaglandin	TMV	Tobacco mosaic virus
PGAL	Phosphoglyceraldehyde	TN-T,I,C	Troponin-T,I,C
pH	$-$Log H^+ concentration	TNBS	Trinitrobenzenesulfonic acid
Phe	Phenylalanine	TRH	Thyrotropin releasing hormone
P:O	High-energy phosphate produced per atom of oxygen consumed	Tris	Tris(hydroxymethyl)aminomethane
PP_i	Inorganic pyrophosphate	tRNA	Transfer ribonucleic acid
PQ	Plastoquinone	Trp	Tryptophan
Pro	Proline	TSH	Thyroid-stimulating hormone (thyrotropin)
Q	Coenzyme Q (ubiquinone)		
R	Gas constant	Tyr	Tyrosine
rDNA	Ribosomal DNA	U	Uracil or uridine
RER	Rough endoplasmic reticulum	UDP-gal	Uridine diphosphate galactose
RIA	Radioimmunoassay	UDP-glc	Uridine diphosphate glucose
RNA	Ribonucleic acid	UMP,UDP,UTP	Uridine 5'-mono-, -di-, -triphosphate
RNAase	Ribonuclease		
rRNA	Ribosomal ribonucleic acid	UV	Ultraviolet
RuBP	Ribulose 1,5-bisphosphate	V	Volt
s	Sedimentation coefficient	v_i	Initial velocity
S	Substrate	V_{max}	Maximal velocity
S	Svedberg unit	Val	Valine
SDH	Succinate dehydrogenase	VLDL	Very-low-density lipoprotein
SDS	Sodium dodecyl sulfate	XMP	Xanthosine monophosphate

Photograph and Illustration Credits

Chapter 1

Figure 1-24 Reproduced with permission from G. Vidal (1984). *Sci. Amer.* 250 (Feb.):48–57.

Figure 1-26 Reprinted from *Molecular Evolution and the Origin of Life*, S. Fox and K. Dose, 2nd ed., copyright 1977, by courtesy of Marcel Dekker, Inc., New York.

Figure 1-27 Reprinted from *Molecular Evolution and the Origin of Life*, S. Fox and K. Dose, 2nd ed., copyright 1977, by courtesy of Marcel Dekker, Inc., New York.

Figure 1-30 (a, b, c) Reproduced from *An Electron Micrographic Atlas of Viruses*, R. C. Williams and H. W. Fisher, copyright 1974, courtesy of Charles C. Thomas, Springfield, IL. *(d)* Reproduced from *Introduction to Virology*, K. M. Smith and D. A. Ritchie, copyright 1980, by permission of Chapman and Hall Publishers, London.

Figure 1-31 Reproduced from Kleinschmidt, A. (1962). *Biochim. Biophys. Acta* 61:857–864, by copyright permission of Elsevier Science Publishers (Biomedical Division), Amsterdam.

Chapter 3

Figure 3-1 Reproduced from V. J. Evans, J. C. Bryant, W. T. McQuilkin, M. C. Fioramonti, K. K. Sanford, B. B. Westfall, and W. R. Earle (1956). *Cancer Res.* 16:87–94, by copyright permission of Cancer Research, Inc., New York.

Figure 3-2 Reproduced from R. P. Perry, K. K. Sanford, V. J. Evans, R. C. Hyatt, and W. R. Earle (1957). *J. Natl. Cancer Inst.* 18:704–714, by courtesy of the National Cancer Institute.

Figure 3-5 (Top four micrographs) Reproduced from *Molecular Biology of the Cell*, B. Alberts, D. Bray, J. Lewis, M. Raff, K. Roberts, and J. D. Watson, copyright 1983, Garland Publishing, New York. *(Bottom two micrographs)* Reproduced from *A Textbook of Histology*, W. Bloom and D. W. Fawcett, 10th ed., copyright 1975, by permission of W. B. Saunders Company, Philadelphia.

Figure 3-6 (Bottom) Reproduced from *An Electron Micrographic Atlas of Viruses*, R. C. Williams and H. W. Fisher, copyright 1974, courtesy of Charles C. Thomas, Publisher, Springfield, IL.

Figure 3-7 (Bottom) Reproduced from *Collagen: The Anatomy of a Protein, Studies in Biology*, J. Woodhead-Galloway, copyright 1980, by permission of Edward Arnold, Ltd., London.

Figure 3-8 (Bottom) Reproduced from *An Electron Micrographic Atlas of Viruses*, R. C. Williams and H. W. Fisher, copyright 1974, courtesy of Charles C. Thomas, Springfield, IL.

Figure 3-9 Reproduced from P. Kalmbach and H. D. Fahini (1978). *Cell Biol. Internat. Reports* 2:389–396, by copyright permission of Academic Press Ltd., London.

Figure 3-12 Reproduced from *Scanning Electron Microscopy in Biology*, R. G. Kessel and C. Y. Shih, copyright 1974, by permission of Springer-Verlag, Heidelberg.

Figure 3-13 (Top) Reproduced with permission from J. C. Kendrew (1961). *Sci. Amer.* 205(Dec.):96–111. *(Middle)* Reprinted with permission from R. E. Franklin and R. G. Gosling, *Nature* 171 (4356):740–741. Copyright 1953, Macmillan Journals, Ltd., London. *(Bottom)* Reproduced from *Modern X-Ray Analysis on Single Crystals*, P. Luger, copyright 1980, by permission of Walter de Gruyter and Company, Berlin.

Figure 3-14 Reproduced from J. G. Torrey, in *Cell, Organism, and Milieu*, D. Rudnick, ed., pp. 189–222, copyright 1959, by permission of John Wiley and Sons, New York.

Figure 3-15 Reproduced from S. H. Blose, D. I. Melton, and J. R. Feramisco (1984). *J. Cell Biol.* 98:847–858, by copyright permission of The Rockefeller University Press, New York.

Figure 3-16 Reprinted by permission of Elsevier Science Publishing Co., Inc., from "A Modification of the Unlabeled Antibody Enzyme Method Using Heterologous Antisera for the Light Microscopic and Ultrastructural Localization of Insulin, Glucagon and Growth Hormone," by S. L. Erlandson, J. A. Parsons, J. P. Burke, J. A. Redick, D. E. VanOrden, and L. S. VanOrden, *J. Histochem. Cytochem.* 23:666–677. Copyright 1975 by The Histochemical Society, Inc.

Figure 3-17 (Middle) Reproduced from D. F. Bainton and M. G. Farquhar (1968). *J. Cell Biol.* 39:286–295, by copyright permission of The Rockefeller University Press, New York. *(Right)* Reproduced from *Atlas of Cell Biology*, J.-C. Roland, A. Szöllösi, and D. Szöllösi, copyright 1977, by permission of Little, Brown and Company, Inc., Boston.

Figure 3-18 (Top right, Bottom right) Reproduced with permission from W. Nagl, in *Mechanisms and Control of Cell Division*, T. L. Rost and E. M. Gifford, Jr., eds., pp. 147–193, copyright 1977, Dowden, Hutchinson, and Ross Publishers, Stroudsburg, PA.

Figure 3-24 (Left) Reproduced from E. M. Croze and D. J. Morré (1984). *J. Cell Physiol.* 119:46–57, by copyright permission of Alan R. Liss, Inc., New York. *(Middle, right)* Reproduced from P. Baudhuin, P. Everard, and J. Berthet (1967). *J. Cell Biol.* 32:181–191, by copyright permission of The Rockefeller University Press, New York.

Figure 3-29 Reproduced from S. H. Blose, D. I. Melton, and J. R. Feramisco (1984). *J. Cell Biol.* 98:847–858, by copyright permission of The Rockefeller University Press, New York.

Figure 3-34 Reproduced from R. Douce and J. Joyard in *Methods in Enzymology*, A. San Pietro, ed., vol. 69, pp. 290–302, copyright 1980, by permission of Academic Press, Inc., Orlando, FL.

Chapter 4

Figure 4-4 Reproduced from J. D. Robertson (1972). *Ann. N.Y. Acad. Sci.* 195:356–371, by copyright permission of New York Academy of Sciences.

Figure 4-7 (Middle) Reproduced from R. B. Park in *Plant Biochemistry*, J. Bonner and J. E. Varner, copyright 1965, pp. 115–145, by permission of Academic Press.

Figure 4-13 Reproduced from P. Sheterline and C. R. Hopkins (1981). *J. Cell Biol.* 90:743–754, by copyright permission of The Rockefeller University Press, New York.

Figure 4-14 Reprinted with permission from C. E. Martin, K. Hiramitsu, Y. Kitajima, Y. Nozseva, L. Skriver, and G. A. Thompson, Jr., *Biochemistry* 15:5218–5227. Copyright 1976, American Chemical Society.

Figure 4-15 Reproduced from J. E. Hainfeld and T. L. Steck (1977). *J. Supramolec. Struc.* 6:301–311, by copyright permission of Alan R. Liss, Inc., New York.

Figure 4-16 Reproduced from T. L. Steck (1974). *J. Cell Biol.* 62:1–19, by copyright permission of The Rockefeller University Press, New York.

Figure 4-18 Reproduced from R. S. Weinstein, J. K. Khodadad, and T. L. Steck (1978). *J. Supramolec. Struc.* 8:325–335, by copyright permission of Alan R. Liss, Inc., New York.

Figure 4-19 Reproduced from J. F. Hainfeld and T. L. Steck (1977). *J. Supramolec. Struc.* 6:301–311, by copyright permission of Alan R. Liss, Inc., New York.

Figure 4-20 Reproduced from C. M. Cohen, J. M. Tyler, and D. Branton (1980). *Cell* 21:875–883, by copyright permission of the Massachusetts Institute of Technology, Boston.

Figure 4-22 Adapted from O. Nilsson and G. Dallner (1977). *Biochim. Biophys. Acta* 464:453–458.

Figure 4-23 Reproduced with permission from W. Stoeckenius (1976). *Sci. Amer.* 234 (June):38–46.

Figure 4-24 Adapted from R. Henderson and P. N. T. Unwin (1977). *Biophys. Struct. Mech.* 3:121–128.

Figure 4-25 Reproduced from T. J. Chai and R. E. Levin (1975). *Applied Microbiol.* 30:450–455, by copyright permission of the American Society for Microbiology.

Figure 4-26 Reproduced from D. J. Politis and R. N. Goodman (1980). *Appl. Environ. Microbiol.* 40:596–607, by copyright permission of the American Society for Microbiology.

Figure 4-28 Reproduced with permission from J. W. Costerton, *Annual Review of Microbiology*, 33:459–479, copyright 1979 by Annual Reviews, Inc., Palo Alto, CA.

Figure 4-31 Reproduced with permission from J. W. Costerton, *Annual Review of Microbiology*, 33: 459–479, copyright 1979 by Annual Reviews, Inc., Palo Alto, CA.

Figure 4-33 (Left, Middle) Reproduced from K. Amako and A. Umeda (1977). *J. Electron Microsc.* 26:155–159, by permission of the Journal of Electron Microscopy.

Figure 4-34 Reproduced from D. Abram (1965). *J. Bacteriol.* 89:855–873, by copyright permission of the American Society for Microbiology.

Figure 4-37 Reproduced from R. Stewart and K. Mühlethaler (1953). *Amer. J. Bot.* 17:295–316, by permission of the Botanical Society of America.

Figure 4-38 Reproduced from *Cells and Organelles*, A. B. Novikoff and E. Holtzman, copyright 1970, Holt, Reinhart and Winston, by permission of H. H. Mollenhauer.

Figure 4-39 Reproduced from *Atlas of Cell Biology*, J.-C. Roland, A. Szöllösi, and D. Szöllösi, copyright 1977, by permission of Little, Brown and Company, Inc., Boston.

Figure 4-40 Reproduced from D. H. Dickson, D. A. Graves, and M. R. Moyles (1982). *Amer. J. Anat.* 165:83–98 by copyright permission of Alan R. Liss, Inc., New York.

Figure 4-41 Reproduced from N. Simionescu and M. Simionescu (1976). *J. Cell Biol.* 70:608–621, by copyright permission of The Rockefeller University Press, New York.

Chapter 5

Figure 5-1 Reproduced from *Scanning Electron Microscopy in Biology*, R. G. Kessel and C. Y. Shih, copyright 1974, by permission of Springer-Verlag, Heidelberg.

Figure 5-6 Adapted from R. LeFevre (1961). *Pharmacol. Rev.* 13:39–49.

Figure 5-12 Adapted from T. Krasne, R. Eisenmann, and S. Szabo (1971). *Science* 174:412–415.

Figure 5-13 (a) Adapted from D. Weinstein, B. A. Wallace, E. R. Blout, J. S. Morrow, and W. Veatch (1979). *Proc. Natl. Acad. Sci. USA* 76:4230–4234.

Figure 5-15 Adapted from J. C. Skow (1957). *Biochim. Biophys. Acta* 23:394–401.

Figure 5-17 Adapted from A. S. Hubbs and R. W. Albers (1980). *Ann. Rev. Biophys. Bioeng.* 9:251–291.

Figure 5-18 Adapted from *Mechanisms and Regulation of Carbohydrate Transport in Bacteria*, M. H. Saier, Jr., copyright 1985, Academic Press, Inc., Orlando, FL.

Figure 5-19 Adapted from I. Bihler and R. K. Crane (1962). *Biochim. Biophys. Acta* 59:79–83 and from *Membranes and Ion Transport*, E. E. Bittar, ed., Vol. 1, copyright 1970, Wiley-Interscience, New York.

Figure 5-22 Adapted from A. L. Hodgkin and R. D. Keynes (1955). *J. Physiol. (London)* 128:61–88.

Figure 5-23 Adapted from B. H. Ginsberg in *Biochemical Actions of Hormones*, G. Litwack, ed., Vol. IV, pp. 313–344, copyright 1977, Academic Press, Inc., Orlando, FL.

Figure 5-26 Reproduced from D. S. Friend and N. B. Gilula (1972). *J. Cell Biol.* 53:758–776, by copyright permission of The Rockefeller University Press, New York.

Figure 5-27 Reproduced from R. S. Decker and D. S. Friend (1974). *J. Cell Biol.* 62:32–47, by copyright permission of The Rockefeller University Press, New York.

Figure 5-28 Reproduced from L. A. Staehelin (1973). *J. Cell Sci.* 13:763–786, by permission of The Company of Biologists, Ltd., Colchester, England.

Figure 5-30 Reproduced from D. E. Kelley (1966). *J. Cell Biol.* 28:51–72, by copyright permission of The Rockefeller University Press, New York.

Figure 5-31 Reproduced from S. Eichenberger-Glinz (1979). *Wilhelm Roux's Archives* 186:333–349, by copyright permission of Springer-Verlag, Heidelberg.

Figure 5-32 (Left) Reproduced from J. Weiner, D. Spiro, and W. R. Loewenstein (1964). *J. Cell Biol.* 22:587–598, by copyright permission of The Rockefeller University Press, New York. *(Right)* Reproduced from N. E. Flower and B. K. Filshie (1975). *J. Cell Sci.* 17:221–239, by permission of The Company of Biologists, Ltd., Colchester, England.

Figure 5-33 Reproduced from I. Simpson, B. Rose, and W. R. Loewenstein (1977). *Science* 195:294–296, by copyright permission of the American Association for the Advancement of Science.

Figure 5-35 Reproduced from E. L. Benedetti and P. Emmelot (1965). *J. Cell Biol.* 26:299–305, by copyright permission of The Rockefeller University Press, New York.

Figure 5-37 Reproduced from W. R. Loewenstein in *Cell Membranes: Biochemistry, Cell Biology, and Pathology*, G. Weissman, ed., pp. 105–114, copyright 1975, by permission of H. P. Publishing Co., Inc., New York.

Chapter 6

Figure 6-1 Reproduced from K. R. Porter (1945). *J. Exp. Med.* 81:233–246, by copyright permission of The Rockefeller University Press, New York.

Figure 6-3 (Top left, Bottom left) Reproduced from K. Tanaka, A. Iino, and T. Naguro (1976). *Arch. Histol. Jap.* 39:165–175, by permission of the Japan Society of Histological Documentation. *(Top right)* Reproduced from *The Cell*, D. W. Fawcett, 2nd ed., copyright 1981, by permission of W. B. Saunders Company, Philadelphia. *(Bottom right)* Reproduced from M. Bielinska, G. Rogers, T. Rucinsky, and I. Boine (1979). *Proc. Natl. Acad. Sci. USA* 76:6152–6156, by courtesy of the National Academy of Sciences.

Figure 6-5 Reproduced from C. deDuve (1971). *J. Cell Biol.* 50:20D–55D, by copyright permission of The Rockefeller University Press, New York.

Figure 6-8 Reproduced from S. Orrenius, J. Ericsson, and L. Ernster (1965). *J. Cell Biol.* 25:625–639, by copyright permission of The Rockefeller University Press, New York.

Figure 6-9 Adapted from L. Ernster and S. Orrenius (1973). *Drug Metab. Dispos.* 1:66–73 and from R. Kuntzman, W. Levin, M. Jacobson, and A. H. Cooney (1973). *Life Sci.* 7:215–224.

Figure 6-10 Adapted from C. M. Redman and D. D. Sabatini (1966). *Proc. Natl. Acad. Sci. USA* 56:608–615.

Figure 6-13 Reproduced from H. W. Beams and R. G. Kessel (1968). *Int. Rev. Cytol.* 23:209–276, by copyright permission of Academic Press, Inc., Orlando, FL.

Figure 6-14 Reproduced from G. T. Cole and M. J. Wynne (1973). *Cytobios.* 8:161–173, by permission of The Faculty Press, Cambridge, England.

Figure 6-15 Reproduced from P. Favard in *Handbook of Molecular Cytology*, A. Lima-de-Faria, ed., pp. 1130–1178, copyright 1969, by permission of Elsevier Science Publishers, Amsterdam.

Figure 6-16 (Left) Reproduced from W. P. Cunningham, D. J. Morré, and H. H. Mollenhauer (1966). *J. Cell Biol.* 28:169–179, by copyright permission of The Rockefeller University Press, New York. *(Right)* Reproduced from J. M. Sturgess, E. Katona, and M. A. Moscarello (1973). *J. Memb. Biol.* 12:367–384, by copyright permission of Springer-Verlag, Heidelberg.

Figure 6-17 Reproduced from J. D. Jamieson and G. E. Palade (1967).

J. Cell Biol. 34:597–615, by copyright permission of The Rockefeller University Press, New York.

Figure 6-18 Adapted from J. D. Jamieson (1967). *J. Cell Biol.* 34:597–615 and from J. D. Castle, J. D. Jamieson, and G. E. Palade (1971). *J. Cell Biol.* 53:290–311.

Figure 6-19 Adapted from J. D. Jamieson and G. E. Palade (1967). *J. Cell Biol.* 34:577–596, 597–615.

Figure 6-20 Reproduced from D. S. Friend (1965). *J. Cell Biol.* 25:563–576, by copyright permission of The Rockefeller University Press, New York.

Figure 6-22 Reproduced from M. Neutra and C. P. Leblond (1966). *J. Cell Biol.* 30:119–136, by copyright permission of The Rockefeller University Press, New York.

Figure 6-23 Reproduced from W. J. Malaisse, F. Malaisse-Lagae, E. Van Obberghen, G. Somers, G. Devis, M. Ravazzola, and L. Orci (1975). *Ann. N.Y. Acad. Sci.* 253:630–652, by copyright permission of the New York Academy of Sciences.

Figure 6-25 Reproduced from D. F. Bainton and M. G. Farquhar (1966). *J. Cell Biol.* 28:277–301, by copyright permission of The Rockefeller University Press, New York.

Figure 6-26 Adapted from J. Berthet, L. Berthet, F. Appelmans, and C. deDuve (1951). *Biochem. J.* 50:182–189.

Figure 6-28 Adapted from C. deDuve, in *Subcellular Particles*, T. Hayashi, ed., pp. 128–159, copyright 1959, Academic Press, Inc., Orlando, FL.

Figure 6-29 Reproduced from P. Baudhuin, P. Evrard, and J. Berthet (1965). *J. Cell Biol.* 32:181–243, by copyright permission of The Rockefeller University Press, New York.

Figure 6-30 (Left) Reproduced from D. F. Bainton and M. G. Farquhar (1968). *J. Cell Biol.* 39:286–298, by copyright permission of The Rockefeller University Press, New York. *(Right)* Reproduced from P. Baudhuin, P. Everard, and J. Berthet (1967). *J. Cell Biol.* 32:181–191, by copyright permission of The Rockefeller University Press, New York.

Figure 6-31 Adapted from J. P. Milson and C. H. Wynn (1973). *Biochem. J.* 132:493–500.

Figure 6-34 Reproduced from R. M. Steinman and Z. A. Cohn (1972). *J. Cell Biol.* 55:616–634, by copyright permission of The Rockefeller University Press, New York.

Figure 6-35 (Top left and right) Reprinted by permission of Elsevier Science Publishing Co., Inc., from "Receptor-Mediated Endocytosis in Cultured Fibroblasts: Cryptic Coated Pits and the Formation of Receptosomes," by M. C. Willingham, A. V. Rutherford, M. G. Gallo, J. Wehland, R. B. Dickson, R. Schlegel, and I. H. Pastan, *J. Histochem. Cytochem.* 29:1003–1013. Copyright 1981 by The Histochemical Society, Inc. *(Bottom left)* Reproduced from T. Kirchhausen and S. C. Harrison (1984). *J. Cell Biol.* 99:1725–1734, by copyright permission of The Rockefeller University Press, New York. *(Bottom right)* Adapted from T. Kirchhausen and S. C. Harrison (1984). *J. Cell Biol.* 99:1725–1734.

Figure 6-36 Adapted from J. L. Goldstein, M. S. Brown, R. G. W. Anderson, D. W. Russell, and W. J. Schneider (1985). *Ann. Rev. Cell Biol.* 1:1–39.

Figure 6-37 Reproduced from D. Zucker-Franklin and J. G. Hirsch (1971). *Lab Invest.* 25:415–421, by copyright permission of Williams and Wilkins Company, New York.

Figure 6-38 Reproduced from E. Essner and H. Haimes (1977). *J. Cell Biol.* 75:381–387, by copyright permission of The Rockefeller University Press, New York.

Figure 6-39 Reproduced from Z. Steplewski, D. Herlyn, G. Maul, and H. Koprowski (1985). *Hybridoma* 2:1–5, by copyright permission of Mary Ann Liebert, Inc., Publishers, New York.

Figure 6-40 Reproduced from J. Boyles and D. F. Bainton (1981). *Cell* 24:905–914, by copyright permission of the Massachusetts Institute of Technology, Boston.

Figure 6-41 Reproduced from M. Locke and A. K. Sykes (1975). *Tissue and Cell* 7:143–158, by permission of Longman Group, Ltd., Essex, England.

Figure 6-42 Reproduced from H. Loeb, G. Jonniaux, A. Resibois, N. Cremes, M. Tondeus, P. E. Gregoire, J. Richard, and P.

Cieters (1968). *J. Pediatr.* 73:860–874, by permission of the C. V. Mosby Co., St. Louis.

Figure 6-43 Reproduced from Ph. Matile, in *Lysosomes in Biology and Pathology*, J. T. Dingle and T. Fell, eds., Vol. 1, pp. 406–430, copyright 1969, by permission of Elsevier Science Publishing Co., New York.

Figure 6-44 Adapted from C. deDuve (1975). *Science* 189:186–194.

Figure 6-45 Adapted from C. deDuve (1975). *Science* 189:186–194.

Figure 6-46 Reproduced from J. M. Barrett and P. M. Heidger, Jr. (1975). *Cell Tiss. Res.* 157:283–305, by copyright permission of Springer-Verlag, Heidelberg.

Figure 6-47 Reprinted by permission of Elsevier Science Publishing Co., Inc., from "Visualization of Peroxisomes (Microbodies) and Mitochondria with Diaminobenzidene," by S. Goldfischer and E. Essner, *J. Histochem. Cytochem.* 17:675–680. Copyright 1969 by The Histochemical Society, Inc.

Figure 6-49 Adapted from C. deDuve and P. Baudhuin (1966). *Physiol. Rev.* 46:323–357.

Figure 6-50 Adapted from C. deDuve (1969). *Proc. Roy. Soc. B* 173:71–83.

Figure 6-52 Reproduced from E. Vigil (1970). *J. Cell Biol.* 46:435–454, by copyright permission of The Rockefeller University Press, New York.

Figure 6-53 Reproduced from B. M. Goldman and G. Blobel (1978). *Proc. Natl. Acad. Sci. USA* 75:5066–5070, by courtesy of the National Academy of Sciences.

Figure 6-55 Adapted from J. E. Rothman and E. P. Kennedy (1977). *Proc. Natl. Acad. Sci. USA* 74:1821–1825.

Chapter 7

Figure 7-10 Reproduced from *Mitochondria*, B. Tandler and C. L. Hoppel, copyright 1972, Academic Press, Inc., Orlando, FL.

Figure 7-11 (Left) Reproduced from K. Blinzinger, M. B. Rewcastle, and H. Hager (1965). *J. Cell Biol.* 25:293–305, by copyright permission of The Rockefeller University Press, New York. *(Right)* Reproduced from T. Samorajski, J. R. Keefe, and J. M. Ordy (1967). *J. Cell Biol.* 28:489–504, by copyright permission of The Rockefeller University Press, New York.

Figure 7-13 Reproduced from H. Fernández-Morán, P. V. Blair, and D. E. Green (1964). *J. Cell Biol.* 22:63–100, by copyright permission of The Rockefeller University Press, New York.

Figure 7-14 Reproduced from J. N. Telford and E. Racker (1973). *J. Cell Biol.* 57:580–586, by copyright permission of The Rockefeller University Press, New York.

Figure 7-15 Reproduced from C. R. Hackenbrock (1968). *J. Cell Biol.* 37:345–369, by copyright permission of The Rockefeller University Press, New York.

Figure 7-16 (Left) Reproduced from B. Sacktor and Y. Shimada (1972). *J. Cell Biol.* 52:465–477, by copyright permission of The Rockefeller University Press, New York. *(Right)* Reproduced from J. D. Jamieson and G. E. Palade (1968). *J. Cell Biol.* 39:589–603, by copyright permission of The Rockefeller University Press, New York.

Figure 7-18 Reproduced from *A Textbook of Histology*, W. Bloom and D. W. Fawcett, 10th ed., by permission of W. B. Saunders Company, Philadelphia.

Figure 7-19 Reproduced from P. M. Andrews and C. R. Hackenbrock (1975). *Exp. Cell Res.* 90:127–136, by copyright permission of Academic Press, Inc., Orlando, FL.

Figure 7-20 (Left) Reproduced from D. F. Parsons, G. R. Williams, W. Thompson, D. Wilson, and B. Chance in *Mitochondrial Structure and Compartmentation*, E. Quagliariello, S. Papa, E. C. Slater, and J. M. Tager, eds., copyright 1967, pp. 29–70, by permission of Adriatica Editrice, Bari, Italy. *(Right)* Reproduced from D. F. Parsons, G. R. Williams, and B. Chance (1966). *Ann. N.Y. Acad. Sci.* 137:643–666, by permission of the New York Academy of Sciences.

Figure 7-23 Reprinted by permission of Elsevier Science Publish-

ing Co., Inc., from "The Ultrastructural Localization of Cytochrome Oxidase via Cytochrome C," by W. A. Anderson, G. Bara, and A. M. Seligman, *J. Histochem. Cytochem.* 23:13–20. Copyright 1975 by The Histochemical Society, Inc.

Figure 7-31 Reproduced from C. R. Hackenbrock (1981). *Trends Biochem. Sci.* 6:151–154, by copyright permission of Elsevier Science Publishers (Biomedical Division), Amsterdam.

Figure 7-32 Reprinted from E. Racker (1967). *Federation Proceedings* 26:1335–1340.

Figure 7-39 Reproduced from A. Asano, N. S. Cohen, R. F. Baker, and A. F. Brodie (1973). *J. Biol. Chem.* 248:3386–3397, by copyright permission of the American Society of Biological Chemists, Inc.

Chapter 8

Figure 8-1 Adapted from R. Emerson and W. Arnold (1932). *J. Gen. Physiol.* 15:391–420.

Figure 8-3 Adapted from R. B. Park (1966). *Int. Rev. Cytol.* 20:67–95.

Figure 8-4 Reproduced from *Plant Anatomy*, A. Fahn, 2nd ed., copyright 1974, Pergamon Press, New York, with permission of Y. Ben-Shaul.

Figure 8-7 Reproduced from M. P. Garber and P. L. Steponkus (1974). *J. Cell Biol.* 63:24–34, by copyright permission of The Rockefeller University Press, New York.

Figure 8-8 Reproduced from J. Rosado-Alberio, T. E. Weier, and C. R. Stocking (1968). *Plant Physiol.* 43:1325–1331, by permission of the American Society of Plant Physiologists.

Figure 8-9 Reproduced from J. Rosado-Alberio, T. E. Weier, and C. R. Stocking (1968). *Plant Physiol.* 43:1325–1331, by permission of the American Society of Plant Physiologists.

Figure 8-10 Reproduced from E. Gantt and S. E. Conti (1966). *J. Cell Biol.* 29:423–434, by copyright permission of The Rockefeller University Press, New York.

Figure 8-12 Reproduced from S. E. Frederick and E. H. Newcomb (1969). *J. Cell Biol.* 43:343–353, by copyright permission of The Rockefeller University Press, New York.

Figure 8-13 Reproduced from J. E. M. Ballantine and B. J. Forde (1970). *Amer. J. Bot.* 57:1150–1159, by copyright permission of the Botanical Society of America.

Figure 8-14 Reproduced from H. Kushida, M. Itoh, S. Izewa, and K. Shibata (1964). *Biochim. Biophys. Acta* 79:201–203, by copyright permission of Elsevier Science Publishers (Biomedical Division), Amsterdam.

Figure 8-16 (Top) Reproduced from A. Douce, R. B. Holtz, and A. A. Benson (1973). *J. Biol. Chem.* 20:7215–7222, by copyright permission of the American Society of Biological Chemists, Inc. *(Bottom)* Reproduced from P. V. Sano, D. J. Goodchild, and R. B. Park (1970). *Biochim. Biophys. Acta* 216:162–178, by copyright permission of Elsevier Science Publishers (Biomedical Division), Amsterdam.

Figure 8-18 Reproduced from D. G. Cran and J. V. Possingham (1974). *Protoplasma* 74:197–213, by copyright permission of Springer-Verlag, Heidelberg.

Figure 8-19 Reproduced from E. Gantt and S. F. Conti (1965). *J. Cell Biol.* 26:365–375, by copyright permission of The Rockefeller University Press, New York.

Figure 8-20 Reproduced from B. E. S. Gunning (1965). *J. Cell Biol.* 24:79–93, by copyright permission of The Rockefeller University Press, New York.

Figure 8-25 Adapted from R. Emerson and W. Arnold (1932). *J. Gen. Physiol.* 16:191–205.

Figure 8-28 Adapted from R. Emerson and C. M. Lewis (1943). *Amer. J. Bot.* 30:165–178 and from R. Emerson, R. Chalmers, and C. Cederstrand (1957). *Proc. Natl. Acad. Sci. USA* 43:133–143.

Figure 8-32 Adapted from J. Neumann and A. T. Jagendorf (1984). *Arch. Biochem. Biophys.* 107:109–119.

Figure 8-36 Adapted from G. Hind and A. T. Jagendorf (1963). *Proc.*

Natl. Acad. Sci. USA 49:715–719 and from A. T. Jagendorf and J. Neumann (1965). *J. Biol. Chem.* 240:3210–3215.

Figure 8-37 Reproduced from M. P. Garber and P. L. Steponkus (1974). *J. Cell Biol.* 63:24–34, by copyright permission of The Rockefeller University Press, New York.

Figure 8-41 Reproduced with permission from K. R. Miller (1979). *Sci. Amer.* 241 (Oct.):102–113.

Figure 8-43 Adapted from U. W. Goodenough, J. J. Armstrong, and R. P. Levine (1969). *Plant Physiol.* 44:1001–1012.

Figure 8-44 Reproduced from A. Melis (1984). *J. Cell. Biochem.* 24:271–285, by copyright permission of Alan R. Liss, Inc., New York.

Figure 8-45 Reproduced from M. Calvin, in *Photosynthesis*, E. Rabinowitch and Govindjee, eds., pp. 220–237, copyright 1969 by permission of John Wiley and Sons, Inc., New York.

Figure 8-47 Adapted from A. T. Wilson and M. Calvin (1955). *J. Amer. Chem. Soc.* 77:5948–5957, and from J. A. Bassham, K. Shibata, K. Steenberg, J. Bourdon, and M. Calvin (1956). *J. Amer. Chem. Soc.* 78:4120–4124.

Figure 8-49 Adapted from H. S. Johnson and M. D. Hatch (1968). *Biochem. J.* 114:127–134.

Figure 8-51 Reproduced with permission from W. M. Laetsch, in the *Ann. Rev. Plant Physiol.* 25:27–52, copyright 1974, by Annual Reviews, Inc., Palo Alto, CA.

Figure 8-53 Reproduced from N. J. Lang and B. A. Whitton in *The Biology of Blue-Green Algae*, N. G. Carr and B. A. Whitton, eds., pp. 66–79, copyright 1973, by permission of Blackwell Scientific Publications, Ltd., Oxford, England.

Figure 8-54 Reproduced from A. R. Varga and L. A. Staehelin (1983). *J. Bacteriol.* 154:1414–1430, by copyright permission of the American Society for Microbiology.

Chapter 9

Figure 9-1 (Left) Reproduced from B. Chailley, A. N'Diaye, E. Boisvieux-Ulrich, and D. Sandoz (1981). *Eur. J. Cell Biol.* 25:300–307, by copyright permission of Wissenschaftliche Verlagsgesellschaft mbH, Stuttgart, FRG. *(Middle)* Reproduced from M. McGill, D. P. Highfield, T. M. Monahan, and B. R. Brinkley (1976). *J. Ultrastruc. Res.* 57:43–53, by copyright permission of Academic Press, Inc., Orlando, FL. *(Right)* Reproduced from L. T. Haimo and B. R. Telzer (1981). *Cold Spr. Harbor Symp. Quant. Biol.* 46:207–217, by copyright permission of Cold Spring Harbor Laboratory, Cold Spring Harbor, New York.

Figure 9-2 Reproduced from S. H. Blose, D. I. Meltzer, and J. R. Feramisco (1984). *J. Cell Biol.* 98:847–858, by copyright permission of The Rockefeller University Press, New York.

Figure 9-3 (Left) Reproduced from L. A. Amos and A. Klug (1974). *J. Cell Sci.* 14:523–549, by permission of The Company of Biologists, Ltd., Colchester, England. *(Right)* Reproduced from C. Pierson, P. R. Burton, and H. R. Hines (1978). *J. Cell Biol.* 76:223–228, by copyright permission of The Rockefeller University Press, New York.

Figure 9-6 Reproduced from R. D. Sloboda, W. L. Dentler, R. A. Bloodgood, B. R. Telzer, S. Granett, and J. L. Rosenbaum in *Cell Motility*, R. D. Goldman, T. D. Pollard, and J. L. Rosenbaum, eds., pp. 1171–1212, copyright 1976, by permission of Cold Spring Harbor Laboratory, Cold Spring Harbor, New York.

Figure 9-7 Reproduced from M. W. Kirschner, R. C. Williams, M. Weingarten, and J. C. Gerhart (1974). *Proc. Natl. Acad. Sci. USA* 71:1159–1163, by courtesy of the National Academy of Sciences.

Figure 9-8 Adapted from J. Bryan (1976). *J. Cell Biol.* 71:749–767.

Figure 9-9 Reproduced from W. L. Dentler, S. Granett, G. B. Witman, and J. L. Rosenbaum (1974). *Proc. Natl. Acad. Sci. USA* 71:1710–1714, by courtesy of the National Academy of Sciences.

Figure 9-10 Reproduced from J. D. Pickett-Heaps (1969). *Cytobios* 3:257–280, by permission of The Faculty Press, Cambridge, England.

Figure 9-11 Reproduced from B. R. Telzer, M. J. Moses, and J. L.

*Rosenbaum (1975). *Proc. Natl. Acad. Sci. USA* 72:4023–4027, by courtesy of the National Academy of Sciences.

Figure 9-12 Reproduced from G. M. Fuller and B. R. Brinkley (1976). *J. Supramolec. Struct.* 5:497–514, by copyright permission of Alan R. Liss, Inc., New York.

Figure 9-13 (Left) Adapted from R. L. Margolis and L. Wilson (1981). *Nature* 293:705–711. *(Right)* Adapted from T. Mitchison and M. Kirschner (1984). *Nature* 312:237–242.

Figure 9-14 Reproduced from R. W. Weisenberg and A. C. Rosenfeld (1975). *J. Cell Biol.* 64:146–158, by copyright permission of The Rockefeller University Press, New York.

Figure 9-15 Adapted from R. D. Sloboda, W. L. Dentler, and J. L. Rosenbaum (1976). *Biochemistry* 15:4497–4501.

Figure 9-16 Reproduced from A. T. Mariassy, C. G. Plopper, and D. L. Dungworth (1975). *Anat. Rec.* 183:13–26, by copyright permission of Alan R. Liss, Inc., New York.

Figure 9-18 Reproduced from W. L. Dentler and E. L. LeCluyse (1982). *Cell Motility* 2:549–572, by copyright permission of Alan R. Liss, Inc., New York.

Figure 9-19 Adapted from P. Satir (1974). *Sci. Amer.* 231 (Oct):45–52.

Figure 9-20 (Top) Reproduced from F. D. Warner and P. Satir (1974). *J. Cell Biol.* 63:35–63, by copyright permission of The Rockefeller University Press, New York.

Figure 9-21 Reproduced from N. B. Gilula and P. Satir (1972). *J. Cell Biol.* 53:494–509, by copyright permission of The Rockefeller University Press, New York.

Figure 9-22 Reproduced from W. L. Dentler (1980). *J. Cell Sci.* 42:207–220, by permission of The Company of Biologists, Ltd., Colchester, England.

Figure 9-23 Reproduced from I. R. Gibbons (1963). *Proc. Natl. Acad. Sci. USA* 50:1002–1010, by courtesy of the National Academy of Sciences.

Figure 9-24 (Right) Reproduced from I. R. Gibbons and A. V. Grimstone (1960). *J. Biophys. Biochem. Cytol.* 7:697–715, by copyright permission of The Rockefeller University Press, New York.

Figure 9-25 Adapted from H. Pitelka in *Cilia and Flagella*, M. A. Sleigh, ed., pp. 437–469, copyright 1974, Academic Press, Inc., Orlando, FL.

Figure 9-27 Reproduced from R. V. Dippell (1968). *Proc. Natl. Acad. Sci. USA* 61:461–468, by courtesy of the National Academy of Sciences.

Figure 9-28 Reproduced from S. P. Sorokin (1968). *J. Cell Sci.* 3:207–230, by permission of The Company of Biologists, Ltd., Colchester, England.

Figure 9-29 Reproduced from J. L. Rosenbaum, J. E. Moulder, and D. L. Ringo (1969). *J. Cell Biol.* 41:600–619, by copyright permission of The Rockefeller University Press, New York.

Figure 9-30 Reproduced from S. F. Goldstein, M. E. J. Holwill, and N. R. Silvester (1970). *J. Exp. Biol.* 53:401–409, by permission of Cambridge University Press, New York.

Figure 9-31 Reproduced with permission from P. Satir (1974). *Sci. Amer.* 231(Oct.):45–52.

Figure 9-32 Reproduced from K. E. Summers and I. R. Gibbons (1971). *Proc. Natl. Acad. Sci. USA* 68:3092–3096, by courtesy of the National Academy of Sciences.

Figure 9-34 (Top) Reproduced from L. E. Roth, D. J. Pihlaja, and Y. Shigemaka (1970). *J. Ultrastruct. Res.* 30:7–37, by copyright permission of Academic Press, Inc., Orlando, FL. *(Bottom left)* Reproduced from M. S. Mooseker and L. G. Tilney (1973). *J. Cell Biol.* 56:13–26, by copyright permission of The Rockefeller University Press, New York. *(Bottom right)* Reproduced from J. B. Tucker (1970). *J. Cell Sci.* 7:793–821, by permission of The Company of Biologists, Ltd., Colchester, England.

Figure 9-35 Adapted from R. D. Berlin, J. M. Oliver, T. E. Ukena, and H. H. Yin (1975). *New Engl. J. Med.* 292:515–520.

Figure 9-36 Adapted from R. D. Berlin, J. M. Oliver, T. E. Ukena, and H. H. Yin (1975). *New Engl. J. Med.* 292:515–520.

Figure 9-38 (Top) Reproduced from R. D. Goldman, E. Lazarides, R.

Pollack, and K. Weber (1975). *Exp. Cell Res.* 90:333–344, by copyright permission of Academic Press, Inc., Orlando. FL. *(Bottom)* Reproduced from R. D. Goldman and D. M. Knipe (1972). *Cold Spr. Harbor Symp. Quant. Biol.* 37:523–534, by copyright permission of Cold Spring Harbor Laboratory, Cold Spring Harbor, New York.

Figure 9-39 (Left) Reproduced from E. Lazarides and K. Weber (1974). *Proc. Natl. Acad. Sci. USA* 71:2268–2272, by courtesy of the National Academy of Sciences. *(Middle)* Reproduced from E. Lazarides (1976). *J. Cell Biol.* 68:202–219, by copyright permission of The Rockefeller University Press, New York. *(Right)* Reproduced from K. Fujiwara and T. D. Pollard (1976). *J. Cell Biol.* 71:848–875, by copyright permission of The Rockefeller University Press, New York.

Figure 9-40 Reproduced from K. B. Pryzwansky, M. Schliwa, and K. R. Porter (1983). *Eur. J. Cell Biol.* 30:112–125, by permission of Wissenschaftliche Verlagsgesellschaft mbH, Stuttgart, FRG.

Figure 9-43 Reproduced from Y. M. Kersey and N. K. Wessells (1976). *J. Cell Biol.* 68:264–275, by copyright permission of The Rockefeller University Press, New York.

Figure 9-44 Reproduced from R. D. Allen (1972). *Exp. Cell Res.* 72:32–45, by copyright permission of Academic Press, Inc., Orlando, FL.

Figure 9-48 Reproduced from G. Albrecht-Buehler (1976). *J. Cell Biol.* 69:275–286, by copyright permission of The Rockefeller University Press, New York.

Figure 9-49 Reproduced from R. Norberg, K. Lidman, and A. Fagraeus (1975). *Cell* 6:507–512, by copyright permission of the Massachusetts Institute of Technology, Boston.

Figure 9-50 Reproduced from R. Rajaraman, D. E. Rounds, S. P. S. Yen, and A. Renbaum (1974). *Exp. Cell Res.* 88:327–339, by copyright permission of Academic Press, Inc., New York.

Figure 9-51 (Left) Reproduced from M. S. Mooseker and R. Tilney (1975). *J. Cell Biol.* 67:725–743, by copyright permission of The Rockefeller University Press, New York. *(Right)* Reproduced from J. Heuser (1981). *Trends Biochem. Sci.* 6:64–68, by permission of Elsevier Science Publishers (Biomedical Division), Amsterdam.

Figure 9-52 Reproduced from P. T. Matsudaira and D. R. Burgess (1981). *Cold Spr. Harbor Symp. Quant. Biol.* 46:845–854, by copyright permission of Cold Spring Harbor Laboratory, Cold Spring Harbor, New York.

Figure 9-53 Reproduced from W. W. Frank, E. Schmid, D. L. Schiller, S. Winter, E. D. Jarasch, R. Moll, H. Denk, B. W. Jackson, and K. Illmensee (1981). *Cold Spr. Harbor Symp. Quant. Biol.* 46:431–453, by copyright permission of Cold Spring Harbor Laboratory, Cold Spring Harbor, New York.

Figure 9-54 Reproduced with permission from K. R. Porter and J. B. Tucker (1981). *Sci. Amer.* 224(Mar.):57–67.

Figure 9-57 Reproduced from M. L. DePamphilis and J. Adler (1971). *J. Bacteriol.* 105:384–395, by copyright permission of the American Society for Microbiology.

Figure 9-58 Adapted from M. L. DePamphilis and J. Adler (1971). *J. Bacteriol.* 105:384–385.

Figure 9-59 (Left) Reproduced from *Fundamental Principles of Bacteriology*, A. J. Salle, 7th ed., copyright 1973, by permission of McGraw-Hill Book Company, New York. *(Right)* Reproduced from A. Lowy and L. Hanson (1965). *J. Mol. Biol.* 11:293–313, by copyright permission of Academic Press, Ltd., London.

Figure 9-60 Adapted from T. Iino (1969). *J. Gen. Microbiol.* 56:227–239.

Figure 9-61 Reprinted by permission from M. Silverman and M. Simon, *Nature* 249(5452):73–74. Copyright 1974, Macmillan Journals, Ltd., London.

Chapter 10

Figure 10-2 Reproduced from L. D. Simon and T. F. Anderson (1967). *Virology* 32:279–297, by copyright permission of Academic Press, Inc., Orlando, FL. Courtesy of Photo Researchers, New York.

Figure 10-3 Adapted from A. D. Hershey and M. Chase (1952). *J. Gen. Physiol.* 36:39–56.

Figure 10-7 Adapted from M. Meselson and F. Stahl (1958). *Proc. Natl. Acad. Sci. USA* 44:671–682.

Figure 10-9 Adapted from R. T. Okazaki, K. Okazaki, K. Sakabe, K. Sugimoto, and A. Sugino (1968). *Proc. Natl. Acad. Sci. USA* 59:598–602.

Figure 10-13 Adapted from G. W. Beadle and E. L. Tatum (1941). *Proc. Natl. Acad. Sci. USA* 27:499–506.

Figure 10-15 Adapted from L. Pauling, H. A. Itano, S. J. Singer, and I. C. Wells (1949). *Science* 110:543–548.

Figure 10-16 Reproduced with permission from V. Ingram (1958). *Sci. Amer.* 198(Jan.):68–74.

Figure 10-18 Reproduced from *Cell Growth and Cell Function* by Torbjoern O. Caspersson, M.D., by permission of W. W. Norton and Company, Inc. Copyright 1950 by W. W. Norton and Company, Inc. Copyright reserved 1978 by Torbjoern O. Caspersson.

Figure 10-19 Adapted from P. Siekevitz (1952). *J. Biol. Chem.* 195:549–565, and from J. W. Littlefield, E. B. Keller, J. Gross, and P. C. Zamecnik (1955). *J. Biol. Chem.* 217:111–124.

Figure 10-20 Reproduced from D. M. Prescott (1964). *Progr. Nucleic Acid Res. Mol. Biol.* 3:33–55, by copyright permission of Academic Press, Inc., Orlando, FL.

Figure 10-21 Adapted from J. Marmur, C. M. Greenspan, E. Palecek, F. M. Kahan, J. Levine, and M. Mandel (1962). *Cold Spr. Harbor Symp. Quant. Biol.* 28:191–199.

Figure 10-22 Adapted from S. Brenner, F. Jacob, and M. Meselson (1961). *Nature* 190:576–581.

Figure 10-23 Adapted from M. B. Hoagland, M. L. Stephenson, J. F. Scott, L. I. Hect, and P. C. Zamecnik (1958). *J. Biol. Chem.* 231:241–257.

Figure 10-47 Adapted from A. M. Maxam and W. Gilbert (1977). *Proc. Natl. Acad. Sci. USA* 74:560–564.

Figure 10-48 Adapted from F. Sanger, S. Nicklen, and A. R. Coulson (1977). *Proc. Natl. Acad. Sci. USA* 74:5463–5467.

Figure 10-49 Reproduced from R. D. Palmiter, G. Norstedt, R. E. Gelinas, R. L. Brinster, and R. E. Hammer (1983). *Science* 222:809–814, by copyright permission of the American Association for the Advancement of Science.

Chapter 11

Figure 11-2 Reproduced from E. W. Daniels, J. M. McNiff, and D. R. Ekberg (1969). *Z. Zellforsch* 98:357–363, by copyright permission of Springer-Verlag, Heidelberg.

Figure 11-3 (Top) Reproduced from R. H. Kirschner, M. Rusli, and T. Martin (1977). *J. Cell Biol.* 72:118–132, by copyright permission of The Rockefeller University Press, New York. *(Middle)* Reproduced from A. C. Fabergé (1974). *Cell Tiss. Res.* 151:403–415, by copyright permission of Springer-Verlag, Heidelberg. *(Bottom)* Reproduced from D. Branton and H. Moor (1964). *J. Ultrastruct. Res.* 11:401–411, by copyright permission of Academic Press, Inc., Orlando FL.

Figure 11-4 Reproduced from U. Scheer, J. Kartenbeck, M. F. Trendelenburg, J. Stadler, and W. W. Franke (1976). *J. Cell Biol.* 69:1–18, by copyright permission of The Rockefeller University Press, New York.

Figure 11-6 Reproduced from C. M. Feldherr (1962). *J. Cell Biol.* 14:65–72, by copyright permission of The Rockefeller University Press, New York.

Figure 11-7 Reproduced from B. J. Stevens and H. Swift (1966). *J. Cell Biol.* 31:55–77, by copyright permission of The Rockefeller University Press, New York.

Figure 11-8 Reproduced from *The Cell*, D. W. Fawcett, 2nd ed., copyright 1981, by permission of W. B. Saunders Company, Philadelphia.

Figure 11-9 Reproduced from J. Olert (1979). *Experientia* 35:283–285, by permission of Birkhäuser Verlag AG, Basel.

Figure 11-10 Reproduced with permission from E. G. Jordan, *The Nucleolus*, 2nd ed., 1978. Carolina Biology Reader Series. Copyright Carolina Biological Supply Company, Burlington, NC.

Figure 11-11 Reproduced from T. S. Ro-Choi, K. Smetana, and H. Busch (1973). *Exp. Cell Res.* 79:43–52, by copyright permission of Academic Press, Inc., Orlando, FL.

Figure 11-13 Reprinted with permission from S. Panyim and R. Chalkley, *Biochemistry* 8:3972–3980. Copyright 1969, American Chemical Society.

Figure 11-15 Reproduced from J. L. Peterson and E. H. McConkey (1976). *J. Biol. Chem.* 257:548–554, by copyright permission of the American Society of Biological Chemists, Inc.

Figure 11-16 Reproduced from D. E. Olins and A. L. Olins (1974). *Science* 183:330–332, by copyright permission of the American Association for the Advancement of Science.

Figure 11-18 Reproduced from R. Kornberg and A. Klug (1981). *Sci. Amer.* 244(Feb.):52–64.

Figure 11-21 Reproduced from F. Thoma, Th. Koller, and A. Klug (1979). *J Cell Biol.* 83:403–427, by copyright permission of The Rockefeller University Press, New York.

Figure 11-22 Reproduced from J. R. Paulsen and U. K. Laemmli (1977). *Cell* 12:817–828, by copyright permission of the Massachusetts Institute of Technology, Boston.

Figure 11-23 Reproduced from J. D. Griffith (1976). *Proc. Natl. Acad. Sci. USA* 73:563–567, by courtesy of the National Academy of Sciences.

Figure 11-24 Reproduced from R. Berezney and D. S. Coffey (1977). *J. Cell Biol.* 73:616–637, by copyright permission of The Rockefeller University Press, New York.

Figure 11-25 Reproduced from J. G. Gall, E. H. Cohen, and M. L. Polan (1971). *Chromosoma (Berl.)* 33:319–344, by copyright permission of Springer-Verlag, Heidelberg.

Figure 11-26 Reproduced from V. Sorsa (1974). *Hereditas* 78:298–302, by copyright permission of Hereditas.

Figure 11-28 (Left) Reproduced from J. M. Amabis and C. Janczur (1978). *J. Cell Biol.* 78:1–7, by copyright permission of The Rockefeller University Press, New York. *(Right)* Reproduced from B. Daneholt (1975). *Cell* 4:1–7, by copyright permission of the Massachusetts Institute of Technology, Boston.

Figure 11-29 Reproduced from H. Sass (1981). *Chromosoma (Berl.)* 83:619–643, by copyright permission of Springer-Verlag, Heidelberg.

Figure 11-30 Reproduced with permission from W. Beermann and U. Clever (1964). *Sci. Amer.* 210(Apr.):50–58.

Figure 11-31 Reproduced from B. Lambert (1973). *Cold Spr. Harbor Symp. Quant. Biol.* 38:637–644, by copyright permission of Cold Spring Harbor Laboratory, Cold Spring Harbor, New York.

Figure 11-32 Reproduced from B. Daneholt (1975). *Cell* 4:1–9, by copyright permission of the Massachusetts Institute of Technology, Boston.

Figure 11-33 Reproduced from Th. K. H. Holt (1970). *Chromosoma (Berl.)* 32:64–78, by copyright permission of Springer-Verlag, Heidelberg.

Figure 11-34 Reproduced from M. Ashburner, in *Developmental Studies on Giant Chromosomes*, W. Beermann, ed., pp. 101–151, copyright 1972, by permission of Springer-Verlag, Heidelberg.

Figure 11-35 Reproduced from A. L. Beyer, O. L. Miller, Jr., and S. L. McKnight (1980). *Cell* 20:75–84, by copyright permission of the Massachusetts Institute of Technology, Boston.

Figure 11-36 Reproduced from A. O. Pogo, V. C. Littau, V. G. Allfrey, and A. E. Mirsky (1967). *Proc. Natl. Acad. Sci. USA* 57:743–750, by courtesy of the National Academy of Sciences.

Figure 11-37 Adapted from R. G. Roeder and W. J. Rutter (1970). *Proc. Natl. Acad. Sci. USA* 65:675–692, and from S. P. Blatti, C. J. Ingles, T. J. Lindell, P. W. Morris, R. F. Weaver, F. Weinberg, and W. J. Rutter (1970). *Cold Spr. Harbor Symp. Quant. Biol.* 35:649–657.

Figure 11-38 Adapted from R. Weinmann and R. G. Roeder (1974). *Proc. Natl. Acad. Sci. USA* 71:1790–1794.

Figure 11-47 Adapted from H. Harris (1959). *Biochem. J.* 73:362–369.

Figure 11-48 Adapted from K. Scherrer, H. Latham, and J. E. Darnell (1963). *Proc. Natl. Acad. Sci. USA* 49:240–248.

Figure 11-51 Reproduced from O. P. Samarina, E. M. Lukanidin, J. Molnar, and G. P. Georgiev (1968). *J. Mol. Biol.* 33:251–263, by permission of Academic Press Ltd., London.

Figure 11-54. Adapted from J. E. Darnell, W. R. Jelinek, and G. R. Molloy (1973). *Science* 181:1215–1227, and from D. Scheiness and J. E. Darnell (1973). *Nature New Biol.* 241:266–268.

Figure 11-55 Reproduced from S. M. Berget, A. J. Berk, T. Harrison, and P. A. Sharp (1977). *Cold Spr. Harb. Symp. Quant. Biol.* 42:523–529, by copyright permission of Cold Spring Harbor Laboratory, Cold Spring Harbor, New York.

Figure 11-58 Reproduced from S. M. Tilghman, P. J. Curtis, D. C. Tiemeier, P. Leder, and C. Weissmann (1978). *Proc. Natl. Acad. Sci. USA* 75:1309–1313, by courtesy of the National Academy of Sciences.

Figure 11-61 Reproduced from E. Sidebottom, in *The Cell Nucleus*, H. Busch, ed., Vol. 1, pp. 439–460, copyright 1974, by permission of Academic Press, Inc., Orlando, FL.

Figure 11-62 Adapted from R. J. Britten and D. E. Kohne (1968). *Science* 161:529–540.

Figure 11-66 Reproduced from R. G. Davidson and H. M. Nitowsky (1963). *Proc. Natl. Acad. Sci. USA* 50:481–485, by courtesy of the National Academy of Sciences.

Figure 11-72 Adapted from A. H.-J. Wang, G. J. Quigley, F. J. Kolpak, J. L. Crawford, J. H. vanBoom, G. vanderMarel, and A. Rich (1979). *Nature* 282:680–686.

Figure 11-73 Reproduced from H. J. Lipps, A. Nordheim, E. M. Lofer, D. Ammermann, B. D. Stollar, and A. Rich (1983). *Cell* 32:435–441, by copyright permission of the Massachusetts Institute of Technology, Boston.

Figure 11-75 Adapted from W. Gilbert and B. Müeller-Hill (1967). *Proc. Natl. Acad. Sci. USA* 58:2415–2421.

Figure 11-77 Adapted from W. Gilbert (1973). *Cold Spr. Harb. Symp. Quant. Biol.* 38:845–855.

Chapter 12

Figure 12-2 Reproduced from J. A. Lake (1976). *J. Mol. Biol.* 105:131–159, by copyright permission of Academic Press, Ltd., London.

Figure 12-5 Reproduced from R. P. Perry (1963). *Exp. Cell Res.* 29:400–406, by copyright permission of Academic Press, Inc., Orlando, FL.

Figure 12-6 Adapted from M. Birnstiel (1966). *Nat. Cancer Inst. Monogr.* 23:431–444.

Figure 12-7 Reproduced from M. L. Pardue, S. Gerbi, R. Eckhardt, and J. Gall (1970). *Chromosoma (Berl.)* 29:268–290, by copyright permission of Springer-Verlag, Heidelberg.

Figure 12-9 Adapted from J. E. Darnell (1968). *Bacteriol. Rev.* 32:262–281, and from K. Scherrer, H. Latham, and J. E. Darnell (1963). *Proc. Natl. Acad. Sci. USA* 49:240–248.

Figure 12-10 Adapted from J. R. Warner and R. Soeiro (1967). *Proc. Natl. Acad. Sci. USA* 58:1984–1990.

Figure 12-14 Reproduced from N. K. Das, J. Micou-Eastwood, G. Ramamarthey, and M. Alfert (1970). *Proc. Natl. Acad. Sci. USA* 67:968–975, by courtesy of the National Academy of Sciences.

Figure 12-15 Reproduced from O. L. Miller, Jr., and B. R. Beatty (1969). *J. Cell Physiol.* 74(Suppl.1):225–232, by copyright permission of Alan R. Liss, Inc., New York.

Figure 12-16 Reproduced from O. L. Miller, Jr., and B. A. Hamkalo (1972). *Int. Rev. Cytol.* 33:1–25, by copyright permission of Academic Press, Inc., Orlando, FL.

Figure 12-19 Reproduced from J. A. Lake (1974). *Proc. Natl. Acad. Sci. USA* 71:4688–4692, by courtesy of the National Academy of Sciences.

Figure 12-20 Adapted from J. A. Lake (1981). *Sci. Amer.* 245(Aug.):84–97.

Figure 12-21 Adapted from H. M. Dintzis (1961). *Proc. Natl. Acad. Sci. USA* 47:247–261.

Figure 12-22 Adapted from R. Kaempfer (1968). *Proc. Natl. Acad. Sci. USA* 61:106–113.

Figure 12-33 Adapted from J. Warner, P. M. Knopf, and A. Rich (1963). *Proc. Natl. Acad. Sci. USA* 49:122–129.

Figure 12-34 Reproduced from J. R. Warner, P. M. Knopf, and A. Rich (1963). *Proc. Natl. Acad. Sci. USA* 49:122–129, by courtesy of the National Academy of Sciences.

Figure 12-36 Adapted from H. Noll, T. Staehelin, and F. O. Wettstein (1963). *Nature* 198:632–638.

Figure 12-37 Reproduced from A. Rich, in *The Neurosciences, A Study Program*, G. Quarton, T. Melnechuk, and F. O. Schmitt, eds., pp. 101–112, copyright 1967, by permission of The Rockefeller University Press, New York.

Figure 12-38 (Left) Reproduced from H. T. Bonnett and E. H. Newcomb (1965). *J. Cell Biol.* 27:423–432, by copyright permission of The Rockefeller University Press, New York. *(Right)* Reproduced from O. Behnke (1963). *Exp. Cell Res.* 30:597–598, by permission of Academic Press, Inc., Orlando, FL.

Figure 12-39 Reproduced from O. L. Miller, Jr., C. A. Thomas, and B. Hamkalo (1970). *Science* 169:392–395, by copyright permission of the American Association for the Advancement of Science.

Figure 12-41 Adapted from M. J. Clemens, E. C. Henshaw, H. Rahaminoff, and I. M. London (1974). *Proc. Natl. Acad. Sci. USA* 71:2946–2950, and from N. J. Gross and M. Rabinovitz (1972). *Proc. Natl. Acad. Sci. USA* 69:1565–1568.

Figure 12-43 Adapted from G. Huez, G. Marbaix, E. Hubert, M. Leclercq, U. Nudel, H. Soreq, R. Salomon, B. Lebleu, M. Revel, and U. Z. Littauer (1974). *Proc. Natl. Acad. Sci. USA* 71:3143–3146.

Figure 12-44 Adapted from J. M. Gilbert and W. F. Anderson (1970). *J. Biol. Chem.* 245:2342–2349.

Figure 12-45 Adapted from D. L. Oxender, G. Zurawski, and C. Yanofsky (1979). *Proc. Natl. Acad. Sci. USA* 76:5524–5528.

Chapter 13

Figure 13-2 Adapted from S. Cooper and C. E. Helmstetter (1968). *J. Mol. Biol.* 31:519–540.

Figure 13-3 Reproduced from J. Cairns (1963). *Cold Spr. Harb. Symp. Quant. Biol.* 28:43–46, by copyright permission of Cold Spring Harbor Laboratory, Cold Spring Harbor, New York.

Figure 13-4 Reproduced from D. M. Prescott and P. L. Kumpel (1972). *Proc. Natl. Acad. Sci. USA* 69:2842–2845, by courtesy of the National Academy of Sciences.

Figure 13-6 Reproduced from *Molecular Genetics*, J. H. Taylor, Part I, copyright 1963, by permission of Academic Press, Inc., Orlando, FL.

Figure 13-7 (Top, Bottom) Reproduced from J. A. Huberman and A. Tsai (1973). *J. Mol. Biol.* 75:5–12, by copyright permission of Academic Press Ltd., London. *(Middle)* Reproduced from D. R. Wolstenholme (1973). *Chromosoma (Berl.)* 43:1–18, by permission of Springer-Verlag, Heidelberg.

Figure 13-10 Reproduced from M. L. Higgins and L. Daneo-Moore (1974). *J. Cell Biol.* 61:288–300, by copyright permission of The Rockefeller University Press, New York.

Figure 13-11 Reproduced from L. M. Santo, H. R. Hohl, and H. A. Frank (1969). *J. Bacteriol.* 99:824–833, by copyright permission of the American Society for Microbiology.

Figure 13-13 Reproduced from *Genetics: Human Aspects*, A. P. Mange and E. J. Mange, copyright 1980. Reprinted by permission of W. B. Saunders Company, Philadelphia.

Figure 13-17 Adapted from M. P. F. Marsden and U. K. Laemmli (1979). *Cell* 17:849–858.

Figure 13-18 Reproduced from R. R. Schreck, D. Warbuster, O. J. Miller, S. M. Beiser, and B. F. Erlanger (1973). *Proc. Natl. Acad. Sci.*

USA 70:804–807, by courtesy of the National Academy of Sciences.

Figure 13-19 Reproduced from C. L. Rieder (1981). *Chromosoma (Berl.)* 84:145–158, by copyright permission of Springer-Verlag, Heidelberg.

Figure 13-21 Reprinted by permission of Elsevier Science Publishing Co., Inc., from "Correlative Immunofluorescence and Electron Microscopy on the Same Section of Epon-Embedded Material," by C. L. Rieder and S. S. Bowser. *J. Histochem. Cytochem.* 33:165–171. Copyright 1985 by The Histochemical Society.

Figure 13-22 Reproduced from R. C. Buck and J. M. Tisdale (1962). *J. Cell Biol.* 13:109–115, by copyright permission of The Rockefeller University Press, New York.

Figure 13-23 Reproduced from T. E. Schroeder (1972). *J. Cell Biol.* 53:419–434, by copyright permission of The Rockefeller University Press, New York.

Figure 13-24 Reproduced from A. M. Lambert and A. S. Bajer (1972). *Chromosoma (Berl.)* 39:101–144, by copyright permission of Springer-Verlag, Heidelberg.

Figure 13-25 Reproduced from J. D. Pickett-Heaps (1967). *Develop. Biol.* 15:206–236, by copyright permission of Academic Press, Inc., Orlando, FL.

Figure 13-28 Reproduced with permission from *The Meiotic Mechanism*, B. John and K. R. Lewis, 2nd ed., 1984. Carolina Biology Reader Series. Copyright Carolina Biological Supply Company, Burlington, NC.

Figure 13-30 Reproduced with permission from *The Meiotic Mechanism*, B. John and K. R. Lewis, 2nd ed., 1984. Carolina Biology Reader Series. Copyright Carolina Biological Supply Company, Burlington, NC.

Figure 13-31 Reproduced from M. Sasaki and S. Makino (1965). *Chromosoma (Berl.)* 16:637–651, by copyright permission of Springer-Verlag, Heidelberg.

Figure 13-33 Adapted from J. A. Taylor (1965). *J. Cell Biol.* 25:57–67.

Figure 13-37 Reproduced from M. S. Valenzuela and R. B. Inman (1975). *Proc. Natl. Acad. Sci. USA* 72:3024–3028 by courtesy of the National Academy of Sciences.

Figure 13-40 (Top) Reproduced from D. E. Comings and T. A. Okada (1972). *Adv. Cell Mol. Biol.* 2:309–382, by copyright permission of Academic Press, Inc., Orlando, FL. *(Middle)* Reproduced from R. Wettstein and J. R. Sotelo (1971). *Adv. Cell Mol. Biol.* 1:109–152, by copyright permission of Academic Press, Inc., Orlando, FL. *(Bottom)* Reproduced from D. E. Comings and T. A. Okada (1971). *Exp. Cell Res.* 65:104–116, by copyright permission of Academic Press, Inc., Orlando, FL.

Figure 13-42 Reproduced from T. F. Roth (1966). *Protoplasma* 61:346–386, by copyright permission of Springer-Verlag, Heidelberg.

Figure 13-43 (Right) Reproduced from O. L. Miller, Jr. (1965). *Natl. Cancer Inst. Monogr.* 18:79–99 by courtesy of the National Cancer Institute.

Figure 13-46 Reproduced from K. R. Porter, D. Prescott, and J. Frye (1973). *J. Cell Biol.* 57:815–836, by copyright permission of The Rockefeller University Press, New York.

Chapter 14

Figure 14-2 Reproduced from D. Thompson, V. Walbot, and E. H. Coe, Jr. (1983). *Amer. J. Bot.* 70:940–950, by permission of the Botanical Society of America.

Figure 14-3 Adapted from B. Singer, R. Sager, and Z. Ramanis (1976). *Genetics* 83:341–353.

Figure 14-4 Reproduced from Y. Yotsuyanagi (1962). *J. Ultrastruct. Res.* 7:121–158, by copyright permission of Academic Press, Inc., Orlando, FL.

Figure 14-5 Reproduced from P. R. Burton and D. G. Dusanic (1968). *J. Cell Biol.* 39:318–331, by copyright permission of The Rockefeller University Press, New York.

Figure 14-6 Reproduced from H. Ris (1962). *J. Cell Biol.* 13:383–391, by copyright permission of The Rockefeller University Press, New York.

Figure 14-7 Reproduced from M. M. K. Nass, S. Nass, and B. F. Afzelius (1965). *Exp. Cell Res.* 37:516–539, by copyright permission of Academic Press, Inc., Orlando, FL.

Figure 14-9 Adapted from D. J. L. Luck and E. Reich (1964). *Proc. Natl. Acad. Sci. USA* 52:931–938.

Figure 14-11 Adapted from R. Wells and M. Birnstiel (1969). *Biochem. J.* 112:777–786.

Figure 14-12 (Left, Bottom) Reproduced from M. M. K. Nass (1966). *Proc. Natl. Acad. Sci. USA* 56:1215–1222, by courtesy of the National Academy of Sciences. *(Right)* Reproduced from M. M. K. Nass (1969). *Science* 165:25–36, by copyright permission of the American Association for the Advancement of Science.

Figure 14-13 Adapted from J. C. Mounolou, H. Jakob, and P. P. Slonimski (1966). *Biochem. Biophys. Res. Commun.* 24:218–224.

Figure 14-14 Adapted from E. S. Goldring, L. I. Grossman, D. Krupnick, D. R. Cryer, and J. Marmur (1970). *J. Mol. Biol.* 52:323–335.

Figure 14-15 Adapted from E. H. L. Chun, M. H. Vaughan, Jr., and A. Rich (1963). *J. Mol. Biol.* 7:130–141, and from P. R. Whitfield and D. Spencer (1968). *Biochim. Biophys. Acta* 157:333–343.

Figure 14-16 Reproduced from J. E. Manning, D. R. Wolstenholme, R. S. Ryan, J. A. Hunter, and O. C. Richards (1971). *Proc. Natl. Acad. Sci. USA* 68:1169–1173, by courtesy of the National Academy of Sciences.

Figure 14-17 (Left) Reproduced from R. J. Rose and J. V. Possingham (1976). *J. Cell Sci.* 20:341–355, by permission of The Company of Biologists, Ltd., Colchester, England. *(Right)* Reproduced from R. Charret and J. André (1968). *J. Cell Biol.* 39:369–381, by copyright permission of The Rockefeller University Press, New York.

Figure 14-18 Adapted from K.-S. Chiang and N. Sueoka (1967). *Proc. Natl. Acad. Sci. USA* 57:1506–1513.

Figure 14-19 (Left) Reproduced from H. Kasamatsu, D. L. Robberson, and J. Vinograd (1971). *Proc. Natl. Acad. Sci. USA* 68:2252–2257, by courtesy of the National Academy of Sciences. *(Right)* Reproduced from K. Koike and D. R. Wolstenholme (1974). *J. Cell Biol.* 61:14–25, by copyright permission of The Rockefeller University Press, New York.

Figure 14-22 Reproduced from M. Wu, N. Davidson, G. Attardi, and Y. Aloni (1972). *J. Mol. Biol.* 71:81–93, by copyright permission of Academic Press Ltd., London.

Figure 14-23 Adapted from L. A. Grivell (1983). *Sci Amer. 248* (Mar.):78–89, and from D. Clayton (1984). *Ann. Rev. Biochem.* 53:573–594.

Figure 14-24 Adapted from P. R. Whitfield and W. Bottomly (1983). *Ann. Rev. Plant Physiol.* 34:279–310.

Figure 14-25 (Top) Reproduced from W. Y. Chooi and C. D. Laird (1976). *J. Mol. Biol.* 100:493–518, by copyright permission of Academic Press Ltd., London. *(Bottom)* Reproduced from H. Falk (1969). *J. Cell Biol.* 42:582–587, by copyright permission of The Rockefeller University Press, New York.

Figure 14-28 Reproduced from W. J. Larsen (1970). *J. Cell Biol.* 47:373–383, by copyright permission of The Rockefeller University Press, New York.

Figure 14-29 Adapted from D. J. L. Luck (1965). *J. Cell Biol.* 24:461–470.

Figure 14-30 Adapted from N. J. Gross, G. S. Getz, and M. Rabinowitz (1969). *J. Biol. Chem.* 244:1552–1562.

Figure 14-31 Reprinted with permission from H. Plattner and G. Schatz, *Biochemistry,* 8:339–343. Copyright 1969, American Chemical Society.

Figure 14-32 Reproduced from J. V. Possingham and R. J. Rose (1976). *Proc. Roy. Soc. London, Ser B.* 193:295–305, by copyright permission of The Royal Society, London.

Figure 14-33 Reprinted by permission from S. M. Ridley and R. M. Leech, *Nature* 227(5257):463–465. Copyright 1970, Macmillan Journals, Ltd., London.

Figure 14-34 Reproduced with permission from U. W. Goodenough and R. P. Levine (1970). *Sci. Amer.* 223(Nov.):22–29.

Figure 14-36 (Top) Reproduced from T. E. Weier and D. E. Brown (1970). *Amer. J. Bot.* 57:267–275, by permission of the Botanical Society of America. *(Bottom)* Adapted from T. E. Weier and D. E. Brown (1976). *Amer. J. Bot.* 57:267–275.

Figure 14-37 Reproduced from S. Klein, G. Bryan, and L. Bogorad (1964). *J. Cell Biol.* 22:433–442, by copyright permission of The Rockefeller University Press, New York.

Figure 14-42 (a) Adapted from R. A. Raff and H. R. Mahler (1972). *Science* 177:575–582. *(b)* Adapted from T. Cavalier-Smith (1975). *Nature* 256:463–468. *(c)* Adapted from L. Reijnders (1975). *J. Mol. Evol.* 5:167–176.

Figure 14-43 Reproduced from S. W. Watson, L. B. Graham, C. C. Remsen, and F. W. Valois (1971). *Arch. Mikrobiol.* 76:183–203, by copyright permission of Springer-Verlag, Heidelberg.

Chapter 15

Figure 15-2 Reproduced from *Scanning Electron Microscopy in Biology,* R. G. Kessel and C. Y. Shih, copyright 1974, by permission of Springer-Verlag, Heidelberg.

Figure 15-4 Reproduced from R. G. Kessel and A. W. Beams (1963). *Exp. Cell Res.* 30:440–443, by copyright permission of Academic Press, Inc., Orlando, FL.

Figure 15-6 Reproduced from R. G. Kessel (1968). *J. Morphol.* 126:211–246, by copyright permission of Alan R. Liss, Inc., New York.

Figure 15-7 (Bottom) Reproduced from R. A. Wallace and J. N. Dumont (1968). *J. Cell Physiol.* 72(Suppl. 1):73–89, by copyright permission of Alan R. Liss, Inc., New York.

Figure 15-9 Reproduced from F. J. Longo and E. Anderson (1969). *J. Exp. Zool.* 172:67–96, by copyright permission of Alan R. Liss, Inc., New York.

Figure 15-12 (Top) Reproduced from G. Schatten and D. Mazia (1976). *Exp. Cell Res.* 98:325–337, by copyright permission of Academic Press, Inc., Orlando, FL. *(Middle, Bottom)* Reproduced from H. Schatten and G. Schatten (1980). *Develop. Biol.* 78:435–449, by copyright permission of Academic Press, Inc., Orlando, FL.

Figure 15-13 Reproduced from W. A. Anderson (1967). *J. Cell Biol.* 32:11–26, by copyright permission of The Rockefeller University Press, New York.

Figure 15-14 Reproduced from D. M. Phillips (1970). *J. Cell Biol.* 44:243–277, by copyright permission of The Rockefeller University Press, New York.

Figure 15-15 Reproduced from D. S. Friend (1982). *J. Cell Biol.* 93:243–249, by copyright permission of The Rockefeller University Press, New York.

Figure 15-16 Reproduced from S. A. Pratt (1968). *J. Morphol.* 126:31–65, by copyright permission of Alan R. Liss, Inc., New York.

Figure 15-18 Reproduced from R. Yanagimachi and Y. D. Noda (1970). *J. Ultrastruct. Res.* 31:465–485, by copyright permission of Academic Press, Inc., Orlando, FL.

Figure 15-19 (Left) Reproduced from G. Schatten and D. Mazia (1976). *Exp. Cell Res.* 98:325–337, by copyright permission of Academic Press, Inc., Orlando, FL. *(Right)* Reproduced from E. Anderson (1968). *J. Cell Biol.* 37:514–539, by copyright permission of The Rockefeller University Press, New York.

Figure 15-20 Reproduced from M. J. Tegner and D. Epel (1973). *Science* 129:685–688, by copyright permission of the American Association for the Advancement of Science.

Figure 15-21 Reproduced from R. D. Grey, P. K. Working, and J. L. Hedrick (1976). *Develop. Biol.* 54:52–60, by copyright permission of Academic Press, Inc., Orlando, FL.

Figure 15-22 Adapted from D. Epel (1977). *Sci. Amer. 237* (Nov.):129–139.

Figure 15-23 Reproduced from F. J. Longo and E. Anderson (1968). *J. Cell Biol.* 39:339–368, by copyright permission of The Rockefeller University Press, New York.

Figure 15-24 Adapted from D. Epel (1967). *Proc. Natl. Acad. Sci. USA* 57:899–906.

Figure 15-26 Reproduced from L. L. Hoefert (1969). *Amer. J. Bot.* 56:363–368, by permission of the Botanical Society of America.

Figure 15-28 Reproduced with permission from J. B. Gurdon (1968). *Sci. Amer.* 219(Dec.):24–36.

Figure 15-29 Reprinted by permission from D. A. Evans, *BioTechnology* 1:253–261. Copyright 1983, Macmillan Journals, Ltd., London.

Figure 15-30 Reproduced from J. L. Marx (1982). *Science* 218:459–460, by copyright permission of The American Association for the Advancement of Science.

Figure 15-31 Adapted from R. T. Schimke (1966). *Bull. Soc. Chim. Biol.* 48:1009–1030.

Figure 15-32 Adapted from R. T. Schimke (1964). *J. Biol. Chem.* 239:3808–3817.

Figure 15-34 Reproduced from M. S. Steinberg, in *Cellular Membranes in Development*, M. Locke, ed., pp. 321–366, copyright 1964, by permission of Academic Press, Inc., Orlando, FL.

Figure 15-35 Adapted from J. E. Morris and A. A. Moscona (1970). *Science* 167:1736–1738.

Figure 15-37 Reproduced from J. Wartiovaara, S. Nordling, E. Lehtonen, and L. Saxén (1974). *J. Embryol. Exp. Morphol.* 31:667–682, by permission of Cambridge University Press, New York.

Figure 15-38 Adapted from L. Hayflick (1975). *BioScience* 25:629–637.

Figure 15-39 Reproduced with permission from L. Hayflick (1980). *Sci. Amer.* 242(Jan.):58–65.

Chapter 16

Figure 16-9 Adapted from W. Y. Ling and J. M. Marsh (1977). *Endocrinology* 100:1571–1578.

Figure 16-13 Adapted from E. V. Jensen and E. R. DeSombre (1973). *Science* 182:126–134.

Figure 16-14 Adapted from J. Gorski, D. Toft, G. Shyamala, D. Smith, and A. Notides (1968). *Recent Prog. Hormone Res.* 24:45–80.

Figure 16-15 Reproduced from G. A. Porter, R. Bogoroch, and I. S. Edelman (1964). *Proc. Natl. Acad. Sci. USA* 52:1326–1337, by courtesy of the National Academy of Sciences.

Figure 16-16 Adapted from G. Shyamala and J. Gorski (1969). *J. Biol. Chem.* 244:1097–1103.

Figure 16-18 Adapted from L. Chan, A. R. Means, and B. W. O'Malley (1973). *Proc. Natl. Acad. Sci. USA* 70:1870–1874, and from S. E. Harris, J. M. Rosen, A. R. Means, and B. W. O'Malley (1975). *Biochemistry* 14:2072–2081.

Chapter 17

Figure 17-3 Reproduced from *A Textbook of Histology*, W. Bloom and D. W. Fawcett, 10th ed., copyright 1975, by permission of W. B. Saunders Company, Philadelphia.

Figure 17-4 Reproduced from *A Textbook of Histology*, W. Bloom and D. W. Fawcett, 10th ed., copyright 1975, by permission of W. B. Saunders Company, Philadelphia.

Figure 17-13 Adapted from N. Hozumi and S. Tonegawa (1976). *Proc. Natl. Acad. Sci. USA* 73:3628–3632.

Figure 17-19 Reproduced from I. Yahara and F. Kakimoto-Sameshima (1977). *Proc. Natl. Acad. Sci. USA* 74:4511–4515, by courtesy of the National Academy of Sciences.

Chapter 18

Figure 18-1 Reproduced from A. Peters in *The Structure and Function of Nervous Tissue*, G. H. Bourne, ed., Vol. 1, pp. 119–186, copyright 1968, by permission of Academic Press, Inc., Orlando, FL.

Figure 18-2 (Bottom) Reproduced from H. Mannen (1966). *J. Comp. Neurol.* 126:75–90, by copyright permission of Academic Press, Inc., Orlando, FL.

Figure 18-3 (Left) Reproduced from *The Fine Structure of the Nervous System*, A. Peters, S. L. Palay, and H. DeF. Webster, eds., copyright 1970, by permission of Harper & Row, Inc., New York. *(Right)* Adapted from *Medical Neurobiology*, W. D. Willis, Jr. and R. G. Grossman, 3rd ed., copyright 1981, C. V. Mosby Company, St. Louis.

Figure 18-4 (Left) Reproduced from A. Peters, in *The Structure and Function of Nervous Tissue*, G. H. Bourne, ed., Vol. 1, pp. 119–186, copyright 1968, by permission of Academic Press, Inc., Orlando, FL. *(Right)* Reproduced from *The Fine Structure of the Nervous System*, A. Peters, S. L. Palay, and H. DeF. Webster, eds., copyright 1970, by permission of Harper & Row, Inc., New York.

Figure 18-5 Reproduced from *The Fine Structure of the Nervous System*, A. Peters, S. L. Palay, and H. DeF. Webster, eds., copyright 1970, by permission of Harper & Row, Inc., New York.

Figure 18-6 Reproduced from R. B. Wuerker (1970). *Tissue and Cell* 2:1–19, by permission of Longman Group, Ltd., Essex, England.

Figure 18-7 Reproduced from Y. J. LeBeux (1973). *Z. Zellforsch* 143:239–272, by copyright permission of Springer-Verlag, Heidelberg.

Figure 18-8 Reproduced from *The Fine Structure of the Nervous System*, A. Peters, S. L. Palay, and H. DeF. Webster, eds., copyright 1970, by permission of Harper & Row, Inc., New York.

Figure 18-9 (Top) Adapted from *Medical Neurobiology*, W. D. Willis, Jr. and R. G. Grossman, 3rd. ed., copyright 1981, C. V. Mosby Company, St. Louis. *(Bottom)* Reproduced from *Medical Neurobiology*, W. D. Willis, Jr. and R. G. Grossman, 3rd ed., copyright 1981, by permission of the C. V. Mosby Company, St. Louis.

Figure 18-10 Reproduced with permission from P. Morell and W. T. Norton (1980). *Sci. Amer.* 242(May):88–117.

Figure 18-11 Adapted from J. H. Schwartz (1980). *Sci. Amer.* 242(Apr.):152–157.

Figure 18-12 Adapted from H. L. Fernandez, P. R. Burton, and F. E. Samson (1971). *J. Cell Biol.* 51:176–192.

Figure 18-14 Reproduced from G. D. Pappas and S. G. Waxman in *Structure and Function of Synapses*, G. D. Pappas and D. P. Purpura, eds., pp. 1–43, copyright 1972, by permission of Raven Press, New York.

Figure 18-15 Reproduced from Y. J. LeBeux (1973). *Z. Zellforsch* 143:239–272, by copyright permission of Springer-Verlag, Heidelberg.

Figure 18-16 Reproduced from Y. J. LeBeux (1973). *Z. Zellforsch* 143:239–272, by copyright permission of Springer-Verlag, Heidelberg. Diagram adapted from Y. J. LeBeux and J. Willemot (1975). *Cell Tissue Res.* 160:37–68.

Figure 18-17 Reproduced from *Synapses and Synaptosomes*, D. G. Jones, copyright 1975, by permission of Chapman and Hall Publishers, London.

Figure 18-18 Reproduced from G. D. Pappas and S. G. Waxman in *Structure and Function of Synapses*, G. D. Pappas and D. P. Purpura, eds., pp. 1–43, copyright 1972, by permission of Raven Press, New York.

Figure 18-19 Reproduced with permission from W. M. Cowan (1979). *Sci. Amer.* 241(Sept.):113–130.

Figure 18-20 Adapted from N. K. Wesells, B. S. Spooner, J. F. Ash, M. O. Bradley, M. A. Luduena, E. L. Taylor, J. T. Wrenn, and K. M. Yamada (1971). *Science* 171:135–143.

Figure 18-21 Reproduced with permission from R. Levi-Montalcini and P. Calissano (1979). *Sci. Amer.* 240(June): 68–77.

Figure 18-22 Adapted from *Principles of Neural Science*, E. R. Kandel and J. H. Schwartz, 2nd ed., copyright 1985, Elsevier, New York.

Figure 18-23 Adapted from A. L. Hodgkin (1985). *Proc. Roy. Soc. (London), Ser. B* 148:1–37.

Figure 18-24 Adapted from A. L. Hodgkin and B. Katz (1949). *J. Physiol.* 108:37–77.

Figure 18-25 Adapted from A. L. Hodgkin (1958). *Proc. Roy. Soc. (London), Ser. B* 148:1–37.

Figure 18-28 Adapted from T. Rushton (1951). *J. Physiol.* 115:101–122.

Figure 18-29 Reproduced with permission from P. Morell and W. T. Norton (1980). *Sci. Amer.* 242(May):88–117.

Figure 18-30 (Left) Reproduced from E. R. Lewis, Y. Y. Zeevi, and F. S. Werblin (1969). *Brain Res.* 15:559–562, by copyright permission of Elsevier Science Publishers (Biomedical Division), Amsterdam. *(Right)* Reproduced from E. Townes-Anderson, P. R. MacLeish, and E. Raviola (1985). *J. Cell Biol.* 100:175–188, by copyright permission of The Rockefeller University Press, New York.

Figure 18-35 Reproduced with permission from H. A. Lester (1977). *Sci. Amer.* 236(Feb.):106–118.

Figure 18-36 Reprinted by permission from P. T. Turner and A. B. Harris, *Nature* 242(5392):57–60. Copyright 1983, Macmillan Journals, Ltd., London.

Figure 18-37 Diagram modified and micrograph reproduced from J. E. Heuser and T. S. Reese (1973). *J. Cell Biol.* 57:315–344, by copyright permission of The Rockefeller University Press, New York.

Figure 18-38 Reproduced from H. C. Fertuck and M. M. Salpeter (1976). *J. Cell Biol.* 69:144–148, by copyright permission of The Rockefeller University Press, New York.

Figure 18-39 Reproduced from W. Schiebler and F. Hucho (1978). *Eur. J. Biochem.* 85:55–63, by copyright permission of Springer-Verlag, Heidelberg.

Figure 18-40 Adapted from E. Ramón-Moliner, in *The Structure and Function of Nervous Tissue*, G. H. Bourne, ed., Vol. I, pp. 205–267, copyright 1968, Academic Press, Orlando, FL.

Figure 18-41 Reproduced from R. Simantor, H. J. Kuhas, G. R. Uhl, and S. H. Snyder (1977). *Proc. Natl. Acad. Sci. USA* 74:2167–2171, by courtesy of the National Academy of Sciences.

Figure 18-42 Adapted from *Principles of Neural Science*, E. R. Kandel and J. H. Schwartz, 2nd ed., copyright 1985, Elsevier, New York.

Chapter 19

Figure 19-1 Reproduced from *Atlas of Descriptive Histology*, E. J. Reith and M. H. Ross, 2nd ed., copyright 1970, by permission of Harper & Row, Inc., New York.

Figure 19-2 Reprinted from F. A. Pepe, in *Biological Macromolecules, Subunits in Biological Systems*, S. N. Timasheff and G. D. Fasman, eds., Vol. V, Part A, pp. 323–353, 1971, by courtesy of Marcel Dekker, Inc., New York.

Figure 19-4 Reproduced from H. E. Huxley (1957). *J. Biophys. Biochem. Cytol.* 3:631–647, by copyright permission of Academic Press, Inc., Orlando, FL.

Figure 19-6 Adapted from L. D. Peachey (1965). *J. Cell Biol.* 25:209–231.

Figure 19-7 Reproduced from L. D. Peachey (1965). *J. Cell Biol.* 25:209–231, by copyright permission of The Rockefeller University Press, New York.

Figure 19-8 Reprinted by permission from H. E. Huxley, *Nature*, 202(4937):1067–1071. Copyright 1964, Macmillan Journals, Ltd., London.

Figure 19-9 Reproduced from C. F. Franzini-Armstrong (1970). *J. Cell Biol.* 47:488–499, by copyright permission of The Rockefeller University Press, New York.

Figure 19-10 (Left) Reproduced from D. A. Fischman, in *The Structure and Function of Muscle*, G. H. Bourne, ed., pp. 75–148, copyright 1972, by permission of Academic Press, Inc., Orlando, FL. *(Right)* Reproduced from D. A. Fischman (1970). *Curr. Top. Develop. Biol.* 5:235–280, by copyright permission of Academic Press, Inc., Orlando, FL.

Figure 19-11 Reproduced from Y. Shimada (1971). *J. Cell Biol.* 48:128–142, by copyright permission of The Rockefeller University Press, New York.

Figure 19-12 Adapted from B. Mintz and W. W. Baker (1967). *Proc. Natl. Acad. Sci. USA* 58:592–598.

Figure 19-13 Micrograph reproduced from F. A. Pepe and B. Drucker (1966). *J. Cell Biol.* 28:505–525, by copyright permission of Academic Press, Ltd., London.

Figure 19-14 Reproduced from A. Elliott, G. Offer, and K. Burridge (1976). *Proc. Roy. Soc. London, Ser. B* 193:45–53, by copyright permission of The Royal Society, London.

Figure 19-17 (Top) Reproduced from H. E. Huxley (1963). *J. Mol. Biol.* 7:281–308, by copyright permission of Academic Press, Ltd., London.

Figure 19-18 Reproduced from F. A. Pepe and B. Drucker (1975). *J. Mol. Biol.* 99:609–617, by copyright permission of Academic Press, Ltd. London.

Figure 19-19 (Left) Reproduced from P. B. Moore, H. E. Huxley, and D. J. DeRosier (1970). *J. Mol. Biol.* 50:279–295, by copyright permission of Academic Press, Ltd., London.

Figure 19-20 Reproduced from J. A. Spudich, H. E. Huxley, and J. T. Finch (1972). *J. Mol. Biol.* 72:619–632, by copyright permission of Academic Press, Ltd., London.

Figure 19-21 Reproduced from H. E. Huxley, in *The Cell*, T. Brachet and A. E. Mirsky, eds., Vol. IV, pp. 365–481, copyright 1960, by permission of Academic Press, Inc., Orlando, FL.

Figure 19-22 Adapted from A. M. Gordon, A. F. Huxley, and F. J. Julian (1966). *J. Physiol.* 184:170–192.

Figure 19-25 Reproduced from E. G. Gray, in *The Synapse*, Oxford Biology Reader, J. J. Head, ed., pp. 2–15, copyright 1973, by permission of Oxford University Press, Oxford, England.

Figure 19-26 Reproduced from A. F. Huxley and R. E. Taylor (1958). *J. Physiol.* (London) 144:426–441, by copyright permission of Cambridge University Press, Cambridge, England.

Figure 19-27 Adapted from C. C. Ashley and E. B. Ridgeway (1970). *J. Physiol.* 209:105–130.

Figure 19-28 Adapted from J. D. Potter and J. Gergely (1974). *Biochemistry* 13:2697–2703.

Figure 19-29 Adapted from *Muscle Physiology*, F. D. Carlson and D. R. Wilkie, copyright 1974, Prentice-Hall, New York.

Figure 19-30 Reproduced from G. Gabella (1976). *Cell Tiss. Res.* 170:187–201, by copyright permission of Springer-Verlag, Heidelberg.

Figure 19-31 Reproduced from R. V. Rice, J. A. Moses, G. M. McMann, A. L. Brady, and L. M. Blasik (1970). *J. Cell Biol.* 47:183–196, by copyright permission of The Rockefeller University Press, New York.

Figure 19-32 Reproduced from P. Cooke (1976). *J. Cell Biol.* 68:539–556, by copyright permission of The Rockefeller University Press, New York.

Figure 19-34 Reproduced from D. J. Scales (1983). *J. Ultrastruct. Res.* 83:1–9, by copyright permission of Academic Press, Inc., Orlando, FL.

Figure 19-35 Reproduced from J. R. Sommer and R. A. Waugh (1976). *Amer. J. Pathol.* 82:192–232, by permission of J. B. Lippincott Company, Philadelphia.

Chapter 20

Figure 20-1 Reproduced from *Textbook of Pathology*, S. L. Robbins, copyright 1957, by permission of W. B. Saunders Company, Philadelphia.

Figure 20-2 Adapted from G. J. Todaro and A. Green (1963). *J. Cell Biol.* 17:299–313.

Figure 20-3 Adapted from S. A. Aaronson and G. J. Todaro (1968). *Science* 162:1024–1026.

Figure 20-4 Reproduced from L. Sachs, in *Molecular Bioenergetics and Macromolecular Biochemistry*, H. H. Weber, ed., pp. 118–128, copyright 1972, by permission of Springer-Verlag, Heidelberg.

Figure 20-5 Reproduced from R. Azarnia and W. R. Loewenstein (1971). *J. Memb. Biol.* 6:368–385 by copyright permission of Springer-Verlag, Heidelberg.

Figure 20-7 Reproduced from K. M. Robinson, L. Maistry, and P. Evers (1981). *Scanning Electron Microscopy* II:213–222, by permission of SEM, Inc., Chicago.

Figure 20-8 Reproduced from *Textbook of Pathology*, S. L. Robbins, copyright 1957, by permission of W. B. Saunders Company, Philadelphia.

Figure 20-9 Reproduced from P. C. Nowell, in *Cancer: A Comprehensive Treatise*, F. F. Becker, ed., Vol. 1, 2nd ed., pp. 3–46, copyright 1982, by permission of Plenum Publishing Corp., New York.

Figure 20-10 Adapted from R. K. Boutwell (1964). *Prog. Exp. Tumor Res.* 4:207–250.

Figure 20-11 Adapted from P. Brooks and P. D. Lawley (1964). *Nature* 202:781–784.

Figure 20-12 Reproduced with permission from W. Henle, G. Henle, and E. T. Lennette (1979). *Sci. Amer.* 241(July):48–59.

Figure 20-15 Reproduced from K. Illmensee, in *Genetic Mosaics and Chimeras in Mammals*, L. B. Russell, ed., copyright 1978, by permission of Plenum Publishing Corp., New York.

Figure 20-16 Adapted from P. Buell and J. E. Dunn, Jr. (1965). *Cancer* 18:656–664.

Figure 20-19 Adapted from S. Meselson and L. Russell in *Origins of Human Cancers, A Book*, H. H. Hiatt, J. D. Watson, and J. A. Winsten, eds., copyright 1977, pp. 1473–1482, Cold Spring Harbor Laboratory, Cold Spring Harbor, NY.

Figure 20-20 Adapted from B. Armstrong and R. Doll (1975). *Int. J. Cancer* 15:617–631.

Index

Boldface page numbers refer to the definition of an entry in the text.

Aaronson, S., 731
A-band of sarcomere, **702**, 713, 716
Abercrombie, M., 323, 730
Abl oncogene, 745
Abscisic acid, **638**
Accessory cell. *See* Follicle cell; Nurse cell; Sertoli cell
Accessory fiber, **592**
Accessory pigment, **264**
Acetabularia, 385
Acetanilide-hydrolyzing esterase, 164
Acetylcholine
  binding to receptor, 694
  and depolarization of sarcolemma, 716
  discovery of, 689
  as neurotransmitter, **689**, 693
  site of action, 690
  structure, 690
Acetylcholine receptor, 693–694
  α-bungarotoxin binding, 693
  acetylcholine binding, **693**
  depolarization of skeletal muscle, 694
  hyperpolarization of cardiac muscle, 694
  location, 693
  membrane potential and, 694
  in myasthenia gravis, 720
  nicotinic (fast) acetylcholine receptor, 693
  subunit composition, 694
  in *Torpedo*, 693
Acetylcholinesterase
  acetylcholine inactivation, **697**
  active site, amino acid sequence, 52
Acetyl CoA. *See* Acetyl coenzyme A
Acetyl coenzyme A
  fatty acid β-oxidation and, 214
  in glyoxylate cycle, 199–200
  in histone acetylation, 395
  hydrolysis, 214
  Krebs citric acid cycle, oxidation in, 214–215
  pyruvate oxidation, **213**
N-Acetyl-D-glucosamine, in bacterial cell wall, 115
N-Acetylmuramic acid, in bacterial cell wall, 115
Acid phosphatase
  cytochemical localization, **71**
  lysosome marker, 182–184
  spherosome marker, 195
cis-Aconitate, in Krebs citric acid cycle, 214
Acquired immune deficiency syndrome (AIDS), 662, **741**
Acridine orange
  cytochemical stain, 69
  as mutagen, 359
Acrosomal process, **593**
Acrosome, **589**, 590
ACTH. *See* Adrenocorticotropin
Actin. *See also* Microfilament; Thin filament
  acrosome process formation, 593

α-actin, **317**
ankyrin association, 109–110, 237
β-actin, **317**
γ-actin, **317**
axonal transport of, 673
binding proteins, 317–318
filamentous (F), **712**, 721
globular (G), **711**
in mitotic spindle, 510
in nonmuscle cells, 317
in red blood cell membrane, 106, 109–110, 327
in skeletal muscle, 709, 711
in smooth muscle, 721
spectrin association, 109, 327
stress fibers, 317, 322
α-Actinin
  actin-binding protein, 318
  calcium ion sensitivity, 318
  in cardiac muscle, 723
  in fibroblasts, 318
  in skeletal muscle, **709**
  in smooth muscle, 721
  in thin filament, 713
β-Actinin
  actin-binding protein, 318
  calcium ion sensitivity, 318
  in skeletal muscle, **709**
  in thin filament, 713
γ-Actinin, 318
  actin-binding protein, 318
  calcium ion sensitivity, 318
Actinogelin, 318
Actinomycin D
  inhibitor of RNA synthesis, **414**, 636
  intercalation into DNA, 414
  structure, 414
Action potential, **682**, 681–698. *See also* Membrane potential
  axon diameter and, 684–685
  cardiac muscle contraction, 723–724
  depolarization of sarcolemma, 716
  discovery of, 681–682
  myelin and, 684–685
  in neurotransmitter release, 691
  in photoreception, 685–688
  and potassium ion membrane permeability, 682–684
  propagation, 684–685
  skeletal muscle contraction, 715–718
  smooth muscle contraction, 722
  and sodium ion membrane permeability, 682–684
  and synaptic transmission, 689–698
Activator protein, **440**
Active site. *See* Enzyme
Active transport, **8, 139**, 140–146
  of amino acids, 144–145
  phosphotransferase system, 142–144
  proton-linked, 145–146
  of sodium and potassium ions, 140–142
  sodium-linked, 144–145
  of sugars, 144–145

Actomyosin, **709**

Adaptor molecule, **362.** *See also* Transfer ribonucleic acid (tRNA)

Adenine, **23**
  base pairing, 344, 367
  structure, 22

Adenosine diphosphate (ADP)
  adenylate kinase and ATP synthesis, 720
  electron transfer in mitochondria, control of, 243
  and mitochondrial ATP synthesis, 229
  mitochondrial transport of, 241–242
  structure, **54**

Adenosine monophosphate (AMP)
  allosteric activation of phosphofructo-kinase, 243
  cyclic AMP hydrolysis, 625
  structure, **54**

Adenosine triphosphatase (ATPase)
  in bacteria, 243–244
  bacterial plasma membrane marker, 106
  CF_1 subunit, **273,** 274, 276
  CF_0 subunit, **273,** 274–275
  dynein, 304
  F_1 subunit, **235,** 236–237, 243–244
  F_0 subunit, **237,** 240
  membrane orientation, red blood cell, 110
  myosin, 710
  oligomycin binding to F_0 subunit, 240
  oligomycin-sensitive ATPase, 557
  oligomycin-sensitivity conferring protein (OSCP), 557
  in oxidative phosphorylation, 235–237, 243–244
  in photophosphorylation, 273–276
  subunit synthesis in chloroplast, 554, 559
  subunit synthesis in mitochondrion, 557–558, 561

Adenosine triphosphatase (ATPase), calcium ion
  in microsomes, 164
  in sarcoplasmic reticulum, **717**

Adenosine triphosphatase (ATPase), potassium ion, 106

Adenosine triphosphatase (ATPase), sodium and potassium ion, **140**
  active transport, sodium and potassium ions, 140–143
  model of, 142
  ouabain binding site, 141
  phosphorylation of, 142
  plasma membrane marker, 106

Adenosine triphosphate (ATP), **53**
  allosteric inhibitor of phosphofructokinase, 243
  in amino acid-transfer RNA binding, 364
  and amoeboid movement, 321
  in axoneme bending, 309–311
  in Calvin cycle, 279–282
  capping of, in primary transcript, 419
  and chromosome movement, 510
  coupled reactions, 53–55
  in glucose oxidation, 217
  in histone phosphorylation, 395
  initiation of RNA synthesis, 413
  in Krebs citric acid cycle, 214
  in muscle contraction, 719–720
  as neurotransmitter, 689
  structure, **54**

Adenosine triphosphate (ATP) synthesis
  in bacteria, 243–245
  in chloroplasts, 269–270
  coupling sites in mitochondria, 235

in mitochondrion, 215–218, 235–241
  proton gradient, 237–241
  regulation by adenosine diphosphate (ADP) and adenosine monophosphate (AMP), 243

S-Adenosylmethionine, in histone methylation, 395

Adenovirus, **742**
  and discovery of introns, 420
  tumors induced by, 740

Adenylate cyclase, **530**
  cyclic AMP synthesis, 625
  in dopamine action, 694–695
  in glycogen breakdown, 625
  in histamine action, 697
  in hormone receptor action, 626–628
  in norepinephrine action, 697
  in octopamine action, 697
  in photoreception, 688–689
  plasma membrane marker, 106
  in serotonin action, 697

Adenylate kinase, **720**

ADP. *See* Adenosine diphosphate

ADP ribosylation, of guanine nucleotide binding protein (G-protein), 628

Adrenal cortex, hormone products, 622

Adrenal medulla, hormone products, 622

Adrenocorticotropin (ACTH)
  and cyclic AMP, 626
  feedback regulation of, 623
  as neurotransmitter, 690
  site of synthesis, **622**
  target organ, 622

Adsorption chromatography, **86**

Aequorin, **597,** 691

Affinity chromatography, **87**
  oligodeoxythymidylate [oligo(dT)], 467
  polyuridylic acid [poly(U)], 467

Affinity labeling
  of membrane transporters, **134**
  and ribosome structure, **463**

Aflatoxin, **752**

a gene, in lactose operon, **437**

Aging, of cells, 613–615

Agonist, 631

AIDS. *See* Acquired immune deficiency syndrome

ALA synthetase. *See* 5-Aminolevulinate synthetase

Aldolase, in Calvin cycle, 281

Aldosterone, 633

Alkaptonuria, **350**

Allen, R., 321, 675

Allen video-enhanced contrast differential interference contrast (AVEC-DIC) microscopy system, 675

Allophenic mouse, **708**

Alloway, J., 341

Alpha cells, in pancreas, 621

Alpha (α) helix, **19**

Altmann, R., 569

α-Amanitin, **410,** 411

Ames, B., 751

Ames procedure, **751**

Amino acid, **15.** *See also individual amino acids*
  binding to transfer RNA (tRNA), 363–364
  chemical structure, 16, 114
  peptide bond formation, 15–18, 473
  transport, 144–145

D-Amino acid oxidase, **196**

Aminoacyl site (A site), **471**

Aminoacyl-transfer RNA (tRNA)

binding to ribosome, 470
  formation of, **364**

Aminoacyl-transfer RNA (tRNA) synthetase
  in cytoplasm, 364
  in mitochondrion and chloroplast, 556
  mechanism of action, **364**

p-Aminobenzoic acid, 45

5-Aminolevulinate, **566**

5-Aminolevulinate synthetase (ALA synthetase), **566**

Amoeboid movement, **320.** *See also* Cell locomotion
  mechanism of, 320–322
  microfilaments and, 320–324

AMP. *See* Adenosine monophosphate

Amp gene. *See* Ampicillin-resistance gene

Amphipathic molecule, **13**

Ampicillin, **379**

Ampicillin-resistance gene, 377–378

Amplification. *See* Gene amplification

Amyloplast, **565**

Amytal, 233

Anaerobic glycolysis, **213**

Anaphase, mitosis, **503**

Anaphase I and II, meiosis, **517, 518**

Anaplasia, **735**

Anesthetics, effect on action potential, 684

Androgen
  effect on α-2-microglobulin synthesis, 636
  site of synthesis, 633
  target tissue, 633

Anfinsen, C., 20

Angiogenin. *See* Tumor angiogenesis factor

Animal pole, **588**

Anion, **83**

Anion transporter, 135. *See also* Band-3 protein

Anisotropic band. *See* A-band of sarcomere

Ankyrin, **106**
  and band 4.1 binding, 109–110
  and mobility of band 3 and glycophorin, 110, 327
  molecular weight, 106
  and spectrin binding, 109–110, 327

Annular material, **387**

Annulate lamellae, **390**

Anther, **600**

Antibiotic-resistance genes, **377**

Antibody, **70, 644**
  antigen recognition, 659–660
  diversity, origin of, 650–652
  genes, 652–655
  in immunocytochemistry, 70–71
  structure, 648–650, 652–655
  synthesis, 645–648, 655–658

Antibody heavy chain, **648**
  gene organization, 654–655
  structure, 650
  types, 654–655

Antibody light chain
  gene organization, 654
  structure, 650
  types, 648–649

Antibody-mediated immunity, **644**

Anticodon, **364.** *See also* Transfer ribonucleic acid (tRNA)
  messenger RNA binding, 364–365
  wobble hypothesis, 471–472

Antidiuretic hormone. *See* Vasopressin

Antigen, **70, 644**
  B-lymphocyte activation, 655–658
  in immunocytochemistry, 70

inactivation by antibody, 659–660
receptors, 656
recognition and innate immunity, 663–664
Thy-1, 661
T-lymphocyte recognition of, 661–662
tumor, 662
Antimycin A, **233**
Antipodal cell, **600**
Antiport, **140**
Antiterminator protein, in RNA synthesis, **414**
Antp mutation, **606**
Anucleolar mutant, **451**
Apoprotein, **43**
Arber, W., 375
Archaebacterial 16S lineage, **575**
Arg. *See* Arginine
Arginine (Arg)
histone content, 394
ionic bond formation, 19–20
in histone methylation, 395
structure, 16
Arginase, turnover of, 607–608
Arm, axoneme. *See* Dynein
Arnold, W., 265
Arnon, D., 250
Arrowhead pattern
actin localization in cell, 317, 712
in ribosomal RNA synthesis, 457
Aryl hydrocarbon hydroxylase, **167**
Ascaris, development of, 603
Ascus, **520**
A site. *See* Aminoacyl site
Asn. *See* Asparagine
Asofsky, R., 661
Asp. *See* Aspartic acid
Asparagine (Asn), structure, 16
Aspartate. *See* Aspartic acid
Aspartate transcarbamoylase, 48
Aspartic acid (Asp)
hydrogen bond formation, 20
ionic bond formation, 19–20
as neurotransmitter, 689–690
structure, 16, 690
Aster, **501**, 508–509
Astrachan, L., 357
Astral microtubules, **507**
ATP. *See* Adenosine triphosphate
ATP synthase. *See* Adenosine triphosphatase (ATPase)
Attardi, G., 552, 553
Attenuation, **484,** 485
A-tubule, **302.** *See also* Axoneme
Aurintricarboxylic acid, 480
Aurovertin, 233
Autocrine pathway, **639**
Autoimmune disease, **663**
Autolysis, **192**
Autophagic lysosome, **192**
Autophagic vacuole, **192**
Autophagy, **192**
Autoradiography, **71,** 72–73
Autotroph, **248**
Auxin, **638**
AVEC-DIC microscopy system. *See* Allen video-enhanced contrast-differential interference contrast microscopy system
Avers, C., 223
Avery, O., 341
Avian leukemia virus, 740
Avidin, 636
Axon, **668.** *See also* Action potential

elongation, 328, 678
growth cone, 678
hillock, 671
morphological features, 671
myelination of, 96, 672, 685
synapse formation, 675–678, 689–697
Axonal transport, 672–675
fast, **674**
slow, **673**
Axoneme, **302.** *See also* Cilium; Flagellum
bending, mechanism of, 309–311
structure, 302–306
subfractionation, 303–306
Axoplasm, **672**
Axoplasmic flow, **672**
Axopod, **312**
Axosome. *See* Basal plate
Axostyle, **312**
5-Azacytidine, 432
Azurophil granule, **181,** 182

Bacilli, **118**
Bacillus, in phospholipid synthesis, 203–204
Bacon, R., 60
Bacterium
active transport in, 142–144, 145–146
ATP synthesis, 243–245
cell division, 499–501
cell wall, 114–119
chemotaxis, 334–335
chromosome replication, 493–495
deoxyribonucleic acid (DNA) synthesis, 345–350
evolutionary orgin, 571
flagellum, 330–335
immune response to, 660
organelles in, 4–10
periplasmic space, 119–120
phospholipid synthesis, 203–204
photosynthetic, 264, 288, 571
plasma membrane, 243–245
restriction endonuclease in, 375
shape, 118
size, 4–5
Bacteriochlorophyll, **264**
Bacteriophage, **30.** *See also* Virus
infection sequence, 32–33
lambda, in cloning, 378, 414
SP8, 358
T-even, 30, 32–33, 357
Bacteriorhodopsin, **112**
light absorption by, 288
membrane orientation, 112
Baker, W., 708
Balbiani ring, **405**
Baltimore, D., 742
BamHI
in cloning, 376
recognition sequence, **375**
Band-3 protein, **106**
ankyrin binding, 109–110, 327
bicarbonate ion–chloride ion transport, 135
freeze fracture analysis, 109
as glycoprotein, 108
membrane orientation, 107–108, 110
mobility, 110, 327
molecular weight, 106
spectrin association, 110
Band-4.1 protein, **106**
actin association, 109–110
membrane orientation, 107, 110
molecular weight, 106

spectrin association, 109–110
Band-4.2 protein, **106**
membrane orientation, 107, 110
molecular weight, 106
Bang, O., 740
Barath, Z., 560
Barr, M., 429
Barr body, **429**
Barski, G., 59
Basal body
in bacterium, **331**
in microtubule assembly, eukaryote, **298**
structure, eukaryote, 306
Basal lamina
in cell migration, 609
location, **123**
Basal plate, **306**
Base. *See* Nitrogeneous base
Basten, A., 647
Bastiani, M., 680
Baur, E., 536
Bayliss, W., 620
B-cell. *See* B-lymphocyte
B-DNA. *See* B-form DNA
Beadle, G., 350
Beerman, W., 405
Beevers, H., 200
Belitzer, V., 235
Belt desmosome. *See* Fascia adherens
Bence-Jones protein, **650**
Bendall, F., 268
Benign tumor, **728**
Berenblum, I., 738
Berezney, R., 402
Berget, S., 420
Berlin, R., 314
Bernstein, J., 682
Berson, S., 623
Berthold, A., 619
Berzelius, J., 35, 36
Beta cells, of pancreas, 621
Beta(β) configuration. *See* Pleated sheet
Beta(β) particle, **82**
B-form DNA, **435**
Bicarbonate ion, transport of, 135
Binary switch gene, **605**
Bindin, **593**
Biological catalyst. *See* Enzyme
Biparental inheritance, **537**
Birefringence
microscopy, **62**
in skeletal muscle, 702–703
Bittner, J., 740
Bivalent chromosome, **516**
Blackman, F., 249
Blastopore, **584**
Blastula, **583**
Blepharoplast, **308**
Blobel, G., 168
B-lymphocyte, **643**
antibody synthesis, 645–648, 655–658
growth factor, 658
immunoglobulin receptors on cell surface, 659–660
interleukin-1 effect, 657
memory cell production, 658, 661
monoclonal antibody production, 658–659
plasma cell and, 644
T-lymphocyte interaction, 647–648
Blue-green algae
evolutionary origin, 571
photosynthesis in, 287–288
Bogorad, L., 559

Bone marrow, in lymphocyte development, 644
Borst, P., 550
Boundary lipid, **103**
Boveri, T., 603
Bowman, W., 702
Brachet, J., 355, 385
Braun, A., 747
Breakage-and-exchange model of genetic recombination, **519,** 520
Brenner, S., 359
Bresslau, E., 540
Brevin, 318
Bridges, C., 404
Briggs, R., 604
Britten, R., 426
Brodie, B., 165
Brown, D., 411, 451
Brown algae, evolutionary origin, 570
Brush border, **326,** 327. *See also* Microvillus
B-tubule, **302.** *See also* Axoneme
B-type virus, 740
Büchner, E., 35
Bundle-sheath cell, 284–286
α-Bungarotoxin, **693**
Burgoyne, L., 397
Burkitt, D., 741
Burkitt's lymphoma, **741**
Burnet, M., 651, 663
Bursa of Fabricius, **644,** 646
Butow, R., 561

C₃ pathway. *See* Calvin cycle
C3a, in complement system, **663**
C3b, in complement system, **663**
C₄ pathways, 283–286, **284**
    discovery of, 283–284
    light-dependent enzyme synthesis, 568
    phosphoenolpyruvate carboxylase in, 283–286
Caenorhabditis elegans, cell lineage in, 605
Caffeine
    in cancer therapy, 732
    phosphodiesterase inhibitor, 625
Cain, D. F., 719
Cairns, J., 346, 493
Calcitonin, 623
Calcium ion
    in amoeboid movement, 321–322
    calmodulin binding, 632
    in cardiac muscle contraction, 724
    cell cycle changes, 531
    cortical response, 594
    cyclic AMP action and, 631
    cytoplasmic concentration, 631–632
    endoplasmic reticulum release, 631
    hormone action, 631–632
    inositol trisphosphate and, 631
    as intracellular messenger, 631–632
    in microtubule formation, 300
    mitochondrial transport, 241–242
    mitotic spindle formation, 508
    neurotransmitter release, 691
    phosphodiesterase activation, 631
    in protein secretion, 178
    in skeletal muscle contraction, 717–718
    in smooth muscle contraction, 722
Calcium pump. *See* Adenosine triphosphatase (ATPase), calcium ion
Callan, H., 527
Callus, **604**
Calmodulin
    allosteric regulation by calcium ion, **632**
    axonal transport, 673

cell cycle changes, 531
in microtubule assembly and disassembly, 300
in microvillus, 326
in mitotic spindle, 508
myosin light-chain kinase activation, 722
in smooth muscle contraction, 722
troponin, comparison with, 722
Calvin, M., 250
Calvin cycle (C₃ pathway), **250**
    ATP requirement, 281–282
    discovery, 279–281
    light-dependent synthesis of enzymes, 568
    NADPH requirement, 281–282
    rate, 282
    ribulose 1,5-bisphosphate, 281
    summary equation, 281–282, 283
    yield of carbohydrate, 282
CAM. *See* Cell adhesion molecule
Cancer, **727**
    causes, 736–748
    genetic and epigenetic views, 747–748
    and heredity, 746–747
    initiation phase, 738
    prevention, 748–752
    promotion phase, 738
    treatments, 752–754
Cancer cell, 729–736
    anchorage-dependent growth, 723
    anchorage-independent growth, 732
    density-dependent growth inhibition, 730–732
    lifespan, 730
    morphological changes, 735–736
    proteolytic enzyme secretion, 734
    restriction point control, 732
    surface alterations, 732–734
    tumorigenicity, 730
Cantor, H., 661
CAP. *See* Catabolite gene activator protein
Cap. *See also* Capping
    in cilium tip, 303
    7-methylguanosine addition to, **419,** 468–469
Capping. *See also* Cap
    B-lymphocytes, **104,** 656–657
    of concanavalin A receptors, 314–315
    and microfilaments, 327
    in protein mobility studies, 103–104
Carbohydrate, **12,** 13. *See also individual carbohydrates*
    addition to protein, 170–172
    in red blood cell plasma membrane, 107–108
    synthesis in smooth endoplasmic reticulum, 163–165
Carbon dioxide
    in C₄ pathway, 283–286
    in Krebs citric acid cycle, 214–215
    pyruvate oxidation, 213
    transport in blood, 135
Carbon dioxide fixation. *See* Calvin cycle
Carbon monoxide, as electron transfer inhibitor, 233
Carbon skeleton, **12**
Carbonyl cyanide m-chlorophenylhydrazone (CCCP)
    chloroplast inhibitor, **273**
    mitochondrial inhibitor, 233, **239**
Carcinoembryonic antigen, **734**
Carcinogenesis, **738.** *See also* Cancer
Carcinogens, chemical, **737,** 736–739

Ames test, 751
    and carcinogenesis, stages of, 738
    environmental sources, 749–752
    examples of, 737
Carcinoma, **727,** 728
Cardiac muscle
    cell structure, **723**
    contraction mechanism, 723
Cardiolipin, **99,** 101, 229
Capsule layer, **113**
Carlson, J. G., 603
Caro, L., 175
β-Carotene, **264**
Carotenoid pigment, 15, **264**
    absorbance spectrum, 263
    species distribution, 264
    structure, 264
    in thylakoid membrane, 101, 259
Carrel, A., 58, 613
Carrier. *See* Transporter
Carrier, in antibody formation, **651**
Carrot cells, totipotency of, 604
Cartilage cells, embryonic derivation of, 608
Caspersson, T., 355, 565
Catabolic enzyme, **440**
Catabolite gene activator protein (CAP), **440,** 441
Catabolite repression, **440**
Catalase, **196,** 197
Catalatic mode, **197**
Catalyst. *See* Enzyme
Catecholamine, **622**
Cation, **83**
Cavalier-Smith, T., 574
C-bands, in chromosome, **505**
CCCP. *See* Carbonyl cyanide m-chlorophenylhydrazone
cDNA. *See* Complementary DNA
Cell, **4**
    chemical composition, 11–24
    evolution, 575
    fractionation by differential centrifugation, 78–80
    lifespan in culture, 613–614
    organelles, 4–10
    origin, 27–30
    photosynthetic, 9
    selection and cloning, 377
    size, 4, 5
    structure, prokaryote versus eukaryote, 10–11
    universal functions, 4
Cell adhesion, 151, 609–610
Cell adhesion molecule (CAM), **610**
Cell aggregation, patterns of, 609, 610
Cell body, of neuron, **668**
Cell–cell recognition
    cell adhesion, 149–151, 610
    glycosyltransferases and, 151
    specialized membrane regions, 149–157
    tissue hierarchy, 609–610
Cell culture
    methods, 58–60
    primary, **59**
    synchrony methods, 490–491
    transformed cell, 59
Cell cycle, 490. *See also individual phases*
    arrest in G₁, 732
    cell differentiation and, 608
    cleavage and, 600
    control of, 528–529
    DNA synthesis in, 493–499
    duration of phases, 491

histone protein phosphorylation and, 507
stimulation after fertilization, 599
synchrony, 490–491
Cell differentiation, 602–608
  cell cycle and, 608
  cytoplasmic influence on, 602–603
  embryonic induction, 611
  gene expression in, 605–606
  protein turnover in, 607–608
  totipotency and, 603–605
Cell division. *See also* Cell cycle;
    Cytokinesis; Meiosis; Mitosis
  in cancer cells, 728–729
  differentiation and, 608
  growth factors and, 529–532
  microfilaments and, 324
  in prokaryotes, 499–501, 529
  regulation of, 528–533
  restriction point control, 732
Cell-free protein synthesis, **358**
Cell fusion, **59**
  monoclonal antibody production, 658–659
  nucleus activation and, 425
  protein mobility studies, 104
Cell-fusion model of myogenesis, **706**
Cell line, **59**, 658–659
Cell locomotion. *See also* Amoeboid
    movement
  chemotaxis and, 323–324
  cytochalasin B and, 323
  microfilaments and, 322–323
  regulation of, 323–324
Cell lysis, virus-induced, 33
Cell membrane. *See individual cell
    membranes*
Cell-mediated immune response, 660–662
Cell-mediated immunity, **644**
Cell migration
  in amoeboid movement, 320–324
  in animal embryogenesis, 608–610
  in plant embryogenesis, 601
Cell organelles
  isolation of, 77–80
  marker substances, 79
Cell plate, **513**
Cell sap, **163**
Cell shape
  changes and morphogenesis, 608–609
  microfilaments and, 324–327
  microtubules and, 315–316
Cell surface
  cancer cell, 732–734
  eukaryote, 120–127
  metabolic functions, 157–158
  morphology and cell cycle, 531
  prokaryote, 113–120
Cell theory, **3**, 667
Cellulase
  auxin effect on, 638
  chloroplast isolation, 257
  protoplast isolation, 604
Cellulose, **13**
  plant cell wall, 120
  structure, 14
Cell wall, bacteria
  cell shape, 118
  gram negative, **115**
  gram positive, **114**
  lipopolysaccharide in, 115
  Omp A protein, 118
  outer membrane, 115
Cell wall, plant, **8**

cellulose in, 13, 14, 120
extensin in, 122
Golgi complex and synthesis of, 123, 182
hemicellulose in, 122
lignin in, 122
microtubules and, 315
pectin in, 122
plasmodesmata and, 122
structure, **120–123**
synthesis, 123
Centrifugal force (g), **73**
Centrifugation
  differential, **74**
  isodensity, **75**, 76–77
  moving-zone, **74**
  principle, 73–74
  velocity, **74**, 75
Cerebroside, 100
Central element, in synaptonemal
    complex, **524**
Central microtubular bridge, **302**
Central sheath, **302**, 311
Central tubules, **302**, 311
Centriole, **306**
  aster and, 501
  cell division, 508
  formation, 306–308, 508–509
  microtubule assembly and, 306
  in mitosis, 501
  mitotic spindle and, 507
  plant cell, 307, 501
  in sperm tail development, 590–592
  structure, 306
Centromere, **495**
  C-bands and, 505
  DNA synthesis in, 506
  heterochromatin, 505, 506
  kinetochore attachment, 506
Centrosome, **298**, 508
CF_0. *See* Adenosine triphosphatase
    (ATPase), CF_0 subunit
CF_1. *See* Adenosine triphosphatase
    (ATPase), CF_1 subunit
C_{H1}, C_{H2}, C_{H3}. *See* Constant region, heavy
    chain
C gene. *See* Constant region gene
Chalone, **754**
Changeux, J.-P., 47
Chapeville, F., 364
Chargaff, E., 341
Chargaff's rule, **343**
Chase, M., 342
Chemical carcinogen. *See* Carcinogen,
    chemical
Chemical synapse, **676**
Chemiosmotic theory, **237**
  lactose transport, 145
  oxidative phosphorylation, 237–241
  photophosphorylation, 269–276
Chemoreceptor, **685**
Chemotaxis, **323**
  bacteria flagella, 334
  cell locomotion, 323–324
  fertilization, 593
  leukocytes, 323
Chemotherapy, **753**
Chiang, K.-S., 549
Chiasma, **516**
Chironomus, 405–406
Chlamydomonas
  chloroplast genetics, 538
  flagellum regeneration, 308
  maternal inheritance, 552
  mutant, stacked thylakoids, 279

Chloramphenicol, **480**, 557
Chlorella, 267
Chloride ion, transport of, 135
p-Chloromercuribenzoate (PCMB), 48
Chlorophyll. *See also individual chloro-
    phyll molecules*
  distribution in chloroplast membranes,
    101, 259
  in photosynthetic unit, 265–266
  synthesis during chloroplast biogenesis,
    566
Chlorophyll *a*
  absorbance spectrum, 263
  in reaction center, 265, 267
  species distribution, **264**
  structure, 263
Chlorophyll *b*
  absorbance spectrum, 263
  in photosystems I and II, 267–268
  species distribution, **264**
  structure, 263
Chlorophyll *c*, **264**
Chlorophyll *d*, **264**
Chloroplast, **9**, 248–292. *See also
    Photosynthesis*
  anatomy, 251–262
  arrangement in cell, 255–257
  ATPase, 273–275, 276–278
  ATP synthesis, 269–270
  CF_0 subunit, 273–275, 276–278
  CF_1 subunit, 273–275, 276–278
  chemiosmotic theory, 269–276
  chlorophyll content, 263–268
  class I and II, 257–258
  developmental origin, 564–568
  electron transfer chain, 266–269
  electron transfer shuttle pathways, 259
  envelope, 258–260
  evolutionary history, 568–575
  genetic code, 556
  genetics, 536–538
  grana, **252**
  growth and division, 564
  inner membrane, **253**, 259
  intermembrane space, **253**
  isolation, 257–258
  lipid distribution, 259
  mitochondrion, proximity to, 255
  outer membrane, **253**, 259
  peroxisome, proximity to, 255
  photophosphorylation, 269–276
  photorespiration, 286
  protein import, 560–561
  protein synthesis, 554–559
  ribosomal RNA (rRNA), 554
  ribosomes, 555–556
  RNA synthesis, 554
  shape, dark and light conditions, 256
  size, 256
  stroma, **9**, **260–261**
  structure, compared with mitochon-
    drion, 255
  subfractionation, 257–258
  transfer RNA (tRNA), 555–557
  thylakoid membrane, **9**, **253**, 262
Chloroplast deoxyribonucleic acid (DNA)
  circularity of, 548
  density of, 546–547
  discovery, 541–543
  displacement loop (D loop), 550–551
  map, *Chlamydomonas*, 538
  map, spinach, 554
  5-methylcytosine, 547
  properties, 546–548

Chloroplast deoxyribonucleic acid (DNA)
  (*Continued*)
    reassociation kinetics, 547
    ribosomal RNA (rRNA) genes, 554
    semiconservative replication, 548–551
    transcription, 552–554
    transfer RNA (tRNA) genes, 553
Chlorpromazine, 695
Choleragen, **627**
Cholesterol, **15**
    distribution in cell membranes, 99, 101
    structure, 100
    transport via low-density lipoprotein, 192
    uptake by cell, 192
Cholesterol esterase, **630**
Cholesterol hydroxylase, 164
Choline, **562**
Choline acetyltransferase, **691**
Chondroid, **8**
Chondroitin sulfate, **126**
Chorion genes, 431
Christensen, H., 144
Chromatid, **495**
    DNA synthesis, 495
    in lampbrush chromosome, 526–528
    separation in meiosis, 518
    separation in mitosis, 503
    in synaptonemal complex, 524–526
Chromatin, **9, 386**
    chemical constituents, 393–396
    condensation in meiosis, 515–516
    condensation in mitosis, 501
    folding, 400–401, 504–505
    isolation, 393
    model, 400
    nuclease digestion, 397
    nucleosome, 396–401
    sperm head, 589
    structural changes after fertilization, 599
Chromatography, **85, 262**
    adsorption, **86,** 87
    affinity, **87,** 467
    gas, **88**
    gas-liquid, **88**
    gel filtration, **85**
    liquid, **85**
    paper, **87**
    thin-layer, **87**
Chromatophore, **8,** 10, 288
Chromomere, **516,** 527
Chromosome, **9, 393**
    banding patterns, 505
    centromeric heterochromatin, 505
    folding and protein phosphorylation, 507
    movement, 503, 509–510
    regulation of formation, 507
    replication, 494–499
    structure, 401–402, 504–507
    translocation, 744–745
Chromosome diminution, **603**
Chromosome puff, 405
    developmental changes, 406
    nonhistone proteins, 407
    ribonucleoprotein formation, 407
    RNA synthesis, 407
Chymotrypsin, active site, with substrate,
    52–53
Chymotrypsinogen, 50
Ciliary necklace, 302
Cilium, **10,** 301–311
    axoneme, 302–306
    basal body, 306
    beat pattern, 302
    cap, 303

centriole and, 306–307
comparison with eukaryotic flagellum,
    301–302
modification in rod cell, 687
motility mechanism, 309–311
Cisterna, **8, 163**
Citric acid cycle. *See* Krebs citric acid cycle
Claman, H., 646
Clathrin, **189**
    axonal transport, 673
    coated vesicle formation, 189
    structure, 188
Claude, A., 77, 160, 161, 218
Cleavage, **583,** 600
Cleavage furrow, **511**
Clever, U., 407
Clonal deletion theory of immune
    tolerance, **663**
Clone
    antibody synthesis, **651**
    cell culture, **59,** 658–659
    recombinant DNA technology, 376–380
Cloning vector, **376**
CoA. *See* Coenzyme A
Coacervation, **27**
Coated pit, **189**
Coated vesicle
    formation, **189**
    synaptic vesicle recycling, 692
Cocci, **118**
Codon, **361**
    anticodon binding, 364
    dictionary, 361
    initiator, 362, 469
    nonsense, 361
    terminator, 362
Coenzyme, **43**
Coenzyme A (CoA), **213**
Coenzyme Q (CoQ). *See* Ubiquinone
Coffey, D., 402
Coffino, P., 629
Cohen, G., 145
Cohesive ends. *See* Sticky ends
Cohn, Z., 187
Colchicine, **295**
    axon elongation, 678–679
    microtubule structure, 295, 312–314
Cole, K., 682
Colicin
    mechanism of action, **244**
    type E$_3$, effect on protein synthesis,
        468, 480
Collagen
    amino acid content, 124
    in basal lamina, 124
    fiber, **124**
    fibril, **124**
    gene, intron number, 421
    triple helix formation, 124–125
    types, 124
Collander, R., 132
Colony hybridization, **374,** 380
Color vision, 687
Competence, **611**
Competition hybridization, **372**
Complementary DNA (cDNA), **350,** 376
Complement system, **648,** 661, 663
Complexes I-IV, mitochondrion, **234**
Conalbumin, 636
Concanavalin A, **314**
Condensation reaction, **26**
Condensing agent, **26**
Condensing vacuole, **176**
Cone cell, **687**

Conjugation, **518**
Connecting piece, sperm flagellum, 592
Connexon, **156**
Constant region
    antibody heavy chain, **650**
    antibody light chain, **650**
    C$_{H1}$, C$_{H2}$, C$_{H3}$, 650
Constant region gene (C gene), **653,**
    654–655
Constitutive heterochromatin, **429**
Contact inhibition, 730
    in cell cycle, 531
    of movement, **323**
Contractile ring, **513**
Contrast, in microscopy, **60**
Cooper-Helmstetter model, **491**
Copy-choice model of genetic recombina-
    tion, **519**
Core complex, photosystems I and II,
    267
Core enzyme, RNA polymerase, **409**
Core oligosaccharide, **171**
Core particle, nucleosome, **399**
Co-repressor, **440**
Correns, C., 536
Cortical fiber, **312**
Cortical granule
    in amphibian oocyte, **587**
    discharge, 595
Cortical reaction, **595**
Cortisol, 621, 633
Cortisone, 607–608
Cot value, **426**
Coupling site, mitochondrion, **235**
    location, 241
    reconstitution, 239–240
Covalent bond, **19**
    peptide bond, 15, 17, 473
    phosphodiester linkage, 23
C-protein, of sarcomere, **709,** 710
Craig, L., 650
Creatine kinase, **711,** 719
Crick, F., 28, 343, 359, 362, 473
Cristae, **9, 218**
Cross-bridge, muscle, **714**
Cross-linking agents, ribosome structure,
    462
Cross-linking protein, 317–318
Cross-wall, **499**
Crown gall disease, **747**
Crystalline gene (α-A$_2$), 442
C-type virus, 740
Curie, M., 739
Curtis, H., 681
Cuticle, **122**
Cyanide ion
    electron transfer inhibitor, mitochon-
        drion, 233
    enzyme inhibitor, 47
3′,5′-Cyclic adenosine monophosphate
    (cyclic AMP)
    adenylate cyclase, 625
    in catabolite repression, 440–441
    in cell cycle, 530
    cellular slime mold chemotaxis, 323
    and dopamine action, 694
    in glycogen breakdown, 625
    hormones and, 625–630
    lactose operon regulation, 440–441
    and opiate action, 696
    phosphodiesterase hydrolysis of, 625
    and protein kinase activity, 482,
        628–630
    synthesis, **625**

Cyclic AMP. *See* 3′,5′-Cyclic adenosine monophosphate
Cyclic GMP. *See* 3′,5′-Cyclic guanosine monophosphate
3′,5′-Cyclic guanosine monophosphate (cyclic GMP)
  hormone action and, 630–631
  and opiate action, 696
  phosphodiesterase hydrolysis of, 688–689
  and photoreception, 688–689
Cycloheximide, **480**, 557
Cyclosis, **319**
Cys. *See* Cysteine
Cysteine (Cys)
  disulfide bond formation, 17, 19, 21
  structure, 16
Cytochalasin B, **319**
  axon elongation, 678
  capping, 327
  endocytosis, 327
  exocytosis, 328
  glucose transport, 135
  microfilament assembly, 319
Cytochalasin D, **327**, 328
Cytochemical techniques
  enzyme cytochemistry, 71
  immunocytochemistry, 70–71
  microscopic autoradiography, 71–73
  staining, 69–70
Cytochrome, **229**. *See also individual cytochromes*
  discovery, 231
  electron transfer chain, chloroplast, 266, 269
  electron transfer chain, mitochondrion, 233–234
  heme group, structure, 230
Cytochrome aa_3. *See* Cytochrome oxidase
Cytochrome *b*
  absorbance spectrum, 230
  electron transfer chain, chloroplast, 266
  electron transfer chain, mitochondrion, 233–234
  gene localization, HeLa mitochondrial DNA, 553, 558
  gene localization, chloroplast DNA, 559
Cytochrome b_5
  mechanism of action, **167**
  microsomal localization, 164
Cytochrome *f*
  electron transfer chain, chloroplast, **266**
  gene localization, chloroplast DNA, 559
Cytochrome oxidase (cyt aa_3)
  absorbance spectrum, 231
  cytochemical localization, 228
  electron transfer chain, mitochondrion, 233–235
  gene localization, HeLa mitochondrial DNA, 553, 558
  import into mitochondrion, 560
Cytochrome P-448, **167**
Cytochrome P-450, **165**
  function, 165–166
  microsomal localization, 164
Cytokinesis, **499**, 504. *See also* Cell division
  in animal cells, 511–513
  mechanism of, 510–513
  in plant cells, 513
  in prokaryotes, 499–501
Cytokinin, **638**
Cytopharyngeal basket, **312**
Cytoplasm, **9**. *See also* Cytosol
Cytoplasmic inheritance

biparental inheritance, **537**
discovery of chloroplast DNA, 541–543
discovery of mitochondrial DNA, 541–543
genetic evidence, 536–540
maternal inheritance, 536
molecular basis, 540–561
Cytoplasmic streaming, **319**, 320
Cytoplast, **635**
Cytosine, **23**
  base pairing, 344, 367
  structure, 22
  uracil, conversion to, 369
Cytosine arabinoside, 732
Cytoskeleton, **10**, 293–337. *See also* Intermediate filament; Microfilament; Microtubule
Cytosol, **9**, 79
Cytotoxic effector T-cell, **661**

DAB. *See* Diaminobenzidine
Dale, H., 689
Dalgarno, L., 468
Danielli, J., 95
Danielli-Davson model of membrane structure, **95**
Darkfield microscopy, **62**
Dark reactions of photosynthesis, **250**
  Calvin cycle, 279–283
  discovery of, 249–251
  summary equation, 281–282, 283
Darnell, J., 453
Davies, R. E., 719
Davson, H., 95
De Duve, C., 160, 182, 570
Degenerate code, **361**
DeLarco, J., 731
Denaturation
  DNA, 371
  protein, 20
Dendrite, **668**
  morphological features, 669–671
  synapse formation, 671
Dendritic spine, **671**
Dense body, **721**
Dense projections, **676**
Density-dependent inhibition of growth, **531**
Density–size relationship, of macromolecules and organelles, 77
Deoxybromouridine, 549
Deoxyribonuclease (DNAase)
  micrococcal nuclease, **397**
  pancreatic DNAase-I, **399**
  rat liver nuclease, **397**
Deoxyribonucleic acid (DNA), **23**, **340**. *See also* Chromatin; Chromosome
  base composition, species comparisons, 341
  base pairing, 344, 367
  chemical synthesis, 381
  chloroplast. *See* Chloroplast DNA
  cloning, 375–380
  colorimetric identification, 81
  complementary, **350**, 376
  conformation and gene expression, 434–437
  cytochemical identification, 69
  deletion, 432
  denaturation, 371, 372
  discovery, 338–339
  DNAase-I sensitivity, 430
  double helix, 343–345
  histones and, 396–401

  linker, **376**, 399
  main band, **428**
  methylation and gene expression, 432
  mitochondrion. *See* Mitochondrial DNA
  nitrogenous bases, 22, 340
  nucleosome, 396–401
  nucleotide, base pairing, 344
  packaging, 401–402
  polarity, 343
  rearrangement, of genes, 433–434, 653–655
  reassociation kinetics, 426–429
  recombinant technology, 375–380
  replication. *See* Deoxyribonucleic acid (DNA) synthesis
  restriction endonuclease cleavage, 375
  sequencing, 380–381
  spacer, **457**
  in sperm head, 590
  steroid–receptor binding, 633–635
  structure, 343
  supercoiling, 434–435
  synthesis. *See* Deoxyribonucleic acid (DNA) synthesis
  template, 346
  viruses, 741–742
Deoxyribonucleic acid (DNA) synthesis, 343–350
  of A- and T-rich sequences, 498
  and bacterial cell cycle, 491
  and bacterial proteins, 350
  base pairing in, 344, 367
  in cell cycle, 493–499
  chromosome replication, 493–499
  cytoplasmic regulation of, 498
  directionality, 493
  DNA polymerase, 345–350
  fertilization and, 599
  heterochromatin, 498
  initiator protein, 494–495
  model, 349
  and nuclear matrix, 497–498
  nucleosomes, 496–497
  Okazaki fragments, 348
  primase and, 349
  rate, 496, 498
  reverse transcriptase and, 350
  rolling-circle mechanism, 452
  semiconservative replication, 344–345
  S phase, 491, 498
  start site, 494
Deoxyribose, 22, **23**
Depolarization, **682**
Depolymerizing proteins, **317**, 318
De Saussure, T., 249
Desensitization, **632**
Desmin
  properties of, 328–329
  in thin filament, **713**
Desmosome, **152**
  keratin content, 329
  types, 152–155
Determination, **584**
Deuterium labeling, 463
Deuterosome, **307**
D gene. *See* Diversity gene
DHAP. *See* Dihydroxyacetone phosphate
Diacumakos, E., 540
Diacylglycerol
  as intracellular messenger, 632
  and platelet-derived growth factor (PDGF), 531
  synthesis, **632**
Diakinesis stage, **517**

Dialysis, **82**
Diaminobenzidine (DAB), **196**
Diatom, 570
Dictyosome, **171**
Dideoxynucleotides, in DNA sequencing, **380–381**
Diet, cancer, 752
Difference spectrum, **232**
Differential centrifugation, **74,** 78–80
Differential interference microscope, **62**
Differential scanning calorimetry, **102**
Diffraction, **67**
Diffusion
  facilitated, **133,** 133–139
  simple, **130,** 130–133
Digalactosyldiglyceride, 259–262
Digitonin, 225
Diglyceride, **13**
Dihydroxyacetone phosphate (DHAP)
  chloroplast electron shuttle pathway, 259
  mitochondrial electron shuttle pathway, 216–217
Dihydrouridine loop (D-loop), **365**
Diisopropylfluorophosphate (DIPF)
  mapping active site, 52
  mechanism of action, 45
2,4-Dinitrophenol, **233,** 239, 675
Dinoflagellate, 570
Dintzis, H., 465
Dipeptide, **15**
DIPF. *See* Diisopropylfluorophosphate
Diphtheria toxin, mechanism of action, **481**
Diploid, **514**
Diplotene stage, **516–517**
Direct filiation theories, **572–575**
Disaccharide, **13**
Displacement loop (D-loop)
  in chloroplast DNA synthesis, 550–551
  in mitochondrial DNA synthesis, **550**
  recA protein and, 523
Dissociation constant (K_d), **148**
Distal centriole, **590,** 592
Disulfide bond, 17, **19,** 21
Diversity gene (D gene), **654,** 655
D-loop. *See* Dihydrouridine loop; Displacement loop
DNA. *See* Deoxyribonucleic acid
DnaB protein, 350
DnaC protein, 350
DNA gyrase, 350, **435**
DNA ligase, **348**
DNA methylase, **432**
DNA polymerase, **345**
  directionality, 347–349
  in DNA repair, 345, 370
  reaction catalyzed by, 345
  DNA synthesis, 345–347
  type I, 345, 350
  type II, 345
  type III, 345–350
DNA repair, 345, **370**
Docking protein, **169**
Doherty, P. C., 661
Dolichol phosphate, **171**
Dopamine
  neurotransmitter, **689**
  and Parkinson's disease, 694
  receptors, 694–695
  and schizophrenia, 689
  site of action, 690
  structure, 690
Dorsal lip of blastopore, **611**
Doty, P., 371

Double-label experiment, **82**
Downstream sequence, **411**
Drosophila, developmental genes, 605–606
Drug detoxification, endoplasmic reticulum, 165–167
Dunant, Y., 691
Dyad, in skeletal muscle, **705**
Dynein
  arm, **302**
  axoneme bending, 311
  localization in axoneme, 304
  in mitotic spindle, 510

Eagle, H., 58
Ebashi, S., 718
Ecdysone, **407**
E. coli. See Escherichia coli
EcoRI
  in cloning, 379
  recognition sequence, **375**
Ectoderm, **584**
Ectoplasm, **319**
  in amoeboid movement, 320–321
  in cytoplasmic streaming, 319–320
Edelmann, G., 314, 610, 648
Edidin, M., 104
EDTA. *See* Ethylenediaminetetraacetate
EF. *See* Elongation factors; Exoplasmic fracture face
EGF. *See* Epidermal growth factor
Egg, **582**
  animal development, 588
  cell coat formation, 588
  depolarization of plasma membrane, 594
  and follicle cell, 585
  and nurse cell, 586
  plant development, 600
  polarity of, 588
  yolk, 585–587
Electrical charge gradient. *See* Membrane potential
Electrical potential. *See* Membrane potential.
Electrical signal, in neuron. *See* Action potential
Electrical synapse, **678**
Electrochemical equilibrium, **146**
Electrochemical gradient, **140**
  in bacterium, 243–245
  in chloroplast, 270–276
  membrane potential component, 238, 272
  in mitochondrion, 237–241
  pH component, 237–241, 270–276
  and protein import into mitochondrion, 560–561
Electrogenic pump, **140**
Electron microscopy, 63–66
  fixation of specimens, 64
  freeze-etch, **65**
  high-voltage, **329**
  scanning (SEM), **66**
  staining of specimens, 64–65
  transmission (TEM), **64**
Electron-spin resonance (ESR), **102**
Electron transfer chain, bacteria, 243–245
Electron transfer chain, chloroplast
  organization into complexes, 274–275
  sequence of carriers in, 268–269
  standard redox potential, 268–269
Electron transfer chain, mitochondrion
  inhibitors, 233
  organization into complexes, 234
  sequence of carriers, 216, 231–235

Electrophoresis, **83**
  isoelectric focusing, 84
  polyacrylamide gel electrophoresis (PAGE), 84–85
  principle, 83–85
  two-dimensional, 84–85, 394, 426
Electroplaque, **694**
Ellerman, V., 740
ELISA. *See* Enzyme-linked immunosorbent assay
Ellis, R. J., 560
Elongation factors (EF), protein synthesis, **470**
  elongation factor-1 (EF-1), 472
  elongation factor-2 (EF-2), 473
  elongation factor-G (EF-G), 475
  elongation factor-Ts (EF-Ts), 472
  elongation factor-Tu (EF-Tu), 472
Embryonal carcinoma, **747**
Embryonic induction. *See* Induction, embryonic
Embryo sac, **600**
Emerson, R., 265, 266
Endergonic reaction, **38**
Endocrine cell, **620,** 639
Endocytosis, **8,** 146
  lysosomes and, 187–192
  membrane recycling, 180
  microfilaments and, 327–328
  receptor-mediated, **189**
  synaptic vesicles, 692
Endoderm, **584**
Endoplasm, **319**
  amoeboid movement and, 320–321
  cytoplasmic streaming and, 319–320
Endoplasmic reticulum (ER), **8,** 160–171, **162**
  calcium ion storage, 631–632
  Golgi complex, relationship to, 175–176
  lipid composition, membrane, 101
  membrane biogenesis, 203–205
  nuclear envelope formation, 504
  oligosaccharide modification of proteins, 170–171
  and phospholipid synthesis, 203–205
  and protein synthesis, 167–171
  rough (RER), **162,** 167–171
  and sarcoplasmic reticulum, 703–705
  smooth (SER), **162,** 163–165
  structure, 161, 162
  yolk synthesis and, 586–587
Endorphin, **639,** 696
β-Endorphin
  site of action, 690
  synthesis, 697
Endosome, **189**
Endosperm, **600**
Endosymbiont theory, **569,** 569–572
Endotoxin, **115**
End plate, **154**
Engelman, D., 463
Engelmann, T., 251
Englehardt, V., 710
Enhancer sequence, **434**
Enkephalin
  binding to opiate receptors, 696
  discovery, 696
  mechanism of action, 696
  site of action, 690
Enterokinase, in trypsinogen modification, 50
Enthalpy (*H*), **37**
Entropy (ΔS), **37**
Enzyme, **36,** 35–56

activation energy, **39**, 40
active site, **51**
allosteric regulation, 47–49, 243
allosteric site, **47**
catalysis and origin of life, 26
catalyst, defining criteria, **36**
catalytic subunit, **48**
classes, 36
cofactor, **43**
competitive inhibitor, **45**
cooperativity, **49**
covalent modification of, 50–51
crystallization of first enzyme, 36
cytochemistry, **71**
discovery, 35–37
feedback inhibition, **47**
induction, **437**
irreversible inhibitor, **45**
mapping active site, 52
maximal velocity (V_{max}), **41**
Michaelis constant (K_m), **41**
multienzyme complex, 50
noncompetitive inhibitor, **46**
pH effect, 42–43
reaction rate, 39–40
regulatory subunit, **48**
reversible inhibitor, **45**
substrate binding, 40
temperature effect, 42–43
Enzyme-linked immunosorbent assay
 (ELISA), **659**
Ephrussi, B., 538
Epidemiology, **749**
Epidermal growth factor (EGF), **529**
 control of cell division, 529
 influence on development, 613
 oncogenes and, 745
 and protein kinase activity, 530
 receptor localization, 530
Epidermis, and embryonic induction, 611
Epinephrine
 cyclic AMP and, 626
 in glycogen breakdown, 625
 site of action, 621
 site of synthesis, **622**
 structure, 621
Epithelium, 584
Epstein-Barr virus (EBV), **741**
Equilibrium constant (K_{eq}), **38**
ER. See Endoplasmic reticulum
erb-B oncogene, **745**
Ergastoplasm, **160**
Erikson, R., 743·
Erythrocyte. See Red blood cell
ES. See Exoplasmic surface
Escherichia coli
 chromosomal replication, 493–494
 DNA Cot curve, 427
 lactose operon, 436–441
 number of genes, 426
 ribosomal protein operon, 481
 tryptophan operon, 484–485
ESR. See Electron spin resonance
Estradiol
 receptor localization, 634
 site of synthesis, **633**
 steroid hormone action, model of, 635
 structure, 633
 target tissue, 633
Estrogen, **15**. See also Estradiol
 ovalbumin messenger RNA half-life, 483
 and ovalbumin synthesis, 636–637
 site of synthesis, 633
 target tissue, 633

Ethidium bromide, 539
Ethylene, **638**
Ethylenediaminetetraacetate (EDTA)
 enzyme inhibitor, 47
 dynein isolation, 304
Etioplast, **278**
 in chloroplast development, 566
 development of, 566
 photosystems I and II appearance in, 278
 prolamellar body in, 566
Eubacterial 16S lineage, **575**
Euchromatin, **429**
Euflavin, 539
Eukaryote, **5**
 evolutionary origin, 571
 organelles, 4–10
eIF-2. See Initiation factors, protein
 synthesis
Exchange diffusion, **137**
Excision-repair. See DNA repair
Excitation-contraction coupling, **715–718**
Excitatory postsynaptic potential, **690**
Exergonic reaction, **38**
Exocrine cell, **620**
Exocytosis, **8**, **146**
 and membrane recycling, 180, 204–205
 microfilaments and, 327–328
 and protein secretion, 178–180
 of synaptic vesicles, 690–692
Exon, **420**
Exoplasmic fracture face (EF), **66**, 276
Exoplasmic surface (ES), **66**, 276
Exotoxin, **115**
Extensin, **122**
Extracellular matrix, **123–127**
 cell migration, 608
 collagens in, 124–125
 fibronectin, 126–127
 laminin, 127
 proteoglycans in, 125–126

F_0. See Adenosine triphosphatase
 (ATPase), F_0 subunit
F_1. See Adenosine triphosphatase
 (ATPase), F_1 subunit
Fab fragment, of antibody, **648**
Fabry's disease, 193
Facilitated diffusion. See Diffusion,
 facilitated
Facultative heterochromatin, **429**
FAD–FADH$_2$. See Flavin adenine dinu-
 cleotide
Familial adenomatosis, 746
Familial hypercholesterolemia, **192**
Farrelly, F., 561
Fascia adherens, **153**, 154
Fascin, 318
Fast axonal transport, **674–675**
Fat. See Neutral fat
Fatty acid, **13**
 acetyl coenzyme A conversion to, 213
 in glyoxylate cycle, 199
 β-oxidation pathway, 199, 214
 saturated, **13**
 structure, 14
 unsaturated, **13**
Fatty acid CoA ligase, microsomal
 localization, 164
Fawcett, D., 524
Fc fragment, of antibody, **648**
Feedback regulation
 hormone release and, **623**
 of phosphofructokinase, **243**
 in ribosomal protein synthesis, 481

Feldherr, C., 387
Feldmann, M., 647
Feline leukemia virus, 740
Feline sarcoma virus, 740
Fermentation. See Anaerobic glycolysis
Fernández-Morán, H., 64
Ferredoxin, **266**
Ferritin, as antibody label, **71**, 104
Fertilization, **582**, 593–600
 and calcium ion, 595–597
 and cortical reaction, 595
 and DNA synthesis, 599
 early responses of, 593–597
 external, 593
 internal, 593
 late responses of, 597–600
 meiosis, completion of, 597
 metabolic changes during, 599
 polyspermy block, 594
 and protein synthesis, 599
 and RNA synthesis, 599
 sperm-egg fusion, 597–598
Fertilization cone, **597**
Fertilization membrane, **582**, 595
Fes oncogene, 745
α-Fetoprotein, 734
Feulgen reaction, **69**
Fgr oncogene, 745
Fibronectin, **126**
 cancer cell, 733
 cell adhesion and, 150
 cell migration and, 609
 in extracellular matrix, 126–127
Fibrous protein, secondary structure, **19**
Filamin, 318
Filopodia
 in cell locomotion, 324
 in growth cone, 678
Fimbrin, 318, 326
Finean, J. B., 96
Fingerprint, peptide, **353**
Fischer, E., 51
Flagellin, **331**, 332
Flagellin genes, phase variation and, **433**
Flagellum, **10**, 301–311
 axoneme structure, 302–306, 309–311
 bacteria, 330–335
 beat pattern of, 302
 cilia, comparison with, 301–302
 in mammalian sperm, 592
 motility, mechanism of, 309–311
 in plant sperm, 600
 regeneration of, 308
Flavin adenine dinucleotide (FAD–
 FADH$_2$)
 absorbance spectrum, 229
 in Krebs citric acid cycle, 214–215
 redox couple, 210, 211
 structure, **211**
 yield of ATP per molecule, 215–216
Flavin mononucleotide (FMN–FMNH$_2$)
 absorbance spectrum, 229
 redox couple, 210
 structure, **211**
Flavoprotein, **210**, 266
Flip-flop, of membrane components. See
 Transverse diffusion
Fluid mosaic model of membrane
 structure, 98–105
Fluorescein, **70**, 104
Fluorescence photobleaching recovery,
 104
Fluorophenylalanine, 332–333
5-Fluorouracil, 539

FMN–FMNH₂. *See* Flavin mononucleotide
Fms oncogene, 745
Focal contact, **322**
Folic acid, 45
Follicle cell, **585**
Follicle-stimulating hormone (FSH),
    622–623
N-Formylmethionine
    in chemotaxis, 323
    in mitochondria and chloroplasts, 556
Formylmethionyl transfer RNA
    (tRNA^fmet), **469**
Fos oncogene, 745
Fox, C. F., 575
Fox, S., 26, 27, 29
Fps oncogene, 745
Fraenkel-Conrat, H., 32
Fragmin, 318
Frameshift mutation, **359**
Franklin, R., 343
Free energy (G), **37**
Free energy of activation, **39**
Free-energy change (ΔG), determination
    of, 37–39
Free radicals, and vitamin E, 614
Freeze-etch method, **65**
Frye, D., 104
FSH. *See* Follicle-stimulating hormone
Fullam, E., 161
Fumarate, 215
Fusidic acid, 480

GABA. *See* Gamma-aminobutyric acid
Galactose oxidase, in membrane protein
    labeling, **107**
β-Galactosidase, **437**, 438–441
Gall, J., 527
Galvani, L., 681
Gamete, **582**
Gametogenesis, **584**, 584–593. *See also*
    Oogenesis; Spermatogenesis
Gamma-aminobutyric acid (GABA)
    mechanism of action, 697
    as neurotransmitter, **689**
    site of action, 690
    structure, 690
Ganglioside, distribution in cell mem-
    branes, 100
Gap junction, **155**
    in cancer cell, 733
    in cardiac muscle, 723
    as electrical synapse, 678
    in Sertoli cell function, 589
    structure, 156
Garrod, A., 350
Gas chromatography, **88**
Gas-liquid chromatography (GLC), **88**
Gastrula, **584**, 608
Gaucher's disease, 193
G-bands, intercalary heterochromatin
    and, **505**
GDP. *See* Guanosine diphosphate
Gelactin, 318
Gelboin, H., 167
Gel-filtration chromatography, **85**
Gelsolin, 318
Gel-sol transition, in amoeboid move-
    ment, 321
Gene amplification
    in cloning DNA, **376**
    in gene regulation, **431**
    oncogenes and, 744
    of ribosomal RNA genes, **452**
Gene conversion, **520**

Gene deletion, **432**
Gene expression. *See also* Ribonucleic
        acid (RNA) synthesis
    bacteriophage proteins and, 436
    chromatin structure and, 429–431
    cytoplasmic influence on, 425–426
    in development, 605–606
    DNA-associated proteins and, 436–442
    DNA conformational changes and,
        434–436
    DNA methylation, 432–433
    DNA rearrangement and, 432–434,
        653–655
    epigenetic changes in, 747
    gene deletion and, 432
    RNA processing and, 442–443
    steroid hormones and, 633–636
Gene injection, **606–607**
Generative nucleus, **600**
Gene rearrangement
    antibody genes, 653–655
    mating-type locus, yeast, 433–434
Genetic code, 359–362
    dictionary of codons, 361
    in cytoplasm, versus in mitochondria
        and chloroplasts, 556
Genetic recombination, **514**
    breakage-and-exchange model, **519**
    conjugation, **518**
    copy choice model, **519**
    molecular basis of, 518–523
    nonreciprocal, **520**
    and recA protein, 523
    synaptonemal complex and, 523–526
    transduction, **518**
    transformation, **518**
Genetic transformation. *See* Transforma-
        tion
Georgiev, G., 418
Gerhart, J., 48
Germ cell. *See* Primordial germ cell
Germinal vesicle, **585**
Germ line theory of antibody diversity, **652**
Ghost, **106**
Gibberellins, **638**
Gibbons, I., 303, 309
Gibor, A., 538
Giemsa stain, in chromosome banding, **505**
Gilbert, W., 380, 437
Glandular secretion, nerve impulse and,
        697
GLC. *See* Gas-liquid chromatography
Glial cell, 608, **672**, 681
Glial fibrilliary acidic protein, 328
Glick, B., 646
Gln. *See* Glutamine
α-Globin, **353**. *See also* Hemoglobin
    control of synthesis, 484
    messenger RNA length, 468
β-Globin, **353**. *See also* Hemoglobin
    gene, 421, 606
    messenger RNA length, 468
Globular protein, secondary structure, **19**
Glu. *See* Glutamic acid
Glucagon
    cyclic AMP and, 626
    glycogen breakdown and, **625**
    site of synthesis, 621, 628
    target organ, 623
β-1,4-Glucan synthetase
    auxin effect on, 638
    in cellulose synthesis, **182**
    in Golgi complex, 172
    plant plasma membrane marker, 106
Glucocorticoid, 633, 637

Gluconeogenesis, peroxisomes and,
    198–199
Glucose, **12**
    regulation of insulin levels, 623
    structure, 13
    transport, 134, 135, 144
Glucose oxidation, 208–218
    efficiency, 217
    energy yield, 215–218
    glycolysis, 211–213
    respiration, 213–215
Glucose-6-phosphatase, **163**, 165
Glucose-6-phosphate dehydrogenase
    gene, 430
β-Glucuronidase, **165**
Glucuronyl transferase, **164**
Glutamate. *See also* Glutamic acid
    in electron shuttle into mitochondrion,
        216–217
    as neurotransmitter, 689–690, 697
    structure of, 690
Glutamic acid (Glu). *See also* Glutamate
    in bacterial cell wall, 115
    ionic bond formation, 19–20
    structure, 16
Glutamine (Gln), structure, 16
Glutamine synthetase, 610
Glutarate, as succinic dehydrogenase
    inhibitor, 45
Gly. *See* Glycine
Glyceraldehyde-3-phosphate, **281**
Glyceraldehyde-3-phosphate dehydrogen-
    ase, **106**
Glycerol, **13**, 14
α-Glycerophosphate shuttle, **216–217**
Glycine (Gly)
    in bacterial cell wall, 115
    collagen, content of, 124
    as neurotransmitter, 689, 690
    structure, 16, 690
Glycocalyx, **123**
Glycogen, **13**
    breakdown, hormonal influence, 625,
        628–629
    energy storage in muscle, 705
    structure, 14
Glycogen phosphorylase
    mechanism of action, **625**
    phosphorylation, 50, 628
Glycogen synthase, **628**
Glycolipid, **13**
    in chloroplast membranes, 101, 259
    distribution in cell membranes, 99, 100,
        101
Glycolysis
    anaerobic, 213
    energy yield, 212
    in muscle contraction, 720
    pathway, 211–213
Glycophorin
    ankyrin binding, 109–110
    blood groups and, 150
    in cell recognition, 150
    mobility of, 110, 327
    red cell membrane localization,
        107–109
    spectrin and, 109–110
    structure, **108**
Glycoprotein, formation of, 170–172
Glycosaminoglycan, **125**
Glycosome, **202**
Glycosyltransferase
    in cell-cell adhesion, **151**
    in Golgi complex, 172
    and protein modification, 170

Glyoxylate cycle, **199**
Glyoxysome, **199**
GnRF. *See* Gonadotropin releasing
    hormone (GnRH)
Goldberg, N., 631
Goldstein, A., 695
Golgi, C., 171, 668
Golgi body. *See* Golgi complex
Golgi complex, **8, 171**
    in acrosome development, 590
    in cell wall synthesis, 123, 182
    cis region, 172
    discovery of, 171
    membrane biogenesis, 203–205
    oligosaccharide addition to protein,
        177–178
    in protein secretion, 171–182
    structure, 171–173
    trans region, 172
    in yolk modification, 586–587
Gomori technique, **71**
Gonadotropin releasing factor (GnRF). *See*
    Gonadotropin releasing hormone
    (GnRH)
Gonadotropin releasing hormone (GnRH)
    neurosecretion and, 698
    target organs, 623
Good, N., 269
Good, R., 645
Goodman, C., 680
Gorski, J., 634, 635
Gorter, E., 94
G_0 phase of cell cycle, **528**
    entry into G_1, 529
    cancer cell, 732
G_1 phase of cell cycle, **491**
    arrest in, 731
    elimination during cleavage, 600
    entry into, 528–529
    regulation of cell division, 528–529
G_2 phase of cell cycle, **491**, 529
G-protein. *See* Guanine nucleotide
    binding protein; Vesicular stomatis
    virus G-protein
Graft-versus-host reaction, **661**
Gram, H. C., 114
Gramicidin, **138**
    chloroplast inhibitor, 273
    mechanism of action, mitochondria,
        233, 239
    structure, 138
Gram stain, **114**
Granick, S., 538
Granum, **252**
Gray crescent, **602**
Green, H., 730
Green algae, evolutionary origin,
    570
Greengard, P., 695
Grendel, F., 94
Griffith, F., 340
Griffith, O. H., 102
Grobstein, C., 612
Growth and division model
    chloroplasts, **564**
    mitochondrion, **561–562**
Growth cone, **678**
Growth factors
    epidermal growth factor (EGF), **529**
    nerve growth factor (NGF), **679**
    oncogenes and, 745
    platelet-derived growth factor (PDGF),
        531
Growth hormone. *See* Somatotropin
Growth hormone gene, injection of, 382

GTP. *See* Guanosine triphosphate
Guanine, **23**
    base pairing, 344, 367
    structure, 22
Guanine nucleotide binding protein
    (G-protein)
    adenylate cyclase and, **626**, 627
    oncogenes and, 745
    in photoreception, 688–689
Guanosine diphosphate (GDP)
    in microtubule formation, 298–300
    in photoreception, 688–689
Guanosine diphosphate (GDP)-mannosyl
    transferase, 164
Guanosine pentaphosphate (pppGpp), **460**
Guanosine tetraphosphate (ppGpp), **460**
Guanosine triphosphate (GTP)
    adenylate cyclase and, 626–628
    in elongation of protein synthesis, 470,
        472
    in initiation of protein synthesis, 467, 470
    in initiation of RNA synthesis, 413
    methylation to form RNA cap, 419
    in microtubule formation, 298–300
    in photoreception, 688–689
    in termination of protein synthesis, 476
Guanylate cyclase, **631**, 688–689
L-Gulonolactone-ascorbate, **165**
Gurdon, J., 425, 451, 604

Hackenbrock, C., 234
HaeIII, **375**
Half-life ($t_{1/2}$)
    messenger RNA, **482**
    protein, **607**
Halobacterium halobium
    photosynthesis in, 288–289
    purple membrane, 112–113
Hamburger, V., 679
Hämmerling, J., 385
Hanson, J., 702, 713
H-2 antigen, **646**
Haploid, **514**
Hapten, **650**
Ha-ras oncogene, 745
Harris, H., 415, 425
Harrison, R., 58, 678
Harvey, E., 95
Hatch, M., 283
Hatch–Slack pathway. *See* C_4 pathway
HAT medium, in monoclonal antibody
    production, **659**
Hayflick, L., 613, 615
Heavy chain
    antibody, **648**, 650, 654–655
    myosin, **709**
Heavy meromyosin, 317, **709**
Heaysman, J., 730
Heilbrunn, L. V., 717
Hela cell
    cancer, 730
    map of mitochondrial DNA, 553
    origin, **59**
    synthesis of mitochondrial DNA, 563
    synthesis of mitochondrial RNA, 552
Helicase, 350
Helix destabilizing protein, 350
Helmholtz, H., von, 681
Helmont, J. B., van, 249
Helper T-cell, 647, **661**
Heme group, structure of, **229**
Hemicellulose, **122**
Hemi-desmosome, **153**, 154
Hemin, **482**

Hemoglobin. *See also* α-Globin; β-Globin
    peptide fingerprint, 352–353
    in sickle-cell anemia, 352–353
    synthesis in red blood cells, 476–478
    three-dimensional structure, 20
Heparin, 126, **413**
Heroin, **695**
Herpes simplex virus, **675**, 740–741
Hers, H., 194
Hershey, A., 342
Hershey-Chase experiment, 342–343
Heterochromatin
    base composition of, 505
    centromeric, **505**, 506
    chromosome banding and, 505
    constitutive, **429**
    facultative, **429**
    intercalary, **505**
    and satellite DNA, 505
Heteroduplex region, **522**
Heterogeneous nuclear RNA (HnRNA),
    416
    discovery of, 415–416
    evidence for, 416
    messenger RNA sequences in, 416–417
    7-methylguanosine cap, 419
    packaging with protein, 417–419
Heterokaryon, 60
Heterophagic lysosome, **192**
Heterotroph, **248**
Hewish, D., 397
Hexose, **12**
HGPRT. *See* Hypoxanthine-guanine
    phosphoribosyltransferase
High-mobility group (HMG) proteins, **394,**
    431
Hill, J., 737
Hill, R., 250, 268
Hill reaction, **250**
Hilschmann, N., 650
Hiscoe, H., 672
His. *See* Histidine
Histamine
    effect on adenylate cyclase, 697
    in innate immunity, **663**
    as neurotransmitter, 689
    site of action, 690
    structure, 690
Histidine (His)
    in chymotrypsin active site, 53
    hydrogen bond formation, 20
    ionic bond formation, 19–20
    methylation of histone, 395
    structure, 16
Histocompatibility antigen, **646, 647**
Histone genes, 431
Histone kinase, **507**
Histone proteins, **394**
    acetylation of, 394, 430
    amino acid content, 394
    H1, 394, 399, 507
    H2A, H2B, H3, H4, 394, 399
    H5, 394
    methylation of, 394
    molecular weight, 394
    in nucleosome, 399–401, 496–497
    octomer, 399
    phosphorylation of, 394, 507
    replacement by protamines in sperm
        DNA, 589
    synthesis in S phase of cell cycle,
        496–497
HIV. *See* Human immunodeficiency virus
HMG protein. *See* High-mobility group
    protein

HnRNA. *See* Heterogeneous nuclear RNA
Hoagland, M., 362
Hodgkin, A., 140, 681
Hoffman, H.-P., 223
Hogeboom, G., 77, 218
Holley, R., 364
Holliday, R., 522
Holtfreter, J., 612
Holtzer, H., 608, 706
Homeo box, **606**
Homeotic gene, **606**
Homogenate, **78**
Homogentisic acid, **350**
Homolog. *See* Homologous chromosome
Homologous chromosome, **514**, 514–518
Hooke, R., 3
Hormone, **619**, 619–642
  calcium ion and, 631–632
  in development, 613
  discovery of, 619–620
  classes of, 621–622
  cyclic AMP and, 626
  and gene transcription, 636–637
  from nonendocrine cells, 638–639
  nucleus, action on, 632–637
  plasma membrane, action on, 623–632
  receptor binding criteria, 624
  secretory pathways, 639–640
  termination of action, 632
Horseradish peroxidase, 71
HPr protein, **143**
HPV. *See* Human papilloma virus
Hozumi, N., 653
HTLV. *See* Human immunodeficiency
    virus; Human T-cell lymphotropic
    virus
Huez, G., 483
Hughes, J., 696
Human immunodeficiency virus (HIV), **741**
Human papilloma virus (HPV), 740
Human T-cell lymphotropic virus (HTLV),
  741
Humoral immunity. *See* Antibody-
    mediated immunity
Hurwitz, J., 357
Huxley, A., 681, 713, 716
Huxley, H., 702, 713
Hyaluronic acid, **126**
Hybridoma, **658**
Hydrocortisone
  and glutamine synthetase activity, **610**
  and tyrosine transaminase activity, **636**
Hydrogen bond
  in DNA, **343**, 371
  in protein, **19**, 20
Hydrogen peroxide, breakdown in
  peroxisomes, **197**
Hydrogen sulfide (H_2S), in photosynthesis,
  288
Hydrolase, 36, 184–185
Hydrophobic interactions, in protein, **19**
Hydroxylapatite chromatography, **87**, 427
Hydroxyl ion transport, in mitochon-
  drion, 241–242
Hyperchromic effect, **371**
Hyperpolarized state, **682**
Hypertonic, **131**
Hypervariable region, antibody molecule,
  650
Hypothalamus, **622**, 690
Hypotonic, **131**
Hypoxanthine-guanine phosphoribosyl-
  transferase (HGPRT), **659**
H-zone, of sarcomere, **702**

I-band, of sarcomere, **702**
Idiotype, **650**
IF. *See* Initiation factor
Ig. *See* Immunoglobulin
i gene, in lactose operon, **437**
Iino, T., 332
Ile. *See* Isoleucine
Illmensee, K., 603, 747
Immune surveillance theory, **734**
Immune system, 643–666. *See also*
    Antibody; Antigen; Immunoglobulin
  antibody diversity, 650–655
  antibody structure, 648–650
  antibody synthesis, 645–648, 655–658
  cell-mediated immune response,
    660–662
  developmental origin, 643–644
  general properties of, 643–644
  immunological tolerance, 662–663
  innate versus acquired immunity,
    663–664
Immunocytochemistry, **70**, 71
Immunoglobulin (Ig), **648**
  A, **648**
  capping, 656–657
  D, **648**
  E, **648**
  G, **648**, 649, 658, 660
  genes, 434, 652–655, 745
  M, **648**, 658, 660
  molecular weight, 648
  number and types of chains, 648
Immunological labeling, and ribosome
  structure, **463**
Immunological tolerance, **662**, 663
Import-receptor protein, **560**
Induced-fit model, **51**
Inducer, in lactose operon, **437**
Induction, embryonic, **584**, 610–612
Induction synchrony, **490**
Inflammation response, **663**
Information transfer, and origin of life,
  26–27
Ingenhousz, J., 249
Ingram, V., 352
Initial velocity (v_i), **41**
Initiation complex, **470**
Initiation factor (IF), protein synthesis
  eukaryote initiation factor 2 (eIF-2), 482
  IF-1, IF-2, IF-3, 467, 470
Initiator codon, **362**, 469
Initiator transfer RNA (tRNA), **466**,
  469–470
Innate immunity, **663**
Inorganic phosphate ion transport, in
  mitochondrion, 241–242
Inosinic acid pyrophosphorylase, **426**
Inositol trisphosphate
  as intracellular messenger, **632**
  in muscle contraction, 718
  platelet-derived growth factor and, 531
  synthesis, 632
Inoué, S., 295, 507, 510
Insertion sequence, **367**
In situ hybridization, **374**, 451
Instructive theory of antibody formation,
  651
Insulin
  feedback regulation, 623
  gene, introns in, 421
  messenger RNA length, 468
  receptor, 147
  site of synthesis, 621
  structure, 17, 621

synthesis and post-translational
    modification, **485**
  target organ, 623
Integral protein
  plasma membrane, **102**
  mitochondrial inner membrane, 234
Interband region, polytene chromosome,
  404
Intercalary heterochromatin, **505**
Intercalated disk, **723**
Intercalating agent, **369**
Intercellular communication
  cell surface and, 147–157
  evolution of, 637–640
  pathways, 639–640
Intercellular junctions. *See* Desmosome;
    Gap junction; Plasmodesmata; Tight
    junction
Interdoublet nexin link, **302**, 311
Interference microscope, **62**
Interferon
  mechanism of action, **482**
  in virus infection, 664
Interleukin-1, 657–658
Interleukin-2. *See* B-lymphocyte, growth
    factor
Intermediate filament, **10**, 328–329
Interneuron, **693**
Interphase. *See also* G_1 phase; G_2 phase; S
    phase
  DNA synthesis and, 491
  meiosis, 518
Interpolar microtubule, **507**
Interspersed repetitive DNA, **429**
Intervening sequence. *See* Intron
Intestinal cell
  glucose transport, 144
  totipotency of, 604
Intracellular messengers. *See* Calcium
    ion; Cyclic AMP; Cyclic GMP;
    Diacylglycerol; Inositol trisphosphate
Intron, **420**
  discovery of, 420–421
  protein domains and, 424
  splicing of, 421–424
Inverted repeat sequence, **367**
Involuntary muscle contraction, **721**
Iodoacetamide, **45**
Iodoacetate
  mapping enzyme active site, 52
  mechanism of action, **45**
Iojap mutation, **537**
Ion exchange chromatography, **85**
Ionic interactions, in proteins, **19**
Ionophore, **137**
  carbonyl cyanide *m*-chlorophenylhy-
    drazone (CCCP), 233, 239
  carrier type, **139**
  channel-former type, **139**
  2,4-dinitrophenol, 233, 239
  gramicidin, 138, 239
  nigericin, 239
  valinomycin, 137, 239
IPTG. *See* Isopropyl-thio-galactoside
Iron-sulfur protein (FeS), **231**, 266
Ishida, M. R., 543
Isocitrate, 214
Isocitrate dehydrogenase, and theories of
  myogenesis, **708**
Isodensity centrifugation, **75**–77
Isoelectric focusing, **84**
Isoelectric point, **84**
Isoleucine (Ile)
  hydrophobic interactions, 20

structure, 16
Isomerase, 36
Isoprene unit, **15**
Isopropyl-thio-galactoside (IPTG), and
  lactose operon, **437**
Isotonic, **131**
Isotopic tracer, 81–82
Isotropic band. *See* I-band, of sarcomere
Isozyme, **51**
Israël, M., 691
Iwanowsky, D., 30

Jacob, F., 47, 357, 437
Jaffe, L., 594
Jagendorf, A., 270
Jamieson, J., 175
Janssen, J., 60
Janssen, Z., 60
Jensen, E., 633
Jerne, N., 651
J region, antibody diversity, **654**
Junctional fold, **716**

Kaback, H., 145
Kaempfer, R., 465
Kappa (κ) chain, in antibody, **649**
Karyotype
  meiotic, 517
  mitotic, **503**
Kasugamycin, 480
Katz, B., 682, 689
Keilin, D., 41, 231
Kendrew, J., 68
Kennedy, E., 203, 218
Keratin, **328–329**
Keratin sulfate, **126**
Keynes, R., 140
α-Ketoglutarate
  and electron shuttle into mitochon-
    drion, 216–217
  in Krebs citric acid cycle, 214
Khorana, H. G., 360, 381
Kinesin, **675**
Kinetics
  enzyme, **39–40**
  reassociation, of DNA, **426–429**
Kinetochore, **506**
  microtubule assembly, 298
  microtubule attachment, 517
  microtubule and mitotic spindle and,
    507
Kinetoplast, **540**, 541
King, T., 604
Ki-ras oncogene, **745**
Kirschner, M., 299
Kishimoto, A., 632
Kleinschmidt, A., 65
Klug, A., 365
Köhler, G., 658
Kohne, D., 426
Kölliker, A., 218
Kornberg, A., 345
Kornberg, R., 138, 399
Kornberg enzyme. *See* DNA polymerase I
Koshland, D., 51
Kramer, F., 381
Krebs, E., 628
Krebs, H., 215
Krebs citric acid cycle, **213–215**
Kuentzel, H., 560

Lactose
  inducer, lactose operon, **437**
  proton-linked transport of, 145–146

Lactose (*lac*) operon, **437–440**
  catabolite repression, **440**
  inducer molecule, **437**
  negative control, **440**
  operator sequence, **437, 440**
  positive control, **440**
  promoter, **437, 440**
  repressor, **437, 440**
Lactate dehydrogenase (LDH), isozymes
  of, **51**
Lactoperoxidase, in membrane protein
  labeling, **107**
Lactose permease
  lactose transport, **145–146**
  γ structural gene, 437
Lambda (λ) chain, in antibody, **649**
Lambda phage. *See* Bacteriophage, lambda
Lambert, B., 406
Lamellipodia, **322**, 735
Lamin proteins, **387**, 507
Laminin, **127**
Lampbrush chromosome, 408, **526,**
  527–528
Landsteiner, K., 650
Langmuir, I., 94
Lateral element, in synaptonemal
  complex, **524**
LDH. *See* Lactate dehydrogenase
LDL. *See* Low-density lipoprotein
L-DOPA. *See* Levo-dihydrophenylalanine
Leblond, C. P., 177
Lecithin. *See* Phosphatidylcholine
Lectin, **150**
  in cancer cell agglutination, 733
  concanavalin A receptors and endocy-
    tosis, 314
  and lymphocyte capping, 658
Leder, P., 360, 653
Leeuwenhoek, A., van, 251
Lehninger, A., 218
Length-regulating proteins, **317**, 318
Leptotene stage, **516**
Leu. *See* Leucine
Leu-5 mutant, **615**
Leucine (Leu)
  hydrophobic interactions, 20
  structure, 16
Leucofuchsin. *See* Schiff reagent
Leukemia, **727–728**
Leukoplast, **565**
Levene, P., 339
Levi-Montalcini, R., 679
Levo-dihydrophenylalanine (L-DOPA),
  695
LH. *See* Luteinizing hormone
LHC. *See* Light-harvesting complex
Ligand, **147**, 149
Ligase, **36**, 348
Light chain
  antibody, **648**, 650, 654
  myosin, **709**
Light-harvesting complex, **267**
Light-harvesting complex 2 (LHC2), **278,**
  279
Light meromyosin, **709**
Light microscope, **60–63**
Light reactions of photosynthesis,
  262–279, **250**
  discovery, 249–251
  electron transfer chain, 266–269
  photophosphorylation, 269–270
  photosynthetic unit, 265–266
  pigments, light-absorbing, 262–265
  P/O ratio, 269

proton gradient and, 270–276
  summary equation, 283
  yield of ATP and NADPH, 269–270,
    272–273, 282–283
Lignin, **122**
Limbic system, **695**
Lineweaver-Burk plot, **42**
Lipid, **13**. *See also* Neutral fat; Phospho-
    lipid; Steroid
  in amphibian oocyte, 587
  bilayer, **95**
  cytochemical identification, 69
  properties, 13–15
  synthesis in endoplasmic reticulum,
    163–165
Lipkin, D., 625
Lipmann, F., 53, 364
Lipofucsin granules, **669**
Lipoplast, **565**
Lipopolysaccharide, **115**
Liposome, **99**
β-Lipotropin, **697**
Lipovitellin, **585**
Liquid chromatography, **85**
Liquid crystal state, of cell membrane, **102**
Liquid scintillation counter, **82**
Liver
  drug detoxification, 165–167
  homogenate in Ames procedure, 751
  hormone products, 622
  proteins synthesized for amphibian
    oocyte, 585
Ljubimowa, M., 710
L-Malate, 214
Lock-and-key theory, **51**
Lodish, H., 204
Loewenstein, W., 157, 733
Loewi, O., 689
Low-density lipoprotein (LDL), **192**
LSD, **697**
Lubrol, 225
Luck, D., 543, 548, 561
Lucké, B., 740
Lucké virus, 740
Lumicolchicine, **315**, 675
Lung cancer, 737, 738, **749**
Luteinizing hormone (LH), **623**
  cyclic AMP and, 626
  site of synthesis, 622
  target organ, 622
Luteinizing hormone-releasing hormone,
  690
Lyase, 36
Lymphatic system, **643**
Lymph node, **644**
Lymphoblast, **661**
Lymphocyte. *See* B-lymphocyte; T-lym-
  phocyte
Lymphocyte inhibitory factor, **662**
Lymphokine, **647**
  B-cell growth factor (interleukin-2), **658**
  discovery of, 647
  interleukin-1, **657**
  lymphocyte inhibitory factor, **662**
  macrophage activating factor, **662**
  macrophage migration inhibition factor,
    662
Lymphoma, **727**, 728
Lyon, M., 429
Lyon hypothesis, **429**, 430
Lys. *See* Lysine
Lysine (Lys)
  acetylation, in histone, 395
  in bacterial cell wall, 115

Lysine (Lys) *(Continued)*
  histone content, 394
  hydroxylation in collagen, 124
  ionic bond formation, 19–20
  methylation, in histone, 395
  structure, 16
Lysosomal storage disease, **193–195**
Lysosome, **8, 184,** 182–195
  in antibody-mediated phagocytosis, 660
  axonal transport of, 674
  digestive functions, 187–191
  discovery of, 182–184
  enzyme constituents, 184–186
  formation, 186–187
  isolation, 184
  membrane fragility, 183
  phospholipid asymmetry in membrane,
    111, 203–204
  primary lysosome, 190
  and residual body, 190
  secondary lysosome, 190
  stimulation by macrophage activating
    factor, 662
Lysozyme
  bacterial cell wall and, **118**
  gene, 33, 421
  in innate immunity, **664**
  synthesis, hormonal control of, **636**

McCarthy, B., 416
McClintock, B., 367
McConnell, H., 102, 138
McKinnell, R., 748
Macrophage, **187**
  activating factor, **662**
  in antibody-mediated phagocytosis, 660
  antibody synthesis and, 647
  in B-lymphocyte activation, 651–658
  chemotactic factor, **662**
  endocytosis and, 187
  migration inhibition factor, **662**
Macromolecule, **11.** *See also* Carbohy-
    drate; Lipid; Protein; Nucleic acid
  colorimetric identification, 81
  methods of analyzing, **80–88**
Magnesium ion (Mg^{2+})
  chlorophyll structure, 264
  ribulose bisphosphate carboxylase
    regulation, 282–283
  RNA polymerase sensitivity, 409–410
Mahler, H., 573
Mahowald, A., 603
Major histocompatibility complex (MHC),
  646, 661–662
Malate-aspartate shuttle, 216–217
Malate dehydrogenase, bacterial plasma
  membrane marker, 106
Malignant tumor, **728,** 729
Malonate, succinate dehydrogenase
  inhibitor, **45,** 215
Mangold, H., 611
Mannose-6-phosphate, in lysosome
  formation, **186**
MAP. *See* Microtubule-associated protein
Marcker, K., 469
Marek's disease virus, 740
Margolis, R., 298
Margulis, L., 570
Marmur, J., 358, 371
Mast cell, **663**
Maternal inheritance, **536,** 551–552
Mating-type locus, yeast, **433–434**
Maturation factor, **658**

Maturation phase, in amphibian develop-
  ment, **588**
Maturation promoting factor (MPF), **507**
Matrix, of mitochondrion, **9, 228**
Matrix porin, **118**
Maxam, A., 380
Maximal velocity (V_{max}), **40**
Mayer, J. R., 249
Mechanoreceptor, **685**
Medawar, P., 663
Megaspore, **600**
Megasporocyte, **600**
Meiosis, **514–528**
  gametogenesis in plants, 600
  genetic recombination in, 518–526
  oogenesis and, 588
  spermatogenesis and, 589
  stages in, 514–518
  and synaptonemal complex, 523–526
Melting temperature (T_m), **372**
Memory cell, **658,** 661
Membrane. *See also individual cell*
  membranes
  biogenesis of phospholipids and
    proteins, 203–205
  structure, 93–105
Membrane disk, in photoreceptor cell, **687**
Membrane potential ($\Delta\psi$), **146.** *See also*
  Action potential
  calculation of, 146–147
  chloroplast, electrochemical potential,
    272
  light-induced changes in, 687–688
  mitochondrion, electrochemical
    potential, **238**
  squid axon, 681
Mendel, G., 339, 350
Menten, M., 40
Mercaptoethanol, **21**
Mereschovsky, C., 569
Meristem, **565**
Meromyosin, 317, **709**
Meselson, M., 344, 519
Mesenchyme, **584**
Mesoderm, **584**
Mesophyll cell, in C_4 pathway, **284–286**
Mesosome, **8,** 494
Messenger ribonucleic acid (mRNA), **357**
  binding proteins, 481–482
  coding region length, 467–468
  criteria for information transfer, 357
  direction of translation, 465
  discovery of, 354–357
  half-life, 482–483
  isolation of, 466–467
  7-methylguanosine cap, 419, 468–469
  polyadenylate sequence, 419, 483
  polycistronic, 414
  processing and export, 442–443
  synthesis and steroid hormones,
    636–637
  translation after fertilization, 599
  translation during cleavage, 600
  untranslated regions, length of, 467–468
Messenger ribonucleoprotein particles,
  482
Met. *See* Methionine
Metabolism, **4, 208**
Metachromatic leukodystrophy, 193
Metal ion
  as cofactor, **43**
  in cytochemical stains, 70
  as enzyme inhibitor, **47**
Metallothionein gene, **382**

Metaphase, mitosis, **501–503**
Metaphase I and II, meiosis, **517, 518**
Metastasis, **729**
Methionine (Met), structure, 16
Methionyl transfer RNA (tRNA^{met}), **469**
Methylation
  in bacterial flagellum motility, **335**
  of DNA, **432**
5-Methylcytosine, **432**
Methyl green, **69**
7-Methyl guanosine, RNA cap, **419**
Mg^{2+}. *See* Magnesium ion
MHC. *See* Major histocompatibility
  complex
Michaelis, L., 40, 218
Michaelis constant (K_m), **41**
Michaelis-Menten equation, **40–42**
Microbody, **8, 196.** *See also* Peroxisome
Micrococcal nuclease. *See* Deoxyribonuclease (DNAase)
Microfibril, in plant cell wall, **121**
Microfilament, **10, 316–328**
  arrowhead pattern, 317
  in brush border, 326–327
  and capping, 327
  and cell division, 324
  cell shape and, 324–327
  in cleavage furrow, 511–512
  cytoplasmic distribution, 317
  in endocytosis and exocytosis, 327–328
  in growth cone, 678
  in micropinocytosis, 327–328
  microtubule interaction, 328
  in morphogenesis, 609
  and terminal web, 326
α-2-Microglobulin, hormonal control of
  synthesis, 636–637
Micromonas, 561
Microscopy, **60–73**
  Allen video-enhanced contrast
    differential interference contrast
    microscopy system (AVEC-DIC), **675**
  darkfield, **62**
  electron, **63–66**
  interference, **62**
  light, **60–63**
  phase contrast, **61–62**
  polarization, **62–63**
Microsome, **163**
  in cell-free protein synthesis, 465
  enzyme orientation in membrane, 164
  fraction, preparation of, 79
  phospholipid asymmetry in membrane,
    111
Microsphere, **27**
Microspore, **600**
Microsporocyte, **600**
Microtome, **61**
Microtrabecular network, **329**
Microtubule, **10, 293, 293–316**
  assembly/disassembly, 295–301
  assembly during mitosis, 501
  in axonal transport, 673
  in cell shape, 315–316
  cell surface connection, 314–315
  directional assembly, 297–298
  in growth cone, 679
  microfilament interaction, 328
  model of, 295
  plant cell wall formation, 315
  properties of, 293
  protofilament, 294
  ring-form, 295, 297
  spindle apparatus structure, 507–509

transport of subcellular components, 312–314, 673
Microtubule-associated proteins (MAPs), **295**, 301
Microtubule-organizing center, **298**
  basal body, **306**
  centriole, **306**
  centrosome, **508**
  kinetochore, **507**
Microvillus, **129.** See also Brush border
  calmodulin, 326
  in cancer cell, 735
  fimbrin, 326
  in follicle cells, 585
  110-kilodalton protein, 326
  structure, 326
  villin, 326
Midbody, **511**
Middle lamella, **120**
Middle piece, sperm flagellum, **592**
Miele, E., 381
Miescher, F., 339
Mil oncogene, 745
Miller, J., 165
Miller, J. F. A. P., 645, 646
Miller, O., 457
Miller, S., 25
Millon reaction, **69**
Mills, D., 381
Milstein, C., 169, 658
Mineralocorticoids, 633
Miniature endplate potential. See Miniature postsynaptic potential
Miniature postsynaptic potential, **690**
Minimal medium, Neurospora, **350**
Mintz, B., 606, 708, 747
Mitchell, G., 646
Mitchell, H., 539
Mitchell, M., 539
Mitchell, P., 145, 237, 270
Mitchison, T., 299
Mitochondrial deoxyribonucleic acid (DNA)
  catenated dimer, **545**
  circular dimer, **543**
  circularity, 543
  comparison of mammalian and yeast, 553–554
  in cytoplasmic inheritance, 546
  density, comparison with nuclear DNA, 543
  discovery of, 541–543
  displacement loop (D-loop) formation, 550
  ethidium bromide binding, 545
  heavy and light strands, 552
  homology with nuclear DNA, 561
  introns in, 554
  map, HeLa cells, 552–553
  open circle, 543
  properties, 543–546
  reassociation kinetics, 546
  ribonucleic acid (RNA) synthesis, 552–554
  ribosomal RNA genes, 553
  semiconservative replication, 548–550
  structure, in yeast, 553
  synthesis and cell cycle, 563
  transfer RNA genes, 553
  turnover, 563
Mitochondrial ribonucleic acid (RNA)
  ferritin labeling, 552
  hybridization to DNA, 552
  synthesis, 552–554

Mitochondrion, **9, 208–247**
  anatomy of, 218–229
  ATP synthesis and, 229–243
  ATP synthesis inhibitors, 233
  in axonal transport, 674
  biogenesis, 561–564
  chloroplast, proximity to, 255
  condensed and orthodox configurations, 222
  cristae, **9,** 218
  electron transfer chain, 229–243
  evolutionary history, 568–575
  fraction, preparation of, 79
  genetic code, 556
  genetics of, 538–540
  growth and division model, 561–562
  inner membrane, **218**
  inner membrane, fluidity, 234
  inner membrane, model, 241
  inner membrane, phospholipid asymmetry, 111
  inner membrane, properties, 228–229
  inner membrane, reconstitution, 235–237
  inner membrane, transport, 241–242
  intermembrane space, **218,** 227–228
  matrix, **9, 218,** 228
  outer membrane, 218, 226–227
  peroxisome, proximity of, 255
  photorespiration and, 287
  protein import, 560–561
  protein synthesis, 554–558
  protein synthesis inhibitors, 557
  ribosomal RNA, 553
  RNA polymerase in, 552
  RNA synthesis, 552–554
  shape, three-dimensional, 223–224
  in skeletal muscle, 705–706
  in sperm flagellum, 592
  spheres, **218**
  structure, compare with chloroplast, 255
  subfractionation of, 225–226
  submitochondrial particle, **226**
  transfer RNA, 555–557
  yolk production and, 587
Mitoplast, **225**
Mitosis, **491, 501–513**
  chromosome movement, 509–510
  chromosome structure, 504–507
  cleavage and, 600
  cytokinesis, mechanism of, 510–513
  mitotic spindle, 507–509
  stages in, 501–504
Mitotic spindle, **501**
  calcium ion and, 508
  cleavage plane location and, 511, 603
  formation, 501–503, 507–508
  microtubule components, 507
  microtubule polarity in, 509
  in mitosis, 501, 504
M-line, of sarcomere, **702**
MMTV. See Mouse mammary tumor virus
Molecular weight–sedimentation coefficient relationship, 74
Molting, chromosome puffs and, 407–408
Monoclonal antibody, **70, 658,** 659
Monod, J., 47, 145, 357, 437
Monogalactosyldiglyceride, 259, 262
Monoglyceride, **13**
Monolayer culture, **58**
Mononucleosis, **741**
Monosaccharide, **12**
Moore, C., 420
Moore, P., 463

Morgan, T. H., 602
Morphine, **695**
Morphogenesis, **608**
  cell adhesion molecules (CAMs), 609–610
  cell–cell interactions and, 608–615
  cell movements and, 608–609
  cell shape and, 608–609
  cell sorting, 609–610
  microfilaments and, 609
Moscona, A., 609
Moses, M., 524
Mos oncogene, 745
Motility, **10.** See Cilium; Flagellum
Motor endplate. See Neuromuscular junction
Mouse mammary tumor virus (MMTV), 637, 740
Moving-zone centrifugation, **74**
Moyle, J., 237
MPF. See Maturation promoting factor
M-protein, in sarcomere, **709,** 711
M-protein, vesicular stomatis virus, **204**
mRNA. See Messenger ribonucleic acid
Mucopolysaccharide. See Glycosaminoglycan
Mueller, G., 165
Müller-Hill, B., 437
Multienzyme complex, **50**
Multiple myeloma, 649–650
Multiple sclerosis, **685**
Murein. See Peptidoglycan
Murine leukemia virus, 740
Murine sarcoma virus, 740
Muscle. See Cardiac muscle; Skeletal muscle; Smooth muscle
Muscular dystrophy, **720**
Mutation, **365**
  chemical induction, 369
  X-ray-induced metabolic mutants in Neurospora, 350–352
Myasthenia gravis, **720**
Myb oncogene, 745
Myc oncogene, 745
Myelin, **96**
  action potential propagation and, 685
  discovery, 672
  sheath, 96, 672
Myeloma protein, **650**
Myoblast, **706**
Myogenesis, in skeletal muscle, 706
Myogenic contraction, 723
Myogenic hypothesis of muscular dystrophy, **720**
Myoglobin, **705**
Myokinase. See Adenylate kinase
Myosin, **708**
  A-band location, 709
  actin binding site, 710
  ATPase activity, 319, 710
  cleavage furrow and, 513
  discovery of, 708
  head, 709
  in mitotic spindle, 510
  nonmuscle cells, 319
  polymerization into filaments, 710
  properties of, 709–710
  structure, 709
Myosin light-chain kinase, **722**
Myosin light-chain phosphatase, **722**

NAD⁺–NADH. See Nicotinamide adenine dinucleotide
NADH-cytochrome b₅ reductase, 164

NADH dehydrogenase
  bacterial plasma membrane marker, 106
  complex I, 234, 241
  coupling site, 1, 235
  electron transfer chain, mitochondrion,
    232, 233
NADP⁺–NADPH. *See* Nicotinamide
  adenine dinucleotide phosphate
NADP reductase, **266**
NADPH-dependent cytochrome P-450
  reductase, 164, **165,** 166
Nageli, C., 94
Nass, M., 541
Nass, S., 541
Nathans, D., 375
Nebenkern, **592**
Negative control, in gene regulation, **437**
Negative staining, **64**
Neoplasm. *See* Tumor
Nernst equation, **146**
Nerve, **672**
Nerve cell. *See* Neuron
Nerve fiber, **672**
Nerve growth factor (NGF), 613, **679**
Nerve impulse. *See also* Action potential
  in excitation-contraction coupling,
    715–718
  physiological responses to, 697–698
  in synaptic transmission, 689–697
Nervous tissue, induction by epidermis, 611
Neu oncogene, 745
Neupert, W., 560
Neural cell adhesion molecule (N-CAM),
  610
Neural crest, cell migration and, **608**
Neurite, **668**
Neuroblast, 603, **678**
Neuroblastoma, **680**
Neurocrine pathway, **639**
Neuroendocrine pathway, **639**
Neuroendocrine system, major pathways,
  622–623
Neurofilament, **671**
  axonal transport, 673
  protein, properties, 328
Neurogenic hypothesis of muscular
  dystrophy, **720**
Neurohormone, **638**
Neuromuscular junction, **716**
  acetylcholine, 689, 690, 716
  acetylcholine receptor, 693–694
  depolarization of sarcolemma, 716–717
  neurotransmitter release, 676–677
  structure, 716
Neuron, **668**. *See also* Action potential;
  Axon
  electrical properties of, 681–689
  formation and cell migration, 608
  growth and development, 678–681
  specificity of connections, 610, 680
  structure, of, 668–671
Neuropeptide, **621**
Neurosecretion, **697**
Neurospora
  leu-5 mutant, 615
  metabolic mutants, 350–352
  poky mutation, 540
Neurotransmitter, **689**. *See also individual
  neurotransmitters*
  axonal transport, 674
  discovery of, 689
  inactivation of, 697
  with inhibitory actions, 697
  receptors, 692–697

release mechanism, 689–692
Neutra, M., 177
Neutral fat, **13**
  breakdown to fatty acids, 213–214
  chemical composition, 13–14
  in muscle, 705
  triglyceride, 13
Neutral *petite* mutation, **539**
Neutrophil. *See* Polymorphonuclear
  leukocyte
Nexin. *See* Interdoublet nexin link
NGF. *See* Nerve growth factor
Nicholson, G., 98
Nicotinamide adenine dinucleotide,
  NAD⁺–NADH, **211**
  absorbance spectrum, 229
  in anaerobic glycolysis, 213
  ATP yield per molecule, 215–218
  cytoplasmic shuttle into mitochon-
    drion, 216–217
  in glycolysis, 211–213
  in glyoxylate cycle, 199
  in Krebs citric acid cycle, 214–215
  redox couple, 210
  regeneration in peroxisome, 198
  structure, 211
Nicotinamide adenine dinucleotide
  phosphate, NADP⁺–NADPH, **211**
  in Calvin cycle, 279–282
  chloroplast electron transfer chain, 266,
    268–269
  redox couple, 210
  requirement for dark reactions, 283
  structure, 211
Nicotinic (fast) acetylcholine receptor. *See*
  Acetylcholine receptor
Niedergerke, R., 713
Niemann-Pick disease, 193
Nigericin
  mechanism of action, **239**
  uncoupler, chloroplasts, 273
  uncoupler, mitochondrion, 233
9 + 2 arrangement, **10,** 302
Nirenberg, M., 358
Nissl, F., 668
Nissl substance, **668**
Nitrogenous base, **22**
  base pairing, 344, 367
  in DNA, 23, 340
  in nucleotide, 23
  peroxisome breakdown, 198
  in RNA, 23, 340
  structure, 22
Nitrous acid, as mutagen, **369**
Node of Ranvier, **672,** 685
Noll, H., 476
Nomura, M., 461, 481
Nonhistone protein, **394**
  in chromosome puff, 407
  enzymes classified as, 396
  heterogeneity of, 394
  phosphorylation, 396
  scaffold, 401, 504–505
Nonreciprocal genetic recombination. *See*
  Gene conversion
Nonsense code. *See* Stop signal
Nonspecific immunity. *See* Innate
  immunity
NOR. *See* Nucleolar organizer region
Norepinephrine
  and adenylate cyclase, 697
  as neurotransmitter, **689**
  site of action, 690
  site of synthesis, 622

structure, 690
  target organs, 622
  and tyrosine hydroxylase activity, 679
Norepinephrine receptor. *See* Dopamine,
  receptors
Northern blotting, **374**
N-ras oncogene, 745
Nucleic acid. *See also* Deoxyribonucleic
  acid (DNA); Ribonucleic acid (RNA)
  absorption spectrum, 80
  chemical composition, 22–24
Nucleic acid hybridization, 357, **371,** 372
Nuclein, **339**
Nuclear body. *See* Nucleoid
Nuclear-cytoplasmic 18S lineage, **575**
Nuclear-division model of myogenesis, **706**
Nuclear envelope, 9, **386,** 386–391
  continuity with endoplasmic reticulum,
    387
  functions, 390–391
  mitosis and, 501, 504
  nuclear lamina and inner membrane,
    387
  nuclear pores, 387–389
  perinuclear space, 387
  structure, 387
  transport pathway, 387–390
Nuclear lamina, **387**
Nuclear matrix, 9, **386,** 403
  chromatin fibers in, 403
  and gene expression, 431
  structure, 402–403
Nuclear *petite* mutation, **538**
Nuclear pore, 9, **387**
  mitosis and, 501, 504
  structure, 387
  transport pathway, 387–389
Nuclear sap. *See* Nucleoplasm
Nuclear totipotency, 603–605
Nucleoid, 9, **386**
Nucleolar organizer region (NOR), **392,** 451
Nucleolus, 9, **386**
  chemical composition, 393
  cortex, **392,** 456–457
  fibrillar core, 456–457
  isolation, 393
  matrix, **393**
  in mitosis, 501, 504
  and ribosome biogenesis, 450–451
  structure, 392–393
Nucleoplasm, 9, **386,** 391
Nucleoplasmin, **389,** 497
Nucleoside, **22**
Nucleoside diphosphatase
  Golgi marker, 172, 182
  microsomal localization, 164
Nucleoside pyrophosphatase, 164
Nucleosome, **397**
  DNA content, 397, 399–400
  histone protein content, 399–401
  model of, 399
  in sperm development, 589
  synthesis of, 496–497
5'-Nucleotidase, 106, 164
Nucleotide, **22**
  base pairing, 344, 367
  codon triplet, 361
  in DNA, 23, 340, 344
  DNA polymerase reaction, 345
  phosphodiester linkage, 23, 347
  in RNA, 23, 340
  RNA polymerase reaction, 357
  structure, 22
Nucleus, **9**

chromatin fibers in, 393–402
crude fraction, preparation of, 79
cytoplasmic influence on, 424–425
gene transcription, mechanisms of, 404–424
gene transcription, regulation, 424–443
histone and nonhistone proteins, 393–396
hormones acting on, 632–637
nuclear envelope, 386–391
nuclear matrix, 402
nucleolus, 392–393
nucleoplasm, 391
nucleosome, 396–401
proteins encoded by oncogenes, 745
in sperm development, 589
subfractionation of, 393
Nude mouse, **661**, 730
Numerical aperature, **60**
Nurse cell, **586**

Ochoa, S., 250
Octomer, in nucleosome, **399**
Octopamine, as neurotransmitter, **697**
O'Farrell, P., 85
Okazaki, R., 347
Okazaki fragments, **348**
Oligoadenylate synthase, **483**
Oligodeoxythymidylate [oligo(dT)] column chromatography, **467**
Oligomycin, **233**, 240
Oligomycin-sensitive ATPase. See Adenosine triphosphatase (ATPase)
Oligomycin-sensitivity conferring protein (OSCP), **557**
Oligosaccharide, **13**
Olins, A., and D., 396
O'Malley, B., 636
Omp A protein, **118**
Oncogene, **743**, 745–746
Oncogenic viruses, **739–746**
Oncologist, **728**
One gene–one enzyme theory, **352**
One gene–one polypeptide chain theory, **354**
Oocyte, **584**
polarity of cytoplasm, 588
primary oocyte, 588
secondary oocyte, 588
Oogenesis, **584**, 585–589
in amphibians, 584–585
growth phase, 588
maturation phase, 588
Oogonium, **584**
Oparin, A., 25, 27
Operator sequence, in lactose operon, **437**, 438–439
Operon
chloroplast ribosomal RNA genes, **554**
lactose (lac), **437–441**
tryptophan (trp), **484–485**
Opiates, **695–697**
Opsin, **687**
Optic tectum, **680**
Organic molecules. See also Carbohydrate; Lipid; Protein; Nucleic acid
origin of life, 25–26
structure, 12–24
Orgel, L., 26, 28
Origin of life, **24–30**
catalysis and, 26
first cells, 27–29
information transfer and, 26–27
nucleic acids before proteins, 27–28

organic molecules, formation of, 25–26
primitive earth conditions, 24–25
proteins before nucleic acids, 29–30
Orly, J., 626
Orthograde movement, in axonal transport, **673**
OSCP. See Oligomycin-sensitivity conferring protein
Osmosis, **118**, 130
Ouabain, **141**
Outer segment, in photoreceptor cell, **687**
Ovalbumin
messenger RNA half-life, 483
messenger RNA length, 468
Ovalbumin gene
intron number, 421
nuclear matrix enrichment, 431
steroid hormone control of transcription, 636–637
Ovary
hormone product, 622
in plants, 600
Overton, C. E., 94, 132
Ovoperoxidase, **595**
Ovum. See Egg
Owen, R. D., 663
Oxalate, 45
Oxaloacetate
in C_4 pathway, 284
electron shuttle into chloroplast, 259
electron shuttle into mitochondrion, 216–217
in Krebs citric acid cycle, **214**
Oxidase, mixed function, **165**
β-Oxidation pathway, **214**
Oxidation-reduction potential. See Redox potential
Oxidation-reduction reaction, **209**
Oxidative phosphorylation, **215**, 229
ATP synthesis, 229, 235
chemiosmotic theory, 237–241
electron transfer chain, 231–235
inhibitors, 233
Oxidized substance, **209**
Oxidoreductase, 36
Oxygen
in photorespiration, **286–287**
production in photosynthesis, **269–270**, 273
uptake in respiration, **233**, 235
Oxytocin
neurosecretion and, 698
site of action, 621
site of synthesis, 622
smooth muscle contraction, **722**
structure, 621

P680, **267**
P700, **267**
Pachytene stage, **516**, 523
Packer, L., 614
Packing ratio, in DNA, **401**
PAGE. See Polyacrylamide gel electrophoresis
Palade, G., 64, 77, 160, 168, 175, 218, 447
Palmiter, R., 382
Pancreas, 620
Pancreatic DNAase-I. See Deoxyribonuclease (DNAase)
Papain, **648**, 709
Paper chromatography, **87**
Papovaviruses, 740, 742
Paracrine pathway, **639**
Parathormone, 623, 626

Pardee, A., 732
Parkinson's disease, **694**
Parsons, D., 225
Parthenogenesis, **589**
Partition coefficient, **132**
Pastan, I., 440
Pasteur, L., 24, 35, 243
Pasteur effect, **243**
Patch clamp technique, **684**
Pauling, L., 19, 68, 352, 651
pBR322, **376**
PC. See Plastocyanin
PCA. See Perchloric acid
PCMB. See p-Chloromercuribenzoate
PDGF. See Platelet-derived growth factor
Pectinase, **257**, 604
Pectins, **120**, 122
Pediculi cristae, **227**
P-element, **606**
Pelling, C., 406
Penicillin, **119**
Penman, S., 330
Pentose sugar, **12**, 22. See also Deoxyribose; Ribose
Pepsin, 50
Pepsinogen, 50
Peptide bond, **15**, 17, 473
Peptide hormone, **621**, 623–632. See also individual hormones
adenylate cyclase interaction, 626–628
calcium ion and, 631–632
and cyclic AMP, 625–626, 628–630
cyclic GMP and, 630–631
examples of, 629
in glycogen breakdown, 628–629
receptor binding criteria, 624–625
termination of action, 632
Peptidoglycan, **115**
Peptidyl site (P site), **471**
Peptidyl transferase, **473**
Perchloric acid (PCA), 82
Perinuclear space, **387**
Periodic acid–Schiff, **69**
Peripheral protein, **102**
Peripheral reticulum, **254**
Periplasmic space, **119**, 120
Perlman, R., 440
Permease. See Transporter
Peroxidatic mode, **197**
Peroxisome, **196**, 195–199
and D-amino acid oxidase, 196
biogenesis, 200
and catalase, 196–197
chloroplast, proximity to, 255
cytochemical identification, 196
evolutionary origin, 202
isolation, 196
metabolic functions, 197–199
mitochondrion, proximity to, 255
photorespiration and, 287
respiration and, 198
and urate oxidase, 195–196
Perry, R., 450
Perutz, M., 68
Petite mutation, **538**
cytoplasmic, 539
mitochondrial structure and, 538
molecular basis of, 538
mutagenic agents and, 538
neutral, 539
nuclear, 538
spontaneous mutation rate, 538
suppressive, 539
PF. See Protoplasmic fracture face

Pfeffer, W., 94
PGA. *See* 3-Phosphoglycerate
pH
  embryonic induction and, 612
  endosome fusion and, 189
  gradient, chloroplast, 272
  gradient, mitochondrion, 237–241
  late responses of fertilization and, 600
  ribulose bisphosphate carboxylase
    regulation by, 282–283
Phagocytosis, 187
  antibody-mediated, 660
  microfilaments and, 327–328
  microtubules and, 314–315
Phalloidin, **319**
Phase contrast microscopy, **61**
Phase plate, **62**
Phase variation, in *Salmonella*, **433**
Phe. *See* Phenylalanine
Phenobarbital, 166–167
Phenylalanine (Phe)
  hydrophobic interactions, 20
  structure, 16
Philadelphia chromosome, **736**
Phorbol ester, **739**
Phosphate group
  alpha-location, 413
  in phosphodiester linkage, 23
  in nucleotide, **22**
Phosphatidic acid phosphatase, 164
Phosphatidylcholine
  asymmetry in cell membrane, 111
  in chloroplast membranes, 259
  distribution in cell membranes, 100, 101
  structure, **100**
Phosphatidylethanolamine
  asymmetry in cell membrane, 111,
    203–204
  in chloroplast membranes, 259
  distribution in cell membranes, 100, 101
  structure, **100**
Phosphatidylglycerol
  in chloroplast membranes, 259
  distribution in cell membranes, 100, 101
  structure, **100**
Phosphatidylinositol
  asymmetry in cell membranes, 111
  in chloroplast membranes, 259
  distribution in cell membranes, 100, 101
  structure, **100**
Phosphatidylinositol 4,5-bisphosphate
  hormone action and, **632**
  platelet-derived growth factor and, **531**
Phosphatidylserine
  asymmetry in cell membranes, 111
  chloroplast membranes and, 259
  distribution in cell membranes, 100, 101
  structure, **100**
Phosphatidylthreonine, **100**
Phosphocreatine, **719**
3′,5′-Phosphodiester linkage, **23, 413**
Phosphodiesterase, **625**
  calcium ion effect on, 631
  cyclic AMP, 625
  cyclic GMP, 688–689
  in glycogen breakdown, 625
  in photoreception, 688–689
Phosphoenolpyruvate carboxylase, **283,**
    284–286
Phosphofructokinase
  allosteric control by ATP and AMP, 243
  effect on ATP synthesis, 243
  in glycolysis, **212**
3-Phosphoglycerate (PGA)

  in Calvin cycle and, **279**
  in chloroplast electron shuttle pathway,
    259
Phosphoglyceride, **13.** *See also individual*
    molecules
  in cell membranes, 99, 100, 101
  structure, 13
Phosphoglycolate, in photorespiration, **287**
Phospholipase C, **632**
  discovery of, 632
  platelet-derived growth factor (PDGP)
    and, 531
  mechanism of action, 632
Phospholipid, **13.** *See also individual*
    molecules
  bilayer, 95
  in chloroplast membranes, 259
  distribution in cell membranes, 100, 101
  structure, 13
  synthesis in cell membranes, 203–205
Phosphorylase. *See also* Glycogen
    phosphorylase
  phosphorylase *a*, **50**
  phosphorylase *b*, **50, 711**
Phosphorylase kinase, **628,** 629–630
Phosphotransferase pathway, **142–144**
Phosvitin, in amphibian development, **585**
Photogene, **559**
Photon, **265**
Photophosphorylation, **269**
  coupling sites and, 269–270
  cyclic, 269
  noncyclic, 269
Photoreceptor, **685**
  cone cell, 219, 687
  cyclic GMP effect, 687–688
  membrane potential changes in,
    687–688
  rod cell, 687
Photorespiration, **286**
  in peroxisome, 198
  photosynthesis and, 286–287
Photosynthesis. *See also* Chloroplast;
    Dark reactions of photosynthesis;
    Light reactions of photosynthesis
  efficiency, 283
  equation describing, 249
  overview of, 249–251
  in prokaryotes and, 287–289
Photosynthetic lamellae, **8,** 10, 288
Photosynthetic pigments. *See* Carotenoid
    pigment; Chlorophylls
Photosynthetic unit, **265.** *See also*
    Photosystems I and II
Photosystems I and II, **267**
  in electron transfer chain, 265–270
  freeze fracture and, 278
  proteins encoded by chloroplast genes,
    554, 559
  reconstitution, 275
  in stacked and unstacked thylakoid
    membrane, 277
  subfractionation, 267
Phragmoplast, **513**
Phycobilin, **265**
Phycobilisome, **254,** 265, 288
Phycocyanin, **263**
Phycocyanobilin, **265**
Phycoerythrin, 263
Phycoerythrobilin, 264
Phytochrome, **568**
Phytol chain, in chlorophyll, **263**
Piericidin, 233
Pigment granule, in amphibian oocyte, **587**

Pinocytosis, **187**
Pittenger, T., 540
Pituitary, 622, 690, 698
Plant
  development, 600–602
  evolutionary origin of, 570
  growth substances, 638
  viruses, 740
Plaque, **153**
Plasma cell, **644**
Plasma membrane, **5**
  active transport, 139–146
  bacterium, 142–144, 145–146, 243–245
  biogenesis of phospholipids and
    proteins, 203–205
  Danielli-Davson model, 95
  diffusion, 130–139
  fluidity, 102–105
  fluid mosaic model, 98–105
  hormones acting on, 623–632
  intercellular junctions, 151–157
  lipid constituents, 99, 100, 101
  membrane potential, 146–147
  molecular organization, 105–113
  protein arrangement, 102
  protein markers, 106
  protein mobility and microtubule,
    314–315
  purple membrane, 112–113
  receptor-ligand affinity, 147–149
  recycling, 187–192
  red blood cell, 106–112
  selective permeability, 8
  unit membrane hypothesis, 95–98
Plasmid, **347,** 376
Plasmin, **734**
Plasminogen, **734**
Plasminogen activator, **734**
Plasmodesmata, **122**
Plasmolysis, **130**
Plastocyanin(PC), **266**
Plastoglobuli, **260**
Plastoquinone(PQ), **266**
  in electron transfer, chloroplast,
    268–269
  redox couple, 267
Platelet-derived growth factor (PDGF)
  in cell cycle, **531**
  oncogenes and, 745
Plaut, W., 541
Pleated sheet, **18**
Pneumococcus pneumoniae, genetic
    transformation in, **340–341**
Poky mutation, **540**
Polar body, **588**
Polar granules, **603**
Polarity
  amphibian oocyte, **588**
  DNA, **343**
  of Golgi membranes, **172**
  water, **12**
Polarization microscopy, **62**
Polar lobe, **602**
Polar nuclei, **600**
Poliovirus, axonal transport of, 675
Pollack, R., 732
Pollen grain, **600**
Pollen tube, **600**
Poly(A). *See* Polyadenylic acid
Polyacrylamide gel electrophoresis
    (PAGE), **84,** 85, 394, 426
Polyadenylic acid [Poly(A)]
  addition to RNA transcript, 419
  cell-free protein synthesis and, **359**

presence in messenger RNA, 419, 483
ribonuclease resistance, 419
Polyadenylic acid [Poly(A)] polymerase,
419
Poly(C). *See* Polycytidylic acid
Polycomplex, **524**
Polycytidylic acid [Poly(C)], in cell-free
protein synthesis, **359**
Polyhedral virus, **30**
Polymorph. *See* Polymorphonuclear
leukocyte
Polymorphonuclear leukocyte, **181**
antibody-mediated phagocytosis, 660
azurophil granule, 181–182, 191
bacterial destruction by, 191
specific granule, 181–182, 191
Polyoma virus, 740
Polypeptide, **15**. *See also* Protein
Polypeptide hormone, **149**. *See also*
Peptide hormone
Polyphosphate, **26**
Polyribosome. *See* Polysome
Polysaccharide, **12**, 12–13
cellulose, 13, 14, 120
colorimetric identification, 81
cytochemical identification, 69
glycogen, 13
Polysome, **478**
discovery of, 476–477
in protein synthesis, 476–479
Polyspermy, **594**
Polytene chromosome, **404**, 404–409
discovery of, 404
DNA synthesis, 406–408
puffing, 405–409
ribonucleoprotein granules in, 407
size, 404
species distribution, 404
tissue distribution, 404
Poly(U). *See* Polyuridylic acid
Polyuridylic acid [Poly(U)]
in cell-free protein synthesis, 359
column chromatography, 467
P/O ratio, **235**
in chloroplasts, 269–270, 273, 282
in mitochondria, 238–239
Porphyrin ring, in chlorophyll, 263
Porter. *See* Transporter
Porter, K., 64, 161, 329
Porter, R., 648
Positive control, in gene regulation, **440**
Positive staining, **64**
Postsynaptic density, **677**
Postsynaptic ending, **676**
Potassium ion channel
acetylcholine receptor, 694
adenosine triphosphatase (ATPase),
sodium and potassium, **140–142**
neuron plasma membrane, 683
Pott, P., 737
$_{pp}60^{c-src}$, **744**
$_{pp}60^{v-src}$, **744**
PQ. *See* Plastoquinone
Precipitation, of macromolecules, 82–83
Preinitiation complex, 467, 470
Pre-messenger RNA (mRNA). *See*
Heterogeneous nuclear RNA (HnRNA)
Preproinsulin, **485**
Prescott, D., 529
Presumptive myoblast, **706**
Presynaptic knob, **676**
Priestly, J., 249
Primary cell coat, **588**
Primary embryonic induction, **611**

Primary lysosome, **190**
Primary oocyte, **588**
Primary structure, of protein, **18**, 19
Primary transcript, **414**
7-methyl guanosine cap, 419
synthesis of, 414
Primary wall, **120**
Primase, **349**
Primordial germ cell
formation, **584**
migration of, 608
Pro. *See* Proline
Proacrosome granule, **592**
Procentriole, **306**
Procollagen, **125**
Proenzyme. *See* Zymogen
Profilin, 318
Progeria, **613**
Progesterone, 633, 636
Progestin, 633
Proinsulin, **485**
Prokaryote. *See* Bacterium; Blue-green
algae
Prolactin, 622–623
Prolamellar body, **566**
Proline (Pro)
hydroxylation in collagen, 124
structure, 16
Promitochondrion, **564**
Promoter sequence, **411**
catabolite gene activator protein
binding, 440
cyclic AMP binding, 440
lactose operon, 437, 440
ribosomal RNA gene, 411–412
Pronucleus, 599
Prophage, **32**, 742
Prophase, mitosis, **501**
Prophase I, meiosis, stages in, **514**, 515–518
Prophase II, meiosis, **518**
Proplastid, **565**
Prostaglandin, **640**
Prosthetic group, **43**
Protamine, **394**, 589
Protease, and ribosome structure, 462–463
Protein, **15**
absorption spectrum, 80
biological functions, 22
colorimetric identification, 81
conformation, 20
cytochemical identification, 69
levels of organization, 18–22
regulatory, 412, 436–441, 441–442
Protein kinase, cyclic AMP-dependent, **628**
dopamine action and, 695
microtubule-associated proteins, and,
300–301
Protein kinase, *start* genes and, **529**
Protein kinase, tyrosine-specific, **530**, 744,
745
Protein kinase C, 531, **632**, 739
Proteinoplast, **565**
Protein phosphatase, **632**
Protein processing
lysosome proteins, 181–182
membrane proteins, 203–204
Protein secretion, 167–171
glycosylation, 170–171
Golgi complex and, 175–180
and rough endoplasmic reticulum
(RER), 168–170
signal hypothesis and, 169–170
Protein synthesis
cell-free, **358**, 464–465

chloroplast, 554–559
coupling to RNA synthesis, 478–479
elongation phase, 470–474
fertilization and, 599
inhibitors, 479–481
initiation of, 466–470
mitochondrion, 554–558
polysomes and, 476–479
posttranslational modifications,
485–486
ribosome association with, 356
ribosome dissociation in, 465–466
termination phase, 474–476
translational control, 481–485
translocation, **473**, 474
Protein turnover
cell differentiation and, 607–608
ubiquitin and, 486
Protenoid, **26**
Proteoglycan, **125**
Protocell, **28**
Protochlorophyllide, **566**, 567
Protoeukaryote, **570**
Protofilament, **294**
Proto-*myc*, **745**
Proton gradient
bacteria, 243–245
bacterial flagella and, 334
chloroplast, 270–276
mitochondrion, 237–241
Proto-oncogene, **744**
Protoplasmic fracture face (PF), **66**, 276
Protoplasmic surface (PS), **66**, 276
Protoplast, **8**
bacterial, 118
plant, 257, 604
Protoribosome, **29**
Provirus. *See* Prophage
Proximal centriole, **590**, 592
PS. *See* Protoplasmic surface
Pseudo-Hurler disease, 193
Pseudopodia, 320–321
Psi factor, **460**
P site. *See* Peptidyl site
Puff. *See* Chromosome puff
Pulse-chase experiment, **82**
Pump. *See* Transporter
Purine, **23**
Puromycin, **471**, 481, 557
Purple membrane, **112–113**, 288–289
Pyrenoid, **260**
Pyridine nucleotide, in electron transfer
chain, **229**
Pyrimidine, **23**
Pyronin G, 69
Pyruvate, **213**, 214
Pyruvate dehydrogenase, 50, **213**

Q-band, **505**
Qβ replicase, **381**
Quanta, in neurotransmitter release, **690**
Quantal cycle, **608**
Quantum. *See* Photon
Quaternary structure, in protein, **20**
Quinacrine mustard, **505**
Quinone, in chloroplast membranes, **259**

Rabies virus, axonal transport of, 675
Racker, E., 235, 273
Radial loop model, **505**
Radial spokes, **302**, 311
Radiation, 739, 753
Radioimmunoassay (RIA), **623**
Radioisotope. *See* Isotopic tracer

Raf oncogene, 745
Raff, R., 573
Rall, T., 625
Ramón y Cajal, S., 171, 678
Ranvier, L.-A., 672
Rate constant
 protein degradation (K_d), **607**
 protein synthesis (K_s), **607**
Raven, P., 570
R-band, in chromosome, **505**
Reaction center, **265**
Reassociation kinetics, of DNA, **426–429**
RecA protein, **523**
Receptor, **147**, 619
 acetylcholine, 693–694
 dopamine, 694–695
 ligand interaction, criteria, 147
 opiate, 695–697
 plasma membrane, 147–149
 T-lymphocyte, 662
Receptor-mediated endocytosis, **189**
Recombinant DNA, **376**, 375–380
Recombinant RNA, **381**
Recombination. *See* Genetic recombination
Red algae, evolutionary origin of, 570
Red blood cell
 metabolic activation of, 425
 plasma membrane structure, 101,
  106–112, 327
Red-drop, **267**
Redman, C., 168
Red muscle, **705**
Redox couple, **209**
Redox potential, **210**
Reduced substance, **209**
Refractive index, **61**
Regulatory gene, **437**
Regulatory protein
 eukaryotic genes, **412**, 441–442
 prokaryotic genes, 436–441
Reich, E., 543, 548
Reijnders, L., 574
Relaxing factor, in muscle contraction, **717**
Release factors (RF), in protein synthesis,
 474
Releasing factors, and hormone secretion,
 622
Repeated DNA sequences, **426**
 centromeric heterochromatin and, 505
 Cot value, 426–427
 reassociation kinetics, 426–428
Replication fork, **496**
Replicon, **493**, 495–496, 498
Repressor protein, lactose operon, 437–439
RER. *See* Rough endoplasmic reticulum
Residual body, **190**, 592
Resolution, **60**
Resolvase, **368**
Respiration, **213**
Respiratory chain. *See* Electron transfer
 chain
Respiratory control, **243**
Resting membrane potential. *See*
 Membrane potential
Restriction endonuclease, **375**
Restriction map, **375**
Restriction point, **528**, 732
Restriction site, **375**
Retina, photoreceptors in, 687
11-*cis*-Retinal, **687**
 bacteriorhodopsin and, 112
 in photoreceptor cell, 687
Retrograde movement, in axonal
 transport, **674**

Retroviruses, **742**, 744
Reverse transcriptase, **350**, 742
RF. *See* Release factors
Rhoades, M., 537
Rhodopsin, **687**, 688
Rho factor, in termination of RNA
 synthesis, 414
Rho⁻ (ρ^-) mutation. *See Petite* mutation
RIA. *See* Radioimmunoassay
Ribonuclease
 and degradation of polysomes, 476–478
 denaturation of, 20–21
 and heterogeneous nuclear RNA-pro-
  tein association, 417–418
 ribosome structure analysis, 462–463
 III, and ribosomal RNA synthesis,
  458–459
Ribonucleic acid (RNA), **23**. *See also*
 Heterogeneous nuclear RNA;
 Messenger RNA; Ribosomal RNA;
 Transfer RNA
 base pairing, 344, 367
 colorimetric identification, 81
 cytochemical identification, 69
 nitrogenous bases, 22–24, 340
 primary transcript, 414
 primer in DNA synthesis, 349–350
 processing, 415, 414–424, 442–443
 recombinant, 381
 splicing mechanism, 423–424
 viruses, 740, 742–745
Ribonucleic acid (RNA) synthesis,
 404–424. *See also individual ribonucleic acids*
 asymmetry, 358
 in chloroplast, 554
 coupling to protein synthesis, 478–479
 elongation phase, 413–414
 initiation, 412–413
 in mitochondrion, 552–554
 termination phase, 414
 visualization of, 404–409, 527–528
Ribonucleoprotein (RNP) particles
 export from nucleus, 389
 lampbrush chromosome and, 527–528
 and messenger ribonucleic acid, **482**
 nucleolar location, 392
 in ribosome subunit formation,
  448–450, 455–456, 459, 461
Ribophorin, **169**
Ribose, **23**
Ribosomal protein
 binding of initiation factors to ribosome,
  470
 nomenclature, 450
 regulation of synthesis, 481
 ribonucleoprotein formation, 448–450,
  455–456, 459, 461–464
 ribosome subunit composition, 450
Ribosomal protein genes
 in chloroplast, 559
 mitochondrion, 558
 operon, in *E. coli*, 481
Ribosomal ribonucleic acid (rRNA)
 amino acid starvation and control of
  synthesis, 460, 484
 in chloroplast, 554
 in situ hybridization, 451
 methylation, 459
 in mitochondrion, 552–554
 processing, 458–459
 protein synthesis and, 357
 ribonucleoprotein formation, 448–450,
  455–456, 459, 461–464
 sequence and cell evolution, 575

synthesis in nucleolus, 450–451
synthesis in prokaryotes, 458–459
5S Ribosomal RNA (rRNA)
 gene, 411–412
 location in ribosome, 449–450, 464
 RNA polymerase III and, 412
 synthesis, 411–412, 455
5.8S Ribosomal RNA (rRNA)
 location in ribosome, 449–450
 synthesis, 453–455
16S Ribosomal RNA (rRNA)
 assembly into ribosome subunit, 461
 3′ end, and messenger RNA binding, 468
 location in ribosome, 449–450, 464
 Shine-Dalgarno sequence, 468
 synthesis, 457–460
18S Ribosomal RNA (rRNA)
 3′ end, and messenger RNA binding, 468
 location in ribosome, 449–450
 synthesis, 453–455
20S Ribosomal RNA (rRNA), synthesis,
 453–455
23S Ribosomal RNA (rRNA)
 location in ribosome, 449–450
 synthesis, 457–460
28S Ribosomal RNA (rRNA)
 location in ribosome, 449–450
 synthesis, 453–455
32S Ribosomal RNA (rRNA), synthesis,
 453–455
41S Ribosomal RNA (rRNA), synthesis,
 453–455
45S Ribosomal RNA (rRNA)
 gene, 456–457
 methylation pattern, 454–455
 processing, 453–455
 synthesis, 453–455
Ribosomal RNA genes
 amplification, 452, 584
 isolation, 451
 operon organization in prokaryotes, 458
 5S RNA transcript, promoter location,
  411–412
 45S RNA transcript, 456–457
Ribosome, **9**, 447–488. *See also* Protein
 synthesis
 biogenesis, 450–460
 chloroplast, 555–556
 80S, composition of, 448–450
 70S, composition of, 448–450
 discovery of, 447–448
 dissociation conditions, 448
 formation, 456–460
 mitochondrion, 555–556
 nucleolus and, 451–452
 polysome, 476–479
 protein. *See* Ribosomal protein
 ribonucleic acid. *See* Ribosomal RNA
 reconstitution of, 461
 ribonucleoprotein particle formation,
  448–450, 455–456, 459, 461–464
 60S subunit, composition of, 448–450
 50S subunit, cell-free assembly, 461
 50S subunit, composition of, 448–450
 40S subunit, composition of, 448–450
 30S subunit, cell-free assembly, 461
 30S subunit, composition of, 448–450
 three-dimensional model, 448–450
Ribulose 1,5-bisphosphate (RuBP), **281**,
 283–284
Ribulose 1,5-bisphosphate carboxylase
 chloroplast gene for, 554
 large subunit, synthesis of, 559–560
 magnesium ion effect, 282
 oxygen and, 387

pH effect on, 282
small subunit, synthesis of, 560–561
structure, **282**
Rich, A., 365, 485
Rickenberg, H., 145
Rifamycin, **413**
Ris, H., 329, 541
RNA. *See* Ribonucleic acid
RNA maturase, **554**
RNA polymerase, **357**
actinomycin D effect, 414
α-amanitin sensitivity, 410–411
core enzyme, **407**
DNA binding and, 411–412, 413
eukaryotic, 410
fractionation, 409
heparin effect on, 413
initiation of RNA synthesis, 412–413
isolation, 410
magnesium ion sensitivity, 409–410
messenger RNA synthesis, 410
modulation of activity by proteins, 436
nuclear location, 410–411
optimum conditions for activity,
409–411
prokaryotic, 409
reaction catalyzed by, 357
ribosomal RNA synthesis, 410
in ribosome biogenesis, 453–456
5S RNA synthesis, 410
sigma factor, **409**
TATA box and, 411
termination of RNA synthesis, 414
transfer RNA synthesis, 410
type I, 410
type II, 410, 411, 412
type III, 410, 412
RNP particles. *See* Ribonucleoprotein
particles
Robertson, J. D., 96
Rod cell, **687**. *See also* Photoreceptor
Roeder, R., 410
Roentgen, W., 739
Rootlet fibers, **306**
Roseman, S., 143, 150
Rosenbaum, J., 297, 308
Ros oncogene, 745
Rosy gene, **606**
Rotenone, **233**
Roth, J., 639, 680
Rothman, J., 203, 204
Rotor, **73**
Rough endoplasmic reticulum (RER), **162**,
167–171. *See also* Protein secretion
Rous, P., 738, 740
Rous sarcoma virus, 740, 743
rRNA. *See* Ribosomal ribonucleic acid
R side chain, in protein, **16**
Rubin, S., 606
RuBP. *See* Ribulose 1,5-bisphosphate
Rutter, W., 410

Sabatini, D., 168
Sager, R., 538, 543, 552, 559
Sakaguchi reaction, **69**
Saltatory conduction, of action potential,
685
Saltatory movement
cytoplasmic streaming and, **320**
of intracellular particles, 675
Sanger, F., 18, 380, 469
Sanger dideoxy method, **380–381**
Sarcolemma, **702**
acetylcholine receptors, 693–694, 716
depolarization of, 716–717

junctional fold, 716
transverse tubules and, 703
Sarcoma, **727**
Sarcomere, **702**
A-band in, 702, 713, 716
cross-bridges, 703, 714
H-zone in, 702
I-band in, 702, 713, 716
M-line in, 702
myofibril and, 702
polarity of, 714–715
sliding filament hypothesis, 713–715
thick filament, 702, 710–711
thin filament, 702, 711–713
Z-line in, 702, 714
Sarcoplasmic reticulum, **703**, 704–705
adenosine triphosphatase (ATPase), cal-
cium ion, 717–718
calcium ion and, 717–718
in cardiac muscle, 723
in smooth muscle, 721
terminal cisternae and, 703
transverse tubules and, 703
Satellite DNA, **428**
in heterochromatin, 505
isolation, 428
ribosomal RNA genes, 451
Satir, P., 309
Saturated fatty acid, **13**
Saturation hybridization, **372**
Scaffold, **401**, 504
Scanning electron microscopy (SEM), **66**
Scatchard plot, **148**
Schachman, H., 48
Schatz, G., 560
Scherrer, K., 453
Schiff reagent, **69**
Schizophrenia, **695**
Schimke, R., 607
Schleiden, M., 3, 667
Schmitt, F. O., 96
Schneider, W., 77, 218
Schramm, M., 626
Schwann, T., 3, 667
Schwann cell, **96**, 672
Scremin, L., 540
Scurvy, and collagen fiber formation, 124
SDS. *See* Sodium dodecyl sulfate
Secondary cell coat, **588**
Secondary constriction, **506**
Secondary embryonic induction, **611**
Secondary lysosome, **190**
Secondary oocyte, **588**
Secondary response, in antibody
synthesis, **660**
Secondary structure, of protein, **19**
Secondary wall, **122**
Second messenger concept, **626**
Secretin
cyclic AMP and, 626
discovery of, **620**
target organ, 623
Sedimentation coefficient (s), **73**, 74
Seed, **601**
Seeley, J., 679
Seidman, J., 653
Selection synchrony, **491**
Self-assembly, 24, 32–33
SEM. *See* Scanning electron microscopy
Semiconservative DNA replication, **344**,
345
Semipermeable membrane, **130**
Sendai virus, **104**
Senebrier, J., 249
Sensory neuron, **679**, 685

Septate junction, **154**
Ser. *See* Serine
SER. *See* Smooth endoplasmic reticulum
Serine (Ser)
in bacterial cell wall, 115
in chymotrypsin active site, 53
hydrogen bond formation, 20
phosphorylation of histone, 395
structure, 16
Serotonin, 689, 690, 697
Serotonin receptor. *See* Dopamine,
receptors
Sertoli cell, **589**
Shadow casting, **65**
Sharp, P., 420
Shearer, R., 416
Shine, J., 468
Shine-Dalgarno sequence, **468**
Shope, R., 740
Shope papilloma virus, 740
Shuttle streaming, **320**
Sickle-cell anemia, 352–353
Siekevitz, P., 447, 464
Sigma factor, **409**
Signal hypothesis, **169**, 170–171
Signal peptidase, 164, **170**
Signal recognition particle (SRP), **169**
Signal sequence, **169**
Silverman, M., 333
Simon, E., 695
Simon, M., 333
Simple diffusion. *See* Diffusion, simple
Singer, S. J., 98
Single-copy DNA sequence. *See* Unique
DNA sequence
Single-strand exchange intermediate, **522**
Sis oncogene, 745
Sitosterol, 99
Sjøstrand, F., 64, 218
Skeletal muscle, **702**, 701–720
abnormal functions, 720
cell structure, 701–706
contraction mechanism, 713–720
developmental origin, 706–708
multinucleate state, 706–708
muscle fiber, 702
myofibril, 702
sarcomere. *See* Sarcomere
sliding filament hypothesis, 713–715
Ski oncogene, 745
Slack, C., 283
Slater, E. C., 237
Sliding filament hypothesis, **714**, 713–715
Sliding mechanism, in motility, 309–311
Sliding-vesicle theory, **675**
Slime layer, in bacteria, **113**
Slonimski, P., 546, 554
Slow axonal transport, **673**
Small nuclear RNAs, **391**
Small nuclear ribonucleoprotein particle
(snurp), **423**, 424
Smith, H., 375
Smith, J., 614
Smooth endoplasmic reticulum (SER),
162, 163–165
Smooth muscle, **721**, 722
S1 nuclease, **372**, 552
Snurp. *See* Small nuclear ribonucleopro-
tein particle
Snyder, S., 695
Sodium butyrate, in gene expression, 430
Sodium dodecyl sulfate (SDS), **84**
Sodium ion channel
in action potential, **682–684**
adenosine triphosphatase (ATPase),

Sodium ion channel *(Continued)*
 sodium and potassium, **140–142**
 block by tetrodotoxin, 684
 rod outer segment membrane, 687–688
Sodium ion transport, mitochondrial
 inner membrane, 241–242
Sodium potassium ATPase. *See* Adenosine
 triphosphate (ATPase), sodium and
 potassium ion
Somatic mutation theory of antibody
 diversity, **653**
Somatic recombination theory of
 antibody diversity, **653**
Somatostatin, **690,** 698
Somatotropin, **621,** 622–623
Southern blotting, **374**
SP8-bacteriophage, **358**
Spacer DNA, **457**
Specific granule, **181,** 182, 191
Spectinomycin, 480
Spectrin, **106–107**
 actin binding, 109, 327
 ankyrin binding, 109–110, 327
 band 4.1 binding, 109–110
 calcium ion sensitivity, 318
 mobility of band 3 and glycophorin,
 110, 327
Spectrophotometric analysis, of macro-
 molecules, 80–81
Spemann, H., 611
Sperm, **582**
 DNA, 590
 formation, 589
 guided movement toward egg, 593
 head, 589–590
 in plants, 307
 sea urchin, 593
 tail, 589, 590–593
Spermatid, **589**
Spermatogenesis, **589,** 589–593
Spermatogonium, **589**
Spermiogenesis, **589**
Sperm nuclei, in plant development, **601**
Sperry, R., 680
S phase, of cell cycle, **491,** 498
Spherosome, **195**
Sphingomyelin, **99,** 100, 101
Sphingophospholipid, **13,** 15
Sphingosine, **13,** 15
Spindle apparatus. *See* Mitotic spindle
Spine. *See* Dendritic spine
Spirilla, **118**
Spirochete. *See* Spirilla
Spirogyra, 251
Spleen, in lymphocyte development, **644**
Splice junction, **423**
Spontaneous chemical reaction, **38**
Spontaneous generation, **24**
Spore, **501,** 499–501
Spore coat, **501**
Sporulation. *See* Spore
Spradling, A., 606
Src oncogene, **743**
SRP. *See* Signal recognition particle
Stahl, F., 344
Standard free-energy change (ΔG°), **37**
Stanley, W., 30
Starch, **13,** 14, 260
Starling, E., 620
Start gene, yeast, **529**
Steck, T., 108
Steinberg, M., 609
Steinman, R., 187
Stem cell, **584,** 643
Steroid, **15**

in cell membranes, 99, 100
 structure, 15
Steroid hormone, **621**
 action on nucleus, 633–636
 mechanism of action, 635
 properties, 621
 transcriptional control, 636–637
Stevens, A., 357
Steward, F., 604
Sticky ends of DNA, **375**
Stigma, **600**
Stigmasterol, 99, 100
Stöffler, G., 463
Stokes formula, **73**
Stomata, **286**
Stop signal. *See* Terminator codon
Strauss, W., 189
Streptomycin
 chloroplast mutagen, 538
 effect on protein synthesis, **461,** 479–480
 resistance, and ribosomal protein, 461
Stress fibers, **317,** 322
Striated muscle. *See* Skeletal muscle
Stringency factor, **484**
Stroma, **9, 252**
 contents, 260
 dark reactions of photosynthesis and,
 250, 279–283
 isolation, 257–258
Stroma center, **260**
Structural gene, **437**
Subcellular fractionation, **77–80**
Submitochondrial particles, **226**
Substance P, **690**
Substrate, **36**
Substrate-level phosphorylation, **229**
Subunits, of protein, **20**
Succinate, 214
Succinate dehydrogenase
 bacterial plasma membrane marker, 106
 competitive inhibitors of, 45, 215
 in Krebs citric acid cycle, **214**
Succinyl coenzyme A, 214
Sudan black, **69**
Sudan red, **69**
Sueoka, N., 549
Sugar, colorimetric identification, 81
Sugar transport. *See* Active transport, so-
 dium-linked; Phosphotransferase
 pathway
Sulfanilamide, 45
Sulfatide, **15,** 100
Sulfoquinovosyldiglyceride, 259, 262
Summers, K., 309
Sumner, J., 36
Supernatant, **74**
Suppressive *petite* mutation, **539**
Suppressor mutation, **476**
Suppressor T-cell, **663**
Surface lamina, **330**
Surgery, **753**
Suspension culture, **58**
Sutherland, E., 624, 625
SV-40 virus, **740**
Svedberg, T., 73
Svedberg unit (S), **74**
Swift, H., 404, 491
Sympathetic neuron, **679**
Symport, **144**
Synapse, **668,** 675–678
 chemical, **676,** 677
 electrical, **678**
 at neuromuscular junction, 676–677,
 716
Synapsis, **516**

Synaptic cleft, **676**
Synaptic vesicle, **676**
Synaptonemal complex
 chemical composition, **524**
 development during meiosis, 524
 formation during meiosis, 516
 and genetic recombination, 526
 structure, 523–526
Synaptosome fraction, **694**
Synergid cell, **600**
Synkaryon, **60**
Szent-Györgi, A., 708

T_3. *See* Triiodothyronine
T_4. *See* Thyroxine
$T_{1/2}$. *See* Half-life
T3 complex, **662**
TψC loop, **365**
Takai, Y., 632
Takebe, I., 604
T-antigen, **742**
TATA box, **411,** 412–413
Tatum, E., 350, 540
Tau proteins, **295**
Tautomer, **367**
Taylor, J. H., 495, 519
Taylor, R., 716
Tay-Sachs disease, 193
TCA. *See* Trichloroacetic acid
T-cell. *See* T-lymphocyte
Teichoic acid, **115**
Telophase, mitosis, **503,** 504
Telophase I, II, meiosis, **517–518**
TEM. *See* Transmission electron micros-
 copy
Temin, H., 742
Temperate virus. *See* Virus, lysogenic
Template, **346**
Teratocarcinoma, **747**
Terenius, L., 695
Terminal cisterna, **703**
Terminalization, **517**
Terminal transferase, **376**
Terminal web, **326**
Termination, in protein synthesis,
 474–476
Termination signal. *See* Terminator codon
Terminator codon, **362**
Terpene, **15**
Tertiary cell coat, **588**
Tertiary structure, of protein, **19,** 20
Tertiary wall, **122**
Testicular feminizing syndrome, **636**
Testis, hormone products, 622
Testosterone, 620, 633
Tetanus toxin, axonal transport, 675
Tet gene. *See* Tetracycline-resistance gene
Tetracycline, **379,** 480
Tetracycline-resistance gene, 377–378
Tetrad, in meiosis, **516**
Tetranucleotide theory, **340**
Tetrodotoxin, **684**
TGF. *See* Transforming growth factors
Theophylline, **625**
Thermodynamics, **37**
Thermoreceptor, **685**
Theta antigen. *See* Thy 1 antigen
Thiamine pyrophosphatase, 172
Thick filament, **702.** *See also* Myosin
 cross-bridges and, **703,** 714
 polarity of, 714–715
 skeletal muscle, 710–711, 713–715
 smooth muscle, 721
 Z-line and, **702,** 714
Thin filament, **702.** *See also* Actin

polarity of, 714–715
skeletal muscle, 711–713, 713–715
smooth muscle, 721
Z-line and, 702, 714
Thin-layer chromatography (TLC), **87**
Thomas, J., 399
Thorazine. *See* Chlorpromazine
Thorn. *See* Dendritic spine
Thr. *See* Threonine
Threonine (Thr)
in histone phosphorylation, 395
structure, 16
Threshold potential, **683**
Thy 1 antigen, **661**
Thylakoid membrane, **9, 253**
adenosine triphosphatase (ATPase), 273–275
CF_0, CF_1, 273–275
electron transfer chain, 266–269
freeze fracture of, 276–277
isolation, 257–258
light-absorbing pigments, 262–266
lipid content, 262
model of, 275
reconstitution, 274
spheres, 253
stacked and unstacked, 253, 257–258, 277–279
Thymidine block, **490**
Thymidine kinase gene, injection of, 606
Thymine, **23**
base pairing, 344, 367
dimer, **369,** 371
enol form, 367
keto form, 367
structure, 22
Thymus, 644–645
Thyroglobulin, **622**
Thyroid hormone, **622, 633,** 636
Thyroid-stimulating hormone (TSH), **622,** 626
Thyrotropin. *See* Thyroid-stimulating hormone (TSH)
Thyrotropin-releasing hormone (TRH)
neurosecretion and, 698
as neurotransmitter, 690
target organ, 623
Thyroxine (T_4), **622**
Tiedemann, H., 612
Tight junction, **151,** 152
Ti plasmid, **382**
TLC. *See* Thin-layer chromatography
T-lymphocyte, **643**
in acquired immune deficiency syndrome (AIDS), 662
antigen receptor, 656
cell-mediated immunity and, 661–662
receptor, 662
sensitized, 661
T_m. *See* Melting temperature
TNBS. *See* Trinitrobenzene sulfonic acid
Tobacco mosaic virus, 31, 32–33
Tobacco smoke, components of, 737
Todaro, G., 730, 731
Toft, D., 634
Tonegawa, S., 653
Tonofilaments, **152**
Torpedo, 693
Totipotency, **604**
Toxin, as antigen, 660
Transcription, **359.** *See also* Ribonucleic acid (RNA) synthesis
5S Transcription factor, **412,** 441
Transduction, **32,** 518
Transfection, **744**

Transfer ribonucleic acid (tRNA), **360**
anticodon, 364
chloroplast, 555–557
discovery of, 362–363
isoaccepting tRNAs, 363
mitochondrion, 554
modified bases in, 365
in protein synthesis, 362–365
and regulation of protein synthesis, 484–485
structure, 365
$T\psi C$ loop, 365
wobble hypothesis, 471–472
Transferase, **36**
Transformation, **340, 376,** 518
Transformed cell, **59,** 730
Transforming growth factors (TGF), **732**
Transformylase, in mitochondria and chloroplasts, **556**
Transitional fibers, **306**
Transitional vesicles, **176**
Transition state, **39**
Transition temperature, **103**
Transition zone, **306**
Transketolase, in Calvin cycle, 281
Translation, **359.** *See also* Protein synthesis
Translocase. *See* Transporter
Translocation, in protein synthesis, **473,** 474
Translocator, in axonal transport, **675**
Transmembrane protein, **102**
Transmission electron microscope, **64**
Transpeptidation reaction, **119**
Transplantation antigen. *See* Histocompatibility antigen
Transporter, **134**
Transposable element, **367,** 606
Transposase, **368**
Transposon, **367**
Transverse diffusion, **137,** 204
Transverse fibers, in synaptonemal complex, **524**
Transverse septum, in bacterial cell division, **499**
Transverse tubule, in muscle, **703,** 716–717
Treadmilling model, **298–299**
Trebst, A., 269
TRF. *See* Thyrotropin releasing factor; Thyrotropin releasing hormone
Triad, in muscle, **705**
Tricarboxylic acid cycle. *See* Krebs citric acid cycle
Trichloroacetic acid (TCA), 82
Triglyceride, **13,** 14
Triglyceride lipase, 629
Triiodothyronine (T_3), 622, 633
Trinitrobenzene sulfonic acid (TNBS), 203
Triose, **12**
Triplet code, **359**
Triton, WR-1339, 196
Tropomyosin, **319, 709**
in muscle contraction, 718
in nonmuscle cells, 319
structure, 712
Troponin, **319, 709**
cardiac muscle, 723
skeletal muscle, 718
structure, 712
T, N, I subunits, 713
tRNA. *See* Transfer ribonucleic acid
Trp. *See* Tryptophan
Trypsin, 50, 709
Trypsinogen, 50
Tryptophan (Trp)

control of synthesis, 440, 484–485
hydrophobic interactions, 20
operon, 440, 484–485
structure, 16
Tryptophan oxygenase, hormonal control of synthesis, 636
Tryptophan pyrrolase, turnover of, **607–608**
Tryptophan synthetase, A subunit mutations, **354**
TSH. *See* Thyroid-stimulating hormone
T-system, in muscle, **705**
Tswett, M., 262
Tubulin, **294.** *See also* Microtubule assembly
α and β subunits, 295
colchicine binding, 295
protofilament formation, 294
Tumor, **728**
Tumor angiogenesis factor, **734**
Tumor-specific cell surface antigens, **734**
Turnover, protein, **607**
Two-dimensional gel electrophoretic separation, **85**
Type II glycogenosis, 194
Tyr. *See* Tyrosine
Tyrosine (Tyr)
hydrogen bond formation, 20
protein kinase, **530, 744–745**
structure, 16
Tyrosine hydroxylase, 679
Tyrosine transaminase, hormonal control of synthesis, 636
Tzagoloff, A., 557

U1 snurp, **423**
U7 snurp, **424**
Ubiquinone, **15**
absorbance spectrum of, 230
electron transfer chain, 232–235
redox couple, 231
Ubiquitin, **430,** 486
UDP-glucose. *See* Uridine diphosphate-glucose
Ultracentrifuge, **73**
Ultraviolet light, as mutagen, **369,** 370, 739–740
Unassigned reading frame (URF), **553**
Unc mutant, 244
Uncoupling agent, **239**
chloroplast, 273
mitochondrion, 233
Unique DNA sequence, **426–427**
Unit membrane hypothesis, **95**
Unsaturated fatty acid, **13**
Upstream sequence, **411**
Uracil, **23**
base pairing, 344, 367
cytosine, conversion to, 369
in RNA, 357
structure of, 22
Urate oxidase, **195,** 196
Urea, 12, 21
Urease, crystallization of, 36
URF. *See* Unassigned reading frame
Uridine diphosphate (UDP)-glucose, 171

Vacuole, **8**
Val. *See* Valine
Valine (Val)
hydrophobic interactions, 20
structure, 16
Valinomycin
as ionophore, **137**
mechanism of action, mitochondria, 239

Valinomycin (Continued)
  uncoupler, chloroplast, 273
  uncoupler, mitochondrion, 233
  structure, 137
Variable region gene (V gene), **650**
  antibody diversity and, 654–655
  in antibody heavy chain, 650
  in antibody light chain, 650
  evidence for, 653
Vasopressin
  cyclic AMP and, 626
  neurosecretion and, 698
  site of synthesis, 622
  target organ, 622
Vector. See Cloning vector
Vegetal pole, **588**
Vegetative nucleus, **600**
Velocity centrifugation, **74,** 75
Vertical transmission, 740
Vesicle hypothesis, of neurotransmitter
    release, **690**
Vesicular stomatis virus (VSV), G-protein,
    204
V gene. See Variable region gene
Villin, 318, **326**
Vimentin, **328–329**
Vinblastine, **312**
Vincristine, **312**
Vinculin
  calcium ion sensitivity, 318
  in cardiac muscle, **723**
  in cell locomotion, **322**
Vinograd, J., 545, 550
Virchow, R., 24, 672
Virus, **30**
  antigen recognition by T-lymphocyte,
      662
  bacteriophage, 30, 32–33, 357, 358, 378,
      414
  defective, **32**
  DNA, **741–742**
  enveloped, **30**
  evolutionary origin, 33
  infection, steps in, 32–33
  lysogenic (temperate), **32**
  oncogenic, 739–746

  reproduction and life cycle, 30–33
  RNA, **742–743**
  virulent, **32**
Vishniac, W., 250
Vitamin
  A₁, 15, 687
  C, 124, 165
  as cofactor, 44
  D, 15
  E, 614
Vitelline envelope, **588**
Vitellogenin, **585**
Vogelstein, B., 431
Volkin, E., 357
Voltage clamp technique, **682**
Voltage-gated channel, **683**

Waksman, B., 645
Wallin, J. F., 569
Walsh, D., 628
Warburg, O., 218
Warren, J., 609
Water
  properties, 11–12
  splitting in photosynthesis, 270
Watson, J., 343
Weigle, J., 519
Weinberg, R., 735
Weintraub, H., 430, 497
Weisenberg, R., 295, 300
Weismann, A., 604
Weiss, P., 672
Weiss, S., 357
Werner's syndrome. See Progeria
Wessells, N., 609, 678
West, I., 145
White, L., 57
White muscle, **705**
Whole-mount preparation, **65**
Wiercinski, F. J., 717
Wilkins, M., 68, 343, 396
Willstätter, R., 36
Wilson, E. B., 339
Wilson, H. V., 149
Wilson, L., 298
Wobble hypothesis, **471–472**

Woese, C., 575
Wöhler, F., 12
Wound tumor virus, 740

Xanthine dehydrogenase, 606
X chromosome, inactivation of, 429–430
Xenopus laevis
  anucleolar mutant, 451
  nuclear transplantation, 425
Xeroderma pigmentosum, 370, 746
X-ray, in cancer, 739
X-ray diffraction, **66,** 67–68

Yalow, R., 623
Yamada, T., 612
Yanofsky, C., 354, 484
Yeast
  cell cycle mutants, 529
  ethidium bromide, as mutagen, 539
  euflavine, as mutagen, 539
  5-fluorouracil, as mutagen, 539
  mating-type locus, 433–434
  petite mutation, 538–540
Yes oncogene, 745
y gene, in lactose operon, **437**
Yin-yang hypothesis, 631
Yolk, **585,** 586–587
Yolk platelet, **585**

Zacharias, E., 339
Zamecnik, P., 464
Z-DNA. See Z-form DNA
Zebra bodies, **194**
Z-form DNA, **435,** 436
z gene, in lactose operon, **437**
Zinkernagel, R. M., 661
Z-line, of sarcomere, **702**
Zona pellucida, **588**
Z-scheme, **268**
Zubay, G., 440
Zygote, **582**
Zygotene stage, **516**
Zymogen, **50**
Zymogen granule, **176**